INTEGRATED PHARMACEUTICS

INTEGRATED PHARMACEUTICS
Applied Preformulation, Product Design, and Regulatory Science

ANTOINE AL-ACHI
MALI RAM GUPTA
WILLIAM CRAIG STAGNER

Campbell University
College of Pharmacy & Health Sciences
Buies Creek, North Carolina

A JOHN WILEY & SONS, INC., PUBLICATION

Library of Congress Cataloging-in-Publication Data:

Al-Achi, Antoine, 1955–
  Integrated pharmaceutics : applied preformulation, product design, and regulatory science /
Antoine Al-Achi, Mali Ram Gupta, William Craig Stagner.
       p. ; cm.
  Includes bibliographical references and index.
  ISBN 978-0-470-59692-0 (cloth)
  I. Gupta, Mali Ram. II. Stagner, William Craig. III. Title.
  [DNLM: 1. Chemistry, Pharmaceutical. 2. Biopharmaceutics. 3. Dosage Forms–standards. 4.
Drug Design. 5. Drug and Narcotic Control. QV 744]

  615.7–dc23

                                                                                          2012025748

Printed in the United States of America

10 9 8 7 6 5 4 3 2 1

To my wife, Pamela Al-Achi, and my sons, Elias Gabriel, Anthony William, and John Peter. To my beloved father, Elias, my mother, Renee, and my sister and brothers, Claudette, Peter, and Kamil and their families.

To my wife, Sulochna Gupta, and my children and grandchildren, Michael, Saijal, Nathan, Maya, and Deepak. To my beloved father, Harchand Rai, my mother, Anachhi, and the entire Mittal family.

To my wife, Nancy Stagner, and my children and grandchildren, Ryan, Beth, Ethan, Tripp, Kelly, Amelia, Justin, and Lauren.

To all our students: past, present, and future.

CONTENTS

PART II

PRODUCT DESIGN

CHAPTER 13   *TABLET PRODUCT DESIGN*   **215**

CHAPTER 14   *CAPSULE PRODUCT DESIGN*   **319**

CHAPTER 15   *DISPERSED SYSTEM PRODUCT DESIGN*   **359**

CHAPTER 16   *AEROSOL PRODUCT DESIGN*   **415**

PART III

REGULATORY SCIENCE

CHAPTER 30   SAFETY, TOXICOLOGY, AND PHARMACOGENOMICS   911

**CHAPTER 31   REGULATORY SCIENCE INITIATIVES FOR ADVANCING
PUBLIC HEALTH**   939

FOREWORD

The wide variety of topics covered by the authors in this book emphasize both the depth and breadth of knowledge needed for pharmaceutical scientists to bring a drug product to the marketplace successfully. The challenge of designing and developing compounds into pharmaceutical products, which are critical to the survival of both the biotech and pharmaceutical industries, will depend largely on the education and extensive training required of young pharmaceutical scientists. This book gives the reader an understanding of the basic and applied sciences involved in the development and approval of a pharmaceutical product through regulatory authorities. Unfortunately, these topics have slowly lost their emphasis over the past several years in graduate courses taught in our colleges of pharmacy.

The eleven chapters in Part I cover preformulation topics, wherein the physical and chemical properties of a drug substance, along with its stability and interactions with excipients and the biological aspects of the formulations, are discussed in detail. These and other preformulation topics covered in this section outline the basic properties that fingerprint both the characteristics of a drug substance and the properties of ingredients that must be considered when formulating a physically and chemically stable dosage form. Part II, Chapters 12 through 20, cover product design topics. The regulatory science aspects of drug development are covered in Chapters 21 through 31. This is yet another area that has received minimal attention in graduate schools.

For both the academic and industrial scientist, as well as graduate students whose research is focused in this field, the authors have emphasized important aspects of materials science and processing that must be addressed for a successful product introduction following approval through regulatory authorities. The chapters in this book flow extremely well and provide very useful information not only to undergraduate and graduate students in pharmaceutics, pharmacy, materials sciences, and engineering, but also to faculty and industrial scientists in these disciplines.

James W. McGinity, Ph.D.

Professor of Pharmaceutics
The University of Texas at Austin

PREFACE

The idea for this book was born out of the authors' desire to create a textbook to be used for several courses in our pharmaceutical science B.S./M.S. curriculum and cooperative B.S./M.S. engineering/pharmaceutical sciences program approved by North Carolina State University, Raleigh, North Carolina and Campbell University, Buies Creek, North Carolina. The book will also be used as part of the College of Pharmacy & Health Sciences Pharm.D. program. The book's theme and scope focus on the application of the principles of physical pharmacy, product design, and regulatory science and how they relate in an intricate web to produce effective dosage forms that deliver drugs to their site of action. Currently, there is a critical shortage of pharmaceutical scientist specialists in product design and related technologies. Historically, most pharmaceutical scientists were educated in pharmacy schools. These programs integrated biology, pharmacology, chemistry, mathematics, physics, and materials science with one overarching goal: drug delivery to treat the human condition. A majority of the Ph.D. pharmaceutical scientists were educated as B.S. pharmacists. Over the past 30 years, pharmacy education has become more drug therapy focused and less drug delivery oriented. Federal funding has also followed this trend, leaving unprecedented shortages of academic pharmaceutical scientists. In addition, the shift in pharmacy education emphasis has required that the pharmaceutical industry hire chemists, biologists, and engineers who do not have the benefit of the integrated educational program that was once offered by schools of pharmacy. These employees are trained (not educated) on the job, although some schools of engineering are trying to fill some of the educational void.

The novel approach of this book is, as much as possible, to integrate international harmonized pharmaceutical development regulatory guidelines and requirements with the science and technology of pharmaceutical product design. New regulatory guidelines, such as quality by design, design space analysis, process analytical technology, polymorphism characterization, blend sample uniformity, stability protocols, and the biopharmaceutical classification system are integrated throughout the text. In Part I, we present the fundamentals of physical pharmacy and preformulation as they apply to pharmaceutical dosage form design. Topics such as thermodynamics, drug solubility, drug stability, rheological aspects of formulation, interfacial science, bioavailability, and others are covered in this part. Other chapters cover basic mathematical, statistical, and design-of-experiment concepts.

In Part II, we elaborate on the complex multifactorial process that brings together drug delivery to treat a human condition with formulation, manufacturing process, and container closure system design. The inextricable interrelationships among the formulation, the process, and the container closure system are emphasized by integrating each of these product design features into a single dosage-form chapter. Unification of appropriate preformulation and regulatory science applications is also highlighted. A similar format is incorporated for most chapters: an introduction that discusses the relevant anatomical and bodily function that affect drug delivery, advantages and disadvantages of the product, formulation design that examines dosage-form-specific preformulation, excipient compatibility, formulation development; process design, relevant process analytical technologies, pertinent scale-up models and practices, container closure system design incorporating critical patient and product considerations, risk management, in-process and final product attribute tests, and new drug application stability assessment programs. Most chapters include extensive reference appendixes of functional excipients, their compendial status, and usage levels. Other reference appendixes include surfactant hydrophile-lipophile balance (HLB) values, oil-required HLB values, sequestering agent stability constants, lyophilization bulking agents, and eutectic and collapse temperatures.

Part III covers regulations as specified by the U.S. Food and Drug Administration (FDA), European Medicines Agency, and other international regulatory agencies. This part provides a broad spectrum of topics from compliance requirements (current good manufacturing and good laboratory practices, and others), International Conference on Harmonization and other global harmonization initiatives, the investigational new drug (IND) and new drug application (NDA) phase-appropriate new drug development process, pre- and postapproval processes [INDs, NDAs, abbreviated NDAs, and drug master files], accelerated approval and initiatives for orphan and pediatric drug development, post–drug approval activities, quality system controls, commissioning and qualification (of facilities, equipment, analytical instruments, and test methods, among others), regulatory requirements for all facets of extemporaneous compounding (from handling of prescription for compounding to patient counseling), recommendations for conducting and reporting results of nonclinical and clinical safety and toxicology studies, barriers and benefits of pharmacogenomics studies, to the most recent FDA initiative on regulatory science.

In summary, the book introduces a fresh approach to presenting industrial pharmacy by combining physical pharmacy, product design, and regulatory science issues in a single compendium. The authors hope that the integrated perspective presented will be useful for undergraduate, graduate, and professional pharmacy students and will provide pharmaceutical scientists with a reference resource. The authors will also greatly appreciate feedback and comments that lead to improvements in the book.

Acknowledgments

The authors wish to thank Bob Stagner for reviewing Chapters 12 through 20, 23, 25, and 26 and for providing constructive suggestions, and Paul Johnson for reviewing and providing helpful suggestions on chapters 23 and 29. The authors are also indebted to Drs. Keith Johnson and Larry Gatlin for their expert advice on

Chapters 16 and 17, respectively. Special thanks go to Scott Staton for his tireless effort in preparing the excellent illustrations.

The authors appreciate the unwavering support of their department chair, Dr. Emanuel Diliberto. We also acknowledge the help of our graduate students in this endeavor.

alachi@campbell.edu   *Antoine Al-Achi*
guptam@campbell.edu   *Mali Ram Gupta*
stagnerw@campbell.edu   *William Craig Stagner*

PART 1

PREFORMULATION

CHAPTER 1

MATHEMATICAL CONCEPTS

1.1 INTRODUCTION

Pharmacy as a profession is art, business, and science. The science of pharmacy, also known as *pharmaceutical science*, requires knowledge of mathematics. Experimentation in pharmaceutical science produces quantitative measures with specific values. Handling these measures mathematically depends on how to apply rules to define them. In turn, these definitions of measures lead to a description of experimental entities. For example, to define a solution's pH, a pH meter is normally used in the measurement. Knowledge of the pH value can define the concentration of hydronium ions present in the solution. The relationship that allows transformation of the pH value to a concentration term is a mathematical expression known as *Sörensen's equation*:

$$\text{pH} = -\log[\text{H}_3\text{O}^+] \tag{1.1}$$

If the pH meter reads pH 10.8 for the solution, equation (1.1) may be used for the determination of $[\text{H}_3\text{O}^+]$:

$$10.8 = -\log[\text{H}_3\text{O}^+]$$

$$[\text{H}_3\text{O}^+] = 1.58 \times 10^{-11} \text{ M}$$

Thus, the concentration of hydronium ions in solution was computed from equation (1.1) by mathematical manipulation employing the rules of logarithms.

Mathematical rules can also aid a pharmaceutical scientist in describing the blood profile following administration of a drug in patients. Following intravenous administration of a drug, the drug is placed in circulation and achieves its highest concentration immediately following injection. The concentration of the drug decreases thereafter through distribution to tissues and via metabolic pathways. The drug disappearance from the circulation over time may be described by an exponential function following the general expression

$$C_{\text{blood}} = C_{\text{initial}}e^{-kt} \tag{1.2}$$

where C_{blood} is the drug concentration at time t, C_{initial} the initial concentration of the drug in the blood immediately following administration, and k the elimination

Integrated Pharmaceutics: Applied Preformulation, Product Design, and Regulatory Science,
First Edition. Antoine Al-Achi, Mali Ram Gupta, William Craig Stagner.
© 2013 John Wiley & Sons, Inc. Published 2013 by John Wiley & Sons, Inc.

rate constant. Equation (1.2) can be made linear by converting it to its logarithmic form:

$$\ln C_{\text{blood}} = \ln C_{\text{initial}} - kt \tag{1.3}$$

The transformation of equation (1.2) to equation (1.3) requires knowledge of the rules of logarithms. *Pharmacokinetics*, which is the study of drug absorption, distribution, and elimination, uses these mathematical manipulations of data to improve patients' therapeutic outcomes. Equation (1.3) describes a linear relationship between the natural logarithm of drug blood concentration and time. This linear relationship is not only important in pharmacokinetics but its applications are well utilized in physical pharmacy applications.

In this chapter we cover the major important mathematical concepts that pharmaceutical scientists utilize in their studies. With the advancement of computer technology, many of these mathematical applications are handled by a computer software program or even by a basic scientific calculator.

1.2 THE SIMPLE LINEAR RELATIONSHIP

When two variables x and y vary with each other linearly, their function may be written as

$$y = a + bx \tag{1.4}$$

where y is the dependent variable and x is the independent variable. The slope of the line is b and the y-intercept is a. The coefficient b can be positive or negative in value. When b is positive, an increase in x results in an increase in y. Conversely, if b is negative, an increase in x produces a decrease in y. Although equation (1.4) can be found manually, the usual method is to input the y and x values into a computer program to generate a linear equation. For example, the following data were obtained from a spectrophotometric experiment measuring the concentration of aspirin in solution:

Concentration (mg/mL)	Absorbance
0.0325	0.003
0.0650	0.006
0.1250	0.011
0.2500	0.023
0.5000	0.049

To obtain the linear relationship between concentration and absorbance, a simple scientific calculator may be used. The following equation is obtained:

$$\text{absorbance} = -0.000771 + 0.098565 \times \text{concentration (mg/mL)} \tag{1.5}$$

Comparing equation (1.4) to equation (1.5), the absorbance value is the dependent variable and the concentration is the independent variable. The y-intercept is negative in this case, and statistically speaking, is not different from zero. The coefficient b is positive, which is expected from relationships that represent

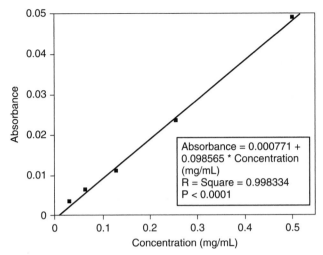

FIGURE 1.1   Positive linear relationship between the concentration of aspirin in solution and absorbance readings. Data points are experimental values, and the solid line is the best-fit line for the data.

Beer's law (Figure 1.1). It is important always to check whether or not the mathematical relationship adheres to the scientific norms. In using equation (1.5), the concentration of aspirin in an unknown solution may be estimated. For example, if the absorbance of an unknown solution of aspirin is 0.015, the estimated concentration of aspirin in solution is

$$0.015 = -0.000771 + 0.098565 \times \text{concentration (mg/mL)}$$

$$\text{concentration (mg/mL)} = 1.6$$

Note that the y-intercept of -0.000771 was used in estimating the concentration.

Based on Beer's law, the absorbance value is the logarithm of the ratio I_0/I, where I_0 and I are the intensities of the incident and emitted light, respectively. The absorbance value is logarithmic; however, the spectrophotometer readily calculates its value and the operator does not need to handle logarithmic calculations. Equation (1.5) follows the general format of *Beer's law:*

$$\text{absorbance} = \text{absorptivity} \times \text{pathlength of light} \times \text{concentration} \qquad (1.6)$$

Comparing equation (1.4) to equation (1.6), the theoretical y-intercept value must be zero, and the coefficient b is absorptivity × pathlength of light. The pathlength of light is predetermined by the instrument's tube holder (normally, 2 cm in length), and thus the slope of line b allows calculation of the absorptivity value, which is an important physical characteristic of a drug. (The absorptivity value varies with the solvent, the temperature, and the wavelength being used in the experiment.) Under the conditions of this experiment, the absorptivity may be calculated as follows,

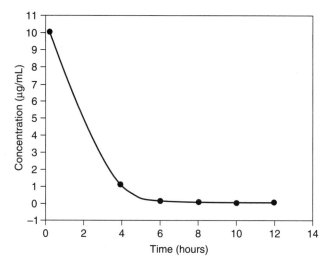

FIGURE 1.2   Exponential decrease in drug blood concentration *vs.* time.

assuming that the pathlength was 2 cm:

$$b = \text{absorptivity} \times \text{pathlength of light}$$
$$0.098565 = \text{absorptivity} \times 2$$
$$\text{absorptivity} = 0.049 \text{ mL}/(\text{mg} \cdot \text{cm})$$

For some linear relationships, the slope of the line is negative. For example, equation (1.3) has a negative slope. The negative slope of equation (1.3) indicates that concentration of the drug in blood decreases with time. It should be emphasized, however, that the linear relationship is between the logarithm of the drug concentration and time, not the concentration of the drug vs. time. Thus, when presented with data such as drug concentration vs. time (Figure 1.2), convert the drug concentration to logarithmic terms (natural or base 10) and then plot ln (drug blood concentration)] vs. time. The resulting graph is a straight line (Figure 1.3).

Time (h)	Concentration (µg/mL)	ln (concentration)
0.25	10	2.30258509
4	1	0
6	0.2	−1.6094379
8	0.1	−2.3025851
10	0.08	−2.5257286
12	0.05	−2.9957323

The equation that relates the drug blood concentration vs. time is presented as

$$\ln(\text{concentration}) = 1.8561797 - 0.4538628 \times \text{time (h)} \tag{1.7}$$

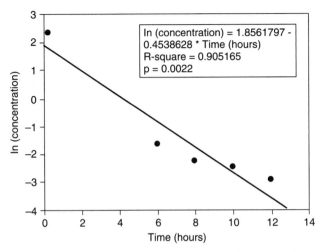

FIGURE 1.3 Linear relationship of the natural logarithm of drug blood concentration *vs.* time.

From this equation, the first-order rate constant for elimination may be calculated from the slope:

$$\text{slope} = -04538628 = -k_{el}$$

Therefore,

$$k_{el} = 0.454 \text{ h}^{-1}$$

The value of k_{el} indicates that 45.4% of the drug concentration remaining is eliminated each hour.

1.3 EXPONENTIAL RULES

In physical pharmacy expressions, many of the calculations require handling terms with exponents. The rules for handling exponents are (Stein, 1977; Anton, 1980):

1. Any number raised to the power of zero results in a value of $1 : x^0 = 1$
2. Any number raised to the power of 1 will equal its value: $x^1 = x$
3. $x^n \times x^m = x^{n+m}$
4. $x^n / x^m = x^{n-m}$
5. $1/x^n = x^{-n}$
6. $(x^n)^2 = x^{2n}$

In preparing buffer solutions, the ability of the resulting solution to resist a change in its pH is known as the *buffer capacity*. In calculating the buffer capacity value, the hydronium ion concentration, the acid dissociation constant, and the total buffer concentration must be known. Assuming that the total buffer concentration was 1 M, $[H_3O^+] = 10^{-4}$ M, and $K_a = 1.47 \times 10^{-4}$, we'll estimate the buffer capacity value. The equation for calculating the buffer capacity is (Martin et al., 1983)

$$\text{buffer capacity} = 2.303C \frac{K_a[H_3O^+]}{\{K_a + [H_3O^+]\}^2}$$

$$= (2.303)(1) \frac{(1.47\text{x } 10^{-4})(10^{-4})}{[(1.47 \times 10^{-4}) + (10^{-4})]^2}$$

$$= 0.56$$

The higher the value of the buffer capacity, the higher the resistance of the buffer is to a change in pH.

1.4   LOGARITHMIC RULES

For most pharmaceutical applications, the *logarithmic function* serves to convert a nonlinear relationship to a linear one. Linearity allows easier calculations for coefficients from a mathematical model. Logarithmic functions are thought of as exponential equations; thus, $y = x^z$ translates into $z = \log_x y$ ($\log_x$ = logarithm of base x). There are two important logarithm symbols: log and ln; *log* is the logarithm to the base 10, whereas *ln* denotes a natural logarithmic function to the base e ($e = 2.71828\ldots$). When handling logarithmic terms in an equation, the following mathematical rules apply (Stein, 1977; Anton, 1980):

1. $\ln x = 2.303 \log x$
2. $\log(x \times z) = \log x + \log z$
3. $\log(x/z) = \log x - \log z$
4. $\log x = z$ or $x = 10^z$
5. $\ln x = z$ or $x = e^z$
6. $\log x^z = z \log x$
7. $\ln e = 1$

For example, consider equation (1.7) and convert the equation to its log form of base 10:

$$\ln(\text{concentration}) = 1.8561797 - 0.4538628 \times \text{time (h)} \qquad (1.7)$$

$$2.303 \log(\text{concentration}) = 1.8561797 - 0.4538628 \times \text{time (h)}$$

$$\log(\text{concentration}) = 0.806 - 0.197 \times \text{time (h)} \qquad (1.8)$$

Equations (1.7) and (1.8) are identical mathematically, and they produce the same value for the elimination rate constant. In using equation (1.8) to calculate k, the slope of the equation is used:

$$\text{slope} = -0.197 = \frac{-k}{2.303}$$

Therefore, $k = (2.303) \times (0.197) = 0.454$ h^{-1}, which is the same value as that obtained using equation (1.7).

In Chapter 8, concepts related to the shelf-life determination of drug products are introduced. One area of concern is the effect of a change in storage temperature on the stability of a drug. One equation uses the logarithm of the ratio of a degradation rate constants at two different temperatures: $\log(k_2/k_1)$. The following calculations illustrate the use of logarithmic rules in solving such equations:

$$k_1 = 0.034 \text{ min}^{-1}$$

$$k_2 = \text{unknown}$$

$$\log \frac{k_2}{k_1} = 0.842$$

To find k_2 we use,

$$\log \frac{k_2}{0.034} = 0.842$$

$$\log k_2 - \log(0.034) = 0.842$$

$$\log k_2 = 0.842 + \log(0.034) = -0.6265$$

$$k_2 = 10^{-0.6265} = 0.236 \text{ min}^{-1}$$

Since drug degradation occurs with faster rates at higher temperatures than at lower ones, k_2 must occur at a temperature much higher than that observed with k_1.

Methods for the sterilization of pharmaceutical units may be divided into thermal and nonthermal. Thermal methods utilize heat as a means of achieving the destruction of microorganisms. The rate at which microbes get killed may be described by

$$M = M_0 e^{-kt} \tag{1.9}$$

where M_0 and M are the initial microbial population and that at time t, respectively. The rate constant for the process of microorganism killing is k. Equation (1.10) can be rearranged to read

$$\frac{M}{M_0} = e^{-kt}$$

Taking the natural logarithm of both sides of the equation results in

$$\ln \frac{M}{M_0} = \ln(e^{-kt}) = -kt \tag{1.10}$$

Equation (1.10) allows calculations of the rate constant if M_0 and M are known. For example, the initial population of spores was 10^4 mL^{-1}; following 60 s of exposure to a temperature of 120°C, the population was reduced to 10 mL^{-1}. Calculate the rate constant:

$$\ln \frac{10}{10^4} = -k(60 \text{ s})$$

$$k = 0.115 \text{ s}^{-1}$$

This means that 11.5% of the remaining population of microorganisms is destroyed every second at 120°C.

1.5 DIFFERENTIAL EQUATIONS

Differential equations may be employed to solve rate-related problems such as when studying drug degradation at a given rate and order of reaction. *Integration* is viewed as summation and is the opposite of differentiation. In pharmaceutical sciences, the notion of *differentiation* is commonly applied to topics that involve rates, such as drug degradation over time, drug diffusion through a membrane over time, and the rate by which a drug disappear's from circulation following administration. The general format for the rate using a differential format is dA/dt, where d indicates a small quantity. Thus, dA/dt indicates that a small change in A occurs for every small change in t. When applied to a differential equation, integration (symbol $= \int$) "sums up" all these small changes, thus, integration is considered to be a summation process.

The solutions for some important differential equations in physical pharmacy can be summarized as follows (Stein, 1977; Anton, 1980):

1. *Zero-order*:

$$\frac{dA}{dt} = -k_0 A^0$$

$$dA = -k_0 dt$$

$$\int_0^A dA = -k_0 \int_0^t dt \qquad (1.11)$$

$$A - A_0 = -k_0(t - t_0)$$

$$\text{given } t_0 = 0$$

$$A = A_0 - k_0 t$$

2. *First-order*:

$$\frac{dA}{dt} = -k_1 A^1$$

$$\frac{dA}{A} = -k_1 dt$$

$$\int_0^A \frac{dA}{A} = -k_1 \int_0^t dt \qquad (1.12)$$

$$\ln A - \ln A_0 = -k_1(t - t_0)$$

$$\text{given } t_0 = 0$$

$$\ln A = \ln A_0 - k_1 t$$

3. *Second-order*:

$$\frac{dA}{dt} = -k_2 A^2$$

$$\frac{dA^2}{A} = -k_2 dt$$

$$\int_0^A \frac{dA^2}{A} = -k_2 \int_0^t dt$$

$$-\frac{1}{A} - \frac{1}{A_0} = -k_2(t - t_0) \qquad (1.13)$$

given $t_0 = 0$

$$\frac{1}{A} = \frac{1}{A_0} + k_2 t$$

4. *Partial derivatives* (Adamson, 1969): When a variable such as y is a function of two or more other variables $(q, x, w, \ldots)$, the notion of partial derivatives (∂) is applicable. Thus, we express y with respect to q, x, as follows (the subscripts indicate that these variables are held constant during the differentiation calculation):

$$\left(\frac{\partial y}{\partial q}\right)_{x,w}$$

$$\left(\frac{\partial y}{\partial x}\right)_{q,w}$$

$$\left(\frac{\partial y}{\partial w}\right)_{q,x}$$

For example, the diffusion coefficient (D) is a function of temperature (T), viscosity of solution (η), and radius of particles (r). A is Avogadro's number and R is the gas constant:

$$D = \frac{RT}{\pi \eta r A} \qquad (1.14)$$

Thus, to write D as a function of T, η, and r, the following expressions may be used:

$$\left(\frac{\partial D}{\partial T}\right)_{\eta,r}$$

$$\left(\frac{\partial D}{\partial \eta}\right)_{T,r}$$

$$\left(\frac{\partial D}{\partial r}\right)_{T,\eta}$$

When all variables change simultaneously,

$$dD = \left(\frac{\partial D}{\partial T}\right)_{\eta,r} dT + \left(\frac{\partial D}{\partial \eta}\right)_{T,r} d\eta + \left(\frac{\partial D}{\partial r}\right)_{T,\eta} dr$$

Assuming that the temperature is held constant, D is a function of viscosity, and the radius of particles, $D = f(\eta, r)$. From equation (1.14),

$$D = \frac{RT}{\pi A}\eta^{-1}r^{-1}$$

$$= (\text{constant})\,\eta^{-1}r^{-1} \qquad (1.15)$$

$$D_\eta = -(\text{constant})\,\eta^{-2}r^{-1}$$

$$D_r = -(\text{constant})\,\eta^{-1}r^{-2}$$

where D_η is the partial derivative of D with respect to the viscosity of the solution, and D_r is the partial derivative of D with respect to the radius of the particles.

5. *Second-order partial derivatives*: With the temperature held constant, equation (1.15) can be written as a function of the viscosity of solution and the particle radius:

$$D = f(\eta, r)$$

$$\frac{\partial(\partial f/\partial \eta)}{\partial r} = \frac{\partial^2 f}{\partial \eta \partial r} \tag{1.16}$$

Equation (1.16) is the form of a second-order partial derivative.

For example, given D at 25°C, we have

$$D = 31.31 \times 10^{-23}\eta^{-1}r^{-1}$$

To find $\partial^2 D/(\partial \eta \partial r)$ we write

$$\frac{\partial^2 D}{\partial \eta \, \partial r} = \frac{\partial(\partial D \partial r)}{\partial \eta} = \frac{\partial(-(\text{constant})\, \eta^{-1}r^{-2})}{\partial \eta}$$

$$= (\text{constant})\, \eta^{-2}r^{-2} = 31.31 \times 10^{-23}\eta^{-2}r^{-2}$$

In thermodynamics (see Chapter 2), a change of volume of an ideal gas at a constant temperature does not result in a change in the internal energy of the system. Thus,

$$\left(\frac{\partial E}{\partial V}\right)_T = 0 \tag{1.17}$$

In addition, from the *gas law expression*, $PV = nRT$, and at a given constant pressure, the following partial differential equation may be written for the change of volume with a change in temperature:

$$\left(\frac{\partial V}{\partial T}\right)_P \tag{1.18}$$

Equation (1.18) and the gas law expression lead to

$$\left(\frac{\partial V}{\partial T}\right)_P = \frac{nR}{P} \tag{1.19}$$

And for 1 mol of an ideal gas ($n = 1$), $(\partial V/\partial T)_P = R/P = \text{constant}$ (Adamson, 1969).

1.6 EXPANDING AND REDUCING FORMULAS

Pharmacists and pharmaceutical scientists are often asked to prepare volumes or quantities different from those given in a prescription or formula. When faced with either expanding or reducing formulas, a simple proportion method is sufficient to solve the problem. For example, a technician was asked to prepare 200 mL of the following buffer solution:

K_2HPO_4	0.50 g
KH_2PO_4	0.35 g
Purified water, enough to make	500 mL

Using 200/500 as a proportion, each value in the formula is multiplied by this ratio:

$$(0.50 \text{ g})(200/500) = 0.20 \text{ g}$$
$$(0.35 \text{ g})(200/500) = 0.14 \text{ g}$$
$$(500 \text{ mL})(200/500) = 200 \text{ mL}$$

In preparing the buffer solution, the technician mixes 0.20 g of K_2HPO_4 and 0.14 g of KH_2PO_4 and dissolves them in enough purified water to make 200 mL of solution. If the volume of the solution to be prepared were 1 L instead of 200 mL, the calculations above would be repeated using the ratio 1000/500 in the calculations.

Another possible use formula expansion or reduction is to be asked to calculate the amount of solute needed for a given volume of solution. Assume that a pharmacist was asked to prepare 500 mL of the following solution:

Na^+	8 mEq
Purified water, enough to make	100 mL

Following the method described above, multiply the quantities in the formula by the ratio 500/100:

Na^+	$(8)(500/100) = 40$ mEq
Purified water, enough to make	$(100)(500/100) = 500$ mL

If the pharmacist were to use NaCl (molecular mass = 58.44 g/mol) as the salt to provide sodium ions in solution, 40 mEq of NaCl equals (40)(0.05844) or 2.34 g. Thus, the pharmacist dissolves 2.34 g of NaCl in enough purified water to make 500 mL of solution.

REFERENCES

Adamson WA. *Understanding Physical Chemistry*, 2nd ed. Menlo Park, CA: W.A. Benjamin; 1969.

Anton H. *Calculus*. New York: Wiley; 1980.

Martin A, Swarbrick J, Cammarata A. Solubility and distribution phenomena. In: *Physical Pharmacy: Physical Chemical Principles in the Pharmaceutical Sciences*, 3rd ed. Philadelphia: Lea & Febiger; 1983. pp. 272–313.

Stein KS. *Calculus and Analytical Geometry*, 2nd ed. New York: McGraw-Hill; 1977.

GLOSSARY

Beer's law

States that the absorbance value of a drug solution is directly and linearly proportional to the concentration of the drug in solution. The Beer's law equation is given as Abs = As $\times$ l $\times$ C, where Abs is the absorbance, As the absorptivity, l the pathlength, and C the concentration of the drug in solution.

Buffer capacity

An indicator of the degree of resistance that a buffer has toward a change in its pH upon challenging it with acids or bases. The higher the value of the buffer capacity, the higher the resistance of the buffer is to a change in pH.

Differential equations

May be employed to solve rate-related problems, such as when studying drug degradation at a given rate and order of reaction. The general format for the rate using a differential format is dA/dt, where d indicates a small quantity. Thus, dA/dt indicates that a small change in A occurs for every small change in t.

Gas law equation for an ideal gas

$PV = nRT$, where P is the pressure, V is the volume, n the number of moles, R the gas constant, and T the temperature in kelvin.

Integration

Viewed as summation; the opposite of differentiation. When applied in a differential equation, integration ($\int$) "sums up" all these small changes.

Logarithmic functions

Thought of as exponential equations; thus, $y = x^z$ translates into $z = \log_x y$ ($\log_x$ = logarithm of base x).

Microbial population estimation for sterilization processes

The rate at which microbes get killed may be described by the equation $M = M_0 e^{-kt}$, where M_0 and M are the initial microbial population and that at time t, respectively. The rate constant for the process of microorganism killing is k.

Partial derivatives

When a variable such as y is a function of two or more other variables $(q, x, w, \ldots)$, the notion of partial derivatives (∂) is applicable.

pH

The pH scale is used to measure the concentration of hydronium ions $[H_3O^+]$ in solution.

CHAPTER *2*

THERMODYNAMICS

2.1 INTRODUCTION

Physical matter is governed by the laws of *thermodynamics*. The science of thermodynamics concerns itself with the transformation of heat into mechanical work and the opposite transformation between the two quantities. It is also concerned with transformations among the various forms of energy (i.e., kinetic, potential, chemical, thermal, nuclear, etc.)

The *gas law* describes an *ideal gas* by the *state equation* $PV = nRT$ (where P is the pressure, V the volume, n the number of moles, R the gas constant, and T the temperature). For the transformation of an ideal gas under constant temperature (*isothermal transformation*), the product PV is constant. The *work of isothermal expansion* of an ideal gas can be written as $W = nRT \log(V_2/V_1) = nRT \log(p_1/p_2)$, where the subscripts 1 and 2 represent the initial and final conditions. The *density of an ideal gas* (ρ) of mass m can be expressed as $\rho = m/V = Mp/RT$, where M is the molecular mass of the gas in grams per mole. For a mixture of gases confined to a rigid container with volume V maintained at a constant temperature, the total pressure exerted by all the gases equals the sum of all the partial pressures of the gases. The *partial pressure* is defined as the pressure exerted by a single gas at the same temperature when it occupies the same volume of the mixture of gases. This statement is known as *Dalton's law of partial pressure* (Fermi, 1936).

2.2 THE ZEROTH LAW OF THERMODYNAMICS

The statement of the *zeroth law of thermodynamics* reflects the recognition that when placed in physical contact with each other, systems reach an equilibrium point when they have the same temperature. It is also understood that based on this law, once the system reaches a point of thermal equilibrium, its other physical properties will remain unchanged as well. This law is the basis of conducting experiments under *isothermal* conditions. For example, immersing a glass beaker containing a solution in a constant-temperature water bath, and observing the temperature of the solution reaching the same temperature as that of the water bath, is a direct application of the zeroth law. Clinical applications of this law include measuring

Integrated Pharmaceutics: Applied Preformulation, Product Design, and Regulatory Science,
First Edition. Antoine Al-Achi, Mali Ram Gupta, William Craig Stagner.
© 2013 John Wiley & Sons, Inc. Published 2013 by John Wiley & Sons, Inc.

a patient's temperature by placing an oral thermometer under the patient's tongue or a heating pad on the patient's skin for a warming effect.

2.3 THE FIRST LAW OF THERMODYNAMICS

The total energy in a system can be thought of as the sum of energies of all the individual particles in the system. If a system is made up of n particles, the total internal energy of the system can be estimated by

$$E = \sum n_i e_i \qquad (2.1)$$

$$n = \sum n_i \qquad (2.2)$$

where E is the total internal energy of a system, n_i the number of particles of the ith component, and e_i the energy of the ith component. Exchanges between the system and its environment (surroundings) can occur in term of molecular or energy exchange, with both exchanges being relatively small. When these exchanges occur, the *change* in the internal energy can be estimated from (Novak, 2003)

$$dE = \sum dn_i e_i + \sum n_i de_i \qquad (2.3)$$

The combined equations (2.1), (2.2), and (2.3) represent the first law of thermodynamics in its molecular format (Novak, 2003).

In general, the statement of the *first law of thermodynamics* is (Wood and Battino, 2001)

$$dE = dQ - dW \qquad (2.4)$$

where dQ is the differential heat absorbed by the system and dW is the differential work done by the system. [Note that the values dQ and dW are dependent on the path the system is taking and therefore their cyclic integral is not zero, whereas that for dE is zero (Wood and Battino, 2001).] Equation (2.4) can also be written as (Manninen, 2004)

$$\Delta E = q - w \qquad (2.5)$$

Equation (2.5) is applicable only to closed systems as opposed to open systems. A *closed system* is one that can exchange heat with its surroundings, but not matter. If a system exchanges matter with its surroundings, the system is an *open system* (Manninen, 2004). The heat exchanged over time for an open system can be calculated by

$$\frac{\partial Q}{\partial t} = \frac{\partial m}{\partial t} \left(\frac{\Delta H}{MW} \right) \qquad (2.6)$$

where $\partial Q / \partial t$ is the change in heat per unit time, $\partial m / \partial t$ the change in mass per unit time, and $\Delta H / MW$ the ratio of enthalpy over the molecular mass of the substance. For example, in the case of water and sublimation, the molecular mass is 0.018 kg/mol and ΔH is the enthalpy for sublimation (Beaty, 2006).

The statement of the first law of thermodynamics is an expression of conservation of energy (and matter) during a process or a transfer. Nothing is created or destroyed during any transfer; rather, it is simply *transformed* from one form into another (e.g., kinetic energy associated with motion is transformed to x-ray) (Bartell, 2001; DeMeo, 2001). In addition, according to *Einstein's equation* ($E = mc^2$, where m is the mass and; c is the speed of light in vacuum), matter itself is a form of energy (Bartell, 2001). It is important to emphasize that the statement of the first law does not consider the *direction* by which the process proceeds.

Another fundamental concept is *heat capacity* (C). It is defined as the amount of heat needed to raise the temperature of 1 mol of a compound by 1 degree. For example, the binding of several drugs to two forms of cyclodextrin (β and γ) was studied by Todorova and Schwarz (2007). The change in heat capacity for the binding process of flurbiprofen (an anti-inflammatory agent) to β and γ-cyclodextrin was found to be in the range -362 ± 48 and -25.1 ± 9.2 J/mol·K, respectively. The solvent water plays a key role in the binding of the sybstance to cyclodextrin structure. A larger reduction in the heat capacity is an indication of a greater rearrangement in water molecule organization during binding reactions. From these data, the authors concluded that the structure of water was reorganized when the drug bound to the β form and not to the γ form of cyclodextrin. Similar applications of these measurements are being used in understanding the binding reactions between DNA and a multitude of drugs, ranging from anticancer to antibiotic to antiviral agents (Haq, 2002). Moreover, the *affinity* of a drug to its receptor may be better explained by studying how a system responde to changes in temperature and the resulting estimations of the enthalpy (ΔH) and entropy (ΔS) changes (see Section 2.4) or "compensations" during the binding process (Raffa, 1999). In this context, Merighi et al. (2010) studied the binding of cannabinoids and their antagonist with their CB1 and CB2 receptors. The affinity of the agents to these receptors was elucidated by estimating the thermodynamic parameters ΔG (change in free energy: the energy available to the system to perform work), ΔH, and ΔS. Knowledge of these values helps define whether the affinity of the drug is enthalpy- or entropy-driven.

2.4   THE SECOND LAW OF THERMODYNAMICS

Entropy is a fundamental expression derived from the *second law of thermodynamics*. The mathematical expression for *entropy* is given by (Wood and Battino, 2001):

$$dS = \frac{dQ_{\text{rev}}}{T} \tag{2.7}$$

where dQ_{rev} is a differential quantity for the amount of heat absorbed by the system during a reversible change and T is the temperature in kelvin. Equations (2.5) and (2.7) may be combined to yield (Wood and Battino, 2001):

$$dE = TdS - PdV \tag{2.8}$$

Equation (2.8) introduces the terms pressure P and volume V, and it describes the change in energy as a function of the two independent variables entropy and volume (Wood and Battino, 2001).

For a natural irreversible spontaneous process, the entropy of the universe increases during that process (Spencer and Lowe, 2003). It can also be seen as a measure for the number of *distinguishable* arrangements in which the available energy can be allotted *fairly* among the various possible arrangements (Spencer and Lowe, 2003). Mathematically, the change of entropy may be estimated from the following expressions (Spencer and Lowe, 2003)

$$\Delta S = nR \ln \frac{V_{\text{final}}}{V_{\text{initial}}} \tag{2.9}$$

Equation (2.9) estimates the change of entropy at a constant temperature for a process involving a volume change (i.e., an expansion).

Whereas, the first law of thermodynamics allows complete conversion of heat into work, the second law prohibits such conversion (Bartell, 2001). [In reality, this may be possible if one considers an ideal gas as being expanded reversibly under isothermal conditions. However, such a system cannot practically produce all work, because as the system is retuning to its original state, some of its work converts to heat (Bartell, 2001).] This is important, because the laws of thermodynamics should be taken together when attempting to define a process.

The statement of the second law of thermodynamics describes the direction by which a process follows. For an irreversible spontaneous and natural process, its direction always proceeds toward a state of *more randomness* or an *increase* in entropy (Manninen, 2004), whereas a reversible process is associated with $\Delta S_{\text{universe}} = 0$. For a given *system*, the change of entropy during any system change can be negative, zero, or positive.

2.5 THE THIRD LAW OF THERMODYNAMICS

The *third law of thermodynamics* is the law of absolute entropy. It states that the entropy of a perfectly crystalline solid is zero at absolute zero temperature (0 K). The reasoning behind this statement is that a perfect crystal has its molecules in perfect arrangement with no possible, molecular movement. Thus, the entropy of this system is at its lowest possible point, which is zero (Fermi, 1936).

2.6 POLYMORPHISM

Polymorphic crystals for the same chemical occur in nature and they share similar chemical properties, but they differ in their physical characteristics (i.e., density, viscosity, melting point, etc.) (Maiwald, 2006). *Polymorphism* arises from the different arrangements and/or confirmations of the molecules within the crystalline lattice. When a polymorph undergoes a phase transformation into a gaseous or liquid state, it loses its *uniqueness* with respect to molecular arrangements. Thus, all polymorphs behave similarly in a gas or liquid state, because polymorphism is strictly a solid-state phenomenon (Giron, 2005). Polymorphic forms for a given substance can be generated by *crystallization* steps, where the substance is dissolved in various solvent systems, and then evaporating the solvent to allow crystallization of the polymorphs. Another method of preparing polymorphs is by *crystallization from the melt*, where the melted material is supercooled to temperatures well below the melting point of the metastable form of the agent. A third method for polymorphic generation involves *removal of the solvent molecules from the solvate form* of the chemical, which can lead to a new crystal for the substance. *Grinding* the crystal form of the drug can result in partial or complete conversion of the crystals to an amorphous state (a noncrystalline state that resembles a supercooled liquid in the arrangement of its molecules; its structure does not have the same rigidity as that seen with crystalline solids; thus, amorphous structures tend to flow if subjected to stress over a prolonged period). Subsequent recrystallization of the amorphous solid can yield metastable polymorphic forms of the drug. Metastable crystals (metastable structures are unstable at a given temperature and pressure and tend to transform *slowly* over time to a stable form; metastable structures can be crystalline or amorphous in nature) can also be generated by *sublimation* provided that the substance is not highly susceptible to degradation at temperatures used during the sublimation process (Maiwald, 2006). Generally speaking, slow crystallization of the material produces thermodynamically stable forms, and faster processes yield metastable crystals. Raman spectroscopy appears to be the method of choice for solid-state analysis of the material for polymorph detection (Anderton, 2007).

Thermodynamically, polymorphs may be distinguished from each other by measuring the difference in their *Gibbs free energy* (Needham and Faber, 2003):

$$\Delta G = \Delta H - T \Delta S \tag{2.10}$$

where ΔG is the difference in Gibbs free energy (i.e., the amount of energy available to perform work; it is negative for a spontaneous natural irreversible processes), ΔH the difference in enthalpy (i.e., the amount of heat absorbed or dissipated), T the temperature in kelvin, and ΔS the difference in entropy. The differences in enthalpy and entropy are those observed between the two polymorphs. In turn, the difference in enthalpy is related to the difference in the internal energy (ΔE) and in the volume (ΔV) under constant pressure (P) (Needham and Faber, 2003):

$$\Delta H = \Delta E + P \Delta V \tag{2.11}$$

The difference in Gibbs free energy between the polymorphs can be estimated *approximately* from their solubility ratio (S_2/S_1) (Needham and Faber, 2003):

$$\Delta G \cong RT \ln \frac{S_2}{S_1} \tag{2.12}$$

where R is the gas constant. Interestingly, hydrate forms of the drug behave similar to the behavior of polymorphic forms because the crystalline structure of both forms depends on the degree of hydrogen bonding occurring within the crystal structure (Needham and Feber, 2003).

Differential scanning calorimetry (DSC) methods are commonly employed to prepare polymorphic forms of a drug using the heating and cooling conditions, or it can be used to monitor the presence of polymorphs in samples (Stodghill, 2010).

Enthalpy measurements are commonly carried out using DSC methodology. According to the Confederation for Thermal Analysis and Calorimetry, DSC is "a technique where the heat flow rate difference into a sample and reference material is measured" (Stodghill, 2010). A signal on the DSC chart (a thermogram) signifies whether the process is exothermic or endothermic. Interpretation of the peaks on the DSC thermogram requires some knowledge of the process. For example, melting of a solid requires absorption of heat by the material; thus, it is an endothermic process. Since a pure crystalline solid melts within a very narrow range of temperatures, the DSC peak is expected to be sharp in this case. In evaporation of a solvent, also an endothermic process, the DSC peak is expected to be broader than that for a crystal melting phenomenon (Stodghill, 2010). Other processes do not necessary show a peak signal on the DSC chart. For example, in the case of glass transition of a sample, a thermogram shows a "stepwise increase in the heat capacity (C_P)" (Stodghill, 2010). Another insturement for differentiating between polymorphs or crystalline material vs. the amorphous state is the *solution calorimeter* (SC). The SC method involves drying a sample and placing it in a 1-mL glass heat-sealed ampoule. The sample is then placed in its compartment and immersed in a calorimeter reaction vessel along with a stirrer, thermistor, and calibration heater (which is a silvered dewar flask containing a cooling solvent). The entire process is performed isothermally under 25°C. Following calibration of the instrument, the ampoule containing the sample is broken, allowing the sample to mix with the cooling solvent. A recorder attached to the instrument registers time vs. temperature plots for determining the type of process, endothermic or exothermic. The SC method relies on the fact that the difference between the heat of solution of two forms $(\Delta H^A - \Delta H^B)$ is independent of the solvent being used and equals the heat of transition (ΔH^T) between the two forms, A and B (Lindenbaum and McGraw, 1985). It should be noted, however, that each heat of solution varies depending on the solvent being used. For exothermic reactions, the more negative the heat of solution, the higher is the solubility of the active pharmaceutical ingredient (API) form in the solvent. Unlike DSC, SC does not require the sample to be heated through the phase transition, which may result in degradation of the API. Another method that is widely acceptable for polymorphic measurement is x-ray diffraction. This method is useful for initial determination of different morphology patterns; however, these differences may be due to variations in orientation rather than being related directly to crystal structure (Lindenbaum and McGraw, 1985). In addition to DSC and SC, polymorphism can be assessed and identified by Raman spectroscopy, Raman microscopy, magic-angle-spinning nuclear magnetic resonance spectroscopy, thermogravimetric analysis, thermomicroscopy, and dynamic vapor sorption methods (Maiwald, 2006).

2.7 PHYSICAL STABILITY OF CRYSTAL FORMS

Pursuant to the discussion above, crystalline forms of the same API can differ in their Gibbs free energy (G). In fact, for two crystalline forms (A and B) of the same API, the one with a smaller G is more stable at a given temperature and pressure (Variankaval and Sheth, 2007):

$$G_A < G_B \tag{2.13}$$

Thus, from equations (2.12) and (2.13), the more stable the crystal, the lower is its solubility.

To the extent that crystals from the same API have the same chemical form, the comparison above remains valid. However, once we begin comparing various crystalline forms of the API that differ in their chemical composition, other factors must be considered in the evaluation. A case in point is the comparison of a hydrous form of the API with an anhydrous crystal. In this situation, the water activity (a_w) is considered for comparison as well as the pressure and temperature (Variankaval and sheth, 2007):

$$API(solid) + n\,H_2O \rightleftharpoons API \cdot n\,H_2O(solid) \tag{2.14}$$

and the equilibrium constant for reaction (2.14) is

$$K_{eq} = \frac{a_w}{[H_2O]^n} \tag{2.15}$$

For example, from the solubility curve conducted at room temperature for two different forms (anhydrous and hemihydrate) of an API in hydroalcoholic solutions with varying ethanol concentrations, it was found that at a concentration of ethanol of 96.8% v/v, the solubility of both forms was 2.8 mg/mL. In theory, this represents thermodynamically stable conditions for the anhydrous and hemihydrate forms at saturation (Variankaval and Sheth, 2007).

2.8   SOLUBILITY

The solubility of APIs is an important parameter that often governs its bioavailability from a dosage form. Thermodynamics can explain whether or not a compound is capable of undergoing dissolution in an aqueous buffer system and how much of it is dissolved given knowledge of its physical structure. *Thermodynamic solubility* is defined as the concentration of the API in solution in equilibrium with a *specific* solid state, not *any* solid state. In this context we refer to the solid state as being stable crystalline, polymorphic, metastable, pseudopolymorphic (solvates or hydrates), or amorphous in nature. The solubility of the API (S) is related to the molar heat of dissolution ($\Delta H_{dissolution} = \Delta H_{solution}$) by the equation (Giron, 2005)

$$\log S = \text{constant} - \frac{\Delta H_{dissolution}}{R}\frac{1}{T} \tag{2.16}$$

where R is the gas constant and T is the temperature in kelvin. The stable crystalline solid form normally has a several fold-lower solubility profile than that of the other forms. It should be noted, however, that a metastable or amorphous form (a noncrystalline form; unlike the crystalline form, it does not exhibit a definite melting point) of the API may not be the best candidate for the final version of the formulation, as these forms are not stable enough to remain in their physical state over the shelflife of the product (Saal, 2010). Analytical methods are available for detecting the presence of unwanted forms of the API. For example, thermal analysis, solid-state nuclear magnetic resonance spectroscopy, or x-ray diffraction is used to detect the amorphous form of a drug. Raman spectroscopy, combined with chemometrics, is particularly useful for detecting very small amounts of amorphous material formed during processing. In tablet manufacturing it was found that exposure to moisture and holding time in the wet granulation were associated with production of the amorphous form of an API (Zhou et al., 2010). Equation (2.16) indicates that a plot of log S vs. $1/T$ is positively linear for exothermic reactions and is negatively linear for endothermic processes. Moreover, the lower the $\Delta H_{\text{solution}}$ value, the higher the predicted affinity of an API to the solvent molecules.

The relative solubility of an API between the crystalline and its amorphous form can be predicted by the following (Hancock and Parks, 2000):

$$\ln \frac{\sigma_T{}^a}{\sigma_T{}^c} = -\frac{RT}{\Delta G_T{}^{a,c}} \tag{2.17}$$

$$\Delta G_T{}^{a,c} = \Delta H_T{}^{a,c} - (T \Delta S_T{}^{a,c}) \tag{2.18}$$

$$\Delta H_T{}^{a,c} = \Delta H_f{}^c - (C_p{}^a - C_p{}^c)(T_f{}^c - T) \tag{2.19}$$

$$\Delta S_T{}^{a,c} = \Delta S_f{}^c - (C_p{}^a - C_p{}^c) \ln \frac{T_f{}^c}{T} \tag{2.20}$$

$$\Delta S_f{}^c = \frac{\Delta H_f{}^c}{T_f{}^c} \tag{2.21}$$

where the subscripts a and c are for the amorphous and crystalline forms of the API, respectively. The subscript f stands for fusion. The term $\sigma_T{}^a / \sigma_T{}^c$ represents the solubility ratio, T the temperature, and R the gas constant. $T_f{}^c$ is the melting point, C_p is the heat capacity at constant pressure, and H and S are the enthalpy and entropy, respectively. The molar heat of fusion ($\Delta H_f{}^c$) is in direct correlation with the bonding strength within the crystal arrangement. The higher value of the $\Delta H_f{}^c$, the higher the melting point of a substance, because stronger forces exist between molecules when the heat of fusion is high. The equations above may be used to estimate the relative solubility of different polymorphs as well (with $\Delta C_p \approx \Delta S_f \approx 0$). Although using these thermodynamic equations can result in overestimating the relative solubility term manyfold, they still point out that the amorphous forms have significantly higher solubility in water than those of their crystalline counterparts (Hancock and Parks, 2000).

REFERENCES

Anderton C. A valuable technique for polymorph screening. Am. Pharm. Rev. 2007;10(3):34, 36–40.

Bartell LS. Apparent paradoxes and instructive puzzles in physical pharmacy. J. Chem. Educ. 2001;78(8):1067–1069.

Beaty N. Lyophilization: heat and mass transfer. Am. Pharm. Rev. 2006;9(2):81–87.

DeMeo S. Making assumptions explicit: how the law of conservation of matter can explain empirical formula problems. J. Chem. Educ. 2001;78(8):1050–1052.

Fermi, E. *Thermodynamics*. New York: Dover; 1936.

Giron D. Polymorphism: thermodynamic and kinetic factors to be considered in chemical development, part 1. Am. Pharm. Rev. 2005;8(4):32, 34–35, 37.

Hancock BC, Parks M. What is the true solubility advantage for amorphous pharmaceuticals? Pharm. Res. 2000;17(4):397–404.

Haq I. Thermodynamics of drug–DNA interactions. Arch. Biochem. Biophys. 2002;403(1):1–15.

Lindenbaum S, McGraw SE. The identification and characterization of polymorphism in drug solids by solution calorimetry. Pharm. Manuf. 1985;2:27–30.

Maiwald M. Examining the range of methods available to derive numerous polymorphic forms. Am. Pharm. Rev. 2006;9(4):95–99.

Manninen AH. Is a calorie really a calorie? Metabolic advantage of low-carbohydrate diets. J. Int. Soc. Sports Nutr. 2004;1(2):21–26.

Merighi S, Simioni C, Gessi S, Varani K, Borea AP. Binding thermodynamics at the human cannabinoid CB1 and CB2 receptors. Biochem. Pharmacol. 2010;79(3):471–477.

Needham F, Faber J. Total pattern analysis using the new organic powder diffraction file: PDF–4/organic. Am. Pharm. Rev. 2003;6(4):10, 12.

Novak I. The microscopic statement of the second law of thermodynamics. J. Chem. Educ. 2003;80(12):1428–1431.

Raffa BR. (Extra) thermodynamics of the drug–receptor interaction. Life Sci. 1999;65(10):967–980.

Rosenthal JA. Demonstration of surface tension. J. Chem. Educ. 2001;78(3):332–333.

Saal C. Optimizing the solubility of research compounds: how to avoid going off track. Am. Pharm. Rev. 2010;13(4):12, 14–18.

Segers DJ, Zatz LJ, Shah PV. In vitro release of phenol from ointment formulations. Pharm. Technol. 1997;21(1):70, 72, 74, 76, 80–81.

Spencer JN, Lowe JP. Entropy: the effects of distinguishability. J. Chem. Educ. 2003;80(12):1417–1424.

Stodghill SP. Thermal analysis: a review of techniques and applications in the pharmaceutical sciences. Am. Pharm. Rev. 2010;13(2):29, 31–36.

Todorova AN, Schwarz PF. The role of water in the thermodynamics of drug binding to cyclodextrin. J. Chem. Thermodyn. 2007;39(7):1038–1048.

Variankaval N, Sheth RA. Physical stability of crystal forms: implications in drug development. Am. Pharm. Rev. 2007;10(6):96, 98–101.

Wood SE, Battino R. Explaining entropy pictorially. J. Chem. Educ. 2001;78(3):311–312.

Zhou Y, LoBello R, Cai C, Crane N, Poshni F, Porter WW III, Cain TY. Measurement of an amorphous form in a pharmaceutical drug product using FT-Raman spectroscopy. Am. Pharm. Rev. 2010;13(5):34–41.

GLOSSARY

API	Active pharmaceutical ingredient.
Dalton's law of partial pressure	For a mixture of gases confined to a rigid container with volume V maintained at a constant temperature, the total pressure exerted by all the gases equals the sum of all the partial pressures of the gases. The *partial pressure* is defined as the pressure exerted by a single gas at the same temperature when it occupies the same volume of the mixture of gases. This statement is known as *Dalton's law of partial pressure*.
Density of an ideal gas	The density of an ideal gas (ρ) of mass m can be expressed as $\rho = m/V = Mp/RT$, where M is the molecular mass of the gas in grams per mole.
DSC	Differential scanning calorimetry.
Einstein's equation	$E = mc^2$ (E = energy; m = mass; c = speed of light in vacuum.

First law of thermodynamics	The statement of the first law of thermodynamics is an expression of conservation of energy (and matter) during a process or a transfer. Nothing is created or destroyed during any transfer; rather, it is simply *transformed* from one form into another and is an expression of conservation of energy (and matter) during a process or transfer. Nothing is created or destroyed during a transfer; rather, it is simply *transformed* from one form into another.
Gas law for an ideal gas	Describes an ideal gas by the *state equation* $PV = nRT$ (where $P =$ pressure, $V =$ volume, $n =$ number of moles, $R =$ gas constant, and $T =$ temperature).
Gibbs free energy	The amount of energy available to perform work; it is negative for a spontaneous natural irreversible process: $\Delta G = \Delta H - T\Delta S$.
Heat capacity	The amount of heat needed to raise the temperature 1 mol of a compound by 1 degree.
Isothermal transformation	Transformation of an ideal gas under constant temperature.
Polymorphism	Polymorphic crystals for the same chemical occur in nature and have similar chemical properties, but they differ in their physical characteristics (i.e., density, viscosity, melting point, etc.) Polymorphism arises from different arrangements and/or confirmations of the molecules within the crystalline lattice.
SC	Solution calorimetry.
Second law of thermodynamics	Describes the direction by which a process follows. For an irreversible spontaneous and natural process its direction always proceeds toward a state of *more randomness* or an *increase* in entropy, whereas a reversible process is associated with $\Delta S_{universe} = 0$. For a given *system*, the change of entropy during any system change can be negative, zero, or positive.
Thermodynamic solubility	The concentration of a drug in solution in equilibrium with a *specific* solid state, not *any* solid state.
Thermodynamics	The science of thermodynamics concerns itself with the transformation of heat into mechanical work and the opposite transformation between the two quantities. It is also concerned with transformations among the various forms of energy (i.e., kinetics, potential, chemical, thermal, nuclear, etc.).
Third law of thermodynamics	States that the entropy of a perfectly crystalline solid is zero at absolute zero temperature (0 K).
Work of isothermal expansion of an ideal gas	The work of an isothermal expansion of an ideal gas can be written $W = nRT \log(V_2/V_1) = nRT \log(p_1/p_2)$, where the subscripts 1 and 2 represent the initial and final conditions.
X-ray diffraction	A useful method for the initial determination of different morphology patterns.
Zeroth law of thermodynamics	When placed in physical contact with each other, systems reach an equilibrium point when they have the same temperature.

CHAPTER 3

SOLUBILITY AND DISSOLUTION

3.1 INTRODUCTION

The primary purpose of *liquid dosage forms* in pharmacy practice is to serve as drug delivery devices for patients. *Solutions*, the main class of liquid preparations, are important dosage forms for all routes of administration. They can be made sterile for *parenteral* administration, *ophthalmic* applications, and for *irrigating* biological tissues during operations. A *solution's pH* can be adjusted to that of a biological fluid. Solutions can be colored, flavored or sweetened, or rendered aromatic by the addition of substances to enhance patient compliance. Perhaps their greatest usefulness is in administering drugs to children and elderly patients. In short, solutions lend themselves to easy manipulation to fit patient needs. From the point of view of a pharmacist, solutions offer a very easy way to deliver medications to patients as well as being simple to prepare and to handle. From the point of view of a compounding pharmacist, this ease translates into an eloquent dosage form that can be prepared with little effort. Many solutions can be made simply by adding a drug to a commercially available vehicle just prior to dispensing.

Other liquid dosage forms have advantages similar to those mentioned above; however, their methods of preparation can be more involved, although they can still be prepared within a reasonable time with moderate effort. Liquid dosage forms contain *pharmaceutical necessities* that may be unique to them; *stabilizing agents* to maintain the pH of the preparation, preservatives to keep the formulation microbial-free, *rheological agents* for ease of pouring and packaging, and substances added to the formulation to render the dosage form *thermodynamically* stable are just a few examples.

As with other solid or semisolid preparations, the *stability* of drugs in liquid dosage forms is an important consideration. When placed in solution, drugs become prone to various chemical reactions that can lead to their degradation. Thus, understanding the issues concerning stability is of utmost importance to formulation scientists. In addition, the scientist needs to determine an expiry date for the formulation. One needs to consider not only the chemical stability but also the *physical stability* of a preparation. Many physical stability problems arise from

Integrated Pharmaceutics: Applied Preformulation, Product Design, and Regulatory Science,
First Edition. Antoine Al-Achi, Mali Ram Gupta, William Craig Stagner.

thermodynamic instability in the preparation or from missing necessary supportive ingredients or from incorrect amounts of these ingredients being added to the formulation.

In this text we discuss various theories and principles that govern the art of preparing pharmaceutical liquid dosage forms (*preformulation*). Various methods for liquid dosage form preparations (*formulation*) are introduced as well.

3.2 CONCENTRATION UNITS

This topic is a necessary "evil" for any discussion concerning a pharmaceutical dosage form. Its importance stems from the fact that the *quantity* of a drug in a preparation is critical to a patient receiving the proper drug *dose*. The Arabs introduced the notion of dosage to medicine. Prior to that, medications were simply given in any quantity desired. Even today, some folklore treatments are given without mentioning the proper dosage. So when a liquid dosage form is being prepared, it is mandatory that its *composition* be known to all parties involved. The concentration units provide scientists with a way to describe the composition of a preparation and the amount of each of its ingredients.

Several concentration units are commonly used in pharmacy practice. In this chapter are cover the most important ones encountered in pharmacy practice. For the sake of simplicity describe a preparation containing two ingredients: a drug substance (D) and a formulation vehicle.

3.2.1 Percent by Weight or Percent Weight per Weight

The *percent weight per weight* (% w/w) is defined for D as the weight of D in grams in 100 g of preparation. Mathematically, this can be written

$$\%\text{w/w} = \frac{W_D}{W_T} \times 100 \tag{3.1}$$

Equation (3.1) defines % w/w as the ratio of the weight of D in grams to the total weight of a preparation (W_T) in grams multiplied by 100.

Example 3.1. Tetracycline hydrochloride (antibacterial, antiamebic, antirickettsial) is sometimes given to patients with milk or yogurt to diminish stomach upset in sensitive persons. The entire content of a tetracycline hydrochloride capsule (contains 250 mg of tetracycline HCl) was emptied and dissolved in a cup of yogurt. The patient was asked to consume the entire yogurt mixture. Express the concentration of tetracycline hydrochloride in yogurt as % w/w. One cup of yogurt measures 120 mL or 250 g.

Solution. Apply equation (3.1):

$$\%\text{w/w} = \frac{0.250 \text{ g}}{250 \text{ g}} \times 100 = 0.1\%$$

So every 100 g of yogurt contains 0.1 g of tetracycline hydrochloride.

Example 3.2. The patient in Example 3.1 consumed only one-third of a cup of yogurt containing tetracycline HCl. How many milligrams of the drug did he get?

Solution. One-third of a cup is (250 g/3 = 83.3 g). Since the yogurt contained 0.1 g of tetracycline HCl in every 100 g, the quantity of drug in 83.3 g is 0.083 g. Thus, the patient received 83.3 mg of tetracycline HCl.

3.2.2 Percent by Volume or Percent Volume per Volume

A *percent volume per volume* (% v/v) concentration is used when the solute drug is a liquid. In many cases a scientist may express a liquid added to another liquid as a percent by volume. The *United States Pharmacopeia* (U.S.P.) states in its General Notices section that this expression of concentration is implied whenever we have a solution containing a liquid in another liquid. As is the case with % w/w, percent by volume indicates the number of milliliters of a substance in 100 mL of final preparation. The mathematical expression for percent by volume is

$$\%v/v = \frac{V_D}{V_T} \times 100 \tag{3.2}$$

where V_D and V_T are the volumes of the substance added and the total volume of the preparation, respectively. Unlike percent by weight, the total volume of a solution cannot be considered to be the algebraic sum of the volumes of the solution's components. Scientists know that mixing 30 mL of alcohol, U.S.P. with 70 mL of water does not result in 100 mL of final solution. There is always some volume contraction when mixing two different liquids. The rule is always to make up the final volume of a preparation by using the more diluted liquid. For example, when mixing alcohol, U.S.P. with purified water, the volumes of each ingredient are first measured using a graduated cylinder, and the two liquids are then mixed together. The mixture is allowed to reach equilibrium at room temperature and then enough water is added to make up the desired final volume.

Example 3.3. A 60-mL preparation contains 2 mL of glycerin as a preservative. Express the concentration of glycerin in the preparation as % v/v.

Solution. Since 2 mL of glycerin is in 60 mL of the preparation, the percent by volume is $2/60 \times 100 = 3.3\%$.

Glycerin is often used in preparations as a preservative, but it can also be used as a solvent, a humectant, or a wetting agent. When added to syrups, glycerin can prevent the cap-lock phenomenon, which occurs due to crystallization of sucrose between the cap and the neck of a bottle.

Example 3.4. A 98-mL emulsion contains 30 mL of oil (the internal phase) and 70 mL of water (the external phase). Express the concentration of olive oil in the emulsion as % v/v.

Solution.   Apply equation (3.2):

$$\frac{30}{98} \times 100 = 30.6\% \mathrm{v/v}$$

In addition to the oily and aqueous phases, emulsions contain stabilizers, known as *emulsifying agents*, for their internal phase.

3.2.3   Percent Weight Per Volume

Percent weight per volume (% w/v) is perhaps the concentration unit most often used in pharmaceutical solutions. Solutions containing solids in liquids are extremely common in pharmacy practice, and based on the U.S.P. recommendations, the concentration of drugs in these solutions is expressed as % w/v. By definition, the amount of the drug (in grams) in 100 mL of a preparation is % w/v.

$$\% \mathrm{w/v} = \frac{W_D}{V_T} \times 100 \tag{3.3}$$

where W_D and V_T are as defined previously.

Example 3.5.   Strong iodine solution, U.S.P., also known as Lugol's solution, contains 50 g of iodine per 1 L of preparation. This solution may be used to prevent goiter development. What is the concentration of iodine in Lugol's solution expressed as % w/v?

Solution.

$$\% \mathrm{w/v} = \frac{50 \text{ g}}{1000 \text{ mL}} \times 100 = 5\%$$

3.2.4   Molarity

A solution containing 1 g molecular weight of a substance in 1 L of preparation is labeled 1 molar (1 M). For example, a 1 M solution of glucose contains 180 g of glucose per liter of preparation, since glucose has a molecular weight of 180 g/mol. By definition, 1 g molecular weight of a substance is 1 mol. Thus, a 1 M solution contains 1 mol of that substance per liter:

$$\text{molarity} = \frac{n_D}{V_T} \tag{3.4}$$

In this case V_T is in liters and n_D is the number of moles of the substance.

Example 3.6.   What is the molar concentration of sulfuric acid (H_2SO_4) in a solution containing 49 g of sulfuric acid in 1 L? The molecular weight of sulfuric acid is 98 g/mol.

Solution.   Since this solution contains ($49/98 = 0.5$ mol) of sulfuric acid in 1 L, its molar concentration is 0.5 M.

Example 3.7.   Diluted hydrochloric acid, U.S.P. is a solution prepared by the addition of 234 mL of concentrated hydrochloric acid solution to water to make 1 L. The usual dose of this preparation is 1 teaspoonful of solution taken diluted with a significant amount of water as a digestive aid. The concentrated hydrochloric acid solution contains 36.5% w/w of acid. Express the concentration of HCl (molecular weight 36.5 g/mol) in the diluted HCl, U.S.P. in mol/L. The specific gravity of the concentrated acid solution is 1.17.

Solution.   Since the specific gravity of the concentrated acid is 1.17, 234 mL of this solution weighs 234 mL × 1.17 = 273.78 g. This weight of the concentrated acid solution contains 99.93 g of HCl (based on 36.5% w/w). Therefore, the diluted hydrochloric acid solution contains 99.93 g of HCl in 1 L of preparation. This is equivalent to (99.93 g/36.5 g/mol)/1 L = 2.7 M.

Example 3.8.   Suppose that a patient diluted 5 mL of the diluted hydrochloric acid, U.S.P. solution with 240 mL of water to make a total volume of 245 mL. Express the concentration of HCl in this solution in molar units.

Solution.   The 5 mL of the diluted acid contains (5 mL × 2.7 mol/1000 mL = 0.0135 mol) of HCl. The 0.0135 mol is present in 245 mL solution. Thus, the molarity of HCl in this solution is 0.0135 mol/0.245 L = 0.055 M.

3.2.5   Normality

Normality is a term commonly used for acid and base solutions, but it is also useful for solutions containing ions. Normality expresses the number of equivalents (Eq) of a solute in 1 L of solution. A solution that contains 1 g *equivalent weight* of a substance in 1 L of solution has 1 Eq of that substance in 1 L. For example, the equivalent weight of sulfuric acid is [98 g/mol/(2 Eq/mol) = 49 g/Eq]. Thus, a solution containing 49 g of sulfuric acid per liter has 1 Eq of sulfuric acid per liter, or 1 normal (1 N).

$$N_D = \frac{(\text{no. Eq})_D}{V_T} \qquad (3.5)$$

where V_T is the total volume of the solution in liters.

Example 3.9.   Express the concentration of HCl in the solution in Example 3.8 in normal units.

Solution.   Since we have 0.0135 mol of HCl (or 0.493 g) in 245 mL of the preparation:

$$\text{equivalent weight} = \frac{36.5 \text{ g/mol}}{1 \text{ Eq/mol}} = 36.5 \text{ g}$$

$$\text{no. Eq} = \frac{0.493 \text{ g}}{36.5 \text{ g}} = 0.0135$$

$$N_{\text{HCl}} = \frac{0.0135}{0.245 \text{ L}} = 0.055$$

Notice that in this case the molarity and the normality are the same since the valence of HCl is 1. To convert from normality to molarity, or vice versa, use the following relationship:

$$\text{normality} = \text{valence} \times \text{molarity} \tag{3.6}$$

Example 3.10. Express the concentration of H_2SO_4 in normal units if the solution is labeled 2 M.

Solution. Apply equation (3.6):

$$\text{normality} = (2 \text{ Eq/mol})(2 \text{ M})$$
$$= 4 \text{ N}$$

Or, a 2 M solution of sulfuric acid contains $(2 \times 98 \text{ g/mole} = 196 \text{ g})$ of H_2SO_4 per liter.

The number of equivalents of H_2SO_4 is $(196 \text{ g}/49 \text{ g/Eq} = 4)$. Therefore,

$$N_{H_2SO_4} = \frac{4}{1} \text{ L} = 4 \text{ N}$$

3.2.6 Milliequivalence

A closely related concentration unit to that of normality is the *milliequivalence*. A milliequivalent weight is one-thousandth its equivalent weight. Thus, 1 Eq = 1000 milliequivalents (mEq). A 1 N solution of hydrochloric acid contains 1 Eq of HCl per liter or 1000 mEq per liter of solution. Since the equivalent weight is in grams, the milliequivalent weight is expressed in milligrams. This concentration unit is particularly useful for solutions containing ions and electrolytes. Its importance stems from the fact that ions always combine milliequivalent with milliequivalent. We cannot make the same statement if we replace milliequivalent by any other term, such as mole or gram.

Example 3.11. An elixir contains 1 mg/mL of calcium ions (40 g/mol). Express the concentration of calcium ions in normal units and as mEq/L. What amount (in grams) of calcium ions is there in 1 L of elixir?

Solution. The equivalent weight of calcium is $40/2 = 20$ g. Its milliequivalent weight is 20 mg. Since the solution contains 1 g/L of calcium ions, the number of equivalents of calcium is $1/20 = 0.05$, and $N_D = 0.05/1 = 0.05$ N. From the value $N_D = 0.05$ we can calculate the concentration in mEq/L from $0.05 \times 1000 = 50$ mEq/L. Each mEq of calcium is 20 mg; therefore $(50 \times 20 = 1000 \text{ mg})$, 1 g of calcium ions exists in 1 L of elixir.

Example 3.12. Suppose that calcium chloride $(CaCl_2; 111 \text{ g/mol})$ was used to provide the amount of calcium ions needed in Example 3.12. How much calcium chloride is needed to make 1 L of solution?

Solution.   Each 111 g of $CaCl_2$ contains 40 g of calcium. Therefore, 1 g of calcium ions is available in 2.78 g of $CaCl_2$.

The *solubility* of $CaCl_2$ in water is 1 g in 1.2 mL of water at 25°C. Thus, 2.78 g of calcium chloride should dissolve easily in 1 L of the aqueous solution. By definition, the solubility of any substance is the concentration of the substance in a saturated solution at a given temperature and pressure. Unless otherwise stated, the pressure is assumed to be 1 atmosphere. Calcium chloride solubility in boiling water is 1 g in 0.7 mL an *endothermic process*; the solubility increases with increasing temperature of the solution). It is always a good idea to check the solubility of compounds in the vehicle to be used for the preparation prior to compounding. A lot of frustration can be avoided if this is done, because that many compounds have low solubility in water, and some compounds exhibit an *exothermic reaction* with respect to solubility (i.e., an increase in the temperature of the solution results in a decrease in the solubility). The pH of the solution should also be considered in this regard, as the solubility of the compound may change upon a change in pH (this is true for solutions of electrolytes, weak acids, and weak bases).

Another issue related to milliequivalence is water of hydration. Many substances can exist as *solvates*, (e.g., *hydrates*). The same compounds can also exist in nonsolvate form (e.g., anhydrous form). When calculating the equivalence weight of hydrous substances, the water of hydration should be included in the weight of the drug.

Example 3.13.   Calculate the equivalent weight of $CaCl_2$ anhydrous (111 g/mol) and $CaCl_2 \cdot 2H_2O$ hydrous (147 g/mole).

Solution.   For the anhydrous form the equivalent weight is $111/2 = 55.5$ g. The hydrous form has an equivalent weight of $147/2 = 73.5$ g.

Example 3.14.   Calculate the amount of $CaCl_2 \cdot 2H_2O$ (in grams) needed to prepare 1 L of the solution in Example 3.12.

Solution.   The equivalent weight of the dihydrous salt is 73.5 g/Eq. Thus, for a solution of 0.05 Eq/L, we need 73.5 g/Eq × 0.05 Eq/L = 3.68 g/L of $CaCl_2 \cdot 2H_2O$. Calcium chloride is a good source of calcium ions as well as being used for its diuretic effect.

A compound such as monobasic sodium phosphate (NaH_2PO_4) may have several equivalent weights, depending on the ion of interest. If we were to use this salt for its potassium ions, the valence is 1. If it is used for its hydrogen content (as in a buffer solution), the valence is 2, whereas for its phosphate ion, 3 is the valence value. Moreover, when placed in water, this salt dissociates, and depending on the pH of the solution, two species are found, the monobasic and the dibasic (Na_2HPO_4):

$$HPO_4{}^{2-} + H^+ \rightleftharpoons H_2PO_4{}^-$$

Because of this variability in milliequivalence for phosphate molecules, phosphate concentration in parenteral preparation is never expressed in milliequivalence units but in millimolar units. The same argument applies to compounds containing carbonate ions, such as sodium bicarbonate ($NaHCO_3$).

3.2.7 Molality

Molality expresses the number of moles of a substance in 1 kg of solvent. Molality is distinguished from molarity in being independent of the volume of the solution. As the total volume of a solution varies with the temperature, the weight of the solvent does not. This is particularly important when comparing similar solution concentrations of the same drug placed at two extreme temperatures (e.g., 40°C vs. 4°C). The concentration of the drug at 40°C expressed as mol/L is different from that at 4°C; however, in molal units (m) both solutions maintain the same concentration value:

$$m_D = \frac{n_D}{W_{FV}} \tag{3.7}$$

where W_{FV} is the weight of the formulation vehicle (the solvent) in kilograms.

Example 3.15. To prepare 100 mL of a glucose solution, a pharmacist added 0.1 mol of glucose (180 g/mol) to 96 mL of water and mixed thoroughly until a solution formed. Express the concentration of glucose in the solution in molal units.

Solution. Apply equation (3.7):

$$m_{glucose} = \frac{0.1}{0.096} = 1.04 \text{ m}$$

3.2.8 Mole Fraction

The *mole fraction* (X), also known as the *number fraction*, is the ratio of the number of moles of one substance to the total number of moles of all the components in the solution. The mole fraction is unitless, since the units of the numerator and the denominator cancel out:

$$X_D = \frac{n_D}{n_T} \tag{3.8}$$

Example 3.16. Express the concentration of glucose in Example 3.16 as a mole fraction.

Solution. The solution contains 0.1 mol of glucose and 96 g of water in every 100 mL of the preparation.

$$X_{glucose} = \frac{0.1}{0.1 + (96/18)} = 0.018$$

If a solution contains only two components (solute and solvent), we can calculate the molality of component D by knowing the mole fraction of D.

$$m_D = \frac{1000X_D}{MW_{solvent}(1 - X_D)} \tag{3.9}$$

Example 3.17. Express the concentration of glucose in Example 3.17 in molal units.

Solution. Apply equation (3.9):

$$m_{glucose} = \frac{(1000)(0.018)}{(18)(1 - 0.018)} = 1.02 \text{ m}$$

3.2.9 Osmolality/Osmolarity

Some of most important concentration units in pharmaceutical practice are osmolarity and osmolality. The difference between these two terms is the same as the difference we discussed earlier between molality and molarity. *Osmolality* is the number of osmols of a substance in 1 kg of solvent (usually, water), and *osmolarity* is number of osmols of a substance in 1 L of solution. In clinical practice, these two terms are almost interchangeable, and thus we are going to focus our discussion on the simpler term, osmolarity. Osmolarity implies a physical parameter; an *iso-osmolar solution* has the same *osmotic pressure* as that of a reference fluid. A more useful term for medical professionals is *tonicity*. Tonicity implies a *biological* similarity (or dissimilarity) with a biological fluid such as blood. An *isotonic solution* is a solution that has the same osmotic pressure as that of a biological fluid and also *maintains the cellular integrity of the biological tissues*. Although an iso-osmolar solution has the same osmotic pressure as the reference fluid, it does not necessary have to be harmless with regard to the cellular structure with which it comes into contact. The key to this point is that if a particular solute is capable of penetrating the cellular membrane of a cell and causing *hemolysis* or other destruction to the membrane, that solution may be called iso-osmotic; however, it cannot be labeled isotonic. A classical example in pharmacy is a 1.9% solution of boric acid. This solution is iso-osmotic with both the lacrimal fluid of the eyes and the blood. However, this solution is isotonic only with regard to the eye tissues, not with the blood. It causes hemolysis when injected into circulation despite its iso-osmolarity.

For nonelectrolytes, osmolarity is identical to molarity. For electrolytes, the situation is quite different. To calculate the osmolarity of a solution of electrolytes from the molarity value, use the following expression:

osmolarity(osmol/L) = (number of ions the molecule dissociates into)(molarity)
$$\tag{3.10}$$

An *osmol* is defined as the weight of a substance in grams that upon addition to a solution yields an Avogadro's number of particles (ions, molecules, etc.)

Example 3.18. A solution was labeled as 1 M for NaCl. Express the osmolarity of sodium chloride in this solution.

Solution.

$$\text{Osmolarity(osmol/L)} = (2)(1) = 2$$

Example 3.19. An isotonic solution of sodium chloride contains 9 g/L. Express the osmolarity of this solution in milliosmols per liter. The molecular weight of NaCl is 58.4 g/mol.

Solution. A milliosmol is 1/1000 of an osmol. Apply equation (3.10):

$$\text{osmolarity(Osmol/L)} = (\text{no. of ions per molecule})(\text{molality})$$
$$= (2)\frac{9 \text{ g}/58.4 \text{ g/mol}}{1 \text{ L}} = 0.308$$

or

$$(0.308 \text{ osmol/L})(1000) = 308 \text{ mOsmol/L}$$

It is interesting to note here that serum osmolarity is considered to be approximately 300 mOsmol/L.

3.2.10 Ratio Strength

Expressing concentrations in *ratio strength* is sometimes used in prescriptions. The physician may write a prescription for hydrocortisone cream 1% as 1 : 100. This ratio strength of 1 : 100 means that there should be 1 g of hydrocortisone in 100 g of cream. A solution labeled as 1 : 50 for sucrose means that there are 1 g of sucrose in every 50 mL of the preparation. Thus, the definition of ratio strength is the number of grams or milliliters of the solute to the number of grams or milliliters of the preparation.

Example 3.20. Find the ratio strength of a solution containing 300 mL of propylene glycol in 1 L of the preparation.

Solution. The ratio strength of propylene glycol in this preparation is 1 : 3.3 (1 mL of propylene glycol in each 3.3 mL of solution).

3.2.11 Parts per Million

Parts per million (ppm) is commonly used for extremely dilute solutions of electrolytes. Parts per million reflects the number of *parts* of the solute in 1 million parts of solution. For example, a fluoride solution is labeled as 1 ppm, 1 g of fluoride ions in 1 million grams of solution. However, in diluted aqueous solutions, it is assumed that their density is 1 and the number of grams of the solution is replaced by the number of milliliters. Thus, the aqueous fluoride solution can be interpreted as having 1 g of fluoride in 1 million milliliters of solution. If, on the other hand, the solution is not dilute, the density (or the specific gravity) of the solution must be known to convert the grams of the solution to the volume in milliliters.

Example 3.21. An aqueous solution contains 0.0025 g of lead in 1 gallon of solution. Express the concentration of lead in parts per million.

Solution. One gallon is 3785 mL. Thus, 0.0025 g of lead is in 3785 mL of solution. By proportion, in 1 million milliliters of solution we have 0.66 g of lead, or 0.66 ppm.

Example 3.22. An aqueous solution contains 4 ppm of fluoride ions (19 g/mol). How much of sodium fluoride (42 g/mol) is needed to make 500 mL?

Solution. A concentration of 4 ppm indicates that there is 4 g of fluoride ions in 1 million milliliters of solution (assuming a dilute solution). Therefore, in 500 mL of solution there is 0.002 g of fluoride ions. To supply this quantity of fluoride ions, we need 0.0044 g ($42 \times 0.002/19 = 0.0044$ g of sodium fluoride), or 4.4 mg. Dentists often use an 8% solution of stannous fluoride as a prophylactic anticariogenic medication in children.

3.3   WHAT SHOULD BE DONE WHEN ALCOHOL IS PRESCRIBED IN A FORMULATION

According to the U.S.P., whenever the term *alcohol* is called for in a prescription, alcohol, U.S.P. should be used in the preparation. Alcohol, U.S.P. is considered to be 95% v/v ethanol, and thus certain calculations may be necessary in those cases.

Example 3.23

Antiemetic	q.s. [as much as suffices]
Alcohol	15%
Purified water, q.s. ad.	50 mL

Solution. The formula above contains 15% v/v alcohol, that is, 15 mL of ethanol in 100 mL of the preparation. Thus, we need 7.5 mL of ethanol. The ethanol has to be provided by alcohol, U.S.P. (95%). Therefore, $7.5 \times 100/95 = 7.9$ mL of alcohol, U.S.P. is needed.

Example 3.24

Antiemetic	q.s.
Alcohol	15 mL
Purified water, q.s., ad.	50 mL

Solution. In this example the quantity of alcohol is 15 mL. Thus, 15 mL of alcohol, U.S.P. must be used. Thus, this preparation contains $15 \times 95/100 = 14.25$ mL of ethanol; or, the concentration of ethanol in percent per volume is 28.5%.

3.4 THE PARTITION COEFFICIENT

Solute molecules possess a different affinity to different solvents which may influence their solubility and partitioning in solvent systems. Thus, depending on their hydrophilicity or lipophilicity, the molecules may or may not *partition* in a solvent system. When drug molecules are placed in a solution and the solution is mixed with another immiscible solvent, the drug molecules can either remain in their solution or partition into the other solvent. The degree to which partitioning occurs is in part a function of the balance between the lipid solubility of the drug and its water solubility. So if a drug is soluble in an aqueous phase initially and then subjected to mixing with another oily phase (immiscible with water), at equilibrium some of the drug is going to be found in the oily layer and some of it will remain in the aqueous phase. The ratio of the concentration of the drug in the oily phase to that in the aqueous phase at equilibrium is known as the *partition coefficient P* (Martin et al., 1983):

$$P = \frac{C_{oil}}{C_{water}} \tag{3.11}$$

It is assumed that equation (3.11) is obtained at constant temperature, or *isothermally*. For example, for a weak acid dissolved in water and then mixed with oil until equilibrium is achieved, equation (3.11) can be written as (Martin et al., 1983)

$$P = \frac{C_{oil}}{([HA] + [A^-])_{water}} \tag{3.12}$$

In equation (3.12) we assume that the weak acid (HA) does not associate (i.e., forms aggregates) in oil. Since the denominator in equation (3.12) is an expression of the total molar concentrations of all the species present, the partition coefficient in equation (3.12) is known as the *apparent partition coefficient*, while the expression in (3.11) is for the *true partition coefficient*. The partition coefficient has a vast number of applications in pharmacy. Drugs or other ingredients can partition between the immiscible phases (e.g., emulsions), and depending on the application, this partitioning may be favorable or unfavorable.

One of these applications is the addition of a preservative to emulsions. Benzoic acid can partition in oil and water phases, and depending on its remaining concentration in the water phase (where microbial contamination is likely to occur) it can be effective or ineffective. In addition, only [HA], the undissociated species, has antimicrobial activity. Thus, for the sake of effectiveness, the pH of the system must remain below 4.5, and definitely below 5.0, where its activity ceases. Another issue with benzoic acid is its association reaction in many nonpolar solvents and oils. Aggregates of benzoic acid in nonpolar solvents contribute further to the complexity of deciding on the amount of benzoic acid needed for emulsion formulations. Benzoate ions are also found to adsorb on surfaces of suspended solid particles. This can also result in reduced antimicrobial activity and can affect the thermodynamic equilibrium of the suspended particles as well (Martin et al., 1983).

In water–oil systems, preservatives added to the formulation should maintain a concentration in the aqueous phase high enough to block any microbial growth. It is also desirable to keep the preservative nonionized so that its effectiveness is maintained. Such a task is accomplished by knowing (1) the pH of the aqueous phase, (2) the true partition coefficient of the preservative in the system, (3) the relative volumes of the aqueous and the oil phase, and (4) the total concentration of the preservative in the aqueous phase at equilibrium. The molar concentration of the preservative remaining in the aqueous phase at equilibrium is (Martin et al., 1983)

$$[HA]_{aqueous} = \frac{C}{qP^n[HA]_{aqueous} + 1 + (K_a/[H_3O^+])} \tag{3.13}$$

where $[HA]_{aqueous}$ is the concentration of the preservative remaining in the aqueous phase in its undissociated form at equilibrium, C the initial concentration of the preservative; q the volume ratio (oil/water), P^n the true partition coefficient, n the number of molecules per aggregates (e.g., for dimmers, n is 2), K_a the acid dissociation constant; and $[H_3O^+]$ the hydronium ion concentration. Equation (3.13) is applicable for preservatives that form aggregates in the oil phase (Martin et al., 1983).

If the preservative does not form aggregates in the oil phase, $[HA]_{aqueous}$ is calculated from (Martin et al., 1983)

$$[HA]_{aqueous} = \frac{C}{qP + 1 + (K_a/[H_3O^+])} \tag{3.14}$$

The value of the true partition coefficient must be obtained before equation (3.13) or (3.14) is employed. In the case of no aggregation in the oil phase, P is calculated from measuring the concentration of preservative remaining in the aqueous phase at equilibrium under different pH conditions. If equal volumes of the aqueous and oil phases are used, a plot of $K_a + [H_3O^+]/C$ vs. $[H_3O^+]$ will yield a straight line with a slope equal to $(P + 1/C)$ and a y-intercept of K_a/C. Then the true partition coefficient can be calculated from the slope value.

In the case of aggregations, the true partition coefficient is obtained from (Martin et al., 1983)

$$P = \frac{(C_{oil})^{1/n}}{C_{aqueous}} \tag{3.15}$$

where C_{oil} is the concentration of preservative in the oil phase remaining at equilibrium, and $C_{aqueous}$ is the concentration of the preservative at equilibrium in water in undissociated form [HA] (this can be achieved for a weak acid by making the solution highly acidic). Examples of acidifying agents used in formulations are acetic acid, glacial acetic acid, citric acid, fumaric acid, hydrochloric acid (concentrated or diluted), lactic acid, malic acid, nitric acid, phosphoric acid (concentrated or diluted), propionic acid, sodium phosphate monobasic acid, sulfuric acid, and tartaric acid. All of these agents are official in the U.S.P. Some are solid at room temperature (e.g., fumaric acid) and others are liquid (e.g., propionic acid) (Allen, 1999).

From a practical point of view, the discussion above is merely academic and to some extent requires a lot of experimentation. For example, if a scientist wishes to determine the amount of preservative needed for a given formulation, the following general guidelines may be appropriate:

1. Consult the literature for information pertaining to a similar formulation.

2. Ask experts in the field (academic centers).

3. If the first two attempts do not result in meaningful information, a simple experiment can be carried out on the formulation. Make several batches of the formulation, each with an increasing concentration of the preservative. Document microbial or fungal growth in batches over a reasonable period of time (the time expected for use), and the formulation with the least amount of preservative that remains free of any growth is the one to choose.

3.5 DISINTEGRATION AND DISSOLUTION

Perhaps the single most important factor that holds back agents from becoming drugs is their solubility. A substance that is not significantly soluble in water would not be a good drug candidate because it would not solubilize in body fluids. In the same vein, an agent without relatively good solubility in lipids cannot easily diffuse through a biological membrane. A balanced solubility in water and lipids is essential for a drug to exhibit its therapeutic action. However, before the substance is absorbed by the biological membrane, it has to be in solution at the site of administration. Efforts to deliver poorly water-soluble substances are focused on improving their dissolution rate and/or on obtaining transient solubilization. Other efforts focus on creating a sustained solubilization state. Techniques such as reduction in the particle size by creating nanoparticles, using a high-energy solid form, or using a salt form are all means of improving the dissolution rate and/or obtaining transient solubilization. However, all of these methods have limitations in vivo and most often are linked to higher variability in the bioavailability of the drug. Sustained solubilization is achieved in vivo by solubilizing the poorly water-soluble agent in a lipid phase that subsequently facilitates the drug's absorption (Giliyar et al., 2006). Lipocine Technologies (Salt Lake City, Utah) has developed delivery systems along this line, in which a facilitated solubilization in a proprietary lipid phase (Lip'ral and Lip'ral-SSR) is capable of producing micelles to maintain sustained solubilization for water-insoluble drugs. The technology is suitable for immediate-release and modified-release delivery systems (Giliyar et al., 2006; Lipocine, Inc.).

For a drug to be released from a solid dosage form such as a tablet, the unit must undergo two steps: disintegration and dissolution. Normally, the disintegrated particles undergo a deaggregation step to convert the larger aggregates produced by the disintegration process into finer particles. Some minor dissolution usually occurs during the disintegration and deaggregation steps; however, for the most part, dissolution occurs following these two steps. Disintegration, being the first step in the process of dissolution, can be a limiting factor for the ultimate absorption of drugs by a biological membrane. (The slowest step in any process is considered to be the rate-limiting step for the entire process.) In designing a dosage form, formulators aim to include in the mixture a *disintegrating agent* that acts between the granules and within the granules of the dosage form to initiate disintegration and finalize it by deaggregation. Most disintegrating agents have the ability to swell and to increase in volume severalfold upon contact with an aqueous environment. Natural cellulosic or carbohydrate materials are usually employed for that purpose. More recently, materials utilized for the purpose of efficiently breaking a solid dosage form apart have been suggested in industry. Examples of these newly introduced disintegrants are Kollidon CL-F and Kollidon CL-SF (BASF, Germany), which are made of insoluble crospovidone and are useful primarily for small tablets and tablets intended to dissolve rapidly in the mouth (Giliyar et al., 2006).

The dissolution rate of solid dosage forms can be estimated by envisioning a solid particle undergoing dissolution in a dissolution medium. Surrounding the solid particle is a stagnant layer of solvent acting as a *diffusion boundary layer* for drug molecules entering a dissolution medium. The thickness of the diffusion layer is h, and the concentration of the drug within the diffusion layer is that of saturation, C_s. According to *Fick's law of diffusion*,

$$J = \frac{dM}{S\,dt} \tag{3.16}$$

where J is the flux, dM/dt the amount of drug "diffused" through the stagnant solvent layer over time, and S the surface area for diffusion. Since

$$J = \frac{KD}{h}\Delta C \tag{3.17}$$

then

$$\frac{dM}{S\,dt} = \frac{KD}{h}\Delta C \tag{3.18}$$

where K is the partition coefficient of drug and in this case is equal to 1, since the "nature" of the diffusion layer is the same as that of the solvent, D is the diffusion coefficient, and $\Delta C = C_s - C$. (C is the concentration of the drug in the dissolution medium outside the stagnant layer.) For all practical purposes, C_s is much greater than C and the equation may be written

$$\frac{dM}{dt} = \frac{DS}{h}C_s \tag{3.19}$$

Replacing the amount of the drug by (concentration $\times$ volume) yields

$$\frac{dC}{dt} = \frac{DS}{hV}C_s \tag{3.20}$$

Equation (3.20) is known as the *Noyes–Whitney expression*. The diffusion coefficient D is a function of the temperature of the solution, the viscosity of the dissolution medium, and the radius of the dissolving particles. Thus, the dissolution rate dC/dt of the drug is influenced by all these factors and the other factors in equation (3.20).

In general, the solubility and dissolution rates for salt forms are much better than those of active pharmaceutical ingredients (API). Thus, one way of enhancing the dissolution rate of APIs is to select a salt of the drug to be included in the formulation instead of the API itself. This is essential, because unless a drug is already in solution, it cannot undergo absorption through the biological membrane (Neervannan, 2004). Several factors must be considered when choosing a counterion for making the salt: (1) its pK_a value, which should be 2 to 3 PH units different from that of the API; (2) whether the final melting point of the resulting salt is planned to be high or low (high with inorganic acids and low with aliphatic organic acids); (3) its physicochemical properties and how they could influence its stability during handling; (4) the ease of handling during manufactur; and (5) its toxicity potential (Neervannan, 2004). Tong and Whitesell (1998) presented data eloquently describing an in situ screening method during the preformulation stage for selecting a salt form with optimal high solubility. They suggested that this method should be applicable to "most basic compounds." This screening method includes four steps: acid selection, solubility study, characterization of the residual solids, and calculation of the solubility product constant K_{sp}. Assuming that the API is a base, the pK_a of the acid selected to form the salt with the API must be 2 pH units lower than the pK_a of the API. During the solubility study, the salt is formed by reacting the API chemically with an acid so that at the end of the reaction an excess of acid remains in the reaction vessel. A measurement of the final solution pH is performed at this point. The resulting salt material is then subjected to a battery of physical testing (i.e., differential scanning calorimetry, thermogravimetric analysis, hot-stage microscopy, x-ray diffraction, etc.) to identify its structure. The final stage includes estimation of K_{sp} and the solubility of the salt formed, S_{salt}:

$$K_{sp} = [BH^+][X^-] \tag{3.21}$$

$$S_{salt} = (K_{sp})^{1/2} \tag{3.22}$$

where [BH^+] and [X^-] are the molar concentration of the protonated species of the base and the counterion, respectively.

In addition to calculating the total solubility, the *common ion effect* (CIE) must also be considered. The CIE reduces the solubility and dissolution rates of the compound to levels even lower than those of the base itself. This effect is due to the presence of an excessive amount of the counterion in solution (e.g., the presence of chloride ions in excess in solutions containing a hydrochloride salt of a basic drug) (Neervannan, 2004). To account for CIE, equation (3.21) can be rewritten as

$$K_{sp} = [BH^+][X^-]_T \tag{3.23}$$

where $[X^-]_T$ is the total concentration of the counterion resulting from the dissociation from the salt and that contributed from an outside source (Tong and

Whitesell, 1998). The following examples illustrate the Tong and Whitesell in situ method for estimating the solubility of a compound.

Example 3.25.   Tong and Whitesell gave an estimate for the solubility (mg/mL) of succinate salt of the basic drug (GW1818X, molecular weight = 571.6 g/mol) 5.60 mg/mL in 0.1 M succinic acid aqueous solution at room temperature. Show calculations for estimating the solubility of the drug given that this solubility of the drug in 0.1 M succinic acid (S) was 3.02 mg/mL, the amount of the drug added (X) was 8.8 mg, the volume of solution (V) was 0.5 mL, the pH of the saturated solution at equilibrium was 3.67, and the pK_a for succinic acid was 4.16.

Solution.   According to Tong and Whitesell, the amount of the drug precipitated can be estimated by the following equation:

$$X_p = X - (SV)$$
$$= (8.8 \text{ mg}) - [(3.02 \text{ mg/mL})(0.5 \text{ mL})] = 7.29 \text{ mg of base} \qquad (3.24)$$

Since succinic acid was present in solution in excess (according to the in situ salt screening method), succinic acid concentration in solution at equilibrium can be estimated by

$$[A_s] = [A] - \frac{X_p}{(V(\text{MW})} \qquad (3.25)$$

where [A] is the concentration of succinic acid used and MW is the molecular weight of the basic drug:

$$[A_s] = 0.1 \text{ M} - \frac{7.29 \text{ mg}}{(0.5 \text{ mL})(571.6 \text{ g/mol})} = 0.0745 \text{ M}$$

The concentration of succinic acid remaining in solution at equilibrium in its ionized form can be estimated from

$$[A_{\text{ionized}}] = \frac{[A_s]}{1 + ([H^+]/K_a)}$$
$$= \frac{0.0745 \text{ M}}{1 + (0.000214/0.000069)} = 0.0182 \text{ M} \qquad (3.26)$$

Expressing the concentration of drug added to solution in molar units, the solubility product constant of the succinate salt is calculated from

$$K_{\text{sp}} = S[A_{\text{ionized}}] = \frac{3.02 \text{ g/L}}{571.6 \text{ g/mol}} (0.0182 \text{ M}) = 0.0000962 \text{ M}^2$$

From equation (3.22),

$$S_{\text{salt}} = (K_{\text{sp}})^{1/2} = (0.0000962 \text{ M}^2)^{1/2} = 0.00981 \text{ M}$$

or,

$$S_{\text{salt}} = (0.00981 \text{ M})(571.6 \text{ g/mol}) = 5.60 \text{ mg/mL}$$

Example 3.26.   The actual solubility of the *authentic* succinic acid salt in Example 3.1 was measured to be 2.6 mg/mL in 0.01 M succinic acid. We assume that the system is at equilibrium and is described by the following scheme (Tong, and Whitesell, 1998):

$$BH^+X^-(solid) \rightleftharpoons BH^+X^-(in\ solution;\ 1\ :\ 1\ salt) \rightleftharpoons B^+ + X^-$$

Recalculate the solubility of the succinate salt of the drug at an equilibrium pH of 3.3 after accounting for the common ion effect.

Solution.

$$K_{sp} = [BH^+][X^-]_T$$

$$[BH^+] = \frac{2.6\ g/L}{571.6\ g/mol} = 0.00455\ M \tag{3.27}$$

$$[X_T] = [BH^+]_{from\ salt} + [HX]_{outside\ source}$$

$$= 0.00455\ M + 0.01\ M = 0.01455\ M$$

$$[X_T] = [X^-]_T + [HX]_{unionized}$$

$$[HX]_{unionized} = \frac{[H^+][X^-]_T}{K_a}$$

$$[X_T] = [X^-]_T + \frac{[H^+][X^-]_T}{K_a} \tag{3.28}$$

$$[X^-]_T = \frac{[X]_T}{1 + ([H^+]/K_a)}$$

$$[X^-]_T = \frac{0.01455\ M}{1 + (0.0000501/0.000069)} = 0.00843\ M$$

$$K_{sp} = [BH^+][X^-]_T = (0.00455\ M)(0.00843\ M) = 0.0000383\ M^2$$

and from equation (3.22),

$$S_{salt} = (K_{sp})^{1/2} = (0.0000383\ M^2)^{1/2} = 0.00619\ M$$

$$= (0.00619\ M)(571.6\ g/mol) = 3.5\ mg/mL$$

Tong and Whitesell (1998) noted that the significant discrepancy between the estimated solubility of the succinate salt by the in situ method (solubility of 5.6 mg/mL) and the solubility of the authentic salt (3.5 mg/mL) was due to the difference in the *type* of salt used in the two experiments. The authentic salt was found to be the *hydrate* form, whereas the one used in the in situ experiment had a different x-ray diffraction profile than that of the hydrated salt. Thus, the results cannot be comparable. We encourage the reader to review the article by Tong and Whitesell and practice estimation of solubility for different salts.

Solid dispersions (SDs) are now commonly used in the pharmaceutical industry to enhance the solubility and dissolution of poorly aqueous soluble APIs. The purpose of this technology is to create a solid dispersion of an API in a solid matrix. The API may be present in this solid matrix as a crystalline, amorphous, or unimolecular form. Subsequently, the solid dispersion may be included as a dry powder component into granules, beads, or tablets (Stodghill, 2010). The solid dispersion is prepared by the fusion, solvent, or melting solvent method. SDs may be used to prepare both immediate- and sustained-release formulations (Monika, and Madon, 2008). For example, felodipine (a calcium-channel blocker and vasodilator; used in the treatment of hypertension) formed a solid dispersion within a poly(vinylpyrrolidone) matrix, which resulted in a marked improvement in its dissolution profile. The glassy or/amorphous state of felodipine within the mixture was confirmed by differential scanning calorimetry methods (Stodghill, 2010).

The release rate of the drug from a tablet matrix made from hydroxypropyl methylcellulose was found to be dependent on the geometry of the tablet. Specifically, the ratio of the tablet surface area to its volume (S/V) was of importance in determining the drug release rate. The higher the S/V ratio, the higher was the release of the drug from the matrix. This was true regardless of the dose, tablet shape, solubility, or the way the drug was released from the tablet matrix (Missaghi et al., 2010).

3.6 CONCLUDING REMARKS

There are many ways to express the concentrations of a substance in formulations: percentage, molality, molarity, normality, mole fraction, and milliequivalence, just to name a few. Scientists must familiarize themselves with ratios and proportions together with some basic algebraic rules. A scientific calculator is sufficient for most of the calculations needed for compounded medications.

When a drug is allowed to partition between two immiscible liquids, the ratio of the resulting concentrations in the two phases at equilibrium is known as the *partition coefficient*. This parameter is useful in determining the extent by which a drug can diffuse through a biological membrane (such as the skin). It can also be used to determine amount of preservative needed for a given formulation containing two immiscible phases (e.g., emulsions). The solubility of the drug in water, the disintegration process, and dissolution are all product characteristics that ultimately can influence an API's therapeutic effectiveness.

REFERENCES

Allen VL. Featured excipients: acidifying agents. Int. J. Pharm. Compd. 1999;3(4):309–310.

BASF, The Chemical Company. http://www.pharma-ingredients.basf.com/Home.aspx. Accessed Feb. 21, 2012.

Giliyar C, Fikstad TD, Tyavanagimatt S. Challenges and opportunities in oral delivery of poorly water-soluble drugs. Drug Deliv. Technol. 2006;6(1):57–63.

Lipocine, Inc. http://www.lipocine.com/technologies.html. Accessed Feb. 19, 2012.

Martin A, Swarbrick J, Cammarata A. Solubility and distribution phenomena. In: *Physical Pharmacy: Physical Chemical Principles in the Pharmaceutical Sciences*, 3rd ed. Philadelphia: Lea & Febiger, 1983. pp. 272–313.

Missaghi S, Patel P, Tiwari SB, Farrell TP, Rajabi-Siahboomi AR. Investigation of the influence of tablet shape, geometry and film coating on drug release from hypromellose extended-release matrices. Drug Deliv. Technol. 2010;10(2):32, 34, 36, 38–41.

Monika, DH, Madan AK. Development of sustained release solid dispersions using factorial design. Drug Deliv. Technol. 2008;8(5):48, 50–53.

Neervannan S. Strategies to impact solubility and dissolution rate during drug lead optimization: salt selection and prodrug design approaches. Am. Pharm. Rev. 2004;7(5):108, 110–113.

Stodghill SP. Thermal analysis: a review of techniques and applications in the pharmaceutical sciences. Am. Pharm. Rev. 2010;13(2):29, 31–36.

Tong W-Q, Whitesell G. In situ salt screening: a useful technique for discovery support and preformulation studies. Pharm. Dev. Technol. 1998;3(2):215–223.

GLOSSARY

API	Active pharmaceutical ingredient.
CIE	Common ion effect.
Equivalent weight (g/Eq)	The ratio of the molecular weight (g/mol) of the solute to its valence (Eq/mol).
Iso-osmolar solution	Has the same *osmotic pressure* as that of a reference fluid.
Isotonic solution	Has the same osmotic pressure as that of a biological fluid and also *maintains the cellular integrity of the biological tissues*.
Milliequivalence	A milliequivalent weight is one-thousandth its equivalent weight; thus, 1 Eq = 1000 mEq.
Molality (m)	Expresses the number of moles of a substance in 1 kg of solvent.
Molarity (M)	A solution containing 1 gram molecular weight of a substance in 1 L of a preparation is labeled as 1 molar (1 M).
Mole fraction	Also known as the number fraction; the ratio of the number of moles of one substance to the total number of moles of all the components in a solution.
Normality (N)	Expresses the number of equivalents of a solute in 1 L of solution.
Osmolality/osmolarity	Osmolality is the number of osmols of a substance in 1 kg of solvent (usually, water), and osmolarity is the number of osmols of a substance in 1 L of solution.
Partition coefficient	The ratio of the concentration of a drug in the oily phase to that in the aqueous phase at equilibrium.
Parts per million	The number of *parts* of a solute in 1 million parts of solution.
Percent by volume (% v/v)	The number of milliliters of a substance in 100 mL of a final preparation.
Percent by weight (% w/w)	Defined for D as the weight of D in grams in 100 g of a preparation.
Percent weight per volume (% w/v)	The amount of a drug (in grams) in 100 mL of a preparation.
Ratio strength	The number of grams or milliliters of a solute in relation to to the number of grams or milliliters of a preparation.
SD	Solid dispersion.
U.S.P.	*United States Pharmacopeia*.

CHAPTER 4

BIOLOGICAL ASPECTS OF FORMULATIONS

4.1 INTRODUCTION

Dosage forms that are equivalents indicate they are similar in certain characteristics; thus, *pharmaceutical equivalents* are preparations of the same drug with the same concentration and the same type of dosage form; *clinical equivalents* produce the same clinical effect in patients suffering from the same condition; *biological equivalents* have the same physiological profile as determined from the concentration in biological fluids following administration. Pharmaceutical, clinical, and biological equivalents are all *chemical equivalents* (i.e., they contain the same chemical agent in the same concentration and in the same dosage form). Chemical equivalents should meet the standards listed in the *United States Pharmacopeia–National Formulary* (U.S.P.–N.F.) (Baweja, 1987).

The *partition coefficient* is an important parameter in determining the ability of a formulation to release a drug to a biological membrane such as the skin. The partition of the drug between the dermatological preparation's vehicle and the skin is of importance to the bioavailability of drugs delivered topically. The higher the partition coefficient, the higher the flux of the drug through the skin. It appears that the closer the value of $\log P_{octanol}$ (the partition coefficient value obtained from an octanol–water experiment) to 2.0, the better the flux of the drug through the skin. One should stress, though, that $\log P_{octanol}$ is only one of the many factors that has to be considered for the precutaneous absorption of drugs.

4.2 BIOAVAILABILITY AND BIOEQUIVALENCE

The U.S. Food and Drug Administration (FDA, 2000) defines the term *bioavailability* as

> The rate and extent to which the active ingredient or active moiety is absorbed from a drug product and becomes available at the site of action. For drug products that are not intended to be absorbed into the bloodstream, bioavailability may be

Integrated Pharmaceutics: Applied Preformulation, Product Design, and Regulatory Science, First Edition. Antoine Al-Achi, Mali Ram Gupta, William Craig Stagner.
© 2013 John Wiley & Sons, Inc. Published 2013 by John Wiley & Sons, Inc.

assessed by measurements intended to reflect the rate and extent to which the active ingredient or active moiety becomes available at the site of action.

When a dose of a drug is given to a patient via a route of administration, and following absorption through the biological membrane (except when intravenous administration is used), the drug establishes a concentration within the blood compartment. Following distribution to various tissues, some amount of the drug reaches the receptor site. It is assumed that the concentration of the drug at the site is only a small fraction of that found in the circulation, however, its rate of change reflects that of the blood compartment (i.e., a decrease in plasma drug concentration by 10% signifies a *similar* decrease at the receptor site). Upon binding to its receptor, the drug exerts a pharmacological response which may translate into a useful clinical response (a therapeutic effect). The science of pharmacokinetics concerns itself with what occurs from the dose to the receptor site, whereas pharmacodynamics describes the steps from the receptor site to the clinical outcome.

Ideally, when two dosage forms for the same drug are bioequivalent, they produce similar biological effects. The FDA (2000) defines *bioequivalence* as

> The absence of a significant difference in the rate and extent to which the active ingredient or active moiety in pharmaceutical equivalents or pharmaceutical alternatives becomes available at the site of drug action when administered at the same molar dose under similar conditions in an appropriately designed study.

In practice, the term *bioequivalent* refers to similar bioavailability, or a similar rate and extent of absorption of a therapeutic agent from its site of administration to the circulation in its pharmacologically active form. The reasons that the blood circulation replaces the clinical outcome as a measure of bioequivalence are that therapeutic outcomes are by their nature subjective, the drug effect is often obscured by a placebo response, and the disease progression is usually unpredictable.

Certain terms are commonly used to describe how drug products compare with each other: Pharmaceutical alternatives and pharmaceutical equivalents for example, term *pharmaceutical alternatives* is The employed to describe drug products containing the same therapeutic ingredient but not necessarily in the same chemical type or amount or dosage form whereas *pharmaceutical equivalents* are drug products that contain exactly as that of the same chemical entity as that of active ingredient and have identical dosage forms and amounts of drug. Pharmaceutical equivalents do not have to contain the same inactive ingredients or have the same expiry date. In addition, pharmaceutical alternatives or equivalents must meet other characteristics as described in reference compendia. The FDA defines these terms as follows (*Code of Federal Regulations*, 21 CFR 320):

> *Pharmaceutical equivalents* means drug products in identical dosage forms that contain identical amounts of the identical active drug ingredient, *i.e.*, the same salt or ester of the same therapeutic moiety, or, in the case of modified release dosage forms that require a reservoir or overage or such forms as prefilled syringes where residual volume may vary, that deliver identical amounts of the active drug ingredient over the identical dosing period; do not necessarily contain the same

inactive ingredients; and meet the identical compendial or other applicable standard of identity, strength, quality, and purity, including potency and, where applicable, content uniformity, disintegration times, and/or dissolution rates.

Pharmaceutical alternatives means drug products that contain the identical therapeutic moiety, or its precursor, but not necessarily in the same amount or dosage form or as the same salt or ester. Each such drug product individually meets either the identical or its own respective compendial or other applicable standard of identity, strength, quality, and purity, including potency and, where applicable, content uniformity, disintegration times and/or dissolution rates.

Furthermore, the FDA states (21 CFR 320) that

Two drug products will be considered bioequivalent drug products if they are pharmaceutical equivalents or pharmaceutical alternatives whose rate and extent of absorption do not show a significant difference when administered at the same molar dose of the active moiety under similar experimental conditions, either single dose or multiple dose. Some pharmaceutical equivalents or pharmaceutical alternatives may be equivalent in the extent of their absorption but not in their rate of absorption and yet may be considered bioequivalent because such differences in the rate of absorption are intentional and are reflected in the labeling, are not essential to the attainment of effective body drug concentrations on chronic use, and are considered medically insignificant for the particular drug product studied.

When conducting bioequivalent studies, dosage forms of the same active ingredient (sometimes a prodrug is also acceptable for comparison) are compared. Generally, healthy volunteers younger in age (18 to 35 years old) and of normal weight are recruited for enrollment in the studies. The purpose of the studies is to compare dosage forms of the same drug moiety in different aspects, such as routes of administration (nasal vs. oral), dosing regimen (twice vs. three times daily), amounts of the drug per unit dose (250 mg vs. 500 mg), vehicles in the dosage form (lactose vs. starch), physical forms of the drug (crystalline vs. amorphous), or manufacturing processes (dry vs. wet granulation in tablet manufacturing). The FDA may require bioavailability studies to be conducted by the manufacturer if the results from clinical studies indicate that different drug products produce different therapeutic results, results from bioavailability studies indicate that different products are not bioequivalent, or a drug has a narrow therapeutic range. In addition, the low solubility of a drug and/or a large dose, and considerably incomplete absorption of the drug, are conditions that may trigger bioavailability studies demanded by the FDA.

The *principal bioavailability parameters* are obtained from drug–blood level vs. time. These are C_{max} (the highest blood concentration of the drug achieved after administration of the formulation), T_{max} (the time when C_{max} occurs), and the *area under the curve* (AUC) (the total area under the blood level–time curve). Terms such as *absolute bioavailability* and *relative bioavailability* are often used to compare the bioavailability of a drug from the route of administration vis-à-vis the intravenous route or another standard route of administration, respectively. For example, oral route vs. intravenous route:

$$F = \frac{(\text{dose}_{i.v.})(\text{AUC}_{oral})}{(\text{dose}_{oral})(\text{AUC}_{i.v.})} \tag{4.1}$$

The bioavailability of drugs is affected by physiological factors, the physical and chemical characteristics of the drugs, and the type of dosage form being used. Among the physiological factors are the age, gender, and disease state of the patient; the degree of metabolism at the site of administration; and the pH prevailing at the administration site. The ability of a drug to permeate the biological membrane, its solubility in biological fluids, whether the drug is an electrolyte or a nonelectrolyte, and the partition coefficient are among the characteristics of a drug affecting its bioavailability. The best oral bioavailability of a drug is generally expected when the drug is given in a solution followed in descending order by suspension, emulsion, capsules, tablets, and coated tablets. Since most drug absorption occurs in the small intestine, increasing the retention time for the drug in this area would improve its bioavailability. It was found experimentally that oral solutions had the least transit time (0.3 hour) in the stomach and the longest transit time (4.4 hours) in the small intestine (compared to tablets, pellets, and capsules). The transit time in the small intestine for pellets, capsules, and tablets was 3.4, 3.2, and 3.1 hours, respectively (Jain et al., 2005). The average pH value in the stomach during fasting is between 1.4 and 2.1, and this increases to about 7.0 when food is present. Average values for the pH in the duodenum, jejunum, and ileum are (4.9 to 6.4), (4.4 to 6.6), and (6.5 to 7.4), respectively (Neervannan, 2004).

Bioavailability parameters (C_{max}, T_{max}, and AUC) are calculated directly from the drug–blood level vs. time. The C_{max} and T_{max} values help characterize the plasma concentration profile by identifying a maximum point; however, the accuracy of these values is dependent on the timing and frequency of blood sampling. Together C_{max} and T_{max} represent a measure of the rate of absorption. The AUC, a measure of the extent of absorption, is usually determined by the equation:

$$\text{AUC}_t = \sum \frac{C_{i-1} + C_i}{2}(t_i - t_{i-1}) \tag{4.2}$$

Example 4.1.   Calculate AUC for the following data:

Time (h)	Concentration (µg/mL)
0	0
1	2
3	15
5	5
7	1

Solution.   Apply equation (4.2):

$$\text{AUC}_t = \sum \frac{C_{i-1} + C_i}{2}(t_i - t_{i-1})$$

$$\text{AUC}_7 = \left(\frac{0+2}{2}\right)(1-0) + \left(\frac{2+15}{2}\right)(3-1)$$

$$+ \left(\frac{15+5}{2}\right)(5-3) + \left(\frac{5+1}{2}\right)(7-5) = 44 \ \mu g \cdot h/mL$$

4.3 PROTOCOLS FOR DETERMINING BIOEQUIVALENCE

To perform bioequivalence studies, the following components must be available: a sensitive drug assay, a study design, and statistical analysis of the data. Chemical determination of the concentration of the drug in biological fluids is of utmost importance for bioequivalence of drug products to be established. These chemical methods must be sensitive enough to detect drug concentration at the µg/mL or ng/mL levels. Study designs can be in either parallel or crossover form. In *parallel designs* every patient receives one and only one of the products being tested, whereas in *crossover designs* every patient receives all the products, but administered at different times. Crossover designs are preferred because they reduce the intersubject variability. Between product administrations, a suitable period of time (a washout period) elapses to ensure that administration of the first product does not influence the data collected from the second product. The last component of bioequivalence protocols is the statistical analysis of the data. It is recognized that bioequivalence cannot be established on the strength of one parameter alone. It is therefore essential that in the process of evaluating bioequivalence, the investigator consider the plasma concentration vs. time curve and each major parameter of bioavailability. It is also important to consider the variability around the mean value for both the plasma concentration and pharmacokinetic parameters. When comparing two products, substantially greater variability for one product suggests that the consistency and reproducibility of plasma concentration curves will be poor when the drug is administered to patients. This may occur despite similar average values in the bioequivalence study. An analysis of variance test is generally used with bioavailability or bioequivalence data.

4.4 BIOEQUIVALENCE PROCEDURE

Suppose that a generic drug company initiates a study to establish the bioequivalence of a brand name product. The general procedure includes the following steps:

1. Twelve to 24 healthy subjects are recruited for the study.
2. Each subject is given the brand-name formulation and the generic product in a randomized crossover design unless a parallel design or other design is more appropriate for valid scientific reasons.
3. The plasma or blood drug levels resulting from each product are monitored over a period of time [at least three times the half-life of the active drug ingredient or therapeutic moiety, or its metabolite(s), measured in the blood or urine]. The sampling times should be identical for comparing oral dosage forms; for studies comparing an intravenous dosage form and an oral dosage form, the sampling times should be those needed to describe both. For other drug delivery systems, the sampling times should be based on valid scientific reasons.

4. The C_{max}, T_{max}, and AUC values for each product are compared statistically.

5. If the difference in the parameters is 20% or less, the products are generally considered bioequivalent. The FDA (2000) "recommend [2] that confidence interval (CI) values not be rounded off; therefore, to pass a CI limit of 80 to 125, the value would be at least 80.00 and not more than 125.00."

6. The FDA lists its judgments of which drug products are bioequivalent based on the data submitted by the manufacturer.

7. Two drug products composed of the same dosage form are *A rated* when no known or suspected bioequivalence problems exist. The FDA's A ratings are assigned as follows:

AA	Products in conventional dosage forms not presenting bioequivalence problems
AN	Solutions and powders for aerosolization
AO	Injectable oil solutions
AP	Injectable aqueous solutions
AT	Topical products
AB	Products in which a study has been submitted demonstrating bioequivalence following an initial A-rating denial by the FDA

4.5   FDA-APPROVED METHODS FOR BIOEQUIVALENCE STUDIES

The *Code of Federal Regulations*, 21 CFR 320, gives the following methods as approved for bioequivalence determinations:

> The following in vivo and in vitro approaches, in descending order of accuracy, sensitivity, and reproducibility, are acceptable for determining the bioavailability or bioequivalence of a drug product.

1. (i) An in vivo test in humans in which the concentration of the active ingredient or active moiety, and, when appropriate, its active metabolite(s), in whole blood, plasma, serum, or other appropriate biological fluid is measured as a function of time. This approach is particularly applicable to dosage forms intended to deliver the active moiety to the bloodstream for systemic distribution within the body; or

   (ii) An in vitro test that has been correlated with and is predictive of human in vivo bioavailability data; or

2. An in vivo test in humans in which the urinary excretion of the active moiety, and, when appropriate, its active metabolite(s), are measured as a function of time. The intervals at which measurements are taken should ordinarily be as short as possible so that the measure of the rate of elimination is as accurate as possible. Depending on the nature of the drug product, this approach may be applicable to the category of dosage forms described in paragraph (b)(1)(i) of this section. This method is not appropriate where urinary excretion is not a significant mechanism of elimination.

3. An in vivo test in humans in which an appropriate acute pharmacological effect of the active moiety, and, when appropriate, its active metabolite(s), are measured as a function of time if such effect can be measured with sufficient accuracy, sensitivity, and reproducibility. This approach is applicable to the category of dosage forms described in paragraph (b) (1) (i) of this section only when appropriate methods are not available for measurement of the concentration of the moiety, and, when appropriate, its active metabolite(s), in biological fluids or excretory products but a method is available for the measurement of an appropriate acute pharmacological effect. This approach may be particularly applicable to dosage forms that are not intended to deliver the active moiety to the bloodstream for systemic distribution.

4. Well-controlled clinical trials that establish the safety and effectiveness of the drug product, for purposes of measuring bioavailability, or appropriately designed comparative clinical trials, for purposes of demonstrating bioequivalence. This approach is the least accurate, sensitive, and reproducible of the general approaches for measuring bioavailability or demonstrating bioequivalence. For dosage forms intended to deliver the active moiety to the bloodstream for systemic distribution, this approach may be considered acceptable only when analytical methods cannot be developed to permit use of one of the approaches outlined in paragraphs (b) (1) (i) and (b) (2) of this section, when the approaches described in paragraphs (b) (1) (ii), (b) (1) (iii), and (b) (3) of this section are not available. This approach may also be considered sufficiently accurate for measuring bioavailability or demonstrating bioequivalence of dosage forms intended to deliver the active moiety locally, e.g., topical preparations for the skin, eye, and mucous membranes; oral dosage forms not intended to be absorbed, e.g., an antacid or radiopaque medium; and bronchodilators administered by inhalation if the onset and duration of pharmacological activity are defined.

5. A currently available in vitro test acceptable to FDA (usually a dissolution rate test) that ensures human in vivo bioavailability.

6. Any other approach deemed adequate by FDA to measure bioavailability or establish bioequivalence.

Concerning item 5 above, the FDA recognizes four different levels to *correlate* a dissolution test with in vivo performance (Uppoor, 2005). Level A is a point-to-point correlation (linear or nonlinear) between the dissolution test and the in vivo input rate, level B utilizes all in vivo and in vitro data; however, it estimates parameters related to statistical moment analysis, level C is a correlation between a pharmacokinetics parameter and a dissolution test estimate (e.g., t_{50}), and in multiple level C correlation several of the pharmacokinetics parameters are correlated with the amount of API dissolved at several time points during the dissolution experiments. Level A and multiple level C correlations are considered to be the best suited for comparing in vitro data with the in vivo studies from point of view of the FDA (Uppoor, 2005).

The FDA also declares that "the basic principle in an in vivo bioavailability study is that no unnecessary human research should be done. An in vivo bioavailability study shall not be conducted in humans if an appropriate animal model exists and correlation of results in animals and humans has been demonstrated. In some situations, an in vivo bioavailability study in humans may preferably and more properly be done in suitable patients."

4.6 APPROACHES TO IMPROVING BIOAVAILABILITY

With the exception of intravenous administration, all other routes require a drug to be able to pass through biological membrane to reach the circulation and ultimately, the site of action. For the drug to diffuse through the biological membrane, it has to be present in solution prior to absorption through the biological membrane. Since all body fluids are aqueous in nature, the solubility of the drug in water is perhaps the most important factor in determining the API biological absorption. The *biopharmaceutical classification system* (BCS) of solutes with respect to their solubility and permeability through biological membranes recognizes four categories: I, highly soluble/highly permeable; II, less soluble/highly permeable; III, highly soluble/less permeable; and IV, less soluble/less permeable (Garad, 2004). For the majority of APIs, class II is perhaps the predominant classification (Yang et al., 2011). It is estimated that about 40% of the drugs utilized in practice suffer from poor water solubility (Liversidge, 2011).

Many techniques have been proposed to improve drug *dissolution rate*, and thus its bioavailability. These techniques include micronizing the drug particles, engulfing the drug within a surface that facilitates its transport through the biological membrane, or by dissolving the drug in a vehicle other than water prior to administration (Yang et al., 2011). Other approaches to improving the dissolution and bioavailability of APIs include hot-melt extrusion, spray drying, solid dispersion, change in physical form, complexation, salt synthesis, micellar solutions, use of a prodrug (Garad, 2004; Yang et al., 2011), and use of a biodegradable carrier system (Wu et al., 2011). In the latter approach, biodegradable porous starch foam material was used to enhance the oral bioavailability of lovastatin (a cholesterol-lowering drug). The dissolution rate and bioavailability of lovastatin, a BCS class II drug, were found experimentally to be improved compared with the crystalline form of lovastatin (available commercially in capsule dosage forms) (Wu et al., 2011).

One of the applications of micronization is the formation of nanoparticles, in particular those of 100 nm or smaller sizes. Nanoparticles can be formulated in nanosuspension products or in the form of *amorphous* nanoparticles that provide greater bioavailability for poorly water-soluble APIs (Yang, et al., 2011). Liquid-filled hard gelatin capsules have been employed to enhance the bioavailability of sparingly water-soluble drugs. Hard gelatin capsules are used primarily to deliver powders; however, they can also enclose nonaqueous solvents. Low-melting-point substances (e.g., docusate sodium), low-dose APIs (e.g., triamterene), materials with known stability problems (e.g., vitamin A), production of a sustained-release effect (e.g., phenylpropanolamine in a semisolid oily vehicle), and enhancement of the bioavailability of drugs are all applications of liquid-filled hard gelatin capsules (Cole, 1989).

The latter application of improving bioavailability has been documented with several drugs, including digoxin in a poly(ethylene global) (PEG) 400, alcohol, and propylene glycol mixture (although bioavailability was enhanced, the drug was unstable in this mixture), nifedipine in a semisolid PEG mixture (PEG 200 and PEG 35,000), and triamiterene in PEG 1500 base. In particular, the use of PEG 6000 as a matrix for many drugs was found to enhance the release rate of the drugs into solution (Cole, 1989). Lipid-based drug delivery systems have been investigated as carrier systems for poorly water-soluble drugs. The lipid-based material (in a form of a lipid-only microemulsion) is liquid at room temperature; however, when mixed with a PEG base (such as PEG 3350) it becomes solid. The solidified mass is delivered in a hard gelatin capsule. Upon contact with the gastrointestinal fluid, the lipid-only microemulsion spontaneously forms an emulsion maintaining the API in a soluble state ready for absorption (Serajuddin et al., 2008). A representative formulation of this type includes (% w/w): API (8%), propylene glycol monocaprylate (27.6%), polyoxyl 35 castor oil (27.6%), and PEG 3350 (36.8%) (Serajuddin et al., 2008). The polyoxyl 35 castor oil is a solubilizing and emulsifying and dispersing agent, while propylene glycol monocaprylate acts as an emulsion stabilizer (both components are miscible with each other and form a single phase presenting a solubilizing medium for the API) (Serajuddin et al., 2008).

Rapid-freezing particle engineering technology is another technique used for improving the bioavailability of poorly water-soluble APIs. Briefly, this method utilizes the formation of a rapidly frozen solution of API on a surface (freezing time normally is between 0.05 and 1 second), with subsequent removal of the solvent by sublimation. The method produces a highly porous amorphous *nanostructure*. Rapid freezing has the potential to produce particulate matter of a poorly water-soluble API suitable for pulmonary delivery (Yang et al., 2011).

The delivery of poorly soluble drugs by parenteral routes can be achieved by a simple change in a solution's pH (very high or very low pH values) or by the inclusion of detergents in the formulation. However, both methods result in undesirable effects on the tissues following injection. In addition, the poorly soluble drug may precipitate out after injection and cause phlebitis or even occlusion of the pulmonary capillaries, resulting in pulmonary infarctions. Other methods for delivery of these drugs include emulsions, cyclodextrin–drug complex formation, and entrapment with liposomes. Whereas a water-insoluble drug is found associated *within* the lipid layering of the liposomes, other systems of delivery incorporate the drug within a core surrounded with lipid layers. This technology combines the highly concentrated micronized drug (or solubilized inside an oily phase), housed inside a core, with lipid modifiers (phospholipids) on the surface to facilitate formation of a homogeneous product. The interaction between the lipid phospholipids and the drug is presumed to be weak. These systems allow the formulation of immediate-release (intravenous administration) or controlled-release delivery (intramuscular, subcutaneous, and intradermal) (Pace et al., 1999).

REFERENCES

Baweja KR. Dissolution testing of oral solid dosage forms using HPLC. Pharm. Technol. 1987;11(1):28, 30–31, 34, 36.

Code of Federal Regulations, 21 CFR 320. Title 21, Food and Drugs; Chapter I, Food and Drug Administration Department of Health and Human Services; Subchapter D, Drugs for Human Use; Part 320; Bioavailability and Bioequivalence Requirements. BP RW. 1, Mer. 2003 Title 21, Vol. 5 Revised Apr. 1, 2009.

U.S. Department of Health and Human Services, Food and Drug Administration Center for Drug Evaluation and Research. http://www.accessdata.fda.gov/scripts/cdrh/cfdocs/cfcfr/CFRSearch. cfm?CFRPart=320&show=1. Accessed Feb. 19, 2012.

Cole TE. Liquid-filled hard-gelatin capsules. Pharm. Technol. 1989;13(9):124, 126, 128, 130, 134, 136, 138, 140.

Garad SD. How to improve the bioavailability of poorly soluble drugs. Am. Pharm. Rev. 2004;7(2):80, 82, 84–85, 93.

FDA. 2000. **Guidance for industry**: Bioavailability and bioequivalence—studies for orally administered drug products; general considerations. http://www.fda.gov/downloads/Drugs/ GuidanceComplianceRegulatoryInformation/Guidances/ucm070124.pdf. Accessed Feb. 19 2012.

Jain KS, Jain NK, Agrawal GP. Gastroretentive floating drug delivery: an overview. Drug Deliv. Technol. 2005;5(7):52, 54–61.

Liversidge G. Controlled release and nanotechnologies: recent advances and future opportunities. Drug Dev. Deliv. 2011;11(1):42, 44–46.

Neervannan S. Strategies to impact solubility and dissolution rate during drug lead optimization: salt selection and prodrug design approaches. Am. Pharm. Rev. 2004;7(5):108, 110–113.

Pace NS, Pace WG, Parikh I, Mishra KA. Novel injectable formulations of insoluble drugs. Pharm. Technol. 1999;23(3):116, 118–120, 122, 124, 126, 128, 130, 132, 134.

Serajuddin ATM, Li P, Haefele TF. Development of lipid-based drug delivery systems for poorly water-soluble drugs as viable oral dosage forms: present status and future prospects. Am. Pharm. Rev. 2008;11(3):34, 36, 38–42.

Uppoor RS. Regulatory considerations for in vitro/in vivo correlations. Am. Pharm. Rev. 2005;8(5):42, 44–45.

Wu C, Wang Z, Zhi Z, Jiang T, Zhang J, Wang S. Development of biodegradable porous starch foam for improving oral delivery of poorly water soluble drugs. Int. J. Pharm. 2011;403(1–2): 162–169.

Yang W, Owens ED, Peters IJ, Williams OR. Improved bioavailability of engineered amorphous nanostructured compositions of poorly water-soluble drugs by rapid-freezing particle engineering technology. Drug Dev. Deliv. 2011;11(1):22, 24–31.

GLOSSARY

API	Active pharmeceutical ingredient.
AUC	Area under the curve.
BCS	Biopharmaceutical classification system.
Bioavailability	The rate and extent to which an active ingredient or active moiety is absorbed from a drug product and becomes available at the site of action.
Bioequivalence	The absence of a significant difference in the rate and extent to which the active ingredient or active moiety in pharmaceutical equivalents or pharmaceutical alternatives becomes available at the site of drug action when administered at the same molar dose under similar conditions in an appropriately designed study.

Bioavailability parameters	C_{max} (the highest blood concentration of the drug achieved after administration of the formulation), T_{max} (the time when C_{max} occurs), and the area under the curve (AUC) (the total area under the blood level–time curve).
CFR	*Code of Federal Regulations.*
Clinical equivalents	Produce the same clinical effect in patients suffering from the same condition; biological equivalents have the same physiological profile as determined from the concentration in biological fluids following administration.
FDA	U.S. Food and Drug Administration.
PEG	Poly(ethylene glycol).
Pharmaceutical alternatives	Drug products containing the same therapeutic ingredient but not necessarily in the same chemical type, amount, or dosage form.
Pharmaceutical equivalents	Drug products that contain exactly the same chemical entity as the active ingredient and have identical dosage forms and amounts of drug.

CHAPTER 5

INTERFACIAL PROPERTIES

5.1 INTRODUCTION

Immiscible phases such as gas–liquid, liquid–liquid, gas–solid, or liquid–solid are frequently encountered in pharmaceutical dosage forms. The boundary that is formed between two immiscible phases is known as the *interface*. At the interface there exists a *tension* that prevents the phases from mixing together. This tension is known as the *interfacial tension* (liquid–liquid or solid–liquid) or the *surface tension* (when one of the phases is air, e.g., liquid–air or solid–air). The higher the tension at the interface, the higher the degree of immiscibility between the phases. A *spreading coefficient* is used to assess the *miscibility* or spreadability of one phase on another phase. For example, one of the issues facing dentures is the ability of the device to adhere to the biological tissues of patients. It is obvious that the material used to help a denture adhere to patient's tissues must have a ability to spread on both surfaces: on the patient's biological tissues and on the denture material. The spreadability of the material used in these types of formulations must produce a positive value for the spreading coefficient (S). That is, the work of adhesion (W_a) between the material (m) and the surfaces (s) is greater in magnitude than the work of cohesion (W_c) of the material:

$$S_{m/s} = W_a{}^{m/s} - W_c{}^m \tag{5.1}$$

$$W_a{}^{m/s} = \gamma_m + \gamma_s - \gamma_{m/s} \tag{5.2}$$

$$W_c{}^m = 2\gamma_m \tag{5.3}$$

$$W_c^s = 2\gamma_s \tag{5.4}$$

It is also understood that the adhesive material being used in dentures should be strong enough to hold the denture in place, yet capable of separating relatively easily from its bonding to tissues and denture material when the patient opts to do so (Ali, 1988).

Equation (5.1) may be written with respect to the surface tension and the interfacial tension for a given system:

$$S_{m/s} = W_a{}^{m/s} - W_c{}^m = (\gamma_m + \gamma_s - \gamma_{m/s}) - 2\gamma_m \tag{5.5}$$

$$S_{m/s} = \gamma_s - (\gamma_m + \gamma_{m/s}) \tag{5.6}$$

Integrated Pharmaceutics: Applied Preformulation, Product Design, and Regulatory Science,
First Edition. Antoine Al-Achi, Mali Ram Gupta, William Craig Stagner.
© 2013 John Wiley & Sons, Inc. Published 2013 by John Wiley & Sons, Inc.

From equation (5.6), for the material to spread on the surface, the sum of its surface tension and the interfacial tension (material/surface) must be smaller than the surface tension.

The applications of surface tension principles are widespread and cover areas such as the formation of crystal nuclei, bubbles, and droplets in liquids. In these systems and others (e.g., colloidal dispersions), it is considered to be a destabilizing force, as in foam systems (Rosenthal, 2001).

5.2 LIQUID–SOLID INTERFACE

A drop of liquid placed on a solid surface forms an angle with the surface known as the *contact angle* (θ). The angle thus formed is a function of three parameters: the surface tension of the liquid (γ_L), the surface tension of the solid (γ_S), and interfacial tension between the solid and the liquid (γ_{SL}) (Emory, 1989):

$$\cos\theta = \frac{\gamma_S - \gamma_{SL}}{\gamma_L} \tag{5.7}$$

Equation (5.7) is known as *Young's equation* (Craig et al., 2001). The measurement of θ is performed at a constant temperature, as all the parameters in equation (5.7) decrease with an increase in temperature. The decrease in surface and interfacial tension with a rise in temperature can be linear (water) or nonlinear (ethanol), depending on the nature of the material (Emory, 1989). The contact angle may be measured using a tensiometer. A simpler way to measure the contact angle using a modified optical microscope was presented by Craig et al. (2001). The contact angle may be measured using a light microscope fitted with a crosshair eyepiece and a protractor positioned around the eyepiece's barrel. Briefly, a drop of liquid (e.g., water) is placed on the surface of the solid and examined under a lowpower (20×) optical microscope. When the crosshair is aligned with the tangent of the liquid droplet, the angle formed between the tangent and the horizontal line (the solid surface) is θ. (The protractor is mounted around the eyepiece barrel such that the crosshair can be moved independent of it.) It is important to standardize the size of the liquid droplet, which is added to the surface using a syringe (Craig et al., 2001).

5.3 DOSAGE-FORM APPLICATIONS

The interfacial phenomenon is a very important concept for various pharmaceutical dosage forms. Following is a summary of some of these applications (Fathi-Azarbayjani et al., 2009).

5.3.1 Solid Dosage Forms

Often during the processing of a product formulation, wetting of the solid ingredients with solvents (usually, aqueous in nature) is needed to produce intermediate structures such as granulations. The morphology of the granules is governed by the work of adhesion (W_a):

$$W_a = 4\left(\frac{\gamma_1{}^d\gamma_2{}^d}{\gamma_1{}^d + \gamma_2{}^d} + \frac{\gamma_1{}^p\gamma_2{}^p}{\gamma_1{}^d\gamma_2{}^p}\right) \tag{5.8}$$

where p and d represent the polar and nonpolar contributions of the surface free energy, respectively, and γ_1 and γ_2 are the surface free energies of the binder and the substrate, respectively. One commonly used binder for tablets and capsules is starch, which may originate from various plant sources (e.g., rice, corn, wheat). Depending on the source of the starch, its surface tension varies with the plant species, and its spread on the surface of the substrate will vary accordingly. Another application in this regard is the coating process of tablets and capsules and the interaction of the coating material with the surface of the solid dosage form. Too high or too low a level of surface tension can hinder application of the coat (high surface tension) or cause the coating material to appear ragged (low surface tension).

5.3.2 Dissolution and Disintegration of Solid Dosage Forms

Dissolution and disintegration are essential for preparing an active pharmaceutical ingredient (API) for absorption. Alteration in the dissolution and disintegration could greatly alter the bioavailability of the API. Although for the most part disintegration is considered to be a first step toward dissolution, in reality some dissolution occurs while disintegration is under way. For a solid dosage form to disintegrate, aqueous media must penetrate the microscopic channels that open on the surface of the unit dose. Entry of water through these channels permits disintegration to initiate. However, for water to enter these "capillary" spaces, the contact angle between the water and the dosage unit material must be below $90°$ for some wetting of the surface to occur. A mathematical expression known as the *Washburn equation* is commonly employed for estimating the volume of liquid (usually, water) (v) that penetrates the capillary channels with a radius r in time t, when the surface tension of the liquid (γ_L), the contact angle (θ), and the viscosity of the penetrating liquid (η) are known:

$$v = \left(\frac{r\gamma_L \cos\theta\, t}{2\eta}\right)^{0.5} \tag{5.9}$$

The product of $\gamma_L \cos\theta$ is known as the *adhesion tension* (AD). If AD is positive, a *spontaneous* penetration of the capillary channels occurs; if the AD is negative, spontaneous penetration does not occur. It should be noted that for AD to be positive, θ must be less than $90°$. The surface tension of several pharmaceutical preparations was estimated using the capillary principles already mentioned. The following equation was used to estimate the surface tension of a liquid (γ_1) when knowing the surface tension of a standard liquid (water) (γ_2) at the same temperature (Al-Achi and Shipp, 2005):

$$\gamma_1 = \frac{\gamma_2\rho_1 h_1}{\rho_2 h_2} \tag{5.10}$$

where ρ is the density and h is the rise of the level of a liquid inside a capillary tube. The subscripts 1 and 2 represent the test liquid and the standard liquid, respectively. This technique provided the surface tension for Listerine, hydrogen peroxide solution, saline solution for soft contact lenses, and Scope as 44.83, 72.35, 73.62, and 37.51 dyn/cm, respectively (Al-Achi and Shipp, 2005). The reference standard for these estimations was water, which had a surface tension value of 72.8 dyn/cm. The presence of alcohol and/or other surface-lowering agents in Listerine and Scope reduced their surface tension well below that of water.

5.3.3   Complexation with Carriers

Many carriers for APIs have been suggested or are being used in pharmaceutical dosage forms. Among these carriers are the liposomes and cyclodextrins (CDs). With CD complexation, the CD–API complex is normally observed when the surface tension of the solution complex decreases. An increase in the surface tension signifies desorption of API from the complex. Liposomes as carriers have recently gained acceptance in the pharmaceutical industry community despite their major drawback as unstable formulations. The stability of the formulation can be improved by *flocculating* the suspension of liposomes, a task that involves adjustment of the interfacial tension existing in the dispersion. Moreover, as the cholesterol content of the lipid bilayer increases, the surface tension decreases. Keeping the lipid composition of the lipid bilayer constant and increasing the drug concentration incorporated in liposomes results in an increase in the surface tension of the carriers. Thus, monitoring the surface tension of the liposomes–API dispersion can serve as a good indicator of the complex stability.

5.3.4   Emulsions

Emulsions are coarse pharmaceutical dispersions, with the dispersed-phase particles averaging larger than $0.5\mu m$ in size. Emulsions are unstable systems, as two immiscible liquids are brought into contact and expected to remain so during the life of the product. The interfacial tension between the two liquid phases ($\gamma^{\alpha\beta}$), along with the large interfacial area of the dispersed phase (ΔA), render the system thermodynamically unstable (due to a very high interfacial free energy, ΔF):

$$\Delta F = \gamma^{\alpha\beta} \, \Delta A \qquad (5.11)$$

Emulsions are stabilized physically either by reducing the interfacial tension between the phases and/or by establishing a barrier at the interface to prevent coalescence. The former may be achieved by the addition of surface-active agents to the formulation. (Surfactants are anionic, cationic, or nonionic. Cationic and anionic surfactants can neutralize each other if they are present in the same formulation.) From the point of view of the physical stability of emulsions, creaming and flocculation are considered reversible processes, whereas coalescence is irreversible. The creaming rate can be estimated in dilute emulsions by *Stokes' equation*:

$$v = \frac{2r^2(\rho_I - \rho_E)g}{9\eta_0} \tag{5.12}$$

From equation (5.12) the rate of creaming (v) is a function of the radius of the droplets, the difference in the densities of internal and external phase ($\rho_I - \rho_E$), the viscosity of the dispersion medium (η_0), and acceleration due to gravity (g) (Anonymous, 2011). Coalescence occurs because the protective film surrounding the droplets of the internal phase breaks, leading to a droplet merging with the eventual complete separation of the two phases. Disproportion or *Ostwald ripening* is a phenomenon where by molecules of the dispersed phase move from smaller droplets to large ones through the external phase. According to the *Laplace equation*, the pressure of the dispersed material (P) is inversely related to the radius of the droplet (r) (Anonymous, 2011):

$$P = \frac{2\gamma}{r} \tag{5.13}$$

where γ is the surface tension. Also note that the pressure of dispersed material is directly proportional to the surface tension for a given size of droplet. Thus, the pressure differential between droplets is the driving force for Ostwald ripening. In addition, the rate of diffusion is directly proportional to the diffusion coefficient (D), which in turn is inversely proportional to the viscosity of the continuous phase and the radius of the droplets. According to the *Stokes–Einstein equation*,

$$D = \frac{RT}{6\pi\eta_0 r \Pi} \tag{5.14}$$

Equation (5.14) relates the diffusion coefficient with the radius of the droplets given, Π being Avogadro's number, R the gas constant, and T the temperature of the dispersion in kelvin (Anonymous, 2011). A special class of emulsions, known as *microemulsions*, exists where the droplets approach the size of molecular dispersions (solutions). Thus, very little interfacial tension is found in these systems. Microemulsions demonstrate superiority over regular emulsions in that the delivery of drugs is accomplished in a more rapid and efficient manner.

It is interesting to note here the importance of equation (5.13) (the Laplace equation) in the formation of ice crystal during freezing. Ice crystals growth must follow a nucleation step, as it does not form *spontaneously* on its own, due to the large pressure (due to the high surface tension) imposed on small crystal formation, which causes the crystal to melt. Nucleation forms easily in *nonhomogeneous systems*, whereas for *homogeneous systems* energy is required following placing the system below its equilibrium freezing temperature (T_f) (i.e., under supercooling conditions with a rate equal to ΔT). The input of energy during this process is necessary to overcome the interfacial tension. For example, in order to freeze biological samples without being damaged, it is important to create small crystals during the process. The radius (r) can be maintained small by increasing the rate of supercooling:

$$r = \frac{2T_f \gamma}{L \Delta T} \tag{5.15}$$

where L is the latent heat (the heat absorbed or released by a system during phase transportation without a change in temperature). It is obvious from equation (5.15) that decreasing the interfacial tension (by the addition of surface-active agents) can also result in a smaller value for r; however, it is not as practical as increasing the rate of supercooling (Rosenthal, 2001).

5.3.5   Suspensions

Suspensions, similar to emulsions, are coarse dispersions in which the dispersed phase is made of solid particles suspended in a liquid medium. Suspensions are also thermodynamically unstable, for the same reasons as those for emulsions. A reduction in the interfacial tension between the phases can achieve physical stability with the formation of flocculated suspensions that do not form a hard cake upon sedimentation. Flocculation is the result of a balance established between the attractive forces existing between suspended particles (van der Waals forces) and repulsive forces. At a given zeta potential, which varies from one system to another, the attractive forces exceed those of repulsion, and flocculation occurs. Suspensions can also be stabilized by using steric macromolecules (e.g., acacia, gelatin, casein, vegetable oils), with a portion of the macromolecule polymer adsorbing on the surface of the dispersed particles and the remaining portion projecting out into the dispersion medium (Anonymous, 2009).

5.3.6   Suppositories

Most APIs administered in the form of suppositories are suspensions made in a suppository base. The relative affinity of the API between the suppository base and the aqueous liquid in the rectal area determines whether the drug would be released at the site of administration. It was found that the API's surface tension plays a predictive role in this context. APIs with a high surface tension would partition into the aqueous liquids easier than would those with a lower surface tension value. High surface tension is associated with lower affinity to the lipophilic base of the suppository.

5.3.7   Topical Preparations

The surface tension of normal human skin (clean and dry) is about 27 to 28 dyn/cm. For a topical preparation to spread on the skin, its surface tension must be equal to or lower than that of the skin. In addition, the interfacial tension between the topical preparation and the stratum corneum may be reduced by the addition of surfactants that act as penetration enhancers.

5.3.8   Aerosols

To produce small droplets of aerosol particles via nebulizers, the surface tension of nebulizing preparations must be decreased by the addition of surface-active agents.

REFERENCES

Al-Achi A, Shipp S. Physical characteristics of selected over-the-counter medications. Int. J. Pharm. Compd. 2005;9(1):75–81.

Ali ES. Denture adhesives: a review and guideline for formulation and evaluation methods. Pharm. Technol. 1988;12(1):44, 46, 50, 52, 54, 56.

Anonymous. Emulsion stability and testing. Tech. Brief 2011;2(Particle Sciences, Inc.):1–2.

_____. Physical stability of dispersed systems. Tech. Brief 2009;1(Particle Sciences, Inc.):1–2.

Craig VSJ, Jones AC, Senden TJ. Contact angles of aqueous solutions on copper surfaces bearing self-assembled monolayers. J. Chem. Educ. 2001;78(3):345–346.

Emory FS. Principles of integrity-testing hydrophilic microporous membrane filters: I. Pharm. Technol. 1989;13(9):68, 70, 74, 76–77.

Fathi-Azarbayjani A, Jouyban A, Chan YS. Impact of surface tension in pharmaceutical sciences. J. Pharm. Pharm. Sci. 2009;12(2):218–228.

Rosenthal JA. Demonstration of surface tension. J Chem Ed 2001;78(3):332–333.

GLOSSARY

AD	Adhesion tension.
Adhesion tension	The product ($\gamma_L \cos \theta$).
API	Active pharmaceutical ingredients.
CD	Cyclodextrin.
Contact angle	An angle with the surface formed when a drop of liquid is placed on a solid surface.
Interface	The boundary formed between two immiscible phases.
Ostwald ripening	A phenomenon where molecules of the dispersed phase move from smaller droplets to large ones through the external phase.
Spreading coefficient	Used to assess the *miscibility* or the spreadability of one phase on another phase.
Surface tension/interfacial tension	Tension at the interface prevents the phases from mixing together; known as surface tension (when one of the phases is air, e.g., example liquid–air or solid–air) or as interfacial tension when one of the phases is liquid, e.g., liquid–liquid or solid–liquid)

CHAPTER 6

ADSORPTION PHENOMENON

6.1 INTRODUCTION

Many drugs, such as human insulin and its analog, lispro (Ling et al., 1999), can undergo adsorption on the surface of containers, such as glass or plastic containers (intravenous tubing). Overcoming the adsorption of insulin on plastic tubing can be achieved by the use of a priming volume of an insulin solution (Goldberg et al., 2006). Glass, in particular, can provide a clean surface where physical adsorption is possible. Pharmaceutical-quality glass is available in four different types. Types I, II, and III are used for sterile and parenteral preparations, and type IV is used for all others. Type I glass is composed of borosilicate and is neutral, while types II, III, and IV are soda glass. In addition, the surfaces of types I and II glass are treated by sulfating or sulfuring to make them more chemically resistant (Shabir, 2008). Thus, in theory, types I and II glass should provide a lower capacity for adsorbing substances. Adsorption of substances can be physical (physisorption) or chemical (chemisorption) in nature. *Chemical adsorption* is considered to be irreversible, whereas *physical adsorption* is a reversible process (i.e., adsorption and desorption). During physical adsorption a single or multiplayer formation can occur; chemisorption is limited to a monolayer formation. Many factors can influence the adsorption of drugs on surfaces, such as the ionic strength, the presence of ionic species (anions and cations), and the presence of organic matter on the surface (Bui and Choi, 2010). For example, the presence of trivalent metal cations (Al^{3+}) in solution produces an intense adsorption of acidic drugs on the surface of silica (Bui and Choi, 2010). The adsorption process of *active pharmaceutical ingredients* (APIs) can be monitored by various analytical methods, such as total internal reflection fluorescence (Mollmann et al., 2006), ellipsometry (Mollmann et al., 2005), and fluorescence correlation spectroscopy (Donsmark et al., 2006), among others.

Models to define the adsorption of gases on solid surfaces are now utilized to describe the adsorption of drugs on solid surfaces such as carrier systems. The Freundlich isotherm and the Langmuir isotherm are two commonly used models for the adsorption phenomenon. A *Freundlich isotherm* (*isotherm* indicates a process proceeding at a constant temperature) is described by the following expression:

$$\log V = \log k + n \log P \tag{6.1}$$

Integrated Pharmaceutics: Applied Preformulation, Product Design, and Regulatory Science,
First Edition. Antoine Al-Achi, Mali Ram Gupta, William Craig Stagner.
© 2013 John Wiley & Sons, Inc. Published 2013 by John Wiley & Sons, Inc.

where V is the volume of gas adsorbed, P the gas pressure, and k and n are constants. The constant k reflects the *affinity* of the substance for the adsorbing surface, and n is a measure of the *capacity* (i.e., maximum adsorption). A plot of $\log V$ vs. $\log P$ yields a straight line with a slope of n and a y-intercept of $\log k$.

The *Langmuir isotherm* expression is

$$\frac{P}{V} = \frac{1}{V_m}P + \frac{1}{V_m b} \tag{6.2}$$

with V_m the maximum volume of gas adsorbed at a monolayer point per gram of adsorbent and b the ratio of adsorption rate constant to desorption rate constant. The relationship between P/V and P is linear with a slope of $1/V_m$ and a y-intercept of $1/V_m b$.

A third model for adsorption, known as the *Braunauer–Emmet–Teller* (BET), *isotherm*, may be used to describe the adsorption process. This model can explain the number of layers of the adsorbing material that can form on the solid surface. The BET expression is

$$\frac{P(P_0 - P)}{V} = \frac{1}{V_m C} + \frac{C - 1}{CV_m}\frac{P}{P_0} \tag{6.3}$$

where P_0 is the saturation pressure and C is a proportionality constant of the difference between the heat of adsorption for monolayer and the heat of condensation for the subsequent layers.

The adsorption of human insulin from solutions onto the surface of soybean solid particles followed the Freundlich [equation (6.4)] and Langmuir isotherms [equation (6.5)] (Al-Achi et al., 2003):

$$\log V = \log k + n \log C \tag{6.4}$$

$$\frac{C}{V} = \frac{1}{V_m}C + \frac{1}{V_m b} \tag{6.5}$$

where C replaces P in equations (6.2) and (6.3). C is the equilibrium concentration and V is the number of units (U) of human insulin adsorbed per gram of soybean powder at a given C. From equations (6.4) and (6.5), V_m was estimated to be 106 U (units) of human insulin per gram of soybean powder. The constant k was 20.9 U/g and n was 0.37. The authors concluded that the adsorption of human insulin to the surface of soybean particles was weak in nature, indicating that soybean particles can act as a carrier system for this hormone if formulated in a suspension dosage form. The interaction of human insulin with erythrocyte membrane was also described as an adsorption phenomenon as well as internalization of human insulin in the red blood cells through a diffusion process (Al-Achi, 1993; Al-Achi, 1998). The chemotherapeutic agent doxorubicin was also shown to adsorb on the surface of human erythrocytes (Al-Achi and Boroujerdi, 1990). The Freundlich adsorption isotherm expression for doxorubicin adsorption was reported as

$$\log V = 2.466 + 1.051 \log C \tag{6.6}$$

Based on equation (6.6), the value of the constant k was 292.42 mg of drug per gram of erythrocytes and that for n was 1.05. A large value for n indicates a greater ratio of adsorbed drug molecules to free molecules. It can also be related to the strength of the bonds between the drug and the adsorbing surface; larger values reflect stronger bonding to the surface (Al-Achi and Boroujerdi, 1990). The authors also showed that disruption of the red blood cell membrane by a mechanical force (sonication) did not alter the nature of adsorption of the drug to the surface. This may suggest that the adsorption was mainly nonspecific rather than being specific to certain adsorbing moieties present on the surface (such as the cellular membrane protein spectrin; it was reported that at most 10% of the binding of doxorubicin to erythrocyte membrane was attributed to binding to spectrin) (Al-Achi and Boroujerdi, 1990).

Thermodynamic parameters for adsorption (free energy ΔG, enthalpy ΔH, and entropy ΔS) may be estimated based on the Langmuir constant b (Akaho and Fukumori, 2001):

$$\Delta G = -RT \ln b \tag{6.7}$$

$$\Delta H = \frac{d(\Delta G/T)}{d(1/T)} \tag{6.8}$$

$$\Delta S = \frac{\Delta H - \Delta G}{T} \tag{6.9}$$

The symbols in equations (6.7), (6.8), and (6.9) are those for temperature in kelvin (T) and the gas constant (R).

The adsorption phenomenon has several applications in pharmacy, including, but not limited to, dosage-form preparations, understanding the stability of APIs in solution in contact with containers or linings, filter retention of APIs on the filter material during processing, purification of proteins, enhancing the solubility of poorly water-soluble APIs, clinical applications, and removal of contaminants from solutions, among others. In the latter case, endotoxin removal from protein solutions represents an important application of adsorption in the area of pharmaceutical applications. Purification of proteins by the *adsorption affinity method* is gaining more and more acceptance in the pharmaceutical industry. This technique relies on the adsorption of the desired protein on an adsorbent, allowing separation of the protein from solution, leaving the unwanted substances behind. For example, purification of insulin from solution was achieved by adsorbing the protein on beads made of poly[ethylene glycol dimethacrylate-N-methacryloyl-(l)-histidine] [poly(EDMA-MAH)]. Langmuir isotherm data showed that maximum insulin adsorption on poly(EDMA-MAH) beads was 24.7 mg insulin/g beads (Sari et al., 2011). Often, activated charcoal may be used to clear solutions from unwanted ingredients by providing an adsorbing surface. The removal of ethylenediaminetetraacetic acid (EDTA) from solution by activated charcoal was studied under varying conditions of pH, EDTA concentration, activated charcoal amount and particle size, and temperature. The adsorption of EDTA on charcoal was found to increase with increasing temperature (an endothermic process) or with an increase in the amount of charcoal added. EDTA adsorption on activated charcoal followed very closely both the Freundlich and Langmuir isotherms (Zhu et al., 2011).

Methylene blue (MB) was successfully removed from aqueous solutions by adsorbing MB on anionic starch microspheres prepared from neutral starch microspheres (NSMs) in the presence of an anionic etherifying agent. The adsorption of MB on the NSMs was documented using Langmuir and Freundlich isotherm models (Yang et al., 2010). Endotoxins can cause harmful manifestations if injected with the protein products, and thus its elimination from the product is very important. For example, endotoxins were totally eliminated (average removal of 99%) from solutions of proteins (hemoglobin, human serum albumin, and lysozymes) by passing the solution through a polymyxin affinity adsorbent. Recovery of the proteins after the endotoxin treatment averaged approximately 86% (Xing et al., 2010).

With respect to poorly water-soluble APIs, the adsorption of the drug on a hydrophilic carrier was shown to improve its solubility manyfold. For example, ketoprofen and griseofulvin, two poorly water-soluble drugs, were adsorbed on the surface of hydrophilic silica aerogels. The physical adsorption of ketoprofen and grieseofulvin on the aerogels resulted in capturing 30% by weight of ketoprofen and 5.4% by weight of grieseofulvin in the aerogel matrix. Due to the high surface area of aerogels and its ability for an immediate collapse in aqueous media, the release rate of the drugs increased by 500% and 450% for ketoprophen and grieseofulvin, respectively (Smirnova et al., 2004). Another issue of importance is the adsorption of plasma proteins on the surfaces of medical devices during treatment. The adsorption of several human plasma proteins, including Apo A-I, Apo A-IV, fibrinogen, haptoglobin, IGHG1 protein immunoglobulin, serum retinal-binding protein, and truncated serum albumin, was reduced significantly by coating the surfaces with dextran. This treatment with dextran can be effective in reducing the incidence of thromboinflammatory responses that may be encountered with the use of medical devices (Tsai et al., 2011). Attempting to lower cholesterol absorption from the gastnointestinal tract, thus reducing the risk of hyperlipidemia, is one approach to combating this disease. A chitosan salt with 2-(4-chlorophenoxy)-2-methylpropionic acid (CMP) was prepared and tested experimentally in rats for its ability to adsorb bile acids. The oral treatment in rats with this salt effectively reduced serum cholesterol and triacylglycerol concentrations for 2 weeks. In vitro, bile acids were exchanged with CMP when chitosan–CMP was added to their solutions, releasing CMP and capturing the bile acids in its place (Murata et al., 2009).

6.2 ADSORPTION ON FILTERS

The filtration process in used extensively as a step during formulation, in particular in the preparation of sterile products. Ophthalmic preparations often contain benzalkonium chloride (BC) as a preservative. During extemporaneous compounding operations of ophthalmic solutions, the method of choice for sterilization is filtration. Normally, a 13-mm filter allows filtration up to 10 mL of sample, and a 25-mm filter permits up to 100 mL of volume filtration. In addition, the preparer has available to him or her several types of filter to use during processing. Various filter types are available from various companies. It was found that the size, the pore size (0.2 μm vs. 0.45 μm), the type, the filter media, and the company all have an effect on the amount of BC adsorbed on the filter material. For BC, the adsorption was a physical type since BC was washed off the filter with water rinsing. The adsorption of BC on filter material was least for Gelman PES (HT Tuffryn) 13 mm in diameter with 0.45-μm pores (Prince and Allen, 1998). Surfactants are commonly used to reduce the interfacial tension between immiscible phases in order to stabilize the formulation. The adsorption of polysorbate 80, a nonionic surfactant, was evidence of whether a sterilizing-grade filter was used in the process, whereas proteins that were present in the same solution with polysorbate 80 adsorbed preferentially on nylon filters (Mahler et al., 2010). It is highly advisable that the adsorption of materials on filters be determined on a case-by-case basis, as this may depend on the type of filter being used, the protein in question, and the specifics of the formulation (Mahler et al., 2010; Tozuka et al., 2010).

6.3   ADSORPTION OF PROTEINS

The delivery of protein presents a new challenge to the pharmaceutical industry, as many new biotechnology products are of this type. The protein content of formulation may be lost to chemical degradation or to physical stress (agitation). The adsorption of protein molecules on various surfaces can also lead to loss of soluble protein from formulation. The nonspecific adsorption of proteins to the surface of glass tubing, for example, can be minimized or overcome by the pretreatment of the surface with bovine serum albumin (Dixit et al., 2011). In certain cases, protein adsorption on the surface can be related to an electrostatic interaction (Zhang et al., 2009; Anand et al., 2010). The adsorption of proteins on an oil–water interface can be observed during processing of the formulation. Upon adsorption on the interface, the protein aggregates, resulting in the formation of a viscoelastic multilayer structure. The viscosity of the multilayer was found to increase with increasing protein concentration, and the rate of adsorption was a factor in the molecular weight of the protein (Baldursdottir et al., 2010). It is rather a common practice in clinical pharmacy to store sterile formulations in prefilled syringes. Silicon oil is used in syringes as a lubricant to facilitate the movement of different parts of a syringe with respect to each other. However, silicon oil was found to provide a surface where various proteins in solution could adsorb on the oil surface, allowing the loss of soluble protein from the solution. Additives present in the solution might enhance or decrease the adsorption. The presence of surfactants decreased the adsorption to silicone oil, whereas sodium chloride enhanced the adsorption process (Ludwig et al., 2010). Long-chain surfactants formed complexes with proteins through electrostatic and hydrophobic interactions, changing the nature by which a protein could interact with the surface (Pradines et al., 2011).

The degree of glycosylation of proteins, a common method to improve the stability and delivery profile of proteins, was found to have no effect on the adsorption of proteins on hydrophobic surfaces; on hydrophilic surfaces, however, a greater degree of glycosylation was associated with a higher adsorption potential (Pinholt et al., 2010).

REFERENCES

Akaho E, Fukumori Y. Studies on adsorption characteristics and mechanism of adsorption of chlorhexidine mainly by carbon black. J. Pharm. Sci. 2001;90(9):1288–1297.

Al-Achi A, Boroujerdi M. Adsorption isotherm for doxorubicin on erythrocyte-membrane. Drug Dev. Ind. Pharm. 1990;16(8):1325–1338.

Al-Achi A, Greenwood R. Human insulin binding to erythrocyte-membrane. Drug Dev. Ind. Pharm. 1993;19(6):673–684.

Al-Achi A, Clark JT, Greenwood R, Sipho Mafu S. Human insulin interaction with soybean powder. Pharm. Eng. 2003;23(1):40–45.

Al-Achi A, Greenwood R. Erythrocytes as oral delivery systems for human insulin. Drug Dev Ind Pharm 1998;24(1):67–72.

Anand G, Sharma S, Dutta AK, Kumar SK, Belfort G. Conformational transitions of adsorbed proteins on surfaces of varying polarity. Langmuir 2010;26(13):10803–10811.

Baldursdottir SG, Fullerton MS, Nielsen SH, Jorgensen L. Adsorption of proteins at the oil/water interface: observation of protein adsorption by interfacial shear stress measurements. Colloids Surf. B 2010;79(1):41–46.

Bui TX, Choi H. Influence of ionic strength, anions, cations, and natural organic matter on the adsorption of pharmaceuticals to silica. Chemosphere 2010;80(7):681–686.

Dixit CK, Vashist SK, Maccraith BD, O'Kennedy R. Evaluation of apparent non-specific protein loss due to adsorption on sample tube surfaces and/or altered immunogenicity. Analyst 2011;136(7): 1406–1411.

Donsmark J, Jorgensen L, Mollmann S, Frokjaer S, Rischel C. Kinetics of insulin adsorption at the oil–water interface and diffusion properties of adsorbed layers monitored using fluorescence correlation spectroscopy. Pharm. Res. 2006;23(1):148–155.

Goldberg PA, Kedves A, Walter K, Groszmann A, Belous A, Inzucchi SE. "Waste not, want not": determining the optimal priming volume for intravenous insulin infusions. Diabetes Technol. Ther. 2006;8(5):598–601.

Ling J, Hu M, Hagerup T, Campbell RK. Lispro insulin: adsorption and stability in selected intravenous devices. Diabetes Educ. 1999;25(2):237–245.

Ludwig DB, Carpenter JF, Hamel JB, Randolph TW. Protein adsorption and excipient effects on kinetic stability of silicone oil emulsions. J. Pharm. Sci. 2010;99(4):1721–1733.

Mahler HC, Huber F, Kishore RS, Reindl J, Rückert P, Müller R. Adsorption behavior of a surfactant and a monoclonal antibody to sterilizing-grade filters. J. Pharm. Sci. 2010;99(6):2620–2627.

Mollmann SH, Bukrinsky JT, Frokjaer S, Elofsson U. Adsorption of human insulin and AspB28 insulin on a PTFE-like surface. J. Colloid Interface Sci. 2005;286(1):28–35.

Mollmann SH, Jorgensen L, Bukrinsky JT, Elofsson U, Norde W, Frokjaer S. Interfacial adsorption of insulin conformational changes and reversibility of adsorption. Eur. J. Pharm. Sci. 2006;27(2–3): 194–204.

Murata Y, Kodama Y, Hirai D, Kofuji K, Kawashima S. Properties of an oral preparation containing a chitosan salt. Molecules 2009;14(2):755–762.

Pinholt C, Fanø M, Wiberg C, Hostrup S, Bukrinsky JT, Frokjaer S, Norde W, Jorgensen L. Influence of glycosylation on the adsorption of *Thermomyces lanuginosus* lipase to hydrophobic and hydrophilic surfaces. Eur. J. Pharm. Sci. 2010;40(4):273–281.

Pradines V, Fainerman VB, Aksenenko EV, Krägel J, Wüstneck R, Miller R. Adsorption of protein–surfactant complexes at the water/oil interface. Langmuir 2011;27(3):965–971.

Prince JS, Allen VL. Sorption of benzalkonium chloride to various filters used in processing ophthalmics. Int. J. Pharm. Compd. 1998;2(3):249–251.

Sari MM, Armutcu C, Bereli N, Uzun L, Denizli A. Monosize microbeads for pseudo-affinity adsorption of human insulin. Colloids Surf. B 2011;84(1):140–147.

Shabir AG. Review of pharmaceutical product stability, packaging and the ICH guidelines. Am. Pharm. Rev. 2008;11(1):139–141.

Smirnova I, Suttiruengwong S, Seiler M, Arlt W. Dissolution rate enhancement by adsorption of poorly soluble drugs on hydrophilic silica aerogels. Pharm. Dev. Technol. 2004;9(4):443–452.

Tozuka Y, Sugiyama E, Takeuchi H. Release profile of insulin entrapped on mesoporous materials by freeze–thaw method. Int. J. Pharm. 2010;386(1–2):172–177.

Tsai IY, Tomczyk N, Eckmann JI, Composto RJ, Eckmann DM. Human plasma protein adsorption onto dextranized surfaces: a two-dimensional electrophoresis and mass spectrometry study. Colloids Surf. B 2011;84(1):241–252.

Xing H, Huang Y, Li Y, Luo J, Zhang L, Ma G, Su Z. Effect of affinity medium and solution conditions on endotoxin removal from protein solutions. Sheng Wu Gong Cheng Xue Bao 2010;26(11):1584–1595.

Yang Y, Wei X, Sun P, Wan J. Preparation, characterization and adsorption performance of a novel anionic starch microsphere. Molecules 2010;15(4):2872–2885.

Zhang S, Tanioka A, Saito K, Matsumoto H. Insulin adsorption into porous charged membranes: effect of the electrostatic interaction. Biotechnol. Prog. 2009;25(4):1115–1121.

Zhu HS, Yang XJ, Mao YP, Chen Y, Long XL, Yuan WK. Adsorption of EDTA on activated carbon from aqueous solutions. J. Hazard. Mater. 2011;185(2–3):951–957.

GLOSSARY

Adsorption	Adsorption of substances can be physical (physisorption) or chemical (chemisorption) in nature.
API	Active pharmaceutical ingredient.
BC	Benzalkonium chloride.
BET	Braunauer, Emmet, and Teller.
Braunauer–Emmet–Teller (BET) isotherm	May be used to describe the adsorption process; can explain the number of layers of the adsorbing material that can form on a solid surface.
Chemical/physical adsorption	Chemical adsorption is considered to be irreversible, whereas physical adsorption is a reversible process (i.e., adsorption and desorption).
CMP	2-(4-Chlorophenoxy)-2-methylpropionic acid.
EDTA	Ethylenediaminetetraacetic acid.
Freundlich and Langmuir isotherms	Two models commonly used for the adsorption phenomenon.
MB	Methylene blue.
NSM	Neutral starch microsphere.
poly(EDMA-MAH)	Poly[ethylene glycol dimethacrylate-N-methacryloyl-(l)-histidine].

CHAPTER 7

RHEOLOGICAL PRINCIPLES

7.1 INTRODUCTION

The science that deals with the flow of liquids is known as *rheology*, and its subject is the parameter *viscosity*. Rheological properties of liquids determine many of their applications; pouring a suspension from its bottle, squeezing a toothpaste gel from its tube, and the flow of a parenteral solution through an infusion feeding tube are just a few examples of these applications. Rendering a liquid thicker (Allen, 1999; Allen, 2000) or reducing another's thickness, dissolving a tablet powder in already prepared syrup, or mixing two different liquids with significantly varied viscosity are common challenges for the formulator. In this chapter we introduce rheology and define its roles in preparing liquid dosage forms.

The viscosity of a liquid at rest is constant under constant conditions of temperature and pressure. The viscosity of a liquid is a reflection of what goes on internally during movement. The various molecules, ions, and aggregates that are dispersed in a liquid undergo *shear* as they slide over each other. The result is internal friction, and *viscosity* is a collective term for this friction during the movement of a liquid. This explains why a more concentrated solution is more viscous than its diluted counterpart. For example, syrups containing 50% w/v sucrose will be less viscous than simple syrup, N.F. (*National Formulary*) (85% w/v). When the temperature of a liquid increases, its molecules, ions, and particles become more energetic and thus they move more easily relative to each other when they come in contact. Thus, the internal friction will be reduced, as will the viscosity. An increase in pressure over the liquid will have the opposite effect on viscosity from that observed with a temperature change.

Certain pharmaceutical systems do not conform particularly to either Newtonian or non-Newtonian flow, and their rheological profile is described with the term *viscoelasticity*. Viscoelastic materials exhibit both the fluid properties of liquids and the elastic properties of solids. Examples of viscoelastic systems in pharmacy are ointments, pastes, and creams (semisolid systems).

Integrated Pharmaceutics: Applied Preformulation, Product Design, and Regulatory Science, First Edition. Antoine Al-Achi, Mali Ram Gupta, William Craig Stagner.

7.2   NEWTONIAN SYSTEMS

When a force (F) is applied to a liquid to move it (shear it), the liquid layers situated in a parallel arrangement begin to move with varying speeds with respect to each other. Due to the existence of molecular bonds between adjacent layers, movement of the layers is restricted. This restriction is known as the *internal friction* of the liquid. The parameter that describes the internal friction is known as the *viscosity* of the liquid. During the shearing movement of the liquid, the difference in the velocity between two adjacent parallel planes of liquid separated by a small distance is known as the *shear rate* (G). *Newton's law of flow* describes the movement of the liquid by the following expression (Allen, 2003):

$$F = \text{constant} \cdot G \qquad (7.1)$$

where the constant is viscosity (η). We could rearrange equation (7.1) in the linear form

$$G = \frac{1}{\eta} F \qquad (7.2)$$

A graph based on equation (7.2) relating G and F (known as a *rheogram*) results in a slope equal to the inverse of viscosity, a term commonly known as the *fluidity* (Figure 7.1). The units for η are poise and centipoise [1 Poise $=$ 100 centipoise (cP)]. One poise is defined as the force per unit area (dyn/cm^2) that is required to bring about a difference in velocity (1 cm/s) between two adjacent parallel planes of liquid separated by a distance of 1 cm and each 1 cm^2 in surface area. Fluidity has units of $poise^{-1}$. Since the viscosity for Newtonian systems is estimated from the slope of the line (F vs. G), the viscosity of Newtonian liquids does not change with a change in G. In other words, as the shearing rate increases or decreases, the viscosity of a Newtonian liquid remains constant (Allen, 2003).

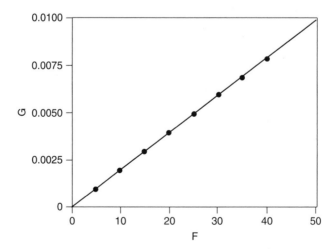

FIGURE 7.1   Newtonian flow. G is the rate of shear and F is the shearing force.

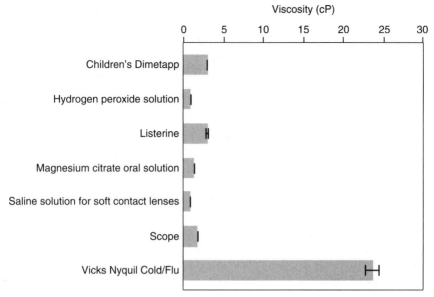

FIGURE 7.2 Viscosity (cP) of some nonprescription products at room temperature. (Adapted from Al-Achi and Shipp, 2005.)

Many pharmaceutical liquid dosage forms behave as Newtonian fluids (Figure 7.2). For example, oral solutions, syrups, and elixirs all exhibit a Newtonian type of flow. Certain liquids, such as mineral oil, glycerin, and purified water, also show a Newtonian rheological profile. Mineral oil exists in two main varieties, heavy and light. These two types have different densities, and to compare them rheologically we use the term *kinematic viscosity*:

$$\text{kinematic viscosity} = \frac{\eta}{\rho} \tag{7.3}$$

where ρ is the density of the liquid. Thus, kinematic viscosity is often used for fluids that exist in different densities and is expressed in units of stokes or centistokes (1 stoke = 100 centistoke). Also, *kinematic viscosity* is adopted by the *United States Pharmacopeia* U.S.P. as the term for viscosity for substances listed in the compendia (Allen, 2003).

7.3 NON-NEWTONIAN SYSTEMS

If the viscosity of a material *changes* with varying shear rate, the material is described as *non-Newtonian*. The overall mathematical expression for these systems may be stated as follows:

$$F^Q = \eta' G \tag{7.4}$$

where Q is the coefficient and relates to the type of flow and η' is the viscosity coefficient. As the value of Q approaches 1, the system becomes more and more

Newtonian in its flow properties. If the value Q is larger or smaller than 1, the material is described as pseudoplastic or dilatant, respectively. *Pseudoplastic materials* demonstrate a decrease in their viscosity (η') (thinning) as the rate of shear on the system increases (Figure 7.3), whereas *dilatant systems* show an increase in their viscosity (thickening) as G increases (Figure 7.4). Emulsions, in particular, demonstrate pseudoplastic flow. For example, emulsions prepared with acacia as an emulsifying agent have shown a clear pseudoplastic rheological profile (Al-Achi et al., 2007). A third non-Newtonian system, known as *plastic*, also shows a reduction in its viscosity with an increase in flow rate; however, its viscosity reaches a minimum constant value at shearing forces exceeding a certain value that varies from one system to another (Figure 7.5). This force is commonly referred to as the *yield value* (f). The constant minimum value for the viscosity of *plastic systems* is known as the plastic viscosity and can be estimated from the latter linear portion of the F vs. G rheogram. The slope of the latter linear portion is the inverse of the plastic viscosity (the slope is known as the *mobility*). For example, Maalox (Novartis Pharmaceuticals) exhibits pseudoplastic flow, whereas Children's Motrin (McNeil Consumer & Specialty Pharmaceuticals) and Dulcolax Milk of Magnesia (Boehringer Ingelheim Pharmaceuticals, Inc.) exhibit plastic flow (Al-Achi and Shipp, 2005).

Both pseudoplastic and plastic materials exhibit *thixotropy*; that is, the system recovers slowly over time after the stress is removed as it returns to its original consistency. Upon applying *constant* shearing (i.e., at a constant G value), pseudoplastic and plastic systems show a decrease in their viscosity value over *time*, and the system *recovers* to its original thickness (i.e., viscosity) when the stress is removed. Thixotropy is a property that is highly desirable and taken advantage of in suspensions, emulsions, and other systems. In these dispersed systems, the formulator aims to have high viscosity for the preparation upon standing and a

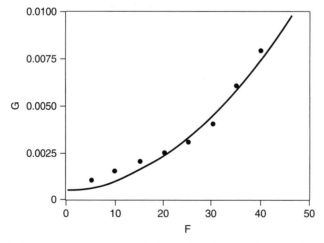

FIGURE 7.3 Rheogram depicting pseudoplastic flow. The y-axis is the shearing rate and the x-axis is the shearing force.

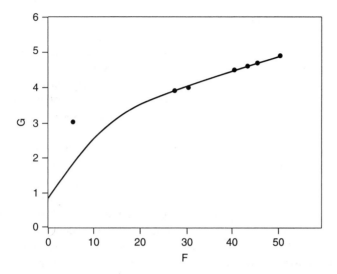

FIGURE 7.4   Rate of shear (G) vs. shearing force (F) for a dilatant material.

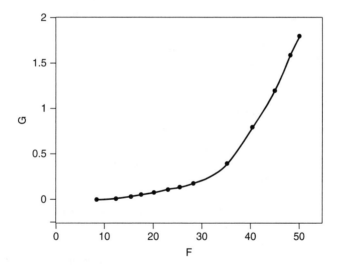

FIGURE 7.5   Plastic flow. G and F are the shearing rate and shearing force, respectively.

reduced viscosity when sheared. This is achieved by employing a pseudoplastic or a plastic material as a vehicle in the formulation (Allen, 2003).

A wide range of instruments are commercially available for measuring the viscosity of Newtonian and non-Newtonian systems. These include capillary type, falling-sphere, cup, and rotational viscometers for measuring Newtonian fluid. A cone and plate viscometer is used for measuring both Newtonian and non-Newtonian systems (Allen, 2003).

7.4 VISCOELASTICITY

Often, rheological properties of semisolids are measured along with textural and mechanical testing of formulations to provide a more complete physical profile of a pharmaceutical product. Textural tests are commonly carried out utilizing a texture analyzer to provide information on the *adhesiveness, hardness*, and *compressibility* of the semisolid formulation. *Adhesiveness* is a term that quantifies the work needed to overcome the adhesive forces existing at the contact surfaces of the instrument probe and the formulation. *Hardness* is a measure of the force required to produce a specified deformation in the structure of the formulation. The amount of work that is used to produce an initial deformation of the material is known as *compressibility* (Jones et al., 2003). Semisolid dosage forms when sheared using non-Newtonian viscometers produce *chaotic* rheograms that have no definite shape. Therefore, instrumentation for measuring the viscoelastic properties of materials is available under the general name *rheometers*. Moreover, conventional viscosity measurements using the non-Newtonian model may not be satisfactory to differentiate between formulations with respect to their rheological properties. For example, Shaw and Parsi (2007) showed that three formulations of an oil-in-water lotion or emulsion showed a similar non-Newtonian profile; however, when tested using rotational rheometry measurements, one of the formulations differed significantly from the other two in its rheological properties. In these rheometry measurements, the system is subjected to constant stress over time. The ratio of strain over stress is known as *compliance* (J). Plotting J (y-axis) vs. time (x-axis) produces a graph known as a *creep curve*. This curve has three distinct regions: a first short region, which is represented as a line perpendicular to the x-axis; followed by a second, curvy-linear segment; and a final, straight-line portion. The first region represents the elastic property of the material, the second segment correlates to both the solid and liquid properties of the system, and the linear segment describes the liquid properties of the material (Davis, 1974). The classical depiction of the creep curve is by a mechanical model composed of a spring and a dashpot (a piston moving inside a viscous Newtonian liquid). The arrangement of these mechanical units is to place the spring and dashpot in series (known as a *Maxwell unit*) with *Voigt unit(s)* (the spring and dashpot are in a parallel arrangement) inserted between the spring and the final dashpot.

The various regions of the creep curve can now be explained by a mechanical model: The initial rise in the creep curve corresponds to the strain in the spring, the curvilinear portion corresponds to the Voight unit(s), and the straight-line segment represents the final dashpot unit. From the slope of the straight-line segment of the creep curve the value of the *ground-state viscosity* (η_0) is found (η_0 is obtained from the reciprocal of the slope value) (Davis, 1974). Because of the nondestructive nature of these measurements, the viscosity value (η_0) remains relatively stable despite subjecting the sample to repeated measurements. The parallel position of the spring and the dashpot in the Voight unit necessitates a parameter to describe how the two units are behaving with respect to each other during the shearing process. This parameter, known as the *retardation time* (τ), is defined as the ratio of the viscosity to the elasticity (remember that the Voight unit is composed of the elastic spring and dashpot placed parallel to each other). For every Voight unit in the model, a retardation time is estimated. The higher the complexity of the material being examined, the higher the number of Voight units needed in the model. The removal of stress from the sample results in a graph similar to that when the stress was applied—however, now in a downward reversal mode. When the sample is highly concentrated (gel-like), the downward curve reaches the x-axis; otherwise, it reaches an equilibrium point above the x-axis (Davis, 1974).

Viscoelastic properties of materials may also be estimated using oscillation stress (Davis, 1974). Several parameters are defined in this process, recognizing the elastic (solid) and viscous (fluid) nature of these materials. Among the viscoelastic parameters being defined are the *elastic (storage) modulus* (G') the *viscous (loss) modulus* (G''), the *dynamic viscosity* (η'), and the *loss tangent* (tan δ) (δ is the phase lag in degrees). G' and G'' are related to δ by the following expressions:

$$G' = \frac{G}{F} \cos \delta \tag{7.5}$$

$$G'' = \frac{G}{F} \sin \delta \tag{7.6}$$

The dynamic viscosity is the ratio of the viscous modulus (G'') to the frequency of oscillation (ω) in rad/s. The *dynamic rigidity* (η'') is the ratio of G' (storage modulus) to the frequency of oscillation (ω) (Jones et al., 2003). The loss tangent (tan δ) is given by

$$\tan \delta = \frac{G''}{G'} \tag{7.7}$$

When the material is perfectly elastic, δ is $0°$ (tan $\delta < 1$), whereas if it is a perfect liquid, the value is $90°$ (tan $\delta > 1$) (Davis, 1974). In general, $\delta \geq 45°$ reflects a viscous system structure, whereas $\delta < 45°$ is associated with elastic systems (Biradar et al., 2009). During cyclic deformation, tan δ is viewed as the ratio of energy lost to the energy stored; a decrease in the loss tangent is indicative that the material is becoming more elastic, whereas an increase in its value is associated with more viscous dominance (Segers et al., 1997).

A *zero-shear-rate viscosity* (*equilibrium viscosity*) may be used to compare formulations made from different concentrations of similar polymeric materials. This viscosity is obtained from the *Cross model*, which is used to define the viscosity under low-shear-rate conditions (Jones et al., 2003). Another parameter to consider is the *complex modulus* ($G*$). This modulus is used to describe the total resistance of a system to strain. The importance of $G*$ is that it describes the system under linear viscosity where the structural integrity of the system is, and remains, intact (Lippacher et al., 2004). In addition, information generated from this region describes the complex viscosity ($\eta*$). The magnitude of $\eta*$ for polymeric dispersions equals the viscosity at a given shearing rate and oscillatory frequency (based on the Cox–Merz rule) (Jones et al., 2003). *The linear viscosity is directly proportional to the frequency of oscillation:*

$$\eta* \alpha \omega^{-\upsilon} \tag{7.8}$$

where v can have values ranging from 0 to 1, with 0 representing liquid and 1 associated with solid (Thorgeirsdóttir et al., 2005).

Carboxymethylcellulose (CMC) is often used in preparations intended for pharmaceutical, cosmetic, or food products. CMC exhibits pseudoplastic flow properties with shear-thinning characteristics as the CMC solution is exposed to an increased shear rate. Non-Newtonian flow behavior is often due to aggregation or disaggregation of the particles as they undergo various orientations, deformations, and disentanglement of their polymeric chains during shearing. For CMC, applying either rotational or oscillatory perturbation instruments results in a similar characterization for its flow properties in solutions (Thurston and Martin, 1978). Oscillatory stress testing was helpful in identifying the best formulation consistency for a bioadhesive chlorhexidine (antimicrobial) formulation targeted to the oropharynx region. The natural oscillatory rate due to ciliary movement was reported to be in the range 0.5 to 3.0 Hz.

The preferred formulation was chosen so that it exhibited a balance between acceptable texture and viscosity characteristics. These were determined by studying formulations containing, in addition to the drug, various combinations of hydroxyethylcellulose (HEC), poly(vinylpyrrolidone) (PVP), and polycabophil (PC). An increase in the concentration of HEC, PVP, and PC in the formulation resulted in an increase in the storage and loss moduli. As the oscillatory frequency increased along with a decrease in the concentration of each of the polymeric components, there was a decrease in the storage modulus (tan δ). Also, the dynamic viscosity of the formulation experienced a reduction in value as the oscillatory frequency increased, opposite to that seen when the concentration of the polymeric materials increased in the formulation. A *plateau region of viscoelasticity* was observed, which is often seen with highly cross-linked polymeric mixtures (creams or gels made of cellulose polymers, PC, or xanthan gum). This plateau region is also observed with systems containing high-molecular-mass uncross-linked polymers (Jones et al., 1998).

Applications of viscoelasticity in pharmacy are numerous and include those pertaining to topical preparations with respect to their spreadability on a skin surface. Viscoelastic monitoring permits identifying changes in the formulation during its storage and shelf life. Filling, mixing, pumping, and stirring are all industrial processes that depend to a certain extent on the viscoelastic properties of material (Lippacher et al., 2004). Thus, viscoelasticity is common during many manufacturing processes that apply forces or pressures on the material during handling. For example, materials under compression forces, such as those in tablet formation, undergo elastic and plastic deformation with a subsequent release of energy when the tablet is formed. During the manufacturing of tablets, powder mixtures or granulations are initially brought together within the narrow space of a die cavity. As compression force is applied, the solid particles change elastically, which leads to fragmentation. Subsequent increase in the compression force initially produces plastic deformation that causes the particles to disintegrate, with the creation of new surfaces. Plastic deformation yields to elastic deformation upon increasing the compression further. When the tablet is released, the elastic deformation energy is dissipated over a period of time that ranges from seconds to hours (Jürgens, 2006).

In tablet compression studies, the rheological properties of the powder mixture were found to have a great effect on tablet formation. For example, lots of powder mixtures with similar particle characteristics as determined by microscopy imaging techniques and laser diffraction methods produced totally different powder compressibility. These lots differed in their rheological properties when tested with a powder rheometer. The rheometer measures the flow energy of the powder as a predictor of powder's flowability (Jürgens, 2006).

7.5 REYNOLDS NUMBER

The *Reynolds number* is an expression of the ratio of inertial to viscous forces in a fluid (Klein, 1999). The usefulness of the Reynolds number (N_{Re}) is in understanding the various phenomena involved in the mixing of fluids and in predicting whether the mixing is laminar or turbulent. N_{Re} relates to several parameters involved in agitating liquids. A direct relationship exists between N_{Re} and the square of the diameter of the agitator blade (D), the density of the solution (ρ), and the agitator speed (B). It is inversely related to the viscosity of the solution (η) (Klein, 1999):

$$N_{Re} = \frac{D^2 \rho B}{\eta} \qquad (7.9)$$

For turbulent mixing, the values for N_{Re} are found to be larger than 0.2. On the other hand, low values of the Reynolds number (high viscosity fluids) were found to be important in estimating the sedimentation rate of particles in suspensions using Stokes' law for sedimentation (Shanebrook, 1978):

$$\upsilon = \frac{d^2(\rho_s - \rho_l)g}{18\eta} \qquad (7.10)$$

where υ is the sedimentation rate, d the average diameter of the suspending particles, g the acceleration due to gravity, ρ_s and ρ_l the densities of the suspended solid particles and the suspending medium, respectively, and η is the viscosity of the suspending medium. Equation (7.10) is applicable only for diluted suspensions, to ensure free settling of particles (Shanebrook, 1978). Unfortunately, pharmaceutical dispersions are much too concentrated for equation (7.10) to be used in estimating υ, even though it may be used *qualitatively* to describe the effect of the various parameters on sedimentation (e.g., higher viscosity indicates lower sedimentation rate).

7.6 CONCLUDING REMARKS

The viscosity of a formulation plays a role during and after a preparation is formulated. A high viscosity for the product can be challenging to prepare in terms of mixing and pouring its ingredients; however, it provides good physical stability for the final product. Substances such as methylcellulose can be added to the formulation to affect its viscosity.

REFERENCES

Al-Achi A, Shipp S. Physical characteristics of selected over-the-counter medications. Int. J. Pharm. Compd. 2005;9(1):75–81.

Al-Achi A, Mosley A, Dembla I. Mineral oil emulsion preparation with acacia. Drug Deliv. Technol. 2007;7(5):58–62.

Allen VL. Featured excipients: viscosity-increasing agents for aqueous systems. Int. J. Pharm. Compd. 1999;3(6):479–486.

———. Featured excipients: stiffening agents. Int. J. Pharm. Compd. 2000;4(1):60–62.

———. Quality-control analytical methods: viscosity measurements. Int. J. Pharm. Compd. 2003;7(4):305–309.

Biradar VS, Dhumal SR, Paradkar A. Rheological investigation of self-emulsification process. J. Pharm. Pharm. Sci. 2009;12(1):17–31.

Davis SS. Is pharmaceutical rheology dead? Pharm. Acta Helv. 1974;49(5–6):161–168.

Jones DS, Lawlor SM, Woolfson DA. Rheological and mucoadhesive characterization of polymeric systems composed of poly(methylvinylether-*co*-maleic anhydride) and poly(vinylpyrrolidone), designed as platforms for topical drug delivery. J. Pharm. Sci. 2003;92(5): 995–1007.

Jürgens K. Influence of particle size and shape on the drug product: aspects of formulation and performance. Am. Pharm. Rev. 2006;9(4):61–63, 65.

Klein FG. A new approach to the scale-up of liquid pharmaceuticals. Pharm. Technol. 1999;23(3):136, 138, 140, 142, 144.

Lippacher A, Müller RH, Mäder K. Liquid and semisolid SLNe dispersions for topical application: rheological characterization. Eur. J. Pharm. Biopharm. 2004;58:561–567.

Segers DJ, Zatz LJ, Shah PV. In vitro release of phenol from ointment formulations. Pharm. Technol. 1997;21(1):70, 72, 74, 76, 80–81.

Shanebrook RJ. Stokes' law, Reynolds number, and the erythrocyte sedimentation rate test. Am. J. Phys. 1978;46(3):306–307.

Shaw S, Parsi S. Use of rheometry in the product development of semi-solids. Drug Deliv. Technol. 2007;7(1):36, 38, 40, 42–45.

Thorgeirsdóttir TO, Kjoniksen A-L, Knudsen KD, Kristmundstottir T, Nystrom B. Viscoelastic and structural properties of pharmaceutical hyfrogels containing monocarpin. Eur. J. Pharm. Biopharm. 2005;59:333–342.

Thurston BG, Martin A. Rheology of pharmaceutical systems: oscillatory and steady shear of non-Newtonian viscoelastic liquids. J. Pharm. Sci. 1978;67(11):1499–1506.

GLOSSARY

CMC	Carboxymethylcellulose.
Fluidity	The inverse of viscosity is a term commonly known as the *fluidity* of Newtonian liquid.
HEC	Hydroxyethylcellulose.
Kinematic viscosity	Often used for fluids that exist in different densities; expressed in units of stokes or centistoke (1 stoke = 100 centistokes). Kinematic viscosity has been adopted by the U.S.P. as the term for viscosity for substances listed in the compendia.
N.F.	*National Formulary.*
Non-Newtonian flow	If the viscosity of material *changes* with varying shear rate, the material is described as exhibiting non-Newtonian flow.
PC	Polycarbophil.
Pseudoplastic materials	Demonstrate a decrease in their viscosity (η') (thinning) as the rate of shear increases on the system.

PVP	Poly(vinylpyrrolidine).
Rheology	The science that deals with the flow of liquids; its subject is the parameter *viscosity*.
Shear rate	The difference in the velocity between two adjacent parallel planes of liquid separated by a small distance during shearing movement of a liquid.
Shearing force	The force per unit area that is applied on a liquid to cause shear.
U.S.P.	*United States Pharmacopeia*.
Viscosity	The parameter that describes internal friction is known as the viscosity of a liquid.

CHAPTER 8

CHEMICAL STABILITY AND SHELF-LIFE DETERMINATION

8.1 INTRODUCTION

Since all pharmaceuticals are chemicals in nature, it is expected that once a preparation is made, chemical degradation of the active pharmaceutical ingredient (API) (and other ingredients) begin, to occur. These degradation reactions reduce the concentration of the API in the product over time and replace it with its by-products, which may be toxic. According to the general rule, the product maintains its *shelf-life* status as long as it contains a minimum of 90% of the labeled API amount (or concentration). [A more exact definition for the shelf life is the time required for a product to lose a quantity of the API beyond that specified by the *United States Pharmacopeia* (U.S.P.).]

From the point of view of chemical stability, APIs may undergo hydrolysis (e.g., esters and β-lactams), oxidation (e.g., aldehydes and nitrite derivatives), decarboxylation (e.g., loss of CO_2 from the carboxyl group with heat), dehydration (e.g., tetracycline), epimerization (e.g., tetracycline drugs), or photochemical reactions (e.g., phenothiazines and nifedipine) (Allen, 2011). Most of the chemical reactions occurring on drug products are either hydrolysis or oxidation. Oxidative reactions can be promoted in solution by the presence of residual trace metals or by exposure to light (about one-third of all drugs are of concern with respect to photostability) (Templeton and Klein, 2007). These reactions can be prevented by using opaque or amber-colored containers and the inclusion of chelating agents in the formulation to trap trace metals. Hydrolysis reactions are common, due to the ubiquitous use of water as a solvent in formulations. Reduction or total replacement of water by other solvents can either eliminate or significantly reduce the hydrolysis reactions. Moreover, the choice of an optimal pH of the preparation can also play a role in reducing both the rate of hydrolysis and oxidation to a significant extent (Shabir and Arain, 2004). Although stability issues can also be physical in nature, they will be discussed in more detail under each specific dosage form, such as pharmaceutical suspensions and emulsions.

Stability issues are not necessarily confined to aqueous pharmaceutical preparations. They can occur in any formulation, including solid products or oily liquids.

Integrated Pharmaceutics: Applied Preformulation, Product Design, and Regulatory Science,
First Edition. Antoine Al-Achi, Mali Ram Gupta, William Craig Stagner.
© 2013 John Wiley & Sons, Inc. Published 2013 by John Wiley & Sons, Inc.

In the latter case, the oils in these formulation undergo lipid peroxidation reactions. This may be controlled in several ways, including trace metal chelation, protection from light exposure, inclusion of antioxidants in the formulation, and preventing hydrolysis reactions for lipids with fatty acid esters (Cannon, 2008).

Traditional stability studies on drug products include forced degradation, accelerated stability testing, and long-term stability determination. Studies have shown that six-month accelerated stability studies are often sufficient to assign a shelf life of 18 to 24 months to a drug product, although certain regulatory agencies (e.g., the International Conference on Harmonization) may require longer durations for stability studies (e.g., 12 months) (Beaman, 2010).

The U.S.P. provides general guidelines for products that are compounded in a pharmacy operation where no other information is available with respect to their stability. According to these guidelines, the *beyond-use date* (BUD) for a compounded solid or nonaqueous prescription is 25% of that listed on the commercial products (when a commercial product is used in its preparation), or six months, whichever is earlier. If the product is prepared from U.S.P. chemicals, the BUD is no more than six months. For water-containing preparations, the BUD is 14 days or the duration of therapy, whichever is earlier, and the product has to be stored under refrigeration. For all other nonsterile preparations the BUD is 30 days or the duration of therapy, whichever comes first. The U.S.P. guidelines for establishing BUD for sterile products are more restrictive and for non-preserved preparations allow a 28-hour BUD under refrigeration. The BUD for multidose preparations containing a preservative can be up to 30 days (the range is between 1 and 30 days), depending on many factors, such as the conditions under which the product is being stored and the number of times the product is used by a patient.

8.2 SHELF-LIFE DETERMINATION

8.2.1 Extraction Concepts

In the determination of the shelf life of a drug in a dosage form, the drug concentration (or amount) is measured over a period of time under accelerated stress conditions of temperature and relative humidity. The dosage form is usually stored under the stress conditions, and periodically a sample is removed for analysis. An extraction process is usually needed to separate the drug from its dosage form to get it ready for chemical analysis. This process of extracting the drug depends on its partition coefficient value between the phases used in the extraction process. Normally, the drug is dissolved in an aqueous solution if it is present in solid dosage form or is extracted directly from its liquid aqueous preparation (such as syrups or elixirs). Once the drug is transferred to an aqueous vehicle, it is extracted with an organic solvent such as chloroform or ether. A multiple extraction procedure is normally done to assure almost complete transfer of the drug from the aqueous solution into the organic phase. Then the organic phase is removed completely by evaporation (normally, under a vacuum), leaving behind a solid residue of the drug. The drug residue is subsequently dissolved in an aqueous vehicle prior to chemical analysis. The following equations govern extraction of the drug from its aqueous vehicle by an organic solvent (Martin et al., 1983):

$$x = \frac{PQ}{PQ + 1} \tag{8.1}$$

$$z = (PQ + 1)^{-1} \tag{8.2}$$

$$x + z = 1 \tag{8.3}$$

$$\text{fraction remaining} = x^n \text{ or } z^n \tag{8.4}$$

$$\text{total fraction extracted} = 1 - x^n \text{ or } 1 - z^n \tag{8.5}$$

where P is the partition coefficient of the drug, Q is a volume ratio (upper/lower), x and z are the fraction of the drug in the upper and lower phases at equilibrium, respectively, and n is the number of extractions. For example, assume that a drug with a partition coefficient value of 0.125 in a water–chloroform system was extracted three times (chloroform was the extractant). What percentage of the drug is extracted from a 25-mL aqueous portion into chloroform (25 mL each)? Since chloroform is heavier than water, it will occupy the lower phase. Calculate for x and z:

$$x = \frac{0.125}{1.125} = 0.111 \text{ or } 11.1\%$$
$$z = 100 - 11.1 = 88.9\%$$
$$\text{fraction remaining} = z^n = (0.111)^3 = 0.00137$$
$$\text{total fraction extracted in three extractions} = 1 - z^n = (1 - 0.00137)$$
$$= 0.9986 \text{ or } 99.86\%$$

Practically speaking, a drug that is 99.86% extracted is considered to be complete. Once the drug is analyzed for its content of the active principle, the results reflect its *actual* concentration in the product (Martin et al., 1983).

8.2.2 Factors Affecting the Stability

Although the factors that can affect the chemical stability of the ingredients in the product are numerous, the most notable are the storage temperature and the pH of the product. (Other factors can be the presence of additives in the formulation, agitation, or the dosage-form type.)

8.2.2.1 Effect of Temperature
Temperature affects the stability of the product by supplying more "free energy" at a higher temperature, allowing the reactions to proceed at a much faster rate (Shabir and Arain, 2004). According to the Arrhenius model, proposed in the 1870s, the rate constant for degradation can be estimated by the following first-order rate expression (Shabir and Arain, 2004):

$$k = Se^{-\Delta H_a/RT} \tag{8.6}$$

where k is the degradation rate constant, S the frequency factor (the number of collisions occurring between the reacting molecules per unit time), ΔH_a the molar heat of activation (the amount of heat needed to drive the reaction forward; for most drugs this value is about 20 kcal/mol, with a normal range between 15 and 60 kcal/mol), R is the gas constant, and T is the temperature of the reaction in kelvin. Equation (8.6) may be linearized by taking the natural logarithm of both sides of the equation:

$$\ln k = \ln S - \left(\frac{\Delta H_a}{R} \frac{1}{T} \right) \qquad (8.7)$$

A plot of $\ln k$ vs. $1/T$ yields a straight line with a slope of $-(\Delta H_a/R)$ and a y-intercept of $\ln S$. The plot affords an estimation of the degradation rate constant k at storage temperatures ($25°C$ or $4°C$) after estimating equation (8.7) at elevated temperatures (i.e., accelerated studies). The value of k at storage temperatures allows estimation of the *shelf life* of a product (usually, it is t_{90}, the time needed for 10% of the drug to degrade) (Shabir and Arain, 2004).

The degradation rate constant that is calculated from equation (8.7) belongs to a rate of reaction of the type

$$\frac{dA}{dt} = k[A]^r \qquad (8.8)$$

where $-(dA/dt)$ is the rate of degradation (the negative sign implies that the drug concentration is decreasing over time), [A] the concentration of the drug at time t, and r the order of the reaction. The symbol $[\cdot]$ signifies a molar concentration. If $r = 0$, the reaction follows a zero-order kinetics; if $r = 1$, the reaction is first-order; when $r = 2$, the reaction is second-order; and so on. Most pharmaceutical APIs follow either zero-, first-, or second-order reactions. A second-order reaction can also be of the type

$$-\frac{dA}{dt} = k[A][B] \qquad (8.9)$$

where B is another substance reacting with component A (Shabir and Arain, 2004).

The degradation of cefazolin sodium in sterile ophthalmic solutions stored in sterile glass dropper bottles was evaluated under various temperatures (7, 17, 25, and 40°C) (Kommanaboyina et al., 2000). Cefazolin sodium undergoes a hydrolysis reaction in solution and thus is susceptible to pH changes. The degradation reaction followed first-order kinetics (or perhaps pseudo-first-order kinetics) (at 7°C the reaction was too slow to reach a definite conclusion as to the reaction order). The energy of activation obtained from an Arrhenius plot averaged 95.7 kJ/mol, with an average frequency factor of 5.6×10^{12} collisions/hour.

When the actual chemical degradation of a drug follows first-order kinetics in solution, it is expected that the same drug behaves as if it follows a zero-order degradation reaction (i.e., an apparent zero-order reaction) in a suspension dosage form (Qasem et al., 2003):

$$k_0 = k_1[A] \qquad (8.10)$$

where k_0 is the zero-order rate constant, k_1 the first-order rate constant, and [A] the solubility of the API in the solvent system. The shelf life of the product can be estimated from

$$t_{90} = \frac{0.1[A_0]}{k_0} \qquad (8.11)$$

with t_{90} being the time for 10% of the drug to degrade, $[A_0]$ the initial concentration of the drug in the formulation, and k_0 the zero-order rate constant for degradation. Equation (8.11) may be used to calculate the beyond-use date for drugs undergoing a zero-order or apparent zero-order reaction. To estimate the shelf life of products that experience first- or second-order reactions, the following equations may be used:

$$t_{90} = \frac{0.105}{k_1} \tag{8.12}$$

$$t_{90} = \frac{0.11}{[A_0]k_0} \tag{8.13}$$

where equation (8.12) is for the first-order reaction and equation (8.13) is for the second-order reaction.

Propylthiouracil is an antithyroid agent used in the treatment of hyperthyroidism. This drug was shown to undergo an apparent zero-order kinetics for its stability in aqueous suspensions with a characteristic energy of activation of 21 kJ/mol and a frequency factor of 12.61×10^3 collisions/hour (Alexander, 2005). The estimated reaction rate constant at room and refrigerator temperatures was 0.078% and 0.04%, respectively. From the rate constants for degradation obtained from the Arrhenius model at various temperatures (4, 30, 50, 60, and 70°C), the calculated shelf life of propylthiouracil in suspensions was 127 and 248 days at 25 and 4°C, respectively (Alexander, 2005). Table 8.1 presents various examples of compounded drug products and their estimated beyond-use dates. Gupta (2007) reported the degradation of desonide (a corticosteroid used for ear inflammation) in an eardrop formulation to follow first-order kinetics at room temperature:

$$\ln[\text{concentration (\%)}] = 4.6087 - 0.0005[\text{time (days)}]$$

From the slope of the line, the first-order rate constant is 0.0005 day^{-1}. The shelf life of the drug is calculated as

$$t_{90} = \frac{0.105}{k_1} = \frac{0.105}{0.0005} = 210 \text{ days}$$

However, the actual reading of the decomposition was 91.8% of the original label claim at day 180. Thus, the author recommended the latter figure for the beyond-use date for desonide at room temperature in the eardrop formulation [50 mg desonide, 2 mL glacial acetic acid, 50 mL glycerin, and q.s. (as much as surfices) to 100 mL with propylene glycol] (Gupta, 2007).

Reaction rate prediction can be problematic when it comes to preparations stored under their *glass transition temperature* (T_g), defined as the temperature above which the material converts from a glass state into a less viscous (more mobile) state, characterized by an increase in flow rate and heat capacity. The kinetics described by the Arrhenius model do not apply in this situation, and to predict the rate constant of reaction, the *empirical Williams–Landell–Ferry* (WLF) *model* is employed instead (Martin and Mo, 2007):

$$\ln k = \frac{-C_1(T - T_g)}{C_2 + (T - T_g)} \tag{8.14}$$

TABLE 8.1 Beyond-Use Date for Compounded Preparations: Selected Examples

Drug Name	Therapeutic Action	Formulation Type	BUD Room Temperature	BUD Under Refrigeration	Comments	Reference
Acyclovir sodium	For herpes virus infection	Sterile solution in 0.9% sodium chloride injection (10 mg/mL)	30 days	Physical instability (precipitations within 5 days)	Stored in polypropylene syringes (Becton, Dickinson Co., NJ)	Ling and gupta, 2001a
Alatrofloxacin mesylate	Antibacterial	Intravenous injection in 5% dextrose injection (1.88 mg/mL)	At least 9 days	—	Sterile preparation in poly(vinyl chloride) plastic containers	Gupta and Bailey, 2000
	Antibacterial	Intravenous injection in 0.45% sodium chloride injection (1.88 mg/mL)	At least 9 days	—	Sterile preparation in poly(vinyl chloride) plastic containers	Gupta and Bailey, 2000

Albuterol sulfate	Bronchodilator/ exercise induced asthma	Nebulizer solution: sterile solution in 0.9% sodium chloride injection (0.2 mg of albuterol/mL)	At least 7 days	At least 7 days	Storage containers: poly(vinyl chloride), polyolefin, polypropylene syringes, borosilicate glass, and polypropylene microcentrifuge tubes	Hunter et al., 1998
Allopurinol	For hyper-uricemia/gout	Suspension (20 mg/mL)	8.3 years	—	Suspension was prepared from tablets and stored in amber-colored glass bottles	Alexander et al., 1997a
Aminocaproic acid	Fibrinolysis inhibitor/for life-threatening hemorrhaging	Sterile solutions in 0.9% sodium chloride injection (10 and 100 mg/mL)	At least 7 days	At least 7 days	Stored in poly(vinyl chloride) bags	Zhang and Trissel, 1997
Amiodarone hydrochloride	Antiarrhythmic	Suspension (20 mg/mL)	193.4 days	677.3 days	—	Alexander, 2003
Amphotericin B	Antifungal	Eyedrop sterile solution in 5% dextrose solution (5 mg/mL)	16 days	120 days	Protected from light exposure	Peyron, et al., 1999

(continued)

TABLE 8.1 (Continued)

Drug Name	Therapeutic Action	Formulation Type	BUD Room Temperature	BUD Under Refrigeration	Comments	Reference
Atenolol	Antihypertensive/for angina pectoris and myocardial infarction	Liquid preparations in simple syrup or Ora-Sweet (2 mg/mL)	14 days	—	Liquid preparations in vehicles of simple syrup and Ora-Sweet (Paddock Laboratories, MN)	Patel et al., 1997
	Antihypertensive/for angina pectoris and myocardial infarction	Liquid preparations in methylcellulose or Ora-Sweet SF (2 mg/mL)	At least 28 days	—	Liquid preparations in vehicles of Ora-Sweet SF (Paddock Laboratories, MN) and methylcellulose	Patel et al., 1997
Atropine sulfate		Sterile solution in 0.9% sodium chloride solution (2 mg/mL)	364 days	364 days	Solution was prepared from atropine sulfate powder dissolved in isotonic saline solution, filtered through 0.2-μm-pore filters, and packaged in polypropylene syringes (Becton, Dickinson Co., NJ)	Donnelly and Cormon, 2008

Bethanechol chloride	For urinary retention (postoperatively)	Oral liquid (concentration unknown)	40 days	—	Making the solution from commercially available tablets was not stable; solutions were made from commercial injections of the drug	Gupta and Maswoswe, 1997b
Brompheniramine	Antihistamine	Oral liquid (0.4 mg/mL)	At least 202 days	—	Stored in amber-colored glass bottles	Gupta and Gupta, 2011
Brompheniramine maleate	Antihistamine	Oral liquid (0.4 mg/mL)	202 days	—	Stored in amber-colored glass bottles	Gupta and Gupta, 2011
Cefazolin sodium	Antibacterial	Ophthalmic sterile solution in 0.9% sodium chloride injection (3.5 mg/mL)	6 days	14 days	Stored in sterile ophthalmic dropper bottles	Kommanaboyina et al., 2000
	Antibacterial/for urinary tract infections/lower respiratory tract infections	Sterile solutions in 0.9% sodium chloride (50 mg/mL)	7 days	22 days	Stored in polypropylene syringes (Becton, Dickinson Co., NJ)	Gupta, 2003a

(continued)

TABLE 8.1 (Continued)

Drug Name	Therapeutic Action	Formulation Type	BUD Room Temperature	BUD Under Refrigeration	Comments	Reference
Cefepime hydrochloride	Antibacterial/urinary tract infections	Sterile solutions in 5% dextrose injection (0.05 mg/mL)	2 days	23 days	Solutions stored in plastic bags (Baxter Healthcare Corp., IL)	Gupta et al., 1997
	Antibacterial/urinary tract infections	Sterile solutions in 5% dextrose injection (0.05 mg/mL)	2 days	23 days	Solutions stored in plastic bags (Baxter Healthcare Corp., IL)	Gupta et al., 1997
	Antibacterial/for urinary tract infections/pneumonia/skin infections	Sterile solution in 0.9% sodium chloride injection (20 mg/mL)	At least 2 days	At least 21 days	Stored in polypropylene syringes (Becton, Dickinson Co., NJ)	Ling and Gupta, 2001b
Cefmetazole sodium	Antibiotic/urinary tract infections	Sterile solution in 5% dextrose injection (0.2 g/mL)	2 days	28 days	Stored in 50-mL plastic bags (Baxter Healthcare Corp., IL)	Gupta and Maswoswe, 1997e
	Antibiotic/urinary tract infections	Sterile solution in 0.9% sodium chloride injection (0.2 g/mL)	2 days	28 days	Stored in 50-mL plastic bags (Baxter Healthcare Corp., IL)	Gupta and Maswoswe, 1997e
Cefotaxime sodium	Antibiotic/urinary tract infections/lower respiratory tract infections	Sterile solution in 0.9% sodium chloride injection (50 mg/mL)	1 day	18 days	Stored in polypropylene syringes	Gupta, 2002b

94

Cesfuroxime sodium	Antibacterial/for urinary tract infections/lower respiratory tract infections	Sterile solution in 0.9% sodium chloride injection (50 mg/mL)	2 days	21 days	Packaged in polypropylene syringes (Becton, Dickinson Co., NJ)	Gupta, 2003b
Clindamycin phosphate	Antibiotics	Sterile solutions in 0.9% sodium chloride (6 and 12 mg/mL)	7 days	30 days	Packaged in AutoDose Infusion System Bags (Tandem Medical, Inc., CA)	Xu and Trissel, 2003
Clonazepam	For epilepsy	Suspension in Hospital for Sick Children (HSC) vehicle	—	At least 60 days	HSC contains simple syrup and methylcellulose; suspensions were stored in poly(vinyl chloride) amber-colored plastic bottles	Roy and Besner, 1997
Desonide	Corticosteroid/for ear inflammation	Ear drops (0.5 mg/mL)	180 days	—	Stored in amber-colored glass bottles (Owens-Illinois Prescription Products, OH)	Gupta, 2007
Dexamethasone sodium phosphate	Adrenocortical steroid	Sterile solutions in 0.9% sodium chloride injection (0.1 and 1 mg/mL)	22 days	—	Stored in polypropylene syringes (Becton, Dickinson Co., NJ)	Gupta, 2002a

(continued)

TABLE 8.1 *(Continued)*

Drug Name	Therapeutic Action	Formulation Type	BUD Room Temperature	BUD Under Refrigeration	Comments	Reference
Diclofenac sodium	Anti-inflammatory	Sterile solution (25 mg/mL)	23 days	Drug precipitates out from solution	Packaged in clear-glass vials and stored in the dark	Gupta, 2006a
Dobutamine hydrochloride	For congestive heart failure	Intravenous injection in 5% dextrose injection (4 mg/mL)	30 days	30 days	Sterile preparation	Webster, 1999
Droperidol	Neuroleptic	Sterile solution in 0.9% sodium chloride injection (0.625 mg/mL)	180 days	—	Stored in polypropylene syringes; solutions were stored protected from light	McCluskey and Lovely, 2011
Ephedrine sulfate	pressor agent during spinal anesthesia	Sterile solutions in 0.9% sodium Chloride injection (5 mg/mL)	60 days	60 days	Solutions were packaged in polypropylene syringes (Becton, Dickinson Co., NJ)	Storms et al., 2001
Ethacrynate sodium	Diuretic	Sterile solution in 0.9% sodium chloride injection (1 mg/mL)	At least 14 days	At least 22 days	Stored in polypropylene syringes (Becton, Dickinson Co., NJ)	Ling and Gupta, 2001c
Ethacrynic acid	Diuretic	Oral aqueous liquid (2.5 mg/mL)	3 days	24 days	The dosage form contained 0.05 M phosphate buffer and 10% mannitol	Ling and Gupta, 2001d

Drug	Use	Formulation	Stability	Stability	Storage	Reference
Fentanyl	Opiate analgesic	Sterile solution (0.05 mg/mL)	28 days (exposed to light)	28 days (protected from light)	Stored in polypropylene syringes or poly(vinyl chloride) bags	Donnelly, 2005
5-Fluorouracil	Antineoplastic agent/for glaucoma/pterygium/retinal detachment/premalignant eye lesions	Ophthalmic sterile solution in 0.9% sodium chloride injection (10 mg/mL)	7 days	7 days	Stored in 1-mL tuberculin syringes (Becton, Dickinson Co., NJ) solutions stored in the freezer (-10°C) also had a beyond-use date of 7 days	Fuhrman et al., 2000
Furosemide	Diuretic	Sterile solutions in 0.9% sodium chloride injection (1.2, 2.4, and 3.2 mg/mL)	84 days	84 days	Solutions were protected from light and stored in polypropylene syringes	Donnelly, 2002
Glycopyrrolate	Preoperative antimuscarinic/for salivary, tracheobronchial, and pharyngeal secretion reduction	Oral liquids prepared from tablets (0.5 mg/mL) in water	25 days	—	Solutions were stored in amber-colored bottles (Owens-Illinois Prescription Products, OH)	Gupta, 2001b
Glycopyrrolate	Preoperative antimuscarinic/for salivary, tracheobronchial, and pharyngeal secretion reduction	Oral liquid (0.5 mg/mL)	At least 129 days	—	Stored in amber-colored bottles	Gupta, 2003c

(continued)

97

TABLE 8.1 *(Continued)*

Drug Name	Therapeutic Action	Formulation Type	BUD Room Temperature	BUD Under Refrigeration	Comments	Reference
	Preoperative antimuscarinic/for salivary, tracheobronchial, and pharyngeal secretion reduction	Sterile solution (0.2 mg/mL)	90 days	90 days	Stored in polypropylene syringes (Becton, Dickinson Co., NJ)	Storms et al., 2003
Heparin sodium	Anticoagulant	Sterile solutions (1000 and 40,000 units/mL)	30 days	—	Polypropylene syringes (CADD-Micro Medication Reservoire, 10 mL, Sims Deltec, Inc.)	Stiles et al., 1997
Hydralazine hydrochloride	For hypertension	Sterile solution in 0.9% sodium chloride injection (0.2 mg/mL)	2 days	—	Packaged in poly(vinyl chloride) bags	Gupta, 2005a
	For hypertension	Sterile solution in 5% dextrose injection (0.2 mg/mL)	1 hour	—	Packaged in poly(vinyl chloride) bags	Gupta, 2005a
	Antihypertensive	Oral solutions (1 and 10 mg/mL)	—	Up to 30 days	Stored in glass and plastic bottles; formulations were either flavored or unflavored	Okeke et al., 2003
Hydrocortisone sodium succinate	Various diseases	Sterile solution in 0.9% sodium chloride injection (10 mg/mL)	7 days	21 days	Stored in polypropylene syringes (Becton, Dickinson Co., NJ)	Gupta and Ling, 2000

Hydromorphone hydrochloride	Analgesic	Sterile solutions in 0.9% sodium chloride injection (1.5 and 80 mg/mL)	At least 60 days	At least 60 days	Packaged in polypropylene syringes (Becton, Dickinson Co., NJ)	Trissel et al., 2002b
Indomethacin sodium trihydrate	Anti-inflammatory/ analgesic	Sterile solution in 0.9% sodium chloride injection (0.1 mg of indomethacin/mL)	At least 10 days	—	Sterile preparations in plastic bags	Gupta and Maswoswe, 1998
Isoniazid	For tuberculosis	Oral aqueous liquid (10 mg/mL)	At least 42 days	—	Stored in amber-colored glass bottles	Gupta and Good, 2005
Ketamine hydrochloride	Anesthetic	Sterile solution in water for injection (10 mg/mL)	At least 30 days	—	Stored in polypropylene syringes	Gupta, 2002c
Ketorolac tromethamine	Nonsteroidal anti-inflammatory	Sterile solution in 5% dextrose injection (0.6 mg/mL)	7 days	At least 50 days	Stored in 50-mL plastic bags (Baxter Healthcare Corp., IL)	Gupta and Maswoswe, 1997c
	Nonsteroidal Anti-inflammatory	Sterile solution in 0.9% sodium chloride injection (0.6 mg/mL)	7 days	At least 50 days	Stored in 50-mL plastic bags (Baxter Healthcare Corp., IL)	Gupta and Maswoswe, 1997c
Levothyroxine sodium	Thyroid hormonal therapy	Syrup (sorbitol 70%) (0.04 mg/mL)	15 days	47 days	—	Alexander and et al., 1997b

(continued)

TABLE 8.1 (*Continued*)

Drug Name	Therapeutic Action	Formulation Type	BUD Room Temperature	BUD Under Refrigeration	Comments	Reference
	For reduced or absent thyroid function	Sterile solution in 0.9% sodium chloride injection (0.1 mg/mL)	—	At least 7 days	Stored in polypropylene syringes (Monoject, MO)	Gupta, 2000c
Lidocaine hydrochloride	Antiarrhythmic/local anesthetic	Sterile solution (20 mg/mL)	90 days	90 days	Stored in polypropylene syringes (Becton, Dickinson Co., NJ)	Storms et al., 2002
Lisinopril	Angiotensin-converting enzyme inhibitor/for hypertension	Syrup (2 mg/mL)	At least 30 days	At least 30 days	Compounded from tablets (Zestril, Zeneca Pharmaceuticals, DE); syrup was stored in amber-colored prescription bottles	Rose et al., 2000
Magnesium sulfate	Tocolytic agent; anticonvulsant	Sterile solution in 0.9% sodium chloride injection (37 g/mL)	3 months	—	Stored in glass bottles (type I) or poly(vinyl chloride) bags	Sarver et al., 1998
	Tocolytic agent; anticonvulsant	Sterile solution in lactated Ringer's injection (37 g/mL)	3 months	—	Stored in glass bottles (type I) or poly(vinyl chloride) bags	Sarver et al., 1998

Mechlorethamine hydrochloride	Anticancer/for mycosis fungoides	Ointment (topical) (0.01%)	7 days	—	Ointment was packaged in screw-cap ointment jars	Zhang et al., 1998
Memantine	For Alzheimer's disease	Oral liquid (0.166 mg/mL)	7 days	28 days	Stored in glass bottles	Yamreudeewong, et al., 2006
Meperidine hydrochloride	Analgesic–pain management	Sterile solutions in 5% dextrose solution (0.25, 1, 10, 20, and 30 mg/mL)	At least 28 days	At least 28 days	Sterile preparations in polypropylene syringes/preservative-free	Donnelly and Bushfield, 1998
	Analgesic–pain management	Sterile solutions in 0.9% sodium chloride solution (0.25, 1, 10, 20, and 30 mg/mL)	At least 28 days	At least 28 days	Sterile preparations in polypropylene syringes/preservative-free	Donnelly and Bushfield, 1998
Methadone	For heroin addiction/analgesic	Oral solutions of methadone powder or concentrate in orange-flavored Tang drink (5 mg/mL)	21 days (chemically stable after 91 days; however, bacterial growth observed after 21 days)	91 days	—	Donnelly, 2004

(*continued*)

TABLE 8.1 (*Continued*)

Drug Name	Therapeutic Action	Formulation Type	BUD Room Temperature	BUD Under Refrigeration	Comments	Reference
Methylprednisolone sodium succinate	Adrenocorticosteroid hormone/for various diseases	Sterile solution in 0.9% sodium chloride injection (10 mg/mL)	4 days	At least 21 days	Stored in polypropylene syringes (Becton, Dickinson Co., NJ, and Monoject, MO)	Gupta, 2001a
Metoclopramide hydrochloride	For stimulating the motility of the upper gastrointestinal tract	Sterile solution in 0.9% sodium chloride injection (0.5 mg/mL)	At least 21 days	—	Packaged in polypropylene syringes (Becton, Dickinson Co., NJ)	Gupta, 2005b
Metoprolol tartrate	Beta-adrenergic blocking agent/for high blood pressure, heart failure, and angina	Aqueous solutions (5 mg/mL)	16 days	—	Solutions were prepared from tablets and stored in amber-colored glass bottles	Gupta and Maswoswe, 1997a
Metronidazole	Antibiotic/antiparasitic	Suspension (15 mg/mL)	90 days	—	Suspensions were prepared from 500-mg tablets; actual shelf life was estimated to be 73 years; However, for practical considerations a 90-day shelf life was proposed	Alexander et al., 1997d

Drug	Use	Formulation			Storage	Reference
Midazolam	For status epileptuss (in dogs)	Intranasal sterile solution (50 mg/mL)	30 days	30 days	Packaged in clear glass vials placed, in turn, in light-resistant containers	Lester Elder et al., 2011
	For status epileptus (in dogs)	Intranasal sterile solution (50 mg/mL)	30 days	30 days	Stored in clear glass vials	Lester Elder et al., 2011
Milrinone lactates	For congestive heart failure and myocardial infarction	Sterile solutions in 0.9% sodium chloride injection (0.4, 0.6, and 0.8 mg/mL)	14 days	14 days	Sterile preparations in poly(vinyl chloride) bags	Nguyen et al., 1998
	For congestive heart failure and myocardial infarction	Sterile solutions in 5% dextrose injection (0.4, 0.6, and 0.8 mg/mL)	14 days	14 days	Sterile preparations in poly(vinyl chloride) bags	Nguyen et al., 1998
	For congestive heart failure and myocardial infarction	Sterile solution in 0.9% sodium chloride injection (0.2 mg/mL)	14 days	14 days	Sterile preparations in poly(vinyl chloride) bags	Wong and Gill, 1998
	For congestive heart failure and myocardial infarction	Sterile solutions in 5% dextrose injection (0.2 mg/mL)	14 days	14 days	Sterile preparations in poly(vinyl chloride) bags	Wong and Gill, 1998
Mitomycin	Antibiotic	Sterile solutions (0.5 mg/mL)	7 days	28 days	Stored in 1-mL tuberculin syringes (Monoject, MO)	Gupta and Maswoswe, 1997d

(continued)

TABLE 8.1 *(Continued)*

Drug Name	Therapeutic Action	Formulation Type	BUD Room Temperature	BUD Under Refrigeration	Comments	Reference
Morphine sulfate	Analgesic	Sterile solution (1 mg/mL)	3 years	—	Terminally sterilized at 120°C for 20 minutes and stored in polypropylene bags (Polimoon Langeskov, (Norway)	Nguyen-Xuan et al., 2006
	Analgesic	Sterile solutions in 0.9% sodium chloride injection (5 and 50 mg/mL)	At least 60 days	At least 60 days (with precipitation developed within 2–4 days)	Packaged in polypropylene syringes (Becton, Dickinson Co., NJ)	Trissel et al., 2002a
	Analgesic	Sterile solutions in sterile water for injection (5 and 50 mg/mL)	At least 60 days	At least 60 days (with precipitation developed within 2–4 days)	Packaged in polypropylene syringes (Becton, Dickinson Co., NJ)	Trissel et al., 2002a
Nafcillin sodium	Antibacterial	Sterile solution in 0.9% sodium chloride injection (10 mg/mL)	At least 7 days	At least 44 days	Stored in polypropylene syringes (Becton, Dickinson Co., NJ)	Ling and Gupta et al., 2000
Naltrexone hydrochloride	For opiate addiction	Sterile solution (1.4 mg/mL)	At least 42 days	—	Stored in clear-glass vials	Gupta, 2008a

Naratriptan hydrochloride	For migraines	Suspension in Syrpalta vehicle (Humco, TX) (0.5 mg/mL)	7 days	90 days	Packaged in amber-colored, plastic, screw-cap prescription bottles	Zhang et al., 2000
	For migrainess	Suspensions made from tablets (0.5 mg/mL)	At least 7 days	At least 90 days	Vehicles used Ora-Plus:Ora-Sweet (1:1) or Ora-Plus:Ora-Sweet SF (1:1) (Paddock Laboratories, MN)	Zhang et al., 2000
Omeprazole	Proton pump inhibitor	Oral liquid (suspension) (2 mg/mL)	4 weeks	—	Content of capsules (20 mg) was emptied and mixed for 3 minutes and suspended in 8.4% sodium bicarbonate solution	Garg et al., 2009
	Proton pump inhibitor	Oral liquid (suspension) (2 mg/mL)	1 week	—	Content of capsules (20 mg) was emptied and turned into a fine powder using a glass mortar; the resulting powder was mixed and suspended in 8.4% sodium bicarbonate solution	Garg et al., 2009

(continued)

TABLE 8.1 *(Continued)*

Drug Name	Therapeutic Action	Formulation Type	BUD Room Temperature	BUD Under Refrigeration	Comments	Reference
Pentobarbital sodium	Sedative/hypnotic/ anticonvulsant	Sterile injections in 0.9% sodium chloride injection (10 mg/mL)	At least 31 days	—	Solutions were packaged in glass or polypropylene syringes (Becton, Dickinson Co., NJ)	Gupta, 2001c
Pentoxifylline	For peripheral vascular diseases	Topical cream (5% w/w) in Dermabase cream	62 days	—	Stored in plastic white opaque ointment jars	Gupta, 2012
Pergolide mesylate	For equine Cushing's syndrome	Oral liquid (suspension) (0.2 mg/mL)	45 days	—	—	Shank and Ofner, 2009
Perphenazine	For schizophrenia/severe nausea and vomiting	Oral liquid in Ora-Sweet syrup (Paddock Laboratories, MN) (0.5 mg/mL)	114 days	—	Final pH = 4.2	Gupta, 2008b
	For schizophrenia/severe nausea and vomiting	Oral liquid in Humco's syrup (Humco, TX) (0.5 mg/mL)	183 days	—	Final pH = 4.5	Gupta, 2008b
	Antiemetic	Oral liquid (0.5 mg/mL)	30 days	—	Vehicle used was Ora-Sweet (Paddock Laboratories, MN)	Gupta, 2005c

Drug	Use	Formulation				Reference
Phenytoin sodium	Antiseizure/topically for open-wound healing	Suspensions in 0.9% sodium chloride solution (20 mg/mL)	14 days	—	Nonsterile suspensions prepared from capsules added to 0.9% sodium chloride solution	Rhodes et al., 2006
Piperacillin sodium	Antibacterial/for urinary tract infections/respiratory tract infections	Sterile solution in 0.9% sodium chloride injection (40 mg/mL)	5 days	28 days	Sterile solutions in polypropylene syringes (Becton, Dickinson and Co., NJ)	Gupta and Ling, 2001
Propylthiouracil	Antithyroid/for hyperthyroidism	Suspension (5 mg/mL)	127 days	248 days	Stored in amber-colored glass bottles	Alexander, 2005
Pyridoxine hydrochloride	B-complex vitamin	Sterile solution (100 mg/mL)	180 days	—	Packaged in polypropylene syringes (Becton, Dickinson Co., NJ)	Gupta, 2006b
Ribavirin	Antiviral	Suspension (40 mg/mL)	At least 28 days	At least 28 days	Suspension was prepared from oral capsules	Chan et al., 2004
Sotalol hydrochloride	For cardiac rhythm disorders	Suspensions in various vehicles (5 mg/mL)	At least 12 weeks	At least 12 weeks	Vehicles used: Ora-Plus:Ora-Sweet (1:1)l Ora-Plus: Ora-Sweet SF (1:1) (Paddock Laboratories, MN); or simple syrup:methylcellulose (1:2.4)	Sidhom et al., 2005

(continued)

TABLE 8.1 (*Continued*)

Drug Name	Therapeutic Action	Formulation Type	BUD Room Temperature	BUD Under Refrigeration	Comments	Reference
Spironolactone	Diuretic	Suspension (5 mg/mL)	90 days	90 days	Suspensions were prepared from 25-mg tablets	Alexander et al., 1997c
Succinylcholine chloride	Skeletal muscle relaxant	Sterile solution (20 mg/mL)	45 days	90 days	Stored in polypropylene syringes (Becton, Dickinson Co., NJ)	Storms et al., 2003
	Skeletal muscle relaxant	Sterile solution (20 mg/mL)	6 months	—	—	Roy et al., 2008
Sufentanil citrate	Opioid analgesic	Sterile solution in 0.9% sodium chloride injection (0.002 mg of sufentanil/mL)	Stable for 24 hours in polypropylene syringes connected to epidural catheters and in entire infusion system using polyethylene tubing; not stable if PVC tubing is present in the system	—	Simulated epidural administration for 24 hours: a polypropylene syringe connected to polyethylene or poly(vinyl chloride) tubing and filter to an epidural catheter	Jappinen et al., 1998

Drug	Category	Formulation			Storage	Reference
Teicoplanin	Antibiotic	Sterile solution in 5% dextrose injection (0.004 mg/mL)	—	6 days	Solutions were stored in poly(vinyl chloride) bags (Baxter, Belgium)	Galanti et al., 2001
Temozolomide	For refractory anaplastic astrocytoma	Suspension in Ora-Plus and Ora-Sweet (1:1) (Paddock Laboratories, MN) (10 mg/mL)	7 days	—	Packaged in amber-colored plastic prescription bottles	Trissel et al., 2006
	For refractory anaplastic astrocytoma	Suspension in Ora-Plus and Ora-Sweet SF (1:1) (Paddock Laboratories, MN) (10 mg/mL)	14 days	—	Packaged in amber-colored plastic prescription bottles	Trissel et al., 2006
Terbutaline sulfate	For bronchospasm/chronic obstructive lung disease	Sterile solution in 0.9% sodium chloride injection (0.1 mg/mL)	23 days	—	Stored in poly(vinyl chloride) bags	Gupta, 2004
Testosterone	Hormonal	Sterile solution in 0.9% sodium chloride solution (0.001 mg/mL)	1 hour	9 hours	Sterile preparation in plastic bags	White and Quercia, 1999

(continued)

TABLE 8.1 *(Continued)*

Drug Name	Therapeutic Action	Formulation Type	BUD Room Temperature	BUD Under Refrigeration	Comments	Reference
Trastuzumab	Anticancer	Sterile solutions in 0.9% sodium chloride injection (0.4 and 4 mg/mL)	28 days	—	Stored in polypropylene bags; solutions were kept at room temperature without protection from light	Kaiser and Krämer, 2011
Treprostinil sodium	For pulmonary arterial hypertension	Sterile solutions (1, 2.5, 5, and 10 mg/mL)	At least 60 days	At least 60 days	Packaged in plastic syringe reservoirs (Medtronic MiniMed, CA)	Xu et al., 2004
Trovafloxacin mesylate	Antibiotic	Oral liquid (10 mg/mL)	14 days	—	Stored in amber-colored glass bottles	Gupta, 2000a
Tubocurarine chloride	Neuromuscular blocking agent	Sterile solution (3 mg/mL)	45 days	90 days	Stored in polypropylene syringes	Stewart et al., 2002

Venlafaxine	Antidepressant	Extended-release suspension in Ora-Plus/Ora-Sweet (Paddock Laboratories, MN) (15 mg/mL)	28 days	28 days	Stored in amber-colored plastic bottles	Donnelly et al., 2011
	Antidepressant	Extended-release suspension in simple syrup (15 mg/mL)	28 days	28 days	Stored in amber-colored plastic bottles	Donnelly et al., 2011
Vinorelbine Tartrate	Anticancer	Sterile solution in 5% dextrose injection (0.5 mg/mL)	120 hours	—	Sterile preparations in poly(vinyl chloride) minibags; exposure to constant fluorescent lighting	Lieu and Gill, 1999
	Anticancer	Sterile solution in 0.9% sodium chloride injection (2 mg/mL)	120 hours	—	Sterile preparations in poly(vinyl chloride) minibags; exposure to constant fluorescent lighting	Lieu and Gill, 1999

111

where C_1 and C_2 are model constants and T is the storage temperature. The WLF model is suitable for systems that exhibit high viscosity with a rubbery consistency upon storage. It was shown that the kinetics in these systems is an order of magnitude higher than that obtained from an Arrhenius liquid. Thus, systems stored at temperatures approaching their T_g value experience a significant increase in reaction rate (Martin and Mo, 2007).

8.2.2.2 Effect of pH Most APIs are either weak acids or weak bases. In solutions, the pH is normally adjusted so that the API is soluble in the preparation. However, the same pH that is helpful in achieving solubility can be detrimental to an API's stability. To achieve an optimum pH for product stability, the degradation rate constant of the API is first obtained under varying pH values. A plot of the logarithm of k vs. pH yields a graph with a nadir at the point of inflection. The corresponding pH to that point represents the pH of optimum stability (Shabir and Arain, 2004). For most applications, it appears that an acidic pH is most favorable for both weak acid and weak base pharmaceuticals, and this acidic pH is also optimal to protect the API from both hydrolysis and oxidation reactions (Shabir and Arain, 2004). In certain cases a slight change in the pH of the formulation can produce a great effect on a drug's stability. For example, the stability of perphenazine (for the control of severe nausea and vomiting; for schizophrenia) in two commercially available syrup vehicles was found to deteriorate rapidly if the pH of the vehicle was reduced from 4.5 to 4.2 (Gupta, 2008b).

8.3 STABILITY OF BIOTECHNOLOGY PRODUCTS

Biotechnology products are increasingly becoming a major component of the pharmaceutical industry, the estimate for marketed products being nearly $50 billion a year (Matejtschuk et al., 2009). Because of their fragile nature, biotechnology products must be stabilized to accommodate handling requirements during processing and by patients. One method for improving the stability of these products is lyophilization (freeze-drying). During the process of lyophilization, the material is subjected first to a freezing cycle, followed by a low-pressure cycle to induce sublimation for initial drying, and a final drying cycle (Blue and Yoder, 2009; Matejtschuk et al., 2009). Optimum conditions (Matejtschuk et al., 2009) for this operation were determined to include an initial rapid-freezing step accompanied by periods of slow cooling, allowing larger ice crystals to materialize. (Larger ice crystals require less time for the sublimation step.) During the sublimation phase, it is desirable to keep the product below its glass transition temperature (T_g) (i.e., the temperature at which noncrystalline components of the product become immobilized). Maintaining the material below T_g was found to be associated with better product stability. Additives in the formulation also play a major role from the point of view of stability. Trehalose, a sugar, was found in many cases to provide a good final stability profile for a product. The final stage of the lyophilization process is a secondary drying step to reduce the content of water "bound" to the API. This is normally achieved by adjusting the temperature to ambient ranges or higher (Blue and Yoder, 2009). The amount of moisture content in the final product can range from below 1% to up to 5% by weight (Blue and Yoder, 2009; Matejtschuk et al., 2009). The closer the moisture content to the T_g value, the poorer is the stability of the preparation (Blue and Yoder, 2009). Ideally, the moisture content should be well below the T_g value and should never exceed it (Blue and Yoder, 2009). Desiccating the product to reduce its moisture content may be necessary as a final step for certain biotechnology materials. Accelerated stability studies are done at relatively high temperatures applying Arrhenius kinetics principles (Matejtschuk et al., 2009).

REFERENCES

Alexander SK. Stability of an extemporaneously compounded propylthiouracil suspension. Int. J. Pharm. Compd. 2005;9(1):82–86.

Alexander SK, Davar N, Parker AG. Stability of allupurinol suspension compounded from tablets. Int. J. Pharm. Compd. 1997a;1(2):128–131.

Alexander SK, Kothapalli RM, Dollimore D. Stability of an extemporaneously formulated levothyroxine sodium syrup compounded from commercial tablets. Int. J. Pharm. Compd. 1997b;1(1):60–63.

Alexander SK, Thyagarajapuram N. Formulation and accelerated stability studies for an extemporaneous suspension of Amiodarone Hydrochloride. IJPC 2003;7(5):389–393.

Alexander SK, Vangala SKSS, White BD, Dollimore D. The formulation development and stability of spironolactone suspension. Int. J. Pharm. Compd. 1997c;1(3):195–199.

———. The formulation development and stability of metronidazole suspension. Int. J. Pharm. Compd. 1997d;1(3):200–205.

Allen VL. Implementing *United States Pharmacopeia* Chapter <795>, Pharmaceutical compounding—Nonsterile preparations, part 3. Int. J. Pharm. Compd. 2011;15(6):488–496.

Beaman VJ. Stability testing: Doing everything or doing the right thing? Am. Pharm. Rev. 2010;13(5):73–76.

Blue J, Yoder H. Successful lyophilization development of protein therapeutics. Am. Pharm. Rev. 2009;12(1):90–96.

Cannon BJ. Chemical and physical stability considerations for lipid-based drug formulations. Am. Pharm. Rev. 2008;11(1):132,134–138.

Chan PJ, Tong HH, Chow LHA. Stability of extemporaneous oral ribavirin liquid preparation. Int. J. Pharm. Compd. 2004;8(6):486–488.

Donnelly FR. Chemical stability of furosemide in minibags and polypropylene syringes. Int. J. Pharm. Compd. 2002;6(6):468–470.

———. Chemical stability of methadone concentrate and powder diluted in orange-flavored drink. Int. J. Pharm. Compd. 2004;8(6):489–491.

———. Chemical stability of fentanyl in polypropylene syringes and polyvinylchloride bags. Int. J. Pharm. Compd. 2005;9(6):482–483.

Donnelly FR, Bushfield LT. Chemical stability of meperidine hydrochloride in polypropylene syringes. Int. J. Pharm. Compd. 1998;2(6):463–465.

Donnelly FR, Corman C. Physical compatibility and chemical stability of a concentrated solution of atropine sulfate (2mg/mL) for use as an antidote in nerve agent casualties. Int. J. Pharm. Compd. 2008;12(6):550–552.

Donnelly FR, Wong K, Goddard R, Johanson C. Stability of venlafaxine immediate-release suspensions. Int. J. Pharm. Compd. 2011;15(1):81–84.

Fuhrman CL, Godwin AD, Davis AR. Stability of 5-fluorouracil in an extemporaneously compounded ophthalmic solution. Int. J. Pharm. Compd. 2000;4(4):320–323.

Galanti ML, Hecq J-D, Jeuniau P, Vanbeckbergen D, Jamart J. Assessment of the stability of teicoplanin in intravenous infusions. Int. J. Pharm. Compd. 2001;5(5):397–400.

Garg S, Svirskis D, Al-Kabban M, Komeshi M, Lee J, Liu Q, Naidoo S, Kairuz T. Chemical stability of extemporaneously compounded omeprazole formulations: a comparison of two methods of compounding. Int. J. Pharm. Compd. 2009;13(3):250–253.

Gupta DV. Stability of an oral liquid dosage form of trovafloxacin mesylate and its quantitation in tablets using high-performance liquid chromatography. Int. J. Pharm. Compd. 2000a;4(3):233–235.

_____. Stability of levothyroxine sodium injection in polypropylene syringes. Int. J. Pharm. Compd. 2000b;4(6):482–483.

_____. Chemical stability of methylprednisolone sodium succinate after reconstitution in 0.9% sodium chloride injection and storage in polypropylene syringes. Int. J. Pharm. Compd. 2001a;5(2):148–150.

_____. Stability of an oral liquid dosage form of glycopyrrolate prepared from tablets. Int. J. Pharm. Compd. 2001b;5(6):480–481.

_____. Stability of pentobarbital sodium after reconstitution in 0.9% sodium chloride injection and repackaging in glass and polypropylene syringes. Int. J. Pharm. Compd. 2001c;5(6):482–484.

_____. Chemical stability of dexamethasone sodium phosphate after reconstitution in 0.9% sodium chloride injection and storage in polypropylene syringes. Int. J. Pharm. Compd. 2002a;6(5):395–397.

_____. Stability of cefotaxime sodium after reconstitution in 0.9% sodium chloride injection and storage in polypropylene syringes for pediatric use. Int. J. Pharm. Compd. 2002b;6(1):234–236.

_____. Stability of ketamine hydrochloride injection after reconstitution in water for injection and storage in 1-mL tuberculin polypropylene syringes for pediatric use. Int. J. Pharm. Compd. 2002c;6(4):316–317.

_____. Chemical stability of cefazolin sodium after reconstituting in 0.9% sodium chloride injection and storage in polypropylene syringes for pediatric use. Int. J. Pharm. Compd. 2003a;7(2):152–154.

_____. Chemical stability of cefuroxime sodium after reconstitution in 0.9% sodium chloride injection and storage in polypropylene syringes for pediatric use. Int. J. Pharm. Compd. 2003b;7(4):310–312.

_____. Stability of oral liquid dosage forms of glycopyrrolate prepared with the use of powder. Int. J. Pharm. Compd. 2003c;7(5):386–388.

_____. Chemical stability of terbutaline sulfate injection after diluting with 0.9% sodium chloride injection when stored at room temperature in polyvinylchloride bags. Int. J. Pharm. Compd. 2004;8(5):404–406.

_____. Chemical stability of hydralazine hydrochloride after reconstitution in 0.9% sodium chloride injection or 5% dextrose injection for infusion. Int. J. Pharm. Compd. 2005a;9(5):399–401.

_____. Chemical stability of metoclopramide hydrochloride injection diluted with 0.9% sodium chloride injection in polypropylene syringes at room temperature. Int. J. Pharm. Compd. 2005b;9(1):72–74.

_____. Chemical stability of perphenazine in oral liquid dosage forms. Int. J. Pharm. Compd. 2005c;9(6):484–486.

_____. Chemical stability of diclofenac sodium injection. Int. J. Pharm. Compd. 2006a;10(2):154–155.

_____. Chemical stability of pyridoxine hydrochloride 100-mg/mL injection, preservative free. Int. J. Pharm. Compd. 2006b;10(4):318–319.

_____. Chemical stability of desonide in ear drops. Int. J. Pharm. Compd. 2007;11(1):79–81.

_____. Chemical stability of naltrexone hydrochloride injection. Int. J. Pharm. Compd. 2008a;12(3):274–275.

_____. Chemical stability of perphenazine in two commercially available vehicles for oral liquid dosage forms. Int. J. Pharm. Compd. 2008b;12(4):372–374.

_____. Chemical stability of phentoxifylline in a topical cream. Int. J. Pharm. Compd. 2012;16(1):80–81.

Gupta DV, and Bailey ER. Stability of alatrofloxacin mesylate in 5% dextrose injection and 0.45% sodium-chloride injection. Int. J. Pharm. Compd. 2000;4(1):66–68.

Gupta DV, Gupta SV. Chemical stability of brompheniramine maleate in an oral liquid dosage form. Int. J. Pharm. Compd. 2011;15(1):78–80.

Gupta DV, and Ling J. Stability of hydrocortisone sodium succinate after reconstitution in 0.9% sodium chloride injection and storage in polypropylene syringes for pediatric use. Int. J. Pharm. Compd. 2000;4(5):396–397.

_____. Stability of piperacillin sodium after reconstitution in 0.9% sodium chloride injection and storage in polypropylene syringes for pediatric use. Int. J. Pharm. Compd. 2001;5(3):230–231.

Gupta DV, Maswoswe J. Quantitation of metoprolol tartrate and propranolol hydrochloride in pharmaceutical dosage forms: stability of metoprolol in aqueous mixture. Int. J. Pharm. Compd. 1997a;1(2):125–127.

———. Stability of bethanechol chloride in oral liquid dosage forms. Int. J. Pharm. Compd. 1997b;1(1):278–279.

———. Stability of ketorolac tromethamine in 5% dextrose injection and 0.9% sodium chloride injection. Int. J. Pharm. Compd. 1997c;1(3):206–207.

———. Stability of mitomycin aqueous solution when stored in tuberculin syringes. Int. J. Pharm. Compd. 1997d;1(1):282–283.

———. Stability of cefmetazole sodium in 5% dextrose injection and 0.9% sodium-chloride injection. Int. J. Pharm. Compd. 1997e;1(3):208–209.

———. Stability of indomethacin in 0.9% sodium chloride injection. Int. J. Pharm. Compd. 1998;2(2):170–171.

Gupta DV, Sood A. Chemical stability of isoniazid in an oral liquid dosage form. Int. J. Pharm. Compd. 2005;9(2):165–166.

Gupta DV, Maswoswe J, Bailey ER. Stability of cefepime hydrocholoride in 5% dextrose injection and 0.9% sodium chloride injection. Int. J. Pharm. Compd. 1997;1(6):435–436.

Hunter AD, Fox LJ, Yoo DS, Michniak BB. Stability of albuterol in continuous nebulization. Int. J. Pharm. Compd. 1998;2(5):394–396.

Jappinen A-L, Kokki H, Rasi SA, Naaranlahti JT. Stability of sufentanil in a syringe pump under simulated epidural infusion. Int. J. Pharm. Compd. 1998;2(6):466–468.

Kaiser J, Krämer I. Physicochemical stability of diluted Trastuzumab infusion solutions in polypropylene infusion bags. Int. J. Pharm. Compd. 2011;15(6):515–520.

Kommanaboyina B, Rhodes C, Lindauer FR, Grady TL. Some studies of the stability of compounded cefaxolin ophthalmic solution. Int. J. Pharm. Compd. 2000;4(2):146–149.

Lester Elder D, Zheng B, White AC, Li D. Stability of midazolam intranasal formula for the treatment of status epilepsus in dogs. Int. J. Pharm. Compd. 2011;15(1):74–77.

Lieu LC, Gill AM. Five-day stability of vinorelbine in 5% dextrose injection and in 0.9% sodium chloride injection at room temperature. Int. J. Pharm. Compd. 1999;3(1):67–68.

Ling J, Gupta DV. Stability of nafcillin sodium after reconstitution in 0.9% sodium chloride injection and storage in polypropylene syringes for pediatric use. Int. J. Pharm. Compd. 2000;4(6):480–481.

———. Stability of acyclovir sodium after reconstitution in 0.9% sodium chloride injection and storage in polypropylene syringes for pediatric use. Int. J. Pharm. Compd. 2001a;5(1):75–77.

———. Stability of cefepime hydrochloride after reconstitution in 0.9% sodium chloride injection and storage in polypropylene syringes for pediatric use. Int. J. Pharm. Compd. 2001b;5(2):151–152.

———. Stability of ethacrynate sodium after reconstitution in 0.9% sodium chloride injection and storage in polypropylene syringes for pediatric use. Int. J. Pharm. Compd. 2001c;5(1):73–74.

———. Stability of oral liquid dosage forms of ethacrynic acid. Int. J. Pharm. Compd. 2001d;5(3):232–233.

Martin A, Swarbrick J, Cammarata A. Solubility and distribution phenomena. In: *Physical Pharmacy: Physical Chemical Principles in the Pharmaceutical Sciences*, 3rd ed. Philadelphia: Lea & Febiger; 1983. pp. 272–313.

Martin WHS, Mo J. Stability considerations for lyophilized biologics. Am. Pharm. Rev. 2007;10(4):31–36.

Matejtschuk P, Malik K, Duru C, Bristow A. Freeze drying of biologicals: process development to ensure biostability. Am. Pharm. Rev. 2009;12(2):12–18.

McCluskey VS, Lovely KJ. Stability of droperidol 0.625 mg/mL diluted with 0.9% sodium chloride injection and stored in polypropylene syringes. Int. J. Pharm. Compd. 2011;15(2):170–173.

Nguyen D, Gill AM, Wong F. Stability of milrinone lactate in 5% dextrose injection and 0.9% sodium chloride injection at concentrations of 400, 600 and 800 µg/mL. Int. J. Pharm. Compd. 1998;2(3):246–248.

Nguyen-Xuan T, Griffiths W, Kern C, Van Gessel E, Bonnabry P. Stability of morphine sulfate in polypropylene infusion bags for use in patient-controlled analgesia pumps for postoperative pain management. Int. J. Pharm. Compd. 2006;10(1):69–73.

Okeke CC, Medwick T, Nairn G, Khuspe S, Grady TL. Stability of hydralazine hydrochloride in both flavored and nonflavored extemporaneous preparations. Int. J. Pharm. Compd. 2003;7(4):313–319.

Patel D, Doshi HD, Desai A. Short-term stability of atenolol in oral liquid formulations. Int. J. Pharm. Compd. 1997;1(6):437–439.

Peyron F, Elias R, Ibrahim E, Amirat-Combralier V, Buès-Charbit M, Balansard G. Stability of amphotericin B in 5% dextrose ophthalmic solution. Int. J. Pharm. Compd. 1999;3(4):316–320.

Qasem GJ, Bummer MP, Digenis AG. Kinetics of placlitaxel 2'-N-methylpyridinium mesylate decomposition. AAPS Pharm. Sci. Tech. 2003;4(2):75–82.

Rhodes SR, Kuykendall RJ, Heyneman AC, May PM, Bhushan A. Stability of phenytoin sodium suspensions for the treatment of open wounds. Int. J. Pharm. Compd. 2006;10(1):74–78.

Rose JD, Webster AA, English AB, McGuire MJ. Stability of lisinopril syrup (2mg/mL) extemporaneously compounded from tablets. Int. J. Pharm. Compd. 2000;4(5):398–399.

Roy JJ, Besner J-G. Stability of clonazepam suspension in HSC vehicle. Int. J. Pharm. Compd. 1997;1(6):440–441.

Roy JJ, Boismenu D, Mamer AO, Nguyen TB, Forest J-M, Hildgen P. Room temperature stability of injectable succinylcholine dichloride. Int. J. Pharm. Compd. 2008;12(1):83–85.

Sarver GJ, Pryka R, Alexander SK, Weinstein L, Erhardt WP. Stability of magnesium sulfate in 0.9% sodium chloride and lactated Ringer's solutions. Int. J. Pharm. Compd. 1998;2(5):385–388.

Shabir GA, Arain SA. The role of R&D in stability of drug development. Am. Pharm. Rev. 2004;7(5):32, 34, 36, 38, 47.

Shank RB, Ofner MC III. Stability of pergolide mesylate oral liquid at room temperature. Int. J. Pharm. Compd. 2009;13(3):254–258.

Sidhom BM, Taft RD, Rivera N, Kirschenbaum LH, Almoazen H. Stability of sotalol hydrochloride in extemporaneously prepared oral suspension formulations. Int. J. Pharm. Compd. 2005;9(5):402–406.

Stewart TJ, Storms LM, Warren WF. Stability of tubocurarine chloride injection at ambient temperature and 4°C in polypropylene syringes. Int. J. Pharm. Compd. 2002;6(4):308–310.

Stiles LM, Allen VL, McLaury H-J. Stability of two concentrations of heparin sodium prefilled in CADD-micro® pump syringe. Int. J. Pharm. Compd. 1997;1(6):433–434.

Storms LM, Stewart TJ, Warren WF. Stability of ephedrine sulfate at ambient temperature and 4°C in polypropylene syringes. Int. J. Pharm. Compd. 2001;5(5):394–396.

———. Stability of lidocaine hydrochloride injection at ambient temperature and 4°C in polypropylene syringes. Int. J. Pharm. Compd. 2002;6(5):388–390.

———. Stability of glycopyrrolate injection at ambient temperature and 4°C in polypropylene syringes. Int. J. Pharm. Compd. 2003;7(1):65–67.

Templeton CA, Klein JL. Practical considerations for effectively conducting pharmaceutical photostability testing. Am. Pharm. Rev. 2007;10(4):124–128.

Trissel AL, Pham L, Xu AQ. Physical and chemical stability of morphine sulfate 5 mg/mL and 50 mg/mL packaged in plastic syringes. Int. J. Pharm. Compd. 2002a;6(1):62–65.

———. Physical and chemical stability of hydromorphone hydrochloride 1.5 mg/mL and 80 mg/mL packaged in plastic syringes. Int. J. Pharm. Compd. 2002b;6(1):74–76.

Trissel AL, Zhang Y, Koontz ES. Temozolomide stability in extemporaneously compounded oral suspensions. Int. J. Pharm. Compd. 2006;10(5):396–399.

Webster AA, English AB, McGuire M, Riggs MR, Lander V, Buckle T. Stability of Dobutamine hydrochloride 4 mg/mL in 5% Dextrose Injection at 5 and 23°C. IJPC 1999;3(5):412–414.

White MC, Quercia R. Stability of extemporaneously prepared sterile testosterone solution in 0.9% sodium chloride solution large-volume parenterals in plastic bags. Int. J. Pharm. Compd. 1999;3(2):156–157.

Wong F, Gill AM. Stability of milrinone lactate 200 µg/mL in 5% dextrose injection and 0.9% sodium chloride injection. Int. J. Pharm. Compd. 1998;2(2):168–169.

Xu AQ, Trissel AL. Stability of clindamycin phosphate in autodose infusion system bags. Int. J. Pharm. Compd. 2003;7(2):149–151.

Xu AQ, Trissel AL, Pham L. Physical and chemical stability of treprostinil sodium injection packaged in plastic syringe pump reservoirs. Int. J. Pharm. Compd. 2004;8(3):228–230.

Yamreudeewong W, Teixeira GM, Mayer EG. Stability of memantine in an extemporaneously prepared oral liquid. Int. J. Pharm. Compd. 2006;10(4):316–317.

Zhang Y, Trissel AL. Stability of aminocaproic acid injection admixtures in 5% dextrose injection and 0.9% sodium chloride injection. Int. J. Pharm. Compd. 1997;1(2):132–134.

Zhang Y, Trissel AL, Johansen FJ, Kimball LC. Stability of mechlorethamine hydrochloride 0.01% ointment in Aquaphor® base. Int. J. Pharm. Compd. 1998;2(1):89–91.

Zhang Y, Trissel AL, Fox LJ. Naratriptan hydrochloride in extemporaneously compounded oral suspensions. Int. J. Pharm. Compd. 2000;4(1):69–71.

GLOSSARY

Beyond-use date (BUD)	The BUD for a compounded solid or nonaqueous prescription is 25% of that listed on the commercial products (when a commercial product is used in its preparation) or six months, whichever is earlier.
API	Active pharmaceutical ingredient.
Shelf life	A product maintains its shelf-life status as long as it contains a minimum of 90% of the labeled API amount (or concentration).
Williams–Landell– Ferry (WLF) model	Used to predict the rate constant of reaction.

CHAPTER *9*

PARTICLE SCIENCE

9.1 INTRODUCTION

Pharmaceutical formulators deal with solid ingredients and incorporate them in preparations. Solid dosage forms are the most common preparations found commercially. About 80% of all drug products are solid dosage forms (Sandler, 2005; Beckmann, 2006). Moreover, the majority of active pharmaceutical ingredients (APIs) are in the form of a crystalline solid material. Strict adherence to the crystallization process parameters is of utmost importance for producing the same type and size of crystals for a given API. A slight variation in the crystallization conditions of an API can produce a wide variety of crystals that do not necessarily share or yield similar bioavailability profiles (Beckmann, 2006). Often, the size of the particles of a solid material can influence the final product's physical and therapeutic benefits. For example, topical applications of particles in the size range 3 to 10 μm have resulted in the accumulation of these particles in hair follicles, while larger particles (>10 μm) remained on the surface of the skin, and particles smaller than 3 μm were able to diffuse through the stratum corneum and the hair follicles (Parikh, 2011).

API particle size can affect not only drug characteristics (dissolution, solubility, bioavailability, content uniformity, product appearance, or stability), but can also influence manufacturing processes (i.e., compactibility, blend uniformity, and flowability) (Sun et al., 2010). Monitoring the size of the solid material *during* processing, as in *online* particle sizing, allows better control of the final product. For example, in tablet formulation it is important to produce larger agglomerations of particles, known as *granules*, for several practical reasons. Among the reasons are improvements in the flowability of the material during processing, in the compressibility of the material, in the dispersibility, in the uniformity, and to reduce dust during operation (Rubino, 1999). Moreover, online monitoring of the granulation during the developmental stage permits transparency of the entire process, allowing process parameters to be changed and tracked so that their effect can be measured immediately, affording easier scale-up transfer from research and development to the production line (Jürgens, 2006).

Integrated Pharmaceutics: Applied Preformulation, Product Design, and Regulatory Science,
First Edition. Antoine Al-Achi, Mali Ram Gupta, William Craig Stagner.

The size of particles of solid mixtures can be small, approaching that of a colloidal range (1 nm to 0.5 μm) or larger, measuring several millimeters in diameter. For example, the production of *pellets* in pharmaceutical and other industries (food and agriculture) is well recognized. Pellets are produced with a great degree of precision in their particle size (several millimeters in diameter), affording means of drug delivery. Four consecutive steps are commonly employed in making pellets as carriers for APIs: (1) granulation, (2) extrusion, (3) spheronization, and (4) drying. The *granulation* step is similar to that employed in the wet granulation process in tablets, except that the resulting mass to be extruded is much more fluid. During the *extrusion* process, the wet mass is forced through openings of a preselected sieve size to form rods. The resulting rods are then spheronized (i.e., the rods are cut into small pieces and then rounded to form a pellet-like shape). As a final step, the pellets are dried by various means before packaging (McConnell et al., 2010).

9.2   PARTICLE SIZE ESTIMATION AND DISTRIBUTION

Instruments used in particle size analysis are expected to be reproducible, accurate, and sensitive (Scarlett, 2003). Particle size influences many aspects of the final pharmaceutical product. To that end, all of the following characteristics of drug products are subject to particle size distribution: dissolution, solubility, bioavailability, stability, content uniformity, appearance, and manufacturing processes (John, 2009; Sun et al., 2010). Not only the average particle size, but also the particle size *distribution* about its mean value, is important, for similar average-size particles for two powder mixtures with totally different distributions can have a significant influence on product performance and/or its manufacturability (Sun et al., 2010). For example, it was found that diluents (e.g., lactose monohydrate and microcrystalline cellulose) added to a tablet blend could influence the flowability of the formulation, with a larger particle size and narrower size distribution producing better control over tablet weight variability during production (Fan et al., 2005). Solid chemicals in the form of a powder represent some what of a challenge to the analyzer, due to the uneven and nonuniform shape of the particles in the mixture. This difficulty in estimating average particle size was resolved by assigning an *equivalent spherical diameter* to the particles, relating them to a sphere with the same surface area, volume, surface-to-volume ratio, settling velocity, aerodynamic behavior, or sieve mesh (Scarlett, 2003; Hubert et al., 2008) that would behave in a manner similar to the powder particles when subjected to the same conditions of testing (Hubert et al., 2008). It is important to note here that "standardizing" the particles allows linearization by fixing the geometric shape of particles to that of a sphere (Scarlett, 2003).

Numerous methods can be employed to measure the average particle size in a collection of powder. Among these methods are sieving, direct imaging, settling, and dynamic light scattering (laser light diffraction) (Scarlett, 2003). For example, the *volume-equivalent diameter* is obtained from a Coulter counter, image analysis produces a *projected area diameter*, and dynamic light scattering yields a *hydrodynamic diameter* (Hubert et al., 2008). The sieving methods allow size analysis in the range 20 μm to 100 mm; imaging techniques can detect particle size on the scale of angstroms and larger; and laser light diffraction analysis detects particle size in the range 0.02 μm to 9 mm. The latter method is increasingly becoming a standard method for measuring the particle size of pharmaceutical powders (John, 2009). Near-infrared spectroscopy (NIR) is also utilized to estimate the particle size of solid materials. The dimension of the particles being examined by NIR is limited by the reference being used during analysis (Kelly and Lerke, 2005). Applications of NIR in the pharmaceutical industry are numerous and in addition to particle sizing, include measurements of moisture content, the content uniformity of dosage forms, blend uniformity, identification of the API, the amorphous-to-crystalline ratio, and the presence of polymorphic forms (Ji et al., 2006). For solid oral dosage forms, the two most useful methods for particle size characterization are laser diffraction and sieving (Sandler, 2005). Both methods have been covered in great detail in the *United States Pharmacopeia* (U.S.P.), which includes specifications of both tests. Several steps are required in performing both tests, including sampling of bulk powder materials and subsampling of bulk powder samples for a specimen, preparation of the specimen, setting up the needed instruments, performing the size measurements and analyzing the data, interpreting the results of the analysis, and writing reports describing the size analysis of the material tested (Sun et al., 2010). Most errors involved in size analysis are not necessarily related to the instruments used but, rather, to the sampling techniques. Caution must be exercised not to distort the sample during handling and to describe in detail the steps taken for sampling (Sun et al., 2010). For most APIs, a size range for their average particle diameter is between 1 and 200 μm (Hubert et al., 2008).

In addition to the particle diameter size, the shape of the particles should be reported to produce a morphological profile of the powder being tested. Most applications in this area of testing rely on microscopy, including optical and/or scanning electron microscopy (Hubert et al., 2008). One of the important descriptive properties of particles is *texture*. Texture is associated with particle size; smaller particles have a finer texture, while large particles are associated with a coarser texture. To define particles well morphologically, a three-dimensional imaging techniques are necessary. Unfortunately, light microscopy is not adequate for measuring the depth of a particle; scanning electron microscopy is better in that aspect (Sandler, 2005). The pharmaceutical industry most often relies on data projection techniques such as *principal components analysis* (PCA) and *self-organizing maps* (SOMs) to provide visualization of particles in a sample. The main difference in utility between the two methods is that PCA is applicable for linear projection, whereas, an SOM is for nonlinear projection. Information obtained from PCA and SOMs is useful for predicting powder *processability*: for example, during tableting (Sandler, 2005).

The determination of particle size by light microscopy is based on examining a minimum of 300 particles and obtaining their diameter at an arbitrary point, usually at the horizontal line crossing the middle of the particle at its point of orientation as seen under the microscope (Parrott, 1984). The following example illustrates the process by which an average particle size for a collection of powder is obtained. Following the determination of individual particle dimensions using a reticle attached to the eyepiece of a light microscope, results are summarized in a

TABLE 9.1   Grouped Frequency Table for Particle Size Determination

Size (μm)	Midsize, d (μm)	Frequency, n	nd	d^2	d^3	nd^2	nd^3
10–60	35	5	175	1,225	42,875	6,125	214,375
60–110	85	12	1,020	7,225	614,125	86,700	7,369,500
110–160	135	405	54,675	18,225	2,460,375	7,381,125	996,451,875
160–210	185	375	69,375	34,225	6,331,625	12,834,375	2,374,359,375
210–260	235	125	29,375	55,225	12,977,875	6,903,125	1,622,234,375
260–310	285	20	5,700	81,225	23,149,125	1,624,500	462,982,500
310–360	335	8	2,680	112,225	37,595,375	897,800	300,763,000

grouped frequency table (Table 9.1).

$$\sum n = 950 \text{ particles}$$

$$\sum nd = 163,000$$

$$\sum nd^2 = 29,733,750$$

$$\sum nd^3 = 5,764,375,000$$

$$\text{arithmetic diameter} = d_{av} = \frac{\sum nd}{\sum n} = \frac{163,000}{950} = 171.58 \ \mu m$$

$$\text{standard deviation} = \sigma = \left\{ \frac{\sum [n(d - d_{av})^2]}{\sum n} \right\}^{0.5}$$

$$\sigma = \left[\frac{(5)(35 - 171.58)^2 + \cdots + (8)(335 - 171.58)^2}{950} \right]^{0.5}$$

$$= \left(\frac{1,766,381.58}{950} \right)^{0.5} = (1,859.35)^{0.5}$$

$$= 43.12 \ \mu m$$

$$\text{mean surface–volume diameter} = d_{vs} = \frac{\sum nd^3}{\sum nd^2} = \frac{5,764,375,000}{29,733,750}$$

$$= 193.87 \ \mu m$$

$$\text{mean surface diameter} = d_s = \frac{\sum nd^2}{\sum nd} = \left(\frac{29,733,750}{163,000} \right)^{0.5}$$

$$= 13.51 \ \mu m$$

$$\text{mean volume diameter} = d_v = \left(\frac{\sum nd^3}{\sum nd} \right)^{1/3} = \left(\frac{5,764,375,000}{163,000} \right)^{1/3}$$

$$= 32.82 \ \mu m$$

TABLE 9.2   Grouped Frequency Table for Particle Size Determination

Size (μm)	Midsize, d (μm)	Frequency, n	log d	n log d
10–60	35	5	1.5440...	7.7203...
60–110	85	12	1.9294...	23.1530...
110–160	135	405	2.1303...	862.7851...
160–210	185	375	2.2671...	850.1893...
210–260	235	125	2.3710...	296.3834...
260–310	285	20	2.4548...	49.0968...
310–360	335	8	2.5250...	20.2003...

When the particle surface is important for a process (e.g., dissolution), diameters that include the surface are employed to estimate the total surface area of a powder. Diameters with a volume term attached may be useful for estimating the total volume of the powder per unit weight (Parrott, 1984). Geometric mean (d_{geo}) (and its geometric standard deviation, σ_{geo}) is employed whenever the distribution of powder particles is suspected to be asymmetrical (Parrott, 1984). The grouped frequecy is shown in Table 9.2.

$$\sum (n \log d) = 2109.52868$$

$$\sum (\log d) = 15.22195$$

$$\log(d_{geo}) = \frac{\sum (n \log d)}{\sum n} = \frac{2109.52868}{950} = 2.2205\ldots$$

$$d_{geo} = 166.17 \ \mu m$$

$$\log(\sigma_{geo}) = \left\{ \frac{\sum [n(\log d - \log d_{geo})^2]}{\sum n} \right\}^{0.5}$$

$$= \left[\frac{5(1.5440\ldots - 2.2205\ldots)^2 + \cdots + 8(2.5250\ldots - 2.2205\ldots)^2}{950} \right]^{0.5}$$

$$= 0.1128\ldots$$

$$\sigma_{geo} = 1.30 \ \mu m$$

Experimental mathematical expressions, known collectively as *Hatch and Choate equations*, are commonly used to convert geometric mean values to other mean diameter values (Parrott, 1984):

$$\log d_{av} = \log d_{geo} + 1.1513 \log^2 \sigma_{geo}$$

$$= 2.2205\ldots + (1.1513)(0.1128\ldots)^2 = 2.2352\ldots$$

$$d_{av} = 171.87 \ \mu m$$

$$\log d_{vs} = \log d_{geo} + 5.7565 \log^2 \sigma_{geo}$$

$$\log d_{s} = \log d_{geo} + 2.3026 \log^2 \sigma_{geo}$$

$$\log d_{v} = \log d_{geo} + 3.4539 \log^2 \sigma_{geo}$$

The reader is encouraged to calculate the values of d_{vs}, d_{s}, and d_{v} from the equations above given the values of $\log d_{geo}$ and $\log \sigma_{geo}$. Similar calculations of mean particle diameters may be obtained from the sieving methods; however, a *weight size distribution* is utilized in the calculation in this case (Parrott, 1984). For example, dried microcapsules containing verapamil and propranolol were evaluated with respect to their size distribution using standard U.S.P. sieves of various mesh sizes (no. 18, 24, and 80) (Khamanga et al., 2009). (There is an inverse relationship between mesh size and opening size; higher mesh sizes indicate smaller openings.) In doing so, the investigators estimated the mean particle size of each fraction from the arithmetic mean size of the apertures of the preceding sieve on which retained microspheres were found. Several other tests were performed on these particles (composed of Eudragit RS and Eudragit RL polymers), including flow properties, in vitro drug release, and determination of drug loading. The *Hausner ratio* (HR; tapped bulk density/bulk density $= \rho_{tap}/\rho_{bulk}$) was estimated as a measure of microsphere flow properties; values of HR greater than 1.25 are indicative of poor flowability. For in vitro release testing, the U.S.P. suggests several apparatus to be used, including the well-known basket (U.S.P. Apparatus 1) and paddle (U.S.P. Apparatus 2) methods. Data obtained from drug release studies may be evaluated using several mathematical models (Khamanga et al., 2009):

Zero-order:	$Q_t = Q_0 + K_0 t$	(9.1)
First-order:	$\ln Q_t = \ln Q_0 + K_1 t$	(9.2)
Higuchi's:	$Q_t = Q_0 + K_H \sqrt{t}$	(9.3)
Korsmeyer–Peppas's:	$Q_t = K_{KP} t^n$	(9.4)
Kopcha's:	$Q_t = A\sqrt{t} + Bt$	(9.5)
Makoid-Banakar's:	$Q_t = K_{MB} t^n e^{(-ct)}$	(9.6)

where Q_0 and Q_t are the initial amount of a drug and the cumulative amount released at time t. The remaining notations are coefficients and constants for the various models. The n parameter in the Korsmeyer–Peppas equation translates into the following diffusion predictions: $n = 0.5$ is associated with a strictly diffusion-controlled type of drug release (it is similar to Higuchi's model in that it assumes pure drug release from the matrix without erosion occurring), $n = 1$ is for zero-order release, and $n > 1.0$ indicates anomalous diffusion (i.e., swelling-controlled release).The Makoid–Banakar model becomes identical to that of Korsmeyer–Peppas when the parameter c is zero. When the parameter A in Kopcha's model is greater than B, the diffusion follows a Fickian profile. The procedure is to fit the drug release data to all of these models, and the best fit is found from the highest value for the *coefficient of determination* (r^2) associated with the models. (The coefficient of determination is a measure of the amount of variability observed with the dependent variable as explained by a given independent variable. The closer the value of r^2 to 1, the better the independent variable *predicts* the changes seen in the dependent variable.) Furthermore, nanoparticle shape is also an important characteristic for powders, which is usually determined by scanning electron microscopy techniques. Also, the percent of drug loading [% loading = (weight of drug/weight of microspheres) × 100], the encapsulation efficiency [EE = (actual drug content/theoretical or assumed drug content) × 100], and the percentage yield [% yield = (weight of microcapsules/total weight of drug and polymer) × 100] are all important parameters to define for every matrix (drug + polymer) formulation (Khamanga et al., 2009).

9.3 MICRONIZATION

It is known that the rate of dissolution of drug particles increases as the particle size decreases. Comminuting the drug to micrometer-size particles also increases its solubility to a certain extent, although for routine pharmacy operations in processing powders, this effect on solubility is seldom achieved or seldom of practical importance. It is interesting to note that the *change* in the solubility of the drug by *micronization* may be procedure dependent. For example, micronizing a poorly water-soluble compound to a particle size of 5 μm using a jet-milling process or mortar and pestle resulted in a different effect on the solubility. Whereas jet-milling the drug did not alter its solubility in aqueous media, the mortar and pestle method resulted in a reduction in its solubility (Dai et al., 2010). The solubility of active chemicals in water is one of the major factors that hinder their candidacy to become a drug. The immediate effect on micronizing the drug is enhancement in its dissolution rate, which consequently improves its bioavailability (Patel et al., 2007). Reduction in particle size may be achieved by milling and grinding, microcrystallization, and spray-drying methods (Dubin, 2005). The milling methods require low energy input (1 to 2%) and include hammer mill, jet mill, and fluid energy mill. Spray-drying methods rely on atomizing energy (vibratory, kinetic, or pressure) to produce particle sizes in the range 5 to 100 μm. For sparingly water-soluble APIs, spray-drying methods are useful for enhancing their solubility in water, as they produce amorphous forms of a drug (Parikh, 2011). Several novel technologies may be used as well: namely, cryogenic, precipitation, and sonication methods (Barot et al., 2006) and supercritical fluid extraction of emulsions (Chattopadhyay and Shekunov, 2006).

Cryogenic technology utilizes the technique of *spray-freezing into liquid* (SFL). In this method a solution of an API is sprayed in the form of very fine particles via a nozzle directly into *liquid nitrogen* (LN_2). Because of the extreme low temperature of liquid nitrogen ($-196°C$), upon contact with LN_2, the fine liquid particles of the API solidify instantaneously. The solidified particles are then separated from LN_2 by sieving, and the residue of LN_2 is subsequently removed under freeze-drying conditions. The resulting dry particles contain the API homogeneously dispersed throughout the porous microdroplets. (Porous channels are created during LN_2 removal.) The use of SFL is widely applicable to small drug molecules as well as to biotechnology products (e.g., peptides, proteins, vaccines)

(Chattopadhyay and Shekunov, 2006). Cryogenic methods utilizing supercritical CO_2 are in particular useful for water-soluble APIs and produce remarkably fine uniform particles useful for pulmonary delivery. *Spray-freeze-drying* (SFD) technology consists of atomizing the API solution (along with additives) through a nozzle along with supercritical CO_2. The resulting frozen solvent is subsequently removed by an atmospheric freeze-drying step. The SFD method may also be applicable to water-insoluble APIs by dissolving the API in an organic solvent or in a form of an oil-in-water emulsion. Powders produced by the SFD have a particulate density of less than 0.5 g/cm^3, a low bulk density of less than 0.1 g/cm^3, and a large aerodynamic shape factor. In comparison to the SFL technology, SFD is much simpler to apply (Chattopadhyay and Shekunov, 2006).

In the precipitation method, the API is dissolved in a solvent with a low boiling point. The solution is then heated at or above the solvent's boiling point and is atomized through a nozzle. During this step, most of the solvent evaporates, leaving behind micronized solid particles of an amorphous form of the API dispersed in supersaturated solution form. (Amorphous solids have a better dissolution profile in water than that of their corresponding crystalline forms.) Subsequently, the dispersion is stabilized and then dried to remove the remaining solvent residue completely (Barot et al., 2006). The particle size produced by this method depends on the nucleation and growth of particles in supersaturated solutions, which may be challenging to control (Chattopadhyay and Shekunov, 2006).

Micronization by sonication involves the dispersion of a solution of the API in the form of aerosols. The resulting aerosols are dispersed in a "nonsolvent" phase (e.g., cyclohexane) to allow crystallization. This dispersion is then subjected to ultrasonication at frequencies between 35 and 45 kHz (Barot et al., 2006).

The supercritical fluid extraction of emulsions is in particular applicable to water-insoluble APIs. It is a closely related method to the precipitation technology, where the oily phase of an oil-in-water emulsion is extracted with supercritical CO_2. The nucleation of particles occurs slowly within the aqueous phase of the droplets, which results in a stable particle size much smaller than that obtained by precipitation methods (Chattopadhyay and Shekunov, 2006).

9.4 PARTICLE SIZE PREPARATION AND REDUCTION FOR PULMONARY DELIVERY

For pulmonary delivery of powders, three techniques have commonly been utilized to prepare powders useful for inhalation: spray-drying, freeze-drying, and spray-freeze-drying (Samaha et al., 2008). For effective systemic effects via the lungs, the preferred particle size is in the range 1 to 3 μm, and smaller particles (less than 1 μm) tend not to be deposited efficiently in the lungs (Panagiotou, 2009). [An ideal particle diameter for pulmonary deposition of dry powders was stated to be 3 μm (Mobley, 1998)]. The most commonly used method for the particulate reduction of powders intended for inhalation is by air jet-milling methods (dry methods). The most common of these methods are spiral jet milling and fluid-bed opposed jet milling. Wet milling processes are also available and are much less complex than the dry methods (Rasenack, 2010). For liposome-containing drugs, freeze-drying techniques are effective in producing solid particles for pulmonary delivery (Desai et al., 2002). Jet milling reduces the particle size further, placing it in the ideal range of about 3 μm; however, the integrity of liposomes is often jeopardized (caused by drug leakage) (Mobley, 1998). Used as a cryoprotectant, sucrose is found to be ideal for liposomes during the lyophilization process (Mobley, 1998; Joshi and Misra, 2001). In addition to milling and freeze-drying techniques, spray-drying methods and supercritical fluid-processing techniques are also available for preparing powders for inhalation; however, their applications in this regard are much more limited. *Pulmosphere technology* utilizes spray drying in which an emulsion of an API containing phospholipids and a propellant are spray-dried to produce large porous particles (Rasenack, 2010).

Assessing the surface morphology of solid particles may be done using scanning electron microscopy. In addition, the particle size can be estimated using a laser diffractometer. A polydispersity index known as the *Span index* is utilized to describe the distribution profile of powder particles. The higher the value of this index, the higher the variability in the size distribution. This index is given by the expression:

$$\text{Span index} = \frac{D_{(v,90)} - D_{(v,10)}}{D_{(v,50)}} \tag{9.7}$$

where $D_{(v,10)}$, $D_{(v,50)}$, and $D_{(v,90)}$ are spherical equivalent volume parameters at 10%, 50%, and 90%, respectively.

The aerodynamic powder particle diameter can be estimated from

$$D_{\text{aer}} = \rho(\text{VMD}) \tag{9.8}$$

where ρ is the tapped bulk density of the powder and VMD is the volume median diameter (Samaha et al., 2008). The aerodynamic particle size distribution is commonly estimated from cascade impaction methods. Briefly, in this method the particles emitted are collected on cascade impactor surfaces or deposition sites. Larger particles strike first and smaller particles are collected at more distant surfaces. Subsequent evaluation of the relative number of particles collected at each deposition site provides a profile for the particle size distribution in a sample (Christopher et al., 2005).

Special considerations for the production of small particles for inhalation products are important to distinguish them from other routes of administration. Although certain organic solvents are appropriate for methods of particle reduction for oral products, residues of these solvents can have a detrimental effect on lung tissues. Thus, high-energy methods for particle size reduction are employed for the pulmonary delivery of drugs: homogenization, ultrasonication, and high-shear fluid processing. Homogenization methods are somewhat limited in their efficiency and reproducibility, due to the choice of valve geometry, orifice size, and pressure. Ultrasonicators are extremely powerful in their action, and when operated at energy levels of 15 to 50 kHz can produce pressures of more than 500 atm. Unfortunately, this intense delivery of energy produces extremely high local temperatures, up to 5000°C. which precludes the practicality of this method for particle size reduction. A favorable method for reducing particle size is high-shear fluid processing technology, which relies on forcing a stream of liquid-containing particles to pass through microchannels at extremely high pressures (100 to 3000 atm) and high shearing rates. This method produces a final product with good uniformity. The high-shear fluid processing technique works well at the laboratory level, and is easy to transfer to a pilot plant or to production (Panagiotou, 2009).

Another problem in delivering solid particles via the pulmonary route is particle aggregation. In particular, in *dry powder inhalers* (DPIs) the fine drug particles tend to adhere to the excipient's larger particles, reducing the amount of drug systemic delivery. However, it appears that the extent to which fine particles aggregate among themselves is not significant enough to cause an alteration in drug delivery (Houzego, 2002). Lactose is commonly used as an excipient for DPIs. Its particle size is in the range 50 to 100 μm in diameter (Houzego, 2002). There are two main solutions to aggregation problems: by the production of fast-moving air over drug–lactose aggregates and by hurling the drug–lactose aggregates onto a surface, thus eliminating aggregation (Houzego, 2002).

9.5   POLYMERIC PARTICULATE MATTER

Polymeric particles intended for parenteral administration must have certain characteristics, most important is that the polymer is biodegradable. This can be achieved simply by employing material similar to that used initially for a resorbable surgical suture. These types of material are biodegradable and are gradually absorbed by body tissues, gradually releasing the drug over time. Other characteristics of polymer particles include a spherical shape, which allows the powder polymer to flour freely; a relatively small particle size, preferably smaller than 125 μm in diameter; and good mechanical properties (ease of handling during manufacturing). Polymers of this type belong to the class of polyesters that includes lactic acid and glycolic acid. Because of the stereospecificity of lactic acid, it can exist as an L, D, or DL type; poly(DL-lactide) has an amorphous structure that produces a better film. One of the major drawbacks of poly(DL-lactide) is that their rate of biodegradation is slow (one year for complete resorbtion). On the other hand, polyglycolides have physical characteristics that prevent them from being practical in formulations; they are crystalline materials that do not readily dissolve in available solvents. Copolymerization of polyglycolides with poly(DL-lactide) in different proportions produces polymers with good solubility in common solvents as well as improving on the rate of biodegradation over that seen with poly(DL-lactide) polymers (reducing the period of degradation from months to weeks) (Tice and Cowsar, 1984).

9.6   NANOPARTICLES

The term *nanoparticles* is commonly used to refer to particulate matter in the size range 1 to 1000 nm (Saleem et al., 2007), with the majority of applications normally having particles in the size range 100 to 1000 nm (Möschwitzer, 2010), where 1 mm is 1 billionth of a meter (10^{-9}m); it is roughly the space occupied by three to six atoms (Parikh, 2011). Nanoparticles may be prepared by precipitation from molecular dispersions, milling methods, or chemical reactions (Möschwitzer, 2010). Potential applications of nanoparticles are in the area of cancer treatment and imaging. For example, the use of up-conversion nanoparticles is increasingly being recognized as a means of delivering chemotherapeutic agents (e.g., doxorubicin) as well as providing contrast material for low-background biomedical imaging (Wang et al., 2011). Other examples of nanoparticulate carriers include liposomes, emulsions, polymers, polymeric micelles, dendrimers, nucleic acid–based carriers, nanoshells, gold shells, nanocrystals for diagnostic imaging techniques, and magnetic nanoparticles composed of iron oxide which are utilized in magnetic resonance imaging, among others (Sadrieh and Tyner, 2008; Patel et al., 2009). When nanoparticles are dispersed in liquid dosage forms, the resulting preparation is referred to as a *nanosuspension*. In turn, nanosuspensions are subjected to drying processes (e.g., lyophilization) to produce dry powders to be incorporated in solid dosage forms (Möschwitzer, 2010). Nanoparticulate carrier drug delivery systems may be incorporated into injectable, implantable, oral, topical and transdermal, and pulmonary technologies (Amin and Shah, 2010).

Potential benefits of nanotechnology include reduction in dosage size due to specific targeting of drugs; improvement in patient compliance; a multitasking approach for the nanoparticles, such as imaging and targeting drug therapy; enhancing the stability of a drug following administration, in particular for peptides and proteins; protecting APIs from degradation; producing a controlled-release delivery mechanism for an API; enhancing drug loading; and improving bioavailability due to an increase in the total surface area available for absorption (Sadrieh and Tyner, 2008; Amin and Shah, 2010). Improvement in oral bioavailability following nanoparticle administration was related to an increase in the intestinal mucosal uptake of the nanoparticles. In addition to improving the oral bioavailability of APIs, percutaneous diffusion of particulate matter through the skin was also improved significantly (Parikh, 2011). As the particle size approaches that of the nanoscale, many of the physical characteristics of the material undergo changes; these alterations in the physical properties of the material include vapor pressure, volatility, volume, conductivity, melting and freezing points, and surface area (Saleem et al., 2007).

For nanoparticle-based dosage forms, in addition to the normal testing procedures performed on conventional dosage forms, a battery of other tests are needed for characterizing them. These include average particle size diameter and size distribution, particle surface morphology, particle aggregation profile, and surface chemical composition, including electric charge (Sadrieh and Tyner, 2008). Manufacturers of nanoparticle-based dosage forms need to define the type and mode of interaction between the API and the nanoparticles (the existence of strong or weak bonds and the type of bond), the release profile of the API from the nanoparticle core, and the safety profile of the nanoparticle–API complex (Sadrieh and Tyner, 2008). For size determination, dynamic light-scattering techniques, small-angle x-ray scattering, small-angle neutron scattering, electron microscopy testing (both scanning and transmission), field-flow fractionation, analytical ultracentrifugation, and atomic force microscopy are all used for diameter size estimation and other morphological characteristics (Saleem et al., 2007; Sadrieh and Tyner, 2008). Care must be taken when sampling for particle size determination, as these systems tend to aggregate readily because of the high internal surface area/volume ratio.

Zeta potential (ζ-potential) measurement permits monitoring of the nanosystems to place them in a nonaggregation zone (usually very high positive or negative ζ-potential values are desirable in this regard) (Saleem et al., 2007). [In an electric double-layer model for suspended particulate systems, the electrical potential decreases linearly from the actual surface of the particles until it reaches the boundary between the tightly bound layer (the Stern zone) and the more diffused layer; it thereafter decreases exponentially to zero when it reaches the electroneutral region of the system. The difference in potential between the outer boundary of the Stern layer and that of the electroneutral region is known as the ζ-potential (Fairhurst and Lu, 2011).] Field-flow fractionation methods combine chromatography and electrophoresis to segregate particles in a collection of particles based on their size or other characteristics (e.g., charge or density) so that they may be subjected to further sizing by other methods (e.g., dynamic light scattering). While optical ordinary microscope can measure particles down to 200 nm in diameter, *electron microscopy* (EM) has the ability to measure dimensions at the angstrom level. EM data cover a wide range of information on particle length, diameter, shape, and area. In addition, atomic force microscopy allows topographical and morphological measurements of a sample (Saleem et al., 2007). Dynamic light scattering methods are utilized widely in sizing particulate matter down to 3 nm, and are useful as long as the particle size does not exceed 3 μm in diameter (distortion of measurement occurs beyond this limit). Similarly, laser diffraction techniques may be used in sizing nanosystems; however, they are limited to particulate matter with a diameter greater than 500 nm. Small-angle x-ray scattering and small-angle neutron scattering methods allow better resolution in particle sizing than do the other methods; however, they are not widely adopted by research laboratories, due to their complex mode of operation (Saleem et al., 2007).

A special type of nanoparticle is the *lipid nanoparticle* (LN) (size range 80 to 1000 nm with a typical range between 200 and 400 nm) (Keck et al., 2007). LNs are divided primarily into two subtypes: *solid lipid nanoparticles* (SLNs) and *nanostructured lipid carriers* (NLCs). LNs are prepared from an oil-in-water (o/w) nanoemulsion in which the oily phase consists of solid lipids at body temperature. The difference between the two subtypes is that an SLN consists of a single solid lipid, whereas an NLC contains a blend of solid lipids with oily liquids. NLC particles possess a higher loading capacity than that of SLNs, due to their "imperfect" internal structure (Keck et al., 2007). Applications of LNs are found for dermal and oral routes, providing an effective way of delivering the drug locally or systematically. Oral applications of this type were found to enhance the bioavailability of drugs such as cyclosporine A. Local skin applications of LNs containing ultraviolet-blocking agents provides prolonged protection against ultraviolet light, due to the ability of the nanoparticles to afford a very slow way to release an API from a matrix, thus maintaining it on the surface of the skin (Keck et al., 2007). Marketed preparations known as *self micro-emulsifying drug delivery systems* (SMEDDSs) are also available, which upon mixing with an aqueous environment form a spontaneous *nanoemulsion*. (The terminology is deceiving in that it uses "micro," whereas in fact the size of the dispersed droplets is in the nano range.) SMEDDSs offer enhanced bioavailability from the gastrointestinal tract (e.g., cyclosporine A) (Mɑllertz, 2007).

Nanotechnology has applications in the production of sustained-release formulations. In 2011, over 1400 products with sustained-release characteristics existed on the market. A general classification of sustained-release technologies includes two types: *extended-release systems* and *delayed-release systems*. An example of the former is the Oros technology by Johnson & Johnson's Alza, and an example of the latter is the SODAS technology produced by Elan Drug Technologies. Both systems are versatile enough to be customized to a variety of medicinal agents (Liversidge, 2011). Top-down approaches (utilizing attrition) and bottom-up methods based on molecular deposition (spray-freezing into liquid, gas antisolvent recrystallization, and rapid expansion from a liquefied-gas solution) have been employed in the production of nanosize particulate matter (Möschwitzer, 2010; Liversidge, 2011). Wet milling technology utilizing high-pressure input is used for the production of NanoCrystal by Elan Drug Technologies. This technique has been incorporated in the production of many marketed formulations, such as Rapamune (immunosuppressant; Pfizer), Emend (prevention of nausea and vomiting associated with chemotherapy; Merck), TriCor (lipid-lowering; Abbott), Megace ES (malnutrition associated with AIDS, Strativa Pharmaceuticals), and Invega Sustenna (schizophrenia; Janssen) (Liversidge, 2011).

9.7 SEGREGATION OF PARTICLES

In any collection of powder particles, differences in size and shape among the particles do exist. If a powder is subjected to shear or shaking during handling, it is expected that particle segregation occurs during the process. The segregation can be serious enough to produce uneven distribution of the API in the final product (e.g., tablets), as it produces demixing of the API with the other ingredients in the formulation. Factors known to contribute to segregation include differences in the particle size and shape between the API and the other ingredients, as well as flow property differences (Lee, 2007). Particle size diversity, the strength of the cohesive forces existing between particles, and mechanical stress during handling and storage also play a role in particle segregation (Deanne and Etzler, 2007). The degree of particle size dispersion within a powder is determined by sampling using scoops, thieves, and riffles, with the latter being the most adequate type for sampling. Important rules to consider during particle size analysis are: (1) samples should be taken while the powder is in motion; and (2) samples should be collected from different locations at different time periods along the entire stream of the powder being tested (Deanne and Etzler, 2007).

REFERENCES

Amin A, Shah M. Nanoparticulate carrier drug delivery systems: a mini review of patented technologies. Drug Deliv. Technol. 2010;10(7):40–45.

Barot M, Modi MD, Parikh RJ. Novel particle engineering technology: a review. Drug Deliv. Technol. 2006;6(9):54, 56, 58–60.

Beckmann W. Particle design of APIs through crystallization. Am. Pharm. Rev. 2006;9(6):110, 112–117.

Chattopadhyay P, Shekunov YB. New enabling technologies for drug delivery. Drug Deliv. Technol. 2006;6(8):64–68.

Christopher D, Pan Z, Lyapustina S. Aerodynamic particle size distribution profile comparisons: considerations for assessing statistical properties of profile comparison tests. Am. Pharm. Rev. 2005;8(1):68–72.

Dai W-G, Dong LC, Li S, Pollock-Dove C, Deng Z. Significantly reduced drug-apparent solubility using a micronization process. Drug Deliv. Technol. 2010;10(2):48–53.

Deanne R, Etzler MF. Particle size analysis and powder sampling. Am. Pharm. Rev. 2007;10(3):132, 134–136.

Desai TR, Wong JP, Hancock REW, Finlay WH. A novel approach to the pulmonary delivery of liposomes in dry powder form to eliminate the deleterious effects of milling. J. Pharm. Sci. 2002;91(2):482–491.

Dubin C. Formulation frustrations. Drug Deliv. Technol. 2005;5(8):30, 32, 34.

Fairhurst D, Lee WR. The zeta potential and its use in pharmaceutical applications: Part 1. Charged interfaces in polar and non-polar media and the concept of the zeta potential. Drug Dev. Deliv. 2011;11(6):60–64.

Fan A, Palleria S, Carlson G, Ladipo D, Dukich J, Capella R, Leung S. Effect of particle size distribution and flow property of powder blend on tablet weight variation. Am. Pharm. Rev. 2005;8(2):73–78.

Houzego P. Deaggregation mechanisms in dry powder inhalers (DPIs). dds&s 2002;2(3):81–85.

Hubert M, Sarsfield B, Shekunov B, Grosso J. Oral solid dosage form: from choice of particle size technique to method development and validation. Am. Pharm. Rev. 2008;11(6):14, 16, 18, 20–21, 23.

Ji Q, Tumuluri V, Wang W, Teelucksingh J, Mecadon M, Vegesna R. Rapid content uniformity determination of low dose TCH346 tablets by NIR. Am. Pharm. Rev. 2006;9(5):20, 22–26.

John E. How to set specifications for the particle size distribution of a drug substance? Am. Pharm. Rev. 2009;12(3):72, 74–77.

Joshi MR, Misra A. Liposomal budesonide for dry powder inhaler: preparation and stabilization. AAPS PharmSciTech 2001;2(4;art. 25): 1–10.

Jürgens K. Influence of particle size and shape on the drug product: aspects of formulation and performance. Am. Pharm. Rev. 2006;9(4):61–63, 65.

Keck CM, Hommoss AH, Müller RH. Lipid nanoparticles for encapsulation of actives: dermal and oral formulations. Am. Pharm. Rev. 2007;10(7):78, 80–82.

Kelly RN, Lerke SA. Particle size measurement technique selection within method development in the pharmaceutical industry. Am. Pharm. Rev. 2005;8(6):72–81.

Khamanga MS, Parfitt N, Nyamuzhiwa T, Haidula H, Walker BR. The evaluation of Eudragit microcapsules manufactured by solvent evaporation using USP Apparatus 1. Dissolut. Tech. 2009;16(2):15–21.

Lee C. Troubleshooting low potency results in solid oral dosage forms: Is it the method or the formulation? Am. Pharm. Rev. 2007;10(6):142–144, 146–147.

Liversidge G. Controlled release and nanotechnologies: recent advances and future opportunities. Drug Dev. Deliv. 2011;11(1):42, 44–46.

Mallertz A. Lipid-based drug delivery systems: choosing the right in vitro tools. Am. Pharm. Rev. 2007;10(4):102, 104, 106–110.

McConnell EL, Pygall SR, Ward R, Seiler C. An industrial insight into pellet manufacture by extrusion spheronization. Drug Deliv. Technol. 2010;10(6):50–54.

Mobley WC. The effect of jet-milling on lyophilized liposomes. Pharm. Res. 1998;15(1):149–152.

Möschwitzer J. Nanotechnology: particle size reduction technologies in the pharmaceutical development process. Am. Pharm. Rev. 2010;13(3):54–59.

Panagiotou T. Achieving optimal particle size distribution in inhalation therapy. Drug Deliv. Technol. 2009;9(2):53–57.

Parikh MD. Recent advances in particle engineering for pharmaceutical applications. Am. Pharm. Rev. 2011;14(2):98–103.

Parrott LE. Characteristics of particles and powders. In: *Pharmaceutical Technology: Fundamental Pharmaceutics*. Delhi, India: Surjeet Publications; 1984. pp. 1–36.

Patel DM, Patel NM, Patel MM. Fast-dissolving rofecoxib tablets: formulation development and optimization using factorial design. Drug Deliv. Technol. 2007;7(2):32, 34–37, 39.

Patel P, Acharya MA, Patel J. Nanoparticles in cancer research: a novel drug delivery and pharmacological approach. Drug Delivery Technol. 2009;9(2):48, 50–52.

Rasenack N. Particle engineering for pulmonary dosage forms. Am. Pharm. Rev. 2010;13(3):76–80.

Rubino PO. Fluid-bed technology: overview and criteria for process selection. Pharm. Technol. 1999;23(6):104, 106, 108, 110, 112–113.

Sadrieh N, Tyner K. Characterization considerations for nanomaterial-based drugs. Am. Pharm. Rev. 2008;11(2):70, 72–74.

Saleem I, Donovan MJ, Smyth HDC. Particle size analysis: from microparticles to nanosystems. Am. Pharm. Rev. 2007;10(2):126, 128–130.

Samaha M, El-Khawas F, El-Maradny H, Mohamed Ragab D. A comparative study of the effect of using different drying techniques for preparation of inhalable protein powders on their aerosolization performance. Drug Deliv. Technol. 2008;8(5):38–43.

Sandler N. New image analysis approach for particle size monitoring of pharmaceutical solids. Am. Pharm. Rev. 2005;8(3):86–91.

Scarlett B. Measuring and interpreting particle size distribution. Am. Pharm. Rev. 2003;6(4):93–101.

Sun Z, Ya N, Adams RC, Fang FS. Particle size specifications for solid oral dosage forms: a regulatory perspective. Am. Pharm. Rev. 2010;13(4):68, 70–73.

Tice TR, Cowsar DR. Biodegradable controlled-release parenteral systems. Pharm. Technol. 1984;8(11):26–30, 32–34, 36.

Wang C, Cheng L, Liu Z. Drug delivery with upconversion nanoparticles for multifunctional targeted cancer cell imaging and therapy. Biomaterials 2011;32(4):1110–1120.

GLOSSARY

API	Active pharmaceutical ingredient.
Coefficient of determination (r^2)	A measure of the amount of variability observed with the dependent variable as explained by a given independent variable. The closer the r^2 value to 1, the better the independent variable *predicts* the changes seen in the dependent variable.
DPI	Dry powder inhalers.
EM	Electron microscopy.
Equivalent spherical diameter	Refers to a sphere with the same surface area, volume, surface/volume ratio, settling velocity, aerodynamic behavior, or sieve mesh as that of particles (Scarlett, 2003; Hubert et al., 2008) that would behave in a manner similar to that of powder particles when subjected to the same conditions of testing (Hubert et al., 2008).
Extrusion	A wet mass is forced through openings of a preselected sieve size to form rods.
Granulation	The granulation step is similar to that employed in the wet granulation process of tablets except that the resulting mass to be extruded is much more fluid.

Hausner ratio	Tapped bulk density/bulk density (ρ_{tap}/ρ_{bulk}) estimated as a measure for microsphere flow properties; values of HR greater than 1.25 are indicative of poor flowability.
HR	However ratio.
LN	Lipid nanoparticle.
LN$_2$	Liquid nitrogen.
Micronization	Comminuting the drug to micrometer- size particles.
Nanoparticle	Commonly refers to particulate matter with a size range of 1 to 1000 nm.
Nanosuspension	The preparation that results when nanoparticles are dispersed in liquid dosage forms.
NIR	Near-infrared spectroscopy.
NLC	Nanostructured lipid carriers.
PCA	Principal components analysis.
Pulmosphere technology	A method utilizing spray-drying in which an emulsion of an API containing phospholipids and a propellant is spray-dried to produce large porous particles.
SFD	Spray-freeze drying.
SFL	Spray-freezing into liquid.
SLN	Solid lipid nanoparticles.
SMEDDS	Self micro-emulsifying drug delivery systems.
SOM	Self organizing maps.
U.S.P	*United States Pharmacopeia.*

CHAPTER 10

BASIC STATISTICS AND DESIGN OF EXPERIMENTAL CONCEPTS

10.1 DESCRIPTIVE STATISTICS

Statistics is the science of organizing and analyzing experimental data. In organizing the data, researchers group their findings in tables and graphs to emphasize trends and/or relationships between variables. In addition to graphical and tabulated methods, variables may be described numerically. These numerical descriptions summarize the data with respect to central location, variability, skewness, and kurtosis. In general, variables are classified into four categories: nominal, ordinal, interval, and ratio. *Nominal* variables are the simplest. For example, "drug type" is a nominal variable. In a nominal scale there is no order among the categories. *Ordinal* variables contain order and hierarchy in their structure. For example, the intensity of tablet color (light, dark, darker) is ordinal in nature. However, the difference between "light" and "dark" cannot be assumed to be the same as that between "dark" and "darker." If the distance between the subcategories within a variable is the same, the scale is called *interval*. However, similar to the ordinal scale, the interval scale does not have a true zero value. When a true zero exists and the distance between the subcategories is the same, the scale is called *ratio* (Elenbaas, 1983a). Nominal and ordinal data are referred to as *categorical*, whereas interval and ratio data are for *continuous* variables (Fenn Buderer, 1997).

The numerical descriptions are called *characteristics* when they are obtained from a sample and are called *parameters* when they refer to a population. *Central location* measures (also known as *central tendency*) include arithmetic mean, median, mode, and percentiles. (In addition to the arithmetic mean, the *geometric mean* may be used more appropriately if the data reflect *growth rate*. The geometric mean is computed by taking the nth root of the product of all the observations. The *harmonic mean* is used as an average value for ratios and rates and is calculated by dividing the number of observations by the sum of the reciprocal of each observation in the sample. In general, the harmonic mean produces the lowest value for the average, while the value of the geometric mean is the largest. Measures of central location are seen as points on a continuous scale; thus, they are used only with continuous variables. The arithmetic mean is sensitive to extreme values in

Integrated Pharmaceutics: Applied Preformulation, Product Design, and Regulatory Science,
First Edition. Antoine Al-Achi, Mali Ram Gupta, William Craig Stagner.
© 2013 John Wiley & Sons, Inc. Published 2013 by John Wiley & Sons, Inc.

a data set, and these values tend to pull the mean to their direction. The median, mode, and percentiles are insensitive to extreme values (Fenn Buderer, 1997). For example, consider the following set of data obtained from a dissolution test (the amount of the drug dissolved in 45 minutes):

$$234, 236, 305, 219, 226, 231, 350 \text{ mg}$$

The value of the arithmetic mean is $(234 + 236 + 305 + 219 + 226 + 231 + 350)/7 = 257.3$ mg; the median value is (219, 226, 231, 234, 236, 305, 350) or 234 mg (50% of the distribution lies above 234 mg). There is no mode value for this set, since the frequency of all the data points is 1. Suppose that we added another measurement to the set above:

$$234, 236, 305, 219, 226, 231, 350, \underline{400} \text{ mg}$$

The new value for the mean is now increased to 275.1 mg, the median value slightly changed to (219, 226, 231, 234, 236, 305, 350, 400) or 235 mg (the average of 234 and 236 mg), and still there is no mode value for this new set. Notice how the mean has moved 400 mg in the direction of the datum. Also notice the insensitivity of both the median and the mode to this new relatively large value.

Measures of variability (Fenn Buderer, 1997) are important in defining a variable's distribution. As opposed to the measures for central location, the measures for variability are considered to be segments on a continuous scale. The single value of the central tendency cannot fully describe the distribution alone. Two distributions may have exactly the same average but differ significantly in their dispersion. The most important measure for variability is the *standard deviation* (SD) (also the *variance*, SD^2), and the simplest measure in this respect is the range. For our original dissolution data set (234, 236, 305, 219, 226, 231, 350 mg), the range is stated as 219 to 350 mg or $350 - 219 = 131$ mg. The SD value for this data set is 50.0 mg. Notice that the SD has the same units as that for the data points. The SD is *thought of* (in a crude way) as the average distance of each data point from the mean. (Not exactly, as mathematically speaking, the average distance of each data point from the mean is always zero.) The variance is the square of the SD, which for our example is 4689.3 mg^2. The units for the variance are squared as well (mg^2). Similar to the mean, all the measures for variability are sensitive to extreme values. Another commonly used measure for variability in the pharmaceutical industry is the *relative standard deviation* (% RSD), defined as the ratio of the standard deviation to the arithmetic mean expressed as a percent age. For the dissolution data, % RSD is $(50.0/257.3) \times 100 = 19.4\%$. % RSD is also known as the *coefficient of variation* (% CV). However, % RSD is adopted by the U.S.P. as the term used for variability. The advantage of % RSD over SD is that the former describes the variability *relative to the mean* of the distribution. Thus, one could declare that one distribution is less or more variable *with respect to its mean* as compared to another distribution.

Let us consider another set of dissolution data for the same drug and the same dosage form, but made by a different manufacturer:

$$222, 220, 224, 225, 233, 300, 229 \text{ mg}$$

For this set of data, the mean is 236.1 mg and the SD is 28.5 mg. Therefore, % RSD for this set is 12.1%. Thus, the second data set is *less* variable than the first set with respect to its mean. By just stating that the SD value of the second set is less than that of the first set is not as powerful as when comparing the % RSD values from both sets.

Statistical software programs provide calculations for measures of skewness and measures of kurtosis. Distributions may be skewed positively toward the right or skewed negatively toward the left. Severe skeweness may affect how the analysis is handled by statisticians, and nonparametric testing is normally employed on data that are significantly skewed. Similarly, measures for kurtosis (i.e., how peaked or flat the distribution looks) are determined by computer software packages, and if kurtosis is relatively significant, nonparametric tests are also used to analyze the data.

The following are equations for estimating some important descriptive statistics:

$$\text{relative frequency} = \frac{\text{frequency}}{\text{Total no.}}$$

$$\text{relative frequency (\%)} = (\text{relative frequency}) \times 100$$

$$\text{arithmetic mean} = X\text{-bar or } Y\text{-bar} = \frac{\sum X_i}{n}$$

$$\text{grouped mean} = \frac{\sum m_i f_i}{N}$$

$$\text{rank percentile} = \text{rank } P\text{th} = \frac{P}{100}(N+1)$$

$$\text{percentile} = P\text{th} = [(Y_{(i+1)} - Y_i)f] + Y_i$$

$$\text{standard deviation} = \text{SD or } S = \left[\frac{\sum(X_i - X\text{-bar}^2)}{N-1} \right]^{1/2}$$

$$\text{variance} = S^2$$

$$\text{grouped SD} = \left[\frac{\sum(m_i - X\text{-bar})^2 f_i}{(N-1)} \right]^{1/2}$$

$$\text{interquartile range} = \text{IQR} = P_{75} - P_{25}$$

$$\text{coefficient of variation (\%)} = \text{relative standard deviation (\%)}$$

$$= \text{CV(\%)} = \text{RSD (\%)} = \frac{\text{SD}}{X\text{-bar}} \times 100$$

$$\text{Pearson's skewness coefficient} = \frac{\text{mean-median}}{\text{SD}}$$

10.2   INFERENTIAL STATISTICS

Having described the data in tables, graphs, and numerical characteristics, the next step is to analyze and compare variables with respect to their characteristics among groups. For examples, two batches of capsules for the same drug manufactured by the same company may differ significantly with respect to their average weight. Or a company may declare that the average weight for its capsules made for one drug must be 300.0 mg. Suppose that a batch of these capsules were made and that on average the weight of 20 capsules selected by *random* from the batch was 296.5 mg. One may argue that 296.5 mg is indeed less than 300 mg and the batch should be rejected. However, another technician weighed a different 20-capsule sample from the same batch and found its average weight to be 304 mg; and 304.0 mg is larger than both 300 mg (the declared average) and 296.5 mg (the average of the first 20-capsule sample). Perhaps you already figured out that repeated sampling from the same batch of capsules would not necessarily yield exactly a 300.0-mg average for all 20-capsule samples selected by random from the entire batch. The question is: How would one ascertain that the average weight for the batch is indeed not *significantly different* from the declared average? The operator in this case has two choices: The first is to weigh out every capsule produced in that batch (not a very sensible way of doing business), or to employ statistics. Statistical analysis provides the answer to this question with a certain degree of confidence (usually, 95% or 99% confidence). Stated differently, what is the probability of finding an average weight for a 20-capsule sample chosen randomly from the batch to be 296.5 mg (or 304.0 mg)? If this probability was a high value, the sample average is no different from the average declared (Cuddy et al., 1983).

Statistical comparison is not limited to comparing one sample average to a declared average but can also be used to compare averages obtained from two or more samples. In this case the operator chooses to compare the average weight (i.e., cost, disintegration time, etc.) of capsules from company A and that from company B. This can also be achieved by calculating the probability that the averages are all equal. If this probability is high, there is no difference between the averages; otherwise, the averages differ.

So far, we have mentioned calculating the probability of an event (equal weights) to occur by chance alone and declaring that this probability is high or low. However, we need to specify the limit by which a probability is high or low. This is done by declaring a value in advance which is commonly known as alpha (α) or the *level of significance*. Thus, α allows us to distinguish between averages that are similar from those that are dissimilar. The value of α is commonly set at 5%; however, other limits, such as the 1% limit, are also acceptable. The confidence by which we declare our conclusion (also known as the *confidence coefficient*) and α are complementary to each other. When α is 5%, the confidence coefficient is 95%. From a statistical point of view, α is the probability of declaring that two averages are not the same when in fact they are the same (also known as the probability of type I error). In addition to type I error, statistical analysis is also associated with type II error. The probability of committing type II error is β. When we declare that two averages are the same when in fact they are not, we commit type II error. In pharmaceutical investigations, β is kept at 20% or less. The complement of β is the *power* of the test, defined as the probability of finding the two averages different when indeed they are. Since power is the complement of β, the value of the power is commonly chosen to be 80 or higher (Elenbaas, 1983b).

Statistical software programs are now available in the pharmaceutical industry to compute all the necessary calculations needed to reach a statistical conclusion. Thus, it is important to understand the fundamental points concerning the statistical analysis and what terms to look for in the statistical report in order to make the final conclusion. Assuming that an investigator is following the scientific methods in his or her investigations, a hypothesis is developed to answer the issue at hand. In scientific investigations the hypothesis is usually tested with an α of 5% and a β of less than 20% (power = 80%). A sample size is then selected that meets the conditions set by an investigator regarding the values of α and β. Sample size selection also requires knowledge of the standard deviation of the variable of interest as well as the *effect size* (Δ) and whether the test is one- or two-tailed. The effect size is based on and extracted from the hypothesis. Suppose that the investigator aims to *increase* the hardness of a tablet formulation by adding more binders to the formulation. (The word *increase* in hardness of tablets as stated in the hypothesis signifies a *direction* for the hypothesis and would make it a one-tailed test. If there were no direction to the hypothesis—for example, the investigator is looking for *any* difference in hardness—the test would be two-tailed) (Elenbaas, 1983b). The increase in hardness of tablets is defined by an investigator as an a minimum increase of 1 kP on average. Also assume that the standard deviation of tablet hardness is 0.25 kP. Therefore, the effect size for this test is (1 kP/0.25 kP = 4). The sample size is therefore selected to detect this increase in tablet hardness of 1 kP with a type I error of 5%, a power of 80%, and an effect size of 4.

After the data are collected, the investigator runs the statistical analysis to see the effect of adding more binders to the formulation on the hardness of tablets. The statistical report includes a term known as the *p value* (Fenn Buderer, 1998a). Based on the *p* value and its relative magnitude compared to α, a decision is made on the hypothesis. Thus, the *p* value is the smallest value for α by which the null hypothesis is rejected (Elenbaas, 1983). In almost all pharmaceutical investigations, one is interested to learn whether a certain factor (or factors) is (are) affecting the outcome. In our example, this certain factor is the amount of binder added, and the outcome is the hardness of the tablet. The hypothesis is rewritten in the form of the *null hypothesis* (H_0), which is simply a statement of no difference (Elenbaas, 1983b). Applying it to our example, H_0 states that the hardness of the tablet is the same regardless of the amount of binder added to the formulation ($H_0 : \mu_1 = \mu_2$). Alternatively, we suggest that adding more binder is going to increase the hardness of the tablets ($H_a : \mu_1 > \mu_2$). By comparing the *p* value from the statistical report to α, one can reach a conclusion as to whether to reject or not to reject H_0. If the *p* value is larger than α, H_0 is not rejected, and vice versa. Assuming that in our example we reach a *p* value of 0.0389, H_0 is rejected since $p < \alpha$. Our conclusion would be that the addition of more binder to the formulation causes the hardness of tablets to *increase* by a minimum of 1 kP. Note that this *p* value of 0.0389 is for a one-tailed test. For a two-tailed test, the *p* value is computed by multiplying it by 2. Thus, for our example, a two-tailed test *p* value would be $2 \times 0.0389 = 0.0778$. With a two-tailed test, since $p > \alpha$, the conclusion would be that the addition of more binders *does not change* the hardness of tablets. Accordingly, a one-tailed test always has *higher power* than a two-tailed test (because H_0 is more likely to be rejected with a one-tailed test than with a two-tailed test). The default, therefore, is to have a two-tailed test for the hypothesis unless a one-tailed test is otherwise justifiable. This is adopted so that the investigator does not use a test with a higher power without proper reasoning.

Another important concept in statistics belongs to the *confidence interval* (CI) (Fenn Buderer, 1998). Unlike the *p* value, the CI provides an estimate of the parameter of interest in the population (usually, the mean value, μ) as *predicted* from the statistics (also known as a *point estimate*) obtained from the sample. The CI utilizes the statistics value from the sample, the standard deviation, and the sample size used to estimate the *true* value of the parameter in the population within a declared confidence level (e.g., 95% or 99%) (Fenn Buderer, 1998). The ratio of the standard deviation to the squareroot of the sample size is known as the *standard error of the mean* (SE) (Fenn Buderer, 1997a). The CI has the following general format:

$$CI = (\text{point estimate}) \pm (\text{table value}) \ (\text{SE}) \tag{10.1}$$

where the table value is the value of the statistics being used (e.g., z, Student's t) and is obtained from standard statistical tables. Most statistical software programs provide estimates of the CI range. Equation (10.1) calculates two values for the parameter, a lower limit and an upper limit. At the 95% confidence level, for example, one can declare *with 95% confidence* that the parameter of the population is within the CI limits. It is important to state the confidence level with the CI statement, because in statistics we cannot declare a 100% confidence in our estimations. A sample obtained from its population inherently contains *random error* due to sampling bias, and estimates obtained from the sample to describe a population's parameters cannot be 100% accurate.

Statistical tests are generally classified into two main types: parametric and nonparametric. Among the *parametric tests* we recognize the Z test, Student's t test, and the analysis of variance test (Elenbaas, 1983; Fenn Buderer, 1998b). *Nonparametric tests* include the chi-squared test, Mann–Whitney U test, Wilcoxon signed-rank test, and Friedman matched samples test (Elenbaas, 1983). The basic hypothesis for all these tests is the same as that outlined earlier. Nonparametric tests are commonly used when the dependent variable (the outcome) is classified at either the ordinal or the nominal scale (Fenn Buderer, 1997). They are also used whenever the sample size is small and there is a considerable deviation from a normal distribution. On the other hand, parametric tests are used for populations that are normally distributed and when the outcome being measured is at either the interval or the ratio scale (Elenbaas, 1983).

The procedures described next summarize the steps needed to run major statistical tests. For all of these tests, a two-sided test is assumed.

10.2.1   One-Sample *Z* Test

1. State the null and alternative hypotheses.

2. State α.

3. Calculate the Z statistic:

$$Z = \frac{Y\text{-bar} - \mu}{\sigma/\sqrt{n}}$$

4. Find the Z tabulated.
5. Compare the Z calculated with the Z tabulated and reach a conclusion.
6. Build a confidence interval on the true value of μ:

$$CI = Y\text{-bar} \pm Z\text{-table } (\sigma/\sqrt{n})$$

10.2.2 Two-Sample Z Test

1. State the null and alternative hypotheses.
2. State α.
3. Calculate the Z statistic:

$$Z = \frac{Y\text{-bar}_1 - Y\text{-bar}_2}{\sigma\sqrt{(1/n_1) + (1/n_2)}}$$

4. Find the Z tabulated.
5. Compare the Z calculated with the Z tabulated and reach a conclusion.
6. Build a confidence interval on the true value for the difference between the means in the population:

$$CI = (Y\text{-bar}_1 - Y\text{-bar}_2) \pm Z\text{-table } [\sigma\sqrt{(1/n_1) + (1/n_2)}]$$

10.2.3 One-Sample t Test

1. State the null and alternate hypotheses.
2. State α.
3. Calculate the t statistic:

$$t = \frac{Y\text{-bar} - \mu}{s/\sqrt{n}}$$

4. Find the t tabulated.
5. Compare the t calculated with the t tabulated and reach a conclusion.
6. Build a confidence interval on the true value of μ:

$$CI = Y\text{-bar} \pm t\text{-table } (s\sqrt{n})$$

10.2.4 Two-Sample t Test (Pooled t Test or Unpaired t Test)

1. State the null and alternative hypotheses.
2. State α.

3. Calculate the t statistic:

$$t = \frac{Y\text{-bar}_1 - Y\text{-bar}_2}{S_p \sqrt{(1/n_1) + (1/n_2)}}$$

$$S_p = \sqrt{\frac{(n_1 - 1)s_1{}^2 + (n_2 - 1)s_2{}^2}{n_1 + n_2 - 2}}$$

4. Find the t tabulated.
5. Compare the t calculated with the t tabulated and reach a conclusion.
6. Build a confidence interval on the true value of the true difference between the means in the population:

$$\text{CI} = (Y\text{-bar}_1 - Y\text{-bar}_2) \pm t\text{-table} \, [S_p \sqrt{(1/n_1) + (1/n_2)}]$$

10.2.5 Paired t Test

1. State the null and alternative hypotheses.
2. State α.
3. Calculate the t statistic:

$$t = \frac{d\text{-bar}}{s/\sqrt{n}}$$

4. Find the t tabulated.
5. Compare the t calculated with the t tabulated and reach a conclusion.
6. Build a confidence interval on the true value of the true mean difference in the population:

$$\text{CI} = d\text{-bar} \pm t\text{-table} \, (s/\sqrt{n})$$

10.2.6 One-Sample Chi-Square Test

1. State the null and the alternative hypotheses.
2. State α.
3. Calculate the χ^2 statistics. For 1 degrees of freedom,

$$\chi^2 = \Sigma \frac{(|\text{observed} - \text{expected}| - 0.5)^2}{\text{expected}}$$

For more than 1 degree of freedom,

$$\chi^2 = \Sigma \frac{(|\text{observed} - \text{expected}|)^2}{\text{expected}}$$

4. Find the χ^2 tabulated.
5. Compare the χ^2 calculated with the χ^2 tabulated and reach a conclusion.

6. Build a confidence interval on the true value for the true proportion in the population:

$$CI = p \pm Z\text{-table } \sqrt{pq/n}$$

10.2.7 Two-Sample Chi-Square Test

1. State the null and alternative hypotheses.
2. State α.
3. Calculate the χ^2 statistics.
   For 1 degree of freedom,

$$\chi^2 = \Sigma \frac{(|\text{observed} - \text{expected}| - 0.5)^2}{\text{expected}}$$

For more than 1 degree of freedom,

$$\chi^2 = \Sigma \frac{(|\text{observed} - \text{expected}|)^2}{\text{expected}}$$

4. Find the χ^2 tabulated.
5. Compare the χ^2 calculated with the χ^2 tabulated and reach a conclusion.
6. Build a confidence interval on the true value of the true difference between the proportions in the population:

$$CI = (p_1 - p_2) \pm Z\text{-table } \sqrt{(p_1 q_1/n_1) + (p_2 q_2/n_2)}$$

10.2.8 Linear Regression

$$y = mx + b$$
$$m = \frac{\Sigma(x_i - x\text{-bar})(y_i - y\text{-bar})}{\Sigma(x_i - x\text{-bar})^2}$$
$$b = y\text{-bar} - m(x\text{-bar})$$
$$r = \frac{\Sigma(x_i - x\text{-bar})(y_i - y\text{-bar})}{v\Sigma(x_i - x\text{-bar})^2 \Sigma(y_i - y\text{-bar})^2}$$

For r:

$$\text{degrees of freedom} = n - 2$$
$$SE = \frac{1}{\sqrt{n - 3}}$$
$$95\% \text{ CI} = Z_r \pm 1.96 \text{ (SE)}$$
$$99\% \text{ CI} = Z_r \pm 2.58 \text{ (SE)}$$

For m:

$$S_{yx} = \frac{\sqrt{SSE}}{n-2}$$

$$SSE = \sum (y_i - y\text{-bar})^2 - m \sum (x_i - x\text{-bar})(y_i - y\text{-bar})$$

$$t = \frac{m}{S_{yx}/v \sum (x_i - x\text{-bar})^2}$$

degrees of freedom $= n - 2$

$$CI_m = m \pm t \frac{S_{yx}}{v \sum (x_i - x\text{-bar})^2}$$

For b:

$$t = \frac{b - b_0}{SE}$$

$$SE = S_{yx} \sqrt{\frac{1}{n} + \frac{(x\text{-bar})^2}{\sum (x_i - x\text{-bar})^2}} \quad \text{degrees of freedom} = n - 2$$

$$CI = b \pm t(SE)$$

10.2.9 One-Way Analysis of Variance

Source	df	SS	MS	F
Between groups	$k - 1$	SS_B	MS_B	MS_B/MS_W
Within groups	$n_{total} - k$	SS_W	MS_W	
Total	$n_{total} - 1$	SS_T		

$$SS_T = \left(\sum X^2\right)_{total} - \frac{\left(\sum X_{total}\right)^2}{n_{total}}$$

$$SS_B = \left[\frac{\left(\sum X_1\right)^2}{n_1} + \frac{\left(\sum X_2\right)^2}{n_2} + \cdots \frac{\left(\sum X_k\right)^2}{n_k}\right] - \frac{\left(\sum X_{total}\right)^2}{n_{total}}$$

$$SS_W = SS_T - SS_B$$

$$MS_B = \frac{SS_B}{k-1}$$

$$MS_W = \frac{SS_W}{n_{total} - k}$$

$$F = \frac{MS_B}{MS_W}$$

degrees of freedom $=$ d.f. numerator $= k - 1$

degrees of freedom $=$ d.f. denominator $= n_{total} - k$

10.2.10   Binomial Probability Estimation

$$p + q = 1$$

$$\Pr(x = k) = \frac{n!}{k!(n-k)!}p^k q^{(n-k)}$$

Recursion rule:

$$\Pr(x = k + 1) = \frac{n-k}{k+1}\frac{p}{q}[\Pr(x = k)]$$

with $k = 0, 1, \ldots, n - 1$

Approximation to normality if $npq \geq 5$ and $np > 5$:

$$\text{mean} = np$$

$$\text{variance} = nq$$

10.2.11   Bayes's Rule

Predicted value positive:

$$\text{PV+} = \frac{(x)(\text{sensitivity})}{x(\text{sensitivity}) + (1-x)(1-\text{specificity})}$$

Predicted value negative:

$$\text{PV-} = \frac{(1-x)(\text{specificity})}{(1-x)(\text{specificity}) + (x)(1-\text{sensitivity})}$$

Likelihood ratio:

$$+ (\text{results}) = \frac{\text{sensitivity}}{1 - \text{specificity}}$$

$$- \text{results} = \frac{1 - \text{sensitivity}}{\text{specificity}}$$

Test Results	Condition	
	Present	Absent
Positive	A	B
Negative	C	D

$$\text{sensitivity} = \frac{\text{true positive}}{\text{true positive} + \text{false negative}}$$

$$= \frac{A}{A + C}$$

$$\text{specificity} = \frac{\text{true negative}}{\text{true negative} + \text{false positive}}$$

$$= \frac{D}{B + D}$$

$$prevalence = \frac{A + C}{A + B + C + D}$$

$$PV+ = \frac{A}{A + B}$$

$$PV- = \frac{D}{C + D}$$

10.2.12 Comparing Two Variances

1. State the null and alternative hypotheses.
2. State α.
3. Calculate the F statistic:

$$F = \frac{S_1^2}{S_2^2}$$

4. Compare the F calculated with the F tabulated and reach a conclusion.
5. Construct a confidence interval on the true variance ratio in the population:

$$lower\ limit = \left[\frac{S_1}{S_2} \frac{1}{\sqrt{F(n_1 - 1), (n_2 - 1)}} \right]^2$$

$$upper\ limit = \left[\frac{S_1}{S_2} \sqrt{F(n_2 - 1), (n_1 - 1)} \right]^2$$

10.3 STATISTICAL APPLICATIONS IN QUALITY CONTROL TESTING

The *validity* of a quality control test can be estimated by precision, accuracy, sensitivity, and specificity. According to general regulations governing current good manufacturing practices, these parameters are essential for assuring that the test is performing at the level expected. *Simple* statistical methods can be used as a quick and easy way to validate the tests used in quality control. Below are descriptions of the statistical methodologies used to obtain the four parameters (Lee et al., 1988).

Accuracy is a measure of the distance between a test value and a "true value." The closer the two values are to each other, the more accurate the instrument. Using samples with known concentrations for a drug, a standard method and the method being tested are both used on the samples to measure the drug concentration. A linear relationship is then developed between the two methods, as described in Chapter 1. Applying statistical testing on the slope of the line and the y-intercept and estimating the confidence interval for both estimates (the slope and the y-intercept) should yield a confidence interval for the slope containing a value of 1 but not containing a zero; for the y-intercept the confidence interval must contain the value zero. If that was the case, the test could be declared to be accurate. Accuracy may be reported as a percent deviation from the true value.

Precision is defined as a measure for the reproducibility of a test. Applying a measure of dispersion such as the relative standard deviation percent (% RSD) serves as a measure of precision. Normally, a minimum of 50 replications are carried out, and the mean and the standard deviation are calculated from the sample. The sample standard deviation is then used to estimate a two-tailed confidence interval on the population standard deviation sigma (S). For a 95% confidence interval on sigma (95% CI) and a two-tailed test, the estimated value for the population parameter is within the range

$$\left[\frac{(n-1)S^2}{\chi_{0.975}^2}\right]^{0.5} < S < \left[\frac{(n-1)S^2}{\chi_{0.025}^2}\right]^{0.5} \tag{10.2}$$

where S^2 is the variance estimate for the sample, n the sample size, and $\chi_{0.975}^2$ and $\chi_{0.025}^2$ are the chi-squared table values at a percentile of 97.5% and 2.5%, respectively, with degrees of freedom of $n-1$. The larger the sample size, the better the value estimated for sigma. Assuming a large enough sample size (>50 replicates), dividing both sides of equation (10.2) by the sample mean and multiplying the ratio by 100 will convert the values to % RSD. The estimated % RSD values are then used as a measure of the precision of the test or the instrument, by comparing them to a predetermined acceptable range for % RSD.

The *sensitivity* of a test is the smallest difference in concentration between two samples of the same drug that an instrument is able to detect. This is accomplished by using a *control* sample containing only the solvent, not the drug being tested. Other samples are then prepared containing an amount of the drug in increasing increments (e.g., 0.01%, 0.02%, 0.03%). The test is applied to all the samples, and the signal obtained from solutions containing the drug are compared to the control solution, using Dunnett's test for comparison. The method is repeated one more time, but this time the control sample is taken as a midpoint concentration for the drug.

Specificity is the ability of a test to distinguish a drug being tested from other substances that may be present in the same solution. This may be done by mixing the drug with the other ingredients present in the formulation, extracting the drug from the formulation mixture with a suitable solvent, and then testing it with the instrument. The test should recognize only the drug, ignoring the other solutes present. In some cases, the signal of the drug may be determined previously at a specified point of the test (e.g., retention time obtained using a high-performance liquid chromatograpic method). If there were no interference with the drug's signal at that point, a conclusion could be drawn based on the presence or absence of the signal. Unfortunately, outside contaminants, not necessarily part of the formulation, may give false-positive results for the test. To affirm the specificity of the test, another test is run in parallel to validate the final results.

10.4   DESIGN OF EXPERIMENT

The *design of experiment* (DOE) procedure is a mathematical method for identifying important factors that can influence an industrial process and pinpointing the values of the *factors* that maximize (or minimize) the *outcomes*. It is intended to optimize a process or a product in the pharmaceutical industry (Bandurek, 2005). It is often useful to identify a method and outcome parameters for a given process. If method parameters are many (i.e., usually more than five), it would be more efficient to run a *screening* design followed by a *modeling* design for the analysis (Hubert et al., 2008). Examples of the types of screening designs are the Taguchi and the two-level fractional factorial. The main objective of these screening designs is to identify and isolate critical method parameters from the important ones (Hubert et al., 2008). Compared to a full factorial design, screening designs have the advantage of producing fewer experimental runs. A modeling design expands the information already learned from the screening design and adds much detail on the method performance (Hubert et al., 2008). The overall purpose of DOE is to identify the best-fit model for a process. For example, a research study investigated the use of cellulose acetate butyrate (CAB) in formulations, with the objective of determining the influence of formulation variables on CAB coating film performance (Price, 1999). A model tablet was used in this experiment containing Polyox (MW 5,000,000), a blue dye, and magnesium stearate. After blending with a V-blender for 3 minutes, the mixture was compressed into 250-mg tablets with a compression force of 400 lb. A film of CAB coating was then applied on the tablets using a pan coater. CAB coating film performance was assessed using the water uptake rate (g/min) by the coated tablets. A 12-run design was created (input variables or factors: CAB, poly(ethylene glycol) (PEG), water, and acetone; outcome: % of water uptake) and the following results were obtained:

$$\text{water uptake rate (g/min)} = (6.504 \times 10^{-5}) - (1.053 \times 10^{-6})$$

$$\text{PEG} - (4.916 \times 10^{-6}) \text{ water} + (2.544 \times 10^{-6})$$

$$\text{PEG} \times \text{water} + (1.020 \times 10^{-5}) \text{ PEG}^2 + (9.627 \times 10^{-7}) \text{ water}^2 \qquad (10.3)$$

Equation (10.3) is a *predictive* expression for estimating the water uptake rate with respect to the amounts of PEG and water in the formulation. The equation takes into account an interaction term (PEG× water) and a polynomial fit of degree 2 with respect to the input variables (PEG^2 and water^2). Moreover, two input factors (CAB and acetone) are not included in the estimation, since their contribution to the mathematical model was insignificant.

In general, DOE can be broken down into six steps: defining the outcome(s), stating the input factors and their levels, identifying the type of run (e.g., full factorial), conducting the experiment and obtaining results, analyzing the results and reaching conclusions, and making changes if necessary to improve the process (Yuan et al., 2008). Orthogonality in DOE is an important concept that is used to locate the sources of variation in an experiment. Orthogonal DOE assures that each independent variable's effect is evaluated *independent* of the other factors (Price, 1999).

10.4.1   Full Factorial Design

Consider the preparation of a gastrointestinal floating device that delivers a drug by gradual degrading in the stomach juices. For the device to remain floating on the surface of the gastric fluid, such a device should manifest a density of less than 1. A gastric floating device is one type of *gastroretentive dosage form*. In this example, the investigators formulated nine preparations with a 3^2 *full factorial design* (HPMC : EC = hydroxypropyl methylcellulose : ethylcellulose 50 : 50, 60 : 40, or 70 : 30 and xanthan gum 0, 10, or 20 mg) (Shah et al., 2009).

The two outcomes measured were t_{50} (the time required for 50% of the drug to be released) and Y60 (the percentage of the drug released in 1 hour). A JMP data file has the following format:

X1	X2	t50	Y60 (%)	X1	X2	t50	Y60 (%)
0	0	232	13.2	1	0	209	19.76
−1	0	274	9.83	0	1	306	11.68
−1	1	315	7.23	1	−1	189	21.34
−1	−1	225	11.53	0	−1	206	16.41
1	1	144	36.9				

t_{50}
Summary of Fit

RSquare	0.86024
RSquare Adj	0.627306
Root Mean Square Error	34.09993
Mean of Response	233.3333
Observations (or Sum Wgts)	9

Y60
Summary of Fit

RSquare	0.884822
RSquare Adj	0.69286
Root Mean Square Error	4.954704
Mean of Response	16.43111
Observations (or Sum Wgts)	9

t_{50}
Analysis of Variance

Source	DF	Sum of Squares	Mean Square	F Ratio
Model	5	21471.583	4294.32	3.6931
Error	3	3488.417	1162.81	**Prob > F**
C. Total	8	24960.000		0.1557

Y60

Analysis of Variance

Source	DF	Sum of Squares	Mean Square	F Ratio
Model	5	565.77641	113.155	4.6093
Error	3	73.64728	24.549	**Prob > F**
C. Total	8	639.42369		0.1192

t$_{50}$

Parameter Estimates

Term	Estimate	Std Error	t Ratio	Prob>|t|
Intercept	253	25.41659	9.95	0.0022
X1	-45.33333	13.92124	-3.26	0.0473
X2	24.166667	13.92124	1.74	0.1810
X1*X2	-33.75	17.04997	-1.98	0.1421
X1*X1	-22	24.1123	-0.91	0.4288
X2*X2	-7.5	24.1123	-0.31	0.7761

Y60

Parameter Estimates

Term	Estimate	Std Error	t Ratio	Prob>|t|
Intercept	11.595556	3.693018	3.14	0.0517
X1	8.235	2.022749	4.07	0.0267
X2	1.0883333	2.022749	0.54	0.6279
X1*X2	4.965	2.477352	2.00	0.1388
X1*X1	4.0016667	3.503505	1.14	0.3363
X2*X2	3.2516667	3.503505	0.93	0.4218

Maximized Desirability
Predicition Profiler

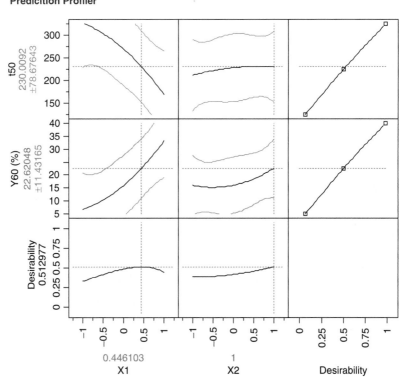

From the results above, maximum values for t_{50} and Y60 are obtained at a HPMC : EC ratio of 0.446 and a xanthan gum amount of 20 mg.

A study was set to optimize fast-dissolving oral tablets made by compression. A randomized full factorial design (3^2) was employed to investigate the effect of two independent variables, the amounts of crospovidone (X1; 4, 8, or 12 mg), and mannitol (X2; 80, 85, or 90 mg), on the disintegration time (DT) and wetting time (WT), measured in seconds. The design also included a center point for each of the input variables. In addition to crospovidone and mannitol, all preparations contained 25 mg of the anti-inflammatory drug rofecoxib, 2% w/w of talc, 1% w/w of magnesium stearate, and 0.5% w/w of sodium lauryl sulfate (Fenn Buderer, 1997). For a full factorial run using JMP, the interactions X1 × X1 and X1 × X2 were found to be insignificant. Thus, they were dropped from the model. The following were the results of the analysis.

For the DT (sec.)

Summary of Fit

RSquare	0.979322
RSquare Adj	0.966916
Root Mean Square Error	0.203134
Mean of Response	4.774444
Observations (or Sum Wgts)	9

Analysis of Variance

Source	DF	Sum of Squares	Mean Square	F Ratio
Model	3	9.7715056	3.25717	78.9361
Error	5	0.2063167	0.04126	**Prob > F**
C. Total	8	9.9778222		0.0001

Parameter Estimates

Term	Estimate	Std Error	t Ratio	Prob>|t|
Intercept	-10.21833	1.424353	-7.17	0.0008
Crospovidone (mg)	-0.124583	0.020732	-6.01	0.0018
Mannitol (mg)	0.1776667	0.016586	10.71	0.0001
(Mannitol (mg)-85)*(Mannitol (mg)-85)	0.0532667	0.005745	9.27	0.0002

For WT (sec.)

Summary of Fit

RSquare	0.965649
RSquare Adj	0.945039
Root Mean Square Error	0.263483
Mean of Response	6.034444
Observations (or Sum Wgts)	9

Analysis of Variance

Source	DF	Sum of Squares	Mean Square	F Ratio
Model	3	9.757906	3.25264	46.8522
Error	5	0.347117	0.06942	**Prob > F**
C. Total	8	10.105022		0.0004

Parameter Estimates

Term	Estimate	Std Error	t Ratio	Prob>|t|
Intercept	-5.113333	1.847516	-2.77	0.0395
Crospovidone (mg)	-0.152917	0.026892	-5.69	0.0023
Mannitol (mg)	0.1333333	0.021513	6.20	0.0016
(Mannitol (mg)-85)*(Mannitol (mg)-85)	0.0622667	0.007452	8.36	0.0004

Response Surface:

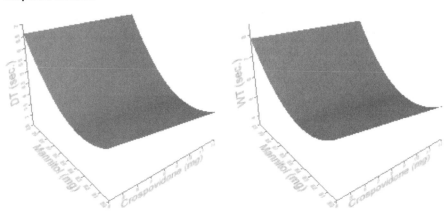

10.4.2 Fractional Factorial Design

To optimize the manufacturing operation for a blow–fill–seal (BFS) technology for sterile solutions, investigators examined factors that affect the presence of particulate matter in the preparation (Price, 1999). This study was developed with the objective of designing a system to eliminate (or greatly minimize) particulate matter contamination in BFS by controlling the critical fill zone where filling was taken place. This system of control was referred to as a *particulate control system*. Six independent variables were included in the design: HEPA (high-efficiency particulate air) height and flow rate (%), damper position (% open), chimney air velocity (ft/min), isolation plate (slotted or holes), and knife cut (double or single). Researchers used a one-fourth fractional factorial design, which reduced the number of runs needed. The experiments were repeated in triplicates. The input factors and their boundaries were chosen so that three factors—HEPA flow rate (20%, 50%, or 80%), damper position (% open) (30%, 55%, or 80%), and chimney air velocity (300, 550, or 800 ft/min)—were at three levels, and three other factors—HEPA height (0 or 0.375 in.), isolation plate [slotted (S) or hole (H)], and knife cut [Double (D) or Single (S)]—were each at two levels (Price, 1999). As an outcome for the analysis, the experimenters chose the natural logarithm for the particulate count. Analysis of the fore going data using JMP statistical analysis software (SAS Institute, Cary, NC) produced the following results:

Summary of Fit

RSquare	0.953762
RSquare Adj	0.933533
Root Mean Square Error	0.720784
Mean of Response	4.792984
Observations (or Sum Wgts)	70

Analysis of Variance

Source	DF	Sum of Squares	Mean Square	F Ratio
Model	21	514.39007	24.4948	47.1480
Error	48	24.93742	0.5195	**Prob > F**
C. Total	69	539.32749		<.0001

Lack of Fit

Source	DF	Sum of Squares	Mean Square	F Ratio
Lack Of Fit	2	2.263692	1.13185	2.2963
Pure Error	46	22.673727	0.49291	**Prob > F**
Total Error	48	24.937419		0.1121
				Max RSq
				0.9580

Term	Prob>|t|
Intercept	<.0001
HEPA Flow Rate (%)[0]	0.5383
HEPA Flow Rate (%)[1]	0.5487
Damper Position (% open)[0]	0.0660
Damper Position (% open)[1]	.
Chimney Air Velocity (ft/min)[0]	0.8849
Chimney Air Velocity (ft/min)[1]	.
HEPA Height (in.)[0]	<.0001
Isolation Plate[H]	<.0001
Knife Cut[D]	0.0117
HEPA Flow Rate (%)[0]*Damper Position (% open)[0]	0.7462
HEPA Flow Rate (%)[0]*Damper Position (% open)[1]	.
HEPA Flow Rate (%)[1]*Damper Position (% open)[0]	.
HEPA Flow Rate (%)[1]*Damper Position (% open)[1]	.
HEPA Flow Rate (%)[0]*Chimney Air Velocity (ft/min)[0]	0.3365
HEPA Flow Rate (%)[0]*Chimney Air Velocity (ft/min)[1]	.
HEPA Flow Rate (%)[1]*Chimney Air Velocity (ft/min)[0]	.
HEPA Flow Rate (%)[1]*Chimney Air Velocity (ft/min)[1]	.
HEPA Flow Rate (%)[0]*HEPA Height (in.)[0]	0.0105
HEPA Flow Rate (%)[1]*HEPA Height (in.)[0]	0.0062
HEPA Flow Rate (%)[0]*Isolation Plate[H]	0.7353
HEPA Flow Rate (%)[1]*Isolation Plate[H]	0.9105
HEPA Flow Rate (%)[0]*Knife Cut[D]	0.0559
HEPA Flow Rate (%)[1]*Knife Cut[D]	0.0212
Damper Position (% open)[0]*Chimney Air Velocity (ft/min)[0]	0.5235
Damper Position (% open)[0]*Chimney Air Velocity (ft/min)[1]	.
Damper Position (% open)[1]*Chimney Air Velocity (ft/min)[0]	.
Damper Position (% open)[1]*Chimney Air Velocity (ft/min)[1]	.
Damper Position (% open)[0]*HEPA Height (in.)[0]	0.0259
Damper Position (% open)[1]*HEPA Height (in.)[0]	.
Damper Position (% open)[0]*Isolation Plate[H]	0.5896
Damper Position (% open)[1]*Isolation Plate[H]	.
Damper Position (% open)[0]*Knife Cut[D]	.
Damper Position (% open)[1]*Knife Cut[D]	.
Chimney Air Velocity (ft/min)[0]*HEPA Height (in.)[0]	.
Chimney Air Velocity (ft/min)[1]*HEPA Height (in.)[0]	.
Chimney Air Velocity (ft/min)[0]*Isolation Plate[H]	.
Chimney Air Velocity (ft/min)[1]*Isolation Plate[H]	.
Chimney Air Velocity (ft/min)[0]*Knife Cut[D]	.
Chimney Air Velocity (ft/min)[1]*Knife Cut[D]	.
HEPA Height (in.)[0]*Isolation Plate[H]	0.9628
HEPA Height (in.)[0]*Knife Cut[D]	0.7154
Isolation Plate[H]*Knife Cut[D]	0.9106

The model had a high value-adjusted r^2 (0.933533), which indicates that there was a good fit for the data. In addition, the lack-of-fit analysis indicates that the model was a good choice for the data at hand ($p = 0.9580$). The values for optimum control of the particulate count occurred with HEPA flow rate = 20%; damper position = 80% open; HEPA height = 0.375 in.; isolation plate = slotted; chimney air velocity = 300 ft/min; and knife cut = single.

The DOE approach also has applications in validation experiments. A *fractional factorial design* may be used in validation experiments if the number of input factors is five or fewer. Often, the number of factors exceeds five and in this case Taguchi L12 arrays are used to limit the number of experimental runs. In Taguchi L12 arrays, each factor is given an upper and a lower limit, and the pattern in the design is balanced so that for each factor an equal number of upper and lower limits is present. The design contains 12 runs and can test for interactions between the factors in pairs or in triplicates. A single component factor effect can also be measured, although it may be confounded by the other input factors present in the design (Bandurek, 2005).

REFERENCES

Bandurek GR. Using design of experiments in validation. BioPharm Int. 2005;18(5):40–52.

Cuddy PG, Elenbaas RM, Elenbaas IK. Evaluating the medical literature: I. Abstract, introduction, methods. Ann. Emerg. Med. 1983;12:549–555.

Elenbaas RM, Elenbaas IK, Cuddy PG. Evaluating the medical literature: II. Statistical analysis. Ann. Emerg. Med. 1983a;12:610–620.

Elenbaas IK, Cuddy PG, Elenbaas RM. Evaluating the medical literature, part III: results and discussion. Ann. Emerg. Med. 1983b;12:679–686.

Fenn Buderer NM. Basic statistical concepts for evaluating pharmaceutical compounding literature: introduction to the series. Int. J. Pharm. Compd. 1997;1(6):395–397.

_____. Basic statistical concepts for evaluating pharmaceutical compounding literature: 2. Confidence intervals and p values. Int. J. Pharm. Compd. 1998a;2(1):42–43.

_____. Basic statistical concepts for evaluating pharmaceutical compounding literature: 4. Student's t-test and one-way analysis of variance. Int. J. Pharm. Compd. 1998b;2(3):216–218.

Hubert M, Sarsfield B, Shekunov B, Grosso J. Oral solid dosage form: from choice of particle size technique to method development and validation. Am. Pharm. Rev. 2008;11(6):14, 16, 18, 20–21, 23.

Lee LM, Taylor M, Kantrowitz J. Statistical evaluation of quality control tests. Pharm. Technol. 1988;12(9):108, 110, 112, 114, 116, 118, 120.

Price J. Blow–fill-seal technology: II. Design optimization of a particulate control system. Pharm. Technol. 1999;23(2):42, 44, 46, 48, 50, 52.

Shah N, Patel R, Amin A, Kanagale P, Venkateshan N. Gastro-retentive dosage forms. Drug Deliv. Technol. 2009;9(1):44–49.

Yuan J, Dunn D, Clipse NM, Newton RJ. Permeability study on cellulose acetate butyrate coating film. Drug Deliv. Technol. 2008;8(2):46–51.

GLOSSARY

Accuracy A measure of the distance between a test value and a "true value." The closer the two values are to each other, the more accurate the instrument.

BFS	Blow–fill–seal.
CI	Confidence interval.
CAB	Cellulose acetate butyrate.
CV	Coefficient of variation.
DOE	Design of experiment.
DT	Disintegration time.
HEPA	High-efficiency particulate air.
Null hypothesis (H_0)	A statement of no difference.
PEG	Poly(ethylene glycol).
Precision	A measure of the reproducibility of a test.
p value	The smallest value which the null hypothesis is rejected.
RSD	Relative standard deviation.
SD	Standard deviation.
Sensitivity	The smallest difference in concentration between two samples of the same drug that an instrument is able to detect.
Specificity	The ability of a test to distinguish the drug being tested from other substances that may be present in the same solution.
Standard error of the mean (SE)	The ratio of the standard deviation to the square root of the sample size.
WT	Wetting time.

CHAPTER 11

FORMULATION DEVELOPMENT CONCEPTS

11.1 PREFORMULATION

Preformulation is a term often used to describe the physical and chemical characteristics of substances and mixtures used in pharmaceutical products. It is used synonymously with *physical pharmacy*. Preformulation is part of a larger science known as *pharmaceutics*. Both preformulation and formulation are parts of pharmaceutics, and their application in formulation development in industry is known as *industrial pharmacy*. Characterization of chemicals can be done on the molecular, particulate, and bulk levels (Brittain, 1999). Nowadays, scientists depend heavily on using instrumentation to characterize or identify a chemical or a final product. Instruments used in research have become extremely complicated, and similar to a piano and a musician's ability to play it, the same instrument can produce bad, good, or excellent data, depending on how it is being run (Scarlett, 2003).

It has been stated that only 1 out of 11 drugs entering phase 1 clinical investigations reaches the market, and only one-third of the drugs that make it to the market cover their development cost (Dubin, 2006). It is also predicted that the market for drug delivery systems will grow at an annual rate of 10%, reaching $132 billion in 2012 (Dubin, 2009). Perhaps the main driving force behind this predicted growth in drug delivery technologies is the extension of a product's life while minimizing cost (Hamilton and Luty, 2005). Development of a new delivery system for a given drug begins in research and development (R&D) departments in the pharmaceutical industry, utilizing the principles of preformulation. However, there is no guarantee whatsoever that a specific formulation that was generated in R&D will be applicable in the production area. In addition, the pharmaceutical industry utilizes what is known as *process analytical technologies* (PATs). The main purpose of PAT is for control and assessment *during* manufacturing. It is important for the R&D department and the production operation facility to coordinate their use of PAT and to allow communication concerning PAT to flow smoothly back and forth. (Arrivo, 2003). The obvious advantage of PAT is that it provides online, in-place analysis for an active pharmaceutical ingredient (API) during the processing steps, eliminating the need for repeated sampling during the manufacturing operation.

Integrated Pharmaceutics: Applied Preformulation, Product Design, and Regulatory Science, First Edition. Antoine Al-Achi, Mali Ram Gupta, William Craig Stagner.
© 2013 John Wiley & Sons, Inc. Published 2013 by John Wiley & Sons, Inc.

Thus, it reduces the exposure of personnel to APIs, a matter of importance when handling highly potent drugs. PAT also reduces both the overall cost and the time of operation (Arrivo, 2003). In September 2004, the U.S. Food and Drug Administration (FDA) issued a Guidance for Industry document concerning PAT. The use of PAT allows the industry to produce a pharmaceutical product of high quality in a *consistent and predictable* manner (Dziki, 2008).

The purpose served by R&D is often questioned, since a considerable amount of information learned there may not be applicable to production. In this chapter we shed some light on issues related to transferring a formulation from R&D to production (i.e., the scale-up process), rate-controlled drug delivery systems, novel drug delivery technologies to enhance oral bioavailability, possible drug–excipient or excipient–excipient interactions, contamination issues, and other issues. To make the point in another way, according to Arthur Calder-Marshall, "out of sight, out of mind" translated into Russian by a computer, then back into English, became "invisible maniac" (Tingstad, 1988). Thus, clear and open communications between a R&D department and production personnel cannot be overemphasized for the ultimate success of the final product. From the point of view of the FDA when evaluating product quality, the FDA believes that "quality cannot be tested into the product, but rather it should be built into the product" during development (Dziki, 2008), through communication among all parties involved.

11.2 SCALE-UP CONSIDERATIONS

Issues related to scaling-up dosage forms from the R&D stage to the pilot stage and thereafter to production are numerous and often lead to failure in producing the planned dosage form unless addressed and resolved. Most problems encountered during scale-up are related to solid dosage forms such as tablets and suppositories (Russo, 1984; Chowhan, 1988) as well as lyophilization techniques (Tchessalov et al., 2007). Variability in the chemicals with respect to their physical characteristics and equipment variability are perhaps the main areas of concern during scale-up (Russo, 1984; Pondell, 1985). For example, during tablet formulation, variations exist during milling and mixing, and in the types of presses used (Russo, 1984). Moreover, a fluid-bed granulation process is often used in industry in tablet manufacturing. The basic fluid-bed granulation unit consists of an air-handling unit, an exhaust air turbine, an exhaust air filter, a processing zone (i.e., dryers, granulators, spray nozzles), and product discharge components (Olsen, 1989). This technique allows the production of granules with ideal features of spherical shape, high porosity, and narrow size range. These features provide excellent flow characteristics and compressibility for the resulting granulation. Production batches are naturally higher in weight and require more time for processing. Fluid beds employed in the production stage produce higher shearing forces to form the desired agglomeration. They utilize warmer fluidizing air than that used with R&D units, and the rate for delivering the binder dispersion is manyfold higher than that of a R&D application. These two factors present in the production unit produce granulation containing a lower moisture content just prior to the drying stage. Production tablets often have higher disintegration times, albeit not significant enough to affect the dissolution profile of tablets. To achieve a smooth transition from R&D to production utilizing fluid-bed technology, factors such as binder delivery rate and moisture content must be considered (Gore et al., 1985). Overall, and for the fluid-bed granulation scaling-up procedure, close attention should be directed to factors such as batch size, type of equipment used, delivery rate of binders, pressure of the atomization air, volume of the fluidization air, and temperature of the inlet air (Gore et al., 1985; Mehta, 1988). Granulations made by high-shear mixtures present another type of challenge for scale-up. There appear to be four areas of concern regarding this scaling-up operation: the first three being variations in the API and the excipients, in-process controls, and failure to recognize the fact that scaling-up formulations from R&D to production is not a linear process. However, the most significant factor affecting the process of transition to large-scale production was found to be the variation encountered with the API (Chowhan, 1988).

Film coating is another area of concern during scale-up. For film coating, two major types of equipment are available: fluid-bed type and pan coaters. Each of these coaters exists in many types, and those used in the R&D stage are obviously smaller than those employed in production. Smaller instruments handle materials in a much gentler way. Large production large equipment could damage a product at any step before and after coating (Pondell, 1985). Therefore, care must be exercised to protect a product during handling in the production areas, despite evidence of "stability" throughout the R&D process. Other factors that may contribute to the quality of the finished coating film include the solid content of the coating dispersions, the number of spray guns used, the pan speed (in the case of pan coating devices), and the coloring agent if used in the coating dispersion. All these factors play a significant role during processing of the final product formulation (Porter and Saraceni, 1988).

Lyophilization is a commonly employed process in industry for the preparation of sterile powdered dosage units. This technique requires several steps that may create challenges during the scale-up process. From the point of view of the scaling-up technique, temperatures during the freezing, primary, and secondary drying steps must be similar for the laboratory and manufacturing processes. This guarantees not only the same product characteristics but also the same stability profile (Tchessalov et al., 2007).

11.3 COMBINATION PRODUCTS

Compliance with taking medications is a huge issue in clinical practice. It is estimated that the cost of noncompliance can be as high as $100 billion annually, accounting for additional health care expenditures and lack of productivity (Dubin, 2005a). Combination products offer one solution to improving patient compliance. The notion of combination products is not a new concept in the pharmaceutical industry. A quick perusal of reference clinical texts reveals their presence since the industry began. However, the old types of combination products were limited to certain drugs that did not show compatibility issues, and their quantities were small enough to accommodate them within a single dosage unit. The new combination products are much more versatile in their ability to include products in larger quantities and regardless of compatibility issues. Multicompartment capsules (e.g., InnerCap Technologies, Inc; Tampa, FL) are made to house several drugs for treating the same or different conditions, and regardless of their compatibilities. The new capsules are made of a suitable material, such as hydroxypropyl methylcellulose, gelatin, starch, or a combination. Coating materials can be applied on the outer walls for special targeted delivery. Another example of multicompartment capsules is BilDil (NitroMed Inc., Lexington, MA), containing the combination vasodilator drugs isosorbide dinitrate and hydralazine hydrochloride, approved by the FDA for the treatment of heart failure in African-American patients (Dubin, 2005b).

In general, combination products can deliver simultaneously, within the same unit, a drug, a biological, and/or a medical device. The majority of these products (90%) are currently being developed for cardiovascular applications and focus on the development of stent drug delivery (Prajapati et al., 2008). Other applications involve areas related to surgical dressing and orthopedics. In addition, an emerging area of combination products technology includes nanotechnology (Ford, 2008). This area has evoleved into a new specialty known as *nanomedicine*. Applications in this area include improved antibacterial agent activity, acting as carriers, improvement in implantable materials, diagnostic applications, and as sensory units in various devices (Prajapati et al., 2008). For example, new technologies are being developed to overcome issues related to post-stent placement (e.g., re-blocking of arteries, thrombosis, blood clots). Among these technologies, Micell Technologies' polymer nanoparticle coating technique allows multiple drug delivery from a single stent as well as controlled delivery of drugs in a sequential manner (Andrews and McClain, 2008).

11.4 RATE-CONTROLLED DRUG DELIVERY

Systems for controlling drug release from a dosage form have been developed to maximize therapeutic benefits while minimizing adverse outcomes of drug therapy. Diffusion systems and activation approaches are the two main classes of rate-controlled drug delivery. Diffusion systems are of several types and include membrane permeation, matrix diffusion, and microreservoir dissolution. The activation technologies utilize osmotic, hydrodynamic, or vapor pressure. Ion or pH activation, ultrasound, and magnetism are all activation delivery techniques (Chien, 1985).

Membrane permeation allows a polymeric barrier to control delivery of a drug housed inside a rigid compartment. Usually, the drug is in the form of a dispersion, and its delivery through the membrane follows a zero-order rate. Matrix diffusion systems depend on dispersing the drug in a biodegradable matrix form. The rate of drug diffusion from the polymeric matrix is a function of the square root of time. The microreservoir dissolution method employs a suspension of the drug particles in an aqueous solution dispersed throughout a hydrophilic polymeric matrix. The suspension may be housed within an additional polymeric membrane to further control the rate of drug release. The rate of drug release from the device is a function of its solubility in the aqueous solution and in the polymeric matrix (Chien, 1985).

Some gastroretentive systems, for example, work via membrane permeation mechanisms. Floating drug delivery systems that contain a drug dispersed in a hydrophilic matrix (e.g., guar gum; hydroxypropyl methylcellulose) exhibit a density below than 1 when placed in an aqueous environment (such as gastric fluid) and thus float on the surface of the aqueous liquid (Jain et al., 2005). The hydrophilic matrix can form a *superporous hydrogel*. Upon absorption of water, the hydrogel material swells with the formation of space within the matrix, known as the *effective pore size*. When the effective pore size is larger than 10 μm, the hydrogel is known as a superporous hydrogel (Amin, 2008a). Some examples of this technology available on the market are Liquid Gaviscon (an antacid; GlaxoSmithKline), which is a floating alginate raft; Baclofen GRS (a bronchodilator; Sun Pharma), which consists of a coated multilayer floating and swelling device; and Glumetza (for blood glucose control; Depomed), which is a polymer-based system (Amin, 2008a). These units release the drug one layer at a time, starting with the most superficial layer first. The outer layer is in contact with aqueous fluid swells and prevents further entry of the fluid inside the matrix. When the outer layer swells, it entraps some air and helps to keep the unit floating on the surface of the fluid (Dubin, 2007).

Other systems contain matrices prepared from swellable polymers and effervescent materials that upon contact with the gastric fluid, generate CO_2 gas; the gas gets entrapped in the polymer structure, helping the unit to float (Dubin, 2007). Once the outer layer is completely degraded and the drug is released from it, a new outer layer is formed, and the process continues until the entire unit is gone. The hydroxypropyl methylcellulose (HPMC), with appropriate hydration rate and viscosity, used in these floatable units is extremely popular in the preparation of controlled-release systems. Its popularity stems from several characteristics, including a high capacity for drug loading, easy compression into tablets, and being nontoxic, among others (Gothoskar, 2005). Floating foam-based systems are also used to provide gastroretentive drug delivery. The matrix of these systems consists of porous (low density) material resembling foam (e.g., Accurel MP; polypropylene foam powder). Drugs with or without added polymers are adsorbed on the foam particles. The entrapped air within the matrix allows the unit to float upon exposure to an aqueous environment. The drug release profile from the system is independent of the foam materials; however, it depends on the type of polymer used with the drug (Amin, 2008a).

Other floating matrices utilize lipid materials as a base. For example, lipid matrices known as Gélucires (Gattefosse, St-Priest, France), composed of mixtures of mono-, di-, and triglycerides with poly(ethylene glycol) esters of fatty acids, have been shown to provide a sustaining effect. Gélucires have *hydrophile–lipophile balance* (HLB) values in the range 1 to 18, with a wide melting-point range between 33 and 65°C. Low-HLB-value Gélucires produce units with floating characteristics gastrointestinal fluids. The absence or low content of PEG esters mixed with Gélucires provides a sustaining effect. [Larger amounts of PEG esters mixed with the lipid matrix accelerate the rate of drug release (Amin, 2008; Patel and

Patel, 2008).] Glyceryl monostearate retards the release of drugs from formulations. When used in combination with Gélucires, the matrix failed to float. Mixtures of ethylcellulose, a release retardant agent, with Gélucires provided both retardation of drug release and floating characteristics. Gélucires combined with HPMC showed an accelerated drug release profile compared to Gélucires alone (Amin, 2008a).

Another oral delivery membrane permeation technology is the Eudramode platform (Evonik Röhm GmbH, Darmstadt, Germany), which is made of layered pellets. The core of the pellets is composed of modulator ions (a salt of weak or strong acids, such as sodium citrate) layered with a modulating film (Eudragit NE 30 D), followed by a layer of the drug, and a covering controlled-release layer of quaternary ammonium ions (Eudragit RL/RS 30 D; ammoniomethacrylate copolymers). The pellets may be placed in a capsule form or compressed into a tablet. The Eudramode platform was shown to deliver the drug in a zero-order release fashion. Three factors play an important role in controlling the drug release from the pellets; these are the salt used in the core and the thickness of both the modulating film and the controlled-release covering outer layer (Ravishankar et al., 2005). Eudragit polymers (Evonik Röhm GmbH, Darmstadt, Germany) are copolymers derived from esters of acrylic and methacrylic acid. Their properties are based on functional groups located on the polymeric chain. Eudragit polymers are available in a wide range of applications, including aqueous dispersions (L50 D-55; NE 30 D; NE 40 D; NM 30 D; RS 30 D; RL 30 D; FS 30 D), powders and granules (L100; L100-55; S 100; RS 100; RL 100; RS PO; RL PO; E-100; E PO), and organic solutions (E12,5; L12,5; S12,5; RS12,5; RL12,5) (Evonik Röhm GmbH). Eudragit polymers can be classified into two main categories: pH-dependent and pH-independent. As pH-dependent films, the polymers become soluble in solution if the pH exceeds a certain value, whereas pH-independent polymer films act as a barrier, allowing diffusion of a drug through the film. When used as a matrix, the polymer acts as a sponge, releasing the drug in a controlled fashion. The drug diffuses out of the pH-independent matrix, whereas it is released by erosion and diffusion when a pH-dependent matrix is used (Álvarez et al., 2007).

Activation technologies that control drug delivery by a pressure mechanism depend on interaction between the device and its environment following administration. Usually, the drug is housed within a physical compartment, and its rate of release through an orifice is controlled by either driving water inside the device through osmosis, absorbing water by a polymeric matrix that upon swelling exerts pressure on the drug reservoir, or by the use of a chemical that vaporizes at body temperature, with the gas thus formed exerting pressure to release drug from its compartment. These systems allow release of the drug from the device by a zero-order process (Chien, 1985; Gupta et al., 2005). An example of these activation technologies is the Oros system (Alza Pharmaceuticals, Mountain View, CA; a subsidiary of Johnson & Johnson). Oros drug delivery is based on osmosis and has wide clinical applications, including drugs used in the treatment of attention-deficit hyperactivity disorder, benign prostatic hyperplasia, hypertension, diabetes (type 2), overactive bladder, pain, and schizophrenia (Conley et al., 2006). Some drug products using Oros include albuterol, chlorpheniramine maleate, doxazosin mesylate, glipizide, ivermectin, nifedipine, and verapamil (Gupta et al., 2005).

Systems that depend on pH activation are covered by a mixture of polymers that deliver the device intact to the intestines. Under the pH conditions available in the intestine, some polymers are soluble and others are not. This causes the polymeric membrane surrounding the device to become porous, allowing diffusion of the drug through the pores. Ion activation systems rely on interaction of the drug with a resin, which upon delivery to the gastrointestinal tract undergoes reactions with available ions to release the drug slowly from its resin–drug complex and to provide controlled delivery (Chien, 1985). Insulin encapsulated in nanoaggregate particles (100 to 230 nm) and later coated with weak polyelectrolytes [(Poly(α, β-1-malic acid) and chitosan] exhibited a pH-dependent release of insulin from the nanoparticles due to the sensitivity of the polyelectrolytes to pH. At physiological blood pH (7.4), insulin release from the matrix was observed, whereas at pH 4 to 5 no insulin release was detected. Furthermore, the addition of a higher number of polyelectrolyte layers to the surface of the nanoaggregates resulted in a reduction of the rate of release of insulin from the nanoparticles (Fan et al., 2006). In addition to the pH, drug lipophilicity as measured by the partition coefficient ($\log P$) was shown to play an important role in drug release (Sawant et al., 2010). For lidocaine ($\log P = 2.6$; lipophilic) and lidocaine hydrochloride ($\log P \leq 0$; hydrophilic) it was observed that when lidocaine was incorporated into a hydrophilic matrix (hydroxypropyl cellulose), the release, rate of the drug followed a slow non-Fickian release rate, whereas the hydrophilic salt of the drug experienced a burst effect for its release from the matrix. When lidocaine was incorporated into a hydrophobic carrier (Eudragit E100), a burst effect was noted. It appears that lidocaine interacted with hydroxypropyl cellulose, which resulted in a slow release of the drug from the matrix. No interaction between Eudragit E100 and lidocaine was noted, which resulted in a burst effect (Sawant et al., 2010).

Activation systems that depend on magnetism employ pellets carrying a tiny magnet within them. The drug–pellet system is coated with impermeable polymeric membrane to the drug, except at the center of the pellet, where no coat is applied. The drug is released from the uncoated center area, and its rate of release is accelerated using an external magnetic field (Chien, 1985). Applications of magnet-containing units may be helpful in situations where an increase in the drug release rate is desirable at a certain point in time. For example, insulin delivery through such a device would be desirable, where the patient can activate delivery of insulin when needed after meals (Langer, 1985). Another possible external activation energy source is ultrasonic waves. A biodegradable polymeric matrix containing the drug may be used, and the rate of polymer degradation is enhanced by an ultrasonic device, thus facilitating release of the drug from its matrix (Chien, 1985).

11.5 DRUG DELIVERY TECHNOLOGIES FOR IMPROVING ORAL DELIVERY

The oral route of administration has been by far the most convenient method of drug administration. Over 50% of the drugs available on the market exist in this form (Jain et al., 2005). The ease of delivery combined with a relatively rapid onset of action, along with lower cost per dose unit, makes this route an ideal way for augmenting the therapeutic effects of drugs. One of the recent innovations in enhancing the delivery of drugs via this route is the use of *orally disintegrating tablets* (ODTs). While oral delivery of drugs by conventional dosage forms (tablets, capsules, etc.) requires the intake of water to facilitate swallowing the solid dosage form, the rapid disintegration and dissolution of ODTs in the mouth cavity eliminate the need of water during administration. One ODT technology uses a specialized lyophilization method (i.e., Zydis, Catalent, New Jersey) in which the entire formulation process is prepared within a blister compartment. The Zydis fast-dissolving technology (within 3 seconds when placed in an aqueous environment) has been shown to improve the bioavailability of drugs as well as the onset of action, comparable to that of an intravenous injection (Catalent Zydis brochure, 2007). ODTs prepared by a direct compression method produce tablets that disintegrate and dissolve rapidly in the mouth cavity; however, the final stage of dissolution and absorption occurs in the gastrointestinal (GI) tract after the patient swallows the resulting "suspension" made of the disintegrated particles (OraSolv, CIMA Labs, Inc., MN). ODT technology can also offer a sustained-release delivery mechanism such as the tablets developed by Eurand N.V. (AdvaTab, Amsterdam, The Netherlands). AdvaTab tablets dissolve quickly within the mouth cavity (15 to 30 seconds) and have the capacity to carry large doses of drug (Eurand N.V.). In general, an ideal ODT unit should dissolve rapidly in a matter of seconds within the mouth cavity and should have a taste-masking property and a smooth mouth feel (Hamilton and Lutz, 2005). In addition to tablets, *oral fast-dissolving films* or *strips* (OFDSs) are also available to enhance the delivery of poorly water-soluble and permeable APIs. The film is composed of cellulose-based polymers or poly(ethylene oxide) containing strong flavoring agents to mask the taste of added ingredients. Examples of these oral strips are Triaminic, Chloraseptic, and Listerine (Ali and Quadir, 2007). The normal size of an OFDS is about the dimension of a postage stamp. Upon contact with the saliva, the film disintegrates instantaneously. Original formulations of OFDSs were introduced on the marker as breath-freshening films. These original breath-fresheners were followed by introducing products for delivering vitamins, minerals, and dietary supplements. The advantages of OFDSs are fast disintegration and dissolution, flexibility of the film, and nonfriablility (Hariharan and Bogue, 2009).

The permeation of drugs through the GI tract after an oral administration is crucial for delivering a sufficient amount of the drug to the circulation in a timely manner. This becomes even more important if the drug has a poor permeability profile through the gut mucosa. A system developed by Merrion Pharmaceuticals, known as GIPET (gastrointestinal permeation enhancement technology), facilitates and enhances the absorption of poorly permeable drugs up to 200-fold. GIPET works via a proprietary micelle–drug membrane transport enhancement method. This technology is applicable to a wide range of drugs, including biotechnology products. Because of its good safety profile, the FDA has included GIPET in its GRAS (generally regarded as safe) listing. Nanosplode, Lip'ral, Lip'ral Synchronized Solubilizer Release, and Hydroance (Lipocine, Inc., Salt Lake City, UT) are other technologies that facilitate the absorption of drugs that suffer from poor absorption and/or a poor solubility profile from the GI tract. [It is estimated that about 40% of all drugs have poor solubility in water (Dubin, 2005a; Chattopadhyay and Shekunov, 2006).] Following rapid disintegration from solid dosage form (e.g., capsules), the drug molecules are engulfed with specialized proprietary lipids that facilitate drug absorption by the GI mucosa (Lipocine, Inc.).

Another important area for delivering drugs is that to the brain and the central nervous system (CNS), which is often limited due to the presence of the blood–brain barrier (BBB). The BBB is due to the presence of tight junctions in the blood vessels of the CNS that do not permit drugs to pass through them, natural protective measure for the brain. LipoBridge (Genzyme Pharmaceuticals, Massachusetts) is a short-chain oligoglycerolipid compound that reversibly opens the tight junctions, thus facilitating the entry of drug molecules into the CNS and brain tissues. This allows an effective concentration of drugs in the CNS and brain to increase by a factor of up to 100 above normal. The system appears to be safe and is excreted unchanged without being metabolized (Genzyme Pharmaceuticals).

11.6   DRUG DELIVERY TECHNOLOGIES FOR IMPROVING TRANSMUCOSAL DELIVERY

Transmucosal delivery is seen as somewhat of an extension to other routes. Via this system, drugs are transported through a transmucosal pathway by placing them in the mouth, nose, or another route. The market for transmucosal delivery is estimated to be in billions of U.S. dollars annually. The transport of drugs through the mucosa of the mouth cavity is reported to be highly efficient, exceeding that of the skin manyfold (4 to 4000 times higher.) An example of such a delivery system is the OraVescent technology (CIMA Labs, Brooklyn Park, MN), which offers mucosal transport of drugs from the mouth cavity (sublingually or between the buccal and gingival tissues). OraVescent is available in compressed tablet form and is dispensed in a PakSolv proprietary blister package (Brown and Hamed, 2007).

Sodium alginate is another type of material found useful through its bioadhesive properties, thus allowing delivery of APIs through the GI tract and nasal mucosa. Sodium alginate is a biodegradable natural polysaccharide that forms viscous solutions upon dissolution in water (20 to 400 cP). Its viscosity is pH-dependent, and it decreases at pH values above 10. It can be used as a matrix in floating beads and tablets of gastroretentive preparations. Research on sodium alginate has also revealed its usefulness to provide or enhance the controlled-release delivery of drugs as well as delivering proteins, DNA, and various other carriers (e.g., liposomes and microspheres) (Patel et al., 1989). With respect to the mucoadhesive properties, various agents may be used to provide such action. Among these agents are chitosan (cationic), carbopol 934 (anionic), and hydroxypropyl methylcellulose (HPMC, nonionic). Majithiya et al. (2011) studied the effect of these agents on the mucoadhesive property of tablets using a rat stomach membrane model. Of all the single agents tested, HPMC had the best mucoadhesive strength. When other agents were mixed with HPMC a synergistic mucoadhesive effect was seen. A combination of chitosan: carbopol $(1:4)$ gave the highest strength for mucoadhesion. Overall, it appears that combination polymeric mixtures provide a stronger mucoadhesive force than that of a single agent alone.

11.7   DRUG DELIVERY TECHNOLOGIES FOR TRANSDERMAL DELIVERY

The skin is the largest organ of the body. It presents a formidable barrier for foreign substances to enter the body and for endogenous ones to escape to the outside. The presence of keratin in the *stratum corneum* prevents water from entering or leaving the skin (except at sweat pores). For normal skin, the pH is slightly acidic (pH 5 to 6), due to the presence of organic acids on the surface of the skin, such as amino acids and fatty acids.

Transdermal patches are often used to deliver drugs through the skin. The solubility of drugs in the adhesive material of the patch can be difficult to estimate. Myatt et al. (2008) presented a practical method for estimating the solubility of solutes in the adhesive matrix. In this method, the amount of drug diffused over time is obtained by diffusion experiments, and the data are then plotted to fit *Higuchi's model*:

$$M = k_0 t^{0.5} \tag{11.1}$$

where M is the cumulative amount of drug released per unit area, t the time, and k_0 is Higuchi's constant. The value of k_0 is calculated at different starting concentrations of the solute added to the adhesive matrix. When k_0 values are plotted against the initial concentration of the drug in the matrix, the graph forms two distinct linear portions that intersect at a point. The point of intersection corresponds to the solubility of the drug in the adhesive matrix read directly from the x-axis (Myatt et al., (2008).

11.8   SPECIAL CONSIDERATIONS FOR BIOTECHNOLOGY AND PROTEIN DELIVERY SYSTEMS

The future of pharmaceuticals will include a great number of biotechnology products that are composed of proteins. Sterility, purity, and stability (physical and chemical) are the obvious challenges in this area of delivery. Each area should be addressed independently since a stable product *does not* guarantee sterility or purity, and vice versa. Considerations for storage conditions (temperature, cleanliness) and handling (shaking the container vs. swirling) are also areas of concern for these delivery systems. It is known in pharmacy practice that vigorous shaking of a protein product produces breakage in the chain of the protein, rendering it denatured (Chan, 2008).

Another area of concern is that proteins in formulation often aggregate in the form of dimers to multimers. The dimers are usually soluble in pharmaceutical solvent systems (generally, aqueous in nature), while larger aggregates may grow to sizes that become visible to the naked eye. Aggregate formation is believed to give rise to immunogenicity as well as to create major challenges in the areas of handling, production yield, and quality (Das and Nema, 2008). Detection of particulate matter due to aggregation of protein molecules can be detected by various measures, including laser diffraction particle analyzers, dynamic image analysis, imaging, and Raman spectroscopy (Das and Nema, 2008). Moreover, the pharmaceutical industry utilizes various techniques for testing the structural integrity of proteins during processing. These include capillary electrophoresis, high-performance liquid chromatography methods, mass spectrometry, glycosylation analysis, amino acid analysis and primary sequencing by Edman chemistry, spectroscopy techniques, differential scanning calorimetry, dynamic and static light-scattering methods, and analytical ultracentrifugation/sedimentation techniques, among others (Chan, 2008). Raman spectroscopy and Raman microscopy are commonly used to monitor the stability of proteins during processing, allowing in situ testing as well as testing of samples in a rapid manner by matching Raman fingerprinting of proteins with that of samples. However, Raman technology does not work well with lyophilized material or when the concentration of protein is too low (Wen et al., 2010).

Routes of administration for biotechnology products are also limited to the parenteral route, although pulmonary and nasal delivery may be alternative options. For example, the delivery of insulin is done primarily by syringes, pens, jet injectors, and pumps (Shankar, 2005), and attempts to market insulin as a buccal spray or in inhalers have been seen recently. Generex Biotechnology is already marketing insulin as a buccal delivery device (Oral-lyn), which is approved for use in several countries (Amin et al., 2008b). Inhaled insulin preparations have been introduced to the market, but for varying reasons (mainly safety issues, efficacy, and cost) they were withdrawn by the manufacturer within months of being offered (e.g., Exubera, Pfizer). Exubera was a fine powder (particle size 1 to 5 μm) of *bioavailable* human insulin. Novo Nordisk's AERx Insulin Diabetes Management System was halted in its development stage before reaching the market. This system was an inhaled insulin system in liquid form. Eli Lilly, in cooperation with Alkermes, has been developing powdered insulin for inhalation that could be delivered using a simple inhaler. The particles are large and have low bulk density, providing them with good aerodynamic properties. Other devices for inhaled insulin are also being developed by several companies (Baxter BioPharma Solutions, BioSante Pharmaceuticals, Kos Pharmaceuticals, and MannKind) (Amin et al., 2008b).

Recent work in the area of oral delivery for peptides and proteins is also showing some hopeful signs of progress, although issues related to the *reproducibility* of delivery remain a big obstacle. The presence of food in the GI tract, the availability of active enzymatic processes, and the challenges facing absorption of high-molecular-mass molecules by the gut mucosa are just a few factors that render oral delivery of proteins and peptides the "holy grail" of drug delivery. Another area of concern is that related to the solubility of API after being released at the site of administration. Improvement in the solubility of peptides and proteins may be facilitated by the inclusion of counterions in the formulation. Acetate and trifluoroacetic acid (TFA) are commonly used as counterions in protein formulations. Higher residual TFA levels are of concern; however, they are tightly regulated because they can cause undesirable effects in patients. Also, modifications made to peptide and protein structure can not only improve solubility but can minimize aggregation and adsorption of the proteins as well (Dubin, 2008).

New approaches, such as Emisphere's Eligen (Cedar Knolls, NJ), are promising technologies for oral delivery of drugs that are otherwise limited to the parenteral route. This technologies has been tested with many important drugs, including calcitonin, growth hormone, heparin, and insulin (Majuru, 2005). It utilizes a proprietary delivery agent that when given along with a drug orally, forms a complex in the GI tract which changes the physical characteristics of the drug, allowing it to be absorbed through the mucosa. Following absorption, the complex, which is held together by weak intermolecular forces, dissociates, releasing the active drug to exert its pharmacological effect.

Emisphere's Eligen technology has been placed on hold by the company with respect to developing it further as a means of delivering insulin orally. In a statement on its website the company declared that "Oral insulin could represent a significant opportunity for Emisphere and, in theory, would meet unmet market needs. However, realistically, at this stage in its development and based on what has been and perhaps what hasnt been done with this product candidate, it presents significant challenges. At this time, the Company is exploring its strategic options for the oral insulin compound." This illustrates the types of challenges facing oral delivery of peptides and proteins.

Delivery of insulin via other routes, besides the parenteral route, has been investigated fiercely for many decades. The main reason for this interest is its importance in managing diabetes, a major health issue worldwide. Diabetic patients are required to self-inject insulin preparations several times a day, a painful and inconvenient way to maintain a normal lifestyle. Thus, developing alternative routes of administration has been a priority in this field. The oral route, in particular, is of greatest interest, as it provides a way that mimics the body's own method of insulin "delivery." Insulin is a high-molecular-mass hormone of 5800 Da. It is synthesized by the beta cells in the islets of Langerhans of the pancreas, and transported via the portal vein to the liver. Similar transport of insulin to the liver occurs via the oral route, which is the natural way for the body to handle insulin. Unfortunately, the GI tract is an extremely hostile environment for insulin, due to the presence of enzymes capable of destroying this regulating hormone. The main enzymes are pepsin, trypsin, and chymotrypsin. Studies performed in vitro have shown that the effect of these three enzymes is rapid (seconds to minutes) and highly effective (80 to 100% degradation). Pepsin requires an acidic environment to act, whereas trypsin and chymotrypsin need a slightly basic pH. This pH requirement corresponds to their location in the gut cavity; pepsin is located in the stomach, and the other two enzymes are found in the small intestine. Another obstacle for the oral delivery of peptides and proteins is the intestinal permeability. By far the most important area for drug absorption is the duodenum. However, the absorbing epithelium is limited in its action to smaller-molecular-mass drugs and more lipophilic solutes. As such, peptides and proteins are not good candidates for absorption in this region, and thus they have a very low oral bioavailability (Cheng and Lim, 2008). Pharmaceutical companies commonly rely on the concept of the *human maximum absorbable dose*

(MAD_{human}) as a measure of drug bioavailability. MAD_{human} measures the oral dose absorbed (mg) (D_{abs}) that reaches the circulation as compared to the oral dose administered (Timpe, 2007):

$$D_{abs} = PSAT \tag{11.2}$$

where P is the apparent permeability coefficient (cm/s), S the saturation solubility of an API (mg/mL), A the surface area available for absorption (cm^2), and T the transit time through the small intestine (seconds). In humans, the estimated values for A and T are 800 cm^2 and 3.3 hours, respectively. The D_{abs} desired is often stated as being 40 to 60% of the efficacious dose (ED_{50}) (Timpe, 2007). It is expected that the value of D_{abs} with respect to the oral absorption for unaided peptides and proteins is very low or nearly zero.

A promising area for delivering proteins orally is the entrapment of the API in a polymeric mesh, from which the drug is released slowly overtime as well as being protected from enzymatic degradation in the gut. An Example of a system that has produced positive *experimental* results in this respect is the polymer system known collectively as a *thiomer*. These polymers are essentially thiolated and provide a zero-order drug release kinetic from their matrix. The matrix itself is made of a three-dimensional structure that offers a tightened tablet matrix for the delivery of peptides and proteins (Langoth and Bernkop-Schnürch, 2005).

One area for protein delivery employs an encapsulated ion of proteins in biodegradable micro- or nanoparticles. The production of these particles involves a double-emulsion method, in which the aqueous phase containing proteins (w_1) is emulsified with the polymeric material dispersed in an organic phase to form a w_1/o emulsion. A double emulsion $(w_1/o/w_2)$ is then formed by the addition of an aqueous solution (w_2) containing 2% poly(vinyl alcohol) (PVAL). The resulting double emulsion is mixed with a PVAL aqueous phase (0.3%), and the organic phase is evaporated to form a suspension of nanoparticles. An example of these microparticles is Novartis's Sandostatin LAR Depot (composed of a biodegradable glucose star polymer with a copolymer of lactic and glycolic acids) delivering the chemotherapeutic agent octreotide acetate, used in the treatment of acromegaly (Bilati et al., 2005).

Delivery of peptides and proteins through the pulmonary route is associated with significant challenges regarding the stability of proteins during transportation, use, and storage. Metered-dose inhalers (MDIs) contain propellants that are not compatible with most proteins. Thus, nebulizer solutions containing unbuffered isotonic solutions with a pH of 5 or higher are preferred over MDIs for delivering protein drugs. Calcium salts (calcium chloride) are often used in protein solutions as stabilizers. Surfactants such as Tween 80 are usually added to protein solutions to minimize the adherence of protein to the container walls and to protect the protein from aggregation and a rise in temperature during nebulization. Instead of solutions, powdered aerosols may be used where the protein is present in freeze-dried form. To facilitate the flow of the protein's powder, an inert coarse powder carrier is usually mixed with the protein particles. Protein products should be protected from high temperatures and humidity during storage, as these conditions lead to denaturing the protein through aggregation reactions. The bioavailability of protein drugs via the lungs can be restricted, as proteolytic enzymes and the presence of macrophages act to eliminate the protein and limit its absorption (Krishnamurthy, 1999).

Other areas of biotechnology products are related to gene delivery. Liposomes are one of the preferred carriers for genes, due to their potential low immunogenic reaction (Felgner et al., 1987). In general, liposomes are classified into three types: *multilamellar vesicles* (MLVs), *small unilamellar vesicles* (SUVs), and *large unilamellar vesicles* (LUVs). With respect to size, MLVs range between 0.1 and 5.0 μm, whereas SUVs and LUVs are 0.02 to 0.05 μm and greater than 0.6 μm in size, respectively. There are two major methods of producing liposomes: by the formation of emulsions and by hydrating a dried lipid film followed by vigorous agitation (sonication or high-shear impeller) (Chrai et al., 2002). As a carrier for genes, liposomes were first suggested by Felgner et al. (1987). The ability of liposomes to travel through cytoplasmic and nuclear membranes is essential for delivering genes to the DNA. In addition, genes thus delivered should be able to integrate with the cell's DNA to become effective. Cationic liposomes are commonly used to deliver the plasmid gene, because these liposomes were found to be most stable in vivo (Bellare and Daga, 2005).

Generally speaking, proteins are prepared as lyophilized dried powders, in solutions, or as frozen preparations. The latter has its own advantages and disadvantages. Frozen protein preparations are desirable during batch manufacturing of a drug product. They reduce the waste associated with the manufacture of these products as well as assuring that the end product meets its specifications during storage and throughout its shelflife. In addition, the low freezing temperature (-20°C) affords a hostile environment for microorganisms and thus helps in contamination control of the product. However, freezing and thawing cycles can cause denaturation of protein. Also, solution heterogeneity, commonly found in the frozen material, can result in physical and chemical alterations that have a detrimental effect on a proteineous drug (Kantor et al., 2011).

11.9 DRUG–EXCIPIENT AND EXCIPIENT–EXCIPIENT INTERACTIONS

The choice of excipients in a formulation is not a marginal matter. Excipients may contain undesirable contaminants that need to be identified prior to incorporating them in formulations. Guidelines for impurities of excipients are traditionally listed in various pharmacopeias. For example, the *United States Pharmacopeia* monograph for poly(ethylene glycol) lists the limits for major and minor contaminants such as ethylene glycol (0.25%) and ethylene oxide (10 ppm) (Erickson, 2005). The functions of excipients in formulations are numerous and include roles as solubilizing agents, stability enhancers, tonicity-adjusting components, preservatives, and release-controlling materials, among others (Dubin, 2005). Interactions among excipients in the same formulation or between excipients and an API can lead to altered or diminished bioavailability and/or product stability (Dubin, 2005b). Isothermal stress methods are often used to test for incompatibilities existing among ingredients in the same formulation. High temperatures with or without moisture are used as storage conditions to determine how the various ingredients may affect the stability of an API in a formulation. However, the use of high temperatures during stress testing has its limitations if the ingredients melt at elevated temperatures (e.g., lipid components). For these formulations, extrapolation from high temperatures to ambient temperatures may be problematic (Holm, 2010). Most pharmaceutical companies that operate on a limited tight budget choose already acceptable excipient blends by the regulatory agencies for their formulations. Otherwise, the company has to justify the excipient's safety before use (Dubin, 2005a; Eatmon and Loxley, 2009). The approved list for excipients for certain formulations may be even more limited, such as those used in parenteral preparations (Dubin, 2005a). Generally speaking, pharmaceutical companies opt to use an excipient blend chosen from an FDA-approved excipients list. Characteristics of a good excipient are compatibility with the API and other added ingredients in a formulation, being a nontoxic substance, and its availability on the market with reasonable purity and an acceptable grade (Eatmon and Loxley, 2009). Some of the excipients used in biopharmaceuticals may require specific handling. For example, human serum albumin (HSA; 66,000 Da) is often used as an excipient in biotechnology formulations. One of requirements of its use is that it be free of any viral contamination. This is accomplished by heating HSA to $60°C$ for 10 hours. This excipient has the advantage of being relatively stable at room or higher temperatures, safe to use due to lack of immunogenicity, and can be used to eliminate the adsorption of other proteins from the surface of containers. HSA is also available from recombinant DNA technology (Alburex, Mitsubushi Welfarma, Japan), which eliminates the problem of virus contamination (Chaubal, 2005).

Other considerations in drug development are related to how excipients and drugs interact with each other in the same formulation. These interactions may affect the physical characteristics of final products, such as the dissolution and disintegration time for tablets. For example, an increase in the mixing time after adding magnesium stearate (a lubricant) to a mixture of ketorolac tromethamine with spray-dried lactose, cornstarch, pregelatinized starch, or a combination, resulted in a decrease in the dissolution rate of the capsules containing the formulation with cornstarch but not with that containing pregelatinized starch. The disintegration time of the capsules with cornstarch increased with an increase in mixing time with magnesium stearate but not with pregelatinized starch. These effects on dissolution time and disintegration time were explained by the interaction between cornstarch agglomerates, magnesium stearate flakes, and drug particles, as documented by scanning electron microscopy. No such interactions were seen with pregelatinized starch (Chowhan and Chi, 1985a). When crospovidone was used in the formula, replacing cornstarch and pregelatinized starch, similar effects were observed on the dissolution and disintegration times. Scanning electron microscopy revealed that drug particles get entrapped in crospovidone, and the resulting agglomerates interact with magnesium stearate flakes as mixing time increased. Larger drug particles had more efficient entrapment with crospovidone particles, resulting in a more pronounced effect on dissolution and disintegration time (Chowhan and Chi, 1985b).

11.10 THE PRESENCE OF CONTAMINANTS IN A FORMULATION

Chemicals used in formulations are available in various purity grades. The pharmaceutical industry utilizes those with characteristics that match those approved by regulatory health agencies. In particular, preparations that are intended for parenteral administration must contain chemicals of the highest possible purity available on the market, with no or an acceptably low content of contaminants as determined by the health authorities. In certain cases, such as the formulation of sterile single-use liquid biopharmaceuticals, the development process requires the use of stainless steel bulk storage tanks for an API that contains several elements, such as aluminum, carbon, chromium, cobalt, copper, iron, manganese, molybdenum, titanium, tungsten, and zirconium. These elements may leach from storage tanks and contaminate the product during the development process (Leveen, 2010). Analytical methods for detecting impurities in APIs or starting materials are based on various chromatography techniques such as high-performance liquid chromatography (HPLC) and supercritical fluid chromatography. The latter method was found to be much superior over HPLC techniques in detecting impurities in APIs and other starting additives (Zelesky, 1988).

Particulate matter contamination of sterile products is another area of concern in the pharmaceutical industry. Particulates can enter sterile products during processing, packaging, or use. Glass, plastic, and rubber materials that are used in vials and ampoules could contribute particulate matter to the products housed within them. Other sources of particulates include personnel entering and working in sterile packaging areas, sealing operation for ampoules, and gas purging of containers without a filtration step. It is always easier to prevent contamination from occurring than to clean a contaminated product. Accordingly, combining multiple steps during manufacturing, such as filling, sealing, and printing (in that order) under one-time operation may minimize the risk of introducing particulates into the product (Dean, 1985). Other forms of contaminants in parenteral sterile products are *leachables* and *extractables*. With increased use of newly synthesized polymers [polyethylene, polypropylene, polyester-laminated polymers, and poly(vinyl chloride)] as container materials for pharmaceutical and biological products, leachables and extractables have become of great concern to the pharmaceutical industry (Northup, 2012). This is true not only for sterile injectables but also for preparations intended for aerosol inhalations, ophthalmic products, and nasal aerosol sprays (Northup, 2012). Examples of leachables in orally inhaled and nasal drug products include the polycyclic aromatic hydrocarbons (PAHs), *n*-nitrosamines, and 2-mercaptobenzothiazole (2-MBT). All of these chemicals are considered to be carcinogens. PAHs and *n*-nitrosamines appear frequently in the environment, including in food products, soil, water, and air. 2-MBT residues are found in rubber products and are known to cause allergic reactions in some people who wear rubber gloves (Norwood and Mullis, 2008). In general, leachables are chemical substances (e.g., zinc ions) that migrate from a container's surface into a product concentrate during normal storage and handling conditions; whereas extractables are chemical substances (e.g., carbon-based compounds) that can be obtained from a container's surface after treatment with solvents under varying conditions of temperature and time (Castner et al., 2004). For example, in April 2010, Johnson & Johnson recalled its children's formulations, citing the presence of the pesticide and flame retardant 2,4,6-tribromophenol in these preparations. This pesticide, which originated from the shipping pellets, had permeated the packaging walls, including the actual container of the product, and reached the formulation (Lynch, 2012). Recommendations to overcome the presence of leachables and extractables are to obtain information on the raw materials and equipment used in making the containers, to utilize various methodologies in their identification, and to be aware of their sources, which may be from outside the container itself (Castner et al., 2004). Various analytical techniques can be utilized in the detection of leachables and extractables. The list includes single-crystal x-ray spectrometry, Fourier transform infrared spectrophotometry, diode-array ultraviolet spectrophotometry, nuclear magnetic resonance spectroscopy, and mass spectrometry. The latter a widely acceptable method for contaminant detection, due to its high sensitivity and its ability to interface with gas and liquid chromatographic techniques (Norwood et al., 2005).

Contamination presents in metered-dose inhalers is another challenge facing the industry. Although contamination can be traced to ingredients used in the product concentrate, the containers themselves may serve as the main source. Inhaling contaminants into the lungs allows easy access of potentially toxic materials to the blood circulation via the alveolar sacs. Thus, proper control of extractables and leachables from containers is an important safety concern (Hansen, 2005). Most important is screening for substances such as nitrosamines and polynuclear aromatic hydrocarbons, as their presence is related to serious adverse health concerns in humans. Quality control methods are important for detecting the presence of these substances at concentrations of nanograms per unit weight of the container material or lower. The leachables profile over the shelf life of the product should be determined, and agreement with the level and type of extractables must be established (Hansen, 2005).

11.11   OTHER CONSIDERATIONS

Certain parameters define the "value" of a product. These are included in the acronym ESCP: efficacy, safety, convenience, and price (Eatmon and Loxley, 2009). It is the efficacy of a drug that makes it useful clinically. Among drugs with similar efficacy, the one with the safest profile receives favorable acceptance among patients and clinicians alike. If efficacy is more crucial than safety (such as the case with chemotherapeutic agents), clinicians are willing to use drugs that are more efficacious, despite pronounced toxic effect. Although patients prefer the use of products that are more convenient to take (e.g., once vs. three times a day), they are not willing to sacrifice efficacy for convenience. Convenience only becomes important when the efficacy and safety characteristics of a drug are similar. Finally, pricing the product the right way by matching it with the consumer's buying capability is an important factor, in particular when efficacy and safety profiles are comparable (Eatmon and Loxley, 2009).

REFERENCES

Ali S, Quadir A. High molecular weight povidone polymer-based films for fast- dissolving drug deliv. applications. Drug Deliv. Technol. 2007;7(6):36, 38–40, 42–43.

Álvarez GD, Esteban E, Abmus M, Skalsky B. Comparison of Eudragit® FS 30 D and Eudragit® L 30 D-55 as matrix formers in sustained-release tablets. Drug Deliv. Technol. 2007;7(7):30–35.

Amin A, Shah T, Parikh D, Shah M. Progress in swellable and low-density excipients for prolonged gastric retention. Drug Deliv. Technol. 2008a;8(2):22, 24–26.

Amin A, Shah T, Patel J, Gajjar A. Annual update on non-invasive insulin delivery technologies. Drug Deliv. Technol. 2008b;8(3):43–48.

Andrews RR, McClain JB. Breaking through the stent-coating process. Drug Deliv. Technol. 2008;8(2):28–31.

Arrivo SM. The role of PAT in pharmaceutical research and development. Am. Pharm. Rev. 2003;6(2):46, 48–49, 51–53.

Bellare J, Daga VK. Liposome-mediated gene delivery. Drug Deliv. Technol. 2005;5(8):60–67.

Bilati U, Allémann E, Doelker E. Protein drugs entrapped within micro- and nanoparticles: an overview of therapeutic challenges and scientific issues. Drug Deliv. Technol. 2005;5(8):40, 42–47.

Brittain GH. What is pharmaceutical physics? Pharm. Technol. 1999;23(9):132, 134, 136, 138.

Brown A-SB, Hamed E. OraVescent® technology offers greater oral transmucosal delivery. Drug Deliv. Technol. 2007;7(7):42–45.

Castner JC, Williams N, and Bresnick M. Leachables found in parenteral drug products. Am. Pharm. Rev. 2004;7(2):70, 72–75.

Catalent Zydis brochure. http://www.catalent.com/drug/oral/zydis/zydis.pdf; 2007:1–6. Accessed Feb. 19, 2012.

Chan PC. Biochemical and biophysical methods currently used for therapeutic protein development. Am. Pharm. Rev. 2008;11(7):48, 50–54.

Chattopadhyay P, Shekunov YB. New enabling technologies for drug delivery. Drug Deliv. Technol. 2006;6(8):64–68.

Chaubal VM. Human serum albumin as a pharmaceutical excipient. Drug Deliv. Technol. 2005;5(8):22–23.

Cheng W, Lim L-Y. Peroral delivery of peptide drugs. Am. Pharm. Rev. 2008;11(4):33–34, 36–41.

Chien WY. The use of biocompatible polymers in rate-controlled drug delivery systems. Pharm. Technol. 1985;9(5):50, 52, 54, 56, 58, 60, 62, 64, 66.

Chowhan TZ. Aspects of granulation scale-up in high-shear mixers. Pharm. Technol. 1988;12(2):26, 28–29, 32, 34, 40, 42, 44.

Chowhan ZT, Chi L-H. Drug–excipient interactions resulting from powder mixing: I. Possible mechanism of interaction with starch and its effect on drug dissolution. Pharm. Technol. 1985a;9(3):84, 86, 90, 92, 94–97.

———. Drug–excipient interactions resulting from powder mixing: II. Possible mechanism of interaction with crospovidone and its effect on in vitro dissolution. Pharm. Technol. 1985b;9(4):28, 30, 32–33, 36, 38–41.

Chrai SS, Murari R, Ahmad I. Liposomes: a review—I: Manufacturing issues. Pharm. Technol. 2002;16: 28, 30, 32, 34.

CIMA Labs, Inc. http://www.cimalabs.com/technology/orasolv. Accessed Feb. 19, 2012.

Conley R, Gupta KS, Sathyan G. Clinical spectrum of the osmotic-controlled release oral delivery system (OROS), an advanced oral delivery form. Curr. Med. Res. Opin. 2006;22(10):1879–1892.

Das T, Nema S. Protein particulate issues in biologics development. Am. Pharm. Rev. 2008;11(4):52, 54–57.

Dean AD. Sources of particulate contamination in sterile packaging. Pharm. Technol. 1985;9(8):32–36.

Dubin CH. Formulation frustrations. Drug Deliv. Technol. 2005a;5(8):30, 32, 34.

_____. Combination products: strategies for a continuous pipeline. Drug Deliv. Technol. 2005b;5(9):40, 42, 44, 46.

_____. Should the molecule dictate drug delivery? Drug Deliv. Technol. 2006;6(1):26, 28, 30–31.

_____. Discussing gastroretentive drug delivery systems. Drug Deliv. Technol. 2007;7(6):26, 28–29.

_____. Formulating proteins and peptides: designing for delivery. Drug Deliv. Technol. 2008;8(2):34, 36–38.

_____. On the rise: drug delivery companies you should know about. Drug Deliv. Technol. 2009;9(1):36, 38–43.

Dziki W. Process analytical technology in product development and its impact on pre-approval inspections. Am. Pharm. Rev. 2008;11(7):12, 14, 16, 18, 20–22.

Eatmon C, Loxley A. Chitosan sources, chemistry, properties and pharmaceutical uses. Drug Deliv. Technol. 2009;9(1):18, 20, 22–23.

Emisphere Technologies. http://www.emisphere.com/contact.html. Accessed Feb. 19, 2012.

Erickson M. Identification and control of impurities in excipients. Am. Pharm. Rev. 2005;8(1):88, 90, 92–94.

Eurand N.V. http://www.aptalispharma.com/en/pharma_technologies. Accessed Feb. 21, 2012.

Evonik Röhm GmbH, Darmstadt, Germany. http://www.pharma-polymers.com/NR/rdonlyres/3D19A053-7628-41B1-BC52-123763B75566/0/080620FinalProduktbrosch%C3%BCre.pdf. Accessed Feb. 19, 2012.

Fan YF, Wang YN, Fan YG, Ma JB. Preparation of insulin nanoparticles and their encapsulation with biodegradable polyelectrolytes via the layer-by-layer adsorption. Int. J. Pharm. 2006;324(2):158–167.

Felgner PL, Gadek TR, Holm M, Roman R, Chan HW, Wenz M, Northrop JP, Ringold GM, Danielsen M. Lipofection: a highly efficient, lipid mediated DNA-transfection procedure. Proc. Natl. Acad. Sci. 1987;84: 7413–7417.

Ford CM. Investment trends driving combination product development. Drug Deliv. Technol. 2008;8(4):36, 38–39.

Genzyme Pharmaceuticals. http://www.genzymepharmaceuticals.com. Accessed Feb. 21, 2012.

Gore YA, McFarland WD, Batuyios HN. Fluid-bed granulation: factors affecting the process in laboratory development and production scale-up. Pharm. Technol. 1985;9(9):114–116, 118, 120, 122.

Gothoskar AV. Study of softening point of swellable matrices and effect on drug-release pattern. Drug Deliv. Technol. 2005;5(2):56–62.

Gupta V, Verma S, Nanda A, Nanda S. Osmotically controlled drug delivery. Drug Deliv. Technol. 2005;5(7):68–76.

Hamilton LE, Lutz ME. Advanced orally disintegrating tablets bring significant benefits to patients and product life cycles. Drug Deliv. Technol. 2005;5(1):34, 36–37.

Hansen JK. Developing a correlation between extractables and leachables in MDIs and DPIs. Drug Delivery Technol. 2005;5(2):66–68.

Hariharan M, Bogue BA. Orally dissolving film strips (ODFS): the final evolution of orally dissolving dosage forms. Drug Deliv. Technol. 2009;9(2):24–29.

Holm R. Enabling pharmaceutical technology lipid based formulations. Am. Pharm. Rev. 2010;13(5):10, 12–15.

Jain KS, Jain NK, Agrawal GP. Gastroretentive floating drug delivery: an overview. Drug Deliv. Technol. 2005;5(7):52, 54–61.

Kantor A, Tchessalov S, Warne N. Quality-by-design for freeze–thaw of biologics: concept and application to bottles of drug substance. Am. Pharm. Rev. 2011;14(4):65–66, 68–72.

Krishnamurthy R. Protein stability in pulmonary delivery formulations: a review. Pharm. Technol. 1999;23(3):48, 50, 52, 54, 56, 58, 60.

Langer R. Biomaterials: new perspectives on their use in the controlled delivery of polypeptides. Pharm. Technol. 1985;9(10):35, 38, 40.

Langoth N, Bernkop-Schnürch A. The use of multifunctional polymers as auxiliary agents in non-invasive peptide delivery. Am. Pharm. Rev. 2005;8(2):80–84.

Leveen L. Pre-sterilized, single use filling systems for liquid bio-pharmaceuticals. Am. Pharm. Rev. 2010;13(5):30, 32, 33.

Lipocine, Inc. http://www.lipocine.com/nanosplode.html. Accessed Feb. 19, 2012.

Lynch PM. Extractables and leachables: quality concerns and considerations for ophthalmic and injectable products. Am. Pharm. Rev. 2011;14(4):93–94, 96.

Majithiya RJ, Raval AJ, Umrethia ML, Ghosh PK, Murthy RSR. Enhancement of mucoadhesion by blending anionic, cationic and nonionic polymers. Drug Deliv. Technol. 2008;8(2):40–45.

Majuru S. Development of an oral insulin solid dosage formulation using emisphere's eligen® technology. Drug Deliv. Technol. 2005;5(9):74, 76–78.

Mehta MA. Scale-up considerations in the fluid-bed process for controlled-release products. Pharm. Technol. 1988;12(2):46–47, 50–52.

Merrion Pharmaceuticals. http://www.merrionpharma.com/content/technology/gipet.asp. Accessed Feb. 19, 2012.

Myatt R, Oladiran G, Batchelor H. Measuring the solubility of a model drug in drug-in-adhesive transdermal patches to validate a theoretical solubility calculator. Drug Deliv. Technol. 2008;8(3):56, 58–59.

Northup SJ. Pharmaceutical containers: safety qualification of extractables/leachables. Am. Pharm. Rev. 2008;11(3):104, 106, 108–109.

Norwood DL, Mullis JO. "Special case" leachables: a brief review. Am. Pharm. Rev. 2009;12(3):78, 80, 82–86.

Norwood DL, Nagao L, Lyapustina S, Munos M. Application of modern analytical technologies to the identification of extractables and leachables. Am. Pharm. Rev. 2005;8(1):78, 80, 82–87.

Olsen WK. Batch fluid-bed processing equipment—a design overview: I. Pharm. Technol. 1989;13(1):34, 36, 38, 40, 42, 44, 46.

Patel VF, Patel NM. Influence of drug-to-lipid ratio and release modifier on dipyridamole release from floating lipid granules. Drug Deliv. Technol. 2008;8(4):22, 24, 26–27.

Patel G, Patel G, Patel R, Patel J, Bharadia P, Patel M. Sodium alginate: physiological activity, usage and potential applications. Drug Deliv. Technol. 2007;7(4):28, 30, 32, 34, 36–37.

Pondell R. Scale-up of film coating processes. Pharm. Technol. 1985;9(6):68, 70.

Porter CS, Saraceni K. Opportunities for cost containment in aqueous film coating. Pharm. Technol. 1988;12(9):62, 64, 66, 68, 70, 74, 76.

Prajapati BG, Patel JK, Patel VM, Prajapati KV. The upcoming era of nanomedicine: a briefing. Drug Deliv. Technol. 2008;8(4):46, 48–49.

Ravishankar H, Patil P, Petereit H-U, Renner G. Eudramode™: a novel approach to sustained oral drug delivery systems. Drug Deliv. Technol. 2005;5(9):48, 50, 52–55.

Russo JE. Typical scale-up problems and experiences. Pharm. Technol. 1984;8(11):46–48, 50–51, 54–56.

Sawant PD, Luu D, Ye R, Buchta R. Drug release from hydroethanolic gels: effect of drug's lipophilicity ($\log P$), polymer–drug interactions and solvent lipophilicity. Int. J. Pharm. 2010;396(1–2):45–52.

Scarlett B. Measuring and interpreting particle size distribution. Am. Pharm. Rev. 2003;6(4):93–101.

Shankar B. Insulin delivery technologies: present and future. Drug Deliv. Technol. 2005;5(9):38–39.

Tchessalov S, Dixon D, Warne N. Principles of lyophilization cycle scale-up. Am. Pharm. Rev. 2007;10(3):88–92.

Tingstad EJ. The key to effective communication. Pharm. Technol. 1988;12(10):30, 32.

Timpe C. Strategic for formulation development of poorly water-soluble drug candidate $\frac{3}{4}$ A recent perspective. Am. Pharm. Rev. 2007;10(3):104, 106–109.

Wen Z-Q, Cao X, Philips J. Application of Raman spectroscopy in biopharmaceutical manufacturing. Am. Pharm. Rev. 2010;13(4):46, 48, 50–53.

Zelesky T. Supercritical fluid chromatography (SFC) as an isolation tool for the identification of drug related impurities. Am. Pharm. Rev. 2008;11(7):56–60, 62.

GLOSSARY

API	Active pharmaceutical ingredient.
BBB	Blood–brain barrier.

CNS	Central nervous system.
FDA	U.S. Food and Drug Administration.
GI	Gastrointestinal.
GIPET	Gastrointestinal permeation enhancement technology.
GRAS	Generally recognized as safe.
HSA	Human serum albumin.
HLB	Hydrophile–lipophile balance.
HPLC	High-performance liquid chromatography.
HPMC	Hydroxypropyl methylcellulose.
Industrial pharmacy	Both preformulation and formulation are parts of pharmaceutics, and their application in formulation development in industry is known as *industrial pharmacy*.
LUV	Large unilamellar vesicles.
MAD	Human maximum absorbable dose.
2-MBT	2-Mercaptobenzothiazole.
MDI	Metered dose inhalers.
MLV	Multilamellar vesicles.
ODTs	Orally disintegrating tablet.
OFDS	Oral fast-dissolving film or strip.
PAH	Polycyclic aromatic hydrocarbon.
PAT	Process analytical technology.
Preformulation	A term often used to describe the physical and chemical characteristics of substances and mixtures used in pharmaceutical products.
Process analytical technology (PAT)	Used for control and assessment *during* manufacturing.
PVAL	Poly(vinyl alcohol).
R&D	Research and development.
SUV	Small unilamellar vesicle.
TFA	Trifluoroacetic acid.

PART II

PRODUCT DESIGN

CHAPTER 12

THE PRODUCT DESIGN PROCESS

12.1 INTRODUCTION

A drug delivery system is designed to meet a patient's need(s) and to provide a specified functionality and performance. Creating this drug delivery system, called *product design*, is a complex, multifactorial process that integrates drug delivery to treat a human condition with formulation, manufacturing process, and container closure system design. The quality of the final product is critically dependent on the integration of the design of these three components. The interrelationship of *formulation design, process design, and container closure system design* is represented in Figure 12.1. Product design is an essential element of the drug development process. Product development can take seven to 10 years. The pharmaceutical industry continues to strive to decrease the overall development time. In general, phases 1, 2, and 3 take about one, two, and three years to complete, respectively. Figure 12.2 shows how the various components of product design fit into the late drug discovery and clinical development process.

Product design activities begin with participation in lead drug candidate selection and end with successful process validation at the commercial manufacturing site. The final process validation effort usually occurs after a new drug application (NDA) has been submitted and before the U.S. Food and Drug Administration (FDA) postapproval NDA commercial site inspection.

Product design can be thought of as a funnel whose top represents the many unknowns that are present at the start of the development process. As the process continues to move toward the narrow opening of the funnel, the number of unknowns shrinks and the variables that have the most impact on product quality are identified. In practice, this smooth learning transition is often interrupted by unexpected findings that cause additional experiments to be undertaken or projects to be abandoned. In general, companies have found that early involvement of cross-functional teams or committees improves an organization's ability to identify and address key issues that can speed up the development process. It is paramount that product design scientists be represented in decision-making cross-function committees.

Integrated Pharmaceutics: Applied Preformulation, Product Design, and Regulatory Science, First Edition. Antoine Al-Achi, Mali Ram Gupta, William Craig Stagner.
© 2013 John Wiley & Sons, Inc. Published 2013 by John Wiley & Sons, Inc.

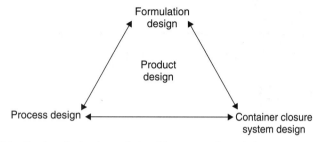

FIGURE 12.1 Product design interrelationships among formulation, process, and container closure system design.

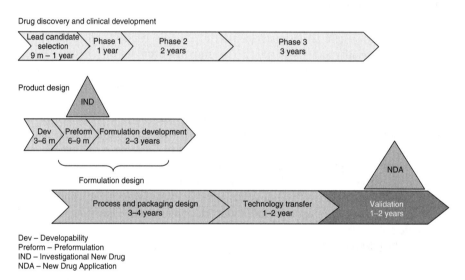

Dev – Developability
Preform – Preformulation
IND – Investigational New Drug
NDA – New Drug Application

FIGURE 12.2 Drug development process.

The goal of product design is to gain timely regulatory approval for a high-quality product that meets the needs of patients, health care providers, and other payers. Product quality should meet the identity, strength, quality, purity, and potency standards set by the FDA. To meet this goal, product design must incorporate the most current regulatory science thinking, which is often provided in the form of FDA and International Conference on Harmonization (ICH) guidance documents. The overarching objectives in product design are to build-in quality from initial lead compound selection to the manufacture of commercial product. This is done to achieve patient acceptance, dose uniformity, dosage form stability (physical, chemical, and microbial), and efficient and robust processes for commercial manufacture.

The product design decision process that follows the selection of the lead candidate can be divided into inputs and outputs as depicted in Figure 12.3. The scope of product design is defined in part by the target product profile (FDA, 2007), quality target product profile (FDA, 2009), and the physicochemical

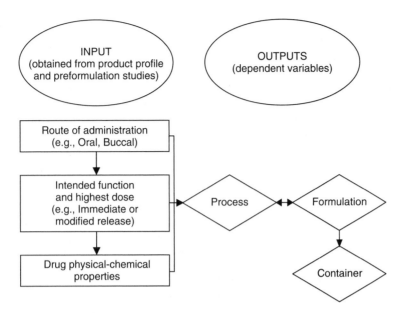

FIGURE 12.3   Product design decision process following lead candidate selection.

properties of the drug. This input information is used by a design scientist to choose the appropriate process, formulation, and container closure system. Note that the manufacturing process is chosen before a formulation is chosen. This is not obvious nor necessarily intuitive. This concept of choosing a manufacturing process before selecting a formulation has very important practical implications in the pharmaceutical industry. Most industrial pharmacy departments are divided into developability, preformulation, formulation, and process development functions or departments. A new scientist will be hired into one of these areas. The significant implication of Figure 12.3 is that the formulation function needs to be vigilant in resourcing the area with scientists who possess a strong understanding of the manufacturing processes. This notion of choosing a process before a formulation is discussed in subsequent chapters.

Product design embraces the FDA and ICH quality by design (QbD) initiatives. QbD concepts such as developing a quality target product profile, building quality into the product at the start by identifying the best lead drug candidate, developing a design space, and identifying critical attributes are also discussed in subsequent chapters.

12.2 FORMULATION DESIGN

Formulation design is the aspect of product design that establishes drug product formulation. Formulation design can be broken down into drug developability, preformulation, and formulation development. These three components of formulation design do not occur at the same stages of drug development, which is illustrated in Figure 12.2 and discussed in more detail at the end of the chapter. Drug developability occurs at the stage of lead candidate identification. Preformulation activities can be broken down into drug physicochemical characterization and drug–excipient compatibility studies. Preformulation physicochemical characterization can start sometime during the lead drug candidate selection process. Drug–excipient compatibility studies typically begin after the lead drug candidate has been chosen and after a manufacturing process has been selected. The excipients that are used in an excipient compatibility study support the process identified and the product design attributes desired. Once the compatibility study is complete, probe formulations are chosen from the compatible excipients.

12.2.1 Developability Assessment

Developability assessment evaluates a potential therapeutic candidate on its in vivo, in vitro, and physical and chemical attributes to determine the level of development risk and commercial viability. This assessment often takes place during lead drug candidate selection at the early stages of product development. Cross-functional or multidisciplinary integration usually consists of representatives from the fields of biology, molecular biology, genetics, medicinal chemistry, bioengineering, pharmacology, safety assessment, metabolism, pharmacokinetics, product design, medicine, and marketing. Each function assesses a compound's developability based on specific criteria that are critical to each function's successful development of the compound. Chemists or bioprocess engineers assess developability based on the ease of manufacture, production cost, and yield. Pharmacologists evaluate proof of concept, the dose–response curve, the concentration of drug required to elicit 50% of the maximum effect ($EC_{50\%}$), and undesired effects on the central nervous and cardiovascular systems. Drug safety personnel evaluate such drug risks as mutagenicity, carcinogenicity, and organ toxicology. Pharmacokinetic scientists study a compound's absorption, distribution, and elimination profiles. Drug metabolism scientists assess a drug candidate's metabolic profile, including the potential for first-pass metabolism. Product design is keen to understand a compound's stability, solubility, and permeability properties, which can significantly affect its developability, depending on the dosage form and route of administration desired. Medical researchers are concerned primarily with a compound's proof of concept, potency, therapeutic index, and route of administration for different intended purposes, such as ambulatory or hospital patient treatment. Marketing personnel differentiate a potential lead compound based on factors such as expected efficacy, safety profile, dose, dosing frequency, route of administration, lack of food effect, metabolic profile, and access to enabling drug delivery systems. The FDA route of administration nomenclature is provided in Appendix 12.1.

The lead drug candidate is evaluated against a number of these parameters that are defined in what is known as a *quality target product profile*, which describes the desired dosage form, probable dosage range, route of administration, intended purpose, patient population, pharmacokinetic and pharmacodynamic profile, safety profile, performance, and quality attributes. The cross-functional team's selection of a lead compound often requires a trade-off analysis that attempts to minimize the development risk while choosing a compound that fulfills most of the quality target product profile's requirements.

The developability assessment criteria are different for each dosage form. Examples of formulation design developability criteria for oral, transdermal, and injectable dosage forms are presented in the next section. The FDA and *United States Pharmacopeia* (U.S.P.) dosage form nomenclature is provided in Appendixes 12.1 and 12.1.

12.2.1.1 Oral Drug Delivery Developability There are three compound properties that tend to have the most impact on oral drug development: therapeutic ratio (therapeutic window or index), solubility, and bioavailability. The *therapeutic ratio* is defined as the ratio of the *median lethal dose* (LD_{50}) to the *median effective dose* (ED_{50}). A larger therapeutic ratio provides the formulator with a wider "operating" range, which increases the likelihood of developing a successful oral product. Animal rights groups have pushed to ban the use of LD_{50}. Because of this, other ratios are starting to replace the therapeutic index as measures of efficacy and toxicity. One such therapeutic index is the ratio of the maximum tolerated concentration and minimum effective concentration. A compound with a narrow therapeutic index seriously constrains a formulation scientist's drug delivery options. So, if all potential lead compound properties are considered equal except their therapeutic index, the candidate with the largest (widest) index would be chosen as the lead candidate. The *Code of Federal Regulations* (21 CFR 320.33) defines a narrow therapeutic ratio as one where "there is less than a 2-fold difference in median lethal dose (LD_{50}) and median effective dose (ED_{50}) values, or have less than a 2-fold difference in the minimum toxic concentrations and minimum effective concentrations in the blood, and safe and effective use of the drug products requires careful dosage titration and patient monitoring."

The ease of developing an oral dosage form is also dependent on a drug substance's solubility and bioavailability. At the lead compound stage of drug development, bioavailability studies are not typically done because of lack of compound, lack of ability to predict the bioavailability in humans from animal data, and time and effort required to perform these studies. A biopharmaceutics classification system (BCS) (Amidon et al., 1995; FDA, 2000) has been developed that attempts to correlate in vitro parameters with in vivo bioavailability. The BCS classifies a drug into one of four classes depending on its level of solubility and permeability: class I, high solubility and high permeability; class II, low solubility and high permeability; class III, high solubility and low permeability; and class IV, low solubility and low permeability. To be considered highly soluble, a compound's highest dose strength must dissolve in 250 mL or less in pH 1 to 7.5 aqueous media at 37°C. To be judged highly permeable, a candidate's extent of absorption must be 90% or greater unless there is evidence that drug degradation occurs in the gastrointestinal track. The FDA has adopted this system to grant biowaivers. The formulation scientist can use this system to help in lead compound selection. For example, a compound that has high solubility and high permeability (class I) and dissolves up to 85% or more in 30 minutes in 900 mL of 0.1 N HCl, 4.5 buffer, and 6.8 using the U.S.P. apparatus I or II at 37°C would be expected to have good absorption and bioavailability. On the other hand, a compound that exhibits high solubility and low permeability (class III) and dissolves rapidly would not be expected to have as good bioavailability. A biopharmaceutics drug disposition classification system (BDDCS) (Wu and Benet, 2005) has been proposed to predict in vivo and in vitro drug disposition. Like the BCS, the BDDCS recognizes solubility as an important factor in drug bioavailability, but it also recognizes the contribution that transport–absorption–elimination interplay has on overall drug disposition. The number of drugs that have been evaluated using the BCS and BDDCS systems is expanding continually. Takagi et al. (2006) and Dahan et al. (2009) discussed the provisional BCS classification for the world's top oral drugs. The FDA–NICHD (2011) has provided a BCS classification for over 350 marketed drugs. Benet et al. (2011) have applied the BDDCS system to over 900 drugs.

A developability classification system (DCS) has recently been proposed, which revises the BCS as an oral product development tool (Butler and Dressman, 2010). The DCS applies biopharmaceutical concepts to product design. BCS class II is divided into dissolution rate–limited absorption and solubility-limited absorption. The dissolution rate is expressed as a target particle size compared to the dose/solubility ratio used in the BCS. The solubility is determined in fasted-state simulated intestinal fluid, and the solubility-limited absorbable dose is based on 500 mL compared to BCS's 250 mL. The DCS is a reasonably simple system that aids the product design scientist in understanding the absorption limitations for a drug. Once these limitations are recognized, strategies to address the limitations can be implemented.

At the end of the day, a compound that has high solubility, fast dissolution, high bioavailability, low or no metabolism, and exhibits low or no facilitated transport is generally easier to develop. This translates to faster development times, lower development costs, and less development risk.

12.2.1.2 Pulmonary Drug Delivery Developability

Decisions to continue development of a potential drug for pulmonary delivery are based primarily on its potency and lack of irritancy and immunogenicity. Potential drugs targeted for lung delivery need to be potent. Generally, less than 20% of an inhaled drug is delivered to the lungs; however, there are reports of greater than 50% lung drug deposition (Leach et al., 2006). Typical doses delivered to the lungs are less than 1 mg. Pulmonary dosing ranges of beta agonists and steroids are 50 to 100 μg and 100 to 400 μg, respectively. Pulmonary insulin was dosed at 1 to 3 mg. Foreign particulates, including drugs, can induce a cough reflex, cause irritancy, or generate an allergic response. There has been a rising concern about inhaled drugs causing immunogenic responses (Ferrari et al., 2003; Cefalu, 2008).

12.2.1.3 Injectable Drug Delivery Developability

The majority of injectable systems are used as solutions. Therefore, solubility, stability, and tissue irritancy are important developability characteristics. The dosage flexibility of subcutaneous (SC) and intramuscular (IM) injection delivery systems are limited by drug solubility, maximum volume of injection, and irritation and pain on injection. Injection volumes typically do not exceed 2.5 and 5 mL for SC and IM injections, respectively. Smaller IM injection volumes are dosed into the smaller deltoid muscle, while larger injection volumes can be given into posterior gluteal and thigh muscles. An open-label double-blind study in patients reported that a single gluteal IM injection ranging from 9.8 to 19.5 mL was well tolerated (Pryor et al., 2001). A number of compounds have been shown to cause concentration-dependent irritation and necrosis at the injection site. Several studies have shown that incorporating the drug in a lipid emulsion can reduce the irritation (Lovell et al., 1994; Wang et al., 1999).

It is desirable to have a stability-indicating method at the developability stage to quantify the stability of the compounds being considered as a lead candidate. At this development stage the method should be capable of monitoring the loss of drug without interference from the degradant peaks to provide a reasonable level of scientific confidence in the results. Potential lead compounds are often tested for light, moisture, pH, oxygen, and temperature stability. Selection of the lead candidate is based on comparing each compound's solubility, stability, and tissue irritancy and choosing the compound that has the best overall attributes.

12.2.2 Preformulation

Preformulation efforts are initiated on the lead candidate and possibly on a backup compound. The purpose of this stage of formulation design is to gain physico-chemical data that support the selection of an acceptable drug delivery system and formulation. Preformulation activities include the development and refinement of a stability-indicating method, characterization of drug solution and solid-state physicochemical properties, and excipient compatibility studies that support drug safety study formulations and earlier formulation development studies. A stability-indicating method is one that has been demonstrated scientifically to detect a change in purity or content over time. The first step in developing a stability-indicating method is to generate degraded product using a forced degradation protocol that employs extreme pH, temperature, oxidative, and light conditions. The second step uses the degraded samples to development a method that separates the degradants from the drug substance. Solution characterization includes determining the drug's pK_a, solubility in various physiological media, such as simulated gastric, small intestine fasted-state, and small intestinal fed-state fluids (U.S.P. 32/*National Formula* (N.F)27; Dressman and Reppas, 2000), and chemical stability. Solid-state characterization includes assessment of a drug's bulk powder properties, such as hygroscopicity; flowability; compressibility; true, bulk, and tapped density; particle size distribution; and specific surface area. Solid-state stability and polymorphism are also evaluated at the preformulation stage. Selection of the dosage form and route of administration involves input from a number of functional areas. Medical and marketing requirements carry the greatest weight in the decision-making process. The formulation scientist provides advice regarding potential development and manufacturing risks and benefits.

Once the drug delivery system has been chosen, excipient compatibility studies are undertaken. The excipients chosen for the compatibility study depend on the manufacturing process, physicochemical properties of the drug, and intended dosage form selected. The aim of an excipient compatibility screening study is to assess a drug's potential stability when exposed to a specific excipient. The study is done under accelerated temperature and moisture conditions for one to three months. Excipients that show the highest level of drug stability are then potential candidates for further formulation studies.

12.2.3 Formulation Development

Formulation development, the final stage of formulation design, is discussed in more detail in subsequent chapters. Basically, the dosage formulation is specific to a given drug delivery system or dosage form. Formulation selection is a combination of art and science, often referred to as the *art and science of formulation development* or simply *formulation knowledge*. The art of formulation is developed from experience and from intuition that is honed over time. The science is based on academic learning, personal research, literature, and science and technology training. Formulation development is typically initiated by a formulation scientist choosing several "probe" formulations, which are selected from the excipients that were shown to be compatible with the drug. Dosage forms are then manufactured and critical product attributes are assessed. As the drug progresses through the clinical phases, the formulation is refined, or in some cases the dosage form is changed, depending on what is learned in the ongoing stability and clinical programs. By the start of phase 3, formulation optimization studies, should be completed and a "final" optimized commercial formulation chosen because any changes in the formulation or process may require bridging safety and bioavailability studies, which can be very costly and potentially risky.

12.3 PROCESS DESIGN

Process design activities include process selection, process characterization, process optimization, scale-up, technology transfer, and validation. Technology transfer and validation are beyond the scope of this book. For each dosage form, there are a number of different processes that can be adopted to manufacture a particular drug delivery system. Similar to formulation design, there is a certain level of art and science associated with process design. At the process development stage it is important to know the working limitations of the pilot plant or manufacturing plant equipment unless a completely new process needs to be developed. Generally, pilot and manufacturing scale equipment have less flexibility regarding operating temperatures, temperature control, speed, airflow, throughput time, and capacity compared to laboratory-scale equipment. Early process design can use *scaling principles* to develop small-scale processes that account for the limitations and scale differences seen in commercial manufacturing. This is called *scaling down*. The use of scaled-down processes in early development is the most efficient way to assure the robust *scale-up* of a unit operation.

Scaling a process usually entails increasing the batch size from laboratory scale to pilot scale to production scale. Pharmaceutical processes involve a number of different processing steps, called *unit operations*. As an example, unit operations can involve milling, blending, granulating, drying, and tableting. It is vital to develop a process that is robust and capable of beind scaled up or down easily. It is desirable to establish a fundamental, scientifically sound approach to scale-down and scale-up. Those processes that can be described by a mathematical equation provide insight into the issues that are crucial to successful scaling. Empirical statistical design approaches can also lead one to a better understanding of the critical process variables. Model equations can be developed that help predict scaling factors that can be used during scaling.

Intuitively, one expects a similarly scaled process to result in a similar product. Several similarities that are important to maintain during scaling are equipment operating principles or concepts governing the operation of a piece of equipment, equipment geometry, formulation, energy input, temperature exposure, and processing time. The FDA (1999a) has provided guidance that assigns class and subclass designations to equipment used in solid-dose unit operations. Equipment of the same class and subclass are considered to have the same design and operating principles. A number of different types of equipment can be used for a given unit operation. Scaling concepts are applicable only if the operating principle remains the same. Once scaling is restricted to equipment with a given operating principle (a tumble V-blender vs. a high-shear powder mixer), the geometric dimension ratios of the equipment need to be held constant. The formulation or ratios of excipients to drug also need to be held constant. Finally, the energy input and temperature exposure need to be similar. Time similarity is referred to as *kinetic similarity*. *Dynamic similarity* maintains similar forces to which that the product is exposed, such as gravitation, centrifugal force, and pressure.

The concept of dimensionless numbers has been used in engineering for many years to scale processes and systems. The idea is to develop dimensionless number(s) that describe a physical process or system. Since by definition the relevant dimensionless number has no dimensions, it is independent of scale. Examples of dimensionless numbers used in scaling pharmaceutical products are Reynolds and Froude numbers. The *Reynolds number* is the ratio of inertial to viscous forces that are important in fluid dynamics and flow. The *Froude number* is the ratio of inertial to gravitational forces. As one scales a process up or down, the Reynolds number or Froude number would remain constant since it is scale invariant. The use of dimensional analysis and derivation of dimensionless numbers is supported by the *pi theorem*, which states that physical relationships can be reduced to independent dimensionless groups. In subsequent chapters we discuss how dimensionless numbers can be used to scale various processes.

The FDA has issued a number of guidances to industry that outline mechanisms for postapproval changes. Under certain specified conditions, formulation and process changes can be made after the NDA has been approved, without prior FDA approval. These changes are referred to as "changes being affected supplement" [21 CFR 314.70(c)]. The guidances are referred to as scale-up and postapproval changes. The guidances define levels of change; recommended chemistry, manufacturing, and controls tests for level of change; in vitro dissolution tests and/or in vivo bioequivalence evaluation; and documentation necessary to support the change (FDA, 1995, 1997a, b). In addition, the FDA has given guidance that classifies equipment based on its operating principles (FDA, 1998, 1999a).

Once a manufacturing process is chosen, it is important to characterize the process and understand which processing variables affect critical product attributes. *Process analytical technology* (PAT) is used to help characterize and control the process. According to the FDA (2004), "PAT is a system for designing, analyzing, and controlling manufacturing through timely measurements (during processing) of critical quality and performance attributes of raw and in-process materials and processes with the goal of ensuring final product quality." An ideal outcome of PAT would be to establish real-time process control and quality assurance. The FDA has defined at-line, online, and in-line guidelines for nondestructive PAT technologies. At-line measurements involve samples that are removed from the process and analyzed near the process line but are not returned to the batch. Online samples are diverted from the process stream, analyzed, and may be returned to the batch. In-line sampling and analysis occur during processing without sample removal. In-line sampling and analysis permit continuous analysis and control of a process.

As a compound moves through phases 1 and 2, the process will be refined, and by phase 3 all processes need to be fully optimized and scaled up. It is most desirable to manufacture at least one phase 3 clinical lot at the commercial production site at a commercial scale. Specific dosage form processes are discussed in detail in later chapters.

The objective of process optimization is to develop a robust process, one that produces a consistent product when processing such factors as excipients, drug substance, manufacturing environment, and operating conditions vary. A robust process such as a robust formulation must be incorporated into the product design process. Process optimization and design of experiments help identify the critical process parameters that significantly affect the critical product attributes.

12.4   CONTAINER CLOSURE SYSTEM DESIGN

The third component of product design is the design of a container closure system (CCS) that provides the necessary functionality for the overall product design. Choosing a CCS depends on a multitude of factors, such as the patient's disease state, dosing regimen, route of administration, patient compliance, physiochemical properties of the drug and dosage form, barrier requirements for proper stability, manufacturing issues and costs, and regulatory considerations such as environmental friendly material that is tamper-evident and child-resistant.

Over the past 25 years there has been a revolution in product design that has affected pharmaceutical packaging especially. The marketing of metered-dose inhalers, powder inhalers, prefilled syringes with automated self-injection systems, transdermal patches, and ionotophoretic drug delivery systems has truly integrated the dosage form and the package. Today the primary pharmaceutical package or CCS consists of those components that are used to contain, protect, preserve, and deliver a safe and efficacious product. In addition to the primary container that is contact with the dosage form, there are secondary packaging components that are also important in conveying critical information and helping with the safe distribution of a product to the warehouse, pharmacies, hospitals, health care professionals, and other locations. As mentioned above, the CCS may ultimately deliver the drug to the patient. The CCS must hold the dosage form as well as protect it from environmental factors such as light, oxygen, and microbes. There is a complex interplay between the material properties of the CCS, the physiochemical properties of the drug, and the ultimate functional needs of the final product. At the end of the day, successful product design results from the full integration of formulation design, process design, and the container closure system design.

CCS design, which depends on the type of dosage form, can range from the selection of container closures that are readily available commercially to designing a system that also acts as a delivery device, as is the case for dry powder inhalers. The selection and design of a CCS is an integral part of the product design process and demands the same level of consideration as formulation and process design. The CCS should protect the dosage form from microbial contamination, moisture, air, and light, all of which can affect the overall product quality. A CCS includes all the packaging components, such as bottle, cap, and cap liner, that together contain and protect the product. The packaging components should be made of safe materials that do not interact with the drug or drug product. Drug products often go to market in a primary and secondary packaging system. The components that are in contact or have the potential for coming into contact with the drug product are referred to as the primary packaging system. Secondary packaging systems add additional protection for the product. The FDA and Consumer Product Safety Commission (CPSC) have regulatory responsibility for human-use packaged products. The FDA requirements for tamper-resistant closures are stated in 21 CFR 211.132. The CPSC requirements for child-resistant and adult-use-effective closures (often referred to as "geriatric friendly" closures) are provided in 16 CFR 1700. The FDA's container closure guidance for industry discusses the likelihood of packaging concerns that are related to different routes of administration and the potential for packaging component–dosage form interactions (FDA, 1999b). The highest level of packaging concern is associated with the parenteral, ophthalmic, and inhalation routes of administration. The integrity of the container is important for maintaining a sterile product, and there are minimal anatomical protective barriers when a drug is delivered by the injectable, ophthalmic, or inhalation route. There is less concern over the container closure system for transdermal patches and nasal delivery. The least concern is associated with packaging for topical and oral products. The likelihood that a drug delivery system will interact with a CCS is highest for solutions. There is a need to evaluate the level of leachables and extractables for liquid systems. The lowest potential for interaction to occur between the formulation and the packaging components is oral solid tablets and capsules. The FDA has also developed five general categories of dosage forms and has recommended that CCS description, suitability, quality control, and stability sections be provided in a regulatory submission. The five general categories are: Inhalation Drug Products, Drug Products for Injection and Ophthalmic Drug Products, Liquid-Based Oral and Topical Drug Products and Topical Delivery Systems, Solid Oral Dosage Forms and Powder for Reconstitution, and Other Dosage Forms. A list of FDA and U.S.P. packaging nomenclature and definitions is given in Appendixes 12.4 and 12.5. U.S.P. packaging fabrication materials and closure types are presented in Appendix 12.6.

REFERENCES

Amidon G, Lenernäs H, Shah V, Crison JA. Theoretical basis for a biopharmaceutical drug classification: the correlation of in vitro and in vivo bioavailability. Pharm. Res. 1995;12:413–420.

Benet LZ, Broccatelli F, Oprea TI. BDDCS applied to over 900 drugs. AAPS J. 2011;13:519–547.

Butler JM, Dressman JB. The developability classification system: application of biopharmaceutics concepts to formulation development. J Pharm Sci. 2000;99:4940–54.

Cefalu WT, Bohannon NJ, Fineberg SE, Teeter JG, Krasner AS. Assessment of long-term immunological and pulmonary safety of inhaled human insulin with up to 24 months of continuous therapy. Curr. Med. Res. Opin. 2008;24:3073–3083.

Code of Federal Regulations. 16 CFR 1700. Poison prevention packaging. http://www.access.gpo.gov/nara/cfr/waisidx_03/16cfr1700_03.html. Accessed Dec. 2011.

———. 21 CFR 211.132. Tamper-evident packaging requirements for over-the-counter (OTC) human drug products. http://www.accessdata.fda.gov/scripts/cdrh/cfdocs/cfcfr/CFRSearch.cfm?fr=211.132. Accessed Dec. 2011.

———. 21 CFR 320.33(b). Criteria and evidence to access actual or potential bioequivalence problems. http://www.accessdata.fda.gov/scripts/cdrh/cfdocs/cfcfr/CFRSearch.cfm?fr=320.33. Revised Apr. 1, 2009. Title 21, Vol. 5. Accessed Dec. 2011.

Dahan A, Miller JM, Amidon GL. Prediction of solubility and permeability class membership: BCS classification of the world's top oral drugs. AAPS J. 2009;11:740–746.

Dressman JB, Reppas C. In Vitro–in vivo correlations for lipophilic, poorly water-soluble drugs. Eur. J. Pharm. Sci. 2000;11(Suppl. 2):S73–S80.

FDA. 1995. Guidance for industry: Immediate release solid oral dosage forms: scale-up and postapproval changes; chemistry, manufacturing, and controls, in vitro dissolution testing, and in vivo bioequivalence documentation. U.S. Department of Health and Human Service,s Food and Drug Administration Center for Drug Evaluation and Research. Nov. 1995. http://www.fda.gov/downloads/Drugs/GuidanceComplianceRegulatoryInformation/Guidances/ucm070636.pdf. Accessed Dec. 2011.

———. 1997a. Guidance for industry: Scale-up and postapproval changes—chemistry, manufacturing, and controls; in vitro release testing and in vivo bioequivalence documentation. U.S. Department of Health and Human Services, Food and Drug Administration Center for Drug Evaluation and Research. May 1997. http://www.fda.gov/downloads/Drugs/GuidanceComplianceRegulatoryInformation/Guidances/ucm070930.pdf. Accessed Dec. 2011.

———. 1997b. Guidance for industry: SUPAC-MR—Modified release solid oral dosage forms; scale-up and postapproval changes; chemistry, manufacturing, and controls; in vitro dissolution testing and in vivo bioequivalence documentation. U.S. Department of Health and Human Services, Food and Drug Administration Center for Drug Evaluation and Research. Sept. 1997. http://www.fda.gov/downloads/Drugs/GuidanceComplianceRegulatoryInformation/Guidances/UCM070640pdf. Accessed Dec. 2011.

———. 1998. Guidance for industry: SUPAC-SS—Nonsterile semisolid dosage forms; manufacturing equipment addendum (draft guidance). U.S. Department of Health and Human Services, Food and Drug Administration Center for Drug Evaluation and Research. Dec. 1998. http://www.fda.gov/downloads/Drugs/GuidanceComplianceRegulatoryInformation/Guidances/ucm070928.pdf. Accessed Dec. 2011.

———. 1999a. Guidance for industry: SUPAC-IR/MR—immediate release and modified release solid oral dosage forms; manufacturing equipment addendum. U.S. Department of Health and Human Services, Food and Drug Administration Center for Drug Evaluation and Research. Jan. 1999. http://www.fda.gov/downloads/Drugs/GuidanceComplianceRegulatoryInformation/Guidances/UCM070637.pdf. Accessed Dec. 2011.

———. 1999b. Guidance for industry: Container closure systems for packaging human drugs and biologics. U.S. Department of Health and Human Services, Food and Drug Administration

Center for Drug Evaluation and Research and Center for Biologics Evaluation and Research. May 1999. http://www.fda.gov./downloads/Drugs/GuidanceComplianceRegulatory Information/Guidances/ucm070551.pdf. Accessed Dec. 2011.

———. 2000. Guidance for industry: Waiver of in vivo bioavailability and bioequivalence studies for immediate-release solid oral dosage forms based on a biopharmaceutics classification system. U.S. Department of Health and Human Services, Food and Drug Administration Center for Drug Evaluation and Research. Aug. 2000. http://www.fda.gov/downloads/Drugs/GuidanceComplianceRegulatoryInformation/Guidances/ucm070246.pdf. Accessed Dec. 2011.

———. 2004. Guidance for industry: PAT—a framework for innovative pharmaceutical development, manufacturing, and quality assurance. U.S. Department of Health and Human Services, Food and Drug Administration Center for Drug Evaluation and Research, Center for Veterinary Medicine, Office of Regulatory Affairs, Pharmaceutical CGMPs. Sept. 2004. http://www.fda.gov/downloads/Drug/GuidanceComplianceRegulatoryInformation/Guidances/UCM0-70305.pdf. Accessed Dec. 2011.

———2007. Guidance for industry and review staff: Target product profile—a strategic development process tool. U.S. Department of Health and Human Services, Food and Drug Administration Center for Drug Evaluation and Research. Mar. 2007. http://www.fda.gov/downloads/Drugs/GuidanceComplianceRegulatoryInformation/Guidances/ucm.080593. Accessed Feb. 2011.

———. 2009. Guidance for industry: Q8(R2)—Pharmaceutical development, revision 2. International Conference on Harmonization. U.S. Department of Health and Human Services, Food and Drug Administration Center for Drug Evaluation and Research and Center for Biologics Evaluation and Research. Nov. 2009. http://www.fda.gov/downloads/Drugs/GuidanceComplianceRegulatory Information/Guidances/ucm.073507. Accessed Feb. 2011.

FDA–NICHD. 2011. Intra-agency agreement between the Eunice Kennedy Shriver National Institute of Child Health and Human Development (NICHD) and the U.S. Food and Drug Administation (FDA) oral formulations platform. Report 1. http://bpca.nichd.nih.gov/collaborativeefforts /initiatives/upload/Formulations_Table_for_Web_11–02-11.pdf. Accessed Jan. 2012.

Ferrari S, Griesenback U, Geddes DM, Alton E. Immunologic hurdles to lung gene therapy. Clin. Exp. Immunol. 2003;132:1–8.

Leach CL, Bethke TD, Boudreau RJ, Hasselquist BE, Drollmann A, Davidson P, Wurst W. Two-dimensional and three-dimensional imaging show ciclesonide has high lung deposition and peripheral distribution: a nonrandomized study in healthy volunteers. J. Aerosol Med. 2006;19:117–126.

Lovell MW, Johnson HW, Hui H-W, Cannon JB, Gupta PK, Hsu CC. Less-painful emulsion formulations for intravenous administration of clarithromycin. Int. J. Pharm. 1994;109:45–57.

Pryor FM, Gidal B, Ramsay RE, DeToledo J, Morgan RO. Fosphenytoin: pharmacokinetics and tolerance of intramuscular loading doses. Epilepsia 2001;42:245–250.

Takagi T, Ramachandran C, Bermejo M, Yamashita S, Yu LX, Amidon GL. A provisional biopharmaceutical classification of the top 200 oral drug products in the United States, Great Britain, Spain, and Japan., Mol. Pharm. 2006;3:631–643.

U.S. Pharmacopeial Convention. 2010. Gastric fluid, simulated, TS. U.S.P. 32–N.F. 27. online, http://www.uspnf.com/uspnf/pub/data/v32272/usp32nf27s2_test-solutions.xml. Accessed Jan. 2010.

Wang Y, Mesfin G-M, Rodriguez CA, Slatter JG, Schuette MR, Cory AL, Higgins MJ. Venous irritation, pharmacokinetics, and tissue distribution of tirilazad in rats following intravenous administration of a novel supersaturated submicron lipid emulsion. Pharm Res. 1999;16:930–938.

Wu C-Y, Benet LZ. Predicting drug disposition via application of BCS: transport/absorption/elimination interplay and development of a biopharmaceutics drug disposition classification system. Pharm. Res. 2005;22:11–23. doi:10.1007/s11095-004-9004-4.

GLOSSARY

BCS	Biopharmaceutics classification system.
BDDCS	Biopharmaceutics drug disposition classification system.

CCS	Container closure systems.
CFR	*Code of Federal Regulations*.
CPSC	Consumer Product Safety Commission.
DCS	Developability classification system.
EC	Effective concentration.
ED	Effective dose.
FDA	U.S. Food and Drug Administration.
ICH	International Conference on Harmonization.
IM	Intramuscular.
IR	Immediate release.
LD	Lethal dose.
MR	Modified release.
NDA	New drug application.
N.F.	*National Formulary*.
PAT	Process analytical technology.
QbD	Quality by design.
SC	Subcutaneous.
SUPAC	Scale-up and postapproval changes.
U.S.P.	*United States Pharmacopeia*.

APPENDIXES

APPENDIX 12.1   FDA Route of Administration Nomenclature

Term	Definition
Auricular (otic)	Administration to or by way of the ear.
Buccal	Administration directed toward the cheek, generally from within the mouth.
Conjunctival	Administration to the conjunctiva, the delicate membrane that lines the eyelids and covers the exposed surface of the eyeball.
Cutaneous	Administration to the skin.
Dental	Administration to a tooth or teeth.
Electroosmosis	Administration of through the diffusion of substance through a membrane in an electric field.
Endocervical	Administration within the canal of the cervix uteri; synonymous with the term *intracervical*.
Endosinusial	Administration within the nasal sinuses of the head.
Endotracheal	Administration directly into the trachea.
Enteral	Administration directly into the intestines.
Epidural	Administration upon or over the dura mater.
Extraamniotic	Administration to the outside of the membrane enveloping the fetus.
Extracorporeal	Administration outside the body.
Hemodialysis	Administration through hemodialysate fluid.
Infiltration	Administration that results in substances passing into tissue spaces or into cells.
Interstitial	Administration to or in the interstices of a tissue.
Intraabdominal	Administration within the abdomen.
Intraamniotic	Administration within the amnion.
Intraarterial	Administration within an artery or arteries.
Intrabronchial	Administration within a bronchus.

APPENDIX 12.1 (*Continued*)

Term	Definition
Intrabursal	Administration within a bursa.
Intracardiac	Administration with the heart.
Intracartilaginous	Administration within a cartilage; endochondral.
Intracaudal	Administration within the cauda equina.
Intracavernous	Administration within a pathologic cavity, such as occurs in the lung in tuberculosis.
Intracavitary	Administration within a nonpathologic cavity, such as that of the cervix, uterus, or penis, or such as that which is formed as the result of a wound.
Intracerebral	Administration within the cerebrum.
Intracisternal	Administration within the cisterna magna cerebellomedularis.
Intracorneal	Administration within the cornea (the transparent structure forming the anterior part of the fibrous tunic of the eye).
Intracoronal, dental	Administration of a drug within a portion of a tooth which is covered by enamel and which is separated from the roots by a slightly constricted region known as the neck.
Intracoronary	Administration within the coronary arteries.
Intracorporous cavernosum	Administration within the dilatable spaces of the corporus cavernosa of the penis.
Intradermal	Administration within the dermis.
Intradiscal	Administration within a disc.
Intraductal	Administration within the duct of a gland.
Intraduodenal	Administration within the duodenum.
Intradural	Administration within or beneath the dura.
Intraepidermal	Administration within the epidermis.
Intraesophageal	Administration within the esophagus.
Intragastric	Administration within the stomach.
Intragingival	Administration within the gingivae.
Intraileal	Administration within the distal portion of the small intestine, from the jejunum to the cecum.
Intralesional	Administration within or introduced directly into a localized lesion.
Intraluminal	Administration within the lumen of a tube.
Intralympanic	Administration within the aurus media.
Intralymphatic	Administration within the lymph.
Intramedullary	Administration within the marrow cavity of a bone.
Intrameningeal	Administration within the meninges (the three membranes that envelope the brain and spinal cord).
Intramuscular	Administration within a muscle.
Intraocular	Administration within the eye.
Intraovarian	Administration within the ovary.
Intrapericardial	Administration within the pericardium.
Intraperitoneal	Administration within the peritoneal cavity.
Intrapleural	Administration within the pleura.
Intraprostatic	Administration within the prostate gland.
Intrapulmonary	Administration within a lung or its bronchi.
Intrasinal	Administration within the nasal or periorbital sinuses.
Intraspinal	Administration within the vertebral column.

(*continued*)

APPENDIX 12.1 (*Continued*)

Term	Definition
Intrasynovial	Administration within the synovial cavity of a joint.
Intratendinous	Administration within a tendon.
Intratesticular	Administration within a testicle.
Intrathecal	Administration within the cerebrospinal fluid at any level of the cerebrospinal axis, including injection into the cerebral ventricles.
Intrathoracic	Administration within the thorax (internal to the ribs); synonymous with the term endothoracic.
Intratubular	Administration within the tubules of an organ.
Intratumor	Administration within a tumor.
Intrauterine	Administration within the uterus.
Intravascular	Administration within a vessel or vessels.
Intravenous	Administration within or into a vein or veins.
Intravenous bolus	Administration within or into a vein or veins all at once.
Intravenous drip	Administration within or into a vein or veins over a sustained period of time.
Intraventicular	Administration within a ventricle.
Intravesical	Administration within the bladder.
Intravitreal	Administration within the vitreous body of the eye.
Iontophoresis	Administration by means of an electric current where ions of soluble salts migrate into the tissues of the body.
Irrigation	Administration to bathe or flush open wounds or body cavities.
Laryngeal	Administration directly upon the larynx.
Nasal	Administration to the nose; administered by way of the nose.
Nasogastric	Administration through the nose and into the stomach, usually by means of a tube.
Not applicable	Routes of administration are not applicable.
Occlusive dressing technique	Administration by the topical route, which is then covered by a dressing that occludes the area.
Ophthalmic	Administration to the external eye.
Oral	Administration to or by way of the mouth.
Oropharyngeal	Administration directly to the mouth and pharynx.
Other	Administration is different from others on this list.
Parentral	Administration by injection, infusion, or implantation.
Percutaneous	Administration through the skin.
Periarticular	Administration around a joint.
Peridontal	Administration around a tooth.
Peridural	Administration to the outside of the dura mater of the spinal cord.
Perineural	Administration surrounding a nerve or nerves.
Rectal	Administration to the rectum.
Respiratory (inhalation)	Administration within the respiratory tract by inhaling orally or nasally for local or systemic effect.
Retrobulbar	Administration behind the pons or behind the eyeball.
Soft tissue	Administration into any soft tissue.
Subarachnoid	Administration beneath the arachnoid.
Subconjuctival	Administration beneath the conjunctiva.

APPENDIX 12.1 (*Continued*)

Term	Definition
Subcutaneous	Administration beneath the skin; hypodermic; synonymous with the term *subdermal*.
Sublingual	Administration beneath the tongue.
Submucosal	Administration beneath the mucous membrane.
Topical	Administration to a particular spot on the outer surface of the body.
Transdermal	Administration through the dermal layer of the skin to the systemic circulation by diffusion.
Transmucosal	Administration across the mucosa.
Transplacental	Administration through or across the placenta.
Transtracheal	Administration through the wall of the trachea.
Transtympanic	Administration across or through the tympanic cavity.
Unassigned	Route of administration has not yet been assigned.
Unknown	Route of administration is unknown.
Ureteral	Administration into the ureter.
Urethral	Administration into the urethra.
Vaginal	Administration into the vagina.
Bead	A solid dosage form in the shape of a small ball.

Source: After http://www.fda.gov/ForIndustry/DataStandards/StructuredProductLabeling/ucm162034.htm. Accessed Feb. 2010.

APPENDIX 12.2   FDA Drug Dosage Form Nomenclature

Term	Definition
Intrabiliary	Administration within the bile, bile ducts, or gallbladder.
Aerosol	A product that is packaged under pressure and contains therapeutically active ingredients that are released upon activation of an appropriate valve system; intended for topical application to the skin as well as local application into the nose (nasal aerosols), mouth (lingual aerosols), or lungs (inhalation aerosols).
Foam	A dosage form containing one or more active ingredients, surfactants, aqueous or nonaqueous liquids, and propellants; if the propellant is in the internal (discontinuous) phase (i.e., of the oil-in-water type), a stable foam is discharged, and if the propellant is in the external (continuous) phase (i.e., of the water-in-oil type), a spray or a quick-breaking foam is discharged.
Metered	A pressurized dosage form consisting of metered dose valves which allow for the delivery of a uniform quantity of spray upon each activation.
Powder	A product that is packaged under pressure and contains therapeutically active ingredients, in the form of a powder, that are released upon activation of an appropriate valve system.
Spray	An aerosol product that utilizes a compressed gas as a propellant to provide the force necessary to expel a product as a wet spray; it is applicable to solutions of medicinal agents in aqueous solvents.
Bar, chewable	A solid dosage form, usually in the form of a rectangle, that is meant to be chewed.
Implant, extended release	A small sterile solid mass consisting of a highly purified drug intended for implantation in the body, which would allow at least a reduction in dosing frequency compared to that drug presented as a conventional dosage form.
Block	Solid dosage form, usually in the shape of a square or rectangle.

(*continued*)

APPENDIX 12.2 (*Continued*)

Term	Definition
Capsule	A solid oral dosage form consisting of a shell and a filling. The shell is composed of a single sealed enclosure or of two halves that fit together and are sometimes sealed with a band; capsule shells may be made from gelatin, starch, or cellulose or other suitable materials, may be soft or hard, and are filled with solid or liquid ingredients that can be poured or squeezed.
Coated	A solid dosage form in which a drug is enclosed within either a hard or a soft soluble container or shell made from a suitable form of gelatin; additionally, the capsule is covered in a designated coating.
Coated pellets	A solid dosage form in which a drug is enclosed within either a hard or a soft soluble container or shell made from a suitable form of gelatin; the drug itself is in the form of granules to which varying amounts of coating have been applied.
Coated Extended release	A solid dosage form in which a drug is enclosed within either a hard or a soft soluble container or shell made from a suitable form of gelatin; additionally, the capsule is covered in a designated coating and releases a drug (or drugs) in such a manner as to allow at least a reduction in dosing frequency compared to that drug (or drugs) presented as a conventional dosage form.
Delayed release	A solid dosage form in which a drug is enclosed within either a hard or a soft soluble container made from a suitable form of gelatin, which releases a drug (or drugs) at a time other than promptly after administration. Enteric-coated articles are delayed-release dosage forms.
Pellets	A solid dosage form in which a drug is enclosed within either a hard or a soft soluble container or shell made from a suitable form of gelatin; the drug itself is in the form of granules to which enteric coating has been applied, thus delaying release of the drug until its passage into the intestines.
Extended release	A solid dosage form in which a drug is enclosed within either a hard or a soft soluble container made from a suitable form of gelatin, which releases the drug (or drugs) in such a manner as to allow a reduction in dosing frequency as compared to that drug (or drugs) presented as a conventional dosage form.
Film coated, extended release	A solid dosage form in which a drug is enclosed within either a hard or soft soluble container or shell made from a suitable form of gelatin; additionally, the capsule is covered in a designated film coating and releases a drug (or drugs) in such a manner as to allow at least a reduction in dosing frequency compared to that drug (or drugs) presented as a conventional dosage form.
Gelatin coated	A solid dosage form in which a drug is enclosed within either a hard or a soft soluble container made from a suitable form of gelatin; through a banding process, the capsule is coated with additional layers of gelatin so as to form a complete seal.
Liquid filled	A solid dosage form in which a drug is enclosed within a soluble gelatin shell which is plasticized by the addition of a polyol, such as sorbitol or glycerin, and is therefore somewhat thicker in consistency than a hard shell capsule; typically, the active ingredients are dissolved or suspended in a liquid vehicle.
Cement	A substance that serves to produce a solid union between two surfaces.
Cigarette	A narrow tube of cut tobacco (or similar material) enclosed in paper and designed for smoking.
Cloth	A large piece of relatively flat absorbent material that contains a drug. It is typically used for applying medication or for cleansing.
Concentrate	A liquid preparation of increased strength and reduced volume which is usually diluted prior to administration.

APPENDIX 12.2 (*Continued*)

Term	Definition
Cone	A solid dosage form bounded by a circular with the surface formed by line segments joining every point of the boundary of the base to a common vertex; a cone (usually containing antibiotics) is normally placed below the gingiva after a dental extraction.
Core, extended release	An ocular system placed in the eye from which the drug diffuses through a membrane at a constant rate over a specified period.
Cream	An emulsion semisolid dosage form, usually containing >20% water and volatiles and/or <50% hydrocarbons, waxes, or polyols as the vehicle; generally for external application to the skin or mucous membrane.
Augmented	A cream dosage form that enhances drug delivery; augmentation does not refer to the strength of the drug in dosage form. *Note*: CDER has decided to refrain from expanding the use of this dosage form, due to difficulties in setting specific criteria that must be met to be "augmented."
Crystal	A naturally produced angular solid of definite form in which the ultimate units from which it is built up are arranged systematically; they are usually evenly spaced on a regular space lattice.
Culture	The propagation of microorganisms or of living tissue cells in special media conductive to their growth.
Diaphragm	A device, usually dome-shaped, worn over the cervical mouth during copulation, for prevention of conception or infection.
Disk	A circular platelike organ or structure.
Douche	A liquid preparation, intended for the irrigative cleansing of the vagina, that is prepared from powders, liquid solutions, or liquid concentrates and contains one or more chemical substances dissolved in a suitable solvent or mutually miscible solvents.
Dressing	The application of various materials for protecting a wound.
Drug delivery system	Modern technology, distributed with or as a part of a drug product that allows for the uniform release or targeting of drugs to the body.
Elixir	A clear, pleasantly flavored, sweetened hydroalcoholic liquid containing dissolved medicinal agents; it is intended for oral use.
Emulsion	A dosage form consisting of a two-phase system comprised of at least two immiscible liquids, one of which is dispersed as droplets (internal or dispersed phase) within the other liquid (external or continuous phase), generally stabilized with one or more emulsifying agents. *Note*: Emulsion is used as a dosage-form term unless a more specific term is applicable (e.g., cream, lotion, ointment).
Enema	A rectal preparation for therapeutic, diagnostic, or nutritive purposes.
Extract	A concentrated preparation of vegetable or animal drugs obtained by removal of the active constituents of the respective drugs with a suitable menstrua, evaporation of all or nearly all of the solvent, and adjustment of the residual masses or powders to the prescribed standards.
Fiber, extended release	A slender and elongated solid threadlike substance that delivers a drug in such a manner as to allow a reduction in dosing frequency compared to that drug (or drugs) presented as a conventional dosage form.
Film	A thin layer or coating.
Extended release	A drug delivery system in a form of a film that releases a drug over an extended period in such a way as to maintain constant drug levels in the blood or target tissue.

(*continued*)

APPENDIX 12.2   (*Continued*)

Term	Definition
Soluble	A thin layer or coating that is susceptible to being dissolved when in contact with a liquid.
For solution	A product, usually a solid, intended for solution prior to administration.
Extended release	A product, usually a solid, intended for suspension prior to administration; once the suspension is administered, the drug will be released at a constant rate over a specified period.
For suspension	A product, usually a solid, intended for suspension prior to administration.
For suspension extended release	A product, usually a solid, intended for suspension prior to administration; once the suspension is administered, the drug will be released at a constant rate over a specified period.
Gas	Any elastic aeriform fluid in which the molecules are separated from one another and so have free paths.
Gel	A semisolid dosage form that contains a gelling agent to provide stiffness to a solution or a colloidal dispersion; a gel may contain suspended particles.
Dentifrice	A combination of a dentifrice (formulation intended to clean and/or polish the teeth, which may contain certain additional agents) and a gel. It is used with a toothbrush for the purpose of cleaning and polishing the teeth.
Metered	A gel preparation, with metered dose valves that allow for the delivery of a uniform quantity of gel upon each activation.
Generator	An apparatus for the formation of vapor or gas from a liquid or solid by heat or chemical action; the term also applies to radioactive columns from which radionuclides are provided.
Globule	Also called pellets or pilules, globules are made of pure sucrose, lactose, or other polysaccharides. They are formed into small globular masses of various sizes, and are medicated by placing them in a vial and adding the liquid drug attenuation in the proportion not less than 1% v/w. After shaking, the medicated globules are dried at temperatures not to exceed 40 degrees Celsius.
Graft	A slip of skin or other tissue for implantation.
Granule	A small particle or grain.
Delayed release	A small medicinal particle or grain to which an enteric or other coating has been applied, thus delaying release of a drug until its passage into the intestines.
Effervescent	A small particle or grain containing a medicinal agent in a dry mixture usually composed of sodium bicarbonate, citric acid, and tartaric acid, which when in contact with water has the ability to release gas, resulting in effervescence.
For solution	A small medicinal particle or grain made available in its more stable dry form, to be reconstituted with solvent just before being dispensed; the granules are prepared to contain not only the medicinal agent but also the colorants, flavorants, and any other desired pharmaceutical ingredient.
For suspension	A small medicinal particle or grain made available in its more stable dry form, to be reconstituted with solvent just before being dispensed to form a suspension; the granules are prepared to contain not only the medicinal agent but also the colorants, flavorants, and any other desired pharmaceutical ingredient.
For suspension extended release	A small medicinal particle or grain made available in its more stable dry form, to be reconstituted with solvent just before being dispensed to form a suspension; the extended-release system achieves slow release of the drug over an extended period of time and maintains constant drug levels in the blood or target tissue.

APPENDIX 12.2 (*Continued*)

Term	Definition
Gum	A mucilaginous excretion from various plants.
Chewing	A sweetened and flavored insoluble plastic material of various shapes which when chewed releases a drug substance onto the oral cavity.
Resin	Natural mixture of gum and resin, usually obtained as exudations from plants.
Implant	A material containing a drug intended to be inserted securely or deeply into a living site for growth, slow release, or formation of an organic union.
Inhalant	A special class of inhalations consisting of a drug or combination of drugs that by virtue of their high vapor pressure can be carried by an air current into the nasal passage, where they exert their effect; the container from which the inhalant generally is administered is known as an inhaler.
Injection	A sterile preparation intended for parenteral use.
Injection, emulsion	An emulsion consisting of a sterile, pyrogen-free preparation intended to be administered parenterally.
Injection, lipid complex	Definition pending. (The FDA is in the process of providing a definition and this is taken verbatim from the FDA website.)
Injectable, liposomal	An injection, which either consists of or forms liposomes (a lipid bilayer vesicle usually composed of phospholipids which is used to encapsulate an active drug substance).
Injection, powder for solution	A sterile preparation intended for reconstitution to form a solution for parenteral use.
Injection, powder for suspension	A sterile preparation intended for reconstitution to form a suspension for parenteral use.
Injection, powder for suspension, extended release	A dried preparation intended for reconstitution to form a suspension for parenteral use which has been formulated in a manner to allow at least a reduction in dosing frequency as compared to that drug presented as a conventional dosage form (e.g., as a solution).
Injection, powder, lyophilized for liposomal suspension	A sterile freeze dried preparation intended for reconstitution for parenteral use which has been formulated in a manner that would allow liposomes (a lipid bilayer vesicle usually composed of phospholipids which is used to encapsulate an active drug substance, either within a lipid bilayer or in an aqueous space) to be formed upon reconstitution.
Injection, powder, lyophilized, for solution	A dosage form intended for the solution prepared by lyophilization ("freeze drying"), a process which involves the removal of water from the products in the frozen state at extremely low pressures; this is intended for subsequent addition of liquid to create a solution that conforms in all respects to the requirements for Injections.
Injection, powder, lyophilized, for suspension	A liquid preparation, intended for parenteral use that contains solids suspended in a suitable fluid medium and conforms in all respects to the requirements for sterile suspensions; the medicinal agents intended for the suspension are prepared by lyophilization ("freeze drying"), a process which involves the removal of water from products in the frozen state at extremely low pressures.
Injection, powder, lyophilized, for suspension, extended release	A sterile freeze dried preparation intended for reconstitution for parenteral use which has been formulated in a manner to allow at least reduction a in dosing frequency as compared to that drug presented as a conventional dosage form (e.g., as a solution).

(*continued*)

APPENDIX 12.2   (*Continued*)

Term	Definition
Injection, solution	A liquid preparation containing one or more drug substances dissolved in a suitable solvent or mixture of mutually miscible solvents that is suitable for injection.
Injection, solution, concentrate	A sterile preparation for parenteral use which, upon the addition of suitable solvents, yields a solution conforming in all respects to the requirements for Injections.
Injection, suspension	A liquid preparation suitable for injection, which consists of solid particles dispersed throughout a liquid phase in which the particles are not soluble. It can also consist of an oil phase dispersed throughout an aqueous phase, or vice-versa.
Injection, suspension, extended release	A sterile preparation intended for parenteral use which has been formulated in a manner to allow at least a reduction in dosing frequency as compared to that drug presented as a conventional dosage form (e.g., as a solution or a prompt drug-releasing, conventional solid dosage form).
Injection, suspension, liposomal	A liquid preparation suitable for injection, which consists of an oil phase dispersed throughout an aqueous phase in such a manner that liposomes (a lipid bilayer vesicle usually composed of phospholipids which is used to encapsulate an active drug substance, either within a lipid bilayer or in an aqueous space) are formed.
Injection, suspension, sonicated	A liquid preparation, suitable for injection, which consists of solid particles dispersed throughout a liquid phase in which the particles are not soluble. In addition, the product is sonicated while a gas is bubbled through the suspension, and results in the formation of microspheres by the solid particles.
Insert	A specially formulated and shaped nonencapsulated solid preparation intended to be placed into a nonrectal orifice of the body, where the medication is released, generally for localized effects.
Extended release	A specially formulated and shaped nonencapsulated solid preparation intended to be placed into a nonrectal orifice of the body, where a drug is released, generally for localized effects; the extended-release preparation is designed to allow for a reduction in dosing frequency.
Intrauterine device	A device inserted in the uterus and left to prevent effective conception.
Irrigant	A sterile solution intended to bathe or flush open wounds or body cavities; they are used topically, never parenterally.
Jelly	A class of gels, which are semisolid systems that consist of suspensions made up of either small inorganic particles or large organic molecules interpenetrated by a liquid—in which the structural coherent matrix contains a high portion of liquid, usually water.
Kit	A packaged collection of related material.
Liner, dental	A material applied to the inside of the dental cavity for protection or insulation of the surface.
Liniment	A solution or mixture of various substances in oil, alcoholic solutions of soap, or emulsions intended for external application.
Lipstick	A waxy solid, usually colored cosmetic, in stick form for the lips.
Liquid	A dosage form consisting of a pure chemical in its liquid state. This dosage-form term should not be applied to solutions.

APPENDIX 12.2   (*Continued*)

Term	Definition
Extended release	A liquid that delivers a drug so as to allow a reduction in dosing frequency compared to that drug (or drugs) presented as a conventional dosage form.
Lotion	An emulsion, a liquid dosage form generally used in external applications to the skin.
Augmented	A lotion dosage form that enhances delivery. Augmentation does not refer to the strength of the drug in the dosage form. *Note:* CDER has decided to refrain from expanding the use of this dosage form, due to difficulties in setting specific criteria that must be met to be considered "augmented."
Lotion/shampoo	A lotion dosage form which has a soap or detergent that is generally used to clean the hair and scalp; it is often used as a vehicle for dermatological agents.
Lozenge	A solid preparation containing one or more medicaments, usually in a flavored, sweetened base, which is intended to dissolve or disintegrate slowly in the mouth. A lollipop is a lozenge on a stick.
Mouthwash	An aqueous solution most often used for its deodorant, refreshing, or antiseptic effect.
Not applicable	The use of a dosage-form term is not relevant or appropriate.
Oil	An unctuous, combustible substance that is liquid, or easily liquefiable, on warming, and is soluble in ether but insoluble in water. Such substances, depending on their origin, are classified as animal, mineral, or vegetable oils.
Ointment	A semisolid dosage form, usually containing <20% water and volatiles and >50% hydrocarbons, waxes, or polyols as the vehicle. This dosage form is generally for external application to the skin or mucous membranes.
Augmented	An ointment dosage form that enhances drug delivery. Augmentation does not refer to the strength of the drug in the dosage form. *Note:* CDER has decided to refrain from expanding the use of this dosage form due to difficulties in setting specific criteria that must be met to be considered "augmented."
Packing	A material usually covered by or impregnated with a drug that is inserted into a body cavity or between the tooth enamel and the gingival margin.
Paste	A semisolid dosage form containing a large proportion (20 to 50%) of solids finely dispersed in a fatty vehicle. This dosage form is generally for external application to the skin or mucous membranes.
Dentifrice	A paste formulation intended to clean and/or polish the teeth, which may contain certain additional agents.
Pastille	An aromatic preparation, often with a pleasing flavor, usually intended to dissolve in the mouth.
Patch	A drug delivery system that often contains an adhesive backing that is usually applied to an external site on the body. Its ingredients either passively diffuse from, or are actively transported from, some portion of the patch. Depending on the patch, the ingredients are delivered either to the outer surface of the body or into the body. A patch is sometimes synonymous with the terms *extended-release film* and *system*.
Extended release	A drug delivery system in the form of a patch that releases a drug in such that there is a reduction in the dosing frequency compared to the drug presented as a conventional dosage form (e.g., a solution or a prompt drug-releasing conventional solid dosage form).

(*continued*)

APPENDIX 12.2   (*Continued*)

Term	Definition
Extended release, electronically controlled	A drug delivery system in the form of a patch which is controlled by an electric current that releases the drug in such a manner that a reduction in dosing frequency compared to that drug presented as a conventional dosage form (e.g., a solution or a prompt drug-releasing conventional solid dosage form).
Pellet	A small sterile solid mass consisting of a highly purified drug (with or without excipients) made by the formation of granules, or by compression and molding.
Coated, extended release	A solid dosage form in which the drug itself is in the form of granules to which varying amounts of coating have been applied, and which releases a drug (or drugs) so as to allow a reduction in dosing frequency compared to that drug (or drugs) presented as a conventional dosage form.
Pill	A small, round solid dosage form containing a medicinal agent intended for oral administration.
Plaster	A substance intended for external application, made of such materials and of such consistency as to adhere to the skin and attach to a dressing. Plaster is intended to afford protection and support and/or furnish an occlusion and macerating action and to bring medication into close contact with the skin.
Poultice	A soft, moist mass of meal, herbs, seed, etc., usually applied hot, in cloth; of gruel-like consistency.
Powder	An intimate mixture of dry, finely divided drugs and/or chemicals that may be intended for internal or external use.
Dentrifice	A powder formulation intended to clean and/or polish the teeth, which may contain certain additional agents.
For solution	An intimate mixture of dry, finely divided drugs and/or chemicals, which upon the addition of suitable vehicles yields a solution.
For suspension	An intimate mixture of dry, finely divided drugs and/or chemicals, which upon the addition of suitable vehicles, yields a suspension (a liquid preparation containing the solid particles dispersed in a liquid vehicle).
Metered	A powder dosage form that is suitable inside a container that has a mechanism to deliver a specified quantity.
Ring	A small circular object with a vacant circular center that is usually intended to be placed in the body by special inserters, where the medication is released, generally for localized effects.
Rinse	A liquid used to cleanse by flushing.
Salve	A thick ointment or cerate (a fat-or wax-based preparation with a consistency between that of an ointment and that of plaster).
Shampoo	A liquid soap or detergent used to clean the hair and scalp; often used as a vehicle for dermatological agents.
Suspension	A liquid soap or detergent containing one or more solid, insoluble substances dispersed in a liquid vehicle that is used to clean the hair and scalp and is often used as a vehicle for dermatological agents.
Soap	Any compound of one or more fatty acids, or their equivalents, with an alkali; soap is a detergent and is widely employed in liniments, enemas, and in making pills. It is also a mild aperient, antacid, and antiseptic.
Solution	A clear homogeneous liquid dosage form that contains one or more chemical substances dissolved in a solvent or mixture of mutually miscible solvents.
Concentrate	A liquid preparation (i.e., a substance that flows readily in its natural state) that contains a drug dissolved in a suitable solvent or mixture of mutually miscible solvents; the drug has been strengthened by evaporation of its nonactive parts.

APPENDIX 12.2   (*Continued*)

Term	Definition
For slush	A solution for the preparation of an iced saline slush, which is administered by irrigation and used to induce regional hypothermia (in conditions such as certain open heart and kidney surgical procedures) by direct application.
Gel forming/drops	A solution which, after (usually) administered in a dropwise fashion, forms a gel.
Gel forming, extended release	A solution that forms a gel when it comes in contact with ocular fluid and which allows at least a reduction in dosing frequency.
Solution/drops	A solution that is usually administered in dropwise fashion.
Sponge	A porous, interlacing absorbent material that contains a drug; typically used for applying or introducing medication, or for cleansing. A sponge usually retains its shape.
Spray	A liquid minutely divided as by a jet of air or steam.
Metered	A nonpressurized dosage form consisting of valves that allow the dispensing of a specified quantity of spray upon each activation.
Suspension	A liquid preparation containing solid particles dispersed in a liquid vehicle and in the form of coarse droplets or as finely divided solids to be applied locally, usually to the nasal–pharyngeal tract or topically to the skin.
Stick	A dosage form prepared in a relatively long and slender, often cylindrical form.
Strip	A long narrow piece of material.
Suppository	A solid body of various weights and shapes, adapted for introduction into the rectal orifice of the human body; suppositories usually melt, soften, or dissolve at body temperature.
Extended release	A drug delivery system in the form of a suppository that allows for a reduction in dosing frequency.
Suspension	A liquid dosage form that contains solid particles dispersed in a liquid vehicle.
Extended release	A liquid preparation consisting of solid particles dispersed throughout a liquid phase in which the particles are not soluble; the suspension has been formulated to allow at least a reduction in dosing frequency compared to that drug presented as a conventional dosage form (e.g., as a solution or a prompt drug-releasing, conventional solid dosage form).
Suspension drops	A suspension that is usually administered in dropwise fashion.
Suture	A strand or fiber used to hold wound edges in apposition during healing.
Swab	A small piece of relatively flat absorbent material that contains a drug. A swab may also be attached to one end of a small stick. A swab is typically used for applying medication or for cleansing.
Syrup	An oral solution containing high concentrations of sucrose or other sugars; the term has also been used to include any other liquid dosage form prepared in a sweet and viscid vehicle, including oral suspensions.
Tablet	A solid dosage form containing medicinal substances with or without suitable diluents.
Chewable	A solid dosage form containing medicinal substances with or without suitable diluents that is intended to be chewed, producing a pleasant-tasting residue in the oral cavity that is easily swallowed and does not leave a bitter or unpleasant aftertaste.
Coated	A solid dosage form that contains medicinal substances with or without suitable diluents and is covered with a designated coating.
Coated particles	A solid dosage form containing a conglomerate of medicinal particles that have each been covered by a coating.

(continued)

APPENDIX 12.2 (*Continued*)

Term	Definition
Delayed release	A solid dosage form that releases a drug (or drugs) at a time other than promptly after administration; enteric-coated articles are delayed-release dosage forms.
Delayed-release particles	A solid dosage form containing a conglomerate of medicinal particles that have been covered with a coating that releases a drug (or drugs) at a time other than promptly after administration. Enteric-coated articles are delayed-release dosage forms.
Dispersible	A tablet that, prior to administration, is intended to be placed in liquid, where its contents will be distributed evenly throughout the liquid. *Note:* The term is no longer used for approved drug products; it has been replaced by the term *tablet, for suspension.*
Effervescent	A solid dosage form containing mixtures of acids (e.g., citric acid, tartaric acid) and sodium bicarbonate, which release carbon dioxide when dissolved in water; it is intended to be dissolved or dispersed in water before administration.
Extended release	A solid dosage form containing a drug that allows at least a reduction in dosing frequency compared to that drug presented in conventional dosage form.
Film coated	A solid dosage form that contains medicinal substances with or without suitable diluents and is coated with a thin layer of a water-insoluble or water-soluble polymer.
Film-coated extended release	A solid dosage form that contains medicinal substances with or without suitable diluents and is coated with a thin layer of a water-insoluble or water-soluble polymer; the tablet is formulated such as to make the medicament contained available over an extended period of time following ingestion.
For solution	A tablet that forms a solution when placed in a liquid.
For suspension	A tablet that forms a suspension when placed in a liquid (formerly referred to as a "dispersible tablet").
Multilayer	A solid dosage form containing medicinal substances that have been compressed to form a multiple-layered tablet or a tablet-within-a-tablet, the inner tablet being the core and the outer portion being the shell.
Multilayer, extended release	A solid dosage form containing medicinal substances that have been compressed to form a multiple-layered tablet or a tablet-within-a-tablet, the inner tablet being the core and the outer portion being the shell, which, additionally, is covered by a designated coating; the tablet is formulated such as to allow at least a reduction in dosing frequency compared to that drug presented as a conventional dosage form.
Orally disintegrating	A solid dosage form containing medicinal substances which disintegrates rapidly, usually within a matter of seconds, when placed on the tongue.
Orally disintegrating, delayed release	A solid dosage form containing medicinal substances which disintegrates rapidly, usually within a matter of seconds, when placed on the tongue, but which releases a drug (or drugs) at a time other than promptly after administration.
Soluble	A solid dosage form that contains medicinal substances with or without suitable diluents and possesses the ability to dissolve in fluids.
Sugar coated	A solid dosage form that contains medicinal substances with or without suitable diluents and is coated with a colored or uncolored water-soluble sugar.
Tampon	A plug made of cotton, sponge, or oakum used variously in surgery to plug the nose, vagina, etc., for the control of hemorrhage or the absorption of secretions.

APPENDIX 12.2 (*Continued*)

Term	Definition
Tape	A narrow woven fabric, or a narrow extruded synthetic (such as plastic), usually with an adhesive on one or both sides.
Tincture	An alcoholic or hydroalcoholic solution prepared from vegetable materials or from chemical substances.
Troche	A discoid-shaped solid containing a medicinal agent in a suitably flavored base; troches are placed in the mouth, where they slowly dissolve, liberating the active ingredients.
Unassigned	A dosage form has yet to be assigned.
Wafer	A thin slice of material containing a medicinal agent.

Source: After http://www.fda.gov/Drugs/DevelopmentApprovalProcess/FormsSubmissionRequirements/Electronic Submissions/DataStandardsManualmonographs/ucm071666.htm. Accessed Feb. 2010.

APPENDIX 12.3   U.S.P. Dosage Form Nomenclature

Term	Definition
Aerosols	Products that are packaged under pressure and contain therapeutically active ingredients that are released upon activation of an appropriate valve system.
Boluses	Large elongated tablets intended for administration to animals.
Capsules	Solid dosage forms in which a drug is enclosed within either a hard or soft soluble container or shell.
Concentrate for dip	A preparation containing one or more active ingredients, usually in the form of a paste or solution.
Creams	Semisolid dosage forms containing one or more drug substances dissolved or dispersed in a suitable base.
Emulsions	Two-phase systems in which one liquid is dispersed throughout another liquid in the form of small droplets.
Extracts and fluid extracts	Concentrated preparations of vegetable or animal drugs obtained by removal of the active constituents of the respective drugs with suitable menstrua, by evaporation of all or nearly all of the solvent, and by adjustment of the residual masses or powders to the prescribed standards.
Gels	Semisolid systems consisting of suspensions made up either of small inorganic particles or large organic molecules interpenetrated by a liquid; sometimes called jellies.
Implants (pellets)	Small sterile solid masses consisting of a highly purified drug (with or without excipients) made by compression or molding.
Infusions, intra-mammary	Suspensions of drugs in suitable oil vehicles.
Inhalations	Drugs or solutions or suspensions of one or more drug substances administered by the nasal or oral respiratory route for local or systemic effect.
Injections	A preparation intended for parenteral administration or for constituting or diluting a parenteral article prior to administration.
Irrigations	Sterile solutions intended to bathe or flush open wounds or body cavities.
Lozenges	Solid preparations intended to dissolve or disintegrate slowly in the mouth.
Ointments	Semisolid preparations intended for external application to the skin or mucous membranes.

(*continued*)

APPENDIX 12.3 (*Continued*)

Term	Definition
Absorption bases	May be divided into two groups: the first group consisting of bases that permit the incorporation of aqueous solutions with the formation of water-in-oil emulsions, and the second consisting of water-in-oil emulsions that permit the incorporation of additional quantities of aqueous solutions.
Hydrocarbon bases	These bases, also known as *oleaginous ointment bases*, are represented by white petroleum and white ointment.
Water-removable bases	Oil-in-water emulsions (e.g., hydrophilic ointment), are more correctly called "creams."
Water-soluble bases	This group of so-called "greaseless ointment bases" comprises water-soluble constituents.
Ophthalmic preparations	Drugs are administered to the eyes in a wide variety of dosage forms, some of which require special consideration.
Ointments	Sterile ointments for application to the eye.
Solutions	Sterile solutions, essentially free of foreign particles, suitably compounded and packaged for instillation into the eye.
Suspensions	Sterile liquid preparations containing solid particles dispersed in a liquid vehicle intended for application to the eye.
Pastes	Semisolid dosage forms that contain one or more drug substances intended for topical application.
Powders	Intimate mixtures of dry, finely divided drugs and/or chemicals that may be intended for internal (oral powders) or external (topical powders) use.
Premixes	Mixtures of one or more drug substances with suitable vehicles.
Solutions	Liquid preparations that contain one or more chemical substances dissolved (i.e., molecularly dispersed) in a suitable solvent or mixture of mutually miscible solvents.
Oral solutions	Liquid preparations, intended for oral administration, that contain one or more substances with or without flavoring, sweetening, or coloring agents dissolved in water or cosolvent–water mixtures.
Syrups	Oral solutions containing high concentrations (near saturation) of sucrose, other sugars, and polyols, such as sorbitol and glycerin. The definition has been extended to include a liquid dosage form prepared in a sweet and viscid vehicle that includes oral suspensions.
Elixirs	Traditionally contain alcohol or cosolvent, such as glycerin and propylene glycol.
Otic solutions	Intended for instillation in the outer ear; are more aqueous, or they are solutions prepared with glycerin or other solvents and dispersing agents.
Spirits	Alcoholic or hydroalcoholic solutions of volatile substances usually prepared by simple solution or by admixture of the ingredients.
Tinctures	Alcoholic or hydroalcoholic solutions prepared from vegetable materials or from chemical substances.
Topical solutions	Solutions, usually aqueous but often containing other solvents, such as alcohol and polyols, intended for topical application to the skin.
Waters, aromatic	Clear, saturated aqueous solutions (unless otherwise specified) of volatile oils or other volatile or aromatic substances.
Suppositories	Solid bodies of various weights and shapes adapted for introduction into the rectal, vaginal, or urethral orifice of the human body.

APPENDIX 12.3 (*Continued*)

Term	Definition
Suspensions	Liquid preparations that consist of solid particles dispersed throughout a liquid phase in which the particles are not soluble.
Ophthalmic	See under "Ophthalmic preparations."
Oral	Liquid preparations containing solid particles dispersed in a liquid vehicle, with suitable flavoring agents, intended for oral administration.
Otic	Liquid preparations containing micronized particles intended for instillation in the outer ear.
Topical	Liquid preparations containing solid particles dispersed in a liquid vehicle, intended for application to the skin.
Systems	
Intrauterine	Self-contained discrete dosage forms intended for release of drug over a long period of time (e.g., one year).
Ocular	Intended for placement in the lower conjunctival fornix, from which the drug diffuses through a membrane at a constant rate.
Transdermal	Self-contained, discrete dosage forms that, when applied to intact skin, are designed to deliver the drugs through the skin to the systemic circulation.
Tablets (molded and compressed)	Solid dosage forms containing medicinal substances with or without suitable diluents.
Chewable	Formulated and manufactured so that they may be chewed, producing a pleasant-tasting residue in the oral cavity that is easily swallowed and does not leave a bitter or unpleasant aftertaste.
Delayed release	Intended to delay the release of the medication until the tablet has passed through the stomach.
Extended release	Formulated in such manner as to make the medicament contained available over an extended period of time following ingestion.

Source: Adapted from U.S.P.–N.F. <1151>, with permission from The United States Pharmacopeial Convention Copyright © 2008. All rights reserved.

APPENDIX 12.4   FDA Package Type Nomenclature

Term	Definition
Ampoule	A container capable of being hermetically sealed intended to hold sterile materials.
Applicator	A prefilled noninjectable pipette, syringe, or tube.
Bag	A sac or pouch.
Blister pack	A package consisting of molded plastic or laminate that has indentations (viewed as "blisters" when flipped) into which a dosage form is placed. A covering, usually of laminated material, is then sealed to the molded part. A strip pack is a specialized type of blister pack where there are no preformed or molded parts; in this case there are two flexible layers that are sealed with the dosage form in between. Suppositories that are strip-packed between two layers of foil are also considered a blister pack.
Bottle	A vessel with a narrow neck designed to accept a specific closure.
With applicator	A bottle that includes a device for applying its contents.
Can	A cylindrical vessel usually made of metal.

(*continued*)

APPENDIX 12.4   (*Continued*)

Term	Definition
Canister	A type of can used to hold a drug product.
Carton	A cardboard box or container is usually considered a secondary packaging component.
Cartridge	A container consisting of a cylinder with a septum at one end and a seal at the other end, which is inserted into a device to form a syringe containing a single dose of a parenteral drug product.
Case	A receptacle for holding something (e.g., that into which some oral contraceptive blister packs are placed).
Cello pack	A plastic "clamshell" (thin plastic preformed structure for a device).
Container	A receptacle designed to hold a specific dosage form.
Cup	A bowl-shaped container.
Unit-Dose	A cup intended to hold a single dose of a nonparenteral drug product.
Cylinder	A container designed specifically to hold gases.
Dewar	A container usually made of glass or metal that has at least two walls, with the space between the walls evacuated so as to prevent the transfer of heat. The inside of the container often has a coating (e.g., silvering) on the inside to reduce heat transfer, and is used especially to store liquefied gases or in experiments at low temperatures. The size can vary from that of a small Thermos bottle up to that which may be mounted on a large truck (also known as a "cryogenic truck").
Dial pack	A dose pack container designed to assist with patient compliance. The patient turns a dial to the correct day, the correct dose is made available, and the container indicates that the dose has been removed.
Dose pack	A container in which a preselected dose or dose regimen of the medication is placed.
Drum	A straight-sided cylindrical shipping container with flat ends, one of which can be opened and closed.
Inhaler	A device by means of which a medicinal product can be administered by inspiration through the nose or the mouth.
Refill	A container of medication intended to refill an inhaler.
Jar	A rigid container having a wide mouth and often no neck, which typically holds solid or semisolid drug products.
Jug	A large, deep container that has a narrow mouth, is typically fitted with a handle, and is used to hold liquids.
Kit	A package that includes a container of drug product(s) and the equipment and supplies used with it.
Not stated	The package type is not stated or is unavailable.
Package	A drug product container with any accompanying materials or components, which may include the protective packaging, labeling, administration device, etc.
Combination	A package in which two or more drug products normally available separately are available together.
Packet	An envelope containing only one dose of a drug product, usually in the form of granules or powder, an example being glassine powder paper containing aspirin. Other examples are aluminum foil packets into which alcohol swabs and pledgets are placed.
Dispensing	A bottle used by a pharmacist to dispense the medication prescribed. It includes preparations for which a dropper accompanies the bottle.

APPENDIX 12.4 *(Continued)*

Term	Definition
Dropper	A bottle that has a device specifically intended for the application of a liquid in a drop-by-drop manner, or a device intended for the delivery of an exact dose (e.g., calibrated dropper for oral medications).
Glass	A glass vessel with a narrow neck designed to accept a specific closure.
Plastic	A plastic vessel with a narrow neck designed to accept a specific closure.
Pump	A bottle that is fitted with a pumping mechanism for the administration of drug product.
Spray	A bottle that is fitted with an atomizer or a device that produces finely divided liquid carried by air.
Unit-dose	A bottle that contains a single whole dose of a nonparenteral drug product.
Box	A square or rectangular vessel usually made of cardboard or plastic.
Unit-dose	A box that contains a single dose of a nonparenteral drug product. *Note*: Boxes that contain 100-unit-dose blister packs should be classified under "blister pack", since this is the immediate container into which the dosage form is placed.
Pouch	A flexible container used to protect or hold one or more doses of a drug product (e.g., a pouch into which oral contraceptive blister packs are inserted, and an overwrap pouch for large-volume parenterals).
Supersack	A multiplayer paper bag for shipping some solid bulk excipients, usually in the form of powder or granules.
Syringe	A device for the administration of parenteral drug products that consists of a rigid barrel fitted with a septum with a plunger at one end and a seal or needle at the other end. The needle assembly may be part of the device or separate.
Glass	A device for the administration of parenteral drug products that consists of a rigid glass barrel fitted with a septum with a plunger at one end and a seal or needle at the other end. The needle assembly may be part of the device or separate.
Plastic	A device for the administration of parenteral drug products that consists of a rigid plastic barrel fitted with a septum with a plunger at one end and a seal or needle at the other end. The needle assembly may be part of the device or separate.
Tank	A large receptacle used for holding, transporting, or storing liquids or gases; often referred to as a *reservoir*.
Tray	A shallow flat receptacle, with a raised edge or rim, used for carrying, holding, or displaying a finished drug product in its primary or market package. A tray and its contents may be encased in shrink-wrapped plastic for shipping, or with a cover or an overwrap as part of a unit-of-use package or kit.
Tube	A flexible container for semisolid drug products, which is flattened and crimped or sealed at one end and has a reclosable opening at the other.
With applicator	A tube provided with a device (the applicator) for administering the dosage form. The applicator may be part of the tube closure or be separate.
Vial	A container designed for use with parenteral drug products.
Dispensing	A vial that is used by the pharmacist to dispense the medication prescribed.
Glass	A glass container designed for use with parenteral drug products.
Multidose	A vial intended to contain more than one dose of a drug product.
Patent delivery system	A vial that has a patented delivery system.

(continued)

APPENDIX 12.4 (*Continued*)

Term	Definition
Pharmacy Bulk Package	A container of a sterile preparation whose contents are intended for use in a pharmacy admixture program and are restricted to the preparation of admixtures for infusion or, through a sterile transfer device, for the filling of empty sterile syringes.
Piggyback	A vial that contains a parenteral preparation that can be attached directly to the tubing of a parenterally administered fluid.
Plastic	A plastic container designed for use with parenteral drug products.
Single-dose	A vial containing a single unit of a parenteral drug product.
Single-use	A vial from which a single dose of a parenteral drug product can be removed, and then the vial and its remaining contents disposed of.

Source: After http://www.fda.gov/Drugs/DevelopmentApprovalProcess/FormsSubmissionRequirements/Electronic Submissions/DataStandardsManualmonographs/ucm071748.htm. Accessed Feb. 2010.

APPENDIX 12.5 U.S.P. Packaging Nomenclature

Term	Definition
Containers	That which holds an article and is or may be in direct contact with the article.
Tamper-evident packaging	Container or individual carton of a sterile article intended for ophthalmic or otic use, except where extemporaneously compounded for immediate dispensing on prescription; sealed so that the contents cannot be used without obvious destruction of the seal.
Light-resistant container	Protects the contents from the effects of light by virtue of the specific properties of the material of which it is composed, including any coating applied to it.
Well-closed container	Protects the contents from extraneous solids and from loss of the article under the ordinary or customary conditions of handling, shipment, storage, and distribution.
Tight container	Protects the contents from contamination by extraneous liquids, solids, or vapors; from the loss of article; and from efflorescence, deliquescence, or evaporation under the ordinary or customary conditions of handling, shipment, storage, and distribution; and is capable of tight reclosure.
Hermetic container	Impervious to air or any other gas under the ordinary or customary conditions of handling, shipment, storage, and distribution.
Single-unit container	Designed to hold a quantity of drug product intended for administration as a single dose or a single finished device intended for use promptly after the container is opened.
Single-dose containers	A single-unit container for articles intended for parenteral administration only.
Unit-dose containers	A single-unit container for articles intended for administration by other than the parenteral route as a single dose, direct from the container.
Unit-of-use container	Contains a specific quantity of a drug product that is intended to be dispensed as such without further modification except for the addition of appropriate labeling.
Multiple-unit container	Permits withdrawal of successive portions of the contents without changing the strength, quality, or purity of the remaining portion.
Multiple-dose container	A multiple-unit container for articles intended for parenteral administration only.

Source: Adapted from U.S.P.–N.F. General Notice, with permission from The United States Pharmacopeial Convention. Copyright © 2008. All rights reserved.

APPENDIX 12.6   U.S.P. Packaging Fabrication Materials and Closure Types

Term	Definition
Packaging fabrication materials	Include substances used to manufacture packaging containers such as glass, plastics [including high-density polyethylene, low-density polyethylene, poly(ethylene terephthalate), poly(ethylene terephthalate G) and polypropylene], other resins, and other materials, as listed in the general test chapter <661>, Containers, and in the FDA Guidance for Industry on Container Closure Systems for Packaging Human Drugs and Biologics.
Glass	The glass packaging material used in the immediate container should meet the glass test requirements for Limits for Glass Types and Chemical Resistance-Glass Containers.
Plastic	Any plastic packaging material used in the immediate container should meet the plastic test requirements for plastic in the general test chapters Containers and Containers—Permeation.
Packaging closure types	Reclosables and nonreclosables may be used for solid, semisolid, and liquid dosage forms.
Reclosables	Containers with suitable closures that may incorporate tamper evidence and child-resistance capabilities.
Nonreclosables	Containers with closures that are nonreclosable, such as blisters, sachets, strips, and other single-unit containers.

Source: Adapted from U.S.P.–N.F. <1136>, with permission from The United States Pharmacopeial Convention. Copyright © 2008. All rights reserved.

CHAPTER 13

TABLET PRODUCT DESIGN

13.1 INTRODUCTION

The oral drug delivery route is generally considered the most preferred route of drug administration for the vast majority of patients. For certain patient populations, tablets are less convenient or inappropriate, such as children under the age of 4 or 5; the elderly, who have trouble swallowing; and patients with specific disease states who find swallowing difficult, such as those with Parkinson's disease. About two-thirds of all pharmaceutical dosage forms are oral products, and about 40% of the approved U.S. Food and Drug Administration (FDA) drug products are tablets. As a result, high-speed high-quality current good manufacturing practice tablet manufacturing is commonplace in the pharmaceutical industry. However, tablet product design is anything but simple and straightforward. Even with highly sophisticated and focused drug discovery groups, well-honed developability drug assessment programs, and data-driven lead candidate selection processes, lead drug candidates rarely (if ever) meet the ideal drug requirements. Consequently, many drug-associated physiochemical problems need to be addressed by product design scientists. In addition, the gastrointestinal tract (GI) is a formidable barrier to drug absorption. The GI tract is characterized by having a bimolecular hydrophilic–lipophilic membrane separated into four segments: esophagus, stomach, small intestine, and large intestine; fluid flux of about 7 to 9 L per day; pH range of about 1 to 8; enterocytes and secreted enzymes with metabolic functions; secreted hormones such as gastrin and motilin; influx and efflux transporters; endocytic cells such as M-cells found in the Peyer's patches that lead to transcytosis of particles; and microbial flora that can metabolize drugs. The GI tract is also known to be where drugs can be absorbed and taken by the portal vein directly to the liver and metabolized before the medicine can be absorbed into systemic circulation and reach the site of action. This shunting of the drug to the liver can lead to presystemic metabolism of the drug, referred to as *first-pass metabolism* which can significantly reduce the oral bioavailability of a drug. GI tract motility under the influence of the hormone motilin can also affect drug absorption. Most drugs are absorbed in the upper GI tract. Peristaltic muscular contractions propel the intestinal contents along the GI tract. Powerful contractions called *housekeeping waves* start in the stomach and sweep undigested food and bacteria from the stomach

Integrated Pharmaceutics: Applied Preformulation, Product Design, and Regulatory Science,
First Edition. Antoine Al-Achi, Mali Ram Gupta, William Craig Stagner.
© 2013 John Wiley & Sons, Inc. Published 2013 by John Wiley & Sons, Inc.

and small intestine into the large intestine. Since most drug absorption occurs in the small intestine, these housekeeping waves can literally sweep the drug beyond these optimum intestinal absorption areas and cause erratic and poor bioavailability. In addition, the residence time of a drug in the GI tract ranges from 10 to 24 hours, which can again lead to bioavailability variation. The interplay between physiochemical drug properties such as intrinsic solubility, pK_a-controlled changes in solubility and percent ionized drug, stability, partition coefficient, and the GI tract factors listed above strongly influence the rate and extent of drug absorption, making tablet product design a complex multifactor endeavor.

Having said this, the tablet drug delivery system is versatile and is one of the most widely used dosage forms because it provides a solid dosage form that is convenient to self-administer and delivers good dose uniformity. The *United States Pharmacopeia* (U.S.P.) and the FDA define tablets as a solid dosage containing medicinal substances with or without suitable diluents. There are a number of ways to classify tablets (Appendixes 12.1 to 12.3). Tablets are classified by the means of manufacture, intended function, and route of administration. Tablets can be manufactured by compression or molding. The U.S.P. and FDA also classify tablets by function, such as chewable, dispersible, tablet for solution, tablet for suspension, delayed release, and extended-release. Routes of administration are also used to describe tablets that may have special requirements, such as subdermal, sublingual, buccal, and vaginal tablets.

Compressed tablets are made by filling a die with dry powder containing a drug. Then, high pressure is applied to the powder by means of an upper and a lower punch that fit inside the die (Figure 13.1). Molded tablets may be made by light compression of wetted powder that is dried in a mold. Molded tablets may also be prepared by freeze-drying a drug solution that is placed in a mold or by pouring a melted matrix into a mold that is subsequently cooled. Figure 13.2 is a photo of an oral disintegrating tablet that has been manufactured by a mold–freeze–drying process. In this case, the liquid tablet components are dispensed into a clear mold. The water is freeze-dried, and the remaining solid takes the shape of the mold. At completion of the freeze-drying process, a foil lidding seals the tablet and gives the unit additional rigidity. Advantages and disadvantages of tablets are listed in Table 13.1.

At first glance, designing a tablet may seem like a simple process. After all, making a tablet does not appear to be much different from making an infrared pellet with a Carver Press. The purpose of product design, in this case tablet design, is to create a drug delivery system that meets specific functional and performance criteria. To better understand the complexity of tablet design we need to start by understanding some of a tablet's functional and performance criteria. Functionally, a tablet is a dosage form that is easily portable. It is small, easily administered, and remains intact through compression, film coating, packaging, storage, distribution, dispensing, and patient administration. Once a tablet is taken orally, it must disintegrate or fall apart, forming smaller aggregates to allow the drug to dissolve in the gastrointestinal contents and to be available for either a local effect or absorption for a systemic effect. The relationship among tablet disintegration, dissolution, and drug availability is illustrated in Figure 13.3.

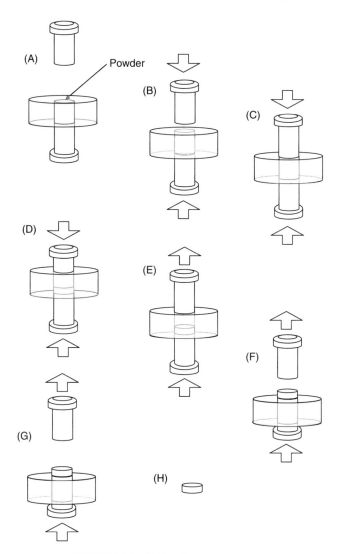

(A)

Powder

(B)

(C)

(D)

(E)

(F)

(G)

(H)

FIGURE 13.1 Tablet formation sequence.

The optimum performance of a tablet depends on a number of criteria that often have competing objectives, which result in complex and significant interaction effects that are not easily predicted or managed. For example, it is desirable to have a tablet remain intact until it is taken by the patient. To remain intact through compression, film coating, packaging, storage, distribution, dispensing, and patient administration, the tablet must have an acceptable degree of *hardness* and *friability* to prevent breakage or abrasion prior to dispensing and patient use. To achieve acceptable hardness and friability, tablets are compressed under specified pressures and speed. *Functional adjuvants* or *excipients* that are therapeutically inactive are often required to provide better powder cohesion and binding. The design criteria

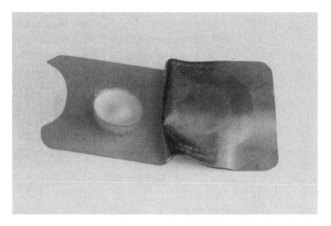

FIGURE 13.2 Molding for freeze-dried tablet.

TABLE 13.1 Advantages and Disadvantages of Tablets

Advantages	Disadvantages
Versatile	Limitations for poorly soluble drugs
Many applications, such as immediate release, controlled release, fast dissolve	Some processes (e.g., wet granulation) require a number of unit operations and equipment
Number of routes of administration	Film coating is generally required for poor-tasting drugs
Wide dose range (micrograms to grams)	Containment requirements for dust exposure and cross-contamination, especially for highly potent and toxic compounds
Convenient to carry and self-administer	
Good dose uniformity	
Solid state provides good physical, chemical, and microbial stability	Bulk handling of powder using scoops or vacuum transfer
Number of processing options readily available	Potential for drug segregation
High-speed production	Dust explosivity
Robust manufacturing process	Limitations for high-dose and poorly compressible drugs
Tamperproof and tamper-evident	Difficult to swallow for young children and the elderly
Product identification color, shape, and logo	
Cost-effective	

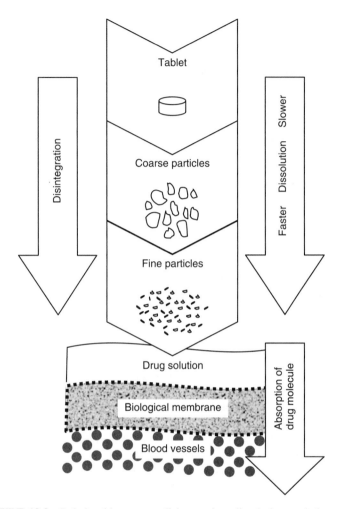

FIGURE 13.3 Relationships among disintegration, dissolution, and absorption.

do not stop here. It is vital that a tablet release the medicinal agent for it to elicit its therapeutic effect. That means that once a patient takes a tablet, it needs to disintegrate or break up and allow the drug to dissolve and be in solution to achieve its therapeutic effect. The formulation scientist needs to incorporate a disintegration mechanism into the design of the tablet. A disintegrant excipient is often chosen to address this design consideration. In the presence of water, disintegrants swell and break up the tablet matrix, causing the tablet to disintegrate. In essence, the function of the disintegrant opposes the function of the compression and the tablet binder. Therefore, the mechanism to predict dissolution performance is complex and difficult to predict.

The design objectives are further complicated by the effect of such environmental conditions as relative humidity. If moisture penetrates the container closure system, the water can react with the tablet matrix, specifically the disintegrant, to cause the tablet to soften and not meet its hardness and friability criteria. There are many other opposing, nonlinear, complex relationships that require development scientists to be highly skilled in the art and science of product design. Statistically based experimental design approaches are now recognized as important tools that can aid formulation scientists to optimize a product. Product design employs a complicated, multifactorial, multidimensional design space. Artificial intelligence and expert systems have also been cited to enable rational product design.

Tablet design is an innovative process in which one starts with a drug substance and eventually creates a tablet that meets vital functional and performance criteria. Human anatomy and physiology, route of administration, intended purpose, and product design are inextricably linked. Human anatomy and physiology must be considered when developing a dosage form. The GI tract provides a hostile environment to drugs. Digestive enzymes, wide pH range, housekeeping waves, and first-pass metabolism can affect a drug's stability and bioavailability. The intended purpose of the drug delivery system may be to control or modify the release of the drug in specific regions of the GI tract. Specified drug release profiles can be achieved by selection of functional excipients and special manufacturing processes. Once a manufacturing process is selected, excipients that support the process and an intended product profile target can be chosen. In general, a tablet formulation will use excipients that provide the desired product attributes.

A pharmaceutical excipient can be defined as any material that is included in the manufacturing process or is in the final product that is not the active pharmaceutical ingredient. Generally, excipients should be inert and have no pharmacological or physiological activity; that is, they should be physically, chemically, and microbiologically stable, compatible with the drug substance and other formulation excipients, odorless, and colorless. However, excipients are used in formulations to provide specific functions to improve a dosage form's performance. In this context, excipients are anything but inert and can be referred to as *functional excipients*. Over the years there has been a growing need for additional functional excipients. There has also been recognition that more attention needs to be paid to the lot-to-lot quality of excipients. In 1995, the International Pharmaceutical Excipients Council was formed. The council, a federation of industrial associations representing the Americas, Europe, Japan, and China, was constituted to ensure that excipients meet appropriate standards for quality, safety, and functionality. Excipients can be categorized as compendial and noncompendial. European, Japanese, and U.S.P.–*National Formulary* (N.F.) pharmacopeial excipients carry regulatory status and must meet standards that are approved in the official monographs. The U.S.P. has several chapters dedicated to excipients: <1059>, Excipient Performance; <1195>, Significant Change Guide for Bulk Pharmaceutical Excipients; and <1074>, Excipient Biological Safety Evaluation Guidelines. Noncompendial excipients can be used, but their inclusion must be supported by a type IV drug master file that is maintained with the FDA and supports an excipient's quality and safety. The FDA's inactive ingredients list gives the name, route of administration, and amount of an excipient used in approved drug products. The U.S.P. is also responsible for setting standards for food ingredients, which are available in the Food Chemicals Codex. Food ingredients have been used successfully as

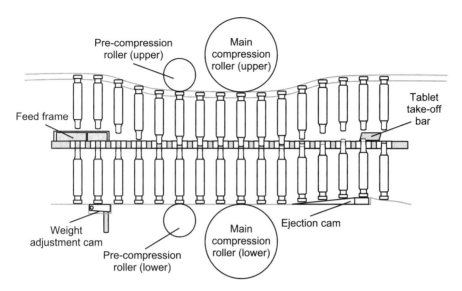

FIGURE 13.4 Rotary tablet press sequence.

pharmaceutical excipients. *The Code of Federal Regulations* (CFR), 21 CFR Parts 182 and 184, provides a list of generally recognized as safe food ingredients.

Tablet performance is dependent on every tablet having the same composition and dose uniformity. This desired state is highly dependent on uniform and consistent powder flow. The automated tablet compression process is illustrated in Figure 13.4. The flowing tableting powder, or *running powder*, fills the tablet die by volume, not by weight. The die fill volume is a function of powder flow rate and the powder's dynamic bulk density. If the flow pulses, one can expect that the die fill volume will not be consistent, leading to tablet weight and drug dose variability. In addition, to achieve a uniform dose and performance, the composition of the running powder must remain constant and not undergo segregation. Segregation may be caused by a number of factors, such as machine vibration, vacuum transfer, hopper design, particle size and shape, and particle density differences.

A number of running powder manufacturing processes use different unit operations to achieve a uniform free-flowing running powder. The choice of the running powder manufacturing process is based primarily on the physiochemical properties of the drug. The active pharmaceutical ingredient (API), dose, compressibility, flowability, stickiness, and stability are some of the factors that are considered. Each running powder manufacturing process has advantages and disadvantages, which are discussed later in the chapter. Table 13.2 summarizes the major processes used to prepare running powders. The four main types of running powder manufacture are dry blend, wet granulation, dry granulation, and hot-melt extrusion. Wet granulation and dry granulation can be broken down further into specific wet and dry granulation processes. It is clear from the table that some running powder manufacturing processes require more steps, which add to their complexity.

The dry blending process for preparing a running powder is generally considered the simplest. The dry blending column shows that this process generally

TABLE 13.2　Running Powder Manufacturing Process Steps

Process	Dry Blend Direct Compression	Wet Granulation High-Shear	Wet Granulation Fluid-Bed	Wet Extrusion	Dry Granulation Roller Compaction	Dry Granulation Slugging	Hot-Melt Extrusion
Screen/mill	×	×	×	×	×	×	×
Mix/blend	×	×	×	×	×	×	×
Solidify					×	×	
Wet mass		×	×	×			
Melt							×
Extrude				×			×[a]
Spheronize				×[b]			×[b]
Mill/size		×		×[c]	×	×	×[d]
Dry		×	×	×			
Mill/size		×	×	×			
Final blend	×	×	×	×	×	×	×

[a] The extrudate can be filled into molds, cooled, and blister-packed.

[b] Spheronizatoin of the extrudate can be used for encapsulation or tableting.

[c] The extruded material can be dried without wet sizing.

[d] Rod-shaped extrudates can be cut into capsule-shaped tablets or pelletized into beads or granules.

requires only three steps or unit operations to prepare running powders. The other running powder processes require at least twice as many steps. The hot-melt process involves four unit operations when the final dosage form is a molded tablet. Slugging and roller compaction use five process steps each. Fluid-bed granulation usually involves six steps, but wet granulation and drying are done in the same granulator–dryer, so the actual number of product-handling steps is five. High-shear wet granulation often requires seven steps. A wet extrusion granulation can involve nine unit operations. Even though the extrusion process involves more steps, it is more amenable to continuous processing and requires significantly less material to undertake experimental runs. This is a major advantage in early development when drug substance is scarce and the product design scientist still has the ability to utilize a design-of-experiments approach because wet extrusion conserves material use. Once the continuous extrusion process is in place—manufacturing efficiencies, costs, and flexibility in capacity—demand planning is improved significantly. The advantages and disadvantages of running powder manufacturing processes are listed in Table 13.3.

13.2　FORMULATION DESIGN

13.2.1　Preformulation

The purpose of a preformulation study is to evaluate a drug's physiochemical properties and identify specific issues that place the development program at risk. For

TABLE 13.3   Advantages and Disadvantages of Running Powder Manufacturing Processes

Process	Advantages	Disadvantages
Dry blend	Less labor, time, equipment, space, and operational energy Absence of solvents Minimizes stability problems due to heat and moisture Generally, better "physical" stability over time: that is, less increase in hardness	Running powder flow and compressibility are very dependent on properties of drug and excipients Difficult to obtain dense, hard tablets for high-dose drugs More prone to have segregation problems, especially with low-dose drugs To minimize segregation, it is desirable to match particle size and density of drug and excipients
High-shear wet granulation	Physical characteristics of drug are minimized Can distribute low concentrations of drug to achieve acceptable content Binding of particles helps ensure blend uniformity Requires less granulating fluid compared to fluid-bed granulation Optimizes bulk density Forms dense granules Wet binders improve compressibility and consolidation Short processing time Dust and segregation are reduced Power consumption curves can be used to determine the granulation endpoint	Instability due to moisture and heat Large number of process steps: each step requires process validation and cleaning validation Longer overall process time, especially drying compared to dry blending High labor and energy manufacturing costs Large capital requirements (equipment and space) Endpoints not an exact science Tablets may tend to increase in hardness over time and result in decreases in dissolution and potentially bioavailability
Fluid bed	Physical characteristics of drug are minimized Binding of particles helps ensure blend uniformity Gentle fluid mixing prevents large agglomeration and eliminates the need for wet milling Granules lead to harder tablets than with high-shear granulation	Instability due to moisture and heat Cleaning is more difficult Large number of process steps, each step requiring process validation and cleaning validation Long process time High labor and energy manufacturing costs Large capital requirements (equipment and space)

(*continued*)

TABLE 13.3 (*Continued*)

Process	Advantages	Disadvantages
	Wet binders improve compressibility and consolidation	Endpoints not an exact science
		Tablets may tend to increase in hardness over time and result in decreases in dissolution and potential bioavailability
	Granulation and drying performed as one unit operation	
		More fines and finer granulation
	Thermodynamic models can assist in endpoint determination	Potential for dust explosion
		Lower-density granulation compared to high-shear wet granulation
Wet extrusion	Lends itself to continuous processing	(See "High-shear wet granulation")
	Significant conservation of API during development trials	
	(See "High-shear wet granulation")	
Roller compaction	Absence of solvents	Drug compressibility may limit compactability
	Minimizes stability problems due to heat and moisture	Overlubrication resulting from two compressional events, roller compacttion and tableting, can lead to poor dissolution
	Denser running powders to attain higher tablet weights compared to blended running powder and wet granulation	
		Binding of powder in the compactor
	Higher output speeds compared to slugging (up to 50,000 to 100,000 kg/h)	Maintaining ribbon hardness and consistency
	Better powder flow than in a dry blend	High levels of material loss (up to 20 to 30% loss)
	Compression step pretableting can increase bonding (basically increases time under compression)	Lose compressibility due to strain or compression hardening
Slugging	Absence of solvents	Drug compressibility may limit the ability to make slugs
	Minimizes stability problems due to heat and moisture	Need to develop two direct tablet formulations and processes
	Denser running powders to attain higher tablet weights compared to powder blend and wet granulation	Slugging results in extended manufacturing time (up to 50 to 60 kg/h)
	Better powder flow compared to dry blend	Overlubrication can lead to poor dissolution
	Pretableting can increase bonding (basically increases time under compression)	Several milling and sizing steps requiring dust control
		Lose compressibility due to strain or compression hardening

TABLE 13.3 (*Continued*)

Process	Advantages	Disadvantages
Hot-melt extrusion	Absence of solvents Excipients and process most amenable to controlled drug delivery systems: pellets, beads, molded tablets Economical number of processing steps Lends itself to continuous processing Uniform dispersion or solid solution Efficient use of API Make molded tablets as part of a continuous process	Drug and excipients need to be thermally stable Drug may be dispersed as a metastable amorphous material Special equipment train that has more constrained uses

example, poor solubility and stability are attributes that may lead to development risks that could slow the development program or cause it to be terminated. Only about one out of 12 compounds that are filed as an investigational new drug (IND) reach a new drug application (NDA) filing. This low success ratio raises important questions as to how much time and money should be spent at the preformulation stage of drug development. Some companies support a minimalistic approach and others undertake more in-depth programs. The aim of an efficient preformulation program should be to provide the necessary information in the shortest possible time at the least cost. Knowledge gained during the developability stage should help highlight the key factors that put a development program at risk. A data-driven preformulation program should leverage this knowledge and emphasize obtaining additional knowledge in key areas. A stability-indicating assay that is selective for the drug substance is required for all preformulation programs. Tablet preformulation activities include physiochemical solid- and solution-state characterization and excipient compatibility. Table 13.4 lists a number of solid-state physical properties that have important tablet formulation and product capability implications.

Once a stability-indicating method has been developed, drug solid- and solution-state stability studies can be initiated. Table 13.5 lists the solid- and solution-state "chemical properties" that have important formulation and product attribute implications. Solubility, permeability, and partition coefficients are generally considered physical properties. These properties are included here for two reasons: first, one needs a chemical assay to measure these properties; and second, the chemical nature of these molecules affects these attributes.

As mentioned in Chapter 12, the decision regarding the route of tablet administration and intended purpose is driven primarily by marketing and medical needs. These decisions immediately constrain the manufacturing process and formulation choices. Once the route of administration, expected highest dose, and intended use have been decided, the formulation scientist will use the physical and chemical

TABLE 13.4 Solid-State Physical Properties and Tablet Implications

Physical Properties	Tablet Implications
Organoleptic properties: appearance and odor	If there is a distinctive color, such as yellow or brown, a colorant may be required to provide a uniform tablet color.
	A disagreeable odor such as a "sulfide" or "rotten egg" smell may require volatile additives such as vanillin or mint to compete with an offensive odor.
	These potential additives should be used in the excipient compatibility screening study.
Morphology and micromeritics: crystal habit, birefringence, particle size, particle shape, particle size distribution, bulk density, tapped density, true density, agglomeration state, and surface area	A cubic crystal will flow better than will a long needle.
	Birefrigence in the absence of anisotropic crystal is indicative of crystalline material compared to amorphous material.
	Large crystals of a highly insoluble drug may have to be milled to increase the surface area to improve dissolution.
	Bulk containers are sized by volume, but powders are handled by weight. Bulk and tapped densities are important in determining what size container and processing equipment are needed.
	Highly agglomerated, cohesive particles will not flow well and are often poorly compressible.
Thermal properties: melting point, enthalpy of fusion, and other thermal transitions	Low-melting compounds can soften or even melt during compression, due to the heat of compression and frictional heat buildup.
	A high enthalpy of fusion suggests that the drug may have poor solubility.
	A high melting point suggests that the drug may have poor solubility.
	Differential scanning calorimetry can pick up other enthalpic transitions, such as polymorphism, dehydration, and decomposition. Polymorphs have different solubility, stability, cohesiveness, density, dissolution rates, and bioavailability.
	Thermal gravimetric analysis can quantitate hydration states and solvent weight loss.
	Loss of water and solvent can occur during milling and micronization processes.

TABLE 13.4   (*Continued*)

Physical Properties	Tablet Implications
Crystallinity and polymorphism	Amorphous material is less stable physically and chemically than its crystalline counterpart. Amorphous material is more soluble, faster dissolving, and more bioavailable than the crystalline state.
Polymorphism tendency	Some drugs have a larger propensity than others to form polymorphs. Since polymorphs can have a significant effect on manufacturability and bioavailability, it is important to be aware of the drug's tendency to form polymorphs.
Hygroscopicity and moisture adsorption–desorption isotherm[a]	The degree of hygroscopicity will dictate the proper storage and handling conditions.
Contact angle	The contact angle indicates how wettable a drug substance is. Wetting or spreading of the dissolution media over the drug is the first step in dissolution.
Static charge and explosivity	The dust clouds that result from handling powders can be lethally explosive. Knowing the explosive potential of powders is crucial to maintaining a safe work environment.
Flow properties	Tablet content uniformity is a function of the flowability of a drug and the running powder. A drug may have such poor flow qualities that it is impractical to use a dry powder blend and direct compression process.
Compressibility	The intrinsic compressibility of the drug substance and dose often dictate what process is used to prepare the running powders.

[a]Callahan et al., (1982, App. 13.1).

properties determined in Tables 13.4 and 13.5 to decide which process will be most likely to provide the desired drug delivery attributes (see Figure 12.3).

13.2.2   Excipient Compatibility Studies

An excipient compatibility study is essentially a screening design whose aim is to identify excipients with which a drug is and is not compatible. The type of tablet (immediate release, controlled release, chewable), route of administration (oral, sublingual, subdermal), and physiochemical properties of the drug will dictate the

TABLE 13.5 Solution-and Solid-State Chemical Properties and Tablet Implications

Chemical Properties[a]	Tablet Implications
Purity and impurities: impurities may include degradation products, materials related to the synthesis, organic volatiles, residual solvents, and heavy metals	This is a quality attribute. These properties can affect safety and efficacy. These properties can affect manufacturability.
Solubility: in water and pH values covering the range of the gastrointestinal tract, fasted simulated intestinal fluid[b] and fed simulated intestinal fluid[b]	High-solubility drugs can be delivered at higher doses with a greater likelihood of good bioavailability. The simulated fluids contain lecithin and the bile acid sodium taurocholate, which are wetting or surface-active agents and form micelles and may be better media to use to predict the solubility of a drug in humans. Solubility data can be applied to various bioavailability classification systems: (e.g., the biopharmaceutics and biopharmaceutics drug disposition classification systems)
Permeability: using cellular absorption models, such as Caco-2 cells	Used in the evaluation of the biopharmaceutics classification system.
Partition or distribution coefficient	Partition or distribution coefficient can affect bioavailability. Application of the rule of 5s,[c] which states that poor bioavailability is expected for drugs that have a partition coefficient above 5.
Acid–base dissociation constant	Drug ionization state affects a drug's solubility, dissolution rate, stability, and bioavailability as a function of pH.
Solution- and Solid-State stability: as a function of room temperature and stressed temperature, pH, oxygen, and light	Stability knowledge affects formulation, process, and container closure decisions.

[a] These properties are included here for two reasons: first, one needs a chemical assay to measure these properties; and second, the chemical nature of these molecules affects these attributes.

[b] Dressman and Repa (2000).

[c] Lipinkski et al. (1997).

choice of the running powder manufacturing process, which in turn drives the selection of functional excipients.

Table 13.6 lists the classes of tablet excipients and their functions. As you can see from the list, the number of excipients in a tablet formulation can be quite large. One tries to minimize the number of excipients used to decrease the complexity and cost of ingredients, manufacture storage and handling costs, and potential for excipient–excipient and excipient–drug physiochemical interactions.

TABLE 13.6   Tablet Excipient Classes and Functions[a]

Excipient Class	Function
Fillers, binders, and bulking agents	Increase the bulk of a tablet to a convenient size.
Binders	Promote the cohesion of powders, making them more compressible.
Disintegrants	Break up a tablet or cause it to disintegrate.
Lubricants	Decrease interfacial friction, especially between a tablet and the die wall.
Antiadherents and antisticking agents	Decrease filming, adherence, and sticking of powder to metal surfaces such as punch faces.
Glidants and flow aids	Decrease interfacial cohesion and friction, especially between the powder particles. Improve running powder flowability.
Wetting agents	Decrease the interfacial energy between a liquid and a powder or solid, which allows the liquid to spread over the solid particles.
Solubilizing agents	Increase the solubility of a drug in the dissolving media.
Drug-release modifiers	Provide delayed or extended drug release.
Stabilizers	Improve the stability of a drug substance through a variety of mechanisms.
Colorants	Add aesthetic appeal and provide product distinction.
Sweeteners and flavors	Improve taste and patient acceptance.
Antimicrobial agents	Decrease microbial bioburden, especially in the case of dispersible tablets and tablets for solution.

[a] U.S.P. Chapter <1059>, Excipient Performance, discusses the functional categories and provides a description of the functional classes, functional mechanisms, and physical and chemical properties of the various excipients.

Generally, a simpler formulation produces a more rugged, reproducible, and better-performing tablet.

Once the manufacturing process has been chosen, the initial set of excipients are selected. Appendixes 13.2 to 13.9 list various excipients by their functional classes and typical use amounts or concentrations. Tablet diluents and direct compression binders are provided in Appendix 13.2. Tablet diluents are used to increase the amount of running powder to make a convenient-size tablet, say 100 mg or greater. Bulking agents are also referred to as *diluents* and *fillers*. Many diluents function as both diluents and direct compression binders. Microcrystalline cellulose, a diluent, is self-lubricating, an antiadherant, and a highly effective binder at concentrations as low as 5 to 10% w/w. The ability of other direct compression excipients to maintain their compressibility in the presence of other excipients and the drug depends on the compressibility of the other components and is determined experimentally. In general, the other diluent–binders require a minimum of 10 to 20% w/w for acceptable tablet hardness. Wet granulation binders for immediate-release tablets are listed in Appendix 13.3. Hot-melt excipients, including diluents and binders, plasticizers, lubricants, and others, are given in Appendix 13.4. As discussed previously, disintegrants are used to break up the tablets. Appendix 13.5 presents typical disintegrants and use levels. A tablet lubricant's primary function is to decrease friction between the tablet and the metal tablet machine components. Lubricants can be classified as hydrophobic and hydrophilic and are listed in Appendix 13.6. Lubricants can also be categorized as boundary, fluid film, and liquid lubricants. Boundary lubricants such as magnesium stearate adhere to solid surfaces such as granules, tablets, and tablet tools and dies. Fluid-film lubricants such as stearic acid soften and may melt under pressure. They resolidify as the temperature and pressure decrease. Liquid lubricants such as mineral oil reduce friction by forming a permanent oily film during compression. Lubricants can also function as antiadherants and antisticking agents. A list of common antiadherants and antisticking agents are provided in Appendix 13.7. Glidants and flow aids improve tablet weight variation by improving the running powder flow quality, including flow rate and flow uniformity. Appendix 13.8 lists frequently used glidants and flow aids. Wetting agents are classified as anionic, nonionic, and cationic and are listed in Appendix 13.9.

Excipient compatibility studies are a screening tool used to find functional excipients that are compatible with the drug. As is evident from the number of excipients listed in Appendixes 13.2 through 13.9, the number of possible excipients tested can be extremely large. Cost-effective studies need to be designed to provide the critical compatibility knowledge required for efficient and productive formulation development. The design of excipient compatibility studies can merge the art and science of protocol development. In the past, statistical designs such as partial factorial and Plackett–Burman techniques have been proposed as cost-effective means to identify the main factors or excipients that are compatible with a drug substance. More recently, the use of $N - 1$ designs (Allen, 2009) has been proposed to provide the main excipient effects as well as interaction among excipients. Other traditional excipient compatibility studies have investigated the binary compatibility of an excipient and a drug. The ratio of the binary mixture has also evolved from using a 1:1 ratio of drug to excipient to a ratio of drug to excipient that reflects the ratio in the final dosage form. Some companies test the stability of the drug in standard formulations as either powder mixes or compressed tablets.

Another, less well documented approach is to study drug compatibility within the various excipient classes. For example, in the case of a direct compression drug formulation, the compound would be blended with groups of excipients from their respective classes, such as binder and bulking agents, disintegrants, lubricants, and antiadherants and glidants. If there are no stability interactions in all four groups of excipients, all the excipients in those groups would be deemed compatible. In the case where degradation was noted in one of the groups, the individual excipients in that group would have to be retested to confirm which excipient or excipients were incompatible with the drug substance. At this stage of formulation design there may be a very limited number of drug substances that require trade-offs between the extent of compatibility knowledge and the amount of drug available.

TABLE 13.7   Typical Excipient Compatibility Stability Protocol

	Initial	1 Month	3 Months	6 Months
25°C	×	×	×	×
40°C/75% RH[a]		×	×	×
ICH visible light[b]				
ICH ultraviolet light[b]				

[a] RH, relative humidity. Closed and open samples are tested for the effect of moisture on the drug and drug–excipient combinations.

[b] Following ICH guidelines (1996).

Solid-state compatibility studies can be conducted using a stress temperature of 60°C, recommended International Conference on Harmonization (ICH) (1996) stress light conditions (visible light at not less than 1.2 million lux hours per meter squared, and ultraviolet light exposure at 200 watthours per meter squared at 365 nm), and exposure to water (20% w/w added water). The dry drug–excipient blends and powder blends with 20% added water are tightly sealed and stored at 60°C for 4 and 8 weeks. All samples are analyzed initially and at the specified exposure times. Many excipient compatibility studies use a 40°C accelerated temperature. These studies are usually carried out for three to six months. The effect of moisture is often evaluated by exposing drug and drug–excipient blends stored in closed and open containers to 40°C/75% relative humidity for three to six months. A typical excipient compatibility stability protocol is shown in Table 13.7.

13.2.2.1 Excipient Compatibility Study: Hypothetical Case Study 1 (Ideal Drug Candidate)
Excipients to be used in an excipient compatibility study are chosen based on which running powder manufacturing process is chosen to be employed. To choose a process, the product design scientist must take into account the drug's physicochemical properties and the quality target product profile. Two hypothetical case studies are used here to demonstrate how the running powder manufacturing process is chosen and how the selection of the process affects the selection of the excipients to be assessed during the study. The hypothetical drug in Case Study 1 is an "ideal" drug candidate, possessing nearly ideal physicochemical properties for an immediate-release tablet dosage form. In Case Study 2, the hypothetical drug's physicochemical properties are those commonly encountered by the development scientist and will be referred to as a "real-life" case. These two cases are continued in the tablet formulation section that immediately follows the excipient compatibility discussion.

In addition, in this section use discuss two types of excipient compatibility design approaches that are applied to the ideal drug. The same two approaches are applied to the real-life drug scenario. The two excipient compatibility approaches will highlight the functional class screening method and the Plackett–Burman statistical design technique, to illustrate how the two methods affect sample preparation and the number of samples required for analysis. As mentioned previously, the excipients chosen for this drug–excipient compatibility study are dictated by the choice of the running powder manufacturing process.

For Case Study 1, the following information has been determined from the hypothetical preformulation physicochemical studies. The drug is a neutral molecule that has good solubility, wettability, and stability over a wide pH range. It is highly permeable, compressible, and free flowing. It has a "true" density of 1.5 g/cm^3 (1500 kg/m^3) and a mean particle size of 150 μm. The quality target product profile states that a stable, immediate-release oral tablet that has a dose range of 25 to 100 mg is desirable. The drug's physiochemical properties and quality target profile lend themselves to a direct compression tablet using a dry blend running powder manufacturing process. The excipients chosen for the drug–excipient compatibility study should be compatible with manufacturing a direct compression tablet.

Table 13.8 is an example of the functional class screening excipient compatibility study approach. This approach tests the compatibility of a functional class of excipients that are blended with the drug and evaluated for stability. If the drug is stable in the mixture, all the excipients are "cleared" for use with the drug. The obvious downside to this approach occurs when a class is found to be incompatible and the development time is extended until the specific source of incompatibility is determined. Table 13.8 lists the proposed excipient functional classes, the specific excipients to be tested in each class, and the drug/excipient ratios that would be used for Case Study 1. The drug/excipient ratios shown in the table represent typical ratios that are employed in a tablet formulation. Let's now consider the thought process for choosing these particular excipients. First, it is important to choose excipients that have a good safety record and are listed in the U.S. European, and Japanese pharmacopeias. The FDA (2011) provides the *Inactive Ingredients List* (IIL), which is updated quarterly. The IIL compiles all

TABLE 13.8   Functional Class Approach with an Ideal Drug Candidate

Excipient Class	Drug/Excipient Ratio	Excipient
Diluents and diluent binders	1 : 4	Microcrystalline cellulose Spray-dried lactose Mannitol
Disintegrants[a]	1 : 9	Croscarmellose sodium Crospovidone Starch glycolate sodium
Wet binders	1 : 9	NA[b]
Lubricants[c]	1 : 19	Magnesium stearate Stearic acid
Antiadherants	1 : 19	NA
Glidants	1 : 49	NA
Wetting agents	1 : 99	NA

[a] Since the drug is a neutral molecule, an ionic interaction with the ionic disintegrants is not expected, so these disintegrants are included in the same group.

[b] NA, not applicable in this case.

[c] Since the drug is stable over a wide pH range, acidic and basic lubricants have been placed in the same group, to decrease the level of effort.

the excipients that have been incorporated in FDA-approved commercial products. The list also notes the dosage form and the amounts or percentages used in the product. In this case, a formulator could go to the IIL and use the "find" facility to determine the excipient amounts that are present in FDA-approved tablets. Selecting an excipient that is found in the major pharmacopeias gives the formulator the most flexibility in developing a tablet that can be marketed worldwide. Manufacturers may also place constraints on what excipients can be chosen for product design. Many manufacturing facilities try to limit the number of excipients in their inventory, and formulators are encouraged to select excipients that have approved vendors and are on the inventory list.

At a minimum, a direct compression tablet formulation will need a diluent–binder, a disintegrant, and a lubricant. A formulation scientist may choose to evaluate only these functional excipients in an initial excipient compatibility study. It may turn out that acceptable tablets can be manufactured without the addition of antiadherants and glidants and that the inclusion of these classes of excipients in an excipient compatibility study would have been wasted effort and cost. In this case study, the preformulation studies show that the drug is easily wetted and soluble so there is little need to study wetting agents or solubilizing agents. The drug was also shown to be flowable, so one could eliminate the need to include flow enhancers in the excipient compatibility study. Again, in this particular case study, the drug is compressible and the dose is in a reasonable dilution range to support the development of a direct compression tablet. On the other extreme, a formulation scientist might choose to study all the functional classes of excipients, including a wet binder and wetting agents, with the idea of having this information available if future manufacturing studies show that a wet granulation or wetting agent is needed. The compatibility information would be available and there would be no delay in the development program waiting for data to be generated.

The ideal drug should be physically stable, with cationic and anionic excipients, since it is a neutral molecule. That is, one would not expect a neutral molecule to have ionic interactions with any of the excipients. On the other hand, if the drug were a positively charged quaternary amine, it is possible that it would participate in ionic interactions with a negatively charged excipient, forming an insoluble ion pair or salt.

Note that the lubricant class contains both magnesium stearate, a basic lubricant, and stearic acid, an acidic lubricant. The separation of the lubricants is a formulator's choice. Grouping acidic and basic compounds together can raise many questions, such as: Will the acidic and basic compounds form a neutral mixture that will confound any true pH degradation effect, or will the compounds react preferentially with themselves instead of the drug? In this case, since the drug is stable over a wide pH range, both magnesium stearate and stearic acid have been grouped together to decrease the number of samples and minimize the amount of drug use. One could also choose to study neutral lubricants such as semisolid poly(ethylene glycol).

Meticulous stability sample preparation is key to getting valid stability data. This study design requires three groups of drug–excipient (diluent–binders, disintegrants, and lubricants) dry blends, three drug–excipient blends with 20% water, drug as a control, and a 20% water–drug blend as a control, for a total of eight different samples to be placed on stability under every test condition. There should be enough samples to perform all testing in duplicate and have at least one sample at each test condition in reserve for any retesting that might be required. The samples are prepared by making a common master blend for the diluent–binder group, disintegrant group, and lubricant group. A portion of each master blend is then evenly moistened with appropriate amount of water to form water-exposed samples. Drug and drug containing 20% w/w added water would serve as controls.

The samples prepared are then placed under the stability test conditions shown in Table 13.7 and analyzed for percent drug remaining at the predetermined time intervals. This particular excipient compatibility approach and stability protocol requires a total of 176 samples to be analyzed (8 initial samples × 2 for duplicate analyses *plus* 8 samples × 2 temperature conditions × 3 stability time points × 2 for duplicate analyses *plus* 8 light samples × 2 light conditions × 2 for duplicates analyses *plus* 8 samples × 4 retest conditions).

A *Plackett–Burman screening design* that incorporates the same excipients that were used in the class function design is shown in Table 13.9. Note that water has been added as a factor and that 12 blends are required for this design. Pure drug is included as a control, which brings the total number of blends to be placed on stability at 13 compared to 8 for the functional class approach. However, these mixtures are much more complex and require significantly more time and effort to prepare and assure blend homogeneity. The percentage of each excipient is indicated by L1, zero percentage excipient, and L2, a predetermined percentage of excipient depending on the excipient class and the typical drug/excipient ratio that is used in a direct compression tablet formulation. Note that Plackett–Burman blend 9 contains the drug and all the excipients that have to be weighed precisely

TABLE 13.9   Plackett–Burman Statistical Design Approach with an Ideal Drug Candidate[a]

Blend	MC	SDL	M	CNa	C	SGS	MgS	SA	H₂O	% Drug	% Degr
1	L1	L2	L2	L2	L1	L1	L1	L2	L1		
2	L1	L1	L2	L1	L1	L2	L1	L2	L2		
3	L1	L1	L2	L1	L2	L2	L2	L1	L1		
4	L1	L2	L1	L1	L2	L1	L2	L2	L2		
5	L2	L1	L1	L2	L1	L2	L2	L1	L1		
6	L2	L1	L1	L1	L1	L1	L1	L2	L1		
7	L1	L2	L1	L2	L2	L2	L1	L1	L1		
8	L2	L2	L2	L1	L1	L1	L2	L1	L1		
9	L2	L2	L2	L2	L2	L2	L2	L2	L2		
10	L2	L2	L1	L1	L2	L2	L1	L1	L2		
11	L2	L1	L2	L2	L2	L1	L1	L1	L2		
12	L1	L1	L1	L2	L1	L1	L2	L1	L2		

[a]L1, absence of excipient; L2, amount of excipient that would provide the ratio suggested in Table 13.8 except water at 20%; MC, microcrystalline cellulose; SDL, spray-dried lactose; M, mannitol; CNa, croscarmellose sodium; C, crospovidone; SGS, starch glycolate sodium; MgS, magnesium stearate; SA, stearic acid; H₂O, water; % Degr, percent degradation.

and mixed uniformly. This particular Plackett–Burman design is a "main effect" design, which can assess the effect of each excipient on the degradation of the drug and percent drug remaining. This particular design will not provide information on any potential excipient–excipient interactions that might affect drug stability.

This particular excipient compatibility approach and stability protocol requires a total of 286 samples to be analyzed (13 initial samples × 2 for duplicate analyses *plus* 13 samples × 2 temperature conditions × 3 stability time points × 2 for duplicate analyses *plus* 13 light samples × 2 light conditions × 2 for duplicates analyses *plus* 13 samples × 4 retest conditions). This is a 62.5% increase in sample analysis effort compared to the functional class excipient compatibility approach, not to mention the increase in time and effort in preparing the more complex samples required in the Plackett–Burman design approach. The advantage of the Plackett–Burman design is the ability to perform statistical analysis on the data and determine which excipients statistically compromise the drug's stability.

In either excipient compatibility approach an excipient would be considered compatible with the drug if 90% or more of the drug remains at the end of six months at 40°C. The excipients can be ranked as to their relative stability, and the highest-ranking excipients in each class would be used in probe formulation studies. This stability criterion is considered to be a conservative assessment of drug–excipient stability. Many scientists consider the drug to be stable with the excipient if 90% or more of the drug remains after three months.

13.2.2.2 Excipient Compatibility Study: Hypothetical Case Study 2 (Real-Life Hypothetical Drug Candidate)

In this case, preformulation studies were conducted on a drug that is a hydrochloride salt of a primary amine. The drug was shown to be stable, but it was poorly soluble and wettable in 0.1 N HCl. The powder was sticky, poorly compressible, agglomerated, and demonstrated poor flow. The hypothetical drug had a true density of 1.3 g/cm^3 ($1300 kg/m^3$) and a mean particle size of 25 μm. The quality product target profile states that an immediate-release oral tablet having a maximum dose of 500 mg is desirable. What running powder manufacturing process best addresses the quality target product profile requirements and the drug's physicochemical properties? The drug, at a maximum dose of 500 mg, has a number of detrimental physicochemical properties (agglomerated sticky powder that has poor solubility, wettability, flow, and compressibility) that need to be overcome by the excipients and the running powder manufacturing process. Since the drug is stable to moisture and heat, a high-shear wet granulation process offers several advantages over dry blend or dry granulation techniques. A tablet binder and wetting agent can be added to the granulating solution and can be sprayed onto the powder mixture in a uniform manner. Higher-shear mixing can deagglomerate the drug and intimately mix the binder solution and wetting agent with the drug and excipients. The granulation process can create a free-flowing, easily wetted, compressible running powder that can overcome the drug's tableting issues.

In this example, the excipient classes that would need to be investigated are diluents, wet binders, disintegrants, lubricants, antiadherants, glidants, and wetting agents. The drug's chemical structure should be taken into account when choosing

TABLE 13.10   Functional Class Approach with a Real-Life Hypothetical Drug Candidate

Excipient Class	Drug/Excipient Ratio	Excipient
Diluents and diluent–binders	1 : 4	Mannitol Microcrystalline cellulose Pregelatinized starch
Disintegrants	1 : 9	Crospovidone
Wet binders	1 : 9	Hydroxypropyl cellulose Hypromellose Poly(ethylene oxide) Povidone
Lubricant	1 : 19	Magnesium stearate Stearic acid
Antiadherants	1 : 19	Talc Starch
Glidants	1 : 49	Colloidal silicon dioxide
Wetting agents	1 : 99	Poloxamer Poly(oxyethylene sorbitan) fatty acid esters

the excipients for the compatibility study. In this case study, the drug is a primary amine hydrochloride salt. Primary and secondary amines are reported to undergo a Maillard reaction with reducing carbohydrates such as lactose, glucose (dextrose), maltose, dextrin, maltodextrin, starch, and cellulose. The aldehyde function of the reducing sugar reacts with the amine, resulting in a "browning reaction." Starch and cellulose excipients have terminal glucose units that can react with primary and secondary amines but to a lesser extent than can lactose or glucose (Roman, 2008). The more hydrolyzed the starch, the larger its reducing capability or ability to undergo a Maillard reaction. The percentage of reducing sugar in the hydrolyzed starch is referred to as the dextrose equivalent.

The drug is an acid salt, so it is positively charged. The cation may form an insoluble salt or ion pair with an anionic compound. One may wish to choose non-ionic excipients for the compatibility study to minimize possible problematic ionic interactions. Tables 13.10 and 13.11 provide the excipient functional class approach and JMP (SAS) Custom Statistical Design excipient compatibility screening study approach, respectively. These excipient studies incorporate the same testing and storage conditions as those presented in Case Study 1. The addition of more classes and excipients increases the size of this study. Lactose has been excluded as an excipient because of its potential to undergo a Maillard reaction with the primary amine function that may exist at pH values near or above the drug's pK_a value. Mannitol, a nonreducing polyalcohol, has replaced lactose as a diluent and direct compression binder. Croscarmellose sodium and starch glycolate sodium were not included in the screening study because the large croscarmellose and starch glycolate anions can potentially form an ionic interaction with hydrochloride drug salt to form an insoluble ion pair. Similarly, one would probably exclude sodium docusate

TABLE 13.11   Custom Design Approach with a Real-Life Hypothetical Drug Candidate[a]

Run	M	MC	PS	HC	H	P	PVP	MgS	SA	T	S	CSD	Tween	H_2O
1	L1	L1	L2	L2	L1	L1	L1	L2	L1	L1	L1	L2	L2	L1
2	L2	L1	L1	L2	L1	L2	L2	L1	L1	L2	L2	L1	L2	L1
3	L2	L2	L1	L1	L1	L1	L1	L1	L2	L1	L1	L1	L2	L2
4	L2	L1	L2	L1	L2	L1	L2	L2	L1	L2	L2	L2	L2	L2
5	L2	L1	L2	L2	L2	L1	L1	L1	L2	L2	L2	L1	L1	L1
6	L1	L2	L1	L1	L2	L2	L1	L2	L2	L1	L2	L1	L2	L1
7	L2	L1	L1	L2	L1	L2	L1	L2	L2	L2	L2	L2	L1	L2
8	L2	L2	L2	L2	L2	L2	L2	L2	L1	L1	L1	L1	L1	L2
9	L2	L1	L1	L1	L2	L2	L2	L1	L2	L1	L1	L2	L1	L1
10	L2	L2	L1	L1	L1	L2	L1	L1	L1	L2	L2	L2	L1	L1
11	L1	L1	L1	L1	L2	L2	L1	L1	L1	L2	L1	L1	L1	L2
12	L1	L2	L2	L2	L2	L2	L2	L1	L2	L2	L1	L2	L2	L2
13	L1	L2	L2	L2	L2	L1	L2	L1	L1	L1	L2	L2	L1	L2
14	L1	L2	L2	L1	L1	L1	L2	L2	L2	L2	L1	L1	L1	L1
15	L1	L1	L1	L1	L1	L2	L2	L1	L2	L1	L2	L1	L1	L2

[a] M, mannitol; MC, microcrystalline cellulose; PS, pregelatinized starch; HC, hydroxypropyl cellulose; H, hypromellose; P, poly(ethylene oxide); PVP, povidone; MgSt, magnesium stearate; SA, stearic acid; T, talc; S, starch; CSD, colloidal silicon dioxide; H_2O, (water).

and sodium lauryl sulfate from the excipient screen and focus on using nonionic wetting agents instead.

Table 13.10 shows the functional class excipient compatibility approach for Case Study 2. There are seven classes of excipients, compared to only three for Case Study 1. In this case there are 14 groups of blends, seven dry and seven with 20% w/w water. Dry drug and 20% w/w water–drug samples are included as controls. This brings the total number of sample groups to 16. A sufficient amount of each sample group needs to be reserved for the possibility of retesting a sample. Using the stability protocol in Table 13.8, a total of 352 assays (16 initial samples × 2 for duplicate analyses *plus* 16 sample groups × 2 temperature storage conditions × 3 time points × 2 for duplicates analyses *plus* 16 light samples × 2 light conditions × 2 for duplicate analysis *plus* 16 × 4 retest conditions) are required for this design. This compares to 176 assays required for the ideal drug in Case Study 1.

A JMP (SAS) Custom Statistical Design was used to create Table 13.11. This design minimizes the number of samples that need to be prepared and analyzed and is capable of evaluating the main excipient effects. It is not designed to assess potential interactions among excipients. In this example, the total number of sample groups is 16: 15 design groups and drug as a control. As in the previous Plackett–Burman design, L1 represents the absence of excipient and L2 represents the presence of the excipient in an amount representative of its concentration in a tablet formulation. In this case, the custom statistical design approach gives the same number of samples as the functional class approach. The functional class approach carries the downside that additional sample preparation and testing would be required if any class exhibited an incompatibility with the drug. On the other hand, the custom statistical design provides a statistical stability analysis for each excipient. However, the statistical design sample preparation is more complex and tedious.

As in for Case Study 1, an excipient would be considered compatible with a drug if 90% or more of the drug remained intact at the end of six months at $40°C$. The excipients can be ranked as to their relative stability, and the highest-ranking excipients in each class would be used in probe formulation studies. The selection of the probe formulations initiates the formulation development process.

13.2.3 Formulation Development

The aim of formulation development is to select excipients that will provide a dosage form that has the proper functionality and performance. The quality of the dosage form is a function of the excipients and the manufacturing processes used to prepare the running powder. Since the running powder manufacturing process was determined prior to the excipient compatibility study, the formulation development stage emphasizes the selection of lead formulation and evaluation of important tablet performance attributes, such as stability, content uniformity, weight uniformity, disintegration, dissolution, hardness, and friability. The impact of a formulation on the tableting process, such as powder flow, tablet punch filming, sticking, and picking, is monitored. Formulation screening is the first step in formulation development. Empirical and statistical screening approaches are used to gain an understanding about which excipients have the most impact on the critical tablet properties.

13.2.3.1 Hypothetical Tablet Formulation Case Study 1 (Continued from Excipient Compatibility Case Study 1 for an Ideal Drug Candidate) The previous hypothetical excipient compatibility cases will be used to illustrate the formulation development process. As a reminder, in Case Study 1, a dry blend direct compression tablet formulation met the quality product target profile criteria. Starting with excipient compatibility case study 1, let's for a moment assume that the excipient compatibility study demonstrated that the drug was compatible with all the excipients tested. Once compatible excipients have been determined, screening or probe tablet formulations can be fabricated and evaluated to determine the influence that various excipients have on the tableting process and important tablet attributes. Statistical screening methods can also be used to evaluate excipient impact.

Table 13.12 lists three direct compression probe formulations. These three formulations will be dry blended, tableted, and evaluated. What is the thought process for choosing these excipients and formulations? The excipient list includes two diluent-direct compression binders, one disintegrant and one lubricant. Microcrystalline cellulose is the most widely used direct compression excipient. It has excellent binding capacity, is self-lubricating, and can act as an antiadherant. Microcrystalline cellulose has reasonable flow properties but often requires a glidant to improve its flow in a formulation. Mannitol is a soluble diluent–binder that is available as a granulated powder that has good compressibility and flow. The mannitol can help improve the overall powder flow. To minimize the chance of drug

TABLE 13.12 Probe Tablet Formulations for an Ideal Drug Candidate

Component	Formulation 1 (% w/w)	Formulation 2 (% w/w)	Formulation 3 (% w/w)
Drug (100-mg dose)	50	50	50
Microcrystalline cellulose	45	—	22.5
Mannitol granules	—	40	18.5
Crospovidone	4	8	8
Magnesium stearate	1	2	1

segregation, it is especially important for a dry blend direct compression formulation that the excipients and drug particle size and density match as closely as possible. The mean particle size of microcrystalline cellulose is approximately 90 μm and its true density ranges from about 1.5 to 1.6 g/cm^3 (1500 to 1600 kg/m^3). Granulated mannitol has a mean particle size of 250 μm and its density is approximately 1.5 g/cm^3. Again, the neutral drug in Case Study 1 had a hypothetical mean particle size of 150 μm and a density of 1.5 g/cm^3. Therefore, the densities of the drug and the two diluent–binders are a match. Matching the density of the excipients with that of the drug will tend to minimize potential segregation due to density differences. It may be necessary to mill the drug and mannitol to get their particle sizes closer to that of microcrystalline cellulose to achieve good tablet weight and content uniformity. Crospovidone is known as a super-disintegrant that can be used in dry and wet running powder processes. It is also a nonionic disintegrant. Magnesium stearate is a very effective lubricant and is the most widely used lubricant. Finally, the likelihood of creating and maintaining a homogeneous drug blend is improved when the formulation contains at least 50% w/w drug. In Table 13.12 the probe tablet formulations are designed to evaluate the diluent–binders separately and in combination. The effect of disintegrant level and lubricant are compared across the three formulations.

A statistical approach to assessing the formulation requirements for hypothetical Case Study 1 is presented in Table 13.13. A JMP constrained custom mixture design with two center points was used to generate the table. The advantage of this mixture design is that it can provide a statistical assessment of the interactions that can occur between excipients that have competing objectives, such as binders and disintegrants. The design also evaluates the effect of drug concentration (33 to 66% w/w), ranges of microcrystalline cellulose (0 to 50% w/w), mannitol (0 to 50% w/w), magnesium stearate (0.25 to 1.0% w/w), and the presence and absence of crospovidone (0 to 8% w/w). A statistically valid mathematical model can be generated that describes how the various excipients affect tablet stability, content uniformity, and dissolution. Tablet characterization is not limited to these three variables. Many other running powder and tablet property attributes can be determined, such as powder flow, compressibility, tablet punch filming, sticking, picking, tablet weight variation, disintegration, capping, lamination, hardness, friability, and more. All these attributes can be modeled. The obvious downside to

TABLE 13.13   A JMP Constrained Custom Mixture Design with Two Center Points for an Ideal Drug Candidate[a]

Run	Drug	MC	M	C	MgS
1	0.66	0.25	0	0.08	0.0025
2	0.48	0.23	0.24	0.04	0.006
3	0.66	0.25	0.00	0.08	0.01
4	0.47	0.28	0.24	0.00	0.0025
5	0.40	0.30	0.24	0.04	0.006
6	0.52	0.00	0.47	0.00	0.01
7	0.33	0.08	0.50	0.08	0.01
8	0.66	0.32	0.00	0.00	0.01
9	0.33	0.50	0.16	0.00	0.01
10	0.33	0.17	0.50	0.00	0.0025
11	0.49	0.00	0.50	0.00	0.0025
12	0.33	0.16	0.50	0.00	0.01
13	0.54	0.00	0.39	0.80	0.0025
14	0.66	0.00	0.25	0.08	0.006
15	0.50	0.50	0.00	0.00	0.0025
16	0.46	0.47	0.01	0.04	0.01
17	0.66	0.00	0.34	0.00	0.0025
18	0.64	0.00	0.26	0.08	0.01
19	0.39	0.50	0.02	0.08	0.008
20	0.48	0.23	0.24	0.04	0.006
21	0.33	0.5	0.14	0.02	0.0025
22	0.33	0.08	0.50	0.08	0.0025
23	0.33	0.50	0.09	0.08	0.0025

[a] MC, microcrystalline cellulose; M, mannitol; C, crospovidone; MgS, magnesium stearate.

implementing the design is the drug substance material requirements, time, and cost associated with the study. If one made 100-g blends, this design would require over 1 kg of drug substance. This quantity of material may not be available at the early stage of formulation design.

13.2.3.2   Hypothetical Tablet Formulation Case Study 2 (Continued from Excipient Compatibility Case Study 2 for a Hypothetical Real-Life Drug Candidate)   In Case Study 2, the product target profile and the drug's physico-chemical properties led to the selection of a high-shear wet granulation process to prepare the running powders. The drug's maximum dose was 500 mg and it had a number of detrimental physicochemical properties, such as poor solubility, wettability, flow, compressibility and stickiness, that need to be overcome by the excipients and the running powder manufacturing process. To continue this hypothetical tablet case, let us say that the pregelatinized starch was shown to be incompatible with the drug and that the drug was compatible with all the other excipients tested in Tables 13.10 and 13.11. In this case, mannitol and microcrystalline cellulose remain

TABLE 13.14 Probe Tablet Formulations

Component	Formulation 1 [mg (% w/w)]	Formulation 2 [mg (% w/w)]	Formulation 3 [mg (% w/w)]
	Intragranular		
Drug	500 (50)	500 (50)	500 (50)
Mannitol	290 (29)	255 (25.5)	125 (12.5)
Microcrystalline cellulose	109 (10.9)	95 (9.5)	95 (9.5)
Crospovidone	60 (6)	100 (10)	60 (6)
Hypromellose	30 (3)	30 (3)	30 (3)
Tween 80	1 (0.1)	10 (1)	0 (0)
Water[a]			
	Extragranular		
Magnesium stearate	10 (1)	10 (1)	10 (1)
Crospovidone	—	—	40 (4)
Mannitol	—	—	140 (14)
Total	1000 (100)	1000 (100)	1000 (100)

[a] Water is used for wet granulation and removed on drying.

as diluents. Table 13.14 shows three high-shear wet granulation probe formulations to be assessed for tablet performance.

What is the thought process for choosing these three probe formulations? First, all the excipients were shown to be stable with the drug and are monographs of the *European Pharmacopoeia*, the *Japanese Pharmacopoeia*, and the U.S.P.–N.F. As diluents, mannitol and microcrystalline cellulose have good compression properties. Mannitol is highly water soluble and contributes to the overall hydrophilic nature of the tablet. However, mannitol can be easily overwetted during the wet granulation process. The addition of microcrystalline cellulose acts to adsorb the granulating fluid. This broadens the range over which the liquid can be added without overwetting the powder bed, which leads to a more robust granulation process.

Hypromellose is a very effective binder that is available in a number of viscosity grades, which provide flexibility in attaining the desired tablet hardness, friability, and dissolution. Hypromellose can be added to the drug–excipient mix as a powder, and water is added as the granulation fluid. Hypromellose can also be added to the granulation liquid to form a solution in water or in mixed-solvent systems. The granulating fluid can be sprayed or poured onto the powder blend.

Tween 80 can be dissolved in the aqueous granulating fluid before or after the hypromellose solution has been prepared. Poloxamer is a commonly used surfactant, but it is not approved in the *Japanese Pharmacopoeia* and therefore was not chosen for these trial formulations.

Crospovidone is a nonionic super-disintegrant that is effective if wetted during wet granulation (intragranular) or when added as a dry powder post-granulation-drying (extragranular). Placing the disintegrant inside the granule may help disrupt the granule that contains the binder as well as disintegrating the tablet. Some formulators prefer to divide the total amount of disintegrant into intragranular and extragranular portions, the argument being that the primary purpose of the extra-granular portion is to disrupt the tablet and that the intragranular portion is present to disrupt the granules.

In the formulations above, water is used as the granulating fluid. The hypromellose and Tween 80 are added to the water and can be sprayed or pumped uniformly onto the moving powder bed during the granulation process.

Hypromellose and magnesium stearate are held at the same concentrations in all three probe formulations. The Tween concentration ranges from 0 to 1% w/w. The crospovidone ranges from 6 to 10% w/w. In formulations 1 and 2, the drug and all excipients are granulated except magnesium stearate, which is added to the extragranular powder after drying. In formulation 3, a portion of the crospovidone and mannitol are incorporated in intragranular and extragranular portions. Magne-sium stearate is added to the extragranular blend. The actual manufacturing process is discussed in more detail in Section 13.3. The primary purpose of selecting these three probe formulations is to discern the effect that the disintegrant level, wet-ting agent level, intra- and extragranular disintegrant, and mannitol levels have on a number of critical quality tablet attributes (ICH, 2009d). Critical quality tablet attributes are discussed in more detail in Section. 13.6.

If one is lucky, one or more of the probe formulations may meet the desired attribute test requirements. Specific difficulties may also be identified which may require additional modifications of the probe formulation. The results from these probe formulations can be used to design statistical screening studies that can look more deeply into excipient–excipient interactions and drug–excipient interactions. As noted on several earlier occasions, the formulation and process are inextricably linked, which always needs to be kept in mind when accessing the effect of different excipients on running powder and tablet attributes.

13.2.4 Formulation Optimization

The formulation screening phase may take several iterations before acceptable tablet attributes are achieved. Tablet punch filming and sticking may have resulted in either Case Study 1 or 2. There are a number of options that could be studied to address this issue. The formulator could try increasing the lubricant level with hopes of not significantly decreasing the tablet hardness or dissolution rate. An antiadher-ant could be added to the formulation, or a combination of increased lubricant level and added antiadherant could be tested. In any case, additional screening studies would need to be performed. Once acceptable tablet attributes have been achieved, the formulation scientist starts formulation optimization. Formulation optimization embraces the design space concepts proposed by the FDA and ICH. "Design Space is a multidimensional combination and interaction of input variables that have been demonstrated to provide assurance of quality" (ICH, 2009d). Design space provides a relationship between formulation variables, process variables, and critical product attributes. Once the design space has been defined and approved by the FDA, the agency has suggested that one can make changes inside the space without those changes requiring regulatory preapproval.

Design of experiments was discussed in detail in Chapter 10. Design of experiments allows analysis of data that are generated from a set of prescribed experiments that include any number of variables. This is counterintuitive to the conventional scientific method, which controls all variables except one. The advantage of statistical designs is that they can identify and quantitate main variable effects and interactions that occur among variables. As mentioned earlier, drug delivery systems are very complex and generally have a number of variables and interactions of variables that affect critical quality attributes. Dissolution is often affected by interactions between binders, lubricants, and disintegrants. Lubricant and binder interactions are known to affect tablet hardness and friability. These interactions are complex, may be nonlinear, and are often not easily or adequately interpreted by the use of conventional scientific experimentation. An example of formulation and process optimization is provided in Chapter 14.

13.3   PROCESS DESIGN

There are a number of process steps or unit operations that can be used to prepare the running powder. The specific process steps or unit operations depend on the running powder process chosen. Several unit operations are common to the preparation of most running powders. Particle sizing and powder blending unit operations are used universally. In this section we discuss running powders prepared from the dry blending, high-shear wet granulation, slugging, and roller compaction methods. Wet granulation extrusion and hot-melt extrusion are discussed in Chapter 20. Tablet compression and coating are also presented in this chapter.

Several FDA guidances can serve as useful scale-up references (FDA, 1995, 1999). The equipment addendum covers the various unit operations and classifies and subclassifies the equipment by operating principles. The unit operations discussed are particle size reduction, blending and mixing, granulation, drying, unit dosing such as tableting and encapsulation, and coating–printing–drilling.

13.3.1   Safety and Industrial Hygiene

Safety and appropriate industrial hygiene consideration are important for all dosage forms. It is appropriate to discuss these important matters briefly during tablet design, since powder processing increases the safety and industrial hygiene risks. There are numerous occupational hazards that need to be considered carefully during pharmaceutical processing and manufacturing. Design of safe operating equipment is paramount. There are a number of moving parts in a tablet press and in high-shear mixers that could cause serious injury. The equipment should have appropriate emergency stops, lockouts, local exhaust ventilation, and containment. Fine aerosolized powder can be very explosive. In special cases, explosion-proof wiring, grounding, and explosion venting should be incorporated into the equipment design. Employees should be properly trained in the use of personal protective equipment and in each piece of manufacturing equipment that they use. Facilities and equipment should be designed to minimize operator exposure to occupational

TABLE 13.15   Occupational Exposure Limits Categorization and Performance-Based Exposure Controls

	Category 1: >1000 μg/m³ᵃ	Category 2: 100 to 1000 μg/m³	Category 3: 10 to 100 μg/m³	Category 4: 1 to 10 μg/m³	Category 5: <1.0 μg/m³
Recommended handling controls	Local exhaust and dust mask	Downflow booth and dust mask	Ventilated closure and dust mask	Ventilated closure and cartridge respiratory	Isolators and powder air purifying respirator

ᵃ Aerosolized particulate occupational exposure level.

particulates. Personal protective equipment should not be relied on as a primary source of personal protection.

A company should establish an industrial hygiene safety (IHS) program that entails compound safety and health assessment for all chemicals used in the manufacture of pharmaceutical dosage forms. Risk assessment should be carried out on solvents, excipients, and drug entities. An IHS program includes hazard identification, compound categorization, exposure control guidelines, and exposure assessment guidelines. Exposure control guidelines are based on occupational exposure limits (OELs) or performance-based exposure control limits.

The no observed effect level (NOEL) is used by the pharmaceutical industry as the basis for establishing OELs. The OEL concept develops a tolerance exposure limit for compounds. In general, a NOEL is identified for the most sensitive pharmacological or toxicological endpoint. The setting of the OEL involves the use of safety factors and uncertainty factors. The OEL is an 8-hour time-weighted average-allowable exposure concentration. If a NOEL is not available, other parameters can be used to establish an OEL, such as the lowest observed effect level or the lowest therapeutic dose. An example of a compound OEL categorization chart is provided in Table 13.15.

The handling controls recommended for each category are based on the physical form (solid, liquid, or gas–vapor) and the quantity of the compound. The greatest concern is material that can be inhaled. Therefore, once a solid compound that can potentially generate dust is in a liquid form, say a solution, the recommended handling controls become less stringent. Equipment manufacturers build compound containment into their designs, such as glove boxes, negative air pressure, controlled venting, and clean-in-place systems. A laminar flow hood can provide an OEL of 50 μg/m³ and above, while a downflow booth can provide an OEL level of 30 μg/m³. Glove box isolator systems provide OELs below 1 μg/m³.

13.3.2   Particle Sizing

13.3.2.1   *Screening and Size Classification*   Typically, the more uniform a drug and excipient particles' size, shape, and densities, the fewer the content uniformity and segregation problems. True density cannot be changed, but sizing techniques can be used to attain a desired particle size and size distribution. Screening is the simplest method of particle sizing if the particles are loosely aggregated

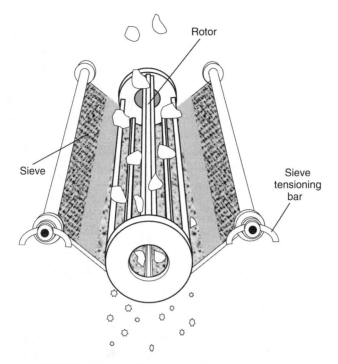

FIGURE 13.5 Oscillating screen/low-shear granulator.

and true particle size reduction is not needed. Small quantities of powder, say less than 1 to 2 kg, can be hand screened in an industrial hygiene-friendly containment system. Vibrating screen assemblies can be used to handle larger quantities. As the industry moves to continuous process and higher dust container systems, the screening system can be placed in-line with powder vacuum transfer systems. Oscillating screening granulators can be used to screen powder agglomerates and reduce the size of particles. Oscillating screening granulators (Figure 13.5) can also be used to size wet granulations prior to drying. The throughput rate depends on the size and type of screen system.

Sieve screen sizes are listed in Appendix 13.10. Sieves are made of wire mesh with openings of known size. The term *mesh* is used to denote the number of openings per linear inch. The opening size of the screen is given in micrometers. A desired particle size range or distribution can be classified by using a series of nested screens. A nest of screens or sieves is placed on a shaker or other appropriate assembly where the top sieve is the largest desired particle size and lower sieve is the smallest acceptable particle size. Material retained on the largest screen and passing through the smallest screen are discarded or reworked. The material is classified as having passed through the largest sieve and been retained on the smallest sieve. Additionally, a smaller nest of sieves can be used to acquire a cumulative weight particle size distribution histogram, which can be useful in assessing and comparing granulations.

13.3.2.2 Milling and Size Reduction Few materials used in the fabrication of pharmaceutical dosage forms exist in the desired or optimum size. Milling is the mechanical process of reducing the particle size of solids to achieve material in the desired size range. Size reduction of APIs, excipients, wet granulations, and dry granulations is a common pharmaceutical unit operation. There are a number of reasons to reduce the particle size. Drug particle reduction can increase the surface area to improve dissolution and bioavailability. Obtaining material in the proper size and shape range can improve such formulation properties as content uniformity, powder flow, and bulk density. The selection of a mill depends on the final particle size and material characteristics desired, such as heat sensitivity, cohesiveness, fibrous nature, and particle hardness. Other considerations include powder explosivity and powder containment. The heat of milling may soften or melt material, making it sticky and difficult to mill. Materials that are highly unstable to heat may undergo degradation during milling. Ball milling up to 240 minutes has resulted in reported temperature increases of more than 32°C (Chieng et al., 2006). Commercial vendors of hammer mills state that these mills exhibit a low temperature rise. Dozier et al. (2005) performed a statistically designed hammer mill experiment that evaluated the effect of mill screen size and production rate on electrical consumption, particle size, and temperature rise. In this study the maximum temperature rise was 3.4°C. These materials may require special cryogenic milling conditions to decrease the product temperature and to make the material harder and more susceptible to brittle fracture. Liquid nitrogen and carbon dioxide are the most common cryogens. When nitrogen or carbon dioxide cryogens are used, work areas need to be monitored actively for oxygen levels. Oxygen levels must not go below 19.5% v/v. Special humidity control may be required for hygroscopic drugs. As little as 5% moisture uptake can decrease the milling efficiency and cause the material to become sticky and clog the mill. Hydrates and other solvates may liberate their solvent of crystallization under mechanical stress or increase in temperature. Higher-energy forms of drug substances may result from increased imperfections that are produced during milling. Higher-energy forms have faster dissolution rates and potentially better bioavailability. It has also been suggested that the formation of amorphous material results from local thermal melting followed by rapid quenching and amorphization. Descamps et al. (2007) suggested an athermal ballistic or dynamic mechanism for amorphization and crystal form transitions based on the position of the glass transition temperature (T_g) of the material and the milling temperature. Ball milling below T_g resulted in the formation of amorphous material while milling above T_g resulted in crystal form changes. The potential for physical form changes needs to be kept in mind when evaluating a milling process. Trasi et al. (2010) studied the thermal behavior of milled griseofulvin. Milling resulted in a decrease in the crystallinity of a compound. Feeley et al. (1998) and Heng et al. (2006) studied the effects of micronization and ball milling on the surface properties of drugs.

The rheology of solids or the study of the flow of solids helps explain milling and particle fracture. For a review of rheology, see Chapter 7. A solid material "engineering" stress–strain profile is provided in Figure 13.6. Engineering stress and strain measurements are based on the original dimensions of material and

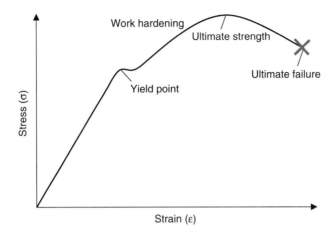

FIGURE 13.6   Plastic deformation engineering curve.

do not take instantaneous changes into account. The profile has two components: *elastic deformation* and *plastic deformation*. The initial profile is linear and is described by *Hooke's law*, which states stress is proportional to strain. Strain is a unitless term that measures deformation as (1) linear strain, change in length per linear unit; (2) angular rotation, in radians; or (3) volumetric strain (change in volume per volume units). The linear stress–strain profile is consistent with an elastic material. As stress is increased, the elastic limit of the material is exceeded and the material is deformed permanently. This area of irreversible plastic deformation is described by a nonlinear stress–strain profile. The yield strength or yield point of the material is the force per unit area (stress) at which strain increases without additional applied stress. This particular stress–strain curve shows an area beyond the yield point, called *strain hardening* or *work hardening*, where there is resistance to plastic deformation with increases in stress until the material reaches its ultimate strength or tensile strength. This strain hardening is similar to the process of "working" steel to make it harder. The tensile strength of a material is determined by the maxima in the stress–strain profile. Ultimate material failure occurs at the material's ultimate breaking point. The stress resulting in breaking is the *breaking strength* of the material.

Milling and subsequent particle size reduction of pharmaceutical materials result from material failure due to brittle or plastic fracture. Brittle fracture occurs during elastic deformation and fractures without yielding, much like the breaking of glass. For brittle material the breaking strength and tensile strength are the same. Plastic fracture occurs for materials that have reached their plastic deformation limit, which usually results for an accumulation of dislocations that build up during strain hardening.

For particles to fracture, the milling stress and force per unit area must exceed the material's breaking strength. All solid materials have minute flaws or dislocations. The flawed areas are most susceptible to forming cracks under stress. The stress increases along the crack and is concentrated as the crack depth increases. As the stress is increased, the breakage strength of the material is exceeded and fracture

TABLE 13.16 Particle Sizes Obtained by Various Mills

Type	Coarse: 1000–5000 μm	Medium: 500–1000 μm	Medium Fine: 150–500 μm	Fine: 50–150 μm	Suprafine: 10–50 μm	Ultrafine: 1–10 μm
Compression (conical/screen)	×	×	×			
Impact (hammer/pin)		×	×	×		
Impact (ball mill)				×	×	×
Fluid energy (jet mill)				×	×	×

Source: Data from Hosokawa Micron Powder Systems.

occurs. Brittle fracture is essentially temperature independent and independent of the stress rate. Plastic material fracture is temperature dependent and increases with increasing temperature. The breakage strength of plastic materials is also dependent on the number of stress cycles. Increases in the number of stress cycles lead to stress fatigue and stress fractures.

Mill selection depends on the type of material being milled and the particle size desired. Fibrous material typically requires a mill that cuts compared to mills that crush by pressure or impact. The most common type of mills can be categorized as compression (cone mills), impact (hammer and pin), and fluid energy (jet) mills. The different mills and the particle sizes that can be obtained from the various mills are shown in Table 13.16.

A compression mill has rotating blades that press the powder through screen openings. There are a number of variables that can be adjusted to achieve the desired particle size and distribution, such as blade shape, blade rotation speed, clearance distance between the blade and the screen, and screen size and shape. A cone mill and several size screens are shown in Figure 13.7, and the blade configuration is shown in Figure 13.8. Cone mills can be used for dry powders or wet granulations. The typical size range obtained by this type of mill is 150 to 5000 μm.

Hammer and pin mills are used to reduce dry powders to 50 to 1000 μm. Particle size and distribution can be controlled by varying the shape of the hammers (blunt hammer or knives) or pins, rotation speed, and screen size. Figures 13.9 and 13.10 show a hammer and a pin mill, respectively. The blunt hammer blade is used for crushing on impact, while the knive blade is more useful in reducing fibrous material with its cutting action.

Fluid energy mills can achieve particle size ranges of 2 to 50 μm. Figure 13.11 is a photograph of a spiral fluid jet mill. The advantage of a jet mill is that it has no moving parts. High-pressure gas (60 to 100 psi) causes particle–particle attrition. Centrifugal force keeps the particles in the mill until they are small enough to leave with the vortex airstream. Nitrogen gas can be used as an inert atmosphere to minimize dust explosion potential. The mill can also be chilled to make the particles more brittle and easier to micronize. self-contained clean-in-place fluid energy mill

FIGURE 13.7 Cone mill.
(Courtesy of Mendel Corporation.)

that can maintain an OEL of less than 1 µg/m^3 is shown in Figure 13.12. Milling waxy or sticky materials is difficult. It is often necessary to cool these types of materials to make them more brittle and amenable to brittle fracture.

13.3.2.2.1 PAT: Milling A focused beam reflectance measurement (FBRM) process analytical technology tool has been used to monitor changes in particle size resulting from crystal growth, milling, and granulation. FBRM instrumentation supplied by Lasentec has been used to measure real-time in-line granulation particle size manufactured by a continuous wet granulation–drying–milling process (Kumar, et al. 2010). The FBRM instrument employs a rotating laser optics system that is projected onto a particle. The instrument measures the chord lengths of the particles, determined by the time that the rotating laser beam remains in contact with the particle. The rotating scan speed can be adjusted from 2 to 8 m/s, so the system can acquire thousands of chord lengths per second. A cord length distribution (CLD) is obtained in real time. A central composite design with two center points, two factors, and three levels was used to compare the CLD obtained from FBRM and particle size distribution data (PSD) obtained from conventional sieve analysis. Milled granule samples were collected at designated times, and sieve analysis was compared to the corresponding real-time CLD data. Both FBRM and conventional sieve analyses resulted in similar bimodal distributions for all 10 batches manufactured. FBRM data of CLD showed a significant correlation with the conventional PSD.

FIGURE 13.8 Cone mill blade configuration. (Courtesy of Hosokawa Micron Powder Systems.)

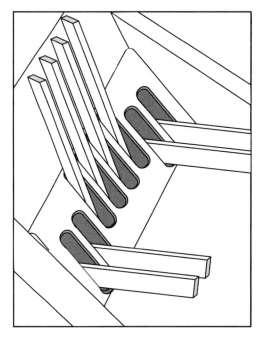

FIGURE 13.9 Hammer mill.

FIGURE 13.10   Pin mill. (Courtesy of Hosokawa Micron Powder Systems.)

A significant relationship (p value < 0.0001) was determined for D20, D50, and D80 between PSD and CLD. The coefficient of determination (r^2) was 0.886.

13.3.2.2.2 Scale-up of the Milling Process   The goal of scale-up is to have the material experience a similar processing environment as the mass of the material or scale is increased. The shape of the equipment, operating principles, forces exerted on the material, and processing time are considered important parameters that affect the scalability. In general, particle *comminution* or size reduction will depend on the tensile strength of the particle, the size and size distribution of the particles, the milling energy, and the particle residence time. As shown in Table 13.16, there are a number of different types of mills that use a variety of operating principles to reduce particle size. Each mill type will have critical operating parameters that specifically affect scaling. Several general precepts apply to comminution. The energy required to achieve a given particle size is inversely proportional to the starting particle size. Smaller particles have proportionally fewer flaws, and the efficiency of size reduction decreases as the particles become smaller. Therefore, smaller desired particle sizes will require higher-energy mills, such as fluid energy mills. The rate of comminution depends on the initial particle size and the strength of the particles. Higher rates of particle size reduction are seen with

FIGURE 13.11   Fluid energy jet mill. (Courtesy of Hosokawa Micron Powder Systems.)

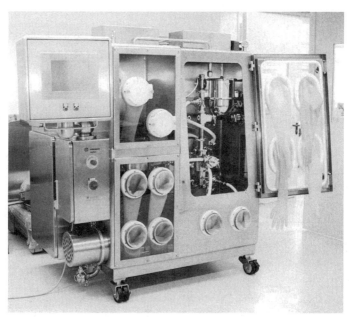

FIGURE 13.12   Fluid energy mill in self-contained clean-in-place isolator. (Courtesy of Hosokawa Micron Powder Systems.)

larger-particle-size starting material, and slower rates of communition occur when the particles are stronger and resist milling.

To have the material experience a similar milling environment as the mass of the material is increased, the "principle of similarities" is evoked. Ideally, the mill operating principle and geometry are the same. For example, both pilot and production mills are fluid energy mills that have geometric point-to-point correspondence and are exact scaled replicas. The scaled process should also have dynamic similarity, which involves the milling force experienced by the particles. Hammer mills reduce the particle size by a combination of impact by the hammer, particle-to-particle attrition due to centrifugal force, and the impact of the particle on the cutting edge of the screen. The tip speed of the hammer is ND_p, where N is the angular velocity or rotation speed and D is the diameter of the hammer. Hammer mill tip speeds can reach 100 m/s, which translates into high kinetic energy. This energy is translated into high impact force. Setting the tip speeds of the pilot- and production-scale mills to the same value should provide similar impact forces on the particles. Another important similarity parameter is kinetic similarity, which accounts for the time that the material is being milled. Kinetic similarity is represented by the milling mean residence time, which is given by $t = M/F$, where M is the mass of powder in the mill and F is the throughput or mass flow rate. The principle of similarities supports the notion that the particle should experience the same impact force and mill residence time.

Thyn (1989) developed a mathematical model that evaluated residence time, product quality, and mill energy consumption as a function of number of hammer rows, number of hammers in a row, mesh screen size, velocity, and mill output. Most of the experimentation was done at a pilot scale with several verifying experiments done at full scale. Radiotracer materials were used to verify the model.

13.3.3 Powder Mixing and Blending

Powder mixing and blending are used synonymously in this chapter. The objective of powder mixing is to generate a uniform mixture with a homogeneous distribution of active ingredient(s) that does not segregate during downstream processing such as granulation, tableting, and encapsulation. As such, blending is an integral part of the overall manufacturing process. The homogeneity of powder blends and dosage units are critical quality attributes. Powder blending is the redistribution or reshuffling of individual particles or groups of particles to create a homogeneous mixture that is usually based on the sample size of a single dosage unit. This critical sample size has been referred to as the *scale of scrutiny*, which is based on the practical need to have a minimum blend weight or unit exhibit an acceptable level of homogeneity. Powder blend homogeneity and the acceptable degree of mixing are discussed in more detail in Chapter 14.

There are four general mechanisms of powder mixing: *diffusion, convection, impact,* and *shear*. Diffusion is described to occur randomly as individual particles or microscale randomization. This random movement by fluidization can be caused by the powder bed expansion during mixing or powders rolling over sloping surfaces, such as in V-blenders or tote tumble blenders. Convection is the movement of

FIGURE 13.13   V-blender. (Courtesy of Mendel Corporation.)

groups of particles within the powder bed, resulting in macrorandomization. Convection mixing can occur in tumble blenders, conical screw blenders, or ribbon blenders. Shinbrot and Muzzio (2000) provide a number of colored illustrations of granular mixing and segregation. Impact mixers are used to disrupt aggregates of cohesive powders. Choppers on high-shear mixers and pin mixers act to deaggregate agglomerates. Shear causes reorganization of particles through the movement of shear planes caused by mechanical stress. High-shear mixer–granulators and cyclomix high-intensity mixers apply significant shear to powder beds. Examples of a V-blender and a cyclomix high-intensity mixer are shown in Figures 13.13 and 13.14. Selection of a specific type of mixer depends on the powder characteristics, such as particle size and distribution, shape, density, moisture, cohesive and adhesive forces, and static charge. Cohesive and agglomerated powders require greater degrees of shearing energy.

The binomial probability distribution can be used to predict the theoretical best-case level of blend uniformity that can be achieved based on the sample size (number of total particles) and the proportion of drug and excipient. The binominal distribution is used to describe the probability of discrete random variables that result in only one of two mutually exclusive outcomes. As mentioned earlier, the objective of powder mixing is to generate a uniform mixture with homogeneous distribution of active ingredient(s) that does not segregate during downstream processing, such as granulation, tableting, and encapsulation. In this case, the sample contains either drug or excipient, which are mutually exclusive. The binomial distribution population mean is $\mu = np$, where n is the total number of particles in the sample and p is the probability of finding a drug particle which is equivalent

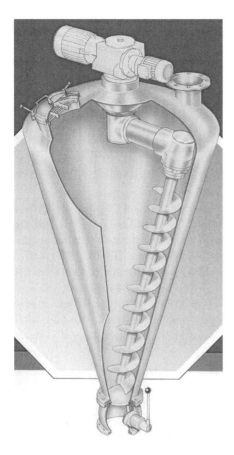

FIGURE 13.14 Conical screw mixer. (Vrieco-Nauta; courtesy of Hosokawa Micron Powder Systems.)

to the proportion of drug in the sample, assuming that the drug and excipient particles have the same density and size and do not interact. The population variance is $s^2 = np(1 - p)$. The powder blend sample size is usually one to three times the target weight of the tablet. One can determine the number of particles in the formulation by knowing the material's volume number diameter (see Chapter 9), density, and the weight of material used in the mixture. Table 13.17 illustrates how the number of particles in the tablet or blend sample and the proportion of drug in the tablet or blend affects the theoretical best-case uniformity.

This illustration shows that one can decrease the theoretical drug % RSD in a powder blend by increasing the total number of sample particles and increasing the proportion of drug. This example also supports one's intuition that the smaller the number and proportion of drug particles in a tablet or blend, the more difficult it is to achieve a uniform blend. Stagner et al. (1991) studied the content uniformity failure of a capsule that contained two sets of beads that were blended and filled into the capsule. The binominal distribution was used to show that even in an ideal case, blending would result in a content uniformity having percent relative standard deviation of 5.6%. The actual mean percent relative standard deviations of

TABLE 13.17 Theoretical Content Uniformity Based on a Binominal Distribution

Number of Particles	Proportion of Drug	Mean	Variance	% Relative Standard Deviation[a]
1000	0.5	500	250	3.16
1000	0.1	100	90	9.49
1000	0.9	900	90	1.05
200	0.5	100	50	7.07

[a] % Relative standard deviation = $\{[(s^2)^{0.5}]/\text{mean}\} \times 100\%$.

five production drug product lots was 5.5%, which confirmed the a priori statistical calculation based on the binominal distribution. The practical implications of this finding suggests that changes in blender design, mixing time, or mixing process would be fruitless and would not lead to improved content uniformity. The authors went on to show that a 3% RSD was possible by increasing the percent of active beads from 60% to 95%.

The random blending process has been described by first-order rate expressions. As discussed above, even ideally blended samples will have a certain level of random mixture variability, which is the best variability expected for real samples. The first-order mixing expression is

$$\text{Ln}(\% \text{ RSD}_t - \% \text{ RSD}_\infty) = \text{Ln}(\% \text{ RSD}_0 - \% \text{ RSD}_\infty) - kt \qquad (13.1)$$

where % RSD_t is the percent relative standard deviation of the blends at time, t, % RSD_∞ is the percent relative standard deviation at time infinity, and % RSD_0 is the relative standard deviation at the initial blend state. The first-order rate constant is k in t^{-1}. Segregation or demixing is often observed for real blends, so the % RDS_8 is taken at the time point where the % RSD is at its minimum.

13.3.3.1 PAT: Powder Mixing and Blending
Near-infrared and light-induced fluorescence (LIF) have been used as PAT tools to monitor the blending process in-line in real time. Recently, LIF was used to measure the blending endpoint for a commercial product (Karumanchi et al., 2011). Two granule blends, each containing an active pharmaceutical ingredient, were added in a 25 : 75 ratio to a 124-L intermediate blending container. Powder blend samples were taken at 10 locations at 0.25, 0.5, 0.75, 1, 2, 5, 10, and 20 minutes using a stainless steel sampling thief. The LIF probe was placed on the surface of the powder, and five fluorescent measurements were taken. Active mixing took place during the first minute of blending. Steady-state blending was achieved at the 2-minute time point.

Near-infrared (NIR) spectroscopy has also been used to monitor the blending process. Moes et al. (2008) used in-line NIR to monitor the blending and content uniformity in a tumble blender using in-line NIR. The drug concentration ranged from 1.98 to 9.25% w/w, and the percent relative standard deviation ranged from 1.0 to 2.5%. Assessing blend and final dosage form content uniformity has been subject to several working party reports (FDA, 2003a; and ICH, 2010a–d).

13.3.3.2 Scale-up of the Blending Process   Effective scale-up of tumble blenders depends on maintaining comparable geometry, rotational speed, and number of rotations. Garcia and prescott (2008) provided an example of scaling a 20-L V-blender to a 200-L V-blender. To maintain the same geometric shape, the blender angles should remain the same and the relative fill height should be kept constant. It is also important to use the same fill pattern during scale-up. The authors used the Froude number, Fr, to calculate the required rpm and blending time for a 200-L vessel that would provide mixing dynamics comparable to those obtained in the 20-L V-blender.

$$Fr = \text{rpm}^2 \frac{r}{g} \qquad (13.2)$$

where rpm is the blender rotations per minute, r the blender's characteristic radius, and g the acceleration of gravity. In this case the characteristic radius will increase $10^{0.33}$ or 2.14 times for the larger blender, where the volume of each blender is volume $\sim r^3$. The Froude number for each size of blender can be used to calculate the comparable rpm for the 200-L blender. Acceptable homogeneity was obtained in the 20-L vessel when the powder was mixed for 20 minutes at 8 rpm.

$$\frac{Fr_{20-L}}{Fr_{200-L}} = \frac{\text{rpm}^2_{20-L} \cdot r_{20-L}/g}{\text{rpm}^2_{200-L} \cdot r_{200-L}/g} \qquad (13.3)$$

Since the goal is to keep the Froude number constant, the ratio of the Froude numbers is 1 and the rpm for the 200-L V-blender is

$$\text{rpm}_{200-L} = \left(\frac{\text{rpm}^2_{20-L} \cdot r_{20-L}}{r_{200-L}} \right)^{0.5} = \left(\frac{8^2}{2.14} \right)^{0.5} = 5.5 \text{ rpm} \qquad (13.4)$$

To achieve a comparable 160 rotations used for the 20-L blender, the 200-L blending time would be 29 minutes. Scaling factors (SFs) for V-blenders are given in Table 13.18, where

$$\text{blending time}_{\text{larger blender}} = \frac{SF_{\text{larger blender}}}{SF_{\text{smaller blender}}} \cdot \text{smaller blender time}$$

TABLE 13.18 Scale-up Factors for V-Blenders

Working Capacity (L)	Scale-Up Factor
28	1.3
85	1.6
142	1.7
566	2.0
1416	2.2
2832	2.8

Source: Harsco Corporation, Harsco Industrial Patterson-Kelley, East Stroudsburg, PA.

13.3.4 Granulation

Granulation is a particle design process that creates larger "agglomerated or granulated" particles from smaller particles to improve the performance characteristics of the running powder. Table 13.3 outlines the two methods of granulation: wet and dry. Wet granulation incorporates a volatile liquid in the powder that is subsequently "wet-massed" to form wet granules which are subsequently sized and dried. Dry granulation does not use a liquid to form the granules. Generally, granulation is used to overcome detrimental physical properties of the drug and excipient lot-to-lot variability. The drug may be needle-shaped, cohesive, poor flowing, and/or poorly compressible. The primary reasons for granulating are to improve running powder flow properties, to improve compressibility, to improve the distribution of binder and wetting agents, to improve the content uniformity, to minimize potential segregation, to increase bulk density, and to decrease dusting, which can lead to unacceptable levels of operator exposure. Improving powder flow and powder flow uniformity can significantly improve tablet, capsule, and powder sachet dose uniformity. There is a limit to how large a tablet or capsule can be and still achieve patient acceptance. This size limitation becomes a serious problem with high-dose drugs. The granulation process causes the powder blend to become denser than the bulk powder. The higher-bulk-density granules can be filled into capsules or tablet dies more readily with higher fill weights.

Wet extrusion spheronization and hot-melt extrusion granulation are discussed in Chapter 20.

13.3.4.1 *Wet Granulation*

The advantages and disadvantages of wet granulation are listed in Table 13.4. Wet granulation incorporates a volatile liquid into the powder with the aim of enlarging the overall particle size distribution. The most common granulating liquid is U.S.P. water because it can be obtained in high quality and quantity; it is nontoxic and is nonflammable or explosive. Since ethanol is flammable, explosive, and expensive, it is reserved for use when its solvency is required to dissolve the drug or excipients, or when the drug is unstable in water. Hydroethanolic solutions decrease the explosion risk and cost but still require explosion-proof-rated equipment and electrical systems.

To design the desired granule size, structure, friability, and compressibility, it is important to understand the mechanisms that result in particle interaction and granule growth. Rumpf (1962) described the mechanisms of interparticle interaction and bonding. During wet massing the amount of liquid added usually produces a mobile liquid film. There are four mobile film states that relate to the amount of liquid and the type of liquid distribution: *pendular, funicular, capillary, and droplet*. The first three states are important in wet granulation. The fourth state results in an overwetted granulation. Several equations help provide an understanding of these stages and the strengths of moist granules. The mobile states are related to how much liquid is in the granule pores. The relative liquid saturation in the granule pores is given by

$$S = \frac{M_l}{M_p}\left(\frac{1-\varepsilon}{\varepsilon}\right)\rho \qquad (13.5)$$

where S is the relative saturation of granule pores, M_l is the mass of granulating solution, M_p is the mass of powder blend, ε the porosity of the granule, and ρ the true density of the powder blend. The first stage of liquid saturation is the pendular state, where liquid bridges cause adhesion through surface tension and capillary force. When the granule void volume is filled with the granulating liquid, the capillary state has been reached. The funicular state is an intermediate state between the pendular and capillary stages where mechanical kneading and consolidation forces increase the density of the granules and force air out of the void spaces. The strength of a moist idealized spherical monodispersed granule is given by

$$\sigma = SP \frac{(1-\varepsilon)}{\varepsilon} \frac{\gamma}{d} \cos\theta \qquad (13.6)$$

where σ is the moist granule strength, S the relative liquid saturation level, P the powder blend material constant, ε the granule porosity, γ the surface tension of the granulating liquid, d the diameter of the capillary wall, or in this case the distance between two particles, and θ the contact angle of the granulating liquid and powder blend. The capillary suction pressure is given by the final two terms in equation (13.6). To increase granule strength the formulator can increase the relative granulating liquid saturation level, (i.e., percent w/w binding liquid), decrease the granule porosity and distance between particles by kneading time and energy, increase the granulating liquid's surface tension, and decrease the contact angle between the powder blend and the granulating liquid. There may be a trade-off between increasing the liquid's surface tension and decreasing the contact angle. Surfactants can be added to the granulating liquid to decrease the contact angle to improve the wetting of the powder. Adding surfactants will also decrease the surface tension of the granulating liquid, which has a negative effect on granule strength. One way to decrease to θ without adding surfactants to the granulating liquid, is to add more hydrophilic excipients, such as mannitol, pregelatinized starch, and microcrystalline cellulose, to the powder blend. Binders also affect the surface tensions of the binder liquid. For example, a 15% w/w povidone solution has a surface tension at 20°C of 57.5 mN/m (viscosity = 15.8 mPa · s), and a 6.3% w/w hypromellose solution has a surface tension of 44.0 mN/m (viscosity = 15.8 mPa · s) compared to water, which has a surface tension of approximately 71.6 mN/m (viscosity = 3 mPa · s) (Pepin et al., 2001).

Immobile liquid films that have higher bond strengths than those of the mobile films mentioned above may be obtained by using higher-viscosity granulating solutions. One issue with using highly viscous granulating liquids is achieving uniform distribution of the granulating solution. The formulation has the best chance of obtaining uniform distribution by using a high-shear granulator and when possible, spraying the liquid onto the powder bed. If the granulating liquid is too viscous to spray, the liquid should be added slowly to the powder bed near the choppers to get as uniform coverage as possible.

The granulation process can be described by three stages or rate processes: the wetting and nucleation stage, the consolidation and growth stage, and the attrition stage. The liquid–powder bonding mechanisms discussed above play a key role in the early stages of granule formation, called the nucleation stage. The nucleation stage is characterized by a wide particle size distribution of capillary granules. The growth of the nucleated granulations occurs through consolidation of the nucleated and small consolidated granules. At some point the bonding forces are overcome by mechanical stress and granule size, resulting in granule attrition. The final granule size and strength are a function of these three rate processes.

The granule drying process may provide additional granule strength that results from the formation of solid or viscoelastic bridges caused by the crystallization of dissolved substances, softening or partial melting of low-melting-point materials, and an increase in viscosity and hardening of the binder.

Higher-shear and fluid-bed granulation are the most commonly used wet granulation techniques employed today. They have superseded the use of low-shear planetary mixers, ribbon mixers, and tumble mixers. High-shear and fluid-bed granulation is also commonly used in tandem with microwave/vacuum drying and fluid-bed drying as a "single process." Single-pot or single-container processes are discussed in Section 13.3.5.

13.3.4.1.1   High-Shear Wet Granulation   A high-shear granulator can be used both for mixing powders as well as preparing a wet granulation. The high-shear mixer granulator has a main impeller and chopper (see Figure 13.15). The rotation speed of the main impeller can be varied depending on the use. The impeller speed can vary from approximately 10 to 20 rpm for wet granulation discharge, to 60 to 90 rpm for powder mixing, to 80 to 160 rpm for wet massing and kneading. The choppers function at much higher speeds and serve to improve the content uniformity of dry powder blends and the wet mass. The choppers are also used

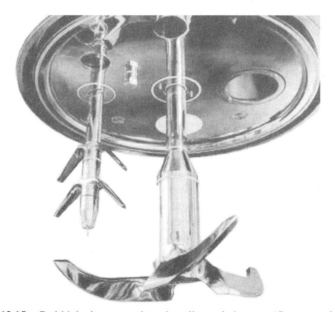

FIGURE 13.15   Gral high-shear granulator impeller and chopper. (Courtesy of GEA Niro Pharma Systems.)

to break up large agglomerates into finer granules. High-shear granulator choppers usually have two settings. A typical chopper speed range is 1000 to 3000 rpm. Powder mixing times range from 1 to 10 minutes. Wet massing endpoints can be determined by mixing time, using power consumption curves, a power measure of main impeller, or a combination of methods.

High-shear wet granulation is especially useful for obtaining acceptable content uniformity for low-dose tablets or capsules. The drug can be solubilized or suspended in the granulating liquid and sprayed onto the powder bed. This granulation technique accomplishes two goals. First, it is easier to obtain a uniform distribution of drug in the granulating powder. Second, once the drug is distributed uniformly onto the powder, it is more likely to stay associated with the granule, resulting in less chance of the drug undergoing segregation. The input mixing energy results in increased product temperature, which can affect the chemical and solid-state stability of drugs and excipients. The increase in product temperature depends on the mass of powder loading, the impeller speed, the percent granulating liquid, the power consumption, and the time. Temperature rises of 20 to 35°C have been reported (Betz et al., 2004; Ghrorab and Adeyeye, 2007).

13.3.4.1.1.1 *PAT: High-Shear Granulation* Leuenberger et al. (1979) monitored the granulation process using power consumption in a low-shear planetary mixer and a Z-mixer. The authors identified five power consumption stages and related those to different granulation phases. The granulation phases that could be distinguished by the power consumption curves were moistening of the powder without perceptible agglomeration, particle agglomeration, acceptable granule formation, overwetting and dough formation, and significant overwetting and suspension formation. Other articles have evaluated the granulation process using a capacitive sensor (Fry et al., 1987), sound and vibration (Briens et al., 2007), and torque (Ghorab and Adeyeye, 2007). Many commercial high-shear granulators provide torque and power consumption readouts to monitor the granulation process and determine the granulation endpoint.

13.3.4.1.1.2 *Scale-up of High-Shear Granulators* There are a number of equipment, formulation, and process parameters that can affect the scale-up of high-shear granulators. Equipment parameters include bowl size and shape, impeller and chopper speeds and design, location of granulating liquid addition ports, powder bed temperature control, and location and size of granulation discharge ports. Other parameters include the powder mass, particle size and solubility of the powder, amount of granulating fluid, type and viscosity of granulating liquid, rate of addition, addition method, and location of granulation liquid addition, to name a few.

Smaller-capacity high-shear granulators are more efficient than larger granulators. This is illustrated in Figure 13.16, which shows the relationship between tip speed and granulator size when the same empirical stress is applied to the granulation mass. It can be seen that as the granulator size increases, there is a corresponding need to increase the granulator tip speed to maintain an equivalent shear stress. Rekhi et al., (1996) evaluated a number of high-shear granulation

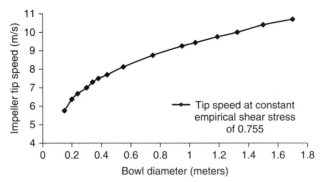

FIGURE 13.16 Granulator bowl size and impeller-tip-speed relationship. (Adapted with permission from a GEA Niro Pharma Systems brochure.)

scale-up parameters and concluded that the granulation process is scalable when the impeller tip speed is kept constant; the volume of granulation liquid is based on the ratio of the increased batch size while liquid addition time is held constant, and the massing timing is the ratio of the impeller rpm. Newer high-speed granulators are designed to maintain constant tip speed when scaling a process.

An empirical equation has been developed for the scale-up of high-shear granulators which maintained the same bowl shape ratio, relative height in the bowl for a given percent fill, and scaled impeller and chopper sizes. The empirical equation allows one to calculate the number of impeller rotations per minute (rpm) that is expected to provide similar blend forces. Previously, the Froude number was used to calculate the change in rpm when scaling from a 20- to a 200-L V-blender. Equation (13.7) is similar to (13.4) except that the power term ranges from 0.80 to 0.85, which was determined experimentally (Tardos et al., 2004):

$$\text{rpm}_2 = \text{rpm}_1 \left(\frac{d_1}{d_2} \right)^x \tag{13.7}$$

where d_1 and d_2 are the operating diameters of the two granulators and x is the empirical stress constant, ranging from 0.80 to 0.85.

Landin et al. (1996) studied the scale-up endpoint for high-shear wet granulators [PMA 25 (7.5 kg), PMA100 (30 kg), and PMA 600 (180 kg)]. The authors used classical dimensionless numbers (the Power, Reynolds, and Froude numbers) to predict the granulation endpoint of scaled batches. The authors concluded that granulation in a high-shear granulator behaves much like a concentrated suspension because of the high state of turbulence.

13.3.4.2 Dry Granulation Dry granulation is used when dry blending is not successful for an API that is poorly compressible, requires a high dose, demonstrates sensitivity to moisture and heat, or exhibits a combination of these factors. As shown in Figure 13.3, dry granulation techniques require fewer manufacturing steps than do wet granulation techniques. This can lead to a more cost-effective method of producing running powders if the material throughput is similar to that of wet granulation methods. The advantages and disadvantages of dry granulation methods are listed in Table 13.4. In general, it is more difficult to develop a robust dry granulation process. Material loss, overlubrication, and loss of compressibility due to multiple compressive stresses are common problems that must be addressed.

Overlubrication of the running powder is a result of multiple processing steps that can detrimentally spread the lubricant that results in softer tablets and potentially poor dissolution performance. The multiple processing steps that involve the lubricant which can lead to overlubrication are lubrication for powder compaction, milling and sizing of the compacts, and lubrication of the dry granulation for tableting. Multiple compressive stresses on plastic material can lead to strain hardening, making additional plastic deformation more difficult (see Section 13.3.2.2) In practical terms, the material becomes less and less compressible and the tablet strength decreases, which results in decreased tablet crushing strength and increased friability.

Consolidation to form a compact or slug typically occurs in several steps. As the powder is subjected to compressional pressure, the powder volume decreases as particles rearrange, increasing the particle–particle interactions until it is difficult for them to slip or move. Additional pressure leads to elastic, viscoelastic, and plastic deformation (irreversible viscous flow). Acceptable compact or slug strength may be achieved during plastic deformation. Particle fracture can also occur, primarily for more brittle materials. This fragmentation increases the surface area of the particles, allowing for more particle–particle contact points, which can lead to "cold fusion or welding" and intermolecular bonding, such as van der Waals forces and hydrogen bonding.

Microcrystalline cellulose undergoes little fragmentation at typical tablet compressions forces and produces tablet strength primarily through plastic deformation. Microcrystalline cellulose, sodium chloride, mannitol, and many polymers exhibit plastic deformation under tablet compression. Dicalcium phosphate and lactose bond primarily through brittle fracture. A number of equations have been used to describe the densification of powders. The Heckel equation (Heckel, 1961a, b) is frequently used to gain insight into the mechanism of consolidation:

$$\ln\left(\frac{1}{1-D}\right) = KP + A \tag{13.8}$$

where P is the compression pressure, the *reciprocal of K* is the *mean compact yield pressure*, A is associated with the initial compact volume and densification prior to bonding, and D is the relative density, which is defined as the ratio of the compact density at P to the compact density at zero porosity. Note that $1/K$ relates to the plasticity of the material. The smaller the value of $1/K$, the more plastic the compact is. An elastic material such as acetaminophen will exhibit a small K or low slope with a comparatively large mean yield, $1/K$. Three characteristic types of Heckel plots have been reported: plastic deformation of different particle size distributions of the same material, resulting in separate independent curves; fracture of different particle size distributions to give a more singular size distribution, which results in a single terminal curve; and plastic deformation with rearrangement.

Wells (1996) described a simple technique that only requires several grams of material to evaluate the compression properties of a drug. Three types of test samples are prepared to evaluate if the drug undergoes plastic, brittle fracture, or elastic deformation. The tablet sample size is 500 mg of drug which is blended with 5 mg of magnesium stearate. The tablet or pellet is made on a conventional Carver press using a 13-mm die. The test procedure uses different blending times and compression dwell times. Each tablet is compressed at 1 ton. The first sample set, A, is blended for 5 minutes and compressed for a dwell time of 2 seconds. Sample set B is blended for 5 minutes and compressed for 30 seconds. Sample C is blended for 30 minutes and compressed for 2 seconds. If the drug or test material undergoes plastic deformation, the crushing strength of sample C will be less than that of sample A, and that of sample A will be less than that of B (C < A < B). The crushing strength of all three samples will be equal if the test material undergoes fracture (A = B = C). If the test material undergoes elastic deformation, one would expect the crushing strengths of A and C to be similar and the samples may cap or laminate. Sample B should exhibit a low crushing strength.

13.3.4.2.1 Roller Compaction Roller compaction is a continuous process that feeds powder between two counterrotating rollers that compress the powder into a ribbon compact that is subsequently sized as a running powder. The roller compactor advantages and disadvantages are provided in Table 13.4. The process is used when the API is sensitive to moisture, heat, or both. Roller compaction is also used to improve the flowability of the running powder and "lock" the drug into the granules that are prepared to decrease segregation. Figure 13.17 is a diagram of the roller region of the compactor.

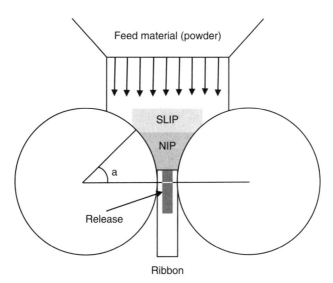

FIGURE 13.17   Defined areas within the roller compaction. (Adapted with permission from Pharmaceutical dosage forms: Tablets, Volume 1, Unit operations and mechanical properties, Augsburger LL, Hoag SW,2008, Informa Healthcare.)

A number of variables can affect the compaction process, such as roll speed, diameter, width, surface, and geometry type (flat, concave, convex); roll gap; fixed or movable rollers; dwell time; type of powder feed system; deareation; and rolls mounted horizontally, vertically, or inclined. Roll pressure, roll gap, and roll speed/screw feed rate typically have the largest impact on a compact's tensile strength and solid fraction. The formulation components also have a major impact on the roller compaction process. It is common practice to have a combination of brittle and plastic materials as part of the formulation. Plastic materials generally deform irreversibly with minimal compression force. However, plastic materials are more sensitive to roll speed and dwell time. They also have a tendency to become overlubricated and can undergo work hardening, which results in lower tablet strength. Brittle materials undergo fracture that creates more surface area and are more resistant to overlubrication and work hardening. Typically, a roller compaction formulation should contain at least 20 to 30% w/w plastic material.

The *solid fraction* or *relative density* is a measure of the extent of the densification process and is:

$$\text{solid fraction} = \frac{\text{ribbon density(mass/ribbon volume)}}{\text{true density}} \tag{13.9}$$

The degree of densification affects the hardness and tensile strength of the ribbon, which can affect the quality of the granules and tablets prepared from the ribbon. The tensile strength can be determined by using a three-point ribbon bending technique (American Society of Testing and Materials C1161, E1820, and D790). The tensile strength is given by

$$\sigma_T = \frac{3Fl}{2wt^2} \tag{13.10}$$

were σ_T is the fracture tensile strength, F the load applied at fracture, l the distance between the two ribbon supports, w the width of the ribbon, and t the thickness of the ribbon.

13.3.4.2.1.1 PAT: Roller Compaction   Near-infrared spectroscopy has been used to provide simultaneous real-time nondestructive roller compactor ribbon content uniformity, moisture content, compact density, tensile strength, and Young's modulus (Gupta et al., 2005). A robust compaction process should maintain the ribbon uniformity of these critical variables. The ability to monitor these variables simultaneously in real time offers the opportunity to have feedback control of various process parameters, such as roll gap, roll pressure, and feed rate.

13.3.4.2.1.2 Scale-up of Roller Compaction   Ideally, as one scales a process it is important to maintain critical equipment angles and shapes as constant as possible. This is difficult to do with roller compaction. As the roll diameter increases, the nip angle, shape, and size change. This makes it difficult to scale the roller compaction process. The roller compactor scale-up of a controlled-release matrix tablet has been reported (Sheskey et al., 2000). The scale-up study evaluated both a laboratory- and a commercial-scale compactor. Throughput rates of 2, 11,

12, 19, 23, 40, 45, 75, 130, 135, 228, and 242 kg/h were studied. Initial testing was conducted on a laboratory scale to determine the optimum processing parameters. Acceptable ribbons and tablets were manufactured regardless of scale when the linear speed was scaled, the screw speed/roll speed ratio was kept constant, and the roll force per linear inches was held constant.

13.3.4.2.2 Slugging The advantages and disadvantages of slugging are outlined in Table 13.4. Slugging has been used routinely to prepare running powders for high-dose heat- and/or moisture-sensitive APIs, such as vitamins. The slugging process requires a reasonably compressible formulation that has excipients similar to those in a dry blend direct compression formulation. The difference in slugging and tableting is that the slugs are about 2.54 cm in diameter, and it requires about 240 MPa of compression pressure to form a slug. This is four to eight times the pressure used in roller compaction. Slugging tablet presses operate at about four times the pressure of a conventional B-tool tablet press. The slugging press also uses a larger punch, called a D tool. The slugs are then milled and screened to the particle size desired. The slugging throughput is about 1000 times less than that of roller compaction. Today, slugging is commonly used for new processes only when there is a need for higher compressional forces to prepare the compact for milling and granule formation. Slugging may also be used if roller compaction equipment or know-how is not available in development and manufacturing. PAT and scale-up are similar to those in tableting and are discussed in more detail in Section 13.3.6.

13.3.5 Drying and Single-Pot Granulation and Drying Processes

Table 13.3 outlines various methods of preparing running powders. Both high-shear and fluid-bed granulation require wetting the powder mass to form granules, which are subsequently dried. In the high-shear granulation method the wet granules must be discharged from the high-shear granulator and transferred to a dryer. The transfer may be part of a batch or continuous process. Granulation and drying for the fluid-bed process are referred to as a *single-pot process* (SPP) and the equipment is referred to as a single-pot processor. The obvious advantage of an SPP is it is more cost-effective. There is less material handling, less operator exposure, fewer training requirements, and fewer required pieces of equipment, saving total equipment costs and space requirements. In general, selection of the dryer depends on the liquid that is to be removed, the maximum allowable product temperature, and need for batch or continuous drying. The selection of an SPP takes into account the granule manufacture and drying requirements. There are a number of different types of dryers: drum, tray, fluid-bed, spray, vacuum tray, tumble and conical, and microwave-vacuum.

Drying of pharmaceutical granulations is based on the removal of granulating liquid by vaporization. Hydroalcoholic and nonaqueous fluids are also used as granulation liquids, and the general drying principles also apply to these liquids. Drying involves heat and mass transfer. Heat can be transferred as convection, conduction, and radiation. Convection drying involves the flow of low-humidity heated air which comes in direct contact with the product and results in vaporization of the water. The vaporized water is removed by the heated air and condensed into its liquid form.

Moisture can be present in a material or drug substance in the form of water of crystallization, bound and unbound water. A material that has a water molecule(s) associated with its crystal structure is often referred to as *water of crystallization*, *water of hydration*, or simply a *hydrate*. This type of water is part of the repeating crystalline lattice structure without being covalently bound to the host molecule. Typically, there are discrete amounts of water associated with each molecule of host material, which can be determined by Karl Fischer techniques, thermal gravimetric analysis, and x-ray diffraction. During differential scanning calorimetery (DSC) analysis, a discrete endotherm(s) will result when this type of water is released from the host material. Other liquids that form this type of crystal lattice association will also exhibit distinct DSC endotherms or x-ray diffraction patterns and are called *solvates*.

Another type of water is *bound moisture*, water associated with the material that has an equilibrium vapor pressure less than that of pure water at a given temperature and pressure. *Unbound* or *free water* is moisture that exhibits an equilibrium vapor pressure equal to that of pure water under the same temperature and pressure conditions. *Equilibrium water* is the amount of associated water that is in equilibrium with a stated vapor pressure of water.

Water or moisture content is expressed as percent water based on wet or dry material weight. The moisture content based on the dry weight or in the case of hydrates, anhydrous weight, is the weight of water divided by the weight of dry or water-free material multiplied by 100%. The wet moisture content is calculated by dividing the weight of water by the weight of water containing material multiplied by 100%. This is the most common way of expressing moisture content. Thermogravimetric analysis, loss on drying, and Karl Fischer techniques provide water content based on wet weight.

During the drying process the material is usually dried to a specified moisture content or moisture content range. This range often reflects the equilibrium moisture content over a reasonable range of relative humidities that do not adversely affect processing or final product attributes. The drying process involves heat and mass transfer and has several different drying stages. After an initial temperature adjustment period, the material will dry at a constant rate while a continuous layer of surface water is available for vaporization. This is referred to as the *constant-rate period*. When the surface of the material is no longer saturated with water and dry spots appear, the drying rate steadily decreases. This is called the *first falling-rate period* or *unsaturated surface drying*. The moisture content at a point which the constant period ends is called the *critical moisture content*. At this content, the drying rate will depend on the transfer of the liquid to the surface and depends on the nature of material. Crystalline or porous granular materials tend to dry faster than amorphous substance or materials that have small pores or capillaries. As the water migrates to the surface, water-soluble materials can be transported preferentially to the surface. This may lead to mottling of dyed material surfaces or case hardening of the dried surface. Depending on the temperature, pressure, and relative humidity of the drying environment, a *second falling-rate period* may be observed where vaporization no longer occurs on the surface of the solid. The vaporization occurs inside the material and is transported to the surface by diffusion. Eventually, the granulation or solid moisture will be in equilibrium with its surroundings and the drying rate will be zero. The solid has reached its equilibrium moisture content and further drying will not result in a change in the solid's moisture content.

During the constant-rate drying period, the rate of drying is a function of heat and mass transfer. The mass transfer is expressed by the first-order equation

$$-\frac{dW}{dt} = k_m A (p_s - p) \tag{13.11}$$

where dW/dt is the rate of drying, k_m a mass transfer coefficient, A the surface area for mass transfer, p_s the solvent vapor pressure at the surface temperature of the solid, and p the partial water vapor pressure in the air environment. The driving force for water removal is the difference in vapor pressure.

Most pharmaceutical drying operations use the terms *absolute humidity* or *dew point* to describe the water content of the process drying air. The interrelationships among temperature, water vapor pressure, heat, and humidity are described by *psychrometry* and psychrometric charts. *Humidity* is the concentration of water vapor in air. Mass-basis absolute humidity is defined as the mass of water vapor per kilogram of dry air. *Relative humidity* (RH) is the ratio of the vapor pressure of water in air to the saturation vapor of water at a given temperature and pressure. *Percent relative humidity* is relative humidity $\times$ 100%. The *dew point* or *dew point temperature* is the temperature at which air becomes completely saturated with water at a given atmospheric pressure. Most manufacturing operations control and report the percent relative humidity of their processing rooms. Processing rooms are usually maintained at $45 \pm 15\%$ RH. Process drying air that has a low relative humidity or low dew point will lead to faster drying than will process air with a higher relative humidity or dew point. The dew point can be determined by knowing the wet and dry bulb temperatures. The *dry bulb air temperature* is the temperature that is not affected by water content and is shielded from radiation. In other words, the dry bulb temperature is similar to using an outdoor thermometer that is placed in the shade to avoid direct radiation from the sun. A wet bulb thermometer is wrapped in wet muslin called a "sock." If the air is not saturated with moisture, the *wet bulb temperature* will be lower than the dry bulb temperature, due to evaporative cooling. If the air is saturated with water (100% RH), the water in the sock will be in equilibrium with the water in the air, and the wet bulb temperature will be the same as the dry bulb temperature.

During the constant-drying-rate period, the moist solid temperature will be the wet bulb temperature, which will be lower than the drying gas temperature. At the first falling-rate period, the solid material temperature will increase until the equilibrium moisture content is reached and drying is complete. Once the solid has come to its equilibrium moisture content, the temperature of the solid will reach the dry bulb temperature of the process air. The solid temperature can be used as an endpoint for drying. The rate of drying during the constant drying period can also be expressed in terms of heat transfer:

$$\frac{dW}{dt} = \frac{hA(T - T_s)}{\Delta H_{vap}} \tag{13.12}$$

where dW/dt is the rate of drying, h the heat transfer coefficient, A the drying surface area, T the temperature of the air, T_s the temperature of the solid surface, and ΔH_{vap} the heat of vaporization. This shows that the temperature differential is also a driving force in drying.

13.3.5.1 Fluid-Bed Granulation and Drying

Fluid-bed granulation and drying are part of an SPP that granulates the powder and uses fluidized air to dry the granules, that have been formed. A fluid-bed granulator dryer is shown in Figure 13.18. The blended powder is added to the bowl of the fluid-bed granulator. At the bottom of the bowl, air is used to fluidize the powder. For the powder to be fluidized, the velocity of the air needs to be high enough to support the weight

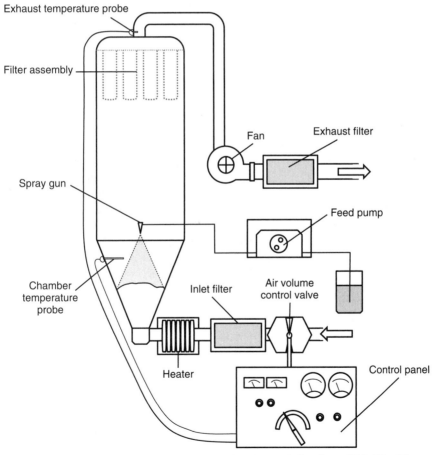

FIGURE 13.18 Fluid-bed granulator. (Adapted with permission from GEA Niro Pharma Systems, 1998.)

of the powder and set the particles in motion. The *minimum fluidization velocity* can be monitored by the pressure drop across the powder bed. As the inlet air velocity is increased, the pressure drop across the bed increases until it reaches a point where frictional drag on the particles in the powder bed equals the force of gravity. As the particles move and become fluidized, the pressure difference across the bed decreases and the minimum fluidization velocity can be determined. The higher the inlet air velocity, the greater the turbulence generated. At even higher velocities, entrainment will occur and the particles with be transported and lost to the exhaust filter. Smaller, less dense particles will be preferentially entrained or transported to the filter, which could lead to a disproportionate loss of drug. Good fluidization can be observed in the sight glass as free downward-flowing granules. Good fluidization is also characterized by a constant exhaust temperature and bed pressure.

13.3.5.1.1 Fluid-Bed Granulation The fluid-bed process is a wet granulation process whose advantages and disadvantages are listed in Table 13.4. The mechanism for wet granule formation was discussed at the beginning of this section. The fluid-bed wet granulation process is more complex than high-shear wet granulation. There are three modes of granulation: top-spray, bottom-spray, and tangential-spray. Most fluid-bed granulations use top-spray granulators because they accommodate the largest batch sizes, are easiest to set up and operate, and cause the least mechanical stress on the granules. A number of process parameters need to be considered in fluid-bed granulation, such as batch size; powder bed pressure drop; bed mixing time; amount of granulation fluid; fluid addition rate; powder bed moisture; granulation time; spray nozzle size, number, and position; air volume and pressure; inlet air distributor plate size, temperature, volumetric flow rate, and dew point; product bed temperature; exhaust air temperature and dew point; filter type; and shaking time and interval. Generally, the most important variables are the spray rate, droplet size or atomization spray pressure, inlet air temperature and dew point, and volume of fluidization air. Control of these parameters can prevent poor fluidization, content inhomogeniety, and excessively course or fine granules.

Excessive fines can be a result of too low inlet air dew point, granulation fluid spray rate, and amount of binder fluid. In other words, the powder bed is not wetted sufficiently to cause particle size enlargement. Excessive fines can also result from too high an inlet air temperature, atomization pressure that produces finer granulating fluid droplets, and airflow rate.

Course granules can result from too low inlet air temperature and spray nozzle pressure that leads to larger granulation fluid droplets. High spray rate and inlet air dew point can also lead to larger granules. In general, the granules that are formed are more porous than those obtained by high-shear granulation. The agglomerates are formed primarily by the surface tension at the air–liquid interface and capillary suction. The granules do not experience the kneading action that occurs during high-shear granulation. The kneading helps densify the powder, which creates smaller capillaries and less porous granules. Therefore, it takes up to three to four times more granulating fluid for fluid-bed granulation than for high-shear granulation. In fluid-bed granulating the drying process occurs concurrently with the granulation, so the granule binder has a tendency to lose moisture, which increases its viscosity and eventually can dry to form solid bridges. In addition, the outside of the granule is sprayed with binder fluid. As a result of these differences, fluid-bed granules tend to form harder tablets.

13.3.5.1.2 Fluid-Bed Drying The fluid-bed granulation drying process is unique since the granulation and drying processes are carried out simultaneously. Since the fluidized granulation is a solid–air suspension, the heat transfer process is very efficient. The drying capacity of the air is dependent on its temperature, humidity, and volume flow. The drying process can be followed by tracking the bed or product temperature and comparing it to the exhaust air temperature. During the constant-rate period, the exhaust air temperature (dry bulb temperature) and product temperature (wet bulb temperature) should remain constant until the first falling-rate period is attained. At the end of drying, the equilibrium moisture content is reached and the exhaust air temperature and product temperature should be the same.

The volume airflow (cubic feet per minute or cubic meters per hour) and drying rate can be adjusted to provide an optimized granule particle size. High-volume airflow will increase the drying rate (decrease the drying time) but will have a tendency to cause granule attrition and potentially create an undesirable level of fines. Depending on the physicochemical properties of the drug and excipients, the nominal inlet air temperature range is 35 to 70°C. The dew point normally ranges from 10 to 20°C.

13.3.5.1.3 PAT: Fluid-bed Granulation and Drying Recently, NIR was used to monitor the drying process of a fluid-bed dryer that was part of a continuous wet extrusion granulation (i.e., fluid-bed drying) milling manufacturing process. Chablani et al. (2011) compared NIR moisture content measurement taken in-line in real time to conventional loss on drying and Karl Fischer moisture analysis. A statistically significant linear relationship was found between all three moisture measurement techniques. Skibsted et al. (2007) discussed examples of NIR-based real-time release. They reported the use of online fluid-bed NIR moisture measurements taken during the granulation process.

13.3.5.1.4 Scale-up of Fluid-bed Granulation and Drying Scale-up of the fluid-bed drying process is common in many industries, including pharmaceuticals. Several articles (Mehta, 1988; Rambali et al., 2003, Parikh and Mogavero, 2005) have explored the scale-up criteria; however, the process remains empirical in nature. To ensure successful product quality, consideration must be given to the geometric similarity of equipment, powder bed moisture content, droplet size (for fluid-bed granulation), and drying endpoint determination. During drying, the bed moisture content depends primarily on batch size, inlet air temperature, dew point, and airflow rate. To achieve similar fluidization characteristics during scale-up, the linear velocity of air (total airflow/cross-sectional area) is usually kept constant. Also, the ratio of batch size to airflow is maintained to have similar drying capacity. Thus, knowing the bed diameter can help calculate batch size and air flow rate.

However, more experimental work may be required if there are significant differences in the air distribution plate and the geometry, as it can change the airflow pattern and bed height, leading to different drying conditions. Inlet air temperature and dew point can be kept constant across different scales since the inlet temperature is normally optimized based on the physical and chemical properties of the material. The dew point control is facilitated by injecting steam into the free airstream or by utilizing a desiccant wheel. It is important to control the drying endpoint to have similar granule compression characteristics. Traditional techniques to detect the end of drying have correlated granule moisture content with product temperature, exhaust temperature, and exhaust dew point or a combination. Engineering models have also been used to monitor the drying process with inputs for product temperature, exhaust temperature, exhaust dew point, and others. A more recent technique for endpoint determination employs near-infrared spectroscopy to monitor moisture content in-line in real time.

13.3.5.2 *Microwave Vacuum Granulation and Drying*   A microwave vacuum granulator dryer marries the concepts of high-shear granulation, microwave heating, and vacuum drying. Microwave vacuum granulation drying (MVGD) is a much less common SPP, which is somewhat surprising because it has been shown to offer significant advantages. Robin et al. (1994) indicated that MVGD offered a reduction in new capital investment since the MVGD could function as a high-shear granulator and a dryer. The authors also noted that the MVGD took less floor space, required fewer material-handling steps, shorter process time, and provided substantial operating cost benefits compared to more conventional technologies. MVDG is especially useful in providing containment of highly potent or toxic compounds, or both, that require wet granulation.

Microwaves result in the direct heating of the wet granulation. The microwave energy is absorbed by water or other polar molecules, which causes rapid orientation and reorientation in an alternating electronic field. Pharmaceutical microwaves operate at 2450 MHz. The water molecules reorient themselves thousands of times per second, which causes friction and the generation of heat. The amount of microwave energy absorbed is (Waldron, 1988)

$$P = 2\pi f v^2 E_0 E_r \tan \delta \qquad (13.13)$$

where P is the power density of the material or energy absorbed (W/m^3), f the frequency (Hz), v the electric field strength (V/m), E_0 the dielectric permissivity of free space, E_r the dielectric constant of the material, and tan d the loss tangent.

Materials that absorb microwave energy more readily exhibit higher loss tangents (loss factor) and higher heating rates. Equation (13.11) is relevant to the drying process. The use of a vacuum removes oxygen- and allows drying to occur at lower temperatures, which is an advantage for oxygen- and temperature-sensitive drugs. Microwave vacuum drying (Pearlswig et al., 1994) has been reported to be scalable and three times faster than vacuum drying alone (30 minutes compared to 90 minutes). The product can be dried without agitation so that there is less granule attrition, dusting, and product loss, which is ideal for potent products.

TABLE 13.19 Loss Tangent and Dielectric Values for Common Pharmaceutical Excipients

Excipient	Loss Tangent (Temperature Dependent)[a]	Dielectric (Temperature Dependent)[b]
Starch	0.41	3–5
Cellulose	0.15	3.2–7.5
Mannitol	0.06	3.0
Calcium carbonate	0.03	6.1–9.1
Lactose	0.048	3.2[c]
Water	12.6	78
Ethanol	8.6	25

[a] Van Vaerenbergh (2011).
[b] Ohmart/VEGA (2011).
[c] Roggenbuck et al. (2010).

Table 13.19 shows the loss tangent for common pharmaceutical excipients. A generic loss tangent factor of 0.001 has been suggested for drug substances (Garcia and Lucisano, 1997). It can be seen that the bulk of the microwave energy is absorbed by the granulating liquids. It should be noted that water of crystallization does not generally absorb microwave energy. Moisture content reproducibility has been reported to be ±0.25%, with a temperature distribution throughout the bed of ±5°C.

Most of the issues seen with earlier equipment, such as temperature control, electrical arcing, local heating, thermal runaway, and electric field breakdown under vacuum conditions, have been addressed in newer designs. The current models can provide variable output and monitor real-time forward power, reflected power, arcing, and product temperature. Microwave leakage safety issues have also been addressed satisfactorily. Dried granule properties tend to be similar to other granulation drying technologies, and product stability has been proven. A number of MVGD products have now been approved around the world.

13.3.5.2.1 PAT: Microwave Vacuum Granulation and Drying Product temperature can be used to monitor the drying endpoint. Reflected energy can also be used as a drying endpoint. As the granulated powder dries, less energy is absorbed into the wet mass and more energy is reflected. The total energy that is absorbed by the product can also be determined in real time and can be used as a process endpoint. The electric field strength increases as the water is removed. This increase in field strength can be used to monitor and control the drying endpoint.

13.3.5.2.2 Scale-up of Microwave Vacuum Granulation Drying The high-shear granulation component of an MVGD system is the same or similar to that of the high-shear granulators that we discussed previously. Therefore, similar scaling factors and techniques can be applied here. Robin et al. (1994) undertook MVGD scale-up trials from 15 kg to 150 kg and found that the drying time of

30 minutes was directly scalable using the same microwave energy per kilogram of granulation. This makes scale-up in an MVGD very attractive. This was not the case for vacuum drying alone at 5 mbar (500 Pa) with intermittent agitation. In this case, the process was not directly scalable and required a drying time of 60 minutes for a 30-kg batch compared to 135 minutes for a 150-kg batch.

13.3.6 Tableting

Formulation design and process design should provide a running powder (powder that flows in the tablet press) that is free-flowing, compressible, and well lubricated, to prevent binding of the tablet press or punches, and maintains content uniformity during the tableting process. Figure 13.1 shows the formation of a tablet where an upper and lower punch enter a die filled with powder. The tablet is formed when pressure is applied to the upper and lower punches, causing compaction of powder that is constrained in the die. Figure 13.4 illustrates how this process is extended to tablet compression on a rotary tablet machine. The compression sequence is described below.

1. The lower punch is lowered inside the die by a cam. The distance that the lower punch is lowered creates the volume for the powder to fill. The powder flows into the die by gravity, and as the fill volume increases, so does the final tablet weight. Force feeders are used with high-speed tablet machines that assist the gravity feed. The final volume and tablet weight are adjusted through a weight adjustment cam that lowers or raises the lower punch. Excess powder in the die is scraped off so that the powder in the die is flush with the top of the die.

2. The upper punch is lowered by a cam until it enters the die and compaction begins. Modern presses often have precompression and main compression rolls (shown in Figure 13.4). The purpose of the precompression rolls is to initiate compaction and force air out of the powder. The main compression roll can be adjusted to provide the appropriate hardness, friability, and tablet surface properties.

3. Once compression is complete, the upper punch is pulled upward out of the die by a cam, and simultaneously, the lower punch is raised by a cam. As the lower punch is raised, it pushes the tablet out of the die. The tablet ejected is removed from the lower punch face by the scrape-off bar.

The upper and lower punch compression time, force, and displacement curves are shown in Figure 13.19. The *compression states* as a function of time are described below (Muzzio et al., 2008).

- *Solidification (consolidation) time*, T_s, is the time that the upper and lower punches are undergoing vertical displacement and compacting the powder.
- *Dwell time*, T_d, is the time period during which the punches are undergoing displacement and remain in contact with the powder compact.
- *Relaxation time*, T_r, occurs during punch decompression.

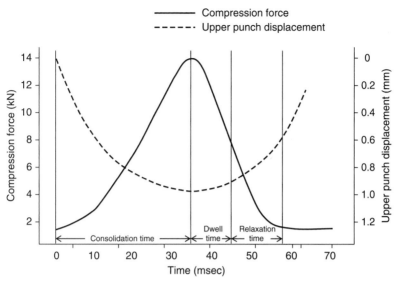

FIGURE 13.19 Compression–time profile. (Adapted with permission from Pharmaceutical dosage forms: Tablets, Volume 3, Manufacture and Process Control, Augsburger LL, Hoag SW,2008, Informa Healthcare.)

- *Contact time*, T_c, is the interval when the punch tips are in contact with the compact and the punch heads are in contact with the compression rolls, where $T_c = T_s + T_d + T_r$.
- *Ejection time*, T_e, is the time required to eject the tablet.
- *Residence time*, R_t, is the time the powder is present in the die; $R_t = T_c + T_e$.

Instrumented tablet press profiles can also record the upper punch precompression force ($\sim$4 to 7 kN), compression force ($\sim$10 to 22 kN), ejection force ($\sim$500 to 1500 N), and tablet take-off force ($\sim$0.1 to 0.4 N). This information can be used to diagnose powder flow problems, nonunifomity of lubricant, and sticking of tablets to the lower punch.

Scale-up is addressed in more detail in Section 13.3.6.2. However, it is important at the tablet process development stage that the scale-up criteria be kept in mind. Scale-up of tablets requires that the unit die fill volume remains constant and that the compression time–force profile is similar for development- and manufacturing-scale equipment. The best way to develop a tableting process is to perform the process design work on a high-speed compaction simulator so that the critical compression times and forces cited above can be matched to the commercial press.

Running powders that bond by undergoing brittle fracture are less sensitive to the compression contact time than are materials that undergo plastic or elastic deformation. As discussed above for dry granulation and roller compaction, plastic material tensile strength can be significantly affected by compaction contact time and compressional force. If the elastic limit is not exceeded during compression, the tablet may have a tendency to cap or laminate during decompression as it is ejected from the die cavity. Elastic recovery may occur days after manufacture, and the tablets may soften over time. In severe cases the tablets may form cracks that can lead to lamination along the tablet's "belly band" or capping near the tablet's land area. Tablets that are underlubricated may adhere to the punch face and cause filming of the punches and a dulling of the tablet surface. The filming may lead to picking and sticking of the tablets to the punch face, causing cratering on the tablet surface or loss of logo and breakline definition.

Temperature-sensitive drugs and excipients may be affected by temperature increases that occur during tableting. Interparticulate friction, die wall friction, fragmentation, elastic and plastic deformation, bonding, and ejection friction can lead to exothermic processes that result in increased tablet temperatures. Using an infrared thermoviewer, Ketolainen et al. (1993) demonstrated that tablet temperatures can rise approximately 1 to 8°C, depending on the excipients and the level of lubricant. Press speed and compressional force are two operational factors that can affect the extent of temperature increase.

Tablet tool and die design and maintenance are central to achieving high-quality tablets. Tablet punch and die sets should be treated as precision instruments. Optimizing the tablet embossing font, font angle, prepicked islands, and tapered areas can minimize sticking and picking of tablet logos. Wider angles allow for better punch surface release. Maintenance of the tools and dies is also critical to overall tablet quality. The tools should be thoroughly cleaned, buffed, and stored in a protective sheath or holder at a controlled temperature and humidity. The tools and dies should be maintained with tight specifications and tolerances. Differences in tool length can cause tablet weight variation and dose nonuniformity.

13.3.6.1 PAT: Tableting Automatic tablet weight feedback control has been available from tablet press manufacturers and instrumentation companies for some time. The simplest weight feedback control principle measures peak compression force, which is related to the amount of material being compressed. Lower and upper compression limits can be set, and when compression forces exceed the limit, a feedback loop is activated and causes the tablet weight to increase or decrease based on the compression force observed. Today, dual process controls are available in which hardness and weight can be adjusted independently in real time. There are also commercial systems that provide online tablet NIR content uniformity measurements and at-line tablet weight, hardness, thickness, and diameter data.

13.3.6.2 Scale-up of Tableting The unit die fill volume is independent of the batch scale-up. Larger batches are manufactured at higher tablet press speeds for longer periods of time with presses that have more punches. The higher compression speeds require the running powder to be free flowing, highly compressible, and resistant to segregation, due to the shear caused by overhead hopper flow, the higher turret speed, and the presence of feeder paddles. To manufacture commercial-scale tablets with the same product attributes as the pilot-scale batches, the running powder should experience a similar time–force compression profile. This means that T_s, T_d, T_r, T_c, T_e, and R_t in the pilot and commercial scales should match. This may be difficult to accomplish with pilot scale presses, and it is generally recognized that high-speed instrumented compaction simulators provide a good alternative. This means that the commercial tablet press needs to be identified as early as in phase 2 clinical development so that the final formulation and process is defined by the start of phase 3 NDA clinical studies.

If a high-speed compaction simulator is not available, several concepts should be kept in mind. At the higher commercial press speeds, T_s and T_d will decrease, which will generally require an increase in the peak compressional force. A compression–hardness profile can be used to demonstrate how the running powder will perform under this increased pressure. One can run a pilot press at higher rpm to match the linear speed of the commercial press at lower rpm. One needs to keep in mind that plastic running powders will be much more sensitive to changes in T_c than running powders that bond through fragmentation and fracture. Compressing at higher speeds and for longer periods of time provides more opportunity for segregation, filming, and sticking of punches. Increases in press and tablet temperatures may cause tableting issues, such as punch filming.

Levin and Zlokarnik (2002) demonstrated the use of dimensional analysis to develop a regression model that was used to predict tablet hardness for viscoelastic and brittle materials using a research press and a midrange production press. The model parameters were depth of fill, tablet thickness, compression roll diameter, maximum applied compression pressure, compression rate, powder compressibility, and geometric dwell time.

13.3.7   Film Coating

Tablets are film coated to enhance the ease of swallowing a tablet; mask the poor taste of drugs by preventing the tablet from disintegrating in the mouth; protect the tablet contents from the environment; mask the color of low levels of impurities or degradation products; provide ready product identification for the patient, pharmacist, and health care provider; create a harder, more durable tablet that can be packaged and handled throughout the distribution channel with less breakage and powdering, which can lead to increased manufacturing cost and unwanted exposure of operators and end users to the drug; and provide specific drug delivery functionality, such as delayed or extended release.

Tablet film coating is a complex multifactorial process that integrates the tablet core characteristics, coating formulation attributes, and the coating process. The quality of the final product is critically dependent on the integration of the design of these three components. Film coating is expensive and time intensive. Unless a continuous coating process is used, it is typically the smallest batch unit operation, which can limit the efficiency of the overall manufacturing process.

The types of coatings are sugar coating, film coating, and press coating or compression coating. Sugar coating is very time consuming and requires a number of steps, including tablet core sealing, subcoating, glossing, coloring, polishing, and printing. In today's economy, sugar coating is too costly and time consuming for new product development. It is still used for older products in the marketplace. Similarly, the product development team or company management team should seriously consider the value added by a conventional film-coated product. In many cases, a well-designed tablet formulation and process can address the reasons typically cited for film coating. It may be prudent to reserve film coating or press coating for those situations that utilize the film coat as an enabling drug delivery technology.

The quality of the finished film coat depends on core tablet quality and the final film-coat formulation and process. The core tablet must exhibit good friability and hardness to withstand the stress experienced during fluid-bed or pan coating. Moderate tablet friability and hardness may limit the batch size or the type of coating process. Embossed tablets, those that have engravings, require more flexible films so that "filling" and "bridging" of the embossment does not occur. As discussed earlier for tableting, selection of the embossing font and cut angle are important for coating performance.

The coating solution or suspension is usually comprised of a polymer(s), plasticizer(s), opacifying agents, colorants, and the solvent system. The solvent system may be aqueous, a water–alcohol cosolvent system, or nonaqueous. Organic and cosolvent systems require significant plant investment to address environment impact, industrial hygiene, and explosion concerns. Recently, colloidal aqueous dispersions of water-insoluble polymers have provided significant advances for aqueous film coating. The polymers used in film coating should be elastic and resistant to mechanical stress; be stable to light, heat, and oxygen; and have excellent dissolution properties.

Aqueous polymer dispersions are now available for water-insoluble polymers. They are classified as true latexes and pseudolatexes. A true latex is prepared by emulsion polymerization, where as a pseudolatex dispersion is produced by mechanical means. Both dispersions have particles in the size range 10 to 10,000 nm. The pseudolatexes have the advantage of being free of the reactants used in the polymerization reaction, such as monomers, initiators, and catalysts. Aquacoat ECD (FMC BioPolymer) and Surelease (Colorcon) are ethylcellulose pseudolatex dispersions. Kollicoat SR 30 D (BASF) is a poly(vinyl acetate) controlled-release polymer dispersion system. Eudragit RL and RS grades are pH-independent sustained-release polymer systems of a copolymer of ethyl acrylate and methyl methacrylate and a low content of methacrylic acid ester with quaternary ammonium groups. These polymer systems are used to provide sustained release.

Acrylate true latexes are provided as Acryl-EZE (Colorcon), Eudragit L30-D55 (Evonik), and Kollicoat MAE 30 DP (BASF). Acrylate polymer systems are designed to offer delayed release and to dissolve at pH values of 5 and higher. Aquacoat CPD (Colorcon) is a cellulose acetate phthalate delayed-release polymer dispersion system that is also pH sensitive. Latex systems use higher concentrations of solids ($\sim$30% w/w) than do aqueous solution polymer systems, which typically contain less than 10% solids. In general, latex systems require higher levels of plasticizer than do aqueous solution polymer systems.

Polymer coatings can also provide moisture protection. Evonik has developed several E-grade dimethylaminoethyl methacrylate polymers that provide excellent protection against moisture ingress. These polymers, which can also be used for taste and odor masking, are soluble in gastric fluid up to pH 5.5. Above pH 5.0, they are swellable and permeable. In the past, most formulators developed their own coating formulation. Today, most scientists use "ready to use" dispersions or "ready for reconstitution" powders provided by a number of coating companies (e.g., BASF, Colorcon, Evonik, FMC BioPolymer).

13.3.7.1 Film-Coating Formulation The film-coating formulation is comprised of functional excipients that need to be compatible with the drug and the tablet core excipients. In addition, the coating excipients should be nontoxic, tasteless, inert, nonhygroscopic, tackfree, and soluble in the coating solvent unless the coating liquid is a dispersion. The excipient classes and functions are listed in Table 13.20. Organic solvents are being used much less frequently than in the past because of environmental and industrial hygiene requirements and restrictions as well as safety and explosion concerns. In this chapter we focus on the use of aqueous-based coating systems.

13.3.7.1.1 Film-Coating and API Compatibility Aqueous film-coating components need to be compatible with the drug substance and excipients. There are various ways to access the compatibility of a drug with a proposed film coat. Dry mixtures of the film-coating excipients and the drug can be vortexed together to prepare intimate mixes that can be placed under accelerated temperature and humidity conditions. Coating formulations that exhibit acceptable compatibility can be used to film-coat core tablets containing APIs and retested for acceptable stability.

13.3.7.1.2 Selection of Plasticizer and Plasticizer Level Once the film-coating system compatibility study has been completed, a simple statistical screening design-of-experiment study can be conducted at low and high levels of different plasticizers. A fractional factorial with three plasticizers at two levels allows one to study the main effects without interactions using four experiments. The level of plasticizer increases the flexibility and elastic modulus of the polymer coat and enhances logo fill-in; decreases the glass transition temperature, T_g, which improves the polymer system's mobility and ability to coalesce; and decreases the minimum film formation temperature (MFT), which allows latex film formation and the coating process to take place at lower manufacturing temperatures. For latex dispersions, coalescence and film formation are problematic above the MFT.

13.3.7.2 Film-Coating Process Aqueous film coating is a complex process that utilizes first principles of evaporative heat and mass transfer. There are significant interactions and interplay between the core tablet and its physiochemical properties; the coating formulation and its physical, chemical, and mechanical properties; and the coating process, which includes the equipment and critical process parameters. Tablet film coating may be done in a pan or perforated pan that causes the tablet bed to tumble, similar to a tumble clothes dryer. Tablets may also be film-coated in a fluid-bed dryer, which introduces the tablets to harsher impact and attrition than does pan coating. The remainder of this discussion focuses on aqueous pan coating. The complexity of pan coating is best illustrated in Table 13.21, which

TABLE 13.20 Film-Coating Excipient Classes and Functions

Excipient Class	Function
Film-formers: 5 to 30% w/w	
Immediate release: hypromellose (HPMC), hydroxyethyl cellulose, hydroxypropyl cellulose, aminoalkyl methacrylate copolymer, poly(vinyl alcohol), poly(vinyl alcohol)–poly(ethylene glycol) graft copolymer	Provide a flexible, hard, crack-free film that enhances the ease of swallowing a tablet, masks the drug taste, prevents tablet disintegration in the mouth, protects the tablet from environmental conditions, and creates a harder, more durable tablet that minimizes breakage and dusting.
Delayed release: cellulose acetate phthalate, cellulose acetate succinate, hydroxypropyl methylcellulose phthalate, poly(vinyl acetate) phthalate, hydroxypropyl methylcellulose acetate succinate, and methacrylic acid copolymers (U.S.P. types A, B, C)	Enteric coating provides delayed release.
Sustained release: ethylcellulose, ammonio-methacrylate copolymers (U.S.P. types A and B), poly(vinyl acetate)	Provide controlled release
Plasticizers: 1 to 20% w/w	Miscible with the polymer and improve film-forming capability, enhance flexibility, reduce the glass transition point, decrease the minimum film-forming temperature, and increase the film toughness
Water soluble: poly(ethylene glycol) (PEG 200, 600 Da), propylene glycol, glycerol, triacetin, triethyl citrate	
Water insoluble: acetylated monoglycerides, acetyltributyl citrated, acetyltriethyl citrate, dibutyl phthalate, dibutyl sebacate, diethyl phthalate, tributyl citrate, water	
Pore-forming agents (incorporated into sustained-release films): 0.25 to 3% w/w	Dissolve and form pores or holes for water to penetrate the insoluble film
Dibasic calcium phosphate, HPMC, mannitol, methylcellulose, sucrose, sodium chloride	
Colorants: 0.5 to 8% w/w	Provide tablet identification, mask the color of low levels of impurities or degradation products, and provide a more uniform appearance
Various iron oxides, FD&C (Food, Drug, and Cosmetic Act) dyes, and aluminum lakes	
Opacifiers: 0.5 to 8% w/w	Mask the color of low levels of impurities or degradation products; provide a more uniform appearance
Titanium dioxide, talc, aluminum silicate, magnesium carbonate	

TABLE 13.20 (*Continued*)

Excipient Class	Function

Glossing agents: 0.1 to 1% w/w
  Carnauba wax, high-molecular-mass
    poly(ethylene glycols), hypromellose
    1 to 4% w/w
	Provide a high gloss and good aesthetic
	appearance

Glidants and antitacking agents: 1 to
  10% w/w
Talc, silicon dioxide, magnesium stearate,
  glyceryl monostearate
	Decrease tackiness of film coat

Antifoaming agents Dimethylpolysiloxane	Decrease foam formed during spraying

Flavors and sweeteners
  Acesulfame potassium, sodium
    saccharine, sucralose (all are reasonably
    heat stable and N.F.)
	Improve aesthetic quality

lists the tablet core, coating formulation, and processing parameters that need to be considered.

The coating formulation and process should provide a smooth and glossy tablet appearance. The coating process requires a fine balance between the coating rate and the drying rate. The coating formulation droplets contact the tablet and need to spread and coalesce to form a smooth coat. In general, the smoother the coat, the glossier the tablet will appear. In addition, the coating needs to adhere to the tablet surface and have enough flexibility that as the film coat dries, it maintains good logo engraving definition. Overwetting of the tablet or spray drying may occur if the balance between the application rate and drying is not adequate. Overwetted tablets may become soft, friable, and sticky. Rowe (1997) has described a number of aqueous film-coating defects, including blistering, chipping, cratering, picking, pitting, orange peel, blooming, blushing, color variation, infilling, and mottling.

Ebbey (1987) derived an aqueous film-coating thermodynamic model employing the first principles of evaporative heat and mass transfer. The author developed the concept of an *environmental equivalency* (EE) *factor*, which allows for adjustment of process variables or scale that provides for an equivalent drying environment. The EE factor is a dimensionless number that represents the relative rate of drying and is equal to the ratio of surface area of heat transfer to surface area of mass transfer. An EE value of 1 results in an overwetting film-coating situation. Campbell et al. (2004) suggested that the optimum EE value for film coating is 4.4, with an operating range of 3.7 to 5.2. The heat and mass transfer equation proposed by Ebbey is

$$\frac{A_h}{A_m} = \frac{(M_w p_{w,wb}/RT_{wb} - M_w p_{w,\infty}/RT_{\infty})h_{fg}}{\rho_{\text{film}} c_p (T_{\infty} - T_B)} \tag{13.14}$$

TABLE 13.21 **Aqueous-Film-Coating Parameters**

Components of the Film-Coating Process	Parameters
Core tablet	Size, shape, surface area, hardness, friability, embossing, excipients (superdisintegrants, highly soluble materials), API stability to heat and moisture, surface roughness and porosity, batch size and loading, tablet bed porosity
Coating formulation	Viscosity, tablet and coating system surface tension, solids content, solution or dispersion state, tackiness, interaction with API or tablet excipients, mechanical properties (tablet adhesion, tensile strength, work of failure, indentation and puncture strength, elastic modulus), spreading, and coalescence
Process	
Spray gun system	Operating design and principle, nozzle design and spray pattern, spray rate, atomization pressure and resulting droplet size, number, spacing, distance and angle of guns to the tablet bed
Coating pan	Operating design and principle, pan size, batch size and loading, rotational speed, baffle design and placement
Air handlers and conditioned air	Inlet air temperature, dew point and moisture content, and volume flow rate; tablet bed temperature; exhaust air temperature, dew point and moisture content, and volume flow rate

where A_h is the surface area of heat transfer, A_m the surface area of mass transfer, M_w the molecular mass of the water, $p_{w,\text{wb}}$ the partial pressure of water vapor under wet bulb conditions, R the ideal gas law constant, T_{wb} the temperature under wet bulb conditions, $p_{w,\infty}$ the partial pressure of water of the free-stream air, T_∞ the temperature of the free-stream air, T_B the temperature of the tablet bed, h_{fg} the latent heat of vaporization, ρ_{film} the density of the moist air at the film temperature T_f, where $T_f = (T_B - T_\infty)/2$, and c_p the specific heat of the air.

Strong (2009) derived an equivalent, yet less complex expression for the EE factor given in terms of the humidity ratio (kilogram of water per kilogram of dry air). This expression is then used to derive the vaporization efficiency, E, which is the reciprocal of EE. E has a finite range of 0 to 1 and is calculated using easily measurable processing parameters:

$$E = \frac{\Delta T}{\Delta T_{\text{wb}}} \tag{13.15}$$

where E is the vaporization efficiency, ΔT the tablet bed temperature minus the free-stream air temperature, and ΔT_{wb} the wet bulb temperature minus the free-stream air temperature.

13.3.7.2.1 PAT: Tablet Film Coating In the past, the film-coating process was determined by tablet weight gain, which reflected the amount of coating solution that was applied to the tablet. Several authors (Pérez-Ramos et al. 2005: Lee et al., 2010) have reported the use of in-line NIR to monitor the film-coating process by measuring film thickness in a pan coater and fluid-bed coater, respectively. Tabasi et al. (2008) reported the use of NIR spectroscopy to monitor the thickness of controlled-release polymer coatings. The objective of this study was to use NIR to prepare calibration models to predict tablet polymer coat thickness and drug release rate. Recently, Müller et al. (2010) used in-line Raman spectroscopy to monitor the application of a drug–polymer film-coating solution to tablets. Spectral results correlated with weight gain for both placebo cores and active core tablets that were pan coated.

13.3.7.2.2 Scale-up of Aqueous Film Coating Aqueous film coating utilizes first principles of evaporative heat and mass transfer. Therefore, to coat at equivalent conditions as one increases the scale of manufacture from grams of tablets to kilograms of tablets, the microenvironment must be the same to assure the same quality of coating. The EE value is a dimensionless number that reflects the rate of drying and has been used routinely to scale film-coated products. Campbell et al. (2004) described a feedback control system that employed the EE factor successfully to scale-up tablet coating processes. The feedback control system is also employed for routine commercial manufacture. The patent also suggests that the target EE value for aqueous film coating should be 4.4, with an operating EE range of 3.7 to 5.2.

13.4 CONTAINER CLOSURE SYSTEM DESIGN

In the case of tablets, the container closure system (CCS) provides a means to transport and dispense a tablet at the same time, maintaining the tablet's physiochemical attributes, such as labeled dose, dissolution and disintegration properties, and lack of microbiological biological contamination. Tablet CCS systems provide for both multiple-unit and unit-dose containers, which are generally bottles and blister packs. These containers are relatively simple CCSs compared to aerosol or transdermal dosage forms. In addition, drug products often go to market in a primary and secondary packaging system.

There are a number of FDA and U.S.P. references that address CCSs and provide information about packaging nomenclature, fabrication materials, drug master files, and performance criteria. Appendix 12.4 notes the FDA package nomenclature, Appendix 12.5 the U.S.P. package nomenclature, and Appendix 12.6 the U.S.P. packaging fabrication materials and closure types. The FDA (1999) has written an industrial guidance, Container Closure Systems for Packaging Human Drugs and Biologics, which is intended to provide guidance for submitting regulatory packaging information. It also provides information regarding the current good manufacturing practices, Consumer Product Safety Commission (CPSC), and U.S.P. requirements for CCSs.

Good Packaging Practices (U.S.P. <1177>) discusses general packaging terminology. The components that are in contact or have the potential to come into contact with the drug product are referred to as the *primary container*. The primary package protects the dosage form from adverse environmental conditions such as light, oxygen, moisture ingress or egress, and microbiological contamination. *Secondary container* systems add additional protection for shipping the product and include labeling information or ancillary devices such as droppers or measuring spoons. The secondary container can also provide protection from light. The secondary container is not included in stability testing and is not included as part of the CCS description. A standard folding carton is most often used as the secondary container. Changes made to the secondary packaging material do not require prior FDA approval. *Critical secondary containers* are not in direct contact with the dosage form but provide essential environmental protection against light, moisture ingress, oxygen ingress, or combination thereof. As an example of a critical secondary container, the GlaxoSmithKline Diskus uses an aluminum pouch to protect inhalation powder from uptake of environmental moisture. *Additional packaging*, such as trays for syringes, droppers for nasal products, and desiccants for oral solids, may also be used. The *final exterior package* is usually corrugated paper or a plastic wrapper.

The FDA and CPSC have regulatory responsibility for human-use packaged products. The FDA requirements for tamper-resistant closures are stated in 21 CFR 211.132. The CPSC requirements for child-resistant and adult-use-effective closures (often referred to as "geriatric friendly") are provided in 16 CFR 1700.

A CCS is specific for each drug and type of dosage form. The interaction of the dosage form with the CCS components is dependent on the physiochemical properties of the dosage form. A cosolvent solution is more likely to interact with polymeric components of the CCS than is a dry powder. Similarly, one would expect more potential for evaporative weight loss for the cosolvent solution than for a dry powder. Additionally, the importance that a CCS plays in overall product design depends heavily on the route of administration. For example, there is more concern associated with CCS integrity and performance characteristics for injectionables and inhalation products than for oral solid dosage forms such as tablets and capsules. Ophthalmics and transdermal dosage forms generally carry an intermediate CCS concern level. The suitability of the container closure system is based on its protection, compatibility, safety, and performance.

CCS selection is based on intended use (multiple- or single-unit use), level of environmental protection required for acceptable product stability, dosage form–container closure compatibility, safety of the packaging components, and performance and functionality. Oral dose multiple-use containers routinely use high-density polyethylene or polypropylene plastic bottles. Single-use containers are often blister packs or blister strips that use a number of plastic films, such as poly(vinyl chloride) (PVC), polypropylene (PP), polyester terephthalate (PET), poly(vinylidene chloride) (PVdC), polystyrene (PS), polychlorotrifluoroethylene, and oriented polyamides. Cyclic olefin copolymer is a newer amorphous halogen-free polymer that has the transparency of glass. It provides a good moisture and oxygen barrier. Aluminum foil is often used in conjunction with polymer films for blister packs. Some fast-dissolve tablets are very friable and rely on the blister pack to provide adequate physical protection from being crushed or broken. Bottles may use U.S.P. purified cotton or purified rayon as a *coiler* to fill the dead space and decrease the chance for tablet attrition or breakage. The CCS should provide the tablet with adequate protection from light and moisture. Preformulation and formulation stability studies are used to understand and drive the CCS environmental protection requirements. Highly moisture-sensitive products may require the use of a glass container with silica gel or molecular sieve desiccants, induction-sealed lidding, and a tightly closing cap. Secondary carton packaging may provide additional light and moisture protection. A critical secondary nitrogen-filled aluminum pouch container can be used for extremely sensitive drugs and tablets. Blister packs may require aluminum–aluminum fabrication to provide the highest level of moisture and light protection. The interaction potential between oral tablets and the CCS is low. The effect of the desiccant on the equilibrium tablet moisture content and on the tablet's performance is the most frequent interaction encountered, and this is rare. The packaging components should be safe, and reference to indirect food additive regulations is sufficient. Leachables and extractables are of special concern for liquid products, but are of minimal concern for oral solids. The CCS should provide product protection, tamper evidence, safety, and be user friendly. The barrier properties for the various film materials are given in Table 13.22.

13.4.1 Multiple-Unit Bottles

13.4.1.1 Plastic Bottles High-density polyethylene (HDPE) is the most commonly used bottle material. Polyethylenes are part of the polyolefin family of polymers. HDPE is crystalline in nature and contains not less than 85% ethylene and not less than 95% total olefins and has a density of about 0.94 to 0.96 g/cm^3 (940 to 960 kg/cm^3). HDPE is chemically inert and has acceptable moisture and oxygen barrier properties for most solid oral dosage forms. Drugs or oral dosage forms that are highly sensitive to moisture or oxygen may need to be packaged in glass or an acceptable blister pack. Polypropylene is also part of the polyolefin family and is one of the least costly polymers. It is highly crystalline and very resistant to chemical attach. Polypropylene is not as good a barrier as HDPE to moisture and oxygen.

13.4.1.2 Glass Bottles Glass bottles provide the greatest protection from the environment. Pharmaceutical glass is classified by the U.S.P. as type I, II, or III (Containers—Glass, <660>). Type I glass is more chemically inert, resistant to water attack, and contains high levels of oxides, such as aluminum and boric oxides as well as alkali metal oxides. Type III glass is soda-lime glass. It is a silica glass that contains alkali earth metal oxides and is very susceptible to water attack that leads to leaching of these oxides and results in "sloughing" of the glass and an increase in the water pH. Acid treatment of the glass, which makes it more resistant to water attack, is referred to as *treated soda-lime glass* or *type II glass*. Type III glass is generally used for solid oral products. To meet the 2015

TABLE 13.22 Permeability Through Selected Packaging Materials

Film Material	Use Thickness (mil)[a]	Water Vapor Transmission Rate[b]	Oxygen Permeability Constant[c]
Cyclic olefin copolymer	10	80	0.017
High-density polyethylene	25	125	2000

Oriented
  polyamide/aluminum/poly
  (vinyl chloride)
	1/1.8/2.4	0	0

Poly(chlorotrifluoroethylene) (Aclar)	3	14[d]	0.025[e]
Poly(ethylene terephthalate)	2	400	45
Polypropylene	25	200	3000
Poly(vinyl chloride), unplasticized	10	1560	1.3
Poly(vinylidene chloride) (Saran)	1.2	20	15
Polyamide (nylon; Usually applied to aluminum foil)	1	3000	40

Source: Mangaraj et al. (2009) unless otherwise stated.

[a] 1 mil = 0.0254 mm.

[b] Test conditions: 37.8°C and 90% RH [$(g \cdot \mu m)/m^2 \cdot day$].

[c] Permeability constant ($cm^3 \cdot \mu m$)/h $\cdot cm^2 \cdot atm$.

[d] Calculated from Massey (2003). Test conditions: 37.8°C and 100% RH [$(g \cdot \mu m)/m^2 \cdot day$].

[e] Calculated from Massey (2003). Standard temperature and pressure.

serialization requirements for ePedigree programs, Roche Diagnostics and Schott have collaborated to develop a laser system to provide a two-dimensional bar code for glass containers.

13.4.1.3 Closures, Desiccants, and Oxygen Scavengers
A bottle closure is typically made up of a HDPE or polypropylene threaded cap that has a liner and inner seal that comes in contact with the product. Metal caps can be used with glass bottles. A flip-cap closure is rarely used for pharmaceutical prescription products. The liner forms a tamper-evident hermetic seal between the cap and the bottle. The liner is often a laminate material. The composition of the liner depends on how the inner seal is attached to the bottle. The inner seal comes in contact with the product, so its composition must meet safety considerations. The inner seal material is usually poly(vinyl chloride) or poly(vinylidene chloride) for moisture and oxygen protection. A food-grade adhesive is used to attach the inner seal to the bottle. Heat induction sealing is a fast and effective method of sealing the inner seal, which can contain aluminum. All the liner layers must use food- or pharmaceutical-grade materials.

Desiccants readily absorb moisture from the environment. They are often included in the CCS to protect the tablet from moisture. The moisture may interact with the tablet excipients, especially disintegrants, and cause the tablet to soften on storage. Desiccants are also added to decrease the presence of moisture and minimize potential hydrolytic drug decomposition. The desiccant materials most commonly used are silica gel and molecular sieves. Silica gel absorbs moisture most efficiently at high relative humidities. Molecular sieves are more efficient at absorbing moisture at low relative humidities. Currently, the use of molecular sieves is approved in the European Union but not in the United States. Plastic canisters or pouches are used to contain the desiccant and prevent it from contaminating the drug product. The materials used for the desiccant containers must meet all the safety criteria for any material that is in direct contact with the drug product.

Often, the manufacture will package bottles in tablet counts that provide a patient with a month's supply of medication. The typical tablet count for bottles is 30 and 100 tablets, which covers once- and twice-daily dosing. The pharmacist will often label this original bottle and dispense it to the patient with the inner seal and desiccant intact. This practice does raise patient awareness and safety concerns. The patient needs to be counseled about the presence of the desiccant and its purpose so they are not surprised and think the product has been adulterated in some way or, even worse, think they should take it.

A key question that needs to be answered is: How much desiccant should be used to provide adequate product protection? One approach is to use the Leeson–Mattock (1958) model, which assumes that hydrolytic solid–water degradation that occurs in tablets can be described by pseudo-first-order kinetics. In this case, the drug degradation rate in the tablet or tablet granulation is determined as a function of moisture content. The moisture-dependent degradation rates are then plotted against moisture content. The intercept on the moisture content axis gives the *critical moisture content* for the solid mixture, the amount of water in the tablet that does not lead to degradation of the drug. This is a useful number that can be used to calculate the amount of desiccant required to protect the product from water that will permeate the container during storage. Let us use the following example to illustrate how the required amount of desiccant can be determined when a set of known conditions is provided.

The Experimental conditions are as follows:

1. Tablets having a total weight of 250 mg are to be packaged in HDPE bottles in counts of 30.

2. The HDPE bottle has a water vapor transmission rate of 5.0 g water $\cdot$ m^{-2} $\cdot$ day^{-1} and has a surface area of 2.77×10^{-2} m^2.

3. The amount of water vapor that permeates the HDPE bottle over a two-year period is 2.8 g:

$$(730 \text{ days})(5.0 \text{ g} \cdot \text{m}^{-2} \cdot \text{day}^{-1})(2.77 \times 10^{-2} \text{ m}^2) = 2.8 \text{ g water over a}$$

two-year period

4. The tablet moisture specification is 2.5% w/w or 6.25×10^{-3} g of water per tablet:

$$(0.25 \text{ g total weight/tablet})(2.5 \text{ g water/100 g total weight})$$
$$= 6.25 \times 10^{-3} \text{ g } H_2O/\text{tablet}$$

5. The critical moisture content is 5% w/w or 1.25×10^{-2} g H_2O/tablet.

6. One gram of desiccant can absorb 1 g of water.

The amount of water that can be taken up by the tablet without any detrimental effects is the difference between the critical moisture content and the tablet moisture specification. In this case, 1.25×10^{-3} g H_2O/tablet -6.25×10^{-3} g H_2O/tablet $= 6.25 \times 10^{-3}$ g H_2O/tablet or 1.88×10^{-1} g water for 30 tablets. Since 2.8 g of water will permeate the container over a two-year storage period and the tablets can absorb up to 1.88×10^{-1} g of water without any detrimental effect, a desiccant needs to be added that has the capacity to absorb the moisture difference, 2.61 g of water. Desiccant canisters are available with different amounts of desiccant. They are typically available in 1, 2, and 3 g of desiccant. Larger packs are also available. In this case, a 3-g desiccant canister should be sufficient to protect the product from moisture permeation if the desiccant adsorbs 1 g of water per gram of desiccant.

Oxygen scavenger packs for bottles are now available. Several companies have also developed oxygen scavenger films that contain iron and cobalt to catalyze oxidation reactions in the film.

13.4.2 Unit-Dose Containers

13.4.2.1 Blister Strips Strips packs predate blister packs. Both are used as unit-dose packs for oral solid dosage forms. Advantages of unit-dose packs compared to multiple-use bottles include individual dose protection until the tablet is removed from the pack, prevention of moisture and oxygen ingress, tamper-evident, child resistance, reduced possibility of accidental misuse, may increase patient compliance, and easily portable when an individual dose is separated from the blister card. Blister packaging is cost-effective compared to bottles. Generally, the break even point for packing in bottles or blisters is considered to be 100 tablets.

Blister strips consist of one or two polymer layers. They are usually formed at lower speeds and require less material than do blister packs. Strip packaging machines are also less complex and smaller than blister pack machines. Tablets are inserted into a heated "pocket area," which is fed by two separate polymer rolls that are sealed in the strip, which takes the shape of the tablet.

13.4.2.2 Blister Packs Compared to blister strips, blister packs have a preformed cavity in which a tablet is placed. Figure 13.20 illustrates the difference between a blister strip and a blister pack. Blister packs may be made by thermoforming or cold forming. PVC, PVC–PVdC, PVC–Aclar, Aclar, PP, PET, and PS are commonly used thermoforming polymers and laminates. Laminates are used to provide the necessary environmental protection and thermoforming properties. See Table 13.23 for moisture vapor and oxygen permeation characteristics for various blister packing materials. PVC is typically the blister material in contact with

(a) Blister strip – top view

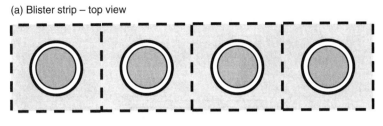

(b) Blister pack – side view

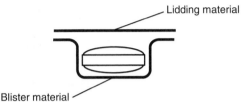

Lidding material

Blister material

FIGURE 13.20 Typical blister strip (a) and blister pack (b). (Adapted from U.S.P.–N.F. Chapter <1146>, with permission from the United States Pharmacopeial Convention. Copyright © 2008. All rights reserved.)

the product. Although PVC is the most commonly used blistering material, it has marginal moisture protection. It has also been shown to produce hydrochloride emissions and highly toxic dioxins under suboptimal conditions during incineration. Germany and Switzerland have banned the incineration of PVC. There has been significant effort to replace PVC with PP.

An aluminum laminate can be used to cold-form the blister cavity. Aluminum–aluminum (foil–foil) laminates can provide complete protection form moisture, oxygen, and light. The aluminum foil is usually composed of several thin layers to help avoid pinholes in the material. Cold-forming involves shaping the foil around a plug to form the cavity. Figure 13.21 shows various types of barrier film laminates used for the blister cavity.

The lidding stock is attached to the blister cavity by applying heat and pressure over time to a heat-seal layer attached to the lidding material. The lidding seal is critical to controlling water vapor transmission and oxygen permeation. The three standard lid stock designs are peel-push, child-resistant peel-push, and push-through. Various types of lidding construction are shown in Figure 13.22. Peel-push lidding is usually constructed of laminated layers of heat-seal material, hard aluminum foil, laminating adhesive, and paper. The paper can be peeled back and the dosage form pushed through the foil. Hard foil is used because of its brittle nature, and the tablet can be pushed easily through the aluminum. The heat-sensitive adhesives are solvent-based or aqueous-based vinyls. Child-resistant peel-push lidding has an additional polymer layer between the aluminum and the paper. The common polymers used are PVP, PET, nylon, and polyester. The foil is a soft aluminum that exhibits high elongation at break and resists puncture if a child tries to bite through the aluminum barrier. Finally, push-through lidding has the simplest design, a laminate of a primer coating on Al foil that is coated with a heat seat. Tablets are

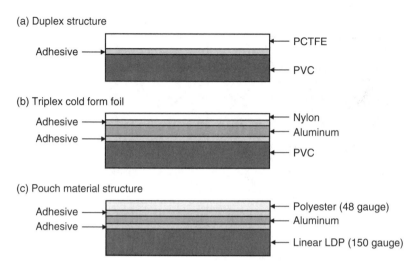

FIGURE 13.21 Laminates for blister cavities. (Adapted from U.S.P.–N.F. Chapter <1146>, with permission from the United States Pharmacopeial Convention. Copyright © 2008. All rights reserved.)

simply pushed through the thin aluminum lidding. Printing materials must be food or pharmaceutical grade and withstand temperatures as high as 300°C. They also need to be resistant to fading, abrasion, and other physical and chemical stresses.

Cold-formed aluminum–aluminum laminates are cold-pressed and provide the highest barrier protection to light, oxygen, and moisture. During the cold-forming process the foil cavity is shaped around a plug. This process is slower than thermoforming. As mentioned above, soft foil has high elongation at break. This is an advantage for child resistance but is a disadvantage for soft tablets

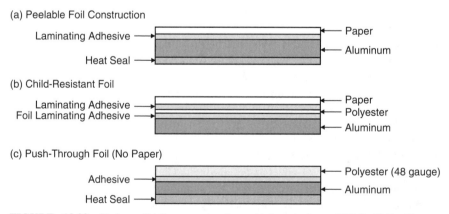

FIGURE 13.22 Various lidding construction. (Adapted from U.S.P.–N.F. Chapter <1146>, with permission from the United States Pharmacopeial Convention. Copyright © 2008. All rights reserved.)

such as lyophilized fast-dissolving tablets and molded tablets. In these cases, less flexible, more brittle hard foil is a more practical choice.

13.4.3 Blister Pack Processing

Blister packaging is an intermittent continuous process that involves thermoforming the cavity, filling the cavity with a tablet, sealing the cavity, printing container information on the lidding, and cartooning the blister sleeves. Control of this process has been enhanced by the use of feedback microprocessors and integrated visioning systems that monitoring and control the blistering process. The general blister pack equipment and process involves thermoform plastic unwinding, heating, forming, cooling, filling, sealing, and final cooling stations (see Figure 13.23).

The lidding material and blister web or forming material are supplied as rolls. The lidding and blister materials are supplied to the downstream processes by an unwinding station. For thermoforming processes the films are heated to provide enough flow to be formed into blister cavities. The various films require different temperatures. PVC films can be heated to 120 to 140°C. PP is heated to higher temperatures, around 140 to 160°C. Once the cavity is formed, the web moves on to a cooling station that cools the cavity before it is filled with a tablet. In the case of aluminum cold forming the aluminum cavity is cold formed around a plug. The lidding material is then matched to the blister web and heat sealed under pressure. The web is then labeled and cut. Typically, a visioning system is used to confirm the presence of a tablet in the blister and that the correct bar code and label have been used.

The sealing step is the most critical process to assure proper product protection, tamper evidence, and child resistance. The sealing step is a function of the

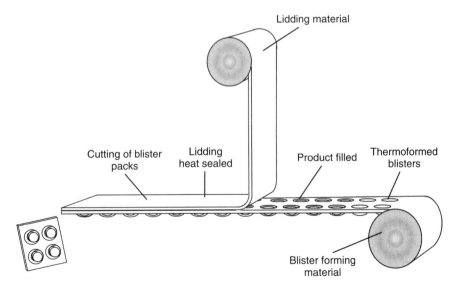

FIGURE 13.23 Typical blister packaging assembly process. (Adapted from Pilchik, 2000.)

blister and lidding construction materials, sealing pressure, temperature, time, drug temperature sensitivity, type of sealing mechanism, sealing width and area, and knurl design. A multivariate process optimization statistical design can be used to document which variables are important to control and over what processing ranges. The sealing width should be at least 3 mm. There are a number of knurl designs: linear, annular rings, diamond, and a combination. The diamond design is thought to provide the best protective seal by creating the most arduous path for air and moisture vapor penetration.

13.4.4 Regulatory Submission Information

The regulatory information needed for a FDA regulatory submission for tablet CCS is nicely summarized in Table 7 of the FDA guidance document Container Closure Systems for Packaging Human Drugs and Biologics (1991). The main areas that need to be covered are a general description and the suitability, quality control, and stability of the CCS. The general description should contain the name product code, manufacture, and materials of construction. Justification of the suitability of the CCS should include a discussion of seal integrity, light protection (U.S.P. <671>), moisture barrier properties (U.S.P. <671>), safety of the materials of construction (U.S.P. <88>), drug–contact material compatibility, and performance on stability. Leak tests should be performed for blister packages. Glass containers should meet the U.S.P. Containers—Glass (<660>) chemical resistance test. Plastic containers should conform to the U.S.P. Containers—Plastics (<661>) requirements. Quality control of the CCS should be supported by a sampling plan, attribute testing, associated acceptance criteria, and quality compliance plan. Stability in commercial CCSs is discussed in more detail in Section 13.7.

The filing of the regulatory information noted above is the responsibility of the sponsor. Often, the manufacturer of the CCS has propriety information that the company does not wish to share with the sponsor. In these cases, the sponsor can request that the CCS manufacturer provide the confidential information in a type III drug master file (DMF), which can be incorporated into the registration document by letter of reference. The letter of reference authorizes the sponsor to reference the DMF in their application. The letter also authorizes the FDA to have access to the DMF for the FDA's review. The CCS manufacturer's letter of reference should specify the sponsor, the component or material in question, and where the information is located in the DMF.

Appendix 12.5 provides the U.S.P. package nomenclature. A U.S.P. reference table, Containers for Dispensing Capsules and Tablets, provides recommended repackaging specifications for all official U.S.P. products. The relevant listed specifications are: tight, well closed, and light resistant.

13.5 RISK MANAGEMENT

Each product will have a set of critical product attributes that have a major influence on the overall quality of the product. ICH (2006) has published a quality risk

TABLE 13.23   Risk Assessment for a Wet Granulated Embossed Film-Coated Tablet[a]

Input Materials	Wet Granulation	Wet Milling	Drying	Milling	Extra-granular Blending	Tableting	Coating	Packaging	Critical Product Quality Attributes
L	L	L	L	L	L	H—picking	H—logo fill-in	L	Description and appearance
H	L	L	L	L	L	L	H	H	Identity
H	M	L	M	M	H	M	M	L	Assay
H	M	L	L	M	H	M	L	L	Content uniformity
H	M	L	M	M	M	H	M	L	Dissolution
L	H	L	H	L	L	L	M	L	Moisture content and uniformity
L	H	M	M	M	M	H	M	L	Hardness
M	M	L	M	H	M	H	L	L	Weight uniformity
M	L	L	M	L	L	L	L	M	Stability

[a] H, high; M, medium; L, low.

management guidance that discusses the general risk management process, risk management methodologies, and tools used to assess risk.

13.5.1   Risk Assessment

Risk is associated with the probability of a critical quality failure and the extent or consequences resulting from product performance failure, keeping in mind that the patient is the ultimate stakeholder. Three basic questions need to be asked in risk assessment: What can go wrong? (risk identification); what is the probability that it will go wrong? (risk analysis); and what are the consequences? (risk evaluation). Risks can be expressed as probabilities or by ranking. An example of a risk-based analysis is provided in Table 13.23.

13.5.2   Risk Control

Ideally, there should be a strategy that explicitly identifies the critical quality attributes and outlines how the attributes are going to be measured and controlled. An example of a control strategy for a wet granulated embossed film-coated tablet is provided in Table 13.24.

13.6   ATTRIBUTE TESTS

There are a number of attributes that can be used to characterize the tablet running powder, tablets, and the film coat. Assessment of the critical nature of these attributes is based on risk assessment, control strategy, and experience with the specific product. Some of the attributes are considered more critical than others,

TABLE 13.24 Risk Control Strategy for a Wet Granulated Embossed Film-Coated Tablet[a]

Input Materials	Wet Granulation	Wet Milling	Drying	Milling	Extra-granular Blending	Tableting	Coating	Packaging	Critical Product Quality Attributes
1		2					3	4	5	Description and appearance
6							7	8	Identity
9					10		11		Assay
					12		13		Content uniformity
14			15	16			17		Dissolution
	18		19				20		Moisture content and uniformity
						21			Hardness
						22			Weight uniformity

[a] 1, color, aggregation state; 2, aggregation state, particle sizing of wet mass; 3, gloss, embossing quality, picking, sticking, lamination/capping; 4, gloss, visual surface texture, color, bridging, chipping, peeling; 5, online visioning system; 6, infrared, certificate of analysis; 7, API HPLC retention time, NIR, IR; 8, NIR or other PAT technology to prevent product mix-up or carryover; 9, HPLC analysis or certificate of analysis; 10, final blend assay; 11, final tablet U.S.P. assay or content; 12, blend uniformity per FDA guidance document, also indication of uniform blend of lubricant; 13, U.S.P. content uniformity; 14, API particle size, moisture content; 15, moisture content; 16, particle size; 17, U.S.P. dissolution of film-coated tablet; 18, granulation endpoint; 19, moisture content, ideally in-line; 20, moisture content, ideally online; 21, in-process control of weight and compression force, online or at-line hardness; 22, online or in-process tablet weight uniformity check and control.

depending on the product performance desired and the overall impact that specific attributes have on performance. Attribute tests for the running powder, tablets, and film coat are presented in this section.

13.6.1 Running Powders

Particle size distribution, flowability, and compressibility are considered important running powder quality attributes.

13.6.1.1 *Particle Size* Particle size distribution of running powders may be determined by a number of techniques. The two most frequently used techniques are light diffraction (U.S.P. Chapter <429>) and sieving (U.S.P. <786>; ICH, 2010d). The light diffraction technique measures the diffraction pattern produced by the powder when irradiated with monochromatic laser light. The pattern is related to the particle size and distribution of the running powder and is transformed into a cumulative distribution that provides a percentile size of interest. Powders ranging from 0.5 to 3500 μm can be dispersed in air and measured by this light diffraction technique. U.S.P. <786> discusses the use of sieve analysis to determine powder particle size distribution. Screen sizes range from 38 to 11,200 μm. Typical granulated running powders will have a size range of 150 to 2000 μm. See Chapter 9 for further details concerning statistical analysis and interpretation of particle size distribution data.

13.6.1.2 Powder Flowability   Powder flowability and its measurement are discussed in U.S.P. <1174>, including angle of repose, compressibility index, flow through an orifice, and shear cell methods as techniques used to measure powder flowability. The powder flow properties for corresponding angles of repose and the compressibility index are provided. Such tables for flow through an orifice or shear cell techniques are not included.

13.6.1.3 Compressibility   The compressibility of running powders was discussed in Section 13.3.4.2.

13.6.2 Tablets

There are a number of tablet attribute tests, such as appearance, color (U.S.P. <631> and <1061>), content uniformity (U.S.P. <905>; ICH, 2010a), weight uniformity (U.S.P. <905>), disintegration time (U.S.P. <701>; ICH, 2009e), dissolution (U.S.P. <711> and <1092>; ICH, 2010b; FDA, 1997), drug release (U.S.P. <724>, thickness, breaking force (hardness) (U.S.P. <1217>), friability (U.S.P. <1216>; ICH, 2010c), moisture content (U.S.P. <731> and <921>), and stability, that define that a tablet's physiochemical and performance characteristics. NDA product stability assessment is discussed in more detail in Section 13.7. Some tablet attributes may be considered more critical than others, depending on the desired product and performance needs. Content uniformity, drug release rate, and hardness are often considered to be critical quality attributes.

13.6.3 Tablet Film Coat

Tablet film coat attributes, such as adhesion to the tablet surface (Wood and Harder, 1970; Fisher and Rowe, 1976), surface roughness and gloss (Reiland and Eber, 1986), logo and break indentation bridging (Rowe, 1997), elastic modulus and tensile strength (Gutierrez-Rocca (and McGinity, 1993), cracking (Rowe, 1997), and puncture strength (Bodmeier and Paeratakul, 1994), are usually assessed and defined during the development stage. The addition of the film coat is monitored during production by tablet weight gain and NIR analysis. The film-coated tablet's appearance and affect on dissolution are typical commercial product attributes that are evaluated.

13.6.4 Excipients

The FDA requires that each excipient be identified by a specific test. Often, a near-infrared or infrared spectrum is used to confirm the identity of an excipient. If the excipient is to be labeled as a pharmacopeia excipient, such as N.F. or U.S.P., the excipient needs to be tested to the pharmacopeia standards. A quality program may approve a specific vendor for a pharmacopeial excipient. Once a vendor is approved for a specific excipient, the company may receive the excipient under the vendor's certificate of analysis. Routine monitoring of the vendor and the certificate of analysis assures that the material meets the U.S.P. criteria.

13.7 NEW DRUG APPLICATION STABILITY ASSESSMENT

The primary purpose of stability assessment is to establish product label storage instructions and a shelf life for the product when it is stored under the labeled storage conditions. To meet a specified shelf life, the product must remain within a set of product specifications that are established for each critical product attribute. A balance must be struck between the time required to establish real-time data to support the labeled storage conditions and shelf life and the need to get important drugs to patients as quickly as possible with the appropriate safety assurances. The regulated process allows the NDA applicant to use appropriately designed stability studies to use real-time data and statistical approaches to extrapolate an estimated shelf life. The stability information that is used to establish the product label storage conditions and shelf life is based on a minimum of three batches of drug product. The three batches of drug product should be manufactured using the proposed market formulation, presentation (color, embossing, bisect), strengths, process, and CCS. The stability program should include tablets packaged in the proposed container closure system in the proposed bottle count sizes and blister pack presentations. These batches should be placed into accelerated and long-term stability testing. It is ideal to have all three batches manufactured and packaged at the proposed commercial production site on a commercial scale. This is often not possible because of cost and lack of available drug substance. In these cases, two of the batches may be manufactured and packaged at one-tenth of full-scale production or 100,000 units, whichever is larger. The third batch may be a smaller pilot scale.

The International Conference on Harmonization guidance documents serve as a starting point for developing a tablet stability protocol (ICH, 1996, 1997, 2003a, b, 2004, 2009a–c). There should be a minimum of 12 months of long-term and three months of accelerated data at the time of the NDA submission unless the FDA has agreed to shorter stability times. The long-term storage condition should be carried out to support the desired shelf life with real-time data. The ICH stability conditions are $25 \pm 2^{\circ}C/60 \pm 5\%$ RH for long-term storage and $40 \pm 2^{\circ}C/75 \pm 5\%$ RH accelerated storage for one and three months. These storage conditions will support the standard FDA label, which states that the product can be stored at $25^{\circ}C$ ($77^{\circ}F$) with excursions permitted to 15 to $30^{\circ}C$ (59 to $86^{\circ}F$). For less stable products that may not achieve 24 months of stability, accelerated testing can be conducted at $30^{\circ}C/65\%$ RH. In addition, the ICH guidance document Q1D Bracketing and Matrixing Designs for Stability Testing of New Drug Substances and Products can significantly reduce the stability commitment if preapproved by the FDA.

Initial testing at time zero typically includes a description of appearance (color, shape, identifying marks such as logo and bisect), two identity methods, assay of active and related products, content uniformity, microbial limits test (unless there is a scientific justification not to include this test), water content, and dissolution. Depending on other quality attributes that are critical for product quality,

TABLE 13.25   NDA Stability Protocol

Product	Container/Closure and Supplier	API Process
Product lot	Package lot	API lot
Batch size	Batch size	Batch size
Manufacturer	Packager	Manufacturer
Manufacturer site	Packaging site	Site of manufacture
Manufacture date	Package date	Manufacture date
Stability start date		

					Time (months)					
Storage ($^\circ$C/% RH)	1	3	6	9	12	18	24	36	48	60
25/60[a]		×	×	×	×	×	×	×	(×)[b]	(×)[b]
40/75[b]		×	×							
Light[c]	×									

[a] $^\circ$C ± 2/% RH ± 5%.
[b] (×), optional testing.
[c] Testing per ICH guidelines.

other tests may be performed as well, such as residual solvents, weight uniformity, hardness, and friability. Ongoing stability would include drug and degradation product assay, microbial limits test (may be done annually), and dissolution. Again, product-specific attributes may be added if deemed appropriate. Validated test methods are required for any test that is used for initial testing and stability testing. Appropriate acceptance criteria should also be included for each test. A typical stability protocol that would support a regulatory submission to the European Union, Japan, and the United States is provided in Table 13.25.

The FDA has categorized such impurities as organic impurities that are process- and drug-related, inorganic impurities, and residual solvents. For drug product stability studies, the FDA (2003d) has given guidance regarding the reporting and listing of degradation levels. The degradation levels are based on the maximum daily dose. Table 13.26 summarizes the reporting and listing requirements for drug product degradation impurities.

The ICH (2004) has issued guidance on the evaluation of stability data. The guidance provides a decision tree that addresses circumstances when statistical analysis is unnecessary, shelf-life extrapolation is appropriate, pooling of data is accepted, and data are not amenable to statistical analysis. The guidance also discusses recommended statistical analysis approaches. In general, statistical analysis is normally considered unnecessary when the data show that the product will remain well within the acceptance criteria for a given product attribute; there is little or no change in the accelerated and long-term condition; and there is little or no variability. Justification for extrapolation of the shelf life can be proposed. When accelerated or long-term (25°C) data show variability or change over time,

TABLE 13.26 Reporting, Identification, and Qualification Thresholds for Degradation Impurities in New Drug Products

Maximum Daily Dose	Threshold
Reporting Threshold	
Less than or equal to 1 g	0.1%[a]
Greater than 1 g	0.05%
Identification	
Less than 1 mg	1.0% or 5 μg TDI,[b] whichever is lower
1 to 10 mg	0.5% or 20 μg TDI, whichever is lower
Greater than 10 mg to 2 g	0.2% or 2 mg TDI, whichever is lower
Greater than 2 g	0.10%
Qualification	
Less than 10 mg	1.0% or 50 μg TDI, whichever is lower
10 to 100 mg	0.5% or 200 μg TDI, whichever is lower
Greater than 100 mg to 2 g	0.2% or 3 mg TDI, whichever is lower
Greater than 2 g	0.15%

[a] Expressed as percent of maximum daily dose.

[b] TDI, total daily intake.

shelf-life extrapolation can be based on statistical analysis or a data-based justification. Statistically analysis can support up to twice the period of real-time 25°C data, but not more than 12 months. An accepted justification for extrapolation of data not amenable to statistical analysis can support a shelf life up to one-and-a-half times the available real-time long-term data, but no more than six months of extrapolation. Other circumstances are also covered in this document and left to the review of the reader.

REFERENCES

Allen TT, Chantarat N, Taslim C. Fractional factorial designs that maximise the probability of identifying the important factors. Int. J. Ind. Syst. Eng.. 2009;4:133–150.

Augsburger LL, Hoag SW, eds. *Pharmaceutical Dosage Forms: Tablets*, Vol. 1, *Unit Operations and Mechanical Properties*, 3rd ed. New York: Informa Healthcare; 2008a.

_____. *Pharmaceutical Dosage Forms: Tablets*, Vol. 2, *Rational Design and Formulation*, 3rd ed. New York: Informa Healthcare; 2008a.

_____ *Pharmaceutical Dosage Forms: Tablets*, Vol. 3, *Manufacture and Process Control*, 3rd ed. New York: Informa Healthcare; 2008c.

Betz G, Junkier P, Leuenberger H. Power consumption measurement and temperature recording during granulation. Int. J. Pharm. 2004;272:137–149.

Bodmeier R, Paeratakul O. Mechanical properties of dry and wet cellulosic acrylic films prepared from aqueous colloidal polymer dispersions used in the coating of solid dosage forms. Pharm. Res. 1994;11:882–888.

Bowman BJ, Ofner CM III, Shott H. Colloidal dispersions. In: Troy DB, ed. *Remington: The Science and Practice of Pharmacy*, 21st ed. Philadelphia: Lippincott Williams & Wilkins; 2006. p. 308.

Briens L, Daniher D, Tallevi A. Monitoring high-shear granulation using sound and vibration measurements. Int. J. Pharm. 2007;331:54–60.

Callahan JC, Cleary GW, Elefant M, Kaplan G, Kensler T, Nash RA. Equilibrium moisture content of pharmaceutical excipients. Drug Dev. Ind. Pharm. 1982;8:355–369.

Campbell DA, Marranca JT, Pope, RE, Stagner RA. Systems for controlling evaporative drying processes using environmental equivalency. U.S. Patent 6770141. Aug. 2004.

Cantor SL, Augsburger LL, Hoag SW, Gerhardt A. Pharmaeutical granulation process, mechanism, and the use of binders. In: Augsburger LL, Hoag HH, eds. *Pharmaceutical Dosage Forms: Tablets*, Vol. 1, *Unit Operations and Mechanical Properties*, 3rd ed. New York: Informa Healthcare; 2008. p. 269.

Chablani L, Tayor MK, Mehrotra A, Rameas P, Stagner WC. Inline real-time near-infrared granule moisture measurements of a continuous granulation-drying-milling process. AAPS PharmSciTech. doi:10.1208/s12249-011-9669-z. Published online Aug. 13, 2011.

Chieng N, Zujovic Z, Bowmaker G, Rades T, Saville D. Effect of milling conditions o the solidi-state conversion of ranitidine hydrochloride Form 1. Int. J. Pharm. 2006;327:36–44.

Code of Federal Regulations. 16 CFR 1700.2. Poison prevention packaging: testing procedure for special packaging. http://www.access.gpo.gov/nara/cfr/waisidx_03/16cfr1700_03.html. Accessed Dec. 2011.

———. 21 CFR 73.1200. Synthetic iron oxide. http://cfr.vlex.com.vid/73-1200-synthetic-iron-oxide-19703771. Accessed Mar. 2011.

———. 21 CFR 211.110. Sampling and testing of in-process materials and drug products. http://www.accessdata.fda.gov/scripts/cdrh/cfdocs/cfcfr/CFRSearch.cfm?fr=211.110. Revised Apr. 1, 2009. Title 21, Vol. 4. Accessed Dec. 2011.

———. 21 CRF 211.132. Tamper-evident packaging requirements for over-the-counter (OTC) drug products. http://edocket.access.gpo.gov/cfr_2001/aprqtr/pdf/21cfr211.132.pdf. Accessed Dec. 2011.

———. 29 CFR 1910.1200(g). Material safety data sheet. http://www.ilpi.com/msds/osha/1910_1200.html#1910.1200(g). Accessed Dec. 2011.

Corvari V, Fry WC, Seibert WL, Augsberger L. Instrumentation of a high-shear mixer: evaluation and comparison of a new capacitive sensor, a watt meter, and a strain-gage torque sensor for wet granulation monitoring. Pharm. Res. 1992;9:1525–1533.

Daniel WW. *Biostatistics: A Foundation for Analysis in the Health Sciences*. New York: Wiley: 1974. p. 64.

Descamps M, Willart E, Dudognon E, Caron V. Transformation of pharmaceutical compounds upon milling and comilling: the role of T_g. J. Pharm. Sci. 2007;96(5):1398–1407.

Dozier WA III, Hanna W, Behnke K. Grinding and pelletizing responses of pearl millet-based diets. J. Appl. Poult. Res. 2005;14:269–274.

Dressman JB, Repa C. In vitro–in vivo correlations for lipophilic, poorly water-soluble drugs. Eur. J. Pharm. Sci. 2000;Suppl. 11:S73–S80.

Ebbey GD. A thermodynamic model for aqueous film-coating. Pharm. Technol. 1987;Apr. 40–50.

FDA. 1991. Guidance for industry: Container closure systems for packaging human drugs and biologics. U.S. Department of Health and Human Services, Food and Drug Administration Center for Drug Evaluation and Research and Center for Biologics Evaluation and Research. May 1999. http://www.fda.gov/downloads/Drugs/GuidanceComplianceRegulatoryInformation/Guidances/ucm070551.pdf. Accessed Dec. 2011.

———. 1995. Guidance for industry: Immediate release solid oral dosage forms—scale-up and postapproval changes; chemistry, manufacturing, and controls, in vitro dissolution testing, and in vivo bioequivalence documentation. U.S. Department of Health and Human Services, Food and Drug Administration Center for Drug Evaluation and Research. Nov. 1995. http://www.fda.gov/downloads/Drugs/GuidanceComplianceRegulatoryInformation/Guidances/ucm070636.pdf. Accessed Dec. 2011.

———. 1997. Guidance for industry: Dissolution testing of immediate release solid dosage forms. U.S. Department of Health and Human Services, Food and Drug Administration Center for Drug Evaluation and Research. Aug. 1997. http://www.fda.gov/downloads/Drugs/GuidanceComplianceRegulatoryInformation/Guidances/ucm070237.pdf. Accessed Dec. 2011.

_____ 1999. Guidance for industry: SUPAC-IR/MR—Immediate release and modified release solid oral dosage forms; manufacturing equipment addendum. U.S. Department of Health and Human Services, Food and Drug Administration Center for Drug Evaluation and Research. Jan. 1999. http://www.fda.gov/downloads/Drugs/GuidanceComplianceRegulatoryInformation/Guidances/UCM070637.pdf. Accessed Dec. 2011.

_____. 2003a. Draft guidance for industry: Powder blends and finished dosage units—stratified in-process dosage unit sampling and assessment. U.S. Department of Health and Human Services, Food and Drug Administration Center for Drug Evaluation and Research. Oct. 2003. http://www.fda.gov./OHRMS/DOCKETS/98fr/03d-0493-gdl0001.pdf. Accessed Dec. 2011.

_____. 2003b. Guidance for industry: Powder blends and finished dosage units—stratified in-process dosage unit sampling and assessment. U.S. Department of Health and Human Services, Food and Drug Administration Center for Drug Evaluation and Research. Revised attachments. Nov. 2003. http://www.fda.gov./downloads/Drug/GuidanceComplianceRegulatoryInformation/Guidances/ucm070314.pdf. Accessed Dec. 2011.

_____. 2003c. Guidance for industry: Q1A—Stability testing of new drug substances and products. U.S. Department of Health and Human Services, Food and Drug Administration Center for Drug Evaluation and Research and Center for Biologics Evaluation and Research. Nov. 2003. http://www.fda.gov/downloads/regulatoryinformation/guidances/ucm128204. Accessed Dec. 2011.

_____. 2003d. Guidance for industry: Q3B(R)—Impurities in new drug products. U.S. Department of Health and Human Services, Food and Drug Administration Center for Drug Evaluation and Research and Center for Biologics Evaluation and Research. Nov. 2003. http://www.bcg-usa.com/regulatory/docs/2003/ICHQ3B(R).pdf. Accessed Dec. 2011.

_____. 2006. Guidance for industry: Q3B(R2)—Impurities in new drug products. U.S. Department of Health and Human Services, Food and Drug Administration Center for Drug Evaluation and Research and Center for Biologics Evaluation and Research. July 2006. http://www.fda.gov/downloads/Drugs/GuidanceComplianceRegulatoryInformation/Guidances/ucm073389.pdf. Accessed Dec. 2011.

_____. 2007. Guidance for industry and review staff: Target product profile—a strategic development process tool. U.S. Department of Health and Human Services, Food and Drug Administration Center for Drug Evaluation and Research. Mar. 2007. http://www.fda.gov/downloads/Drugs/GuidanceComplianceRegulatoryInformation/Guidances/ucm.080593. Accessed Feb. 2011.

_____. 2009. Guidance for industry: Q8(R2)—pharmaceutical development, revision 2. International Conference on Harmonization. U.S. Department of Health and Human Services, Food and Drug Administration Center for Drug Evaluation and Research and Center for Biologics Evaluation and Research. Nov. 2009. http://www.fda.gov/downloads/Drugs/GuidanceComplianceRegulatoryInformation/Guidances/ucm.073507. Accessed Feb. 2011.

_____. 2011. Inactive ingredients list. http://www.accessdata.fda.gov/scripts/cder/iig/index.cfm. Accessed Dec. 2011.

Feeley JC, York P, Sumby BS, Dicks H. Determination of surface properties and flow characteristics of salbutamol sulphate, before and after micronisation. Int. J. Pharm. 1998;172:89–96.

Fisher DG, Rowe RC. The adhesion of film coatings to tablet surfaces: instrumental and preliminary evaluation. J. Pharm. Pharmacol. 1976;28:886–889.

Fry WC, Stagner WC, Wichman KC. Computer-interfaced capacitive sensor for monitoring the granulation process. J. Pharm. Sci. 1984;73:420–421.

Fry WC, Wu PP, Stagner WC, Wichman KC, Anderson NR. Computer-interfaced capacitive sensor for monitoring the granulation process: II. System response to process variables. Pharm Technol. 1987;Oct.: 30–41.

Garcia TP, Lucisano LJ. Single-pot processing. In: Parikh DM, ed. *Handbook of Pharmaceutical Granulation Technology*. Drugs and the Pharmaceutical Sciences, Vol. 81. New York: Marcel Dekker; 1997. p. 317.

Garcia TP, Prescott JK. Blending and blend uniformity. In: Augsburger LL, Hoag SW, eds. *Pharmaceutical Dosage Forms: Tablets*, Vol. 1, *Unit Operations and Mechanical Properties*, 3rd ed. New York: Informa Healthcare; 2008. pp. 111–174.

GEA Niro Pharma Systems. Current issues and troubleshooting fluid bed granulation. Pharm. Technol. Eur. 1998; May. http://www.niroinc.com/html/pharma/phpdfs/niroreprint2.pdf.

Ghorab M, Adeyeye MC. High shear mixing granulation of ibuprogen an β-cyclodextrin: effects of process variables on ibuprofen dissolution. AAPS PharmSciTech 2007;8(4):art. 84. doi:10.1208/pt0804084.

Gupta A, Peck, GE, Miller, RW, Morris, KR. Real-time near-infrared monitoring of content uniformity, moisture content, compact density, tensile strength, and Young's modulus of roller compacted powder blends. J. Pharm. Sci. 2005;94:1589–1597.

Gutierrez-Rocca JC, McGinity JW. Influence of aging on the physical–mechanical properties of acrylic resin films cast from aqueous dispersions and organic solutions. Drug Dev. Ind. Pharm. 1993;19:315–332.

Heckel RW. Density–pressure relationships in powder compaction. Trans. Metal. Soc. AIME 1961a;221:671–675.

———. An analysis of powder compaction phenomena. Trans. Metal. Soc. AIME 1961b; 221:1001–1008.

Heng JYY, Thielmann F, Williams DR. The effects of milling on the surface properties of form I parectamol crystals. Pharm. Res. 2006;23:1918–1927.

Hogan J, Shue P-I, Podczek F, Newton MJ. Investigations into the relationship between drug properties, filling, and release of drugs from hard gelatin capsules using multivariate statistical analysis. Pharm. Res. 1996;13:944–949.

ICH. 1996. Q1B: Photostability testing of new drug substances and products. Nov. 1996. http://www.fda.gov/Drugs/GuidanceComplianceRegulatoryInformation/Guidances/ucm065005. Accessed Dec. 2011.

———. 1997. Q1C: Stability testing for new dosage forms. June 1997. http://www.fda.gov/Drugs/GuidanceComplianceRegulatoryInformation/Guidances/ucm065005. Accessed Dec. 2011.

———. 2003a. Q1D: Bracketing and matrixing designs for stability testing of new drug substances and products. Jan. 2003. http://www.fda.gov/Drugs/GuidanceComplianceRegulatoryInformation/Guidances/ucm065005. Accessed Dec. 2011.

———. 2003b. Q1A(R2): Stability testing of new drug substances and products. Nov. 2003. http://www.fda.gov/Drugs/GuidanceComplianceRegulatoryInformation/Guidances/ucm065005. Accessed Dec. 2011.

———. 2004. Q1E: Evaluation of stability data. June 2004. http://www.fda.gov/Drugs/GuidanceComplianceRegulatoryInformation/Guidances/ucm065005. Accessed Dec. 2011.

———. 2006. Q9: Quality risk management. June 2006. http://www.fda.gov/Drugs/GuidanceComplianceRegulatoryInformation/Guidances/ucm065005. Accessed Dec. 2011.

———. 2009a. Q4B, Annex 4A: Microbiological examination of non-sterile products—microbial enumeration tests. General chapter. Apr. 2009. http://www.fda.gov/Drugs/GuidanceComplianceRegulatoryInformation/Guidances/ucm065005. Accessed Dec. 2011.

———. 2009b. Q4B, Annex 4B: Microbiological examination of non-sterile products—tests for specified micro-organisms. General chapter. Apr. 2009. http://www.fda.gov/Drugs/GuidanceComplianceRegulatoryInformation/Guidances/ucm065005. Accessed Dec. 2011.

———. 2009c. Q4B, Annex 4C: Microbiological examination of non-sterile products—acceptance criteria for pharmaceutical preparations and substances for pharmaceutical general use. Apr. 2009. http://www.fda.gov/Drugs/GuidanceComplianceRegulatoryInformation/Guidances/ucm065005. Accessed Dec. 2011.

———. 2009d. Q8(R2): Pharmaceutical development—international harmonized tripartite guide-line. Nov. 2009. http://www.fda.gov/RegulatoryInformation/Guidances/ucm128003. Accessed Dec. 2011.

———. 2009e. Q4B, Annex 5: Disintegration test. General chapter. Dec. 2009. http://www.fda.gov/Drugs/GuidanceComplianceRegulatoryInformation/Guidances/ucm065005. Accessed Dec. 2011.

———. 2010a. Q4B, Annex 6: Uniformity of dosage units. General chapter. Apr. 2010. http://www.fda.gov/Drugs/GuidanceComplianceRegulatoryInformation/Guidances/ucm065005. Accessed Dec. 2011.

_____. 2010b. Q4B, Annex 7: Dissolution test. General chapter. Apr. 2010. http://www. fda.gov/Drugs/GuidanceComplianceRegulatoryInformation/Guidances/ucm065005. Accessed Dec. 2011.

_____. 2010c. Q4B Annex 9: Tablet friability. General chapter. Apr. 2010. http://www.fda. gov/Drugs/GuidanceComplianceRegulatoryInformation/Guidances/ucm065005. Accessed Dec. 2011.

_____ 2010d. Q4B Annex 12: Analytical sieving. General chapter. Sept. 2010. http://www.fda. gov/downloads/Drugs/GuidanceComplianceRegulatoryInformation/Guidances/ucm194501. Accessed Dec. 2011.

Jarkko K, Jukka I, Paronen P. Temperature changes during tableting measured using infrared thermoviewer. Int. J. Pharm. 1993;92:157–166.

JMP 8. Design of experiments software. SAS Institute, Cary, NC; 2008.

Karumanchi V, Taylor MK, Ely KJ, Stagner WC. Monitoring powder blend homogeneity using light induced fluorescence, AAPS PharmSciTech [online early access]. doi:10.1208/s12249-011-9667-7. Published online Aug. 13, 2011.

Ketolainen J, Ilkka J, Paronen P. Temperature changes during tableting measured using infrared thermoviewer. Int. J. Pharm. 1993;92:157–166.

Kumar V, Mehrota A, Taylor, M, Stagner W. Real-time particle size analysis using focused beam reflectance measurement (FBRM) as a process analytical technology (PAT). American Association of Pharmaceutical Scientists Annual Meeting, New Orleans, LA, Nov. 2010.

Landin M, York P, Cliff MJ, Rowe RC, Wigmore AJ. Scale-up of a pharmaceutical granulator in fixed bowl mixer–granulators. Int. J. Pharm. 1996;133:127–131.

Lee M-J, Park C-R, Kim A-Y, Kwon B-S, Bang K-H, Cho Y-S, Jeong M-Y, Choi G-J. Dyanamic calibration for the in-line NIR monitoring of film thickness of pharmaceutical tablets processed in a fluid-bed coater. J Pharm Sci. 2010;99:325–335.

Leeson LJ, Mattock AM. Decomposition of aspirin in the solid state. J. Am. Pharm. Assoc. Sci. Educ. 1958;47:329–338.

Leuenberger H, Bier H-P, Sucker HB. Theory of the granulating-liquid requirement in the conventional granulation process. Pharm. Technol. 1979;June: 61–68.

Levin M, Zlokarnik M. Dimensional analysis of the tablet process. In: Levin M, ed. *Pharmaceutical Process Scale-up*. Drugs and the Pharmaceutical Sciences, Vol. 118. New York: Marcel Dekker; 2002. pp. 253–258.

Lipinski CA, Franco L, Dominy BW, Feeney PJ. Experimental and computational approaches to estimate solubility and permeability in drug discovery and development settings. Adv. Drug Deliv. Rev. 1997;23:3–25.

Liversidge GG, Cundy K, Bishop JF, Czekai DA. Surface modified drug nanoparticles. U.S. Patent 5145684. 1992.

Managaraj S, Goswami TK, Mahajan PV. Applications of plastic films for modified atmosphere packaging of fruits and vegetables: a review. Food Eng. Rev. 2009;1:133–158.

Massey LK. *Permeability Properties of Plastics and Elastomers: A Guide to Packaging and Barrier Materials*, 2nd ed. Norwich, NY: Plastics Design Library/William Andrew Publishing; 2003. p. 96.

McGinity JW, ed. *Aqueous Polymeric Coatings for Pharmaceutical Dosage Forms*, 2nd ed. Drugs and the Pharmaceutical Sciences, Vol. 79. New York: Marcel Dekker; 1997.

Mehta AT. Scale-up considerations in the fluid-bed process for controlled-release products. Pharm. Technol. 1988;Feb. 46–52.

Moes JJ, Ruijken MM, Gout E, Frijlink HW, Ugwoke MI. Application of process analytical technology in tablet process development using NIR spectroscopy: blend uniformity, content uniformity, and coating thickness measurements. Int. J. Pharm. 2008;357:108–118.

Müller J, Knop K, Thies J, Uerpmann C, Kleinebudde P. Feasibility of Raman spectroscopy as PAT tool in active coating. Drug. Dev. Ind. Pharm. 2010;36:234–243.

Muzzio FJ, Marinthi I, Portillo P, Llusa M, Levin M, Morris KR, Soh JLP, McCann RJ, Alexander A. A forward-looking approach to process scale-up for solid dose manufacturing. In: Augsburger LL, Hoag SW, eds. *Pharmaceutical Dosage Forms: Tablets*, Vol. 3, *Manufacture and Process Control*, 3rd ed. New York: Informa Healthcare; 2008. pp. 136–139.

Ohmart/VEGA. Listing of dielectric constants. http://www.ohmartvega.com/en/DK_CD.htm. Accessed Dec. 2011.

Parikh DM, Mogavero M. Batch Fluid Bed Granulation. In: Parikh DM, ed. *Handbook of Pharmaceutical Granulation Technology*, 2nd ed. Drugs and the Pharmaceutical Sciences, Vol. 154. Boca Raton FL: Taylor & Francis; 2005. pp. 282–286.

Pearlswig DM, Lucisano LJ, Robin P. Simulation modeling applied to the development of a single-pot process using microwave/vacuum drying. Pharm. Technol. 1994;June: 44–60.

Peck GE, Soh JLP, Morris KR. Dry granulation. Augsburger LL, Hoag SW, ed. *Pharmaceutical Dosage Forms: Tablets*, Vol. 1, *Unit Operations and Mechanical Properties*, 3rd ed. New York: Informa Healthcare; 2008. p. 316.

Pepin X, Blanchon S, Gouarraze G. Powder consumption profiles in high-shear wet granulation: I. Liquid distribution in relation to power and binder properties. J. Pharm. Sci. 2001;90: 322–331.

Pérez-Ramos JD, Findlay WP, Peck G, Morris KR. Quantitative analysis of film coating in a pan coater based on in-line sensor measurements. AAPS PharmSciTech. 2005;6:E127–E136.

Pilchik R. Pharmaceutical blister packaging: II. Machinery and assembly. Pharm. Technol. 2000; Nov.: 56–60.

Rambali B, Baert L, Massart DL. Scaling up of fluidized bed granulation process. Int. J. Pharm. 2003;252:197–206.

Reiland, TL, Eber AC. Aqueous gloss solutions, formula and process variables, effect on on the surface texture of film coated tablets. Drug Dev. Ind. Pharm. 1986;12:231–245.

Rekhi GS, Caricofe RB, Parikh DM, Augsburger LL. A new approach to scale-up of a high-shear granulation process. Pharm. Technol. 1996;Oct.: 2–10.

Robin P, Lucisano LJ, Pearlswig DM. Rationale for selection of a single-pot manufacturing process using microwave/vacuum drying. Pharm. Technol. 1994;May: 28–36.

Roggenbuck A, Schmitz J, Deninger A, Mayorga IC, Hemberger J, Güsten R, Grüninger M. Coherent broadband continuous-wave terahertz spectroscopy on solid-state samples. New J. Phys. 2010. doi:12:043017 (http://www.nip.org/).

Roman R. Stability kinetics. In: Augsburger, LL, Hoag SW, eds. *Pharmaceutical Dosage Forms: Tablets*, Vol. 1, *Unit Operations and Mechanical Properties*, 3rd ed. New York: Informa Healthcare; 2008. pp. 485–517.

Rowe RC. Defects in aqueous film-coating tablets. In: McGinity JW, ed. *Aqueous Polymer Coatings for Pharmaceutical Dosage Forms*. 2nd ed. Drugs and the Pharmaceutical Sciences, Vol. 79. New York: Marcel Dekker; 1997. pp. 419–440.

Rowe RC, Sheskey PJ, Owen SC, eds. *Handbook of Pharmaceutical Excipients*, 5th ed. London: Pharmaceutical Press; Washington, DC: American Pharmacists Association; 2006.

Rumpf H. The strength of granules and agglomerates. In: Knepper W, ed. *Agglomeration*. New York: Interscience; 1962. pp. 379–414.

Sheskey P, Pacholke K, Sackett G, Maher L, Polli J. Roll compaction granulation of a controlled-release matrix tablet formulation containing HPMC: effect of process scale-up on robustness of tablets, tablet stability, and predicted in vivo performance. Pharm. Technol. 2000;Nov.: 30–52.

Shinbrot T, Muzzio FJ. Nonequilibrium patterns in granular mixing and segregation. Phys. Today 2000;53:25–30.

Skibsted ETS, Westerhuis JA, Smilde AK, Witte DT. Examples of NIR based real time release in tablet manufacturing. J. Pharm. Biomed. Anal. 2007;43:1297–1305.

Srinivasan A, Iser R, Devinder SG. Common deficiencies in abbreviated new drug applications: 2. Description, composition, and excipients. Pharm. Technol. 2010;34:45–51.

Stagner WC, Cerimele BJ, Fricke GH. Content uniformity: separation and quantitation of sources of dose variation. Drug Dev. Ind. Pharm. 1991;17:233–244.

Strong JC. Psychrometric analysis of the environment equivalency factor for aqueous tablet coating. AAPS PharmSciTech 2009;10:303–309.

Tabasi SH, Fahmy R, Bensley D, O'Brien C, Hoag SW. Quality by design: II. Application of NIR spectroscopy to monitor the coating process for a pharmaceutical sustained release product. J. Pharm. Sci. 2008;97:4052–4066.

Tardos GI, Hapgood KP, Ipadeola OO, Michaels JN. Stress measurements in high-shear granulators using calibrated "test" particles: application to scale-up. Powder Technol. 2004;140: 217–227.

Thyn J. Optimization of hammer mills. Isotopenpraxis isotopes in environmental and health studies. 1989;25:151–156.

Trasi NS, Boerrigter SX, Bryrn Sr. Investigation of the milling-induced thermal behavior of crystalline and amorphous griseofulvin. Pharm. Res. 2010;27:1377–1389.

U.S. Pharmacopeia. U.S.P.–N.F. online, First Supplement U.S.P. 33–N.F. 28 Reissue. Accessed Oct. 2010.

Van Scoik KG, Zoglio MA, Carstensen JT. Drying and driers. In: Swarbrick J, Boylan JC, eds. *Encyclopedia of Pharmaceutical Technology*, Vol. 4. New York: Marcel Dekker; 1991. p. 495.

Van Vaerenbergh G. Microwave drying. http://www.gea-ps.com/npsportal/cmsdoc.nsf/webdoc/webb87bhnb. Accessed Dec. 2011.

Waldon MS. Microwave vacuum drying of pharmaceuticals: The development of a process. Pharm. Eng. 1988;8:9–13.

Wells JW. Tablet testing. In: Swarbrick J, Boylan JC, eds. *Encyclopedia of Pharmaceutical Technology*. Vol 14. New York: Marcel Dekker; 1996. pp. 401–416.

Wood JA, Harder SW. The adhesion of film coatings to the surfaces of compressed tablets. Can. J. Pharm. Sci. 1970;5:18–23.

GLOSSARY

AMG	Acetylated monoglycerides.
API	Active pharmaceutical ingredient.
ATBC	Acetyltributyl citrate.
ATEC	Acetyltriethyl citrate.
BCS	Biopharmaceutics Classification System.
BDDCS	Biopharmaceutics Drug Disposition Classification System.
CAP	Cellulose acetate phthalate.
CAS	Cellulose acetate succinate.
COC	Cyclic olefin copolymers.
CCS	Container closure system.
CFR	*Code of Federal Regulations*.
CLD	Cord length distribution.
CPSC	Consumer Product Safety Commission.
DBP	Dibutyl phthalate.
DBS	Dibutyl sebacate.
DE	Dextrose equivalent.
DEP	Diethyl phthalate.
DMF	Drug master file.
DSC	Differential scanning calorimetry.
EE	Environmental equivalency.
E.P.	*European Pharmacopoeia*.

EU	European Union.
FaSSIF	Fasted simulated intestinal fluid.
FBG	Fasting blood glucose.
FBRM	Focused-beam reflectance measurement.
FDA	U.S. Food and Drug Administration.
FeSSIF	Fed simulated intestinal fluid.
GMP	Good Manufacturing Practices.
GRAS	Generally recognized as safe.
GT	Gastrointestinal
HDPE	High-density polyethylene.
HEC	Hydroxyethyl cellulose.
HPC	Hydroxypropyl cellulose.
HPMC	Hydroxypropyl methylcellulose.
HPMCAS	Hydroxypropyl methylcellulose acetate succinate.
HPMCP	Hydroxypropyl methylcellulose phthalate.
ICH	International Conference on Harmonization.
IHS	Industrial hygiene safety.
IIL	Inactive Ingredients List.
IR	Immediate release.
J.P.	*Japanese Pharmacopoeia.*
LIF	Light-induced fluorescence.
LOEL	Lowest Observed Effect Level.
LTD	Lowest Therapeutic Dose.
MFT	Minimum film formation temperature.
MR	Modified release.
MVGD	Microwave vacuum granulation drying.
NDA	New drug application.
N.F.	*National Formulary.*
NIR	Near-infared.
NOEL	No observed effect level.
OEL	Occupational exposure limits.
RSD	Relative standard deviation.
PAT	Process Analytical technology.
PCTEF	Poly(chlorotrifluroethylene) or Aclar.
PEG	Poly(ethylene glycol).
PET	Polyester terephthalate.
PP	Polypropylene.
PS	Polystyrene.
PSD	Particle size distribution.
PVAP	Polyvinyl acetate phthalate.
PVC	Poly(vinyl chloride).
PVdC	Poly(vinylidene chloride).
RH	Relative humidity.
SPP	Single-pot process.
SUPAC	Scale-up and postapproval changes.
TA	Triacetin.
TBC	Tributyl citrate.
TEC	Triethyl citrate.
U.S.P.	*United States Pharmacopeia.*
WVTR	Water vapor transmission rate.

APPENDIXES

APPENDIX 13.1 Hygroscopicity Classification System

Class I: Nonhygroscopic. Essentially no moisture increases occur at relative humidities below 90%. (Figure App. 13.1). Furthermore, the increase in moisture content after storage for 1 week above 90% relative humidity is less than 20%.

Class II: Slightly hygroscopic. Essentially no moisture increases occur at relative humidities below 80%. The increase in moisture content after storage for 1 week above 80% relative humidity in less than 40%.

Class III: Moderately hygroscopic. The moisture content does not increase above 5% after storage at relative humidities below 60%. The increase in moisture content after storage for 1 week above 80% relative humidity is less than 50%.

Class IV: Very hygroscopic. Moisture increase may occur at relative humidities as low as 40 to 50%. The increase in moisture content after storage for 1 week above 90% relative humidity may exceed 30%.

Source: Adapted from Callahan et al. (1982), with permission from Informa Healthcare.

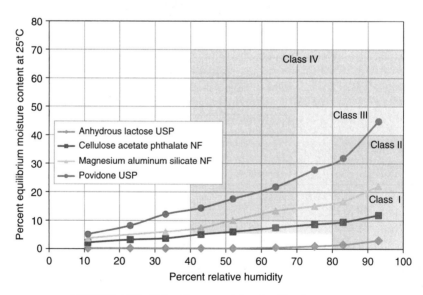

FIGURE APP. 13.1 Hygroscopicity classification system.

APPENDIX 13.2   Diluents and Direct Compression Binders[a]

Excipient	Trade Names
Calcium phosphate, dibasic, anhydrous, E.P./J.P./U.S.P.–N.F.	A-Tab, Emcompress Anhydrous
Calcium phosphate, dibasic dihydrate, E.P./U.S.P.–N.F.	Di-Tab, Emcompress
Calcium phosphate, tribasic, E.P./U.S.P.–N.F.	Tri-Cal, Tri-Cafos, Tri-Tab
Calcium sulfate dihydrate, E.P./U.S.P.–N.F.	Compactrol, Destab
Cellulose	
Microcrystalline cellulose, E.P./J.P./N.F.	Avicel, Celex, Celphere, Ceolus, Emcocel, Ethispheres, Fibrocel, Pharmacel, Tabulose, Vivapur
Powdered cellulose, E.P./J.P./N.F.	Arbocel, Elcema, Solka Floc
Dextrose, E.P./J.P./U.S.P.–N.F.	Caridex, C*PharmDex, Dextrofin, Lycadex PF, Tabfine D-100 (glucose)
Coprocessed/combination excipients	
Calcium carbonate 95%/maltodextrins 5%	Cal-Carb
Calcium carbonate/sorbitol	ForMaxx
Calcium sulfate 93%/vegetable gum 7%	Cal-Tab
Dextrose and ∼5% maltose	Emdex
Dextrose and ∼6% maltodextrin	Unicex
Dextrose, fructose, maltose, isomaltose, and higher polysaccharides	Celubtab, Sweetex
Dextrose, di-, tri-, and higher polysaccharides	Cantab
Lactose α-monohydrate 75%/cellulose 25%	Cellactose, Microelac
Lactose monohydrate 50%/hydroxypropyl methylcellulose 50%	RetaLac
Lactose monohydrate 95%/povidone 3%/crospovidone 3%	Ludipress
Lactose anhydrous/5% lactitol	Pharmatose DCL40
Lactose/15% cornstarch	StarLac
Microcrystalline cellulose/guar gum	Avicel CE-15
Microcrystalline cellulose/silicone dioxide	ProSolve
Sucrose/dextrin 3%	Di-Pac
Sucrose ∼95%/invert sugar ∼4%/magnesium stearate 0.5%	Nu-Tab
Sucrose ∼90%/invert sugar ∼10%	Sugartab
Lactitol, E.P./U.S.P.–N.F.	Finlac DC, Lacty
Lactose	
Lactose, β-anhydrous, E.P./J.P./U.S.P.–N.F.	Anhydrous lactose, Lactopress, Pharmatose DCL21 and DCL22, Super-Tab Anhydrous
Granulated lactose, E.P./J.P./U.S.P.–N.F.	Pharmatose DCL 15
Lactose, monohydrate, U.S.P.–N.F.	Generic (diluent/filler)
Spray-dried lactose, α-monohydrate, U.S.P.–N.F.	Fast-Flo Lactose
Spray-dried lactose, α-monohydrate, E.P./J.P./U.S.P.–N.F.	FlowLac, Lactopress, Pharmatose DCL11 and DCL14 (spray-dried), Super-Tab spray dried lactose

(*continued*)

APPENDIX 13.2 (*Continued*)

Excipient	Trade Names
Maltodextrin, E.P./U.S.P.–N.F.	Lycatab DS, Maltrin, Star-Dri
Maltose, J.P./U.S.P.–N.F.	Advantose (spray-dried), Maltose, Sunmalt
Mannitol, E.P./J.P./U.S.P.–N.F.	
Mannitol α	Pearlitol SD (spray-dried)
Mannitol β	Partek, Pearlitol DC (direct compression)
Sorbitol, E.P./J.P./U.S.P.–N.F.	Liponic, Nerutol, Neosorb, Sorbogem
Starch	
Dextrates, U.S.P.–N.F. (hydrolyzed)	Celutab (hydrolyzed starch)
Pregelatinized starch, E.P./J.P./U.S.P.–N.F.	Starch 1500 G (global), Swelstar
Starch, E.P./J.P./U.S.P.–N.F. (corn, potato, wheat)	C*PharmGel, Melojel, Meritena, Paygel 55, Perfectamyl D6PH,Pure-Bind, Pure-Cote, Pure-Dent, Pure-Gel, Pure-Set, Purity 21, Purity 826, Tablet White
Sucrose, E.P./J.P./U.S.P.–N.F.	Generic
Xylitol, E.P./J.P./U.S.P.–N.F.	Klinit, Xylifin, Xylisorb, Xylitab, Xylitolol

[a] A comprehensive list of products and suppliers is provided in Rowe et al. (2006).

APPENDIX 13.3 Wet Granulation Binders[a]

Excipient	Trade Names	Common Use Range (% w/w of formulation)	Use Range[b, c] (mg)
Immediate Release			
Acacia, E.P./J.P./U.S.P.	Generic	2–5	5.00–15.6
Alginic acid, E.P./U.S.P.–N.F.	Kelacid, Protacid, Satialgine	3–6	22.2–400
Alginate sodium, E.P./U.S.P.–N.F.	Kelcosol, Keltone, Protanal	3–5	35.0–300
Gelatin	Byco, Cryogel, Instagel, Solugel	1–4	1.00–45.3
Cellulose derivatives			
Carboxymethylcellulose calcium, E.P./J.P./U.S.P.–N.F.	Nymcel ZSC	5–10	13.3–242
Carboxymethylcellulose sodium, E.P./J.P./U.S.P.–N.F.	Akucell, Aquasorb, Blanose, Finnfix	1–6	2.20–155
Hydroxyethyl cellulose, E.P./U.S.P.–N.F.	Nymcell, Tylose CB	2–10	11.8–150
Hydroxypropyl cellulose, E.P./J.P./U.S.P.–N.F.	Cellosize HEC, Klucel, Natrosol, Tylose PHA	2–10	1.00–240
Hypromellose, E.P./J.P./U.S.P.–N.F.	Benecel MHPC, Methocel E, F, K; Metolose Tylopur	1–10	0.75–445
Methylcellulose, E.P./J.P./U.S.P.–N.F.	Benecel, Culminal MC, Methocel A, Metolose	1–6	2.75–183

APPENDIX 13.3   (Continued)

Excipient	Trade Names	Common Use Range (% w/w of formulation)	Use Range[b] (mg)
Poly(ethylene oxide), U.S.P.–N.F.	Polyox	2–8	57.8–544
Polydextrose, U.S.P.–N.F.	Generic	2–6	3.8–7.7
Povidone[c,d] E.P./J.P./U.S.P.–N.F.	Kollidon, Plasdone	1–15	0.27–80
Poly(vinyl alcohol), E.P./U.S.P.–N.F.	Elvanol, Mowiol, Polyvinol	1–6	0.69–20.0
Sucrose, E.P./J.P./U.S.P.–N.F.	Generic	2–15	12.0–1200
Starch (see Appendix 13.5)	Generic	1–5	14.1–615
Pregelatinized starch, E.P./J.P./U.S.P.–N.F.	Starch 1500G	2–20	22–435
Tragacanth, E.P./J.P./U.S.P.–N.F.	Generic	1–3	4.00–7.50
Controlled Release			
Carbomer, E.P./U.S.P.–N.F.	Acitamer, Carbopol, Pemulen, Ultrez	5–30	1.50–90.0
Polymethacrylates, E.P./U.S.P.–N.F.	Acryl-EZE, Eudragit	—	25.1–117
Poly(vinyl alcohol), E.P./U.S.P.–N.F.	Airvol, Alcotex, Elvanol, Gelvatol, Gobsenol, Lemol, Mowiol, Polyvinol	—	Up to 20.0
Povidone (see above)		5–30	0.27–80

[a] A comprehensive list of products and suppliers is provided in Rowe et al. (2006).

[b] FDA Inactive Ingredient List, http://www.accessdata.fda.gov/scripts/cder/iig/index.cfm.

[c] Higher concentrations are often used as direct compression blend.

[d] Lack of justification for peroxide impurity level may lead to an FDA deficiency letter (Srinivasan et al., 2010).

APPENDIX 13.4   Hot-Melt Excipients[a]

Diluents and Binders	Trade Names	Common Use Range (% w/w)	T_g or T_m (°C)
Hydrophilic/Water Soluble			
Chitosan, EP		30–50	203
Cellulose			
Hydroxypropyl cellulose (see Appendix 13.2)	Klucel	16–24	130
Hydroxypropyl methylcellulose (see Appendix 13.2)	Methocel	8–12	175
Microcrystalline cellulose (see Appendix 13.2)	—	32–48	265
Pectin, U.S.P.–N.F.	Generic	16–24	
Polycarbophil, U.S.P.–N.F.	Noveon AA-1	4–6	102

(*continued*)

APPENDIX 13.4 (*Continued*)

Diluents and Binders	Trade Names	Common Use Range (% w/w)	T_g or T_m (°C)
Poly(ethylene glycol)[b] (see Appendix 13.5)	PEG 1500 (30 cSt at 99°C)	10–25	40
	PEG 4000 (134 cSt at 99°C)	—	55
	PEG 6000 (320 cSt at 99°C)	1–5	59
Poly(ethylene oxide) (see Appendix 13.2)	Polyox	20–35	−67, 73[c]
Polymethacrylates, E.P./U.S.P.–N.F.	Eudradit S 100	50–65	50–60
Povidone[b] (see Appendix 13.2)	—	15–25	168
Pregelatinized starch (see Appendix 13.2)	—	10–20	
Xanthan gum, E.P./U.S.P.–N.F.	Keltrol, Rhodigel, Vanzan, Xantural	10–25	Chars at 270
	Lipophilic/Insoluble		
Cellulose			
Cellulose acetate butyrate, U.S.P.–N.F.	CAB 381-0.5	40–60	125,157[c]
Ethylcellulose, E.P./U.S.P.–N.F.	Ethocel	10–25	131
Hypromellose acetate succinate, U.S.P.–N.F.	Aqoat	—	113
Hypromellose phthalate, E.P./J.P./U.S.P.–N.F.	Mantrocel HP-55	10–25	137,150[c]
Waxes			
Cetyl esters, U.S.P.–N.F.	Crodamol SS, Cutina CP, Liponate SPS, Protachem MST, Ritaceti, Starfol Wax CG, Synaceti 116	—	45
Microcrystalline, U.S.P.–N.F.	Generic	15–30	67
Nonionic emulsifying, U.S.P.–N.F.	Collon NI, Crodex N, Emulgade 1000NI, Permulgin D, Polawax, Ritachol 2000, T-Wax	—	50
White wax	Generic, Lunaceram, Lunacera P	10–25	63
	Plasticizers, Solubilizers, and Lubricant		
Citrate esters			
Acetyltributyl citrate, E.P./U.S.P.–N.F.	ATBC, Citroflex A-4	—	Liquid at room temperature

APPENDIX 13.4 (*Continued*)

Diluents and Binders	Trade Names	Common Use Range (% w/w)	T_g or T_m (°C)
Acetyltriethyl citrate, U.S.P.–N.F.	ATEC, Citroflex A-2	12–18	Liquid at room temperature
Dibutyl sebacate, U.S.P.–N.F.	Kodaflex DBS	—	−10
Glycerin, E.P./J.P./U.S.P.–N.F.	Croderol, Glycon-100	—	Liquid at room temperature
Glyceryl esters			
Glyceryl monooleate, E.P./U.S.P.–N.F.	Aldo MO, Atlas G-695, Capmul GMO, Kessco GMO, Ligalub, Stepan GMO	—	35
Glyceryl monostearate, E.P./J.P./U.S.P.–N.F.	Capmul GMS-50, Cutina GMS, Kessco GMS, Stepan GMS	2–5	≥55
Glyceryl palmitostearate, U.S.P.–N.F.	Precirol ATO 5, Captex 500	—	54
Medium-chain triglycerides, E.P./U.S.P.–N.F.	Bergabest, Captex 300, Captex 355, Miglyol 810, Miglyol 812, Neobee M5	24–36	−5
Triacetin, E.P./U.S.P.–N.F.	Captex 500	10–15	−78
Fatty acid alcohols			
Cetyl alcohol, E.P./J.P./U.S.P.–N.F.	Avol, Cachalot, Crodacol, Hyfatol, Kessco CA, Lanette 16, Lipocol C, Tita CA, Tego Alkanol 16	—	50
Stearyl alcohol, E.P./J.P./U.S.P.–N.F.	Cachalot, Crodacol S95, Hyfatol 18-98, Rita SA, Stenol, Tego Alkanol 18	—	59
Oils			
Castor oil, E.P./J.P./U.S.P.–N.F.	EmCon CO, Lipovol CO	—	−12
Mineral oil, E.P./J.P./U.S.P.–N.F.	Citation	—	Liquid at room temperature
Phthalate esters			
Cellulose acetate phthalate, E.P./J.P./U.S.P.–N.F.	CAP	—	165,192[c]
Dibutyl phthalate, E.P.	Araldite 502, Kodaflex DBP, Staflex DBP, Witcizer 300	—	−35
Diethyl phthalate, E.P./U.S.P.–N.F.	Kodaflex DEP	—	−40
Poly(ethylene glycols) (see above)			

(*continued*)

APPENDIX 13.4 *(Continued)*

Diluents and Binders	Trade Names	Common Use Range (% w/w)	T_g or T_m (°C)
Propylene glycol, E.P./J.P./U.S.P.–N.F.	Generic	—	−59
Tocofersolan, U.S.P.–N.F.	Vitamin E TPGS, Tocopherol, poly(ethylene glycol) 1000 succinate	20–35	Liquid at room temperature
Vegetable oil Hydrogenated, J.P./U.S.P.–N.F.	Lubritab, Sterotex, Softisan 154, Lipovol HS-K, Sterotex HM	—	61
	Antioxidants		
Butylated hydroxyl anisole, E.P./U.S.P.–N.F.	Nipanox BHA, Nipantiox 1-F, Tenox BHA	0.05–5	47
Butylated hydroxyltoluene, E.P./U.S.P.–N.F.	Agidol, Dalpac, Impruvol, Nipanox BHT, Tenox BHT, Vianol	0.05–5	70

[a] A comprehensive list of products and suppliers is provided in Rowe et al. (2006).

[b] Lack of justification for peroxide impurity level may lead to an FDA deficiency letter (Srinivasan et al., 2010).

[c] T_g, T_m (°C).

APPENDIX 13.5   Disintegrants[a]

Excipient	Trade Names	Common Use Range[b] (% w/w, relative swelling volume)	Use Range[c] (mg)
Alginic acid (see Appendix 13.3)	—	1–5 (dry running powders)	22.5–400
Alginate sodium (see Appendix 13.3)	—	1–5 (dry running powders)	35–350
Croscarmellose sodium, E.P./U.S.P.–N.F.	Ac-Di-Sol, Explocel, Nymcel ZSX, Pharmacel XL, Primellose, Solutab, Vivasol	1–5, 600% (wet or dry granulation)	2–180
Crospovidone,[d,e] E.P./U.S.P.–N.F.	Kollidon, Polyplasdone XL	2–5, 150% (wet or dry granulation)	4.5–792
Microcrystalline cellulose (see Appendix 13.2)	—	5–20 (wet or dry granulation)	
Starch (see Appendix 13.2)	Generic	3–15, 110% (dry running powders)	14.9–615
Pregelatinized starch, E.P./J.P./U.S.P.–N.F.	Starch 1500 G	5–10 (dry running powders)	43–165

APPENDIX 13.5 (*Continued*)

Excipient	Trade Names	Common Use Range[b] (% w/w, relative swelling volume)	Use Range[c] (mg)
Starch glycolate sodium, E.P./U.S.P.–N.F.	Explosol, Explotab, Glycolys, Primojel, Tablo, Visatar P	2–8, 1680% (dry running powders)	2–876

[a] A comprehensive list of products and suppliers is provided in Rowe et al. (2006).

[b] Hogan et al. (1996).

[c] FDA Inactive Ingredient List, http://www.accessdata.fda.gov/scripts/cder/iig/index.cfm.

[d] Lack of justification for peroxide impurity level may lead to an FDA deficiency letter (Srinivasan et al., 2010).

[e] Newer versions of crospovidone have lower peroxide specifications of 30 and 50 ppm.

APPENDIX 13.6 Lubricants[a]

Excipient	Trade Names	Common Use Range (% w/w of formulation)	Use Range[b] (mg)
	Hydrophobic		
Glyceryls			
Glyceryl behenate, E.P./U.S.P.–N.F.	Compritol 888	0.5–3	2.50–60.1
Glyceryl behenate and poly(ethylene glycol) Behenate	Compritol HD5	0.5–3	
Glyceryl monostearate E.U./U.S.–N.F.	Precirol ATO5	1–3	
Magnesium trisilicate, E.P./U.S.P.–N.F.	Generic	1–5	15–76.9
Starch (see Appendix 13.2)		5–10	4.90–6.15
Stearates (magnesium is most common lubricant)			
Calcium stearate, E.P./J.P./U.S.P.–N.F.	HyQual	0.5–2	0.70–47.2
Magnesium stearate, E.P./J.P./U.S.P.–N.F.	Generic	0.25–2	0.32–400
Zinc stearate, E.P./U.S.P.–N.F.	Generic	0.5–2	4.61–36.0
Steric acid, E.P./J.P./U.S.P.–N.F.	Crodacid, Edenor, Emersol, Hystene, Industrene, Kortacid 1895, Pearl Steric, Pristerene, Tegostearic	1–5	0.90–180
Stearyls			
Stearyl alcohol, E.P./J.P./U.S.P.–N.F.	Cachalot, Crodacol S95, Hyfatol 18	1–6	25–244

(*continued*)

APPENDIX 13.6 (*Continued*)

Excipient	Trade Names	Common Use Range (% w/w of formulation)	Use Range[b] (mg)
Stearyl fumarate sodium, E.P./U.S.P.–N.F.	Lanette 18, Lipocol S, Rita SA, Stearol, Stenol, Tego Alkanol 18, Pruv	0.5–2	2–27
Talc, E.P./J.P./U.S.P.–N.F.	Altac, Magsil Star, Superiore	1–5	1.5–91.2
Vegetable oil, hydrogenated, J.P./U.S.P.–N.F.	Cottonseed oil: Akofine, Lubritab, Sterotex	1–6	0.6–402
	Palm oil: Softisan 154	1–6	
	Soybean oil: Lipovol HS-K, Sterotex HM	1–6	3.00–13.5
	Water-Soluble Lubricants		
Fumaric acid, U.S.P.–N.F.	Generic	2–5	10–55.6
Sodium lauryl sulfate, E.P./J.P./U.S.P.–N.F.	Elfan 240, Texapon, K12P	1–3	0.02–51.6
Leucine, E.P./J.P./U.S.P.–N.F.	Generic	1–5	3.60
Polyethylene glycol[c], E.P./U.S.P.–N.F.	Carbowax, Kipoxol, Lutrol E, PEG, Pluriol E	2–5	0.3–454
	(4000–6000 MW)		
Sodium benzoate, E.P./J.P./U.S.P.–N.F.	Generic	3–6	0.34–60
Sodium chloride, E.P./J.P./U.S.P.–N.F.	Generic	3–6	7.50–335

[a] A comprehensive list of products and suppliers is provided in Rowe et al. (2006).

[b] FDA Inactive Ingredient List, http://www.accessdata.fda.gov/scripts/cder/iig/index.cfm.

[c] Lack of justification for peroxide impurity level may lead to an FDA deficiency letter (Srinivasan et al., 2010).

APPENDIX 13.7 Antiadherants and Antisticking Agents[a]

Excipient	Trade Names	Common Use Range (% w/w of formulation)	Use Range[b] (mg)
Colloidal silicon dioxide, E.P./U.S.P.–N.F.	Aerosil, Cab-O-Sil, M-5P, Wacker HDK	1–3	0.65–99
Microcrystalline cellulose (see Appendix 13.2)	—	5–20	16–1120
Starch (see Appendix 13.2)	Generic	1–5	14.1–615
Pregelatinized starch, E.P./J.P./U.S.P.–N.F.	Starch 1500G	2–20	22–435
Talc (see Appendix 13.6)	—	1–5	1.5–91.2

[a] A comprehensive list of products and suppliers is provided in Rowe et al. (2006).

[b] FDA Inactive Ingredients List, http://www.accessdata.fda.gov/scripts/cder/iig/index.cfm.

APPENDIX 13.8   Glidants and Flow Aids[a]

Excipient	Trade Names	Common Use Range (% w/w of formulation)	Use Range[b] (mg)
Colloidal silicon dioxide (see Appendix 13.7)	—	0.1–2.0	0.65–99
Magnesium oxide, E.P./J.P./U.S.P.–N.F.	Destab, Magcal, Magchem 100, Maglite, Magnyox, Marmag, Oxymag	0.5–2	25.7–63
Magnesium silicate, J.P./U.S.P.–N.F.	Generic	0.5–2	10–30
Starch (see Appendix 13.2)	Generic	0.2–10	14.1–615
Pregelatinized starch, E.P./J.P./U.S.P.–N.F.	Starch 1500	5–10	22–435
Talc (see Appendix 15.6)	—	0.2–10	1.5–91.2

[a] A comprehensive list of products and suppliers is provided in Rowe et al. (2006).

[b] FDA Inactive Ingredients List, http://www.accessdata.fda.gov/scripts/cder/iig/index.cfm.

APPENDIX 13.9   Wetting Agents[a]

Excipient	Trade Names	Common Use Range (% w/w of formulation)	Use Range[b] (mg)
		Anionic	
Deoxycholate sodium	Generic	0.1–1	
Docusate sodium, E.P./U.S.P.–N.F.	Generic	0.01–1	0.002–11
Lauryl sulfate sodium (see Appendix 13.6)	—	0.01–1	0.02–51.69
		Nonionic	
Octylphenol ethoxylate, U.S.P.–N.F.	Triton X-100 Surfactant	0.01–0.1	
Poloxamer, E.P./U.S.P.–N.F.	Lutrol, Monolan, Pluronic, Supronic, Symperonic	0.1–5	5.61–106.7
Polyoxyethylene alkyl ethers (polyoxyl 20 cetostearyl ether), E.P./U.S.P.–N.F.	Atlas G-3713, Cremophor A 20, Valpo CS20	0.1–1	
Polyoxyethylene castor oil derivatives (polyoxyl 35 and 40 castor oil) E.P./J.P./U.S.P.–N.F.	Cremophor ELP, Etocas 35, PEG-35, Cremophor RH 40, Jeechem CAH-40, PEG-40	35: 0.01–0.1 40: 0.05–0.5	2 25
Polyoxyethylene sorbitan fatty acid esters (Polysorbate 80) E.P./J.P./U.S.P.–N.F.	Atlas E, Capmul POE-O, Cremophor PS 80, Tween 80	0.1–3	0.07–21.25

(*continued*)

APPENDIX 13.9 (*Continued*)

Excipient	Trade Names	Common Use Range (% w/w of formulation)	Use Range[b] (mg)
Polyoxyethylene stearate (polyoxyl 40 stearate) J.P./U.S.P.–N.F	Crodet S40, Emerest 2672, Lipo-PEG 39S, Myrj 52, PEG-40, Ritox 52	0.1–3	2–8.48
Tocofersolan [vitamin E TPGS; tocopherol poly(ethylene glycol) 1000 succinate], U.S.P.–N.F.	Generic	0.05–2	1.03
	Cationic		
Benzalkonium chloride, E.P./J.P./U.S.P.–N.F.	Hyamine 3500, Pentonium, Zephiram	0.001–0.01	
Benzethonium chloride, E.P./J.P./U.S.P.–N.F.	Hyamine 1622	0.1–0.1	
Cetylpyridinium chloride, E.P./U.S.P.–N.F.	Generic	0.05–0.5	
Lecithin, U.S.P.–N.F. (isoelectric point ~3.5)	Egg lecithin, soybean lecithin	0.25–10	48.0

[a] A comprehensive list of products and suppliers is provided in Rowe et al. (2006).

[b] FDA Inactive Ingredients List, http://www.accessdata.fda.gov/scripts/cder/iig/index.cfm.

APPENDIX 13.10 Nominal Screen Mesh Number and Sieve Opening

U.S. Sieve Opening Size (μm)	Mesh Number
5660	3.5
4760	4
4000	5
3360	6
2830	7
2380	8
2000	10
1680	12
1410	14
1190	16
1000	18
840	20
710	25
590	30
500	35
420	40
350	45

APPENDIX 13.10 (*Continued*)

U.S. Sieve Opening Size (μm)	Mesh Number
297	50
250	60
210	70
177	80
149	100
125	120
105	140
88	170
74	200
62	230
53	270
44	325
37	400

CHAPTER 14

CAPSULE PRODUCT DESIGN

14.1 INTRODUCTION

Capsules are a versatile solid dosage form. Capsule nomenclature is provided in Appendixes 12.1 and 12.2. Capsules are divided into two types: hard-shell and soft-shell. Both types of capsules can be clear or opaque. Like tablets, they provide a solid dosage form that is convenient to self-administer and provide good dose uniformity. A wide variety of materials can be filled into hard-shell capsules, including solids (powders, granules, pellets, tablets, and capsules), semisolids (gels, pastes, and thermosetting polymers), and liquids (solutions, suspensions, emulsions, and microemulsions). Hard-shell capsules are filled with running powder that can be prepared by the same technologies as those used to prepare running powder for tablets without the necessary criteria of being highly compressible. The running powder is filled into the capsule either by a direct volumetric fill into the capsule body or by creating a volumetric plug in a tube or disk outside the capsule that is subsequently filled inside the capsule. Semisolids and liquids are added to the capsules using a volumetric dosing type of syringe. Extended- and delayed-release beads or pellets can also be filled without being compressed. In addition, hard-shell capsules are uniquely suited to overencapsulate tablets for blinded clinical trials.

Hard-shell capsules are comprised of two parts: a body and a cap. They are less flexible than soft-shell capsules. Hard-shell capsules are oblong and the cap fits over the longer body. Hard-shell capsules can be prepared from a number of polymers: gelatin, hypromellose or hydroxypropyl methylcellulose, hydroxypropyl starch and starch modifications, and pullulan. The most commonly used polymer is gelatin. The alternative polymers have been developed to overcome issues associated with gelatin, such as cross-linking and decrease in bioavailability; concerns regarding potential disease caused by pions, leading to bovine spongiform encephalopathy; and religious and dietary restrictions. The alternative polymers have advantages and disadvantages as well and are discussed in more detail in subsequent sections.

Soft-shell capsules are more elastic than their hard-shell counterparts. They are molded in a single form–fill–seal process that allows many shapes to be formed, such as spheres, ovals, and oblong capsules. Soft-shell capsules are prepared from gelatin and hypromellose. The form–fill–seal process generally

Integrated Pharmaceutics: Applied Preformulation, Product Design, and Regulatory Science,
First Edition. Antoine Al-Achi, Mali Ram Gupta, William Craig Stagner.
© 2013 John Wiley & Sons, Inc. Published 2013 by John Wiley & Sons, Inc.

limits the soft-shell capsule fill to liquids that can be added at the filling stage of the continuous form–fill–seal process.

14.2 HARD-SHELL CAPSULES

Capsule manufacture has always been considered the expensive solid oral dosage form alternative. In the past, semiautomated capsule dosage-form manufacture and the relatively high cost of hard-shell gelatin capsules made capsule manufacturing significantly more expensive than tablet manufacture. Today, automated high-speed manufacturing capsule machines are available, although they are still three to five times slower than modern tablet machines. In addition, capsule-filling innovations have significantly affected the use of capsules in early development programs. Very accurate semiautomated filling–weighing systems (Xcelodose by Pfizer and Powdernium by Autodose) have been developed that can speed the early formulation development process. These systems can fill capsules at a rate of 120 capsules per hour over a range of 50 µg to several grams with a percent relative standard deviation of 1 to 5%. As the development costs escalate, the time to market must decrease. It is possible to develop directly into human studies a simple capsule formulation that progresses to a new drug application filing with little or no formulation changes. This would significantly accelerate the time to market. Obviously, this is easier said than done. New drug molecules tend to require significant formulation interventions, but for those molecules that can be formulated in simple powder or liquid hard-shell formulation, the benefits can be very significant.

Cole (1998) evaluated the development and product costs of tablets and capsules. The cost analysis included the costs of excipients, equipment, total product time, labor, good manufacturing practices space requirements, in-process controls, analytical testing, cleaning, and process validation. The total manufacturing cost for a powder-filled capsule was found to be more than that for a direct-compression film-coated tablet and less than that for a wet granulated film-coated tablet. This reemphasizes the need to evaluate the need for a film-coated tablet. The production costs for a direct-compression non-film-coated tablet would provide a corporation with a significant cost savings over the lifetime of the product. A careful and thorough case-by-case evaluation needs to be made for each active pharmaceutical ingredient (API) that incorporates the specific physiochemical implications and consequences.

The advantages and disadvantages of hard-shell capsules are listed in Table 14.1.

The various components of hard-shell capsules and their functions are listed in Table 14.2.

14.2.1 Manufacture of Hard-Shell Capsules

14.2.1.1 Hard-Shell Capsule Dip Molding In this process, stainless steel pins are dipped into a heated hard-shell capsule solution formulation. The capsule shell formulation wets and coats the stainless steel pins. The pins are then removed

TABLE 14.1 **Advantages and Disadvantages of Hard-Shell Capsules Compared to Tablets**

Advantages	Disadvantages
Elegant appearance	Greater chance of sticking to the esophagus than with a tablet
Ease of use	
If taken with water: smooth, slippery, easy to swallow	Slower manufacturing speeds than those for tablets; up to five times slower
Taste masking delivery system	Powder density–dose limitations
Range of colorants	Less tamper resistant than a tablet
Ideal for blinding for clinical trials	Decreased stability with moisture-sensitive drugs
Good dose uniformity	Harder to interpret accelerated testing data; higher temperature/% RH may change capsule moisture and formulation content moisture; potential for cross-linking at accelerated conditions
Highly versatile "container system"; can be filled with many types of materials; does not require a compressible running powder	
Versatile functionality; oral, chewable, delayed release, extended release, vaginal capsule	More difficult to enteric-coat
	Hygroscopic drug may adsorb moisture from capsule shell
Faster development time compared to a tablet	Higher weight variation than with tablets
Fewer manufacturing steps than those for a direct-compression film-coated tablet	Generally higher manufacturing costs
	Need to rely on a capsule manufacturer for source of capsules
Can fill low-melting-point API	
Improved environmental and occupational safety using a liquid-fill system	
The thickness of capsule shell and addition of opacifying agents and colorants can provide additional light protection	
Relatively high-speed manufacturing process	
Reasonably robust manufacturing process	
Less stringent manufacturing and storage controls compared to manufacture of soft-shell capsules	
Lower capital investment in plant and equipment than for tablets	

TABLE 14.2   Hard-Shell Capsule Component Classes and Functions

Component Class	Function
Film-forming biocompatible polymers, 85 to 90% w/w	Forms a soluble, edible, biocompatible container
Gelatin, E.P./J.P./U.S.P.–N.F.	
Hypromellose, E.P./J.P./U.S.P.–N.F.	
Pullulan, J.P./N.F.	
Starch (potato), E.P./J.P./U.S.P.	
Water equilibrium moisture content at 35 to 55% relative humidity	
Gelatin (13 to 16% w/w, bound with low water activity)	For gelatin: water acts as plasticizer to maintain flexibility; brittle below 13% moisture
Hypromellose (4 to 6% w/w)	For hypromellose: moisture does not act as a plasticizer
Pullulan (10 to 15% w/w)	
Starch	
Plasticizers	
Poly(ethylene glycol) (PEG) 4000, E.P./U.S.P.–N.F. (3 to 6% w/w)	Gelatin: decreases brittleness caused by hygroscopic fills
Processing aids	
Carrageenan, U.S.P.–N.F. (0.1 to 0.4% w/w)	Improve gelling and setting system
Cations such as KCl (1 to 1.5% w/w)	Improve gelling and setting system
Surfactants such as sodium lauryl sulfate, E.P./U.S.P.–N.F. (0.05 to 0.25% w/w)	Wetting agent to provide stainless steel pin wetting and lubrication; improve wetting and dissolution
Colorants	Provide product identification; light protection; conceal contents
Iron oxide pigments (0.01 to 1.5% w/w)	
Soluble dyes (0.05 to 1.5% w/w)	
Opacifying agents	Light protection; conceal contents
Titanium oxide (0.01 to 1.0% w/w)	
Preservatives	Decrease bioburden and prevent microbiological growth
Methylparaben, E.P./J.P./U.S.P.–N.F. (0.25 to 0.75% w/w)	
Propylparaben, E.P./E.P./U.S.P.–N.F. (0.06 to 0.2% w/w)	
Flavors	Mask poor-tasting drugs

from the heated capsule shell solution formulation. The capsule shell formulation quickly gels during a cooling and drying process. Once the capsule shell gel is set, the hard shell is removed from the pins, trimmed, and the body and cap are joined. The ability of the capsule shell formulation to wet the stainless steel pins, to form a film, to gel without dripping or thinning, and to release from the pins without sticking is critical. The dipping method is used for the manufacture of hard-shell gelatin, hypromellose, and pullulan capsules.

14.2.1.1.1 Gelatin Hard-Shell Capsules Gelatin capsules have been used in the pharmaceutical industry for over 80 years. Gelatin is a natural polypeptide prepared by acid (type A) or base (type B) hydrolysis of collagen. The isoelectric point of acid-treated gelatin ranges from 6.0 to 9.4 and for base-treated gelatin ranges from 4.8 to 5.2. Gelatin has unique properties that make it ideal as a drug delivery container system. It is biocompatible, nontoxic, a good film-former and gelling agent, relatively stable and nonreactive, and is soluble in gastrointestinal fluids at 37°C. Gelatin capsules rupture in less than 2 minutes in simulated gastric and in intestinal fluids at 37°C (El-Malah and Nazzal, 2007). It is interesting to note that gelatin's solubility in aqueous fluids decreases rapidly below 37°C, and capsules will swell and not disintegrate.

At 37°C, a gelatin capsule usually starts to dissolve and break up at the ends of the capsules in 1 minute. Cross-linking, or pellicle formation, of gelatin can lead to a decrease in dissolution and bioavailability (Digenis et al., 1994; Ofner et al., 2001). Cross-linked gelatin forms a pellicle that swells and becomes a rubberlike mass that is insoluble in the dissolution media. The primary factors that contribute to cross-linking are prolonged exposure to high temperature, humidity, aldehydes, and light. Li et al. (2006) has determined the aldehyde concentration in 31 excipients that are commonly used in capsule formulation. Appendix 14.1 lists the excipients studied and their respective aldehyde concentrations. Poly(ethylene glycol) had the highest level of aldehydes. Appendix 14.2 lists the aldehyde levels of 30 excipients that are commonly used for liquid-fill formulations (Li et al., 2007). A number of these commonly used excipients have reasonably high levels of aldehydes. The presence of aldehydes in a container closure system also needs consideration. Schwier et al. (1993) demonstrated that rayon, which is used as a coiler to keep capsules from rattling inside a bottle, contained furfural as a reactive aldehyde.

In the early 1990s, a number of gelatin capsule products failed to meet dissolution specifications. It was speculated that the dissolution failures were the result of gelatin cross-linking. Since then, a two-tier dissolution methodology has been developed that incorporates the use of pepsin and pancreatin to cleave the gelatin when cross-linking is suspected (Gelatin Capsule Working Group, 1996; FDA, 1997).

Serious concern has been raised about contaminated gelatin and gelatin capsules, specifically their role in causing *bovine spongiform encephalopathy* (BSE) or *mad cow disease*. In 2008, the U.S. Food and Drug Administration (FDA) strengthened previous regulations to protect public safety. The European Commission Health and Consumer Protection Scientific Steering Committee (2003) issued findings that addressed transmittable spongiform encephalopathy infectivity. The committee presented a number of gelatin-processing steps that can provide 4.5 log inactivation.

Gelatin is more prone to microbiological contamination than other capsule-forming materials, and gelatin capsules generally are formulated with antimicrobial agents. A surfactant is often added as a wetting agent to wet the stainless steel pins to provide even film formation, as a lubricant to decrease gelatin adhesion to the pins postdrying to aid capsule stripping, and as a wetting agent to improve capsule dissolution. A typical base capsule formulation (i.e., Torpac is provided in Table 14.3.) Colorant and opacifiers can be added to make it distinctive. Also, Qualicaps sells a gelatin capsule that contains 5% w/w PEG 4000, which acts as a plasticizer for hygroscopic formulations that might cause gelatin capsules to become brittle. The water that is present in the capsule serves as a plasticizer. Tightly bound water leads to low water activity and makes it more difficult for microbes to grow and for capsule moisture to be gained or lost. Gelatin capsules will become brittle after exposure to low relative humidity.

14.2.1.1.2 Hypromellose Hard-Shell Capsules Hypromellose capsules were developed to overcome a number of the gelatin capsule drawbacks. Hypromellose capsules are resistant to cross-linking and will not cross-link with aldehydes. Hypromellose is a neutral molecule, and capsule dissolution is pH-independent. Hypromellose is derived from nonanimal origin, has benefits for patients with dietary or religious restrictions, does not carry a BSE risk, and is preservative-free. Hypromellose has about half the amount of water associated with a hard shell as with a gelatin capsule. In addition, this water is not tightly bound and is not present as a plasticizer. The capsule moisture can be decreased without affecting the capsule's physical properties. The lower water content provides beneficial stability to moisture-sensitive drugs.

The hypromellose capsules are less likely to build up static charge, which improves the handling of the capsules during the manufacturing process. The decreased static charge can lead to improved dose delivery of micronized powder from dry powder inhalers. Hypromellose capsules are soluble at room temperature. However, there is a lag time in capsule rupture of approximately 5 minutes in

TABLE 14.3 Typical Gelatin Hard-Shell Capsule Formulation

Material	Weight (as % w/w)
Gelatin	85.64
Methylparaben	0.57
Propylparaben	0.14
Sodium lauryl sulfate	0.14
Water	13.49

Source: After http://www.torpac.com/Reference/capsule_ingredients.htm. Accessed Dec. 2011.

TABLE 14.4   Typical Hypromellose Hard-Shell Capsule Formulation

Material	Weight (as % w/w)
Hypromellose	91.25
Potassium chloride	1.85
Carrageenan	1.90
Water	5.00

Source: Yamamoto et al. (2002).

simulated gastric fluid and 10 minutes in simulated intestinal fluid (El-Malah and Nazzal, 2007; Groshens et al., 2009). A new hypromellose capsule that does not contain a gelling agent has recently been reported. It is claimed that the newly formulated hypromellose has superior dissolution performance characteristics (Ku et al., 2010).

A typical hypromellose capsule composition is provided in Table 14.4 (Yamamoto et al., 2002). Carrageenan and gellan gum are often used as gelling agents, and salts are used as gelling aids.

14.2.1.1.3   Pullulan Hard-Shell Capsules   Pullulan capsules have been commercially available in Japan for more than 25 years. Pullulan provides the clearest and most elegant capsule base. It is NF and JP monograph material and has been generally recognized as safe in the United States. It is a polysaccharide that is derived from a nongenetically engineered fermentation process. Pullulan capsules have good physical and chemical stability and do not undergo cross-linking. Pullulan is starch-free, gluten-free, preservative-free, kosher, and halal. A typical pullulan capsule composition is provided in Table 14.5.

14.2.1.2   Hard-Shell Capsule Injection Molding   Injection molding offers significant manufacturing flexibility compared to the dipping process. Capsules can be molded into specific shapes that are associated with a particular therapeutic use, such as a heart shape for blood pressure–lowering medication. Capsules can

TABLE 14.5   Typical Pullulan Hard-Shell Capsule Formulation

Material	Weight (as % w/w)
Pullulan	87.50
Potassium chloride	1.05
Carrageenan	0.25
Sodium lauryl sulfate	0.20
Water	11.00

Source: Scott, et al. (2005).

be molded to provide different compartments within the capsule to separate drugs that are incompatible. The logo and capsule identification can be embossed into the capsule. Compared to gelatin capsules, the body and cap fit flush and are sealed with a thin layer of a water–alcohol solution during the filling process. Compared to the lip on a gelatin capsule, the flush fit makes the molded capsule ideal for coating with functional polymers. The sealing also makes them tamper-evident.

It has been reported that molded capsule composition can include polymers that give a capsule enteric properties that prevent dissolution in the stomach or upper intestinal tract, which can provide targeted drug delivery to the lower intestine and colon (Wittwer et al., 1988). The manufacturing process involves melting and extruding of the capsule formulation into molds under high temperatures (80 to $240°C$) and pressures (1×10^5 to 3.0 to 10^8 Pa). The molds are cooled and the capsule is removed from the mold without elastic deformation. Commercial hard-shell starch capsules are made by ejection molding.

14.2.1.2.1 Injection-Molded Starch Hard-Shell Capsules As mentioned above, ejection molding of hard-shell capsules offers greater flexibility and advantages compared to the dip-molding process. In this section, starch and modified starch are used interchangeably, because both materials have been shown to make acceptable capsules. Starch hard-shell capsules are rigid and dense with a flush cap and body. This makes the starch capsule amenable to coating.

Starch hard-shell capsules are used exclusively in the drug delivery system Targit, which emphasizes site-specific delivery to the gastrointestinal track, specifically to the lower small intestine and colon. Enteric coatings were discussed in Chapter 13. Eudragit L and Eudragit S pH-sensitive polymers have been found to be particularly well suited for coating starch capsules and achieving the desired targeting. Eudragit L dissolves at a pH value above 6, and Eudragit S dissolves at pH 7. A 3 : 1 blend by weight of L100 to S100 is preferred. The coating solution can contain talc as an antitacking agent and dibutyl sebacate as a plasticizer (Watts and Smith, 2005).

The composition of a typical hard-shell starch capsule is shown in Table 14.6. There is about 12 to 14% water in the starch capsule. Approximately 50% is tightly bound.

The hard-shell sizes and their liquid and powder capacities are provided in Table 14.7. Special capsules have also been developed for preclinical animal studies.

TABLE 14.6 Typical Hard-Shell Starch Capsule Composition

Material	Weight (as % w/w)
Potato starch	80
Glycerin	6
Water	14

Source: Wittwer, et al. (1988).

TABLE 14.7 Hard-Shell Capsule Sizes and Capacities

Capsule Size	Capsule Volume (mL)	Capacity at the Stated Specific Apparent Densities (g)			
		0.6 g/mL	0.8 g/mL	1.0 g/mL	1.2 g/mL
000	1.37	0.882	1.096	1.370	1.644
00el[a]	1.02	0.612	0.816	1.020	1.224
00	0.91	0.546	0.728	0.918	1.092
0el[a]	0.78	0.468	0.624	0.780	0.936
0	0.68	0.408	0.544	0.680	0.816
1	0.5	0.300	0.400	0.500	0.600
2	0.37	0.222	0.296	0.370	0.444
3	0.30	0.180	0.240	0.300	0.360
4	0.21	0.126	0.168	0.210	0.252
5	0.10	0.078	0.104	0.130	0.156

Source: Stegemann (2002).

[a] Elongated shell.

14.2.2 Hard-Shell Capsule Formulation Design

14.2.2.1 *Preformulation of Hard-Shell Capsules* Capsule preformulation activities include physiochemical solid- and solution-state characterization and excipient compatibility. The physiochemical characterization is as discussed for tablets in Chapter 13. Tables 13.5 and 13.6 provide lists of the solid and solution-state properties that serve to guide the formulation design efforts. Most of the implications listed in these tables also apply to capsules. In some cases, the physiochemical properties may lead the formulator to choose a capsule as the preferred solid dosage form. For example, APIs with a disagreeable odor may be formulated in a banded or sealed hard-shell capsule that can serve as an effective odor barrier. Encapsulated material does not experience the level of friction or temperature rise that is observed with tableting. Therefore, encapsulation may be better suited than tableting for low-melting-point APIs. Good API compressibility is not as important a physical property for encapsulation as it is for tableting. However, flowability of the API for direct filling is equally or more important.

On the other hand, an API may have physiochemical properties that are not compatible with a capsule dosage form. An API that is highly hygroscopic may pull water from a hard-shell matrix and cause the capsule to become brittle. Loss of capsule moisture is more critical for gelatin capsules, where the water serves as a plasticizer for the shell. Hypromellose is not as sensitive to moisture loss and may be a better alternative in this particular case. APIs that are unstable to low levels of moisture are not good candidates for capsules since the water in the capsules may interact with the drug.

14.2.2.2 *Drug–Excipient Compatibility Studies of Hard-Shell Capsules*
Drug–excipient compatibility studies are initiated after the route of administration,

dosage form, and manufacturing processes have been decided. Excipient compatibility studies are a screening design whose aim is to identify excipients that are compatible with the drug. The type of capsule fill (dry running powder, liquid or semisolid, pellets, minitablets, or a combination) and physiochemical properties of the drug will dictate the choice of excipients selected for the compatibility study. The API will need to be compatible with fill excipients and the hard-shell capsule. The excipient compatibility can be done in steps where the stability of the fill excipients and the API are determined first. Once this stability is determined, the stability of the mixture of fill excipients, API, and hard-shell capsule is assessed. Most capsule fills use dry running powders. Before proceeding, a brief review of the manufacture of dry running powders is recommended. Seven different methods of preparing dry running powders are listed in Table 13.3. Table 13.4 outlines the advantages and disadvantages of each of dry running powder manufacturing methods. The excipient classes used to make running powders are provided in Table 13.2 and Appendixes 13.1 to 13.8.

A drug–excipient compatibility study generally requires a number of excipient–drug blends and test conditions that can require a relatively large number of test samples. In Section 13.2.2 we discuss several approaches to designing efficient and informative excipient compatibility studies. Preformulation studies for other types of capsule fills, such as liquids or semisolids, incorporate similar precepts, but the excipients that are evaluated are specific to the type of capsule fill. The drug–excipient compatibility is evaluated at 25°C, under accelerated conditions of 40°C/75% relative humidity (RH), International Conference on Harmonization (ICH) visible light, and ICH ultraviolet light. The temperature samples are assayed at one, three, and six months. Many companies only test at three months and use the 25°C condition as a control.

14.2.2.3 Liquid-Fill Hard-Shell Capsule Excipient Compatibility Studies

In addition to evaluating the compatibility of a drug with the fill excipients, it is necessary to determine the physiochemical compatibility of the fill material with the specific type of capsule of interest. Hard-shell capsules can be filled with a number of different materials, including liquids and semisolid materials. These excipients can affect the overall integrity and performance of the capsule. Appendix 14.1 lists excipients containing aldehydes that can cause gelatin cross-linking. Gelatin is more prone than hypromellose to becoming brittle when exposed to liquid or semisolid excipients. There is little excipient compatibility information for starch and pullulan capsules. Nagata and Tochio (2002) showed that gelatin softened when stored for one month at 45°C in the presence of propylene glycol. Under the same storage conditions, gelatin capsules filled with PEG 400 became brittle and broke. Hard-shell gelatin capsules filled with soybean oil, cottonseed oil, sesame seed oil, and medium-chain triglycerides demonstrated increased brittleness without breakage or change in appearance when exposed to 45°C for one month. Triacetin and triethyl acetate showed the best compatibility with hard-shell gelatin capsules. On the other hand, hypromellose hard-shell capsules were compatible with all the excipients tested except propylene glycol and PEG 400. In the case of propylene glycol, the capsules softened. PEG 400 capsules "sweated." None of the excipients tested with the hypromellose capsules demonstrated breakage due to brittleness. Liquid-fill hard-shell capsule excipients are listed in Appendix 14.2. The liquid-fill compatibility study could be set up to assess the affect of pure excipients or typical liquid-fill carrier compositions on capsule integrity and performance when stored under accelerated conditions of 40°C/75% RH for one, three, and six months.

14.2.2.4 *Formulation Development of Hard-Shell Capsules*

The formulation and process development times are often shorter with hard-shell capsules. A formulation scientist does not have to be concerned about designing a compressed tablet that needs to have acceptable hardness and friability to prevent abrasion and breakage during film coating or packaging. Nor does the formulator need to worry about tablet disintegration. Often, there are fewer excipients and processing steps compared to those for tablets unless a granulated running powder is needed for encapsulation.

The technologies used to prepare tablet running powders are exactly the same for capsules. Direct-compression running powders could be used for tablets and capsules. Similarly, techniques used to improve API flowability or to increase the density of the API for tablets can be used for capsules. Similarly, many of the excipients used for tablets are also used for capsules and can be reviewed in Table 13.2 and Appendixes 13.1 through 13.9. The objective of formulation development is to select excipients that enable proper capsule function and performance. The capsule quality is a function of the excipients and the robustness of the manufacturing process. The excipients are chosen based on an API's physiochemical properties (i.e., solubility, potency, particle size, bulk density, flow characteristics, melting point, hygroscopicity, stability, bioavailability, and more), type of fill, intended capsule function (immediate release, delayed release, extended release), site of delivery, proposed manufacturing process, literature examples, and the formulator's knowledge and skill in the art of capsule formulation. In this section we focus on immediate-release formulations.

14.2.2.4.1 Immediate-Release Hard-Shell Capsule Fills Prepared from Running Powders Selection of the type of running powder manufacturing process (dry blend, high-shear granulation, fluid-bed granulation, extrusion, roller compaction, slugging, or hot-melt) and excipients to use for the capsule fill depends on many factors that were discussed for tablets. A critical component of the selection process is determined by a drug's physiochemical properties. In this section we concentrate on immediate-release formulations. The relationship between the drug formulation and capsule attributes is complex and results from nonlinear responses and interactions among the drug and the excipients. As a result of this complexity, each drug formulation must be developed case by case. However, there are some general findings that can be used as starting points or general rules of thumb.

Using a combination of statistical design and a multivariate nonparametric canonical analysis, Hogan et al. (1996) investigated how capsule weight variation and drug dissolution were affected by such drug properties as solubility, particle size, and drug concentration. Formulation considerations such as type and concentration of filler, disintegrant, concentration of magnesium stearate, and concentration of colloidal silicon dioxide were also evaluated. The authors evaluated five drugs having a 1000-fold difference in solubility, five fillers with different solubilities, five disintegrants having a 15-fold difference in relative swelling capacity, and magnesium stearate and colloidal silicon dioxide at five levels. The study showed that filling performance was affected by drug particle size, API shape, drug concentration, type of filler, and concentration of glidant. Interestingly, lubricant and disintegrant type had less impact on weight variability. A statistically significant relationship between formulation factors and drug release was not found. However, data trends suggest that drug solubility, type of disintegrant, and type of filler have the most influence on drug dissolution and disintegration. A reasonably strong relationship was show between disintegrant degree of swelling and rate of drug release. Lubricant level (range from 0 to 2% w/w) did not demonstrate a major influence on the dissolution rate.

Soluble excipients tend to increase the dissolution performance of soluble and insoluble drugs compared to excipients that are less soluble (Botzolakis, et al. 1982; Koparkar et al., 1990). This general rule of thumb was highlighted during the period from mid-1967 through 1969 when a pharmaceutical manufacturer produced a batch of 100-mg phenytoin sodium capsules with lactose as the excipient instead of calcium sulfate, and 87% of the patients ended up having blood levels higher than the therapeutic range, which resulted in anticonvulsant intoxication. When the excipient change was identified as the cause of the outbreak of serious side effects, phenytoin blood levels were measured. The lactose capsules resulted in a two- to four fold increase in phenytoin blood levels compared to the calcium sulfate capsules (Tyrer et al., 1970). Table 14.8 provides the intrinsic dissolution rates for excipients studied by Koparkar et al. (1990).

TABLE 14.8 Intrinsic Dissolution Rates of Common Hard-Shell Capsule Excipients at 37°C

Excipient	Dissolution Medium	Rate $(\mathrm{mg \cdot min^{-1}\ cm^{-2}})$
Anhydrous lactose	Purified water	21.9
Hydrous lactose	Purified water	12.4
Dicalcium phosphate	0.1 M HCl, $I = 0.25$ M[a]	6.37
dihydrate	0.01 M HCl, $I = 0.25$ M	0.90
Anhydrous dicalcium	0.1 M HCl, $I = 0.25$ M	5.75
phosphate	0.01 M HCl, $I = 0.25$ M	0.69
Tricalcium phosphate	0.01 M HCl, $I = 0.25$ M	0.30
Calcium sulfate	0.1 M HCl, $I = 0.25$ M	1.36
dihydrate	0.01 M HCl, $I = 0.25$ M	0.75

Source: Koparkar et al. (1990).

[a] Ionic strength.

Dahl et al. (1991) showed that high levels of croscarmellose sodium (10 and 25% w/w) can overcome the decrease in capsule drug release that is caused by capsules exposed to high relative humidity.

With the advent of new capsule matrices, there are few restrictions on the types of fill materials. Drugs or excipients that interact adversely with the hard shell should be excluded from use. In the case of gelatin, substances that are aldehydes or contain aldehydes should be avoided to minimize the risk of cross-linking. Drugs or formulations that are hygroscopic and are capable of adsorbing moisture from the gelatin shell should be used with caution or avoided altogether to prevent capsule brittleness, cracking, or breaking. Fill material that is high in moisture may give up its water to the capsule, which may cause the capsule to soften or become distorted. The dose is limited by the physical size of the capsule. Low-bulk or tapped-density drugs may require too large a volume to accommodate a high-dose product. Using a size 0 capsule, the highest capsule fill weight is about 0.82 g if the apparent powder bulk density is 1.2 g/mL.

14.2.2.4.2 Hard-Shell Capsule Fills Prepared from Beads, Pellets or Minitablets, Tablets, and Capsules Bead formulation and processing are discussed in Chapter 19. Recently, a delayed-release capsule containing enteric-coated minitablets of fenofibric acid (Triplex) has been commercialized. Methylacrylic acid copolymer provides the enteric release coating.

Capsules are ideally suited for blinding clinical trials (Goodson and Stagner, 1998). Blinding is used to render different drugs and placebos indistinguishable from each other and to conceal commercial identifying marks. Beads, pellets, minitablets, tablets, and capsules can all be filled into an opaque capsule that blinds the contents to the clinical trial subject. This technique is especially useful when the purpose of the study is to compare a commercial product to a new chemical entity or to compare one commercial product to another. The only manipulation of the commercial product is overencapsulating the commercial tablet or capsule and possibility topping off the capsules with an inert excipient.

14.2.2.4.3 Hard-Shell Capsule Filling of Liquids and Multicomponents of Multiphases Liquid-filled capsules are finding increased application for those compounds that are classified as low solubility–high permeability and low solubility–low permeability biopharmaceutics classification system (BCS) compounds. The ability to seal hard-shell capsules at manufacturing speeds to prevent the liquid formulation from leaking was a major technological breakthrough. Liquid fills include a broad range of materials that are flowable at temperatures ranging from room temperature to 70°C. In the case of filling thermosoftened or melted materials, viscous pastes, and thixotropic formulations, the fill is flowable during filling and thickens and sets postfilling. Other liquids, such as emulsions, microemulsions, suspensions, and solutions, can be filled into hard-shell capsules that are *banded*. Liquid-filled capsule formulations of poorly water-soluble drugs are on the rise. Liquid-fill formulations are especially useful for BCS class II compounds that are poorly soluble and highly permeable. Liquid fills are also useful for low-melting-point drugs. Solutions of low-dose and highly potent drugs reduce concerns about content uniformity. Liquid fills also decrease the occupational exposure to highly potent APIs.

As more and more poorly soluble drugs enter development, there is an increasing demand to explore enabling technologies that have not been fully utilized in the past. Formulators also take advantage of significant advances made by capsule equipment manufacturers, which have overcome the limitations that previously existed with hot-melt and liquid filling. Appendix 14.3 lists a number of excipients that can be used in liquid formulations.

A number of different types of nonaqueous solution-fill systems have been reported. Since they are nonaqueous, drug stability is usually greater than that seen in aqueous solution. The liquid drug delivery system is also typically dispersible in gastrointestinal fluids. Pouton (2006) proposed the lipid formulation classification system shown in Table 14.9. The type I system is a nondispersing lipophilic system that requires digestion. These formulations have low solvent capacity unless the drug is highly lipophilic. The type II system is a water-insoluble self-emulsifying drug delivery system. Type II formulations are much more likely to maintain their solvent capacity when it dispersed. Types IIIA and IIIB are self-emulsifying or self-microemulsifying formulations that contain water-soluble surfactants. The type IIIA system may or may not incorporate hydrophilic cosolvents. Type IIIB scaffolds contain less than 20% w/w oil and 20 to 50% hydrophilic cosolvents. Type III systems deliver the drug without digestion, but type IIIB, which contains less oil, may be more prone to losing its solvent capacity following dispersion. Type IV systems are oil-free and may be more stable in hard-shell capsules than the other liquid-fill formulation types. The principal excipients are water-soluble surfactants and hydrophilic cosolvents. Type IV systems disperse into micelles.

Meinzer et al. (2004) patented a cyclosporine formulation which is an oil-free liquid that was stable in hard-shell gelatin capsules. Liquid formulations previously proposed became brittle when stored for two to three years. The

TABLE 14.9 Lipid Formulation Classification System

Excipients	Content of Formulation (% w/w)				
	Type I	Type II	Type IIIA	Type IIIB	Type IV
Oils: triglycerides or mixed mono- and diglycerides	100	40–80	40–80	<20	—
Water-insoluble surfactants (hydrophile–lipophile balance <12)	—	20–60	—	—	0–20
Water-soluble surfactants (hydrophile–lipophile balance >12)	—	—	20–40	20–50	30–80
Hydrophilic cosolvents (e.g., PEG, propylene glycol, Transcutol)	—	—	0–40	20–50	0–50

Source: Poulton (2006), with permission from Elsevier.

commercial formulation (Gengraf) hard-shell capsule is an oil-free type IV lipid system that contains cyclosporine A (ciclosporin) in a mixture of PEG, a hydrophilic base; a surfactant mixture of polyoxyl 35 castor oil N.F., polysorbate 80 N.F., and sorbitan monooleate N.F.; propylene glycol; and 12.8% v/v alcohol. The Meinzer patent indicates that the PEG preferred molecular mass range is 200 to 600 Da. PEG 300 is cited in the patent examples.

In the case of filling thermosoftened or melted materials, viscous pastes, and thixotropic formulations, the fill is flowable. Once the capsule is filled, the material generally stiffens and becomes more viscous on setting. Hot-melt formulations have been shown to increase the dissolution rate and bioavailability of poorly soluble drugs. Joshi et al. (2004) used a hydrophilic 3 : 1 hot-melt mixture of poly(ethylene glycol)–polysorbate 80 to improve the bioavailabilty of a poorly soluble drug. The solubility of the drug above pH 5.5 was below the detection limit of the analytical method, which was 0.02 µg/mL. The percent absolute bioavailability of the drug was increased 21-fold, from 1.7% absorbed for a micronized drug–capsule formulation to 35.8% absorbed for the hydrophilic hot-melt formulation. A solution was used as the oral dosage-form comparison. The oral solution had an absolute bioavailability of 59.6%. The dissolution profile showed that the capsules remained unchanged at room temperature for six months. There was a decrease in the dissolution rate of capsules after storage at six months and 40°C/75% RH.

Schamp et al. (2006) developed an in vitro–in vivo correlation for hot-melt lipid formulations of a poorly soluble lipophilic drug. Gélucire 44/14 (lauroyl polyoxylglyceride) and Soluphor (2-vinylpyrrolidone) were used as a hot melt to solubilize the drug completely. Solubility and dissolution studies were carried out in several dissolution media. These media included simulated gastric fluid containing 0.1% Triton X 100 surfactant, fasted-state simulated intestinal fluid, and fed-state simulated intestinal fluid. The dissolution rate profiles reached a maximum dissolution concentration that was two to three times the intrinsic solubility of the drug. This supersaturated concentration was maintained for over 3 hours. The authors concluded that maintaining the drug in solution and preventing it from precipitating once it is dissolved is important for absorption of BCS class II compounds.

Hot-melt capsule formulations prepared from low-melting excipients have been exploited. S.M.B. of Belgium has developed a eutectic system (Fenofibrate Lidose) which melts at 37°C. This is an ideal system for a compound such as fenofibrate, which is insoluble and has a low melting point of 79 to 82°C. The eutectic excipient mixture and drug are melted and filled into hard-shell capsules. The formulation solidifies on cooling. When a patient takes the capsule, the eutectic matrix melts at body temperature, forming a liquid drug delivery system. Lipofen by Kowa Pharmaceuticals is also a hot-melt formulation of fenofibrate that is marketed in the United States. The hot-melt excipients used in the formulation are lauroyl polyoxyglyceride, poly(ethylene glycol) 20,000 and 8000, and hypromellose.

So far we have discussed the usefulness of using liquid fills to improve drug bioavailability. Liquid fills can also be used to improve drug stability. Bowtle et al. (1988) showed that vancomycin, a hygroscopic drug that is unstable in solution, could be formulated into a stable immediate-release hot-melt capsule formulation.

The drug itself picked up 10% moisture when exposed to 50% RH over two consecutive manufacturing runs. PEG 4000, 6000, and 8000 were studied because they have acceptable low moisture uptake. PEG 8000 had poor performance characteristics compared to the 4000- and 6000- Da molecular mass material. Hot-melt formulations were shown to provide adequate stability and dissolution profiles, with 75% of the drug dissolved in 60 minutes. It is noteworthy that the commercial vancomycin formulation does contain PEG.

14.2.3 Hard-Shell Capsule Process Design

14.2.3.1 Running Powder Blend Uniformity

Process development should include blend uniformity testing to determine optimum blending times, the extent of blend variability expected, and identification of potential dead spots within the blender. The basic tenet of content uniformity assessment is stratified sampling of a sufficient number of samples done in triplicate. Manufacturers are legally bound by 21 CFR 211.110 to establish methods to monitor and validate the adequacy of mixing to assure blend uniformity. The U.S. Food and Drug Administration (FDA 2003a,b) has issued a guidance document to assist companies in meeting federal regulations. The guidance recommends a stratified sample approach that involves developing an a priori sampling plan that identifies the sampling interval, sample replicates, sample quantity, sampling methodology, and sample locations. The guidance recommends sampling at least 10 sample locations for V-blenders and double-cone blenders. The samples should be taken in triplicate, and powder samples should not be larger than three times the dosage-form unit weight unless larger sample amounts can be justified. The guidance recommends that one sample for each location be assayed, that the percent relative standard deviation across all locations should be less than or equal to 5%, and that all individual results be within 10.0% (absolute) of the mean of the results. A decision tree is used to relate the results of the blend data to dosage units.

Blend variability may be due to one or more factors, such as analytical error, sampling error, or inadequate blending. The Parenteral Drug Association (Berman et al., 1997) provided another, separate approach to evaluating blend uniformity. Occasionally, discrepancies are uncovered between blend uniformity and dosage form uniformity. It is puzzling and difficult to explain how blends that exhibit unacceptable uniformity can lead to tablets or capsules that have acceptable content uniformity. Prescott and Garcia (2001) has developed a troubleshooting approach that can be used to investigate these types of discrepancies.

14.2.3.2 Filling Running Powders into Hard-Shell Capsules

Dry powder fills using running powder made from dry blends, high-shear wet granulation, fluid-bed wet granulation, extrusion, roller compaction, and slugging are discussed in detail in Chapter 13. Interrelated processes such as particle sizing, blending, and drying were also discussed.

Once the dry powder capsule fill, *running powder*, has been prepared and the capsule body has been removed from its cap, encapsulation can begin. The encapsulation process has evolved over time from manual to semiautomatic to automatic filling processes. There are two general types of powder filling processes. In one, the capsule itself serves as the chamber for a volumetric fill. Called *direct filling*, this method requires that the running powder's tapped density be adjusted to provide the correct volumetric fill. In the second case, *indirect filling*, the fill is measured in a chamber that is external to the capsule body and involves preparing a compacted plug in a dosating tube or disk.

14.2.3.2.1 Direct Hard-Shell Capsule Auger Filling Process   An auger direct-filling semiautomated Model 8 capsule filling machine was the production workhorse for many years. The manufacturing capacity of the Model 8 is about 15,000 capsules per hour. Newer models using the same operating principle are rated at up to 30,000 capsules per hour. The Model 8 is still commonly used for manufacturing clinical materials and small-volume commercial products. The encapsulation process starts by *rectifying* the capsules into stainless steel rings that have openings that are sized to hold the capsules. Rectifying the capsules is the process by which all the capsules are oriented with their bodies entering the ring first. Once the rectified capsules are in the ring, a vacuum is applied to the capsule, which separates the body and cap. The ring is separated and the portion of the ring holding the bodies is placed on the capsule machine. A hopper equipped with an auger is positioned over the top of the ring. As the auger forces the powder into the capsule bodies, the ring is also automatically rotated under the hopper, allowing all the capsules to be filled. The capsules are filled by volume. Once all the capsules have been filled, the hopper is removed from the ring manually. The capsule-holding ring is pulled off the capsule machine and rejoined to the half-ring holding the caps. The ring is placed on a *stop plate* and then positioned in front of a set of pegs that are used to join the caps and bodies. The stop plate serves as a backstop to aid in the joining process. Once the capsule is rejoined, the stop plate is removed and the pegs are inserted farther into the ring to eject the capsules from the ring. This process is illustrated in Figure 14.1. Powder flow is important even though the auger assists in moving the powder into the capsules. Highly compressible cohesive powders may actually plug the hopper foot and increase weight variation.

14.2.3.2.2 Indirect Hard-Shell Capsule Dosator Filling Process   Dosator encapsulators are automatic fillers that can run intermittently or continuously, depending on the speed and manufacturer. The most common intermittent filling machine is the Zanasi, which has an operating capacity of up to 40,000 capsules per hour. Continuous dosator fillers are MG2 and Matic. The Macophar, Pedini, and Farmatic also use the dosator filling principle. These automatic machines can fill upward of 120,000 capsules per hour.

The dosator principle determines the fill quantity by forming a "plug" or "loose pellet" inside a stainless steel tube called the dosator. Lowering the dosator tube into a preformed powder bed forms the powder plug. The volume of the plug is controlled by the internal volume of the dosator, which can be adjusted by the piston or plunger that is inside the dosator tube. The piston serves two additional functions: Once the dosator has entered the powder bed and a specific volume of

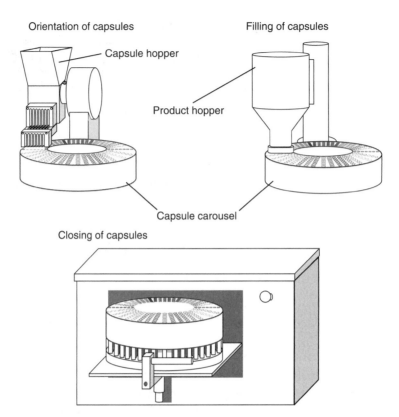

FIGURE 14.1 Encapsulation using a semiautomated capsule machine.

powder has entered the dosator, the piston can be tamped to provide precompression and compression to compact and strengthen the plug. The dosator is removed from the powder bed, aligned over the capsule body, and the plug is transferred into the capsule by the piston or plunger (see Figure 14.2). To minimize fill weight variation, the plug needs to have enough strength to remain intact throughout this process. The fill weight is also controlled by the height of the powder bed and the ability to flow and backfill the hole created by the plug. The running powder needs sufficient lubrication to aid plug ejection.

14.2.3.2.3 Indirect Hard-Shell Capsule Dosing Disk Filling Process
Dosing disk encapsulators are intermittent automatic filling machines. The models are the Bosch GFK (formerly Höfliger-Karg, commonly known as the H&K), Harro Höfliger, Impressa, and Index. Machine output of 180,000 capsules per hour can be achieved by these fillers. Development encapsulators such as the In-CAP from Bonapace can manufacture up to 3000 capsules per hour.

The dosing disk operating principle uses a large stainless steel plate or disk that has holes drilled into it which serve as die cavities to hold the powder that is filled into the cavities. Powder is tamped into the dies by a series of pistons

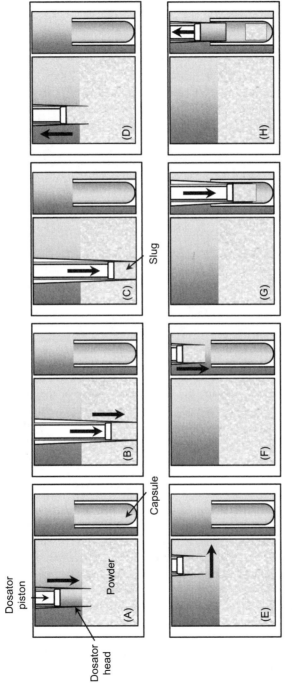

FIGURE 14.2 Dosator sequence.

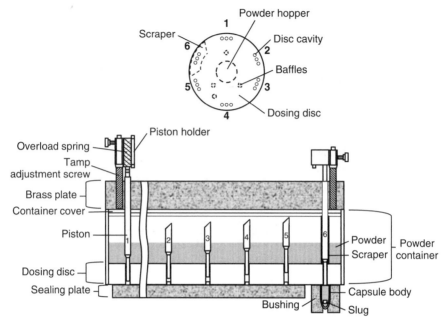

FIGURE 14.3 Dosing disk filling principle. (Adapted with permission from G.S. Banker and C.T. Rhodes, *Modern Pharmaceutics*, Informa Healthcare, New York, 1996.)

or plungers to a preset displacement. The disk is then positioned over an empty capsule body and a plunger pushes the plug into the capsule body. The dose is controlled by the volume of the die cavity, the volume of the powder added, and the thickness of the plug (see Figure 14.3). Powder compressibility and flow is not as critical with disk dosing as are dosator fillers. The powder needs to have enough flow to fill the dies uniformly. The running powders should have enough lubrication to allow smooth ejection of the plug into the capsule and prevent powder buildup in the die cavity.

14.2.3.3 Filling Beads, Pellets, Tablets, and Capsules into Hard-Shell Capsules
Today's automatic fillers can encapsulate other types of solids, such as beads, pellets, and tablets into capsule bodies by direct or indirect filling operations. Direct fills are accomplished by feeding the material into the capsule until the chamber is full. Tablets and capsules are typically delivered through tubes directly into the capsule. It is also possible to fill several types of solid materials into one capsule. This may be desirable for combination products when there is an incompatibility between the actives. Filling different functional material may provide controlled-release by filling an immediate release powder with a controlled-release bead. Indirect filling provides increased dosing flexibility. In this case, the filling material will be dispensed into a separate volumetric metering chamber. The accurately measured volume will then be transferred to the capsule body. Some encapsulators use vacuum to transfer the fill material accurately to the volumetric

filling chamber. Once the chamber is full, the material is transferred to the capsule. Particle size and shape can affect fill weight variability. Electrostatic charge can also affect content uniformity and flow properties. To maintain good content uniformity for blends of beads or pellets, it is especially important to use the largest capsule size possible and maintain a narrow particle size distribution.

14.2.3.4 Filling Liquids and Multicomponents of Multiphases into Hard-Shell Capsules
With advances in hard-shell filling equipment, it is now possible to fill room-temperature or hot (60 to 70°C) liquids at a laboratory to commercial scale. Production speeds range from 3000 to 120,000 capsules per hour. Hard-shell capsules, can withstand higher filling temperatures than can soft-shell capsules, ranging up to approximately 60 to 70°C. The filling material can be dosed accurately as long as the metering operation has a clean break between filling each unit and there is no resulting carryover. Typical filling viscosities range from 0.01 to 1 Pa · s (10 to 1000 cP).

Preventing leakage from the hard-shell capsules has been a major technology breakthrough. Several laboratory- and production-scale technologies have been developed. Scale-up and validation of sealing technologies is minimal since the development- and commercial-scale processes are the same. One sealing method uses a gelatin banding technique in which a layer of liquid gelatin is used to seal the junction of the capsule cap and body to prevent leakage. Another process employs a liquid fusion process whereby a hydroalcoholic solution is applied to the cap–body junction. The fusing solution is drawn into the space between the capsule cap and body by capillary action where the two are sealed together.

The ability to seal liquids has recently expanded the filling flexibility to include multiphase multicomponent capsules (Miller, 2010). One claim involves filling a capsule with a combination of ibuprofen in solution and celecoxib in its solid form. Examples are available showing four components in different physical states in one capsule.

14.2.3.5 Capsule Identification
The primary method of capsule identification is accomplished by printing capsules with pharmaceutical-grade inks that contain food-grade FDA-approved pigments and lakes. Printing can be performed on the capsule axially or radially. Distinctive cap, body, and banding colors help to identify a particular strength, product, or combination. Injection-molded capsules offer additional identification flexibility. The molds can contain product and strength identification that is embossed into the capsule shell. NanoInk has developed and received FDA approval for NanoEncryption. The technology employs overt and covert nanocodes directly on the capsule to thwart counterfeiting.

14.2.3.6 Scale-up of the Hard-Shell Capsule Encapsulation Process
Scaling-up of direct and indirect capsule-filling operations should pose fewer risks than scaling-up tablet processes. Tablet scale-up is achieved primarily by increasing the output rate. Increasing the tableting speed leads to a number of scale-up concerns, such as decreased compression and dwell times. Encapsulation scale-up, on the other hand, is generally accomplished by increasing the

number of manufacturing stations, not the encapsulation speed. For auger direct capsule-filling operations, larger volumes of capsules are typically obtained by increasing the production time or by adding more machines. For dosator or disk dosing encapsulators, increases in capsule output are accomplished by adding more dosators or tampers, disks that operate at about the same speed as the laboratory models. Some equipment modifications are required to accommodate the larger scale, such as larger powder bowls and dosing disks. This may require additional measures to maintain uniform bed height and density for uniform filling of the disk cavities.

Even though the output is increased with scale-up, the manufacturing time is also typically increased. There is more time for running powders to consolidate in the bulk hopper and become less flowable, which might lead to content uniformity problems. The powder may reside in the dosing chamber longer and be exposed to extended mixing, which may lead to overlubrication. Overlubrication may lead to decreased dissolution and bioavailabilty. Since dosator and dosing disk encapsulators have different design characteristics, bridging studies would need to be done to compare capsule batches manufactured with separate design characteristics.

14.2.3.7 Process Capability Index   A number of capability indexes have been developed to describe how capable a process is in producing a defect-free product. Any process can be evaluated by process capability analysis. Process capability examines the variability of the process and how well the process operates inside a set of specifications of upper and lower limits. The *potential capability (Cp)*, also known as the *process capability index*, *production capability index*, and *production capability ratio*, is (Hill and Lewicki, 2006)

$$Cp = \frac{\text{upper limit} - \text{lower limit}}{6 \text{ standard deviations}} \tag{14.1}$$

This assumes that the production process can be described by normal distribution and that the ratio describes the proportion of the normal curve that falls within the specification range. This assumes that the manufacturing process is centered at the mean. Capability analysis can also take into account a process that is not centered at the mean of the specification range. This is called the *actual capability index* or *demonstrated excellence* (Hill and Lewicki, 2006) and is abbreviated *Cpk*. This index combines *Cp* and *k*, where *k* is the ratio of the difference between the process mean and the specification mean:

$$k = \frac{\text{process mean} - \text{specification mean}}{\text{specification mean}} \tag{14.2}$$

Cpk accounts for where the actual distribution lies within the specification range and is expressed as

$$Cpk = (1 - k)Cp \tag{14.3}$$

If the process mean is centered within the upper and lower limits of the process specifications, *k* is zero and *Cpk* and *Cp* are equal. As the process shifts from the mean specification or target mean, *k* increases and *Cpk* becomes smaller than *Cp*. The actual process capability index, *Cpk*, can never be larger than *Cp*. *Cpk* is calculated as the ratio of the difference between the process mean and the nearest specification limit to three times the standard deviation.

$$Cpk = \frac{\text{closest specification limit} - \text{process mean}}{3\text{SD}} \tag{14.4}$$

where SD is the standard deviation of the process. A process that demonstrates a *Cpk* of 1 is generally considered satisfactory. A *Cpk* of 2 is a "six-sigma" process and would have 3.4 failures in 1 million attempts. In the case of weight variation, this would mean that 3.4 tablets or capsules would fall outside the upper and lower limits of the weight specification.

Capability analysis was carried out on the content uniformity of 5- and 10-mg capsules as part of a validation program. The percent label claim (LC) for the 5-mg capsules was determined for 178 capsules that were collected during a 6-hour encapsulation run. The % LC for the 10-mg capsules was determined for 77 capsules that were collected over a 3-hour run. The results are shown in Figures 14.4 and 14.5. The target % LC was set at 100% LC, and the lower and upper specification limits were set at 95% LC, and 105% LC, respectively. The data were analyzed using JMP (2008). The *Cpk* analysis for the 5-mg capsule is shown in Figure 14.4. The top right-hand table shows that the capsule mean percent label

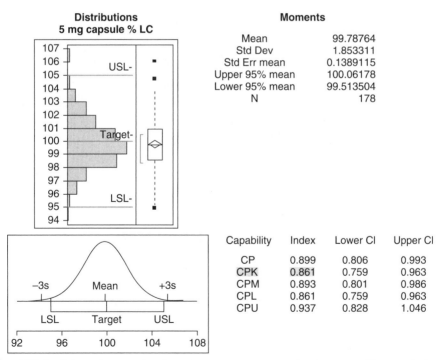

FIGURE 14.4 *Cpk* evaluation for a 5-mg capsule % LC. LSL (lower specification limit), USL (upper specification limit), and s (standard deviation). Analysis of the data used JMP statistical analysis software (SAS Institute, Cary, North Carolina).

claim was 99.8%, and the percent relative standard deviation can be calculated to be 1.85% LC. Most formulation and process scientists would be happy with a mean % LC of 99.8% and relative standard deviation of 1.85%. However, the *Cpk* value of 0.861 suggests that the process is marginal at best. The top graph shows that of the 178 capsules assayed, three capsules (indicated by the three squares) fell outside the 95 to 105% LC specification range. The lower graph shows the % LC distribution vs. the target and upper and lower specification limits and compares these limits to the actual mean and 3 standard deviations of a normal disturibution. This figure highlights the fact that the % LC at 3 standard deviations falls outside the lower and upper specifications.

The mean percent label claim for the 10-mg capsules was 100.8% and the relative percent standard deviation was 1.14% LC. These results seem very comparable to the 5-mg capsule data. However, in this case the 10-mg capsules have a *Cpk* value of 1.224, which suggests that the encapsulation process is satisfactory. Both graphs in Figure 14.5 illustrate this point. All the data fit inside the upper and lower specification limits. The bottom graph shows that a normal distribution curve set at 3 standard deviations fits easily inside the specification limits. This is just one example of the use of capability analysis. Capability analysis can be applied to any unit operation.

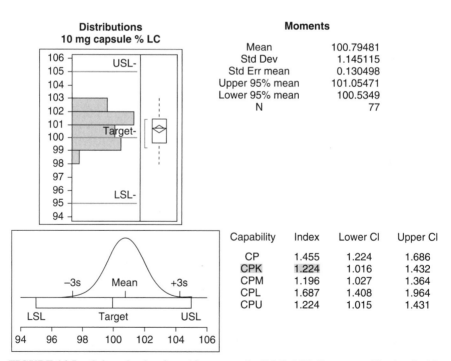

FIGURE 14.5 *Cpk* evaluation for a 10-mg capsule % LC. LSL (lower specification limit), USL (upper specification limit), and *s* (standard deviation). Analysis of the data used JMP statistical analysis software (SAS Institute, Cary, North Carolina).

TABLE 14.10   Advantages and Disadvantages of Soft-Shell Capsules

Advantages	Disadvantages
Ideally suited for liquid filling	Typically, development and manufacturing are contracted out
Accurate dosing of liquid fill	to a specialty soft-shell manufacturing company
Solution fills provide excellent dose uniformity for low-dose and potent drugs	Expensive manufacturing process; drying the soft shell may take 3 to 10 days
	Potential for leakage
Liquid filling provides less exposure risk to operators	Cross-linking of gelatin, resulting in slower dissolution and potential decrease in bioavailability
Good for low-melting drugs that can be formulated into a liquid	High water content ($\sim$40%) in shell before drying
	Softens and becomes tacky in hot, humid climate zones
Molding allows for a wide variety of shapes	Potential for cross-diffusion of drug, fill excipients, and soft-gel-shell excipients affecting shell integrity and drug stability
Rapid release of contents	
Chewable dosage form	
More portable than stand-alone liquids	

14.3   SOFT-SHELL CAPSULES

As the name implies, soft-shell capsules refer to capsules that are softer and more flexible than hard-shell capsules. The walls of the shell are also thicker than hard-shell capsules. The advantages and disadvantages of soft-shell capsules are listed in Table 14.10.

The various components of the soft-shell formulation and their functions are listed in Table 14.11.

14.3.1   Manufacture of Soft-Shell Capsules

Soft-shell capsules are manufactured by a continuous form–fill–seal process (see Figure 14.6). First, a casting film is produced by adding film formers and processing aids to water. This composition is then heated under vacuum to form a molten mass. The molten mass is applied to two cooled casting drums to form a casted film or ribbon on the drum. This film is called a wet film, because it can contains around 40% w/w water. The two films are fed independently to two opposing rotating dies, which form the capsule half-shells. The films are heated below their melting point and are formed to the die shape. The capsules are formed simultaneously from the half-shells, filled, cut, and sealed under temperature and pressure. The filled capsules are then placed on dying conveyors that keep the capsules from touching each other, and capsule drying is initiated. The capsules are then placed on drying trays and placed in forced-air drying ovens. Drying may take 3 to 10 days at 20 to 30°C and 25% RH to achieve a moisture level of 6 to 12% w/w water. This water

TABLE 14.11   Soft-Shell Capsule Component Classes and Functions

Component Class	Function
Film-forming biocompatible polymers (40 to 60% w/w)	Form a soluble, edible, biocompatible container
Gelatin	
Modified Starch (hydroxypropylated starch)	
Plasticizers	Improve flexibility
Glycerin	
Sorbitol	
Propylene glycol	
Poly(ethylene glycol)	
Hydrogenated saccharides and polysaccharides	
Sorbitans	
Processing aids	
Carrageenan (kappa and iota)	Elasticizing agents for modified starch
Buffer system (phosphate is preferred)	Enhances the stability of carrageenans
Colorants	Identification; light protection; concealing contents
Iron oxide pigments (0.01 to 1.5% w/w)	
Soluble dyes (0.05 to 1.5% w/w)	
Opacifying agents	Identification; light protection; concealing contents
Titanium oxide (0.01 to 1.0% w/w)	
Preservatives	Decrease bioburden and prevents microbiological growth
Methylparaben	
Propylparaben	
Flavors	Mask poor-tasting drugs, especially for chewable capsules

is bound water and is not easily removed. Once dried, the capsules can be printed and packaged. A typical soft-shell gelatin capsule shell formulation is shown in Table 14.12.

A typical soft-shell modified starch capsule formulation is shown in Table 14.13.

14.3.2   Soft-Shell Capsule Formulation Design

Although soft-shell capsules have been filled with beads and thixotropic semisolids, commercial soft shells are available primarily as liquids. The liquid forms can be solutions, suspensions, emulsions, microemulsions, self-emulsifying,

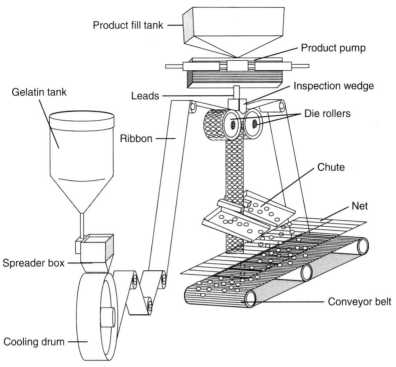

FIGURE 14.6 Rotary die process for soft gelatin capsules. (Adapted with permission from G.S. Banker and C.T. Rhodes, *Modern Pharmaceutics*, Informa Healthcare, New York, 1996.)

TABLE 14.12 Typical Soft-Shell Gelatin Capsule Shell Composition

Material	Weight (% w/w)
Gelatin	61.0
Glycerin	32.0
Methylparaben	0.6
Propylparaben	0.2
Water	6.2

Source: Brox (1988).

and self-microemulsifying drug delivery systems. Pouton (2006) has discussed the advantages and disadvantages of various types of liquid formulation systems as they relate to a drug's physiochemical properties and the physiology of the gastrointestinal tract.

TABLE 14.13 Typical Soft-Shell Starch Capsule Shell
Composition

Material	Weight (% w/w)
Modified starch	34.00
iota-Carageenan	15.20
Glycerin	42.00
Disodium phosphate	2.10
Methylparaben	0.55
Propylparaben	0.15
Purified water	6.0

Source: Tanner et al. (2002).

Soft-shell capsules are more permeable than hard-shell capsules, which can result in a greater potential for diffusion of shell material out of the shell and diffusion of fill contents into the shell. The fills may be lipophilic or hydrophilic in nature and are listed in Appendix 14.3. Low-molecular-mass excipients such PEG 400 and propylene glycol tend to migrate into a soft shell. Volatile solvents such as alcohol can be lost from the fill by permeating through the shell and evaporating. Soft-gelatin capsules face the same cross-linking issues as those of hard-shell gelatin capsules. Aldehyde levels of many liquid-fill excipients are provided in Appendix 14.2.

14.3.2.1 Soft-Shell Capsule Preformulation Refer to Chapter 13 for a detailed discussion of the purpose of preformulation and physiochemical property assessment.

14.3.2.2 Drug–Excipient Compatibility A drug's physiochemical, permeability, and drug disposition properties drive excipient choices for drug–excipient compatibility studies, as discussed above for hard-shell capsules and earlier for tablets.

14.3.2.3 Liquid-Fill Soft-Shell Compatibility The formulation scientist needs to be keenly aware of the increased potential for fill materials to permeate a capsule's shell, affecting capsule and drug-release performance. Similarly, shell materials can diffuse into the liquid fill and interact with components of the fill. Once the drug and excipients have been chosen, these can be studied with particles of the soft shell or in actual soft shells. Soft shells can be manufactured empty and the liquid fill can be injected directly into the empty shell and sealed with a hydroalcoholic solution. Accelerated studies exceeding 40°C/75% RH may cause the capsules to become too soft or cause excessive leakage.

14.3.2.4 Soft-Shell Capsule Formulation Development   Soft-shell capsule formulation design strategies need to take into account the potential interactions that can occur between the fill and the capsule shell composition. There are a number of steps in the manufacturing process where physical and chemical interactions might occur. The liquid fill may contain surfactants or oils that can interfere with the sealing of the shell during the form–fill–seal process. During capsule drying, the water or glycerin may be driven into the capsule fill and may react with the drug or filling material. Depending on the pH of the fill, the water may accelerate hydrolysis reactions, or glycerin may undergo esterification or transesterification reactions. The potential for aldehyde cross-linking of gelatin may be increased for soft-gelatin capsules during the drying phase since the water content of the capsules is around 40% w/w and formaldehyde and acetaldehyde are water soluble. Extra attention needs to be paid to excipients such as Gélucires; PEG 200, 400, and 600; and Labrasol, which have been shown to have high levels of these aldehydes. There is also a potential for hygroscopic drugs to interact with the high-moisture-content shell.

Hydrophilic fills are more problematic than lipophilic or water-free fills. In general, fills should not contain more than 5% water, which can make the shell too soft. On the other hand, hydrophilic fills containing PEG can interact with the water in the shell, causing the shell to become hard and brittle.

Some examples of commercial liquid-fill soft-gelatin capsule products will be discussed. Ciclosporin (cyclosporine A) is an important immunosuppressant agent that is used in organ transplant. The first oral formulation of ciclosporin was an emulsion-filled soft-gelatin capsule branded Sandimmune. It was a lipid formulation classification type III self-emulsifying lipid formulation system containing corn oil, linoleoyl poly(ethylene glycol), and ethanol that had significant inter- and intrasubject variability. A neoral soft-gelatin capsule was developed later as a type III self-emulsifying microemulsion. The microemulsion fill contained corn oil, polyoxy 40 hydrogenated castor oil, tocopherol propylene glycol, and ethanol. Compared to Sandimmune, the microemulsion demonstrated significant increase in blood levels and decreased inter- and intrapatient variability (Novartis; Kovarik et al., 1994).

14.3.3 Soft-Shell Capsule Process Design

Soft-shell capsules are more elastic than their hard-shell counterparts. They are molded in a single form–fill–seal process that allows many shapes to be formed, such as spheres, ovals, and oblong capsules. Soft-shell capsules are prepared from gelatin and hypromellose. The form–fill–seal process generally limits the soft-shell capsule fill to liquids that can be added at the filling stage of the continuous form–fill–seal process. The filling temperature is lower for soft-shell capsules, with sealing occurring from 37 to 40°C. The soft-shell manufacturing process was developed as a proprietary process and continues to be primarily proprietary. There are now equipment suppliers that provide soft-shell encapsulators. The nutraceutical industry is a leader in developing in-house soft-shell manufacturing capabilities. This encapsulation process continues to be held as proprietary, and few research articles have been written concerning process design, scale-up, and process analytical technologies applied to soft-shell manufacturing.

14.4   FORMULATION AND PROCESS OPTIMIZATION

Advances in statistical analysis and design-of-experiment software have placed sophisticated tools and platforms in the hands of the product design scientist. The purpose of product design is to create a drug delivery system that meets specific functional and performance criteria. There are many opposing complex nonlinear relationships that require the development scientist to be highly skilled in the art and science of product design. Statistical analysis and experimental design approaches are now recognized as important tools that can aid the design scientist in optimizing a product in this complicated, multifactorial, multidimensional design space. Formulation and process screening and optimization designs can be used to enhance the understanding of the formulation composition and process factors that affect the overall function and performance of a drug delivery system.

The final drug product attributes are highly dependent on the interplay between the formulation and the process. Statistical programs are now available that allow the design scientist to integrate formulation and process optimization. The next example illustrates the use of the "custom design" platform in the JMP statistical program. This program allows the design scientist to optimize for formulation factors and processing factors for any number of key product attributes. In this particular case, a dry fill capsule containing three powder excipients was filled on an H&K encapsulator. The effect of excipient level, number of tamps, and encapsulation speed on percent dissolved in 30 minutes was evaluated.

A "custom" JMP design of experiments incorporated a mixture design for the excipients: pregelatinized starch (0 to 50% w/w), fast-flow lactose (0 to 100% w/w), and microcrystalline cellulose (0 to 100% w/w). High, low, and midpoint values were used for the number of tamps and encapsulator speed. The capsule lubricant level was held constant. Two midpoints were chosen to provide a measure of variability and nonlinearity. The effect of these factors on capsule dissolution is shown in Table 14.14.

All three excipients had a significant effect on dissolution ($p < 0.05$). The interaction between lactose and tamping and microcrystalline cellulose and tamping were also significant at $p < 0.05$. The model equation describing the effect of these factors on dissolution was

$$\% \text{ label claim dissolved in 30 min} = 109.2(S) + 75.7(L)$$
$$+ 75.5(C) - 7.1(L)(T) - 6.4(C)(T)$$

where S is the pregelatinized starch concentration, L the fast-flow lactose concentration, C the microcrystalline concentration, and T the number of tamps. The tamping interaction caused a slowing in the dissolution with the increased number of tamps. Encapsulation speed was not found to have a statistically significant impact on the dissolution rate. Figure 14.7 shows a contour plot that illustrates

TABLE 14.14   Formulation and Process Optimization of a Capsule Formulation

Pregelatinized Starch	Fast-Flow Lactose	Microcrystalline Cellulose	Number of Tamps	Speed	% Label Claim Dissolved in 30 min
0	0	1	1	1	69
0.5	0	0.5	1	−1	81
0.5	0.5	0	1	1	78
0.25	0	0.75	1	−1	77
0	1	0	1	1	69
0.20	0.38	0.42	0	0	82
0	0.5	0.5	1	−1	70
0.25	0.75	0	1	1	75
0	1	0	1	−1	67
0.5	0	0.5	1	1	82
0.5	0.5	0	−1	1	88
0	0	1	−1	−1	85
0	0	1	1	−1	66
0.25	0.75	0	−1	−1	77
0.5	0	0.5	−1	−1	90
0	0.5	0.5	−1	−1	84
0	0.5	0.5	−1	1	83
0.5	0.5	0	−1	−1	82
0.5	0.5	0	1	−1	78
0	0	1	−1	1	84
0	1	0	−1	−1	82
0.20	0.38	0.42	0	0	83
0.25	0	0.75	−1	1	72
0	1	0	−1	1	86
0	0	1	1	1	69
0.5	0	0.5	−1	1	88

the effect of microcrystalline cellulose and pregelatinized starch on % label claim dissolved in 30 minutes.

The statistical program has the capability of finding a formulation composition and process conditions that maximize the % LC dissolved in 30 minutes. Using this statistical facility, the maximum % LC dissolved in 30 minutes was predicted to be 88%. The formulation composition predicted was 50% pregelatinized starch and 50% microcrystalline cellulose. The number of tamps was set at the lowest setting. At 50% microcrystalline cellulose, the percent dissolved is reasonably stable and robust, ranging from 31 to 50% pregelatinized starch or 19 to 0% fast-flow lactose. Finally, the model equation indicates that starch has the largest positive impact on drug dissolution. To attain higher % LC values in 30 minutes, a second design could be undertaken using only pregelatinized starch (100%) as part of the mixture design.

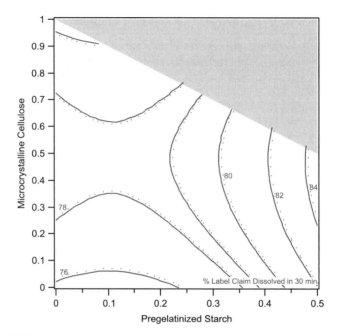

FIGURE 14.7 Contour plot of % label claim dissolved in 30 minutes.

14.5 CONTAINER CLOSURE SYSTEM

Most of the tablet container closure discussion in Section 13.4 also pertains to capsules. Capsules can easily be bent or crushed, so single-use containers need to provide more physical protection than is typically needed for tablets. Peal-back lidding is typically used instead of punch-through lidding. The use of desiccants and their amount needs careful consideration because it is possible to dry out the capsule shell and cause it to become brittle and break easily on dispensing or in a patient's hands. Additionally, the use of any packing component, such as a rayon coiler, which may expose capsules to aldehydes should be evaluated carefully. The equilibrium moisture of a capsule shell may also affect the capsule's dissolution and bioavailability performance.

14.6 RISK MANAGEMENT

Risk management assessment and control should be developed as discussed in Chapter 13.

14.7 ATTRIBUTE TESTS

A number of attribute tests can be used to characterize the running powder, particulates, liquid fill, and capsules.

14.7.1   Running Powder

Running powder particle size, powder flow, compressibility, and content uniformity testing was discussed Chapter 13. In addition, running powder bulk and tapped density are important factors affecting capsule fill (U.S.P. Chapter <616>; ICH, 2010c).

Manufacturers are mandated by law (21 CFR 211.110) to establish methods to monitor and validate the adequacy of mixing and to assure blend uniformity. The FDA (2003a,b) has issued a guidance to assist companies in meeting federal regulations. The guidance recommends a stratified sample approach that involves developing an a priori sampling plan that identifies the sampling interval, sample replicates, sample quantity, sampling methodology, and sample locations.

14.7.2   Liquid Fill

Capsule weight uniformity is dependent on the viscosity of the liquid fill and the ability to get a clean break in the liquid to prevent stringing before dispensing liquid into the next capsule. Depending on the type of liquid fill, a rheometer or viscometer (U.S.P. <911>) may be used to characterize the viscosity, thixotropic, and viscoelastic properties of the liquid.

The visual appearance of the liquid should be documented. If the liquid fill is a dispersed system such as an emulsion or suspension, it is important to determine dispersion homogeneity in-process. The pH of the liquid may have an impact on the physical and chemical stability of the liquid fill.

The congealing temperature (U.S.P. <651>) of a hot-melt fill is critical to maintaining capsule weight and content uniformity. Characterization of the crystallinity of the congealed melt may be important if a drug is designed to be a molecular dispersion or amorphous solid in the congealed matrix (U.S.P. <941>). Precipitation, cloudiness, and particle size of the fill are also tests that may be important to undertake, depending on the specific fill characteristics.

14.7.3   Capsules

The elasticity or hardness of soft-shell capsules can be measured by using an Instron or Barreiss tester. These instruments can be used to measure resistance to deformation. The resistance force can be used to quantify how hard or soft the shell is. This test can be used for in-process and final product assessment, similar to tablet hardness testing.

The water content of the shells, especially that of gelatin hard shells, is an important determinant for shell flexibility and brittleness. The water content of the shells can be determined by a number of methods (U.S.P. <921>).

A number of capsule attributes, such as appearance, color (U.S.P. <631>,<1061>), brittleness, content uniformity (U.S.P. <905>, ICH, 2010a), weight uniformity (U.S.P. <905>), disintegration time (U.S.P. <701>, ICH, 2009d), dissolution (U.S.P. <711>, <1092>, ICH, 2010b; FDA, 1997), and drug release (U.S.P. <724>) are typically evaluated. Capsule content uniformity, dissolution, and microbiological limits are usually considered critical capsule attributes.

Since gelatin and starch are natural products, their capsules can support microbiological growth and contamination. The microbiological testing requirements are provided by the U.S.P. (Chapters <61> and <62>) and ICH (2009a,b).

14.7.4 Excipients

The FDA requires that each excipient be identified by a specific test. Often a near-infrared or infrared spectrum is used to confirm the identity of an excipient. If the excipient is to be labeled as a pharmacopoeia excipient, such as N.F. or U.S.P., it needs to be tested to the pharmacopoeia's standards. A quality program may approve a specific vendor of pharmacopoeia excipients, which may allow a company to accept the U.S.P. testing provided by the vendor.

14.8 NEW DRUG APPLICATION STABILITY ASSESSMENT

Risk assessment should be used to identify critical product attributes that are evaluated as part of the new drug application stability assessment program. Initial testing at time zero typically includes description and appearance (color, shape, identifying marks such as logo, brittleness), two identity methods, assay of the active ingredients, identification and quantitation of degradation products, content uniformity,

TABLE 14.15   Typical Capsule Stability NDA Protocol

Product	Container/Closure and Supplier					API Process				
Product lot	Package lot					API lot				
Batch size	Batch size					Batch size				
Manufacturer	Packager					Manufacturer				
Manufacturer site	Packaging site					Site of manufacture				
Manufacture date	Package date					Manufacture date				
Stability start date										

Storage	Time (months)									
(°C/% RH)	1	3	6	9	12	18	24	36	48	60
25/60[a]		×	×	×	×	×	×	×	(×)[b]	(×)[b]
40/75[a]		×	×							
Light[c]	×									

[a] °C ± 2/% RH ± 5%

[b] optional testing.

[c] Testing per ICH guidelines.

microbial limit test (unless there is scientific justification not to include the test), and dissolution. Depending on other critical quality attributes that are important for product quality, other tests may be performed as well, such as moisture content. Ongoing stability would include assay, degradation products, microbial limits (may be done annually) test, and dissolution. Again, product-specific attributes may be added if deemed appropriate. Validated test methods are required for any test that is used for initial testing and stability testing. Appropriate acceptance criteria should also be included for each test. The reporting, identification, and quantitation of degradation impurities are discussed in detail in Section 13.3.4. A typical stability protocol that would support a regulatory submission to the European Union, Japan, and the United States is provided in Table 14.15.

REFERENCES

Berman J, Elinski DE, Gonzales CR, Hofer JD, Jimenez PJ, Planchard JA, Tlachac RJ, Vogel PF. Blend uniformity analysis: validation and in-process testing. Technical Report No. 25. PDA (Parenteral Drug Association). PDA J Pharm Sci Technol.. 1997;51:Suppl 3:i–iii, S1–99.

Botzolakis JE, Small LE, Augsburger LL. Effect of disintegrants on drug dissolution from capsules filled on a dosator-type automatic capsule-filling machine. Int. J. Pharm.. 1982; 341–349.

Bowtle WJ, Barker NJ, Wodhams J. A new approach to vancomycin formulation using filling technology for semisolid matrix capsules. Pharm. Technol. 1988;June: 86–97.

Brox W. Soft gelatin capsules and methods for their production. U.S. Patent 4744988. 1988.

Code of Federal Regulations. 21 CFR 211.110. Sampling and testing of in-process materials and drug products. http://www.accessdata.fda.gov/scripts/cdrh/cfdocs/cfcfr/CFRSearch.cfm?fr = 211.110. Revised Apr. 1, 2009.Title 21, Vol. 4. Accessed Dec. 2011.

Cole G. Evaluating development and production costs: tablets versus capsules. Pharm. Tech. Eur. 1998;5:17–26.

Dahl TC, Sue IT, Yum A. The influence of disintegrant level and capsule size on dissolution of hard gelatin capsules stored in high humidity conditions. Drug Dev. Ind. Pharm. 1991;17: 1001–1006.

Digenis GA, Gold TB, Shah VP. Crosslinking of gelatin capsules and its relevance to their in vitro-in vivo performance. J. Pharm. Sci. 1994;83:915–21.

El-Malah Y, Nazzal S. Hard gelatin and hypromellose (HPMC) capsules: estimation of rupture time by real-time dissolution spectroscopy. Drug Dev. Ind. Pharm. 2007;33:27–34.

European Commission Health and Consumer Protection Scientific Steering Committee. Updated opinion on the safety with regard to TSE risks of gelatin derived from ruminant bones or hides. 2003; 1–15.

FDA. 1997. Guidance for industry: Dissolution testing of immediate release solid dosage forms. U.S. Department of Health and Human Services, Food and Drug Administration Center for Drug Evaluation and Research. Aug. 1997. http://www.fda.gov/downloads/Drugs/GuidanceComplianceRegulatoryInformation/Guidances/ucm070237.pdf. Accessed Dec. 2011.

———. 2003a. Draft guidance for industry: Powder blends and finished dosage units—stratified in-process dosage unit sampling and assessment. U.S. Department of Health and Human Services, Food and Drug Administration Center for Drug Evaluation and Research. Oct. 2003. http://www.fda.gov./OHRMS/DOCKETS/98fr/03d-0493-gdl0001.pdf. Accessed Dec. 2011.

——— 2003b. Guidance for industry: Powder blends and finished dosage units—stratified in-process dosage unit sampling and assessment. U.S. Department of Health and Human Services, Food and Drug Administration Center for Drug Evaluation and Research. Revised attachments. Nov. 2003. http://www.fda.gov./downloads/Drug/GuidanceComplianceRegulatoryInformation/Guidances/ucm070314.pdf. Accessed Dec. 2011.

Gelatin Capsule Working Group. Collaborative development of two-tiered dissolution testing for gelatin capsules and gelatin-coated tablets using enzyme-containing media. Pharm. Forum 1996;24:7045–7050.

Goodson GW, Stagner WC. Clinical supply manufacture. In: Monkhouse DC, Rhodes CT, eds. *Drug Products for Clinical Trials: An International Guide to Formulation, Production, Quality Control*. New York: Marcel Dekker; 1998. pp. 127–147.

Groshens E, He X, Cade D. Hypromellose hard capsules: reproducibility of the in vitro dissolution performance in USP SGF media. Capsugel Library BAS 410, Poster, AAPS Annual Meeting, Los Angeles, CA, Nov. 2009.

Hill T, Lewicki P. Process analysis. In: *Statistics: Methods and Applications—a Comprehensive Reference for Science, Industry, and Data Mining*. Tulsa OK: StatSoft; 2006. pp. 419–442.

Hogan J, Shue P-I, Podczeck F, Newton JM. Investigations into the relationship between drug properties, filling and the release of drugs from hard gelatin capsules using multivariate statistical analysis. Pharm. Res. 1996;13:944–949.

ICH. 2009a. Q4B, Annex 4A: Microbiological examination of non-sterile products—microbial enumeration tests. General chapter. Apr. 2009. http://www.fda.gov/Drugs/GuidanceComplianceRegulatory Information/Guidances/ucm065005. Accessed Dec. 2011.

———. 2009b. Q4B, Annex 4B: Microbiological examination of non-sterile products—tests for specified micro-organism. General chapter. Apr. 2009. http://www.fda.gov/Drugs/Guidance ComplianceRegulatoryInformation/Guidances/ucm065005. Accessed Dec.2011.

——— 2009c. Q4B, Annex 4C: Microbiological examination of non-sterile products—acceptance criteria for pharmaceutical preparations and substances for pharmaceutical general use. Apr. 2009. http://www.fda.gov/Drugs/GuidanceComplianceRegulatoryInformation/Guidances/ucm065005. Accessed Dec. 2011.

——— 2009d Q4B, Annex 5: Disintegration test. General chapter. Dec. 2009. http://www.fda.gov /Drugs/GuidanceComplianceRegulatoryInformation/Guidances/ucm065005. Accessed Dec. 2011.

——— 2010a Q4B, Annex 6: Uniformity of dosage units. General chapter. Apr. 2010. http://www.fda. gov/Drugs/GuidanceComplianceRegulatoryInformation/Guidances/ucm065005. Accessed Dec. 2011.

——— 2010b Q4B, Annex 7: Dissolution test. General chapter. Apr 2010. http://www.fda.gov/ Drugs/GuidanceComplianceRegulatoryInformation/Guidances/ucm065005. Accessed Dec. 2011.

——— 2010c. Draft Annex 13: Bulk density and tapped density of powders. General chapter. July 2010. http://www.fda.gov/Drugs/GuidanceComplianceRegulatoryInformation/Guidances/ucm065005. Accessed Dec. 2011.

JMP®. Design of experiments software. SAS Institute, Cary, NC; 2008.

Joshi HN, Tejwani RW, Davidovich M, Sahasrabudhe VP, Jemal M, Bathala MS, Varia SA, Serajuddin ATM. Bioavailability enhancement of a poorly water-soluble drug by solid dispersion in polyethylene glycol–polysorbate 80 mixture. Int. J. Pharm. 2004;269:251–258.

Koparkar AD, Augsburger LL, Shangraw RF. Intrinsic dissolution rates of tablet filler–binders and their influence on the dissolution of drugs from tablet formulations. Pharm. Res. 1990;7: 80–86.

Kovarik JM, Mueller EA, van Bree JB, Tetzloff W, Kutz K. Reduced inter- and intra-individual variability in cyclosporine pharmacokinetics from a microemulsion formulation. J. Pharm. Sci. 1994;83:442–444.

Ku MS, Li W, Dulin W, Donahue F, Cade D, Benameur H. Performance qualification of a new hypromellose capsule: 1. Comparative evaluation of physical, mechanical, and processability quality attributes of Vcaps Plus®, Quali-V® and gelatin capsules. Int. J. Pharm. 2010; 386:30–41.

Li Z, Jacobus LK, Wuelfing WP, Golden M, Martin GP, Reed RA. Detection and quantification of low-molecular-weight aldehydes in pharmaceutical excipients by headspace gas chromatography. J. Chromatogr. A 2006;1104:1–10.

Li Z, Kozlowski BM, Chang EP. Analysis of aldehydes in excipients used in liquid/semi-solid formulations by gas chromatography—negative chemical ionization mass spectrometry. J. Chromatogr. A3 2007;1160:299–305.

Meinzer A, Buuggingen DE, Haeberlin B, Riehen CH. Oil-free pharmaceutical compositions for cyclosporine A. U.S. Patent 6723339. 2004.

Miller FH. Multi-phase, multi-compartment capsular delivery apparatus and methods for using same. U.S. Patent 7670612. 2010.

Nagata S, Tochio S. Compatibility between liquid and semi-solid excipients and HPMC capsules. Poster, C.R.S. Annual Meeting, Seoul, Korea, 2002.

Neoral® package insert. Novartis. http://www.pharma.us.novartis.com/product/pi/pdf/neoral.pdf. Accessed Dec. 2011.

Ofner CM III, Zhang Y-E, Jobeck VC, Bowman BJ. Crosslinking studies in gelatin capsules treated with formaldehyde and in capsules exposed to elevated temperature and humidity. J. Pharm. Sci. 2001;90:79–88.

Pouton CW. Formulation of poorly water-soluble drugs for oral administration: physiochemical and physiological issues and the lipid formulation classification system. Eur. J. Pharm. Sci. 2006;29:278–87.

Prescott JK, Garcia TP. A solid dosage and blend uniformity troubleshooting diagram. Pharm Technol. 2001; March: 68–87.

Rowe RC, Sheskey PJ, Owen SC, eds. *Handbook of Pharmaceutical Excipients*, 5th ed. London: Pharmaceutical Press; Washington, DC: American Pharmacists Association; 2006.

Sandimmune® Package insert. Novartis. http://www.pharma.us.novartis.com/product/pi/pdf/sandimmune.pdf. Accessed Dec. 2011.

Schamp K, Schreder S-A, Dressman J. Development of an in vitro/in vivo correlation for lipid formulations of EMD 50733, a poorly soluble, lipophilic drug substance. Eur. J. Pharm. Biopharm. 2006;62:227–234.

Schwier JR, Cooke GG, Hartauer KJ, Yu L. Rayon: a source of furfural—a reactive aldehyde capable of insolubilizing gelatin capsules. Pharm. Technol. 1993;May: 78–79.

Scott R, Cade D, He X, Cole ET. Pullulan capsules. U.S. Patent Application 20050249676. 2005.

Stegemann SS. Hard gelatin capsules today—and tomorrow. Capsugel Library. 2002; 3–23.

Tanner KE, Draper PR, Getz JJ, Burnett SW, Youngblood E. Film forming compositions comprising modified starches and iota-carrageenan and methods for manufacturing soft capsules using same. U.S. Patent 6340473. 2002.

Torpac®. http://www.torpac.com/Reference/capsule_ingredients.htm. Accessed Dec. 2011.

Tyrer JH, Eadie MJ, Sutherland JM, Hooper WD. Outbreak of anticonvulsant intoxication in an Australian city. Br. Med. J. 1970;4:271–273.

U.S. Pharmacopoeia U.S.P.–N.F. Online, First Supplement U.S.P. 33–N.F. Reissue. Accessed Dec. 2011.

Watts P, Smith A. Targit™ technology: coated starch capsules for site-specific drug delivery into the lower gastrointestinal tract. Expert Opin. Drug Deliv. 2005;2:159–167.

Wittwer F, Tomka I, Bodenmann H-U, Raible T, Gillow LS. Method for forming pharmaceutical capsules from starch compositions. U.S. Patent 4738724. 1988.

Yamamoto T, Nagata S, Matsuura S. Process for producing hard capsule. U.S. Patent 6413463. 2002.

GLOSSARY

API	Active pharmaceutical ingredient.
BCS	Biopharmaceutics classification system.
BSE	Bovine spongiform encephalopathy.
E.P.	*European Pharmacopoeia*.
FDA	U.S. Food and Drug Administration.
GRAS	generally recognized as safe.
HPMC	Hydroxypropyl methylcellulose.
ICH	International Conference on Harmonization.
J.P.	*Japanese Pharmacopoeia*.
LC	Label claim.

LFSC	Lipid formulation classification system.
N.F.	*National Formulary*.
PEG	poly(ethylene glycol).
RH	Relative humidity.
U.S.P.	*United States Pharmacopoeia*.

APPENDIXES

APPENDIX 14.1   Aldehyde Levels in Commonly Used Capsule Excipients

Excipient	Aldehydes[a] Detected (µg/g)
Butylated hydroxyanisole	ND[b]
Castor oil	ND
Corn oil	ND
Croscarmellose sodium	0.5, 0.3
Crospovidone	2.4, 4.5
Dibasic calcium phosphate	ND
Ethylcellulose	2.5, >10[c]
Hydroxypropyl cellulose	0.6, 0.5
Hypromellose	1.6, 0.1
Lactose	0.1
Magnesium stearate	0.1
Mannitol	0.2
Medium-chain triglycerides	ND
Microcrystalline cellulose	0.4, 0.1
Mono- and diglycerides	ND
Partially pregelatinized cornstarch	2.5, 0.1
Poloxamer 407	0.1, 2.0
Poly(ethylene glycol) 400	>10[c], 8.2
Poly(ethylene glycol) 600	>10[c], 4.1
Poly(ethylene glycol) 1450	0.4, 0.7
Poly(ethylene glycol) 4000	0.4, 2.1
Polyoxyl 35 castor oil	1.1
Polysorbate 80	1.6, 0.6
Povidone	4.7, 4.2
Propyl gallate	ND
Sodium chloride	ND
Sodium starch glycolate	0.1, 2.7
Sodium stearyl fumarate	0.1
Sucrose	0.1
Titanium dioxide	ND
Triethyl citrate	0.2, 0.6

Source: After Li et al. (2006), with permission from Elsevier.

[a] Formaldehyde and acetaldehyde.

[b] ND not detected.

[c] Peak signal exceeded the linearity range.

APPENDIX 14.2   Aldehyde Levels in Commonly Used Liquid-Fill Capsule Excipients

Excipient	Aldehyde[a] Detected (μg/g)
α-Tocopherol poly(ethylene glycol) succinate	0.2, ND[b]
Corn oil	ND
Diethylene glycol monoethyl ether	1.0, 2.8
Glycerol monolinoleate	0.2, 0.6
Glycerol monooleate	ND, 0.9
Glyceryl monostearate	0.4, 0.1
Hydroxypropyl-β-cyclodextrin	ND
Lauroyl polyoxyglycerides	28.2, 4.7
Medium-chain mono- and diglycerides	0.2
Medium-chain triglycerides, brand A[c]	ND
Medium-chain triglycerides, Brand B[c]	ND
Oleoyl polyoxyglycerides	ND, 2.0
Poloxamer 407	0.7, 0.6
Poly(ethylene glycol) 200	107.0, 12.5
Poly(ethylene glycol) 400	102.5, 2.7
Poly(ethylene glycol) 600	65.2, 12.2
Poly(ethylene glycol) 4000	0.4, 2.3
Poly(ethylene glycol) 4600	ND
Polyoxyethylene 400 caprylic/capric glycerides, brand A	3.9, 0.8
Polyoxyethylene 400 caprylic/capric glycerides, brand B	22.9, 30.7
Polyoxyl 35 castor oil	3.5, ND
Polyoxypropylene stearyl ether	11.2, 5.4
Polysorbate 80	5.2
Propylene glycol	0.3, 0.5
Propylene glycol dicaprylate/dicaprate	ND
Propylene glycol monocaprylate, brand A	ND, 0.1
Propylene glycol monocaprylate, brand B	0.8, 1.8
Propylene glycol monolaurate	0.2, 0.6
Sorbitan monooleate	0.3
Sulfobutyl ether β-cyclodextrin	ND

Source: After Li et al. (2007), with permission from Elsevier.

[a] Formaldehyde and acetaldehyde.

[b] ND not detected.

[c] Brands A and B refer to different companies.

APPENDIX 14.3   Capsule Liquid-Fill Excipients[a]

Excipient	Trade Names
Colloidal silicon dioxide, E.P./U.S.P–N.F.	Aerosil, Cab-O-Sil, M-5, Waker HDK
Diethylene glycol monethyl ether, E.P./U.S.P.–N.F.	Transcutol
Ethanol, 8–10% w/v E.P./J.P./U.S.P.–N.F.	Generic
Glyceryl monolinoleate, E.P./U.S.P.–N.F.	Maisine 35-1
Glyceryl monooleate, E.P./U.S.P.–N.F.	Aldo MO, Atlas G-95
Glyceryl monosterate, E.P./U.S.P.–N.F.	Capmul GMS-50, Cutina GMS, Kessco GMS
Glyceryl palmitostearate, EP/USPNF	Precirol ATP 5
Hydrogenated coco-glycerides	Softisan 100, Softisan 142
Medium-chain triglycerides, E.P./U.S.P.–N.F.	Captex 355, Miglyol 810, Miglyol 812, Neobee
Triacetin, E.P./U.S.P.–N.F.	Captex 500
Fatty acids	
Lauric acid	Hydrofol Acid, Nino AA62 Extra
Oleic acid, E.P./U.S.P.–N.F.	Crossential 094, Metaupon
Stearic acid, E.P./J.P./U.S.P.–N.F.	Crodaci, Pearl Steric
Fatty acid alcohols	Avol, Cahalot, Crodacol
Cetyl alcohol, E.P./J.P./U.S.P.–N.F.	Cachalot, Crodacol S95
Stearyl alcohol, E.P./J.P./U.S.P.–N.F.	
Lecithin, U.S.P.–N.F.	Egg lecithin, soybean lecithin
Oils (natural)	
Corn oil, E.P./J.P./U.S.P.–NF	Maize oil, Majsao CT
Cottonseed oil, U.S.P.–N.F.	Generic
Olive oil, E.P./J.P./U.S.P.–N.F.	Generic
Soybean oil, E.P./J.P./U.S.P.–N.F.	Calchem IVO-114, Lipex 107, Shogun CT
Sunflower oil, E.P./U.S.P.–N.F.	Generic
Oils (hydrogentated)	
Cottonseed oil, J.P./U.S.P.–N.F.	Akosol 407, Sterotex, Lubritab
Palm Oil, J.P./U.S.P.–N.F.	Softisan 154
Soybean oil, J.P./U.S.P.–N.F.	Lipovol HS-K, Sterotex HM
Poly(ethylene glycol), E.P./J.P./U.S.P.–N.F.	Generic
Molecular mass 200 to 600 Da for liquid fills at up to 30% w/v	
Molecular mass 1540 Da and greater for hot melt	
Polyoxyethylene castor oil derivatives, E.P./J.P./U.S.P.–N.F.	Cremophor ELP, Etocas 35, PEG-35, Cremophor RH 40

[a] A comprehensive list of products and suppliers is provided in Rowe et al. (2006).

CHAPTER 15

DISPERSED SYSTEM PRODUCT DESIGN

15.1 INTRODUCTION

For purposes of this chapter, a dispersed system has one or more components that are dispersed in a dispersion medium. Dispersed systems may be classified based on the particle size of the component dispersed. Pharmaceutically relevant dispersions have been classified as *coarse dispersions* (1 μm to several hundred micrometers), *colloid dispersions* (1 nm to less than 1 μm), *nanodispersions* or *ultrafine particle dispersions* (one or more dimensions are 100 nm or less), and *molecular dispersions* or *true solutions*, where the dispersed particle is a molecule of material. In this chapter we discuss only coarse, colloidal, and nanodispersions. In 2003 the 21st Century Nanotechnology Research and Development Act and the National Nanotechnology Initiative were established. As part of the initiative, *nanoparticles* were defined as materials that had one or more dimensions less than 100 nm. There are situations where large molecular compounds with molecular masses of approximately 30,000 Da or greater form true solutions, yet the size of the solute molecules places them in the nanoparticle and colloid size range. Examples of such molecules are proteins (e.g., albumin), natural products (e.g., gelatin), and polymers (e.g., povidone). With the exception of molecular dispersions, dispersed systems consist of at least two phases, known as the *dispersed* or *discontinuous* or *internal phase* and the *supporting* or *continuous* or *external phase*. The continuous phase is also known as the *dispersion medium* or *vehicle*.

Dispersed systems are also classified by the physical state of the dispersed and continuous phase, which is shown in Table 15.1.

As shown in the table, pharmaceutical dispersed systems cover a wide array of dosage forms, including oral, topical, and parenteral suspensions; oral, topical, and parenteral emulsions; pulmonary and topical aerosols; topical semisolids; and solid dispersion tablet, capsule, and suppository matrices. Here *topical* is used in a very broad sense to mean dosage forms that are applied externally to the body (including organs such as the skin and eye and mucous membranes such as the nose, rectum, and vagina) to achieve a local therapeutic effect. There are also a number of enabling dispersion technologies, such as micelles, microemulsions, liposomes, niosomes, drug delivery vectors, and dendrimers.

Integrated Pharmaceutics: Applied Preformulation, Product Design, and Regulatory Science, First Edition. Antoine Al-Achi, Mali Ram Gupta, William Craig Stagner. © 2013 John Wiley & Sons, Inc. Published 2013 by John Wiley & Sons, Inc.

TABLE 15.1 Dispersed System Classification Based on the State of Matter

Dispersed Phase	Dispersion Medium	Type of System	Example
Liquid	Gas	Aerosol	Nebulizer solution
Solid	Gas	Aerosol	Propellant inhaler
Gas	Liquid	Carbonation	Effervescent tablet
Liquid	Liquid	Coarse emulsion (macroemulsion)	Mineral oil emulsion, United States Pharmacopeia (U.S.P.)
Liquid	Liquid	Microemulsion	Capsule, liquid-fill
Solid	Liquid	Coarse suspension	Hydroxyzine pamoate oral suspension, U.S.P.
Solid	Liquid	Sols and gels	Aluminum hydroxide gel, U.S.P.
Liquid	Solid	Absorption base	Hydrophilic petrolatum, U.S.P.
Solid	Solid	Solid matrix	Hot-melt dispersion suppositories

A cursory overview of the chapter is presented to help orient the reader. The formulation design section begins with a general review of important physical pharmacy considerations that affect the design of physically stable coarse liquid dispersions. Liquid or flowable coarsely dispersed systems such as coarse suspensions and macroemulsions rely on a number of common physical pharmacy formulation design principles. This discussion is followed by general preformulation and excipient compatibility study sections. The organization of the formulation development section is based on the states of matter classification presented in Table 15.1. The dosage forms presented in this chapter are suspensions and other flowable solid–liquid medium dispersions, macroemulsions and other flowable liquid–liquid medium dispersions, semisolid dispersions, and solid dispersions. The sections are subdivided further according to the size of the material dispersed. The process design section consolidates process discussions where possible. Dosage-form-specific processes, attribute testing, and stability testing are discussed as needed.

15.2 FORMULATION DESIGN

Suspensions and emulsions are liquid or flowable dispersed system dosage forms that have many therapeutic applications. Some dispersed systems are semisolids, used primarily for topical therapy. Solid dispersions are used in tablets, capsules, and suppositories. Hot-melt dispersion matrices are discussed in chapter 13 and 14. Aerosol dispersed systems are reviewed in Chapter 16. Chapter 17 presents dispersed systems used in parenteral drug delivery. Enabling dispersion technologies are addressed in Chapters 19 and 20. The advantages and disadvantages of liquid coarse dispersed systems are listed in Table 15.2.

Suspensions and emulsions (*macroemulsions*) make up the largest percentage of dispersed system dosage forms. *Macroemulsion* is used synonymously with

TABLE 15.2   **Advantages and Disadvantages of Oral Liquid Coarse Dispersed Systems**

Advantages	Disadvantages
Ease of swallowing	Thermodynamically unstable system
Liquid dispersions are easier to swallow than tablets or capsules	Particle growth, polymorphic conversion, and sedimentation can lead to nonuniform or less bioavailable doses
Target populations are the young and the elderly	Less chemical stability than with tablets and capsules
Enhanced chemical stability for drugs that degrade in the presence of water	Less patient convenience than with tablets and capsules
Reconstituted suspensions increase product shelf life prior to preparation	Susceptible to particle size growth and resulting increase in sedimentation rate
Insoluble suspended drug is easier to taste-mask than are highly concentrated drug solutions	
Better bioavailability than with tablets and capsules	

emulsion to distinguish coarse emulsions from *microemulsions*. Microemulsions have particle sizes in the nanometer range, often 30 nm or less, and are thermodynamically stable. *Nanoemulsions*, on the other hand, are submicrometer (usually, 50 nm or larger) turbid emulsions that are not thermodynamically stable. Like nanoemulsions, suspensions and macroemulsions, are thermodynamically unstable dosage forms that can be stabilized to provide pharmaceutically acceptable drug delivery systems. The proper use of functional excipients plays an important role in stabilizing the dispersion to achieve a shelf life of several years. Coarse, colloidal, and nanodispersions are designed to have optimum particle sizes and particle size ranges that impart critical product attributes. Solutes and small colloidal particles remain dispersed indefinitely through Brownian motion. As the dispersed phase increases in size beyond 1 μm or so, Brownian motion is no longer sufficient to keep the particles suspended, and sedimentation or creaming will result due to the density differences in the dispersed and internal phases. *Sedimentation* results in settling of the dispersed phase, and *creaming* results in the dispersed phase floating to the top of the liquid. As sedimentation or creaming occurs, the dispersed particles move closer together, which may result in irreversible coagulation for solid particles and coalescence for liquid dispersed material. Sedimentation and creaming rates are predicted by Stokes' law:

$$v = \frac{2r^2(\rho_d - \rho_m)g}{9\eta_0} \tag{15.1}$$

where v is the terminal velocity of sedimentation or creaming (in cm/s); r the radius of the particle (cm); ρ_d and ρ_m the dispersed phase and dispersion medium densities, respectively; g the acceleration due to gravity (981 cm/s^2), and η_0 the Newtonian viscosity [poise (g · cm^{-1} · s^{-1}). The viscosity standard international unit is pascal-seconds (kg · s^{-1} · m^{-1}), where the acceleration due to gravity is 9.81 m/s^2.

The dispersed particle's solubility in the vehicle also significantly affects the dispersion's physical stability. The particle or drug solubility in the dispersion medium should be less than 100 parts per million (0.01% w/w or ~0.1 mg/mL). It is even more desirable to control the solubility to less than 10 parts per million. The equilibrium concentration of solute in the dispersion medium is equal to the drug's equilibrium solubility at a specified temperature and pressure. The equilibrium solubility is a dynamic process where particles dissolve and molecules in solution crystallize or precipitate out of the solution phase. The recrystallizing material will generally deposit on the larger particles because this lowers the free energy of the dispersion and leads to particle growth. The smaller the particle, the greater its surface area per volume (specific surface area) and thus the greater its surface free energy. The surface free energy of the particles is

$$\Delta F = \gamma \Delta A \tag{15.2}$$

where ΔF is the change in the surface free energy, γ the interfacial tension between the dispersed material and the dispersion medium, and ΔA the change in the surface area of the dispersed phase. A smaller ΔF is the more thermodynamically stable dispersion, and a negative ΔF results in a spontaneous change. The surface free energy can be reduced by decreasing the interfacial tension of the dispersion. The surface free energy can also be decreased by decreasing the particle surface area. A decrease in particle surface area results from an increase in the particle size of the dispersion. Equation (15.2) predicts that particle growth (decrease in surface area) is a spontaneous process (a negative ΔF). For example, a nanodispersion is prepared that has an average particle size of 10 nm. The specific surface area of this dispersion is approximately 1×10^6 cm^2/cm^3. If the particles grow to 10 μm, the specific surface area is approximately 1×10^3 cm^2/cm^3 and the change in surface area (final state−initial state) is -9.99×10^5 cm^2/cm^3, indicating that the increase in particle size is a spontaneous process. Particle growth is accelerated by temperature fluctuations and polydispersed particle size range.

Particle growth also occurs by a phenomenon called *Ostwald ripening*, a thermodynamically driven process that lowers the free energy of the dispersed system. In this case, small colloidal particles preferentially dissolve and deposit onto larger particles. These very small particles actually have a higher level of solubility than the larger particles because of their greater specific surface. The greater specific surface area creates a greater particle free surface energy, which increases the escaping tendency of individual molecules from the surface of the particle. The relationship between particle size and solubility is given by the *Kelvin and Ostwald−Freundlich equation:*

$$\ln \frac{S}{S_\infty} = \frac{2\gamma M}{r \rho RT} \tag{15.3}$$

where S is the solubility of colloidal particles of radius r, S_∞ the solubility of a particle with an infinitely large radius, γ the solid and dispersion medium interfacial energy, M the molecular mass of the colloid particle, ρ the particle density, R the gas constant, and T the temperature in kelvin. The effect of particle size, r, on solubility becomes a factor at very small colloidal and nanoparticle size ranges. The interfacial tension cannot be measured directly. The interfacial tension has been determined indirectly by applying equation (15.4) and measuring the solubility of particles at difference particle sizes. M/ρ, the molar volume of the particle, has a direct effect on a particle's solubility.

Drug solubility is also affected by a drug's physical state. High-energy thermodynamically unstable amorphous or polymorphic materials will have higher solubility than will the thermodynamically stable form of the drug. Crystal imperfections resulting from the crystallization process or from milling lead to local high-energy sites that can result in higher solubility than expected. These physical states need to be considered, evaluated, and controlled.

As mentioned, the size of the dispersed particles is critical to their product performance, such as suspendability, reaching target sites in the lungs, or dissolving quickly into solution. Aggregation of particles can have a detrimental effect on product function. Van der Waals and ionic forces can affect aggregation and the dispersion particle size. The *DLVO theory* (named after Derjaguin and Landau of Russia and Verwey and Overbeek of the Netherlands) describes the interplay between electrostatic repulsion and van der Waals attractive forces as a function of particle distance (Figure 15.1). The DLVO theory predicts that at very close particle distances, relatively strong short-range van der Waals attractive forces cause irreversible *coagulation* of the particles. At midrange distances, the particles can experience electrostatic repulsion. If the electrostatic repulsive potential energy

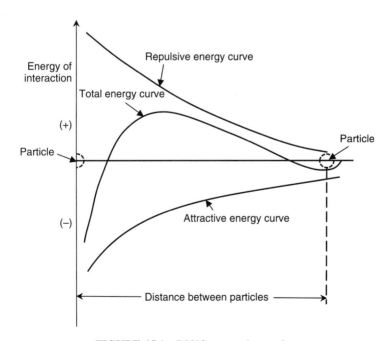

FIGURE 15.1 DLVO energy interaction.

is greater than the kinetic energy of the approaching particles, the particles are repelled. At even greater particle distances, a second, longer-range weak van der Waals attractive forces cause a loose aggregation of the particles, which is also called *flocculation*.

The electrical charge carried by dispersed particles affects repulsion and attraction forces. The electrical charge on the particles creates an *electric double layer*, comprised of the Stern layer and the Gouy–Chapman layer. The *Stern layer* is made up of counterions that are highly bound to the surface of the charged particle. At distances beyond the Stern layer, a more diffuse layer, the *Gouy–Chapman layer*, of counterions is present. The Stern layer and the Gouy–Chapman layer make up what is referred to as the *electric double layer*. As the concentration of the counterions approaches the concentration of the bulk, the electric potential goes to zero. Neither the particle surface potential nor the potential of the Stern and Gouy–Chapman layers can be measured directly. The *zeta* (ζ) or *electrokinetic potential* can be measured. The ζ-potential measures the potential between the bulk dispersion medium and a bound liquid layer that is associated with the dispersed particle. Particles in the double layer bind the dispersion medium, usually water, and form a bound liquid layer or shell of bound water. When the particle moves, the bound fluid and the charged particles inside the bound layer move as one. This movement creates a shear plane with the bulk dispersion medium. The electric potential at this shear plane is the ζ-potential, a relative measure of the potential at the surface of the particle. The ζ-potential is related to the particle's electrophoretic mobility:

$$\zeta = \frac{4 \sqcap \eta}{D} \mu \tag{15.4}$$

where ζ is the zeta potential, η the viscosity of the double layer, D the dielectric of the double layer, and μ the electrophoretic mobility, which is the particle velocity caused by a potential gradient of 1 V/cm. Equation (15.4) shows that decreasing the dielectric of the dispersion medium by adding lower-dielectric water-miscible solvents such as alcohol, glycerin, and propylene glycol increases the ζ-potential. Hydrophobic colloidal particles with a ζ-potential of less than ± 30 mV tend to coagulate irreversibly as a result of the short-range van der Waals interactions or form large aggregates called *flocs*, due to weak long-range attractive forces. ζ-Potentials of hydrophobic colloid particles greater than ± 60 mV tend to repel each other and form *peptized* or *deflocculated dispersions*. In the case of coarse dispersions (particles 1 μm or larger), the primary attractive forces are due to weak long-range van der Waals forces that result in loose aggregates or flocs. Flocs are generally formed at ζ-potentials ranging from plus to minus 30 mV.

Sedimentation volume, F, and *degree of flocculation*, β, are measures of a coarse suspension's state of deflocculation and flocculation. The sedimentation volume is the ratio of the equilibrium volume of sediment to the total volume of the suspension:

$$F = \frac{V_{\text{sed}}}{V_{\text{sup}}} \tag{15.5}$$

where V_{sed} is the sedimentation volume and V_{sup} is the suspension volume. The sedimentation volume can range from 0 to 1. A sedimentation volume of 1 is considered an *ideal* suspension since the dispersion medium is completely occupied by the dispersed phase, so no sedimentation should occur. The *degree of flocculation* is the ratio of sedimentation volume of a flocculated suspension to the sedimentation volume of a completely deflocculated suspension:

$$\beta = \frac{F_f}{F_d} \tag{15.6}$$

where F_f is the sedimentation volume of a flocculated suspension and F_d is the sedimentation volume of a completely deflocculated suspension. As the degree of flocculation approaches unity, the suspension approaches a deflocculated state.

15.2.1 Preformulation

The purpose of a preformulation study is to evaluate a drug's physiochemical properties and identify specific issues that place the development program at risk. The determination of the dispersed material's particle size, shape and particle size distribution, surface area, and true density are of particular importance in accessing the particle's tendency to grow, sediment, or cream in the dispersion medium. The implications of the active pharmaceutical ingredient's physical properties on dispersed system dosage forms are presented in Table 15.3.

The chemical properties that are investigated during preformulation are cited in Table 15.4.

15.2.2 Excipient Compatibility Studies

Once the preformulation information is collected, a physiochemical profile can be developed for a drug. This profile of the drug's physical and chemical nature serves as a knowledge base upon which dosage form, formulation, and process decisions are based. Once the dosage form is chosen, typical excipients associated with that dosage form will be evaluated for physical and chemical compatibility. Different excipient compatibility methodologies are presented in Chapter 13. The excipient studies will also vary depending on the type of dispersion system. Tables 15.5 and 15.6 give the various classes of excipient that are employed in the formulation of suspensions and emulsions, respectively. The dispersed system dosage forms covered in this chapter can be included in two broad categories: suspensions and emulsion. The sizes and types of particles will differ for the individual dosage forms, but many of the excipients will remain common for coarse suspensions; lyophilic colloids of large molecules, gels and particulates such as nanoparticles, vectors, and dendrimers; lyophobic colloids; and association colloids.

Sucrose is the most common density-adjusting agent. The addition of viscosity-enhancing agents also affects the liquid density. Viscosity and structure enhancers are given in Appendix 15.1. Wetting, sweetening, flocculating, buffering,

TABLE 15.3 Solid-State Physical Properties and Dispersed System Implications

Physical Properties	Implications
Organoleptic properties: appearance and odor	If there is a distinctive color such as yellow or brown, a colorant may be required to provide a uniform color to the product.
	A disagreeable odor such as a "sulfide" or "rotten egg" smell may require volatile additives such as vanillin or mint to compete with the offensive odor.
	These potential additives should be used in the excipient compatibility screening study.
Morphology and micromeritics: crystal habit, birefringence, particle size, particle shape, particle size distribution, bulk density, tapped density, true density, state of agglomeration, and surface area	A cubic crystal will sediment more ideally than will a long needle.
	Birefrigence in the absence of anisotropic crystal is indicative of crystalline material compared to amorphous material. Amorphous material will have a higher thermodynamic energy (higher solubility, higher surface energy), and these materials will readily undergo undesirable particle growth and exhibit increased sedimentation.
	Large particles will need to be milled to the desirable particle size range for good suspendability.
	Differences in the true density of the drug and the suspending media result in sedimentation or creaming. Matching the density of the drug and dispersion media prevents separation from occurring.
	Highly agglomerated particles may require special dispersion steps to achieve separate particles.
	Bulk containers are sized by volume, but powders are handled by weight. Bulk and tapped densities are important in determining what size container and processing equipment are needed.
	High-surface-area materials increase the level of thermodynamic instability for dispersed systems.
Thermal properties: melting point, enthalpy of fusion, and other thermal transitions	Low-melting compounds can soften or even melt during high-energy dry powder milling.
	A high enthalpy of fusion suggests that the drug may have low solubility, which is an advantage for designing dispersed systems.
	Differential scanning calorimetry can pick up other enthalpic transitions, such as polymorphism, dehydration, and decomposition. Polymorphs have different solubility, stability, cohesiveness, density, dissolution rates, and bioavailability. High-energy polymorphs have higher solubility and may lead to undesirable particle growth.

TABLE 15.3 (*Continued*)

Physical Properties	Implications
	Thermal gravimetric analysis can quantitate hydration states and solvent weight loss. Loss of water and solvent can occur during the milling and micronization processes.
Crystallinity and polymorphism	Amorphous material is less physically and chemically stable than is its crystalline counterpart. Amorphous material is less stable, more soluble, faster dissolving, and more bioavailable than is the crystalline state.
Polymorphic tendency	Some drugs have a larger propensity than others to form polymorphs. Since polymorphs can have a significant effect on manufacturability and bioavailability, it is important to be aware of a drug's tendency to form polymorphs. Can affect solubility and particle growth.
Hygroscopicity and moisture adsorption/desorption isotherm[a]	The degree of hygroscopicity will dictate proper storage and handling conditions.
Contact angle	The contact angle indicates how wettable a drug substance is. In preparing suspensions the material must first be wetted to form a uniform, well-dispersed suspension.
Static charge and explosivity	The dust clouds that result from handling powders can be lethally explosive. Knowing the explosive potential (minimum ignition energy) of powders is crucial to maintaining a safe work environment.
Flow properties	Flow properties are important for vacuum transfer of bulk materials.

[a]Callahan et al. (1982, App.13.1)

and preserving agents are provided in Appendixes 13.9, 15.2, 15.3, 17.3, and 17.4, respectively. Humectants are generally nonvolatile materials that interact with water to decrease its vaporization. Common humectants are glycerin and propylene glycol.

For the purposes of classifying excipients, coarse emulsions and *association colloids* share a number of excipients. Association colloids include drug delivery systems such as micelles, microemulsions, liposomes, and niosomes, which are discussed later in this chapter. The classification of excipients for these types of dispersed drug delivery systems are presented in Table 15.6.

Many of these excipients are the same as listed in Table 15.5 for suspensions. Emulsifying agents and their hydrophile–lipophile balance values are listed in Appendix 15.4. The required HLB values for many oils are given in Appendix 15.5. Antioxidants are listed in Appendix 17.5.

TABLE 15.4 Solution- and Solid-State Chemical Properties and Dispersed System Implications

Chemical Properties	Implications
Purity and impurities: impurities may include degradation of products, materials related to the synthesis, organic volatiles, residual solvents, and heavy metals	This is a quality attribute. These properties can affect safety and efficacy. These properties can affect manufacturability.
Solubility: in water and oils	For stable solid–water dispersions, water solubility should be lower than 0.1 mg/mL. For oil-in-water emulsions, where the drug is expected to be dissolved in the oil, oil solubility sufficient to deliver the required dose is essential.
Permeability: using cellular absorption models such as Caco-2 cells	Application of the biopharmaceutical classification system to bioavailability.
Partition–distribution coefficient	Affects how much drug will be in the oil and water phases. It is important for the drug to distribute preferentially to the oil phase for drugs that are unstable or have poor taste in water.
Acid–base dissociation constant	Drug ionization state affects a drug's solubility, dissolution rate, stability, and bioavailability as a function of pH.
Solid- and solution-state stability as a function of room temperature and stressed temperature, pH, oxygen, and light	Stability knowledge affects formulation, process, and container closure decisions.

15.2.3 Formulation Development and Dosage-Form-Specific Processes

15.2.3.1 Coarse Suspensions or "Suspensions" The dispersed particles in suspensions are solids. As is the case with all drug delivery systems, dose accuracy, the dose administered, is a critical product attribute. Tablet and capsule doses are "locked in" once the tablet is compressed or the capsule encapsulated. This is not the case with a suspension. For suspensions, each patient dose is at risk because it is not locked in at the time of manufacture. Why is this? First and foremost, a suspension is thermodynamically unstable. Suspended particles will settle as a result of gravity. They will tend to grow, which in turn makes them more susceptible to sedimentation. Temperature fluctuations not only affect a drug's solubility and the rate of particle growth, they also affect the viscosity of the suspending agents and the drug's chemical stability. Suspension dosage forms are one of the most difficult product design challenges faced by the pharmaceutical scientist. Suspension design requirements are quite daunting. An accurate dose must

TABLE 15.5 **Excipient Classes and Functions for Liquid Suspensions and Solid Colloid Dispersions**

Class	Function
Density modifiers	Increase the density of the dispersion medium to match the density of the dispersed particles as closely as possible.
Viscosity modifiers and structure enhancers	Increase in viscosity and structure decreases the sedimentation.
Wetting agents and surfactants	Decrease the interfacial energy between a liquid and a solid allows the liquid to spread over the solid particles and disperse them. Wetting agents also decrease the free energy of the dispersed system and decrease the tendency for particle growth.
Flocculating agents	Control the level of loose aggregate or floc formation.
Humectants	Nonvolatile hydrophilic liquids that decrease the rate of product moisture loss. For oral suspensions, humectants also form a thin lubricant film of nonvolatile liquid on the bottle neck and cap, which decreases "cap lock" when an unopened bottle is stored for a prolonged period of time.
Preservatives	Decrease bioburden and prevent growth of microbiological organisms.
Buffers	Maintain optimum pH for enhanced stability. In the case of parenterals and ophthalmic dispersion, buffers can maintain physiological pH compatibility. Buffers can be used to adjust the pH to decrease the solubility of weak acid and base drugs and lower the chances of crystal growth and Ostwald ripening.
Sweeteners	Improve the taste and palatability of the oral vehicle.
Flavors	Improve the taste and palatability of the oral vehicle.
Colorants	Mask color of dispersion and help provide product distinction.

be poured from the container closure system into a measuring–delivery device (say, a calibrated measuring spoon) between the time of resuspending the suspension and the completion of dose measurement. The suspension should also be smooth, nongritty, easily redispersed, and readily pourable. The suspension must also be designed to deliver an accurate dose for every dose in the bottle. For example, if a bottle of suspension contains a month's supply of medication, the last dose should contain the labeled amount of drug. It is not uncommon that there be bias in the dose delivered, leading to each successive or series of doses a little higher or lower than the preceding dose. This can lead to significant over- or underdosing of the last doses delivered. Therefore, it is important to design a suspension that delivers an accurate dose until the product is depleted. This means that "patient

TABLE 15.6 Excipient Classes and Functions for Macroemulsions and Association Colloids

Class	Function
Emulsifying agents (surface-active agents that act to emulsify the emulsion)	Stabilize emulsion droplets. It is desirable that emulsifiers be chemically stable; exhibit surface activity and adsorb rapidly to the dispersed particles to form a condensed film; impart an electrical charge to the dispersed particles; increase the viscosity of the system; are nonirritant, nontoxic, and compatible with the drug; are reasonably odorless and tasteless; are effective at low concentrations; and are low in cost.
Flocculating agents	Control the level of loose aggregate or floc formation.
Viscosity agents	Increase the viscosity and decrease creaming.
Preservatives	Prevent growth of microbiological organisms.
Antioxidants	Prevent oxidation of oils and the drug.
Buffers	Adjust pH, control flocculation, and improve drug stability.
Sweeteners	Improve the taste and palatability of the oral vehicle.
Flavors	Improve the taste and palatability of the oral vehicle.

simulation use tests" need to be devised that evaluate the product uniformity and dose accuracy for the first, middle, and end doses.

To achieve a stable suspension that has a two- to three-year shelf life, practical formulation design incorporates the concepts of equations (15.1) through (15.4). Drug solubility, particle size, and particle size distribution are critical drug attributes that affect dosing accuracy and the aesthetic and physical stability of the suspension. The drug solubility in the dispersion medium should be 0.1 mg/mL or less. The practical upper particle size limit for an oral suspension is 50 to 75 μm. Particles larger than 75 μm tend to give a gritty mouth feel. They also settle rapidly, and once settled at the bottom of the container, the particles tend to sinter and form hard-to-disperse cakes. A narrow particle size range of 5 to 10 μm is highly desirable. This narrow range can be achieved using modern classifying mills.

Application of Stokes' law, $v = [2r^2(\rho_d - \rho_m)g]/9\eta$, gives the formulation scientist several levers that can be used to optimize a suspension's resistance to sedimentation. First, it is important to note that Stokes' law predicts that the sedimentation velocity can be brought to *zero* if the dispersed particle density, ρ_d, is the same as the density of the dispersion medium. ρ_m. That is, when $\rho_d - \rho_m = 0$, the right-hand side of the equation goes to zero and the sedimentation velocity is zero. In addition to viscosity modifiers, which can increase the density of the dispersion medium, sucrose, glycerol, and propylene glycol can also be used to modify the density of the dispersion medium. The density of most drugs ranges from 1.1 to 1.7 g/cm^3 (1100 to 1700 kg/m^3). A 25% w/w sucrose solution has a density of 1.1 g/cm^3 at 25°C. A 65% w/w sucrose solution has a density of 1.3 g/cm^3 at 25°C, so it is possible to match the densities of the dispersed phase and the dispersion medium. Matching densities of the dispersed phase and the dispersion medium is the master design lever; all other issues that affect dose uniformity, such as changes in solubility due to temperature fluctuations, particle growth due to Ostwald ripening or temperature fluctuations, caking due to sedimentation, and decisions to create a deflocculated or flocculated suspension, are minimized or eliminated. If particle and medium densities cannot be matched, attempts should be made to minimize the differences. Stokes' law also predicts that particle settling terminal velocity is directly related to the particle radius (r) squared. The settling rate for a 5-μm-diameter particle is approximately 100-fold slower than a 50-μm-diameter particle. Moreover, Stokes' law predicts that a 10-fold increase in the suspension viscosity (η) results in a 10-fold decrease in the sedimentation rate. Structured vehicles that are shear thinning and thixotropic in nature (see Chapter 7) may provide adequate viscosity on standing to minimize sedimentation and still be sufficiently resuspendable and pourable when shaken to provide an accurate dose.

Viscosity modifiers can be classified as carbomers, colloidal silicon dioxide; hydrated silicates (bentonite, hectorite, kaolin, and magnesium aluminum silicate); hydrophilic celluloses (methyl cellulose, hypromellose, microcrystalline cellulose, and sodium carboxymethyl cellulose); and polysaccharides [acacia, alginates, chondroitin (animal origin), pectins, tragacanth, and starch]. Appendix 15.3 gives the trade names of viscosity-enhancing and structure-modifying agents. The type of rheological system and common use ranges are also provided.

Equations (15.2) and (15.3) indicate that decreasing the interfacial tension between the dispersed particle and the dispersion medium should lead to a decrease in the suspension's free energy and make it more thermodynamically stable. Surface-active agents and their use concentrations are listed in Appendix 13.9.

Another way to prevent sedimentation is to create a flocculated system that has a sedimentation volume of unity. When it is not possible to achieve a sedimentation volume of 1, the goal is to maximize F and the degree of flocculation in a structured shear-thinning thixotropic viscosity modifier. Optimizing the extent of floc formation is called *controlled flocculation*. The flocs that are formed should be weakly aggregated so that they break up readily under the shear stress of shaking. The structured vehicle's viscosity should also decrease under shear stress to allow easy redispersion of any cake that might have formed and breakup of the flocs that are present. After the dose has been measured and the container has been placed at rest, floc formation should again occur. Since the flocs are aggregates of individual particles, their larger size will tend to cause them to settle more readily than the individual particles. To decrease or prevent rapid sedimentation, the viscosity and structure formation of the thixotropic vehicle should increase faster than floc formation.

Flocculating agents include salts, surfactants, and hydrophilic polymeric materials. Flocculating salts often include sodium salts (sodium chloride, sodium phosphate, sodium citrate, sodium tartrate, and sodium sulfate), potassium salts (the same counterions as sodium), calcium chloride, and magnesium chloride. The divalent counterions tend to be more efficient flocculators of hydrophobic particles, but they are also more prone to form insoluble complexes with other ionized substances that may be present in the suspension, such as viscosity modifiers and surfactants.

The overarching goal of successful suspension design is to develop a patient-acceptable dosage form that delivers an accurate, bioavailable dose of a physically, chemically, and microbiologically stable product. Another suspension design aim is to achieve zero sedimentation. If sedimentation cannot be prevented, secondary aims target a suspension that employs a thixotropic structured vehicle that has a high sedimentation volume, a relatively slow settling rate, good resuspendability, and good drainage from the container closure system.

15.2.3.1.1 Scale-up of Coarse Suspensions The general suspension unit operation steps include:

1. Drug milling and particle sizing
2. Dispersion of the sized drug in a prehydrated premanufactured structured suspending vehicle that contains a wetting agent and protective colloid if needed
3. Addition of water-soluble excipients as a separate solution to the drug suspension
4. Addition of a flocculating agent to the suspension mixture while stirring to obtain a uniform mixture
5. Passage of the flocculated suspension through a colloid mill to disperse the drug completely and uniformly
6. Batch processing through deaerating equipment

The selection of dry powder comminution mills and milling scale-up was discussed in Chapter 13. Thorough wetting, mixing, and dispersion of the drug in the structured suspending vehicle are discussed in this section. Mixing is often broken down into *macromixing* and *micromixing*. *Macromixing* requires that adequate material flow occurs to prevent stratification and sedimentation of the bulk components. *Micromixing* reflects the status of individual particles and their state of aggregation and homogeneous dispersion. Agitation, mixing, and blending are often used synonymously. Some authors distinguish the use of these terms. Mixing and blending refer to the intermingling and distribution of separate components. *Mixing* is the more generic of the two terms and can be used to refer to the combination of miscible or immiscible phases. *Blending*, on the other hand, is frequently reserved for the intermingling of miscible phases, such as water and alcohol. Agitation maintains a state of mixedness, but it does not necessarily result in macro- or micromixing. The forces encountered during fluid mixing are inertial, viscous, and gravitational. Inertial forces are produced by the mixer impeller. The viscosity of the system resists material flow and is related directly to the shearing stress or force per unit area. The gravitational force must also be accounted for when the weight of the liquid is being moved. To attain acceptable macro- and micromixing, a mixer must delivery adequate *pumping* or material flow for macroscale mixing and *shear* to provide desired particle dispersion. The work done by a mixer can be calculated using the following equation, which separates the work done to attain micromixing and work done to realize macromixing:

$$W = [C_1 N^2 D^2 (\text{shear})] \cdot [C_2 N D^3 (\text{pumping})] \qquad (15.7)$$

where W is work done by the mixer, C_1 and C_2 are constants, N is the rotation speed, and D is the impeller diameter. Equation (15.7) shows that the work required to pump the liquid is affected primarily by the diameter of the impeller, not by the rpm. The micromixing due to shear is affected by the impeller diameter and rotational speed (rpm). Three major flow patterns are created by standard prop or marine impellers. The flow patterns are described as axial, radial, and tangential. A vortex is created when tangential flow predominates. Little macro- or micromixing occurs when a vortex is formed. Tangential flow can be used to pull solids at the surface down toward the mixer impeller to break up agglomerates and wet the solid. The vortex may also pull air into the vehicle, which is undesirable, so care needs to be taken to minimize air entrapment if it is necessary to use tangential flow. The most efficient mixing occurs when turbulence results in bulk flow and shear. The *Reynolds number* has been used to describe laminar, irregular, and turbulent flow in stirred vessels:

$$Re = \frac{D^2 N \rho}{\eta} \qquad (15.8)$$

where Re is the Reynolds number, which is the ratio of inertia and viscous forces; D the impeller diameter; N the impeller speed; ρ the density of the fluid; and η the viscosity of the fluid. The Reynolds number has been shown to be useful when scaling from a propeller of one diameter to a propeller of a second diameter when the propeller types are the same. The goal would be to maintain the same ratio of inertia and viscous flow by adjusting the impeller speed. Using a Rushton turbine impeller, turbulent flow is achieved in a stirred tank when the Reynolds number is 10,000 and greater. A low Reynolds number indicates that the flow is dominated by the viscosity of the system and laminar flow is expected to occur. Equation (15.8) indicates that macroscopic flow needs to overcome the viscosity of the system to become turbulent to increase the efficiency of the mixing process.

The dimensionless *Power number* is related to power consumption, the density of the media, the impeller rotational speed, and the media:

$$N_p = \frac{Pg}{\rho N^3 D^5} \qquad (15.9)$$

where N_p is the power number, P the power applied, g the gravitational constant, ρ the density of the media, N the impeller rotational speed, and D the diameter of the impeller. Equation (15.9) shows that the power number is very sensitive to the third power of rotation speed and the fifth power of the impeller diameter. For a given impeller, a correlation is frequently developed between the Reynolds number and the Power number. This relationship is used during scale-up mixing operations to estimate power requirements or mixing speeds.

In scaling from volume$_1$ to volume$_2$, the ratio of the power consumption to volume can be used to determine impeller diameter and mixing speed to achieve turbulent flow:

$$\frac{P_1/V_1}{P_2/V_2} = \left(\frac{N_1}{N_2}\right)^3 \left(\frac{D_1}{D_2}\right)^2 \qquad (15.10)$$

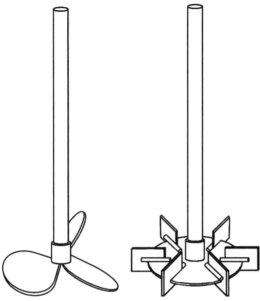

Marine prop impeller   Rushton turbine impeller   **FIGURE 15.2**   Mixer impellers.

where V is the vessel volume and the other terms are as in equation (15.9). Different mixers and processors may be needed, depending on the suspension formulation and desired particle size. For micronized drug, it may only be necessary to mix the powder uniformly into the structured suspension vehicle. Uniform dispersions may be accomplished using a prop mixer or turbine mixer (Figure 15.2). However, in most cases, it is a good idea to run the dispersed suspension through a piston homogenizer, rotor–stator disperser (Figure 15.3), or colloid mill (Figure 15.4) to make sure that all potential agglomerates are dispersed. The piston homogenizers generally operate at 4000 to 6000 psi (27.6 to 41.4 MPa).

It is also often advantageous to operate the disperser under a vacuum to decrease the amount of air that could potentially be trapped in the suspension.

15.2.3.2 Colloidal Solid–Liquid Medium Dispersions

15.2.3.2.1 Vectors Since the human genome has been sequenced, there has been an increasing interest in delivering genes to treat a multitude of targets, such as cancer. The gene must get into the nucleus of the target cell. Vectors are used to carrier the gene to the target cell so that it can be recognized and translocated into the cytoplasm and into the nucleus. There are two general classes for vectors: viral (or viral associated) and nonviral. Early vectors were viruses and are still under active research. The viral vectors can be further categorized into integrating vectors and nonintegrating vectors. *Integrating vectors* can provide lifelong expression and are retroviruses and adeno-associated viruses. *Nonintegrating vectors* are part of the adenovirus family that has more than 50 serotypes (Pfiefer and Verma, 2001). The nonviral vectors involve the incorporation of plasmids (*naked DNA*), DNA,

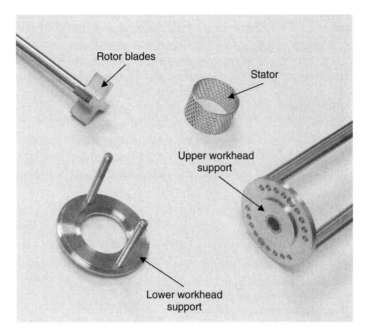

FIGURE 15.3 High-shear mixer components (Silverson Model SL2T).

RNA, and synthetic oligonucleotides with cationic lipids (*lipoplex*) and cationic polymers (*polyplex*). The nonviral vectors may be modified by covalently bonding poly(ethylene glycol) to increase the circulation time. Vectors range in size. The size of recombinant adenovirus has been shown to be approximately 65 to 85 nm (Bondoc and Fitzpatrick, 1998). As with other colloidal particles, viruses may aggregate. This aggregation may be irreversible, leading to large aggregates that precipitate. Once the virus has precipitated, it is no longer an effective vector. The aggregation of dimers, trimmers, and oligomers can affect transfection. In this case, where the virus is a colloidal particle, deflocculation is desired. Adjusting the ζ-potential to increase particle repulsion can be helpful. The nonviral vectors are generally charged, and the charge is referred to as the *charge ratio*, which is the ratio of the positive charge equivalents of the cationic vector delivery component to the negative charge equivalents of the nucleic acid component.

15.2.3.2.2 Dendrimers Dendrimers are a novel class of emerging polymers first synthesized by Donald Tomalia in 1979. *Dendrimer* is derived from the Greek word *dendron*, which means "tree." Dendrimers are spherical three-dimensional polymers that have branches, which have internal void spaces. Dendrimers are distinctly different from linear polymers and have a number of advantages. They are generally monodisperse, the size and number of branches can be controlled as well as their porosity and surface characteristics, and they are less viscous. Dendrimers are synthesized by multistep processes each step called a *generation*. For polyamidoamine (PAMAM) dendrimers, the first-generation particle

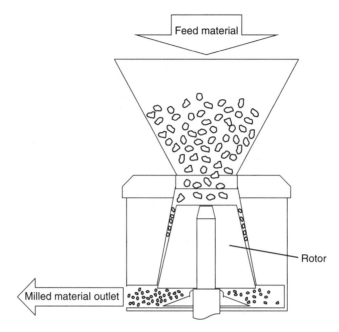

FIGURE 15.4 Colloid mill.

has four surface groups, a molecular mass of 517 Day and a diameter of 1.5 nm. A 10-generation PAMAM dendrimer has 4096 surface groups, a molecular mass of 934,720 Day and a diameter of 13.5 nm (Nanjwade et al., 2009). Controlled synthesis can create particles in the micrometer range. Dendrimers have been made from a number of starting materials and have been studied as drug delivery systems for oral, ocular, transdermal, and pulmonary drug delivery. They have been fabricated to provide controlled release and to deliver genes. A review article is available that details the emerging status of dendrimers (Nanjwade et al., 2009).

15.2.3.2.3 Nanoparticles As noted in Section 15.1, in 2003 the 21st Century Nanotechnology Research and Development Act and the National Nanotechnology Initiative were established. As part of the initiative, nanoparticles were defined as materials with one or more dimensions of less than 100 nm. Since then, significant attention has been given to nanoparticle research as drug delivery systems. The U.S. Food and Drug Administration (FDA) (2010) has issued a reporting format for nanotechnology-related drug submissions. It is interesting that they define *nanotechnology* and *nanoparticle* as objects having a size range from 1 to 100 nm; at the same time the FDA defines *nanomaterial/nanoscale material* as any material with at least one dimension of less than 1000 nm, which is the definition of a colloid. The FDA goes on to define a *nanomedicine* as the use of nanoscale material for medical applications. They also define dendrimer, liposomes, micelles, nanoemulsions, nanocrystal, primary particle, and metal colloids, all of which can be developed as nanomaterials. Biodegradable nanoparticles fabricated from poly(lactic acid) (PLA) and poly(lactic-*co*-glycolic acid) polymers have been studied extensively. PLA–poly(ethylene glycol) nanoparticles tend to increase the circulation time of the material. Drug loading and final particle size are key development parameters. Kim and Dobson (2009) has written a review on nanomedicine for targeted drug delivery.

A relatively new and extremely interesting nanotechnology is the Print (particle replication in nonwetting templates) technology, which has adopted techniques used by the microelectronics industry to fabricate micro- and nanoscale particles with controlled shape, size, composition, and surface functionality. The technology employs an imprint lithographic process that uses nanoscale molds made from low-surface-energy perfluoropolyether matrices. Kelly and DeSimone (2008) have described the nanomolding, filling, lyophilization, and harvesting of nanoscale monodispersed albumin particles and albumin particles that contain siRNA and paclitaxel. Vaccine manufacture is also being investigated.

15.2.3.3 Coarse Emulsions or Macroemulsions

The term *coarse emulsion* or *macroemulsion* is used to distinguish this liquid dispersion system from microemulsions, which are discussed in Section 15.2.3.4.3, and other association colloids. Macroemulsions are in the size ranges 75 to 125 μm. Like suspensions, a narrower size range (5 to 10 μm) gives more stable emulsions. Emulsions of particle sizes greater than 1 μm look white or milky. Colloidal emulsions of particle size 100 to 1000 nm appear blue-white. Colloidal systems below 100 nm become increasingly transparent as the particle size decreases. Emulsions are dispersions, so there is a dispersed or internal phase and a continuous or external phase. The common components of an emulsion are oil and water. When the oil is the dispersed phase and water is the continuous phase, the emulsion system is called an *oil-in-water* (o/w) *emulsion*. Similarly, when water is the dispersed phase and oil is the continuous phase, the emulsion system is called a *water-in-oil* (w/o) *emulsion*. To fabricate an emulsion, the interfacial tension between the oil and water needs to be lowered so that the materials can spread over each other before globules can be formed (see Chapter 5). Surfactants or surface-active agents are used to decrease the interfacial tension. Once the interfacial tension has been sufficiently lowered, energy in the form of shear needs to be applied to the system to form the globules of the dispersed phase. Since these systems do not form spontaneously, they are thermodynamically unstable. Pharmaceutically acceptable emulsions that have shelf lives of two to three years can be designed by using the proper excipients and manufacturing processes. This section focuses primarily on oral drug delivery. The U.S.P. has three official monographs for oral emulsions: Castor Oil Emulsion, Mineral Oil Emulsion, and Simethicone Emulsion.

A number of terms are used to describe the physical instability of macroemulsions. Complete irreversible separation of the emulsions often happens in steps. The first instability step is called *creaming*. Like suspensions, there are typically density differences between the dispersed phase and the continuous phase. Over time, the two phases will separate as a result of these density differences. Similar to suspension sedimentation, emulsion density separation, *creaming*, results in two emulsion phases. For o/w emulsions, creaming occurs upward, with the upper emulsion phase rich in the dispersed oil and the lower emulsion phase deficient in the internal oil. Water-in-oil emulsions cream downward, with the lower emulsion phase rich in the internal water phase and the upper emulsion phase deficient in dispersed water. If the two emulsions can be redispersed into a single uniform emulsion, the system is considered stable, although it is desirable to minimize creaming. The enriched phase will have a greater tendency for globule–globule interaction. This increased potential for interparticle interaction can lead to aggregation or flocculation, which is the next step to irreversible emulsion separation. If the floccules are weakly aggregated, they can be redispersed by gentle shaking to achieve a uniform dispersion. However, if the cohesive forces are high enough, the flocs will aggregate more strongly, and finally, separate globules will coalesce into a single larger droplet. This coalescence is irreversible and leads to complete phase separation.

Phase inversion describes the process by which the phases reverse. That is, the internal phase becomes the external phase, and vice versa. This can be a sudden change that happens during the manufacture of emulsions. Generally, in making an o/w emulsion, a heated water phase is added to the heated oil phase with stirring. Initially, the oil phase constitutes the largest volume and the water is incorporated or dispersed into the oil. At some point during the addition and mixing process, the water phase can no longer be incorporated in the oil, and an apparent change in viscosity and color is accompanied by a phase inversion where the oil becomes dispersed or incorporated into the water phase. For o/w emulsions, a phase inversion temperature (PIT) can be reached where the emulsion inverts to a w/o emulsion. Phase inversion occurs primarily because the surfactant becomes preferentially more soluble in the oil at higher temperatures. At lower temperatures, the surfactant can hydrogen bond with water. As the temperature is raised, the hydrogen bonds become weaker, which decreases the surfactant's solubility in water. The decrease in hydrogen bonding leads to a more favorable solubility in the oil. This shift in solubility causes emulsion inversion. This phenomenon is very important in o/w parenteral emulsions, where injection of an inverted o/w emulsion could cause a pulmonary embolism. The PIT can be used as a measure of the emulsion's stability. In other words, the higher the PIT, the more stable the emulsion. It has been shown that the PIT is inversely proportional to the solubility of the surfactant in the oil. As a general rule of thumb, a commercial emulsion should have a PIT of 50°C, which is 20°C above the upper end of controlled room temperature (30°C), Eccleston (1992). Unfortunately, the phase inversion technique cannot be used to optimize the emulsion stability for w/o emulsions because as the test temperature is raised, the emulsifiers become more soluble in the oil phase and continue to support a w/o emulsion.

The type of emulsion and its stability depend on the *phase volume* of the system, the tendency to cream and flocculate, the surfactant/emulsifier *hydrophile–lipophile balance* (HLB), the chemical nature of the emulsifier, the concentration of emulsifier, the presence of mixed emulsifiers, and the strength and type of interfacial film. The phase volume is

$$\varphi = \frac{\text{volume of the internal phase}}{\text{volume of the total emulsion}} \tag{15.11}$$

where φ is the phase volume. Oil-in-water emulsions are typically more stable than w/o emulsions because the polar groups are better barriers to coalescence than are the hydrocarbons. The optimum phase volume for an o/w emulsion is 0.4 to 0.6 (40 to 60%). The phase volume for w/o emulsions is often lower, around 0.3 to 0.45. The hydrocarbon tails that stick out into the continuous oil phase of an w/o emulsion are a weaker barrier to coalescence. The theoretical maximum phase volume is 0.74, which represents a rhombohedral packing of spherical globules that touch. Above a phase volume of about 0.5 or 50%, it becomes more difficult to design a physically stable emulsion. As a side note, many lipophilic materials used in emulsions are solids. As a result, emulsion compositions are often given by weight. In this case, an estimate of the volume of the oil phase can be determined by taking the total weight of the lipophilic components and dividing by a density value of 0.85 g/mL. The estimated phase volume can then be determined by dividing the volume of the internal phase by the total emulsion volume.

Emulsifiers should be surface active and reduce the interfacial tension between the water and oil phases. Since the success of forming dispersed droplets is thought to be the result of the competing rate processes of forming droplets and coalescence of droplets, it is important that the emulsifier adsorb quickly to the formed dispersed phase before it has a chance to coalesce. The physical stability of an emulsion is affected by its tendency to cream and flocculate. Stokes' law is at play, and the ramifications of density differences between phases, globule size, and viscosity all affect the emulsion's physical stability. In addition, the surface charge as measured by the ζ-potential affects the tendency for the creamed material to flocculate and coalesce irreversibly.

The hydrophile–lipophile balance is an empirical concept that is based in part on the solubility and dispersability preference of the emulsifier in water and oil (Griffin, 1949). An emulsifier that interacts more strongly at the water interface will tend to form an o/w emulsion. Water-in-oil emulsion formation is favored with lipophilic emulsifiers. Griffin (1949) established an *HLB scale* ranging from 1 to 50. The more lipophilic the surfactant is, the lower the HLB number. Empirically, HLB values of 3 to 8 form w/o emulsions, HLB values of 9 to 16 form o/w emulsions, and compounds that have HLB values of 7 to 9 act as good spreading agents. Since Griffin first established the HLB scale, advances have been made in understanding the relationship between a surfactant's HLB and a number of other physical parameters, such as dielectric constant, partition coefficient, heat of interfacial adsorption, solubility parameters, PIT, emulsion inversion point, and interfacial tension (Becher and Schick, 1987). In addition to Griffin's method, these experimental methods can be used to determine a surfactant's HLB. Shinoda and Harai (1964) have shown that temperature, concentration of emulsifier, and type of oil can affect the HLB. It has been suggested that the term *effective HLB* be used to indicate that the HLB varies depending on the physical and chemical conditions of the particular emulsion system. The HLB values for a number of pharmaceutically acceptable emulsifiers are shown in Appendix 15.4.

To make an emulsion, it is also necessary to know the *required HLB number of the oil*. Different oils will have different required HLB values, depending on their chemical structure and lipophilicity. The required HLB for any specific oil is different to form o/w and w/o emulsions. For example, the required HLB values for a heavy mineral for a w/o and an o/w emulsion are 4 and 10.5, respectively. The required HLB values for pharmaceutically acceptable oils are listed in Appendix 15.5. It may be necessary to design an oil system emulsion for which the required HLB value has not been reported. The required HLB can be determined using the Griffin method. A number of emulsions are prepared with the oil, using emulsifiers that have a fairly broad range of HLB values. The most stable emulsions are determined from this initial experiment and the second series of emulsions are prepared using emulsifiers with a narrower range of HLB values. This iterative process is continued until an optimum narrow range of HLB values is established and the required HLB can be reported. This method can be used to find the o/w or w/o required HLB value for oil.

Another method for determining the required HLB for an oil has been reported by Becher and Schick (1987). In this method the surface tension of the oil can be used to determine its required HLB. The equation that relates the o/w required HLB for an oil is

$$\text{required HLB} = \frac{20\gamma}{\gamma + 2.49V_o^{0.33}} \tag{15.12}$$

where γ is the surface tension of the oil and V_o is the molar volume of the oil. Orafidiya and Oladimeji (2002) reported the use of turbidity measurements to establish the required HLB for several essential oils.

In addition, the o/w emulsion's stability is affected by temperature, the emulsifier, and the emulsifier concentration. The effectiveness of various emulsifiers and their concentration can be evaluated using the *emulsion inversion point* (EIP), the composition of water, oil, and emulsifier that results in a phase transition from an o/w to a w/o at a constant temperature. The EIP is the value of the ratio of milliliters of water to the milliliters of oil determined at the time of the emulsion inversion.

The simplest emulsion will have three components: water, oil, and emulsifying or surfactant agent. The selection of the surfactant and its HLB and the phase ratio will depend on the type of emulsion. The amount of surfactant required depends heavily on the particle size desired. It is important to have the surfactant cover the complete surface area of the particle dispersed. The following example illustrates an approach that can be employed to determine the amount of emulsifier to use in an o/w emulsion. Let us determine the amount of emulsifier that is needed to prepare 200 mL of an o/w emulsion with a phase volume of 0.50 and a desired particle size of 5.0 μm.

Example 15.1. Determine the amount of emulsifier required given that the surface area of a spherical droplet is πd^2 and the volume is $d\pi^3/6$. The molecular mass of the surfactant is 1000 Da and each surfactant molecule covers 30 Å^2. The volume of oil is 100 mL and its density is 1 g/cm^3. The 5.0-μm globule surface area needs to be completely covered by the emulsifier. Use centimeters as the common unit for all lengths, surface area, and volume dimensions.

Unit conversions:

$$\text{diameter of 5.0 } \mu m = 5.0 \ \mu m (1 \text{ mm}/1000 \ \mu m)(1 \text{ cm}/10 \text{ mm})$$
$$= 5.0 \times 10^{-4} \text{ cm}$$
$$\text{surface area of emulsifier} = (3.0 \times 10^{1} A^{2}/\text{molecule})(1 \times 10^{-8} \text{ cm/A})^{2}$$
$$= 3.0 \times 10^{-15} \text{ cm}^{2}/\text{molecule}$$
$$\text{volume of 5-}\mu m \text{ globule} = \frac{\pi(5.0 \times 10^{-4} \text{ cm})^{3}}{6} = 6.5 \times 10^{-11} \text{ cm}^{3}/\text{globule}$$
$$\text{surface area of 5.0-}\mu m \text{ globule} = \pi(5.0 \times 10^{-4} \text{ cm})^{2} = 7.9 \times 10^{-7} \text{ cm}^{2}/\text{globule}$$

Solution.

Step 1. Determine how many 5.0-μm globules can be formed from 100. mL of oil.

Step 2. Determine the total globule surface area that needs to be covered by the emulsifier.

Step 3. Calculate the number of surfactant molecules that are needed to cover the total globule surface area.

Step 4. Determine the amount of surfactant by knowing its molecular mass.

Example calculations:

Step 1. The total number of 5-μm globules is

$$\frac{(100 \text{ mL oil})(1 \text{ cm}^{3}/\text{mL})}{6.5 \times 10^{-11} \text{ cm}^{3}/\text{globule}} = 1.5 \times 10^{12} \text{ globules}$$

Step 2. The total globule surface area is

$$(1.5 \times 10^{12} \text{ globules})(7.9 \times 10^{-7} \text{ cm}^{2}/\text{globule})$$
$$= 1.2 \times 10^{6} \text{cm}^{2} \text{total}$$

Step 3. Calculate the number of surfactant molecules needed to cover the oil droplet surface area:

$$\frac{1.2 \times 10^{6} \text{ cm}^{2}}{3.0 \times 10^{-15} \text{ cm}^{2}/\text{molecule surfactant}}$$
$$= 4.0 \times 10^{20} \text{ molecules of surfactant}$$

Step 4. Calculate the amount of surfactant needed to cover the oil globule surface:

$$\left(\frac{4.0 \times 10^{20} \text{ surfactant molecules}}{6.023 \times 10^{23} \text{ molecules/mole}} \right) \cdot 1000 \text{ g/mol} = 6.6 \times 10^{-1} \text{ g}$$

Therefore, in this particular case, the o/w emulsion would require 0.66 g of emulsifying agent to completely cover the surface area of the 5-μm oil globules.

Now the question is: What surfactant or emulsifying agent should be used? The emulsifying agent's HLB number should match the o/w *required HLB value for the oil*. For purposes of this example, let's use light mineral oil as the oil phase. From Appendix 15.5, the light mineral RHLB value required is 10 to 12 for an o/w emulsion. Polyoxyethylene (5) sorbitan monooleate (polysorbate 81) has an HLB value of 10.0. Therefore, 0.66 g of polysorbate 81 could be used to emulsify the light mineral oil to manufacture an emulsion that would have a phase volume of 0.5 and an oil droplet size of 5 μm. Mineral oil, U.S.P. is manufactured using acacia as the emulsifying agent. Acacia has an HLB value of 11.9, which is within the reported required HLB of 10 to 12. The final selection of the emulsifier and the HLB number should be supported by stability studies designed to evaluate the chemical and physical stability of the product.

A number of research reports have shown that a mixture or *blend* of emulsifiers that span or bracket the required HBL value result in more stable emulsions. At this point it is important to note that both emulsifier HLB numbers and required HLB values are experimental values that cannot be measured to the accuracy to which many experimental responses are measured. Therefore, pragmatically, it makes sense to bracket a required HLB with emulsifying or surfactant agents that have lower and higher HLB values compared to the oil's stated required HLB. In addition, it has been shown that mixed surfactants can have a synergistic effect on an emulsion's physical stability. It has been shown that mixed or blended surfactants can form liquid crystals that accumulate at the oil–water interface (Friberg, 1969; Sagitani, 1981). It is thought that the presence of these liquid crystals reduces the cohesion potential between dispersed particles, which increases the stability of the emulsion. Surfactant blends are also thought to produce mixed interfacial films that allow more dense packing and higher surface coverage than those of individual components. The individual amounts of the surfactant blend can be determined by calculating the composite weighted average of the individual HLB values of each emulsifier. It is not uncommon for an emulsion to have several oils in the formulation. In this case, the composite *weighted* required HLB is calculated. An example of using a surfactant blend for an o/w emulsion that contains several oils or lipids is presented below. Table 15.7 gives the emulsion composition and each of the oil's required HLB values for an o/w emulsion. For this example, the amount of emulsifier blend is given as 2.00 g.

TABLE 15.7   Topical Emulsion Formulation

Component	Amount (g)	Required HLB (o/w)
Cetyl alcohol (solid)	5.00	15.0
Lanolin (solid)	10.0	12.0
Beeswax (solid)	15.0	9.0
Mineral oil, light	20.0	10.0
Emulsifier blend	2.00	
Preservative (water miscible)	0.200	
Water, add up to	100.	

Several things need to be determined before an emulsion can be manu-factured. We need to know the required HLB for the oil mixture, including the lipophilic solids (cetyl alcohol, lanolin, and beewax). Once this *composite required HLB* for the total "oil" component is determined, the appropriate emulsifiers can be chosen. For this example, nonionic emulsifiers with HLB values that bracket the composite required HLB are used. Determination of the total oil or *composite required HLB* can be done in the following steps.

Example 15.2. Determine the weighted average oil composite required HLB.

Solution.

Step 1. Establish the oil components and calculate their total amount.

Step 2. Assign to each lipophilic component the required HLB value.

Step 3. Calculate the weighted average composite required HLB (RHLB) for the oil system using the following equation:

$$\text{composite } \text{RHLB}_{\text{oil system}} = f_A \cdot \text{RHLB}_{\text{component A}} + f_B \cdot \text{RHLB}_{\text{component B}}$$
$$+ f_C \cdot \text{RHLB}_{\text{component C}} + f_D \cdot \text{RHLB}_{\text{component D}} \tag{15.13}$$

where f is fractional weight of the specific oil component to the total weight of all the oil components in the emulsion. In this example, the oil phase weighs 50 g, which includes cetyl alcohol, lanolin, beeswax, and mineral oil.

$$\text{RHLB}_{\text{oil system}} = \frac{5.00}{50.0} \cdot (15.0) + \frac{10.0}{50.0} \cdot (12.0) + \frac{15.0}{50.0} \cdot (9.0)$$
$$+ \frac{20.0}{50.0} \cdot (10.0) = 10.6$$

Now that the composite required HLB values for the oil components have been determined, the emulsifiers need to be chosen. For this example, sorbitan monooleate (Span 80, HLB = 4.3) and poly(oxyethylene sorbitan monooleate) (Tween 80, HLB = 15) will be used for the surfactant blend since they bracket the RHLB value 10.6. The total amount of emulsifier blend is 2.00 g. The emulsifier blend amounts are calculated using the weighted average amounts necessary to achieve an HLB of 10.6. Let x = the grams of Span 80 and $2 - x$ = the grams of Tween 80. The weighted average HLB for the surfactant blend can be calculated by

$$\frac{x}{2.00} \cdot 4.3 + \frac{2-x}{2} \cdot 15 = 10.6 \quad \text{and} \quad 2.15x - 7.5x + 15 = 10.6$$

Solving for x, $-5.35x = -4.4$ and $x = 0.82$ g Span 80, so $2 - x = 1.18$ g Tween 80.

In this example, the emulsifier blend that would bracket the composite oil required HLB value, 10.6, would be made up of 0.82 g of Span 80 and 1.18 g of Tween 80. The emulsion can be prepared by melting the solid lipids and adding the mineral oil and Span 80 to the lipophilic liquid. This oil phase is stirred and maintained at 70°C. The water-miscible preservative and Tween 80 are added to the water with mixing and heating to 72°C. The 72°C water phase is then added slowly to the heated oil phase while the oil phase is stirred vigorously at 1000 rpm. It is desirable to have the temperature of the water phase higher than that of the oil phase so that the water does not cause the oil phase to temperature-congeal. As the water phase is added, the temperature of the two phases can be decreased, and at some point during water addition there will be a phase inversion at which the water phase becomes the external continuous phase. At this point, the o/w emulsion is formed and the product becomes more viscous and turns white and creamy.

As stated previously, to make an emulsion, the interfacial tension between the oil and water needs to be lowered significantly. Once the emulsion has been formed, the stability of the emulsion is achieved principally by preventing the dispersed particle from flocculating and coalescing. Droplet-to-droplet contact can be prevented by charge repulsion or by creating physical films around the particles. Increasing the viscosity and continuous phase structure can also decrease the kinetic motion and decrease the chances of collision. Emulsifiers can stabilize macroemulsions by forming monomolecular films, multimolecular films, and solid colloidal interfacial barriers. *Auxiliary emulsifiers* can increase the viscosity and structure of an emulsion system. Anionic and cationic surfactants can form charged monomolecular films around the dispersed droplet. The charges provide a repulsive electrostatic barrier. Nonionic surfactants can also form flexible monomolecular films and are not susceptible to pH changes or interactions with other charged compounds in the emulsion. Water-loving natural colloids such as acacia, gelatin, and tragacanth can form strong rigid multimolecular interfacial films. Colloidal solids such as veegum and bentonite are wetted by both water and oil phases and form a colloidal steric barrier. Auxiliary agents such as fatty acids (palmitic and stearic acid), fatty alcohols (cetyl and steryl alcohol), fatty esters (glyceryl mono-, di-, and tristearate), and sterols (cholesterol) can increase the viscosity and structure and viscocity of the emulsion, which can increase the stability of the emulsion.

A statistical experimental design study can help determine the optimum emulsifier(s) and auxilliary emulsifier(s) concentrations. The effect of temperature-dependent emulsifier solubility can also be explored. As discussed previously, blend surfactants may also contribute to enhanced emulsion stability. The preservatives used in emulsions are shown in Appendix 17.4. The amount of preservatives used in emulsions is usually higher than that used in other dosage forms because some of the antimicrobial agent will partition into the oil phase. It has also been well documented that alkyl hydroxybenzoates (methyl, ethyl, and propyl parabens) interact with emulsion and suspension excipients such as Tween 80, Tween 20, Mryj 52, povidone, poly(ethylene glycol) (PEG), methylcellulose, and gelatin. This interaction results in the binding of the parabens, which leads to a decrease in the amount of "free" antimicrobial agent in the water phase and a decrease in the antimicrobial preservative effectiveness of the product.

15.2.3.3.1 Multiple Emulsions Multiple emulsions are designated $w_1/o/w_2$ (water-in-oil-in-water) and $o_1/w/o_2$ (oil-in-water-in-oil). This designation indicates that a *primary* emulsion (w_1/o or o_1/w) is the dispersed phase. This phase

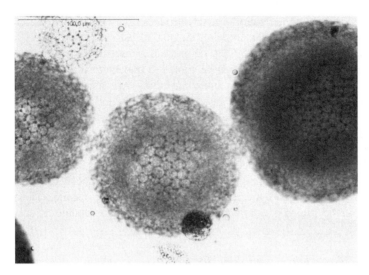

FIGURE 15.5   Multiple emulsion.

forms a *secondary emulsion* with the external phase (w_2 or o_2). It is noted that the component of the internal phase of the primary emulsion is the same component as that used for the external phase of the secondary emulsion. Figure 15.5 shows a multiple emulsion. Multiple emulsions can be used to taste-mask bitter-tasting drugs, to provide controlled drug release, and to improve drug stability. As can be imagined, the level of design complexity is even higher than that for suspensions or emulsions. The chief difficulty is stabilizing three thermodynamically unstable phases. Coalescence can occur in the internal or secondary emulsion. Two methods are used to fabricate multiple emulsions: the one-step or two-step process. The two-step process is used more commonly and is discussed here. The process involves manufacturing the primary emulsion first and then preparing the secondary emulsion from the primary stock. A w/o/w multiple emulsion is produced by first preparing a w/o primary emulsion. A lipophilic surfactant is added to the heated oil phase and heated water is added to oil slowly under higher-speed shear for approximately 15 minutes. The emulsion is allowed to cool slowly during the high-shear mixing. The w/o emulsion that is formed is then added slowly to water that has a hydrophilic surfactant. The primary emulsion is added for 5 to 10 minutes under a lower stirring shear rate to prevent rupture of the primary emulsion. Although multiple emulsions are interesting and provide some advantages over single emulsions, to date there are no commercial multiple emulsions.

15.2.3.3.2 Scale-up of Macroemulsions   The scale-up of macroemulsions can employ dimensionless numbers in a manner similar to that used for suspensions. The final particle sizing of the dispersed droplets uses high pressures and microfluidics to form small, relatively uniform dispersed droplets in the size range 5 to 10 μm. The Microfluidizer introduces the emulsion to microchannels some 50 μm in diameter at pressures of 30,000 psi (207 MPa). The split fluid, which has a velocity of 400 m/s, is rejoined in a reaction chamber, creating high impact and shear. The material can be cooled to prevent heat buildup and possible drug degradation. Other manufacturers, including Avestin and Bee International, also use high pressure, but the reaction chambers function differently than they do in the Microfluidizer. The Bee high-pressure homogenizer offers several different mechanisms to create high shear. In one case the reaction chamber is comprised of ceramic cylinders that have various openings. The material requiring final emulsification is forced through the openings at pressures as high as 45,000 psi (310 MPa). Droplet size reduction is achieved by liquid cavitation and shear. Cavitation occurs when the material reaches a ceramic cylinder with a larger opening. These homogenizers can be run in a continuous mode and are also capable of recirculation cycles.

15.2.3.4 Association Colloidal Dispersions *Association colloids* result from the association of *amphiphilic compounds*, which form aggregates of colloidal dimensions. Amphiphilic compounds contain distinct structural regions that are hydrophilic (water loving) and lipophilic (oil loving). Amphiphilic compounds are also called *amphiphiles*. The association is driven by the common idea that "likes associate with likes." In thermodynamic terms, the association or aggregation is primarily an entropy-driven process and is referred to as the *hydrophobic effect*. The association of amphiphiles occurs in water, which has a relatively high degree of structure, resulting from hydrogen bonding. When an ionic or polar solute is added to the native water, its structure is disrupted. In the case of ion or polar solutes, electrostatic interactions with the water can "compensate" for the initial structural disruption. When hydrophobic or lipophilic solutes are added to water, there is no such structural compensation. In fact, the addition of nonpolar solutes to water causes the water to become more structured around hydrophobic regions. This structuring of water decreases the entropy or increases the order of the system, which gives a large negative entropy change (entropy$_{final}$ − entropy$_{initial}$ is negative). The negative entropy change leads to an increase in free energy:

$$\Delta G = \Delta H - T \Delta S \qquad (15.14)$$

where ΔG the change in Gibbs free energy, ΔH is the change in enthalpy, T the temperature in kelvin, and ΔS the entropy change. Since ΔS is negative in this case, the negative entropy change multiplied by the negative sign gives a positive value for the second term $(T \Delta S)$ and ΔG is increased. To counteract this decrease in entropy and increase in free energy, the hydrophobic groups will tend to remove themselves from contact with the water and associate with each other. Another way of looking at this is that the increasing structure constrains the freedom of movement of the hydrocarbon chains. When the water pushes or forces the hydrocarbon chains to interact with one another, the hydrocarbon tails have more freedom of motion. The hydrophobic effect is a response to the structuring of water. By removing the hydrophobic chains from contact with the water, the free energy of the system is reduced and a state of minimum free energy is achieved,

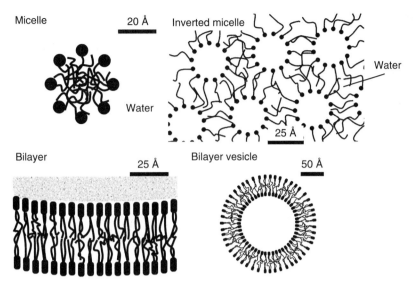

FIGURE 15.6 Amphiphilic associations. (Adapted with permission from H. A., Lieberman, M. M., Rieger, and G. S., Banker, *Pharmaceutical Dosage Forms: Disperse Systems*, Informa Healthcare, New York, 1998.)

causing the formation of association colloids. The transfer free energy of 1 mol of dipalmitoylphosphatidylcholine surfactant from water to a micelle is approximately 15.3 kcal/mol. This large thermodynamically favorable energy change due to the hydrophobic effect also explains why the micelles form spontaneously.

Figure 15.6 illustrates amphiphilic association vesicles. Association colloids have found many uses in the pharmaceutical industry. They have been used to increase the solubility of poorly soluble drugs and to improve drug stability. They have been used as drug delivery systems for oral and parenteral use, including controlled drug delivery.

15.2.3.4.1 Micelles Surfactants are amphiphiles. Above a critical concentration, *critical micelle concentration* (CMC) surfactants self-associate spontaneously to form *micelles*. About 30 to 100 surfactant molecules will form a micelle. This is called the *aggregation number*. Micelles have a nominal diameter of approximately 3 to 8 nm. Table 15.8 provides the CMC values for selected surfactants.

Compared to solid nanoparticles, the surfactant monomers are in dynamic equilibrium with the micelles. The exchange time is in microseconds. The CMC decreases with increasing hydrocarbon chain length. Ionic surfactant micelles are generally smaller than nonionic surfactant micelles because of the electrostatic repulsion that occurs at the polar head groups. This charge can be modulated by adding salt. Increasing the salt concentration can decrease the CMC of ionic surfactant micelles and increase the size and aggregation number. The salt effect is more significant than the longer hydrocarbon chain. Micelles may take the shape of spheres, cylinders, and lamellar bilayers. Cylinder or rodlike aggregates and

TABLE 15.8 Critical Micelle Concentration Values for Selected Surfactants

Surfactant	Trade Name	Molecular Mass (Da)	CMC[a]	Aggregation Number	Cloud Point	HLB[b]
		Nonionic				
Poloxamer[c], (E.P.)/U.S.P.-(N.F.)	Pluronic F-68	8350	0.033	—	—	29
Polyoxyl 23 lauryl ether[c], E.P./*Japanese Pharmacopocia* (J.P.)/U.S.P.–N.F.	Brij 23	1200	0.013	20–40	>100	16.9
Polysorbate 80[c], E.P./J.P./U.S.P.–N.F.	Tween 80	1310	0.012	60	65	15
Sorbitan monooleate[d], E.P./U.S.P.–N.F.	Span 80	429	0.005–0.013	—	—	4.3
		Anionic				
Sodium deoxycholate[d]	Generic	415	0.17	3–12	—	16
Sodium lauryl sulfate[c], E.P./J.P./U.S.P.–N.F.	Elfan 240	288.38	0.23	62	>100	40
		Cationic				
Benzalkonium chloride[e], E.P./J.P./U.S.P.–N.F.	Zephiran	Alkyl mixture	0.009			
Cetrimide[c], E.P.	Cetralol	337	0.14	80		

[a] CMC, critical micelle concentration as % w/v measured at 20 to 25°C.

[b] HLB, hydrophile–Lipophile balance.

[c] http://www.sigmaaldrich.com/etc/medialib/docs/Sigma/Instructions/detergent_selection_table.Par.0001.File.tmp/detergent_selection_table.pdf.

[d] Grant et al. (2006).

[e] Kopeck et al. (2007).

lamellar micelles usually form at higher surfactant concentrations. The lamellar micelles can form liquid crystals that are birefringent and highly viscous. Low HLB surfactants that are oil soluble form *reverse micelles* when dissolved in oil. In this case the polar head groups orient themselves inward and the hydrocarbon tails are directed into the oil. Water and water-soluble drugs can be carried inside the core of the reverse micelle. Reverse micelles are relatively small because the large polar head groups are in the core. Aggregations numbers range up to the low 20s and 30s.

The *Kraft temperature* is a characteristic temperature for each surfactant. It is the temperature at which the surfactant solubility equals the critical micelle concentration. Below this unique Kraft temperature, micelles will not form. Nonionic surfactants undergo a separate phase transition at a higher temperature called the *cloud point* or *cloud temperature*. At the cloud point, the kinetic energy overcomes the hydrogen bonds formed between the surfactant and water, and the surfactant begins to precipitate. At high concentrations of nonionic surfactants (25 to 50%), liquid crystals may form, increasing the viscosity and forming gel-like structures.

Above the CMC, the surface tension no longer decreases and the formation of micelles can lead to significant increases in the solubility of lipophilic drugs by being solubilized in the lipophilic core of the micelle. The solubility of semipolar drugs may be increased by the drug intercalating between the polar heads and nonpolar hydrocarbon chains. The stability of drugs may also be improved once they are taken up into the micelle and protected by the nonaqueous hydrocarbon environment.

Several micelle-containing formulations are on the market. Micelles are used to increase the solubility of amphotericin B using sodium deoxycholate micelles (Fungizone). An aqueous colloidal dispersion of vitamin K (AquaMephyton) uses a polyoxyethylene fatty acid surfactant to form micelles to increase phytonadione's solubility. Soybean lecithin is used to solubilize an intravenous multivitamin mixture (Cernevit).

15.2.3.4.2 Liposomes and Niosomes Liposomes are spherical vesicles consisting of one or more concentric spheres of lipid or phospholipid bilayers that are separated by water. Niosomes are neutral liposomes made from nonionic surfactants and auxiliary agents such as cholesterol. Liposomes and niosomes can range in size from 50 nm to hundreds of micrometers. The nonionic surfactants that are used to prepare niosomes offer the advantages of generally being more chemically stable and less expensive than phospholipids. Liposomes are classified as multilamellar vesicles, small unilamellar vesicles, and large unilamellar vesicles, which are illustrated in Figure 15.7.

Although liposomes are constructed of amphiphiles, there are a number of differences between micelles and liposomes. The lamellar organization of liposomes was noted above. A high degree of surface activity does not itself lead to bilayer formation. Israelachvili et al. (1977) showed that the aggregate structure depended on a number of factors related to the *surfactant parameter*, N_s:

$$N_s = \frac{V_c}{L_c \sigma_A}$$
(15.15)

where N_s is the surfactant parameter, V_c the volume of the hydrophobic portion of the surfactant, L_c the length of the hydrocarbon chain(s), and σ_A the effective area per head group, which depends on the interaction with water and neighboring molecules. Bilayers form when N_S ranges from 0.5 to 1, and spherical micelles form when the surfactant parameter is approximately 0.33. Lipids that form bilayers have bulky alkyl chains, especially those with two fatty acid chains. This creates a relative large V_c compared to the effective area of the head group. This makes it difficult for the chains to fit inside a micelle. On the other hand, the bulky hydrocarbon chains can form bilayers more readily. Sodium lauryl sulfate has a large effective head group area compared to the volume of the hydrocarbon group, producing a lower N_S value, and micelle formation is spontaneous.

Compared to micelles, liposome formation commonly requires the expenditure of energy. Liposomes often require sonication or shear to form. This suggests

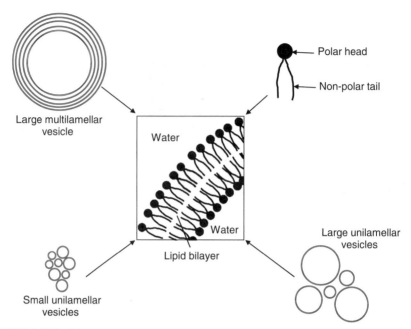

FIGURE 15.7 Liposome structures. (Adapted with permission from H. A., Lieberman, M. M., and Rieger, G. S., Banker, *Pharmaceutical Dosage Forms: Disperse Systems*, Informa Healthcare, New York, 1998.)

that liposomes are kinetically stable. That is, they are stable over a finite period of time, but they are not thermodynamically stable. Once the liposome is prepared, it tends to maintain its structure when diluted.

Liposomes have been fabricated from a number of amphiphilic compounds that commonly have two alkyl hydrocarbon chains instead of the single hydrocarbon-chain surfactants used to form micelles. The expanded V_c value resulting from the two alkyl chains leads to an N_S value of approximately 1. This surfactant parameter value supports a lamellar structure. The use of phospholipids, especially phosphatidylcholine (PC), also known as lecithin is a frequent practice. At physiological pH, phosphatidylcholines have a neutral charge. Other phospholipids that have been used in liposomes are phosphatidylethanolamine (PE), phosphatidylglycerol (PG), phosphatidylinositol (PI), and phosphatidylserine (PS). PE and PS are also neutral at physiological pH, whereas PG and PI are negatively charged. The fatty acid chains that attach to the phosphtidyl group can be saturated or unsaturated. The unsaturated fatty acids can have one, two, three, and up to six double bonds. The double bonds can readily undergo oxidation reactions, so most commercial products use semisynthetic lecithins that have been saturated. A number of synthetic saturated phospholipids are available: dimyristoylphosphatidylcholine, dimyristoylphosphatidylethanolamine, dimyristoylphosphatidylglycerol, dipalmitoylphosphatidylcholine, dipalmitoylphosphatidylethanolamine, dipalmitoylphosphatidylglycerol, dipalmitoylphosphatidylserine, and distearoylphosphatidylcholine. The structures of several phospholipids are shown in Figure 15.8.

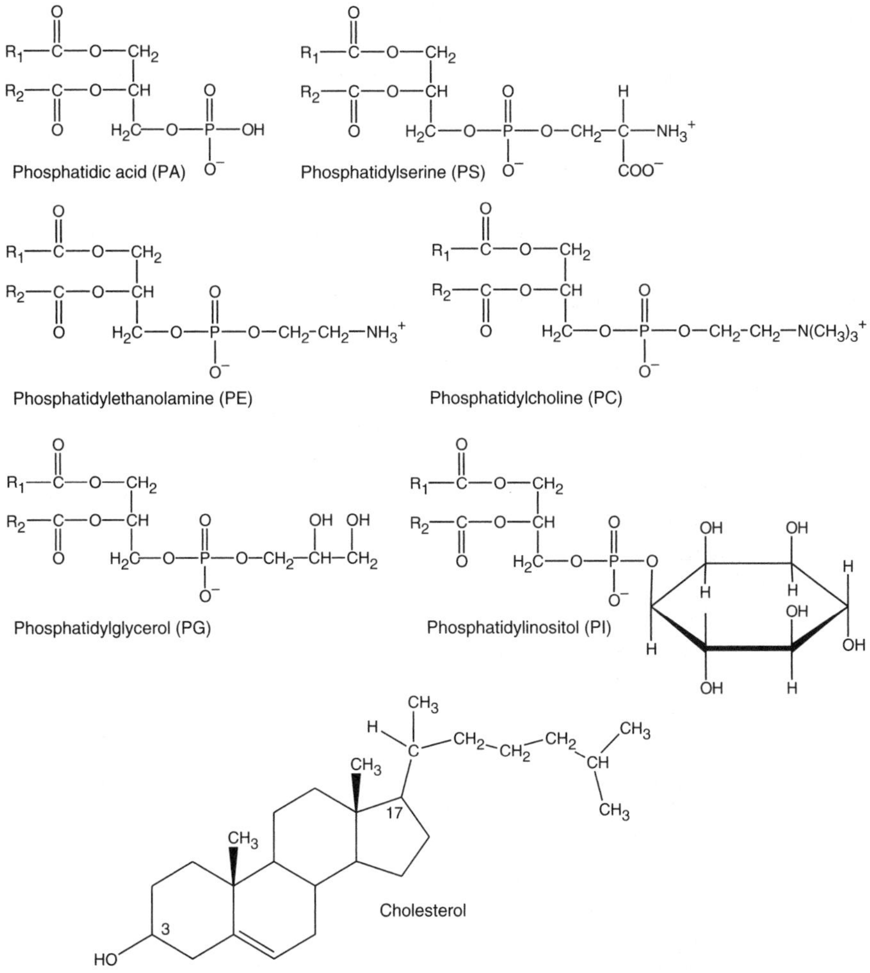

FIGURE 15.8 Phospholipids and cholesterol structures.

Semisynthetic and saturated phospholipids are used in commercial liposome preparations (see Chapter 17). AmBisome, a commercial amphotericin B liposome product, incorporates hydrogenated soy lecithin as one of the bilayer components. The naturally occurring double- and triple-bonded fatty acids are hydrogenated to create saturated fatty acids, which are more chemically stable than their unsaturated counterparts. The liposome product also contains saturated distearoylphophotidyl-glycerol and cholesterol. The phospholipids are rapidly hydrolyzed by endogenous phospholipases. "Stealth" liposomes have been designed to prolong the cir-culation time of the phospholipid liposomes by coating the liposomes with poly(ethylene glycol).

Cationic liposomes have also found some success in delivering DNA into cells. The more commonly used cationic lipids are dioctadecyldimethylammonium

bromide and dioctadecyldimethylammonium chloride, dioleoyloxypropyltrimethy-
lammonium chloride, and dioleoyloxytrimethylammoniopropane. The cationic
phospholipids often contain colipids such as dioleoylphosphatidylethanolamine.

15.2.3.4.3 Microemulsions A *microemulsion* is a colloidal mixture of
oil and water that forms a thermodynamically stable isotropic liquid dispersion.
Similar to oil–water dispersions, the microemulsion is stabilized using surfactants.
Microemulsions are also called *swollen micelles*. The microemulsions may be o/w,
w/o, or *bicontinuous* systems. A bicontinuous phase occurs when microdomains of
water and oil are interdispersed within the system. The size of the microemulsion
droplets are 10 to 200 nm. Like macroemulsions, the HLB value of the surfactant has a significant impact on the type of oil–water emulsion that is prepared.
Microemulsions will form only at ultralow water–oil interfacial tensions. The interfacial tension for microemulsions is estimated to be on the order of 1×10^{-3} mN/m,
compared to emulsion interfacial tensions, which range from 1 to 10 mN/m. This
ultralow interfacial tension is what compensates [equation (15.2)] for increases
in surface areas as high as 200 m^2/g. The overall process is entropy driven.
The large increase in the number of dispersed particles significantly increases the
entropy of the system. The key to reaching these very low interfacial tensions is
to create a highly condensed interfacial film between the water and oil. In contrast to macroemulsions that require 2 to 4% w/w surfactant, microemulsions can
require 10 to 30% w/w surfactant. Sometimes a cosurfactant such as ethanol or
longer-chain alcohols are used to enhance the condensed film. Short-chain alcohols tend to increase the head group compared to the hydrocarbon tails, which
leads to an N_s value [see equation (15.10)] below 1, which favors the interfacial
curvature toward the water. An $N_s = 1$ results in a lamellar phase as discussed
with liposomes or a bicontinous structure. An N_s packing value greater than 1
results in a w/o microemulsion that has the interface curve toward the oil phase.
Temperature affects the solubility of the surfactants, so it can affect the type of
microemulsion that is prepared. Understanding the temperature effects on surfactant
solubility and microemulsion stability should be evaluated early in microemulsion design. The advantages and disadvantages of microemulsions are given in
Table 15.9.

Microemulsions are most suitable for dermatological applications because the
high levels of surfactant make them poorly tolerated for oral use. Small quantities
of microemulsions may be used in intravenous lines and other parenteral uses.
Low-irritancy microemulsion formulations have been evaluated for ocular drug
delivery.

15.2.3.5 Semisolid Dispersions A semisolid has structure, is not pourable;
and does not conform to its container at room temperature. A semisolid deforms
readily under low pressure or shear stress and generally exhibits plastic flow.
Typically, semisolids are used for topical drug application. There are a number of
definitions for such topical dosages as lotions, gels, creams, and ointments. FDA

TABLE 15.9   Advantages and Disadvantages of Microemulsions

Advantages	Disadvantages

Improved physical stability
    Thermodyamically stable
    Microemulsion formation is
    reversible under thermal
    stress; microemulsions will
    reform on return from the
    thermal stress condition
Increased solubilization
Improved bioavailability and
    efficacy
	Thermal stress can break the microemulsion
	High surfactant levels are required
	Higher cost
	Higher risk of adverse events such as irritation
	    and/or poor taste
	Cosolvents may be costly and have unwanted
	    toxicities
	Affected by the ionic strength of the
	    gastrointestinal tract

scientists, Buhse et al. (2005) and others, have provided a topical classification system and decision tree to support the topical dosage-form nomenclature more systematically. Desired properties of a topical semisolid are that it have a smooth texture and be nonirritating, nonsensitizing, easy to apply, nongreasy, nonstaining, and washable. A large number of topical semisolid preparations are used to decrease dry flaky skin or to decrease wrinkles. Other vehicles are used to treat such topical conditions as eczema, itching, and infections. Still other preparations are designed to deliver drugs systemically, such as nitroglycerin and nicotine.

The U.S.P. classifies semisolid preparations intended for external application to the skin or mucous membranes as *ointments*. The U.S.P. further subdivided ointments into *absorption, hydrocarbon* or *oleaginous, water-removable* or *creams*, and *water-soluble bases* or *greaseless ointment bases*. The U.S.P. subdivisions are more or less based on how the bases interact with water. The U.S.P. term *ointment* is somewhat confusing, because most people think of an ointment as being greasy and not washable. The FDA (2006) defines an ointment as a semisolid dosage form that usually contains less than 20% water and volatiles and greater than 50% hydrocarbons, waxes, or polyols as the vehicle. Absorption bases as defined by the U.S.P. are divided into two classes. The first class of absorption bases allows water to be absorbed into the base to form a w/o semisolid emulsion. The second class is w/o emulsions that permit additional water to be incorporated into the base. Absorption bases often contain lanolin, which is a natural wax secreted from lambs' sebaceous glands. It is composed of long-chain fatty acid esters, lanolin alcohols, and hydrocarbons. Cholesterol is also used in absorption bases. Hydrocarbon bases are made of hard and soft paraffin. Water-removable bases are o/w emulsions, better known as *creams*. The FDA (2006) defines creams as semisolid emulsions that usually contain greater than 20% water and volatiles and/or less than 50% hydrocarbons, waxes, and polyols as the vehicle. Water-soluble bases are often prepared from low- and high-molecular-mass poly(ethylene glycol)s. Two groups of excipients seem to be omitted from both the U.S.P. and FDA definitions of bases:

fats and fixed oils, and silicones. Fats and fixed oils are mono-, di-, and triglycerides of saturated and unsaturated fatty acids from plant and animal origin. Silicones are water repellent and have low surface tension, so they can be incorporated readily into creams.

Hydrocarbon bases protect the skin from irritants and tend to hold moisture in the skin by forming an occlusive barrier. They are also smooth, easy to apply, nonirritating, and nonsensitizing. Silcones have similar benefits and can be incorporated readily into creams. Glycerin and propylene glycol are hygroscopic and can "soak up" water from the skin to decrease its hydration level. Water-soluble bases have the least hydration effect and are nonocclusive. O/w creams and water-washable bases will absorb skin fluid exudates that may result from weeping lesions.

15.2.3.5.1   Gels   A *gel* is a semisolid dosage form that contains a gelling agent to provide structure. The gel vehicle contains drug that may be in the form of a solution, suspension, or colloidal dispersion. A *hydrogel* is a dispersion of a polymer or long-chain molecules that is interpenetrated by water, which forms a three-dimensional, semisolid structural network. Hydrogels can be amorphous or semicrystalline in nature. Hydrogels have been prepared from poly(vinyl alcohol), carbopol, hydroxyethyl methacrylate, and polyoxyethylene. Gels are also referred to as *gelled solutions*. Highly concentrated polymers such as gelatin (1 to 15%), methylcellulose (1 to 5%), and starch (5 to 10%) can form clear to translucent gels. Gelatin and starch are, to some extent, *elastic gels*. Elastic gels deform to a force, but rebound when the force is removed. Other pharmaceutical gelling agents are alginic acid, guar gum, xanthan gum, carrageenans, nonionic surfactants, colloidal silicon dioxide, and other semisynthetic cellulose derivatives. Gels that are formed by insoluble particulates are referred to as *jellies*. Even though the three-dimensional structure spans the volume of the liquid, the liquid makes up the majority of the gel structure. Reversible gels are structured semisolid viscous dispersions at low temperature and unstructured low-viscosity dispersions or solutions at higher temperatures. Some polymers, such as carbopol, form gels (0.5 to 2%) with a change in pH. Carbopol polymers are formed by cross-linking acrylic acid and ally ethers of pentaerythritol. Carbopols form acidic dispersions in water, and when they are neutralized or made slightly basic, they form a gel structure. A number of different carbomers are listed in the U.S.P. This gelation process is reversed by adding acid to the gel. Gels can be made in a number of fluids such as water (hydrogels), oil (organogel), and air (aerogels). Structured foams use air as the continuous medium. Organogels can be formed with colloidal silicon dioxide, veegum, low-HLB surfactants, or magnesium stearate. Another nonaqueous gel contains 5% low-molecular-mass poly(ethylene glycol) (PEG) in mineral oil. The PEG is dissolved in mineral oil above 90°C and allowed to cool. As the PEG cools, it forms a three-dimensional gel structure where the mineral interpenetrates the PEG. Semisolid oil gels have been used for intramuscular depot injections. Extended-release systems employ pharmaceutically acceptable gelling agents that form a semisolid matrix in situ when given orally.

15.2.3.5.2 Creams Creams are semisolid o/w or w/o emulsions. The primary difference between liquid emulsions and creams is the higher amount of solid lipophilic materials that are present in creams. Excipients that increase the solid nature of creams are fatty alcohols (cetyl, palmitoyl, and stearoyl alcohol), stearic acid, cholesterol, and white wax. Creams are widely used for application to the skin or mucous membranes and are listed as a subcategory in the U.S.P. under ointments and water-removable bases. The FDA (2006) defines creams as semisolid emulsions that usually contain greater than 20% water and volatiles and/or less than 50% hydrocarbons, waxes, and polyols as the vehicle. More than 70 cream monographs are listed in the U.S.P. The majority of the creams are used for topical delivery of antifungal agents, antibiotics, and anti-inflammatory agents. There are a handful of vaginal creams that have U.S.P. monographs.

An interesting commercial cream formulation is EMLA, which exploits the use of a *eutectic mixture* of the two active topical analgesics, lidocaine and prilocaine. The eutectic mixture is a *single* chemical composition that has the lowest melting point compared to another mixture composition. In this case, lidocaine and prilocaine are both solids at room temperature. When the two anesthetics are mixed in equal weights, the resulting mixture melts at room temperature to form an oil. The melting point of the eutectic mixture is $16°C$. Product design scientists took advantage of this eutectic mixture to design an *oil*-in-water cream for use as a topical anesthetic. The formulation also includes polyoxyl 40 stearate (Myrj 52) as an emulsifier and carbomer (Carbopol) as a thickening agent. The pH of the emulsion is adjusted to approximately 9. Interestingly, the high pH value allows the product to pass the U.S.P. preservative effectiveness test without the need to add antimicrobial preservatives.

15.2.3.5.3 Pastes The FDA (2006) defines a paste as a semisolid dosage form that contains a large proportion (20 to 50%) of solids finely dispersed in a fatty vehicle. Pastes are shear thickening, that is, as the shear is increased, the viscosity or resistance to flow of the paste increases. Pastes are usually prepared for oleaginous bases such as white petrolatum. There are several U.S.P. monograph pastes. Probably the best known paste is zinc oxide paste, U.S.P. which is used in treating and preventing diaper rash. It is composed of 25% w/w zinc oxide, 25% w/w starch, and 50% w/w white petrolatum. Zinc oxide is also used to help prevent sunburns. It is one of the broadest ultraviolet A (315 to 400 nm) and ultraviolet B (280 to 315 nm) spectrum reflectors. Another listed U.S.P. paste is triamcinolone acetonide dental paste.

15.2.3.5.4 Ointments The FDA (2006) defines an ointment as a semisolid dosage form that usually contains less than 20% water and volatiles and greater than 50% hydrocarbons, waxes, or polyols as the vehicle. The U.S.P. defines an ointment as a semisolid preparation intended for external application to the skin or mucous membranes. The U.S.P. also defines a hydrocarbon base as white petrolatum or white ointment. The U.S.P. lists more than 60 topical and 20 ophthalmic ointments. Mupirocin (bactroban) nasal ointment is listed in a U.S.P.

monograph as a topical antibiotic. A number of ophthalmic ointments contain antiviral agents.

15.2.3.5.5 Scale-up of Semisolid Dispersions Semisolids demonstrate the dual rheological properties of liquids (viscosity) and solids (elasticity). The dual rheological nature is referred to as demonstrating viscoelastic properties. Semisolid products tend to "age" postmanufacuring, which results in changes in the viscoelastic properties of the semisolid over time. It is a good idea to let the semisolid material age before packaging to reach a viscoelastic steady state. The FDA has issued two guidance documents concerning scale-up and postapproval changes (SUPAC) (FDA, 1997, 1998). The 1997 document also discussed in vitro release testing and bioequivalence documentation of semisolid products. As with all the SUPAC guidances, changes are categorized by levels, depending on the extent and impact of the change. The guidance does recommend that the minimum size for an NDA pivotal clinical batch or biobatch be 100 kg or 10% of the production scale batch, whichever is larger. The 1998 guidance is an equipment addendum that classifies typical semisolid manufacturing equipment by unit operation and operating principle.

Special equipment is required to handle the higher-viscosity semisolid products. Anchor and scrap-surface mixers are used to manufacture semisolids. As the name indicates, an anchor mixer has an anchor-shaped impeller. Anchor mixers operate at variable low speeds (below 50 rpm) and high torque. The impeller has tight tolerances to the vessel sidewall, so it not only mixes the semisolid bulk material but also mixes the material on the sidewalls. In some cases it is necessary to add a scraper to remove the material adhering to the sidewalls. The scraper material is often made of a polymeric material such as propropylene.

A roller mill is frequently used as the final dispersion step to make sure that product is smooth and free of agglomerates. The shear rate for a three-roll mill is given by

$$\text{shear rate} = \frac{105r(\Delta\text{rpm})}{\text{nip clearance}} \tag{15.16}$$

where r is the roll radius and Δrpm is the difference in the roll speed between the center roller and the final roller. Equation (15.16) shows that the shear rate in seconds increases with the radius of the roll and roll speed. The shear rate is inversely proportional to the nip clearance. Equation (15.16) can also be used to adjust the roll mill parameters to achieve the same shear rate for a scaled process.

15.2.3.6 Solid Dispersions

15.2.3.6.1 Hot-Melt Solid Dispersions Solid dispersions can be formed as molecular, colloidal, or coarse drugs dispersed in a solid matrix. The colloidal and coarse drug can be in the amorphous or crystalline solid state. Hot-melt technology was discussed briefly in Chapter 13 as a means of forming running powder. Filling of molten drug matrices into hard capsules was discussed in Chapter 14. Hot-melt extrusion processes can efficiently use small quantities of drug at the early clinical development stage. Increasingly, drug discovery is identifying poorly soluble drug candidates for development. A significant amount of research has expanded the use of solid dispersions to enhance the solubility and bioavailability of poorly soluble drugs. Effervescent immediate-release hot-melt dispersions have been patented (Robinson and McGinity, 2000). Melt dispersions have found a number of applications in the last decade. As discussed in Chapter 14, technological advances have made it possible to fill hot-melt dispersions into hard-shell capsules on a commercial scale. The use of hot-melt extrusion technologies to form kinetically stable amorphous solid dispersed systems has also been exploited to enhance the bioavailability of poorly soluble, poorly absorbed drugs. Novel stabilizing systems that maintain the amorphous nature of the drug have been reported. Several companies have focused on stabilizing amorphous solids in a kinetically "trapped" metastable state that imparts a robust and acceptable product shelf life. Several review articles are recommended for further reading (Crowley et al., 2007; Repka et al. 2007)

Controlled-release hot-melt extrusion has found a number of applications. These are discussed in more detail in Chapter 20.

15.2.3.6.2 Spray-Dried Solid Dispersions Solid drug material can be dispersed in an aqueous or nonaqueous polymer suspension and spray-dried. The drug can also be dissolved in water or an other solvent system and spray-dried to form a solid dispersion of the drug in the polymer system. The drug can be in an amorphous or crystalline form. Amorphous solids are stabilized by polymer systems that are not hygroscopic and have high glass transition temperatures (T_g). Hydroxypropyl methylcellulose acetate succinate spray-dried dispersions (SDDs) of amorphous drugs have been shown to provide kinetically stable drug–polymer amorphous dispersions, to increase the observed drug solubility compared to the crystalline form, and to inhibit the crystallization of the drug in solution. These systems can be used to increase the solubility of poorly soluble drugs, their dissolution rate, and their bioavailability. Drug dissolution from SDD systems show that supersaturated drug levels are achieved rapidly and maintained for over 4 hours without the drug crystallizing out of solution (Friesen et al., 2008). Aqueous soluble stabilizing excipients have also been studied. Dextran, trileucine, and albumin are promising candidates for inhalation and parenteral SDDs. These spray-dried solid dispersions can be formulated into a number of different dosage forms, such as tablets, capsules, suspensions, and aerosols.

15.2.3.6.3 Suppositories The U.S.P. defines suppositories as solid bodies of various weights and shapes that are adopted for introduction into the rectal, vaginal, and urethral orifices of the human body. After insertion, suppositories soften, melt, disperse, and dissolve in the cavity fluids. The suppository should soften just below the cavity temperature. Suppository bases are divided into two general categories: fatty bases and water-soluble or dispersible bases.

Fatty bases are used primarily for rectal suppositories. Fatty bases are composed predominantly of triglycerides. Theobroma oil or cocoa butter has been used traditionally as a fatty suppository base. The use of hydrogenated triglyceride bases improves the base's resistance to oxidation. Both the naturally occurring theobroma oil and synthetic triglycerides are polymorphic. That is, they can exist in different crystalline forms which have different melting or softening transition temperatures. The α and β polymorphic forms of theobroma oil have been studied extensively. The α crystalline form melts below 30°C, which is the upper end of controlled room temperature. The α polymorph can be formed by heating the theobroma above 60°C and cooling it rapidly. This form converts slowly to the β polymorph, which has a melting point around 35°C. The manufacture of threobroma oil suppositories involves melting the base material at 40 to 50°C, dispersing or dissolving the drug in the molten liquid, pouring the liquid into suppository molds, cooling the suppositories to allow them to solidify, and collecting the suppositories as they are released from the molds.

Synthetic triglycerides exhibit less lot-to-lot variability than do natural triglycerides such as threobroma oil. The synthetic suppository bases are listed in the U.S.P. has *hard fat bases*. The commercial bases also have other additives to adjust the softening and melting points. Some have surfactants such as lecithin and polysorbates, which can help wet the drug.

Preformulation studies need to establish the drug's solubility in the suppository base as a function of temperature. Drugs that go into solution at the suppository manufacturing temperature and then crystallize on cooling may result in expansion of the suppository in the mold. The recrystallization process may also result in drug polymorphs that can affect dissolution and bioavailability. High-melting bases may be required for an oil-soluble drug that lowers the melting point of the base. To ensure proper dose per suppository, the dispersed drug's displacement volume needs to be determined and accounted for in the final volume fill. The relationship between Stokes' law and drug particle size needs to be understood to make sure that the drug is uniformly distributed in the suppository during cooling. Nonuniform distribution may result in suppository structural defects, causing unsuitable mold release and poor appearance.

Water-soluble or dispersible suppository bases do not depend on melting or softening to function. They are soluble or dispersible in the body cavity fluids. Water-soluble or dispersible bases use glycerinated gelatin, glycerin–sodium stearate, PEG, poly(oxyethylene stearate)s, and poly(oxyethylene sorbitan) fatty acid esters. Molded glycerinated vaginal gelatin suppositories are made by dispersing or dissolving the drug in about 10% w/w water and gelatin ($\sim$20% w/w) plasticized with glycerin ($\sim$70%). Higher levels of gelatin, up to 30%, form firmer suppositories, used for rectal insertion. Glycerinated gelatin suppositories are often used for vaginal drug delivery. They are hygroscopic and need to be protected from moisture. PEG suppositories are prepared by melting mixtures of different molecular mass to give varying degrees of solubility and melting range. The drug is incorporated in the molten liquid and the mixture is added to molds, which on cooling form the solid suppositories. Water-soluble or water-dispersible poly(oxyethlyene stearate) nonionic surfactants (e.g., polyoxyl 40 stearate, U.S.P.) are used to prepare melt–cool–mold suppositories. Similarly, nonionic polyoxyethylene sorbitan fatty acid esters (e.g., polysorbates) are also incorporated into suppositories.

Suppositories can be made on semiautomated or automated suppository equipment where the melt is dispensed into preformed foil or plastic molds that serve as the final primary package container. The filled molds are then conveyed on a chilling belt to a cooling chamber, where the cooling rate is tightly controlled. Suppositories can also be compression molded. This process is available for drugs that are heat sensitive. The suppository mixture is screw-fed into chilled molds. Pressure is applied to the mixture to fill the mold and form the suppository.

15.3   PROCESS DESIGN

Process specific design and scale-up issues have been discussed for various dispersed systems. For more detailed discussions of equipment selection, see the article by (Scott and Tabibi 1998), and for the scale-up of dispersed systems, see the articles by Block (1998) and Cherian and Portnoff (1998). Literature examples of PAT and scale-up applications are discussed below.

15.3.1   Coarse Suspensions

15.3.1.1   PAT: Coarse Suspensions   Focused-beam reflectance has been used to study flocculation (Blanco et al., 2002) and to report mean particle size and particle size distribution of suspended aggregates in-line and in real time. Although the example cited is from the paper industry, it can be applied to pharmaceutical suspensions as well. A commercial focused-beam reflectance measure system, which rotates a highly focused laser beam across particles at a known rotation speed and measures the backscattered light, was used in the study. Particle chord lengths are determined by knowing the laser beam rotational speed and the time the particle backscattered the light. Thousands of chord lengths are determined each second. In this study, deflocculation and reflocculation were studied as a function of shear rate and shear time. Particle size growth can also be measured. A focused-beam system has been used to follow the crystal growth that occurs during crystallization processes.

15.3.1.2   Scale-up of Suspensions   Process characterization and optimization studies have been carried out for the reconstitution of cefuroxime axetil powder (Jozwiakowski et al., 1990; Stagner, 1998). Process scale-up studies involved spray-coating a melted wax onto the drug to mask the exceptionally bitter taste. The primary trade-off was to maintain a bioequivalent product while providing an acceptable-tasting suspension. Taste attributes such as overall taste assessment, time to first appearance of bitterness, duration of maximum bitterness, and grittiness and consistency were evaluated on an ordinal scale. The score sheet for four of the 17 trials is shown in Figure 15.9.

Suspension dissolution and stability 14 days postreconstitution were also evaluated. The dominant process variables were powder bed temperature, molten wax temperature, atomization air pressure, and coating spray rate. The contour plots of overall taste and dissolution showed that these responses could be optimized at high atomization pressure, high temperature, and low spray rate. It is interesting to note that subjective product attributes such as taste can be modeled and used successfully in statistical designs. The bioavailability of the optimized

Participant :_____A_____

P.M.

Evaluate the following by checking (✓) your preference:

Definitions:

		SAMPLE	20-6	20-9	20-3	20-19
1. Overall taste assessment						
Score	Meaning	Feeling				
1	horrible unpalatable	☹	✓			✓
2	Very poor, some aversion to taking the product	☹				
3	Minimally acceptable, some bad taste	😐		✓		
4	Fair-pleasant, "not bad for medicine"	🙂			✓	
5	Very good, I like this stuff!	😊				
2. Time of first appearance of bitterness						
Score	Time					
1	Immediate					
2	0 – 5 sec.					✓
3	5 – 15 sec.		✓	✓		
4	15 – 30 sec.				✓	
5	>30 sec. (none)					
3. Duration of maximum bitterness						
Score	Time					
1	>5 min.					
2	2 – 5 min.					✓
3	1 – 2 min.		✓	✓		
4	15 – 60 sec.				✓	
5	0 – 15 sec. (none)					
4. Grittiness/consistency						
Score	Definition					
1	"Sandpaper in mouth"		✓			
2	not very smooth			✓		✓
3	neutral					
4	Smooth, easy to swallow				✓	
5	"melts in your mouth"					

FIGURE 15.9 Taste score sheet. (With permission from The Pediatric Pharmacy Advocacy Group, from Stagner, 1998)

process compared to that of a slow-dissolving powder was determined in six dogs. The optimized formulation provided significantly higher bioavailability than that of the slower-dissolving formulation.

15.3.2 Spray-Dried Solid Dispersions

15.3.2.1 Scale-up of Spray-Dried Solid Dispersions Dobry et al. (2009) proposed a model-based spray-drying development methodology that creates a data-driven knowledge-based approach to maximize efficient scale-up. As part of the process development scheme, critical spray-drying parameters were evaluated and a relationship between a specific drying ratio and the inlet dry gas temperature was plotted. The specific drying ratio is the ratio of solution feed rate to the drying-gas flow rate. Other critical process parameters that are defined at the process development stage are inlet dry gas temperature, outlet gas temperature, and relative saturation of the solvent in the gas at the spray-drying outlet conditions. Physiochemical properties of the drug and polymer system and the desired production output rate constrain these parameters. The maximum input gas temperature is restricted to temperatures that do not cause stickiness or drug degradation. Similarly, the maximum outlet temperature is based on product stability and other critical product attributes, such as particle morphology, density, dissolution, and bioavailability. The minimum outlet temperature is determined to some extent by the relative saturation limit placed on the drying gas. The resulting plot of specific drying ratio to inlet gas temperature provides the thermodynamic operating space and process design space. The scale-up of the spray-drying process would maintain the same critical process levels that were identified during the process design program.

15.4 CONTAINER CLOSURE SYSTEM DESIGN

The container closure system design requirements are different for liquids such as suspensions and emulsions than for semisolid and solid dispersions. The reader is directed to Sections 12.3, 13.3, 13.3.1, and 17.4.1 for further discussion of container closure systems. Suspensions and emulsions are usually provided in multiple-use plastic bottles with child-resistant caps, although clear plastic polyester terephthalate bottles have gained popularity. Clear bottles can be supplied in amber to provide additional light protection. Glass bottles provide the greatest protection from water vapor transmission loss, but they are breakable, which can be a disadvantage. Glass and high-density polyethylene wide-mouth jars are available for semisolid gels, creams, and ointments.

Most semisolid pharmaceutical products are filled into tubes. Tube materials acceptable for pharmaceutical packaging are aluminum, laminate, and plastic. The tubes need to be deformable or collapsible so that the semisolid can be squeezed readily out of the tube. Many tubes are made of high-purity aluminum (99.5%) that is recyclable. When the tubes are sealed properly, they are leak-free, airtight, impermeable to light, and can be printed or labeled. Laminate tubes can be made of purely plastic laminates (plastic barrier laminates) or a combination of plastics and aluminum foil (aluminum barrier laminates). A typical plastic barrier laminate may have as many as four or five polymer layers. Ethylvinyl alcohol is commonly used as a moisture and oxygen barrier. Polyethylenes of various densities are typically used in the other layers. Aluminum barrier laminates are comprised of polyethylene layers and an aluminum foil barrier layer. The aluminum barrier laminate is a good compromise compared to the cost of a pure aluminum tube and to the superior barrier properties compared to those of an all-plastic laminate tube. The tubes need to conform to the following U.S.P. chapters: <671>, Moisture Barrier Test; <88>, Safety of Material of Construction; <661> Extractable Plastics; and <660> Plastics. The reader is also directed to the FDA (1999) container closure guidance for additional regulatory considerations.

15.5 RISK MANAGEMENT

Risk management, which involves risk assessment and risk control, was discussed in Section 13.4. Risk identification is part of the risk assessment process. An Ishikawa or fishbone diagram can be used to visually identify which factors can affect product quality. Figure 15.10 is an example of an Ishikawa diagram for an emulsion product that identifies five main factors that might affect critical emulsion product attributes, such as particle size, creaming rate, and viscosity. The five main factors are surfactant, oil–water phase, other additives or excipients, emulsification process, and cooling process. Each of these factors have subfactors. For a given emulsion, some factors will have a greater impact on the final product quality. A semiquantitative ranking of factor impact can be accomplished by developing an impact score, indicating which factor has the greatest impact on the critical product qualities. A scoring scheme can be developed that assesses the probability and magnitude of impact. The probability and magnitude scores are then multiplied to give an impact score, which can be ranked to identify critical parameters that may need further confirmation through additional research.

15.6 ATTRIBUTE TESTS

15.6.1 Universal Tests

A number of tests are performed routinely on dispersed systems. In addition, there are specific tests that apply to different types of dispersions, and these are discussed under their respective products. All products at the initial testing time point will require a description or appearance (texture, phase separation, ability to redisperse easily, color, and odor), two drug identification tests, dose uniformity, pH, and an assay for the drug. Depending on the propensity of the drug to undergo polymorphic transitions, solid-state polymorphic purity characterization may be warranted. All dispersions containing water will require the U.S.P. Microbial Enumeration Test (Chapter <61>), Antimicrobial Effectiveness Test (<51>), and Antimicrobial

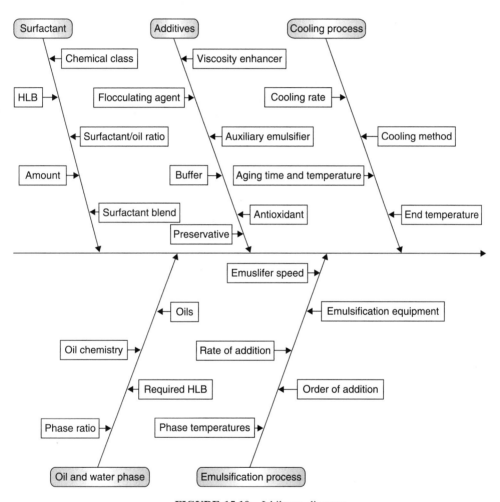

FIGURE 15.10   Ishikawa diagram.

End of the Stability Test Agents—Content (<341>). Identification and assay of preservatives are also done at the start of the stability program, at yearly intervals, and at end of life. The microbial limits are generally set at total viable aerobic count: not more than 10^2 aerobic bacteria and fungi colony-forming units (cfu) per milliliter or gram; total yeast and mold count: not more than 10 fungi cfu/mL or g; Enterobacteriaceae: not more than 10 cfu/mL or g; and absence of *Escherichia coli*, *Pseudomonas aeruginosa*, *Staphylococcus aureus*, and *Salmonella* species. Products should meet Minimum Fill (<755>) for liquids and semisolids.

The FDA requires that each excipient be identified by a specific test. Often a near-infrared or infrared spectrum is used to confirm the identity of an excipient. If the excipient is to be labeled as a pharmacopeia excipient such as N.F. or U.S.P., the excipient needs to be tested to the pharmacopeia standards. A quality program may approve a specific vendor of pharmacopeia excipients, which may allow a company to accept the U.S.P. testing provided by the vendor company under their certificate of analysis.

15.6.2 Suspensions and Macroemulsions

Typical suspension attribute tests include particle size, sedimentation or creaming rate, viscosity, resuspendability (suspensions), and redispersibility, caking (for suspensions), dissolution, pH, ζ-potential, specific gravity, and weight loss on stability. For constitutable products, the reconstitution dose, dose uniformity, and reconstitution time should be reported. *Simulated use tests*, including the stability of reconstituted powder and dose through exhaustion, also need to be undertaken. The protocol should simulate actual use of the reconstituted protocol, including the storage temperature, number of doses, and constituted shelf life.

15.6.3 Nanoparticle and Colloidal Systems

Particle size and dissolution are two critical parameters that should be determined at the initial product release and on stability. Specific gravity, weight loss on stability, and pH are other parameters that are routinely evaluated.

15.6.4 Semisolids

Product fill weight, viscosity, particle size, water content, phase separation, tube leaking, and weight loss on stability are customarily performed on gels, creams, and ointments. The minimum fill is based on the target fill weight (Minimum Fill, <755>). If the target fill weight is less than 60 g, the net content of any single container should not be less than 90% of the labeled amount. If the net content of any single container is between 60 and 150 g, the net content of any single container should not be less than 95% of the labeled amount. The content of gels, creams, and ointments is determined from the top, middle, and bottom of the tube. A leakage test for semisolids is provided in Chapter <771>, Ophthalmic Ointments.

The leak test is performed by placing 10 containers horizontally on blotting paper at $60 \pm 3\,^{\circ}$C for 8 hours. The test passes if none of the tubes show leakage. If one or more tubes leak, the test is repeated with 20 additional tubes. A test for metal particles, Metal Particles in Ophthalmic Ointments (<751>), involves transfer of the semisolid into a petri dish that is heated until the semisolid melts completely. The molten mass is allowed to resolidify and is examined for metal particles using an optical microscope. In vitro drug release testing methodology has been suggested by the FDA (1997).

15.6.5 Solid Dispersions

The solid-state form of the dispersed phase needs to be confirmed at initial release testing and on stability. In vitro release is also required. For suppositories, the softening and melting points are determined along with some measure of hardness. Depending on the type of suppository, the moisture content may also be measured.

15.7 NEW DRUG APPLICATION STABILITY ASSESSMENT

The primary purpose of stability assessment is to establish product label storage instructions and a shelf life for the product when it is stored under the labeled storage conditions. To meet a specified shelf life, the product must remain within a set of product specifications that are established for each critical product attribute. A balance must be struck between the time required to establish real-time data to support the labeled storage conditions and shelf life and the need to get important drugs to a patient as quickly as possible with the appropriate safety assurances. The regulated process allows an applicant for a new drug application (NDA) to use appropriately designed stability studies to use real-time data and statistical approaches to extrapolate an estimated shelf life. Extra samples should be placed on stability to allow for full retesting at one or two time points. NDA stability assessment is discussed in detail in Section 13.7, which also covers the FDA's guidance concerning the reporting, identification, and quantification for degradation impurities.

Stability protocols vary with dosage form and the storage condition proposed. Table 15.10 provides a generic stability protocol for a product to be stored at controlled room temperature.

A general freeze–thaw cycle for suspensions and macroemulsions is 2 days at -10°C and 2 days at 25°C for three cycles. Cycling is often 5 to 25°C for 6 hours for 12 days. Microemulsions can be studied under more strenuous conditions. A typical freeze–thaw cycle is -20 to $+30^{\circ}$C at 7-day cycles for 28 days (one month). A usual cycle study involves temperature cycling from 5 to 40°C every 6 hours for a month.

TABLE 15.10 NDA Stability Protocol

Product	Container/Closure and Supplier	API Process
Product lot	Package lot	API lot
Batch size	Batch size	Batch size
Manufacturer	Packager	Manufacturer
Manufacturer site	Packaging site	Site of manufacture
Manufacture date	Package date	Manufacture date

Storage (°C/% RH)	Time (months)									
	1	3	6	9	12	18	24	36	48	60
$25/60^{a,b}$		×	×	×	×	×	×	×	$(×)^c$	$(×)^c$
$40/75^{a,b}$	×	×								
Light^d	×									
Freeze–thaw^e	×									
Cycle^f	×									

^{a}°C ±2/% relative humidity ±5%.

^{b}Liquid samples are stored upright and horizontal to expose the liquid to the cap.

^{c}(×), optional testing.

^{d}Testing per International Conference on Harmonization guidelines.

^{e}Two days at −10°C, 2 days at 25°C for three cycles.

^{f}5 to 25°C every 6 hours for 12 days.

REFERENCES

Becher P, Schick MJ. Macroemulsions. In: Schick MJ, ed *Non-ionic Surfactants Physical Chemistry*. Surfactant Sciences Series, Vol. 23. New York: Marcel Dekker; 1987. pp. 435–491.

Blanco A, Fuente E, Negro C, Tijero J. Flocculation monitoring: focused beam reflectance measurement as a measurement tool. Can. J. Chem. Eng. 2002;80:1–7.

Block LH. Scale-up of disperse systems: theoretical and practical aspects. In: Lieberman HA, Rieger MM, Banker GS, eds. *Pharmaceutical Dosage Forms: Disperse Systems*, 2nd ed., Vol. 3. New York: Marcel Dekker; 1998. pp. 363–394.

Bondoc LL Jr, Fitzpatrick S. Size distribution analysis of recombinant adenovirus using disc centrifugation. J. Ind. Microbiol. Biotechnol. 1998;20:317–322.

Bowman BJ, Ofner CM III, Schott H. Colloidal dispersions. In: Troy DB, ed. *Remington: The Science and Practice of Pharmacy*, 21st ed. Philadelphia: Lippincott Williams & Wilkins; 2006. pp. 293–318.

Buhse L, Kolinski R, Westenberger B, Wokovich A, Spencer J, Chen CW, Turujman S, Gautam-Basak M, Kang GJ, Kibbe A, Heintzelman B, Wolgang E. Topical drug classification. Int. J. Pharm. 2005;295:101–112.

Callahan JC, Cleary GW, Elefant M, Kaplan G, Kensler T, Nash RA. Equilibrium moisture content of pharmaceutical excipients. Drug Dev. Ind. Pharm. 1982;8:355–369.

Cherian M, Portnoff JB. Scale-up of dispersed parenteral dosage forms. In: Lieberman HA, Rieger MM, Banker GS, eds. *Pharmaceutical Dosage Forms: Disperse Systems*, 2nd, ed., Vol. 3. New York: Marcel Dekker; 1998. pp. 395–422.

Crowley MM, Zang F, Repka MA, Thumma S, Upadhye SB, Battu SK, McGinnity JW, Martin C. Pharmaceutical applications of hot-melt extrusion: 1. Drug Dev. Ind. Pharm. 2007;33:909–926.

Dobry DE, Settell DM, Baumann JM, Ray RJ, Graham LJ, Beyerinck RA. A model-based methodology for spray-drying process development. J. Pharm. Innov. 2009;4:133–142.

Eccleston GM. Emulsions. In: Swarbrick J, Boylan JC, eds. *Encyclopedia of Pharmaceutical Technology*, Vol. 5. New York: Marcel Dekker; 1992. pp. 137–188.

FDA. 1997. Guidance for industry: Nonsterile semisolid dosage forms—scale-up and postapproval changes; chemistry, manufacturing, and controls; in vitro release testing and in vivo bioequivalence documentation. U.S. Department of Health and Human Services, Food and Drug Administration Center for Drug and Evaluation and Research. May 1997. http://www.fda.gov/downloads/Drugs/GuidanceComplianceRegulatoryInformation/Guidances/ucm070930.pdf. Accessed Aug. 2011.

————. 1998. Guidance for industry: SUPAC-SS—Nonsterile semisolid dosage forms: manufacturing equipment addendum. U.S. Department of Health and Human Services, Food and Drug Administration Center for Drug and Evaluation and Research. Dec. 1998. http://www.fda.gov/downloads/Drugs/GuidanceComplianceRegulatoryInformation/Guidances/ucm070928.pdf. Accessed Aug. 2011.

————. 1999. Guidance for industry: Container closure systems for packaging human drugs and biologics. U.S. Department of Health and Human Services, Food and Drug Administration Center for Drug and Evaluation and Research and Center for Biologics Evaluation and Research. May 1999. http://www.fda.gov/downloads/Drugs/GuidanceComplianceRegulatoryInformation/Guidances/ucm070551.pdf. Accessed Feb. 2012.

————. 2006. CDER data standards manual definitions for topical dosage forms. Center for Drug Evaluation and Research. http://www.fda.gov/ohrms/dockets/ac/06/briefing/2006-4241B1-02-30-FDA-Topical%20Dosage%20Forms%20Definitions%20Old%20%20.pdf. Accessed Feb. 2012.

————. 2010. Reporting format for nanotechnology-related information in CMC review. Manual of Policies and Procedures. Office of Pharmaceutical Science, Center for Drug Evaluation and Research. MAPP 5015.9. http://www.fda.gov./downloads/AboutFDA/CentersOffices/CDER/ManualofPoliciesProcedures/UCM214304.pdf. Accessed Feb. 2012.

Friberg S. Mesomorphous phases, a factor of importance for the properties of emulsions. J. Colloid Interface Sci. 1969;29:155–156.

Friesen DT, Shanker R, Crew M, Smithey DT, Curatolo WJ, Nightingale JAS. Hydroxypropyl methylcellulose acetate succinate–based spray-dried dispersions: an overview. Mol. Pharm. 2008;5:1003–1019.

Grant J, Cho J, Allen C. Self-assembly and physicochemical and rheological properties of a polysaccharide–surfactant system formed from the cationic biopolymer chitosan and nonionic sorbitan esters. Langmuir. 2006;22:4327–4335.

Griffin WC. Classification of surface-active agents by "HLB." J. Soc. Cosmet. Chem. 1949;1:311–326.

Griffin WC. Calculation of HLB values of nonionic surfactants. J. Soc. Cosmet. Chem. 1954;5:249–256.

Guess WL. Determination of the hydrophile–lipophile balance of acacia using various surfactants. J. Pharm. Sci. 1961;50:238–239.

Israelachvili JN, Mitchell DJ, Ninham BW. Theory of self-assembly of lipids bilayers and vesicles. Biochim. Biophys. Acta. 1977;470:185–201.

Jozwiakowski MJ, Jones DM, Franz RM. Characterization of a hot-melt fluid bed coating process for fine granules. Pharm Res. 1990;7:1119–1126.

Kelly JY, DeSimone JM. Shape-specific, monodisperse nano-molding of protein particles. J. Am. Chem. Soc. 2008;130:5438–5439.

Kim DK, Dobson J. Nanomedicine for targeted drug delivery. J. Mater. Chem. 2009;19:6294–6307.

Kopecký F, Fazekaš T, Kopecká B, Kaclík P. Hydrophobicity and critical micelle concentration of some quaternary ammonium salts with one or two hydrophobic tails. Acta Faculty Pharm. Univ. Comenianae 2007;54:84–94.

Nanjwade BK, Bechra HM, Derkar GK, Manvi FV, Nanjwade VK. Dendrimers: emerging polymers for drug-delivery systems. Eur. J. Pharm. Sci. 2009;38:185–96.

Nash RA. Suspensions. In: Swarbrick J, Boylan JC, eds. *Encyclopedia of Pharmaceutical Technology*, 2nd ed., Vol. 3. New York: Marcel Dekker; 2002. pp. 2654–2668.

Orafidiya LO, Oladimeji FA. Determination of the required HLB values of some essential oils. Int. J. Pharm. 2002;237:241–249.

Pfeifer A, Verma IM. Gene therapy: promises and problems. Annu. Rev. Genom. Hum. Genet. 2001;2:177–211.

Repka MA, Battu SK, Upadhye SB, Thumma S, Crowley MM, Zhang F, Martin C, McGinnity JW. Pharmaceutical applications of hot-melt extrusion: II. Drug Dev. Ind. Pharm. 2007;33:1043–1057.

Robinson JR, McGinity JW. Effervescent granules and methods for their preparation. U.S. Patent 6071539. June 6, 2000.

Rosoff M. Specialized pharmaceutical emulsions. In: Lieberman HA, Rieger MM, Banker GS, eds. *Pharmaceutical Dosage Forms: Disperse Systems* 2nd ed; Vol. 3. New York: Marcel Dekker, 1998. pp. 1–42.

Rowe RC, Sheskey PJ, Owen SC, eds. *Handbook of Pharmaceutical Excipients*. 5th ed. London: Pharmaceutical Press; Washington; DC: American Pharmacists Association; 2006.

Rowe RC, Sheskey, PJ, Quinn, ME, eds. *Handbook of Pharmaceutical Excipients*, 6th eds. London: Pharmaceutical Press; Washington; DC: American Pharmacists Association; 2009.

Sagitani H. Making homogeneous and fine droplet o/w emulsions using nonionic surfactants. J. Am. Oil. Chem. Soc. 1981;58:738–743.

Scott RR, Tabibi SE. A practical guide to equipment selection and operating techniques. In: Lieberman HA, Rieger MM, Banker GS, eds. *Pharmaceutical Dosage Forms: Disperse Systems*, 2nd ed., Vol. 3. New York: Marcel Dekker; 1998. pp. 291–362.

Shinoda K, Arai H. The correlation between phase inversion temperature in emulsion and cloud point in solution of nonionic emulsifier. J. Phys. Chem. 1964;68:3485–3490.

Stagner WC. Pediatric drug formulation challenges: technical and regulatory perspectives. 1998; J. Pediatr. Pharm. Pract. 1998;3:203–211.

Swarbrick J, Rubino JT, Rubino OP. Coarse dispersions. In: Troy DB, ed. *Remington: the science and practice of pharmacy*. 21st ed. Philadelphia: Lippincott Williams & Wilkins; 2006. pp. 319–337.

U.S. Pharmacopeia. U.S.P.–N.F. online, First Supplement U.S.P. 33–N.F. 28 Reissue. Accessed Oct. 10, 2010.

GLOSSARY

API	Active pharmaceutical ingredient.
CMC	Critical micelle concentration.
DMPC	Dimyristoylphosphatidylcholine.
DMPE	Dimyristoylphosphatidylethanolamine.
DMPG	Dimyristoylphosphatidylglycerol.
DLVO theory	Named after Derjaguin and Landau of Russia and Verwey and Overbeek of the Netherlands, this theory describes the interplay between electrostatic repulsion and van der Waals attractive forces as a function of particle distance.
DODAB	Dioctadecyldimethylammonium bromide.
DODAC	Dioctadecyldimethylammonium chloride.
DOTMA	Dioleoyloxypropyltrimethylammonium chloride.
DPPC	Dipalmitoylphophatidylcholine.
DPPE	Dipalmitoylphophatidylethanolamine.
DPPG	Dipalmitoylphophatidylglycerol.
DPPS	Dipalmitoylphophatidylserine.
DSC	Differential Scanning Calorimeter.
DSPC	Distearolyphosphatidylcholine.
EIP	Emulsion inversion point.
E.P.	*European Pharmacopoeia*.
FDA	U.S. Food and Drug Administration.
HLB	Hydrophile–lipophile balance.
HPMCAS	Hydroxypropylmethylcellulose acetate succinate.
J.P.	Japanese Pharmacopoeia.
LUV	Large unilamellar vesicles.
MLV	multilamellar vesicles.
NDA	New drug application.
N.F.	*National Formulary*.
o/w	Oil-in-water emulsion.
o/w/o	Oil-in-water-in-oil multiple emulsion; the primary emulsion (o/w) is the dispersed phase and the secondary emulsion (o) is the external phase.

PAMAM	Polyamidoamine.
PAT	Process analytical technology.
PC	Phosphatidylcholine.
PE	Phosphatidylethanolamine.
PEG	Poly(ethylene glycol).
PG	Phosphatidylglycerol.
PI	Phosphatidylinositol.
PIT	Phase inversion temperature.
PLA	Poly(lactic acid).
PS	Phosphatidylserine.
RHLB	Required hydrophile-lipophile balance.
SDD	Spray-dried dispersion.
SUPAC	Scale-up and postapproval changes.
SUV	Small unilamellar vescicles.
U.S.P.	*United States Pharmacopeia.*
w/o	Water-in-oil emulsion.
w/o/w	Water-in-oil-in-water multiple emulsion; the primary emulsion (w/o) is the dispersedphase and the secondary emulsion (w) is the external phase.

APPENDIXES

APPENDIX 15.1 Suspending, Viscosity-Enhancing, and Structure-Modifying Agents

Excipient	Trade Names[a]	Rheology	Use[b](% w/v)
Acacia, E.J./J.P./U.S.P.–N.F.	—	Pseudoplastic	64.80
Alginate sodium, E.P./U.S.P.–N.F.	Kelcosol, Keltone, Protanal	Newtonian/ pseudoplastic	0.12
Bentonite, E.P./J.P./U.S.P.–N.F.	Albagel, hydrated aluminum silicate, polargel, Veegum HS	Plastic/thixotropic	1.3^b–6.5^c
Carbomer, E.P./U.S.P.–N.F.	Acrypol, Carbopol, Pemulen	Plastic	0.45^b–5.0^c
Carboxymethylcellulose sodium, E.P./J.P./U.S.P.–N.F.	Aquasorb, Cethylose, Nymcel ZSB, SCMC	Pseudoplastic	0.1 —40.0, 50 mg
Carrageenan, U.S.P.–N.F.	Gelcarin, Genu, Viscarin	Newtonian/pseudoplastic	1.5, 20.15 mg
Dimyristoyl- phosphotidylcholine	—	—	0.15–0.34, 7.05^d
Dioleophosphotidylcholine	—	—	0.42
Dipalmitoyl- phosphotidylcholine	—	—	0.045^b–0.09^b
Guar gum, E.P./U.S.P.–N.F.	Galactosol, Meyprodor	Plastic/thixotropic	0.099^b–0.49^c
Hectorite	Hectabrite DP, Laponite	Plastic/thixotropic	
Hydroxyethyl cellulose, E.P./U.S.P.–N.F.	Cellosize HEC, Natrosol, Tylose H	Pseudoplastic	0.25^b–15.0^c
Hydroxypropyl cellulose, E.P./J.P./U.S.P.–N.F.	Klucel, Nisso HPC	Pseudoplastic	0.0004–6.7

(*continued*)

APPENDIX 15.1 (*Continued*)

Excipient	Trade Names[a]	Rheology	Use[b](% w/v)
Hypromellose, E.P./J.P./U.S.P.–N.F.	Hydroxypropyl methylcellulose, Methocel, Metolose, Tylopur	Pseudoplastic	0.4–3.0
Magnesium aluminum silicate, E.P./U.S.P.–N.F.	Gelsorb, Pharmasorb, Veegum	Plastic/thixotropic	0.85[b]–60[c]
Methylcellulose, E.P./J.P./U.S.P.–N.F.	Cellacol, Viscol	Plastic/pseudoplastic	0.025–1.19
Microcrystalline cellulose, E.P./J.P./U.S.P.–N.F.	Avicel PH, Celex, Pharmacel, Vivapur	Plastic/thixotropic	1.45–58.65
Microcrystalline cellulose and carboxymethylcellulose sodium, E.P./U.S.P.–N.F.	Avicel CL-611, Avicel RC-591	Plastic/thixotropic	1.90–11.25
Poloxamer, E.P./U.S.P.–N.F.	Lutrol, Pluronic, Supronic	—	0.009–12.6
Poly(vinyl alcohol), E.P./U.S.P.–N.F.	Elvanol, Mowiol, Polyvinol	Newtonian	0.05–1.4
Povidone, E.P./J.P./U.S.P.–N.F.	Kolloidon, Plasdone, Poly(vinylpyrrolidone)	Newtonian/ pseudoplastic	0.26—0.6, 0.27–75 mg
Tragacanth, E.P./J.P./U.S.P.–N.F.	—	Pseudoplastic	0.266[b]–24[c]
Xanthum gum, E.P./U.S.P.–N.F.	Keltrol, Vanzan NF, Xantural	Plastic/thixotropic	0.05–18.68

[a] A comprehensive list of products and suppliers is provided in Rowe et al. (2009).

[b] FDA Inactive Ingredients List. http://www.fda.gov/Drugs/InformationOnDrugs/ucm113978.htm. (Accessed Jan. 2012.

[c] FDA Inactive Ingredients List. http://www.fda.gov/Drugs/InformationOnDrugs/ucm113978.htm. (Accessed June 2011.

[d] Listed for solutions.

APPENDIX 15.2 Sweetening Agents

Excipient	Trade Names[a]	Use[b](% w/v, mg)
Acesulfame potassium, E.P./U.S.P.–N.F.	Sunett, Sweet One	0.12–0.9, 3.0—117.0 mg
Aspartame, E.P./U.S.P.–N.F.	Equal, Nutra Sweet, Sanecta	0.0125–45.0, 0.8–233 mg
Mannitol, E.P./J.P./U.S.P.–N.F.	Emprove, Mannogem, Pearlitol	2.94[b]–29.36[c], 8.6–1035.17 mg
Neotame, U.S.P.endash N.F.	—	0.0011
Saccharin sodium, E.P./J.P./U.S.P.–N.F.	Crystallose	0.05–32, 0.4—20 mg
Sorbitol, E.P./J.P./U.S.P.–N.F.	Neosorb, Sorbitab	3.0–90.0, 3.11—337.28 mg

APPENDIX 15.2 (*Continued*)

Excipient	Trade Names[a]	Use[b](% w/v)
Stevia	PureVia, Reb-A, Rebiana, Sun Crystals, SweetLeaf, Truvia	$0.04-0.31^d$, $0.04-4.0$ mg[d]
Sucralose, U.S.P.–N.F.	Splenda	$0.03^a-8.0^b$, $0.018-5.75$ mg
Sucrose, E.P./J.P./U.S.P.–N.F.	—	$12.0-93.24$, $12.0-1200.0$ mg

[a] Rowe et al. (2009).

[b] FDA Inactive Ingredients List, http://www.fda.gov/Drugs/InformationOnDrugs/ucm113978.htm. (Accessed Jan. 2012).

[c] FDA Inactive Ingredients List, http://www.fda.gov/Drugs/InformationOnDrugs/ucm113978.htm. (Accessed June 2011.

[d] Stevia is 300 to 400 times sweeter than cane sugar, which is sucrose. Stevia in the raw, http://www.steviaextractintheraw.com/FAQs.aspx?gclid=CPmDnOTza0CFQpV7AodCXLziw. Accessed on Jan. 2012.

APPENDIX 15.3   Flocculating Agents and Buffers

Electrolytes	Nonelectrolytes
$CaCl_2$, E.P./J.P./U.S.P.–N.F.	Benzyl alcohol, E.P./J.P./U.S.P.–N.F.
KCl, E.P./J.P./U.S.P.–N.F.	Phenylethyl alcohol, U.S.P.–N.F.
NaCl, E.P./J.P./U.S.P.–N.F.	Lecithin, U.S.P.–N.F.
NaH_2PO_4, E.P./U.S.P.–N.F.	
Na_2HPO_4, E.P./J.P./U.S.P.–N.F.	
Na_2SO_4, E.P./U.S.P.–N.F.	

APPENDIX 15.4   Emulsifying Agents

Emulsifier	Trade Names[a]	HLB	Use[b](% w/v)
	Anionic		
Acacia, E.P./J.P./U.S.P.–N.F.	Generic, gum arabic	11.9	12.5
Ammonium lauryl sulfate	Akyposalals 33	31^c	39.75
Sodium lauryl sulfate, E.P./J.P./U.S.P.–N.F.	Elfan 240, Texapon K12P	40^d	$0.33-2.5$
Trolamine lauryl sulfate	Maprofix TLS 500, TEA-lauryl sulfate	34^c	$0.13-10.8$
	Nonionic		
Glyceryl monostearate, E.P./J.P./U.S.P.–N.F.	Capmul GMS-50, GMS, Kessco GMS, Tegin 515	3.8^d	7^b-20^e
Hexylene glycol	—	—	12^e
Lanolin alcohols, E.P./U.S.P.–N.F.	Agrowax, Ritawax, Super Hartolan	1^c	6
Oleic acid, E.P./U.S.P.–N.F.	Crodolene, Emersol, Priolene	1^f	25

(*continued*)

APPENDIX 15.4 (*Continued*)

Emulsifier	Trade Names[a]	HLB	Use[b](% w/v)
Polysorbate 40, E.P./U.S.P.–N.F.	Liposorb P-20, Montanox 40, polyoxyethylene 20 sorbitan monopalmitate, Tween 40	15.6[c]	6.0
Polysorbate 60, E.P./U.S.P.–N.F.	Liposorb S-20, Montanox 60, polyoxyethylene 20 sorbitan monostearate, Tween 60	14.9[c]	0.42–8.0
Polysorbate 80, E.P./J.P./U.S.P.–N.F.	Liposorb O-20, Montanox 80, polyoxyethylene 20 sorbitan monooleate, Tween 80	15.0[c]	0.5–5.0
Propylene glycol monostearate, E.P.	Atlas G-922, Atlas G-2158	3.4[c]	7.0–9.3[e]
Sorbitan monolaurate, E.P./U.S.P.–N.F.	Arlacel 20, Liposorb L, Montane 20, Span 20	8.6[d]	4.74
Sorbitan monooleate, E.P./U.S.P.–N.F.	Arlacel 80, Liposorb O, Montane 80, Span 80	4.3[d]	0.2–3.5
Sorbitan monopalmitate, E.P./U.S.P.–N.F.	Arlacel 40, Liposorb P, Montane 40, Span 40	6.7[d]	2.0
Sorbitan monostearate, E.P./U.S.P.–N.F.	Arlacel 60, Liposorb S, Montane 60, Span 60	4.7[d]	2.0–8.0
Sorbitan sesquioleate, E.P./J.P./U.S.P.–N.F.	Protachem SQI	3.7[d]	2.0[g]
Sorbitan trioleate, E.P./U.S.P.–N.F.	Aracel 85, Liposorb TO, Montane 85, Span 85	1.8[d]	
	Cationic		
Dimyristoyl-phosphotidylcholine	—	—	0.15–0.34[h]7.05[i]
Dioleophosphotidylcholine	—	—	0.42[h]
Dipalmitoyl-phosphotidylcholine	—	—	0.045[b,h]—0.09[e,h]
Lecithin, U.S.P.–N.F.	Phospholipon 100H, ProKote LSC, Sternpur	8.0[c]	0.33–1.2

[a] A comprehensive list of products and suppliers is provided in Rowe et al. (2009).

[b] FDA Inactive Ingredients List, http://www.fda.gov/Drugs/InformationOnDrugs/ucm113978.htm. Accessed Jan. 2012.

[c] Becher and Schick (1987).

[d] Griffin (1954).

[e] FDA Inactive Ingredients List, http://www.fda.gov/Drugs/InformationOnDrugs/ucm113978.htm. Accessed June 2011.

[f] Griffin (1949).

[g] Listed for ointments.

[h] Listed for suspensions.

[i] Listed for solutions

[j] Guess (1961).

[k] U.S.P.

APPENDIX 15.5   Required HLB for Oils and Waxes

Oils and Waxes	Emulsion Type	Required HLB	Use (% w/v)[a]
Alcohol			
Cetyl	o/w	13[b]	3.23–15[c]
Decyl	o/w	15[d]	
Hexadecyl	o/w	11–12[d]	
Isodecyl	o/w	14[d]	
Lauryl	o/w	14[d]	
Oleyl	o/w	14[d]	10
Stearyl	o/w	15–16[d]	1.0–42.5
	w/o	7[e]	
Tridecyl	o/w	14[d]	
Beeswax	w/o	5[e]	3.5–5
	o/w	10–16[b]	
Butyl stearate	o/w	11[d]	3.7
Candelilla wax	o/w	14.5[e]	
Carnauba wax	o/w	14.5[e]	
Castor oil	o/w	14[d]	5–12.5
Ceresin wax	o/w	8[d]	7
Cocoa butter	o/w	6[d]	
Corn oil	o/w	8[d]	
Cottonseed oil	w/o		
	o/w	7.5[b]	
Decyl acetate	o/w	11[d]	
Diisooctyl phthalate	o/w	13[d]	
Dimethyl silicone	o/w	9[d]	
Ethyl aniline	o/w	13[d]	
Ethyl benzoate	o/w	13[d]	
Gelatin	o/w	17[e]	
Isopropyl myristate	o/w	12[d]	1–10
Isopropyl palmitate	o/w	12[d]	1.8–5.5
Isostearic acid	o/w	15–16[d]	25
Lanolin	o/w		2
Lanolin anhydrous	w/o	8[b]	0.2–2
	o/w	15[b]	
Lauric acid	o/w	16[d]	
Lauryl amine	o/w	12[d]	
Linoleic acid	o/w	16[d]	
Maize oil	o/w	8[d]	
Methyl silicone	o/w	11[d]	
Methylphenyl silicone	o/w	7[d]	
Microcrystalline wax	o/w	9.5[b]	0.45
Mineral oil, heavy	w/o	4[b]	15–50.62
Mineral oil, light	w/o	4[b]	5–25
Nonyl phenol	o/w	14[d]	

(*continued*)

APPENDIX 15.5 (*Continued*)

Oils and Waxes	Emulsion Type	Required HLB	Use (% w/v)[a]
Oleic acid	o/w	17^d	25
Palm oil	o/w	7^e	
Paraffin	w/o	4^b	4.5–15
	o/w	9^b	
Petrolatum	w/o	4^b	5.3–58.2
	o/w	10.5^e	
Polyethylene wax	o/w	15^d	
Propylene, tetramer	o/w	14^d	
Rapeseed oil	o/w	7^d	
Ricinoleic acid	o/w	16^d	
Safflower oil	o/w	7^d	
Silicone oil	o/w	10.5^e	
Soybean oil	o/w	6^d	10
Stearic acid	w/o	—	3.0–22.6
	o/w	17^e	

[a] FDA Inactive Ingredients List, http://www.fda.gov/Drugs/InformationOnDrugs/ucm113978.htm. Accessed Jan. 2012.

[b] Griffin (1954).

[c] FDA Inactive Ingredients List, http://www.fda.gov/Drugs/InformationOnDrugs/ucm113978.htm. Accessed June 2011.

[d] Becher and schick (1987).

[e] Griffin (1949).

CHAPTER 16

AEROSOL PRODUCT DESIGN

16.1 INTRODUCTION

An aerosol is the dispersion of a liquid or solid particle in a gas. Pharmaceutical aerosols are developed to deliver drugs topically to body organs such as the skin and to mucous membranes such as the mouth and nose. Foams are a specialized form of a three-phase aerosol system that has a gas or propellant in the internal phase of an oil-in-water emulsion that is used for application to the skin or vagina. The terms *inhalation* and *pulmonary aerosol delivery* are often used synonymously for aerosolized drugs that are delivered to the lungs. In this chapter we focus on the product design of inhalation and nasal aerosols.

Inhalation aerosols have significantly improved the treatment of asthma, chronic obstructive pulmonary disease, and other lung diseases, such as cystic fibrosis. Bronchodilators and anti-inflammatory drugs have been targeted directly to the lung receptors, which has resulted in increased efficacy and reduced side effects. Similarly, deoxyribonuclease and the antibiotic aztreonam lysine have also been approved for inhalation drug therapy. The lung is a reasonably sensitive membrane, and this sensitivity can be exacerbated by lung disease. Instillation of drugs and even normal saline can cause reactions ranging from bronchospasms to hyperreactive lungs. Drugs and excipients can cause irritation and immunological responses, so special care and awareness are required in drug and excipient selection. As an additional level of safety, aqueous pulmonary inhalation products are sterilized.

As has been mentioned on several occasions, product design is a complex multifactorial process that integrates drug delivery to treat a human condition with formulation, manufacturing process, and container closure system design. The quality of the final product is critically dependent on the integration of the design of these three components. This is especially true for aerosol products. The marketing of metered-dose inhalers, nasal aerosols, and powder inhalers has required the integration of the formulation, process, and container closure into what is commonly called the *inhalation drug delivery system*. A drug delivery system has been defined "as a sophisticated dosage form, which, by its construction, is able to modify and control the availability of the drug substance to the body by temporal and spatial considerations" (Cambridge Healthtech Institute, 2011). In the case of inhalation

Integrated Pharmaceutics: Applied Preformulation, Product Design, and Regulatory Science,
First Edition. Antoine Al-Achi, Mali Ram Gupta, William Craig Stagner.
© 2013 John Wiley & Sons, Inc. Published 2013 by John Wiley & Sons, Inc.

and nasal aerosols, all aspects of the drug delivery system—formulation, process, container, and accessories such as spacers—collectively affect the delivered dose.

The three primary inhalation systems discussed in this chapter are nebulizers, pressurized metered-dose inhalers (MDIs), and metered-dose dry powder inhalers (DPIs). With the advent of newer nebulization technologies, the use of *metered-dose liquid inhalers* is now used to distinguish them from the more conventional nebulizers. To be effective, these systems must deliver particles that are typically 5 μm or smaller. This particle size range is often referred to as the *respirable dose* or *fine-particle fraction*. Fluid-mill micronization with a size classifier is frequently used to manufacture drug particles in the respirable range. Other methods of preparing micronized particles include spray drying and solvent or supercritical fluid precipitation. The extent of hygroscopic particle growth that can occur in the lungs when the drug is exposed to almost 100% relative humidity needs to be considered when determining the size requirements for the processed drug substance.

A number of issues must be addressed to achieve dose reproducibility through the useful life of a product. The most critical product attributes are dose reproducibility, particle size, and particle size distribution. Nebulizers and pressurized inhalers provide energy to disrupt aggregate particles and generate droplets or solid particles that have the correct particle size. DPIs, on the other hand, often depend on the patient's inspiration velocity to accomplish deaggregation by forcing the powder through baffles and screens. In some cases, deaggregation energy is provided by compressed gas or motor-driven propeller dispersers. More details regarding the specific fabrication designs are provided in Section 16.3.

A general category of nasal and oral cavity aerosols is also discussed briefly. Delivery of aerosols to the buccal cavity and sublingual membrane has increased over the past few years. The anatomy and physiology of the nasal cavity influences drug delivery and particle impact. The human nasal cavity has an approximate 90° bend, a volume of 16 to 19 mL, a mucous lining containing a number of enzymes, and cilia cells that move particles toward the nasopharynx. The particles are then sent to the stomach by swallowing. The 90° angle causes almost 100% of the particles to have a greater than 10-μm aerodynamic diameter to deposit on the back of the nasal cavity. Almost 50% of aerosols containing 3-μm particles are deposited in the nasal cavity, and 80% deposition is observed for aerosols having 5-μm droplets or particles (Chien et al., 1989).

16.2 FORMULATION DESIGN

16.2.1 Inhalation Aerosol Formulation Design

Inhalation aerosols must traverse the oral pharyngeal area and the *pulmonary tree*, which consists of the trachea, bronchi, bronchioles, terminal bronchioles, respiratory bronchioles, alveolar ducts, and finally, the alveolar sacs, to reach the highly vascularized alveoli. Along the way, the aerosolized particles are exposed to a high relative humidity of around 99.5% at 37°C, mucus, cilia, and enzymes. The areas from the trachea to the terminal bronchioles are often referred to as *conducting airways* because gas exchange does not occur in this part of the pulmonary tree. This part of the lung anatomy is the primary location of airway obstruction caused by asthma and chronic obstruction pulmonary disease. The respiratory bronchioles, alveolar ducts, and alveolar sacs are called *respiratory airways*. Gas exchange occurs in the respiratory airways. As one goes down the pulmonary tree, the diameter of the passages narrows, with the trachea being about 18,000 μm in diameter and the alveoli having a diameter of approximately 400 μm.

Particles deposit in the airways as a function of their aerodynamic particle size. What is an aerodynamic particle size? The aerodynamic particle size takes into account how the particle's size, shape, and density affect how it settles in air compared to a spherical particle that has the same terminal settling velocity and a density of 1 g/cm^3 or 1 kg/L. Particles with an aerodynamic size of 5 to 10 μm or greater will be deposited on the back of the throat and upper airways. This deposition is due to particle inertia that cannot be overcome when the airflow bends at the back of the throat. Smaller particles can compensate or relax and follow the airflow farther down the pulmonary tree. Powders having aerodynamic particles sized between 0.5 and 5 μm are deposited in the tracheobronchial and respiratory airways. Aerodynamic particle sizes smaller than 0.5 μm are deposited by Brownian motion and diffusion. The advantages and disadvantages of inhalation drug delivery are provided in Table 16.1.

Portable delivery devices that target drugs directly to the lungs have significantly improved the treatment of asthma and chronic pulmonary disease. Targeted drug delivery to the lungs has increased the drug's efficacy while decreasing its side effects. Albuterol is a good example of a drug that has a narrow therapeutic window. The side effects can result in serious cardiovascular arrhythmias, dizziness, headaches, and vomiting. Introduction of the pressurized metered-dose inhaler created a paradigm shift in the treatment of asthma. The dose of albuterol and side effects were meaningfully reduced. The MDI dose was 200 μg every 4 to 6 hours, compared to the tablet dosage regimen of 4000 to 8000 μg and the syrup regimen of 2000 to 4000 μg every 3 to 4 hours. Significant advances are now being made with the local treatment of pulmonary infections (Le et al., 2010). Taylor and Kellaway (2001) also provide an overview of pulmonary drug delivery.

16.2.1.1 *Preformulation of Inhalation Aerosols*   The purpose of a preformulation study is to evaluate a drug's physiochemical properties and identify specific issues that place the development program at risk. For example, drug solubility is more important in developing nebulizing solutions and metered-dose inhalers compared to dry powder inhalers. In the case of nebulizers, good aqueous drug solubility is desired. On the other hand, MDIs require low drug solubility (parts per million) in the liquid propellants unless the drug is soluble in the liquid propellant or cosolvents and a solution aerosol can be designed. As is the case for all dosage-form development projects, acceptable drug stability is required for a successful product. The aim of an efficient preformulation program should be to provide the necessary information in the shortest possible time and at the least cost. Knowledge gained during the developability stage can help highlight the key factors

TABLE 16.1   Advantages and Disadvantages of Inhalation Aerosols

Advantages	Disadvantages
Rapid onset of action	Expensive delivery systems that need to target the drug to various areas in the lung
Noninvasive	Performance can deteriorate over time
Can treat the diseased tissue or organ directly, which generally requires less drug to achieve efficacy	Limited to potent low-dose drugs
Low doses generally result in lower systemic toxicity	Complex physiochemical constraints Critical particle size distribution Dose uniformity Associated delivery device often has complex components and moving parts
More convenient and less painful than injection	Effectiveness may be limited to patient's disease state and inspiratory flow rate or the ability of the patient to coordinate actuation with breathing
Less harsh absorption environment than that of the gastrointestinal tract	The lungs in diseased patients are generally hyperreactive and may be sensitive to the drug or carriers, which may cause bronchospasms
Absence of first-pass metabolism	Drugs and excipients may cause irritation, bronchospasm, and allergenic response
For systemic action, the lung membranes have a highly vascular surface area that is often more permeable to drugs than are other membranes	Mucus may decrease drug bioavailability

that put the development program at risk. A data-driven preformulation program should leverage this knowledge and emphasize obtaining additional knowledge in key areas. A stability-indicating assay that is selective for the drug substance is required for all preformulation programs. Aerosol preformulation activities include physiochemical solid- and solution-state characterization and excipient compatibility. Once a stability-indicating method has been developed, drug solid- and solution-state stability studies can be initiated. Table 16.2 lists the solid properties that have important formulation and product attribute implications.

Table 16.3 outlines the solution and chemical properties that are evaluated during preformulation.

16.2.1.2   Excipient Compatibility of Inhalation Aerosols   Compared to the other dosage forms, only a modest number of excipients are typically used in aerosol

TABLE 16.2   Solid-State Physical Properties and Aerosol Implications

Physical Properties	Aerosol Implications
Organoleptic properties: appearance, taste, and odor	Odor may elicit an olfactory response when a drug is being inhaled.
	Taste can result in an adverse event, especially if poor taste leads to vomiting. Taste is not usually evaluated until an investigational new drug (IND) is available. It may be possible to perform a taste trial under an exploratory IND if precautions are taken to minimize exposure. Extraordinarily poor taste can "kill" a potential drug candidate.
Morphology and micromeritics: crystal habit, birefringence, particle size, particle shape, particle size distribution, bulk density, tapped density, true density, state of agglomeration, and surface area	The particle's shape, size, and density will affect its aerodynamic size.
	The particle's aerodynamic size will affect its deposition in the lungs.
Thermal properties: melting point, enthalpy of fusion, and other thermal transitions	Melting point is a reflection of a drug's purity and solubility.
Crystallinity and polymorphism	Amorphous material is less stable physically and chemically than its crystalline counterpart. Amorphous material is generally more hygroscopic than its crystalline counterpoint. In the 99.5% relative humidity (RH) of the lung, hygroscopic amorphous drugs may sorb a significant amount of water, which can increase its particle size and affect its lung deposition.
Polymorphism tendency	Some drugs have a greater propensity than others to form polymorphs. Since polymorphs can have a significant effect on solubility, stability, hygroscopicity, manufacturability, and bioavailability, it is important to be aware of the drug's tendency to form polymorphs.
	Can affect the fraction of drug that is emitted as the fine particle fraction.
Hygroscopicity and moisture adsorption–desorption isotherm[a]	The degree of hygroscopicity will dictate the proper storage and handling conditions.
	Moisture sorption in the lung can affect the drug's deposition.

(*continued*)

TABLE 16.2 (*Continued*)

Physical Properties	Aerosol Implications
Contact angle	The contact angle indicates how wettable the drug substance is. To suspend the drug in the liquid propellants, it is necessary to wet the drug with the propellant to achieve an adequate suspension.
Static charge and explosivity	The dust clouds that result from handling powders can be lethally explosive. Knowing the explosive potential of powders is crucial to maintaining a safe work environment.
	Static charge can cause agglomeration of the particles and affect their aerodynamic particle size.

[a] Callahan et al. (1982).

formulations. Safety concerns have limited the number of excipients available to the pharmaceutical design scientist. As mentioned in Section 16.1, the lung can become hyperactive when exposed to foreign substances, and this is especially true for patients who suffer from asthma. Foreign substances can be irritating to the lungs and lead to bronchial constriction and possible unwanted immunological responses. The U.S. Food and Drug Administration (FDA) guidance (1998, 2002a,b) for nasal and inhalation products makes it clear that the sensitive nature of the patient population places additional burdens on efforts to completely characterize and control the quality of the excipients. The guidance implies that the *United States Pharmacopeia* (U.S.P.)–*National Formulary* (N.F.) excipient monographs will probably not provide an acceptable level of characterization and control. The 1998 draft guidance discussed the characterization and control expectations for a number of common aerosol excipients, such as dehydrated alcohol, lecithin, oleic acid, lactose monohydrate, and propellants. A draft acceptance criterion was given for HFA-134a. A rather detailed discussion of the physical, chemical, and microbiological assessment of lactose monohydrate was also included. α-Lactose monohydrate is the most frequent inert carrier for dry powder inhalers. Lactose is a reducing sugar and can react with primary and secondary amines and form degradation products. Physiochemical factors that can influence drug–excipient adhesion and redispersion during inhalation were discussed, such as surface roughness, surface energy, crystallinity, level of hydration, and purity. The microbiological purity was also addressed, with emphasis on microbial limits and bacterial endotoxins. Analysis of a specific and quantitative level of proteins was mentioned. The absence of transmissible spongiform encephalopathy infectious agents should also be determined or guaranteed. The types of inhalation excipients are given in Table 16.4.

Appendix 16.1 provides a list of the excipients that is available through the FDA Inactive Ingredients List. The appendix lists the function and the use

TABLE 16.3 Solution and Chemical Properties and Their Aerosol Implications

Chemical Properties[a]	Aerosol Implications
Purity and impurities: Impurities may include degradation products, materials related to the synthesis, organic volatiles, residual solvents, and heavy metals	This is a quality attribute. These properties can affect safety and efficacy. These properties can affect manufacturability.
Solubility: in hydrofluoroalkane (HFA) propellants HFA-134a and HFA-227; water; 5, 10, 15% ethanol–HFA cosolvent mixtures; 5, 10, 15% glycerol–HFA cosolvent mixtures; HFA saturated with water; and pH values ranging from 4 to 8.	Solubility in these solvents affects the developability of the drug in a particular dosage form. If the drug is highly soluble in water, it is a good candidate for nebulizer development. If the drug is highly soluble in an HFA, it would be a good candidate for pressurized MDI development. The solubility of water in HFAs is reasonably high and can lead to solubilization and crystal growth of water-soluble drugs that are developed as suspensions. (Water solubility: HFA 134a: 20°C, ~800 ppm; 30°C, ~1200 ppm and HFA 227: 20°C, ~620 ppm: 30°C, ~690 ppm.)[b]
Partition–distribution coefficient	Partition–distribution coefficient can affect bioavailability.
Acid–base dissociation constant	Drug ionization state affects a drug's solubility, dissolution rate, stability, and bioavailability as a function of pH.
Solid- and solution-state stability: as a function of room temperature and stressed temperature, pH, oxygen, and light	Stability knowledge affects formulation, process, and container closure decisions.

[a] These properties are included here for two reasons: (1) one needs a chemical assay to measure these properties, and (2) the chemical nature of these molecules affects these attributes.

[b] E.I. du Pont de Nemours (2011).

concentrations that are reported. The design of the excipient compatibility is similar to that discussed in Chapters 13 and 15. Since the number of excipients is so limited for the various inhalation technologies, it is possible to combine the excipient compatibility study with early probe formulations studies that evaluate the effect of the level of the excipients on the stability and performance of the inhaler system.

16.2.1.3 *Inhalation Aerosol Formulation Development*

The three primary inhalation systems discussed in this chapter are nebulizers, pressurized MDIs and metered-dose DPIs. The advantages and disadvantages of these inhalation systems are presented in Table 16.5.

TABLE 16.4 Excipient Classes and Function

Excipient Class	Function
Nebulizer	
Water	Solvent for drug.
Tonicity-adjusting agents	Used to adjust the tonicity of the nebulizing solution.
Antimicrobial agents	Used to maintain the sterility of multiple-use containers. It should be noted that the FDA and European Medicines Agency (EMEA) discourage the use of aqueous multiuse containers because of concern over the use of antimicrobial agents and risk of lack of sterility.
pH-adjusting agents	Maintain stability and solubility.
Pressurized metered-dose inhaler	
Propellants	Provides the energy to deliver and disperse the aerosol without relying on the patient's inspiration velocity and energy.
Cosolvent	Used to solubilize the drug.
Surfactants	Used to wet, disperse, and suspend the drug particles.
Dry powder inhaler	
Inert carrier material	The carrier serves to improve the flowability of the active powder mixture.
	The drug is distributed uniformly in the carrier.
	The carrier has a large enough particle size that it does not deposit significantly in the lungs.
	The weak attractive forces between the drug and the carrier are overcome during the inhalation step, which allows the micronized drug to be separated from the carrier material and deposited in the lungs.

16.2.1.3.1 Nebulizer Formulation Development and Specific Manufacturing Processes Nebulizers are devices that generate droplet aerosols from aqueous or hydroalcoholic solutions. Nebulizers are approved by the Center for Devices and Radiological Health under the submission of what is called a 510(k). Solid particles can also be generated from a suspension such as budesonide inhalation suspension. The trade-off between drug dose and solubility generally dictates whether a nebulizing solution or suspension is developed. A large dose may require an extended nebulization time, say longer than 15 minutes, if the solution concentration of drug is too low. In this case, to achieve a higher concentration of drug that can be delivered in a shorter period of time, a suspension of an insoluble salt form of the drug might be chosen for development.

Nebulizing liquids need to be manufactured as sterile products (see Chapter 17 for a discussion of pharmaceutical sterilization processes). Many of the

TABLE 16.5   Comparative Advantages and Disadvantages of Conventional Nebulizers, Pressurized MDIs, and Passive (Breath-Actuated) DPIs

Conventional Nebulizer	Pressurized MDI	Passive DPI
	Advantages	
Simple solution: less potential for formulation problems (suspensions are more complex)	Independent of patient's inspiratory flow rate	Simple formulation
Deliver higher doses	Portable	Breath-actuated requires less need for patient inhalation coordination
High dose uniformity	Greater problems than nebulizer with dose uniformity	Propellant-free
Independent of patient's inspiratory flow	Ready for patient use	Less potential for extraction of contaminants from the device components
Less need for patient inhalation coordination	No sterility requirement	No sterility requirement
Propellant-free		Can use with higher-dose drugs ($\sim$2 to 5 mg)
		Minimal concern about extractables
	Disadvantages	
Requires longer treatment times	Requires propellant	Respirable dose is dependent on the patient's inspiratory flow rate and pattern (including device air-flow resistant design issues)
Aqueous system needs to be sterilized	Needs patient coordination with actuated dose	Dose non uniformity
Less portable device	Dose nonuniformity can arise from device complications	Complexity and robustness of the delivery device design
Requires a new drug application (NDA) and possibly a device 510(k) approval	Extractable adulterating material	Large variability between performance of different devices
FDA requires testing with different nebulizer devices if the device is included in the product label	Temperature dependent: emitted pressure and effect on aerodynamic particle size	Powder carrier may cause a cough reflex

(continued)

TABLE 16.5 *(Continued)*

Conventional Nebulizer	Pressurized MDI	Passive DPI
Limited to water-soluble drugs or highly water-insoluble drugs (formulation becomes more complex)	Limited to low-dose drugs	Some devices are less protected from environmental moisture
Low-density polyethylene (LDPE) containers are prone to diffusional contamination of label-adhesive materials and environmental permeants such as naturally occurring vanillin from cartoning products	High exit velocity may require the use of a spacer	Complex interactions between the drug and the inert carrier
LDPE containers can require protective foil pouches. Once the container(s) is removed from the protective foil, the product must be used within a specified time (~20 to 30 days)	Crystal growth and crystalline adhesion to the can walls and valve components	
Drugs are typically less stable in aqueous solution or suspension	Accuracy of the metered valve is limited	

nebulizing liquids are prepared as unit-dose products. **Antimicrobial preservatives** (see Appendixes 16.1 and 17.4) should be used for multiuse containers to maintain product sterility. As noted above, it is very unlikely that multiuse aqueous inhalation containers will be approved by the FDA and the EMEA. The product should also be isotonic and have a pH value around 7. When other pH values need to be used for solubility or stability reasons, the buffer capacity of the buffer system should be relatively low (0.01 or so). Surfactants and antioxidants may also be incorporated into the design of the formulation.

Four main types of nebulizers are used to deliver active pharmaceutical ingredients: jet nebulizers (DeVilbiss Model 646-T; Invacare Reusable Jet Nebulizer; Omron 9973 Jet Air II Nebulizer; Pari-LC-Jet Plus Nebulizer), ultrasonic nebulizers (Aeroneb, AeroGen; Beurer Nebulizer IH50; Omron U22; Pari eFlow; Respironics i-Neb), high-pressure microspray nebulizers (Respimat, Boehringer Ingleheim; AERx, Aradigm), and electrohydrodynamic nebulizers (Mystic EHD, Battelle-Pharma). The AeroGen pulmonary technology is discussed in more detail by Uster (2007).

Jet nebulizers form respirable drug particles by forcing high-velocity air through a nebulizing liquid, which is forced through a small orifice that creates a small particle fraction. Significant advances have been made to reduce the size and noise of the compressor used to generate the high-velocity air or oxygen. The nebulizer liquid volume is usually 1 or 2 mL. Nebulization times can be as short as 3 to 5 minutes at mean airflow rates of 5.5 L/min. Actual doses at the mouthpiece vary from nebulizer and liquid formulation. Budesonide Inhalation Suspension has a 17% nominal dose at the mouth when it is nebulized with the Pari-LC-Jet Plus Nebulizer and Pari Master Compressor System (Budesonide Inhalation Suspension package insert). Air-jet nebulizers have been discussed by Niven and Brain (1994), who evaluated a variety of nebulizers and studied the effect of nebulization pressure on particle size. The recombinant human deoxyribonuclease protein is nebulized with a number of recommended jet nebulizers and compressor combinations. The protein is provided in a single-dose sterile ampoule that needs to be stored at 2 to 8°C and protected from light. The drug solution tonicity is adjusted with sodium chloride.

Ultrasonic nebulizers use an electronic generator that creates an ultrasonic frequency that causes a piezoelectric cell to vibrate at that frequency. The piezoelectric cell is brought into contact with the nebulizing liquid. The energy from the ultrasonic frequency is transmitted to the liquid, which is aerosolized into a respirable aerosol. The main advantage of ultrasonic nebulizers is their ability to fabricate portable handheld devices that can operate on batteries. Ultrasonic nebulizers are also much quieter than jet nebulizers. The major disadvantage is that the ultrasonic energy raises the temperature of the nebulizing liquid, which can denature proteins and other thermally unstable molecules. It is more difficult to achieve effective nebulization with high viscosity and suspension formulations.

Another ultrasound technology uses a vibrating mesh membrane, which is activated by an ultrasonic vibrating piezoelectric cell, creating an aerosol from a porous mesh membrane that has thousands of laser-drilled pores. This type of nebulizer is often referred to as *vibrating mesh technology* (VMT). A liquid sample of a predetermined volume is applied to the rear of the membrane. Ultrasonic energy is applied to the membrane and a liquid droplet is ejected from the front of the membrane (Pillai and Rolland, 2001; Ivri and Wu, 2007; Pfichner and Pumm, 2009). The Omron 14 VMT has attempted to overcome the earlier design problems related to hole clogging and mesh durability. The AeroGen, AeroNeb, and the Pari e-Flow use a curved vibrating porous membrane technology that increases the dosing flexibility. These systems are capable of delivering small and large volumes of low- and high-dose drugs. They are also capable of delivering suspensions.

High-pressure microsprays have been patented and commercialized. These systems are also referred to as *soft mist inhalers*. The Respimat, a portable handheld multidose metered-dose nebulizer, is shown in Figure 16.1. This device uses a spring to create tension which when released squeezes a flexible container that holds the nebulizing liquid. The pressure forces the liquid to flow through nozzles that atomize the nebulizing fluid. An atomizing nozzle and spray-generating device has been described by Bartels et al. (1999). Another high-pressure nebulizer has recently been patented (Kelliher et al., 2011). The drug is aerosolized by forcing the

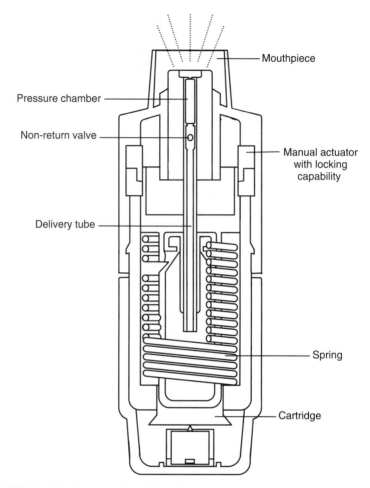

FIGURE 16.1 Respimat Soft Mist inhaler. (Adapted from Kunze et al., 2011.)

liquid formulation through a porous membrane. The particle size is controlled by the properties of the nebulizing liquid and the size and arrangement of the pores. The AERx is a portable handheld device that expels drug through a single-use nozzle that contains small laser-drilled holes that deliver particles in the respirable range. Multiuse metered doses are delivered from a collapsible container. The AERx particle size distribution and percent dose delivery were featured in articles by Schuster et al. (1997) and Schustr and Farr (2007).

The Battlelle Pharma Mystic EHD nebulizer uses electrohydrodynamic dispersion to produce respirable aerosols (Davies et al., 1997; Davies et al., 2011). The device works by passing the nebulizing liquid through a capillary. An electric field causes the liquid in the capillary to form a cone, and small respirable droplets form at the apex of the cone.

The Phillip Respironics i-Neb AAD (adaptive aerosol delivery) system is portable, small (6 inches high), relatively lightweight (~7.5 ounces), quiet, and has a long battery life (40 treatments per charge). It uses a VMT that delivers respirable drug particles with each breath. The device includes a breath analyzer which accesses a patient's breathing pattern during drug delivery. There is a feed-back loop that provides breathing information to the patient and adjusts the dosing accordingly. Denyer et al. (2008) have described a patented AAD system. To learn more about nebulizers, the reader is directed to Knoch and Finlay (2007) and Watts et al. (2008).

Nebulizing solutions are often prepared and filled using a blow–fill–seal process that fills and packages a sterilized nebulizing solution. This specialized sterile fill process is an aseptic process. The packaging material used in this process is low-density polyethylene (LDPE). Pellets of pure LDPE are poured into a hopper and melted. The melted fluid is then injected into a mold and air pushes the melted fluid against the mold. The LDPE cools and hardens to form the walls of the container. Once the walls have been formed, the nebulizing solution is filled into the mold and the two halves of the mold are heat-sealed. The molds can have identifying names, strength, and shape that are embossed or debossed on the nebule. LDPE is highly permeable to volatile chemicals from the local environment and from label adhesives. Product recall has resulted when contaminants from adhesive and parts per million of vanillin (which originated from the natural components of the carton material) were identified in the LDPE nebulizing solution. The FDA found that a majority of non-overwrapped LDPE nebules contained chemical contaminants from secondary packaging and labeling components. Therefore, it is highly recommended not to use paper labels with adhesive on LDPE vials and to overwrap the ampoules with an impervious material such as aluminum foil or other impervious flexible packaging.

16.2.1.3.2 Pressurized MDI Formulation Development and Specific Manufacturing Processes

Historically, MDI referred to an aerosolized drug that was delivered by a unit-dose metering chamber that used liquefied gas pressure and expansion postactuation to form an aerosol. The accurate unit-dose metering chamber distinguished MDIs from nebulizers. Now that there are a number of inhalation systems that provide a single accurately measured dose without the use of liquefied propellants, the term *pressurized MDI* more accurately describes this inhalation system. *Pressurized MDI* and *MDI* are used synonymously in this chapter.

The MDI formulation, manufacturing process, and container closure system are intimately interrelated and interdependent. The performance of an MDI inhala-tion system is directly dependent on the interrelationships between the formulation, manufacturing process, and the container closure system. The complexity of this dosage form is acknowledged by the FDA. The FDA considers an MDI a device. Sponsors of an MDI should discuss with the FDA whether the MDI should be submitted as a device in a 510(k) submission or as a drug in an NDA. FDA inter-center agreements between the Center for Devices and Radiological Health and the Center for Drug Evaluation and Research (CDER) provide guidance as to which center will review the submission. In some cases, both centers may be involved with a review. In this case, one of the centers will take the primary lead and will coordinate the review with the sponsor and the other center. In general, an aerosol delivery device is considered a drug product and regulated by the CDER "when the primary purpose of the device is delivering or aiding the delivery of a *specific* [italicized by the author for emphasis] drug and the device is distributed with the drug" (FDA, 1993).

The inhaler system consists of three major components: a propellant–drug formulation, a container closure system (container and metering valve), and an actuator–mouthpiece, which is referred to as an *inhalation system* (see Figure 16.2). Each of these components has a number of important parameters that affect the dose delivered and the aerodynamic particle size. The metering value (Figure 16.3) is comprised of a number of critical components that control dose delivery. The complexity of the MDI also results from the multiple physical processes that occur during actuation and aerosolization. Upon actuation, the aerosol propellant formulation is released at high velocity (∼15 m/s), which decelerates to about 3 m/s after 10 ms. It undergoes rapid expansion and flash evaporation. As discussed above, the clinical efficacy of the aerosol is critically dependent on the aerodynamic diameter of the drug. The size and shape of the exit actuator nozzle, propellant pressure, surface tension of the formulation, and suspension or solution concentration can affect the plume geometry and the size of the liquid droplets. The initial droplet sizes typically range from 20 to 50 μm and generally decreases to 5 μm or less. The rate of evaporation and droplet size reduction depend on the initial size of the droplet. It takes a 50-μm droplet 24 times longer to vaporize than it takes a 10-μm droplet. The liquid droplets may contain solid drug in the case of a suspension formulation or a drug in solution. As the volatile propellant evaporates,

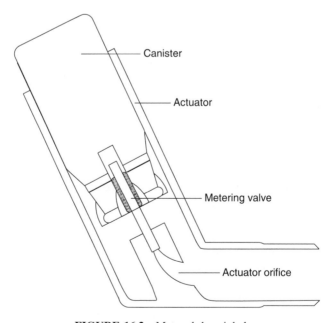

FIGURE 16.2 Metered-dose inhaler.

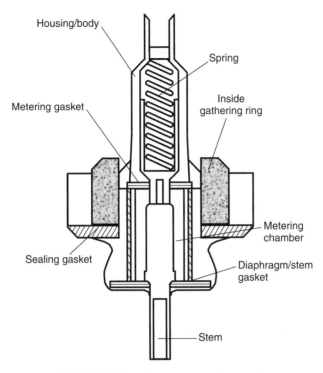

FIGURE 16.3 Metering valve (inverted).

the droplets become more concentrated with drug. Hygroscopic drugs may adsorb moisture in the high humidity of the lungs, which may cause particle growth. All these complex physical processes happen in milliseconds.

The primary formulation components of a pressurized aerosol are the propellant and the drug. Several commercial MDI formulations contain only liquefied propellant and drug. The early liquefied gases that served as *propellants* were butane, isobutane, and chlorofluorocarbons. Butane is highly explosive, and filling butane inhalers was dangerous and required extreme caution and safety measures. More recently, the use of chlorofluorocarbons has been banned over concern that they cause long-term environmental effects, such as depletion of the atmospheric ozone. Today, the hydrofluoroalkanes HFA-134a and HFA-227 are used as liquefied propellants. Liquefied gases are preferred to compressed gases such as carbon dioxide and nitrogen because they maintain a constant emitted pressure as the product is used and the propellant is exhausted. In addition, liquefied propellant flash evaporates, yielding a respirable aerosol. The physical properties of HFA-134a and HFA-227 are vastly different from those of water. Several key physical properties of the HFAs are listed in Table 16.6.

MDIs are formulated as solutions or suspensions. Liquefied propellant suspension formulations utilize the principles discussed in Chapter 15. Dispersed system principles, such as Stokes' law, describe the effect of particle density, vehicle

TABLE 16.6 Physical Properties of HFA-134a and HFA-227

Property	HFA-134a (Tetrafluoroethane)	HFA-227 (Heptafluoropropane)	Water
Formula[a]	CF_3CH_2F	CF_3CHFCF_3	H_2O
Boiling point[a] (°C) at 101.3 kPa	−26.7	−17.0	100
Vapor pressure[a] at 21°C at 101.3 kPa (Pa, atm)	5.59×10^5, 5.51	3.99×10^5, 3.94	3.25×10^3, 0.321
Liquid density at 21°C at 101.3 kPa (g/mL, kg/L)	1.22, 1.22×10^{-3}	1.41, 1.41×10^{-3}	0.998, 9.98×10^{-4}
Heat of vaporization[b], kJ/mol	19.56	23.23	4.07×10^4

[a] Hallworth (1994).

[b] Williams and Liu (1998).

density, particle size, and viscosity on the sedimentation or creaming rate [equation (15.1)], the influence of surface free energy and surface area on particle growth [equation (15.2)], the affect of particle size on solubility [the Kelvin equation (15.3)], and the control of the ζ-potential on flocculation [equation (15.4)] all apply to the formulation of liquefied propellant suspensions. To obtain low solubility in propellants (0.01% w/w or less), highly polar salts of the drug are frequently used. Unfortunately, such salts have a tendency to be soluble in water, so even low moisture levels in the propellants can lead to solubilization of the drug, which can lead to Ostwalt ripening and crystal growth and adhesion to the walls and surfaces of the container closure system. The crystal growth and adhesion leads to loss of the drug, which affects the dose and dose uniformity. Moisture levels of excipients need to be tightly controlled. It is preferable to use non-hygroscopic excipients and excipients that do not contain waters of hydration. Environmental moisture needs to be tightly controlled during manufacture. Postmanufacturing, moisture ingress can occur via environmental water vapor diffusing through the valve and rubber seal materials. In the case of pharmaceutical aerosols, the drug size range approaches that of colloidal particles, so that the Derjaguin–Landau–Verwey–Overbeek theory of agglomeration can be applied to stabilizing aerosol suspensions. Measuring charges on particles in a highly volatile nonpolar liquid is technically challenging. Sandstrom et al. (1994) reported measuring the electrophoretic mobility of albuterol suspension in mixed propellants.

Matching the drug density and the propellant formulation can prevent density gradients and minimize dose variability. The propellant densities can be greater than those of the drug, so in such cases the drug will cream to the top of the inhaler canister, where the metering value is located. Rapid creaming can lead to a higher concentration of drug entering the metering chamber, which results in superpotent doses. Over time, the amount of drug becomes depleted and subsequent doses will be subpotent. Alcohol or glycerin can be added to the suspending vehicle to reduce the density of the propellant vehicle and help match the density of the drug and the vehicle. If alcohol or glycerin is added to provide a density match between the drug and the propellant vehicle, it is important that the addition of these excipients not increase the drug solubility above 0.01% w/w. Alcohol and glycerin can also be used to increase the solubility of surfactants that might be required to stabilize the drug suspension. The HFA propellants can also be mixed to match the density of the drug.

Oleic acid, lecithin, and sorbitan trioleate have been used as stabilizing surfactants for aerosol suspensions. The surfactant concentrations can affect evaporative rates and aerodynamic particle size. Higher surfactant concentrations have a tendency to delivery larger particle sizes, so it is necessary to determine the effect of surfactant concentration and type on the respirable fraction (Clark, 1996). Similarly, alcohol and glycerin are inclined to provide larger aerosolized droplets. Suspension concentration can also affect metering uniformity and aerodynamic particle size. The metering value mechanism can become clogged or drug may agglomerate and impede free movement of the valve. Suspension concentrations greater than 0.01% w/w are apt to form emitted droplets that include multiple particles, which have a propensity to aggregate, forming clusters of aerosolized particles having a larger aerodynamic particle size than that of the primary suspended particles. Gonda (1985, 1988) studied the effects of aggregation on aerodynamic behavior.

Mixing propellants or adding alcohol or glycerin will change the headspace and exit pressure, which can affect the particle size and clinical efficacy. The vapor pressure of the mixtures can be measured directly by a pressure gauge attached to a cannula. The cannula is used to puncture the pressurized aerosol canister. Williams (1998) showed that mixing HFA-134a and HFA 227 follows Raoult's law reasonably well over the temperature range 6 to 42°C. In this study, the total vapor pressure of the mixture was shown to be equal to the sum of the partial vapor pressures of the individual propellants. The partial vapor pressure of a single propellant is

$$P_i = P_i{}^\circ x_i \tag{16.1}$$

where P_i is the partial pressure of constituent i, $P_i{}^\circ$ the vapor pressure of the pure liquid ($x_i = 1$), and x_i the mole fraction of constituent i in solution. For a two-component system, the total pressure above the solution is the sum of the partial pressures of the two components:

$$P_t = P_1 + P_2 \tag{16.2}$$

The addition of alcohol to the propellant mixtures was also studied at 25°C. As might be expected, the addition of ethanol over the range 2.37 to 11.85% w/w showed a decrease in the total vapor pressure of the propellant mixture with an increase in the alcohol concentration. This HFA–alcohol solution system also obeyed Raoult's law relatively well. The authors also studied the effect that the addition of insoluble or suspended solids had on the propellent's vapor pressure. The data were somewhat mixed: In several cases the suspended solids tended to decrease the vapor pressure slightly; in other cases the suspended solids did not affect the vapor pressure.

The heats of vaporization for HFA-134a and HFA-122 were also determined using the integrated form of the *Clausius–Clapeyron equation*:

$$\ln P = \frac{\Delta H_{vap}}{RT} + C \tag{16.3}$$

where P is the vapor pressure for the pure propellant at a given temperature T (kelvin), R the gas constant, ΔH_{vap} the heat of vaporization of the propellant, and C the integration constant. By applying Raoult's law and the Clausius–Clapeyron equation to propellant mixtures, the propellant mixture's vapor pressure can be estimated. This is an important benefit to the formulation design scientist because it is well know that propellant exit pressure can affect aerosol droplet size and respirable dose fraction. Generally, lower exit pressures create larger droplets than can higher exit aerosolization pressures (Clark, 1996). Using mathematical models or design of experiments, a relatively small number of experiments can be used to model the effect of temperature on the respirable dose. Since lifesaving medicinal aerosols are used in all types of temperature conditions, it is important to know the impact of use temperature on the respirable dose so that appropriate product labeling can be provided to the patient. Ethanol or glycerin may improve drug solubility and surfactant wetting. The effect of these excipients on the respirable dose and aerodynamic particle size needs to be evaluated. Both ethanol and glycerin will reduce the propellant's vapor pressure, so the trade-off between increased solubility and reduction in vapor pressure needs to be weighed. Ethanol up to 30% v/v and glycerin up to 7.8% v/v have been used in aerosol formulations (Appendix 16.1). Ventolin HFA is a very simple formulation that contains a suspension of microcrystalline albuterol sulfate and HFA-134a. The Atrovent HFA oral inhalation aerosol is a solution of ipratropium bromide, HFA-134a, sterile water, dehydrated alcohol, and anhydrous citric acid. Xopenax HFA is an example of a more complicated suspension formulation of levalbuterol tartrate. This oral inhalation aerosol suspension contains micronized drug, dehydrated alcohol, and oleic acid. Additional discussions of pressurized metered-dose inhalers are available: Hickey (1992, 2004), Hallworth (1994), and Sciarra and Sciarra (2006).

Drug particle engineering has led to major improvements in the percentage of drug that can be delivered to the lungs. Engineering a drug particle's size, aerodynamic density, and morphology has led to severalfold increases in lung deposition and respirable fraction. Particle engineering is discussed in more detail in Section 16.2.1.3.3. Engineered particles can also be used in MDIs. The majority of MDIs are filled by a single-step pressure-filling process. In this one-step process, the total aerosol can contents, propellant, and formulation are filled through the metering valve. Suspension and solution aerosol products can be filled using this technology. Maintaining adequate suspension homogeneity is a critical process parameter. In some cases a two-step process is used to fill the aerosol canisters. A *product concentrate* that contains a low-volatile portion of the formulation is filled into an open, cooled canister. The product concentrate usually contains the drug, surfactants, and high-boiling cosolvents such as alcohol and glycerin. The metering valves are then crimped onto the aerosol cans and the propellant is filled through the metering valve. In either the one- or two-step process it is important to maintain a low-humidity environment so that moisture condensation or environmental moisture is not introduced to the final product.

Once the cans are filled, they are tested for propellant leaks. The standard method for checking leaks is to immerse the filled cans in a hot-water bath to ensure that they are pressure resistant and do not leak propellant. A number of companies have now developed alternative methods to detect leaks. Machines perform individual leak detection online at speeds up to 500 cans per minute. The testers are very sensitive and can detect the presence of propellants at very low leak rates. The leak-tested cans are then placed in quarantine to allow the seals to set. The quarantine time varies, but it is typically 3 to 4 weeks. At the end of the quarantine period, a weight check is made as a final leak test check. The cans are then sent off for packaging, where the can is placed in the actuator, boxed, and cartoned. Pressurized aerosol filling of pharmaceutical products is discussed in detail by Sirand et al. (2004).

16.2.1.3.3 Dry Powder Inhaler Formulation Development and Specific Manufacturing Processes Dry powder inhalers (DPIs) are classified as single-dose factory-metered, multidose factory-metered, and multidose patient-metered. There are several dozen DPI devices on the market or in development. All dry powder inhalers attempt to provide a predetermined dose per inhalation. In the late 1960s, the earliest DPI (Spinhaler by Fisons) was a single-dose factory-metered capsule dosage form, with the amount of drug placed in each capsule controlled through in-process testing and control. The hard gelatin capsule was punctured to release the dose prior to the patient inhaling the aerosolized powder. The earlier devices depended on *passive* breath actuation. The Turbuhaler (Figure 16.4, AstraZeneca) was the first (1988) multidose patient-metered breath-actuated device to deliver pure aerosolized drug to a patient's lungs. In the mid-1990s the Diskus (Figure 16.5, GlaxoSmithKline) inhaler was introduced and has become the most widely used passive multidose factory-metered powder inhaler. The device uses lactose powder larger than 10 μm as a carrier. The lactose carrier improves the flowability of the powder blend, which improves blister fill weight uniformity. The degree of adhesion of the 1- to 5-μm drug particles is critical to maintaining blend and weight uniformity during the filling process. Similarly, the degree of adhesion affects the degree of deaggregation or dispersion of the drug that occurs during a patient's inspiration. In the case of passive DPIs, it is the inspiratory energy and device design that causes the drug to disperse or separate from the carrier excipient. The carrier excipient is large (30 to 60 μm) compared to the drug (1 to 5 μm), and inertial forces will cause the carrier to impinge on the back of the throat while the smaller drug particles will be deposited in the lower lung. A number of physiochemical particle properties affect powder dispersibility, such as van der Waals forces, electrostatic forces, surface energy, particle size and size distribution, density, morphology (crystalline habit and porosity), surface roughness (rugosity), surface tension of adsorbed water, acidity and basicity, and surface hydrophobicity.

To overcome some of the disadvantages of passive DPIs, *active DPIs* have now been developed and marketed. The Inhale Therapeutic Systems (now Nektar) inhalation device system has many unique features, which were codeveloped with

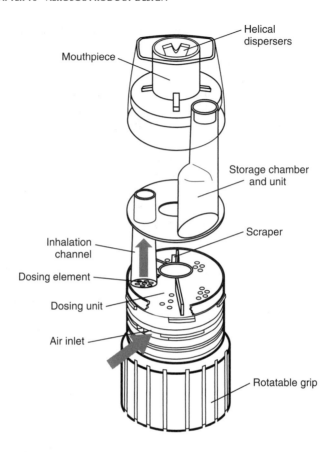

FIGURE 16.4 Turbohaler. (Adapted from Dagsland, 2001.)

Pfizer, for pulmonary insulin drug delivery. The delivery system was approved by the FDA under the trade name Exubera and was later taken off the market by Pfizer, stating marketing reasons. The inhaler incorporated a mechanical "cocking mechanism" that compressed air that was used to actively suspend the insulin drug particles in a built-in spacer. This *aerosol cloud's* characteristic was independent of the patient's inspiratory rate or profile. The 1- and 3-mg doses were filled into blisters using a patented filling technology to achieve a low-weight variation of 2 to 3% relative standard deviation (Parks et al., 1998).

Other very sophisticated active devices are in development. One such device is from Oriel Therapeutics, Inc., which was purchased by Sandoz, a division of Norvatis. The system is a multiple-dose computer-controlled dry powder inhaler device that provides adjustable energy output that takes into account the patient's inspiratory rate and the powder's characteristics. The computer system produces an inhalable dispersed powder using a vibrating piezoelectric membrane (Crowder et al., 2006; Hickey et al., 2008). Atkins (1994) and Hickey, 1992 and 2004 provide additional details regarding dry powder inhaler systems.

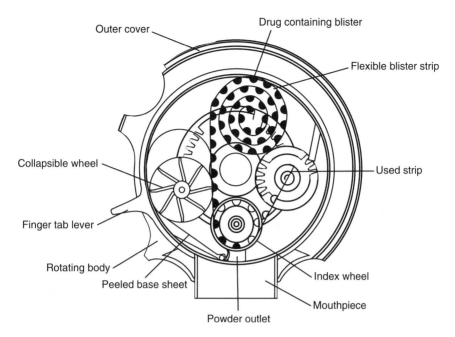

FIGURE 16.5   The Diskus Inhaler. (Adapted from Davies et al., 1997.)

16.2.1.4   Inhalation Aerosol Process Design   The specific processes for manufacturing nebulizer ampules, pressurized oral inhalers, and dry powder inhalers were discussed above with the associated oral inhalation dosage form. In this section we emphasize spray drying as a means of intentional particle design to achieve desired aerodynamic particle size, particle distribution, surface energy and adhesion properties, surface morphology, hygroscopicity, and flow. This area of particle engineering is a rapidly growing field. Early efforts evaluated the effect of particle shape (Chan and Gonda, 1989), hygroscopic growth of a water-soluble compound at 97% relative humidity (Hickey et al., 1990), surface roughness (Ganderton, 1992), and surface energy (Feeley et al., 1998). In 1997, the concept of engineering large porous particles for aerosol pulmonary delivery was published (Edwards et al., 1997). This work took advantage of the fundamental property of the equivalent aerodynamic dynamic particle size, which is the diameter of a unit-density sphere that has the same settling velocity as that of the sphere measured. A simplified equation describes a particle's aerodynamic diameter, d_a, in terms of the true density, ρ_t, the apparent particle density, ρ_a, and the measured geometric mean diameter, d_g:

$$d_a = d_g \left(\frac{\rho_a}{\rho_t} \right)^{0.5} \tag{16.4}$$

where the particle's true density can be measured by helium pycnometry, and its apparent density is calculated from the mass of the particle divided by the volume of a sphere of diameter d_g. A number of porous particles that are large and have the advantage of better flow properties and less adhesion and agglomeration have been reported and have been given a number of trade names: AIR (Edwards et al., 1999), spray-dried large porous particles based on dipalmitoylphosphatidylcholine albumin formulation, and PulmoSpheres (Weers et al., 2001), based on solid foam particles prepared by spray drying as an emulsion of perfluorooctylbromide in the dispersed phase and drug and excipients such as hydroxyethylstarch in the continuous phase. The AIR particles have apparent densities of less than 0.1 g/mL. These particles have been shown to provide very high emitted doses from passive DPIs. Emitted drug doses as high as 90 to 100% have been reported.

Insulin particles were engineered with a corrugated folded-surface glassy particle that demonstrated good physical and biological stability (Platz et al., 2000, 2003; Foster et al., 2001). The rugosity was thought to improve particle dispersibility and reduce aggregation. The technology was trademarked PulmoSol. The glassy particles are stabilized with mannitol, glycine, and sodium citrate. Microencapsulation using leucine and trileucine as particle engineering excipients has resulted in apparent aerosol particle densities as low as 0.1 g/mL. These materials act as surface-active agents and lubricants that affect spray-drying surface properties and decrease aggregation. The materials are capable of encapsulating large biomolecules such as immunoglobulin, growth hormone, and plasmid DNA.

Vehring (2008) has written a comprehensive review covering pharmaceutical spraying, drying, and particle engineering. Reviews of other particle engineering technologies, such as sonocrystallization, supercritical fluid-assisted atomization, nebulization and electrostatic collection, and spray-freeze-drying in liquid and gas, include the following: Chan and Chew (2003), Tong and Chow (2006), Shoyele and Cawthorne (2006) and Chow and Tong (2007).

16.2.1.4.1 PAT: Spray Drying Real-time monitoring and control of the spray-drying process should lead to a more efficient and robust manufacturing process. The real-time monitoring and control of critical process parameters such as drying airflow rate, inlet drying air temperature, outlet air temperature, and percent solvent saturation are widely used process analytical technologies. Particle atomized droplet size is another critical process parameter. Several laser-based analytical techniques are available to provide real-time in-line or online particle size analysis that can be used to determine when the process is at equilibrium and give the particle size and distribution. A focused-beam reflectance (FBRM) technique supplied by Lasentec has been used to measure granule particle size in-line and in real time. A laser beam that uses a rotating laser optics system is projected onto a particle. The particle reflects the laser beam until the rotating laser reaches the opposite side of the particle. The rotational speed can be adjusted from 2 to 8 m/s, so the system can acquire thousands of chord lengths per second. A cord length and cord length distribution is obtained in real time by software that calculates the cord length by knowing the rotational speed and the time the particle reflected the beam (Kumar et al., 2010). The lower absolute sizes (1 to 5 μm) are at the sensitivity limit of this technique. However, this FBRM method is acceptable for engineered particles that have an actual physical size of 5 to 15 μm or larger whose aerodynamic size is in the range 1 to 5 μm. The FBRM technology's operation sensitivity and accuracy are a function of the droplet's ability to reflect the laser beam. This technology performs better as the particle density and reflective ability increase. One other laser technique that has been developed for spray drying is the Insitec system from Malvern. This system uses Mie laser-light-scattering technology to determine the particle size and distribution. Laser beam diameters are available from 3 to 10 mm. This system can measure droplet size ranges from 0.1 to 1000 µm. A focal length of 100 mm is used to measure droplet sizes of 0.1 to 200 µm. This system can make particle size measurements of droplets from aerosol concentrations of 2 to 95%. The advantage of the Mie system is that it is model independent. Both systems can suffer from "foiling or coating" of the optical lens. A purge of clean air or pulses of clean air are used in an attempt to maintain a clean lens. Heng et al. (2009) discussed the use of the Insitec system to measure real-time and in- or at-line particle size of oil-encapsulated spray-dried microspheres. The in- and at-line measurements were made in different places in the process. The particle size for the in-line location was installed in the exit line of the spray drier. The at-line location was installed after the spray drier and cyclone size selection apparatus. As expected, the two locations gave different particle sizes and particle size distributions. The in-line measure gave a bimodal distribution, and the at-line measurement, which was made after a sizing operation, was monomodal. The process analytical technologies discussed above can be instrumental in rapidly scaling the spray-drying process.

16.2.1.4.2 Scale-up of Spray Drying Dobry et al. (2009) have discussed the use of engineering and kinetic models to develop and scale-up spray-dried formulations. Process constraints in conjunction with formulation constraints are identified early in process development. The drying gas inlet temperature is an important processing parameter that may lead to formulation discoloration or melting. Lower temperatures may cause the material to become sticky or undergo other undesirable physical or chemical changes. The drying gas inlet temperature is identified early and is incorporated into the spray-drying process. The authors also discuss the use of mass and energy thermodynamic modeling and design space that involves five key spray-drying process parameters: spray-drying solution feed rate, drying gas flow rate, drying gas inlet and outlet temperature, and relative saturation of solvent at the outlet drying conditions. Drying kinetics and atomization process conditions were also discussed. Both these process parameters affect the droplet size. Computational fluid dynamics models can be used to evaluate the drying kinetics with small amounts of material that can incorporate findings from off-line analytical techniques. The models can also be used to simulate the effect of various drying parameters on droplet particle sizes. This approach is purported to permit efficient optimization of the drying kinetics, which can result in rapid and robust scale-up and commercialization. The drying kinetics can also affect the morphology of the particle. A hot fast-drying process may lead to a particle with a smooth surface, whereas a cool slow-drying process may result in a "dimpled or ridged" particle. As discussed above, the surface morphology can affect the tendency to aggregate, the flowability, and the aerodynamic density. Vehring and Foss (2007) and Vehring (2008) have discussed the use of the dimensionless Péclet number to model and gain an understanding of how drying rate and formulation factors can affect particle morphology. The *Péclet number* is the ratio of the drying rate constant to the diffusional coefficient of the solute:

$$Pe_i = \frac{k}{8D_i} \tag{16.5}$$

where Pe_i is the dimensionless Péclet number for the ith solute or dispersed particle, k the evaporation rate constant, and D_i the diffusional coefficient of the ith species. The surface and surface morphology have been shown to be a function of the mechanisms that influence the distribution of components in the drying droplet. A low Péclet number, say below 1, indicates that the evaporation rate occurs more slowly than the diffusion of the solute, so there is little solute enrichment throughout the dried particle. This is especially true if the solute is soluble in the solvent and the time to reach surface saturation is close to the lifetime of the droplet. For a Péclet number greater than 1, the drying front moves faster than the diffusion of the solute or dispersed particle, and the surface is enriched with solid or dispersed material and shell formation occurs. The shells that are formed can take on different physical characteristics. They may be crystalline or amorphous. The difference in drying rate and particle diffusion may lead to hollow spheres, collapsed spheres, or particles with high rugosity or corrugated-ridged appearance.

16.2.2 Nasal, Buccal, Lingual, and Sublingual Aerosol Formulation Design

Drug delivery to the nasal and oral mucosa for local and systemic delivery has been on the rise. Buccal, sublingual, and lingual drug delivery dosage form for systemic therapeutic effects have exploded over the last decade. New companies have been formed that develop fast-dissolving tablets, dissolvable films, and non-dissolvable patches. In this section we focus on the design of nasal, buccal, lingual, and sublingual aerosols. Local-acting nasal sprays (aerosols) have been available as decongestants, antihistamines, and corticosteroids. Local drug delivery to the nose provides rapid and convenient treatment that minimizes systemic side effects. Use of the nasal mucosa for systemic protein and peptide drug delivery became well established in the 1970s and 1980s. Although sublingual nitroglycerin tablets have been used since the early twentieth century for the treatment of angina attacks, few other drugs have been developed that use this route for systemic drug delivery. Exploitation of the nasal and oral mucous membranes as routes for systemic drug delivery has expanded rapidly in the last two decades. The advantages and disadvantages of systemic aerosol drug delivery to these mucosal membranes are listed in Table 16.7.

The anatomy and physiology of the nose provide a number of levels of defense against foreign particles entering the lungs. The initial defense is hairs that guard the entrance to the nasal cavity. The anterior vestibular portion of the nose is also lined with squamous epithelium. The posterior portion of the nose forms folded turbinates that are lined with secretory columnar and goblet cells which form mucin. The upper or superior turbinate is lined with olfactory cells. The

TABLE 16.7 **Advantages and Disadvantages of Aerosol Drug Delivery to the Nose and Oral Mucosa for Systemic Therapy**

Advantages	Disadvantages
The nasal or oral cavities are convenient locations for drug delivery	Limited to relatively potent molecules
Delivery option for patients with nausea and vomiting	Absorption surface area is only about 0.5% that of the small intestine
These mucous membranes offer reasonably high absorption surface areas: nasal ($\sim$160 cm^2) and oral mucosa ($\sim$100 cm^2)	Irritation, allergenic, or immunogenic reaction to the drug or excipients needs consideration
Mucosal permeability is higher than that of skin but lower than intestinal cell permeability	Thick mucus can be a barrier to absorption
Delivery devices are easy to use	"Runny nose" can cause dose variability and excess drug clearance
Good blood flow for drug absorption	Mucus and salivary clearance; about 0.5 to 2 L of saliva is produced in the oral cavity per day
Generally, less metabolism seen compared to gastrointestinal (GI) tract	Nasal clearance is about 5 to 6 mm/min[a]
Bypasses first-pass metabolism	
Less variability in the absorption environment compared to the GI tract (pH, enzymes, food effect, GI housekeeping waves, and motility)	

[a] Chien et al. (1989 p. 9).

posterior columnar cells are covered with hairlike cilia that beat at a rate of about 10 Hz, which leads to mucus and periciliary fluid clearance. This clearance serves as the body's defense mechanism. The nasal fluids contain a number of lysozymes, IgG, and IgA, which also provide additional protection against foreign matter. The nasal fluid does contain peptidases, which can metabolize peptides and proteins. The nasal mucosa is highly vascularized with superficial and deeper arterioles. About 1.5 to 2 L of mucus is secreted by the goblet cells per day. The pH of the mucus is around 5.5 to 6.5.

Molecular mass, lipophilicity, and the extent of ionized–nonionized permeant affect the drug's permeability. As is seen with many membranes, including the lungs and intestinal membranes, there is an inverse relationship between molecular mass and amount permeated or absorbed. Also, the extent of absorption has been related to the lipophilic octanol–water and nasal mucosal–water partition coefficient. Having said this, studies have shown that reasonable levels of ionized molecules are absorbed, which may be due to paracellular passive diffusion or solvent drag transport or active ion transport mechanisms. Usually, nasal drug absorption is higher than oral absorption. The nasal cutoff is roughly 20,000 Da, compared to oral absorption of 200 to 500 Da.

The oral epithelium contains keratinized and non-keratinized stratified squamous cells. The sublingual and buccal portions for the oral mucosa are non-keratinized and form a flexible relatively permeable membrane. These cells are bathed with mucus and saliva. As stated above, about 0.5 to 2 L of saliva are produced per day. Saliva is hypotonic and contains varying amounts of mucus and enzymes such as lysozyme and amylase, and antibodies. The permeability of the mucosal membrane is intermediate compared to the skin and the small intestine. The buccal membrane is roughly two to eight times thicker than the sublingual mucosa. These membranes are highly vascularized and are also supplied by a lymphatic network. The blood flow to the buccal mucosa is around twice the sublingual blood flow. The pH of saliva is approximately 6.2 to 7, which is higher than the nasal mucus and periciliary fluid. Sublingual drug absorption is affected by drug molecular mass, lipophilicity, and extent of ionization.

The delivery of drugs directly on the tongue has evolved. Since the taste buds are located on the tongue, taste is one the largest formulation design issues that needs to be addressed. In addition to taste buds, the tongue is composed of keratinized stratified squamous cells which decrease the permeability of drugs compared to the sublingual membrane. In the case of fast-dissolving tablets, the dosage form quickly dissolves on the tongue without the need for large quantities of water and the drug is swallowed with the saliva that is present in the mouth. Several products designed for lingual delivery are now also commercially available.

16.2.2.1 Nasal, Buccal, Lingual, and Sublingual Spray Aerosols

The FDA defines a *spray* as an "aerosol product which utilizes a compressed gas as the propellant to provide the force necessary to expel the product as a wet spray; it is applicable to solutions of medicinal agents in aqueous solvents." Nasal spray products are solutions and suspensions of drugs. The dose can be delivered by squeeze bottles or a metered-dose pump. Both mechanical systems depend on creating dispersive energy by forcing the formulation through a nasal orifice under applied pressure. Squeeze bottles are normally used for topical local-acting drugs where highly accurate doses are not required. Squeeze bottle container closure systems suffer not only from high-dose variability, but there is potential for product contamination from "vacuum suck-back" of foreign material that occurs once the squeeze pressure is released. Systemic nasal delivery products use a metered pump spray or are premetered during manufacture. Premetered devices are often single-use containers. Powder nasal sprays are also mostly premetered. The metered pumps are designed to prevent or minimize vacuum suck-back. The mechanism for nasal deposition is similar to that described for the lungs and occurs by impaction, sedimentation, and diffusion. Most nasal aerosols are nasal sprays that have particle sizes between 5 and 10 μm. The liquid dispersed phase is most commonly water. The primary product attributes for a nasal spray are: dose reproducibility, particle size distribution for suspensions, dispersed droplet size distribution, spray pattern, and plume geometry. Nasal sprays do not have to be manufactured as sterile products, but they must conform to U.S.P. Chapter <61>, Microbiological Examination of Non-sterile Products—Microbial Enumeration Tests. The product must also maintain its microbiological quality throughout the expiry period.

Functional excipients used for nasal sprays and their use percentages are listed in Appendix 16.1. Nasal sprays will contain antimicrobial preservatives and tonicity-adjusting agents. They may also contain viscosity-inducing agents, absorption enhancers, solubilizing agents, antioxidants, and humectants such as glycerin. To improve systemic nasal drug delivery, several strategies can be employed. The primary ways to improve absorption are to increase the permeability and residence time in the nasal cavity. Often, the primary mechanism of nasal drug absorption is by passive diffusion. Therefore, the absorption rate will be concentration dependent. A higher initial concentration of drug will cause a larger concentration gradient, which increases the diffusional driving force, resulting in faster absorption. The available drug contact or residence time is affected by the mucociliary clearance, so absorption rate and residence time are crucial to good drug bioavailability.

Lee et al., 1991, Illum and Davis (1992), and Davis and Illum (2003) have reviewed absorption enhancers for nasal drug delivery. Some of the general classes of absorption enhancers are surfactants, phospholipids, bile salts, fatty acid salts, cyclodextrins, chitosan, glycols, soybean-derived steryl glucosides, alkylglycosides and disaccharides, sucrose esters of fatty acids, glycyrrhetinic acid derivatives, and chelators. A number of mechanisms of action have been proposed. It has been speculated that enhancers increase the fluidity of the membrane by disrupting the phospholipid structure or by extracting lipids and proteins from the membrane. Other enhancers may operate by forming intercellular aqueous pores in the paracellular junctions. Some enhancers may act by inhibiting critical metabolic enzymes, such as peptidases and proteolytic enzymes. It has been suggested that steryl glucosides affect tight junction-associated calcium. The fatty acids and bile salts may form lipophilic ion pairs with positively charged drug molecules. Penetration enhancers should not be irritating, toxic, allergenic, or immunogenic. Any changes they cause to the membrane or ciliary apparatus should be rapidly reversible. A number of studies have evaluated the effect of absorption enhancers and their concentrations on ciliary beat frequency, nasal cell integrity, and morphology. Cyclodextrins and glycholate bile salts have shown minor inhibitory effect on ciliary beat frequency and membrane morphology.

Increasing the viscosity of the nasal spray can increase the drug residence time and bioavailability. Cellulose derivatives, carbomers, chitosan, poloxamers, starch derivatives, pectin, alginate derivatives, and pectin have been used to increase the aerosol viscosity. Although the increased viscosity may slow down ciliary transport, increased viscosity also decreases the drug diffusion in formulation. To improve the bioavailability the viscosity needs to be optimized through trade-off analysis. Thermogelling polymers such as Pluronic F127 (polaxamer) become more viscous and form gels at physiological temperatures (32 to 37°C).

Many viscosity-enhancing polymers can also act as bioadhesives. Carbomers can interact with mucus and decrease its clearance. Drug dry powder formulations that contain viscosity-inducing polymers or bioadhesives have been shown to improve the bioavailability of drugs. Colloidal microspheres that contain bioadhesives and viscosity-enhancing agents have also increased the bioavailability of drugs. Liposomes have been shown to provide sustained release of rapidly absorbed drugs and to improve the absorption of poorly absorbed drugs. Delivery of drug powders in the absence of bioadhesives and viscosity-inducing agents has been successful in increasing bioavailability. Additionally, cyclodextrin complexes have shown beneficial absorption increases.

There are a number of non-complex and complex molecules (using U.S.P. terminology for small and biotechnology drugs) that are available commercially as systemic nasal drug delivery systems. Sumatriptan (a non-complex drug) nasal spray was approved in 1997 for migraine headaches. This was a breakthrough therapy for patients who suffered nausea and vomiting with their migraines and wanted to use an injection. The formulation is buffered with a phosphate buffer at a pH 5.5. The osmolality of the solution is 372 or 742 mOsmol for those products of strength 5 and 20 mg per 100 μL. The product is supplied as a single-dose compression pump nasal spray. Controlled clinical studies demonstrated that drug administration into a single nostril provided effective treatment for an acute migraine. Fentanyl nasal spray was approved for the treatment of cancer pain in 2011. It is available as a 100- and 400-μg per 100 μL of spray. This product is sprayed into one or two nostrils, depending on the patient's dose titration level. The pH is adjusted to 3 to 4.2. The formulation includes mannitol and sucrose, which are possibly used as tonicity-adjusting agents and humectants. The product also contains phenylethyl alcohol and propylparaben as microbiological preservatives. Pectin is included as a patented excipient to provide a rapid and prolonged effect (Birch et al., 2010). Desmopressin and calcitonin are polypeptide nasal sprays. Both spray formulations include benzalkonium chloride as a microbiological preservative and sodium chloride as the tonicity-adjusting agent. The reader is directed to the following references for more detailed discussions of nasal spray drug delivery: Chien et al. (1989), Su (1993), Merkus and Verhoef (1994), Lansley and Martin (2001), Lim et al. (2007), and Pillion et al. (2007).

The development of oral sprays is expanding. In 2008, Zolpimist (zolpidem), an oral spray, was approved by the FDA. Zolpimist affords fast onset of action while providing equivalent area-under-the-curve and maximum blood concentration as the reference listed product (Ambien). Zolpimist delivers 5 mg of zolpidem tartarte in 100 μL of spray. The excipients include artificial cherry flavor and neotame to modulate the taste. Citric acid and hydrochloric acid are used to adjust the pH. Propylene glycol and purified water are the solvents. The product is sprayed directly into the mouth onto the tongue. Sativex is an anhydrous buccal spray of Δ^9-tetrahydrocannabinol used as adjunctive therapy for neuropathic pain in multiple sclerosis. The product is approved in Canada and the UK. The excipients include ethanol, propylene glycol, and peppermint oil. Mucosal irritation is high, 22.3%. Hoogstraate et al. (2001) and Smart (2007) have addressed oral transmucosal drug delivery and the use of bioavailability enhancers for buccal and sublingual absorption.

There are several pressurized nitroglycerin products on the market. NitroMist was approved by the FDA in 2006 for the prophylactic and acute treatment of angina pectoris. NitroMist has a longer shelf life than that of sublingual tablets and can be sprayed on (lingual) or under (sublingual) the tongue. It is a metered-dose pressurized spray that delivers 400 μg of nitroglycerin per actuation. The excipients include caprylic–capric diglycerol succinate, which probably acts as a solubilizer and penetration enhancer. It also contains peppermint oil and menthol, which are added as flavors. Menthol has been reported to be a penetration enhancer, so it could also serve in this capacity. Butane is the propellant. Glytrin Spray is a sublingual spray of nitroglycerin. It contains peppermint oil as a flavor, ethanol as a cosolvent, and HFA-134a as the propellant. A pressurized buccal insulin has received commercial approval in Ecuador and is in clinical trials in the United States and the U.K. The patented technology trademarked RapidMist uses HFA-134a as the propellent (Modi, 2007). Dugger et al. (2007) have reviewed the clinical results of lingual and buccal spray formulations.

16.3 CONTAINER CLOSURE SYSTEM DESIGN

The FDA (1999) has written an industrial guidance, Container Closure Systems for Packaging Human Drugs and Biologics, which is intended to provide guidance for submitting regulatory packaging information. It also provides information regarding the current good manufacturing practices, Consumer Product Safety Commission (CPSC), and U.S.P. requirements for container closure systems (CCSs). Good Packaging Practices (U.S.P. <1177>) discusses general packaging terminology. The components that are in contact or have the potential for coming into contact with the drug product are referred to as a *primary container*. The primary package protects the dosage form from adverse environmental conditions such as light, oxygen, moisture ingress or egress, and microbiological contamination. *Secondary container* systems add additional protection for shipping the product and include labeling information or ancillary devices such as dropper or measuring spoons. The secondary container can also provide protection from light. The secondary container is not included in stability testing and is not included as part of the CCS description. A standard folding carton is most often used as the secondary container. Changes made to the secondary packaging material do not require prior FDA approval. *Critical secondary containers* are not in direct contact with the dosage form but provide essential environmental protection against light, moisture ingress, oxygen ingress, or a combination thereof. The GlaxoSmithKline Diskus uses an aluminum pouch to protect the inhalation powder from uptake of environmental moisture. *Additional packaging*, such as trays for syringes, droppers for nasal products, and desiccants for oral solids, may also be used. The *final exterior package* is usually a corrugated or plastic wrapper.

The FDA and CPSC have regulatory responsibility for human-use packaged products. The FDA requirements for tamper-resistant closures are stated in the *Code of Federal Regulations* (21 CFR 211.132). The CPSC requirements for child-resistant and adult-use-effective closures (often referred to as "geriatric friendly") are provided in 16 CFR 1700. The CCS should provide product protection, tamper evidence, safety, and be user friendly.

The FDA document lists the packaging of inhalation aerosols as one of the routes of administration having the highest degree of concern and having a high likelihood of packaging component–dosage form interaction. Nasal aerosols and sprays are also listed as a route of administration having a high degree of concern and a high likelihood of packaging component–dosage form interaction. Lingual (including sublingual) and buccal aerosols were listed as a route of administration having a low degree of concern and a high likelihood of packaging–dosage formulation interaction. As such, the suitability requirements concerning protection, compatibility, safety, and performance for inhalation nebulizing solutions, pressurized inhalation aerosols, and nasal sprays are more stringent than those for inhalation powders, which are more stringent than those for lingual aerosols. (For details the reader is referred to Tables 2 and 3 of the FDA 1999 guidance document Container Closure Systems for Packaging Human Drugs and Biologics.

Leachables and extractables are of special concern for inhalation products, but are also a concern for nasal dosage forms. The concern is decreased significantly for oral transmucosal dosage forms. Leachable material is a substance or substances that are removed from the packaging material by the solvency of the formulation. Extractable studies are performed to identify the substances that can be removed from the packaging material under harsh or accelerated conditions. The extracted substances are identified and become part of a list of potential leachable substances that might appear in the formulation over longer common storage conditions. The drug product or control (formulation without the drug) is a suitable extracting solvent system. A stronger extraction solvent than the drug formulation is recommended to determine a qualitative extraction profile to establish quality control criteria.

Nebulizing solutions are often packaged in low-density polyethylene ampoules or *nebules*. The FDA has issued a guidance for inhalation drug products packaged in semipermeable container closure systems (FDA, 2002b). The document provides guidance on appropriate secondary packaging materials, embossing or debossing on the primary container, the acceptable number of unit-dose containers in the secondary package, and time limits on unprotected nebules.

It is well documented that the reproducibility of the dose administered is critically dependent on the interplay between the aerosol formulation, the performance of the container closure system, and the appropriate patient self-administration. The performance of the container closure system is also directly dependent on the design and quality control of the device components. Selection of the nasal pump and inhalation valve components requires consideration of the interaction between the formulation and the CCS components. Formulation parameters such as viscosity and other rheological properties, such as thixotropy, liquid density, surface tension, vapor pressure, and solvents and cosolvents, affect the selection of the device components. Actuator parameters such as force and speed also need to be understood and controlled. The need for reproducible and accurate delivery volume is a critical product attribute. The CCS needs to be designed to prevent partial metering. The importance of precise metering of the dose is also a critical quality parameter. Doses are limited to microliter quantities. The typical metered volumes range between 25 and 100 μL. Vendor quality control of all components is critical. The large number of attribute tests listed in Section 16.5 is a testament to the complexity of the interrelationships among formulation, container closure device performance, and the critical quality attributes.

16.3.1   Regulatory Submission Information

The regulatory information needed for an aerosol CCS FDA regulatory submission is nicely summarized in Table 7 of the FDA's guidance document, Container Closure Systems for Packaging Human Drugs and Biologics (1999). The main areas that need to be covered are general description, suitability, quality control, and stability of the CCS. The general description should contain the name, product code, manufacturer, and materials of construction. Justification of the suitability of the CCS should include a discussion of seal integrity, light protection (U.S.P. <671>), moisture barrier properties (U.S.P. <671>), safety of the materials of construction (U.S.P. <88>), drug–contact materials compatibility, and performance on stability. Leak tests should be performed on pressurized aerosols. Plastics, including LDPE, should comply with the U.S.P. Containers—Plastics (<661>) requirements. Quality control of the CCS should be supported by a sampling plan, attribute testing and the associated acceptance criteria, and a quality compliance plan. Stability in the commercial CCS is discussed in Section 16.6.

Other CCS information that should be included as outlined by the FDA (2002a) includes the fabricators of the CCS (including the pump and valve); unique identifiers of the CCS (including the pump and valve); precise dimensions of CCS components, control extraction methods, and extractable profile data on the elastomeric components; toxicological evaluation for the extractables; acceptance criteria, test procedures, and sampling plans; and performance characteristics of the pump and valve.

The can, valve, and actuator form the basic package container. The can considerations are important. The can interior should be smooth, press-molded, and free of seams. If coatings are used, a drug master file that describes the coating material and coating process should be provided. The value is a critical component of the container closure system and the inhalation device. The value development should be discussed in the regulatory submission. A valve development report should be prepared that discusses the quality and robustness of its performance. Relevant performance attributes, such as actuation force, simulated use testing, drain-back, leakage, and lot-to-lot variation are often reported. A link needs to be made between the value, formulation, and actuator used in drug safety studies and the pivotal clinical studies. The actuator development should also be presented in the regulatory document. Critical actuator design and performance elements should be reported.

Several other important specific device studies for pressurized MDIs are valve priming, simulated use, and dose exhaustion tests. The *valve priming study* evaluates how many priming shots are required to achieve the target dose. Typically, the priming study uses six inhalers that have been allowed to stand upright for a day to a week. The storage time may also reflect the actual use of the product and mirror the proposed dosing regimen. Intra- and interdose variability as a function of the number of actuations or priming shots are also determined. The actual test procedure requires that the inhaler be shaken prior to the actuation. The inhaler is then fired and the dose determined. Generally, 8 to 10 actuations are studied. The data are then used to support the labeled patient-use instructions, which recommend the number of priming shots required before an actual dose is taken.

The *simulated use test* generally evaluates six inhalers that are actuated throughout the end use to evaluate the dose from the beginning of use to the end of use. To simulate patient use, two actuations are fired in the morning and two in the evening. The dose per actuation is analyzed. In some cases, the *dose exhaustion test* is done in conjunction with the simulated use test. Ideally, the inhaler's last dose should exhaust the container so that no additional partial doses are left. The purpose of the dose exhaustion test is to determine just how quickly the inhaler is exhausted of doses and if any drop-off in shot weight or dose is observed at the end of actuation life. Particle size distribution is also assessed by a simulated use test and through the exhaustion phase of dosing. Spacers are often used with inhalers. Typical studies include simulated use that measures emitted dose, dose variability, particle size distribution analysis, drug deposition, and clearability.

Filing the information described above is the responsibility of the sponsor of the regulatory submission. Often, the manufacturer of the CCS has proprietary information that the company does not wish to share with the sponsor. In these cases, the sponsor can request the CCS manufacturer to provide the confidential information in a type III drug master file (DMF), which can be incorporated into the registration document by a letter of reference, which authorizes the sponsor to cite the DMF in its application. The letter also authorizes the FDA to have access to the DMF for the FDA's review. The CCS manufacturer's letter of reference should specify the sponsor, the component or material in question, and where the information is located in the DMF.

Acceptance criteria, test procedures, and analytical sampling plans should be provided for the critical components. A detailed list of information required is provided in the FDA's Nasal Spray and Inhalation guidance document (2002a).

16.4 RISK MANAGEMENT

Each product will have a set of critical product attributes that have a major influence on the overall quality of the product. The International Conference on Harmonization (ICH, 2006) has published a quality risk management guidance that discusses the general risk management process, risk management methodologies, and tools used to assess risk.

16.4.1 Risk Assessment and Control

Risk is associated with the probability of a critical quality failure and the extent of consequences resulting from product performance failure, keeping in mind that the patient is the ultimate stakeholder. Three basic questions need to be asked in risk assessment: What can go wrong? (risk identification), what is the probability that it will go wrong? (risk analysis), and what are the consequences? (risk evaluation). Risks can be expressed as probabilities or by ranking. Ideally, there should be a strategy that explicitly identifies the critical quality attributes and outlines how the attributes are going to be measured and controlled.

16.4.2   Risk Management Matrix

The Risk Management Working Group of the International Pharmaceutical Aerosol Consortium on Regulation and Science presented an interactive risk management matrix poster for orally inhaled and nasal drug products (Horhota et al., 2011). An adapted version of the poster is shown in Table 16.8. The table identifies four important performance considerations: assurance of a reproducible therapeutic dose, prevention of product adulteration, ease of use, and need to minimize unintended effects. These performance considerations need to be managed to minimize patient risk arising from use of the product. Four major areas that affect the performance targets were also identified: the aerosol formulation; the manufacturing process for making the product; the inhalation systems, including the canister, valve, and actuator; and the patient. In this case, the areas that can affect the overall product performance are also called *risk factors*. This risk management matrix can be used in aerosol product design to address the risks that affect the four highlighted performance targets.

TABLE 16.8   Risk Management Matrix for Orally Inhaled and Nasal Drug Products

Risk Factor	Performance Target			
	Assure Reproducible Therapeutic Dose[a]	Prevent Product Adulteration	Enable Proper Use	Minimize Unintended Effects
Formulation	Control drug purity	Control drug purity	Poor taste or smell may decrease compliance	Be aware of the significant level of lung hypersensitivity that can exist in a patient population
	Maximize drug stability	Control excipient purity	Eliminate or minimize need to shake	
	Control excipient properties such as surface energy and surface roughness of lactose for DPIs	Maintain drug stability	Minimize need to prime	Consider tonicity, particulate load, irritation, allergenic, and immunogenic potential for the drug and excipients
		Choose formulation excipients that do not leach container components	Eliminate or minimize the effect of resting time on resting position	
	Maintain suspension uniformity through use life	Aqueous products require proper preservation	Ensure accurate dose emitted throughout the product life	
	Control excipient purity and performance characteristics			

(*continued*)

TABLE 16.8   (*Continued*)

Risk Factor	Performance Target			
	Assure Reproducible Therapeutic Dose[a]	Prevent Product Adulteration	Enable Proper Use	Minimize Unintended Effects
Product process	Control environmental moisture levels Maintain uniform suspension during the manufacturing process Minimize fill weight variability Eliminate leaks	Produce reliably sterile product Eliminate microbial ingress Qualify cleaning of CCS parts Prevent particle generation from manufacturing equipment		Prevent cross-contamination at all process steps
CCS/Device	Design precise and robust valves Prevent moisture ingress Required aerosolization energy should be independent of disease or patient inspiratory rate Power source should be reliable and robust Overall design should be rugged Should be reliable and robust over wide temperature ranges	Prevent "suck-back" to prevent microbial contamination Select materials that eliminate or minimize leachables Nitrosamines and polynuclear aromatics are of particular concern Minimize foreign non-viable particles Design device to minimize area susceptible for microbiological growth Design devices to be cleaned easily and thoroughly	Provide friendly, intuitive patient design Have a dose counter Warn patient when the last two or three days' supply has been reached	Breath-actuated or self-actuated systems should not misfire

TABLE 16.8 (*Continued*)

Risk Factor	Performance Target			
	Assure Reproducible Therapeutic Dose[a]	Prevent Product Adulteration	Enable Proper Use	Minimize Unintended Effects
CCS–formulation interactions	Repeatable acceptable simulated patient use Dose proportionality with different strengths Consistent spray pattern and plume geometry Reliable and robust over a wide user temperature range Minimize changes in static charge, adhesion forces that lead to loss of drug to the CCS Create proper formulation lubrication to prevent valve sticking	Eliminate leachables Eliminate or minimize adsorption or absorption of preservatives to the CCS		
Patient factors	Initiate dose actuation with inspiration Provide appropriate use training to patient Allow for harsh abuse of device by patient	Instruct patient on appropriate cleaning techniques	Remove dose buildup Eliminate a required use orientation	Eliminate excess drug doses when dose counter is at "zero" Minimize confusion that may arise when taking two or more products that have different CCS and patient instructions

(*continued*)

TABLE 16.8 (*Continued*)

	Performance Target			
Risk Factor	Assure Reproducible Therapeutic Dose[a]	Prevent Product Adulteration	Enable Proper Use	Minimize Unintended Effects
	Allow the patient to use the device in harsh weather conditions			Consider how to avoid patient taking two or more drugs with different dosing regimens, leading to mix-ups

[a] As measured by aerodynamic particle size distribution and dose content uniformity.

16.5 ATTRIBUTE TESTS

As mentioned in Section 16.3, the complexity of the interrelationships between the aerosol formulation, container closure system and device, and performance has required the development of aerosol-specific attribute tests. The number of product attributes that are evaluated for a given aerosol has grown over the years to what is now an extensive array of tests. The attribute tests for nebulizer solutions, pressurized MDIs, DPIs, nasal aerosols, and lingual aerosols have a significant amount of commonality. However, each dosage form has specific critical quality attributes that are not common to aerosol dosage forms. The U.S.P. performance tests for aerosols, nasal sprays, metered-dose inhalers, and dry powder inhalers are outlined in Chapter <601> and the pH chapter (<791>). There are also a number of other applicable U.S.P. tests: Elastomer Closures for Injections, <381>; Uniformity of Dosage Units, <905>; Containers—Plastics, <661>; Biological Reactivity Tests, In Vitro, <87>; and Biological Reactivity Tests, In Vivo, <88>. Nebulizing solutions, which are sterilized products, must comply with U.S.P. Sterility Tests (<71>) and Bacterial Endotoxin Test (<85>). Nebulizing solutions and nasal aerosols may also be tested according to the chapters on Osmolality and Osmolarity (<785>), Minimum Fill (<751>), and Color and Achromicity (<631>). Nasal sprays are also tested for Particulate Matter (<788>), Antimicrobial Effectiveness Testing (<51>), Antimicrobial Agents—Contents (<341>), and Microbial Enumeration Test (<61>). A number of water content methods are discussed in U.S.P. (<921>), Water Determination. The Karl Fischer direct titration and coulometric titration methods are the most commonly used. U.S.P. <610> has methods for propellants, leak testing, delivered-dose uniformity, delivered-dose uniformity over the entire contents, particle size analysis, and aerodynamic size distribution.

The FDA has provided several guidances related to attribute testing of aerosols (1998, 2002a). Table 16.9 summarizes the tests outlined by the FDA for

TABLE 16.9   Summary of FDA Attribute Tests for Aerosols

Attribute Test	Nebulizer Solutions	Pressurized MDIs	DPIs	Nasal Sprays
Drug Product Specifications				
Description	+	+	+	+
Appearance of the CCS				
Color and clarity of the formulation (need appropriate acceptance criteria for color)				
Identification	+	+	+	+
Two independent methods to identify the drug				
Assay	+	+	+	+
Employ a stability-indicating method for drug content				
Impurities and degradation products	+	+	+	+
Employ validated analytical methods for analysis				
Acceptance criteria should be set for individual and total impurities and degradation products				
Preservatives and other stabilizing excipients	[a]	[a]	[a]	[a]
Chemical content				
Microbiological preservative effectiveness testing				
Valve delivery/pump delivery	−	+	−	+
Assess pump-to-pump reproducibility				
Spray content uniformity (SCU)	−	+	−	+
Spray discharged through the actuator for drug content from beginning to end (for an individual container, among containers, and across batches)				
Spray pattern	−	+	−	+
Evaluated on a routine basis				
Should provide information on the size and shape of the plum (density of plum)				
Droplet size distribution	+ (from nebulizer)	−	−	+
Particle size distribution	+	+	+	+

(*continued*)

TABLE 16.9 (*Continued*)

Attribute Test	Nebulizer Solutions	Pressurized MDIs	DPIs	Nasal Sprays
Particulate matter	+	[b]	[b]	+
Microbial limits	[c]	[d]	+	+
Weight loss	+	+	−	+
Net content	+	+	−	+
U.S.P. <751>, Minimum Fill				
Number of doses	−	+	+	−
Leachables	+	+	−	+
PH	+	−	−	+
Osmolality	+	−	−	+
Viscosity	−	−	−	+
Aerodynamic particle size distribution	+	+	+	−
Acceptance criteria are based on the mass median aerodynamic diameter and the geometric standard deviation				
Leak rate	−	+	−	−

Drug Product Characterization Studies

Attribute Test	Nebulizer Solutions	Pressurized MDIs	DPIs	Nasal Sprays
Priming and repriming (multiple-use products)	−	+	−	+
Instructions should be developed for priming and repriming after different periods of non-use in the upright and horizontal configurations				
Effect of resting time (multiple-use products)	−	+	−	+
Determine the effect of increased resting time on the first spray of unprimed units (over a period of 2 days)				
Temperature cycling	+	+	+	+
Use to simulate shipping conditions 12-hour cycles: freezer (−10 to 20°C) and 40°C for at least 4 weeks				
Effect of moisture	−	+	+	−
Determine effect of low and high humidity on SCU and particle size distribution				
In vitro dose proportionality (for multiple strengths)	+	+	+	+

TABLE 16.9 (*Continued*)

Attribute Test	Nebulizer Solutions	Pressurized MDIs	DPIs	Nasal Sprays
Drug deposition on mouthpiece	+	+	+	−
Determine amount of drug deposited on the mouthpiece, adapters, and other accessories				
Cleaning instructions	+	+	+	+
In-use studies need to determine the frequency of cleaning				
Cleaning instruction should be provided				
Device (as part of the CCS) robustness	−	+	+	+
Performance should be studied for in-use factors (shaking, high temperature, low temperature, low humidity, high humidity), inadvertent use (dropping), and transportation (vibration and environmental conditions)				
Effect of dosing orientation	−	+	+	+
Profiling of sprays near container exhaustion (trail-off characteristics)	−	+	+	+
Effect of varying inspiratory flow rates	−	+	+	−
Determine the effect of different flow rates that can be generated by children, adults, patients with severe lung disease, and when spacers and other accessories are used				
Effect of storage on particle size	−	+	+	+
Primary concern is for suspensions				
Plume geometry	−	+	−	+
Complementary to the spray pattern test				
Determines the shape of the entire plume				
Plume geometry can be evaluated by high-speed flash photography to allow monitoring of the plume development over time; not required as a routine test				
Preservative effectiveness and sterility maintenance	−	−	−	+

(*continued*)

TABLE 16.9 *(Continued)*

Attribute Test	Nebulizer Solutions	Pressurized MDIs	DPIs	Nasal Sprays
Preservation must be maintained at a lower-limit preservative level				
Nebulizing solutions must maintain their sterility throughout the life of the product				
Microbial challenge	−	+	+	+
Confirm that product will not support the growth microbes				
Characterization of nebulizer specified in the labeling	+	−	−	−
Nebulizer conditions must provide a labeled dose				
Photostability	+	+	+	+
Performed for products whose primary package allows light exposure				
Stability of primary (unprotected) package	+	−	+	−
Applies to products that have a protective secondary package (foil for LDPE nebulizer solutions, or foil for dry powder inhalers)				
Data should confirm use time for unprotected product in the primary container				

Source: Adapted from FDA guidances (1998 and 2002[a]) The guidance is not directed for dry powder oral inhalation products. General concepts have also been applied to nebulizer solutions and dry powder inhalers.

[a] Applicable for all excipients used to stabilize the product, such as antioxidant and antimicrobial preservatives.

[b] Particle size distribution of drug substance; for pressurized MDIs where the drug is suspended.

[c] Nebulizer solutions should be sterile.

[d] HFAs do not support microbial growth.

nebulizer solutions, pressurized MDIs, DPIs, and nasal sprays. All test methods need to be validated. The critical quality attributes for lingual and sublingual sprays should follow similar testing, keeping in mind that the aerodynamic particle size distribution and spray patterns are not as critical for these drug products. In addition, lingual and sublingual products are not required to be sterile.

16.6 NEW DRUG APPLICATION STABILITY ASSESSMENT

The primary purpose of stability assessment is to establish product label storage instructions and a shelf life for a product when it is stored under the labeled storage conditions. To meet a specified shelf life, the product must remain within a set of product specifications that are established for each critical product attribute. A balance must be struck between the time required to establish real-time data to support the labeled storage conditions and shelf life and the need to get important drugs to the patient as quickly as possible with the appropriate safety assurances. The regulated process allows the NDA applicant to use appropriately designed real-time stability studies and statistical approaches to extrapolate an estimated shelf life. The stability information that is used to establish the product label storage conditions and shelf life is based on a minimum of three batches of drug product. The three batches of drug product should be manufactured using the market formulation and process proposed and should be packaged in the inhalation system proposed, comprised of a canister, valve, and actuator. These batches should be placed on accelerated and long-term stability testing. It is ideal to have all three batches manufactured and packaged at the commercial production site on a commercial scale. This is often not possible because of cost and lack of available drug substance. In these cases, two of the batches may be manufactured and packaged at one-tenth of full-scale production or 100,000 units, whichever is larger. The third batch may be a smaller pilot-scale batch.

The International Conference on Harmonization guidance documents serve as a starting point for developing an aerosol stability protocol (ICH, 1996, 1997, 2003a, b, 2004, 2009a–c). There should be a minimum of 12 months of long-term and three months of accelerated data at the time of an NDA submission unless the FDA has agreed to shorter stability times. The long-term storage condition should be carried out to support the desired shelf life with real-time data. The ICH stability conditions are $25 \pm 2°C/60 \pm 5\%$ relative humidity (RH) for long-term storage and $40 \pm 2°C/75 \pm 5\%$ RH accelerated storage for one and three months. These storage conditions will support the standard FDA label, stating that the product can be stored at $25°C$ ($77°F$) with excursions permitted to 15 to $30°C$ (59 to $86°$ F). For less stable products that may not achieve 24-month stability, accelerated testing can be conducted at $30°C/65\%$ RH. In addition, the ICH (2003a) guidance document Q1D: Bracketing and Matrixing Designs for Stability Testing of New Drug Substances and Products can significantly reduce the stability commitment if preapproved by the FDA.

Initial testing at time zero typically includes a description of appearance, two drug identity methods, an assay of active product and degradation of products, content uniformity, microbial limits test (unless there is a scientific justification not to include the test), and water content. Both external and internal appearance should be included. For example, the pressurized MDI suspension should be white to off-white. The internal surface of the canister should be smooth without pitting or indication of crystal growth and adherence to the can. Depending on other critical quality attributes that are important for product quality, other tests may be performed as well, such as valve delivery, spray content and content through life, aerodynamic particle size, leak rate, number of actuations, and others (see Table 16.9). Again, product-specific attributes may be added if deemed appropriate. Validated test methods are required for any test that is used for initial testing and stability testing.

A typical stability protocol that would support a regulatory submission to the European Union, Japan, and the United States is provided in Table 16.10. For nebulizer solutions and pressurized aerosols, a typical freeze–thaw cycle is −20 to +30°C at 7-day cycles for 28 days (one month). A usual cycle study involves temperature cycling from 5 to 40°C every 6 hours for a month. A powder inhaler stability program would be more analogous to that of other solid dosage forms, such as tablets and capsules, and would not have freeze–thaw or cycling test conditions.

The FDA has categorized impurities as organic impurities that are process- and drug-related, inorganic impurities, and residual solvents. For drug product

TABLE 16.10   NDA Stability Protocol

Product					Container/Closure and Supplier				API Process	
Product lot					Package lot				API lot	
Batch size					Batch size				Batch size	
Manufacturer					Packager				Manufacturer	
Manufacturer site					Packaging site				Site of manufacture	
Manufacture date					Package date				Manufacture date	
Stability start date										

Storage					Time (months)					
°C/% RH	1	3	6	9	12	18	24	36	48	60
25/60^{a,b}		×	×	×	×	×	×	×	(×)	(×)^c
40/75^{a,b}	×	×								
Light^d	×									
Freeze–Thaw^e										
Cycle^f										

^{a}°C ± 2/%RH ± 5%.

^{b}Liquid samples are stored upright and horizontal to expose the liquid to the valve.

^{c}(X), optional testing.

^{d}Testing per ICH guidelines.

^{e}Seven days at −20°C, 7 days at 30°C, 7 days at −20°C, and 7 days at 30°C.

^{f}5 to 40°C for 6 hours for a month.

TABLE 16.11   Reporting, Identification, and Qualification Thresholds for Degradation Impurities in New Drug Products

Reporting Thresholds	Threshold
Maximum Daily Dose	
Less than or equal to 1 g	−0.1%[a]
Greater than 1 g	−0.05%
Identification	
Maximum Daily Dose	
Less than 1 mg	−1.0% or 5 μg TDI[b], whichever is lower
−1 to 10 mg	−0.5% or 20 μg TDI, whichever is lower
Greater than 10 mg to 2 g	−0.2% or 2 mg TDI, whichever is lower
Greater than 2 g	−0.10%
Qualification	
Maximum Daily Dose	
Less than 10 mg	−1.0% or 50 μg TDI, whichever is lower
−10 to 100 mg	−0.5% or 200 μg TDI, whichever is lower
Greater than 100 mg to 2 g	−0.2% or 3 mg TDI, whichever is lower
Greater than 2 g	−0.15%

[a] Expressed as percent of maximum daily dose.
[b] TDI, total daily intake.

stability studies, the FDA (2002, b) has given guidance regarding the reporting and listing of degradation levels. The degradation levels are based on the maximum daily dose. Table 16.11 summarizes the reporting and listing requirements for drug product degradation impurities.

The ICH (2004) has issued a guidance on the evaluation of stability data. The guidance provides a decision tree that addresses circumstances when statistical analysis is unnecessary, shelf-life extrapolation is appropriate, pooling of data is accepted, and data are not amenable to statistical analysis. The guidance also discusses recommended statistical analysis approaches. In general, where the data show that the product will remain well within the acceptance criteria for a given product attribute, there is little or no change in the accelerated and long-term condition, and there is little or no variability, statistical analysis is normally considered unnecessary. Justification for extrapolation of the shelf life can be proposed. The FDA recommended shelf-life extrapolation in this case would be twice as long as the real-time data, but not longer than 12 months beyond the available long-term data. Where accelerated or long-term data show variability or change over time, shelf-life extrapolation can be based on statistical analysis or a data-based justification. Statistical analysis can support up to twice the period of long-term data, but not more than 12 months. An accepted justification for extrapolation of data not amenable to statistical analysis can support a shelf life of up to one-and-a-half times the available real-time long-term data, but no more than six months of extrapolation.

REFERENCES

Atkins PJ. Metered-dose inhaler: nonpressurized systems. In: Swarbrick J, Boylan JC, eds. *Encyclopedia of Pharmaceutical Technology*, Vol. 9. New York: Marcel Dekker; 1994. pp. 287–298.

Bartels F, Bachtler W, Dunne ST, Eicher J, Freund B, Hart WB, Lessmoellmann C. Atomizing nozzle and filter and spray generating device. U.S. Patent 5911851. 1999.

Batheja P, Thakur R, Michniak B. Basic biopharmaceutics of buccal and sublingual absorptions. In: Touitou E, Barry BW, eds. *Enhancement in Drug Delivery*. Boca Raton, FL: CRC Press; 2007. pp. 175–202.

Budesonide Inhalation Suspension package insert. Teva Pharmaceuticals, Sellersville, PA. http://www.drugs.com/pro/budesonide-inhalation-suspension.html. Accessed Aug. 20, 2011.

Callahan JC, Cleary GW, Elefant M, Kaplan G, Kensler T, Nash RA. Equilibrium moisture content of pharmaceutical excipients. Drug Dev. Ind. Pharm. 1982;8:355–369.

Cambridge Healthtech Institute. Drug delivery and formulation glossary and taxonomy. http://www.genomicglossaries.com/content/drugdelivery.ASP. Accessed Aug. 20, 2011

Chan HK, Chew NYK. Novel alternative methods for the delivery of drugs for the treatment of asthma. Adv. Drug Deli. Rev. 2003;55:793–805.

Chan HK, Gonda I. Respirable form of crystals of cromoglycic acid. J. Pharm. Sci. 1989;78;176–180.

Chien YW, Su KSE, Chang S-F. *Nasal Systemic Drug Delivery*. Drugs and the Pharmaceutical Sciences, Vol. 39. New York: Marcel Dekker; 1989.

Chow AHL, Tong HHY, Chattopadhyay P, Shekunov BY. Particle engineering for pulmonary drug delivery. Pharm. Res. 2007;24:411–437.

Clark AR. MDIs: physics of aerosol formation. J. Aerosol Med. 1996;9(Suppl. 1):S19–S26.

————. Dry powder inhalation systems from Inhale Therapeutic Systems. In: Rathbone MJ, Hadgraft J, Roberts MS, eds. *Modified-Release Drug Delivery Technology. Drugs and the Pharmaceutical Sciences,*. Vol. 126. New York: Informa Healthcare; 2007. pp. 903–912.

Crowder TM, Hickey AJ, Boekestein V. Apparatus, systems and related methods for dispensing and / or evaluating dry powders. U.S. Patent 7118010. 2006.

Dagsland A. Inhalation device. U.S. Patent 63325061. 2001.

Davis SS, Illum L. Absorption enhancers for nasal drug delivery. Clin Pharmacokinet. 2003;13:1107–28.

Davies LA, Davies DN, Wan MSK, Coffee RA. Devices and formulations. U.S. Patent 7883032. 2011.

Davies MB, Hearne DJ, Rand PK, Walker RI. Inhalation device. U.S. Patent 5590645. 1997.

Denyer JSH, Dyche A, Marsden R. Controlling drug delivery apparatus. U.S. Patent 7451760. 2008.

Dobry DE, Settell DM, Baumann JM, Ray RJ, Graham LJ, Beyerinck RA. A model-based methodology for spray-drying process development. J. Pharm. Innov. 2009;4:133–142.

Dugger HA, Cleaver KE, Cox DP. Immediate–immediate release (I^2R) lingual or buccal spray formulations for transmucosal delivery of drug substances. In: Rathbone MJ, Hadgraft J, Roberts JS, eds. *Modified-Release Drug Delivery Technology*. Drugs and the Pharmaceutical Sciences, Vol. 126. New York: Informa Healthcare; 2007. pp. 431–446.

Edwards D, Hanes J, Caponetti G, Hrkach JS, Ben-Jebria A, Eskew ML, Mintzes J, Deaver D, Lotan N, Langer R. Large porous aerosols for pulmonary drug delivery. Science 1997;276:1868–1871.

Edwards DA, Langer RS, Vanbever R, Mintzes J, Wang J, Chen D. Preparation of particles for inhalation. U.S. Patent 5985309. 1999.

E.I. du Pont de Nemours and Company, Wilmington, DE. Mutual solubilities of select HCFCs and HFCs and water. http://www2.dupont.com/Refrigerants/en_US/assets/downloads/k06028_Suva_mutual_solubility.pdf. Accessed Aug. 2011.

FDA. 1987. Guidance for industry: guideline for submitting documentation for the stability of human drugs and biologics. Food and Drug Administration Center for Drugs and Biologics. U.S. Department of Health and Human Services, Feb. 1987. http://www.seoho.biz/GMP_Quick_Search/Data/1.%20FDA%20Documents/1.6.31.pdf. Accessed Sept. 2011.

_____. 1993. Reviewer guidance for nebulizers, metered dose inhalers, spacers, and actuators. Office of Device Evaluation, Division of Cardiovascular and Respiratory Devices, Anesthesiology and Defibrillator Devices Branch. Oct. 1993. http://www.fda.gov/MedicalDevices/DeviceRegulationandGuidance/GuidanceDocuments/ucm081282.htm. Accessed Sept. 2011.

_____. 1998. Guidance for industry (draft): Metered dose inhaler (MDI) and dry powder inhaler (DPI) drug products—chemistry, manufacturing, and controls documentation. U.S. Department of Health and Human Services, Food and Drug Administration Center for Drug Evaluation and Research. Oct. 1998. http://www.fda.gov/downloads/Drugs/GuidanceComplianceRegulatoryInformation/Guidances/ucm070573.pdf. Accessed Sept. 2011.

_____. 1999. Guidance for industry: Container closure systems for packaging human drugs and biologics. U.S. Department of Health and Human Services, Food and Drug Administration Center for Drug Evaluation and Research and Center for Biologics Evaluation and Research. May 1999. http://www.fda.gov./downloads/Drugs/GuidanceComplianceRegulatoryInformation/Guidances/ucm070551.pdf. Accessed Dec. 2011.

_____. 2002a. Guidance for industry: Nasal spray and inhalation solution, suspension, and spray drug products—chemistry, manufacturing, and controls documentation. U.S. Department of Health and Human Services, Food and Drug Administration Center for Drug Evaluation and Research. July 2002. http://www.fda.gov/downloads/Drugs/GuidanceComplianceRegulatoryInformation/Guidances/ucm070575.pdf. Accessed Sept. 2011.

_____. 2002b. Guidance for industry: Inhalation drug products packaged in semipermeable container closure systems. U.S. Department of Health and Human Services, Food and Drug Administration Center for Drug Evaluation and Research. July 2002. http://www.fda.gov/OHRMS/DOCKETS/98fr/02d-0254-gdl0001-vol1.PDF. Accessed Oct. 2011.

Feeley J, York P, Sumby B, Dicks H. The use of inverse gas chromotagraphy (IGC) to highlight batch-to-batch variations in powders used for dry powder inhalers (DPIs). Int. J. Pharm. 1998;172:89–96.

Foster LC, Kuo M-C, Billingsley Sr. Stable glassy state powder formulations. U.S. Patent 6258341. 2011.

Ganderton D. The generation of respirable clouds from coarse powder aggregates. J. Biopharm. Sci. 1992;3:101–105.

Gonda I. Development of a systematic theory of suspension inhalation aerosols: I. A framework to study the effects of aggregation on the aerodynamic behaviour of drug particles. Int. J. Pharm. 1985;27:99–116.

_____ Development of a systematic theory of suspension inhalation aerosols: II. Aggregates of monodisperse particles nebulized in polydisperse droplets. Int. J. Pharm. 1988;41:147–157.

Hallworth, GW. Metered-dose inhalers: non-pressurized systems. In: Swarbrick J, Boylan JC, eds. *Encyclopedia of Pharmaceutical Technology*. Vol. 9. New York: Marcel Dekker; 1994. pp. 287–329.

Heng PWS, Chan LW, Tan LH. Laser Diffraction as a PAT tool for spray drying: an investigation into the capabilities of at- and in-line laser diffraction analysis for improved process monitoring and control. PharmaManufacturing, 2009. http://www.pharmamanufacturing.com/articles/2009/096.html?page = print. Accessed Sept. 2011.

Hickey AJ, Crowder T, Warden JA, Johnson KA, Ketner ME, Fording JK, Garten MD, Riley WM, Anderson SD, Ferris BS, Rockwell PG. Dry powder inhalers, related blister package indexing and opening mechanisms, and associated methods of dispensing dry powder substances. U.S. Patent. 7451761. 2008.

Hickey AJ, ed. *Pharmaceutical Inhalation Aerosol Technology*. Drugs and the Pharmaceutical Sciences, Vol. 54. New York: Marcel Dekker, 1992.

_____. *Pharmaceutical Inhalation Aerosol Technology*. 2nd ed. Drugs and the Pharmaceutical Sciences. Vol. 134. New York: Marcel Dekker, 2004.

Hickey AJ, Gonda I, Irwin WJ, Fildes FJ. Effect of hydrophobic coating on the behavior of a hygroscopic aerosol powder environment of controlled temperature and relative humidity. J. Pharm. Sci. 1990;79:1009–1014.

Hoogstraate J, Benes L, Burgaud S, Horriere F, Seyler I. Oral trans-mucosal drug delivery. In: Hillery AM, Lloyd AW, Swarbrick J, eds. *Drug Delivery and Targeting for Pharmacists and Pharmaceutical Scientists* Boca Raton, FL: CRC Press; 2001. pp. 185–206.

Horhota S, Leiner S, Berger R, Purrington A, Kaerger S, Lyapustina L, Butterworth N, Patel R, Davidson B, Johansson E, Grant A, Smith MA. International Pharmaceutical Aerosol Consortium on Regulation and Science (IPAC-RS), Risk Mangagement Working Group. http://www.ipacrs.come/PDFs/Posters/RM interative poster.pdf. Accessed Oct. 2011.

ICH. 1996. Q1B: Photostability testing of new drug substances and products. Nov. 1996. http://www.fda.gov/Drugs/GuidanceComplianceRegulatoryInformation/Guidances/ucm065005. Accessed Dec. 2011.

———— 1997. Q1C: Stabiligy testing for new dosage forms June 1997. http:www.fda.gov/Drugs/GuidanceComplianceRegulatoryInfomation/Guidances/ucm065005. Accessed Dec. 2011.

———— 2003a. Q1D: Bracketing and matrixing designs for stability testing of new drug substances and products. Jan. 2003. http://www.fda.gov/Drugs/GuidanceComplianceRegulatoryInformation/Guidances/ucm065005. Accessed Dec. 2011.

————. 2003b. Q1A(R2): Stability testing of new drug substances and products. Nov. 2003. http://www.fda.gov/Drugs/GuidanceComplianceRegulatoryInformation/Guidances/ucm065005. Accessed Dec. 2011.

————. 2004. Q1E: Evaluation of stability data. June. 2004. http://www.fda.gov/Drugs/GuidanceComplianceRegulatoryInformation/Guidances/ucm065005. Accessed Dec. 2011.

————. 2006. Q9: Quality risk management. June 2006. http://www.fda.gov/Drugs/GuidanceComplianceRegulatoryInformation/Guidances/ucm065005. Accessed Dec.2011.

————. 2009a. Q4B, Annex 4A: Microbiological examination of non-sterile products—microbial enumeration tests. General Apr. 2009. http://www.fda.gov/Drugs/GuidanceComplianceREgulatoryInformation/Guidances/ucm065005. Accessed Dec. 2011.

————. 2009b. Q4B, Annex 4B: Microbiological examination of non-sterile products—tests forspecified micro-organisms. General chapter. Apr. 2009. h8ttp://www.fda.gov/Drugs/GuidanceComplianceRegulatoryInformation/Guidances/ucm065005. Accessed Dec. 2011.

————. 2009c. Q4B, Annex 4C: Microbiological examination of non-sterile products—acceptancecriteria for pharmaceutical preparations and substances for pharmaceutical general use. Apr. 2009. http://www.fda.gov/Drugs/GuidanceComplianceREgulatoryInformation/Guidancesucm065005. Accessed Dec. 2011.

Illum L, Davis SS. Intranasal insulin: clinical pharmacokinetics. Clin Pharmacokinet. 1992;1:30–41.

Ivri Y, Wu C. Liquid dispensing apparatus and methods. U.S. Patent 7,174,888. 2007.

Kelliher G, Fujimoto D, Drish J. Nozzle pore configuration for intrapulmonary delivery of aerosolized formulations. U.S. Patent 7958887. 2011.

Knoch M, Finlay W. Nebulizer Technologies. In: Rathbone MJ, Hadgraft J, Roberts MS, eds. *Modified-Release Drug Delivery Technology*. Drugs and the Pharmaceutical Sciences, Vol. 126. New York: Informa Healthcare; 2007. pp. 849–856.

Kunze H, Hausmann M, Besseler J, Henning C, Kladders H, Witte F, Geser J, Mast M, Lanci A, Dworzak C, Mock E. Nebuliser and container. U.S. Patent 7950388. 2011.

Kumar V, Ely K, Taylor MK, Stagner WC. Real-time particle size analysis using focused beam reflectance measurement (FBRM) as a process analytical technology (PAT). American Association of Pharmaceutical Scientists Annual Meeting, New Orleans, Nov. 2010.

Lansley AB, Martin GP. Nasal drug delivery. In: Hillery AM, Lloyd AW, Swarbrick J, eds. *Drug Delivery and Targeting for Pharmacists and Pharmaceutical Scientists*. Boca Raton; FL: CRC Press; 2001. pp. 237–268.

Lee HLL, Yamamoto A, Kompella UB. Mucosal penetration enhancers for facilitation of peptide and protein drug absorption. Critical Reviews in Therapeutic Drug Carrier System. 1991;8:91–192.

Le J, Ashley ED, Neuhauser MM, Brown J, Gentry C, Klepser ME, Marr AM, Schiller D, Schwiesow JN, Tice S, VandenBussche HL, Wood GC. Reviews of Therapeutics—Consensus summary of aerosolized antimicrobial agents: application of guideline criteria. Pharmacotherapy 2010;30:562–84.

Lim ST, Forbes B, Brown MB, Martin GP. Physiological factors affecting nasal drug delivery. In: Touitou E, Barry BW, editors. Enhancement in drug delivery. Boca Raton: CRC Press; 2007. pp. 355–372.

Merkus FWHM, Verhoef JC. Nasal drug delivery: trends and perspectives. In: Swarbrick J, Boylan JC, eds. *Encyclopedia of pharmaceutical technology*. Vol. 10. New York: Marcel Dekker; 1994. pp. 191–221.

Modi P. Metered dose spray device for use with macromolecular pharmaceutical agents such as insulin. U.S. Patent 7255102. 2007.

Niven R, Brain J. Some functional aspects of air-jet nebulizers. Int. J. Pharm. 1994;104:73–85.

Parks DJ, Rocchio MJ, Naydo K, Wightman DE, Smith AE. Powder filling systems, apparatus and methods. U.S. Patent 5826633. 1998.

Pfichner A, Pumm G. Aerosol generation device and inhalation device therewith. U.S. Patent 7472701. 2009.

Pillai RS, Rolland A. Sonic nebulized nucleic acid/cationic liposome complexes and methods for pulmonary gene delivery. U.S. Patent 6271206. 2001.

Pillion DJ, Arnold JJ, Meezan E. Nasal delivery of peptide drugs. In: Touitou E, Barry BW, eds. *Enhancement in drug delivery*. Boca Raton, FL: CRC Press; 2007. pp. 373–392.

Platz RM, Brewer TK, Boardman TD. Dispersible macromolecule compositions and methods for their preparation and use. U.S. Patent 6051256. 2000.

Platz RM, Patton JS, Foster L. Spray drying macromolecules to product inhaleable dry powders. U.S. Patent 6582728. 2003.

Rowe RC, Sheskey PJ, Owen, SC. eds. *Handbook of Pharmaceutical Excipients*, 5th ed. London: Pharmaceutical Press; Washington; DC: American Pharmacists Association; 2006.

Sandstrom KB, Eriksson PM, Rosenholm JB. Electrophoretic mobility of salbutamol drug powder in mixed propellant solvents. J. Pharm. Sci. 1994;83:1380–1385.

Schuster J, Farr SJ. The AER_X pulmonary drug delivery system. In: Rathbone MJ, Hadgraft J, Roberts MS, eds. *Modified-Release Drug Delivery Technology*. Drugs and the Pharmaceutical Sciences, Vol. 126. New York: Informa Healthcare; 2007. p. 825–834.

Schuster J, Rubsamen R, Lloyd P, Lloyd J. The AER_X aerosol delivery system. Pharm. Res. 1997;14:354–57.

Sciarra JJ, Sciarra CJ. Aerosols. In: Troy DB, ed. *Remington: The Science And Practice of Pharmacy*. 21st ed. Philadelphia: Lippincott Williams & Wilkins; 2006. pp. 1000–1017.

Shoyele SA, Cawthorne S. Particle engineering techniques for inhaled biopharmaceuticals. Adv. Drug Deliv. Rev. 2006;58:1009–1029.

Sirand C, Varlet J-P, Hickey AJ. Aerosol-filling equipment for preparation of pressurized-pack pharmaceutical formulations. In: Hickey AJ, ed. *Pharmaceutical Inhalation Aerosol Technology*. 2nd ed. Drugs and the Pharmaceutical Sciences, 2nd ed. Vol. 134. New York: Marcel Dekker; 2004. pp. 311–343.

Smart JD. Chemical enhancers in buccal and sublingual absorptions. In: Touitou E, Barry BW, editors. *Enhancement in Drug Delivery*. Boca Raton, FL: CRC Press; 2007. pp. 203–213.

Su KSE, Intranasal drug delivery. In: Swarbrick J, Boylan JC, editors. *Encylopedia of Pharmaceutical Technology*. Vol. 8. New York: Marcel Dekker; 1993. p. 175–201.

Taylor G, Kellaway I. Pulmonary drug delivery. In: Hillery AM, Lloyd, AW, Swarbrick J, eds. *Drug Delivery and Targeting for Pharmacists and Pharmaceutical Scientists*. Boca Raton, FL: CRC Press; 2001. pp. 269–300.

Tong HHY, Chow AHL. Control of physical forms of drug particles for pulmonary delivery by spray drying and supercritical fluid processing. KONA 2006;24:27–40.

U.S. Pharmacopeia. U.S.P.–N.F. online, First Supplement U.S.P. 33–N.F. 28 Reissue. Accessed Oct. 10, 2011.

Uster PS. Aerogen pulmonary delivery technology. In: Rathbone MJ, Hadgraft J, Roberts MS, eds. *Modified-Release Drug Delivery Technology*. Drugs and the Pharmacentical Sciences, Vol. 126. New York: Informa Healthcare; 2007. pp. 817–823.

Vehring R. Pharmaceutical particle engineering *via* spray drying. Pharm. Res. 2008;25:999–1022.

Vehring R, Foss WR, Lechuga-Ballesteros D. Particle formation in spray drying. J. Aerosol Sci. 2007;38:728–746.

Watts AB, McConville JT, William RO III. Current therapies and technological advances in aqueous aerosol drug delivery. Drug Dev. Ind. Pharm. 2008;34:913–922.

Weers JG, Schutt EG, Dellamary LA, Tarara TE, Kabalnov A. Stabilized preparations for use in metered dose inhalers. U.S. Patent 6309623. 2001.

Williams RO III, Liu J. Influence of formulation additives on the vapor pressure of hydrofluoroalkane propellants. Int. J. Pharm. 1998;166:99–103.

GLOSSARY

AAD	Adaptive aerosol delivery.
CCS	Container closure systems.
CDER	Center for Drug Evaluation and Research.
CFR	*Code of Federal Regulations.*
CGMP	*Current Good Manufacturing Practice.*
CLD	*Cord length distribution.*
CPSC	Consumer Product Safety Commission.
DMF	Drug master file.
DPI	Dry powder inhaler.
EHD	Electrodydrodynamic.
EMEA	European Medicines Agency.
E.P.	*European Pharmacopoeia.*
FBRM	Focused-beam reflectance measurement.
FDA	U.S.Food and Drug Administration.
GI	Gastrointestinal.
HFA	Hydrofluoroalkane.
ICH	International Conference on Harmonization.
IND	Investigational new drug.
IPAC-RS	International Pharmaceutical Aerosol Consortium on Regulation & Science.
J.P.	*Japanese Pharmacopoeia.*
LDPE	Low-density polyethylene.
MDI	Metered-dose inhaler.
MDLI	Meter-dose liquid inhaler.
NDA	New drug application.
N.F.	*National Formulary.*
PAT	Process analytical technology.
RH	Relative humidity.
SCU	Spray content uniformity.
U.S.P.	*United States Pharmacopeia.*
VMT	Vibrating mesh technology.

APPENDIX

APPENDIX 16.1   Aerosol Excipients

Excipient	Trade Names[a]	Use[b] (% w/v)
Absorption Enhancers (Nasal, Buccal, and Lingual Delivery)		
Alkylgylcosides[c,d]		
Benzalkonium chloride,[c] E.P./J.P./U.S.P.–N.F.	Hyamine 3500, Pentonium, Zephiran, Tetranyl	0.025–1
Benzethonium chloride, E.P./J.P./U.S.P.–N.F.	Hyamine1622	0.02
Carbomer, E.P./U.S.P.–N.F.[e]	Acritamer, Carbopol, Pemulen	0.1–1
Cetylpyridinium chloride,[c] E.P./U.S.P.–N.F.	Cetamiun	0.02–1.5
Chitosan, E.P.[c,e]	Chitin D, Chitosan 100, Daichitosan 100D	0.1
Cyclodextrin,[c] E.P./U.S.P.–N.F.	Captisols, Cavitron	1–5
Disodium edetate,[c] E.P./J.P./U.S.P.–N.F.	Ronacare	0.02
Ethanol,[d,e] E.P./J.P./U.S.P.–N.F.	Generic	0.7–2
Glycine,[d] E.P./J.P./U.S.P.–N.F.	Generic	0.06–0.05%
Lecithin,[c,d] U.S.P.–N.F.	LSC 5050, LSC 6040, Phosal 53 MCT, Phospholipon 100	0.1
Poly(methyl methacrylate)[e]	Darvan 7, Good-Rite 765, Lubras DS	
Polycarbophil,[e] U.S.P.–N.F.	Noveon AA-1	1.37 mg
Buccal film	—	3.25 mg
Buccal tablet	—	
Saponins[e]		
Soybean steryl glucosides[c,d]		
Sucrose esters of fatty acids[c,d]	Gaotong	
Surfactants		
Nonionic		
Polyoxyethylene-9-lauryl ether, U.S.P.–N.F.	Laureth-9, Volpo L9	
Polysorbate 80, E.P./J.P./U.S.P.–N.F.	Tween 80, Atlas E, Armotan PMO 20, Capmul POE-O, Cremophor PS 80, Crillet 4, Liposorb O-20	
Bile salts		
Sodium deoxycholate	Generic	0.004–10
Sodium taurocholate	Generic	
Antimicrobial Preservatives		
Benzalkonium chloride, E.P./J.P./U.S.P.–N.F.	Hyamine 3500, Pentonium, Zephiran, Tetranyl BC-80	0.002–0.02
Cetylpyridinium chloride, E.P./U.P.S.–N.F.	Cepacol, Cetamiun, Pristacin, Pyrisept	0.02–1.5
Chlorobutanol, E.P./J.P./U.S.P.–N.F.	—	0.5

(*continued*)

APPENDIX 16.1 (*Continued*)

Excipient	Trade Names[a]	Use[b] (% w/v)
Edetate disodium, E.P./J.P./U.S.P.–N.F.	Ronacare	0.02–5
Ethanol, E.P./J.P./U.S.P.–N.F.	Generic	0.4–0.5
Methylparaben, E.P./J.P./U.S.P.–N.F.	Nipagin M, Uniphen P-23	0.025–0.07
Phenylethyl alcohol, U.S.P.–N.F.	—	0.2–0.254
Propylparaben, E.P./J.P./U.S.P.–N.F.	Nipasol M, Propagin,	0.015–0.03
	pH-Adjusting Agents	
Citric acid, E.P./J.P./U.S.P.–N.F.	Generic	0.1–0.2
Hydrochloric acid, E.P./J.P./U.S.P.–N.F.	Generic	1.72–3.50
Sodium citrate, E.P./J.P./U.S.P.–N.F.	Generic	0.3–2.0
Sodium hydroxide, E.P./J.P./U.S.P.–N.F.	Generic	8.0
Sulfuric acid, E.P./U.S.P.–N.F.	Generic	12.50
	Powder Carriers	
Albumin, E.P./U.S.P.–N.F.	Albuconn, Albuminar, Bumin, Proserum	0.5%
Dextran	Gentran, TRITC, Dextran 40000 "Ebewe," Rheomacrodex	1%
Lactose monohydrate, E.P./J.P./U.S.P.–N.F.	Lactochem, Pharmatose	20–25 mg
Mannitol,[d] E.P./J.P./U.S.P.–N.F.	C*PharmMannidex, Mannogem, Pearlitol	0.05%
	Propellants	
Heptafluoropropane	Dymel 227 EA/P, R-227, Solkane 227, Zephex 227	
Tetrafluoroethane	Dymel 134a/P, Genetron134a	0.86–51.52
	Solvents	
Ethanol, E.P./U.S.P.–N.F.	Generic	0.8–33 (95.89)[f]
Glycerol, E.P./J.P./U.S.P.–N.F.	Croderol, Glycon G-100, Kemstrene Optim, Pricerine	7.30
Poly(ethylene glycol) 1000, E.P./J.P./U.S.P.–N.F.	Carbowax, Carbowax Sentry, Lipoxol, Lutrol E, Pluriol E	0.0224
Propylene glycol, E.P./J.P./U.S.P.–N.F.	ProGlyc 55, Adeka PG Dowfrost	20–25
	Surfactants	
Lecithin,[d,e] U.S.P.–N.F.	LSC 5050, LSC 6040, Phosal 53 MCT, Phospholipon 100	0.0002–0.28

APPENDIX 16.1 *(Continued)*

Excipient	Trade Names[a]	Use[b] (% w/v)
Oleic acid, E.P./U.S.P.–N.F.	Crodolene, Crossential 094, Emersol, Glycon, Groco, Hy-Phi, Industrene, Metaupon, Neo-Fat	0.0003–0.26
Polysorbate 20, E.P./U.S.P.–N.F.	Tween 20, Armotan PML 20, Capmul POE-L, Crillet 1, Drewmulse, Durfax 20, Glycosperse L-20	2.5
Polysorbate 80, E.P./J.P./U.S.P.–N.F.	Tween 80, Atlas E, Armotan PMO 20, Capmul POE-O, Cremophor PS 80, Crillet 4, Liposorb O-20	
Sorbitan trioleate, E.P./J.P./U.S.P.–N.F.	Span 85, Ablunol S-85, Arlacel 85, Crill 45, Glycomul TO, Hodag STO, Liposorb TO, Montane 85, Tego STO	0.0175–0.0694
	Tonicity-Adjusting Agent	
Sodium Chloride, E.P./J.P./U.S.P.	Generic	0.7–1.9

[a] A comprehensive list of products and suppliers is provided in Rowe et al. (2006).

[b] FDA Inactive Ingredients List, http://www.fda.gov/Drugs/InformationOnDrugs/ucm113978.htm. Accessed Oct. 15, 2011.

[c] Loosen tight junctions.

[d] Affect membrane fluidity.

[e] Mucoadhesion.

[f] Listed in the FDA Inactive Ingredients List for inhalation, but seems unusually high.

CHAPTER 17

STERILE INJECTABLE PRODUCT DESIGN

17.1 INTRODUCTION

Since sterile injectable products are delivered through the primary defense tissues of the human body, they must be free of any viable microorganisms. *Pyrogens* or materials that elicit a febrile response, such as bacterial endotoxins, are highly controlled to extremely low levels. Injectables must also be free of visible particulate matter. As with other dosage forms, product design is multifactorial and must take into consideration the human condition as well as formulation design, process design, and container closure design. Sterile injectable product design involves an interplay between a number of factors, such as the condition of the patient and the status of the disease, the highest drug dose and intended purpose, the route of administration, the drug's physiochemical properties, the maximum injection volume, the pH, the tonicity, the absence of microorganisms, and the minimal acceptable level of bacterial endotoxin pyrogens and particulates. Bacterial endotoxin pyrogens cause the body to undergo a febrile reaction that results in an increase in body temperature. The gram-negative bacterial endotoxin lipopolysaccharide that causes pyrogenic reactions is very potent. As little as 1 ng/kg can initiate a pyrogenic response (Elin et al., 1981). Bacterial endotoxins can cause septic shock and death.

The term *parenteral* is often used interchangeably with *injectable*. The U.S. Food and Drug Administration (FDA) defines parenteral as administration by injection, infusion, or implantation. The FDA's dosage form nomenclature list mentions 17 different types of injectable products (see Appendix 12.2). The most common injectables are solutions, suspensions, emulsions, and dry solid for solution or suspension. The intradermal (ID), intramuscular (IM), intravenous (IV), and subcutaneous (SC) injection sites are the primary routes of administration. Injection volumes typically do not exceed 0.2, 2.0, and 5 mL for ID, SC, and IM injections, respectively. Smaller IM injection volumes are administered in the smaller deltoid muscle, while larger injection volumes can be given in the posterior gluteal and thigh muscles. An open-label double-blind study in patients reported that a single gluteal IM injection ranging from 9.8 to 19.5 mL was well

Integrated Pharmaceutics: Applied Preformulation, Product Design, and Regulatory Science,
First Edition. Antoine Al-Achi, Mali Ram Gupta, William Craig Stagner.
© 2013 John Wiley & Sons, Inc. Published 2013 by John Wiley & Sons, Inc.

tolerated (Pryor et al., 2001). A number of compounds have been shown to cause concentration-dependent irritation and necrosis at the injection site. Several studies have shown that incorporating the drug in a lipid emulsion can reduce irritation (Lovell et al., 1994; Wang et al., 1999). Injection pH ranges of 6.8 to 7.9, 3 to 7, 2.5 to 8.5, and 3 to 10 have been reported for commercial ID, SC, IM, and IV product administration, respectively. Other less common injection sites are intraabdominal, intraarticular, intrabursal, intraocular, and intrathecal. Appendix 12.2 lists a number of additional injection routes of administration. Each route of administration requires special consideration regarding acceptable injection volume, pH, vehicle constituents, tonicity, and constituents such as preservatives, buffer system, and type of dosage form (solution, suspension, emulsion).

The *United States Pharmacopeia* (U.S.P.) also categorizes injectables into "large-" and "small-volume" injectables or parenterals. Large-volume intravenous parenterals contain more than 100 mL of sterile solution and are single-use delivery systems. A small-volume injectable or parenteral contains 100 mL or less of sterile solution. Large-volume parenterals are used to provide fluid, expansion of plasma volume, electrolytes, a source of nutrition, or a combination of these uses. A large-volume intermittent infusion system can be used to deliver small-volume injectables. Large- and small-volume parenterals have specific product attributes that need to be met. For example, there are different acceptance levels for particulate matter and aluminum content. Large-volume parenterals are single-dose containers. Small-volume parenterals can be single- or multiple-dose containers. In addition, large-volume parenterals are given only by intravenous or hypodermoclysis infusion. Small-volume injectables can be given by the other routes noted above. According to the FDA, a large-volume parenteral needs to be a terminally sterilized aqueous single-dose container that is labeled to contain 100 mL or more (21 CFR 310.509) and intended to be used intravenously in a human. The advantages and disadvantages of injectable dosage forms are listed in Table 17.1.

17.2 FORMULATION DESIGN

The FDA injectable nomenclature includes solutions, emulsions, suspensions, and liposome dosage forms. The nomenclature also recognizes reconstitutable powders for solution, suspension, and liposomes. The U.S.P. has classified drug products as noncomplex active drug products, biotechnology-derived drug products, vaccines, blood and blood products, and gene and cell therapy products. The U.S.P. also lists compendia requirements for identification, assay, characterization, physiochemical evaluation, functionality, process tests, product tests, and safety tests. In this chapter we discuss the product design for noncomplex active drug products, commonly referred to as *small molecules* and biotechnology-derived drug products, often referred to as *therapeutic proteins*. As we discussed the formulation design, process design, packaging design, and attribute testing for disperse systems in Chapter 15, in this chapter we concentrate on the special differences and requirements that exist for injectable products. The individual emulsion, liposome,

TABLE 17.1   Advantages and Disadvantages of Injectable Dosage Forms

Advantages	Disadvantages
Dose drug directly into the bloodstream	Sterility, pyrogen, and particulate matter requirements add significant formulation and manufacturing constraints
Ability to dose a patient who cannot take a drug orally	
Quick onset	
Can control discontinuation of administration	Cost of manufacture
	Inconvenient dosing
Can maintain accurate control of blood levels	Pain on administration
	Solubility and stability properties of the drug can severely constrain formulation approaches
Absolute bioavailability	
Avoids harsh gastrointestinal environment	
	Potential biohazard due to blood-borne pathogens
Generally, the dosage form of choice for peptide and proteins	Microbiological contamination, especially of multiple-dose containers
Controlled drug delivery is possible	Facilities, personnel, equipment, excipients including water; container closures must adhere to microbiological control requirements

and suspension sections combine formulation design and process design that are specific to each of these particular parenteral dosage forms.

17.2.1   Preformulation

The route and intended purpose of injection are determined primarily by medical and marketing needs. The feasibility of a route of administration is limited primarily by the dose and the drug's solubility. The drug's solubility and stability also affect selection of the injectable dosage form and sterilization method. There may be a medical need for an extended-release formulation. This would require a different formulation approach that uses oils, extended-release microparticles, and implants. Preformulation efforts are designed to give the formulator an appreciation for the physiochemical properties of the drug. An understanding of the drug's physiochemical properties lays the framework for identifying major development risks. Once the risks are assessed, a formulation and process development plan can be prepared.

Injectable preformulation studies are not that dissimilar from other dosage forms. Special attention needs to be made with stopper and elastomeric compatibility with the drug and other excipients, such as preservatives. Tables 17.2 and 17.3 outline the preformulation studies that are typically done for small-volume injectables. There is overlap between those studies done for noncomplex molecules and therapeutic proteins and other biopharmaceutical products. Specific studies are highlighted.

TABLE 17.2 Solid-State Physical Properties and Injectable Dosage Implications

Physical Properties	Injection Dosage Implications
Noncomplex Molecules (U.S.P. Terminology)	
Organoleptic properties: color, appearance, and odor	If there is a distinctive color, such as yellow or brown, lot-to-lot color may affect product release. Patient complaints may be registered if they detect differences.
	A disagreeable odor such as a "sulfide" or "rotten egg" smell may alarm a patient as to the acceptable quality of a product.
Morphology and micromeritics: crystal habit, birefringence, particle size distribution, bulk density, tapped density, true density, and state of agglomeration	Birefrigence in the absence of anisotropic crystal is indicative of crystalline material compared to amorphous material.
	Large crystals of a highly insoluble drug may have to be milled to increase the surface area to improve dissolution.
	Bulk containers are sized by volume, but powders are handled by weight. Bulk and tapped densities are important in determining what size container and processing equipment are needed.
	True density is important for suspension formulation.
Thermal properties: melting point, enthalpy of fusion, and other thermal transitions	A high enthalpy of fusion suggests that the drug may have poor solubility.
	Differential scanning calorimetry can pick up other enthalpic transitions, such as polymorphism, dehydration, and decomposition. Polymorphs have different solubility, stability, cohesiveness, density, dissolution rates, and bioavailability.
	The solid-state form of the drug is especially important in developing injectable suspensions.
	Thermal gravimetric analysis can quantitate hydration states and solvent weight loss.
	Loss of water and solvent can occur during milling and micronization processes. The quantity of drug used in a formulation is corrected for the amount of water and other volatiles present.
Crystallinity and polymorphism	Amorphous material is less stable physically and chemically than its crystalline counterpart. Injectable suspensions should use the most physically stable form of the drug.

TABLE 17.2   (*Continued*)

Physical Properties	Injection Dosage Implications
Polymorphism tendency	Some drugs have a larger propensity to form polymorphs than others. Since polymorphs can have a significant effect on manufacturability and bioavailability, it is important to be aware of the drug's tendency to form polymorphs.
Hygroscopicity and moisture sorption isotherm	The degree of hygroscopicity will dictate proper storage and handling conditions.
	Degree of hygroscopicity can affect the physiochemical nature of lyophilized materials.
Contact angle	Indicates how wettable the drug substance is. Wetting or spreading of the dissolution media over the drug is the first step in dissolution, which can affect solution preparation time.
	Wettability of a drug can affect suspension formulation.
	Solids need to be adequately wetted to form a proper suspension.
Static charge and explosivity	The dust clouds that result from handling powders can be lethally explosive. Knowing the explosive potential of powders is crucial to maintaining a safe work environment.
Flow properties	Good flow allows for easier weighing and bulk transfer.

Complex Molecules (Therapeutic Proteins—More-Specific Studies)

Noncovalent reactions Aggregation Precipitation Adsorption Denaturation	Proteins are surface active and tend to be adsorbed on glass and plastic, which can result in denaturation, aggregation, and precipitation.
	Shaking or shear can lead to protein denaturation and other noncovalent reactions.
	Denaturation and alteration in the three-dimensional conformation of the protein may result from these reactions. Denaturation can lead to loss of specificity and potency. The effect of freezing, freeze–thaw cycling, temperature, shaking, pH, high drug concentration, salt concentration and type of salt, and dehydration that occurs during lyophilization can also cause denaturation.
	Aggregates have been shown to cause immunogenic reactions.

TABLE 17.3   Solution- and Solid-State Chemical Properties and Injection Dosage-Form Implications

Chemical Properties[a]	Injection Dosage Implications
Noncomplex Molecules	
Purity and impurities: Impurities may include degradation products, materials related to the synthesis, organic volatiles, residual solvents, and heavy metals	This is a quality attribute. These properties can affect safety and efficacy. These properties can affect manufacturability.
Solubility: in water and pH values covering the acceptable range for injections—ID (pH 6.8 to 7.2), SC (pH 3 to 7), IM (pH 2.5 to 8.5), and IV (pH 3 to 10)[b]	Dose may be limited by the drug's aqueous solubility. For very low-solubility drugs, suspension injectables may be acceptable. However, suspensions cannot be administered intravenously.
Solubility in cosolvents and other solubilizing agents	Dose may be limited by the drug's solubility.
Partition/distribution coefficient	For poorly soluble drugs that are lipophilic, it may be possible to use an emulsion to obtain an acceptable emulsion injection concentration.
Acid–base dissociation constants	Drug ionization state affects the drug's solubility, dissolution rate, and stability. Fast dissolution of the drug into the injection media is an important manufacturing consideration.
Solid- and solution-state stability: as a function of room temperature and stressed temperature, pH, oxygen, and light[c]	Stability knowledge affects formulation, process, and container closure decisions.
Stability to steam autoclave at the optimum pH in the presence and absence of different rubber closures[d]	All injectable products should be terminally sterilized when possible. Steam autoclaving is the preferred terminal sterilization technique. It is important to know how stable the drug is under steam sterilization conditions.
Drug adsorption or absorption to plastic tubing, filters, syringes, rubber closures, glass, plastic containers, IV fluid containers, and hospital administration sets	Adsorption or absorption of the drug can result in low assay or content product values. Drug loss to rubber closures during manufacture or on storage can lead to low assay or content values. Drug losses to filters and plastic hospital administration sets can lead to therapeutic failure.

TABLE 17.3   (*Continued*)

Chemical Properties[a]	Injection Dosage Implications
Complex Molecules (Therapeutic Proteins—More-Specific Studies)	

Additional structure verification
  Amino acid sequence
  Secondary structure: α-helix and
    β-sheet
  Tertiary structure: three-dimensional
  Quaternary structure: dimers, x-mers
	All structure levels are critical to the molecule's specificity and activity.

Check for adventitious viruses and microorganisms	Adventitious organisms can cause patient illness and death. Adventitious organisms include mycoplasma, cells without walls that cause pneumonia; cytomegalovirus which is usually a harmless herpes virus that causes fever and chills; chickenpox; human immunodeficiency virus, human T-lymphotropic virus, which causes T-cell leukemia and T-cell lymphoma; hepatitis B virus, which causes liver cirrhosis and liver cancer; hepatitis C virus, which causes cirrhosis and liver cancer; parvovirus or "fifth disease" or "slapped cheek," which causes fever; adenovirus, which causes the common cold and pneumonia.
Determine isoelectric point	The isoelectric point is the pH where a molecule's net charge is zero. Generally, the solubility of a protein is at its minimum at the isoelectric point.
Degradation reactions associated with proteins: deamidation, disulfide exchange, hydrolysis of the peptide (amide) bond, racemization, and Maillard (browning) reaction	Asparagine is more reactive than glutamine to deamidation. Peptide bond hydrolysis or fragmentation is more stable than ester hydrolysis.
Functional assay or bioassay	These assays measure the potency of the biomolecule. They can be developed in animals, cell cultures, or other biomimetric systems.

[a] These properties are included here for two reasons: (1) one needs a chemical assay to measure these properties, and (2) the chemical nature of these molecules affects these attributes.

[b] Appendix 17.3.

[c] FDA (1996).

[d] If the drug is not stable to steam autoclave conditions, the stability to other sterilizing methods, such as ethylene oxide, gamma irradiation, electron-beam irradiation, and x-ray, should be evaluated.

17.2.2 Excipient Compatibility Studies

The preformulation studies should have identified formulation and process risks and the classes of excipients that would need to be evaluated for excipient compatibility. Excipients used for parenteral products require additional scrutiny. They need to be evaluated for the presence of bacterial endotoxin pyrogens, microorganisms, and insoluble particulates. Injectable excipients should also be nonhemolytic, nonirritating, nonantigenic, and stable to heat sterilization conditions. Some excipients may need to be sterilized by filtration or may be sterilized by a final product filtration process. An excipient vendor program should have bioburden specifications as part of its qualification criteria. Once the route of administration, treatment intent, preformulation physiochemical assessment, and dosage form selection has been completed, an excipient compatibility study can be initiated. Table 17.4 lists the injectable excipient classes and functions. Specific excipients can exhibit multiple functions.

Appendixes 17.1 to 17.7 provide an extensive list of excipients and the use percentages that have been incorporated in injectable formulations (Nema et al., 1997; Powell et al., 1998; Strickley, 1999, 2000a,b, 2004; FDA, 2011). Many injectable formulations are primarily pH- and tonicity-adjusted aqueous solutions that involve minimal excipient compatibility assessment. As solubility and stability limitations increase, the level of excipient compatibility assessment grows exponentially. See Section 13.2.2 for a more detailed discussion of excipient compatibility methodologies.

Typically, the primary concerns for injectable dosage form design is the dose, drug solubility, and stability. There are a number of technologies that can be used to enhance a drug's solubility (Liu, 2008). Often, solubility-enabling technologies can provide a level of stability enhancement as well. In Sections 17.2.1.1.1 to 17.2.1.1.8 we discuss various methods that can be evaluated during the preformulation process to increase the solubility and stability of an active pharmaceutical ingredient (API). Depending on the dose and the aqueous solubility of the drug, additional effort may be required to enhance the drug's solubility by use of cosolvents, surfactants, micelles, oils, emulsions, liposomes, and complexing agents. Appendixes 17.1, 17.2, and 17.3 provide lists of solvents, cosolvents, surfactants, solubilizing agents, buffers, and component percentages that have been used in commercial products and listed in the FDA Inactive Ingredients List. Apte and Ogwu (2003) published a review of emerging excipients that could be used for complex drug solution formulations and injection systems to deliver genes, therapeutic proteins, and other biopharmaceutical APIs.

17.2.2.1 Solubilization and Stabilization

17.2.2.1.1 Salt and Prodrug Solubilization and Stabilization The drug's solubility and stability should be addressed at the developability stage of drug development. At this stage it is possible to select a soluble salt (Stahl and Wermuth 2002) or synthesize a soluble prodrug (Christrup et al., 1996) that can address

TABLE 17.4   Injectable Excipient Classes and Functions

Excipient Class	Function
Antimicrobial preservatives	Maintain a sterile product and prevent growth of microorganisms
Antioxidants	Reduce the oxidative degradation of a drug by being preferentially oxidized, impeding an oxidative chain reaction, or a combination.
Buffers	Maintain a desired pH range, typically to improve drug solubility or stability.
Bulking agents	Used in freeze-drying. Fillers improve the dried cake attributes and provide adequate cake mass.
Chelating agents	Complex trace metals that can catalyze oxidation reactions of drugs.
Inert gases	Provide an inert environment in the liquid and headspace to eliminate or reduce the potential oxidation of drugs.
Protectants	Stabilize protein and peptide solutions. Stabilize and decrease denaturation during the freezing and drying processes that occur during lyophilization.
Solubilizing agents	Used to increase the solubility of a drug.
Suspending agents	Add structure, decreases the settling rate of injectable suspensions, and aids in achieving a uniform dose on resuspension.
Surfactants	Act to wet insoluble hydrophobic drugs. Aid in making emulsions by decreasing the interfacial tension between the oil and water phases.
Thickening agents	Increase viscosity and decrease the sedimentation rate of suspensions or creaming rate of emulsions.
Tonicity-adjusting agents	Provide an isotonic solution with cells and biological fluids.
Water for injection	Primary vehicle used in injectable formulations

a solubility issue very early in development without affecting the development timeline. In fact, selection of a soluble and stable API at the earliest stage of drug development can significantly accelerate product design. Stahl showed that salts can increase the solubility of a drug about 4000-fold. The author has seen solubility increases as high as 24,000-fold for amorphous salts or when using vitro salt preparation methods. Leppänen et al. (2000) prepared a phosphate ester of entacapone and increased the drug's solubility 1700- and 20-fold at pH 1.2 and 7.4, respectively. Adding an acid or base to the injection solution is a common method of achieving increased solubility.

Insoluble salts may be synthesized or prepared in situ to improve the physical and chemical stability of suspensions. Insoluble salts may also be useful in preparing lyophilized powder for suspension. Insoluble salts are generally less hygroscopic and can improve the physiochemical stability of the suspension. Prodrugs can also improve the stability of APIs. The prodrug hetacillin is prepared by reacting acetone with the ampicillin amino group side chain. This prevents the autoaminolysis that occurs with aqueous sodium ampicillin solutions. The stabilized hetacillin is readily converted to ampicillin in vivo (Christrup et al., 1996).

17.2.2.1.2 Cosolvent Solubilization and Stabilization Cosolvents can be used to increase the solubility of a drug. A number of commercial formulations have used cosolvents to provide adequate solubility and stability. Solubility increases associated with the use of cosolvent concentrations acceptable for human injection can range from a severalfold increase to a several-thousandfold increase in solubility (Jouyban, 2010). Appendix 17.1 lists commonly used injectable cosolvents. As with all functional excipients, cosolvents can exhibit concentration-dependent irritation and toxicity. Also, cosolvents can cause muscle toxicity (Brazeau and Fung, 1989) and subcutaneous irritation (Radwan, 1994). Medium-chain triglycerides generally exhibit less inflammation than does alcohol or propylene glycol. Different suppliers of peanut oils have been shown to elicit different levels of irritation. Sesame oil had a level of inflammation comparable to that of peanut oil, which was about a fivefold increase compared to normal saline. A 172-, 45-, 38-, and 31-fold increase in skinfold thickness was observed for 10% benzyl alcohol, 10% ethyl oleate, 15% lecithin, and 4% ethanol, respectively. Increases in percentage increase in skinfold thickness were used as a surrogate measure of inflammation. Reed and Yalkowsy (1985) studied the hemolysis of several solvents used as cosolvents. The ranking of least cytotoxic to most cytotoxic was reported to be N,N-dimethylacetamide, poly(ethylene glycol) (PEG) 400, ethanol, propylene glycol, and dimethyl sulfoxide. Cadwalladar and co-workers also evaluated the effect of cosolvent concentration on hemolysis. They showed that hemolysis can be minimized when alcohol, propylene glycol, and poly(ethylene glycol) are used in concentrations of less than 10% (Cadwallader, 1978), 32% (Cadwallader et al., 1964) and 40% (Smith and Cadwallader, 1967), respectively. The use of cosolvents for SC injection is bound by restrictions caused by low blood and fluid flow.

Nonaqueous solvent systems have been developed that require dilution before use. Rajagopalan et al. (1988) observed an 840-fold increase in amphotericin B solubility using a 50:50 propylene glycol/poly(ethylene glycol) 400 solvent mixture. The presence of sodium deoxycholate in a nonaqueous system resulted in an additional 20-fold solubility increase due to sodium deoxycholate ion-pair formation. In general, an excess level of cosolvents is required to decrease the rate and extent of precipitation that commonly occurs during IV infusion or dilution with other injectable drugs or IV fluids.

The dielectric constant of the solvent system can increase or decrease the degradation rate of a drug (Sinko, 2006). This has important pharmaceutical implications since cosolvents are often used to increase drug solubility. The inclusion of cosolvents to water will decrease the bulk dielectric, which can affect drug stability. The medium effect depends on the charges on the drug and other reactants.

If a drug and reactant have opposite charges, a decrease in the solvent dielectric constant will generally result in an increase in drug degradation. Conversely, the degradation rate for like charged reactants will decrease in lower-dielectric-constant solvents. If the drug is a neutral dipolar molecule that reacts with an ion such as hydronium, a decrease in the solvent dielectric constant leads to an increase in the reaction rate. A 2.25-fold increase in acid-catalyzed glucose degradation was seen with a 2.5-fold decrease in the dielectric constant (Heimlich and Martin, 1960).

17.2.2.1.3 Surfactant and Micelle Solubilization and Stabilization A list of surfactants used in injectables is provided in Appendix 17.2. Surfactants and micelles have also been used to increase the solubility of drugs for injectables. In a number of cases the drug is formulated as a concentrate that is diluted further. Seedher and Kanojia (2008) studied the solubilization of four sulfonylureas and two glitazone antidiabetic agents in nonionic, anionic, and cationic surfactants. In a number of cases the mixed micelles of the ionic and nonionic surfactants provided a significant solubility increase compared to the single surfactants alone. Solubility increases of about 380-fold were observed. In many cases, a 7.4 buffer provided additional solubility to the mixed micelle system. Fugizone Intravenous uses 0.0082% w/v sodium deoxycholate to solubilize Amphotericin B 40-fold. Ren et al. (2008) evaluated the effect of five neutral surfactant molecules (Brij 58, Poloxamer 188, Cremophor RH40, Gélucire 44/14, and PEG 6000) on the stability of rabeprazole sodium. Brij 58 increased the aqueous stability of the drug 3.4- and 5.2-fold at 37 and $60^{\circ}C$, respectively. The improved stability was thought to be due to micellar solubilization of the principal degradation product, preventing it from precipitating out of solution. Hamid and Parrott (1971) studied the alkaline hydrolysis of benzocaine and homatropine. The degradation rate constants for benzocaine and homatropine were inversely proportional to the concentration of surfactant. It was also shown that the degradation rate constant was linearly related to the reciprocal of the apparent solubility of benzocaine in the surfactant. Micellar catalysis of organic reactions has also been reported extensively by Broxton and co-workers. Broxton (1982) showed that the hydrolysis of aspirin was inhibited by micelles in the pH range 6 to 8 but was accelerated by the presence of micelles at a pH in excess of 9.

17.2.2.1.4 Oil Solubilization and Stabilization Water-immiscible parenteral vehicles are primarily refined fixed oils and esters of saturated and unsaturated fatty acids. Fixed oils are nonvolatile triglyceride esters of fatty acids of vegetable origin. There are a number of fixed oils that have *National Formulary* (N.F.) monographs, such as almond oil, corn oil, cottonseed oil, olive oil, palm kernel oil, palm oil, sunflower oil, and sesame oil. Safflower oil has a U.S.P. monograph. These oils vary in their fatty acid and sterol composition. These differences can be used to solubilize the drug and provide a more stable injectable. For example, palm kernel oil has 40 to 52% w/w dodecanoate (laurate) ester compared to palm oil, which contains about 40% hexadecanoate (palmitate) and 40% octadecanoate (stearate) esters. Also, the level of unsaturation can affect the stability of the oil itself. The more unsaturated the oil, the more likely it will be sensitive to thermal and oxidation reactions. Sesame oil has been the most frequently used oil, primarily because it is more stable to oxidation due to intrinsic antioxidants. Fixed oils should be free of hydrocarbon oils and waxes since these cannot be metabolized by the body. Recently, there has been a significant increase in allergic reactions to peanuts, peanut powder, and other peanut components. Therefore, other fixed oils should be given preferential consideration. Even though there are increased concerns about peanut allergies, propyliodone in peanut oil is still listed in the online U.S.P. 33–N.F. 28 S1 reissue monograph. There are a number of U.S.P. SVPs that contain fixed oils. As of this writing, fixed oils are undergoing major N.F. revisions concerning their tests and labeling. Antioxidants that are added to the oils are now required to be named as well as the quantity defined. Oils that are intended to be used for injectables need to be labeled as such. The oil should also be free of herbicides and pesticides.

Fatty acid esters are also classified as water-immiscible injectable vehicles. Esters are generally less viscous than fixed oils and are more fluid at lower temperature, which aids in the ease of injection. Ethyl oleate (unsaturated C_{18}), ethyl myristate (saturated C_{14}), and ethyl palmitate (C_{16}) have N.F. monographs. Saturated shorter-chain-length esters (C_8 to C_{12}) have good stability and solubilizing capability. Wate-immiscible injectables are given IM and should never be given IV. These oils can be used in emulsions and microemulsions. Larsen et al. (2002) showed that there was a linear correlation between log molar solubility in 2 : 1 fractionated coconut oil/castor oil and the melting point of the solutes studied. Benzocaine is 24-fold more soluble in isopropyl myristate than in water. Prankerd and Stella (1990) found that rhizoxin's solubility in safflower oil was 125-fold greater than in water and penclomdine's water solubility was increased 492,000-fold in safflower oil. New and Kirby (1997) presented a Macrosol technology for solubilizing hydrophilic molecules including proteins into oil vehicles in the absence of water. A suspension of dapiprazole HCl in a hydrocarbon–oil mixture was shown to be stable for 16 months compared to an aqueous solution, which required use within 21 days of reconstitution (Hanna, 2010).

17.2.2.1.5 Emulsion Solubilization and Stabilization Prankerd and Stella (1990) developed a 10-mg/mL emulsion of penclomedine, which demonstrated a 25,000-fold increase in solubility compared to the drug's aqueous solubility. Akkar et al. (2004) solubilized itraconazole in Lipofundin MCT 20% to the extent of 10 mg/mL using a solvent-free high-pressure homogenization. This was about a 5600-fold increase in solubility. Repta (1981) discussed the use of oil-in-water emulsions to increase the stability of drugs that are highly unstable in water and exhibit high oil stability and solubility. He showed that the larger the o/w partition coefficient and phase volume, the greater the stabilization expected. Theoretically, if the primary route of drug degradation is in water, an emulsion containing a drug that has an o/w partition coefficient of 50 and a phase volume of 0.22 would be expected to show a 12-fold improvement in drug stability. An emulsion formulation example was provided which showed that an o/w emulsion increased the stability 360-fold compared to the stability of an 2 : 1 acetone–water solution of the drug.

17.2.2.1.6 Liposome Solubilization and Stabilization Ambisome is a reconstituted lyophilized liposomal product of amphotericin B. The reconstituted product concentration is 40-fold its aqueous solubility. The increased solubility is thought to be due to an ion-pair association between the liposome hydrogenated soy phosphatidylcholine and amphotericin B. Additionally, Ambisone has been shown to exhibit increased stability in human plasma and decreased toxicity (Adler and Proffitt, 1993; Proffitt et al., 2004).

17.2.2.1.7 Complexation Solubilization and Stabilization The solubility of drugs has been enhanced by forming soluble complexes. Cyclodextrins have been shown to increase drug solubility over 90,000-fold (Loftsson and Brewster, 1996). Modi and Tayade (2007) undertook a comparative solubility enhancement study of valdecoxib. The authors compared the solubilities obtained from the use of surfactant micelles, cosolvents, and cyclodextrins. The largest valdecoxib solubility increases observed were 80-, 70-, and 30-fold increases obtained from a 25% v/v PEG 400/water cosolvent system, Cremophor EL micelles, and sulfobutyl ether-7β-cyclodextrin, respectively. Other solubilizer systems provided lower enhancements. Devarakonda et al. (2005) compared the complex solubilization capability of polyamidoamine (PAMAM) dendrimers and cyclodextrins. The niclosamide–PAMAM complex aqueous solubilities increased from generation 0 to generation 4. The fourth-generation complex exhibited about a 6100-fold increase in solubility compared to hydroxypropyl-β-cyclodextrin, which gave almost a 10-fold increase. Green and Guillory (1989) showed that a chlorambucil/heptakis-β-cyclodextrin complex provided a 40-fold increase in solubility and a 20-fold increase in stability.

17.2.2.1.8 General Considerations The solubility of a drug can be enhanced by using a combination of solubilization approaches, such as using a salt and a complexing agent. Ziprasidone's intrinsic solubility was increased approximately 67,000-fold using a mesylate salt and sulfobutyl ether β-cyclodextrin soluble complexing agent. When evaluating these formulation aids it is important to take into account not only the solubility enhancement but also the total systemic exposure of the excipient in 24 hours per kilogram of body weight or a combination thereof. For example, Ativan (lorazepam) is a 100% nonaqueous IM injection. The recommended IM dose is a 1-mL injection that contains 2% v/v benzyl alcohol (0.02 mL), 18% poly(ethylene glycol) 400 (0.18 mL), and 80% propylene glycol (0.8 mL). On the other hand, Valium (diazepam) contains 40% v/v propylene glycol and 10% ethanol. If the maximum IV dosage regimen for severe recurrent seizures is used for a day, propylene glycol and ethanol exposure would be 28.8 mL of propylene glycol and 7.2 mL of alcohol, respectively. The maximum daily IV exposures for a number of commonly used cosolvents and solubilizing agents has been calculated from the product package inserts and are provided in Table 17.5.

17.2.2.2 Buffers Buffers are used to resist changes in pH caused by interacting with container closure components or degradants or a combination. The target pH

TABLE 17.5 Maximum Daily Exposure of Selected Solubilizing Agents

Product	Solubilizing Agent	Maximum Exposure/Day
Intralipid, Liposyn[a]	Egg yolk lecithin	24 g
Valium	Ethanol	7.2 mL[b]
Intralipid, Liposyn[a]	Glycerin	50 mL[a]
Intralipid, Liposyn[a]	Long-chain triglycerides	250 g
Lipofundin[a]	Medium-chain triglycerides	200 g
Robaxin	PEG 300	15 mL[b]
Valium	Propylene glycol	28.8 mL[b]
Vfend	Sulfobutyl ether β-cyclodextrin	9.6 g[b]
VePesid	Polysorbate 80 (Tween 80)	0.9 g[b]

[a] Depending on the size and maturity of the patient.

[b] IV infusion rate was not stated explicitly.

is usually based on the pH that is close to the physiological pH of blood, 7.4, and where the drug is most soluble and stable. The most common buffers used for injectable formulations are provided in Appendix 17.3. The buffers should have sufficient buffer capacity to maintain the desired pH and adjust readily to pH 7.4 when injected. Buffer concentrations generally range between 0.01 and 0.1 M.

17.2.2.3 Antimicrobial Agents

Antimicrobials (Appendix 17.4) are added to maintain the sterility of multiple-dose containers unless the drug or vehicle has antimicrobial activity sufficient to prevent inadvertent contamination. Because of their nature and function, they exhibit cell irritation and toxicity at relatively low concentrations. Using antimicrobial agents to provide an additional level of assurance to aseptically processed injectables is controversial. Many argue that adding preservatives to aseptically processed products can hide poor manufacturing techniques. Obviously, the use of antimicrobial agents is not a substitute for good manufacturing practices. The benefit of their presence must be weighed against their potential toxicity. Antimicrobial preservative activity is dependent on their concentration, formulation composition, and the pH of the system (Akers et al., 1984). A number of commonly used parenteral antimicrobial preservatives, their use ranges, antimicrobial spectrum, and effective pH ranges are presented in Table 17.6.

Methyl and propylparaben are effective against a wide range of microorganisms. Although they are effective over a relatively wide pH range they are susceptible to ester hydrolysis. The pH of optimum stability is 4 to 4.5 (Kamada et al., 1973; Sunderland and Watts, 1984). The paraben loss under autoclave conditions of 121.5°C for 20 minutes was predicted to be 0.09% at a pH of 3 and 0.05% at a pH of 4. Raval and Parrott (1967) determined that autoclaving methyl paraben at 121.5°C for 30 minutes resulted in 5.5% degradation at pH 6 and 42.0% paraben loss at pH 9. Parabens can undergo transesterification reactions with cosolvents such as ethanol and propylene glycol. Methylparaben has been shown to undergo specific buffer catalysis with phosphate buffers (Blaug and Grant, 1974). There was

TABLE 17.6   Antimicrobial Preservatives

Type	Concentration (% w/v)s	Antimicrobial Activity[a]				pH[b]
		Gram +	Gram −	Fungi	Yeasts	
Benzyl alcohol	0.1–3.0	+ + +	+ + +	−	−	3–6
m-Cresol	0.1–0.3	+ +	+ +	+ +	+ +s	4–10
Methylparaben	0.08–0.1	+ +	+ +	+ +	+ +	3–9
Propylparaben	0.001–0.02	+ + +	+ + +	+ + +	+ + +	3–9
Phenol	0.2–0.5	+ +	+ +	+ +	+ +	4–10
Thimersol[c]	0.1–0.4	+ +	+ +	+ +	+ +	4–8

Source: Akers and DeFelippis, (2000), with permission from Taylor & Francis.

[a] + + +, most effective; + +, moderately effective; −, poorly effective.

[b] Effective pH range.

[c] Thimersol is not acceptable in Japan.

a 2.3-fold increase in the stability rate constant when the phosphate concentration was increased from 0.02 M to 0.20 M. Parabens lose their activity by adsorbing to macromolecules such as cetomacrogol or poly(ethylene) glycol monocetyl ether (Crooks and Brown, 1974) and polysorbate 80 (Pisano and Kostenbauder, 1959). Parabens can also be absorbed into rubber stoppers, tubing, and other plastic material used during manufacturing or in IV administration sets.

The *European Pharmacopoeia* (E.P.) has much more stringent antimicrobial effectiveness criteria than those of the U.S.P. The E.P. requires a 3 log reduction of bacteria after 6 hours of testing compared to the U.S.P. which requires a 1 log reduction after 7 days. There are additional differences during the testing period. Additionally, after 28 days of testing the E.P. requires that there be no recoverable bacteria and no increase in fungi. The U.S.P. requires that there be no increase in bacteria or fungi.

17.2.2.4 Antioxidants

Oxidative degradation is a common drug degradation mechanism that is identified in preformulation studies. Drugs undergo oxidation by removal of hydrogen or the addition of oxygen. Antioxidants act by preferentially giving up hydrogen or accepting an oxygen molecule. Oxidative degradation is complex because many of the initiating species have very short half-lives and are difficult to detect and quantitate. In addition, reactions can be initiated by trace amounts of oxygen, metals, bases, and impurities such as peroxides. Oxidation reactions also occur with excipients such as unsaturated oils, phospholipids, colorants, and flavors. Degradation results not only in drug loss but can lead to discoloration, precipitation, changes in drug release rate, flavor, and odor. Oxidative decomposition is generally described as an autoxidation process that is mediated by a free-radical chain reaction that involves three stages: initiation, propagation, and termination. Antioxidants and sequestering agents are targeted to delay or prevent the initiation stage. A simply schematic of the stages follows:

$$\text{Initiation:} \quad RH \xrightarrow{\text{heat, light, metal, base}} R\cdot + H\cdot \qquad (17.1)$$

Propagation: $R\cdot + O_2 \rightarrow RO_2\cdot$ (peroxy radical)   (17.2)

$RO_2\cdot + RH \rightarrow ROOH$ (hydroperoxide) $+ R\cdot$ (17.3)

Termination: 2 free radicals $\rightarrow$ molecular products   (17.4)

Hydrated metal ions can catalyze reduction or oxidation of ROOH in equation (17.3) to form $RO\cdot$ or $ROO\cdot$, respectively. Ultraviolet light and visible light have sufficient energy to dissociate organic covalent bonds, especially peroxy bonds. Photodecomposition pathways have been discussed in detail by Greenhill (1995) and Asker (2007). The autoxidation reaction can continue until the entire drug is degraded or the oxygen is depleted. Antioxidants are readily oxidizable molecules that possess lower oxidation potentials than those of the drug and are therefore preferentially oxidized. The antioxidants are oxidized by giving up a hydrogen atom or an electron. The antioxidant reaction that undergoes preferential oxidation is:

$AntioxH + R\cdot \rightarrow RH + Antiox\cdot$ (free radical)   (17.5)

$AntioxH + ROOH \rightarrow RO\cdot$ (free radical) $+ Antiox\cdot$ (free radical) $+ H_2O$ (17.6)

The antioxidant radical may react with hydroperoxide compounds and generate free radicals. To inhibit or delay the chain reaction, it is not only important for the antioxidant to be preferentially oxidized because of its relative lower oxidation potential but, it is also desirable that the resulting antioxidant radical be stabilized and therefore less likely to propagate the chain reaction. The butylated hydroxyltoluene free radical stabilizes itself by resonance and, in some cases, dimerization. The effectiveness of the antioxidant also depends on the extent of free-radical formation in equations (17.5) and (17.6). Excess antioxidants can actually result in a prooxidative effect. Therefore, there is an optimum antioxidant concentration.

Just as antioxidants are in their reduced form, so are many drugs. Drugs undergo oxidation by removal of hydrogen or the addition of oxygen. Oxidative degradation can be described by the simplified Nernst equation, which relates the oxidative potential and the oxidation reaction. The relationship between a compound's oxidation potential and the oxidation–reduction reaction is illustrated using hydroquinone. Equation (17.7) shows the chemical and charge balance for the oxidation reaction of hydroquinone and the related Nernst equation (17.8):

hydroquinone (reduced state, Rd) $\leftrightarrow$ quinone (oxidized state, Ox)

$+ 2H^+ + 2$ electrons$^-$ (17.7)

$$E = E^0 + \frac{RT}{F2} \ln \frac{[Ox] \cdot [H]^{+2}}{[Rd]}$$ (17.8)

where E is the oxidation potential, E^0 the standard oxidation potential, R the gas constant, T the temperature in kelvin, F the Faraday constant, 2 the number of electrons involved in the reaction, [Ox] and [H$^+$] the concentration of the products of the reaction, and [Rd] the concentration of the reactant.

The effect of pH on this oxidation–reduction reaction (17.7) can be described by Le Châtelier's principle. As the hydrogen-ion concentration is increased, the reaction equilibrium is shifted to the left in favor of the hydroquinone (in the reduced form). Addition of base will remove the H^+ and "pull" the equilibrium to the right in favor of quinone, the oxidized form. The effect of pH on the hydroquinone oxidation reaction can also be understood from the Nernst equation. As the hydrogen-ion concentration increases, the oxidation potential of the reaction increases, which makes the oxidation reaction more difficult. As the hydrogen-ion concentration decreases, as is the case at a more basic pH, the oxidative potential is decreased and the oxidation reaction is favored. Both Le Châtelier's principle and the Nernst equation predict that basic pH conditions favor oxidation. Having said this, there are literature examples where basic solutions were more stable than oxidation reactions (Asker and Ferdous, 1996; Chinnian and Asker, 1996).

A number of antioxidants and their use concentrations are listed in Appendix 17.5. Very low concentrations of the antioxidants have been shown to be effective. Theoretically, the selection of an antioxidant could be made based on the difference in the oxidative potential of the drug and the antioxidant. However, the impact of trace metals, pH, and other excipients make even the simplest injectable formulations reasonably complex when it comes to predicting antioxidant effectiveness. Akers (1979) wrote one of the few research reports that discusses in detail the measurement of oxidation potentials and its application to preformulation screening. Several antioxidants alone and in combination were evaluated. Akers reported that measuring the decrease in equivalence volume with time was a better indicator of antioxidant capability than standard oxidation potentials. It was also shown that in the case of epinephrine, a combination of antioxidants provided better stability than did single compounds. The major limitation of the method was the nonspecificity of the method for the antioxidants and the drug. Considering the various complexities of antioxidant screening, knowing the difference between the oxidation potential for the drug and antioxidant is still a good starting point for selecting an effective antioxidant. Table 17.7 lists the oxidation potentials of several antioxidants. The compound with a higher (more positive) oxidation potential would be expected to be preferentially oxidized as opposed to a compound with a lower oxidation potential. A review article of antioxidants used in pharmaceutical products is also available (Akers, 1982).

In addition to oxidative potential, there a number of physiochemical properties that should be considered when selecting antioxidants. By their nature, antioxidants are reactive. Antioxidants should be stable to sterilizing autoclave conditions, be stable over a fairly wide range of pH, form colorless degradation products, have good compatibility with other formulation components and container closure systems, and be in solution with the drug. Important physiochemical properties of some antioxidants are discussed below. Sodium sulfite salts are often used in aqueous systems (Schroeter, 1961, 1963). The concentration and pH govern which sulfite species predominates. Generally, the metabisulfite, bisulfite, and sulfite are used at low, intermediate, and high pH values, respectively. Bisulfites undergo reactions with alkene functions and are inactivated. Their antioxidant activity has also been inhibited by mannitol, aldehydes, and ketones, so special attention needs to be made with drugs that have these functional groups. Many oxidized compounds discolor and turn yellow or brownish. This is also true for antioxidants such as ascorbic acid, where oxidized ascorbic acid solutions turn brownish in color. Formulation scientists also need to be aware of physical incompatibilities, such as

TABLE 17.7 Standard Oxidation Potentials for Various Antixoxidants[a]

Compound	E^0 V	pH	Temperature (°C)
Dithiothreitol[b]	+0.208	7.0	30
Sodium thiosulfate[b]	+0.050	7.0	30
Thiourea[b]	+0.029	7.0	30
Ascorbic acid	+0.003[b]	7.0	25
	−0.390[c]		
Sodium metabisulfite[b]	−0.114	7.0	25
Sodium bisulfite[b]	−0.117	7.0	25
Propyl gallate[b]	−0.199	7.0	25
Acetyl cysteine[b]	−0.293	7.0	25
Cysteine[b]	−0.34		
Vitamin K[b]	−0.363		
Glutathione[b]	−0.430		
Hydroxyquinone[c]	−0.601		
Pyrogallol[c]	−0.661		
Hydroquinone[c]	−0.673		
α-Tocopherol[c]	−0.684		
p-Cresol[c]	−1.038		
m-Cresol[c]	−1.080		
Phenol[c]	−1.089		

[a] Sulfites may cause allergic-type reactions, and a warning to that effect must be on the label of any prescription drug, regardless of the quantity (21 CFR 201.22).

[b] Akers (1979).

[c] Nash (1958).

absorption of butylated hydroxytoluene, butylated hydroxyanisole, and propyl gallate into rubber closures. Borosilicate glass contains calcium and barium oxide that can be extracted from the glass. The sulfites are oxidized to sulfates, which may precipitate as insoluble calcium and barium sulfate salts.

17.2.2.5 Metal Ion Sequestrants Oxidation and electron transfer between drugs and hydrated trace transition metal ions are well known. Hydrated metal ions serve as oxidation catalysts. Appendix 17.6 lists a number of sequestrants and their stability constants for metal ions, including some transition metals. The stability constants are also pH dependent (Taylor, 1956). Ethylenediaminetetraacetic acid (EDTA, edetic acid) is one of the most commonly used sequestrants and has been used alone or in combination with antioxidants. Concentrations as low as 0.005% w/v EDTA have stabilized discoloration formation resulting from oxidation reactions. Biocompatible amino acids; alanine, cysteine, glycine, leucine, and valine can be used alone or in combination with EDTA as sequestrants.

17.2.2.6 Other Compatibilities and Considerations Drug compatibility with the major IV fluids should also be determined for 0.9% w/v sodium chloride,

5% w/v dextrose, lactated Ringer's, and 10% fructose and 10% invert sugar. Drug compatibility with filters and IV lines can be expanded if issues were identified during preformulation.

A critically important excipient is water, which has a number of U.S.P. monographs. Several of the water monographs that are relevant to injectables are listed in Table 17.8.

The particulate matter test can be performed by two different methods: light obscuration or microscopic determination. There are different compliance requirements for each test and for small- and large-volume parenterals (SVP and LVP). The compliance requirements are summarized in Table 17.9.

17.2.3   Formulation Development and Formulation-Specific Processes

There are a number of different injectable dosage forms that are ready-to-use injectables or require reconstitution. Reconstitution may result in a solution or dispersed system such as a suspension or liposome. Injectable dispersed systems are covered later in the chapter. A more comprehensive discussion of injectable dispersed systems is also available (Burgess, 2005). In the case of a parenteral solution or dispersed system, it is desirable to formulate a product that matches the tonicity and pH of human plasma and interstitial fluids. The desired tonicity range and pH is 270 to 320 mOsmol/L and 7.4, respectively. The tonicity can be checked by evaluating the solution's effect on red blood cells. Hypotonic solutions will cause the red blood cells to swell and rupture irreversibly if the injectable is too hypotonic. Hypertonic solutions will cause the red blood cells to shrink reversibly and become scallop shaped. Typically, hypotonic solutions are adjusted to make them isotonic. High drug or electrolyte doses may make it impossible to formulate an isotonic solution. In this case, slow IV infusion will allow the hypertonic solution to be diluted rapidly. On the other hand, the cerebral spinal fluid is only about 100 to 160 mL, and there is very limited ability for this fluid to adjust the pH or tonicity of an intrathecal injection.

Table 17.1 lists the general classes of excipients that are used in injectable formulations. The simplest SVP formulation is a ready-to-use solution containing drug, water for injection, and a tonicity-adjusting agent. Solubility and stability problems may require the use of functional excipients to improve solubility and stability. Large-volume parenterals are primarily electrolytes such as sodium chloride, potassium chloride, calcium chloride, and sodium lactate; carbohydrate solutions such as dextrose (d-glucose), fructose, and invert sugar (a mixture of glucose and fructose); and hyperalimentation fluids containing fat emulsions and carbohydrates. LVPs typically do not contain added antimicrobial agents, and they rarely contain antioxidants. LVPs range in pH from about 4 (carbohydrate solutions) to 8.3 (hyperalimentation fluids). The buffering capacity is relatively low for LVPs. Lactated Ringer's injection, on the other hand, has a high buffer capacity at a pH of about 6.5.

17.2.3.1   *Small-Volume Parenteral Solutions*   A stable isotonic ready-to-use aqueous solution is the injectable dosage form of choice. Sodium chloride,

TABLE 17.8   U.S.P. Water Monographs

Monograph and Preparation	Tests	Limits
Purified Water[a]: The feed water for purified water is national primary drinking feed water. This water is then treated by filtration, softening, dechlorination, deammoniafication, and deionization by reverse osmosis, distillation, or ultrafiltration. This may be treated with ultraviolet light. *Note:* No microbiology test is required. However, the U.S.P. recommends that the maximum action level be set at 100 cfu[b,d]/mL.	Total organic carbon, water conductivity, bulk water (three-stage process)	0.50 mg carbon/L^3; stage 2: 2.1 µS/cm^3
Water for Injection (WFI)[c]: purified water is feed water that is distilled or processed by reverse osmosis or an equivalent process to remove microorganisms and chemicals. WFI is used for cleaning equipment and as an excipient for the manufacture of injectable products. It is typically stored and distributed between 65 and 85°C to inhibit microbial growth. A final or terminal sterilization process is required before a product can be released to commerce.	Bacterial endotoxins Total organic carbon Water conductivity	<0.25 endotoxin unit per mL See above See above
Note: No microbiology test is required. However, the U.S.P. recommends that the maximum action level be set at 10 cfu/100 mL.		

TABLE 17.8   (Continued)

Monograph and Preparation	Tests	Limits
Sterile water for injection[d]: water for injection is the feed water that is sterilized. It contains no antimicrobial agents or added substances. It is used to reconstitute powders and dilute sterile solutions.	Sterility Particulate matter Oxidizable substances Bacterial endotoxin Water conductivity	No evidence of microbial growth See Table 17.9 No pink color See above See above
Pure Steam (clean steam)[c]: Water heated above 100°C, prepared from primary drinking water. It contains no added substance.	Condensate tests Bacterial endotoxin Total organic carbon Water conductivity	See above See above See above

[a] 1S (U.S.P. 33),

[b] 2S (U.S.P. 32),

[c] U.S.P. 32–N.F. 27,

[d] cfu, colony-forming unit.

TABLE 17.9   U.S.P. Particulate Requirements for Injectables

Test	Particle Size (μm)	Number of Particles per mL
Light obscuration		
SVP	>10	Does not exceed 6000
	>25	Does not exceed 600
LVP	>10	Does not exceed 25
	>25	Does not exceed 3
Microscopy		
SVP	>10	Does not exceed 3000
	>25	Does not exceed 300
LVP	>10	Does not exceed 12
	25	Does not exceed 2

potassium chloride, and glucose are commonly used excipients to adjust the tonicity of injectables. WFI is the vehicle of choice. Water is the most biocompatible solvent and it raises the least concerns about safety and unwanted side effects, such as injection-site irritation and cytotoxicity. Pharmaceutically acceptable salts, buffers, cosolvents, surfactants and micelles, oils, emulsions, liposomes, and complexing agents can provide acceptable solubility and stability.

Cosolvents are added to increase the solubility and stability of injectable solutions of noncomplex molecules. As discussed in Section 17.2.2.1.2, the potential irritation, hemolysis, and toxicity are dose dependent and are specific to each cosolvent system. The daily exposure needs to be taken into account when formulating with nonaqueous solvents. In addition, if flammable cosolvents such as ethanol are used, the flash point American Society of Testing and Materials (ASTM) D93-10a) of the final formulation needs to be tested to meet U.S. Department of Transportation requirements. Using the ASTM D93-10a flash point method, a flash point of 37.8°C (100°F) or lower must be labeled as flammable material for transportation purposes, which has significant implications concerning the product label and transportation restrictions. For example, Torisel Injection (temsirolimus) contains 25 mg/mL temsirolimus, 55 to 60% v/v propylene glycol, 39.45% v/v ethanol, and 1% v/v dl-α-tocopherol, has a flash point of 21.1°C, and is labeled as a flammable liquid.

Injectable solutions are the easiest parenteral systems to manufacture and sterilize terminally. Preformulation studies are designed to determine how stable a drug is to steam sterilization conditions. Other terminal sterilization processes, such as gamma and electron-beam irradiation may be considered if moist heat sterilization leads to unacceptable levels of drug degradation. Sterile filtration is held as the last sterilization method if other processes cannot be used. The effect of cosolvents on the filter integrity is an important consideration in the proper selection of the sterilizing filter.

Hydrolysis and oxidation of noncomplex and biotechnology-derived molecules are commonly encountered stability problems, especially when the molecules are exposed to moist-heat sterilization conditions. Hydrolysis and oxidation reactions are sensitive to pH conditions, and optimum stability pH ranges can be determined by establishing a degradation rate vs. pH profile.

As discussed in Sections 17.2.2.4 and 17.2.2.5, oxidation reactions are complex and involve minute quantities of free radicals, oxygen, and trace metals. A basic understanding of oxidation, autoxidation, and the factors that affect these reactions is critical to developing formulation stabilization strategies. More detailed discussions of oxidation, autoxidation, antioxidants, and sequestrants may be found in Johnson and Gu (1988), Pezzuto and Park (2002), and Asker (2007). Oxidation reactions are complex and multifactorial, so it is difficult to compare stabilization conditions across different matrices and delivery systems. Since there are a number of factors as well as a complex interaction of factors involved in oxidation reactions, design-of-experiment (DOE) approaches are ideally suited for developing stabilizing formulations. The effects of light, oxygen, trace metals, antioxidants, sequestrants, temperature, pH, ionic strength, dielectric constant, and drug concentration can be studied through a screening design that is followed by confirmation and optimization designed experiments.

From a practical point of view, it makes good sense to remove oxygen, decrease the storage temperature, minimize the source of trace metals, minimize the source of peroxides, and protect the product from light. Oxygen can be removed from WFI by boiling. Many WFI loops often store and circulate water at 80°C. Since oxygen is not soluble in water at this temperature, circulating WFI loops are effectively free of oxygen. WFI water can also be cooled under vacuum to maintain an oxygen-free product. Very low levels of oxygen can be obtained by purging water with purified nitrogen. During manufacturing, the headspace of vials and ampoules can be purged with nitrogen to maintain a low oxygen concentration in the final product. As a security measure, an antioxidant such as the bisulfites can be added. Bisulfites scavenge oxygen by a complex free-radical reaction that is catalyzed by metal ions. The formulation design may incorporate an antioxidant system that contains two antioxidants and one or two sequestrants. The antioxidants should complement each other by working through different mechanisms. One antioxidant could be an oxygen scavenger while the second molecule could be preferentially oxidized compared to the drug.

A similar approach can be used for sequestrants. For example, the level of EDTA can be reduced by including glycine, gluconic acid, or cysteine. If the DOE studies show that trace metals are a significant factor affecting drug oxidation, it may be worthwhile to determine the source(s), types, and levels of trace metals. The search for the source of metal ions should not be limited to the formulation excipients. WFI water can be very corrosive to 316 stainless steel, which contains approximately 63% iron, 20% chromium, 15% nickel, and 2% manganese and may be a source of trace metals in the WFI. Curing agents for the rubber closures can contain metal oxides that can be leached from the rubber component into the injectable solution. Knowing the source of the metal ions and what trace metals are most prevalent may help to refine the choice of sequestrant agents. Atomic absorption spectroscopy has been used to determine the content of trace metals in herbal tea extracts (Slaveska et al., 1998).

Poly(ethylene glycol)s and nonionic surfactants can contain reasonably high levels of peroxide, so these materials deserve special consideration before being used with drugs that are prone to oxidation. Kumar and Kalonia (2006), Ray and Puvathingal (1985), and Segal et al. (1979) have reported procedures that can be used to remove peroxides from these excipients. The best ways to protect the product from light are to use an oversized label on the glass container and through secondary packaging. Amber glass decreases Ultraviolet light transmission significantly compared to flint glass, and it can decrease visible light transmission by 40% or more. However, ferric oxide is used to achieve the amber color. The formulator needs to weigh the advantage of decreasing light transmission with the potential of providing a source of transition metal ions that can result from extraction or erosion of the amber glass. While being stored and prior to use, secondary packaging can provide excellent light protection.

Oils have been used for immediate and controlled-release IM injections. There are a few reports in the literature that discuss the formulation and evaluation of oil injectables. Radd et al. (1985) evaluated corn oil, seasame oil, mineral oil, methy oleate, ethyl oleate, propyl oleate, methyl myristate, and isopropyl myristate oils that were mixed with acetic, propanoic, butanoic, pentanoic, hexanoic, octanoic, linoleic, and oleic liquid acids. The addition of the acids increased the haloperidol solubility in the oils 40- to 60-fold. Drug release was varied some 30-fold for different combinations of oil and acid. Oils cannot be sterilized by autoclaving since they do not contain water. Without the presence of superheated water there is no moisture available to kill the microorganisms by denaturing or hydrolyzing their proteins. A specification for the amount of water in oils and other nonaqueous injectables should be provided in the regulatory documentation.

Therapeutic proteins have made significant inroads as drug therapies used to treat human conditions that were not treatable with noncomplex molecules such as growth hormone replacement and multiple sclerosis. These molecules have special physiochemical characteristics that need to be considered carefully during formulation development. In addition to the effect of pH and temperature on deamidation, dimerization, oxidation, proteolysis, disulfide exchange, β-elimination, Maillard reactions, and racemization reactions, therapeutic proteins require additional attention to maintaining their three-dimensional, native, folded, active state. Physical changes such as unfolding or denaturation, aggregation, and precipitation can occur due to shear or shaking, adsorption, heating, freezing, dehydration, and pH changes. Peptides and proteins are surface active and therefore migrate to the air–vehicle and container closure plastic or glass interface with the solution vehicle. Adsorption may lead to protein unfolding, aggregation, and precipitation. Also, as surface-active materials, peptides and proteins tend to self-associate or aggregate and lose activity. The level of aggregation increases with increased peptide or protein concentration. Aggregation can lead to significant drug loss. There is also a growing association between the presence of aggregates and the stimulation of immunogenic responses to the aggregates.

Table 17.10 lists the typical peptide and protein degradation reactions. These reactions are more likely to occur if the reactive amino acid is on the surface of the protein or accessible to the aqueous environment and not buried in the interior or the more lipophilic folded structure.

Preformulation studies should identify the most critical physical and chemical stability issues that are specific to a particular therapeutic protein. A protein-specific stabilization strategy can then be developed. The following discussion presents an array of stabilizing excipients and methods that can be used to increase the shelf life of a therapeutic protein. Human serum albumin is a well-known stabilizer that has a molecular mass of approximately 65,000 Da, 584 amino acids, and 17 disulfide bonds. It is well known as a binding molecule for noncomplex drugs that circulate in the plasma. Albumin has some interesting physiochemical properties that make it well suited as a stabilizing excipient. Even though it is a protein, it is reasonably stable to changes in temperature and pH. Albumin is stable to pasteurization conditions of 60°C for 10 hours. At a pH of 1 to 2, albumin elongates but returns reversibly to its native state as the pH is increased to its isoelectric point of 4.8. Unlike many proteins, albumin is soluble at its isoelectric point. It is also soluble in dilute base. Albumin has been used as a stabilizer for a number of commercial proteins (Wang and Hanson, 1988). At 0.003% w/v, albumin decreased the adsorption of erythropoietin to glass and plastic surfaces. At 0.003% w/v, albumin decreased the dissociation of alcohol dehydrogenase subunits. It has been used at concentrations ranging from 0.1 to 15% w/v as a cryoprotectant, an agent that protects a protein from denaturing during the freezing step of freeze-drying (Wang and Hanson 1988). Albumin may be a carrier or solubilizer for peptides and proteins in the same way that it is a carrier for bilirubin.

TABLE 17.10   Typical Peptide or Protein Degradation Reactions

Type of Chemical Reaction	Affected Amino Acid	pH of Most Reactivity (Generalized)	Generalized Optimal Stable pH
β-Elimination	Cysteine, serine, and threonine	>7	3–6
Deamidation	Asparagine, glutamine	<3 and >7.5	3–5
Dimerization (covalent bond)	Reaction between the N- and C-terminal ends. Other reactions between amino acids, such as disulfide formation and amide formation		
Oxidation	Methionine, cysteine, histidine, tryptophan, tyrosine, phenylalanine, and proline	~5–7 and greater	2–4 (in general, oxidation reactions increase with increasing pH)
Proteolysis	All amide linkages, and aspartic acid–proline linkages, are more susceptible	Hydroylysis (<3 and >8) Diketopiperazine (5–8 and greater)	Hydrolysis 4–6 Diketopiperazine 3–4
Maillard reaction	Lysine (primary and secondary amines react with aldehydes, and more specifically, reducing sugars: the Maillard reaction)	Acid catalyzed	4–6
Disulfide exchange	Cysteine	~5–7 and greater	3–4

Amino acids such as aspartic acid and glutamic acid have been shown to inhibit protein adsorption to silicone-treated glass. Glycine and alanine have been shown to stabilize proteins undergoing heat treatment. Glycine has been used as a stabilizer for anti-Rh, α-interferon, and antithymocyte globulin at concentrations in the range 0.015, 2, and 2.25% w/v, respectively. Phospholipids and surfactants such as polysorbate 80 (Tween 80), polysorbate 20 (Tween 20), and poloxamer (Pluronic) have been shown to be stabilizers (Wang and Hanson, 1988). They may preferentially adsorb to surfaces and compete for interfaces with the proteins. They may also decrease the surface tension at the interface and make proteins less likely to unfold during shaking or other forms of shear.

Polyols such as poly(hydric alcohol) (glycerin, erythritol, xylitol, and sorbitol), nonreducing disaccharides (sucrose, trehalose, raffinose, and maltose), poly(ethylene glycol), and hydroxymethyl cellulose have been used to decrease aggregation by affecting water structure. Polymers such as poly(vinylpyrrolidone), gelatin, poly(ethylene glycol), and hydroxymethyl cellulose may act as protective colloids that prevent the proteins from interacting to self-associate.

Peptides have been shown to be stable for several years in solution at room temperature. Oxytocin has a five-year shelf life when stored at room temperature. Most solution therapeutic proteins require refrigerated storage. For example; humanized growth hormone, insulin, and natalizumab have shelf lives of 18 months, 18 to 24 months, and four years, respectively, when stored at 2 to 8°C. These products require a cold storage distribution chain from manufacturing, to the distribution warehouse or wholesaler, to the pharmacy, to the patient, and in the home in the case of insulin. Many product labels also warn against shaking the product, which may result in loss of potency. For more extensive discussions concerning protein formulation development, see Wang and Hanson (1988), Pearlman and Wang (1996), Wang (2000), and McNally and Hastedt, (2008).

17.2.3.1.1 Parenteral Solution Scale-up Carstensen and Mehta (1982) and Gorsky (2006) have discussed solution scale-up in detail. The manufacturing of parenteral drug solutions involves a number of processing steps. All processing equipment needs to be rendered sterile before manufacturing can be initiated. Liquids have to be pumped from dispensing vessels into manufacturing vessels. Dissolution of the API and the functional excipients in water or another solvent generally requires heating and agitation. Cosolvent systems will require mixing to efficiently achieve a uniform solution. A solution clarification step using a 5- to 10-μm micron filter eliminates most foreign particulates. The filtration rate depends on the filtering pressure, effective filter surface area and porosity, and amount of particulate material.

Heat–cooling cycles take longer with increasing scale. It may be necessary to heat a solvent to improve dissolution. Applying first principles, the degradation rate constant for an API or functional excipient such as methylparaben is a function of temperature, which is also a function of the heating–cooling times. The agitation rate in linear liquid velocity (cm/s) at the impeller tip decreases with scale. The impeller's shape, diameter, and distance from the wall of the mixing vessel affect the mixing efficiency. These factors need to be considered during scale-up. The solution is then filled into an appropriate container closure system. Pumping and filling rates need to be scaled. Excessive turbulence during pumping and filling may lead to foaming and poor fill uniformity.

Compared to production-scale equipment, laboratory- and pilot-scale equipment are often designed with more operating flexibility, such as rpm and temperature ranges. To improve the chances of a smooth process transfer from laboratory to production, it is often desirable to confirm the *scalability* of a potential process by first assessing the key operating principles and limitations of the production equipment. Geometric similarity and dimensionless numbers can then be used to investigate scale-down requirements for laboratory and pilot-plant equipment. This is referred to as *scaling down* or "working backward" while keeping in mind the ultimate need to successfully scale-up the process. Effective scale-down depends on maintaining comparable impeller, vessel geometries, and rotational speeds. Generally, it is important to maintain equivalent fluid dynamics or average fluid velocity as the process is scaled. It is important to keep geometric ratios such as d_i/d_v and h_l/d_t where d_i, d_v, h_l, and d_t are impeller diameter, vessel diameter, height of the liquid in the vessel, and tank diameter, respectively, as similar as possible. For geometrically similar equipment, the Froude number can be used to work backward, that is, to scale-down the process for mixing speed and time. An example of the use of the Froude number to scale-up a V-blending process was detailed in Section 13.3.3.2. The goal is to have an equivalent Froude number for the production- and smaller-scale equipment. The production mixer's midrange rpm can be used to calculate the laboratory- or pilot-scale rpm that is expected to provide equivalent liquid motion by using the *Froude equation:*

$$F = \text{rpm}^2 \left(\frac{d}{g}\right) \tag{17.9}$$

where F is the Froude number, rpm the impeller's rotations per minute, d the diameter of the impeller, and g the acceleration due to gravity.

$$\frac{F_P}{F_L} = \frac{\text{rpm}_P^2 \cdot d_P/g}{\text{rpm}_L^2 \cdot d_L/g} \tag{17.10}$$

where F_P, F_L, rpm_P, d_P, rpm_L, and d_L are the production Froude number, laboratory Froude number, production impeller rotations per minute, diameter of the production impeller, laboratory impeller rotations per minute, and diameter of the laboratory impeller, respectively. For geometrically similar production- and laboratory-scale equipment, one can set the Froude number ratio to 1 and solve for the laboratory-scale rpm:

$$\text{rpm}_L = \left(\text{rpm}_P^2 \frac{d_P}{d_L}\right)^{0.5} \tag{17.11}$$

The laboratory process would then use this mixing rpm. The optimum mixing time at this rpm would be established by appropriate experimentation. Once the appropriate laboratory mixing time is defined, the production mixing time can be determined by calculating the equal number of rotations at the mixer's rpm midrange.

17.2.3.2 *Powder Fills for Reconstitution*   Sterile powder fills provide a dry storage environment for drugs that are not stable in aqueous-based solutions. The reconstituted powders may form solutions or suspensions. Many injectable penicillin and cephalosporin antibiotics are sterile-filled powders for reconstitution. Pentothal (sodium thiopental) is sterile-filled with anhydrous sodium carbonate as a buffer. Sterile powder filling is done by aseptic processing, which means that a practical level of sterility assurance is maintained at all times. Acceptable aseptic processing starts with adequate facility design that addresses proper cleaning surfaces, air classifications, humidity and temperature control, airlocks, water and

process systems, personnel, equipment, and material flow. Personnel have been found to be the biggest cause of sterile manufacturing failures, so aseptic training and retraining is critical to successful aseptic processing.

Sterile powder filling begins with the manufacture of a sterile API. This is typically done by dissolving the drug and filtering the solution through a 0.22-μm sterilizing filter system. The sterile filtered material may be crystallized, lyophilized, or spray-dried. The API is crystallized using sterile seed crystals, an antisolvent, or a combination of techniques. The recrystallized sterile powder may need to undergo an aseptic sizing or milling operation. The drug is then isolated in an aseptic environment. In the case of *powders for solution*, it may be necessary to aseptically blend the drug with other sterile excipients, such as buffers, soluble flow aids, lubricants, and tonicity-adjusting agents. Sterile sodium chloride can serve multiple functions as a flow aid, lubricant, and tonicity-adjusting agent. *Powders for suspension* will require the addition of appropriate functional excipients. Spray-drying or lyophilization can eliminate this blending step. The API and other soluble excipients can be sterile filtered. This solution can then be spray-dried to produce a flowable powder without the need for dry blending. The spray-drying process controls the particle size by maintaining the solution concentration, the atomization pressure, atomization temperature, spray rate, drying air temperature, and nozzle size. The aerosolized plume is dried using hot sterile air. Lyophilization requires an aseptic sizing or milling step before it can be powder filled. The final powder is then filled into vials, which are stoppered, capped, rinsed, and dried for labeling and packaging. Ideally, the final package product should be terminally sterilized. Dry heat, electron beam, gamma irradiation, and x-ray should be considered as means of terminally sterilizing the final product. These sterilization techniques are discussed more extensively Section 17.3.

17.2.3.3 Emulsions

Examples of emulsions that are commercially available are Diazemul, diazepam; Dipivan, propofol; Vitalipid, vitamins A, D, E, and K_1; Oncosol, perfluorocarbon; and Intralip and Liposyn, fat emulsions. Intravenous emulsions are oil-in-water emulsions. Pai et al. (2006) evaluated the percent hemolysis caused by the commercial micellar amphotericin B product compared to a novel amphotericin B emulsion as a function of dose. The emulsion caused significantly less hemolysis than did the commercial product. The emulsion formulation also achieved higher amphotericin B concentrations in body organs than did the conventional formulation. Hansrani et al. (1983) and Mikrut (2005) have discussed the formulation and manufacturing aspects of intravenous fat emulsions. The Swedish, American, and Japanese formulations use egg lecithin, whereas the German and French formulations use soy lecithin as the emulsifying agent. According to Mikrut, soy lecithin carries more toxicity risk than does egg lecithin. Egg lecithin is used at 1.2% w/v, whereas soy lecithin ranges from 0.38 to 1.5% w/v. Glycerol (2.5% v/v), xylitol (5.0 and 10.0% v/v), and sorbitol (5.0 to 10.0% v/v) are used as tonicity-adjusting agents. Glucose and dextrose are not used as tonicity-adjusting agents because they react with the phospholipid primary amine functions, resulting in a browning or Maillard reaction. The pH is adjusted to approximately 9.5, which is reported to increase the stability of the emulsion by creating a net negative charge that causes the lipid particles to repel each other. This electrostatic repulsion minimizes coalescence. The alkaline pH also causes hydrolysis of the fatty acids of the phospholipids. The hydrolyzed fatty acids form sodium soaps, which act as auxiliary emulsifiers.

Lyons (2005) and Mikrut (2005) provided case studies that discuss the formulation, development, and scale-up of intravenous perfluorocarbon (PFC) and fat emulsions, respectively. The key issues that are addressed in these reports are reduction of emulsion droplet size for safe IV injection and development of emulsions that are stable to terminal steam sterilization and resist flocculation when added to plasma. In both cases, high-pressure homogenization was used to attain the desirable oil droplet size, and nitrogen was used throughout the process to protect the phospholipid emulsifying agents from oxidation. After homogenization at 10,000 psig ($\sim$69,000 kPa) and 35 to 40°C, the emulsion was passed through a 10-μm stainless steel mesh filter before filling. Two types of batch sterilizers are used for emulsions: an air overpressure shaking cycle and a rotating cycle. The shaking cycle autoclave provides an oscillatory agitation frequency of 70 cycles per minute along the center axis of the container lying on its side. The rotary cycle autoclave rotates horizontally fixed containers at approximately 10 rpm. The PFC emulsion was terminally sterilized in a "rotating" steam autoclave. The final particle size was 200 to 300 nm. During scale-up, terminally sterilized product had free or poorly emulsified PFC that sedimented due to its density being greater than that of water. It was determined that a portion of PFC would vaporize during steam sterilization, and this vapor would then condense during the cooling cycle. To address this issue, the headspace was minimized.

The fat emulsion was prepared in a fashion similar to that for the PFC emulsion. The heated oil phase was added to a heated aqueous phase under high-speed mixing. In this case, all components were prefiltered through a 0.45-μm filter. The fat emulsion was again filtered through an 0.8-μm filter after high-pressure homogenization. The pH of this emulsion was adjusted to 9.5 and the emulsion was passed through a 5-μm stainless steel filter under vacuum-nitrogen gassing. The final emulsion was steam-sterilized and had a final particle size of about 0.45 μm.

17.2.3.4 Liposomes
Liposome formulation and manufacture are discussed in Chapter 15. More than a dozen parenteral liposome products are commercially available, and the number of clinical studies continues to grow. For example, AmBisome (amphotericin B) and Doxil (pegylated liposomal doxorubicin) are two well-known commercial parenteral liposomal products.

Additional parenteral requirements include the need for the liposome to be sterile-filtered, and the product, including the excipients, needs to be safe and well tolerated for IV use. Adler and Proffitt (1993) describes the development and characterization of unilamelar AmBisome. Hydrogenated soy phosphatidylcholine (HSPC), distearoylphosphatidylglycerol (DSPG), and cholesterol were carefully chosen liposomal excipients that resulted in the liposomes binding specifically to the fungal cell surface. The liposome formulation also reduced acute and chronic toxicity by minimizing the amount of free drug in equilibrium with the drug associated with the liposome. A stable ion pair was formed between DSPG and amphotericin B (Proffitt et al., 2004). The ion pair was soluble in small amounts of organic solvent and added to the high level of liposomal amphotericin B associated with the liposome. It is thought that the ion pair is intercalated into the lipid membrane layer of the liposome, giving it good stability even in plasma. The dissociation of amphotericin B from the liposome was studied in 50% human plasma at $37°C$ for 72 hours. After incubation for 72 hours, 95% amphotericin B was still associated with the liposome particle. The saturated phospholipid is not only less susceptible to oxidation but its presence in the formed liposome results in a gel-to-liquid transition temperature when above $37°C$. This also improves the liposome's physical stability. The manufacturing steps involved preparing the ion-pair complex, isolating the complex, and preparing the liposome with the ion complex. The liposome formulation contained HSPC, DSPG, cholesterol, sucrose, and disodium succinate. The liposome formed was sized using high-pressure homogenization to produce particles less than 100 nm. The liposome product was sterile filtered and freeze-dried. The reconstituted liposome composition was 0.40, 0.42, 0.672, 1.7, and 7.2% amphotericin B, cholesterol, HSPC, DSPG, and sucrose, respectively. Low-ionic-strength disodium succinate was used as a buffer and provided a final dispersion pH of 5 to 6. Sucrose was used to stabilize the liposome dispersion by inhibiting aggregation and promoting drug–liposome association. It also served as the bulking agent for the freeze-drying process. The lyophilized powder is reconstituted with sterile water for injection and filtered through a 5-μm filter that is provided with the product before being added to 5% dextrose. The shelf life of the freeze-dried liposome formulation is three years.

Doxorubicin is an encapsulated liposome for IV administration. In this case the drug is entrapped in the liquid core of the liposome compared with the intercalation of amphotericin B in the lipid membrane layer. The liposome was fabricated in 250 mM ammonium sulfate solution. The external ammonium sulfate was replaced with sucrose solution by cross-flow filtration. Doxorubicin HCl was loaded inside the inner aqueous core by allowing the ampiphillic doxorubicin to permeate the liposome bilayer. Entrapment was accomplished by the formation of a gel-like doxorubicin sulfate precipitate. The final liposome is 35 to 65 nm and is sterile-filtered. The shelf life of the liquid dispersion is 18 months. To achieve longer systemic circulating times, methoxy poly(ethylene glycol) (MPEG) was surface grafted to the liposome (Mayer et al., 1998; Martin, 2005). The composition of the liposomal dispersion system is 0.2% doxorubicin, 0.3% MPEG-distearoylphosphotidylethanolamine, 0.958% HSPC, and sucrose to maintain isotonicity. The pH was adjusted to 6.5 with 0.2% histidine and hydrochloric acid.

17.2.3.5 Coarse and Colloidal Solid Dispersions

There are a number of coarse parenteral suspensions on the market: for example, Celestone, betamethasone sodium phosphate; Soluspan, betamethasone acetate; Depo-Medrol, methylprednisone acetate; Depo-Provera, medroxyprogesterone acetate; Bicillin C-R, penicillin G benzathine/penicillin G procaine; and NPH insulins. Most of the injectable suspensions are ready-to-use formulations. Parenteral suspensions are typically chosen for drugs that have very low aqueous solubility, that require controlled release, for which local drug delivery is desirable, or for a combination of these. Product design for an injectable suspension carries a high level of difficulty. The parenteral administration route places additional restrictions on suspension formulation. The number of acceptable excipients is much more limited, which complicates formulation development. The suspension concentration is limited to provide acceptable syringeability and injectability. The viscosity of an injectable suspension is also constrained to meet syringeability and injectability requirements. In addition, particulate counts and size must meet the requirements of the U.S.P. injectable particulate monograph. Moreover, terminal steam sterilization is complicated by the potential for Ostwald ripening because the drug's solubility will usually increase with an increase in temperature. Furthermore, if sterile filtration is the only acceptable sterilization process, the low-micrometer particle size cutoff results in additional process complexity. The solids content of parenteral suspension is generally around 0.5 to 5%, reaching as high as 30% for higher-dose drugs. The API particle size is required to be smaller, with a narrower size range, to minimize needle plugging and pain on injection. The particle size and distribution depend on the method of preparation. A typical size range is 1 to 5 μm, with 10 μm approaching the maximum size. Parenteral suspensions should be easily injectable through a 25-gauge needle. No particle should be greater than one-third the size of the needle's internal diameter. For example, the internal diameter of a 24-gauge needle is 300 μm, so the largest suspension particles should not be larger than 100 μm, which is significantly larger than the desired upper size of 10 μm. A flocculated suspension is desirable because the flocs are low-viscosity easily broken assemblies, which results in good syringeability. The types of functional formulation excipients used for injectables are the same as those used for nonparenteral suspensions. These typically include a dispersing or suspending agent, a wetting agent, a buffer, a tonicity-adjusting agent, and an antimicrobial agent for multidose containers. However, the list of acceptable parenteral excipients is much more restricted.

There are several compendial (ampicillin, dimercaprol, epinephrine, penicillin G procaine) and commercial (aurothioglucose) parenteral oil suspensions. Aluminum monostearate has been used as a suspending agent for oils. Also, fixed oils and medium-chain triglycerides have been used as oil vehicles. Sims and Worthington (1985) reported the process for preparing a parenteral oil suspension. Aluminum monostearate was dispersed in the oil and heated to 130°C with continuous stirring. Above 100°C the dispersion went through a phase transition and became a semitransparent gel. Several sterililization methods were evaluated, including dry heat at 170°C for 2 hours and gamma irradiation at 25 kGy from [60]Co.

In general, two different processes are used to prepare injectable suspensions. In one method, sterile powder and sterile vehicle are combined aseptically to form the dispersion. In a second method, separate sterile solutions are prepared and mixed to form the dispersed phase in situ. Akers et al. (1987) provided examples for both methods. Penicillin G procaine suspension is prepared by the addition of sterile penicillin powder to a sterile suspending vehicle. The suspension vehicle contained sodium citrate, povidone, and parabens. The excipients were dissolved in heated water for injection. Lecithin was then dispersed slowly with agitation, and benzyl alcohol was added last to the suspending vehicle. This suspending vehicle was transferred to a tank outfitted with a variable-speed mixer and steam jacket. The vehicle was sterilized in place. The sterile penicillin powder was obtained from a sterile crystallization process. The crystalline powder was then isolated and micronized aseptically. The sterile micronized penicillin powder was added to the cooled sterile suspending vehicle with constant mixing to form an injectable suspension. Once all the powder was incorporated, high-speed mixing homogeneously dispersed the suspension. The suspension was then coarse-filtered through a stainless steel mesh screen to a dispensing tank that was mixing the suspension continuously to maintain homogeneity. The suspension was filled from this dispensing tank into an appropriate container closure system, which was stoppered and crimped with an aluminum seal. The final suspension contained penicillin G procaine 300,000 units/mL, 4% sodium citrate (buffer), 1% lecithin (wetting agent and flocculating agent), 1% benzyl alcohol (preservative), 0.15% methylparaben (preservative), 0.1% povidone (viscosity agent, protective polymer, and suspending agent), and 0.02% propylparaben (preservative).

The second method that was reported involved the preparation of isophane insulin by the in situ precipitation technique. Isophane insulin is a neutral suspension of equal molar concentrations of insulin and protamine. The in situ process involved the preparation of two sterile solutions, solution A and solution B. Sterile solution A contained zinc insulin, glycerin (tonicity-adjusting agent), phenol (preservative), and cresol (preservative) at pH 3 to 4. Solution B contained protamine sulfate (precipitating agent), phosphate buffer, glycerin, phenol, and cresol. Both solutions are sterile filtered prior to mixing. The two solutions are then added together. The crystallization process and particle size are controlled by the concentration of excipients, pH, temperature, and stirring rate.

Producing a sterile micronized crystalline powder for injectable suspensions is not a trivial task. It is also often difficult to produce the right particle size and polymorphs by the in situ method of preparing a parenteral suspension. O'Neill (1976) disclosed a simple and clever process to produce an autoclaved sterilized suspension of the API that maintains its particle size and polymorph character. The underlying principle of the method uses a saturated sodium chloride solution whose solubility does not change significantly over a wide temperature range. This maintains the water activity nearly constant throughout the steam sterilization process. The patentee presented dexamethasone acetate to illustrate the value of this methodology when other sterilization techniques were problematic. When dexamethasone acetate was dry heat–sterilized at 100 to 120°C, it discolored. Aseptic in situ recrystallization resulted in needle-shaped crystals which could not be formulated into an acceptable suspension. Lyophilization of a sterile solution and aseptic glass bead milling resulted in a polymorphic crystalline change. Dexamethasone acetate was also micronized by jet milling to give an initial particle size distribution of 90% less than 10 μm. The dexamethasone acetate particles grew to 300 to 400 μm when this powder was added to water or an unsaturated sodium chloride solution and sterilized by autoclaving.

The use of wet milling techniques to stabilize nanosized solids of poorly soluble drugs (less than 0.1 mg/mL) has proven to be a major breakthrough. The technology has made a number of advances over the last 20 years (Liversidge et al., 1992; De Castro, 1996; Ryde and Ruddy, 2002). This technology overcomes the solubility issue by creating a high drug-loading system of colloid particles in an acceptable dose volume using generally recognized as safe materials. As discussed in Chapter 15, colloid dispersions are commonly considered to range from 1 to 1000 nm. The U.S. government (2011) has defined *nanotechnology* as the "understanding and control of matter at dimensions between approximately 1 and 100 nm." Wet milling involves reducing the particle of insoluble API to 100 to 500 nm using high-energy media mills to produce a stable colloid dispersion of API, water, and a noncovalent adsorption stabilizer. The colloid product can be further processed for oral, inhalation, topical, and parenteral uses. For parenteral use the colloid material can be sterilized by aseptic filtration, steam autoclaving, and gamma irradiation. The ability to sterilize a nanodispersion aseptically depends on the particle size of the dispersion, its viscosity, and the concentration of the drug and excipients. Steam autoclaving generally results in particle size increases. The magnitude of particle size increase depends on many factors, including the impact of temperature on the solubility of the API, the concentration of the stabilizer, the strength of the noncovalent adsorptive forces between the API and the stabilizer, and the water activity. Most wet-milled dispersions designed for parenteral use are lyophilized and terminally sterilized by gamma radiation at 25 kGy. The wet milling processes are proprietary, so there is little information regarding there scalability. According to Lee (2005), the proprietary NanoMill process is directly scalable from 100 mg to 500 kg of API.

17.2.3.6 Controlled Release Injectable controlled-release products offer a variety of advantages over oral controlled-release dosage forms. Prolonged release decreases significantly the concern about product failure resulting from poor compliance. The use of controlled-release antibiotics and contraceptives fall into this category. Many proteins and peptides need to be injected to circumvent contact with the harsh environment of the gastrointestinal system. Long-acting injectable formulations can significantly decrease the number of injections to three or four times a year. The longest-acting oral formulation is limited to the gastrointestinal transit time. In many cases, controlled-release injectables improve drug utilization, minimize dose, and decrease side effects. The largest and most serious disadvantage of long-acting injectables is dose dumping or having a potentially large dose of drug in the body if a drug-associated adverse reaction occurs.

A number of the products that have been discussed previously, such as oils, coarse dispersions, and colloidal dispersions, can be designed to provide controlled release. Some of the oldest products that are still on the market are aqueous suspensions of antibiotics (Bicillin C-R and L-A), steroids, and insulin. Depo-Medrol is methylprednisone acetate (MPA) single-use injectable suspension, U.S.P. It can be administered by intramuscular, intraarticular, or intralesional injection. Each milliliter of suspension contains: 40 mg of MPA, 29 mg of PEG 3350, and 0.195 mg of myristyl-γ-picolinium chloride. Depending on the use, the injection interval can range from daily for systemic use to 5 weeks for intraarticular delivery. Depo-Provera is medroxyprogesterone acetate multiple-use injectable suspension, U.S.P. Each milliliter contains 150 mg of drug, 28.9 mg of PEG 3350, 2.41 mg of polysorbate 80, 1.37 mg of methylparaben, and 0.150 mg of propylparaben. This contraceptive injection is administered every three months.

Controlled release of the products noted above was dependent on the drug being insoluble and injected as a suspension which dissolved slowly over time. The breakthrough development of injectable microcapsules created the ability to design prolonged release of highly soluble parenteral drugs. There are several other advantages of microcapsules. As much as 80% drug loading can be achieved with microcapsules. On the other hand, controlled drug delivery can be achieved with very low drug loads of 2% or less. Some commercial products have as long as four-month dosing intervals. One of the main polymers that has been used in commercial microcapsule products is poly(lactic-*co*-glycolic acid). This copolymer has been used for biocompatible biodegradable sutures for years. The injection volume is the primary limiting factor and main disadvantage of these products. To offset the volume-dose limitation, the drug needs to be potent enough to maintain its effect over a prolonged period.

The first approved injectable microsphere product that is still on the market was Lupron Depot. Lupron (leuprolide) is a nonapeptide analog of gonadrotropin-releasing hormone. The product is a lyophilized biocompatible biodegradable lactic acid and glycolic acid copolymer microsphere that comes in a dual-chamber pre-filled syringe. One chamber contains 3.75 mg of leuprolide acetate, 0.65 mg of gelatin, 33.1 mg of copolymer, and 6.6 mg of mannitol. The second chamber contains the diluents: 5 mg of sodium carboxymethylcellulose, 50 mg of mannitol, 1 mg of polysorbate 80, glacial acetic acid, and water for injection. There are several formulations that allow IM injections every three to fours months.

The formulation design and manufacturing process for preparing water-soluble peptide controlled-release parenteral beads has been described by Okada et al. (1987). The controlled-release microsphere is produced by preparing a water-in-oil (w/o) emulsion in which the aqueous phase contains the soluble peptide and a drug-retaining agent, and a hydrophobic organic solvent phase that contains poly(lactic-*co*-glycolic acid). A w/o/w emulsion is formed by emulsification of another water phase. The solvent in the hydrophobic phase is removed and the microcapsule wall is formed. Gelatin is used as the drug-retaining agent. It can be gelled by lowering the temperature of the emulsion or adding a cross-linking agent such as glutaraldehyde or acetaldehyde. The microspheres are isolated aseptically by centrifugation or ultracentrifugation and then lyophilized in the dual syringe.

The next major breakthrough was the development of protein microcapsules. The obvious development challenge was to maintain the natural three-dimensional protein structure and activity while fabricating the microcapsules using methods similar to those discussed above, which are not necessarily conducive to protein physiochemical stability. Nutropin Depot (somagtropin, recombinant human DNA origin, for injectable suspension) is a reconstitutable controlled-release microcapsule dosage form of recombinant human growth hormone (rhGH). The 191 amino acid hormone is embedded in a biocompatible, biodegradable poly(lactide-*co*-glycolide) microsphere. In this case, the rhGH was stabilized by a zinc–GH insoluble complex. The nonaqueous encapsulation process embedded the zinc–rhGH complex in the polymer by atomizing the polymer–drug suspension through a sonicating nozzle onto a bed of frozen ethanol which was overlaid with liquid nitrogen. This cryogenic process eliminated the need for surfactants or emulsification that was required to fabricate the leuprolide microcapsules. In addition to the zinc–rhGH complex, zinc carbonate was incorporated separately into the microcapsule matrix, which was found to affect the stability of the growth hormone and its release profile (Johnson et al., 1997a; Johnson et al., 1997b). A 5- and a 10-mg product are currently available. The 10-mg product is a lyophilized powder containing 10 mg of GH, 90 mg of mannitol, 3.4 mg of sodium phosphates, and 3.4 mg of glycine that is intended for subcutaneous injection after reconstitution with bacteriostatic water for injection. The 10-mg vial is diluted with 1 to 10 mL of bacteriostatic water for injection. The reconstituted solution should be clear and stored at 2 to 8°C and should not be shaken or frozen. The dose can be given at weekly intervals.

In 2003, Plenaxi (abarelix), a decapeptide–carboxymethylcellulose complex controlled-release injectable aqueous suspension was approved by the FDA. Gefter et al. (2001) discuss the formulation and processing of this controlled-release product, which is designed for once-a-month administration.

Several reports have discussed the quality and performance criteria for modified-release parenterals. A summary report of a workshop organized by the European Federation of Pharmaceutical Scientists in association with the American Association of Pharmaceutical Scientists, the European Agency for the Evaluation of Medicinal Products, the *European Pharmacopoeia*, the U.S. Food and Drug Administration, and the *United States Pharmacopeia* discussed a number of relevant topics. Discussion topics included in vitro/in vivo correlation for controlled-release parenterals, accelerated in vitro release testing, stability, particle size assessment, foreign particulate requirements, qualification of new biopolymers, residual solvents, reconstitution time, syringeability and injectability, resuspendability, bioavailability, and dosage-form-specific topics (Burgess et al., 2004). A summary of the American Association of Pharmaceutical Scientists and the Controlled Release Society discussions concerning the critical variables associated with the in vitro and in vivo performance of parenteral controlled-release products has also been published (Martinez et al., 2010).

17.2.3.7 *Lyophilized Powder for Reconstitution*   The terms *lyophilization* (LYO) and *freeze-drying* (FD) will be used interchangeably throughout this chapter. As more unstable chemotherapeutic and biotechnology injectable drugs are discovered, there has been an increasing need to stabilize these products by FD. There are now more than 100 freeze-dried medicinal products on the U.S. market. The therapeutic benefit to society has supported the significant manufacturing costs that are required to manufacture a lyophilized sterile injectable product. Moreover, many drugs would not be on the market if it were not for LYO technology. FD is the technology of choice for compounds that are extremely unstable to moisture and oxygen. It also allows the drying of heat-sensitive drugs. The dried solid matrix that is left in the vial after FD is referred to as *freeze-dried cake, lyophilization cake, cake*, or *plug*. Drying takes place at low product temperature and low pressure. The low pressure creates an environment for sublimation, which also decreases the number of gas molecules present. The overall decrease in the amount of air in the system leads to a decrease in the amount of oxygen present. Purging the lyophilized

cake with nitrogen can be performed before the vials are stoppered to decrease the oxygen to even lower levels. The lyophilized powder cake is designed to have high porosity and surface area, which leads to rapid reconstitution by adding sterile water for injection. The vial contents of small molecules are shaken to decrease the reconstitution time. Proteins typically require gentler reconstitution methods to minimize denaturation and aggregation. The reconstituted solid cake can form a solution, suspension, or other dispersed systems, such as liposomes.

The interrelationship and interdependence between formulation design, process design, and container closure system design is paramount to successful lyophilization. It is almost impossible to discuss one without discussing the other, so it might be helpful to provide a brief overview of lyophilization before we discuss formulation design. Nail and Gatlin (1993) provide an in-depth discussion of FD, a three-step drying process that involves freezing the liquid containing the drug and the cake-forming excipients; primary drying at low temperatures (below the freezing point of water) and low pressure (vacuum), and secondary drying at higher temperatures (typically at room temperature or higher) and low pressure. Akers (2006) has provided a typical freeze-drying cycle when there is no formulation or product history. The FD cycle is illustrated in Figure 17.1. FD takes advantage of ice sublimation, which involves the transformation of ice directly to water vapor without going through the liquid (water) phase. Since ice sublimation occurs below the triple point of water ($0.0098°C$ and 4.580 torr or 4580 µmHg or 610.6 Pa), LYO affords the benefit of low-temperature drying for highly moisture- and heat-sensitive drugs and helps assure product quality throughout its shelf life.

Prewashed presterilized tube glass vials are filled aseptically with the solution or dispersed liquid. Special prewashed siliconized presterilized slotted stoppers are then partially inserted into the vial. The presence and shape of the slot are important, to allow ice sublimation to occur. The shape of the slot also helps to align the

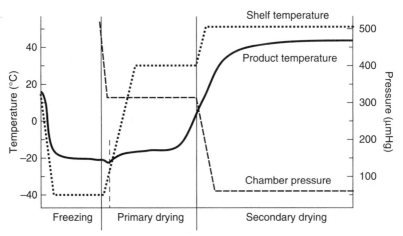

FIGURE 17.1 Freeze-drying cycle. (Adapted with permission from Pharmaceutical Dosage Forms: Parenteral Medications, Vol. 2.)

stopper in the vial. The vials are placed aseptically on trays which are aseptically transferred to the shelves in the freeze-dryer chamber. The liquid is solidified by lowering the material to below -30 to $-50°C$. Once the product is frozen, the freeze-drying chamber is evacuated to pressures that are lower than the ice vapor pressure at the product temperature. Typical primary drying chamber pressures range from 75 to 300 μmHg (0.0750 to 0.300 torr or 10 to 39.9 mPa). Primary drying is initiated at this point by raising the temperature of the freeze-drier shelves. The primary product drying temperature is a critical process parameter. Primary drying is successful when all the frozen water is sublimated and only sorbed water remains in the solid matrix.

At the completion of primary drying, secondary drying is initiated to decrease the amount of sorbed water that might still be present in the solid matrix. Secondary drying is accomplished by raising the product temperature and in some cases decreasing the chamber pressure. Decreasing the chamber pressure decreases the conductive heat, so again the interplay between chamber pressure and temperature needs to be optimized. Secondary product drying temperatures and chamber pressures usually range from 30 to 60°C and 25 to 200 μmHg (0.025 to 0.200 torr or 3.33 to 26.6 Pa), respectively. At the end of secondary drying, the stoppers are inserted completely into the vial under partial vacuum. For extremely moisture- or oxygen-sensitive products, dry nitrogen may be introduced into the vials before final stopper insertion. An aluminum band is then crimped onto the seated stopper and glass vial. The seal must prevent moisture, air, and microbes from penetrating the product. Product stability as a function of residual moisture must be determined on a case-by-case basis. Hopefully, this brief description of the FD process helps to illustrate the extensive interplay between the liquid formulation that contains the cake-forming excipients, the FD process, and the container closure system.

A typical FD cake-forming formulation (product) will contain drug, bulking or cake-forming agents, buffers, tonicity-adjusting agents, and sterile water for injection. The product may also be a dispersed system such as a suspension or liposome. If the product is going to be used as a multiuse vial, a preservative would also be required. Appendixes 17.4 and 17.7 provide a list of preservatives, bulking, and tonicity-adjusting agents. Selection of the particular LYO excipients is based on the results of the drug–excipient compatibility study, desired moisture content of the cake, product stability predicted, and cake properties desired. The cake should be easy to reconstitute and maintain its quality, shape, and integrity throughout shipping, handling, and shelf life. A cake that is crystalline in nature tends to be porous, dries quickly during primary and secondary drying, reconstitutes quickly, and has good physical structure and durability. Mannitol, glycine, and sodium chloride form crystalline cakes that are acceptable for small molecules. The degree of cake crystallinity not only affects its physical attributes, but can also affect the stability of the drug substance. Complex molecules can undergo denaturation during the freezing and drying steps. Functional excipients that protect the protein from denaturation during freezing, drying, or both can have a negative effect on cake quality. Most of the excipients that stabilize complex molecules are mono-, di-, and polysaccharides, albumin (a protein with interesting stabilizing properties); and polymers such as poly(ethylene glycol) and poly(vinylpyrrolidone) form amorphous cakes. For the remainder of this discussion, *amorphous* and *glass* are be used interchangeably. Amorphous cakes have a tendency to shrink over time, are less porous, and reconstitute more slowly. It is speculated that the formation of an amorphous cake allows the stabilizer to remain in more intimate contact with the protein as compared to excipients that crystallize during cake formation. Crystallization of the excipient during freezing will separate the excipient from the complex molecule and diminish its stabilizing influence. So a trade-off is generally made between cake quality and protein stability.

The solid matrix structure or cake, which is formed during freezing and subsequent drying, is the key to providing a robust and efficient lyophilization process. The physical properties of the cake also affect reconstitution time, cake friability, and product stability. It is imperative to thoroughly characterize the thermal and physical behavior of the frozen formulation and the FD cake. Freezing causes a microseparation of the cake-forming formulation into a two-phase mixture of ice and a solute-rich phase. Some excipients are more prone to crystallize during this separation, whereas other excipients tend to separate as an amorphous or glassy material. If all the formulation components crystallize during freezing and microseparation, a *eutectic mixture* is formed which has a *eutectic composition* and *eutectic freezing and melting point*, also called the *eutectic temperature*, T_e. The eutectic temperature is a property of the material and is independent of the solution concentration. The eutectic composition is an intimate mixture of a specific ratio or percentage of two or more crystalline materials that melt as if the mixture were a single pure compound. It has been shown that the frozen crystalline mixture will collapse as a result of melting at its eutectic temperature. In FD, the eutectic melting point is also referred to as the *maximum allowable product temperature, eutectic collapse temperature*, T_{ec}, or simply the *collapse temperature*. The subscript "ec" is used here to distinguish the collapse of the eutectic crystalline mixture. Crystalline cakes will melt, shrink, or collapse at or above the eutectic temperature or collapse temperature: thus the use of the descriptive term *collapse temperature*. The final FD product that undergoes partial shrinkage or collapse is unsightly, reconstitutes poorly, and can affect product stability. Appendix 17.8 provides a list of excipients commonly used in FD, together with their eutectic temperatures.

In the majority of cases, not all the formulation components will crystallize during freezing and microseparation. These systems form ice and a concentrated amorphous dispersion of solute and unfrozen water. This concentrated amorphous dispersion, or glass, is highly viscous and has low mobility. It has been shown that the *glass collapse temperature*, T_{gc}, is principally dependent on the viscosity of the glass. The glass collapse temperature phase transition involves the conversion of a low-mobility high-viscosity glass to a viscous liquid dispersion.

The freezing process for a product that microseparates by crystallization is shown in Figure 17.2. Generally, the cake-forming formulation is a solution that is filtered aseptically before it is filled into vials. This solution, by design, has a low particulate level, and when it is cooled it is common for the solution to supercool before it begins to crystallize. This supercooling occurs from point *a* to point *b*. Once nucleation begins (point *b*) and ice crystals form, the temperature rises due to the exothermic crystallization process. This temperature rise slows

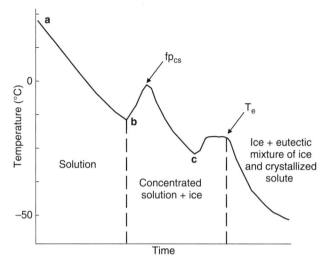

FIGURE 17.2 Freezing of a crystallizing system cycle. (Adapted with permission from Pharmaceutical Dosage Forms: Parenteral Medications, Vol. 2.)

the crystallization process and the solute concentrates in the supercooled water, which does not form ice crystals. The temperature will rise until it reaches the freezing point of concentrated solution, fp_{cs}. As the product continues to freeze from fp_{cs} to point c, nucleation and ice crystal formation continue along with the concentration of solute in the unfrozen solution. At point c, crystallization of the eutectic mixture of ice and other crystallizing solute components begins to occur. As eutectic crystallization occurs, the temperature rises to the eutectic temperature, T_e. Eutectic crystallization is complete when all the material has crystallized at the eutectic composition. Under continued cooling the eutectic mixture temperature will approach the temperature of the cooling source. During primary drying, the product temperature should not exceed the eutectic temperature. Typically, primary drying is performed 2 to 3°C below the T_e.

The freezing process for materials that form amorphous or glassy matrices is provided in Figure 17.3. Supercooling from a to b is seen with the amorphous system, as was the case with the crystallizing system. At b the supercooled water begins to nucleate and forms ice crystals. The solutes begin to concentrate in the unfrozen solution. The water crystallization process raises the product temperature to the freezing point of the concentrated solution, fp_{cs}. As cooling continues, the solutes *freeze concentrate* and the concentrate becomes more viscous until the glass transition temperature, T_g', is reached at the maximum freeze concentration, C_g'. At this temperature and solute concentration, the viscosity of the system prevents further ice formation or solute concentration for the time scale of the lyophilization process. The glass dispersion contains the amorphous solutes and unfrozen water. T_g' and C_g' for glassy cakes are somewhat analogous to T_e and the eutectic composition for crystalline cakes. In either case, if the product drying temperature reaches or exceeds T_g' or T_e, the solid matrix will start to collapse.

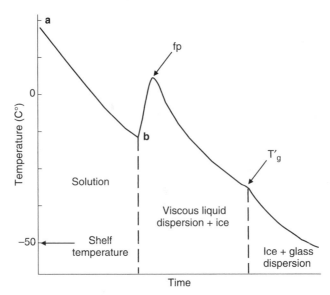

FIGURE 17.3 Freezing of a noncrystallizing system. (Adapted with permission from Pharmaceutical Dosage Forms: Parenteral Medications, Vol. 2.)

It should be noted, however, that there are some differences. In the case of the crystalline matrix, T_{ec} and T_e are the same temperature. The general distinction between the two is based on how they are measured. Typically, T_{ec} is determined by visually observing matrix collapse using microscopy. T_e is often measured by differential scanning calorimetry or thermoelectric analysis. Unlike the eutectic temperature, the measured glass transition, T_g', and the glass collapse temperature, T_{gc}, are composition dependent and vary to some extent by the method of measurement. The glass transition temperature is often measured in a closed system such as a differential scanning calorimeter, where the thermal transition is measured on a sample of known composition that has been hermetically sealed. The glass collapse temperature is often measured by microscopy, which is an open system. In an open system the composition of the cake changes over time as ice water is removed by sublimation. It has also been speculated that the decrease in viscosity at the glass transition temperature is not always sufficient to affect the structural collapse that can be observed with a microscope. Blue and Yoder (2009) observed a 17.7°C difference between T_g' and the glass collapse temperature. In this study, as the protein concentration was increased, there was an increase observed in T_{gc} but T_g' remained constant or decreased slightly. To account for product temperature variation, the upper limit of the primary drying temperature is set at 2 to 3°C below the T_g'. Secondary drying starts when all the ice has been sublimated. Similar care needs to be taken so that the product temperate during secondary drying does not exceed the *glass transition, T_g, of the cake*. T_g is the glass transition temperature for an amorphous solid in the *absence of ice*. As the cake is dried, the viscosity increases and T_g also increases. Franks (1990) has shown that T_g is very sensitive to the level of water content. A 6% increase in water content depressed T_g

some $40°C$. Appendix 17.9 lists T_{gc}, T'_g, and T_g for a selected number of LYO excipients.

So far we have discussed crystalline and glassy systems. A *crystalline matrix* is often developed for proteins that contain crystallizing agents and amorphous stabilizers. The crystallizing agents can provide mechanical strength and a reasonably porous matrix that aids in LYO and reconstitution. The crystalline matrix has a higher collapse temperature, T_{cm}, and the crystalline structure can serve as a support for the amorphous material to collapse on. Crystalline matrix collapse can result in a microcollapse or partial collapse of the product that has acceptable functional and aesthetic attributes. The cake-forming formulation for a crystalline matrix usually incorporates the crystalline component at two to four times the amount of amorphous freezing material. The solids content ranges from 5 to 30%, with a target of 7 to 10%. The amount of amorphous material should be held to a minimum. The goal is to develop a formulation that has a high T_c values, minimal amorphous-forming excipients, and dries to form a porous, nonfriable, nonhygroscopically, stable cake that reconstitutes readily. Pikal (1985) showed that a $1°C$ increase in the primary drying temperature resulted in at least a 13% reduction in drying time. For protein formulation development, the crystalline matrix offers the advantage of being able to include protein stabilizers that form glassy cakes and rely on the crystalline components to provide a structural matrix that enables efficient and reproducible FD. The crystalline matrix cake should also provide good physiochemical stability and rapid reconstitution.

17.2.3.7.1 Effects of Freeze Concentration As water crystallizes to form ice, the solutes will concentrate in the unfrozen, slushy water mixture (freeze concentration). As the solute concentrates, it may exceed its solubility and precipitate as crystalline or amorphous material. Normal saline at its eutectic temperature is 24-fold more concentrated than it is at room temperature. This highly concentrated solution could have a significant detrimental effect on protein solubility and stability. The differential solubility of buffer components can dramatically affect the pH of the freeze concentrate. Disodium hydrogen phosphate is less soluble than sodium dihydrogen phosphate at low temperatures. Anchordoqu and Carpenter (1996) showed that the differential solubility of the phosphate buffer resulted in a drop in pH from 7.5 to 4.5 and resulted in protein denaturation. To minimize buffer effects, buffer concentrations in the millimolar range are recommended, and phosphate, succinate, and tartrate buffers should be avoided. The freeze concentration can increase the bimolecular degradation reactions in partially frozen solutions (Pikal, 2004). Oxygen solubility in water increases more than a 1000 times in partially frozen water (Wisniewski, 1998), which could increase bimolecular oxidative reactions. It is well know that concentrated solutions of proteins are generally less stable than are more dilute solutions. Freeze concentration of proteins may lead to loss of activity. Proteins are surface active and tend to accumulate and denature at such interfaces as air–liquid and liquid–glass. During FD, ice is formed, which provides another interface upon which a protein can interact and denature. As can be seen in this brief discussion, freezing during LYO is associated with many complex and interrelated physiochemical processes and requires thoughtful consideration.

17.2.3.7.2   Freezing and Metastable Solid Phases   Freezing conditions may give rise to metastable glass and crystalline phases for a drug and its excipients. Both mannitol and glycine are known to be polymorphic. Depending on the freezing rate, glasses can exhibit different transition temperatures that relate to the energetic nature of the glass. Similarly, drugs may crystallize into metastable crystalline or amorphous form during the freezing process, which can affect the long-term stability of the drug. It is desirable to have all the components in the matrix in their most stable thermodynamic state. In the next section we discuss the use of thermal treatment to convert metastable solid phases into their more stable form.

17.2.3.7.3   Thermal Treatment: Annealing Before Primary Drying   A number of factors can lead to unwanted metastable solids in the frozen matrix. Nucleation is random, and individual vials may experience different freezing times, depending on when nucleation was initiated. Normally, the greater the degree of supercooling, the more rapidly crystallization occurs once nucleation is initiated. Shock cooling or solidification is a technique used to create high-energy glasses and crystals. The vials near the edges of the lyophilizer experience different temperatures because of radiant energy from the chamber wall. Vials closer to the chamber wall will experience a higher temperature. The other components of the formulation may inhibit crystallization.

Thermal treatment or annealing is used to convert high-energy solids into more stable thermodynamic forms. Annealing results when material, amorphous or crystalline, is held above its transition temperature and a phase change is initiated. The use of annealing and a thermal treatment cycle is determined during FD cycle development. Annealing can take place at one or more times during freezing. However, thermal treatment is most often done postfreezing, once T'_g or T_e has been reached. Refer to Figures 17.2 and 17.3 to better understand what happens when the annealing temperature is set above the T'_g, or T_e and held for up to several hours to assist in product uniformity. In either case, the increased temperature will move the matrix composition to the left of the glass or eutectic mixture between b and T'_g or T_e. At temperatures above T'_g or T_e, the smaller ice crystals melt and the freeze concentrate is diluted. During annealing, Ostwald ripening can occur: The small ice crystals melt and then recrystallize into larger ice particles. Over time, the ice particle size distribution increases. During drying, these larger ice crystals can leave larger pores, which provide less resistance to the sublimation process and secondary drying. The increased temperature may provide enough thermal energy to transform high-energy amorphous excipients such as mannitol and glycine, or a drug substance, into lower-energy crystalline forms. It is well documented that high-energy forms of a drug are less chemically stable than are their low-energy counterparts. The crystalline forms can provide structural integrity and protection against collapse. The crystalline matrix also provides a porous structure that aids primary and secondary drying. In addition, the crystalline forms of mannitol and glycine provide good mechanical cake strength, which decreases cake friability and breakage during labeling, packaging, shipping, distribution, and use. Annealing can minimize vial cracking and breaking due to the crystallization of mannitol.

Vial breakage during freezing can be minimized by decreasing the concentration of mannitol to less than 5% w/v, using wider vials to achieve decreased fill height, filling to less than 30% of the vial height, and freezing at the slow cooling rate of $1°C$/min, and cooling to no lower than $-25°C$ until the mannitol has had a chance to crystallize (annealing as a waypoint during freezing). Annealing of mannitol can also be done once T_e or T'_g is reached. The frozen matrix can be reheated to -20 to $-25°C$ for 2 or more hours if the matrix fill depth is 1 cm or more. In any case, annealing should take place before primary drying is initiated to minimize vial cracking and breakage. Finally and very importantly, annealing can increase T'_g, which allows primary drying to take place at a higher temperature, which can improve the speed and efficiency of the drying process.

Thermal treatment can also have negative consequences. The most likely issue is increased drug degradation as a result of higher mobility at the higher temperature and being exposed to a higher temperature for an extended period of time. Protein phase separation may occur during annealing, which could also lead to increased degradation. It may be possible to obtain the desired annealing with a shorter exposure to elevated temperatures and mobile water by using annealing cycles that do not use as high a temperature or hold time.

17.2.3.7.4 Cake Characterization   The dried cake is often characterized for visual appearance (color, cracking, shape, indication of collapse, or microcollapse), moisture content, hygroscopicity, extent of crystallinity, reconstitution time, pH, extent of solution and solution particulate level, friability, and drug stability as a function of moisture content. In addition to visual appearance, scanning electron microscopy can be used to characterize the cake morphology and bulk porosity. The moisture content of the cake can have a significant effect on the cake's physical stability as well as drug stability. The final cake moisture content is a function of the cake components, primary and secondary drying conditions, the crystalline and amorphous nature of the cake, the use of inert gases and vacuum on sealing, and the moisture content of the stoppers. Sorption studies that evaluate moisture uptake and loss are undertaken to define the hygroscopic nature of the cake. The hygroscopicity classification system developed by Callahan et al. (1982) and presented in Appendix 13.1 can be used to define the cake's hygroscopic nature. The level of moisture content can lower the cake's transition temperature (T_g), which can affect cake's structural integrity and the drug's stability. The extent of the cake's amorphous and crystalline nature can be determined by x-ray diffraction or differential scanning calorimetry. Amorphous material is generally more hygroscopic than crystalline matter, and the water in an amorphous cake is more mobile and reactive, which can lead to poorer product stability. Crystalline or crystalline matrix cakes will commonly have some amorphous character. The level of amorphous content and moisture will affect the cake's T_g. The reproducibility of the cake's T_g can be used as a surrogate measure of the reproducibility of the FD process. If the process is well controlled, there should be little deviation in T_g. LYO products exhibit much better stability when they are stored below their T_g value. It is desirable to have a cake T_g above 60 or $70°C$ to allow for room-temperature storage and shipping-temperature excursions. The maximum acceptable reconstitution time is somewhere around 3 minutes. Rapid and complete dissolution is desired. The reconstituted solution must pass the U.S.P. particulate requirements. There are no standard cake friability tests. The cake crushing strength may be investigated using an Instron or comparable instrument. Cake friability can be assessed by simulated shipping tests or other types of standardized in-house shaking or dropping tests. Accelerated stability testing can be undertaken during the development stage to help predict product physical and chemical stability.

17.2.3.7.5 Formulation of Lyophilized Complex Molecules   Proteins tend to denature when subjected to extremes in temperature and pH; mechanical stress such as filtration, shaking, and aeration; exposure to air–liquid, ice–liquid, glass–liquid, and stopper–liquid interfaces; increased ionic strength; increased protein concentration; classical protein chemical reactions; and dehydration. LYO creates a number of stresses that can denature a protein. The two major LYO stresses are due to freezing and drying. A great deal of research has been published over the last 20 years with regard to protein denaturation and chemical degradation. However, protein formulation development remains a fairly empirical process. The literature is replete with examples of excipients that provide stabilization in the case of one protein, only to be shown to accelerate the loss of activity for another protein. This discussion will provide a general lyophilized protein formulation development framework that focuses on protein stabilization during the freezing and drying stages of LYO. A number of references discuss protein formulation in general and LYO specifically in much greater depth than is presented here (e.g., Wang and Hanson, 1988; Nail and Gatlin, 1993; Pearlman and Wang, 1996; Carpenter et al., 1997; Wang, 2000; Costantino, 2004; Rey and May, 2004; and McNally and Hastedt, 2008).

The physiochemical changes that occur during freezing were discussed in Section 17.2.3.7.1. These changes need to be kept in mind when developing a formulation that is designed to provide cryoprotection. This discussion focuses on the use of functional excipients to stabilize the native protein structure during freezing. Multimeric proteins seem to be particularly unstable to freezing. The *solute excluded* hypothesis has been proposed to explain solute stabilization of proteins in solution and during freezing. Polar solutes such as sugars and polyols are excluded from the unfolded protein hydrocarbon motifs. The solutes excluded increase the surface tension of the water surrounding the protein, which leads to *preferential hydration* of the protein's hydrophilic surface and stabilization of its native conformation. Pikal (2004) and Carpenter et al. (2004) discuss the implications and shortcomings of the excluded solute model for stabilizing the native protein structure during the freezing stage of LYO.

A number of different types of molecules have been shown to provide nonspecific cryostabilization of proteins. The most common cryostabilizers are monosaccharides (glucose), disaccharides (sucrose, trehalose, lactose, maltose), polysaccharides (dextran), amino acids (proline, glutamic acid, glycine), polyols (inositol, mannitol, sorbitol, dithioritol), other proteins (gelatin, egg albumin, bovine serum albumin, human serum albumin), nonionic surfactants [Brij 30, poly(oxyethylene lauryl ether); Tween 80, poly(ethylene sorbitan monooleate);

Pluronic, poloxamer], zinc, and polymers [PVP, poly(vinylpyrrolidine); PEG, poly(ethylene glycol)]. Glucose, lactose, maltose, and dextran, to a much lesser extent, are reducing sugars that undergo a Maillard, or browning, reaction with proteins. These reducing sugars should be avoided if at all possible. As a rule, polymers and proteins are much more potent cryoprotectants than sugars and polyols. The stabilizers are usually formulated on molar or w/v% basis. Excellent protein stabilization has been accomplished with 1% w/v PEG and PVP. On the whole, higher-molecular-mass PEGs seemed to be more effective. Low concentrations (0.1 to 1% w/v) of gelatin and albumin have also been effective cryoprotectants. Amino acids and sugars, on the other hand, must be used at relatively high concentrations. For glycine and mannitol, which can readily crystallize during freezing, it is only the amorphous portion that is thought to impart protection. Sucrose has been used at 10 to 30% w/v to provide cryostabilization. This level of amorphous material can have significant negative consequences on the matrix's mechanical strength, primary drying, and secondary drying times and efficiency.

Stabilization of proteins during drying has been explained to varying degrees by the *water substitute concept* or *immobilization mechanism*. These have been discussed by Carpenter et al. (2004) and Pikal (2004). Very simply, the water substitute concept hypothesizes that functional excipients that are capable of hydrogen bonding protect proteins by replacing the water hydrogen bonds that are vacated during drying. It is thought that the substitute excipient hydrogen bonds will stabilize the protein. The immobilization mechanism proposes that the protein is immobilized in the glass, restricting unfolding and stabilizing the native state. It is also speculated that immobilization may keep proteins from interacting with each other, decreasing the chance for aggregate formation during storage or upon reconstitution. Glass-forming excipients that are capable of hydrogen bonding can have a synergistic stabilizing effect with cryoprotectant. The amount of dehydration stabilizer chosen for a formulation is usually based on the weight ratio of the protein. A 1 : 1 ratio of amorphous stabilizer to protein is usually a good starting point for developmental studies. Marketed lyophilized protein products range from 0.125 to 25 mg/mL. Some development formulations have evaluated hundreds of mg/mL. At higher protein concentrations, a 1 : 1 weight ratio of an amorphous stabilizer results in a fairly high solids concentration in the cake. Since many proteins are formulated at concentrations of less than 5 to 10 mg/mL of protein, the ratio of dehydration stabilizer can be several times greater than the 1 : 1 ratio without affecting primary drying or cake quality. Sucrose and trehalose are commonly used drying stabilizers.

17.3   PROCESS DESIGN

17.3.1   Elements of Design and Qualification

The product design codependency of formulation design, process design, and the container closure system was discussed earlier in this chapter. To manufacture an FDA-approved product commercially it must be manufactured under current good manufacturing practices (CGMPs). The most stringent CGMPs are applied to sterile product manufacture. Although clean room design and qualification, facility design and qualification, water processing design and qualification, and personnel qualification are not within the scope of this book, these areas are the CGMP underpinnings for successful sterile product manufacture and are an essential part of a quality-by-design philosophy. The FDA guidance document Sterile Drug Products Produced by Aseptic Processing—Current Good Manufacturing Practice (2004) details the building and facilities, personnel training, qualification, and monitoring; components and container closures; endotoxin control; validation of aseptic processing and sterilization; and more. U.S.P. Chapter <1116>, Microbiological Evaluation of Clean Rooms and Other Controlled Environments, can also serve as an introduction to the compendial expectations.

The equipment cleaning and sterilization processes need to be qualified and validated. The time frame between cleaning, sterilization, and use needs to be defined and validated. A room-cleaning protocol, schedule, and lapsed time between cleaning and use need to be defined. Routine testing schedules and test specifications need to be assigned to facility systems, such as high-efficiency particulate air, facility good manufacturing process air, clean steam, chilled water, U.S.P. purified water, U.S.P. water for injection, vacuum, compressed gas, and nitrogen. In the remainder of this section we highlight the lyophilization and sterilization processes.

17.3.2 Lyophilization

Lyophilization generally involves:

1. Dissolving the drug and excipients in water for injection
2. Aseptically sterilizing the solution through two (for redundancy) 0.22-μm filters
3. Aseptically filling the sterile drug solution into sterile containers
4. Carrying out online weight checks
5. Partially stoppering with washed, sterilized, siliconized vented stoppers
6. Aseptically transferring the partially stopper vials onto cooled temperature shelves, located in a presterilized freeze-drying chamber
7. Freezing the solution under controlled cooling conditions
8. Annealing the frozen matrix
9. Initiating primary drying by adjusting the chamber to low pressure (vacuum)
10. Heating the shelves to sublime the ice
11. Initiating secondary drying by lowering the pressure further and initiating a controlled temperature ramp to a specific drying temperature to decrease the cake moisture content
12. Nitrogen-purging the chamber if necessary and stoppering the vials completely by bringing the shelves together while the chamber is at a slight vacuum

13. Aseptically transporting the stoppered vials to a capping line
14. Capping the vials with an aluminum crimp
15. Carrying out online weight, crack, and leak checks
16. Labeling the vials and checking the labels online with a visioning system
17. Placing the vials in a carton with the package insert

The standard FD temperatures and pressures were discussed earlier. As mentioned previously, lowering chamber pressure during secondary drying may not help decrease drying time or decrease the amount of residual water because conductive heating is also decreased at higher degrees of vacuum. The interplay between chamber pressure, temperature, and cake porosity and composition must be studied carefully to determine the optimum secondary drying conditions.

The LYO process presents many more opportunities for microbiological contamination than are presented during aseptic solution manufacturing. The filled vials are only partially stoppered and the vials remain partially stoppered for days during LYO, allowing access to microbiological contamination. Because the vials are partially stoppered, it is expected that all product handling and transfer steps are done under laminar airflow using high-efficiency filters. The drying process is operated under vacuum, so there is potential for microbes to be drawn into the LYO chamber. Leak tests need to be performed routinely. Leaks may arise from around door seals; from pinholes in the thermal fluid lines to the shelves; from poor-fitting sterile gas filter connections; and from deficient condenser refrigerant and vacuum pump seals. The LYO chamber and shelves need to be sterilized, preferably by steam under pressure. Biological indicators and temperature sensors are used to support the effectiveness of the sterilization procedure. The integrity of the gas filters needs to be checked routinely. Redundant filters are often used to minimize the risk of filter failure. Once the readiness of the lyophilizer is confirmed, manufacturing can begin.

The FD process steps were described briefly in Section 17.2.3.7. Some practical guidance is offered in the following sections. Figure 17.4 shows the various components of a small freeze-dryer capable of manufacturing clinical supplies. Bindschaedler (2004) has discussed critical process parameters and independent and dependent process parameters.

17.3.2.1 Freezing Process

Once the formulation has been chosen and critical freezing transition temperatures have been determined, a freezing cycle can be developed. A statistical experimental design approach can be used to establish the relationship between product temperature, shelf temperature, and chamber pressure. The predictive model equations can be used to create a design space that delivers products of acceptable quality. A freezing cycle can be defined within the design space that sets a target shelf temperature and chamber pressure along with acceptable upper and lower ranges. A representative freezing cycle involves the following conditions. Once the vials have been filled and transferred aseptically to the lyophilizer, the vials, with partially seated slotted stoppers, are loaded onto the shelves. The shelves can be precooled to about $5°C$, or the product can be loaded onto the shelves at room temperature. In the latter case, once all the vials have

Kinetics lyophilizer chamber & thermocouples

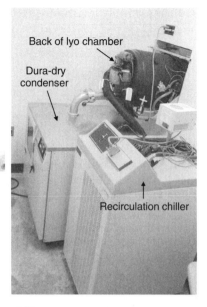

FIGURE 17.4 Lyophilizer.

been loaded, the shelf temperature is decreased to 5°C. Shelf precooling is done to decrease the potential for supercooling. Keep in mind that when the product is loaded onto precooled shelves, there is a risk that condensation may build up on the shelves, where it can freeze and affect the heat transfer and increase the sublimation load on the drier. Whether loading is done at room temperature or at 5°C, the load is conditioned at 5°C for 15 to 30 minutes. This initial hold improves the prefreeze vial temperature uniformity. After the 5°C hold, the target shelf temperature is set and the shelves are cooled at 1 to 2°C/min. The final shelf temperature is frequently set to −40 to −50°C. The shelf temperature must be 2 to 3°C below T_e or T_g' to ensure that the product is below these temperatures.

A lower cooling rate may be required for formulations containing high concentrations of mannitol to ensure complete crystallization and a decrease in the potential for vial breakage. Normally, slow freezing will result in lower supercooling, larger ice crystals, larger pores, and lower resistance during the sublimation stage. Slower freezing rates may lead to more freeze separation from stabilizers, resulting in poorer stability. The freezing cycle may include an annealing step as a waypoint to induce bulking agents to crystallize. Depending on the design-of-experiments (DOE) cycle development study, the temperature is held constant for 30 to 60 minutes before cooling is resumed. Once the desired product temperature has been reached, the vials will again be conditioned at this temperature for 1 to 2 hours, depending on the depth of fill. Customary depths of fill range from 1 to 2 cm. Postfreezing annealing is achieved by increasing the product to 10 to 20°C above T_e or T_g'. The heat treatment causes the smaller ice crystals to melt and refreeze into larger crystals (Ostwald's ripening). The larger ice crystals sublime and leave larger pores, which decreases the mass transport resistance during primary and secondary drying and decreases the reconstitution time. The heat treatment can also cause amorphous material to crystallize, which increases the mechanical strength of the cake. Postannealing, the product temperature is reestablished below T_e or T_g'.

17.3.2.2 *Primary Drying*

Primary drying is the longest phase of FD. DOE can identify key factors and levels that can reduce the primary drying time significantly. In addition, a DOE approach can be used to establish the relationship of product temperature as a function of the shelf temperature and chamber pressure. The predictive model equations can be used to create a design space that delivers products of acceptable quality. Primary drying must take place below the product's collapse temperature, T_c, which may be significantly higher than T_g' (Blue and Yoder, 2009). The upper limit of the target product primary drying should provide a reasonable safety margin, to prevent collapse. For this discussion, the *margin of safety* is defined as the difference between the collapse temperature and the upper limit of the target product primary drying temperature. Tang and Pikal (2004) recommended different margins of safety, depending on the length of the primary drying cycle. A $2°C$ safety margin was recommended for primary drying cycles of more than 2 days. For cycles lasting less than 10 hours, a large safety margin of $5°C$ was proposed. A $3°C$ safety margin was suggested for primary drying cycles that have intermediate drying time requirements. If the product temperature gets dangerously close to T_c, adjusting the chamber pressure has a nearly immediate effect on the product temperature. This is another reason to keep the chamber pressure between 150 and 300 μmHg, so there is flexibility to change the chamber pressure to affect the product temperature quickly if necessary. On the other hand, adjusting the shelf temperature is a sluggish process, taking 20 to 40 minutes to reach an equilibrium temperature. The authors also state that the product temperature should not be higher than $-15°C$ because the heat and mass transfer capabilities of the freeze-dryer may be overloaded. The primary drying chamber pressure usually ranges from 75 to 300 μmHg, with 150 to 250 μmHg commonly used. During sublimation, the product undergoes cooling as a result of the endothermic sublimation process. At the end of primary drying, the product temperature rises and approaches the shelf temperature. This temperature rise can be used as primary drying endpoint. Vials near the outside of the chamber dry faster than do vials in the center, due to uneven heat distribution caused by primary differences in thermal radiation at the sides of the lyophilizer. Additional drying time is required to guarantee that the inner vials are dried adequately. Product temperature can be measured by a number of methods. An excellent discussion of various methods appears in Nail and Gatlin's article. The drying endpoint can also be determined by monitoring the vapor state in the chamber. Primary drying endpoint determination is discussed in more detail in Section 17.3.2.4.

17.3.2.3 *Secondary Drying*

The third step in FD is secondary drying. Secondary drying is initiated when all the ice has been removed during primary drying. The purpose of secondary drying is to remove the sorbed moisture from the cake. Amorphous material tends to have relatively high levels of residual moisture (20 to 30% w/w), which can significantly lower T_g. Significantly lowering T_g also increases the chance of inadvertently exceeding T_g, which can result in softening or partial collapse. Therefore, at the early stage of secondary drying a low heating rate of around 0.1°C/min has been suggested. As the product reaches room temperature, higher heating rates can be used. Crystalline products should not collapse after primary drying, so heating rates of about 0.5°C/min can be safely employed. Secondary drying is accomplished by raising the product temperature and maintaining or decreasing the chamber pressure. Secondary product drying temperatures usually range from 30 to 60°C. The chamber is usually set at around 25 to 150 μmHg. Process development studies should support the final target drying conditions based on product stability data. Cake moisture contents of 1% w/w or less is not uncommon.

17.3.2.4 PAT: Lyophilization

In many ways, LYO process analytical technology has been ahead of most other technologies. Techniques to measure the product temperature, shelf temperature, condenser temperature, and chamber pressure are critical to manufacturing a robust lyophilized product. These critical quality attributes have been measured for many years using a variety of techniques. Resistance thermometers, thermocouples, and to a lesser extent, thermistors, have been used to measure product temperature during freezing and drying. Resistance thermometers measure the change in electrical resistance of metals as a function of temperature and are known as *resistance temperature detectors* (RTDs). Platinum RTDs are well suited for freeze-drying applications.

A thermocouple is formed when the ends of two dissimilar metals are connected to form an electrical junction. It has been shown that a conductor will generate voltage if subjected to a thermal gradient. This is known as the *Seebeck effect*. The thermocouple will produce an electromotive force that is related to the temperature of the object. The thermocouple is often standardized using an electronic bridge circuit that provides a standard electromotive force (EMF) reference temperature or *cold junction*. Copper–constantan and chromel–constantan are commonly used for FDs. The thermocouple EMF vs. temperature plot is not linear. Thermocouples are also not as stable as RTDs.

Thermistors are semiconductors. A plot of relative resistance vs. temperature is plotted and a steep negative slope is generated. This indicates that resistance increases sharply with a decrease in temperature, which is the opposite of what is seen for RTDs. Thermistors have a ±20% variation and are not typically used in FD applications.

Product temperature probes are used routinely in process development, but there are problems that must be dealt with. Accurate probe placement and probe placement maintenance are common problems. Placement of the sample vials with thermocouple wires can also be problematic. The temperature probes can serve as heterogeneous nucleation sites not representative of the batch. Handling the probes and vials raises microbiological contamination concerns. For this reason, clinical trial supply manufacturing and commercial product manufacturing do not typically use temperature probes to measure the product temperature and the drying process. Shelf temperature and chamber pressure data are used to monitor the clinical and commercial manufacturing process. Some lyophilizers are equipped to collect *pressure rise* data to help determine the primary drying endpoint.

The chamber pressure can be measured in a number of ways. A Pirani gauge is a thermal conductivity gauge that uses a Wheatstone bridge to measure pressure at a constant current. At low pressures (100 µmHg or less) the gas energy loss by conduction increases linearly with pressure. Pirani gauge output is a function of gas composition, so if the gas composition changes over the course of the FD process, which occurs as water is removed from the matrix, measurement errors may be encountered.

Capacitance manometers measure pressure that causes deformation of a thin diaphragm. Capacitance is directly proportional to the relative static permittivity, (dielectric) of the medium, the vacuum permittivity, and a geometric factor. A flexible metal diaphragm, which serves as the medium, is placed between two fixed electrodes. The reference side of the diaphragm is evacuated to 9×10^{-5} µmHg and sealed. The diaphragm deflection is proportional to the gas pressure applied, which is the total chamber pressure, independent of composition. FD capacitance manometers have gained broad acceptance because of their wide operating range (0.1 to 1000 µmHg) and accuracies (from 0.05 to 3%). Many development driers use a combination of Pirani and capacitance sensors, since as the water vapor is removed, the Pirani output changes and eventually reaches that of the capacitance gauge when primary drying is complete. Some commercial freeze driers come equipped with the ability to measure the *pressure rise* that results during sublimation when the lyophilization chamber is closed to the condenser. The primary drying process can be monitored by intermittently closing a special fast-acting valve that connects the chamber to the condenser and vacuum pump. Sublimation will result in a pressure increase, and as the sublimation process reaches completion, the pressure rise reaches an equilibrium value. For more details concerning temperature and pressure measurement devices, see the article by Nail and Gatlin (1993).

These process analytical technology (PAT) techniques can interfere or affect the FD process. It is well known that thermocouples that are inserted into vials are a site for heterogeneous nucleation. In practice, this means that vials that contain the thermocouples will not supercool to the same extent as vials that do not contain the thermocouples and, hence, begin to freeze sooner. As a general rule, freezing times are extended 1 to 2 hours. This is called *soaking time* or *conditioning time* and is required to make sure that all vials have been completely frozen. The pressure rise method has the potential to cause collapse during measurement by allowing the pressure to rise, which decreases evaporative cooling and leads to an increase in product temperature and possible collapse. Again, pressure is a master key that can almost instantaneously affect product temperature.

Tang et al. (2006a–c) and Gieseler et. al., 2007b have reported using a manometric temperature measurement system as a process analytical tool. A tunable diode laser spectroscopy (Gieseler et al., 2007a) and Raman and near-IR spectroscopy (De Beer et al., 2009) have been reported to provide in-line real-time freeze-drying process monitoring.

17.3.2.5 Scale-up of the Lyophilization Process

In general, laboratory lyophilizers are not as well insulated as commercial units and have more radiant heat. It is not uncommon to use laboratory FD ramp rates of 2 to 4°C/min, compared to 0.5 to 1°C/min for commercial lyophilizers. For secondary drying, the commercial drying temperature is often set 3°C higher than the laboratory setting. Over the past two decades there have been a number of mathematical models developed to describe the FD process (Pikal, 1985; Liapis and Bruttini, 1994; Liapis et al., 1996; Sadikoglu and Liapis, 1997). The fundamental drying principles developed in these model equations were used successfully to scale up a labile freeze-dried product from a laboratory-scale to a pilot-scale lyophilizer (Tsinontides et al., 2004). Data from the three scale-up runs were used to evaluate the robustness of the cycle and simulate the operational conditions for the pilot-scale demonstration batch. The cycle set points and LYO cycle were demonstrated successfully.

17.3.2.6 Lyophilization Design Optimization Each LYO stage has a large number of factors that affect final product quality. Screening studies can be undertaken at the early process development stages to determine the most significant factors. A freezing screening study may investigate a number of factors, such as the vial fill volume, initial loading shelf temperature, initial conditioning temperature and time, cooling and freezing rate, waypoint annealing temperature and time, postfreezing annealing temperature and time, and postfreezing conditioning temperature and time. T'_g, percent crystallinity, vial breakage, and postfreezing drug stability are responses that are usually evaluated. Primary and secondary drying factors usually include ramp heating rate and primary and secondary drying temperatures and pressures. Responses include product temperature, collapse and degree of collapse, cake moisture content, cake appearance and friability, reconstitution time, and drug stability.

The information from these screening designs identifies the significant factors and helps narrow the number of factors that are incorporated in the response surface designs used to optimize the LYO process. The optimization study operating ranges often employed are ramp rates, 0.5 to 1.5°/min; primary and secondary drying temperatures, ±3°C; and chamber pressure, ±25 μmHg. Product attributes would include the responses listed above. An expert system using manometric temperature measurement has been reported that allows the development of an optimized freeze-drying process during a single laboratory lyophilization experiment (Tang et al., 2007; Gieseler et al., 2007a).

17.3.3 Sterilization

Sterilization is a process that renders a product free of viable microorganisms. By definition, a product is sterile when there is a complete absence of viable microorganisms. A number of acceptable sterilization processes are used in the pharmaceutical industry, including moist heat, dry heat, chemical (cold), radiation (cold), and filtration. All the sterilization methods except filtration sterilize the product by the destruction of microorganisms. Filtration eliminates microorganisms from a product by size exclusion. An aseptic sterilization filtration process requires that processing times between bulk preparation, sterile filtration, and filling be minimized and highly controlled. Process qualification and validation efforts need to support the optimal processing time. The process may be validated for extended processing times that could account for unexpected delays. Moist heat autoclaves include saturated steam, superheated water, and air-over-steam methods. Moist heat and dry heat sterilizers can be operated as a batch or a continuous process. Chemical sterilization processes involve ethylene oxide and vaporized hydrogen peroxide–steam. Radiation sterilization involves electromagnetic waves (gamma rays and x-rays) and particles (electron beam). Filtration uses polymer membranes made from a variety of polymers. U.S.P. Chapter <1222>, Terminally Sterilized Pharmaceutical Products—Parametric Release, discusses the use of test parameters and thorough validation to release product without performing final product sterility testing. Parametric release uses biological, chemical, and physiochemical indicators to confirm that sterilization conditions were achieved during the sterilization process. U.S.P. Chapter <55>, Biological Indicators—Resistance Performance Tests; <1035>, Biological Indicators for Sterilization; and <1211>, Sterilization—Chemical and Physiochemical Indicators and Integrators, cover these areas. For further reading, Hagman (2006) has written a chapter on sterilization that provides a good introduction to pharmaceutical sterilization.

The quality of a sterile product begins with the appropriate design; construction; qualification; cleaning; bioburden control; and maintenance of building, facilities, and equipment. There must also be proper control of components, container closure systems, manufacturing processes, laboratory procedures, packaging, and labeling of the final product. Proper hygiene and gowning training of personnel are also critical for minimizing the product bioburden. Despite this enormous quality-by-design effort, product sterility cannot be assured by end product sterility testing. The sterility test itself has certain limitations. For example, the media may not stimulate the growth of all the microorganisms present. The number of test samples is relatively small, so low levels of microorganism contamination may go undetected. In general, only major contamination is detected. There is also the practical problem of false-positive results of sterility tests. That is, microbes are inadvertently introduced into the test sample during the testing procedure. The sample is then reported to be contaminated when it was actually sterile.

In the case of moist heat, the microorganism inactivation process denatures a microorganism's proteins. Moist heat sterilization is a first-order or exponential process which means theoretically that the number of microbes destroyed never reaches zero. This conundrum has led to the concept of a *sterility assurance level* (SAL) or *probability of nonsterility* (PNS). These terms are used interchangeably. PNS is the probability of a product bioburden surviving the sterilization process. The SAL or PNS is set to minimize the risk of releasing a nonsterile product, knowing that it is impossible to reduce the risk to zero. SAL or PNS is most commonly set at 1×10^{-6}. This PNS limit is interpreted to mean that one out of 1 million units produced would not be sterile. Another way of stating this PNS is that the probability of the product bioburden surviving the sterilization process for any manufactured unit is 1 in 1 million.

The inability to confirm the sterility of the end product and the implementation of a PNS has led to the regulatory acceptance of the parametric release of terminally sterilized products. Terminally sterilized products are sterilized after they have been sealed in their primary container. In principle, the monitoring and control of critical in-process *parameters* can provide a higher level of sterility assurance than that of end-product sterility testing. As one can imagine, to be able to release a sterile product without final product sterility testing requires an extraordinary level of qualification and documentation. In addition, a *sterilization load monitor* is evaluated with each product load. The monitors can be a physical, chemical, or biological indicator and are used to confirm that the sterilization process has met certain critical parameters and there is a state of process control. The FDA has accepted parametric release for moist heat terminal sterilization since 1985. A guidance for industry (FDA, 2010) and the Pharmaceutical Inspection Convention Pharmaceutical Inspection Co-operative Scheme (2007) provide more detailed information on parametric release. The different sterilization processes are discussed in subsequent sections. The choice of sterilization process depends primarily on the physiochemical properties of the drug, which is determined at the preformulation stage of product design. The container closure system also affects the sterilization process chosen. For example, an irradiation sterilization method may be required for a suspension product of a drug that melts at 97°C. Moist heat, dry heat, gamma radiation, and electron-beam radiation are used to sterilize products terminally. Aseptic filtration is used to produce sterile products when terminal sterilization is not possible. Ethylene oxide, vapor hydrogen peroxide, and steam–formaldehyde are used to sterilize surfaces, equipment, utensils, parts, and packaging materials. Product sterilization methods are discussed below.

17.3.3.1 *Moist Heat Sterilization*

Saturated steam is the terminal sterilization method of choice. A steam autoclave is pictured in Figure 17.5. Only heat-sensitive or moisture-sensitive drugs and container closure systems should use other sterilization methods. If other methods are used, the FDA requires that a data-based justification be available for their review. When steam condenses on an object, it releases about 2200 kJ/kg of heat, and the temperature of the object and its contents rises. When the temperature of the drug product reaches above 80°C, significant microbiochemical effects are elicited, which leads to cell death. Common moist-heat sterilization temperatures are 121 and 130°C. The vapor pressure of saturated steam at 121°C is 2.05 bar or 205 kPa and is a thermodynamic function of the water vapor-pressure curve (Figure 17.6). Once the water in the vial or ampoule reaches 121°C, the vapor pressure of the water inside the container closure system is 2.05 kPa. Therefore, saturated steam sterilization is done in a pressure vessel called an autoclave. Each autoclave has a pressure rating that must not be exceeded. Autoclaves should be inspected annually to maintain pressure certification. The container closure system also experiences an increase in internal pressure, due to the expansion of the liquid and the pressure increase in the air of the headspace. This increase in internal pressure is more than 1 bar or 1 kPa greater than the chamber pressure. Small-volume sealed ampoules can withstand this pressure increase. However, the overpressure can expel prefilled syringe plungers, deform plastic or semirigid bags, and raise safety and quality concerns for glass stoppered vials. In the case of stoppered vials, the stopper may be partially lifted or distorted, which may allow ingress of contaminant poststerilization.

Primus autoclave
(non-sterile side)

Primus autoclave
(sterile side)

FIGURE 17.5 Steam autoclave.

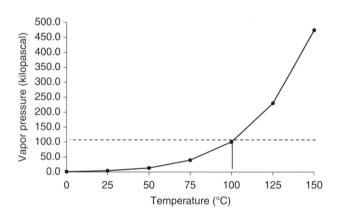

FIGURE 17.6 Water vapor pressure–temperture profile.

Superheated water spray and air-over-steam autoclaves are used to furnish chamber counter pressure that is independent of the autoclave temperature. The counter pressure is controlled by compressed air. The presence of air in a saturated steam autoclave can cause steam concentration and temperature gradients because of the density differences between steam and air. The counter pressure methods utilize special design features to overcome this issue. The superheated water spray method introduces the superheated water at the top of the autoclave through spargers. The superheated spray reaches the lower load through perforated shelves. Counter pressure is maintained by compressed air. Controlled cooling is achieved by sparging the load with the recovered condensed sterile water using a programed cooling rate to prevent glass breakage from temperature shock. When the product temperature reaches 70°C, the load is removed from the chamber to dry spontaneously. Air-over-steam autoclaves use steam that is introduced through spargers at the bottom of the autoclave. Compressed air is again used to provide chamber counter pressure. Magnetically driven ceiling fans are used to maintain a homogeneous mixture of air and steam. Cooling is achieved by providing cooling air, which is distributed by the fans. Both of these methods maintain chamber temperature uniformity of ±0.5°C.

People involved with moist heat sterilization will need to become familiar with D_{10}, z, and F_0 values. Destruction of microorganisms by moist heat has been shown to follow first-order kinetics and is described by

$$N = N_0 \cdot 10^{-(t/D_{10})} \tag{17.12}$$

where N is the number of microorganisms at time t, N_0 the initial number of microbes in the product or product bioburden, and D_{10} the *decimal decay time*. D_{10} is inversely proportional to the first-order constant. As with other first-order rate constants, it is a function of temperature. D_{10} also varies with changes in pH and ionic strength. Decimal decay time, D_{10}, is defined as the time required to reduce the number of surviving microorganisms by 0.1, or 1/10. The D_{10} value is also equal to the time required to reduce the number of surviving organisms by one order of magnitude or a 1 logarithm reduction. The D_{10} value is the reciprocal of the slope of a plot of the logarithm of N vs. time. Equation (17.12) can be rearranged to solve for t, the sterilization time:

$$t = D_{10} \frac{\log N_0}{N} \tag{17.13}$$

where $\log N_0/N$ is called the *inactivation factor or sterility criteria* (Denyer et al., 2004) and N in this case is the PNS, 1×10^{-6}. Nabla (∇) is used as the symbol for the inactivation factor or sterility criterion. If the product bioburden is 10 and the PNS is 1×10^{-6}, the sterility criterion, $\log N_0/N$, is 7 ($\log 1 \times 10^1/1 \times 10^{-6}$). This means that the product will be considered sterile when there is a 7-logarithmic unit reduction or a seven-order decrease in the number of microorganisms. Remember that D_{10} is the time to reduce the number of organisms by 1 logarithmic unit. The sterilization time required to meet the sterility criteria is then given by

$$t = D_{10} \cdot \nabla \tag{17.14}$$

For example, if $D_{10,121.1°C}$ (decimal reduction time) is 1.5 minutes (the commonly reported $D_{121.1°C}$ value for endospores of *Bacillus stearothermophilis*) and ∇ (sterility criterion) is 7, the product would need to be sterilized for 10.5 minutes at 121.1°C to meet the PNS of 1×10^{-6}. *Bacillus stearothermophilis* D_{10} and z values are commonly used to set conservative sterilization cycles.

The z *value* is a measure of the microbial inactivation sensitivity to temperature changes. It is called the *temperature coefficient of microbial destruction*. The

z value is the number of degrees Celsius that leads to a 10-fold change in the D_{10} value. The z value can be expressed by

$$D_{10} = D_{T=0} \cdot 10^{-(T/z)} \tag{17.15}$$

where D_{10} is the D value at a given temperature T in Celsius, and $D_{T=0}$ is the D_{10} value at $0°C$. The z value is the reciprocal of the slope of a plot of the logarithm of D_{10} vs. T. An organism with a larger z value is less sensitive to changes in temperature than an organism with a smaller z value.

A steam sterilization cycle involves loading the product in the autoclave, securing the pressure rated door, and introducing steam or superheated water into the autoclave. The product temperature rises until it reaches the target temperature. The product is then held at the target temperature for a predetermined length of time, at which point cooling is initiated. Biological damage and microbial inactivation begin to occur at about $80°C$, which means that cell destruction is actually occurring during the heat ramping, hold, and cooling phases. There are a number of reasons that it is desirable to include the destruction that occurs during all phases of the sterilization cycle. There is often a trade-off between the need to sterilize a product and to minimize drug degradation during the sterilization cycle. The hold time at the target temperature can be decreased when the microbial inactivation is accounted for during the heating and cooling stages. The sterilization time can be decreased, which over a year's time can increase manufacturing efficiencies. The concept of F_0 is used to account for the microbe destruction that occurs throughout the autoclave cycle. F_0 or the equivalent sterilization time at $121.1°C$ at a z value of $10°C$ is calculated as:

$$F_0 = \Delta t \sum 10^{(T - 121.1°C)/z} \tag{17.16}$$

where F_0 is the equivalent sterilization time in minutes, Δt the incremental time interval between two consecutive measurements of T, T the product temperature in Celsius, and z the temperature coefficient of microbial destruction, which is set to $10°C$. Today's autoclaves are equipped with programmable logic controllers that can monitor the autoclave temperature at given time intervals and control the autoclave cycle to achieve a set F_0 value. Previously, a sterilization time of 10.5 minutes was calculated using equation (17.14), the $D_{10,121.1°C}$ for *Bacillus stearothermophilus*, and the sterility criterion. To realize a conservative sterilization cycle of an equivalent lethality, one would set F_0 to 10.5 minutes. U.S.P. <1222> defines an overkill cycle "as a process that would deliver a minimum F_0 of 12 minutes."

17.3.3.1.1 PAT: Moist Heat Sterilization Parameteric release of drug product is the ultimate goal of PAT, which has been realized for moist heat sterilization. U.S.P. <1222> defines parametric release as "the release of terminally sterilized batches or lots of sterile product based upon compliance with defined critical parameters of sterilization without having performed the requirements under Sterility Tests <71>." Temperature, pressure, and moisture sensors have been used since 1985 to attain a level of quality assurance that has allowed product to be released using primarily PAT to monitor and control the sterilization process. The sterilization process is also monitored by sterilization load monitors that can measure the level of lethality by using biological indicator organisms, temperature monitors, pressure monitors, and steam monitors. Six different types of chemical load monitors are discussed in (ISO International Standards Organisation) 11140-1 (2005). A class 5 steam integrating chemical monitor is designed to react to time, temperature, pressure, and steam. The gold standard for load monitors is biological indicators that contain a number of resistant spore-forming organisms that confirm the adequacy of the sterilization process. These units only monitor the sterilization process; they are not used as control devices. The load monitors are placed in predetermined challenge locations and are evaluated at the completion of the process.

17.3.3.1.2 Scale-up of Moist Heat Sterilization The load configuration is typically evaluated during scale-up. The autoclave temperature distribution should be within $\pm 0.5°C$ with different load configurations, and the same F_0 is maintained at each scale.

17.3.3.2 Dry Heat Sterilization Dry heat sterilization has only two critical process variables, temperature and time. A depyrogenation/dry heat sterilizer is shown in Figure 17.7. It is well known that dry heat is less lethal than moist heat, which requires dry heat sterilization temperatures to be 40 to 60°C degrees Celsius higher than moist heat sterilization temperatures. However, compared to moist heat, dry heat can be used to depyrogenate equipment and containers. Depyrogenation dehydrates, oxidizes, and decomposes the lipopolysaccharide endotoxins, which can be pyrogenic and cause febrile reactions. The dry heat sterilization and depyrogenation cycle temperature and times are given in Table 17.11.

It is generally accepted that if an object is depyrogenated, it is also considered sterilized. Dry heat can be used to depyrogenate container closure systems, equipment, glass and stainless steel vessels, and utensils. The limulus amebocyte lysate test can detect the presence of endotoxins, which is a telltale sign that once-living microorganisms were present in the test sample.

F_H is the dry heat equivalent sterilization time at 170°C and a z value of 20°C. F_d has been used to represent the dry heat equivalent depyrogenation time at 250°C and a z value of 46.4. Akers et al. (1982) suggested that endotoxin

TABLE 17.11 Dry Heat Sterilization and Depyrogenation Cycles

Dry Heat Method	Temperature (°C)	Cycle Time[a] (min)
Sterilization	160	120–180
	170	90–120
	180	45–60
Depyrogenation	230	60–90
	250	30–60

Source: Modified from Hagman (2006).

[a] Depending on load configuration.

Gruenberg depyrogenation
oven (non-sterile side)

Gruenberg depyrogenation
oven (sterile side)

FIGURE 17.7   Depyrognation oven.

inactivation or depyrogenation followed a second-order, not a first-order, process. U.S.P. <1035> states that a decrease in the endotoxin titer of 3 or more logs will result in a process that achieves a PNS that is substantially lower than 1×10^{-6}. A 3-log reduction of endotoxin is generally accepted as a sufficient level of eradication. As noted previously, moist heat sterilization conditions are not sufficient to cause depyrogenation, so glassware and other manufacturing components need to be depyrogenated.

Dry heat transfer can be achieved by all three mechanisms of heat transfer: conduction, convection, and irradiation. Conductive heat transfer is by direct physical contact, which results in heat flowing from a high temperature to a low temperature. Convective or radiant heat transfer is through a fluid medium such as air. It is important to have adequate air circulation in a dry heat oven to prevent variations in air density and temperature. Radiation is the transfer of heat via electromagnetic radiation such as infrared or microwave. Radiation is absorbed, reflected, or transmitted. Only absorbed radiation transfers the energy to heat. The dry heat process can be a batch process that uses a hot-air oven or a continuous tunnel process that can use hot air or infrared radiation as the source of heat. Tunnels operate continuously and are ideal for large-capacity requirements. The tunnel sterilization process conditions are typically 20 minutes at 280°C or 3 to 4 minutes at 300°C. This condition is not harsh enough to eradicate the endotoxins. Depyrogenation cycles can be run for 20 minutes at 350 ± 15°C. The tunnels are fed sterile laminar-flow air that is heated. The tunnels need to have a flexible design that can accommodate a variety of materials of varying shapes and sizes. There needs to be ample control of conveyor belt speed, air velocity, and air temperature. All construction materials must be highly resistant to large temperature changes and meet CGMP cleaning and validation requirements.

Batch sterilizers are used routinely to sterilize and depyrogenate objects. They are ideal for varying load types, small loads, and infrequent use. Oils cannot be sterilized by moist heat because they are free of water. Oily drug products that are not heat labile can be dry heat sterilized.

17.3.3.2.1 PAT: Dry Heat Sterilization and Depyrogenation Parametric release of sterile product counts on sterile control at each unit operation. Time and temperature are critical dry heat sterilization and depyrogenation parameters. Datalogger technologies which can measure time and temperature have made significant advances. Dataloggers with thermal shields have been developed that operate with good accuracy and precision at 160 to 400°C. Dataloggers can be used with the actual production batch to confirm the actual conditions. Temperature load monitors can also be used to confirm the processing conditions.

17.3.3.3 Gamma-Radiation Sterilization

Gamma radiation offers a number of benefits that have resulted in its increasing use. Gamma radiation is a cost-effective continuous process that can provide a precise uniform dose, high penetration capability (making a flexible sterilization alternative), and rapid processing. It can sterilize the product in its final container so that the entire system is sterilized. It is also approved for dosimetric final product release. Importantly, there is no radioactive residual. Another potential advantage of sterilization by ionizing radiation in general, and particularly gamma radiation, is the growing body of data that support the inactivation of viruses. There are more than 200 commercial gamma-radiation facilities around the world. The number and types of pharmaceutical products that are gamma-irradiated is increasing constantly. In addition, the restriction of daily exposure of ethylene oxide to 1 ppm has made gamma radiation a more attractive sterilization technique for packaging container closure systems. The approval of gamma irradiation of foods has decreased the concerns generally associated with radiation exposure. Pharmaceutical products that have been gamma sterilized are antibiotics such as cephalosporins and penicillins, ophthalmic solutions, suspensions, creams and ointments, and aseptically filled sterile products such as prefilled syringes.

Gamma radiation interacts with matter by forming ion pairs and reactive free radicals. The effects may be direct or indirect. A matter-photon excitation or ionization results from a direct interaction. A primary indirect interaction involves the radiolysis of water, which forms a number of activated species, including the hydroxyl radical and hydrogen peroxide. These reactive species then react directly with other matter. Free-radical reactions can cause a number of chemical reactions, such as cross-linking, oxidation, abstraction, dissociation, double-bond formation, and exchange. Microorganism cell death is due primarily to DNA damage. Damage to organelles and the cell membrane may also lead to cell destruction. Interactions with the drug can lead to instability. Drug stability can also be affected adversely by ionizing radiation. In addition, interactions with container closure systems have been problematic, so a trade-off is often required to maximize cell death and minimize undesirable drug and CCS effects. Radiation can cause polymer grafting, oxidation, cyclization, scission, isomerization, amorphization, and crystallization. These reactions can lead to softening, stiffening, embrittlement, odor formation, and discoloration. Most materials have satisfactory compatibility. Poly(vinyl chloride), poly(tetrafluoroethylene), and acetal polymers are considered marginally incompatible. The extent of the reactions is a function of the oxygen concentration, moisture concentration, and temperature. Decreasing these three factors can lead to a decrease in the extent of the detrimental CCS reactions, but it will also decrease the lethality of the irradiated dose.

Cobalt-60 is the commercial source of ionizing radiation. Cobalt-60 has a half-life of 5.27 years and it emits two photons which have 1.17 and 1.33 MeV of energy. A generally accepted sterilization dose is 25 kGy. The dose is measured by dosimeters that are placed with the product undergoing sterilization. Dosimeters measure the chemical changes induced by radiation exposure. One such dosimeter is a ceric–cerous sulfate dosimeter, which is based on the reduction of ceric to the cerous oxidation state on exposure to ionizing radiation. A G value is the number of chemical species formed per 100 eV. The cerous ^{60}Co G value is 2.4 The ^{60}Co radiation dose can be quantified by spectroscopy or electrochemical methods and the G value.

As the desire to terminally sterilize products has intensified, more effort has been placed in determining the optimal radiation dose. The kGy sterilization requirement for an aseptically filtered product is significantly lower because the initial bioburden is much lower. The Stumbo–Murphy–Cochran equation can be used to determine the D_{10} value for an aseptically sterilized prefilled syringe. The D_{10} value, decimal reduction dose, is the radiation dose that elicits a 10-fold or 1 logarithm unit reduction in microbial count and is described by equation (17.17). The Strumbo–Murphy–Cochran dose survival method can be used in conjunction with equation (17.17) to determine the D_{10} value.

$$D_{10} = \frac{D}{\log(N_0/N)} \tag{17.17}$$

where D is the radiation dose delivered, N_0 the initial number of microorganisms in colony-forming units (cfu), and N is log (n/r) with n being the total number of samples irradiated at D and r the number of samples that show no growth. The sterilization dose can be calculated by

$$D = D_{10} \cdot \nabla \tag{17.18}$$

where D is the sterilization dose, D_{10} the decimal reduction dose, and ∇ the inactivation factor or sterility factor, log $N_0/$PNS. For sake of discussion, let D_{10} for the challenge organism be 2.0 kGy. The sterilizing radiation dose calculated for a prefilled syringe having an N_0 of 1×10^{-3} and a PNS of 1^{-6} would be

$$D = 2.0 \text{ kGy} \cdot \frac{\log 1 \times 10^{-3}}{1 \times 10^{-6}} = 6.0 \text{ kGy} \tag{17.19}$$

This ionizing radiation dose is much less than the 25 kGy that is routinely used. Many pharmaceutical products may be compatible at this lower radiation dose. For more information, see Jacobs (1992, 2007).

17.3.3.3.1 PAT: Gamma Radiation Exposure time is the most important variable that is controlled. In addition to measuring exposure time, dosimeters are used to confirm the delivered radiation dose.

17.3.3.4 Other Forms of Ionizing Radiation Sterilization Currently, the commercial use of electron-beam and x-ray technologies for pharmaceutical sterilization is minimal. As commercial cobalt-producing reactors age and require expensive upgrades and global radioactive transportation requirements become more costly, interest in electron-beam and x-ray sterilization has increased. The ultimate source for commercial electron-beam and x-ray sterilization is electricity, so the source of the ionizing radiation is much easier to control. It can be turned on or off compared to transporting radioactive material or dispensing and handling radioactive material such as ^{60}Co, which has a 5.27-year half-life.

Electron-beam sterilization uses high-energy electrons that travel through matter. The high-energy electrons cause the target material to eject electrons that form ion pairs and excited matter. The electron-beam energy ranges up to around 5 MeV at 6 kW of beam power vs. gamma photon radiation energy of 1.17 and 1.33 MeV. Similar to gamma irradiation, the ion pairs form reactive free radicals that can destroy the DNA of microbes and react with the drug and container closure system. In fact, one of the primary uses of electron-beam processing is to cross-link a number of polymers used in the pharmaceutical industry, such as poly(vinyl chloride), poly(vinylidene fluoride), and low- and high-density polyethylene. Electron-beam irradiation does not leave reaction residues or radioactive contaminants. Sterilization is a fast and continuous process. Compared to gamma radiation there are a number of critical process parameters that have to be optimized and controlled to maximize the kill efficiency while minimizing deleterious effects on the drug and packaging materials. The critical parameters that must be qualified and controlled are the electron-beam energy in MeV, beam power in kW, dose rate, beam distribution and angle across the conveyor, and product density and configuration. Rey (2004a) studied the effect of electron-beam, gamma-radiation, and x-ray sterilization on lyophilized vaccines and other biological products. The radiation dose of 25 kGy was delivered in seconds (electron beam) and hours (gamma radiation and x-ray) at liquid nitrogen and room temperature. The results varied depending on the drug tested. For more detailed discussions of electron-beam sterilization, see Cleland and Beck (1992) and ISO 11137 (2006a–c).

X-rays have been used for a number of years to sterilize matter. Sterilization processes use high-energy x-rays, 5 to 7.5 MeV. The electromagnetic photons emitted exhibit excellent penetration capability. The 7.5-MeV energy offers a number of advantages over the 5-MeV emission, such as a two-fold increase in production capacity at the same beam current, better penetration, and improved dose uniformity. The regulations for the higher dose are more stringent, as one would expect. X-rays can deliver a directional ionizing beam compared to the gamma-ray panoramic emission. This creates more flexibility in sterilizing varying load sizes and local shielding. Typical irradiators employ a multipass multilayer system. Pallets are treated for 30 to 60 minutes, depending on the energy and current levels used for the sterilization exposure. Radiation sterilization is discussed in detail in ISO 11137 (2006 a–c).

17.3.3.5 Filtration Sterilization Sterile filtration involves the removal of microorganisms from a fluid stream, such as air, gas, or a liquid product. Filtration is the only method discussed that is not designed to destroy the microorganisms. Filtration with a 0.2- or 0.22-μm filter removes microbes by size exclusion and entrainment by adsorption. It should be kept in mind that only a presterilized or sterile 0.2-μm filter can be used for aseptic processing. The FDA (2004) has developed performance criteria for sterilizing filters. Filter manufactures typically qualify their filters according to this guidance. It is the company's responsibility to validate the filter as a sterilizing filter. The FDA recommends direct inoculation of the test organism, *Brevundimonas diminuta* (ATCC 19146), into the drug formulation to provide an assessment of the effect of the drug product on the filter matrix and the challenge organism. Factors that can affect filter efficiency and performance are surfactants, which can affect the absorptive nature of the filter, compatibility of the formulation components with the filter, pH, filter pressure, flow rate, maximum use time, liquid temperature and viscosity, osmolality, and hydraulic shocks, to name a few. The filter efficacy can also be affected by the prefiltration bioburden and nonviable particulate level. Moist-heat filter sterilization may also affect a filter's retentive characteristics. Filter extractables and nonviable particulate levels need to be quantitated and identified where possible. Adsorption of functional excipients and the drug should be evaluated. Once a filter has been validated as a sterilizing filter, a correlation between a bubble point integrity test and the filter's microbe retention efficiency can be established. Once the correlation has been established, a bubble point integrity test can be used in the production setting to assure the filter's performance. The bubble point is based on the amount of pressure that is required to displace water from the water-saturated filter pores with air. The bubble point is based on the Laplace capillary rise equation,

$$P = \frac{4\gamma \cos \theta}{d} \tag{17.20}$$

where P is the pressure required to expel the water from the pore, γ the surface tension of the liquid in the pores (in this case, water), $\cos \theta$ the contact angle associated with the adhesive forces between the water and the membrane pore material, and d the effective diameter of the filter membrane pore.

According to equation (17.20), smaller-diameter pores develop greater hydrostatic pressure; that is, the pressure inside the pore is inversely proportional to the pore diameter. This means that greater counterpressure is required to force water out of the smaller pores. Once the air pressure forces the water out of the pore, the air flows through the pore and forms bubbles on the opposite side of the filter. The pressure at which the bubbles are free flowing (also called *viscous flow*, which is described by the Poiseulle equation) is called the *bubble point*. It should be noted equation (17.20) also predicts that *P* will decrease as the surface tension is decreased. Surfactants in the liquid product can decrease the bubble point. This is an example of why is it important to perform the correlation of bubble point with filter efficacy using the product formulated. The bubble point needs to be distinguished from diffusive airflow bubbles that result from air that dissolves under the test pressure. A pressure-hold test is routinely incorporated to check the integrity of the filter system, which includes the filter housing and the O-rings, which make an airtight fit between the filter membrane or cartridge and the filter housing. A rapid pressure drop is indicative of a leak in the filter system. The bubble point integrity test should be performed after moisture sterilization and before the product is sterile filtered. Once the sterile filtration process is completed, the filter should be retested for integrity. As a general rule, as an extra precautionary measure, redundant filters are used for the sterile filtering process. In addition, the liquid product is prefiltered to decrease bioburden and nonviable particulates. The size of the prefilter depends on the liquid being sterile filtered. For solutions, it is common to use a 0.45-μm prefilter.

Since filters sterilize by removing organisms from the fluid product, it is possible that microorganisms can grow in the pores and eventually *grow through* and penetrate the filter, causing downstream microbial contamination (Jornitz, 2006). This possibility must be kept in mind when validating a manufacturing time, filter changeover, and bioburden. As a precaution to prevent such contamination, the FDA has suggested that filling operations last no longer than 8 hours.

17.4 CONTAINER CLOSURE SYSTEM DESIGN

Injectable container closure systems (CCSs) must be sterile and protect a product from the environment and from microbiological contamination. Glass and elastomeric closures (rubber stoppers) are frequently used as part of injectable CCSs. Seal integrity is a critical product attribute. Lam and Stern (2010) have discussed techniques to assess the design factors that affect the interaction between the vials and stoppers. In addition to rubber-stoppered vials, elastomers are also used as a seal for prefilled syringes and lyophilized dual chamber vials. In these cases, the elastomer, which is in constant contact with the formulation, must maintain seal integrity but also needs to have sufficient lubrication and flexibility to move under pressure when used.

Glass and elastomers have limitations. Glass allows light transition, and elastomers are not impervious to moisture and oxygen. These materials are not completely inert, so they can interact with the formulation. The proper selection of CCSs is linked inextricably to the formulation and process. Injectable products are liquids (solutions, emulsions, liposomes, and suspensions) and dry solids are for reconstitution. APIs or critical functional excipients such as antimicrobial agents may adsorb or absorb to the elastomeric stopper or plunger material, which could lead to unacceptable losses of these compounds. Similarly, extensive extractable and leachable tests need to be performed to identify and predict the level of leachables that could become present in an injectable during storage. Extractable testing is carried out under stress temperature and solvent conditions. The purpose of an extractable test is to identify what material could possibly migrate into the product. Once the extractable materials have been identified, leachable studies can be executed. Leachable studies use more moderate test conditions and should represent those types of materials that would leach into the product during long-term storage. The placebo solution should be used as one of the "solvent" systems in the extractable study. The product formulation includes a number of excipients that could interact with the CCS. If the drug is stable under the extraction test conditions, the product formulation can be used. Typically, this is not the case, and the placebo solution is used to investigate specific interactions between the formulation and the CCS.

In each section below we discuss the U.S.P. test requirements that are often used for regulatory submissions. A general discussion of CCS regulatory requirements is presented in Chapter 13.

17.4.1 Glass Containers

U.S.P. type I, II, and III glass was discussed in Section 13.4.1.2. Most parenterals use type I glass. Molded glass and tubular glass typically have small differences in their overall composition. Silicon dioxide makes up some 70 to 80% of type I glass. Other major components are boron oxide, aluminum oxide, calcium oxide, and sodium oxide (Sanga, 1979). Molded glass dimensions are more variable than tubular glass, which is why tubular glass is preferred for lyophilization. Molded glass comes in larger volumes than does tubular glass. Glass has many advantages compared to other sterile product CCSs. It is impervious to water vapor and other gases, such as oxygen and ozone. It is not as prone to drug sorption as are many polymers. Glass typically has lower levels of extractables and leachables than those of polymeric CCSs. Like all material, glass does have drawbacks. In 2011, the FDA issued an advisory (FDA, 2011b) to drug manufacturers about a well-know problem call *glass lamellae*. Glass lamellae are thin flexible fragments of glass that are shred from the glass containers. The concern is that these lamellae may cause embolic or other cardiovascular incidents when given IV. When administered SC, there is the possibility of developing a foreign-body granuloma, local injection-site irritation, and increased risk of immunogenicity. The advisory cites a number of situations that can lead to increased glass lamellae. For example, tube glass is processed at higher temperatures and is less resistant than molded glass to high pH, to citrate and tartrate buffers, to storage time and storage temperature, and to terminal sterilization.

Other glass reactions can include leaching of such ions as sodium, aluminum, and calcium. This is most likely to occur in acidic pH solutions. As a result of leaching, the pH of the sterile solution may increase, which may accelerate degradation reactions. Leached calcium may form insoluble salts such as calcium carbonate. Aluminum can also be leached. Accumulated levels of aluminum can lead to central nervous system disorders and bone toxicity. The *Code of Federal Regulations* (21 CFR 201.323) sets an upper aluminum level limit of 25 μg/L for large-volume parenterals. Small-volume parenterals that are used in the preparation of total parenteral nutritional products also has stated on the label that the container contains no more than the labeled amount. Acid-treated glass decreases the leaching. The glass may be treated with sulfuric acid, sulfur trioxide, or sulfur dioxide and is called *dealkalinization*. The treatment creates a thin surface that contains a lower concentration of alkali oxides, especially sodium, by forming soluble alkali sulfate salts. The soluble sulfate salts are washed off the glass surface and the newly generated alkali oxide deficient surface is more chemically resistant to acid attack. A recent surface treatment involves the coating of 100% SiO_2 at a thickness of 40 to 200 nm, depending on the particular application. The technology, *plasma impulse chemical vapor deposition*, imparts a thin relatively inert hydrophobic surface that can be advantageous in protein and lyophilization applications. At high pH, the etching of the glass can occur as the alkali reacts with the glass to form soluble sodium silicate or *water glass*. The etching is visible and is a sign of a reaction with silica.

For drugs that undergo photooxidation, iron oxide can be added to the glass to provide additional ultraviolet and visible light protection. However, it should be kept in mind that leached iron can accelerate or catalyze oxidation reactions for drugs that are susceptible to oxidation. The amber glass not only has to pass the U.S.P. chemical resistance test (U.S.P. <660>, Containers—Glass), it also has to pass the U.S.P. light transmission test (U.S.P. <671>, Containers—Performance Testing). There are different transmission limits for flame-sealed and closure-sealed containers as well as container size. Flame-sealed containers are usually thinner than their closure-sealed counterpart to allow flame sealing. The larger the container size, the lower the maximum percentage transmittance allowed at *any* wavelength between 290 and 450 nm. There is a limit to the level of light protection that can be provided because parenteral containers need to be 100% visually inspected for particulates.

Glass vials, syringes, and ampoules are available. Lyophilized powders also are packaged in dual-chamber vials and dual syringes in which the freeze-dried powder is packaged with the reconstituting diluent in the same CCS. The diluent is separated by an elastomeric plug. For a dual vial, the sterile reconstituting liquid is in the top portion of the vial and the lyophilized powder is in the lower portion. To reconstitute the lyophilized powder, an activation button is pressed down, which displaces the center stopper allowing the diluent to enter the lower chamber, which starts the reconstitution process. A dual-chamber syringe operates under a similar principle. Usually, the syringe plunger is attached to the dual chamber and the plunger is gently pressed to force the diluent to enter the chamber containing the freeze-dried powder.

The rapid growth of therapeutic proteins and other parenteral products has led to the increased use of injection devices and prefilled syringes. An association between tungsten that is used in making the prefilled syringes and protein aggregation has been identified (Liu et al., 2010). Particles isolated from a prefilled syringe were determined to contain protein and tungsten. Tungsten was traced to tungsten pins used to manufacture syringe barrels. Staked needles have the needle permanently attached to the syringe barrel and are separated from the product by elastomeric material. As opposed to Luer-tip needle systems, staked needles are thought to have less tungsten contamination. The disadvantage of the staked needle system is that the user does not have a choice of needle size.

17.4.2 Plastic Containers

New cyclic olefin copolymers (Topas) and cyclic olefin polymer (Crystalline Zenith) have been developed as alternatives to glass. These plastics are clear, have no ion or metal leachables, are break resistant, have low particulate matter, and have low water vapor permeability. They readily transmit light. Crystalline Zenith has been used to develop a silicon-free plastic syringe with a prefixed needle. Plastics used for sterile products are gaining in popularity. These plastics are subject to several U.S.P. chapters: <88>, Biological Reactivity Tests, In Vivo; <661>, Containers—Plastics; and <671>, Containers—Performance Testing.

17.4.3 Elastomeric Closures

Elastomer closures for parenteral products are stoppers for vials, tips on plungers for syringes, and plugs used in dual-chamber vials and prefilled syringes. Selection of the stopper is part of the formulation development process since it can affect product stability. *Rubber stopper* is a generic term that is synonymous with "elastomer" and generally refers to a vulcanized (thermoset) or a thermoplastic elastomer or rubber. Vulcanized or thermoset rubbers undergo chemical cross-linking via the addition of sulfur compounds, peroxides, and amines as vulcanizing or curing agents. This process is not reversible, so stoppers that are molded from the thermoset process cannot be remolded. Thermoplastic material, on the other hand, forms physical cross-links in which the polymer chains intertwine to form an elastic and resilient rubber. Thermoplastics find limited use in sterile products because they undergo deformation during moist-heat sterilization. There are a number of polymers that are used in making parenteral elastomeric closures. These include butyl and halobutyl rubber (isobutylene copolymer); natural rubber (*cis*-1,4-polyisoprene); neoprene polychloroprene); nitrile (butadiene–acrylonitrile copolymer); silicone [poly(dimethylsiloxane)]; fluroelastomers [poly(tetrafluoroethylene)]; urethane (polyester isocyanate); ethylene propylenediene monomer; and polybutadiene (*cis*-polybutadiene). As a rule of thumb, highly unsaturated polymers have excellent physical properties, such as compression recovery, sealing after needle penetration, lack of coring caused by needle penetration, and resilience.

Conversely, highly unsaturated polymers have relatively poor chemical properties, including resistance to moisture vapor transmission, gas transmission, and ozone transmission. They are more reactive and less stable to light, heat, and organic cosolvents. Saturated polymers have relatively poor physical attributes, but demonstrate excellent resistance to chemical exposure. As might be expected, moderately unsaturated elastomers generally have good physical and chemical properties. Fluroroelastomers and butyl/halobutyl rubbers have good all-around physiochemical properties but are more costly than polyisoprene, which is the most commonly used stopper. Polyisoprene is very resistant to coring and is a good choice for multidose vials. Butyl rubber gives excellent barrier protection to moisture or gas transmission. Butyl, neoprene, and nitrile rubbers have good compatibility with oils.

As with other CCS components, designing elastomers is a science in and of itself. A thermoset rubber has a number of functional ingredients, such as the curing or vulcanizing agent (cross-linking agent), accelerator (reduces the cure time and can act as a cross-linking agent), activator (increases the rate of cross-linking by reacting with accelerators to form more efficient cross-linkers), antioxidants–antiozonants, plasticizers–lubricants, and fillers. Many of these components can be leached or extracted and need to meet the requirements of U.S.P. <381>, Elastomer Closures for Injections. Examples of curing agents are sulfur, peroxides, and amines. Accelerators are amines such as hexamethylene amine, thiazoles, thiurams, and sulfonamides. Stearic acid and zinc oxide act as activators when sulfur is used as a curing agent. Carbon black, calcined aluminum silicate, and talc are used as fillers. Butylated hydroxytoluene and butylated hydroxyanisole are used as antioxidants.

Pikal and Lang (1978) reported that lyophilized powders can absorb volatile sulfur and paraffin wax components of rubber closures, which can lead to haze formation when the lophilized cake is reconstituted. The authors showed that the rate of adsorption increases sharply with a decrease in air pressure. They also showed that haze formation was also a function of the stopper nature, concentration of reconstituted product, surface area, cake composition, and polymorphic form of the product. Franke et al. (1994) repoted haze formation of antibiotic powders for reconstitution. They demonstrated that slightly polar volatile compounds such as unsaturated hydrocarbons and aromatic substances were adsorbed and subsequently caused haze formation on reconstitution. On the other hand, nonpolar molecules did not adsorb to the powder mixture.

Elastomers should conform to the U.S.P. physiochemical (<381>), biological (<87>, <88>), and functional (<381>) test chapters. The *Code of Federal Regulations* has a number of applicable sections: 21 CFR: 175, Indirect Food Additives; 177 and 178, Indirect Food Additives Polymers; 182, 184, and 185, Generally Recognized as Safe; and 177.2600, Rubber Articles Intended for Repeated Use. Additional CCS information is available in articles by Hopkins (1965), Keim (1975), and Smith (2002).

17.4.4 Autoinjector and Needle-free Devices

There is an ever-increasing need to provide injectable devices that can be used by the patient at home. The pressure on hospitals to decrease hospital stays means that health care has to be transferred to the home. New therapies that require parenteral administration are also driving the need for convenient, painless, and cost-effective hand held patient injection devices. Autoinjectors and needle-free injection devices have been developed to allow self-administration of drugs outside the hospital setting. A number of advances have been made to achieve these goals. Spring and air pressure–powered injection devices have been developed that control the force and depth of needle penetration for intradermal, subcutaneous, and intramuscular injection. Newer devices now have automatic guards that prevent accidental sticks. Devices have automated retraction mechanisms as well. There are reusable and disposable systems. Reusable devices minimize the environmental impact and are gaining in popularity.

In 2010 the first needle-free injection device was approved by the FDA. This approval was soon followed by a second pediatric approval of another needle-free device for growth hormone that is given subcutaneously three times a week. Needle-free injection devices are designed for subcutaneous and intramuscular self-injection. The subcutaneous delivery is achieved by delivering a stream of liquid drug under pressure through the epidermis and into the subcutaneous tissue. The active drug reaches the subcutaneous tissue in several seconds.

17.5 RISK MANAGEMENT

Each product will have a set of critical product attributes that have major influence on the overall quality of the product. The International Conference on Harmonization (ICH, 2006) has published a quality risk management guidance that discusses the general risk management process, risk management methodologies, and tools used to assess risk. Risk management assessment and control should be developed as discussed in Chapter 13. Risk management assessment and control strategy will identify critical product attributes that need to be tested. The critical product attributes are specific to each drug product. Parenteral solutions, dispersed systems, and lyophilized products will each have their own set of critical product attributes.

17.6 ATTRIBUTE TESTS

17.6.1 Universal Tests

In addition to the specific tests listed for glass, plastics, and elastomers, there are a number of additional tests related to sterile products that can be found in U.S.P. Chapter <1>, Injections and <1211>, Sterilization and Sterility Assurance of Compendial Articles. Label claim content and impurity limits are critical product attributes. Sterility is a critical product attribute and is covered under U.S.P. <71>, Sterility Tests. The sterile integrity of the container closure system is discussed in U.S.P. <1207>, Sterile Packaging Integrity Evaluation. Terminal sterilization is reviewed in U.S.P. <1222>, Terminally Sterilized Pharmaceuticals.

For multiple-use sterile products that contain antimicrobial agents, U.S.P. <51>, Antimicrobial Effectiveness Testing <341>, and Antimicrobial Agents—Content, are applicable. The U.S.P. sets limits on the levels of preservatives for products exceeding 5 mL. The limits are 0.5% (chlorobutanol, cresol, phenol, and similar types of antimicrobial agents), 0.01% (agents containing mercury and cationic surfactants), and 0.2%, sulfur dioxide or an equivalent amount of sulfite, bisulfite, or metabisulfite).

A test for the *determination of the volume of injection in containers* is included in U.S.P. <1>. This test is done to assure that there is enough "extra fill volume" to allow removal of the appropriate dose volume—volumes in the case of a multiple-use container. Injections in cartridges and prefilled syringes are discussed. Other general U.S.P. tests include <788>, Particulate Matter in Injections, which places different particulate restrictions on small- and large-volume parenterals; <785>, Osmolality and Osmolarity; <85>, Bacterial Endotoxins Test; <791>, pH; and <631>, Color and Achromicity.

17.6.2 Constituted Products

Requirements for clarity of solution and completeness of solution are discussed in U.S.P. Chapter <1>, Constituted Solutions, and <641>, Completeness of Solution. The appearance and reconstitution time of lyophilized products should be determined. The water content of a lyophilized product is critical to its reconstitution time and shelf life. Several water content methods are discussed in U.S.P. <921>, Water Determination. The Karl Fischer direct titration and coulometric titration methods are the most common.

17.6.3 Dispersed Products

Suspension particle size distribution and globule size distribution in lipid injectable emulsions (covered in U.S.P. <729>) are important attributes to measure and control. Other product attributes are discussed in Chapter 15.

17.6.4 Biotechnology-Derived Products

Since organisms are used to produce biotechnology-derived products, there is a higher risk of microbial and viral contamination than with most noncomplex drug products. General microbiological testing is found in U.S.P. <61>, <62>, and <1111>. Specific tests: Mycoplasma Tests (<63>) and Viral Safety Evaluation of Biotechnology Products Derived from Cell Lines of Human and Animal Origin are included for official articles. The pyrogen test (<151>) and the bacterial endotoxins test are also listed.

17.7 NEW DRUG APPLICATION STABILITY ASSESSMENT

The primary purpose of stability assessment is to establish product label storage instructions and a shelf life for the product when it is stored under the labeled storage conditions. To meet a specified shelf life, the product must remain within a set of product specifications that are established for each critical product attribute. A balance must be struck between the time required to establish real-time data to support the labeled storage conditions and shelf life and the need to get important drugs to the patient as quickly as possible with the appropriate safety assurances. The regulated process allows the new drug application applicant to use appropriately designed stability studies to use real-time data and statistical approaches to extrapolate an estimated shelf life. This is discussed more extensively in Chapter 13.

The ICH guidance documents serve as a starting point for developing a stability protocol (ICH, 1996, 1997, 2003a,b, 2004, 2009a–c). Data from three batches of drug product manufactured using the proposed market formulation and process, packaged in the proposed CCS in the proposed count sizes and presentations, are placed on accelerated and long-term stability testing. It is ideal to have all three batches made at the proposed commercial production site on a commercial scale. Other FDA submission requirements are discussed in Chapter 13. Parenteral solutions, dispersable systems, and powder for reconstitution have universal stability requirements and specific dosage-form requirements.

17.7.1 Universal Tests

At initial testing, all parenteral products should meet specifications for: product description (color, odor, precipitate, clarity), two identity tests, assay for active and degradation products, sterility, bacterial endotoxins, visible particulates, U.S.P. particulate matter, osmolality, and pH. Container closure integrity testing is also required. Microbiological and chemical preservative assays are necessary for multidose products that contain preservatives. Multidose containers also need a volume of injection in containers performed at the zero stability time point. If alcohol is used as a cosolvent or preservative, it also needs to be assayed. Stability and end-of-life tests would include description, assay of active and degradation products, sterility, bacterial endotoxins, visible particulates, U.S.P. particulate matter, pH, color, and clarity. Microbial and chemical preservative testing is usually done annually and at the end of the stability program.

17.7.2 Dispersed Systems

In addition to the initial time-zero universal tests, dispersed system initial tests would include particle size distribution, dose uniformity, dissolution, syringeability, viscosity, sedimentation/creaming rate, specific gravity, caking, and redispersability. Stability and end-of-life tests should include all the tests mentioned above.

17.7.3 Constituted Solids for Solution

Initial and stability tests should deal with moisture content, constituted strength, constitution time, and constitution pH. Simulated-use tests should also be considered. The stability of the constituted product would also have to be confirmed. The length of the stability study and the particular conditions would depend on the intended use of the product.

17.7.4 Prefilled Syringes

Functional testing of prefilled syringes should be performed at the initial time point, throughout the stability program.

17.7.5 Stability Protocols

Stability protocols (FDA, 2003) vary with dosage form and the proposed storage condition. Table 17.12 provides a generic parenteral stability protocol for a product to be stored at controlled room temperature. A typical freeze–thaw cycle is −20 to 30°C at 7-day cycles for 28 days (one month). A usual temperature cycle study involves temperature cycling from 5 to 40°C every 6 hours for a month. The freeze–thaw cycles and cycling studies are adjusted for type of dosage form and proposed storage conditions. The freeze–thaw cycles for disperse systems are generally less harsh. A general freeze–thaw cycle for emulsions is 2 days at −10°C, then 2 days at 25°C for three cycles. Cycling is often 5 to 25°C for 6 hours for 12 days.

The stability of constituted powder also needs to be addressed. The protocol should simulate actual use of the reconstituted protocol, including the storage temperature, number of doses, and constituted shelf life. Extra samples should be kept

TABLE 17.12 NDA Stability Protocol

Product	Container/Closure and Supplier	API Process
Product lot	Package lot	API lot
Batch size	Batch size	Batch size
Manufacturer	Packager	Manufacturer
Manufacturer site	Packaging site	Site of manufacture
Manufacture date	Package date	Manufacture date

Storage (°C/% RH)	Time (months)									
	1	3	6	9	12	18	24	36	48	60
25/60[a,b]		×	×	×	×	×	×	×	×[c]	(×)[c]
40/75[a,b]	×	×								
Light[d]	×									
Freeze–thaw[e]	×									
Cycle[f]	×									

[a] °C ± 2/% RH ± 5%

[b] Liquid samples are stored upright and horizontal to expose of the liquid to the stopper.

[c] (×), optional testing.

[d] testing per ICH guidelines.

[e] Seven days at −20°C, 7 days at 30°C, 7 days at −20°C, and 7 days at 30°C.

[f] 5 to 40°C every 6 hours for a month.

stable to allow for full retesting at one or two time points. Shelf-life estimation and limits for impurities are discussed in Chapter 13. The API shelf life is generally expected to be 95 to 105% of the label claim.

REFERENCES

Adams GDJ, Ramsay Jr. Optimizing the lyophilization cycle and the consequences of collapse on the pharmaceutical acceptability of *Erwinia* L-asparaginase. J. Pharm. Sci. 1996;85:1301–1305.

Adler JP, Proffitt RT. Development, characterization, efficacy and mode of action of AmBisome, a unilamellar liposomal formulation of amphotericin B. J. Liposome Res. 1993;3:429–450.

Akers MJ. Preformulation screening of antioxidant efficiency in parenteral solutions. J. Parenter. Drug Assoc. 1979;33:346–356.

———. Antioxidants in pharmaceutical product. J. Parenter. Sci. Technol. 1982;36:222–227.

———. Parenteral preparations. In: Troy DB, ed. *Remington: The Science and Practice of Pharmacy*, 21st ed. Philadelphia: Lippincott Williams & Wilkins; 2006. pp. 802–836.

Akers MJ, DeFelippis MR. Peptides and proteins as parenteral solutions. In: Frokjaer S, Hovgaard L, eds. *Pharmaceutical Formulation: Development of Peptides and Proteins*. London: Taylor & Francis; 2000. p. 167.

Akers MJ, Ketron KM, Thompson BR. *F* value requirements for the destruction of endotoxin in the validation of dry heat sterilization/depyrogenation cycles. J. Parenter. Sci. Technol. 1982;38:23–27.

Akers MJ, Boand A, Brinkley D. Preformulation method for parenteral preservative efficacy evaluation. J. Pharm. Sci. 1984;73:903–907.

Akers MJ, Fites AL, Robison RL. Formulation design and development of parenteral suspensions. J. Parenter. Sci. Technol. 1987;41:88–98.

Akkar A, Namsolleck P, Blaut M, Müller RH. Solubilizing poorly soluble antimycotic agents by emulsification via a solvent-free process. AAPS PharmSciTech [serial online] 2004;5(1): article 24. http://www.aapspharmscitech.org.

Anhordoquy TJ, Carpenter JF. Polymers protect lactate dehydrogenase during freeze-drying by inhibiting dissociation in the frozen state. Arch. Biochem. Biophys. 1996;332:231–238.

Apte SP, Ogwu SO. A review and classification of emerging excipients in parenteral medications. Pharm. Technol. 2003;Mar.: 46–60.

Asker AF. Photostability studies of solutions and methods of preventing photodegradation. In: Piechocki JT, Thoma K, eds. *Pharmaceutical Photostability and Stabilization Technology*, Vol. 163. New York: Informa Healthcare; 2007. pp. 345–378.

Asker AF, Ferdous AJ. Photodegradation of furosemide solutions. PDA J. Pharm. Sci. Technol. 1996;50(3): 158–162.

ASTM D93-10a. Standard test methods for flash point by Pensky–Martens closed-cup tester. http://www.astm.org/Standards/D93.htm. Accessed Dec. 2011.

Bell LN, Hageman MJ, Muraoka LM. Thermally induced denaturation of lyophilized bovine somatotropin and lysozyme as impacted by moisture and excipients. J. Pharm. Sci. 1995;84:707–712.

Bellows RJ, King CJ. Freeze-drying of aqueous solutions: maximum allowable operating temperature. Cryobiology 1972;9:559–561.

Bindschaedler C. Lyophilization process validation. In: Rey L, May JC, eds. *Freeze-Drying/Lyophilization of Pharmaceutical and Biological Products*. 2nd ed. Drugs and the Pharmaceutical Sciences, Vol. 137. New York: Marcel Dekker; 2004. pp. 535–573.

Blaug SM, Grant DE. Kinetics of degradation of the parabens. J. Soc. Cosmet. Chem. 1974;25:495–506.

Blue J, Yoder H. Successful lyophilization development of protein therapeutics. Am. Pharm. Rev. 2009;12(1): 90–94.

Brazeau GA, Fung H-L. Physicochemical properties of binary organic cosolvent–water mixtures and their relationships to muscle damage following intramuscular injection. J. Parenter. Sci. Technol. 1989;43:144–149.

Broxton TJ. Micellar catalysis of organic reactions: VIII. Kinetic studies of the hydrolysis of 2-acetyloxybenzoic acid (aspirin) in the presence of micelles. Aust. J. Chem. 1982;35:1357–1363.

Burgess DJ, ed. *Injectable Dispersed Systems: Formulation, Processing, and Performance*. Drugs and the Pharmaceutical Sciences, Vol. 149. Boca Raton, FL: Taylor & Francis; 2005.

Burgess DJ, Crommelin DJA, Hussain AS, Chen M-L. Assuring quality and performance of sustained and controlled release parenterals: EUFEPS Workshop Report. AAPS PharmSci 2004;6(1): art. 11. http://www.aapspharmsci.org.

Cadwallader DE. Erytrocyte stability in ethanol–saline solutions. Br. J. Anaesth. 1978;50:81.

Cadwallader DE, Wickliffe BW, Smith BL. Behavior of erythrocytes in various solvent systems: II. Effect of temperature and various substances on water–glycerin and water–propylene glycol solutions. J. Pharm. Sci. 1964;53:927–931.

Callahan JC, Cleary GW, Elefant M, Kaplan G, Kensler T, Nash RA. Equilibrium moisture content of pharmaceutical excipients. Drug Dev. Ind. Pharm. 1982;8:355–369.

Campbell AN, Smith NO. *Alexander Findlay: The Phase Rule and Its Applications*, 9th ed. New York: Dover; 1951. p. 141.

Carpenter JF, Pikal MJ, Chang BS, Randolf TW. Rational design of stable lyophilized protein formulations: some practical advice. Pharm. Res. 1997;14:969–975.

Carpenter JF, Izutsu K, Randolph TW. Freezing- and drying-induced perturbations of protein structure and mechanisms of protein protection by stabilizing additives. In: Rey L, May JC, editors, *Freeze-Drying/Lyophilization of Pharmaceutical and Biological Products*, 2nd ed. Drugs and the Pharmaceutical Sciences, Vol. 137. New York: Marcel Dekker; 2004. pp. 147–186.

Carstensen JT, Mehta A. Scale-up factors in the manufacture of solution dosage forms. Pharm. Technol. 1982;June: 64–77.

Chang BS, Randall CS. Use of subambient thermal analysis to optimize protein lyophilization. Cryobiology 1992;29:632–656.

Chinnian D, Asker AF. Photostability profiles of minoxidil solutions. PDA J. Pharm. Sci. Technol. 1996;50(2): 94–98.

Christrup LL, Moss J, Steffansen B. Prodrugs. In: Swarbrick J, Boylan JC, eds. *Encyclopedia of Pharmaceutical Technology*, Vol. 13. New York: Marcel Dekker; 1996. pp. 45–46.

Cleland MR, Beck JA. Electron beam sterilization. In: Swarbrick J, Boylan JC, eds. *Encyclopedia of Pharmaceutical Technology*, Vol. 5, New York: Marcel Dekker; 1992. pp. 105–136.

Code of Federal Regulations. 21 CFR 201.22. Prescription drugs containing sulfites; required warning statements. http://www.accessdata.fda.gov/scripts/cdrh/cfdocs/cfcfr/CFRSearch.cfm?fr = 201.22. Revised Apr. 1, 2010. Accessed Dec. 2011.

———— 21 CFR 201.323. Aluminum in large and small volume parenterals used in total parenteral nutrition. http://www.accessdata.fda.gov/scripts/cdrh/dfdoscs/cfcfr/CFRSearch.cfm?fr = 301.33. Revised Apr. 1, 2010. Accessed Apr. 2011.

———— 21 CFR 310.509(b). Parenteral drug products in plastic containers. http://edocket.access.gpo.gov/cfr_2007/aprqtr/pdf/21cfr310.509.pdf. Apr. 1, 2007 edition. Accessed Dec. 2011.

Costantino HR. Excipients for use in lyophilized pharmaceutical peptide, protein, and other bioproducts. In: Costantino HR, Pikal MJ, eds. *Lyophilization of Biopharmaceuticals*. Arlington, VA: American Association of Pharmaceutical Scientists; 2004. pp. 139–228.

Costantino HR, Pikal MJ, eds. *Lyophilization of Biopharmaceuticals*. Arlington, VA: American Association of Pharmaceutical Scientists; 2004.

Crooks MJ, Brown KF. Competitive interaction of preservative mixtures with cetomacrogol. J. Pharm. Pharmacal. 1974;26:235–242.

Crowe JH, Crowe LM, Carpenter JF. Preserving dry biomaterials: the water replacement hypothesis. Biopharm 1993;6(2): 40–43.

Crowe JH, Carpenter JF, Crowe LM. The role of vitrification in anhydrobiosis. Annu. Rev. Physiol. 1998;60:73–103.

De Beer TRM, Vercruysse P, Burggraeve A, Quinten T, Ouyang J, Zhang X, Vervaet C, Remon JP, Baeyens WRG. In-line and real-time process monitoring of a freeze drying process using Raman and NIR spectroscopy as complementary process analytical technology (PAT) tools. J. Pharm. Sci. 2009;98:3430–3446.

De Castro L. Method of preparing stable drug nanoparticles. U.S. Patent 5534270. 1996.

Denyer S, Hodges NA, Gorman SP. *Hugo and Russell's Pharmaceutical Microbiology*. Oxford, UK: Blackwell Science; 2004. pp. 50–62.

Devarakonda B, Hill RA, Liebenberg W, Brits M, de Villiers MM. Comparison of the aqueous solubilization of practically insoluble niclosamide by polyamidoamine (PAMAM) dendrimers and cyclodextrins. Int. J. Pharm. 2005;304:193–209.

Duddu SP, Dal Monte PR. Effect of glass transition temperature on the stability of lyophilized formulations containing a chimeric therapeutic monoclonal antibody. Pharm. Res. 1997;14; 591–595.

Elin RJ, Wolff SM, McAdam KPWJ, Chedid L, Audibert F, Bernard C, Oberling F. Properties of reference *Escherichia coli* endotoxin and its phthalylated derivative in humans. J. Infect. Dis. 1981;144:329–336.

FDA 1996. Guidance for industry Q1B—Photostability testing of new drug substances and products.

———. 2003. Guidance for industry Q1A—Stability testing of new drug substances and products. U.S. Department of Health and Human Services, Food and Drug Administration Center for Drug Evaluation and Research and Center for Biologics Evaluation and Research. Nov. 2003. http://www.fda.gov/downloads/regulatoryinformation/guidances/ucm128204.pdf. Accessed Dec. 2011.

———. 2004. Guidance for industry: Sterile drug products produced by aseptic processing—current good manufacturing practice. U.S. Department of Health and Human Services, Food and Drug Administration Center for Drug Evaluation and Research and Center for Biologics Evaluation and Research. Sept. 2004. http://www.fda.gov/downloads/Drugs/GuidanceComplianceRegulatoryInformation/Guidances/ucm070342.pdf. Accessed Dec. 2011.

———. 2010. Guidance for industry: Submission of documentation in applications for parametric release of human and veterinary drug products terminally sterilized by moist heat process. U.S. Department of Health and Human Services, Food and Drug Administration Center for Drug Evaluation Research, Center for Veterinary Medicine, and Center for Biologics Evaluation and Research. Feb. 2010. http://www.fda.gov/downloads/Drugs/GuidanceComplianceRegulatoryInformation/Guidances/ucm072180.pdf. Accessed Dec. 2011.

——— 2011a. Inactive ingredients in FDA approved drugs. FDA Center for Drug Evaluation and Research, Office of Generic Drugs. Jan. 12, 2011. http://www.accessdata.fda.gov/scripts/cder/iig/index.cfm. Accessed Dec. 2011.

———. 2011b. Advisory to drug manufacturers: Formation of glass lamellae in certain injectable drugs. http://www.fda.gov/Drugs/DrugSafety/ucm248490.htm. Accessed Dec. 2011.

Franke H, Hencken P, Ross G, Kreuter J. Influence of volatile rubber stopper components on the clearness of parenteral antibiotics stored in multiple or single dose containers after reconstitution with water. Eur. J. Pharm. Biopharm. 1994;40:379–387.

Franks F. Freeze drying: from empiricism to predictability. Cryo-Letters 1990;11:93–110.

Gefter ML, Barker N, Musso G, Molineaux CJ. Pharmaceutical formulations for sustained drug delivery. U.S. Patent 6180608. Jan. 2001.

Gieseler H, Kessler WJ, Finson M, Davis SJ, Mulhall PA, Bons V, Debo DJ, Pikal MJ. Evaluation of tunable diode laser absorption spectroscopy for in-process water vapor flux measurements during freeze drying. J. Pharm. Sci. 2007a;96:1776–1793.

Gieseler H, Kramer T, Pikal MJ. Use of manometric temperature measurement (MTM) and SMART™ freeze dryer technology for development of an optimized freeze-drying cycle. J. Pharm. Sci. 2007b;96:3402–3418.

Gorsky I. Parenteral drug scale-up. In: Levin M, ed. *Pharmaceutical Process Scale-up*, 2nd ed. Drugs and the Pharmaceutical Sciences, Vol. 157. New York: Marcel Dekker; 2006. pp. 71–87.

Green AR, Guillory JK. Heptakis(2,6-di-O—methyl)-β-cyclodextrin complexation with the antitumor agent chlorambucil. J. Pharm. Sci. 1989;78:427–431.

Greenhill JV. Photodecomposition of drugs. In: Swarbrick J, Boylan JC, eds. *Encyclopedia of Pharmaceutical Technology*, Vol. 12. New York: Marcel Dekker; 1995. pp. 105–135.

Hagman DE. Sterilization. In: Troy DB, ed. *Remington: The Science and Practice of Pharmacy*, 21st ed. Philadelphia: Lippincott Williams & Wilkins; 2006. pp. 776–797.

Hamid IA, Parrott, EL. Effect of temperature on solubilization and hydrolytic degradation of solubilized benzocaine and homatropine. J. Pharm. Sci. 1971;60:901–906.

Hanna C. Non-aqueous oil delivery system for ophthalmic drugs. U.S. Patent Application US2010/0323978A1, 2010.

Hansrani PK, Davis SS, Grove MJ. The preparation and properties of sterile intravenous emulsions. J. Parenter. Sci. Technol. 1983;37:145–149.

Heimlich KR, Martin A. A kinetic study of glucose degradation in acid solution. Am. Pharm. Assoc. Sci. Educ. 1960;49:592–597.

Hopkins GH. Elastomeric closures for pharmaceutical packaging. J. Pharm. Sci. 1965;54: 138–143.

ICH. 1996. Q1B: Photostability testing of new drug substances and products Nov. 1996. http://www.fda.gov/Drugs/GuidanceComplianceRegulatoryInformation/Guidances/ucm065005. Accessed Dec. 2011

_____. 1997. Q1C: Stability testing for new dosage forms. June 1997. http://www.fda.gov/Drugs/ GuidanceComplianceRegulatoryInformation/Guidances/ucm065005. Accessed Dec. 2011.

_____. 2003a. Q1D: Bracketing and matrixing designs for stability testing of new drug substances and products Jan. 2003. http://www.fda.gov/Drigs/GuidanceComplianceRegulatoryInformation/ Guidances/ucm065005. Accessed Dec. 2011.

_____. 2003b. Q1A (R2): Stability testing of new drug substances and products. Nov. 2003. http://www.fda.gov/Drigs/GuidanceComplianceRegulatoryInformation/Guidances/ucm065005. Accessed Dec. 2011.

_____. 2004. Q1E: Evaluation of stability data. June 2004. http://www.fda.gov/Drigs/Guidance ComplianceRegulatoryInformation/Guidances/ucm065005. Accessed Dec. 2011.

_____. 2006. Q9: Quality risk management. June 2006. http://www.fda.gov/Drigs/GuidanceCompliance RegulatoryInformation/Guidances/ucm065005. Accessed Dec. 2011.

_____. 2009a. Q4B, Annex 4A: Microbiological examination of non-sterile products— microbial enumeration tests. General chapter. Apr. 2009. http://www.fda.gov/Drigs/Guidance ComplianceRegulatoryInformation/Guidances/ucm065005. Accessed Dec. 2011.

_____. 2009b. Q4B, Annex 4B: Microbiologial examination of non-sterile products— tests for specified micro-organisms. General chapter. Apr. 2009. http://www.fda.gov/Drigs/ GuidanceComplianceRegulatoryInformation/Guidances/ucm065005. Accessed Dec. 2011.

_____. 2009c. Q4B, Annex 4C. Microbiological examination of non-sterile products—acceptance criteria for pharmaceutical preparations and substances for pharmaceutical general use. Apr. 2009. http://www.fda.gov/Drigs/GuidanceComplianceRegulatoryInformation/Guidances/ucm065005. Accessed Dec. 2011.

ISO/DIS 11140-1. 2005. Sterilization of health care products—chemical indicators. Part 1: General requirements.

_____. 11137. 2006a. Sterilization of healthcare products: radiation, Part 1. Requirements for development, validation, and routine control of sterilization process for medical devices. Association for the Advancement of Medical Instrumentation.

_____. 2006b. Sterilization of healthcare products: radiation, Part 2. Establishing the sterilization dose. Association for the Advancement of Medical Instrumentation.

_____. 2006c. Sterilization of healthcare products: radiation, Part 3. Guidance on dosimeteric aspects. Association for the Advancement of Medical Instrumentation.

Jacobs GP. Gamma radiation sterilization. In: Swarbrick J, Boylan JC, eds. *Encyclopedia of Pharmaceutical Technology*, Vol. 6, New York: Marcel Dekker, 1992. pp. 305–332.

_____. Radiation sterilization of parenterals. Pharm Technol. May 2007. http://pharmtech. findpharma.com/pharmtech/Aseptic+processing/Radiation-Sterilization-of-Parenterals/Article Standard/Article/detail/423948. Accessed Dec. 2011.

Johnson D, Gu LC. Autoxidation and antioxidants. In: Swarbrick J, Boylan JC, eds. *Encyclopedia of Pharmaceutical Technology*, Vol. 1. New York: Marcel Dekker, 1998. pp. 415–449.

Johnson OL, Ganmukhi MM, Berstein H, Auer H, Khan MA. Composition for sustained release of human growth hormone. U.S. Patent 5654010. 1997a.

Johnson OL, Jaworowicz W, Cleland JL, Bailey L, Charnis M, Duenas E, Wu C, Shepard D, Magil S, Last T, Jones AJS, Putney SD. The stabilization and encapsulation of human growth hormone into biodegradable microspheres. Pharm. Res. 1997b;14:730–735.

Jornitz MW, Meltzer TH. Grow-through and penetration of 0.2/0.22 "sterilizing" membranes. Pharm. Technol. 2006; 3. http://pharmtech.findparma.com/pharmtech/Aseptic+processing/Grow-Through-and-Penetration-of-the-02022-Sterliz/ArticleStandard/Article/detail/31124. Accessed Apr. 2011.

Jouyban A. *Handbook of Solubility Data for Pharmaceuticals*. Boca Raton, FL: CRC Press; 2010.

Kamada A, Yata N, Kubo K, Arakawa M. Stability of p-hydroxybenzoic acid in an acidic medium. Chem. Pharm. Bull. 1973;21:2073–2076.

Keim FM. Design and development of elastomeric closure formulation. Bull. Parenter. Drug Assoc. 1975;29:46–53.

Kumar V, Kalonia DS. Removal of peroxides in polyethylene glycols by vacuum drying: implications in the stability of biotech and pharmaceutical formulations. AAPS PharmSciTech 2006;7:art. 62. doi:10.1208/pt070362.

Lam P, Stern A. Visualization techniques for assessing design factors that affect the interaction between pharmaceutical vials and stoppers. PDA J. Pharm. Sci. Technol. 2010;64:182–187.

Larsen DB, Parshad H, Fredholt K, Larsen C. Characteristics of drug substances in oily solutions: drug release rate, partitioning, and solubility. Int. J. Pharm. 2002;232:107–117.

Lee RW. Case study: development and scale-up of NanoCrystal® particles. In: Burgess DJ, ed. *Injectable Dispersed Systems: Formulation, Processing, and Performance*. Drugs and the Pharmaceutical Sciences, Vol. 149. Boca Raton, FL: Taylor & Francis; 2005. pp. 355–370.

Leppänen J, Huuskonen J, Savolainen J, Nevalainen T, Taipale H, Vepsäläinen J, Gynther J, Järvinen T. Synthesis of a water-soluble prodrug of entacapone. Bioorg. Med. Chem. Lett. 2000;10: 1967–1969.

Levine H, Slade L. Another view of trehalose for drying and stabilizing biological materials. Biopharm 1992;5(5): 36–40.

Liapis AI, Bruttini R. A theory for the primary and secondary drying stages of the freeze-drying stages of the pharmaceutical crystalline and amorphous solutes: comparison between experimental data and theory. Sep. Technol. 1994;4:144–155.

———. Freeze-drying of pharmaceutical crystalline and amorphous solutes in vials: dynamic multi-dimensional models of the primary and secondary drying stages and qualitative features of the moving interface. Dry. Technol. 1995;13:43–72.

Liapis AI, Pikal MJ, Bruttini R. Research and development needs and opportunities in freeze drying. Dry. Technol. 1996;14:1265–1300.

Liu R, ed. *Water-Insoluble Drug Formulation*. 2nd ed. Boca Raton; FL: CRC Press; 2008.

Liu W, Swift R, Torracca G, Nashed-Samuel Y, Wen Z-Q, Jiang Y, Vance A, Mire-Sluis A, Freund E, Davis J, Narhi L. Root cause analysis of tungsten-induced protein aggregation in prefilled syringes. PDA J. Pharm. Sci. Technol. 2010;64:11–19.

Liversidge GG, Cundy KC, Bishop JF, Czekai DA. Surface modified drug nanoparticles. U.S. Patent 5145684. Sep. 1992.

Loftsson T, Brewster ME. Pharmaceutical applications of cyclodextrins: 1. Drug solubilization and stabilization. J. Pharm. Sci. 1996;85:1017–1025.

Lovell MW, Johnson HW, Hui H-W, Cannon JB, Gupta PK, Hsu CC. Less-painful emulsion formulations for intravenous administration of clarithromycin. Int. J. Pharm. 1994;109:45–57.

Lyons RT. Case study: formulation,development, and scale-up production of an injectable perfluorocarbon emulsion. In: Burgess DJ, ed. *Injectable Dispersed Systems: Formulation, Processing, and Performance*. Drugs and the Pharmaceutical Sciences, Vol. 149. Boca Raton, FL: Taylor & Francis; 2005. pp. 371–392.

MacKenzie AP. The physico-chemical basis for the freeze-drying process. Dev. Biol. Stand. 1977;36:51–67.

Martin FJ. Case study: Doxil, the development of pegylated liposomal doxorubicin. In: Burgess DJ, ed. *Injectable Dispersed Systems: Formulation, Processing, and Performance*. Drugs and the Pharmaceutical Sciences, Vol. 149. Boca Raton, FL: Taylor & Francis; 2005. pp. 427–480.

Martinez MN, Rathbone MJ, Burgess D, Huynh M. Breakout session summary from AAPS/CRS joint workshop on critical variables in the in vitro and in vivo performance of parenteral sustained release products. J Control. Release 2010;142:2–7.

Mayer LD, Bally MB, Cullis PR, Ginsberg RS. Liposomal antineoplastic agent compositions. U.S. Patent 5795589. 1998.

McNally EJ, Hastedt JE, eds. *Protein Formulation and Delivery*, 2nd ed. Drugs and the Pharmaceutical Sciences, Vol. 175. New York: Marcel Dekker; 2008.

Mikrut B. Case study: formulation of an intravenous fat emulsion. In: Burgess DJ, ed. *Injectable Dispersed Systems: Formulation, Processing, and Performance*. Drugs and the Pharmaceutical Sciences, Vol. 149. Boca Raton, FL: Taylor & Francis; 2005. pp. 415–425.

Modi A, Tayade P. A comparative solubility enhancement profile of valdecoxib with different solubilization approaches. Indian J. Pharm. Sci. 2007;69:274–278.

Nail SL, Gatlin LA. Freeze drying: principles and practice. In: Avis KE, Lieberman HA, Lachman L. eds. *Pharmaceutical Dosage Forms: Parenteral Medications*, Vol. 2. New York: Marcel Dekker; 1993. pp. 163–233.

Nash RA. The oxidation potentials of antioxidants in drug stabilization. Am. J. Pharm. 1958;130:152–164.

Nema S, Washkuhn RJ, Brendel RJ. Excipients and their use in injectable products. PDA J. Pharm. Sci. Technol. 1997;51:166–171.

New RRC, Kirby CJ. Solubilisation of hydrophilic drugs in oily formulations. Adv. Drug Deliv. Rev. 1997;25:59–69.

Okada H, Ogawa Y, Yashiki T. Prolonged release microcapsule and its production. U.S. Patent 4652441. 1987.

O'Neill JL. Sterilization of solid non-electrolyte medicinal agents employing sodium chloride. U.S. Patent 3962430. 1976.

Overcashier DE, Patapoff TW, Hsu CC. Lyophilization of protein formulations in vials: investigation of the relationship between resistance to vapor flow during primary drying and small-scale product collapse. J. Pharm. Sci. 1999;86:455–459.

Pai SA, Rivankar SH, Sudhakar K. Amphotericin B structure emulsion. U.S. Patent 7053061 B2. 2006.

Pearlman R, Wang YJ, eds. *Formulation, Characterization, and Stability of Protein Drugs: Case Studies*, Vol. 9. New York: Plenum Press; 1996.

Pezzuto JM, Park EJ. Autoxidation and antioxidants. In: Swarbrick J, Boylan JC, eds. *Encyclopedia of Pharmaceutical Technology* 2nd ed. Vol. 1. New York: Marcel Dekker; 2002. pp. 97–113.

Pharmaceutical Inspection Convention Pharmaceutical Inspection Co-Operative Scheme. Recommendation on guidance on parametric release. http://www.picscheme.org/publication.php?id = 8. Accessed Sept. 25, 2007.

Pikal MJ, Use of laboratory data in freeze drying process design: Heat and mass transfer coefficients and the computer simulation of freeze drying. J. Parenter. Sci. Technol. 1985;39:115–138.

⸺⸺⸺. Mechanisms of protein stabilization during freeze-drying and storage: the relative importance of thermodynamic stabilization and glassy state relaxation dynamics. In: Rey L, May JC, eds. *Freeze-Drying/Lyophilization of Pharmaceutical and Biological Products*, 2nd ed. Drugs and the Pharmaceutical Sciences, Vol. 137. New York: Marcel Dekker; 2004. pp. 63–107.

⸺⸺⸺. Lang JE. Rubber closures as a source of haze in freeze dried parenterals: test methodology for closure evaluation. J. Parenter. Drug Assoc. 1978;32:162–173.

⸺⸺⸺. Shah S. The collapse temperature in freeze drying: Dependence on measurement methodology and rate of water removal from the glassy phase. Int. J. Pharm. 1990;62:165–186.

Pisano FD, Kostenbauder HB. Interaction of preservatives with macromolecules: II. Correlation of binding data with required preservative concentrations of *p*-hydroxybenzoates in the presence of Tween 80. J. Am. Pharm. Assoc. Sci. Ed uc. 1959;48:310–314.

Powell MF, Nguyen T, Baloian L. Compendium of excipients for parenteral formulations. PDA J. Pharm. Sci. Technol. 1998;52:238–311.

Prankerd RJ, Stella VJ. The use of oil-in-water emulsions as a vehicle for parenteral drug administration. J. Parenter. Sci. Technol. 1990;44:139–149.

Pretrelski SJ, Pikal KA, Arakawa T. Optimization of lyophilization conditions for recombinant human interleukin-2 by dried-state conformation analysis using Fourier-transform infrared spectroscopy. Pharm. Res. 1995;12:1250–1259.

Proffitt RT, Adler-Moore J, Chiang S-M. Amphotericin B liposome preparation. U.S. Patent 6770290. Aug. 2004.

Pryor FM, Gidal B, Ramsay RE, DeToledo J, Morgan RO. Fosphenytoin: pharmacokinetics and tolerance of intramuscular loading doses. Epilepsia 2001;42:245–250.

Radd BL, Newman AC, Fegely BJ, Chrzanowski FA, Lichten JL, Walkling WD. Development of haloperidol in oil injection formulations. J. Parenter. Sci. Technol. 1985;39:48–51.

Radwan M. In vivo screening model for excipients and vehicles used in subcutaneous injections. Drug Dev. Ind. Pharm. 1994;20:2753–2762.

Rajagopalan N, Dicken CM, Ravin LJ, Sternson LA. A study of the solubility of amphotericin B in nonaqueous solvent systems. J. Parenter. Sci. Technol. 1988;42:97–102.

Raval NN, Parrott EL. Hydrolysis of methylparaben. J. Pharm. Sci. 1967;56:274–275.

Ray WJ Jr, Puvathingal JM. A simple procedure for removing contaminating aldehydes and peroxides from aqueous solutions of polyethylene glycols and of nonionic detergents that are based on the polyoxyethylene linkage. Anal. Biochem. 1985;146:307–312.

Reed KW, Yalkowsy SH. Lysis of human red blood cells in the presence of various cosolvents. J. Parenter. Sci. Technol. 1985;39:64–68.

Ren S, Park M-J, Sah H, Lee B-J. Effect of pharmaceutical excipients on aqueous stability of rabeprazole sodium. Int. J. Pharm. 2008;359:197–204.

Repta AJ. Formulation of investigational anticancer drugs. In: Breimer DD, Speiser P, eds. *Topics in Pharmaceutical Sciences*. Elsevier/North-Holland Biomedical Press; 1981. pp. 131–151.

Rey L, May JC. A new development: irradiation of freeze-dried vaccine and other selected biological products. In: Rey L, May JC, eds. *Freeze-Drying/Lyophilization of Pharmaceutical and Biological Products*, 2nd ed. Drug and the Pharmaceutical Sciences, Vol. 137. New York: Marcel Dekker; 2004. pp. 575–185.

Ryde NP, Ruddy SB. Solid dose nanoparticle compostions comprising a synergistic combination of a polymeric surface stabilizer and dioctyl sodium sulfosuccinate. U.S. Patent 6375986. 2002.

Sadikoglu H, Liapis AI. Mathematical modeling of the primary and secondary drying stages of bulk solution. Drying Technol. 1997;15:791–810.

Sanga SV. Review of glass types available for packaging parenteral solutions. J. Parenter. Drug Assoc. 1979;33:61–67.

Schroeter LC. Sulfurous acid salts as pharmaceutical antioxidants. J. Pharm. Sci. 1961;50:891–901.

––––––. Kinetics of air oxidation of sulfurous acid salts. J. Pharm. Sci. 1963;52:559–563.

Seedher N, Kanojia M. Micellar solubilization of some poorly soluble antidiabetic drugs: a technical note. AAPS PharmSciTech 2008;9:431–436.

Segal R, Azaz E, Donbrow M. Peroxide removal from non-ionic surfactants. J. Pharm. Pharmacol. 1979;31:39–40.

Sims EE, Worthington HEC. Formulation studies on certain oily injection products. Int. J. Pharm. 1985;24:287–296.

Sinko PJ. Chemical kinetics and stability. In: *Martin's Physical Pharmacy and Pharmaceutical Sciences: Physical Chemical and Biopharmaceutical Principles in the Pharmaceutical Sciences*. Philadelphia: Lippincott Williams & Wilkins; 2006. pp. 415–416.

Slade I, Levine H, Finley JW. Protein–water interactions: water as a plactisizer of gluten and other protein polymers. In: Philips RD, Finley JW, eds. *Protein Quality and the Effects of Processing*. New York: Marcel Dekker; 1989. pp. 9–124.

Slaveska R, Spirevska I, Stafilov T, Ristov T. The content of trace metals in some herbal teas and their aqueous extracts. Acta Pharm. 1998;48:201–209.

Smith BL, Cadwallader DE. Behavior of erythrocytes in various solvent systems: III. water–polyethylene glycols. J. Pharm. Sci. 1967;56:351–355.

Smith EJ. Elastomeric components for the pharmaceutical industry. In: Swarbrick J, Boylan JC, eds. *Encyclopedia of Pharmaceutical Technology*, 2nd ed., Vol. 1. New York: Marcel Dekker; 2002. pp. 1050–1065.

Stahl PH, Wermuth CG, eds. *Handbook of Pharmaceutical Salts: Properties, Selection, and Use*. Zürich; Switzerland: Verlag Helvetica Cimica; Wienheim; Germany: Acta Wiley-VCH; 2002.

Strickley RG. Parenteral formulations of small molecules therapeutics marketed in the United States (1999): I. PDA J. Pharm. Sci. Technol. 1999;53:324–349.

_____. Parenteral formulations of small molecules therapeutics marketed in the United States (1999): II. PDA J. Pharm. Sci. Technol. 2000a;54:69–96.

_____. Parenteral formulations of small molecules therapeutics marketed in the United States (1999): III. PDA J. Pharm. Sci. Technol. 2000b;54:152–169.

_____. Solubilizing excipients in oral and injectable formulations. Pharm. Res. 2004;21:201–230.

Sunderland VB, Watts DW. Kinetics of the degradation of methyl, ethyl, and n-propyl 4-hydroxybenzoate esters in aqueous solution. Int. J. Pharm. 1984;19:1–15.

Tang XC, Pikal MJ. Design of freeze-drying processes for pharmaceuticals: practical advice. Pharm. Res. 2004;21:191–200.

Tang X, Nail SL, Pikal MJ. Freeze-drying process design by manometric temperature measure: design of a smart freeze-dryer. Pharm. Res. 2005;22:685–700.

_____. Evaluation of manometric temperature measurement, a process analytical technology tool for freeze-drying: I. Product temperature measurement. AAPS PharmSciTech 2006a;7(1): Art. 14. http://www.aapspharmscitech.org.

_____. Evaluation of manometric temperature measurement, a process analytical technology tool for freeze-drying: II. Measurement of dry-layer resistance. AAPS PharmSciTech 2006b;7(4): Art. 93. http://www.aapspharmscitech.org.

_____. Evaluation of manometric temperature measurement, a process analytical technology tool for freeze-drying: III. Heat and mass transfer measurement. AAPS PharmSciTech 2006c;7(4): Art. 97. http://www.aapspharmscitech.org.

Taylor A. Organic chelating agents as aids to industry. Chem. Ind. 1956; Oct. 20: 1131–1135.

Taylor LS, Zografi G. Sugar–polymer hydrogen bond interactions in lyophilized amorphous mixtures. J. Pharm. Sci. 1998;87:1615–1621.

Torisel MSDS. http://media.pfizer.com/files/products/material_safety_data/WP00085.pdf. Rev. 6.1. Revision date Jan. 22. 2010. Accessed Dec. 2011.

Tsinontides SC, Rajniak P, Hunke WA, Placek J, Reynolds SD. Freeze drying: principles and practice for successful scale-up to manufacturing. Int. J. Pharm. 2004;280:1–16.

U.S. Government. 2011. United States National Technology Initiative. http://www.nano.gov/html/facts/whatIsNano.html. Accessed Feb. 2011.

U.S. Pharmacopeia. U.S.P.–N.F. online, <1222>, Terminally sterilized pharmaceutical products: parametric release. Second Supplement U.S.P. 33–N.F. 28 Reissue. Accessed Mar. 2011.

Wang Y. Lyophilization and development of solid protein pharmaceuticals. Int. J. Pharm. 2000;203:1–60.

Wang W, Mesfin G-M, Rodriguez CA, Slatter JG, Schuette MR, Cory AL, Higgins MJ. Venous irritation, pharmacokinetics, and tissue distribution of Tirilazad in rats following intravenous administration of a novel supersaturated submicron lipid emulsion. Pharm. Res. 1999;16:930–938.

Wang Y-CJ, Hanson MA. Technical report no. 10 parenteral formulations of proteins and peptides: stability and stabilizers. J. Parenter. Sci. Technol. 1988;42:S3–S26.

Willemer H. Experimental freeze-drying: procedures and equipment. In: Rey L, May JC, eds. _Freeze-Drying/Lyophilization of Pharmaceutical and Biological Products_, 2nd ed. Drugs and the Pharmaceutical Sciences, Vol. 96. New York: Marcel Dekker; 1999. pp. 79–121.

Wisniewski R. Developing large-scale cryopreservation systems for biopharmaceutical products. Biopharm 1998;11:50–60.

GLOSSARY

API	Active pharmaceutical ingredient.
ASTM	American Society of Testing and Materials.
CCS	Container closure system.
CFR	_Code of Federal Regulations_.
CFU	Colony forming units
CGMP	Current good manufacturing practices.

COC	Cyclic olefin copolymers
COP	Cyclic olefin polymer
CZ	Crystalline Zenith®
DOE	Design of experiment.
DOT	Department of Transportation
DSC	Differential Scanning Calorimeter
DSPG	Distearoylphosphotidylglycerol.
EDTA	Ethylenediaminetetraacetic acid.
EMF	Electromotive force.
E.P.	*European Pharmacopoeia*.
EPDM	Ethylene propylenediene monomer
FD	Freeze-drying.
FDA	U.S. Food and Drug Administration.
HSPC	Hydrogenated soy phosphatidylcholine.
ICH	International Conference on Harmonization.
ID	Intradermal.
IM	Intramuscular.
IV	Intravenous.
J.P.	*Japanese Pharmacopoeia*.
LVP	Large-volume parenterals.
LYO	Lyophilization.
MPA	Methylprednisone acetate.
MPEG	methoxy poly(ethylene glycol).
N.F.	*National Formulary*.
PAMAM	Polyamidoamine.
PAT	Process analytical technology.
PEG	Poly(ethylene glycol).
PFC	Perfluorocarbon.
PICVD	Plasma impulse chemical vapor deposition
PLG	Polylactide-coglycolide.
PLGA	Poly-lactic-co-glycolic acid
PNS	Probability of nonsterility; the probability that the product bioburden will survive the sterilization process.
PVP	Poly(vinylpyrorlidone), povidone.
rhGH	Recombinant human growth hormone.
RTD	Resistance temperature detectors.
SAL	Sterility assurance level; the probability that the product bioburden will survive the sterilization process.
SC	Subcutaneous.
SVP	Small-volume parenterals.
U.S.P.	*United States Pharmacopeia*.
w/o	Water-in-oil emulsion.
w/o/w	Water-in-oil-in-water multiple emulsion; the primary emulsion (w/o) is the dispersed phase and the secondary emulsion (w) is the external phase.
WFI	Water for injection.
z **value**	Temperature coefficient of microbial destruction, which is set to $10°C$ and is the inactivation sensitivity to temperature change.

APPENDIXES

APPENDIX 17.1   Injectable Solvents and Cosolvents[a]

Excipient	Route of Administration[b]	Use (% w/v)[c]
Water Soluble		
N,N-Dimethylacetamide, E.P.	IM	
	IV	1.8
	IV (infusion)	6.0
Ethanol, E.P./J.P./U.S.P–N.F.	IM	0.6–10
	IV (bolus)	6–10
	IV (infusion)	5–80
	IVS	50
	SC	6 or less
Glycerin, E.P./J.P./U.S.P.–N.F.	ID	5.0
	IM	15 or less
	IV (bolus)	15
	IV (infusion)	2.5 or less
	IVS	
	SC	32.5
Poly(ethylene glycol) (PEG), E.P./J.P./U.S.P.–N.F.		
PEG 200	IM	30
PEG 300	IM	50
	IV (bolus)	50
	IV (infusion)	65
PEG 400	IM	20–30
	IV (bolus)	20–30
	IV (infusion)	11.25
PEG 3350	IAR	3.0
	IL	3.0
	IM	3.0
	Intrasynovial	3.0
PEG 4000	IAR	3.00
	IL	3.00
	Intrasynovial	3.00
PEG 6000	IV	5.00
Poly(propylene glycol)	IM	40
	IV	40
Propylene glycol, (E.P./J.P./U.S.P.–N.F.)	IM	20–75.2
	IV (bolus)	20–40
	IV (infusion)	2.5–51.80

APPENDIX 17.1 *(Continued)*

Excipient	Route of Administration[b]	Use (% w/v)[c]
	SC	1.40
Sorbitol, E.P./J.P./U.S.P.–N.F.	IL	45
	IAR	50
	IM	2.0
	Intrasynovial	45
	IV	4.5–48
	IV (infusion)	7.14–45
Oil Soluble		
Benzyl benzoate, E.P./J.P./U.S.P.–N.F. (soluble in ethanol 95%, fatty acids, essential oils)	IM	0.01–46
Castor Oil, E.P./J.P./U.S.P.–N.F. (miscible in ethanol 95%)	IM	Up to 100
Cottonseed oil, U.S.P.–N.F. (miscible in ethanol 95%)	IM	Up to 100
Ethyl oleate, E.P./U.S.P.–N.F. (miscible in ethanol 95%, fixed oils, organic solvents)	IM SC	
Long-chain triglycerides		
Peanut oil, E.P./J.P./U.S.P.–N.F. (slightly soluble in ethanol 95%, miscible with oils)	IM	
Poppyseed oil	—	1
Safflower oil, E.P./U.S.P.–N.F. (miscible in ethanol 95%, lipid solvents, oils)	IV (infusion)	5–10
Sesame oil, E.P./J.P./U.S.P.–N.F.	IM	Up to 100
Soybean oil, E.P./J.P./U.S.P.–N.F.	IM	5–20
Triolein	Epidural	0.01
Medium-chain triglycerides, E.P./U.S.P.–N.F. (miscible in ethanol, 95%, ethyl acetate)	IV (infusion)	10–20
Vegetable oil, J.P./U.S.P.–N.F.	IM	

[a] Solvents and cosolvents were found in the following references: Costantino and Pikal, (2004); FDA Inactive Ingredients List, http://www.fda.gov/Drugs/InformationOnDrugs/ucm113978.htm accessed Nov. 10, 2010; Nema et al. (1997); Powell et al. (1998); and Strickley (1999, 2000a,b).

[b] IAR, Intraarticular; ID, intradermal; IL, intralesional; IM, intramuscular; IV, intravenous; IVS, intravescial; SC, subcutaneous. FDA Inactive Ingredient List, http://www.accessdata.fda.gov/scripts/cder/iig/index.cfm.

APPENDIX 17.2 Injectable Surfactants, Solubilizing, Emulsifying, and Thickening Agents[a]

Excipient	Route of Administration[b]	Use (% w/v)[c]
Acacia, E.P./J.P./U.S.P.–N.F.	ID	7.0
Aluminum monostearate,	IM	≤0.034–2
J.P./U.S.P.–N.F.	IV	≤ 0.0001
Carboxymethylcellulose,	IM	1%
E.P./J.P./U.S.P.–N.F.		
Carboxymethylcellulose sodium,	IM	0.05–0.75
E.P./J.P./U.S.P.–N.F.		
Cremophor E.P., U.S.P.–N.F.	IV (infusion)	0.02–10
	Intravesical	18
Cremophor RH 80, U.S.P.–N.F.	IV (infusion)	≤ 0.08
Desoxycholate sodium	IV	0.4
Egg yolk phospholipid	IV	1.2
Gelatin, E.P./J.P./U.S.P.–N.F.	IM	16
	IV	1.40
	SC	16
Lecithin, U.S.P.–N.F.	IM	0.4–2.3
Polyoxyethylated fatty acid	IV	7
Polysorbate 20, E.P./U.S.P.–N.F.	IM	0.01
	IV	0.40
	IV (infusion)	2.40
	SC	0.01–0.2
Polysorbate 80, E.P./J.P./U.S.P.–N.F.	IAR	0.40
	Intrabursal	0.04
	IL	0.2
	IM	0.01–12
	IV	0.001–10
	IVS	0.015
	IQ	0.0005
	SC	0.01–0.02
Sodium dodecyl sulfate,	IV	0.018
E.P./J.P./U.S.P.–N.F.		
Sorbitol, E.P./J.P./U.S.P.–N.F.	IL	45
	IAR	50
	IM	2.0
	Intrasynovial	45
	IV	48
	IV (infusion)	7.14–45

[a] Injectable surfactants, solubilizing, emulsifying, and thickening agents were found in the following references: Costantino and Pikal (2004); FDA Inactive Ingredients List, http://www.fda.gov/Drugs/InformationOnDrugs/ucm113978.htm, accessed Nov. 10, 2010; Nema et al. (1997); Powell et al. (1998); and Strickley, (1999; 2000a,b).

[b] IAR, intraarticular; ID, intradermal; IL, intralesional; IM, intramuscular; IQ, intracutaneous, IV, intravenous; IVS, intravescial; SC, subcutaneous.

APPENDIX 17.3   Injectable Buffers and pH-Adjusting Agents[a]

Excipient	pH	Route of Administration[b]	Use (% w/v)[c]
Acetate			
Sodium, E.P./J.P./U.S.P.–N.F.	4.2–5.2	IM	—
	7.0–7.5	ITO	0.39
	3.0–5.5	IV	0.005–0.2
	Neutral	SC	0.14–0.65
Acetic Acid, E.P./J.P./U.S.P.–N.F.	2.5	IM	
	3.0–4.5	IV	0.1–0.46
Ammonium	7.0	IM	0.4
Benzenesulfonic acid	3.25–3.65	IV	
Bicarbonate sodium, E.P./J.P./U.S.P.–N.F.	6.5–9.0	IV	0.01–1.64
Carbonate sodium, E.P./J.P./U.S.P.–N.F.	5.0–8.0	IV	0.12–3.3
	6.0–8.5	IM	0.063–3.3
Citrate			
Acid, E.P./J.P./U.S.P.–N.F.	3.0–6.9	IV	0.006–0.075
	5.0	IM	0.02
	3.3–5.5	Various	0.02
Anhydrous	4.0	IV	0.21–2.2
	6.5–6.9	LI	1.0
	3.3–5.5	Various	0.02
Sodium	3.0–8.0	IV	0.23–1.2
	4.5–8.5	IM	0.05–2.85
	6.0–8.0	IAR	0.1
	7.0–7.5	ITO	0.17
Trisodium, E.P./J.P./U.S.P.–N.F.	6.1	IV	0.98
Diethanolamine, U.S.P.–N.F.	10	IV	0.3
Glycine, E.P./J.P./U.S.P.–N.F.	3.6–7.5	IM	0.5–2.25
	6.4–8.5	IV	1.58–2.25
	6.7–7.4	SC	0.034–2.44
Hydrochloric acid, E.P./J.P./U.S.P.–N.F.	3.5–7.0	IM	
	3.5–7.0	IV	
	4.5–6.5	IAR	
	6.0	IL	
	3.0–5.0	SC	
Lactic acid, E.P./J.P./U.S.P.–N.F.	4.0	IM	0.724–1.158
	3.2–4.0	IV	0.0282–0.129
Maleic acid, E.P./U.S.P.–N.F.	3.0	IM	1.6
	5.0	IV	
Methanesulfonic acid	3.6±0.4	IM	
Monoethanol amine		IM	1.7–4.2
Phosphate			
Acid (phosphoric), E.P./U.S.P.–N.F.	6.0–7.3	IM	

(*continued*)

APPENDIX 17.3 (*Continued*)

Excipient	pH	Route of Administration[b]	Use (% w/v)[c]
	7.3	IV	1.0
Monobasic sodium,	3.5–8.0	IM	0.04–0.69
E.P./U.S.P.–N.F.	7.0–7.8	IV	0.017–0.45
	6.8–7.9	ID	0.34
Dibasic sodium,	3.5–8.0	IM	0.0226–1.75
E.P./J.P./U.S.P.–N.F.			
	6.7–7.9	IV	0.072–0.18
	6.8–7.2	ID	0.71
Tribasic sodium	—	IM	0.014
Sodium hydroxide,			
E.P./J.P./U.S.P.–N.F.			
Sodium tartarate	—	IM	0.475–1.2
Tromethamine	7.4±0.3	IV	0.12

[a] Injectable buffers were found in the following references: Costantino and Pikal (2004); FDA Inactive Ingredients List, http://www.fda.gov/Drugs/InformationOnDrugs/ucm113978.htm, accessed Nov. 10, 2010; Nema et al. (1997); Powell et al. (1998); and Strickley (1999, 2000a,b).

[b] IAR, Intraarticular; ID, intradermal; IL, intralesional; IM, Intramuscular; ITO, intraocular; IV, intravenous; LI, local injection; SC, subcutaneous.

APPENDIX 17.4   Injectable Antimicrobial Preservatives[a]

Excipient	Route of Administration[b]	Use (% w/v)[c]
	Lipophilic	
Benzyl alcohol, E.P./J.P./U.S.P.–N.F.	IV	0.104–3.0
(miscible in fixed and volatile oils)	IAR, IL, SC, IA	0.9
	ICN	0.84
	IM	0.5–8.92
Chlorobutanol, E.P./J.P./U.S.P.–N.F.	IM	0.25–0.5
(miscible in volatile oils)	IV	0.5
m-Cresol, J.P./U.S.P.–N.F. (miscible in fixed and volatile oils)	SC	0.1–0.3
Paraben methyl, E.P./J.P./U.S.P.–N.F.	IM	0.018–0.18
(soluble in glycerin, propylene	IV	0.02–0.18
glycol)	SC	0.18
	IF	0.1
	LI	0.15
Paraben propyl, E.P./J.P./U.S.P.–N.F.	IM, IV	0.01–0.02
(soluble in propylene glycol)	LI	0.02

APPENDIX 17.4 *(Continued)*

Excipient	Route of Administration[b]	Use (% w/v)
Phenol, E.P./J.P./U.S.P.−N.F. (miscible in fixed and volatile oils, glycerin)	SC	0.02
	IM	0.002−0.5
	IV	0.5
	ID	0.5
	IQ	0.28
Phenoxyethanol, E.P./U.S.P.−N.F. (soluble in glycerol)	SC	0.45−0.5
	IM	0.5
	SC	0.5−1.0
Hydrophilic		
Benzalkonium chloride, E.P./J.P./U.S.P.−N.F.	ID	0.02
Benzethonium chloride, E.P./J.P./U.S.P.−N.F.	IM	0.01
Benzyl alcohol, E.P./J.P./U.S.P.−N.F. (4% w/v[c])	IV	0.104−3.0
	IAR, IL, SC, IA	0.9
	ICN	0.84
Paraben methyl, E.P./J.P./U.S.P.−N.F. (0.5% w/v[c])	IM	0.018−0.18
	IV	0.02−0.18
	SC	0.18
	IF	0.1
	LI	0.15
Phenol, E.P./J.P./U.S.P.−N.F. (6.6% v/v[c])	IM	0.002−0.5
	IV	0.5
	ID	0.5
	IQ	0.28
	SC	0.45−0.5
Phenoxyethanol, E.P./U.S.P.−N.F. (2.3% w/v)[c]	IM	0.5
	SC	0.5−1.0
Thiomersal, E.P./U.S.P.−N.F. (50% w/v)[c]	IV	0.0002−0.01
	IM	0.003−0.01
	ID	0.01
	SC	0.007−0.01

[a] Injectable antimicrobials were found in the following references: Costantino and Pikal (2004); FDA Inactive Ingredients List, http://www.fda.gov/Drugs/InformationOnDrugs/ucm113978.htm, accessed Nov. 10, 2010; Nema et al. (1997); Powell et al. (1998); and Strickley (1999, 2000a,b).

[b] IA, Intraarterial; IAR, intraarticular; ICN, intracavernosal; ID, intradermal; IF, infiltration; IL, intralesional; IM, intramuscular; IQ, intracutaneous; IV, intravenous; LI, local injection; SC, subcutaneous.

[c] % w/v solubility of compound in water.

APPENDIX 17.5   Injectable Antioxidants[a]

Excipient	Route of Administration[b]	Use (% w/v)
Lipophilic		
Butyl hydroxyanisole (BHA), E.P./U.S.P.–N.F.	IM	0.00028–0.03
Butyl hydroxytoluene (BHT), E.P./U.S.P.–N.F.	IM	0.00116–0.03
Propyl gallate, E.P./U.S.P.–N.F.	IM	0.02
Hydrophilic		
Acetone sodium bisulfite	ISP, IT	0.2–0.4
Ascorbic acid, E.P./J.P./U.S.P.–N.F. (28.5% w/v[c])	IM	0.1–4.8
Cysteine/cysteinate HCl	IM, SC	0.07–0.1
Dithionite sodium	SC, IM	0.1
Glutamate monosodium, U.S.P.–N.F.	SC	0.1
Metabisulfite potassium, E.P./U.S.P.–N.F. (45.45% w/v)[c]	IV, IM	0.1
Metabisulfite sodium, E.P./J.P./U.S.P.–N.F. (52.6% w/v)[c]	IV	0.02-1
Monothioglycerol, U.S.P.–N.F.	IM	0.1–1
Propyl gallate, E.P./U.S.P.–N.F. (0.1% w/v[c])	IM	0.02
Sodium bisulfite (28.5% w/v[c])	IM, IV	0.02–0.66
Sodium formaldehyde sulfoxylate, U.S.P.–N.F.	IM	0.005–0.5
Sodium thiosulfate, E.P./J.P./U.S.P.–N.F.		0.1–0.5
Sodium sufite, E.P./J.P./U.S.P.–N.F. (31.25% w/v[c])	IV	0.2
	IM	0.05–0.1
Thioglycolate sodium	IM, SC	0.66

[a] Injectable antioxidants were found in the following references: Costantino and Pikal (2004); FDA Inactive Ingredients List, http://www.fda.gov/Drugs/InformationOnDrugs/ucm113978.htm, accessed Nov. 10, 2010; Nema et al. (1997); Powell et al. (1998); and Strickley (1999, 2000a,b).

[b] IM, intramuscular; ISP, intraspinal; IT, intrathecal; IV, intravenous; SC, subcutaneous.

[c] % w/v solubility of compound in water.

APPENDIX 17.6 Logarithms of Stability Constants for Selected Sequestering Agents

Sequestering Agent	Metal										
	Al^{+3}	Ca^{2+}	Co^{2+}	Co^{3+}	Cr^{2+}	Cu^{2+}	Fe^{2+}	Fe^{3+}	Mn^{2+}	Ni^{2+}	Zn^{+2}
Alanine	—	1.24	4.82	—	—	8.18	—	—	3.24	5.96	5.16
Citric acid, E.P./J.P./ U.S.P.–N.F.	11.7	3.5	4.4	—	—	6.1	3.2	11.85	3.2	4.8	4.5
Cysteine	—	—	—	9.3	—	19.2	6.2	—	4.1	10.4	9.8
EDTA[a] (edetate disodium), E.P./J.P./ U.S.P.–N.F.	16.13	10.7 / 10.69[b]	16.21 / 16.3[b]	41.4[b]	13.6[b]	18.8 / 18.8[b]	14.3 / 14.3[b]	25.7 / 25.1[b]	13.56 / 13.8[b]	18.56 / 18.6[b]	16.5
Fumaric acid, U.S.P.–N.F.	—	2.00	—	—	—	2.51	—	—	0.99		
Gluconic acid	—	—	—	—	—	18.29	—	—	—	—	1.7
Glycine, E.P./J.P./ U.S.P.–N.F.	—	1.43	5.23	—	—	8.22	4.3	10.0	3.2	6.1	5.16
Leucine, E.P./J.P./ U.S.P.–N.F.	—	—	4.49	—	—	7.0	3.42	9.9	2.15	5.58	4.92
Pyrophosphate	—	5.0	—	—	—	6.7	—	22.2	—	5.8	8.7
Valine	—	—	—	—	—	7.92	3.39	9.6	2.8	5.37	5.0

Source: T. E. Furia, Sequestrants in foods, in: *CRC Handbook of Food Additives*, Vol. 1, CRC Press, Boca Raton, FL, 1972, pp. 275–277.

[a] FDA Inactive Ingredients List, http://www.fda.gov/Drugs/InformationOnDrugs/ucm113978.htm. Accessed Jan. 1, 2010. Percentage range for intravenous use is from 0.01 to 10%.

[b] Formation constants K_{MY} for EDTA–Metal ion complexes. http://www.cem.msu.edu/~cem333/EDTATable.html. Accessed Dec. 27, 2010.

APPENDIX 17.7 Injectable Bulking Agents, Protectants, and Tonicity-Adjusting Agents[a]

Excipient	Route of Administration[b]	Use (% w/v)
Alanine (16.65 % w/v[c])	IV	0.668–1.11
Albumin (40 % w/v[c])	IV	0.25
	IM	<10.0
Albumin, human, E.P./U.S.P.–N.F. (40% w/v[c])	IV	0.25–5.0
	IM	0.0063–0.5
	SC	0.5–1.25
	IL	0.1
L-Arginine	IM	1.56–26.0
Asparagine (3.53% w/v[c])	IVS	
Calcium chloride, E.P./J.P./U.S.P.–N.F.	IV	0.022–0.056
	IM	0.004
	ITO	0.048
Citric acid, E.P./J.P./U.S.P.–N.F.	IM	0.02–1.0

(*continued*)

APPENDIX 17.7 (*Continued*)

Excipient	Route of Administration[b]	Use (% w/v)
	IV	0.006
	IVS	
Dextrose, E.P./J.P./U.S.P.−N.F.	IM	3.75−4.4
	IV	3.2−5
	SC	1.25
	Spinal anesthesia, SA	8.25
	ID	3.0
	IU	1.0
Gelatin, E.P./J.P./U.S.P.−N.F.	IV	0.5
	SC	0.05
	IM	
Glucose, E.P./U.S.P.−N.F.	IV	<2.0
Glycerin, E.P./J.P./U.S.P.−N.F.	IV	2.25
	IM	1.5−15
	IVS	1.6
	SC	32.5
Glycine, E.P./J.P./U.S.P.−N.F.	IV	0.1−3.3
(25% (w/v[c])	IM	1.5−2.4
	ID	0.09
	SC	0.034−2.44
	IL	2.0
Histidine (4.19% w/v[c])	IV	0.018−0.85
Imidazole	IV	< 0.005
Lactose, E.P./J.P./U.S.P.−N.F.	ID	8.5
	IM	0.4−4.0
	ICN	17.2
	IV	1.0−10.0
Magnesium chloride	ITO	0.03
(54.3% w/v[c])	IM	2.5−6.0
Magnesium sulfate	IVS	
Mannitol, E.P./J.P./U.S.P.−N.F.	IV	0.1−5.0
(18.2% w/v[c])	SC	0.158−10.0
	IM	0.005−8.0
Poly(ethylene glycol),	IV	0.0005−65.0
E.P./J.P./U.S.P.−N.F.	IM	2.03−3.0
	IL	3.0

APPENDIX 17.7 *(Continued)*

Excipient	Route of Administration[b]	Use (% w/v)
Polysorbate 80,	IV	0.001–10.0
E.P./J.P./U.S.P.–N.F.	IM	0.01–12.0
	IAR	0.4
	IL	0.2
	IQ	0.0005
	IVS	0.025
	SC	0.002
Potassium chloride,	SC	0.16
E.P./J.P./U.S.P.–N.F.	ITO	0.075
(35.7% w/v[c])		
Povidone, E.P./J.P./U.S.P.–N.F.	IM	0.55–0.6
	IV	0.2
Sodium chloride,	IV	0.17–1.0
E.P./J.P./U.S.P.–N.F.	IM	0.0017–0.9
(35.7% w/v[c])	IAR, IPC	0.9
	IL, IVS	0.85
	IA	0.26–0.49
	SC	0.35–0.9
	IF	0.33–0.67
	ID	
	IT	0.625–0.9
	ITO	0.32–0.9
	Spinal anesthesia	0.67
Sodium succinate	SC	0.072
Sodium sulfate	IM	1.1
Sorbitol, E.P./J.P./U.S.P.–N.F.	IV	0.698–4.8
	IM	2.0
	SC	
Sucrose, E.P./J.P./U.S.P.–N.F.	IM	8.5
	IV	1.0–20.1
	SC	5.0

[a] Bulking agents, protectants, and tonicity-adjusting agents were found in the following references: Costantino and Pikal (2004); FDA Inactive Ingredients List, http://www.fda.gov/Drugs/InformationOnDrugs/ucm113978.htm, accessed Nov. 10, 2010; Nema et al. (1997); Powell et al. (1998); and Strickley (1999, 2000a,b).

[b] IA, Intraarterial; IAR, intraarticular; ICN, intracavernosal; ID, intradermal; IF, infiltration; IL, intralesional; IM, intramuscular; IPC, intrapericardially; IT, intrathecal; ITO, intra ocular; IV, intravenous; IVS, intravescial; SA, subarachniod; SC, subcutaneous.

[c] % w/v solubility of compound in water.

APPENDIX 17.8 Eutectic Temperatures for Selected Compounds

Compound	Eutectic Temperature ($^\circ$C)
Citric acid[a], E.P./J.P./U.S.P.−N.F.	−12.2
Glycine[b], E.P./J.P./U.S.P.−N.F.	−3.5
Mannitol[a], E.P./J.P./U.S.P.−N.F.	−1.0
Sodium chloride[a], E.P./J.P./U.S.P.−N.F.	−21.5
Sodium sulfate[a]	−1.1

[a] Campbell and Smith (1951).
[b] Nail and Gatlin (1993).

APPENDIX 17.9 Glass Collapse Temperature T_{gc}, Maximally Freeze-Concentrated Glass Transition Temperature T_g', and Material Glass Transition Temperature T_g for Selected Excipients

Compound	T_{gc} ($^\circ$C)	T_g', ($^\circ$C)	T_g ($^\circ$C)
Albumin			
Bovine serum albumin	—	−13[a]	
Egg albumin	−10[b]	−11[c]	
Human serum albumin, E.P./U.S.P.−N.F.	−9.5[d]	−11[c]	
Dextran	−9[b]	−10[e]	91
Gelatin, E.P./J.P./U.S.P.−N.F.	−8[b]		
	−10[f]		
	−8[d]		
Glucose, E.P./U.S.P.−N.F.	−40[b]	−43, 29.1%[g]	4[h]
	−41.5[i]	—	32[j]
			39, 29.1%[g]
Hydroxyethyl starch, E.P./U.S.P.−N.F.	> −5[d]	−12[k]	> 110[l]
Inositol	−27[b]	—	−27[b]
Lactose, E.P./J.P./U.S.P.−N.F.	−32[b]	—	101[j]
	−30.5[m]	—	114[n]
Maltose, J.P./U.S.P.−N.F.	−32[b]	−30, 20%[g]	43, 20%[g]
			82[n]
Methylcellulose, E.P./J.P./U.S.P.−N.F.	−9[b]		
Poly(ethylene glycol), E.P./J.P./U.S.P.−N.F.	−13[e]		
Poly(vinylpyrrolidone), E.P./J.P./U.S.P.−N.F.	−23[b]	—	100[o]
	−24[f]		

APPENDIX 17.9 (*Continued*)

Compound	T_{gc} (°C)	T'_g, (°C)	T_g (°C)
Sorbitol, E.P./J.P./U.S.P.–N.F.	-45^b	-44, $18.7\%^g$	-3, $18.7\%^g$
	-51^f		
Sucrose, E.P./J.P./U.S.P.–N.F.	-32^b	-32, $35.9\ \%^g$	57, $35.9\%^g$
	-34^f		
Trehalose, U.S.P.–N.F.	-29^m	$-29.5, 16.7\ \%^g$	78^j
	-34^p		46^h
			117^q
			118^n

^{a} Slade et al. (1989).

^{b} MacKenzie (1977).

^{c} Chang and Randall (1992).

^{d} Pikal and Shaw (1990).

^{e} Willemer (1999).

^{f} Nail and Gatlin (1993).

^{g} Franks (1990); T_g, w/w water.

^{h} Prestrelski et al. (1995).

^{i} Bellows and King (1972).

^{j} Levine and Slade (1992).

^{k} Crowe et al. (1993).

^{l} Crowe et al. (1998).

^{m} Adams and Ramsay (1996).

^{n} Taylor and Zografi (1998).

^{o} Bell et al. (1995).

^{p} Overcashier et al. (1999).)

^{q} Duddu and Dal Monte (1997).

CHAPTER 18

OPHTHALMIC PRODUCT DESIGN

18.1 INTRODUCTION

The eye anatomy, physiology, and biochemistry present a formidable challenge for ophthalmic product design. The structure of the eye (Figure 18.1) is designed to provide the unique function of connecting the body to the world through clear, undistorted vision. The fragile nature of this unique organ is highlighted by the fact that critical tissues such as the cornea, retina, and optical nerve lack the ability to regenerate. Nature has enlisted a number of protective mechanisms to minimize trauma and maintain the eye's highly specialized function. The eyeball or *globe* sits inside a four-sided pyramidal boney structure called the *orbit*. The orbit, forehead, brow, eyelids, eyelashes, tears, and tear film are anatomical structures that protect the eye from the environment and physical trauma. The globe is composed of three layers; the sclera (the white of the eye), choroid, and retina. At the anterior of the eye a transparent structure called the *cornea* forms the outer layer. The cornea provides a transparent path to the retina and imparts 60% of eye refraction. The globe is spherical except for a slight bulge produced by the cornea. To focus light onto the retina accurately, the dimension of the path needs to remain constant within fine tolerances. The sclera and cornea form a tough outer layer that is resistant to deformation caused by the aqueous and vitreous humors that fill the internal eye cavity. The eye has the ability to focus, adjust light intensity, and respond to depth of field. The photoreceptors of the retina, in conjunction with the optic nerve, send the image information to the brain, which provides an enhanced stereo image (Wilson et al., 2007).

The human eye blinks about 15 times a minute and has a tear turnover rate of 0.5 to 2.2 μL/min. The tear volume is 7 to 30 μL, so about 15 to 30% of the tear volume is turned over every minute. The eye blinks at a high speed of tens of centimeters per second and has a duration of 300 to 400 ms. The tear film experiences a shear rate of 10,000 to 40,000 s^{-1}. This compares to a shear rate of approximately 20,000 s^{-1} when a liquid is squeezed from a collapsible aerosol plastic bottle. The purpose of eye blinking is to spread the tear film over

Integrated Pharmaceutics: Applied Preformulation, Product Design, and Regulatory Science, First Edition. Antoine Al-Achi, Mali Ram Gupta, William Craig Stagner.

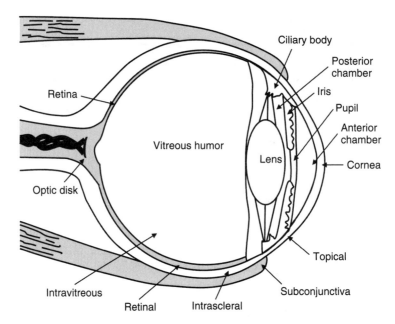

FIGURE 18.1 Eye anatomy.

the eye to keep it lubricated and to remove foreign particles to the *conjunctiva* and *nasolacrimal apparatus*. Blinking also prevents foreign material from entering the eye. The nasolacrimal apparatus has two small openings in the corner of the eyelids which drain tears and tear film to the nasolacrimal duct, which joins the nasal turbinates. This all leads to a short drug contact time. Depending on the viscosity of the formulation and other product attributes, such as mucoadhesion agents, the drug contact time generally ranges from only 3 to 6 minutes. It has been noted that only 5% of a topically applied ophthalmic drug reaches the intraocular portions of the eye (Wilson et al., 2001).

Tear film keeps the anterior portion of the eye lubricated and moist. Tear film exhibits shear-thinning pseudoplastic flow with a yield value of 32 cP (0.032 Pa·s) at 33°C. Once shear is initiated, the tear film has a viscosity of 1.3 to 5.9 cP (0.0015 to 0.0059 Pa·s). The purpose of tears is to bath the eye and, together with blinking, to wash out foreign matter.

The pH of human tears ranges from 7.14 to 7.82 (mean of 7.4) and has a milliosmolarity of 306 mOsmol/L. The buffer capacity of tears is low. Butty et al. (1996) reported the buffer capacity of mice tears to be 0.001 to 0.01. They showed that when acidic and alkaline solutions with a buffer capacity below 0.001 were instilled in the mouse eye, the instilled solution behaved like an unbuffered solution. This suggested that the tears were capable of reestablishing the physiological pH of 7.4. This was not the case for instilled solutions having buffer capacities greater than 0.01. In these cases, the surface pH of the cornea was the same as that of the instillation solution. The Hind–Goyan phosphate buffer system (Hind and Goyan 1947) has a buffer capacity of approximately 0.04.

The tear surface tension changes for patients who experience dry eye. Normal tear surface tension ranges from 43.6 to 46.6 mN/m (or dyn/cm), compared to 49.6 mN/m for patients suffering from dry eye (Wilson et al., 2001).

Although the tear film can be a barrier to drug delivery, it can also be used to enhance drug residence time and absorption through the use of mucoadhesives. The tear film is described as a two- or three-layered structure. In the three-layer model the inner layer that is in contact with the eye is referred as the *precorneal layer*. This layer contains *mucin*, which adheres to the cornea and is about 0.02 to 0.05 μm thick. Mucin is produced by the globlet cells located in the conjunctiva and is made up of primarily negatively charged mucopolysaccharides such as chrondroitin sulfate and hyaluronan and sialic acid–rich glycoproteins. Sialic acid has a pK_a of 2.6 and is negatively charged at the pH of the tears. These negatively charged molecules serve as sites of interaction for positively charged mucoadhesives. The middle aqueous layer is the thickest tear film layer (4 to 7 μm) and is composed of electrolytes, proteins, and phospholipids. Phospholipids accumulate at both the mucin–aqueous layer and the air–tear film interface. The outer layer of the tear film is primarily a cholesterol- and lipid-rich region with some polar lipids. It has also been suggested that a lipid-binding protein, lipocalin, transverses through the upper and lower lipid layers and helps interconnect corneal glycocalyx (a glycoprotein found on the surface of the cornea) and the tear film. The corneal microvilli are also thought to aid in the interaction between the mucus layer and the corneal surface to help anchor the tear film to the cornea (Johnston et al., 2003).

The conjunctiva is another important external anatomical feature. It is thin mucous membrane that extends from the eyelid margin to corneoscleral junction. The conjunctiva is not keratinized and plays a crucial role in maintaining tear-film stability. The conjunctiva is highly vascularized and can absorb a drug before it can elicit its topical effect or be absorbed across the cornea. On the other hand, intraocular noncorneal drug absorption must occur across the conjunctiva and sclera. The conjunctival sac, the space formed between the lower eyelid and eyeball, is used as a location to place eyedrops. The lower eyelid is gently pulled away from the globe which creates a pouch or sac. The volume of this sac is approximately 30 μL, but when the eyelid is retracted, the volume decreases. Chrai et al. (1973) studied the effect of instilled volume on the rate of precorneal drainage. The authors evaluated 5, 10, 25, and 50 μL of instilled solution. They found a positive linear relationship between instilled volume and the precorneal drainage rate. That is, the larger the instilled volume, the more rapid the loss of drug solution. An instilled drop of 8 to 15 μL has been suggested as an optimal volume. Typical commercial eyedroppers deliver 35 to 56 μL, so a majority of the drug is lost to the exterior of the eyelid. It should also be pointed out that volumes larger than 3 μL destabilize the tear film (Wilson et al., 2001).

The internal eye structures include the uveal tract, lens, anterior aqueous humor, posterior vitreous humor, retina, and optic nerve. The *aqueous humor* is produced by ciliary epithelium and it baths the cornea, iris, and lens. The aqueous humor is isomolar with plasma and creates a modest intraocular pressure that normally ranges from 13 to 19 mmHg (1733 to 2522 Pa). The aqueous humor drains from the *canal of Schlemm* and *episcleral veins*. This directional fluid flow creates a barrier to passive diffusion through the aqueous humor for intraocular delivery. In adults, the volume of the aqueous humor and vitreous humor is 250 µL and 4 mL, respectively. The viscosity or the *vitreous humor* is about 2 to 4 cP (0.002 to 0.004 Pa·s). It is a viscoelastic gel that contains collagen fibers and hyaluronic acid. Drugs may be targeted to reach any of these internal structures. In some cases, drug localization in these structures is undesired, such as drugs that bind to melanin, which is located in the iris (Wilson et al., 2007).

Drugs and excipients can cause ocular irritation, allergenicity, toxicity, and photocarcinogenicity. Great care must be taken to prevent irreversible damage to the eye, including the cornea and internal tissues and structure. Even isotonic saline can be toxic to the cornea, conjunctiva, and other internal tissues compared to a solution that is more representative of the composition of tears, which contains sterol esters, glycoproteins (e.g., mucin), carbonate buffers, electrolytes, and other constituents. Albino rabbits are often used to evaluate a drug or formulation's toxicity. The rabbit eye exhibits less blinking and tearing than does the human eye. This tends to increase the drug and formulation contact time with the eye, which generally results in a conservative or overestimate of ophthalmic toxicity. The rabbit cornea is also more sensitive to irritants. *United States Pharmacopeia* (U.S.P.) chapter <1074>, Excipient Biological Safety Evaluation Guidelines, covers the types of ocular safety tests that are recommended for excipients to be used in the eye. Additional care needs to be taken for functional excipients that are used with intraocular drug delivery systems. Excipients used inside the eye (intraocular) have a better chance to accumulate in eye tissues and structures, which could lead to long-term safety issues.

Ocular drug delivery development challenges have resulted in only a comparatively small number of drug approvals over the past 15 years. There were no ocular drug approvals between 1995 and 2003. Fortunately, there has been a renaissance of drug delivery systems that have taken advantage of eye penetration enhancers, mucoadhesives, transporter-targeted prodrugs, microneedles, ultrasound, and iontophoretic drug delivery. *Soft drugs*, which are metabolically inactivated after exerting their therapeutic effect, have been developed that improve a drug's therapeutic index. Adaprolol is a soft beta-blocker that reduces the intraocular pressure with reduced cardiovascular and pulmonary side effects. Another example is oteprednol etabonate, which is a soft steroid that is used to treat topical eye inflammation. Oteprednol etabonate has an improved safety profile and exhibits lower cataractogenic effects without significantly raising intraocular pressure. In this chapter we focus on topically applied ophthalmic drug delivery to treat local periocular conditions such as dry eye, itching and inflammation, and intraocular diseases such as infections and glaucoma. Topical ocular drug absorption into the intraocular tissues occurs through the cornea or across the conjunctiva and sclera. The latter is known as the *noncorneal route*. The noncorneal route has also been referred to as the *nonproductive route* since drug may be picked up by capillaries of conjunctiva and other tissues and shunted to systemic circulation. Direct intravitreal delivery by injection or implant will also be mentioned in this chapter. Systemic, transscleral (subconjunctival, peribulbar, retrobulbar, and sub-tenon), therapeutic contact lens, and punctual plug drug delivery systems are not

TABLE 18.1   Advantages and Disadvantages of Topical Ocular Drug Delivery

Advantages	Disadvantages
Minimizes the side effects associated with systemic delivery of potent compounds	Five percent or less of the drug is absorbed
	Systemic side effects for highly potent drugs
	Patient discomfort, tearing, eyelids stick together
Reasonably quick onset of action, especially for local anesthetics, infections, and pupil dilation	Blurred vision
	Potential for bacterial contamination and overgrowth, leading to permanent cornea damage
Reasonably easy to self-administer drops	
Administration is generally painless	Inadvertent scratching or damage of cornea
Possible to achieve controlled release	Frequent dosing

covered. Even though only about 5% of the instilled drug is absorbed, there are a number of advantages of ocular drug delivery. Table 18.1 lists some of the major advantages and disadvantages of ocular drug delivery.

18.2   FORMULATION DESIGN

The key formulation challenge for periocular or intraocular drug delivery is maintaining a desired residence time. The precorneal factors discussed above, such as blinking, tearing, tear turnover, and nasolacrimal drainage, lead to a drug residence time of 3 to 6 minutes. The trilaminar avascular cornea poses a significant barrier to intraocular drug delivery. The cornea tissue is a barrier to hydrophilic and hydrophobic drugs. Generally, highly potent drugs with low molecular mass that balance solubility and hydrophobicity are the most likely to demonstrate acceptable absorption properties. The strategies that are used to optimize topical ocular delivery are also first-line approaches for topical intraocular delivery. Optimization starts with the proper placement of eyedrops in the eye.

Care should be taken to prevent possible product contamination by avoiding contact of the dropper with the eyelid. The drop is placed in the lower eyelid cul-de-sac by gently pulling lid away from the eyeball or globe to create a pocket for the drop. Manual nasolacrimal or punctual occlusion slows drainage and systemic absorption. Typical commercial eyedroppers deliver 35 to 56 µL, which exceeds the capacity of the cul-de-sac and can cause tearing. The total dose of smaller drop sizes, 8 to 15 µL, may be retained better. For lower-permeability drugs, smaller drop sizes of 5 to 10 µL can result in a twofold increase in drug concentration (Keister et al., 1991). Drug solubility or irritancy may prohibit the use of smaller droplet sizes. Precorneal residence time may be improved by using dispersed systems (suspensions, emulsions, microemulsions, microparticles, nanoparticles, liposomes), viscosity-enhancing agents, bioadhesives, ion-exchange polymers, and biodegradable and nonbiodegradable inserts. These retention techniques can be coupled with strategies to enhance ocular drug absorption. There has been a resurgence of research directed at prolonging and improving the bioavailability of ophthalmic drugs. Formulation approaches are highlighted in this chapter.

Ophthalmic products must be sterile, contain low levels of particulate matter, and be nonirritating, nonallergenic, noncarinogenic or nonphotocarinogenic. It is extremely important to note that formulations for intraocular drug delivery, especially intraocular injections and implants, cannot incorporate many of the already limited number of ophthalmic excipients. Antimicrobial preservative agents have been shown to be toxic to tissues in the anterior eye segment. While dispersed system dosage forms can be used for topical ophthalmic delivery, only aqueous solutions should be used for intraocular use.

18.2.1 Preformulation

The reader is directed to Section 17.2.1, as ophthalmic drug preformulation tests mirror closely those performed for drugs intended for sterile products, described in Chapter 17. The ocular drug's solubility and stability will have the largest impact on selection of the ophthalmic dosage form and sterilization method. Most ophthalmic drops are packaged in plastic bottles, which cannot withstand moist-heat sterilization conditions. Therefore, if terminal product sterilization is desired, gamma or electron irradiation sterilization is recommended and the preformulation studies should include an evaluation of a drug's stability when subjected to these sterilization methods.

18.2.2 Excipient Compatibility Studies

Preformulation studies should have identified formulation and process risks and the classes of excipients that would need to be evaluated for excipient compatibility. Excipients used for ophthalmic products require additional scrutiny; they need to be evaluated for the presence of microorganisms and insoluble particulates. Ophthalmic excipients should also be nonirritating, nonallergenic, and stable to heat or gamma sterilization conditions. Some excipients may need to be sterilized by filtration or may be sterilized by a final product filtration process. An excipient vendor program should have bioburden specifications as part of its qualification criteria. Once the type of dosage form and manufacturing process is selected, an excipient compatibility study can be initiated.

Table 18.2 lists the ophthalmic excipient classes and functions. Specific excipients can exhibit multiple functions.

Appendix 18.1 is a list of excipients that have been used in topical ophthalmic formulations. See Section 13.2.2 for a more detailed discussion of excipient compatibility methodologies.

18.2.3 Topical Ophthalmic Formulation Development

It is important in ophthalmic product design to minimize stinging and burning upon drug application. Stinging and burning can lead to excessive tearing and blinking,

TABLE 18.2 Ophthalmic Excipient Classes and Functions

Excipient Class	Function
Absorption and Permeation enhancers	Improve the ability of the drug to be absorbed by a number of different mechanisms, such as loosening tight junctions and increasing membrane fluidity
Antimicrobial preservatives	Reduce bioburden and prevent the growth of microorganisms
Antioxidants	Reduce the oxidative degradation of a drug by being preferentially oxidized, impeding an oxidative chain reaction or a combination
Buffers	Maintain a desired pH range, often to enhance drug stability or solubility
Chelating agents	Complex trace metals that can catalyze oxidation reactions of drugs
Ion-exchange agents	Control drug release by ion exchange of drug with ions present in the tears and tear film
Solubilizing agents	Increase the solubility of a drug
Suspending and thickening agents	Increase liquid structure to decrease the settling rate of dispersed material and aid in achieving a uniform dose on redispersion; increase viscosity to increase the drug's retention and residence time in the eye; decrease creaming rate of emulsions
Surfactants and wetting agents	Act to wet insoluble hydrophobic drugs; aid in making emulsions by decreasing the interfacial tension between the oil and water phases; surfactant irritancy is ranked from most irritating to least irritating: cationic > anionic > nonionic
Tonicity-adjusting agents	Provide an isotonic solution with eye and ocular fluids
Vehicle	Provides bulk for proper drug delivery

which can significantly reduce drug residence time in the eye. The intrinsic irritancy property of the drug and excipients, pH, buffer capacity, and isotonicity can affect patient comfort. The issue of drug irritancy is a critical attribute and should be addressed at the time the lead compound is selected. The number of excipients used in commercial ophthalmic drug products is a relatively small, limiting number, which increases the challenges faced in ophthalmic product design.

The average tear pH is 7.4. When possible, it is best to adjust the product pH to within the normal tear pH of 7.14 to 7.80. A drug is often too unstable or insoluble to be formulated within this pH range. In those cases, a compromised pH is used that provides acceptable drug stability or solubility and minimizes burning or irritation when administered to the eye. The buffer capacity of the buffer system should be low enough to allow the tears to neutralize the formulation pH.

Commercial epinephrine hydrochloride, dipivefrin hydrochloride, and atropine sulfate ophthalmic formulations have pH values as low as 3. Most commercial products have pH values between 5 and 7. Dexamethasone sodium phosphate has a pH range of 6 to 7.8. Borate, phosphate, and citrate buffers are most commonly used.

The normal osmolality of tears ranges from 290 to 310 mOsmol/kg. A wide range of 100 to 640 mOsmol/kg seems to be well tolerated (Wilson et al., 2001). The osmolarity of most commercial products range from 260 to 330 mOsmol/kg. A commercial 4% pilocarpine ophthalmic solution lists the osmolarity range to be from 550 to 600 mOsmol/kg.

Normal tear fluid has a surface tension of 43.6 to 46.6 mN/m. Matching the tear fluid surface tension is ideal. Formulations that have a surface tension much below this normal range tend to destabilize the tear film. Disruption or destabilization of the tear film may affect the adhesive characteristics of the mucin, which interacts with the corneal microvilli and could affect the effectiveness of drug-retentive mucoadhesive formulations.

A number of topical ophthalmic drug delivery strategies have been studied to improve the efficacy of topically applied ocular drugs. Some of these strategies are discussed below.

18.2.3.1 *Increased Formulation Viscosity*   Increasing the viscosity of ocular solutions tends to increase the drug contact time with the eye before the drug is removed by tears and blinking. Optimization studies can be executed to determine the optimum viscosity, after which an increase in viscosity results in marginal or no improvement. Linn and Jones (1968) evaluated the effect of viscosity on the rate of lacrimal excretion in humans by measuring the onset of dye appearance in the nasolacrymal duct. The formulation used 0.25 to 2.5% (6 to 30,000 cP, or 6 to 30,000 mPa·s) methylcellulose as the viscosity-inducing agent. The control dye appearance time in the nasolacrimal duct was 60 seconds. The dye appearance time was increased 1.5 to 4.25 times for 0.25 and 2.5% methylcellulose, respectively. Adler et al. (1971) studied the effect of vehicle viscosity on the absorption of fluorescein into the human eye and showed that there were marginal increases in penetration. Three viscosities were studied: 5 and 10 cP (or mPa·s) buffered poly(vinyl alcohol) and 2500 cP (or mPa·s) buffered methylcellulose solutions. It was mentioned that subjects could not discern the difference between saline and polymer solutions when the viscosity was below about 10 cP. Solution viscosities greater than 30 cP felt uncomfortable and sticky or gritty. Putting the viscosities in perspective, the commercial "liquid tears" products range from Liquifilm [1.4% poly(vinyl alcohol) and 4 to 6 cP] to Isopto (0.5% hypromellose and 10 to 30 cP).

Several minor disadvantages of viscosity-inducing agents need to be considered. Viscosity-enhancing polymers can cause momentary blurred vision. They also tend to form a dried film of polymer on the eyelids and eyelashes, which is readily removed by a clean moistened cloth.

18.2.3.2 *Ophthalmic Suspensions*   Ophthalmic suspensions are dispersed systems that should have particle sizes less than 10 μm to prevent eye irritation. It is generally accepted that suspended particles should reside in the cul-de-sac

for a longer time than a solution and thus provide a more prolonged effect and better bioavailability. However, the complexity of competing kinetic processes requires that each drug and suspension formulation be evaluated on a case-by-case basis. The competing kinetic processes involve drug diffusion across the cornea, drug dissolution to maintain a saturated solution, and drug elimination by various routes. Drug diffusion is controlled by the drug's physiochemical properties, such as water–lipid solubility, molecular size, and solubility, that affect the concentration gradient. Drug dissolution is described by the Noyes–Whitney equation, which states that the dissolution rate is directly proportional to the drug solubility, particle surface area, and the diffusion coefficient. The diffusion coefficient is directly dependent on temperature and is inversely proportional to the drug's molecular size and system viscosity. Just from this brief discussion it should be apparent that formulation–drug interactions add significant complexity to ophthalmic formulation design. One such competing or antagonistic interaction involves the system viscosity, where increased viscosity can improve the drug contact time but can also decrease the drug dissolution rate and decrease drug diffusion in the vehicle, which can lead to less drug being available for absorption. Schoenwald and Stewart (1980) studied the effect of dexamethasone suspension particle size on ophthalmic bioavailability. Three suspensions were manufactured with mean dexamethasone particles sizes of 5.75, 11.5, and 22.0 μm. A statistically significant rank-order correlation of increased dexamethasone concentration in the cornea and aqueous humor was observed with decreased particle size. An example formulation of a commercial nepafenac ophthalmic suspension is given in Table 18.3.

Ali et al. (2000) described a sterile manufacturing process for brinzolamide ophthalmic suspension. An adapted schematic of the process is shown in Figure 18.2. The schematic shows that carbomer solution and other water-soluble materials (benzalkonium chloride, edetate sodium, sodium chloride, and mannitol)

TABLE 18.3 Composition of Nepafenac Ophthalmic Suspension

Component	Percent w/v	Function
Nepafenac	0.1	Active
Benzalkonium chloride	0.005	Antimicrobial preservative
Tyloxapol	0.01	Surfactant and wetting agent
Edetate sodium	—	Combination antimicrobial preservative and absorption enhancer
Sodium chloride	0.4	Tonicity agent
Carbomer	0.5	Viscosity enhancer
Mannitol	2.4	Tonicity agent
Sodium hydroxide	Adjust to a pH of ~ 7.4	pH-adjusting agent
Purified water, U.S.P.	Add up to volume	Vehicle

Source: Wong (2010).

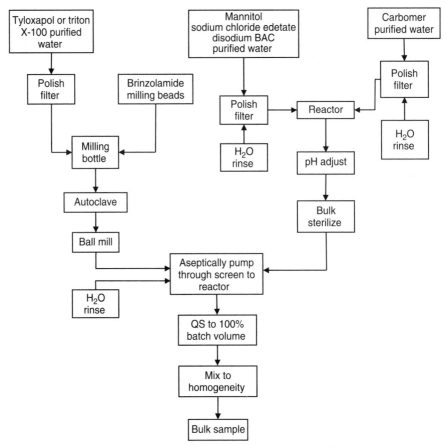

FIGURE 18.2 Sterile manufacturing process for brinzolamide opthalmic suspension. (Adapted from Ali et al., 2000.)

are polish-filtered, combined, and bulk-sterilized. Brinzolamide is added to the milling media along with a solution of tyloxapol. This mixture is autoclaved prior to ball-milling. Once autoclaved, brinzolamide is ball-milled to create a uniform drug suspension. The patent states that the use of tyloxapol is critical to obtaining the desired milled particle size. Any concern about particle size growth following autoclaving is eliminated in this scheme since milling occurs after the sterilization process has been completed. Once the ball milling is complete, brinzolamide is pumped aseptically through a screen reactor and mixed with the carbomer polymer and other soluble components.

18.2.3.3 Ophthalmic Ointments

Ophthalmic ointments are used when enhanced drug residence time is crucial. Ophthalmic ointments are typically preservative-free anhydrous dispersions for drug in a mineral oil–white petrolatum base. This ophthalmic system is useful for drugs that are highly unstable to moisture. Ocular ointments are usually administered at night to minimize the effect of blurred vision.

18.2.3.4 Gel-Forming Ophthalmic Solutions

There are basically three methods of in situ gelling: temperature change, pH change, and ion activation. Poloxamers are temperature-sensitive polymers that gel above room temperature. The surface temperature of the eye is about $34°C$, and a solution of Pluronic F127 will gel when administered to the eye. However, a rather high concentration of polymer is required for gel formation. Xanthan gum and gellan gum (Gelrite) have been used to form in situ gels once administered to the eye. The cation concentrations present in the tears cause these gums to gel in the eye. Timolol maleate ophthalmic gel forming solution (Timoptic-XE) uses gellan gum to extend the contact time, which allows the dosing interval to be reduced from twice a day for the solution formulation to once a day for the gel-forming solution. The use of gellan gum has been described by Mazuel and Friteyre (1989). The patent suggests a range of 0.1 to 2% gellan gum with a preferred concentration of 0.6%. Missel et al. (1993) describe the use of xanthan gum, locust bean gum, gellan gum, and carrageenan as gel-forming solutions intended for ocular administration.

18.2.3.5 Ophthalmic Ion-Exchange Formulation

Ion-exchange resins have been used for a number of dosage forms to provide controlled release. Jani and Rhone (2007) documented the history of the first U.S. Food and Drug Administration (FDA)-approved ion-exchange ophthalmic suspension. In 1989, Betoptic S 0.25% ophthalmic suspension was approved, which had the same clinical activity as that of the 0.5% solution. One of the major disadvantages of the betaxolol hydrochloride 0.5% solution was significant stinging that resulted from initial rapid and high localized drug concentrations which stimulated cornea sensory nerves. A betaxolol hydrochloride suspension was formulated with Amberlite (polacrilin potassium), a copolymer of methacrylic acid and divinylbenzene with sulfonic acid functional groups. As a cationic-exchange resin, polacrilin formed an ion pair with betaxolol hydrochloride. The monovalent and divalent ions in the tears and tear film exchange with betaxolol and release the drug for ocular absorption. The solution and ion-exchange formulations were studied using an in vitro system that simulates the precorneal tear fluid. The solution showed rapid absorption with drug release, which is completed in 30 minutes. The ion-exchange formulation showed sustained drug release over a 2-hour period with a maximum peak drug concentration about 30% less than the solution formulation. In a clinical study, statistically fewer patients experienced stinging and burning. Animal studies demonstrated that the half-strength ion-exchange suspension formulation was bioequivalent to the full-strength solution.

The betaxolol formulation can be surmised by using the Betoptic S package insert, FDA Inactive Ingredients List (FDA, 2011), and U.S. Patent 4911920 (Jani and Harris, 1990). The postulated formulation contains 0.25% w/v betaxolol as the base; 0.25% w/v Amberlite (polacrilin), 4% mannitol, 0.2% w/v carbomer 934P, and 0.01% benzalkonium chloride. Mannitol is used to adjust the tonicity of the formulation. Salts such as sodium chloride which are commonly used as tonicity-adjusting agents cannot be used in this formulation because the ions would compete with the drug-Amberlite binding. The carbomer is an anionic polymer that forms a plastic structured vehicle. This vehicle was shown to keep the drug-resin particles uniformly suspended up to four weeks.

The manufacturing process has been described by Jani and Rhone (2007). Betaxolol as the hydrochloric acid salt, mannitol, benzalkonium chloride, and disodium edetate are dissolved in purified water. The carbomer suspending agent is also dispersed in purified water. Separately, the Amberlite is acidified and milled so that not less than 100% of the particles are less than 90 μm and not less than 99.5% of the particles are less than 25 μm. The drug, mannitol, carbomer, and milled Amberlite liquids are added together and mixed. The pH of this mixture is adjusted to about 7. At this pH the positively charged betaxolol electrostatically binds to the negatively charged sulfonic acid groups of the Amberlite.

DuraSite is a ophthalmic drug delivery system developed by InSite Vision, DuraSite is comprised of polycarbophil ($\sim 0.8\%$ w/v), disodium edetate ($\sim 0.1\%$ w/v), and sodium chloride ($\sim 0.4\%$ w/v). This ophthalmic drug delivery system (Roy et al., 2004) was used to formulate Besivance (besifloxacin hydrochloride) to achieve sustained ocular drug release. In this case, the polycarbophil, which is an acrylic acid polymer cross-linked with divinyl glycol, forms a polymer–drug complex that is suspended using poloxamer 407.

18.2.3.6 Ophthalmic Penetration Enhancers

Penetration or absorption enhancers increase drug absorption. Drug transport across the cornea occurs by two primary routes: transcellular and paracellular. Penetration enhancers can affect one or both routes. Transcellular absorption can occur through passive concentration gradient diffusion, active transport, endocytosis, and transcytosis. The permeation rate is largely dependent on a drug's size, aqueous solubility, and lipophilicity. Drugs that are absorbed by permeating between or around cells follow the paracellular route. Drug transport often occurs by both routes, with one of the routes being the primary pathway for absorption. Cells are held together by four major anatomical features: tight junctions or zonula occludens, zonula adherens, gap junctions, and desmosomes. At the tight junctions the neighboring cells join to form a barrier to fluid. The tight junction is formed by transmembrane proteins that have extracellular domains that interpenetrate through attractive interactions such as van der Waals forces and hydrogen bonding. The tight junction properties are influenced by secondary messengers such as tyrosine kinases, calcium ion, G proteins, and others. The zonula adherens junction is formed by the transmembrane protein cadherin. The tightness of this junction is modulated by the presence of calcium ions. The gap junction connects the cytoplasm of two cells directly via connections. The gap junctions do not pose a barrier to drug penetration. The desmosomes bind the cells together by cell adhesion proteins. Absorption enhancers can act by a number of mechanisms. They can interact with the phospholipid bilayer and make it more fluid. Similarly, they can interact with the tranmembrane proteins between cells to loosen the junctions and increase drug permeability. Calcium chelation can affect the nature of the tight junctions and zonula adherens. A comprehensive review on penetration enhancer mechansims is that of Lee et al. (1991).

A number of penetration or absorption enhancers are listed in Appendix 18.1. A number of these compounds are commonly used ophthalmic excipients that also have the ability to enhance the absorption of drugs. This is especially true of benzalkonium chloride and disodium edetate, which are well-known antimicrobial preservatives that can also act as permeation enhancers. There are safety concerns regarding the long-term use of molecules that disrupt and change the natural protective and vision properties of the eye. Furrer et al. (2002) studied the ocular tolerance of nine permeation enhancers compared to normal saline. At the 1% concentration, the irritancy rank order was dimethyl sulfoxide $\sim$ decamethonium $<$ Tween 20 (polyoxyethylene sorbitan monolaurate) $<$ Brij 35 [poly(oxyethylene lauryl ether)] $\sim$ edetate $<$ glycocholate $\sim$ cholate $<<$ saponin $<<$ fusidate. The 0.5% solutions for Tween 20, Brij 35, and cholate were statistically less irritating than the 1% solutions. All the absorption enhancers showed a statistically significant greater degree of irritation when compared to normal saline. Permeation studies are carried out almost exclusively in in vitro or animal models. In animal studies, penetration enhancers can increase the apparent permeability 1.5- to as much as 60-fold (Lee and Robinson, 2003). Typically, the penetration enhancer concentrations evaluated for approved ophthalmic excipients are studied at their approved levels.

When possible, it is best to optimize drug permeation at the lead drug candidate selection stage. This may mean using the traditional approach of optimizing drug solubility and lipophilicity. In a number of cases this has meant the design of prodrugs and soft drugs to optimize drug absorption. More recently, researchers are looking to exploit influx transporters that have been identified.

18.2.3.7 Ophthalmic Mucoadhesives   A number of mechanisms have been proposed specifically for bioadhesives and mucoadhesives. Physical entanglement and interpenetration have been suggested as a mechanism for bioadhesion. Other attractive forces, such as van der Waals, hydrogen, electrostatic, and hydrophobic bonding, have been cited. Repulsive forces include steric hinderance and electrostatic repulsion. Other factors that affect interaction include chain mobility, branching, chain length, cross-linking density, swelling, and hydrated porosity.

As mentioned earlier, mucin is a major component of the tear film. Mucoproteins, which are rich in sialic acid, bathe the cornea membrane and attach to the cornea microvilli. The pK_a value of sialic acid is 2.6, so for practical purposes the acid is nearly 100% negatively charged at pH 4.6 and higher. This means that at the pH of tears, 7.4, the surface of the cornea is negatively charged. Similarly, most commercial ophthalmics are buffered at a pH of 5 and higher, so once again the cornea surface will be negatively charged. According to Johnston et al. (2003), carbomer ($pK_a \sim 6$), sodium carboxymethylcellulose ($pK_a \sim 4.3$), polycarbophil (pKa ~ 6), and sodium alginate ($pK_a \sim 3.5$) are excellent mucoadhesives. Interactions between mucin and these polymers are reported to be strongest in acidic pH. This suggests that the acid or protonated form of the polymer acts as a mucoadhesive by bonding with the anionic mucin. One would expect that carbomer and polycarbophil should interact more strongly since they would be more protonated and more capable than carboxymethylcellulose or alginate of interacting, which would have significant negative charges at pH 4 to 5.

Chitosan, poly(D-glucosamine), is a cationic polyamine that has a high positive charge density below 6.5. Chitosan is a well-known mucoadhesive that has been studied extensively for oral controlled drug delivery. One would expect the electrostatic attraction between the anionic sialic acid and cationic amines to produce good mucoadhesion. Recently, Singh and Shinde (2011) reported the preparation of brimonidine tartrate–loaded chistosan nanoparticles that furnished significant in vivo sustained effect compared to conventional eyedrops. In another study, chitosan-coated nanoparticles interacted with a mucin dispersion over a 6-hour period. This interaction was attributed to the electrostatic interaction of chitosan and mucin (Yoncheva et al., 2011).

A bioadhesive ophthalmic drug insert (BODI) sustained drug delivery system has been reviewed by Felt-Baeyens and Gurny (2007). This system contains hydrophilic, hydrophobic, and bioadhesive polymers that are extruded into soluble rod-shaped inserts which can release drug over days and weeks. Carbomer was reported as the bioadhesive used to keep the insert positioned in the inferior lateral conjunctival cul-de-sac. The insert was shown to be well tolerated. High extrusion temperatures of 140 to 160°C and the use of gamma-radiation sterilization limit the number of drugs that can be formulated in this system.

18.2.3.8 Other Dosage Formulations

Ophthalmic emulsions have been studied widely and shown to reduce drug-induced irritation, to increase bioavailability, and to offer controlled release. However, emulsion stability and irritation caused by surfactants and emulsifiers has limited emulsion commercial use. Restatis is the only commercially available ophthalmic emulsion available in the United States. Restatis (cyclosporine ophthalmic emulsion, 0.05%) is used to increase tear production in patients suffering from inflammation associated with keratoconjuctivitis sicca. Abdulrazik et al. (2007) describe this product as an anionic *submicron emulsion* (SME) or *nanoemulsion*. Restatis is a white opaque to slightly translucent emulsion. The Restatis formulation composition is proprietary but is speculated to be cyclosporine (0.05%), polysorbate 80 (0.05%) as an emulsifier, glycerin (2.2%) as a tonicity-adjusting agent, carbomer (4%) as a viscosity structure-forming agent and emulsifier, and castor oil (5%) as a solvent for cyclosporine. The product is available as sterile 0.4-mL single-use vials. Notably, the very low phase volume ratio of 5% and the use of the more polar castor oil are important factors that lead to the ability to manufacture an SME (Ding et al., 1995).

Many other drug delivery systems have been studied and have yet to reach the marketplace. Microspheres, microcapsules, nanoparticles, nanocapsules, micelles, liposomes, niosomes, dendrimers, cyclodextrin complexes, and other ophthalmic drug delivery dosage forms have been reported. For more detailed discussions the reader is directed to published work of Wilson et al. (2001), Lang et al. (2002), Kothuri et al. (2003), and Abdulrazik et al. (2007).

18.2.3.9 Nondegradable and Erodible/Biodegradable Ocular Inserts

The first ocular insert was the result of pioneering inventions by Ness (1971) and Higuchi (1971). The nonerodible, nonbiodegradable ocular insert was commercialized by the Alza Corporation to sustain the release of pilocarpine to treat glaucoma. The Ocusert Pilo-20 and Pilo-40 (pilocarpine ocular system, U.S.P.) are elliptical flexible laminate patches which are placed in the conjunctival sac and deliver 20 or 40 µg of pilocarpine per hour for 7 days. After the 7 days the insert is removed and replaced with a new system. The flexible laminate system contains a pilocarpine–alginic acid drug reservoir and rate-controlling ethylene–vinyl acetate copolymer membrane. Some disadvantages of the system have been reported, such as insertion and retrieval of the insert, and inadvertent expulsion of the insert especially during sleeping.

The use of drug-containing contact lenses has been explored since the 1960s. There continue to be novel technologies that integrate drug delivery and contact lenses, such as molecularly imprinted hydrogels, liposome-loaded lenses, immobilization of nanoparticles onto the contact lens surface, and gas-permeable scleral drug delivery lenses. Erodible ocular inserts have the advantage that they do not need to be retrieved from the eye once they are inserted. Lacrisert is a very small 3.5×1.27 mm cylindrical rod of hydroxypropyl methylcellulose (hypromellose) that is inserted into the cul-de-sac to treat dry eye. The insert can be manufactured by extrusion and terminally sterilized by gamma irradiation. The insert is designed to provide lubrication for 1 day. The primary disadvantage of erodible or biodegradable inserts is designing zero-order drug release, which is problematic and creates a number of challenges for the design scientist.

Bioadhesive controlled-release ocular inserts have also been studied (Felt-Baeyens and Gurny, 2007). The bioadhesive inserts were composed of approximately 3% carbomer (Carbopol 934P), 30% ethylcellulose (Ethocel N50), and 67% hydroxylpropyl methylcellulose (Klucel HXF). The inserts were extruded as rods at 140 to 160°C and 200 to 300 kPa of pressure. The bioadhesive inserts were placed in the cul-de-sac to provide sustained release. The high-temperature extrusion condition limits the number of drugs that can be developed with this process.

18.2.3.10 Device-Assisted Ocular Penetration: Iontophoresis, Ultrasound, Electroporation, and Microneedles

Since the early 1980s, continual progress has been made regarding the safety and efficacy of ocular iontophoretic drug delivery. Iontophoresis uses direct current to push topically applied, charged drug molecules through the cornea or transsclera by exploiting the electrostatic principle that like charges repel. In general, an electrode carrying the same charge as the topically applied ionized drug is used to drive the drug across the eye membranes. The ground, which carries a charge opposite to that of the drug, is placed on the forehead or in reasonably close proximity to complete the circuit. Iontophoresis typically operates at a low voltage of 10 V or less. During the electrophoresis procedure, the resistance of the cell membrane decreases and the voltage is adjusted to maintain about 0.5 mA/cm^2 of direct current. In 2010, EyeGate Pharma initiated a pivotal phase 3 clinical trail using its proprietary

EyeGate II ocular iontophoretic delivery system to treat dry eye syndrome with dexamethasone phosphate. A successful phase III study could bolster additional ocular iontophoretic drug delivery research. Ocuphor (Iomed, Inc.) and Vsulex (Acient, Inc.) are also evaluating transscleral iontophoresis. More information concerning ocular iontophoresis may be found in articles by Myles et al. (2003) and Eljarrat-Binstock et al. (2007).

Interest in using ultrasound or sonophoresis to deliver drugs through the eye has lagged iontophoresis by a decade or so and there are a limited number of reports. A number of mechanisms have been offered to explain how sonophoresis increases drug penetration. It is thought that the ultrasound frequency disrupts the lipid bilayer. The formation and subsequent collapse of gas bubbles may lead to cavitation and the formation of holes that decrease diffusional resistance. An increase in heat resulting from increased vibration due to exposure to the ultrasound frequency can also increase the diffusional coefficient. It is also speculated that acoustic shear rising from unequal distribution of pressure forces may enhance diffusion and penetration. Zderic et al., (2004) employed ultrasound to deliver sodium fluorescein across the cornea of rabbit eyes. The study was performed at 880 kHz and at intensities ranging from 0.19 W/cm^2 to 0.56 W/cm^2. The permeability increased 2.1 to 4.2 times as the ultrasound intensity was increased from 0.19 to 0.56 W/cm^2. The authors reported that the surface corneal epithelium appeared swollen and lighter in color, which suggested membrane rupture. The surface epithelium also had 3- to 10-μm holes. No structural changes were observed in the stroma. Studies continue to elucidate the mechanisms of enhanced penetration and impact of sonophoresis on body tissues.

Similarly, there are few reports that study ocular electroporation, and most of the studies are related to gene therapy. Electroporation typically employs short bursts of electric pulses that range between 10 and 30 V. The electrical pulses generally have a duration of 20 to 50 ms with about a 180-ms interval between pulses.

Transscleral microneedle drug delivery has also been reported recently. The use of microneedle ocular drug delivery has shown significant increases in drug permeation, intraocular drug effects, and drug concentrations to the back of the eye. Patel et al. (2011) showed that hollow microneedles can deliver nanoparticle and microparticle suspensions into the suprachorodial space. Donnelly et al. (2010) reviewed microneedle drug delivery systems, including transscleral drug delivery.

18.2.4 Intravitreal Injections and Implants

Even with the significant advances in ocular drug delivery, there are cases in which insufficient drug concentrations are achieved for posterior segment ophthalmic diseases. Additionally, many posterior eye diseases are chronic and necessitate chronic therapy. Intravitreal injections are generally used in such cases. There are a number of limitations to intravitreal injections. Preservative and other excipients should be used with a great deal of caution because they can have negative effects on many eye tissues. Moreover, drug half-lives in the vitreous are relatively short, ranging from 5 to 30 hours. Additionally, the injection volume is limited to 0.1 to 0.2 mL.

Two intraocular products, Miostat (carbachol), a miotic, and Miochol (lyophilized acetylcholine), use *balanced salt solution* (BSS) as the vehicle. Each milliliter of BSS is composed of 0.64% sodium chloride, 0.075% potassium chloride, 0.048% calcium chloride dihydrate, 0.03% magnesium chloride hexahydrate, 0.39% sodium acetate trihydrate, and 0.17% sodium citrate dihydrate adjusted to a pH of 7.5. The osmolality of the BSS is 300 mOsmol/kg.

The first FDA-approved antisense product was Vitravene (fomovirsen sodium), which is an intravitreal injection of a 21-phosphorothioate oligonucleotide that is indicated for cytomegalovirus retinitis. The formulation is a sterile preservative-free bicarbonate buffer isotonic injection solution that has a pH of 8.7 and an osmolality of 290 mOsmol/kg. The injeciton volume is only 0.05 mL.

There are currently three intravitreal inserts: Vitrasert (ganciclovir), Retisert (fluocinolone acetonide), and Ozurdex (dexamethasone). Vitrasert and Retisert are nondegradable. Vitrasert is a tablet containing drug and 0.25% magnesium stearate that is coated with poly(vinyl alcohol) and ethylene–vinyl acetate. Retisert is a tablet comprising drug, microcrystalline cellulose, poly(vinyl alcohol), and magnesium stearate. The formulation is designed to deliver 0.3 to 0.4 µg/day for about 30 months. Ozurdex is a biodegradable D,L-lactide-*co*-glycolide polymer system. Six months of clinical data have been reported.

18.3   PROCESS DESIGN

Ophthalmic products are required to be sterile [21 *Code of Federal Regulations* (CFR) 200.50(a)(1)]. Since many topical eyedrops are dispensed in low-density polyethylene plastic bottles, many opththalmic products are terminally sterilized using gamma radiation. Some unit-dose eyedrops are manufactured using an aseptic form–fill–seal process. The specific processes for manufacturing dispersed systems and sterile formulations may be reviewed in Chapters 16 and 17.

18.4   CONTAINER CLOSURE SYSTEM DESIGN

A high level of concern is associated with ophthalmic container closure system protection, integrity, and inertness. The need to evaluate leachables and extractables is the same as that required for sterile products. The American Academy of Ophthalmology has recommended a uniform color-coding system for the caps and labels of all topical ocular medications (FDA, 1999). The system recommended is shown in Table 18.4.

18.5   ATTRIBUTE TESTS

Ophthalmic ointments should pass the U.S.P. criteria set out in Chapter <771>, Ophthalmic Ointments, which describes a tube leakage test and determination of metal particles. Microscopic examination and particle size determinations are

TABLE 18.4   Color Coding Recommended for Topical Ophthalmic Medications

Therapeutic Class	Color[a]
Anti-infectives	Tan
Anti-inflammatories and steroids	Pink
Mydriatics and cycloplegics	Red
Nonsteroidal anti-inflammatories	Gray
Miotics	Green
Beta-blockers	Yellow
Adrenergic agonists	Purple
Carbonic anhydrase inhibitors	Orange
Prostaglandin analogs	Turquoise

[a] Pantone numbers are also provided in the FDA (1999) guidance.

often done for ophthalmic ointments as well. The compendial sterility requirements should also be met. U.S.P. <1211>, Sterilization and Sterility Assurance of Compendial Articles and <71>, Sterility Tests are critical product attributes. The sterile integrity of the container closure system is discussed in U.S.P. <1207>, Sterile Packaging Integrity Evaluation, and terminal sterilization is reviewed in U.S.P. <1222>, Terminally Sterilized Pharmaceuticals. For multiple-use containers that employ microbiological preservatives, preservative identification, preservative assay, and preservative efficacy- (the Antimicrobial Effectiveness Test (U.S.P. <51>) needs to be performed.

18.6   NEW DRUG APPLICATION STABILTY ASSESSMENT

The new drug application stability program for ophthalmic products should incorporate the timing and testing that is noted for sterile products in Chapter 17. Other dosage-form specific testing should also be incorporated and is described in earlier chapters.

REFERENCES

Abdulrazik M, Behar-Cohen F, Benita S. Drug delivery systems for enhanced ocular absorption. In: Touitou E, Barry BW, eds. *Enhancement in Drug Delivery*. Boca Raton, FL: Taylor & Francis; 2007. pp. 489–526.

Adler CA, Maurice DM, Paterson ME. The effect of viscosity of the vehicle on the penetration of fluorescein into the human eye. Exp. Eye Res. 1971;11:34–42.

Ali Y, Beck RE, Sport RC. Process for manufacturing ophthalmic suspensions. U.S. Patent 6071904. 2000.

Butty P, Mayer JM, Kalin P, Etter JC. In situ determination of pH and buffer capacity of the mouse corneal surface using epi-microscopic fluorescence. Eur. J. Pharm Biopharm. 1996;42: 399–401.

Chrai SS, Patton TF, Mehta A, Robinson Jr. Lacrimal and instilled fluid dynamics in rabbit eyes. J. Pharm. Sci. 1973;62:1112–1121.

Code of Federal Regulations. 21 CFR 200.50(a)(1). Ophthalmic preparations and dispensers. http://law.justia.com/cfr/title21/21-4.0.1.1.1.3.1.1.html. Accessed Dec. 2011.

Ding S, Tien WL, Olejnik O. Nonirritating emulsions for sensitive tissue. U.S. Patent 5474979. 1995.

Donnelly RF, Singh TRR, Woolfson AD. Microneedle-based drug delivery systems: microfabrication, drug delivery, and safety. Drug Deliv. 2010;17:187–207.

Eljarrat-Binstock E, Frucht-Pery J, Domb AJ. Iontophoresis for ocular drug delivery. In: Touitou E, Barry BW, eds. Enhancement in Drug Delivery. Boca Raton, FL: Taylor & Francis; 2007. pp. 549–71.

FDA. 1999. Guidance for industry: Container closure systems for packaging human drugs and biologics. U.S. Department of Health and Human Services, Food and Drug Administration Center for Drug Evaluation and Research and Center for Biologics Evaluation and Research. May 1999. http://www.fda.gov./downloads/Drugs/GuidanceComplianceRegulatoryInformation/Guidances/ucm070551.pdf. Accessed December 2011.

FDA. 2011. Inactive ingredients list. http://www.fda.gov/Drugs/InformationOnDrugs/ucm113978.htm. Accessed Jan. 2012.

Felt-Baeyens O, Gurny R. Bioadhesive ophthalmic drug inserts (BODI). In: Rathbone MJ, Hadgraft J, Roberts MS, eds. Modified-Release Drug Delivery Technology. Drugs and the Pharmaceutical Sciences, Vol. 126. New York: Informa Healthcare; 2007. pp. 341–347.

Furrer P, Mayer JM, Plazonnet B, Gurny R. Ocular tolerance of absorption enhancers in ophthalmic preparations. AAPS PharmSci 2002;4(1): art. 2, 6–10. doi:10.1208/ps04102.

Gaudana R, Ananthula HK, Parenky A, Mitra AK. Ocular drug delivery. AAPS J. 2010;12: 348–360.

Higuchi T. Ocular insert. U.S. patent 3630200. 1971.

Hind HW, Goyan FM. A new concept of the role of hydrogen ion concentration and buffer system in the preparation of ophthalmic solutions. J. Am. Pharm. Assoc. 1947;36:33–41.

Jani R, Harris RG. Sustained release, comfort formulation for glaucoma therapy. U.S. Patent 4911920. 1990.

Jani R, Rhone E. Ion exchange resin technology for ophthalmic applications. In: Rathbone MJ, Hadgraft J, Roberts MS, eds. Modified-Release Drug Delivery Technology. Drugs and the Pharmaceutical Sciences, Vol. 126. New York: Informa Healthcare; 2007. pp. 315–328.

Johnston TP, Dias CS, Mitra AK, Alur H. Mucoadhesive polymers in ophthalmic drug delivery. In: Mitra AK, ed. Ophthalmic Drug Delivery Systems, 2nd ed. Drugs and the Pharmaceutical Sciences, Vol. 130. New York: Marcel Dekker; 2003. pp. 409–435.

Kaur IP, Batra A. Ocular penetration enhancers. In: Touitou E, Barry BW, eds. Enhancement in Drug Delivery. Boca Raton, FL: CRC Press; 2007. pp. 527–548.

Keister JC, Cooper ER, Missel PJ, Lang JC, Hager DF. Limits on optimizing ocular drug delivery. J. Pharm. Sci. 1991;80:50–53.

Kompella UB, Kadam RS, Lee VHL. Recent advances in ophthalmic drug delivery. Ther. Deliv. 2010;Sept. doi:10.4155/TDE.10.40.

Kothuri MK, Pinnamaneni S, Das NG, Das SK. Microparticles and nanoparticles in ocular drug delivery. In: Mitra AK, ed. Ophthalmic Drug Delivery Systems, 2nd ed. Drugs and the Pharmaceutical Sciences, Vol. 130. New York: Marcel Dekker; 2003. pp. 437–466.

Lang JC, Roehrs RE, Rodeheaver DP, Missel PJ, Jani R, Chowhan MA. Design and evaluation of ophthalmic pharmaceutical products. In: Banker GS, Rhodes CT, eds. Modern Pharmaceutics, 4th ed. Drugs and the Pharmaceutical Sciences, Vol. 121. NewYork: Marcel Dekker; 2002. pp. 415–478.

Lang JC, Roehrs RE, Jani R. Ophthalmic preparations. In: Troy DB, ed. *Remington: The Science and Practice of Pharmacy*, 21st ed. Philadelphia: Lippincott Williams & Wilkins; 2006. pp. 850–870.

Lee TWY, Robinson Jr. Ocular penetration enhancers. In: Mitra AK, ed. *Ophthalmic Drug Delivery Systems*, 2nd ed. Drugs and the Pharmaceutical Sciences, Vol. 130. New York: Marcel Dekker; 2003. pp. 281–308.

Lee VH, Yamamoto A, Kompella UB. Mucosal penetration enhancers for facilitation of peptide and protein drug absorption. Crit. Rev. Ther. Drug Carrier Syst. 1991;8:91–192.

Linn ML, Jones LT. Rate of lacrimal excretion of ophthalmic vehicles. Am. J. Ophthalmol. 1968;65:76–78.

Mazuel C, Friteyre M-C. Ophthalmic composition of the type which undergoes liquid–gel phase transition. U.S. Patent 4861760. 1989.

Missel PJT, Lang JC, Jani R. Use of combination gelling polysaccharides and finely divided drug carrier substrates in topical ophthalmic compositions. U.S. Patent 5212162. 1993.

Myles ME, Loutsch JM, Higaki S, Hill JM. Ocular iontophoresis. In: Mitra AK, ed. *Ophthalmic Drug Delivery Systems*, 2nd ed. Drugs and the Pharmaceutical Sciences, Vol. 130. New York: Marcel Dekker; 2003. pp. 365–408.

Ness RA. Ocular insert. U.S. Patent 3618604. 1971.

Patel SR, Lin AS, Edelhauser HF, Prausnitz MR. Suprachoroidal drug delivery to the back of the eye using hollow microneedles. Pharm. Res. 2011;28:166–176.

Rowe RC, Sheskey PJ, Quinn ME, eds. *Handbook of Pharmaceutical Excipients*, 6th ed. London: Pharmaceutical Press; Washington, DC: American Pharmacists Association; 2009.

Roy S, Chandrasekaran SK, Imamori K, Asaoka T, Shibata A, Takahashi M, Bowman LM. Quinolone carboxylic acid compositions and related methods of treatment. U.S. Patent 6685958. 2004.

Schoenwald RD, Stewart P. Effect of particle size on ophthalmic bioavailability of dexamethasone suspension in rabbits. J. Pharm. Sci. 1980;69:391–394.

Singh KH, Shinde UA. Chitosan nanoparticles for controlled delivery of brimonidine tartrate to the ocular membrane. Pharmazie 2011;66:594–599.

Wilson CG, Zhu YP, Kurmala P, Rao LS, Dhillon B. Ophthalmic drug delivery. In: Hillery AM, Lloyd AW, Swarbrick J, eds. *Drug Delivery and Targeting for Pharmacists and Pharmaceutical Scientists*. Boca Raton, FL: CRC Press; 2001. pp. 329–354.

Wilson CG, Semenova EM, Hughes PM, Olejnik O. Eye structure and physiological functions. In: Touitou E, Barry BW, eds. *Enhancement in Drug Delivery*. Boca Raton, FL: Taylor & Francis; 2007. pp. 473–487.

Wong W. Topical nepafenac formulations. U.S. Patent 7834059. 2010.

Yoncheva K, Vandervoort J, Ludwig A. Development of mucoadhesive poly(lactide-*co*-glycolide) nanoparticles for ocular application. Pharm. Dev. Technol. 2011;16:29–35.

Zderic V, Clark JI, Martin RW, Vaezy S. Ultrasound-enhanced transcorneal drug delivery. Cornea 2004;23:804–811.

GLOSSARY

BODI	Bioadhesive ophthalmic drug insert.
BSS	Balanced salt solution.
CFR	*Code of Federal Regulations*.
E.P.	*European Pharmacopoeia*.
FDA	U.S. Food and Drug Administration.
J.P.	*Japanese Pharmacopoeia*.
N.F.	*National Formulary*.
SME	Submicron emulsion.
U.S.P.	*United States Pharmacopeia*.

APPENDIX

APPENDIX 18.1   Ophthalmic Excipients

Excipient	Trade Name[a]	Use (% w/v)[b]
Absorption and Permeability Enhancers		
Antimicrobial Preservatives[c,d]		
Benzalkonium chloride, E.P./J.P./U.S.P.–N.F.	Hyamine 3500; Zephiran	0.004–2
Benzyl alcohol, E.P./J.P./U.S.P.–N.F.	Phenyl carbinol; α-toluenol	0.5[e]
Cetylpridinium chloride, E.P./U.S.P.–N.F.	Pristacin	
Chlorhexidine, E.P./J.P./U.S.P.–N.F.	Nolvasan; Tubulicid	0.5[e]
Chlorobutanol, E.P./J.P./U.S.P.–N.F.	Methaform; Sedaform	0.2–0.65
Paraben methyl, E.P./J.P./U.S.P.–N.F.	Solbrol M; Tegosept M	0.05
Paraben propyl, E.P./J.P./U.S.P.–N.F.	Solbrol P; Tegosept P	0.01–0.015
Phenylethyl alcohol, U.S.P.–N.F.	Benzyl carbinol; PEA	0.25–0.5
Calcium chelator[c]		
Edetate disodium, E.P./J.P./U.S.P.–N.F.	Disodium EDTA; edathamil disodium; versene disodium	0.01–10.00
Fatty acids[c,d]		
Capric acid, E.P./U.S.P.–N.F.	Decanoic acid	0.5[e]
Oleic acid, E.P./U.S.P.–N.F.	Crodolene; Emersol	
Glycosides[d]		
Digitonin	Digitin	
Saponin	Sapogenin glycosides	0.01–0.5[e]
Mucoadhesives		
Carbomer, E.P./U.S.P.–N.F.	Acrypol; Acritamer; carbopol; Pemulen; polyacrylic acid	0.05–4.0
Carboxymethylcellulose sodium, E.P./J.P./U.S.P.–N.F.	Akucell; Aquasorb; Nymcel ZSB; Tylose CB	0.5
Polycarbophil, U.S.P.–N.F.	Noveon AA-1; acrylic acid polymer cross-linked with divinyl glycol	0.86–0.9
Surfactants[c,d]		
Nonionic surfactants		
Octoxynol-40	Triton X-405	0.01–0.05
Nonoxynol-9	—	0.125
Polyoxyl lauryl ether, E.P./J.P./U.S.P.–N.F.	Laureth-9; polyoxyethylene-9-lauryl ether; Volpo L9	

(*continued*)

APPENDIX 18.1 *(Continued)*

Excipient	Trade Name[a]	Use (% w/v)[b]
Polysorbate 80, E.P./J.P./U.S.P.–N.F.	Atlas E; Capmul POE-O; Tween 80	0.05–4
Sorbitan monstearate, U.S.P.–N.F.	Arlacel 60; Atlas 110K; Capmul S; Span 60	
Anionic surfactants	Desoxycholic acid	0.025–0.1[e]
Deoxycholic acid	Cholaic acid	1[6]
Taurocholic acid	—	0.075–0.1[e]
Taurodeoxycholic acid		
Other		
Chitosan[c,d]	Deacetylated chitin	
Cyclodextrins[f]	Captisol; Cavitron; Encapsin	
Laruocapram[c,d]	Azone	0.025–5[e]
Antimicrobial Preservatives		
Benzalkonium chloride, E.P./J.P./U.S.P.–N.F.	Hyamine 3500; Zephiran	0.004–2
Edetate disodium, E.P./J.P./U.S.P.–N.F.	Disodium EDTA; edathamil disodium; versene disodium	0.01–10.00
Benzododecinium bromide	BDAB; benzyldodecyldimethyl ammonium bromide	0.012
Ethanol, E.P./J.P./U.S.P.–N.F.	Absolute alcohol; anhydrous alcohol; grain alcohol	0.5–1.4
Paraben methyl, E.P./J.P./U.S.P.–N.F.	Solbrol M; Tegosept M	0.05
Paraben propyl, E.P./J.P./U.S.P.–N.F.	Solbrol P; Tegosept P	0.01–0.015
Perborate sodium	Dexol; GenAqua	
Phenylethyl alcohol, U.S.P.–N.F.	Benzyl carbinol; PEA	0.25–0.5
Phenylmercuric acetate, E.P./U.S.P.–N.F.	Agrosan GN 5; Mersolite 8; Unisan	0.0008
Phenylmercuric nitrate, E.P./U.S.P.–N.F.	Merpectogel; Phe-Mer-Nite	0.002
Polyaminopropyl biguanide	Dymed	0.00005[g]
Polyhexamethylene biguanide	PHMB	
Polyquaternium-1	Polidronium chloride; Polyquad	0.001[h]
sofZia		
Boric acid, E.P./J.P./U.S.P.–N.F.	Borofax	0.06
Propylene glycol, E.P./J.P./U.S.P.–N.F.	Sirlene; Solargard P; Ucar 35	0.75
Sorbitol, E.P./J.P./U.S.P.–N.F.	Neosorb; Sorbitab	0.25
Zinc chloride	Butter of zinc; Zintrace	0.0025
Sorbic acid, E.P./U.S.P.–N.F.	Sorbistat K	0.1–0.2
Stabilized oxyborate complex	Dissipate[h]	

APPENDIX 18.1 *(Continued)*

Excipient	Trade Name[a]	Use (% w/v)[b]
Stabilized oxychloro complex	Purite	0.005
Thimerosal, E.P./U.S.P.–N.F.	Thimerosal Sigmaultra	0.001–1
	Antioxidants	
Creatinine	TEGO Cosmo C 250	0.2–0.5
Bisulfite sodium	Sodium hydrogen sulfite	0.06–0.1
Sulfite sodium, E.P./J.P./U.S.P.–N.F.	—	0.2
Thiosulfate sodium, E.P./J.P./U.S.P.–N.F.	Ametox; Sodothiol; Sulfothiorine	0.31–5
	Buffers	
Acetate sodium, E.P./J.P./U.S.P.–N.F.	Plasmafusin; Thomaegelin	0.05–1.28
Acetic acid, E.P./J.P./U.S.P.–N.F.	Glacial acetic acid; vinegar acid	0.09–0.2
Borate sodium, E.P./J.P./U.S.P.–N.F.	Borax	0.028–1.1
Boric acid, E.P./J.P./U.S.P.–N.F.	Borofax	0.06–37.2
Citrate sodium, E.P./J.P./U.S.P.–N.F.	Trisodium citrate dihydrate	0.3–2.2
Citric acid monohydrate, E.P./J.P./U.S.P.–N.F.	Acetonum	0.05–0.2
Hydrochloric acid, E.P./J.P./U.S.P.–N.F.	Chlorohydric acid; hydrogen chloride	0.17–1.06
Phosphate potassium monobasic	Potassium dihydrogen orthophosphate; Sorensen's potassium phosphate	0.065–0.44
Sodium hydroxide, E.P./J.P./U.S.P.–N.F.	Aetznatron; Rohrputz	0.1–0.4
Sulfuric acid, E.P./U.S.P.–N.F.	—	0.02
Sulfate sodium	Glauber's salt; Mirabitite	0.15–1.2
Tromethamine	TRIS	0.75–0.94
	Chelating Agents	
Edetate disodium, E.P./J.P./U.S.P.–N.F.	Disodium EDTA; edathamil disodium; versene disodium	0.01–10.00
	Ion-Exchange Agents	
Polacrilin potassium, U.S.P.–N.F.	Amberlite IRP-88; methacrylic acid with divinylbenzene, potassium salt	
Polycarbophil, U.S.P.–N.F.	Noveon AA-1; acrylic acid polymer cross-linked with divinyl glycol	0.86–0.9
	Solubilizing Agents	
Castor oil, E.P./J.P./U.S.P.–N.F.	EmCon CO; Lipovol CO	5.0

(continued)

APPENDIX 18.1 (*Continued*)

Excipient	Trade Name[a]	Use (% w/v)[b]
Glycerin, E.P./J.P./U.S.P.–N.F.	Croderol; Kemstrene; Optim; Pricerine	2.2–3
Poly(ethylene glycol), E.P./J.P./U.S.P.–N.F.	Carbowax; Lipoxol; PEG	
Ophthalmic ointment	PEG 400	5.0
Ophthalmic solution	PEG 8000	2.0
Poly(propylene glycol)	—	15
Propylene glycol, E.P./J.P./U.S.P.–N.F.	Sirlene; Solargard P; Ucar 35	0.75–10.00
Poloxamer, E.P./U.S.P.–N.F.	Lutrol; Pluronic; Supronic; block copolymers of ethylene oxide and propylene oxide	0.1–0.2
Suspending and Thickening Agents		
Carbomer, E.P./U.S.P.–N.F.	Acrypol; Acritamer; Carbopol; Pemulen; polyacrylic acid	0.05–4.0
Carboxymethylcellulose sodium, E.P./J.P./U.S.P.–N.F.	Akucell; Aquasorb; Nymcel ZSB; Tylose CB	0.5
Gellan gum	Gelrite; Phytagel	0.6
Hydroxyethyl cellulose, E.P./U.S.P.–N.F.	Cellosize HEC; Natrosol; Tylose H	0.25–0.5
Hypromellose, E.P./J.P./U.S.P.–N.F.	Benecel MHPC; HPMC; Metolose; Pharmacoat; Tylose MO	0.5–2.25; 50[i]
Methylcellulose, E.P./J.P./U.S.P.–N.F.	Cellacol; Mapolose; Tylose; Viscol	0.164–0.5
Polycarbophil, U.S.P.	Noveon AA-1; acrylic acid polymer cross-linked with divinyl glycol	0.86–0.9
Poly(vinyl alcohol), E.P./U.S.P.–N.F.	Alcotex; Elvanol; Lemol; Polyvinol; PVA	1.4
Povidone	Kollidon; Plasdone; poly(vinylpyrrolidone); PVP	
	Povidone K30	0.6–2.0
	Povidone K90	1.2
Surfactants/Wetting Agents		
Poloxamer, E.P./U.S.P.–N.F.	Lutrol; Pluronic; Supronic; block copolymers of ethylene oxide and propylene oxide	0.1–0.2
Polyoxyl 40 stearate, JP/U.S.P.–N.F.	Crodet S40; Lipal 395; Myrj 52; PEG-40 stearate; poly(oxyethylene stearate)	0.5–7
Polysorbate 80, E.P./J.P./U.S.P.–N.F.	Atlas E; Capmul POE-O; Tween 80	0.05–4
Tyloxapol	Alevaire; Macrocyclon; Triton A-20	0.1–0.3

APPENDIX 18.1   *(Continued)*

Excipient	Trade Name[a]	Use (% w/v)[b]
	Tonicity-Adjusting Agents	
Glycerin, E.P./J.P./U.S.P.–N.F.	Croderol; Kemstrene; Optim; Pricerine	2.2–3
Mannitol, E.P./J.P./U.S.P.–N.F.	Emprove; Mannogem; Pearlitol	2.4–5.6; 23[i]
Potassium chloride, E.P./J.P./U.S.P.–N.F.	Chloride of potash	0.075–0.14; 22.2[i]
Sodium chloride, E.P./J.P./U.S.P.–N.F.	Alberger	0.41–0.9; 55[i]
	Vehicles	
Mineral oil, E.P./J.P./U.S.P.–N.F.	Avatech; Drakeol; Sirius	0.1–59.5
Purified water, E.P./J.P./U.S.P.–N.F.		
Water for injection, E.P./J.P./U.S.P.		
White petrolatum, E.P./J.P./U.S.P.	Snow white; Soft white	85–89

[a] A comprehensive list of products and suppliers is provided in Rowe et al. (2009).

[b] FDA Inactive Ingredient List, http://www.fda.gov/Drugs/InformationOnDrugs/ucm113978.htm. Accessed Jan. 2012.

[c] Loosen tight junctions.

[d] Affect membrane fluidity.

[e] Reported by Lee and Robinson (2003), but not listed in the FDA Inactive Ingredient List.

[f] Solubilization of lipophilic drug.

[g] Renu sensitive multipurpose solution, http://www.bausch.com/en/Our-Products/Contact-Lens-Care/Soft-Contact-Lens-Multipurpose-Solutions/Renu-sensitive-multi-purpose-solution. Accessed Dec. 2011.

[h] The Eye Digest, http://www.agingeye.net/dryeyes/dryeyesdrugtreatment.php. Accessed on Dec. 2011.

[i] Use uncertain.

CHAPTER 19

TRANSDERMAL PRODUCT DESIGN

19.1 INTRODUCTION

As with all dosage forms, there are advantages and disadvantages specific to transdermal drug delivery. Most disadvantages are associated with overcoming the well-evolved barrier function provided by the skin. The advantages and disadvantages of transdermal drug delivery are listed in Table 19.1.

The skin's structure and numerous functions are critical for life. The skin serves as the first barrier to harmful chemical, microbiological, radiation, thermal, and other environmental assaults. It is also the key to regulating body temperature; containing body fluids and tissues; providing tactile, pain, and thermal sensory functions; and handling synthesis, metabolic, and excretion roles. The skin is generally described as a trilayer organ divided into the *epidermis, dermis*, and *hypodermis* or *subcutaneous fatty tissue* (see Figure 19.1). Transdermal drug delivery is affected primarily by the epidermis, which is the primary barrier to drug absorption, and the dermis, which is composed of connective tissue that gives the skin its elasticity. The dermis is also supplied with blood vessels that carry a drug into systemic circulation. The epidermis is only 100 to 150 μm thick and is subclassified into the viable epidermis and stratum corneum. The viable epidermis is stratified into layers that result from progressive keratinocyte differentiation. The keratinocytes make up approximately 90% of the epidermis. Desmosomes or maculae adherens are specialized structures that form "spot welds" that create a junction complex between adjacent keratinocytes. The stratum corneum or horny layer or cornified layer is the nonviable layer of the skin and is the major barrier to both drugs and water permeation. The corneocytes are comprised mainly of keratin fibers and a cell envelope. The corneocytes are embedded in lipids comprised of about equimolar amounts of ceramides, cholesterol, and fatty acids. The stratum corneum is 10 to 25 μm thick and is made up of about 20 layers of corneocytes. The stratum corneum can take up to 15 to 20% of dry weight at modest relative humidities. When the skin is occluded, it will imbibe water, making the stratum corneum more pliable and increasing drug permeability through the corneocytes. In general, most

Integrated Pharmaceutics: Applied Preformulation, Product Design, and Regulatory Science,
First Edition. Antoine Al-Achi, Mali Ram Gupta, William Craig Stagner.
© 2013 John Wiley & Sons, Inc. Published 2013 by John Wiley & Sons, Inc.

TABLE 19.1   Advantages and Disadvantages of Transdermal Drug Delivery

Advantages	Disadvantages
Provide predictable controlled delivery	Passive diffusion generally limited to:
Can maintain potent drugs within narrow therapeutic windows	Highly potent drugs; highest daily dose is about 20 mg/day (effective plasma concentrations in ng/mL)
Can maintain steady-state blood levels for short half-life drugs	Drug log partition coefficient range of 0.8 to 3.5
Obtain constant blood level over the delivery system lifetime	Molecular mass below 500 Da
Controlled delivery can be maintained for as long as 3 to 7 days in some cases	Solubility of drug in formulation of at least 100 μg/mL
Enabling systems can provide programmed delivery	Develop contact dermatitis with chronic use
Avoids first-pass metabolism	Patches come off accidentally without being noticed
Less hostile environment than that in the gastrointestinal tract	Exposure to moisture (sweating, showering, swimming, etc.) reduces patch adhesion or washes topical drug away
Can discontinue treatment immediately	
Large surface area for varying dosing site	
Less inter- and intrapatient variability	
Good patient acceptance-and compliance	
Pain-free self-administration	
Reasonably cost-effective, especially for longer-acting systems that last 3 to 7 days	

passive diffusion occurs through the tortuous intercellular lipid pathway. The pH of the skin surface ranges from about 4 to 7.

The dermis is 1 mm thick or about 10 times thicker than the epidermis. Its connective tissue serves as a support for the epidermis, appendages such as hair follicles, sweat glands, eccrine glands, arrectores pilorum muscles, blood and lymphatic vessels, and nerve endings. Passive and physical permeation enhancers strive to balance the desire to increase drug absorption and minimize damage to the viable epidermis and dermis.

Another major barrier to transdermal drug delivery is the skin's reactivity to foreign material, including drugs and excipients, as a form of external assault. The skin's reaction is often a function of assault time, intensity, or concentration. The skin is a very immunologically active organ, with *Langerhams cells* present throughout the epidermis and dermis. Langerhams cells are antigen-presenting immune dendritic cells. Their purpose is to elicit a defensive immunogenic response to foreign material. Common skin reactions include contact irritancy, allergenicity, photoirritation, and photoallergenicity. There are well established in vivo safety protocols that should be performed as early as possible in the drug selection process to minimize drug development risks arising from skin safety issues. Many

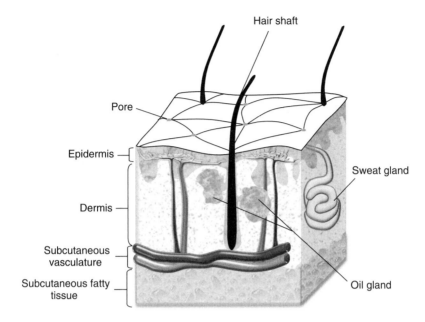

FIGURE 19.1 Skin anatomy.

potential drug candidates have been abandoned because of skin toxicity issues. Similar screening safety studies are also carried out in human volunteers using the drug formulation. Typically, bridging safety studies are done with any change in the formulation because additive or synergistic toxicity issues cannot currently be predicted.

Skin structure and physiology can also affect transdermal bioavailability. Certain areas of the body are more permeable to drugs, such as the axilla, the scrotum, behind the ear, the face, and the scalp. The skin condition, such as level of hydration or disease state, can affect the tightness of the living epidermal cell junctions or lipid content of the lipid matrix in the stratum corneum, which can affect the rate and extent of drug permeation. For example, hydrated skin is more permeable than dry skin. In the elderly, the skin becomes thinner, especially in sun-exposed areas, and it becomes more friable. This patient population's skin may be more sensitive to the strength of the adhesives used in transdermal patches. Irritated or debrided skin is more permeable to drugs, and except for treatment of such areas, application of transdermal delivery systems should be avoided for these areas. Finally, the skin is a living organ and has synthesis and metabolic functions. Presystemic metabolism can affect transdermal bioavailability. The skin has a number of enzymes, including cytochrome P450 enzymes. Drug metabolism is generally much less extensive that that seen in the liver, and the transdermal route does avoid first-pass metabolism. For more information concerning the skin's structure, function, and physiology, the reader is directed to an article by de Jager et al. (2007).

19.2 FORMULATION DESIGN

Even though transdermal drug delivery affords a relatively large number of important and distinguishing advantages, no pharmaceutical company has developed a first-in-class new chemical entity specifically for transdermal drug delivery. In fact, in the United States there are fewer than 20 drugs available in the more than 50 U.S. Food and Drug Administration (FDA)-approved transdermal products. All the drugs were first marketed by some other route. Since scopolamine was introduced into the marketplace in 1979, there has been continued and growing interest in transdermal drug delivery. The *United States Pharmacopeia* (U.S.P.) now lists five transdermal systems: clonidine, estradiol, fentanyl, nicotine, and nitroglycerin. Major breakthroughs in technology and in thinking about the skin have led to enormous advances in the ability to delivery large molecules, such as proteins, vaccines, oligonucleotides, and DNA constructs. Recently, product design thinking has expanded again to considering the use of multiple techniques in coupled, parallel, or staged ways.

19.2.1 Preformulation

Since other dosage forms are typically developed prior to the development of transdermal dosage forms, preformulation data are often available. Additional studies would include solubility determinations with a number of the excipients used specifically in transdermal products.

19.2.2 Excipient Compatibilities Studies

Generally, excipient compatibility data should be available from previous excipient compatibility studies performed for previous dosage form(s). Further, excipient studies should be undertaken for those excipients used specifically in transdermal products that were not evaluated previously. Excipients used in transdermal products are listed in Appendix 19.1.

19.2.3 Formulation Development

Transdermal drug delivery requires that a drug permeate the stratum corneum, the dermis, and then permeate into the dermal microvasculature to enter the systemic circulation. These permeation processes are governed by Fick's laws of diffusion. Typically, when a transdermal formulation is applied to the skin, a passive diffusional process occurs where a concentration gradient is established between the drug concentration in the formulation (C_f) and the drug concentration in the dermis (C_d). Figure 19.2 provides an illustration of a steady-state situation that assumes that C_f remains constant and C_d is much smaller than the concentration in the formulation, due to efficient uptake into the dermal microvascular circulation. At

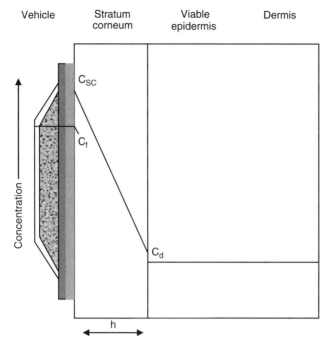

FIGURE 19.2   Steady-state flux.

steady state, the concentration gradient across the stratum corneum is linear and *Fick's first law* applies where the flux across the stratum corneum is constant:

$$J = \frac{M}{At} = \frac{D_a K}{h} C_f = P C_f \tag{19.1}$$

where J is the drug flux [having units of mass (M) per surface area (A) per time (t)], D_a the drug's apparent diffusivity in the stratum corneum, K the drug's stratum corneum–formulation partition coefficient, and h the thickness of the stratum corneum. The DK/h terms are defined as the drug's skin permeability coefficient, P, which is formulation dependent. The discontinuity at the formulation–stratum corneum interface in Figure 19.2 shows that the drug prefers to be in the stratum corneum. The drug's partition coefficient, K, is

$$K = \frac{C_{sc}}{C_f} \tag{19.2}$$

where C_{sc} is the drug concentration in the stratum corneum and C_f is the concentration in the formulation. The $\log_{octanol/water}$ partition coefficient for marketed transdermal drugs ranges from about 0.5 to 4.3.

From equation (19.1), flux can be manipulated by changing the formulation variables K and C_f. It is important to realize that K and C_f are not independent of each other, as shown in equation (19.2). It is also important to note the K is not controlled solely by the formulation. The formulation–stratum corneum partition coefficient is also a function of the skin's affinity for the drug. To reach accepted steady-state plasma concentrations there needs to be constant and sufficient drug concentration in the formulation. However, if formulation adjustments are made to increase the saturation solubility in the formulation, this may adversely affect K, and the drug may favor residing in the vehicle to residing in the stratum corneum. This tension between K and C_f requires optimization of these two formulation parameters to achieve the desired steady-state blood levels.

Also at steady state, the drug blood concentration is constant. This means that the rate of drug entering systemic circulation is equal to the rate of the drug being eliminated from the body. The drug's flux affects the rate at which a drug enters systemic circulation. Since the application surface is known or can be set, the amount of drug entering the bloodstream per unit time (M/t) can be found by a simple rearrangement of equation (19.1) to give $JA = M/t$. In other words, the rate of drug entering the central circulation is given by JA. It is also known that at steady state the drug elimination rate is controlled by the drug clearance (Cl) and the steady-state plasma concentration (C_{ss}). The rate of drug exiting the body is given by $Cl \cdot C_{ss}$. Therefore, at steady state there is a direct relationship between drug flux and plasma concentration:

$$(\text{drug rate in} = JA) = (\text{drug rate out} = Cl \cdot C_{ss}) \qquad (19.3)$$

Equations (19.1) and (19.3) can also be used to estimate a range of required drug formulation concentrations, C_f, for preformulation and formulation studies. Since other dosage forms are typically developed prior to the development of transdermal dosage forms, pharmacokinetic data are usually available at the time that preformulation and formulation studies are initiated. By knowing the plasma clearance and steady-state drug plasma concentration and placing an upper practical limit on the application surface area (A), an estimated flux (J) value can be calculated from a simple rearrangement of equation (19.3) to solve for flux. Knowing the flux, a range of required concentrations of drug in the formulation (C_f) to achieve the steady-state plasma concentration can then be estimated a priori by solving equation (19.1) for C_f and substituting the flux value calculated and estimating parameter ranges for D_a, K, and h.

Equation (19.1) can also be used to understand how formulation design changes can improve drug permeability. Equation (19.1) can be rearranged to

$$\frac{M}{t} = \frac{D_a K A}{h} C_f \qquad (19.4)$$

This equation can help the product design scientist understand what factors can affect drug permeation. As we will see later, the implications of this equation are used in a variety of ways to enhance drug absorption. In the case of passive diffusion, equation (19.4) supports the intuitive notion that increasing the applied surface area should increase the drug input rate. However, there are practical limitations to application area size, especially in the case of patches. Microabrasion techniques are being developed that decrease the thickness of the stratum corneum (h) that result in an increased delivery rate through the skin. The apparent drug diffusivity is affected by a number of factors, including the tortuosity of the diffusional path. The diffusion coefficient of the drug is given by

$$D = \frac{kT}{6\pi r \eta} \tag{19.5}$$

where D is the drug's diffusion coefficient, k is Boltzmann's constant, T is the temperature in kelvin, r the radius of the drug, and η the viscosity of the diffusional medium. The term kT represents the kinetic energy of the drug. The collected terms, $6\pi r \eta$, are sometimes referred to as the *frictional coefficient, f*. Equation (19.5) suggests that increases in the temperature of the applied dosage form should result in an increase in D. Furthermore, according to equation (19.4), an increase in D should result in a proportional increase in the amount of drug delivered per unit of time. Equation (19.5) also explains why larger drugs are expected to have smaller diffusion coefficients, which would lead to proportional decreases in the drug input rate. The currently marketed transdermal drugs have molecular masses of less than 350 Da. The drug's molecular size is a major hurdle for passive transdermal delivery. A number of drug companies are developing enabling technologies to overcome this significant limitation to transdermal drug delivery. Equation (19.5) also indicates that the diffusional coefficient is inversely related to the viscosity of the diffusional media. Permeability enhancers that decrease the viscosity and increase the fluidity of the stratum corneum lipid matrix have been used for many years to improve the permeability rate of drugs.

The magnitude of the diffusional coefficient also affects the lag time that can be expected before steady state is achieved. The diffusional lag time is defined as

$$t_l = \frac{h^2}{6D_a} \tag{19.6}$$

where t_l is the diffusional lag time and the other terms are as defined earlier. Figure 19.3 is a plot of cumulative penetrated drug per unit area vs. time which shows the lag time before steady state is reached. At early time points, very little drug has penetrated. Eventually, steady-state diffusion occurs. The lag time can be determined graphically by extrapolating the linear steady-state line to the x-axis. Equation (19.6) states that the lag time is inversely proportional to the apparent diffusivity of the drug. Therefore, a larger D_a value should give a shorter lag time, and steady-state blood levels should be achieved sooner.

19.2.3.1 *Formulation-Based Transdermal Drug Delivery*

19.2.3.1.1 Chemical Penetration Enhancers The skin is designed to be a formidable barrier to the penetration of foreign matter, and the primary limitations to transdermal drug delivery are the result of this barrier function. Without the use of enhancing drug delivery technologies, only a few drugs have acceptable physical and chemical properties to be delivered transdermally. Currently, most of the marketed transdermal drugs have a molecular mass of less than 350 Da, log $P_{octanol/water}$ ranging from 0.5 to 4.3, and are highly potent drugs that have doses of less than 20 mg/day and often have doses in the microgram and very low

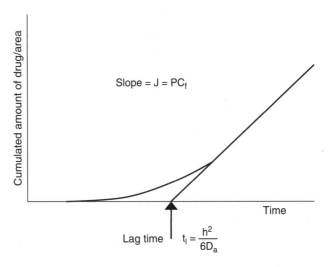

FIGURE 19.3 Cummulative drug permeation.

milligram ranges. This illuminates the degree of constraint placed on this route of drug delivery. Moreover, many drugs demonstrate skin irritation, allergenicity, photoirritation, or photoallergenicity or a combination of these toxicity-limiting issues.

Chemical penetration enhancers are commonly used to increase drug delivery and to expand the number of drugs that can be delivered transdermally. It has been stated that there have been 300 to 400 compounds evaluated as penetration enhancers. Many of the chemicals studied are listed as GRAS (generally recognized as safe) materials. Water has been shown to increase the skin permeability of drugs. Occlusion of the skin can increase the moisture content of the skin from 15 to 20% to 400%, based on stratum corneum dry weight. The mechanism for observed increases in drug absorption is unclear. It is known that water and other penetration enhancers can function by a variety of mechanisms that include a complex interplay between mechanisms. Penetration enhancer mechanisms include fluidization of the lipoidal intercellular matrix, change in solubility in the corneocytes and lipid matrix, phase separation, lipid extraction, solvent drag, and formation of aqueous pores at corneodesmosomes. Since many chemical accelerants have been shown to operate by several potential mechanisms, these promoters are typically classified by chemical class. Williams and Barry (2007) has classified chemical enhancers into 13 or so classes. Although many chemicals have been found to provide large increases in permeability, ranging upward to about a 1500-fold increase (Barry and Williams, 1995), it has been extraordinarily difficult to separate the beneficial increases in absorption and concomitant skin toxicity. In addition to having very low skin toxicity and toxicity in general, chemical enhancers should be effective quickly and at low concentrations, chemically and physically compatible with a multitude of drugs, pharmacologically inactive, odorless, and colorless.

Alcohol, glycerin, propylene glycol, and saturated fatty alcohols such as cetyl and stearyl alcohol have been used extensively in conventional topical formulations to provide solubility and humectant capability. It has been shown that these compounds also improve skin permeability. Examples of their use in commercial transdermal compounds are provided in the next section. Androgel (testosterone) contains 67% alcohol, and the Duragesic (fentanyl) transdermal patch also contains alcohol. Glycerol and propylene glycol esters have also been shown to accelerate percutaneous absorption.

Dimethly sulfoxide (DMSO) has been studied extensively as an absorption enhancer. It is a slightly yellow polar aprotic solvent (Figure 19.4). It has been studied extensively as a topical analgesic and anti-inflammatory. Its effects on skin permeability have been shown to be highly concentration dependent, requiring concentrations as high 30 to 40% to be effective when used with cosolvents. DMSO has been shown to cause the drying of skin, irritation, stinging and burning, erythema, and contact urticarial. It is known that DMSO "dries the skin," probably

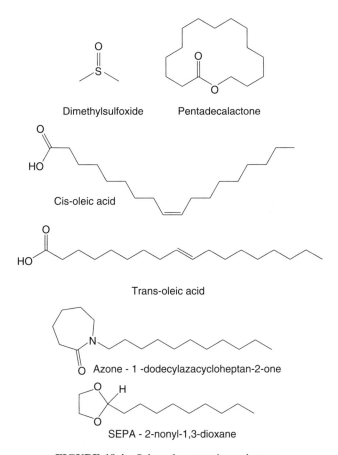

Dimethylsulfoxide   Pentadecalactone

Cis-oleic acid

Trans-oleic acid

Azone - 1 -dodecylazacycloheptan-2-one

SEPA - 2-nonyl-1,3-dioxane

FIGURE 19.4   Selected permeation enhancers.

by its hygroscopic nature and ability to extract lipids. It also can denature proteins and alter the structure of keratin. It has also been speculated that DMSO can interact with polar headgroups of the phospholipids to distort the bilayer lipids. It may also alter the polarity of the lipid matrix and affect drug partitioning. Pennsaid (diclofenac sodium) incorporates 44.5% DMSO in a mixed solvent system to delivery diclofenac sodium solution transdermally.

Pentadecalactone (Figure 19.4) is a GRAS material that is approved for direct food contact and is used as a flavoring agent. It has been shown to enhance percutaneous absorption (Gyurik, 2011) and is present in Testim 1% gel (testosterone) at a concentration of 8% w/w. Pentadecalactone is a macrocyclic lactone that is insoluble in water and glycerin. It has a melting range of 30 to 37°C.

Oleic acid (Figure 19.4) at the 1 to 10% level has been shown to be an effective absorption aid. It also appears that the *cis* form is more effective than *trans*-oleic acid. Shorter-chain saturated fatty acids such as capric acid (*n*-decanoic acid, C_{10}) and lauric acid (*n*-docecanoic acid, C_{14}) at about 10% have led to 100-fold increases in permeability.

Azone (laurocapram) was the first molecule specifically designed as a penetration enhancer. Laurocapram (Figure 19.4) is a colorless and odorless liquid that has an estimated $\log P_{\text{octanol/water}}$ value of 6.28. It has low irritancy and low toxicity (oral LD_{50} in rat of 9 g/kg). Unlike DMSO, laurocapram is more effective than neat oil at lower concentrations. Most studies have shown enhanced permeation at concentrations ranging from 0.1 to 5% concentration. The mechanism of action is still under investigation. There is good evidence that it interacts with lipids to elicit phase changes that lead to accelerated drug absorption.

SEPA (soft enhancer of percutaneous absorption), 2-nonly-1,3-dioxane, is a compound patented by Macro Chem (Samour and Daskalakis, 1989). It has been used as a penetration enhancer with a number of classes of compounds that have reached phase 3 clinical trials (Figure 19.4). In phase 3 minoxidil hair growth studies, formulations containing 10% SEPA significantly outperformed commercial Rogaine. Additionally, SEPA is highly volatile and is a much more aesthetically acceptable for a hair application product than propylene glycol, which leaves a "sticky" feel.

Using a high-throughput screen, Karande et al. (2004) selected and assessed over 5000 binary combinations of enhancers in 50% ethanol buffer and evaluated the most efficient combinations for irritation potential. Less than 1% of the combinations were tested further for flux enhancement. One of the most potent mixtures was a combination of sorbitan monolaurate and *n*-lauroyl sarcosine (Figure 19.5). Based on this original work, Karande et al. (2005) has also reported strategies that are important for the design of new chemical penetration enhancers.

19.2.3.1.2 Conventional Topical Dosage Forms   A broad array of topical dosage forms is marketed for transdermal drug delivery, including solutions, sprays, gels, creams, and ointments. Details concerning these dosage forms have been discussed in previous chapters. Selected examples of marketed products are discussed here.

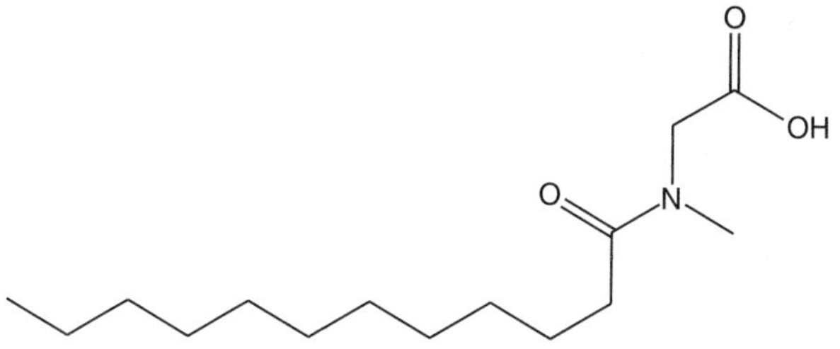

n-Lauroyl sarcosine

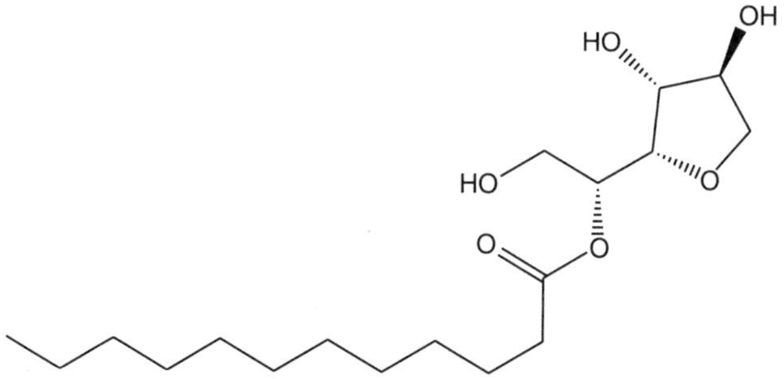

Sorbitan monolaurate

FIGURE 19.5 Potent combination of penetration enhancers.

Pennsaid (diclofenac sodium) is administered as a 1.5% topical solution for the treatment of joint pain. Four 10-drop applications of solution are applied and rubbed into the skin around the knee daily. The formulation excipients are: 45.5% dimethyl sulfoxide, propylene glycol, alcohol, glycerin, and purified water. Diclofenac's molecular mass is 318 Da, log octanol–water partition coefficient is 1.13 (Sallmann, 1986), and daily dose is 100 to 150 mg/day.

Evamist (estradiol) is a 1.7% transdermal metered-pump spray that delivers 1.53 mg of estradiol per actuation. The dose regimen is one 90-μL spray application on the forearm daily. It is indicated for the treatment of moderate to severe vasomotor symptoms of menopause. The excipients are alcohol, U.S.P. and octisalate, U.S.P. (68.85%). Octisalate, 2-ethylhexyl salicylate, is used in sunscreens to absorb ultraviolet B radiation. It may be used here as a solvent and to assist in drug transport across the skin. Estradiol has a molecular mass of 272 Da, log octanol–water partition coefficient of 2.49, a plasma clearance of about 700 L/h, a target steady-state blood level of 0.04 to 0.06 ng/mL, and a half-life of 0.05 hour.

Testim (testosterone) gel is provided in unit-dose tubes of 30 tubes per carton. Information from the product monograph, patent (Gyurik, 2011), and information found in the FDA Inactive Ingredients List suggests that the Testim formulation is testosterone 1%, pentadecalactone 8%, propylene glycol 5%, carbomer 1.5%, ethanol 73.6%, glycerin 5%, poly(ethylene glycol) 1000 0.5%, tromethamine, 0.1% and purified water, 5%. Testosterone has a molecular mass of 288 Da, a log octanol–water partition coefficient of 3.31, and a plasma concentration of about 50 ng/mL.

AndroGel (testosterone) is a 1% gel that is packaged in a metered-dose pump and is indicated for testosterone replacement therapy in males. The metered-dose pump delivers 1.25 g of testosterone per actuation. The starting dose is 5 g of product, which contains 50 mg of drug. It is applied once daily to dry skin on the shoulder, upper arms, or abdomen or a combination of locations. Only about 10% of the applied dose is absorbed for 24 hours. The formulation excipients are carbomer, ethanol 67.0%, isopropyl myristate, water, and sodium hydroxide. The product information stated that in clinical trials where male and female partners had vigorous skin-to-skin contact for 15 minutes after a 10-g application of product had been applied to the male for 2 to 12 hours, the unprotected female partner had serum testosterone concentrations greater than two times baseline.

Trolamine salicylate, U.S.P. is available as an over-the-counter 10% cream that is used for local arthritis, simple muscle strains, and ankle sprains. The excipients include alcohol, glycerin, mineral oil, cetyl alcohol, steric acid, triethanolamine, methyl- and propylparaben, and sodium edetate.

Nitroglycerin ointment, U.S.P. is available from Fourgera. Each inch of ointment contains 15 mg of nitroglycerin.

19.2.3.1.3 Patch Transdermal Delivery Systems   There are a number of U.S.P. transdermal systems: clonidine, estradiol, nicotine, nitroglycerin, and testosterone. The U.S.P. defines *systems* as "preparations of drugs in carrier devices that are applied topically or inserted into body cavities." There are three primary types of transdermal patches: reservoir or rate-controlling membrane, layered, and adhesive patches, which are shown in Figure 19.6. All the patches have some common features. They have an impermeable backing which provides occlusion and

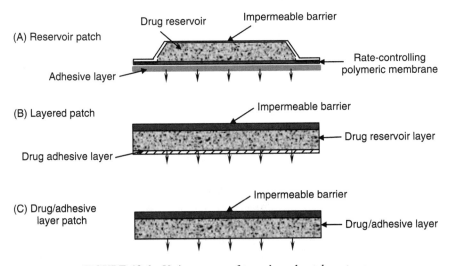

FIGURE 19.6   Various types of transdermal patch systems.

maintains unidirectional diffusion. They also have a protective polymer film that protects the pressure-sensitive adhesive layer until it is peeled off prior to use. The reservoir or rate-controlling membrane patch has a drug reservoir that may be a liquid or gel. A microporous rate-controlling membrane separates the drug reservoir from the adhesive. This was the first patch system approved in the United States. The layered patches are less complex and use different polymer systems to form a drug–polymer matrix and a drug–adhesive layer. The adhesive patch system, in which there is only the drug–adhesive layer, is the simplest.

The impermeable backing is usually a laminate material that includes an aluminum foil layer. The polymers typically used are polyester, polyolefin, poly(vinyl chloride), polyethylene, polyisobutylene, and polypropylene. The pressure-sensitive adhesives are usually acrylate, isobutylene, or silicone-based. The pressure adhesives provide adhesion under slight pressure. The vehicles used to solubilize the drug or form polymer matrices include alcohol, glycerin, propylene glycol, mineral oil, silicone oil, poly(ethylene glycol), isopropyl myristate, methyl laurate, glycerol monooleate, povidone, silicon dioxide, and polyisobutylene.

It is now generally accepted that the skin is generally the rate-determining barrier, including patches that incorporate a rate-controlling membrane. Selected examples of transdermal systems are discussed here. Padmanabhan et al. (2007) has summarized the development of a number of the Alza transdermal patch systems.

Transderm Scop (scopolamine) for treatment of nausea was the first transdermal patch approved for use in the United States, in 1979. The patch is 0.2 mm thick, 2.5 cm^2, and has four layers. The impermeable backing is an aluminized polyester film. The drug reservoir contains scopolamine, light mineral oil, and polyisobutylene. The rate-controlling membrane is a microporous polypropylene. The adhesive layer is mineral oil (12.4 mg), polyisobutylene (11.4 mg), and scopolamine. The patch is to be placed in the posturicular area. The system has a pharmacodynamics lag time of about 4 hours and delivers approximately 1 mg of drug per day for a period of 3 days. Scopolamine has a molecular mass of 303 Da, a log octanol–waer partition coefficient of 1.24, a plasma clearance of 67.2 L/h, a target steady-state blood level of 0.04 ng/mL, and a half-life of 2.9 hours.

Three types of nitroglycerin patches for the treatment of angina pectoris were approved in the early 1980s: Transderm-Nitro (reservoir with rate-controlling membrane), Deponit (layered), and Nitro-Dur (adhesive) patches. All the systems deliver nitroglycerin at about 0.4 mg/h. They are available in various sizes to provide flexible dosing. The Transderm-Nitro patch has an aluminized impermeable backing. The drug reservoir contains nitroglycerin that is adsorbed to lactose, silicon dioxide, and silicone medical fluid. The rate-controlling membrane is a nitroglycerin-permeable ethylene–vinyl acetate copolymer and the hypoallergenic silicone serves as the pressure-sensitive adhesive. The Nitro-Dur patch contains nitroglycerin in an acrylic-based polymer adhesive. Nitroglycerin has a molecular mass of 227 Da, a log octanol–water partition coefficient of 2, a plasma clearance of 966 L/h, a target steady-state blood level of 1.2 to 11 ng/mL, and a half-life of 0.04 hour.

Nicotine has a molecular mass of 162 Da, a log octanol–water partition coefficient of 1.2, a plasma clearance of 77.7 L/h, a target steady-state blood level of 10 to 30 ng/mL, and a half-life of 2 hours. There are a number of nicotine patches available for the treatment of smoking cessation. These patches deliver up to 21 mg/day, which is the highest level of drug delivery from a patch to date. Habitrol is a reservoir patch that contains nicotine and fractionated coconut oil. Acrylate vinyl acetate copolymers and methacrylic acid ester copolymers serve as the rate-controlling membrane. Nicoderm CQ is a matrix patch that contains nicotine, polyisobutylene, and ethylene–vinyl acetate copolymer. The backing is comprised of high-density polyethylene that is sandwiched between polyester films.

The Clonidine Trandermal System by Mylan is a layered patch whose backing is a polyethylene and polyester film. The matrix contains clonidine, mineral oil, polyisobutylene, and colloidal silicon dioxide. An adhesive formulation is layered onto the matrix. Clonidine has a molecular mass of 230 Da, a log octanol–water partition coefficient of 0.5, a plasma clearance of about 13 L/h, a target steady-state blood level of about 1 ng/mL, and a half-life of about 13 hours.

Duragesic (fentanyl) is a potent opioid analgesic. It is a reservoir patch system that contains drug, alcohol, and gelled hydroxyethyl cellulose. The rate-limiting polymer is ethylene–vinyl acetate copolymer. The backing is a polyester film and the adhesive layer contains a fentanyl–silicone adhesive mixture. The Mylan fentanyl patch is a matrix patch that has a polyolefin film and a drug-in-adhesive layer. Fentanyl has a molecular mass of 337 Da, a log octanol–water partition coefficient of 2.9, a plasma clearance of about 50 L/h, a target steady-state blood level of 1 ng/mL, and a half-life of about 7 hours.

19.2.3.1.4 Other Transdermal Formulation Approaches Liposomes, niosomes (nonionic surfactant vesicles), *elastic* niosomes (using laurate esters of poly(ethylene glycol) and sucrose to create more flexible membranes) and microemulsions have been shown to concentrate drug in the epidermal layer with minimal systemic drug delivery. This is targeted delivery, ideal for treating topical skin diseases such as atopic dermatisis with corticosteroids. However, these vesicles do not significantly improve transcutaneous drug delivery. Two formulation approaches, Transfersomes and ethosomes, have been developed that have shown promise for augmented transdermal drug delivery.

Transfersomes were invented and patented by Cevc (1996). Transfersomes are typically composed of phospholipid, a surfactant edge activator that controllably destabilizes the lipid bilayers, and alcohol at approximately 5 to 15%. The destabilizer makes the membrane more flexible and deformable, which allows the vesicle to "squeeze" through lipid pores. Destabilizers include bile salts, polysorbates, glycolipids, and polyoxyethylene alkyl esters. The final formulation is buffered to yield a lipid content of about 10% by weight. Transfersomes have been shown to be able to deliver drugs through the skin in a number of human clinical trials. The ability of Tranferinsulin to lower blood glucose in human volunteers was compared to an equivalent to subcutaneous insulin injection. The Transferinsulin glucose levels were about 35% of the subcutaneous administration. In animal studies comparing Transfenac to the commercial Voltaren topical gel, Transfenac delivered 2 to 10 times more drug than was delivered by the commercial product.

The Transfersome manufacturing process is fairly straightforward. Lecithin is mixed in ethanol with sodium cholate or other acceptable surfactants. This mixture is added to a buffer to bring the total lipid concentration to about 10 % by weight. The suspension is sonicated and frozen. The formulation is processed through a freeze–thaw cycle two or three times to initiate vesicle formation and growth. The vesicles are then sized to a nanometer size range using pressure homogenization or some other method of sizing. According to Cevc (2007), the average particle size is about 120 nm and contains 8.7% w/v lecithin, 1.3% w/v sodium cholate, and 8.5% v/v ethanol.

Ethosomes which were invented and patented by Touitou (1996), have also been shown to deliver drugs transdermally. Alcohol was deemed to have a destabilizing effect on the bilipid membrane until the discovery of ethosomes. An ethosome is composed of a phospholipid, a C_2 to C_4 aliphatic alcohol, optionally propylene glycol, and water. The alcohol content generally ranges from 20 to 50%. When the ethosome contains both alcohol and propylene glycol, the total percentage of the two can reach 70%. Ethosomes have been shown to give entrapment efficiencies as high as 80 to 90%. The physical structure allows for efficient entrapment of hydrophilic and lipophilic drugs. The vesicles themselves are more fluid than conventional liposomes. The melting transition for ethosomes is about 20 to 35°C lower than liposomes having the same composition without ethanol. It is also thought that the ethanol in ethosomes affects the fluidity of the lipid matrix surrounding the corneocytes, which leads to more efficient transport through these intercellular spaces. Ethosomal formulations have also shown clinical efficacy. An ethosome formulation of acyclovir was compared to the commercial formulation, Zovirax, in a two-armed double-blind randomized study. The time to crust formation and healing period were both decreased significantly with the ethosomal formulation.

For further reading, Touitou and Godin (2007) have reviewed vesicular transdermal carriers in detail.

19.2.3.2 Physical–Mechanical Technologies for Mediated Transdermal Drug Delivery

The term *physical–mechanical technologies* is used here to distinguish those technologies that rely on physical and mechanical means, such as temperature, iontophoresis, electrophoresis, sonophoresis, microneedles, thermal ablation, laser ablation, needle-free injection, and microabrasion, to achieve enhanced permeability from the more conventional formulation approaches that use chemical penetrants or patch systems.

19.2.3.2.1 Thermal Penetration Enhancement The Synera patch, commercialized by Zars Pharma, is an emulsion of the eutectic mixture of lidocaine and tetracaine in a reservoir that uses heat to decrease the lag time [equation (19.6)] for local anesthetic effect by increasing the drug's diffusion coefficient [equation (19.5)]. Equation (19.5) shows that the diffusion coefficient is directly proportional to temperature. As the temperature is increased, the diffusional coefficient is increased. Equation (19.6) shows that the lag time is inversely proportional to the magnitude of the diffusional coefficient, so as the diffusional coefficient increases, the lag time to reach steady state and maximal drug delivery is decreased.

Moreover, the total amount of drug delivered can also be increased. The drug reservoir component contains poly(vinyl alcohol), sorbitan monopalmitate (Span 40), water, methylparaben, and propylparaben. The active portion of the patch is 10 cm^2. The overall patch, which measures 50 cm^2, which contains a heating component that is activated by oxygen. The compounds in the heating component are iron powder, activated charcoal, sodium chloride, wood flour, water, and filter paper. This component of the patch is protected from the atmosphere, especially oxygen, by an impervious removable film. At the time of use, the impervious film is removed, which exposes air vents and allows an exothermic oxidation reaction to take place to generate heat. This is called the CHADD (controlled heat-assisted drug delivery) system. The CHADD system increases the local skin temperature by 5°, but the maximum skin temperature will not exceed 40°C. One of the first patents to reveal this approach was that of Konno et al. (1987).

19.2.3.2.2   Iontophoresis Penetration Enhancement   Iontophoresis is an electrotransport technology that facilitates ionizable transdermal drug delivery by applied electrical potential. The driving force is electrostatic repulsion. Since most drugs are ionic in nature, there is a sizable opportunity for iontophoretic transcutaneous drug delivery. Indeed, some 75% of all marketed drugs are basic molecules, while about 20% are acidic in nature. This suggests that about 95% of these drugs may be amenable to electrophoresis as ionized molecules. Another advantage of iontophoresis is that the skin is *permselective* to cations. Since the majority of drugs are weak bases, they can be converted to cations by selecting the appropriate solution pH or using acid salts for the weak base drug. The isoelectric point of the epidermis is pH 4 to 5. At a physiological pH of 7.4, the skin is negatively charged and is more permeable to cations. Acid salts of basic drugs will carry a positive charge and are potential candidates for permselective iontophoresis. Iontophoresis also induces convective fluid flow, which produces flux for neutral and charged molecules, increasing the utility of iontophoresis. This induced fluid flow is called *electroosmosis*. Iontophoretic flux is then the sum of electromigration, electroosmotic migration, and passive diffusion. The electron migration of cationic drugs is responsible for the large majority of drug that is transported across the skin.

A schematic of iontophoretic drug delivery is provided in Figure 19.7. The skin serves to complete the electrical circuit. Commercial systems use consumable electrodes. The silver/silver chloride consumable electrode is most common in commercial products. The active electrode matrix is usually composed of an ionized drug-loaded hydrogel in a nonionizing polymer. Another type of matrix uses a hydrophilic fabric or hydrophilic porous film that is filled with an ionized drug solution. Positively charged drugs such as the acid salts of a basic drug are delivered from the anode. This is often referred to as *anodal delivery*, and the anode is called the *active electrode*. Negatively charged drugs such as sodium salts of weak acids are pushed into the skin at the cathode, which is called *cathodal delivery*. The ion flux is determined by the electric field strength and the electrical resistance of the skin. The electric field strength is governed by the current density (mA/cm^2). Current density can be controlled by adjusting the current intensity and modifying the electrode size. Increasing the electrode size will decrease the current density.

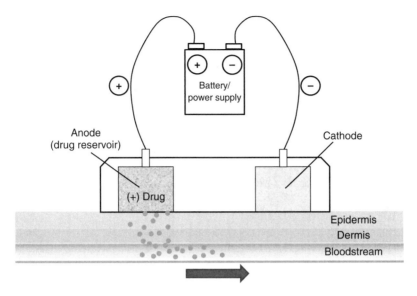

FIGURE 19.7 Schematic of an iontophoretic drug delivery system.

The quantity of ionized drug electrophoresed is proportional to the applied current, duration of treatment, and concentration of ionized drug in solution. It has been shown that at constant current, *Faraday's law* reasonably describes drug flux during steady-state electrophoresis:

$$J_i = \frac{t_i I}{z_i F} \qquad (19.7)$$

where J_i is the electromigration flux of the ith species, t_i the ion transport number, I the current applied, z the valence of the ith species, and F is Faraday's constant. Electrical currents of 3 to 5 mA are typically used for transdermal iontophoresis. It has also been shown that iontophoretic percutaneous drug delivery is primarily via skin appendages such as hair follicles and sweat glands. Iontophoresis exhibits two other primary differences from other physical enhancing technologies: It does not affect the skin permeability per se and flux is independent of surface area. At a fixed current, iontophoretic fluxes do not increase proportionally to the application surface area. That is, at a 1 mA, the same amount of drug would be transported over 10-cm^2 and 20-cm^2 application areas. Once the flux vs. current relationship is developed for any given drug and formulation, drug flux at any other applied current can be predicted relatively well.

Studies have shown that the skin's resistance changes quickly and dramatically during iontophoresis. *Ohm's law* states that voltage is directly proportional to the product of resistance times current:

$$E = IR \qquad (19.8)$$

where E is the electromotive force in volts, R the electrical resistance in ohms, and I the current in amperes. To maintain constant current and drug delivery as the skin's resistance changes, there is a need for the voltage to be adjusted by the electrical circuitry.

Chemical burns can result from the accumulation of counterions at the active electrode if the current density is too high or the treatment duration is too long. For example, positive ions will build up at the negatively charged cathode, forming sodium hydroxide, which can cause chemical irritation and chemical burns. The skin is more sensitive to sodium hydroxide toxicity than the buildup of hydrochloric acid at the positively charged anode.

Figure 19.8 shows a schematic of the E-Trans iontophoretic system (Phipps et al., 2007). The electronic controller is comprised of a printed circuit board assembly, power source, and user interface. In the newer iontophoretic devices, the

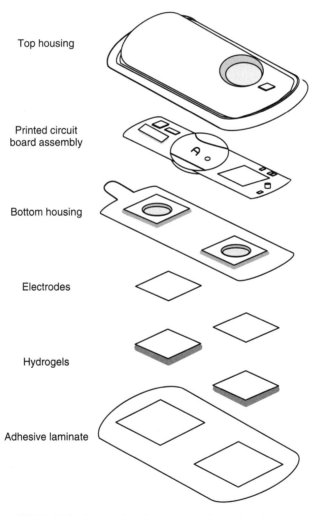

Top housing

Printed circuit board assembly

Bottom housing

Electrodes

Hydrogels

Adhesive laminate

FIGURE 19.8   Schematic of the E-Trans iontophoretic system.

power source consits of commercial batteries that can be placed in series to provide a maximum voltage of about 10 V to achieve a maximum acceptable current density close to 0.5 mA/cm². A field-effect transistor with a feedback resistor can create a constant-current electrical system that adjusts to changes in the skin's electrical resistance. It is also possible to design controllers that adjust to a patient's need for more or less drug and to vary the current accordingly within a dosing protocol range.

Applying Faraday's law and equation (19.7), formulation design efforts try to maximize the drug's transport number, t_i. The transport number is the fraction of the total charge transported by a specific ion during electrophoresis. Therefore, formulation design efforts seek to maximize the fraction of ionized drug that is transported by manipulating pH, ionic strength, and drug concentration. Transport numbers, which have values between 0 and 1, measure the efficiency of drug transport. Equation (19.7) also indicates that a drug with a valence of 1 will be transported twice as fast as a drug that has a valence of 2. Sodium chloride can be thought of as an ideal model salt for iontophoresis because its ions are relatively small, mobile, and carry a single charge. During cutaneous saline electrophoresis the Na^+ and Cl^- transport numbers are approximately 0.6 and 0.4, respectively. Drug transport numbers are expected to be one-tenth to one-third these values. It is also interesting to point out that the higher transport number for a sodium ion compared to a chloride ion reflects the difference in charge density and the skin's permselectivity to cations. Therefore, single-valence small-cation drug molecules are expected to give the best transport results. Using equation (19.7), the drug's transport number for various formulations can be assessed by plotting flux vs. applied current and determining the slope of the line. Typical iontophoretic drug formulations contain buffers to adjust the pH to maximize the ionic character of the drug, keeping in mind the isoelectric point of the epidermis. The formulation may also contain penetration enhancers, antimicrobial preservatives, and antioxidants. As more ions are added to a formulation, the competition to carry the charge increases since the sum of the transport numbers is 1. For a drug to compete effectively for electrophoresis, its ion concentration must favor transportation. Therefore, organic buffers of large molecular mass are preferred to smaller inorganic buffers. Increasing the mole fraction of a drug compared to other charged molecules also favors drug transport.

There is interplay between the choice of a drug salt and the choice of solvent. It is most desirable to have a drug in its completely ionized state. This is generally accomplished by using water as the solvent. Water has a high dielectric constant, which supports ion dissociation and minimizes ion pair formation. It is also highly biocompatible. On the other hand, water's high dielectric constant has a greater impact on decreasing the electric field strength of the electric field surrounding a charged drug molecule. Other cosolvents, such as glycerin, alcohol, propylene glycol, and poly(ethylene glycol), may be added to increase drug solubility, improve drug stability, decrease the dielectric solvent effect, or any combination thereof. Similarly, it is desirable to pick a salt that dissociates readily in the solvent, so there may be trade-offs associated with solubility, stability, cosolvents, and degree of salt dissociation. The drug counterion must also be biocompatible and compatible with the electrodes and the electrochemical reactions that occur at the electrodes. The electrolytes in the counter reservoir must be biocompatible and compatible with the electrode. They should provide adequate conductivity so as to minimize the system's voltage requirements.

Iontophoresis is a proven transdermal technology. The Companion 80 wireless ionotophoresis system (Chattanooga), LidoSite system (Vyteris), and Iontopatch WEDD (wearable lectronic disposable drug delivery) system (Teikoku) are used to deliver lidocaine for local anesthesia. The Dupel system (Empi) uses iontophoresis for localized delivery of dexamethasone. There are now some dozen companies developing iontophoretic epicutaneous devices.

The Ionsys fentanyl iontophoretic transdermal system was a novel patient-controlled device that provided on-demand systemic delivery of fentanyl, an opioid analgesic. The device was approved by the FDA in 2006 and marketed for use in a hospital setting. As of 2008, the FDA has listed the product as discontinued. The product was recalled by the European Medicines Agency because in some cases, component corrosion led to inadvertent leakage of fentanyl. Nevertheless, Ionsys has paved the way for future feedback-controlled iontophoretic systems. Ionsys incorporated a number of safety features to prevent inadvertent overdosing. Once activated for the first time, the system had an operating time of 24 hours and allowed a maximum of 80 doses to be given in 13 hours. The device incorporated a flashing-light algorithm that permitted the patient and health care provider to know approximately how many doses have been supplied by the iontophoretic patch. The system could be placed on the chest or upper outer arm. It was shown that about 20% less drug was delivered at the arm location than of the chest location. Intersubject variability comparing intravenous and transdermal fentanyl administration was 28 and 33%, respectively. The Ionsys hydrogel contained fentanyl hydrochloride and the inactive excipients cetylpyridinium chloride, citric acid, polacrilin (Amberlite IRB-88 potassium), poly(vinyl alcohol), sodium citrate, and sodium chloride. Fentanyl hydrochloride hydrogel was positioned at the anode. The sodium chloride was placed in the cathode hydrogel.

For further details regarding iontophoretic skin penetration enhancement, the reader is directed to the work of Mudry et al. (2007) and Hu et al. (2010).

19.2.3.2.3 Electroporation or Electropermeabilization Enhancement Electroporation has been used primarily to deliver DNA through bilayer cell membranes into cells. The use of electroporation to delivery drugs transcutaneously has been studied since the early 1990s. Compared to iontophoresis, electroporation uses short bursts (milliseconds to microseconds) of high-voltage pulses (above 50 to 1000 V) to create what is believed to be extraordinarily small aqueous pores ($\sim$10 nm) in the stratum corneum. Pulsing protocols have shown that five 100-V pulses with a pulse interval of 100 ms gives permeation comparable to that of fifty 100-V pulses having a 10-ms pulse (Bonner and Barry, 2007a). The data also suggested that more pulses of shorter duration may cause less membrane damage.

Research also continues to evaluate the use of electroporation in combination with other enhancing techniques, such as iontophoresis and chemical penetration enhancers. Good reviews of electroporation are those of Bonner Barry (2007a,b).

To date, no commercial products use electroporation for transdermal drug delivery. Recently, Dolter et al. (2010) reported that several integrated electroporation devices are in clinical testing for delivery of vaccines. The reported devices are Elgen and Cellectra by Inovio Biomedical and TriGrid by Ichor Medical. The TriGrid device consists of three components: a pulse simulator, an integrated applicator, and a sterile single-use applicator cartridge. The electroporation process is completed in a few seconds.

19.2.3.2.4 Ultrasound or Sonophoresis Permeation Enhancement
Ultrasound is a mechanical energy wave that can oscillate at different frequencies and propagates in the direction of the oscillations. The oscillation of pressure is transmitted through air, liquids, and solids. The ultrasound frequency is 18 to 20 kHz and higher. *Acoustic impedance* is directly related to the medium density and speed of propagation:

$$Z = \rho C \tag{19.9}$$

where Z is the acoustical impedance, ρ the medium density, and C the propagation speed. The impedance of air and skin are 400 and 1.6×10^6 kg·m^{-2}·s^{-1}, respectively. The intensity of the ultrasound pressure wave is given by

$$I = \frac{A^2}{2Z} \tag{19.10}$$

where I is the ultrasound intensity (W/cm^2), A the wave amplitude, and Z the impedance. Equation (19.10) states that the ultrasound intensity or concentration of power is directly proportional to the amplitude of the pressure wave squared and inversely proportional to the impedance. As the skin is exposed to ultrasound energy, the impedance of the skin decreases, which results in increasing intensity over exposure time. Therefore, care needs to be taken to account for the change in impedance during ultrasound transcutaneous drug delivery.

Sonophoresis is often used to describe the use of ultrasound to achieve percutaneous absorption of drugs. Frequencies ranging from 20 kHz to 16 MHz have been employed for sonophoresis. Most of the earlier studies used medium-frequency ultrasound (1 to 3 MHz) for sonophoresis studies. Typical transport improvements were on the order of 10-fold or less. This is not a noteworthy improvement compared to chemical enhancers, by whose use drug absorption can be increased 100- to 1000-fold. The largest advantage of the use of ultrasound in these early studies was a general lack of skin toxicity, which is often seen with chemical enhancers. More recently it has been shown that low-frequency ultrasound (20 to 100 kHz) elicits significantly higher orders of transcutaneous absorption. Two different methods have been used to evaluate low-frequency sonophoresis: continuous ultrasound and pretreatment ultrasound. Continuous sonophoresis applies ultrasound throughout the application process. Low-frequency ($\sim$ 20 kHz), low-intensity (1 W/cm^2) ultrasound is applied to a coupling medium that contains the drug. The ultrasound causes temporary alterations in the skin barrier and induces convection through the pressure waves, which leads to increased drug penetration. The low-frequency pretreatment protocol usually uses higher-intensity ultrasound (7 W/cm^2) for up to

10 hours prior to applying a patch or other applicable dosage form. Continuous treatment has led to a 1000-fold increase in permeation, and pretreatment can be expected to deliver up to 100-fold improvement in percutaneous absorption.

Ultrasound has thermal, cavitation, and acoustical streaming effects. Different materials possess different abilities to absorb the ultrasound energy, which is referred to as the *absorption coefficient*. Materials with a high absorption coefficient will absorb more ultrasound energy. The absorption coefficient of a material increases directly with the ultrasound frequency, resulting in a temperature increase. The increased temperature may increase the fluidity of the lipid bilayer and increase the drug diffusion coefficient. The *time to threshold* is the safe ultrasound exposure time. Exposure times beyond the time to threshold are expected to increase the risk of tissue damage. Ultrasound causes pressure variations in the tissue, which leads to cavitation, which results in the rapid growth and collapse of gaseous bubbles. The collapse of the bubbles is thought to lead to the formation of shock waves, which can trigger such structure changes as fluidization of the lipid bilayer in the skin barrier. Other speculated forms of bubble collapse that lead to inertial cavitation are microjet penetration and microjet impact. Ultrasound treatment can also produce *acoustic streaming*, which is unidirectional fluid flow generated by the ultrasound pressure wave. It is not clear what impact acoustic streaming has on the overall effect of drug absorption. Ultrasound seems to be well tolerated by the skin. The toxic skin effects are affected by the application duration, frequency, and intensity. Skin changes are typically rapidly reversible.

SonoPrep by Echo Therapeutics is a sonophoretic device that delivers transdermal lidocaine. The device was approved by the FDA in 2004. It is a portable battery-operated ultrasound device that runs at 55 kHz and 12 W/cm^2. The application probe is 0.8 cm^2. The system also provides real-time skin conductivity that is part of a feedback loop that turns the unit off at a preset threshold. The electrical feedback loop is designed to prevent overexposure to the ultrasound energy. The application time takes 5 to 30 seconds. The ultrasound housing is filled with a coupling phosphate buffer that contains saline solution and 1% sodium lauryl sulfate. Although this system has been discontinued, the development and approval provide a known pathway to approval of future sonophoresis systems. The ability to self-administer a wearable device is one of the major commercial hurdles.

Reviews regarding sonophoresis may be found in work by Kost and Wolloch (2007), Mitragotri and Kost (2007), Escobar-Chávez et al. (2009), and Meidan and Michniak-Kohn (2010).

19.2.3.2.5 Permeation Enhancement of a Stratum Corneum Circumvented or Removed Recently, paradigm-shifting research has shown that controlled puncturing, ablation, or removal of the stratum corneum does not result in irreversible changes or clinically important risks of prolonged irritation or infection potential. This is especially true when these technologies are used for vaccinations that are done once or twice over a span of weeks. A number of technologies have been advanced based on the nominal safety risks associated with these minimally invasive methodologies. In general, these technologies are designed to either circumvent or remove the stratum corneum. These new transdermal delivery systems include microneedles; thermal, laser, and radio-frequency ablation; microabrasion; and ballistic propulsion. A number to these technologies have reached phase 3 clinical trials and FDA New Drug Application submission. These systems are handled by the FDA in a manner similar to inhalation systems. The complexity of this dosage form is acknowledged by the FDA. The FDA considers many of these technologies to be a device. Sponsors of these transdermal systems should discuss with the FDA whether the system should be submitted as a device in a 510(k) submission or as a drug in a new drug application. The FDA has issued an intercenter agreements document between the Center for Devices and Radiological Health and the Center for Drug Evaluation and Research (CDER) that provides guidance as to which center will review the submission. In some cases, both centers may be involved with a review. In this case, one of the centers will take the primary lead and will coordinate the review with the sponsor and the other center. In general, a transdermal drug delivery device is considered a drug product and regulated by CDER "when the primary purpose of the device is delivering or aiding the delivery of a specific drug and the device is distributed with the drug" (FDA, 1993).

19.2.3.2.5.1 Microneedle Penetration Enhancement The use of microneedles for transdermal drug delivery is a minimally invasive technique that was described over 40 years ago. However, it was not until 1998 that the first research paper described the fabrication and testing of microfabricated microneedles (Henry et al., 1998). Microneedles range from 100 to 1000 μm in length and 20 to 600 μm in diameter. They are attached to a base that can contain arrays of 1000 microneedles. These arrays are used to puncture the skin to create micropores in the various layers of the epidermis. The goal is to create micropores without penetrating the dermis and eliciting a pain response. By comparison, the epidermis is about 100 to 150 μm thick and the dermis is about 1000 μm thick. Therefore, the frequency and extent of microporation can be controlled by the number and length of the microneedles.

Microneedles are often classified as solid or hollow constructs. Recent advances in microneedle construction have led to the use of a number of different materials and shapes. Microneedles have been fabricated from elemental silicon, glass, metal (NiFe, stainless steel, and titanium), biodegradable polymers [poly(glycolic acid), poly(lactic acid), and polyl(actic-*co*-glycolic acid)], soluble maltose, carboxymethlycellulose, and others. The solid needles can be coated with drug so that microneedle poration simultaneously circumvents the epidermis and delivers the drug through the epidermis. A dipping method is typically used to coat the drug onto the needles. It has been shown that insertion force increases linearly with needle tip cross-sectional area. The penetration force has been determined to range from 0.1 to 3 N. A microneedle roller has been developed which requires less insertion pressure and is said to be less painful. In another study, the penetration force was reduce by greater than 70% when insertion was accompanied by a vibratory actuator operating in the kilohertz range (Yang and Zahn, 2004). The amount of drug delivered by the coating method is limited to 1 to 2 mg. To overcome this limitation, hollow microneedles have been manufactured. In this case, liquid formulations containing a drug can be infused through the hollow needles, which allow for larger doses and controlled delivery for longer periods of time, in some cases longer than 6 hours.

A number of safety studies have been conducted that have looked at potential irritation, microbiological invasion, and healing or pore closure. In general, these studies confirm that the use of microneedles is a minimally invasive technology. Skin function often returns to its normal function after 3 to 4 hours and reseals completely in less than 24 hours. Occlusion tends to slow the resealing process up to 40 hours, depending on the needle geometry. This may be desirable in the case of controlled release. Since the microneedles bypass the body's natural defense system, they should be sterile and the necessary precautions that are used for injections should be applied to the use of microneedles. Donnelly et al. (2010) have written a review of microneedle-based drug delivery that gives a more detailed summary of the current state of this microporation technique.

There are about a dozen companies that are developing microneedle intradermal systems. A few selected products will be discussed. The solid microstructured transdermal system (sMTS) and hollow microstructered transdermal system (hMTS) has been developed by 3M Corporation. The sMTS delivers up to 0.3 mg of highly potent proteins or vaccines. There are about 375 to 1300 needles/cm^2, which are about 150 to 700 μm long and spaced around 550 μm distance from each other. The dry-coated solid active ingredient is likely to be more stable than material formulated in solutions or suspensions. The application contact time ranges from 30 seconds to 10 minutes. The hMTS has a delivery capacity of 2 mL and has 18 hollow needles that are approximately 700 μm long and have an outer diameter of about 280 μm. The contact time ranges from 2 to 20 minutes. For additional information we refer the reader to Burton et al. (2010) and 3M (2012).

MacroFlux by Zosana uses the patented Alza technology, which employs an array of titanium microneedles or microprojections approximately 175 to 430 μm in length. The arrays are 5 or 10 cm^2 and have up to 320 microprojections/cm^2. MacroFlux is an integrated design that can be used for short-duration effects such as vaccination or allergen testing. It can also be used in conjunction with passive and electrotransport transdermal systems (Cormier and Daddona, 2007; Cormier et al., 2009).

19.2.3.2.5.2 Thermal Ablation Permeation Enhancement   Altea's Passport system uses an array of heating elements that when activated for microseconds heats the stratum corneum to temperatures sufficient to vaporize the tissue and create microchannels for intracutaneous drug transport. The idea is to maintain a steep thermal gradient that creates microchannels in the stratum corneum 30 to 50 μm in depth and 50 to 200 μm in width while not causing deleterious effects to the viable portions of the skin. The system has a reusable handheld thermal ablation applicator and single-use disposable patch. Once the ablation poration process is complete, a patient-friendly fold-over patch is placed directly over the newly formed pores or microchannels. Lilly had a partnership with Altea to develop a transdermal insulin system; Amylin had signed an agreement to develop an exenatide delivery system for type 2 diabetes; and Hospira sought the Altea technology for transdermal delivery of the anticoagulant enoxaparin. In late 2011, these partners and Hospira halted funding and Altea was forced to abandon its projects and sell its technology.

19.2.3.2.5.3 Laser Ablation Permeation Enhancement P.L.E.A.S.E. (painless laser epidermal system), developed by Pantec Biosolutions AG, utilizes an erbium-doped yttrium aluminum garnet (Er/YAG) laser to create micropores 20 to 150 μm deep and about 200 μm wide to circumvent the barrier layer of the skin. Norwood Abbey has developed Epiture Easytouch to deliver 4% lidocaine. Miniaturation and significant cost reduction are needed to make this device more competitive with microneedles and thermal ablation.

19.2.3.2.5.4 Radio-frequency Ablation Permeation Enhancement Over the past decade, the use of radio-frequency heating of the skin to form micro channels has been undertaken. The use of radio-frequency waves at 100 to 500 kHz can cause vibrational heat that induces water evaporation and cell ablation. The process leads to microchannels about 50 μm in depth and 30 to 50 μm wide. Levin (2008) has reported design features of the ViaDerm RF system from TransPharma Medical. The system has a device that pretreats the skin to create RF microchannels and a unique printed patch that is placed over the RF application site. The device is a hand held battery-operated reusable electronic-controlled microelectrode array that adapts the heating energy to the skin texture. The microelectrode array is a single-use low-cost disposable unit that contains hundreds of electrodes. The array is manufactured in three sizes: 1, 2.5, and 5 cm^2. A printed-patch technology is used to manufacture highly potent and costly proteins and oligonucleotides. Printing is accomplished by dispensing and drying very small droplets of concentrated protein liquid on the patch. The printed patch is then placed over the application area, where the interstitial skin fluids migrate through the micropores and dissolve the drug, which diffuses passively through the skin. The amount of drug and rate of drug delivery are affected by the drug concentration on the patch and the microelectrode density pattern. Lilly and TranPharma are codeveloping human parathyroid hormone for osteoporosis.

19.2.3.2.5.5 Needle-free Jet Injector Permeation Enhancement Jet injectors exploit a high-speed jet to porate the skin for intracutaneous drug delivery without the use of a needle. Much of the physics of jet injectors is the domain of ballistics, which is the study of the motion of projectiles. In this case the projectile is a drug in the form of a liquid or dry powder. The power source is usually a spring or compressed gas. The release of the spring or gas impinges on a piston or pressure diaphragm that causes a surge in pressure and release of a drug from its container through a confined nozzle. The speed in the nozzle can be 100 m/s and higher. The pressure and speed generated must overcome frictional and inertia forces of the drug formulation. Once the drug exits the nozzle, it must have enough force to penetrate the skin to the desired depth. The depth of penetration is primarily a function of the size of the drug, formulation, orifice diameter, and jet exit velocity. There are a number of material science issues that must be addressed with the device itself. The material of construction needs to withstand the pressure and shock waves that are generated during activation. The cost of materials, disposal costs, and sterility requirements are also important considerations in the design. A number of major advances have been made in jet injectors to decrease the pain of injection. The introduction of reusable injectors decreases the environment impact of these devices. Antares was an early pioneer in needle-free injectors. They marketed an insulin injector in 1979 and have manufactured and distributed a spring-loaded jet injector for growth hormone since 1994. A number of jet injectors have since been approved: the PowderJect (used by Novartis and Pfizer for vaccines) and Intraject (used in Sumavel, sumatriptan subcutaneous injection from Zogenix) systems. Additional information regarding needle-free product development is available in the work of Bellhouse and Kendall (2007), Levy (2007), and Pass et al. (2007).

19.2.3.2.5.6 Microdermal Abrasion Permeation Enhancement Microdermal abrasion is being developed by Intercell as a vaccine enhancement patch system. The system consists of a microabrasion strip and a patch containing dry immunostimulating adjuvants used in conjunction with conventional vaccination by injection. The adjuvant, IC31, is comprised of an oligodeoxynucleotide containing inosine/deoxycytosine (ODN1a) and an antimicrobial peptide, KLKL(5)KLK. The vaccination is completed in three steps. The microabrasion strip unit is first pressed down on the skin at the intended injection site and the strip is pulled from the unit under pressure. This step microabrades the skin. The intradermal or intramuscular vaccine is then injected at or near this site and the immunostimulating patch is placed over the abraded skin area. The interstitial fluids dissolve the immunostimulating adjuvants and diffuse into the skin, interacting with immunomodulating dendritic cells to enhance the immunogenicity of the vaccine.

19.3 CONCLUSIONS

Significant advances in transdermal drug delivery have been made over the many years since the scopolamine patch was approved by the FDA in 1979. A wide variety of formulation-based and physical–mechanical-based technologies have been commercialized. Most physical–mechanical systems are minimally invasive or noninvasive and carry very low safety risks. Most safety evaluation studies have shown that skin function returns quickly to normal when the skin barrier is circumvented or removed, while the ability to deliver hydrophilic and large peptides and oligonucleotides has increased significantly. As noted is a number of the examples, the use of a combination of technologies is also becoming more prevalent. Many studies are now looking at the synergistic effect of coupling technologies. The use of chemical penetration enhancers, iontophoresis, electroporation, ultrasound, and stratum corneum circumvention and removal in combination is being explored. Significant augmentation has been observed when enabling technologies are coupled. With the continual growth of biotechnology and oligonucleotide drug development, transdermal drug delivery should have a bright future. The transdermal route offers a number of advantages: a large site of administration; convenient patient pain-free dosing, significant opportunity for controlled delivery of highly potent compounds for up to 7 or more days, avoidance of hepatic first-pass metabolism, a less harsh metabolic and pH environment than that with oral drug delivery, ability to terminate medication abruptly, dosing on demand, and others. All these advantages lead to improved patient acceptance and compliance. There are safety considerations that need to be well understood, such as skin irritation, allergenicity, photoallergenicity, carcinogenicity, and photocarcinogenicity. On the whole, transdermal drug delivery provides exciting research and commercial opportunities. Some useful review articles are those of Prausnitz and Langer (2008), Paudel et al. (2010), Singh et al. (2010), and Kalluri and Banga (2011).

REFERENCES

3M. http://solutions.3m.com/wps/portal/3M/en_WW/3M-DDSD/Drug-Delivery-Systems/Transderml-Microneedle-Directory?WT.sv1=TransdermalBody&amp%3bWT.ac+keymatch. Accessed Jan. 2012.

Barry BW, Williams A. Permeation enhancement through skin. In: Swarbrick J, Boylan JC, eds. *Encyclopedia of Pharmaceutical Technology*, Vol. 11. New York: Marcel Dekker; 1995. pp. 449–493.

Bellhouse BJ, Kendall MAF. Dermal powderject device. In: Rathbone MJ, Hadgraft J, Roberts MS, eds. *Modified-Release Drug Delivery Technology*. Drugs and the Pharmaceutical Sciences, Vol. 126. New York: Informa Healthcare; 2007. pp. 607–617.

Birchall JC. Stratum corneum bypassed or removed. In: Touitou E, Barry BW, eds. *Enhancement in Drug Delivery*. Boca Raton, FL: Taylor & Francis; 2007. pp. 337–351.

Bonner MC, Barry BW. Electroporation as a mode of skin penetration enhancement. In: Touitou E, Barry BW, eds. *Enhancement in Drug Delivery*. Boca Raton, FL: Taylor & Francis; 2007a. pp. 303–315.

———. Combined chemical and electroporation methods of skin penetration enhancement. In: Touitou E, Barry BW, eds. *Enhancement in Drug Delivery*. Boca Raton, FL: Taylor & Francis; 2007b. pp. 331–336.

Burton SA, NG C-Y, Simmers R, Moeckly C, Brandwein D, Gilbert T, Nathan J, Brown K, Alston T, Prochnow G, Siebenaler K, Hansen K. Rapid intradermal delivery of liquid formulations using a hollow microstructured array. Pharm. Res. Published online June 26, 2010. doi: 10.1007/s11095-010-0177-8.

Cevc G. Preparation for drug application in minute droplet form. EU 0475160, 1996.

———. Transfersomes: innovative transdermal drug carriers. In: Rathbone MJ, Hadgraft J, Roberts MS, eds. *Modified-Release Drug Delivery Technology*. Drugs and the Pharmaceutical Sciences, Vol. 126. New York: Informa Healthcare; 2007. pp. 533–546.

Cormier M, Daddona PE. Macroflux technology for transdermal delivery of therapeutic proteins and vaccines. In: Rathbone MJ, Hadgraft J, Roberts MS, eds. *Modified-Release Drug Delivery Technology*. Drugs and the Pharmaceutical Sciences, Vol. 126. New York; Informa Healthcare: 2007. p. 589–598.

Cormier MJN, Young WA, Johnson JA, Daddona PE, Armeri M. Transdermal drug delivery devices having coated microprotrusions. U.S. Patent Application US 2009/0186147 A1. 2009.

de Jager MW, Ponec M, Bouwstra JA. The lipid organization in stratum corneum and model systems based on ceramides. In: Touitou E, Barry BW, eds. *Enhancement in Drug Delivery*. Boca Raton, FL: Taylor & Francis; 2007. pp. 217–232.

Dolter KE, Evans CF, Hannaman D. In vivo delivery of nucleic acid-based agents with electroporation. Drug Deliv. Technol. 2010;10:37–41.

Donnelly RF, Singh TRR, Woolfson AD. Microneedle-based drug delivery systems: microfabrication, drug delivery, and safety. Drug Deliv. 2010;17:187–207.

Escobar-Chávez JJ, Bonilla-Martinez D, Billegas-González MA, Villegas-Gonzalez MA, Rodriquez-Cruz IM, Dominguez-Delgado CL. The use of sonophoresis in the administration of drugs throughout the skin. J. Pharm. Pharm. Sci. 2009;12:88–115.

FDA. 1993. Reviewer guidance for nebulizers, metered dose inhalers, spacers, and actuators. Office of Device Evaluation, Division of Cardiovascular and Respiratory Devices, Anesthesiology and Defibrillator Devices Branch. Oct. 1993. http://www.fda.gov/MedicalDevices/DeviceRegulationandGuidance/GuidanceDocuments/ucm081282.htm. Accessed Feb. 2012.

Gyurik RJ. Pharmaceutical composition. U.S. Patent 8063029. 2011.

Henry S, McAllister DV, Allen MG, Prausnitz MR. Microfabricated microneedles: a novel approach to transdermal drug delivery. J. Pharm. Sci. 1998;87:922–925.

Hu L, Batheja P, Meidan V, Michniak-Kohn BB. Iontophoretic transdermal drug delivery. In: Kulkarni VS, ed. *Handbook of Non-invasive Drug Delivery Systems*. Oxford, UK: Elsevier; 2010. pp. 95–118.

Kalluri H, Banga AK. Transdermal delivery of proteins. AAPS PharmSciTech 2011;12:431–441.

Karande P, Jain A, Mitragotri S. Discovery of transdermal penetration enhancers by high-throughput screening. Nat. Biotechnol. 2004;22:192–197.

Karande P, Jain A, Ergun K, Kispersky V, Mitragotri S. Design principles of chemical penetration enhancers for transdermal drug delivery. Proc. Natl. Acad. Sci. 2005;10:4688–4693.

Konno Y, Kawata H, Aruga M, Sonobe T, Mitomi M. Patch. U.S. Patent 4685911. 1987.

Kost J, Wolloch L. Ultrasound in percutaneous absorption. In: Touitou E, Barry BW, eds. *Enhancement in Drug Delivery*. Boca Raton, FL: Taylor & Francis; 2007. pp. 317–330.

Levin G. Advances in radio-frequency transdermal drug delivery. Pharm. Technol. 2008;Apr. S12–S19.

Levy A. Intraject: prefilled, disposable, needle-free injection of liquid drugs and vaccines In: Rathbone MJ, Hadgraft J, Roberts MS, eds. *Modified-Release Drug Delivery Technology*. Drugs and the Pharmaceutical Sciences, Vol. 126. New York: Informa Healthcare; 2007. pp. 619–631.

Meidan V, Michniak-Kohn BB. Ultrasound-based technology for skin barrier permeabilization. In: Kulkarni VS, ed. *Handbook of Non-invasive Drug Delivery Systems*. Oxford, UK: Elsevier; 2010. pp. 119–134.

Mitragotri S, Kost J. Ultrasound-mediated transdermal drug delivery. In: Rathbone MJ, Hadgraft J, Roberts MS, eds. *Modified-Release Drug Delivery Technology*. Drugs and the Pharmaceutical Sciences, Vol. 126. New York: Informa Healthcare; 2007. pp. 561–569.

Mudry B, Guy RH, Delgado-Charro MB. Iontophoresis in transdermal delivery. In: Touitou E, Barry BW, eds. *Enhancement in Drug Delivery*. Boca Raton, FL: Taylor & Francis; 2007. pp. 279–315.

Padmanabhan R, Gale RM, Phipps JB, van Osdol WW, Young W. D-Trans technology. In: Rathbone MJ, Hadgraft J, Roberts MS, eds. *Modified-Release Drug Delivery Technology*. Drugs and the Pharmaceutical Sciences, Vol. 126. New York: Informa Healthcare; 2007. pp. 481–498.

Pass F, Hayes J. Needle-free drug delivery. In: Rathbone MJ, Hadgraft J, Roberts MS, eds. *Modified-Release Drug Delivery Technology*. Drugs and the Pharmaceutical Sciences, Vol. 126. New York: Informa Healthcare; 2007. pp. 599–606.

Paudel KS, Milewski M, Swadley CL, Brogden NK, Ghosh P, Stinchcomb AL. Challenges and opportunities in dermal/transdermal delivery. Ther. Deliv. 2010;1:109–131.

Phipps JB, Southam M, Bernstein KJ. Device for transdermal electrotransport delivery of fentanyl and sufentanil. WO9639224(A1). 1996.

Phipps JB, Padmanabhan RV, Young W, Panos R, Chester AE. E-Trans technology. In: Rathbone MJ, Hadgraft J, Roberts MS, eds. *Modified-Release Drug Delivery Technology*. Drugs and the Pharmaceutical Sciences, Vol. 126. New York; Informa Healthcare, 2007. pp. 499–511.

Prausnitz MR, Langer R. Transdermal drug delivery. Nat. Biotechnol. 2008;26:1261–1268.

Prausnitz MR, Ackley DE, Gyory Jr. Microfabricated microneedles for transdermal drug delivery. In: Rathbone MJ, Hadgraft J, Roberts MS, eds. *Modified-Release Drug Delivery Technology*. Drugs and the Pharmaceutical Sciences, Vol. 126. New York; Informa Healthcare, 2007. pp. 513–531.

Rowe RC, Sheskey PJ, Quinn, ME, eds. *Handbook of Pharmaceutical Excipients*, 6th ed. London: Pharmaceutical Press; Washington, DC: American Pharmacists Association; 2009.

Sallmann AR. The history of diclofenac. Am. J. Med. 1986;80:29–33.

Samour CM, Daskalakis S. Percutaneous absorption enhancers, compositions containing same and method of use. U.S. Patent 4861764. 1989.

Singh TRR, Garland MJ, Cassidy CM, Migalska K, Demir YK, Abdelghany S, Ryan E, Woolfson D, Donnelly RF. Microporation techniques for enhanced delivery of therapeutic agents. Recent Pat. Drug Deliv. Formulation 2010;4:1–17.

Touitou E. Compositions for applying active substances to or through the skin. U.S. Patent 5540934. 1996.

———. Compositions for applying active substances to or through the skin. U.S. Patent 5716,638, 1998.

Touitou E, Godin B. Vesicular carriers for enhanced delivery through the skin. In: Touitou E, Barry BW, eds. *Enhancement in Drug Delivery*. Boca Raton, FL: Taylor & Francis; 2007. pp. 255–278.

U.S. Pharmacopeia. U.S.P.–N.F. online, First Supplement U.S.P. 34–N.F. 29 Reissue. Accessed Jan. 2012.

Williams AC, Barry BW. Chemical Permeation Enhancement. In: Touitou E, Barry BW, eds. *Enhancement in Drug Delivery*. Boca Raton, FL: Taylor & Francis; 2007. pp. 233–254.

Yang M, Zahn JD. Microneedle insertion force reduction using a vibratory actuation. Biomed. Microdevices 2004;6:177–82.

GLOSSARY

CDER	Center for Drug Evaluation and Research.
CHADD	Controlled heat-assisted drug delivery.
DMSO	Dimethyl sulfoxide.
E.P.	European Pharmacopoeia.
FDA	U.S. Food and Drug Administration.
GRAS	Generally recognized as safe.
hMTS	Hollow microstructured transdermal system.
sMTS	Solid microstructured transdermal system.
J.P.	*Japanese Pharmacopoeia*.
N.F.	*National Formulary*.
SEPA	Soft enhancers of percutaneous absorption.
U.S.P.	*United States Pharmacopeia*.
WEDD	Wearable electronic disposable drug delivery.

APPENDIX

APPENDIX 19.1   Transdermal Excipients

Excipient	Trade Names[a]	Use[b] (% w/v or mg)
	Absorption Enhancers	
Alcohol, E.P./J.P./U.S.P.–N.F.	Absolute alcohol; anhydrous alcohol; ethanol; grain alcohol	74.0–358.7 mg, 67.0–74.0% w/v
Dimethyl sulfoxide, E.P./U.S.P.–N.F.	Deltan; DMSO; Kemsol; Procipient; Rimso-50	16.5 mg, 45.5% w/v[c]
Laurocapram	Azone	
n-Lauroyl sarcosine	n-Dodecanoyl sarcosine	0.75–7.5% w/v[c]
n-Methyl pyrrolidone	NMP; Pharmasolve; m-Pyrol	
Octylsalicylate, U.S.P.–N.F.	Dermoblock OS; Escalol 587; Neo Heliopan OS	
Oleic acid, E.P./U.S.P.–N.F.	Crodolene; Emersol	5.51–22.00 mg
Pentadecalactone	Cyclopentadecanolide; Exaltolide; Muskalactone	8.0% w/v

(*continued*)

APPENDIX 19.1   (*Continued*)

Excipient	Trade Names[a]	Use[b] (% w/v or mg)
Propylene glycol, E.P./J.P./U.S.P.–N.F.	Sirlene; Solargard P; Ucar 35	58.13 mg, 6.0% w/v
Sorbitan monolaurate, E.P./U.S.P.–N.F.	Arlacel 20; Liposorb L; Montane 20; Span 20	4.74% w/v[c]
Adhesives		
Silicone	Silicone 4102	228.23 mg
	Silicone 4502	57.14 mg
Acrylic-based polymer	Acrylic Adhesive 2287	121.1 mg
	Acrylic Adhesive 788	20.08 mg
Polybutene	Amoco H-15; Indopol	3.25–221.25 mg
Antimicrobial Preservatives		
Alcohol, E.P./J.P./U.S.P.–N.F.	Absolute alcohol; anhydrous alcohol; ethanol; grain alcohol	74.0–358.7 mg, 67.0–74.0% w/v
Cetylpyridinium chloride, E.P./U.S.P.–N.F.	Pristacin	1.2 mg
Paraben methyl, E.P./J.P./U.S.P.–N.F.	Solbrol M; Tegosept M	0.35 mg,[c] 0.108–70% w/v[c]
Paraben propyl, E.P./J.P./U.S.P.–N.F.	Solbrol P; Tegosept P	0.02 mg,[c] 0.011–30% w/v[c]
Propylene glycol, E.P./J.P./U.S.P.–N.F.	Sirlene; Solargard P; Ucar 35	58.13 mg, 6.0% w/v
Antioxidants		
Ascorbyl palmitate, E.P./U.S.P.–N.F.	Vitamin C palmitate	0.0044–0.02% w/v[c]
Sodium metabisulfite, E.P./J.P./U.S.P.–N.F.	Disodium pyrosulfite	0.5 mg,[c] 0.03–0.3165% w/v[c]
Vitamin E, E.P./J.P./U.S.P.–N.F.	Copherol F1300	0.002% w/v
Buffers		
Citric acid monohydrate, E.P./J.P./U.S.P.–N.F.	Acetonum	1.4 mg
Citrate sodium, E.P./J.P./U.S.P.–N.F.	Trisodium citrate	2.2 mg
Sodium hydroxide, E.P./J.P./U.S.P.–N.F.	Aetznatron; Rohrputz	0.85–4.20 mg, 4.72% w/v
Strong ammonia solution, E.P./U.S.P.–N.F.	Spirit of Hartshorn	1.2% w/v[c]
Trolamine, E.P./U.S.P.–N.F.	TEA; Tealan; triethanolamine	0.35% w/v
Tromethamine	TRIS	0.10% w/v
Chelating Agents		
Edetate disodium, E.P./J.P./U.S.P.–N.F.	Disodium EDTA; edathamil disodium; versene disodium	0.06% w/v
Tartaric acid, E.P./J.P./U.S.P.–N.F.	Dihydroxysuccinic acid	

APPENDIX 19.1 (*Continued*)

Excipient	Trade Names[a]	Use[b] (% w/v or mg)
	Ion-Exchange Agents	
Polacrilin potassium, U.S.P.–N.F.	Amberlite IRP-88; methacrylic acid with divinylbenzene, potassium salt	1.10 mg
	Matrix Formers	
Carbomer, E.P./U.S.P.–N.F.	Acrypol; carbopol; Pemulen; polyacrylic acid	
	Carbomer 940	1.2% w/v
	Carbomer 980	1.5% w/v
	Carbomer 1342	24.3 mg, 0.3% w/v
Colloidal silicondioxide, E.P./J.P./U.S.P.–N.F.	Aerosil; Cab-O-Sil; SAS; Wacker HDK	9.94–49.00 mg
Crospovidone, E.P./U.S.P.–N.F.	Crospopharm; Kollidon CL; PVPP; Polyplasdone XL; poly(vinylpolypyrrolidone)	60.0 mg
Fractionated coconut oil, E.P./J.P./U.S.P.–N.F.	Pureco 76	20% w/v[c]
Gelatin, E.P./J.P./U.S.P.–N.F.	Byco; Cryogel; Instagel; Kolatin; Solugel; Vitagel	
Hydroxypropyl cellulose, E.P./J.P./U.S.P.–N.F.	Klucel; Nisso HPC	19 mg
Lactose, E.P./J.P./U.S.P.–N.F.	Aletobiose	675.0 mg, 18.9% w/v
Light mineral oil, E.P./J.P./U.S.P.–N.F.	Citation	74.0–162.0 mg
Methacrylic acid esters copolymers, E.P./U.S.P.–N.F.	Acryl-EZE; Eastacryl; Eudragit; Kollicoat MAE	
Mineral oil, E.P./J.P./U.S.P.–N.F.	Avatech; Drakeol; Sirius	1.52–11.80 mg
Poly(methyl methacrylate)	Korad; PMMA	
Povidone, E.P./J.P./U.S.P.–N.F.	Kollidone; Plasdone; poly(vinylpyrrolidone)	
	Povidone K29/32	7.266 mg
Silicone oil	Baysilon	353.51 mg
Sodium polyacrylate	Polyco	
	Rate-Controlling Membrane	
Poly(ethylene-*co*-vinyl acetate)	CoTran; EVA; EVM; acetic acid; ethylene ester polymer with ethane	735.0 mg
Polyethylene	Petrothene	85 mg
Polyester film	—	3.61–96.00 mg
Polypropylene	—	13.5–19.3 mg
Poly(vinyl acetate)	Rhodopas; Sovial; Vinac	3.99–16.00 mg
Poly(vinyl chloride)	Bakelite; Flocor	927.0 mg
Poly(vinyl chloride-*co*-vinyl acetate)	—	899.88 mg

APPENDIX 19.1 (*Continued*)

Excipient	Trade Names[a]	Use[b] (% w/v or mg)
	Solubilizing Agents	
Alcohol, E.P./J.P./U.S.P.–N.F.	Absolute alcohol; anhydrous alcohol; ethanol; grain alcohol	74.0–358.7 mg 67.0–74.0% w/v
Diethylene glycol monoethyl ether	Poly-Solv	5.0% w/v
Ethyl acetate, E.P./U.S.P.–N.F.	Acetidin	36138.0 mg
Glycerin, E.P./J.P./U.S.P.–N.F.	Croderol; Kemstrene; Optim; Pricerine	5.0% w/v, 306.2 mg
Glyceryl monooleate, E.P./U.S.P.–N.F.	Atlas G-695; Kessco GMO; Peceol; Stephan GMO; Tegin	18.8 mg
Isopropyl alcohol, E.P./J.P./U.S.P.–N.F.	IPA	4–78% w/v[c]
Isopropyl myristate, E.P./U.S.P.–N.F.	HallStar IPM-NF; Rita IPM; Stepan IPM; Tegosoft M	0.86% w/v, 58.08 mg
Lauryl lactate	Ceraphyl 31; Crodamol LL	12.0 mg
Methyl laurate	—	17.6 mg
Polyethylene glycol, E.P./J.P./U.S.P.–N.F.	Carbowax; Lipoxol; PEG	0.3–84% w/v[c]
Polyoxyl 20 cetostearyl ether, E.P./U.S.P.–N.F.	Atlas G-3713, Ceteareth 20; Volpo CS20	1–10% w/v[c]
Triacetin, E.P./U.S.P.–N.F.	Captex 500	22.1 mg
	Viscosity-Inducing Agents	
Hydroxyethyl cellulose, E.P./U.S.P.–N.F.	Cellosize HEC; Natrosol; Tylose H	20 mg
Polyvinyl alcohol, E.P./U.S.P.–N.F.	Elvanol; Mowiol; Polyvinol; PVA	119.0 mg
Sodium carboxy methyl cellulose, E.P./J.P./U.S.P.–N.F.	Akucell; Aquasorb; Nymcel ZSB; Tylose CB	3.5% w/v[c]

[a] A comprehensive list of products and suppliers is provided in Rowe (2009).

[b] FDA Inactive Ingredient List, http://www.fda.gov/Drugs/InformationOnDrugs/ucm113978.htm. Accessed Feb. 2012.

[c] Listed for topical formulations in the FDA Inactive Ingredient List.

CHAPTER 20

ORAL MODIFIED-RELEASE PRODUCT DESIGN

20.1 INTRODUCTION

Modified-release drug delivery systems are available for most routes of administration. The most popular routes of modified release are oral, parenteral, implantable, and transdermal. In this chapter we focus on oral modified-release products. The scope of the chapter does not include oral modified-release technologies that increase dissolution and bioavailability or targeted delivery to specific sites of action such as the Peyer's patches or the colon.

The modified-release nomenclature has developed over time. *Modified-* and *controlled-release drug delivery* are terms often used synonymously and refer to dosage forms that release drug in a modified or controlled manner compared to a conventional dosage form. The U.S. Food and Drug Administration (FDA) and *United States Pharmacopeia* (U.S.P.) have brought more standardization to this class of dosage forms. The FDA (1997a) has defined *modified-release dosage forms* as "dosage forms whose drug-release characteristics of time course and/or location are chosen to accomplish therapeutic or convenience objectives not offered by conventional dosage forms such as a solution or an immediate release dosage form. Modified-release solid oral dosage forms include both *delayed* and *extended release* drug products." The U.S.P. (2012) and FDA (2010) also define delayed- and extended-release products. The U.S.P. discusses the use of *enteric coatings*, used to delay the release of drug so that it is not degraded or inactivated in the gastric fluid. The U.S.P. also discusses circumstances when delayed-release products are used to prevent gastric mucosa irritation. The U.S.P. uses the term *delayed release* for pharmacopeial purposes when enteric coatings are used, and specific drug-release specifications are provided in U.S.P. Chapter <724>, Drug Release. The U.S.P. also recognizes that a number of terms have been used for formulations that provide extended release. Other terms mentioned in the U.S.P. are *sustained release, prolonged release*, and *repeat action*. The U.S.P. uses *extended release* for pharmacopeial purposes, and each monographed drug usually has its own drug release specifications. The FDA nomenclature is presented in Appendix 20.1. Other common descriptive terms are used to name other types of modified oral release

Integrated Pharmaceutics: Applied Preformulation, Product Design, and Regulatory Science, First Edition. Antoine Al-Achi, Mali Ram Gupta, William Craig Stagner.

dosage forms, such as *multifunctional release, pulsatile* or *pulsed release*, and *chronorelease*. Multifunctional release is typified by providing immediate release and sustained action. This type of release strategy might be useful for pain management where there is a need for rapid onset of action followed by a prolonged pharmacological effect. Pulsatile release design may find utility when there is a desire or need to simulate multiple daily doses in a more convenient single-dose tablet or capsule. For example, pulsed drug delivery may be desirable for a once-a-day drug regimen involving two drugs that have different half-lives requiring pulsed doses of the drugs at different times during the day. It is known that many of the body's physiological functions have a certain chronicity or pattern, called *circadian rhythm* or the *body clock*. In general, there are certain broad times during the day when blood pressure and hormone levels are at their high or low levels. The largest spike in blood pressure is in the morning around 6 to 7 A.M., and the highest level is around 7 to 8 P.M. A pulsed-release dosage form could be designed to generate maximum drug blood levels at these two times. The highest incidents of cardiovascular events are in the morning. A chronorelease product could be designed to be taken at bedtime to provide a blood-level profile that gives maximum protection against a cardiovascular event in the morning hours. Similarly, hormone replacement therapy may use a chronorelease dosage form to mimic the hormonal biorhythms of the body. Advantages and disadvantages of modified-release dosage forms are listed in Table 20.1.

The literature abounds with additional nomenclature that has been used to describe different types of systems and mechanisms of release. Hybrid systems and specialty mechanisms make the taxonomy of classifying modified-release systems that much more difficult. Modified-release dosage forms are often classified as monolithic and multiparticulate systems. In *monolithic systems* the drug is usually distributed throughout the dosage form, like a tablet or single unit or monobloc. *Multiparticulate systems*, as the name implies, are dosage forms that contain multiple particles, such as coated beads or minitablets. Multiparticulate systems are thought to have several advantages over monolithic systems. Catastrophic failure of the rate-controlling mechanism, resulting in dose dumping, is more likely to occur with a monolithic system than with a multiparticulate system, where the likelihood of the control mechanism failing for all particles is significantly lower. Multiparticulate drug release is less affected by food and there is more predictable first-order emptying from the stomach. There is also less patient-to-patient and intrapatient release variability. In addition, the chances of gastrointestinal irritation caused by high local drug concentrations is reduced when using multiparticulate dosage forms.

The monolithic and multiparticulate systems can both have reservoir or matrix structures. The *reservoir structure* has a core–shell arrangement in which the drug is located inside the bead or tablet core and the shell consists of a rate-controlling membrane. Ethylcellulose (Ethocel) and copolymers of methacrylic acid and methyl methacrylic acid (Eudragit) are commonly used as rate-controlling membranes. In the *matrix structure*, the matrix composition provides the release rate control. Oral modified-release product design can also involve a hybrid approach using a matrix controlled-release core that is coated

TABLE 20.1 Advantages and Disadvantages of Modified-Release Drug Delivery

Advantages	Disadvantages
Reduction in dosing frequency	Dose dumping
Reduction in fluctuation of circulating drug levels	Reduce potential for dosage adjustment
Maintain the drug level in the *therapeutic range* or *window* where the drug level remains above the minimum effective concentration and below the minimum toxic concentration	May not be able to break tablets
Should result in a uniform and predictable therapeutic effect	Limitations caused by the gastrointestinal (GI) tract
Avoid nighttime dosing or the need to get up in the night	Limited GI residence time
Mimic circadian rhythm	GI *housekeeping waves*
Deliver drug at therapeutic levels when the patient is most vulnerable to a pathological or disease event	pH, enzymes, changes in surface area and liquid contents
Provide the ability to dose several drugs in one dosage form that have very different pharmacokinetic profiles	Influence of food or alcohol consumption on drug levels
Improve patient compliance	Concern about alcohol-related dose dumping
	Potential for decreased absorption
	May bypass an absorption window
	Unpredictability of the coatings or matrices
	Unpredictable in vivo–in vitro correlation
	Manufacturing complexity
	Laser drilling
	Precise uniform cutting of rate-controlling membrane
	Precise and uniform coating of beads or matrices such as capsules and tablets
	Accurate and uniformly blending of rapid-release and controlled-release components
	Impact of compressional force on coated pellets placed inside a tablet core
	Breakage or cracking of the controlled-release membrane that could lead to dose dumping
	Overdose if patient inadvertently double-doses

with a controlled-release membrane. Similarly, there can be particulate systems that consist of matrix particles that do not have a controlled-release coating but rely on the particle matrix to provide the modified release.

Modified-release systems can be subdivided further by the predominant mechanism of drug release. The main mechanisms for modified drug release include *diffusion, dissolution, swelling, erosion*, and *degradation*. Diffusion, which is described by Fick's first and second laws of diffusion, occurs in all modified-release dosage forms where the distribution of water, drug, polymers, and electrolytes occurs across concentration gradients. However, the other mechanisms listed above may be the release rate–controlling mechanism. Diffusion-controlled drug delivery depends on the drug's solubility and dosage form's material design and fabrication. In a diffusion-controlled monolithic matrix system, the drug is dispersed in a nonswelling, nondissolving, noneroding, and nondegrading polymer matrix such as ethylcellulose. The dispersion can be a molecular dispersion (solution) or a homogeneous dispersion of solid drug particles. In the case of a tablet monolithic molecular dispersion, the dissolution media forms liquid channels through which the drug molecules can diffuse. The diffusional process is controlled by the drug's solubility in the gastrointestinal fluid, its diffusional coefficient in the liquid, and the porosity and tortuosity of the liquid channels. The monolithic solid particulate dispersion requires an additional drug dissolution step before the drug can diffuse through the dissolution fluid channels. Depending on the solubility of the drug in the penetrated liquid and how fast the liquid permeates the matrix, excess drug may exist in the matrix, which provides a saturated solution of drug and a constant concentration gradient until the entire amount of drug is dissolved.

In a diffusion-controlled monolithic reservoir system, a rate-controlling membrane is employed to control the diffusion rate of the drug across the membrane. The gastrointestinal fluid diffuses across the membrane, dissolves the drug, and the dissolved drug diffuses out of the core across the membrane. The concentration gradient established within the core depends on the drug solubility in penetrating liquid in the core and the amount of drug load. If a saturated drug solution can be maintained for a given time interval, a constant concentration gradient will be achieved along with a constant zero-order drug release. If no excess drug is available to maintain a saturated drug solution, the release profile will be first order. Multiparticulate monolithic and reservoir diffusional systems can also be designed. Similarly, modified-release systems can leverage a combination or hybrid of particulate and monolithic systems.

Dissolution-controlled modified release relies on the dissolution process to regulate the rate of drug delivery. Many tablet excipients dissolve or partially dissolve when exposed to the gastrointestinal contents. The dissolution process for low-molecular-mass drugs and excipients is described by the Noyes–Whitney equation. The dissolution process for polymers and other high-molecular-mass materials is based on a *reptation model* proposed by Herman and Edwarde (1990). The concept of reputation is based on the snakelike motion of long entangled macromolecules in solution. Polymer dissolution control typically involves dissolution of the polymer, disentanglement of polymer chains, decrease in an entangled polymer network, increase in localized polymer mobility, and erosion at the surface of the dosage form. A number of factors affect polymer dissolution, such as polymer molecular mass, diffusion coefficient, and the strength of the water–polymer interaction. The release of the drug also depends on the solubility of the drug and its diffusional coefficient in the polymer milieu. Poly(vinyl alcohol) (Polyviol) is a popular dissolution controlled-release soluble polymer.

Polymer swelling can be used to slow down drug release by competing for the gastrointestinal dissolution fluid, decreasing the drug's diffusional coefficient compared to the dissolution fluid alone, and increasing the structural volume, which increases the diffusional distance the drug must travel before reaching the bulk dissolution media. Hydroxypropyl methylcellulose (Methocel) is known to swell significantly when hydrated. Other materials that form hydrogels are hydropropyl cellulose (Klucel), xanthan gum, guar gum, alginate, and carrageenan. Different grades of hydroxypropyl methylcellulose can be chosen that hydrate at different rates and exhibit different viscosities once hydrated. Faster hydrating grades are required for drugs that have higher solubilities and dissolve fast. The rate of drug release can also be controlled by the polymer's hydrated viscosity. The water penetration and swelling occurs from the surface to the center of the matrix. Different release rates may also be achieved by incorporating different-grade polymers in bi- or trilayered tablets.

Erosion is a physical process that results in the loss of polymer from a fabricated dosage form. The erosion results from polymer dissolution, hydration, lowering of the glass transition temperature, or a combination of these and other events that increase the polymer's mobility and loss of entanglements. The loss of polymer can occur at the surface of the dosage form, which results in overall shrinkage of the dosage form over time, known as *surface erosion*. By using a mixture of different types of polymers, polymer loss can be designed to occur within the dosage form called *bulk erosion*. Bulk erosion can occur when more soluble polymers are embedded in an ethylcellulose matrix. For example, poly(ethylene glycol) and poly(vinyl alcohol) can be hot-melt-extruded with an ethylcellulose matrix. The soluble polymers will dissolve, leaving the interior of the matrix less dense and more porous.

Polymer degradation is a chemical process which may involve hydrolysis or other chain scission mechanisms that lead to the formation of lower-molecular-mass oligomers and monomers. Surface and bulk erosion can result from polymer degradation. Reactive polymers tend to degrade rapidly at the surface of the tablet and undergo surface erosion. Predominantly, polyanhydrides undergo degradation surface erosion. Polifeprosan 20 is a polyanhydride copolymer that consists of a 20 : 80 ratio of poly[bis(*p*-carboxyphenoxy)propane : sebacic acid] and is used in the Gliadel wafer implant. Poly(lactic acid) and poly(lactic-*co*-glycolic acid) are slower-hydrolyzing polymers and typically show bulk polymer loss.

The best drug candidates for modified release are biopharmaceutical system classes I and II, which are drugs that have high permeability. The low solubility of class II compounds complicates modified-release product design and may exploit special technologies to enhance drug solubility. Drugs that undergo first-pass metabolism or demonstrate significant fasted–fed absorption can also pose substantial design issues that must be kept in mind.

In vitro–in vivo correlations that correlate in vitro drug release to in vivo plasma-level profiles are highly desirable. In vitro–in vivo correlations that are established early in the product design process can significantly reduce development time and cost, decrease development risk, and reduce the number of human bioavailability studies required during the design process. Optimization efforts that require formulation and process changes can be carried out rapidly with a relatively high degree of confidence that the changes will not affect the drug's bioavailability negatively. In vitro–in vivo correlations also support the establishment of more robust dissolution methods, specifications, and regulatory drug applications. The benefit of an in vitro–in vivo correlation continues throughout a product's life cycle. The FDA (1997a) guidance document Scale-up and Postapproval Changes: Chemistry, Manufacturing and Controls; In Vitro Dissolution Testing and In Vivo Bioequivalence Documentation notes specific postapproval changes that can be made if an in vitro–in vivo correlation exists. The guidance also requires that the "before" and "post-change" product dissolution profiles pass the f_2 dissolution similarity test. The in vitro f_2 dissolution profile comparison (FDA, 1997a) is a model-independent approach that defines the similarity factor, f_2, as

$$f_2 = 50 \text{Log} \left\{ \left[1 + \frac{1}{n \sum_{t=1}^{n} (R_t - T_t)} \right]^{0.5} \right\} \times 100 \qquad (20.1)$$

where log is the base 10 logarithm, n the number of sample points, S the summation over all the sample time points, R_t the dissolution at time t for the reference product before the change, and T_t the dissolution at time t for the test product after the change. It is recommended that only one point past the dissolution plateau value be used in calculating f_2 and that a lag-time correction prior to similarity testing not be performed unless justified. The average difference at any dissolution time point should not be greater than 15% between the unchanged and changed product. Using equation (20.1) and the criteria above, an f_2 value between 50 and 100 indicates that the two dissolution profiles can be considered similar.

U.S.P. Chapter <1088>, In Vitro and In Vivo Evaluation of Dosage Forms, and the FDA (1997b) guidance document Extended Release Oral Dosage Forms: Development, Evaluation, and Application of In Vitro/In Vivo Correlations speak to the growing importance and expectation for developing an in vitro–in vivo correlation. The U.S.P. states that the "concept of correlation level is based on the ability of the correlation to reflect the entire plasma drug concentration–time curve that will result from the administration of the given dosage form. It is the relationship of the entire in vitro dissolution curve to the entire plasma level curve that defines the correlation." The U.S.P. presents three levels of correlation: levels A, B, and C, which are categorized in descending order of usefulness. Level A correlation is a point-to-point relationship between the dissolution profile and the in vivo dosage-form drug input rate. The FDA defines an in vitro–in vivo correlation as "a predictive mathematical model describing the relationship between an in vitro property of an extended release dosage form (usually rate or extent of drug dissolution or release) and a relevant in vivo response, e.g., plasma drug concentration or amount of drug absorbed." The FDA guidance describes four levels of correlation categories with level A being the most useful and having the highest level of regulatory acceptance and level B the least useful for regulatory purposes. The FDA (2011) has released an excellent illustrative example of how to prepare a regulatory submission using the quality-by-design principle. The example presented is based on a modified-release tablet that contains controlled-release beads. The section "Dissolution Method Development and Bioequivalence Studies" advances the FDA's thinking on how to develop an in vitro–in vivo correlation.

The reader is highly encouraged to read the FDA's (2011) quality-by-design modified-release tablet example. The 161-page example provides an excellent chronological development project design template for a modified-release dosage form. The template can also be extended easily to other dosage forms. The example discusses the quality target product profile, drug substance characterization, excipients, excipient compatibility, drug product formulation development, manufacturing process development, dissolution method development, in vitro–in vivo correlation development, design of experiments, and more. It is a "must read" for those actively involved with product design.

In the remainder of the chapter we discuss various examples of modified-release products. There are a multitude of these products on the market and in development. An attempt has been made to provide examples that cover a number of different types of product design concepts.

20.2 COATINGS

Compounds used in film coating and their different functional properties are discussed in Chapter 13. A brief review is provided here, followed by examples of their use in modified-release products. Most delayed-release products use pH-sensitive polymers that are stable and insoluble at the pH of the stomach and dissolve at more alkaline pH. Table 13.21 lists a number of polymers suitable for delayed-release dosage forms. The acrylate latex dispersions provided under the trade names Acryl-EZE, Eudragit, and Kollicoat are commonly used. These polymer systems are designed to dissolve at such designated pH values as 5.5, 6.0, and 7.0, depending on the most desirable intestinal site for drug release, absorption, and action. Aquacoat CPD is a cellulose acetate phthalate delayed-release dispersion that is also pH sensitive. Common sustained-release polymers use ethylcellulose aqueous dispersions (Aquacoat ECD and Surelease). Kollicoat SR 30 D is a sustained-release poly(vinyl acetate) polymer. Eudragit RL and RS grades are pH-independent sustained-release polymer systems containing a copolymer of ethyl acrylate and methyl methacrylate and a low content of methacrylic acid ester with quaternary ammonium groups.

20.2.1 Bupropion HCl (Wellbutrin XL): Monolithic Modified-Release Example

Bupropion hydrochloride is an antidepressant. Bupropion has a molecular mass of 239.1 Da, a melting point of 233 to 234°C, a water solubility value of about 31.2% w/v at room temperature, and an experimental log P of 3.6 (Drug Bank, 2012). Bupropion hydrochloride has a biphasic elimination where there is rapid distribution over 2 to 3 hours and a terminal half-life of 21 hours. Food increases the absorption 11 to 35%, and the drug is about 84% protein bound (Wellbutrin XL product monograph, 2010). Bupropion is unstable at pH values above 2.5 (Laizure and DeVane, 1985; O'Byrne et al., 2010) and it is sensitive to moisture.

The FDA (2011) has classified bupropion hydrochloride in biopharmaceutical system class I. Oberegger et al. (2008) leveraged the use of two coating systems to design a once-a-day extended-release tablet that was stable to moisture ingress. Wellbutrin XL is a monolithic extended-release tablet. The core tablet is prepared by wet granulation of drug and binder. The dry granulation contains about 94% w/w drug and 3.3% w/w poly(vinyl alcohol) as binder. The dried granulation is lubricated with 2.9% w/w glyceryl behenate. The granulation is compressed, and the core tablet is coated with a sustained-release ethylcellulose nonaqueous coating system. The dry film contained 57% w/w ethylcellulose 100, 31.33% w/w poly(vinylpyrrolidone) as a water-soluble rate-modifying polymer, and 11.6% w/w poly(ethylene glycol) as a plasticizer. This dried film coat was then coated with a moisture-barrier outer coat containing a moisture barrier polymer, plasticizer, and permeation enhancer. In this case, the moisture barrier polymer was Eudragit L 30 D-55, which made up 65.6% of the dry film. The plasticizer was a combination of poly(ethylene glycol) at 6.6% w/w and triethyl citrate at 3.3% dry film weight. A permeation enhancer was incorporated into the film to allow drug release at early time points in the stomach. The permeation enhancer used here was silicon dioxide at 24.6% dry weight. The stability and bioavailability data show that the two coatings functioned as designed. Tablets with a moisture barrier coating gained significantly less moisture, and the controlled-release barrier did achieve an acceptable once-daily dose regimen with suitable drug and metabolite levels.

20.2.2 Multiparticulate Modified-Release Example

The compound in this example is a biopharmaceutical system class I drug. It was developed as an extended-release bead formulation. The relative high dose excluded the possibility of spray coating the active drug on an excipient bead. A wet granulation–extrusion–spheronization process was designed to create a matrix bead multiparticulate system. The matrix beads were then coated with an aqueous ethylcellulose controlled-release coating and then encapsulated. The dissolution rates were modeled to aid setting dissolution specifications and to evaluate how processing changes affected the release rate. Seven hundred and twenty dissolution samples were evaluated using zero- and first-order dissolution models. As discussed above, depending on the drug's solubility, the diffusion rate through the controlled-release membrane, and the presence of a constant concentration gradient, a zero- or first-order release profile may occur. Figures 20.1 and 20.2 show the zero- and first-order plots of the dissolution data. The plot of percent drug dissolved vs. time is curvilinear and does not fit the zero-order linear dissolution model ($r^2 = 0.849$). The first-order dissolution model explains 99.9% ($r^2 = 0.999$) of the dissolution data when the natural logarithm of the fraction of drug remaining to be dissolved is plotted vs. time. This excellent fit significantly enhances the product design scientist's ability and confidence to develop a multipoint robust dissolution specification. Knowledge of the dissolution mechanism can also be used to identify critical process and in-process material quality attributes. Guy et al. (1982) provided equations that modeled the drug release rates for spherical particles. Although introduction

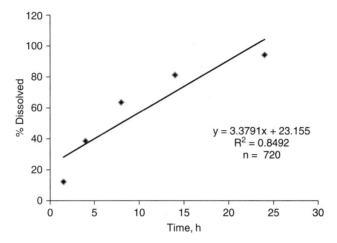

FIGURE 20.1 Zero-order dissolution model.

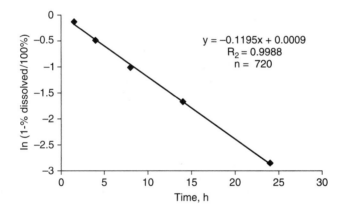

FIGURE 20.2 First-order dissolution model (fraction of drug remaining to be dissolved).

of a rate-controlling membrane takes liberties with the assumptions developed by Guy and co-workers, rearrangement of equation (36) in the original manuscript and substitution for M_8 gives a first-order equation (20.2) similar to the one used to plot Figure 20.2. Equation (20.2) states that a plot of the natural logarithm of the fraction of drug remaining to be dissolved vs. time should be linear:

$$\ln\left(\frac{1 - M_t}{M_\infty}\right) = \frac{-3\kappa Dt}{r_0^2} = -kt \tag{20.2}$$

where ln is the natural logarithm, M_t the amount of drug dissolved at time t, M_∞ the amount of drug when it is all released, κ describes the interfacial transfer process, D is the integral diffusional coefficient, t is the time, and r_0 is the radius of the spherical particle at the initial time. From equation (20.2) the slope of the first-order plot is $-k$ (the apparent first-order rate constant), which also equals $-3\kappa D/r_0^2$. This relationship indicates that the first-order apparent rate constant is inversely proportional to the square of the radius extruded coated bead. This supports the intuitive thought that smaller beads should lead to a faster dissolution process. The fact that the inverse relationship follows the bead radius squared may not be as intuitive and amplifies the need to control the bead particle size meticulously throughout the manufacturing process. The predictive power of this model equation can be verified by changing the bead particle size and comparing the actual first-order rate constant and dissolution values to the values predicted.

20.3   MATRIX SYSTEMS

20.3.1   Polymer Advances

Over the past decade, the companies Dow Chemical and Ashland, Inc. have created subsidiaries, Dow Wolff Cellulosics and Ashland Specialty Ingredients, to meet the product design and regulatory needs of the pharmaceutical, food, and cosmetic industries. The common modified-release polymers that are used for monolithic matrix particles or tablets are ethylcellulose, methylcellulose, hypromellose (hydroxypropyl methylcellulose), hydroxypropylcellulose, and polyethylene oxide. Ethylcellulose is nonswelling and insoluble in water. Direct compressible forms of ethylcellulose, Aqualon T 10 and Ethocel Standard FP Premium, have recently been developed by these specialty companies. Enhanced compressional properties expand the use of ethylcellulose to layered and press-coated modified-release tablets. In addition, faster-hydrating direct compression forms of hypromellose and hydroxpropylcellulose have also been commercialized, which has significantly expanded the ability to design zero- and first-order monolithic dosage forms.

Poly(oxyethylene oxide) is a thermoplastic water-soluble polymer that can be calendered (pressed between rollers), extruded, and injection molded. Poly(ethylene oxide) affords modified-release through controlled dissolution and swelling. Some other polymers commonly used for hot-melt processing are copovidone [poly(vinylpyrrolidone-*co*-vinyl acetate], poly(ethylene glycol), ethylcellulose, and poly(vinylpyrrolidone). Several examples of monolithic matrix systems are discussed below.

20.3.2   Meltrex

Meltrex is a melt-extrusion calendaring process used to manufacture modified-release tablets or caplets having delayed- and extended-release properties. Meltrex is a registered trademark of Soliqs, which is a drug delivery business of Abbott GmbH & Co. Two commercial products are available that have used the Meltrex technology: Kaletra (lopinavir/ritonavir) and Calan SR. Both lopinavir and ritonavir are considered biopharmaceutical system class II compounds which are low-solubility high-permeability drugs. The Kaletra Meltrex product offers a number of advantages over the soft gelatin capsule that was first approved and introduced into the market. The soft gelatin capsule formulation contained 42.4% alcohol, required refrigerated storage, and needed to be used within two months if stored at room temperature. In addition, the dosing regimen involved taking six capsules daily at 133 mg lopinavir/33 mg ritonavir per capsule. The design of a bioequivalent hot-melt dispersion of 200 mg lopinavir/50 mg ritonavir per capsule eliminated the need for alcohol, afforded long-term room-temperature storage, and decreased the daily dosing requirement to two tablets twice daily. In addition, the tablets did not show the food effect seen with soft gelatin capsules.

The Kaletra core tablet excipients and approximate % w/w concentrations (Berndl et al., 2011) are copovidone [poly(vinylpyrrolidone-co-vinyl acetate], 70.7%; sorbitan monolaurate, 7.0%; sodium stearyl fumarate, 1.0%; and colloidal silicon, 0.7%. Lopinavir and ritonavir are mixed with the copovidone and sorbitan monolaurate in a high-shear mixer. This mixture is then added to a twin-screw extruder and melted at 119°C. The exudate is cut into pieces, allowed to solidify, and milled with an impact mill to a particle size of approximately 250 μm. The milled material is then blended with colloidal silicon dioxide and sodium stearyl fumarate. This blend is then compressed and film-coated. Copovidone serves as the thermoplastic polymer that forms the solid dispersion with the drugs. Sorbitan monolaurate (Span 20), a surfactant, was found to improve the bioavailability of lopinavir and ritonavir. Sodium stearyl fumarate is the lubricant, and colloidal silicon dioxide acts as a flow aid.

There is growing concern over the effect of ingestion of alcohol when taken in conjunction with modified-release products. Roth et al. (2009) evaluated the effect of 0 to 40% alcohol on the dissolution profiles of three commercial verapamil sustained-release products and a verapamil Meltrex modified-release dosage. Only the Meltrex product showed invariant dissolution profiles over the alcohol range studied.

20.3.3 DiffCORE

DiffCORE technology, which has been commercialized by GlaxoSmithKline, employs orifice technology that penetrates an impermeable membrane that otherwise coats a tablet (Staniforth, 1991; Buxton et al., 2009). The drug solubility; type of impermeable membrane or film coat; type of core formulation components; and number, size, and shape of the orifice(s) control the release rate of the drug. The aperture(s) generally cover 30 to 60% of the tablet surface, so this physical opening is much larger than the laser-drilled "holes" utilized in osmotic pump dosage forms discussed in Section 20.5. Zero-order release, first-order release, delayed release, and combinations thereof are possible. Lamictal XR extended-release tablets use the DiffCORE technology in combination with an enteric coat and a polymer system that swells and erodes to control the release rate of lamotrigine. Lamital XR tablets are drilled on two sides of the tablet, and this modified-release system is designed to deliver drug for 12 to 15 hours.

The release-modifying agent may consist of swelling agents, osmagents, effervescent couples, eroding or degrading polymers, and others. The lamotrigine % w/w core excipients consist of 15.7% hydroxypropyl methylcellulose K100 LV controlled-release grade, 11.3% hydroxypropyl methylcellulose E4M controlled-release grade, 22.6% lactose monohydrate, and 0.4% magnesium stearate. The tablet core running powder is prepared by wet granulation. The impermeable membrane is prepared from a film coat of Eudragit L30 D-55 that controls the release of the drug in the stomach. The matrix methylcellulose polymer retards release by a swelling and erosion mechanism.

20.3.4 Three-Dimensional Printing

Three dimensional printing is a solid free-form fabrication technology that controls the drug release through flexible three-dimensional positioning of the drug and the rate-controlling polymer, and control of the matrix porosity, shape, and functional design. Cima et al. (1993) reported this technique, and it was soon licensed for the design and manufacture of drug delivery systems. The modified-release dosage form is fabricated by preparing layers of powders that contain drug and excipients. The exact position of the various components employs a computer-controlled ink-jet printing assembly. The technology has been used to fabricate various types of systems that exhibit a variety of modified-release mechanisms, such as zero-order release, immediate or extended release, breakaway tablet portions, enteric dual pulses, and dual pulses (Rowe et al., 2000; Monkhouse et al., 2003). Lin et al. (2001) developed an in vitro–in vivo correlation for an ethinyl estradiol biodegradable implant fabricated using three-dimentional printing technology. A detailed mechanical description of this technology is given in Iskra's patent (2005).

20.3.5 NRobe

NRobe entails the compacting of a drug containing powder that is *enrobed* with a film of rate-controlling polymer (Teckoe et al., 2009). Enrobbing involves the wrapping of lightly consolidated material by preformed films to provide a coated nonfriable dosage form. Figure 20.3 shows the mechanism of the basic steps of powder compaction and enrobement via steps a to l: a, a rate-controlling film is laid upon a platen; b, a vacuum port in the lower piston pulls the film against the lower piston to form a pocket; c, a quantity of running powder is introduced into the pocket and the upper punch compresses the powder (d); e and f, the excess film is cut with a cutting tool; g, the lower punch partially ejects the plug; h, a second film is layered over the partially ejected plug; i, a vacuum is again applied, drawing the second film against the partially ejected plug; j, the second film is cut and trimmed; k, the fully enrobed plug is completely ejected from the die; l, the loose ends of the films are heat-sealed (Teckoe et al., 2009). Hypromellose is the preferred thermoplastic polymer, although hydroxypropyl cellulose, poly(vinyl

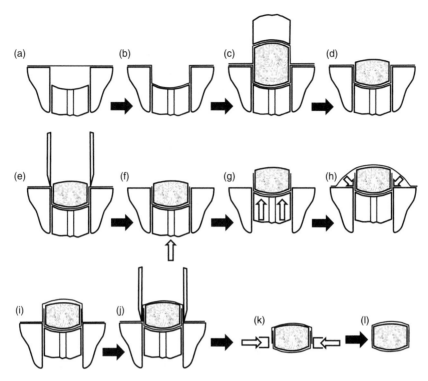

FIGURE 20.3   Powder compaction and enrobing. (From Teckoe et al., 2009.)

alcohol), poly(ethylene oxide), and others are mentioned as possible polymers. Named plasticizers are lactic acid, triacetin, propylene glycol, glycerin, and more. The novel dosage form is excellent for compacting beads that have brittle coatings or coatings that cannot withstand the compressional forces required to make conventional tablets. A hypromellose-enrobed dosage form containing a high drug load of a water-soluble compound and 5% HPMC K 100M gave a zero-order release profile (FMC, 2012). It was speculated that the zero-order dissolution profile was the result of the combination of the high porous nature of the compact and the hypromellose-enrobed film, which allowed time for the powder compact to form a release-retarding gel.

20.4   GASTRORETENTIVE DEVICES

A number of gastroretentive mechanisms have been evaluated, including mucoadhesion, floating dosage forms that use low-density matrices or gas generation, swelling matrices, and combinations thereof. Generally, the most effective gastroretentive devices are those that swell to a size that prevents them from exiting the stomach or floating systems. Only two devices of the many in the marketplace or in development are covered here.

20.4.1 AcuForm

The AcuForm (Depomed) gastroretentive technology has been used in two FDA-approved products, ciprofloxacin (Proquin XR) and metformin hydrochloride (Glumetza), which are both biopharmaceutical class III drugs. Shell and Louie-Helm (1999) discusses the use of poly(ethylene oxide) for gastroretention. Gusler et al. (2004) describe the use of poly(ethylene oxide), hypromellose, and poly(vinylpyrrolidone) for gastroretention and controlled release of ciprofloxacin and metformin hydrochloride. Poly(ethylene oxide) swells in the gastric fluid and erodes over time. It is typically used in the range of 30 to 40% w/w. Poly(ethylene oxide) is also used in combination with hypromellose in a 1 : 2 to 2 : 1 ratio. These polymers swell and prevent discharge of the swollen tablet into the small intestine. Over time the swollen polymers erode and eventually exit the stomach.

It is recommended that Glumetza be taken once daily with the evening meal. According to the product package insert, "Glumetza tablets must be swallowed whole and never split, crushed or chewed. Occasionally, the inactive ingredients of Glumetza 500 mg may be eliminated in the feces as a soft, hydrated mass, while the 1000 mg may leave an insoluble shell that may resemble the original tablet." Both low- and high-fat meals prolong metformin T_{max} by approximately 3 hours, but C_{max} was not affected.

20.4.2 Gires

The Gires gastroretention device is both expandable and floating. Gires is comprised of a tablet that contains a bicarbonate carbon dioxide–generating couple that is enclosed in a poly(vinyl alcohol) (PVA) film pouch which retains the carbon dioxide that results in the expansion of and decrease in density of the polymer pouch (Blackshields et al., 2012). The pouch that holds the tablet is placed in a capsule which the patient takes. A number of patents have been issued that address the automation process to seal the effervescent tablet inside the PVA pouch and then place the pouch inside a capsule (Moodley, 2011). In vivo studies have shown that the device gastric emptying time ranges from 16 to 24 hours. Pouch retrieval times range from 28 to 53 hours (Coughlan et al., 2008).

20.5 OSMOTIC CONTROLLED RELEASE SYSTEMS

The concept of an osmotic controlled-release system is based on designing a core matrix that contains drug and functional excipients that create an osmotic pressure gradient when the matrix is coated with a semipermeable membrane and subsequently exposed to the aqueous environment of the gastrointestinal tract. The water in the gastrointestinal tract is pulled into the core matrix across the semipermeable membrane. The rate of water movement is governed by the thickness of the semipermeable membrane and the composition of the core. Controlled drug release is affected by creating orifices in the semipermeable membrane which allow release of the core drug solution or suspension at the same rate as water is taken into the core. The advantage of this controlled delivery system is that the release can be tightly controlled. Drug release is not dependent on food, pH, gastrointestinal motility, the presence of enzymes, or other physiological factors. Osmotic systems can deliver drug payloads as high as 750 mg of drug over a 24-hour period with nearly zero-order release.

Osmotic controlled release was pioneered by the Alza Corporation, which was founded in 1968. Alza's Theeuwes and Higuthi (1974) are credited with the first osmotic dispensing device patent. In 1989, the FDA-approved Alza's Procardia XL (nifedipine) Oros as the first osmotic controlled-release drug delivery system. Since then more than a dozen osmotic drug delivery systems have been commercialized.

Osmotic systems can be designed to give a wide range of release profiles. They can be designed to give zero-order delayed combinations of pulse and delayed, and an increasing release profile. The Oros Push-Pull device incorporates a layered tablet where one layer contains the drug and osmotically active excipients. The second layer, the *push layer*, is composed of osmotic agents and polymers that expand upon hydration. This second layer helps to expel the drug solution or suspension that resides in the first layer of the orifice. Nifedipine (Wong et al., 1988) and verapamil (Jao et al., 1992), another calcium-channel ion flux inhibitor, employ the Oros Push-Pull technology. The Covera-HS (verapamil hydrochloride) extended-release tablet is designed to be taken at bedtime. In addition to the push-pull technology, the tablet was designed to provide delayed release. The delayed release is accomplished by coating the core tablet with a delayed-release subcoating and overcoating the delayed-release subcoating with a semipermeable membrane. The composite drug release profile was designed to achieve peak drug levels that coincide with the rise in blood pressure in the early morning. Drug release designed to occur at a predesignated time of day is often referred to as *chronotherapy*. The drug-containing matrix contains poly(ethylene oxide), sodium chloride, poly(vinylpyrrolidine), and magnesium stearate. The push matrix contains poly(ethylene oxide), hypromellose, sodium chloride, and magnesium stearate. The delayed-release coating contains hydroxyethylcellulose, hydroxypropyl cellulose, and polysorbate 80. This delay coating is solubilized and released. The semipermeable membrane contains cellulose acetate and poly(ethylene glycol).

20.6 CONCLUSIONS

Oral modified-release product design has evolved over the past half century and a number of controlled-release products have reached the marketplace, giving patients more convenient dosing and, in many cases, improved drug therapy employing specific release profiles, such as delayed release, pulsed release, and chronorelease. Theophylline modified-release product design is a classic example of maintaining drug levels successfully within a narrow therapeutic range. Literally hundreds of novel modified-release systems have been commercialized or are in development.

It is beyond the scope of this chapter to cover this vast array of delivery systems in development, so only a few novel uncommercialized modified-release devices were presented as examples of reasonably advanced systems that will probably be commercialized in the future.

REFERENCES

Berndl G, Rosenberg J, Liepold B, Breitenbach J, Reinhold U, Alani L, Ghosh, S. Solid pharmaceutical dosage form. U.S. Patent 8025899. 2011.

Blackshields C, McHugh C, Leonard T, Cullen A, Madigan K, Duffy P, Crean A. Impact of film properties and drug physicochemical properties on diffusion through a poly(viny alcohol) film for the development of Gires™. http://www.google.com/#pq=merrion+pharmaceuticals+ireland &hl=en&cp=29&gs_id=3d&xhr=t&q=merrion+pharmaceuticals+ Gires&pf=p&sclient=psy-ab& source=hp&pbx=1&oq=merrion + pharmaceuticals + Gires&aq=f&aqi=&aql=&gs_sm=&gs_upl =&bav=on.2,or.r_gc.r_pw.,cf.osb&fp=f2900e4a30b61916&biw=1218&bih=561. Accessed Feb. 2012.

Buxton IR, Currie R, Dela-Cruz MA, Goodson GW, Karolak W, Maleki M, Iyer VM, Muppirala G, Parr AF, Sidhu JS, Stagner RA, Vijay-Kumar AV. Sustained release formulations comprising lamotrigine. EP 1524981 B1. 2009.

Cima MJ, Haggerty JS, Sachs EM, Williams PA. Three-dimensional printing techniques. U.S. Patent 5204055. 1993.

Coughlan DC, O'Toole E, Leonard TW. Enhanced gastroretentive dosage form with minimal food effects: GIRES™. In: Rathbone MJ, Hadgraft J, Roberts MS, Lane ME, eds. *Modified-Release Drug Delivery Technology*, 2nd ed., Drugs and the Pharmaceutical Sciences, Vol. 183. New York: Informa Healthcare; 2008. pp. 231–237.

Drug Bank. http://www.drugbank/ca/drugs/DB01156/. Accessed Jan. 2012.

FDA. 1997a. Guidance for industry: SUPAC-MR—modified release solid oral dosage forms; scale-up and postapproval changes; chemistry, manufacturing, and controls; in vitro dissolution testing and in vivo bioequivalence documentation. U.S. Department of Health and Human Services, Food and Drug Administration Center for Drug Evaluation and Research. Sept. 1997. http://www.fda.gov/downloads/Drugs/GuidanceComplianceRegulatoryInformation/Guidances/UCM070640.pdf. Accessed Dec. 2011.

———. 1997b. Guidance for industry: Extended release oral dosage forms: development, evaluation, and application of in vitro/in vivo correlations. U.S. Department of Health and Human Services, Food and Drug Administration Center for Drug Evaluation and Research. Sept. 1997. http://www.fda.gov/downloads/Drugs/GuidanceComplianceRegulatoryInformation/Guidances/ucm070640.pdf. Accessed Jan. 2012.

———. 2010. Dosage. http://www.fda.gov/Drugs/DevelopmentApprovalProcess/FormsSubmission Requirements/ElectronicSubmissions/DataStandardsManualmonographs/ucm071666.htm. Accessed Jan. 2012.

———. 2011. Quality by design for ANDAs: an example for modified release dosage forms. Dec. 2011. http://www.fda.gov/downloads/Drugs/DevelopmentApprovalProcess/HowDrugsareDeveloped andApproved/ApprovalApplications /AbbreviatedNewDrugApplicationANDAGenerics/UCM286595. pdf. Accessed Jan. 2012.

FDA–NICHD. 2011. Intra-agency agreement between the Eunice Kennedy Shriver National Institute of Child Health and Human Development and the U.S. Food and Drug Administration oral formulations platform, report 1. http://bpca.nichd.nih.gov/collaborativeefforts/initiatives/upload/ Formulations_Table_for_Web_11-02-11.pdf. Accessed Jan. 2012.

FMC Application Bulletin. NRobe™ technology: a simplified approach to controlled release formulation. http://www.fmcmagenta.com/Portals/FMCMagenta/Content/Docs/NRobe%20Application%20Bulletin_Matrices%20010808.pdf. Accessed Feb. 2012.

Gusler G, Berner B, Chau M, Padua A. Optimal polymer mixtures for gastric retentive tablets. U.S. Patent 6723340, 2004.

Guy RH, Hadgraft J, Kellaway IW, Taylor MJ. Calculations of drug release rates from spherical particles. Int. J. Pharm. 1982;11:199–207.

Herman MF, Edwards SF. A reptation model for polymer dissolution. Macromolecules 1990;23:3662–3671.

Iskra JP. Apparatus, systems and methods for use in three-dimensional printing. U.S. Patent 6905645. 2005.

Jao F, Wong PS, Huynh HT, McChesney K, Wat PK. Verapmil therapy. U.S. Patent 5160744. 1992.

Laizure SC, DeVane CL. Stability of bupropion and its major metabolites in human plasma. Ther. Drug Monit. 1985;7:447–450.

Lin S, Chao P-Y, Chien YW, Sayani A, Kumar S, Mason M, West T, Yang A, Monkhouse D. In vitro and in vivo evaluations of biodegradable implants for hormone replacement therapy: Effect of system design and PK-PD relationship.

Monkhouse D, Yoo J, Sherwood JK, Cima MJ, Bornancini E. Dosage forms exhibiting multi-phase release kinetics and methods of manufacture thereof. U.S. Patent 6514518. 2003.

Moodley J. Process and machine for automated manufacture of gastro-retentive devices. U.S. Patent 8033083. 2011.

Oberegger W, Eradiri O, Zhou F, Maes P. Modified-release tablet of buproprion hydrochloride. Canadian Patent 2524300. 2008.

O'Byrne PM, Williams R, Walsh JJ, Gilmer JF. The aqueous stability of bupropion. J. Pharm. Biomed. Anal. 2010;53:376–381.

Roth W, Setnic B, Zietsch M, Burst A, Breitenbach J, Sellers E, Brennan D. Ethanol effects on drug release from verapamil Meltrex® innovative melt extruded formulation. Int. J. Pharm. 2009;368:72–75.

Rowe CW, Katstra WE, Palazzolo RD, Giritlioglu B, Teung P, Cima MJ. Multimechanism oral dosage forms fabricated by three dimensional printing. J. Control. Release 2000;66:11–17.

Shell JW, Louie-Helm J. Gastric-retentive, oral drug dosage forms for the controlled-release of sparingly soluble drugs and insoluble matter. U.S. Patent 5972389. 1999.

Staniforth JN. Controlled release device with an impermeable coating having an orifice for release of drug. U.S. Patent 5004614. 1991.

Teckoe J, Merwood C, Dann M, Kessel SR, Povey I, Good M. Powder compaction and enrobing. U.S. Patent. 7625622. 2009.

Theeuwes F, Higuchi T. Osmatic dispensing device for releasing beneficial agent. U.S. Patent 3845770. 1974.

U.S. Pharmacopeia. U.S.P.–N.F. online, First Supplement U.S.P. 34–N.F. 29 Reissue. Accessed Jan. 2012.

Wellbutrin XL product monograph. http://www.rxlist.com/wellbutrin-xl-drug.htm. Accessed Sept. 2010.

Wong PSL, Barclay BL, Deters JC, Theeuwes F. Osmotic device for administering certain drugs. U.S. Patent 4765989. 1988.

GLOSSARY

FDA	U.S. Food and Drug Administration.
GI	Gastrointestinal.
PVA	Poly(vinyl alcohol).
U.S.P.	*United States Pharmacopeia*.

APPENDIX

APPENDIX 20.1 FDA Nomenclature for Oral Modified-Release Dosage Forms

Dosage Form	Definition
Capsule Coated, extended release	A solid dosage form in which the drug is enclosed within either a hard or a soft soluble container or from a suitable form of gelatin; additionally, the capsule is covered in a designated coating which releases a drug (or drugs) in such a manner to allow at least a reduction in dosing frequency as compared to that drug (or drugs) presented as a conventional dosage form.
Delayed release	A solid dosage form in which the drug is enclosed within either a hard or a soft soluble container made from a suitable form of gelatin, and which releases a drug (or drugs) at a time other than promptly after administration. Enteric-coated articles are delayed-release dosage forms.
Delayed-release pellets	A solid dosage form in which the drug is enclosed within either a hard or a soft soluble container or "shell" made from a suitable form of gelatin; the drug itself is granules to which enteric coating has been applied, thus delaying release of the drug until its passage into the intestines.
Extended Release	A solid dosage form in which the drug is enclosed within either a hard or a soft soluble container made from a suitable form of gelatin, and which releases a drug (or drugs) in such a manner to allow a reduction in dosing frequency as compared to that drug (or drugs) presented as a conventional dosage form.
Film coated, extended release	A solid dosage form in which the drug is enclosed within either a hard or a soft soluble container or "shell" made from a suitable form of gelatin; additionally, the capsule is covered in a designated film coating and releases a drug (or drugs) in such a manner to allow at least a residue in the oral cavity that is easily swallowed and does not leave a bitter or unpleasant after-taste.
Tablet Delayed release	A solid dosage form which releases a drug (or drugs) at a time other than promptly after administration. Enteric-coated articles are delayed-release dosage forms.
Delayed-release particles	A solid dosage form containing a conglomerate of medicinal particles that have been covered with a time other than promptly after administration. Enteric-coated articles are delayed-release dosage forms.
Extended release	A solid dosage form containing a drug which allows at least a dosing frequency as compared to that drug presented in conventional dosage form.
Film coated, extended release	A solid dosage form that contains medicinal substances with or without suitable diluents and is coated with a thin layer of a water-insoluble or water-soluble polymer; the tablet is formulated in such a manner as to make the contained medicament available over an extended period of time following ingestion.
Multilayer, extended release	A solid dosage form containing medicinal substances that have been compressed to form a multiple-layered tablet or a tablet within a tablet, the inner tablet being the core and the outer portion being the shell, which, additionally, is covered in a designated coating; the tablet is formulated in such a manner as to allow at least a reduction in dosing frequency as compared to that drug presented as a conventional dosage form.

Source: FDA (2010).

PART III

REGULATORY SCIENCE

CHAPTER 21

REGULATORY PRACTICES AND GUIDELINES

21.1 WORLDWIDE REGULATORY AGENCIES

Each country has an advisory and/or regulatory agency to promote and protect public health. A list of most of these agencies along with their websites may be found at the U.S. Food and Drug Administration (FDA) website: International Organizations and Foreign Government Agencies (http://www.fda.gov/InternationalPrograms/Agreements/ucm131179.htm). A brief outline of key agencies along with the roles and responsibilities of their major departments (centers) is given below.

21.1.1 U.S. Food and Drug Administration

The FDA is an agency within the U.S. Department of Health and Human Services (HHS). The Bureau of Chemistry, the predecessor of the FDA, began in 1862 under the Department of Agriculture. In 1940, the FDA was transferred to the Federal Security Agency, which later became the Department of Health, Education, and Welfare (HEW) in 1953. The Department of Education Organization Act, signed into law in 1979, divided HEW into two departments: the Department of Education and the Department of HHS.

Most federal laws concerning the FDA are part of the Food, Drug, and Cosmetic (FD&C) Act, passed in 1938 and amended extensively since. The key milestones in the food and drug law history are listed below and can be found in Title 21 of the *Code of Federal Regulations* (CFR). To issue regulations, the FDA follows a process known as *notice and comment rulemaking*, which allows for public input on a proposed regulation before the FDA issues a final regulation.

21.1.1.1 Key Milestones (FDA, 2010a)
- 1862—President Lincoln appoints a chemist, Charles M. Wetherill, to serve in the new Department of Agriculture. This was the beginning of the Bureau of Chemistry, the predecessor of the FDA.
- 1902—Biologics Control Act; ensures purity and safety of serums, vaccines, and similar products.

Integrated Pharmaceutics: Applied Preformulation, Product Design, and Regulatory Science, First Edition. Antoine Al-Achi, Mali Ram Gupta, William Craig Stagner. © 2013 John Wiley & Sons, Inc. Published 2013 by John Wiley & Sons, Inc.

- 1906—Food and Drugs Act; prohibits interstate commerce in misbranded and adulterated foods, drinks, and drugs.

- 1912—Sherley Amendment Act; in *U.S.* v. *Johnson*, the Supreme Court ruled (in 1911) that the 1906 Food and Drugs Act did not prohibit false therapeutic claims, only false and misleading statements about the ingredients or identity of a drug. To overcome this ruling, Congress enacted the Sherley Amendment Act to prohibit labeling medicines with false therapeutic claims intended to defraud the purchaser.

- 1938—Food, Drug, and Cosmetic (FD&C) Act; the basic food and drug law of the United States. This law is intended to assure consumers that foods are pure and wholesome, safe to eat, and have been produced under sanitary conditions; that drugs and devices are safe and effective for their intended uses; that cosmetics are safe and made from appropriate ingredients; and that all labeling and packaging is truthful, informative, and not deceptive. This act was a complete revision of the obsolete 1906 Food and Drugs Act, and Congress passed it in response to the death in 1937 of 107 people, many of whom were children, resulting from ingestion of elixir of sulfanilamide containing the poisonous solvent diethylene glycol. Congress authorized new provisions in this act:

  - Extending control of cosmetics and therapeutic devices

  - Requiring new drugs to be shown safe before marketing; began a new system of drug regulation

  - Eliminating the Sherley Amendment requirement to prove defraud in drug-misbranding cases

  - Providing that safe tolerances be set for unavoidable poisonous substances

  - Authorizing standards of identity, quality, and fill of container for foods

  - Authorizing factory inspections and adding the remedy of control injunctions to previous penalties of seizures and prosecutions

- 1951—Durham–Humphrey Amendment; defines the types of drugs requiring medical supervision and restricts their sale to prescription by a licensed practitioner.

- 1962—Kefauver–Harris Drug Amendments; ensures drug efficacy and greater drug safety; drug manufacturers are required to prove effectiveness of their products to the FDA before marketing.

- 1966—Fair Packaging and Labeling Act; requires all consumer products in interstate commerce to be labeled honestly and informatively.

- 1976—Medical Device Regulation Act; ensures safety and effectiveness of medical devices, including diagnostic products; manufacturers to register with the FDA and follow quality control procedures; some products must have premarket approval by the FDA.

- 1984—Drug Price Competition and Patent Term Restoration Act (Hatch–Waxman Act); designed to encourage the development of generic versions of prescription drugs.

- 1988—Prescription Drug Marketing Act; bans the diversion of prescription drugs from legitimate commercial channels.
- 1988—Food and Drug Administration Act of 1988; officially establishes the FDA as an agency of HHS with a commissioner appointed by the President and spells out the responsibilities of the HHS secretary and the commissioner for research, enforcement, education, and information.
- 1992—Prescription Drug User Fee Act (PDUFA); requires drug and biologic manufacturers to pay fees for product applications and other services and for the FDA to hire more reviewers to assess applications.
- 1994—Dietary Supplement Health and Education Act; establishes specific labeling requirements, provides a regulatory framework, and authorizes the FDA to promulgate good manufacturing practices regulations for dietary supplements.
- 1997—Food and Drug Administration Act; reauthorizes the PDUFA and mandates the most wide-ranging reforms in agency practices since 1938. The provisions include:
  - Measures to accelerate review of devices
  - Regulation of advertising of unapproved uses of approved drugs and devices
  - Regulation of health claims for foods
- 2002—Best Pharmaceuticals for Children Act (BPCA); improves safety of patented and off- patent medicines for children.
- 2002—Medical Device User Fee and Modernization Act (MDUFMA); sponsors to pay fees for services; inspections by accredited third parties.
- 2003—Animal Drug User Fee Act; permits the FDA to collect subsidies for the review of certain animal drug applications from sponsors, similar to the Prescription Drug User Fee Act of 1992.
- 2007—Food and Drug Administration Amendments Act of 2007; this new law represents a very significant addition to FDA authority and provides additional resources necessary for comprehensive reviews of new drugs and devices; the PDUFA and MDUFMA were reauthorized and expanded, and the BPCA and Pediatric Research Equity Act were reauthorized.
- 2009—Family Smoking Prevention and Tobacco Control Act; the FDA establishes the Center for Tobacco Products; establishes bans on cigarettes with flavors characterizing fruit, candy, or clove.
- 2009—Biologics Price Competition and Innovation (BPCI) Act of 2009 (FDA, 2011a); creates an abbreviated approval pathway for biological products that are demonstrated to be "highly similar" (biosimilar) to or "interchangeable" with an FDA-approved biological product. This became a part (as Title VII, Subtitle A) of the broader legislation, the Patient Protection and Affordable Care Act of 2010 (FDA, 2011b).

21.1.1.2 Public's Expectations of Drug Regulations   According to Janet Woodcock [Director, FDA Center for Drug Evaluation and Research (CDER)] (FDA, 2011c), the U.S. medical product regulatory system has been evolving over the course of the last 100 years, and the basic public expectations from the agency have been that:

- All marketed drugs are safe and effective.
- Unsafe or ineffective drugs are removed from the market.
- Drugs should be of high quality.
- The claims and advertising made for drugs are policed, and false, misleading, and flagrant ones are prohibited.
- Generic competition is allowed to help maintain reasonable prices and to help control health care costs.
- Seriously ill patients have access to investigational drugs.
- Information is shared on how to use approved drugs: for example, use of drugs in children, the elderly, and women.
- Development and availability of approved drugs for children and pediatric formulations are promoted and mandated as needed.
- A robust and flourishing drug development research program is established to make new therapies available to patients rapidly, and to ensure that all studies on human subjects are ethical and safe.

21.1.1.3 FDA Responsibilities and Centers (FDA, 2010b)   The mission of the FDA is to enforce laws enacted by the U.S. Congress and regulations established by the agency to protect consumers' health, safety, and pocketbooks. The FDA regulates products, accounting for about a quarter of each dollar that Americans spend (FDA, 2012), and has the responsibility:

- To protect public health by assuring the safety, efficacy, and security of human and veterinary drugs, biological products, medical devices, food supply, tobacco products, cosmetics, and products that emit radiation
- To advance public health by helping to speed innovations that make medicines and foods more effective, safer, and more affordable
- To help public get the accurate, science-based information they need to use medicines and foods to improve their health

  Some of the agency's specific responsibilities include (FDA, 2009a):

- Biologics
    - Product and manufacturing establishment licensing
    - Safety of the nation's blood supply
    - Research to establish product standards and develop improved testing methods

- Cosmetics
  - Safety
  - Labeling
- Drugs
  - Product approvals
  - Over-the-counter (OTC) and prescription drug labeling
  - Drug manufacturing standards
- Foods
  - Labeling
  - Safety of all food products (except meat and poultry)
  - Bottled water
- Medical devices
  - Premarket approval of new devices
  - Manufacturing and performance standards
  - Tracking reports of device manufacturing and serious adverse reactions
- Radiation-emitting electronic products
  - Radiation safety performance standards for microwave ovens, television receivers, diagnostic x-ray equipment, cabinet x-ray systems (such as baggage x-rays at airports), laser products, ultrasonic therapy equipment, mercury vapor lamps, and sunlamps
  - Accrediting and inspecting mammography facilities
- Tobacco products
- Veterinary products
  - Livestock feed
  - Pet foods
  - Veterinary drugs and devices

The FDA regulates almost all facets of prescription drugs (except for the prescribing and dispensing activities, which are relegated to state agencies), including manufacturing, labeling, testing, advertising, efficacy, and safety; however, it has limited authority on other types of products and activities, which are either controlled by other federal agencies or regulated as a joint effort (FDA, 2009b). For example:

- Federal Trade Commission: all advertising, excluding prescription drugs and medical devices
- Department of Justice's Bureau of Alcohol, Tobacco, Firearms, and Explosives: labeling and quality of alcoholic beverages
- Consumer Product Safety Commission: safety of consumer goods such as household products (excluding radiation-emitting products), paint, child-resistant packages, and baby toys

- Drug Enforcement Administration: illegal drugs with no approved medical use
- Department of Agriculture's Food Safety and Inspection Service: safety and labeling of traditional meats and poultry
- Environmental Protection Agency (EPA): shares pesticide regulation with the FDA and with the U.S. Department of Agriculture
- EPA: develops national standards for drinking water from municipal water supplies
- Local county health departments: inspect and license restaurants and grocery stores

The FDA carries out its responsibilities through the following key product centers and offices (FDA, 2011d):

- Office of Medical Products and Tobacco
  - Center for Drug Evaluation and Research
  - Center for Biologics Evaluation and Research
  - Center for Devices and Radiological Health
  - Center for Tobacco Products
  - Office of Special Medical Programs
  - Office of Good Clinical Practice
  - Office of Orphan Product Development
  - Office of Pediatric Therapeutics
  - Office of Combination Products
- Office of Foods
  - Center for Veterinary Medicine
  - Center for Food Safety and Applied Nutrition
- Office of Global Regulatory Operations and Policy
  - Office of Regulatory Affairs
  - Office of International Programs
- Office of the Chief Scientist
  - National Center for Toxicological Research
  - Office of Counterterrorism and Emerging Threats
  - Office of Regulatory Science and Innovation
- Office of Critical Path Programs
- Office of Women's Health
- Office of Minority Health
- Office of Ombudsman

21.1.1.3.1   Center for Drug Evaluation and Research (CDER) (FDA, 2011e)   The CDER is responsible to ensure that safe and effective drugs are available to improve the health of people in the United States. The center:

- Regulates all drugs (prescriptions, generics, OTCs, and most genetically engineered biotechnology therapeutics)
- Reviews and approves new drug applications
- Ensures that consumers have access, as quickly as possible, to new treatments
- Provides information to doctors and patients on proper use of medicines
- Evaluates benefits and risks of drugs
- Monitors marketed drugs for unexpected health risks and takes appropriate actions, including removal of a product from the market
- Ensures truth in advertising

21.1.1.3.2   Center for Biologics Evaluation and Research (CBER) (FDA, 2011f)

- Ensures the safety, purity, potency, and effectiveness of biological products, including vaccines, blood and blood products, cells, tissues, and gene therapies
- Approves new biologics
- Defends the public against threats of emerging infectious diseases and bioterrorism

21.1.1.3.3   Center for Veterinary Medicine (CVM) (FDA, 2011g)

- Regulates the manufacture and distribution of food, food additives, and drugs (excluding vaccines, which are handled by the U.S. Department of Agriculture) that are given to animals, including food animals and pets
- Ensures that the medications used in food animals do not affect the human food supply

21.1.1.3.4   Center for Food Safety and Applied Nutrition (CFSAN) (FDA, 2010c)

- Promotes and protects public health by ensuring that the nation's food supply, including dietary supplements, is safe, sanitary, wholesome, and honestly labeled
- Ensures that cosmetic products are safe and properly labeled

21.1.1.3.5   Center for Tobacco Products (CTP) (FDA, 2011h)

- Oversees implementation of the Family Smoking Prevention and Tobacco Control Act (passed in 2009)
- Sets performance standards
- Reviews premarket applications for new and modified-risk tobacco products, requiring new warning labels
- Establishes and enforces advertising and promotion restrictions

21.1.1.3.6 Office of Regulatory Affairs (ORA) (FDA, 2011i) This office leads all FDA field activities by inspecting products and manufacturers, including reviewing imported products, and develops the FDA-wide policy on compliance and enforcement and executes FDA's import strategy and food protection plans.

21.1.1.3.7 Office of Combination Products (OCP) (FDA, 2011j)
- Is responsible for coordinating and facilitating the application, review, and approval of drug–device, drug–biologic, and device–biologic combination products

21.1.1.3.8 Office of Regulatory Science and Innovation (FDA, 2011k) To advance regulatory science through its new Regulatory Science Initiative, Margaret Hamburg, FDA commissioner, announced establishment of this new office in 2010. Its mission is to provide strategic leadership, coordination, infrastructure, and support for excellence and innovation in FDA's science, thus further advancing the agency's ability to protect and promote the health of the public (refer to Chapter 31 for more details on regulatory science initiatives).

21.1.1.3.9 Office of Pediatric Therapeutics (OPT) (FDA, 2011l)
- Assures access of innovative, safe, and effective medical products to children

21.1.1.3.10 Office of International Programs (OIP) (FDA, 2011m)
- A focal point for all international matters concerning the FDA
- Advances the mission of the FDA and public health worldwide in partnership with other FDA components, other U.S. agencies, foreign governments, and international organizations

21.1.2 European Medicines Agency (EMA, 2011a–c)

The European Medicines Agency (EMA, formerly the EMEA), created in 1995, is a decentralized body of the European Commission, with headquarters in London. The main responsibility of the agency is to protect and promote public and animal health through scientific evaluation of European Marketing Authorization applications (a centralized procedure) and supervision of medicines for human and veterinary use. Under the centralized procedure, companies submit one single marketing authorization application to the EMA for the following medicinal product categories. Once granted, a centralized (or "community") marketing authorization is valid in all European Union (EU) and European Economic Area–European Free Trade Association (EEA–EFTA) states (Iceland, Liechtenstein, and Norway).

- If derived from biotechnology and other high-technology processes
- If needed for the treatment of HIV/AIDS, cancer, diabetes, or neurodegenerative diseases
- All designated orphan medicines intended for the treatment of rare diseases
- All veterinary medicines intended for use as performance enhancers to promote growth of treated animals or to increase yields from treated animals

- Other categories, provided that the medical product constitutes a significant therapeutic, scientific, or technical innovation or the product is in any other respect in the interest of patient or animal health

21.1.2.1 EMA Responsibilities/Committees Working with the member states and the European Commission (EC) as partners in a European medicines network, the EMA:

- Establishes a pool of multinational scientific expertise to achieve a single evaluation via the centralized or mutual recognition marketing authorization procedures
- Reinforces the supervision of existing medicinal products (by coordinating national pharmacovigilance and inspection activities)
- Creates databases and electronic communication facilities as necessary to promote the rational use of medicines
- Provides independent, science-based recommendations on the quality, safety, and efficacy of medicines, and on more general issues relevant to public and animal health that involve medicines
- Applies efficient and transparent evaluation procedures to help bring new medicines to the market by means of a single, EU-wide marketing authorization granted by the EC
- Implements measures for continuously supervising the quality, safety, and efficacy of authorized medicines to ensure that their benefits outweigh their risks and where appropriate, organize withdrawal of medicinal products from the EU
- Provides scientific advice and incentives to stimulate the development and improve the availability of innovative new medicines
- Recommends safe limits for residues of veterinary medicines used in food-producing animals, for the establishment of maximum residue limits by the EC
- Involves representatives of patients, health care professionals, and other stakeholders in its work, to facilitate dialogue on issues of common interest
- Publishes impartial and comprehensive information about medicines and their use
- Develops best practice for medicine evaluation and supervision in Europe, and alongside the member states and the EC, contributes to the harmonization of regulatory standards at the international level
- Contributes to the EU's international activities through its work with the *European Pharmacopoeia*, the World Health Organization (WHO), the International Conference on Harmonization of Technical Requirements for Registration of Pharmaceuticals for Human Use, the International Cooperation on Harmonization of Technical Requirements for Registration of Veterinary Products, and with other international organizations and initiatives

- Provides scientific mediation of issues related to medicines submitted for a decentralized market approval process

The following six scientific committees, composed of members from all EU and EEA–EFTA states, including patients' and doctors' representatives conduct the main scientific work of the agency:

- Committee for Medicinal Products for Human Use (CHMP)
- Committee for Medicinal Products for Veterinary Use (CVMP)
- Committee for Orphan Medicinal Products (COMP)
- Committee on Herbal Medicinal Products (HMPC)
- Committee for Advanced Therapies (CAT)
- Paediatric Committee (PDCO)

21.1.2.1.1 Committee for Medical Products for Human Use The CHMP was established in accordance with EC Regulation 726/2004 and is responsible for preparing the agency's opinions on all questions concerning medicinal products for human use.

21.1.2.1.2 Committee for Medical Products for Veterinary Use The CVMP is responsible for preparing the agency's opinions on all questions concerning veterinary medicinal products, in accordance with EC Regulation 726/2004.

21.1.2.1.3 Committee for Orphan Medical Products The COMP, established in 2001, is responsible for all matters related to orphan drugs, including reviewing applications from persons or companies seeking "orphan medicinal product designation" for products they intend to develop for the diagnosis, prevention, or treatment of life-threatening or very serious conditions that affect no more than 5 in 10,000 persons in the EU.

21.1.2.1.4 Paediatric Committee The PDCO, established in 2004, is responsible for assessing the content of pediatric investigation plans and adopt opinions on them in accordance with EC Regulation 1901/2006 as amended. This includes the assessment of applications for a full or partial waiver and assessment of applications for deferrals.

21.1.2.1.5 Committee on Herbal Medicinal Product The HMPC, established in 2004 in accordance with EC Regulation 726/2004 and Directive 2004/24/EC, provides scientific opinions on questions relating herbal medicinal products and is responsible for registration of traditional herbal medicinal products in EU member states.

21.1.2.1.6 Committee for Advanced Therapies The CAT was established in accordance with EC Regulation 1394/2007 to assess the quality, safety, and efficacy of advanced-therapy medicinal products and to follow scientific development in the field.

21.1.3   Ministry of Health, Labour and Welfare of Japan
(MHLW, 2011a,b)

21.1.3.1   Responsibilities of the Ministry of Health, Labour and Welfare (MHLW)   To protect and promote public health and welfare, the MHLW is responsible for a wide range of activities administered through its various bureaus and their departments.

- Welfare for the elderly
- Self-reliance and rehabilitation for the disabled
- Healthy upbringing for children
- Public assistance
- Medical care insurance and pensions
- Medical care supply system
- Sickness prevention and treatment
- Quality control of medical supplies
- Food hygiene
- Water supply
- Waste treatment
- Relief for survivors of the war dead

These activities are mutually related and are aimed at securing the livelihood of the Japanese people in a comprehensive manner.

21.1.3.2   Major Bureau: Pharmaceutical and Medical Safety Bureau   Of the several bureaus under the secretariat, this bureau deals with assurance of efficacy and safety of pharmaceuticals, quasi-pharmaceuticals, cosmetics, and medical devices. Also, it deals with assurance of blood supplies, regulation of poisonous and deleterious substances, promotion of antinarcotic measures, and similar concerns.

21.1.4   World Health Organization (WHO, 2011a,b)

The World Health Organization (WHO), established in April 1948, is the directing and coordinating authority for health within the United Nations system. The World Health Assembly, the supreme decision-making body for the WHO, meets every year in May in Geneva, Switzerland, and is attended by delegates from all 193 member states. The WHO is responsible for providing leadership on global health matters, shaping the health research agenda, setting norms and standards, articulating evidence-based policy options, providing technical support to countries, and monitoring and assessing health trends.

21.1.5   Global Harmonization

Globalization, a fact of twenty-first century economic life, is providing considerable synergy by reducing unwarranted contradictory regulatory requirements and redundant application of similar requirements by multiple regulatory agencies. This harmonization effort should enhance public health protection, improve government efficiencies, and reduce costs substantially to bring new therapies to numerous regions of the world. Following is a brief introduction to four such harmonization task forces.

21.1.5.1 *International Conference on Harmonization* (ICH, 2011) The International Conference on Harmonization of Technical Requirements for the Registration of Pharmaceuticals for Human Use (ICH) was established as a joint regulatory and industry project to improve the efficiency of the process for developing and registering new medicinal products in Europe, Japan, and the United States. Since its inception in 1990, the ICH has evolved, through its Global Cooperation Group, to respond to the increasingly global face of drug development, so that the benefits of international harmonization for better global health can be realized worldwide. The mission of the ICH is to achieve greater harmonization to ensure that safe, effective, and high-quality medicines are developed and registered in the most resource-efficient manner.

The six founding members include the EU, the European Federation of Pharmaceutical Industries and Associations, the MHLW, the Japan Pharmaceutical Manufacturers Association, the FDA, and the Pharmaceutical Research and Manufacturers of America. In addition to these members, another important group of nonvoting members (WHO, the European Free Trade Association, Health Canada, and the International Federation of Pharmaceutical Manufacturrs and Associations) acts as a link between the ICH and non-ICH countries and regions.

The Global Cooperation Group, formed in 1999 as a subcommittee of the ICH Steering Committee in response to a growing interest in ICH guidelines beyond the three ICH regions, has now been expanded to include representatives from the Asia-Pacific Economic Cooperation group, the Association of Southeast Asian Nations, the Pan American Network for Drug Regulatory Harmonization, the Southern African Development Community, and regulators from Australia, Brazil, China, Chinese Taipei, India, Republic of Korea, Russia, and Singapore.

The ICH harmonization activities fall into four categories (formal ICH procedures, maintenance, revision, and Q&A procedures). The development of a new harmonized guideline and its implementation (the formal ICH procedure) involves five steps:

1. Consensus building. This step begins when the steering committee adopts a concept paper as a new topic.
2. Confirmation of six-party consensus. Once consensus is reached among all six expert working group members, the consensus text is approved and signed off by the steering committee.
3. Regulatory consultation and discussion. At this stage, the approved guideline embodying the scientific consensus goes through a normal process of wide-ranging regulatory consultation in the three regions. In the United States it is published as draft guidance in the *Federal Register*, in the EU it is published

as a draft CHMP guidance, and in Japan it is translated and issued by MHLW for internal and external consultation.

4. Adoption of an ICH harmonized tripartite guideline. This step is reached when the Steering Committee agrees that there is sufficient consensus on the technical issues and that the three regulatory parties to ICH have affirmed that the guideline is being recommended for adoption for their respective regions.

5. Implementation. This step is carried out according to the same national/regional procedures that apply to other regional regulatory guidelines and requirements in these three regions. Information on the regulatory action taken and implementation dates are reported back to the Steering Committee and published by the ICH Secretariat on the ICH website (www.ich.org/home.html).

21.1.5.2 International Cooperation on Cosmetic Regulation (ICCR)
(ICCR, 2011) The ICCR maintains the highest level of global consumer protection while minimizing barriers to international trade. The ICCR members include the FDA, EU (Directorate—General Enterprise, EC), MHLW, and Health Canada.

21.1.5.3 Global Harmonization Task Force (GHTF)
(GHTF, 2012) The GHTF fosters harmonization in the regulations to ensure safety, effectiveness, and quality of medical devices. It is a collaborative effort of the regulatory and industry authorities from Europe, the Asia-Pacific, and North America.

21.1.5.4 International Cooperation on Harmonization of Technical Requirements for Registration of Veterinary Products (VICH)
(VICH, 2011) The VICH, modeled after the ICH, harmonizes technical requirements for veterinary product registration among regulatory and industry authorities from Europe, Japan, and United States.

21.2 GOOD MANUFACTURING PRACTICE

There are several dozen different official (national and supernational) statements on good manufacturing practice (GMP) worldwide, which are published as guides, codes, and/or regulations. The ICH guidelines have a strong influence on several of these GMPs. Three GMP documents (from the FDA, EU, and Japan) are covered in this chapter. Detailed regulatory requirements for the FDA GMP are provided in Section 21.2.1, the key features from the EU GMP in Section 21.2.2, and the Japan GMP in Section 21.2.3.

- The *Code of Federal Regulations*, 21 CFR Parts 210 and 211, U.S. Food and Drug Administration, referred to here as *current good manufacturing practice* (CGMP) (FDA, 2011n,o). A more detailed title would include:

○ *Code of Federal Regulations*, Title 21 (Volume 4), Food and Drug Administration, Department of Health and Human Services, Subchapter C, Drugs: General

■ Part 210: Current Good Manufacturing Practice in Manufacturing, Processing, Packing, or Holding of Drugs; General

■ Part 211: Current Good Manufacturing Practice for Finished Pharmaceuticals

The following references would provide more in-depth understanding of these regulations:

- Preamble to Title 21, Subchapter C, Human and Veterinary Drugs, Current Good Manufacturing Practice in Manufacture, Processing, Packing, or Holding (110 pages long) (http://www.fda.gov/cder/dmpq/preamble.txt).

- Citations from various FDA inspection deficiency observations (typically referred to as *483 observations*) and/or warning letters issued to manufacturers. The nickname "483" derives from U.S. government form FDA 483, and is issued to a firm if the FDA inspection team observes situations, practices, and/or an activity not in compliance (but not necessarily violations) with the CGMP requirements and/or the company's own policies or procedures. Although the "483" reflects the opinion of the inspection team, the observations listed in the form should be taken seriously. Otherwise, this could result in the company receiving a *warning letter*, a more serious regulatory action, and would require the company to respond within 15 days.

  ○ A small number of 483s are posted at the FDA's office of Regulatory Affairs (ORA) FOIA Electronic Reading Room (http://www.fda.gov/AboutFDA/CentersOffices/OfficeofGlobalRegulatoryOperationsandPolicy/ORA/ORAElectronicReadingRoom/default.htm).

  ○ Warning letters are posted on the FDA website, http://www.fda.gov/ICECI/EnforcementActions/WarningLetters/default.htm.

- EU Guidelines to Good Manufacturing Practice for Medicinal Products for Human and Veterinary Use (ECGMP), Volume 4 of the Rules Governing Medicinal Products in the European Union, by the Commission of the European Community. This guide was first published in 1989; the second edition, published in 1992, implemented Commission Directives 91/356 of June 13, 1991 and 91/412 of July 23, 1991. The second edition also included 12 (from a total 20, at this writing) additional annexes. Requirements unique to the ECGMP, along with the Principles from its nine chapters, are presented in Section 21.2.2 (EudraLex, 2011).

- MHLW Ministerial Ordinance No. 136 (MHLW, 2004a), on the Standards for Quality Assurance of Drugs, Quasi-drugs, Cosmetics and Medical Devices, and Ministerial Ordinance No. 179 (MHLW, 2004b), on the Standards for

Manufacturing Control and Quality Control for Drugs; again, unique features from these ordinances are listed in this chapter.

21.2.1 Current Good Manufacturing Practice (FDA, 2011n,o)

The CGMP for human pharmaceuticals affects every American. Consumers expect that each batch of medicines they take is safe and effective. The CGMP requirements and adherence to the CGMP regulations assure the identity, strength, quality, and purity of drug products. The CGMPs provide systems that assure proper design, monitoring, and control of manufacturing processes and facilities. This formal system of controls helps to prevent possible contamination, mix-ups, deviations, failures, and errors and assures that drug products meet their quality standards. Adherence to these formal systems is enforced by the FDA.

The CGMPs are minimum but flexible requirements that allow each manufacturer to decide individually how best to implement the necessary controls and to use modern technologies and innovative approaches to achieve higher quality through continual improvements. Accordingly, the "C" in "CGMP" stands for "current," requiring companies to use technologies and systems that are up to date in order to comply with the regulations. A breakdown of the following CGMP regulations as they relate to the various aspects of a quality systems is listed in Appendixes 29.5 to 29.8.

Status of Current Good Manufacturing Practice Regulations [Sec. 210.1] and Subpart A — General Provisions: Scope [Sec. 211.1]

- The regulations set forth in 21 CFR Parts 210 to 226 contain the minimum CGMPs for methods, facilities, or controls used for the manufacture, processing, packing, or holding of a drug product to assure that it meets the requirements of the Food, Drug, and Cosmetic Act for safety, identity, strength, quality, and purity (SISQP) characteristics that it is represented to possess.

- According to Sec. 501(a)(2)(B) of the act, failure to comply with any of these regulations will render a drug adulterated, and the person responsible for the failure to comply will be subjected to regulatory action. "A drug or device shall be deemed to be adulterated—if it is a drug and the methods used in, or the facilities or controls used for, its manufacture, processing, packing, or holding do not conform to or are not operated or administered in conformity with current good manufacturing practice to assure that such drug meets the requirements of this Act as to safety and has the identity and strength, and meets the quality and purity characteristics, which it purports or is represented to possess" [Sec. 501(a)(2)(B)] (FDA, 2009c).

- These requirements are not enforced for OTC drug products if these products and all their ingredients are ordinarily marketed and consumed as human foods, and are considered drugs by virtue of their intended use as determined by the regulations under 21 CFR Part 110 and, where applicable, Parts 113 to 129.

Subpart B — Organization and Personnel

Responsibilities of the Quality Control Unit [Sec. 211.22]
- The quality control unit (QCU) has the responsibility and authority to:
  - Approve or reject all materials (components, drug product containers, closures, in-process materials, packaging materials, labeling, and drug products)
  - Review the production record to assure that records are error-free and that the errors have been fully investigated
  - Approve or reject contracted items
  - Approve or reject all procedures or specifications affecting drug product identity, strength, quality, and purity
- The QCU has an adequate laboratory facility.
- All QCU responsibilities and procedures are written and followed.

Personnel Qualifications [Sec. 211.25]
- Personnel and supervising staff have the necessary education, training, and/or experience to perform assigned functions to assure that the resultanting drug products have the required SISQP characteristics.
- Ongoing training with sufficient frequency is provided in the personnel's particular areas of responsibilities and in CGMP requirements related to their specific operations.
- Training records must be maintained.
- A sufficient number of qualified personnel are available to perform and supervise all assigned functions.

Personnel Responsibilities [Sec. 211.28]
- Wear appropriate clean clothing and protective apparel (such as head, face, hand, and arm coverings) to protect drug products from contamination.
- Practice good sanitation and health habits.
- Do not enter into areas designated as limited-access areas unless authorized by supervisory personnel.
- A person with a health condition (illness or open lesion) that may adversely affect the safety or quality of a drug product should report to a supervisor and avoid coming in direct contact with any drug items until the condition is corrected or competent medical personnel determined it not to jeopardize the safety or quality of drug products.

Consultants [Sec. 211.34]
- Consultants must have appropriate qualifications to advise on the manufacture, processing, packing, or holding of drug products.
- Records stating the name, address, and qualifications of all consultants and the types of services they provide must be maintained.

Subpart C — Buildings and Facilities

Design and Construction Features [Sec. 211.42]

- Buildings used in drug product processing must be of suitable size, construction, and location to facilitate cleaning, maintenance, and proper operations.

- Design each building to have adequate space for the orderly placement and flow of all equipment and materials to prevent mix-ups between different materials.

- Perform each operation within specifically defined area of adequate size.

- Provide separate or defined areas or other control systems to prevent contamination or mix-ups during the course of the following procedures:

  ○ Receipt, identification, storage, and withholding from use of incoming materials pending sampling, testing, or examination by the QCU before release for manufacturing or packaging

  ○ Holding of rejected materials

  ○ Storage of released material

  ○ Storage of in-process materials

  ○ Manufacturing and processing operations

  ○ Packaging and labeling operations

  ○ Quarantine storage before release of drug products

  ○ Storage of drug products after release

  ○ Control and laboratory operations

  ○ Aseptic processing, which includes as appropriate:

    ▪ Floors, walls, and ceilings made of smooth hard surfaces that are easily cleanable

    ▪ Temperature and humidity controls

    ▪ An air supply filtered through high-efficiency particulate air filters under positive pressure, regardless of whether flow is laminar or nonlaminar

    ▪ Monitoring of environmental conditions

    ▪ Cleaning and disinfecting the room and equipment to produce aseptic conditions

    ▪ Maintaining any equipment used to control aseptic conditions

- Operations relating to the manufacture, processing, and packing of penicillin must be performed in facilities separate from those for other nonpenicillin drug products used for humans.

Lighting [Sec. 211.44]

- Provide adequate lighting in all areas.

Ventilation, Air Filtration, and Air Heating and Cooling [Sec. 211.46]

- Provide adequate ventilation.

- When appropriate, provide equipment adequate to control air pressure, microorganisms, dust, humidity, and temperature.

- Use air filtration systems, when appropriate, on air supplies to production areas.

- If air is recirculated to production areas, control recirculation of dust from production.

- In areas where air contamination occurs during production, provide adequate exhaust systems or other systems adequate to control contaminants in those areas.

- Air-handling systems for the manufacture, processing, and packing of penicillin must be completely separate from those for other drug products for human use.

Plumbing [Sec. 211.48]

- Supply potable water under continuous positive pressure in a defect-free plumbing system to avoid contamination to any drug product.

- If the potable water does not meet the standards prescribed in the EPA's primary drinking water regulations set forth in 40 CFR Part 141, do not permit it in the potable water system.

- Drains must be of adequate size, and where connected directly to a sewer must be provided with an air break or other mechanical device to prevent back-siphonage.

Sewage and Refuse [Sec. 211.50]

- Dispose of sewage, trash, and other refuse in and from the building and immediate premises in a safe and sanitary manner.

Washing and Toilet Facilities [Sec. 211.52]

- Provide adequate washing facilities, including hot and cold water, soap or detergent, air driers or single-service towels, and clean toilet facilities easily accessible to working areas.

Sanitation [Sec. 211.56]

- Maintain buildings used in the manufacture, processing, packing, or holding of a drug product in a clean and sanitary condition

- Buildings must be free of infestation by rodents, birds, insects, and other vermin (other than laboratory animals).

- Dispose of trash and organic waste matter in a timely and sanitary manner.

- Write and follow the procedures for all the tasks listed below.
  - Assigning responsibility for sanitation
  - Cleaning schedules
  - Methods, equipment, and materials to be used in cleaning the buildings and facilities

○ Selection and proper use (to avoid contamination of equipment and drug materials) of suitable rodenticides, insecticides, fungicides, fumigating agents, and cleaning and sanitizing agents and ensuring that these materials are registered and used in accordance with the Federal Insecticide, Fungicide, and Rodenticide Act [7 *United States Code* (U.S.C.) 135].

Maintenance [Sec. 211.58]
- Maintain all buildings in a good state of repair.

Subpart D — Equipment

Equipment Design, Size, and Location [Sec. 211.63]
- Equipment used in processing drug products is appropriately designed, of adequate size, and suitably located to facilitate operations for its intended use and for its cleaning and maintenance.

Equipment Construction [Sec. 211.65]
- Equipment surfaces that come in contact with any drug material must not be reactive, additive, or absorptive so as to alter the SISQP of the drug product beyond the official or other established requirements.
- Any substances required for operation, such as lubricants or coolants, must not come in contact with drug product materials so as to alter the SISQP of the drug product beyond the official or other established requirements.

Equipment Cleaning and Maintenance [Sec. 211.67]
- Clean, maintain, and sanitize, as appropriate, equipment and utensils to prevent malfunction or contamination that would alter the SISQP of the drug product beyond the official or other established requirements.
- Establish and follow written procedures for cleaning and maintenance of equipment and utensils. Include the following in these procedures:
  ○ Assignment of responsibility for cleaning and maintaining equipment
  ○ Maintenance and cleaning schedules, including, where appropriate, sanitizing schedules
  ○ Methods, equipment, and materials used in cleaning and maintenance operations
  ○ Methods of disassembling and reassembling equipment to assure proper cleaning and maintenance
  ○ Removal or obliteration of previous batch identification
  ○ Protection of clean equipment from contamination prior to use
  ○ Inspection of equipment for cleanliness immediately before use
- Keep maintenance, cleaning, sanitization, and inspection records as specified in Secs. 211.180 and 211.182.

Automatic, Mechanical, and Electronic Equipment [Sec. 211.68]

- To assure proper performance, routinely calibrate, inspect, and check equipment, including computers, according to written procedures.

- Maintain written records of those calibration checks and inspections.

- Exercise strict controls over computer and related systems to assure that changes in master production and other control records are instituted only by authorized personnel and are checked for accuracy.

- Maintain a backup file of data entered into these systems. When this is not feasible (e.g., chromatographic charts, etc.), maintain a written record of the program used along with appropriate validation data.

- Maintain hard-copy or alternative systems, such as duplicates, tapes, or microfilm, designed to assure that backup data are exact, complete, and secure from alteration, inadvertent erasures, or loss.

- A second check, as required by Secs. 211.101, 211.103, and 211.182 or 211.188, is not needed if automated equipment used is in conformance and if one person has checked that the equipment performed the operation properly.

Filters [Sec. 211.72]

- Filters used for filtering injectable drug products must not release fibers into such products.

- If use of a fiber-releasing filter becomes necessary, an additional non-fiber-releasing filter having a maximum nominal pore size rating of 0.2 μm (0.45 μm if the manufacturing conditions so dictate) must subsequently be used to reduce the content of particles in the injectable drug product.

- The use of an asbestos-containing filter is prohibited.

Subpart E — Control of Components and Drug Product

General Requirements [Sec. 211.80]

- Write and follow detailed procedures on the receipt, identification, storage, handling, sampling, testing, and approval or rejection of all components and drug product containers and closures

- Handle and store these items in a manner to prevent contamination.

- Store bagged or boxed items off the floor and suitably spaced to permit cleaning and inspection.

- Identify each container or grouping of containers of these items with a distinctive code for each lot in each shipment received.

- Use this code in recording the disposition (quarantine, approved, or rejected) of each lot.

- Identify each lot (each code) as to its status (i.e., quarantined, approved, or rejected).

Receipt and Storage of Untested Components, Drug Product Containers, and Closures [*Sec. 211.82*]

- Upon receipt and before acceptance, visually examine each container or grouping of containers of each item for appropriate labeling as to contents, container damage or broken seals, and contamination.

- Quarantine untested items until they have been tested or examined, and released.

Testing and Approval or Rejection of Components, Drug Product Containers, and Closures [*Sec. 211.84*]

- Do not use a quarantine lot until the lot has been sampled, tested or examined, and released for use by the QCU.

- Collect representative samples from each lot for testing or examination; consider the following criteria when selecting the number of containers to be sampled and the amount of material to be taken from each container:
  - Component variability
  - Confidence levels, and degree of precision desired
  - Past quality history of the supplier
  - Quantity needed for analysis and reserve where required by Sec. 211.170

- Collect samples in accordance with the following procedure:
  - Clean the containers selected for sampling in a manner to prevent introduction of contaminants into the component.
  - Carefully open, sample, and reseal containers to prevent contamination of their contents and contamination of other items.
  - Use sterile equipment and aseptic sampling techniques when necessary.
  - If it is necessary to sample a component from the top, middle, and bottom of its container, do not make a composite of the subdivisions for testing.
  - Identify sample containers with the following information:
    - Name of the material sampled
    - Lot number
    - Container from which the sample was taken
    - Date on which the sample was taken
    - Name of the person who collected the sample
  - Mark the containers from which samples were taken to show that samples have been removed from them.

- Examine and test the samples as follows:
  - Conduct at least one test to verify the identity of each component; use the specific identity tests, if available.
  - Test each component for conformity with all appropriate written specifications for purity, strength, and quality; in lieu of such testing, a report

of analysis (a certificate of analysis) may be accepted from the supplier provided that:

■ At least one specific identity test is conducted.

■ The reliability of the supplier's analyses is established through appropriate validation of the supplier's test results at appropriate intervals.

○ Test containers and closures for conformity with all appropriate written specifications; in lieu of such testing, a certificate of testing may be accepted from the supplier, provided that:

■ At least a visual identification is conducted.

■ The reliability of the supplier's test results are established through appropriate validation of the supplier's test results at appropriate intervals.

○ When appropriate, examine the components microscopically.

○ Examine each lot of each item if liable to contamination with filth, insect infestation, or other extraneous adulterant against established specifications for such contamination.

○ Test each lot for microbiological contamination that is objectionable in view of its intended use prior to releasing it for use.

○ Approve and release a lot of an item if it meets the appropriate written specifications of identity, strength, quality, and purity (ISQP); otherwise, such a lot is rejected.

Use of Approved Components, Drug Product Containers, and Closures [Sec. 211.86]

• Rotate approved items to assure that the oldest approved stock is used first; typically referred to as *first-in-first-out*.

• Deviation from this requirement is permitted if such deviation is temporary and appropriate.

Retesting of Approved Components, Drug Product Containers, and Closures [Sec. 211.87]

• Retest or reexamine a lot if it is stored for long periods or exposed to adverse environmental conditions; based on the results from the evaluation for the ISQP, the lot is approved or rejected by the QCU.

Rejected Components, Drug Product Containers, and Closures [Sec. 211.89]

• Identify the rejected items and control them under a quarantine system to prevent their use in manufacturing or processing operations.

Drug Product Containers and Closures [Sec. 211.94]

• Ensure that drug product containers and closures are not reactive, additive, or absorptive so as to alter the SISQP of the drug beyond the official or established requirements.

- The container closure systems provide adequate protection against foreseeable external factors in storage and use that can cause deterioration or contamination of the drug product.
- Drug product containers and closures are cleaned and, if required, sterilized and processed using the validated depyrogenation processes to remove pyrogenic properties to assure they are suitable for their intended use. Follow written standards or specifications, methods of testing, and methods of cleaning, sterilizing, and processing to remove pyrogenic properties.

Subpart F — Production and Process Controls
Written Procedures; Deviations [Sec. 211.100]
- Write the procedures for production and process controls to assure that the drug products have the ISQP required.
- These written procedures, including any changes, are drafted, reviewed, and approved by an appropriate organizational unit and reviewed and approved by the QCU.
  - Follow these procedures during execution of various control functions and document at the time of performance.
  - Record and justify any deviation from the written procedure.

Charge-in of Components [Sec. 211.101] Include the following in the written production and control procedures to assure that the resulting drug products have the ISQP required:

- Formulate each batch with the intent to provide no less than 100% of the labeled or established amount of active ingredient.
- Appropriately weigh, measure, or subdivide components for drug product manufacturing.
- If a component is removed from the original container to another, identify the new container with the following information:
  - Component name or item code
  - Receiving or control number
  - Weight or measure in the new container
  - Batch for which the component was dispensed, including its product name, strength, and lot number
- Adequately supervise weighing, measuring, or subdividing operations for components.
- Each container of component dispensed to manufacturing is examined by a second person to assure that:
  - The component was released by the QCU
  - The weight or measure is correct as stated in the batch production records
  - The containers are identified properly

- ▪ If the weighing, measuring, or subdividing operations are performed by automated equipment under Sec. 211.68, only one person is needed to assure compliance to these requirements.
- Addition of each component is verified by a second person unless aided by automated equipment.

Calculation of Yield [Sec. 211.103]

- Determine actual yields and percentages of theoretical yield at the conclusion of each appropriate phase of manufacturing, processing, packaging, or holding of the drug product.
- The yield calculations must be verified independently by a second person unless:
  - ○ Calculated by automated equipment under Sec. 211.68; then independent verification is required by the person who performed this function.

Equipment Identification [Sec. 211.105]

- Properly identify all compounding and storage containers, processing lines, and major equipment used during the production at all times to indicate their contents and, when necessary, the processing phase of the batch.
- Identify major equipment by a distinctive identification number or code and record it in the batch production record to show its use in the manufacturing process.

Sampling and Testing of In-Process Materials and Drug Products [Sec. 211.110]

- Establish and follow written procedures describing in-process controls and tests (or examinations) to be performed on appropriate samples of in-process materials of a batch to assure batch uniformity and integrity of drug products.
- Establish control procedures to monitor the output and to validate the performance of those manufacturing processes that may be responsible for causing the variability of in-process material and the drug product. Include the following in these control procedures where appropriate:
  - ○ Tablet or capsule weight variation
  - ○ Disintegration time
  - ○ Adequacy of mixing to assure uniformity and homogeneity
  - ○ Dissolution time and rate
  - ○ Clarity, completeness, or pH of solutions
  - ○ Bioburden testing
- Valid in-process specifications are consistent with drug product final specifications and derived from previous acceptable process average and process variability estimates; employ suitable statistical procedures where possible.
- Assure that the drug product and in-process material conform to specifications through examination and testing of samples.

- Test in-process materials for the ISQP and approved or rejected by the QCU.
- Identify materials rejected in-process and quarantine them to prevent their use in manufacturing or processing.

Time Limitations on Production [Sec. 211.111]
- Establish appropriate time limits for completion of each phase of production to assure drug product quality.
- Deviation from established time limits may be acceptable if such deviation does not compromise the quality of the drug product; such deviations must be justified and documented.

Control of Microbiological Contamination [Sec. 211.113]
- Establish and follow appropriate written procedures designed to prevent objectionable microorganisms in drug products.
- For sterile products, validate procedures used for all aseptic and sterilization processes.

Reprocessing [Sec. 211.115]
- Establish and follow written procedures to ensure that the batches reprocessed will conform with all established standards, specifications, and characteristics.
- Reprocessing is not performed without the review and approval of the QCU.

Subpart G — Packaging and Labeling Control
Materials Examination and Usage Criteria [Sec. 211.122]
- Write and follow detailed procedures on the receipt, identification, storage, handling, sampling, examination, and/or testing of labeling and packaging materials
- Collect representative samples from each receipt of these materials and examine or test before use.
- Approve and release a receipt if it meets appropriate written specifications; otherwise, reject it.
- Maintain records for each shipment received, indicating receipt, examination or testing, and whether accepted or rejected.
- Store labels and other labeling materials separately for each drug product, strength, dosage form, or quantity of contents, with suitable identification.
- Limit access to the storage area to authorized personnel.
- Destroy all obsolete and outdated labels, labeling, and other packaging materials to prevent their use.
- Use of gang-printed labeling for different drug products, or different strengths or net contents of the same drug product, is prohibited unless the labeling from gang-printed sheets is differentiated adequately by size, shape, or color.
- If cut labeling is used, packaging and labeling operations must include one of the following special control procedures:

- ○ Dedication of labeling and packaging lines to each strength of each drug product

- ○ Use of appropriate electronic or electromechanical equipment to conduct a 100% examination for correct labeling during or after completion of finishing operations

- ○ Use of visual inspection to conduct a 100% examination for hand-applied labeling; such examination must be performed by one person and verified independently by a second person

- Monitor printing devices on manufacturing lines to assure that all imprinting conforms to the print specified in the batch production record.

Labeling Issuance [Sec. 211.125]

- Exercise strict control over labeling issued for use in drug product labeling operations.

- Carefully examine labeling materials issued for a batch for identity and conformance to the labeling specified in the master or batch production records.

- Establish procedures to reconcile the quantities of labeling issued, used, and returned.

- Investigate if a discrepancy is found outside the narrow preset limits (established based on historical operating data) in accordance with Sec. 211.192.

- ○ Labeling reconciliation is waived for cut or roll labeling if a 100% examination for correct labeling is performed in accordance with Sec. 211.122.

- Destroy all excess labeling bearing lot or control numbers.

- Maintain and store returned labeling in a manner to prevent mix-ups and with proper identification.

- Write and follow procedures describing the controls employed for the issuance of labeling.

Packaging and Labeling Operations [Sec. 211.130]

- Write a procedure by incorporating the features listed below and follow them to assure that correct labels, labeling, and packaging materials are used.

- ○ Prevent mix-ups and cross-contamination by physical or spatial separation from the operations used for other drug products

- ○ Properly identify filled drug product containers that are set aside and held in unlabeled condition for future labeling operations.

- ○ Identify the drug product with a lot or control number

- ○ Examine packaging and labeling materials to assure that they are suitable and correct before beginning packaging operations.

- ■ Document such examination in the batch production record.

- ○ Inspect packaging and labeling facilities immediately before use to assure that all drug products, packaging, and labeling materials not suitable for this planned operation have been removed from the previous operations.

■ Document the results of such inspection in the batch production records.

Tamper-Evident Packaging Requirements for Over-the-Counter Human Drug Products [Sec. 211.132]

- Under the FD&C Act, the FDA has the authority to establish a uniform national requirement for tamper-evident packaging for OTC drug products to improve package security and help assure the safety and effectiveness of OTC drug products.

- An OTC drug product (except a dermatological, dentifrice, insulin, or lozenge product) for retail sale is adulterated under Sec. 501 of the act and/or misbranded under Sec. 502 if not packaged in a tamper-resistant package or not properly labeled as required in this section.

- Requirements for tamper-evident packages are as follows:
  - If an OTC product is accessible to the public while held for sale, it must be packaged with tamper-evident package (see the Glossary for more information).
  - Design the tamper-evident feature to ensure that it remains intact when handled in a reasonable manner during manufacture, distribution, and retail display.
  - In addition to tamper-evident packaging, seal any two-piece hard gelatin capsule using an acceptable tamper-evident technology.

- Labeling
  - To alert consumers to the specific tamper-evident feature(s), each package is required to bear a statement that:
    ■ Identifies all tamper-evident feature(s) and any capsule-sealing technologies used to comply with this section.
    ■ Is placed prominently on the package at a place where it will be unaffected if the tamper-evident feature of the package is breached or missing; for example, the labeling statement on a bottle with a shrink band could say: "For your protection, this bottle has an imprinted seal around the neck."

- Request for exemptions from packaging and labeling requirements
  - A manufacturer or packer may request an exemption from the packaging and labeling requirements in the form of a citizen petition under 21 CFR 10.30; the petition is required to contain the following:
    ■ The name of the drug product or, if the petition seeks an exemption for a drug class, the name of the drug class and a list of products within that class
    ■ The reason for compliance to these requirements is unnecessary or cannot be achieved

- ■ A description of alternative steps that are available, or that the petitioner has already taken, to reduce the likelihood that the product or drug class will be the subject of malicious adulteration
- ■ Other information justifying an exemption
- The OTC drug products subject to approved new drug applications
  - ○ The holders of approved new drug applications for OTC drug products are required to notify the agency of any changes made in the packaging and labeling to comply with the requirements of this section unless the changes (except for capsule sealing) were made prior to FDA approval [according to 21 CFR 314.70 (c)].

Drug Product Inspection [Sec. 211.134]

- Examine packaged and labeled products during finishing operations to assure that the containers and packages in a lot have the correct label.
- Collect representative samples of labeled units at the completion of finishing operations, and examine visually for correct labeling.
- Record the results of these examinations in the batch production or control records.

Expiration Dating [Sec. 211.137]

- Determine the expiration date using an appropriate stability testing protocol (as described in Sec. 211.166) to assure that the drug product, bearing this date, would meet applicable standards of ISQP until that date. Criteria for imprinting expiration dating on labeling:
  - ○ The date relates to any storage conditions stated on the labeling.
  - ○ If reconstituted at the time of dispensing, the labeling must have expiration information for both the reconstituted and unreconstituted drug products.
  - ○ Expiration dates are imprinted on the labeling in accordance with the requirements of Sec. 201.17.
- Homeopathic drug products are exempted from this requirement.
- Allergenic extracts are also exempted if labeled "No U.S. Standard of Potency."
- New drug products for investigational use are exempted provided that they meet appropriate standards or specifications as demonstrated by stability studies during their use in clinical investigations.
  - ○ If these drugs are reconstituted at the time of dispensing, their labeling must bear expiration information for the reconstituted drug product.
- These requirements are not enforced for human OTC drug products if their labeling does not bear dosage limitations and they are stable for at least three years as supported by appropriate stability data.

Subpart H — Holding and Distribution
Warehousing Procedures [Sec. 211.142]
- Establish and follow written procedures on the warehousing of drug products to include:
  - Quarantine of drug products before release by the QCU.
  - Storage of drug products under appropriate conditions of temperature, humidity, and light so that the ISQP of the drug products are not affected.

Distribution Procedures [Sec. 211.150]
- Establish and follow written procedures for the distribution of drug products.
  - Distribute the oldest approved stock of a drug product first (first approved is shipped first).
    - Deviation from this requirement is permitted if such deviation is temporary and appropriate.
- Establish a system by which the distribution of each lot of each drug product is readily determined to facilitate its recall if necessary.

Subpart I — Laboratory Controls
General Requirements [Sec. 211.160]
- Specifications, standards, sampling plans, test procedures, or other laboratory control mechanisms, including any change, are drafted by the appropriate organizational unit and reviewed and approved by the QCU.
- Follow these requirements and document at the time of performance.
- Record and justify any deviation from these requirements.
- Establish these laboratory controls scientifically to assure that all items will conform to applicable ISQP standards.
- The laboratory controls include:
  - Determination of conformity to applicable written specifications (listed below) for the acceptance of each lot within each shipment of all items
    - Description of the sampling and testing procedures used
    - Representative samples adequately identified
    - Appropriate retesting of an item that is subject to deterioration
  - Similar conformance specifications applied to in-process materials and drug products
  - Calibration of instruments, apparatus, gauges, and recording devices at suitable intervals in accordance with an established written program; the calibration program should include:
    - Specific directions, schedules, limits for accuracy and precision
    - Provisions for remedial action in the event that accuracy and/or precision limits are not met

Do not use any instrument, apparatus, gauge, or recording device if it fails to meet established specifications.

Testing and Release for Distribution [Sec. 211.165]

- Prior to releasing a batch, determine conformance to the final specifications of the drug products, including the identity and strength of each active ingredient.

- Where appropriate, test each batch to ensure that it is free of objectionable microorganisms.

- Write and follow sampling and testing procedure, including the method of sampling and the number of units per batch to be tested

- Establish statistically based criteria for the sampling and testing specifications for approval or rejection of batches; the statistical quality control criteria should also include appropriate acceptance levels and/or appropriate rejection levels.

- Establish and document validation (accuracy, sensitivity, specificity, and reproducibility) of test methods in accordance with Sec. 211.194.

- Rejected batches may be reprocessed and approved provided that these reprocessed batches meet appropriate standards, specifications, and any other relevant criteria.

Stability Testing [Sec. 211.166]

- Write a testing program designed to assess the stability characteristics of drug products.

- Use the results of stability testing to determine appropriate storage conditions and expiration dates.

- Include the following in the written stability program:
  - Sample size and test intervals based on statistical criteria for each attribute examined to assure valid estimates of stability
  - Storage conditions for samples retained for testing
  - Reliable, meaningful, and specific test methods
  - Testing of the drug product in the same container and closure system as that in which the drug product is marketed
  - Testing of drug products for reconstitution at the time of dispensing (as directed in the labeling) as well as after they are reconstituted

- Test an adequate number of batches of each drug product to determine an appropriate expiration date and maintain a record of such data.

- Accelerated studies, combined with basic stability information on the components, drug products, and container and closure system, may be used to support tentative expiration dates provided that full shelf-life studies are not available and are being conducted.
  - Where data from accelerated studies are used to project a tentative expiration date (that is beyond a date supported by the actual shelf-life studies),

the stability studies, including drug product testing, must be continued at appropriate intervals until the tentative expiration date is verified or the appropriate expiration date is determined.

- Allergenic extracts that are labeled "No U.S. Standard of Potency" are exempt from the requirements of this section.

Special Testing Requirements [Sec. 211.167]

- Establish test procedures and conformance specifications for the following special testing requirements:
  - A batch purporting to be sterile and/or pyrogen-free
  - The presence of foreign particle and harsh or abrasive substances in an ophthalmic ointment batch
  - The rate of release of each active ingredient for each batch of controlled-release dosage form

Reserve Samples [Sec. 211.170]

- Retain a properly identified reserve sample representative of each lot in each shipment of each active ingredient.

- The reserve sample consists of at least twice the quantity necessary for all tests required, to determine whether the active ingredient meets its established specifications, except for sterility and pyrogen testing.

- The retention time for an active ingredient is as follows:
  - For an active ingredient in a drug product, except for radioactive and OTC drug products, retain for one year after the expiration date of the last lot of the drug product containing the active ingredient.
  - For an active ingredient in an OTC drug product that is exempt from bearing an expiration date under Sec. 211.137, retain for three years after distribution of the last lot of the drug product containing the active ingredient.

- Retain a properly identified reserve sample representative of each lot or batch of drug product and store it under conditions consistent with product labeling.
  - Store the reserve sample in the same immediate container-closure system in which the drug product is marketed or in one that has essentially the same characteristics.
  - The reserve sample consists of at least twice the quantity necessary to perform all the tests required, except those for sterility and pyrogens.
  - Except for radioactive drug products, examine visually reserve samples from representative sample lots or batches selected by acceptable statistical procedures at least once a year for evidence of deterioration unless visual examination would affect the integrity of the reserve sample.
  - Investigate if there is any evidence of reserve sample deterioration in accordance with Sec. 211.192; record the results of the examination and maintain them with other stability data on the drug product.

○ The retention time for a finished product is as follows:

■ For a drug product, except for radioactive and OTC drug products, retain for one year after the expiration date of the drug product and for three years after the distribution of the lot or batch of an OTC drug product if the product is exempted for bearing an expiration date under Sec 211.137.

Laboratory Animals [Sec. 211.173]

• Maintain and control animals if used in testing components, in-process materials, or drug products for compliance with established specifications to assure their suitability for their intended use.

• Identify and maintain adequate records showing the history of their use.

Penicillin Contamination [Sec. 211.176]

• If there is a reasonable possibility that a nonpenicillin drug product has been exposed to cross-contamination with penicillin, test the nonpenicillin drug product for the presence of penicillin.

• Do not ship such drug product to market if detectable levels of penicillin are found when tested according to procedures specified in the document A Review of Procedures for the Detection of Residual Penicillin in Drugs, which can be downloaded from http://www.fda.gov/downloads/AboutFDA/Centers Offices/OfficeofMedicalProductsandTobacco/CDER/UCM095812.pdf.

Subpart J — Records and Reports
General Requirements [Sec. 211.180]

• Maintain all compliance related, production, control, or distribution records associated with a batch of a drug product for at least one year after the expiration date of the batch or for three years after distribution of a batch of certain OTC drug products lacking expiration dating.

• Maintain records for all components, drug product containers, closures, and labeling for at least one year after the expiration date or, in the case of certain OTC drug products lacking expiration dating, three years after distribution of the last lot of drug product incorporating the component or using the container, closure, or labeling.

• All records required under this part, or copies of such records, must be readily available for authorized inspection during the above-stated retention period at the establishment where the activities described in such records occurred.

○ Records that can be retrieved immediately from another location by computer or other electronic means are considered as meeting the requirements of this paragraph.

○ Records may be retained as originals or true copies, such as photocopies, microfilm, microfiche, or other accurate reproductions of the original records.

• These records or copies thereof should be suitable for photocopying or other means of reproduction as part of such inspection.

- Where reduction techniques, such as microfilming, are used, suitable reader and photocopying equipment must be readily available.
- Maintain written records for their use in the evaluation (at least annually) of the quality standards of each drug product to determine the need for changes in drug product specifications or manufacturing or control procedures.
- Write (include the following) and follow procedures for such evaluations:
  - A review of a representative number of batches, whether approved or rejected, and, where applicable, records associated with the batch
  - A review of complaints, recalls, returned, or salvaged drug products, and investigations conducted under Sec. 211.192 for each drug product
- Establish procedures to assure that the responsible officials of the firm are aware of:
  - Any investigations conducted under Sec. 211.198, 211.204, or 211.208 of these regulations
  - Any recalls
  - Reports of inspectional observations issued by the FDA
  - Any regulatory actions relating to good manufacturing practices brought by the FDA

Equipment Cleaning and Use Log [*Sec. 211.182*]
- Keep a written record of major equipment cleaning and maintenance (except for routine maintenance requiring lubrication and adjustments)
- Keep individual equipment use logs that show the date, time, product, and lot number of each batch processed.
- If equipment is dedicated to the manufacture of one product, individual equipment use logs are not required, provided that:
  - Lots or batches of such product follow in numerical order
  - Lots or batches are manufactured in numerical sequence
  - The records of cleaning, maintenance, and use are part of the batch record
- The persons performing and double-checking the cleaning and maintenance (or, if the cleaning and maintenance is performed using automated equipment under Sec. 211.68, just the person verifying the automated operation) must date and sign or initial the log indicating that the work was performed.
- Entries in the log must be in chronological order.

Component, Drug Product Container, Closure, and Labeling Records [*Sec. 211.184*]
- Include the following in these records:
  - The identity and quantity of each shipment of each lot of components, drug product containers, closures, and labeling
  - The name of the supplier
  - The supplier's lot number(s), if known

○ The receiving code as specified in Sec. 211.80 and the date of receipt

○ The name and location of the prime manufacturer, if different from that of the supplier

○ The results of any test or examination performed [including those performed as required by Sec. 211.82(a), 211.84(d), or 211.122(a)] and conclusions derived

- Maintain an individual inventory record of each incoming material and reconciliation of the use of each lot of each component

- Include sufficient information in these inventory records to allow determination of any batch or lot of drug product associated with the use of each material.

- Document the examination and review of labels and labeling for conformity with established specifications in accord with Sec. 211.122 and 211.130.

- Document the disposition of rejected materials.

Master Production and Control Records [Sec. 211.186]

- To assure batch-to-batch uniformity, master production and control records for each drug product, including each batch size thereof, is prepared, dated, and signed (full signature, handwritten) by one person and independently checked, dated, and signed by a second person.

- Describe the preparation of master production and control records in writing and follow them.

○ The name and strength of the product and a description of the dosage form

○ The name and weight or measure of each active ingredient per dosage unit or per unit of weight or measure of the drug product

○ A statement of the total weight or measure of any dosage unit

○ A complete list of components designated by names or codes sufficiently specific to indicate any special quality characteristic

○ An accurate statement of the weight or measure of each component, using the same weight system (metric, avoirdupois, or apothecary) for each component.

■ Reasonable variations, in the amount of components necessary for the preparation of the dosage form, may be permitted provided that they are justified in the master production and control records.

○ Any calculated excess of component

○ Theoretical weight or measure at appropriate phases of processing

○ Theoretical yield, including the maximum and minimum percentages of theoretical yield beyond which investigation is required according to Sec. 211.192

○ A description of the drug product containers, closures, and packaging material

- A signed and dated (by the person or persons responsible for approval of such labeling) specimen or copy of each label and all other labeling materials

- Complete manufacturing and control instructions, sampling and testing procedures, specifications, special notations, and precautions to be followed

Batch Production and Control Records [Sec. 211.188]
- Batch production and control records are prepared for each batch of drug product produced.

- An accurate reproduction of the appropriate master production or control record is checked for accuracy, dated, and signed.

- Documentation that each significant step in the manufacture, processing, packing, or holding of the batch was accomplished, including:
  - Dates
  - Identity of individual major equipment and lines used
  - Specific identification of each batch of component or in-process material used
  - Weights or measures of components used in the course of processing
  - In-process and laboratory control results
  - Inspection of the packaging and labeling area before and after use
  - A statement of the actual yield and a statement of the percentage of theoretical yield at appropriate phases of processing
  - Complete labeling control records, including specimens or copies of all labeling used
  - Description of drug product containers and closures
  - Any sampling performed
  - Identification of the persons performing and directly supervising or checking each significant step in the operation, or if a significant step in the operation is performed by automated equipment under Sec. 211.68, the identification of the person checking the significant step performed by the automated equipment
  - Any investigation made according to Sec. 211.192
  - Results of examinations made in accordance with Sec. 211.134

Production Record Review [Sec. 211.192]
- The quality control unit must review and approve all drug product production and control records, including those for packaging and labeling, for:
  - Compliance with all established and approved written procedures before a batch is released or distributed
- The QCU must thoroughly investigate any unexplained discrepancy or failure of a batch or any of its components to meet any of its specifications, whether or not the batch has already been distributed.

- ○ Extend the investigation to other batches of the same drug product and other drug products that may have been associated with the specific failure or discrepancy.
- ○ Maintain a written record of the investigation conducted, including the conclusions and follow-up.

Laboratory Records [Sec. 211.194]
- Laboratory records consist of complete data derived from all tests necessary to assure compliance with established specifications and standards, including examinations and assays. The records should include:
  - ○ A description of the sample received for testing with:
    - ▪ The identification of the source (i.e., the location where the sample was obtained)
    - ▪ The quantity
    - ▪ The lot number or other distinctive code
    - ▪ The date the sample was taken
    - ▪ The date the sample was received for testing
  - ○ A statement of each method used in the testing of the sample; this statement should include the location of data which used to establish that the method met proper standards of accuracy and reliability as applied to the product tested
    - ▪ If the method is taken from an approved new drug application, the *United States Pharmacopeia–National Formulary*, or other recognized reference, indicating the method and reference would suffice.
    - ▪ Verify the suitability of all testing methods used under actual conditions of use.
  - ○ A statement of the weight or measure of sample used for each test, where appropriate
  - ○ A complete record of all data secured in the course of each test, including all graphs, charts, and spectra from laboratory instrumentation, with proper identification to show the specific item and the lot tested
  - ○ A record of all calculations performed in connection with the test, including units of measure, conversion factors, and equivalency factors
  - ○ A statement of the results of tests and how the results compare with established standards of the ISQP for the item tested
  - ○ The initials or signature of the person who performed each test and the date(s) the tests were performed
  - ○ The initials or signature of a second person, showing that the original records have been reviewed for accuracy, completeness, and compliance with established standards
- Maintain the following records:
  - ○ If any modification is made to an established test method:

- Include the reason for the modification and data to verify that the results from the modification test method are at least as accurate and reliable as the established method.
  - Testing and standardization of laboratory reference standards, reagents, and standard solutions
  - Periodic calibration of laboratory instruments, apparatus, gauges, and recording devices required by Sec. 211.160
  - Stability testing performed in accordance with Sec. 211.166

Distribution Records [Sec. 211.196]
- Distribution records should contain the following:
  - The name and strength of the product and description of the dosage form
  - The lot or control number
  - The name and address of the consignee
  - The date and quantity shipped

Complaint Files [Sec. 211.198]
- Establish and follow procedures for handling of all written and oral complaints regarding a drug product; include the following in such procedures:
  - Provisions for the QCU to review any complaints of drug product not meeting its established specifications and to determine if an investigation is needed in accordance with Sec. 211.192
  - Provisions for a review to determine whether the complaint represents a serious and unexpected adverse drug experience that must be reported to the FDA in accordance with the 21 CFR 310.305 and 514.80.
- Keep a written record of each complaint at the location where the drug product involved was manufactured, processed, or packed. If records are kept at another facility, they must be readily available for inspection at that other facility.
- Maintain these records for at least one year after the expiration date of the drug product, or one year after the date that the complaint was received, whichever is longer.
- For OTC drug products lacking expiration dating, maintain records for three years after distribution of the drug product.
- Include the following information in the written record:
  - Name and strength of the drug product
  - Lot number
  - Name of complainant
  - Nature of complaint
  - Reply to complainant
  - Findings of the investigation and follow-up conducted according to Sec. 211.192

    ◦ If an investigation is not conducted, include the reason and the name of the person responsible for making the determination

Subpart K — Returned and Salvaged Drug Products

Returned Drug Products [Sec. 211.204]

- Properly identify and store all returned drug products in a separate area.
- If there is any doubt that the product returned has been exposed to adverse conditions which casts doubt on the SISQP of the drug product, it must be destroyed unless through examination, testing, or other investigations prove that the product meets appropriate standards of SISQP.
- A drug product may be reprocessed provided that the resulting drug product meets appropriate standards, specifications, and characteristics.
- Maintain records of the following information on drug products returned:
  ◦ Name and label potency of the drug product dosage form
  ◦ Lot number (or control number or batch number)
  ◦ Reason for the return
  ◦ Quantity returned
  ◦ Date of disposition
  ◦ Ultimate disposition of the drug product returned
  ◦ If the reason for a drug product being returned implicates associated batches, conduct an appropriate investigation of these batches in accordance with the requirements of Sec. 211.192
- Write and follow the procedures for the holding, testing, and reprocessing of drug products returned.

Drug Product Salvaging [Sec. 211.208]

- Do not salvage and distribute a drug product if it has been subjected to improper storage conditions, including extremes in temperature, humidity, smoke, fumes, radiation due to natural disasters, fires, accidents, or equipment failures.
- If there is a reason to salvage such a product, salvaging operations may be conducted only if there is evidence:
  ◦ From laboratory tests and assays (including animal feeding studies where applicable) that the salvaged drug products meet all applicable ISQP standards
  ◦ From inspection of the premises that the drug products and their associated packaging were not subjected to improper storage conditions as a result of the disaster or accident
- Maintain all records associated with a salvaged product, including name, lot number, and disposition of the salvaged drug products.

21.2.2 EU Good Manufacturing Practice (EudraLex, 2011)

Two European Commission Directives (2003/94/EC and 91/412/EEC) provide principles and guidelines of common GMPs for the manufacture of medicinal products of both human and veterinary use, published as the Guide to Good Manufacturing Practice. The guide is presented in three parts of basic requirements and annexes on specific areas of activity. Part I covers GMP principles for the manufacture of medicinal products, Part II covers GMPs for active substances used as starting materials, and Part III covers GMP-related documents. Throughout the guide it is assumed that the requirements of marketing authorization relating to the safety, quality, and efficacy of products are systematically incorporated into all of the manufacturing, control, and release-for-sale arrangements by the holder of the manufacturing authorization. A system of marketing and manufacturing authorizations ensures that all products authorized for the European market are manufactured only by authorized manufacturers and that their activities are regularly inspected by competent authorities. Manufacturing authorizations are required by all pharmaceutical manufacturers in the European Community, whether the products are sold within or outside the community.

Each of the nine chapters in the guide begin with the "Principle," followed by detailed and clearly stated guidelines for achieving compliance to the contents of that chapter. For the most part, these guidelines are similar to those required by U.S. GMPs but are described more clearly and may provide better understanding of the overall GMP requirements. For example, a statement such as "Pipe work should be labeled to indicate the contents and direction of flow" (in Chapter 3/3.42) clearly states the requirements for the EU GMPs. The principle and key requirements from each chapter are described below.

Chapter 1: Quality Management

Principle Senior management of a manufacturing authorization holder is responsible to ensure that medicinal products are manufactured to fit their intended use, to comply with the requirements of the marketing authorization, and do not place patients at risk due to inadequate safety, quality, or efficacy. To achieve the quality objective reliably there must be a comprehensively designed and correctly implemented system of quality assurance incorporating good manufacturing practice, quality control, and quality risk management. It should be fully documented and its effectiveness monitored. All parts of the quality assurance system should be resourced adequately with competent personnel and suitable and sufficient premises, equipment, and facilities. The basic concepts of quality assurance, good manufacturing practice, quality control, and quality risk management are interrelated; the quality assurance and quality risk management concepts are described here in more detail.

Quality assurance is a wide-ranging concept that individually or collectively influences the quality of a product and should ensure that:

- Medicinal products are designed and developed in a way that takes account of the requirements of GMP.

- Production and control operations are clearly specified and GMP adopted.
- Managerial responsibilities are clearly specified.
- Arrangements are made for the manufacture, supply, and use of the correct starting and packaging materials.
- All necessary controls on intermediate products, any other in-process controls, and validations are carried out.
- The finished product is processed and checked correctly according to the procedures defined.
- Medicinal products are not sold or supplied before a qualified person has certified that each production batch has been produced and controlled in accordance with the requirements of the marketing authorization and any other regulations relevant to the production, control, and release of medicinal products.
- Satisfactory arrangements exist to ensure, as far as possible, that the medicinal products are stored, distributed, and subsequently handled so that quality is maintained throughout their shelf life.
- There is a procedure for self-inspection and/or quality audit, which regularly appraises the effectiveness and applicability of the quality assurance system.

Quality Risk Management   Quality risk management is a systematic process for the assessment, control, communication, and review of risks to the quality of medicinal products. It can be applied both proactively and retrospectively. The system should ensure that:

- Evaluation of the risk is based on scientific knowledge, process experience, and ultimately links to the protection of the patients.
- The level of effort and documentation is commensurate with the level of risk.

Chapter 2: Personnel

Principle   To establish and maintain a satisfactory quality assurance system and proper manufacture of medicinal products, a manufacturer must have sufficient qualified personnel who are well trained and clearly understand their individual responsibilities. All personnel should be aware of the GMP principles that affect them and receive initial and continuing training, including hygiene instructions, relevant to their needs.

In this chapter, key full-time personnel, such as the head of production and the head of quality control, are identified as being independent and with specific and joint responsibilities. Other key personnel include a qualified person who ensures that each batch produced within the EC or imported has been tested and checked properly in accordance with the directives and marketing authorization. According to Directive 75/319/EEC, such a controlled batch does not have to be recontrolled in any other member state of the EC.

Shared or joint responsibilities of heads of production and quality control departments are as follows:

- Authorization of written procedures and other documents, including amendments
- Monitoring and control of the manufacturing environment
- Plant hygiene
- Process validation
- Training
- Approval and monitoring of suppliers of materials
- Approval and monitoring of contract manufacturers
- Designation and monitoring of storage conditions for materials and products
- Retention of records
- Monitoring of compliance with the GMP requirements
- Inspection, investigation, and tracking of samples to monitor factors that affect product quality

Chapter 3: Premises and Equipment

Principle   Premises and equipment must be located, designed, constructed, adapted, and maintained to suit the operations. Their layout and design must minimize the risk of errors and permit effective cleaning and maintenance to avoid cross-contamination, buildup of dust or dirt, and, in general, any adverse effect on the product quality.

Chapter 4: Documentation

Principle   Good documentation is an essential part of the quality system. Clearly written documentation prevents errors from spoken communication and permits tracing batch history. Manufacturing formulas and instructions, procedures, and records must be error-free and available in writing. The legibility of documents is a prerequisite.

Chapter 5: Production

Principle   Production operations must follow clearly defined procedures and must comply with the principles of the GMP and relevant manufacturing and marketing authorizations requirements. Validation is emphasized in the EU GMP and consists of:

- Validation studies are conducted according to defined procedures, and the results and conclusions are recorded.
- For a new manufacturing formula or method preparation to be adopted, its suitability for routine processing is demonstrated.
- Amendments to the manufacturing process that may affect product quality and reproducibility should be validated.
- Processes and procedures should undergo periodic critical revalidation to ensure that they are capable of achieving the results intended.

Chapter 6: Quality Control

Principle Quality control (QC) personnel are responsible for sampling, specifications, testing, documentation, and release procedures to ensure that relevant tests are carried out and judged satisfactory prior to releasing materials for use and drug products for sale or supply. QC personnel are not confined to laboratory operations, but must be involved in all decisions that may concern the quality of a product. The independence of QC from production personnel is considered fundamental to the satisfactory operation of quality control.

Chapter 7: Contract Manufacturing and Analysis

Principle Contract manufacture and analysis must be correctly defined, agreed upon, and controlled to avoid misunderstandings, which could result in a product of unsatisfactory quality. A written contract between the contract giver and the contract acceptor must clearly state the duties of each party and the way in which the qualified person releasing each batch of product for sale will exercise his or her full responsibility.

Chapter 8: Complaints and Product Recall

Principle All complaints and other information concerning potentially defective products must be reviewed according to written procedures. A system should be designed to recall from the market promptly and effectively products known or suspected to be defective.

Chapter 9: Self-Inspections

Principle Self-inspections should be conducted to monitor implementation and compliance with GMP principles and to propose necessary corrective measures.

- All GMP-related activities should be examined at suitable intervals to verify conformance with the principles of quality assurance.
- Self-inspections should be thorough and conducted independently by the designated person(s) from the company; an external audit may also be useful.
- All self-inspections should be recorded.
- Reports should contain all the observations made during the inspections and proposals for corrective measures.
- Actions taken subsequent to the audit should also be recorded.

21.2.3 Japan MHLW Ministerial Ordinances on Standards

The following two ordinances provide standards for quality assurance for drugs (Ministerial Ordinance No. 136, 2004) and standards for manufacturing control and quality control for drugs and quasi-drugs (Ministerial Ordinance No. 179, 2004).

MHLW Ministerial Ordinance No. 136 on Standards for Quality Assurance for Drugs, Quasi-drugs, Cosmetics, and Medical Devices (MHLW, 2004a)

Ministerial Ordinance No. 136 provides the standards in accordance with item (1) of Article 12-2 of the Pharmaceutical Affairs Law (Law No. 145, 1960). It comprises

25 articles divided into four chapters. Requirements from Chapters 1 and 2, as they pertain to "drugs," are included in this section; readers should refer to the MHLW website for the requirements for all other types of products, including drugs from biological (cell/tissue-based drugs and medical devices) origins.

Chapter 1   General Provisions (Articles 1 and 2)

    Article 1 Purpose

    Article 2 Definitions (refer to the Glossary section)

- Quality assurance duty
- Market release
- Lot

Chapter 2   Standards for Quality Assurance for Drugs (Articles 3 to 16)

    Article 3 Duties of General Marketing Manager

    Article 4 Organization and Personnel for Quality Assurance Duties

    Article 5 Quality Standard Code

    Article 6 Quality Assurance Duty Procedure Document

    Article 7 Contracts with Manufacturers, etc.

    Article 8 Duties of Quality Assurance Manager

    Article 9 Control of Market Release

    Article 10 Ensuring Proper Manufacturing Control and Quality Control

    Article 11 Handling Information on Quality, etc. and Quality Defects, etc.

    Article 12 Handling Recall

    Article 13 Self-Inspection

    Article 14 Training

    Article 15 Control of Drug Storage, etc.

    Article 16 Control of Documents and Control

Chapter 3   Standards for Quality Assurance for Quasi-drugs and Cosmetics (Articles 17 to 20)

Chapter 4   Standards for Quality Assurance for Medical Devices (Articles 21 to 25)

Supplementary Provision

Unique features of this ordinance:

- The quality assurance (QA) manager (responsible for QA activities) has a minimum of three years' experience in quality assurance or similar duties.
- The general marketing manager, who supervises the QA manager, has the following key duties:
  - Must respect the opinions of the QA manager.

- ○ Based on information from the QA manager, decide necessary measures and instruct the department and other departments responsible for quality assurance duties to implement such measures.

- ○ Ensure that the QA department collaborates closely with the safety control management department and with other departments concerned with quality assurance duties.

- The marketing authorization holder (MAH) of drugs must establish documents that describe marketing approval and other items related to the quality of each drug.

- Documents related to the quality standard code and QA duties should be placed in the offices of the general marketing manager and the QA department.

- The MAH must establish a contract document describing the details of the agreement as to quality assurance duties to ensure that the manufacturing and quality controls are conducted properly and efficiently by the manufacturer (or by the contract manufacturer).

- The MAH should appoint the following persons in advance of the execution of activities:

- ○ A person at the contract manufacturing site to establish procedures to control market release conducted by the manufacturer

- ○ A person in the quality assurance department to verify periodically that the contract manufacturer conducts the market release duties properly and efficiently and keeps records concerned with the results of the verification properly

- ○ A person other than the QA manager who would periodically conduct self-inspections on quality assurance duties, establish records of inspection results, and report them to the QA manager in writing

- ○ A person to establish and implement a plan of training for the personnel engaged in the quality assurance duties, to keep the training records, and to report progress of training to the QA manager in writing

- Maintain all records and documents and retain for five years for drugs other than biological-origin products or cell/tissue-based drugs.

- At a manufacturing site, the site manufacturing manager supervises both independent departments (manufacturing control and quality control departments).

- Establish and maintain a quality-approved *Seihin Hyojun Sho* (drug master file).

- Establish and maintain a protocol for conducting validation.

- A product that can be dispersed easily and cause hypersensitive reactions in a minute amount must be processed in exclusive work areas or rooms with a separate air-handling system.

- Conduct sanitation control of personnel, and establish and maintain records thereof.
- Designate person(s) to conduct all relevant duties (establish procedures, execute protocols, evaluation, investigation, follow-up, keep required documents, and report results to the QC manager, as necessary) associated with the following tasks:
  - Validation of manufacturing processes, manufacturing control, and quality control of all products
  - Change control: changes made to manufacturing processes that could affect product quality
  - Deviation control: when any deviation from the manufacturing process has occurred
  - Handling information on quality and quality defects
  - Handling recalls
  - Self-inspections: written reports are issued to the manufacturing site manager
  - Training
  - Control of documents and records: maintain the documents and records (excluding training records) for five years from the date of establishment or one year plus the shelf life, whichever is longer

MHLW Ministerial Ordinance No. 179 on Standards for Manufacturing Control and Quality Control of Drugs and Quasi-drugs (MHLW, 2004b) This ordinance was established in accordance with the provisions of item (4) of paragraph 2 of Article 14 and item (4) of paragraph 2 of Article 14 applied *mutatis mutandis* (necessary changes) under paragraph 5 of Article 19-2 of the Pharmaceutical Affairs Law (Law No. 145, 1960), MHLW ministerial ordinance to revise the Drugs and Quasi-drugs Manufacturing Control and Quality Control Regulations (MHLW Ministerial Ordinance No. 16, 1999). The contents of this ordinance are as follows:

Chapter 1   General Provision (Articles 1 to 3)

   Article 1 Purpose

   Article 2 Definitions

   Article 3 Scope

Chapter 2   Manufacturing Control and Quality Control in Manufacturing Sites of Drug Manufacturers, etc.

Section 1 General Rules (Articles 4 to 20)

   Article 4 Manufacturing Department and Quality Department

   Article 5 Manufacturing Manager

   Article 6 Personnel

   Article 7 *Seihin Hyojun Sho*

Supplementary Provisions:

> Article 1 Enforcement Date
>
> Article 2 Transitional Measures
>
> Article 3 MHLW Ministerial Ordinance No. 62, 1999, to expire March 31, 2005

Unique features of this ordinance (limited to Chapter 2, Section 1; selected definitions from Article 2 are listed in the Glossary section)

- The manufacturing manager at each manufacturing site supervises both the manufacturing control and quality control departments.

- The manufacturing manager must verify that necessary actions are taken promptly in cases where quality defects or potential risk could affect product quality.

- A quality-approved *Seihin Hyojun Sho* (drug master file) is available for each product.

- Provides specifics on validation, change control, and deviation control.

- The manufacturer must designate personnel beforehand to conduct duties of validation, change control, deviation control, handling of information on quality and quality defects, handling of recalls, self-inspections, training, and control of documents and records.

21.3 FDA INSPECTION AND REGULATORY ACTIONS
(IOM, 2006, 2010)

The FDA inspects pharmaceutical manufacturing facilities worldwide, using scientifically and CGMP-trained investigators to assess if a company is complying with the federal Food, Drug, and Cosmetic Act and the CGMP regulations. This inspection authority is discussed in the Investigations Operations Manual 2.2 (IOM, 2010). If a company is not in compliance, any drug made by that company would be considered adulterated and possibly misbranded under the law (FD&C Act, Sec. 501 or 502, respectively); this does not necessarily mean that there is a direct connection to a product defect or failure. Therefore, the purpose of a drug inspection is to assess:

- If a firm is distributing drugs without FDA approval, including counterfeit or diverted drugs (Sec. 505 of the act requires that new drugs are approved by the FDA)

- A firm's adherence to good manufacturing practice

- That production and control procedures are in place to ensure the identity, strength, quality, and purity of finished drug products

- If the deficiencies (e.g., nonconformance with official compendia, super/subpotency, substitution, etc.) that could lead to the manufacturing and distribution of products in violation of the act are identified and corrective actions implemented

- If the new drugs are manufactured by the procedures and formulas specified in the new drug application documents
- The drug labeling and promotional practices of the firm
- Compliance with the requirements of the Prescription Marketing Act, post-market adverse drug experience reporting, or other appropriate requirements and regulations

Types of inspection may include:

- Quality system/good manufacturing practice inspections
  - Biennial routine inspections
  - Initial inspections of new facilities or newly registered establishments
  - Initial inspections under new management and/or ownership
- Premarket inspections
- Foreign inspections

Based on the seriousness of the compliance violation, the FDA would take appropriate regulatory actions against the company and on the drug products involved. The range of enforcement actions could include:

- To correct and prevent violations
- To remove violative products from market
- To punish offenders

So, depending on the nature of the violation, the FDA could issue a warning letter (Form 483) notifying an individual or firm of a violation and requesting correction to criminal prosecution of the person or firm. Persons responsible for adulteration or misbranding may be liable for a violation of the act and, if found guilty, be subjected to the penalties specified by the law.

21.3.1 Regulatory Actions taken by the FDA

Upon completion of the inspection and before leaving the premises, the FDA investigator(s) will provide inspectional findings on an FDA 483 form (Inspectional Observations) to the highest management official at the facility. The FDA will review a firm's response to a Form 483 (if received within 15 days) before taking regulatory action. The firm is given about two weeks to respond in writing with plans to address the observations reported. Depending on the progress the firm has made in correcting deficiencies and on the seriousness of the deficiencies reported, the FDA may take the following additional actions:

- *Warning letters*: sent to individuals or firms requesting them to respond in writing (usually within 15 days) as to the steps that will be taken to correct a violation. If there is no response or if the response is inadequate, the FDA will take further action, including delays of new product approvals, import denials, or product recall. Warning letters are reviewed by higher-level FDA officials and reflect FDA's current thinking.

- *Recalls*: removing from the market a product that is either defective or potentially harmful or correcting the problem is the most effective means to protect the public. Recalls are almost always voluntary; in rare cases (e.g., infant formula) the FDA will formally request a recall. If a firm refuses to recall a product in violation, the FDA may issue press releases to warn the public and/or pursue seizure or injunction remedies through federal courts. The FDA's role is to oversee a company's strategy and assess the adequacy of the recall. Recalls may be categorize into one of the following three classes based on the level of hazard involved.
  - Class I: dangerous or defective products that could cause serious health problems or death (e.g., label mix-up on a lifesaving drug, a defective artificial heart valve)
  - Class II: products that might cause a temporary health problem or pose only a slight threat of a serious nature (e.g., a subpotent drug not used to treat life-threatening situations)
  - Class III: products that are unlikely to cause any adverse health reaction but that violate FDA labeling or manufacturing laws (e.g., a minor container defect, lack of English labeling in a retail food
- *Seizure*: to remove specific goods from commerce
- *Injunction*: a court order against individuals and/or corporations to prevent them from violating or causing violations of the act.
- *Criminal prosecution*: may be recommended in appropriate cases for violation of Section 301 of the act. Misdemeanor convictions, which do not require proof of intend to violate the act, can result in fines and/or imprisonment up to one year. Felony convictions, which apply in the case of a second violation or intent to defraud or mislead, can result in fines and/or imprisonment for up to three years.

REFERENCES

EMA. 2011a. . http://www.ema.europa.eu/ema/index.jsp?curl = /pages/home/Home_Page.jsp&jsenabled = true. Accessed Dec. 2011.

———. 2011b. About us. http://www.ema.europa.eu/ema/index.jsp?curl = pages/about_us/general/general_content_000235.jsp&murl = menus/about_us/about_us.jsp&mid = WC0b01ac058001ce7d& jsenabled = true. Accessed Dec. 2011.

———. 2011c. Agencies and decentralized bodies. http://europa.eu/agencies/community_agencies/emea/index_en.htm. Accessed Dec. 2011.

EudraLex. 2011. Good Manufacturing Practice Guidelines, Vol. 4. Feb. 2011. Updated July 7, 2011. http://ec.europa.eu/health/documents/eudralex/vol-4/index_en.htm. Accessed Mar. 18, 2011, and Dec. 2011.

FDA. 2009a. What FDA regulates. Dec. 2, 2009. Updated Dec. 2, 2009. http://www.fda.gov/AboutFDA/WhatWeDo/WhatFDARegulates/default.htm. Accessed Dec. 2011

———. 2009b. What FDA doesn't regulate. Apr. 27, 2009. Updated Apr. 27, 2009. http://www.fda.gov/AboutFDA/WhatWeDo/WhatFDADoesntRegulate/default.htm. Accessed Dec. 2011.

———. 2009c. *U.S. Code*, Sec. 501 [21 USC 351]. Adulterated drugs and devices. Updated Apr. 30, 2009. http://www.fda.gov/RegulatoryInformation/Legislation/FederalFoodDrugandCosmeticActFDC

Act/FDCActChapterVDrugsandDevices/ucm108055.htm?utm_campaign = Google2&utm_source = fdaSearch&utm_medium = website&utm_term = 501(a)(2)(B)&utm_content = 5 Accessed Apr. 2012

———. 2010a. Significant dates in U.S. Food and Drug Law History. Oct. 14, 2010. Updated Oct. 14, 2010. http://www.fda.gov/AboutFDA/WhatWeDo/History/Milestones/ucm128305.htm. Accessed Dec. 2011.

———. 2010b. What does FDA do? Dec. 17, 2010. Updated Dec. 17, 2010. http://www.fda.gov/AboutFDA/Transparency/Basics/ucm194877.htm. Accessed Dec. 2011.

———. 2010c. CFSAN—What we do. Updated May 12, 2010. http://www.fda.gov/AboutFDA/CentersOffices/OfficeofFood/CFSAN/WhatWeDo/default.htm. Accessed Dec. 2011.

———. 2011a. Implementation of the Biologics Price Competition and Innovation Act of 2009. Mar. 10, 2011. Updated Mar. 10, 2011. http://www.fda.gov/Drugs/GuidanceCompliance RegulatoryInformation/ucm215089.htm. Accessed Dec. 2011.

———. 2011b. Guidance for industry: Questions and answers regarding the effect of Section 4205 of the Patient Protection and Affordable Care Act of 2010 on state and local menu and vending machine labeling laws. Jan. 31, 2011. Updated May 23, 2011. http://www.fda.gov/Food/Guidance ComplianceRegulatoryInformation/GuidanceDocuments/FoodLabelingNutrition/ucm223408.htm. Accessed Dec. 2011.

———. 2011c. A conversation about the FDA and drug regulation with Janet Woodcock, M.D., Deputy FDA Commissioner for Operations. May 28, 2009. Updated Aug. 12, 2011. http://www.fda.gov/Drugs/ResourcesForYou/Consumers/ucm143467.htm. Accessed Dec. 2011.

———. 2011d. FDA Organisations. Updated Nov. 23, 2011. http://www.fda.gov/AboutFDA/CentersOffices/default.htm. Accessed Dec. 2011.

———. 2011e. CDER: the consumer watchdog for safe and effective drugs. May 21, 2009. Updated Aug. 12, 2011. http://www.fda.gov/Drugs/ResourcesForYou/Consumers/ucm143462.htm. Accessed Dec. 2011.

2011f. About the Center for Biologics Evaluation and Research. Updated Sep. 30, 2011. http://www.fda. gov/AboutFDA/CentersOffices/OfficeofMedicalProductsandTobacco/CBER/default.htm. Accessed Dec. 2011.

———. 2011g. Structure and responsibilities. Updated Nov. 1, 2011. http://www.fda.gov/AboutFDA/CentersOffices/OfficeofFoods/CVM/WhatWeDo/ucm072570.htm. Accessed Dec. 2011.

———. 2011h. About the Center for Tobacco Products. Mar. 3, 2011. Updated Oct. 4, 2011. http://www.fda.gov/AboutFDA/CentersOffices/AbouttheCenterforTobaccoProducts/default.htm. Accessed Dec. 2011.

———. 2011i. About the Office of Regulatory Affairs. Updated Nov. 22, 2011. http://www.fda.gov/AboutFDA/CentersOffices/ORA/default.htm. Accessed Dec. 2011.

———. 2011j. Office of Combination Products. Updated Oct. 21, 2011. http://www.fda.gov/AboutFDA/CentersOffices/OC/OfficeoScienceandHealthCoordinationOfficeofCombinationProducts/default.htm. Accessed Oct. 2011.

———. 2011k. Office of Science and Innovation. Updated Aug. 11, 2011. http://www.fda.gov/AboutFDA/CentersOffices/OC/OfficeofScientificatinMedicalPrograms/ucm197864.htm. Accessed Dec. 2011.

———. 2011l. Office of Pediatric Therapeutics. Updated Oct. 21, 2011. http://www.fda.gov/AboutFDA/CentersOffices/OfficeofMedicalProductsandTobacco/OfficeofScienceandHealthCoordination/ucm2018186.htm. Accessed Feb. 2012

———. 2011m. Office of International Programs. Mar. 2, 2011. Updated Dec. 9, 2011. http://www.fda. gov/AboutFDA/CentersOffices/OC/OfficeofInternationalPrograms/default.htm. Accessed Dec. 2011.

———. 2011n. *Code of Federal Regulations*. 21 CFR 210. Apr. 2011. Updated Apr. 1, 2011. http://www.accessdata.fda.gov/scripts/cdrh/cfdocs/cfcfr/CFRSearch.cfm?CFRPart = 210&showFR = 1. Accessed Dec. 2011.

———. 2011o. *Code of Federal Regulations*. 21 CFR 211. Apr. 2011. Updated Apr. 1, 2011. http://www.accessdata.fda.gov/scripts/cdrh/cfdocs/cfcfr/CFRSearch.cfm?CFRPart = 211. Accessed Dec. 2011

———. 2011p. Recall—firm press release. Feb. 6, 2011. Updated Feb. 6, 2011. http://www.fda.gov/Safety/Recalls/ucm242398.htm. Accessed Dec. 2011.

————. 2012. FDA news release. Feb. 13, 2012. http://www.fda.gov/NewsEvents/Newsroom/ PressAnnouncements/ucm291691.htm. Accessed Mar. 2012

GHTF. 2012. Welcome to the Global Harmonization Task Force website. Updated Mar. 23, 2012. http://www.ghtf.org/. Accessed Apr. 2012.

ICCR. (International Cooperation on Cosmetic Regulation). 2011. Updated Aug. 5, 2011. http:// www.fda.gov/InternationalPrograms/HarmonizationInitiatives/ucm114513.htm. Accessed Dec. 2011.

ICH. 2011. Welcome to the ICH official website. http://www.ich.org/home.html. Accessed Dec. 2011.

IOM 2006. Chapter 7 - Recall Activities. June 2006. http://www.fda.gov/ICECI/Inspections/IOM/ucm 123363.htm. Accessed Dec. 2011

————. 2010. SUBCHAPTER 2.2 - STATUTORY AUTHORITY. March 2010 http://www.fda.gov/ ICECI/Inspections/IOM/ucm122510.htm. Accessed Dec. 2011.

MHLW. 2004a. Ministerial Ordinance 136, 2004. Tentative translation ver. 4.6 (as of Sep. 5, 2005). Sep. 22, 2004. http://www.emergogroup.com/fils/japan-mhlw-ordinance-136-english.pdf. Accessed Dec. 2011.

————. 2004b. Ministerial Ordinance 179, 2004. Tentative translation ver. 1.24. Dec. 2004. https://tuv-sud.jp/english/infoservice/pdf/english_gmp_2.pdf. Accessed Dec. 2011.

————. 2011a. Pharmaceutical and Medical Safety Bureau. http://www.mhlw.go.jp/english/org/policy/ p13-14.html. Accessed Dec. 2011.

————. 2011b. Organization of the Ministry of Health, Labour and Welfare. http://www.mhlw.go.jp/ english/org/detail/index.html. Accessed Dec. 2011.

VICH. 2011. International Cooperation on Harmonization of Technical Requirements for Registration of Veterinary Products. Dec. 2010. Updated Aug. 5, 2011. http://www.fda.gov/ InternationalPrograms/HarmonizationInitiatives/ucm114624.htm. Accessed Dec. 2011.

WHO. 2011a. About. http://www.who.int/about/en/. Accessed Dec. 2011.

————. 2011b. The role of WHO in public health. http://www.who.int/about/role/en/index.html. Accessed Dec. 2011.

GLOSSARY

Acceptance criteria The product specifications and acceptance/rejection criteria, with an associated sampling plan, that are necessary for making a decision to accept or reject a lot or batch (or any other convenient subgroups of manufactured units).

Active ingredient Any component that is intended to furnish pharmacological activity or other direct effect in the diagnosis, cure, mitigation, treatment, or prevention of disease, or to affect the structure or any function of the body of man or other animals. The term includes those components that may undergo chemical change in the manufacture of the drug product and be present in the drug product in a modified form intended to furnish the specified activity or effect.

Actual yield The quantity that is actually produced at any appropriate phase of manufacture, processing, or packing of a particular drug product.

Air lock An enclosed space with two or more doors which is interposed between two or more rooms (e.g., of differing class of cleanliness) for the purpose of controlling the airflow between those rooms when they need to be entered. An air lock is designed for and used by either people or goods.

Batch A specific quantity of a drug or other material that is intended to be of uniform character and quality, within specified limits, and is produced according to a single manufacturing order during the same cycle of manufacture.

Batch number A distinctive combination of numbers and/or letters that identifies a batch specifically.

BPCA   Best Pharmaceuticals for Children Act.

BPCI Act   Biologics Price Competition and Innovation Act of 2009.

Bulk product   Any product that has completed all processing stages up to, but not including, final packaging.

Calibration   A set of operations that establish, under specified conditions, the relationship between values indicated by a measuring instrument or measuring system, or values represented by a material measure, and the corresponding known values of a reference standard.

CAT   Committee for Advanced Therapies.

CDER   Center for Drug Evaluation and Research.

CFR   *Code of Federal Regulations*.

CGMP   Current good manufacturing practice.

CHMP   Committee of Medicinal Products for Human Use.

Clean area   An area with defined environmental control of particulate and microbial contamination, constructed and used in such a way as to reduce the introduction, generation, and retention of contaminants within the area.

Clean/contained area   An area constructed and operated in a manner that will achieve the aims of both a clean area and a contained area at the same time.

Code of Federal Regulations   An annual publication that contains regulations of executive departments and agencies of the U.S. federal government. The CFR is divided into 50 titles that represent broad areas subject to federal regulations. Each title is divided into chapters that bear the name of the issuing agency. Each chapter is further subdivided into parts covering specific regulatory areas. FDA's regulations are in Title 21, Parts 1 to 1271.

COMP   Committee for Orphan Medicinal Products.

Compliance policy guides   Compliance policy guides explain the FDA's policy on regulatory issues related to the FDA laws and regulations.

Component   Any ingredient intended for use in the manufacture of a drug product, including those that may not appear in such a product.

Computerized system   A system that includes the input of data, electronic processing, and the output of information to be used either for reporting or automatic control.

Contained area   An area constructed and operated in such a manner (and equipped with appropriate air handling and filtration) as to prevent contamination of the external environment by biological agents from within the area.

Containment   The action of confining a biological agent or other entity within a defined space.

Primary containment: A system of containment that prevents the escape of a biological agent into the immediate working environment. It involves the use of closed containers or safety biological cabinets along with secure operating procedures.

Secondary containment: A system of containment that prevents the escape of a biological agent into the external environment or into other working areas. It involves the use of rooms with specially designed air handling, the existence of airlocks and/or sterilizers for the exit of materials and secure operating procedures. In many cases it may add to the effectiveness of primary containment.

Controlled area	An area constructed and operated in such a manner that some attempt is made to control the introduction of potential contamination and the consequences of accidental release of living organisms. The level of control exercised should reflect the nature of the organism employed in the process. At a minimum, the area should be maintained at a pressure negative to the immediate external environment and allow for the efficient removal of small quantities of airborne contaminants.
Cross contamination	Contamination of a material or of a product with another material or product.
CVMP	Committee of Medicinal Products for Veterinary Use.
Depth of recall	The level in the distribution chain to which the recall is to extend. This will depend on the product's degree of hazard and the extent of distribution. Levels are as follows:
	Consumer or user level: may vary with product, including any intermediate wholesale or retail level. Consumer or user may include the individual consumer, patients, physicians, restaurants, and hospitals.
	Retail level: recall to the level immediately preceding the consumer or user level. It includes retail groceries, pharmacies, hospital pharmacies, dispensing physicians, institutions such as clinics and nursing homes, and any intermediate levels.
	Wholesale level: all distribution levels between the manufacturer and the retailer. This level may not been encountered in every recall situation (i.e., the manufacturer may sell directly to the retailer).
	Note: In some cases, the user level and the retail level may appear to refer to the same group. Certain devices, radiological products, biologics, and drugs are supplied directly to physicians or hospital-type environments, which will, in turn, will use the product in the prescribed manner on or for the consumer. This type of special environment will be considered the user level.
Drug product	A finished dosage form (e.g., tablet, capsule, solution, etc.,) that contains an active drug ingredient generally, but not necessarily, in association with inactive ingredients. The term includes a finished dosage form that does not contain an active ingredient but is intended to be used as a placebo.
EC	European Commission.
EEA–EFTA	European Economic Area–European Free Trade Association.
Effectiveness checks	Actions taken to verify that all consignees at the recall depth specified by the strategy have received notification about the recall and have taken appropriate action. The method for contacting consignees may be accomplished by personal visits, telephone calls, letters, or a combination of both. These checks are to be conducted by the recalling firm as part of its recall strategy. In the determination of which consignees should be contacted by the firm for effectiveness checks, the FDA may consider as acceptable for a portion of the verification of recall effectiveness: The firm's receipt and retention of response cards, e-mails, or letters from consignees in accordance with instructions issued in the recall notification; or signed records or reports of actions accomplished by the firm's own representatives or signed records or reports from direct or subaccounts.
Effectiveness check levels	An alphabetical term representing the extent to which effectiveness checks will be made within the distribution chain, including consumers or patients where appropriate.

Level A: 100% percent of the total number of consignees to be contacted (at the recall depth specified).

Level B: some percentage of the total number of consignees to be contacted, to be determined on a case-by-case basis, but greater that 10% and less than 100% of the total number of consignees to be contacted (at the recall depth specified).

Level C: 10% of the total number of consignees to be contacted (at the recall depth specified).

Level D: 2% of the total number of consignees to be contacted (at the recall depth specified).

Level E: no checks.

EMA (formerly, EMEA)	European Medicines Agency.
EPA	Environmental Protection Agency.
EU	European Union.
FDA	U.S. Food and Drug Administration.
FD&C Act	Food, Drug, and Cosmetic Act.
Fiber	Any particulate contaminant with a length at least three times greater than its width.
Finished product	A medicinal product that has undergone all stages of production, including packaging in its final container.
Gang-printed labeling	Labeling derived from a sheet of material on which more than one item of labeling is printed.
GHTF	Global Harmonization Task Force.
GMP	Good manufacturing practice.
Herbal medicinal product	Medicinal product containing, as active ingredients, exclusively plant material and/or vegetable drug preparations.
HEW	Department of Health, Education, and Welfare.
HHS	Department of Health and Human Services.
HMPC	Committee for Herbal Medicinal Products.
ICCR	International Cooperation on Cosmetic Regulations.
ICH	International Conference on Harmonization of Technical Requirements for the Registration of Pharmaceuticals for Human Use.
Inactive ingredient	Inactive ingredient means any component other than an active ingredient.
Infected	Contaminated with extraneous biological agents and therefore capable of spreading infection.
In-process control	Checks performed during production to monitor and, if necessary, to adjust the process to ensure that a product conforms to its specification. The control of the environment or equipment may also be regarded as a part of in-process control.
In-process material	Any material fabricated, compounded, blended, or derived by chemical reaction that is produced for, and used in, preparation of a drug product.
Inspection	For-cause inspection An inspection carried out in response to specific information that raises questions, concerns, or problems associated with an FDA-regulated firm or commodity. This information could come to the attention of the FDA from any source, and includes, but is not limited to, the following: the results of a sample analysis, observations made during prior inspections, recall or market withdrawal, consumer or employee complaint, adverse reaction report, or suspicion of fraud.

Intermediate product	Partly processed material that must undergo further manufacturing steps before it becomes a bulk product.
ISQP	Identity, strength, quality, and purity.
Lot	A batch, or a specific identified portion of a batch, having uniform character and quality within specified limits; or, in the case of a drug product produced by continuous process, a specific identified amount produced in a unit of time or quantity in a manner that assures its having uniform character and quality within specified limits.
Lot number, control number, or batch number	Any distinctive combination of letters, numbers, or symbols, or any combination of them, from which the complete history of the manufacture, processing, packing, holding, and distribution of a batch or lot of drug product or other material can be determined.
MAH	Marketing authorization holder.
Market withdrawal	A firm's removal or correction of a distributed product that involves a minor violation for which the FDA would not initiate legal action, or which involves no violation. These include normal stock rotation practices, routine equipment adjustments and repairs, and product improvement. Replacement of device components that fail or wear out after a reasonable life span are considered a market withdrawal unless a violation of the FD&C Act has occurred and can be supported. Recovery of investigational products is normally considered a market withdrawal unless the product has been sold in domestic commercial distribution or a significant health hazard is involved, necessitating classification and publication of the action as a recall. The removal of products from the market as a result of actual or alleged tampering with individual units, where there is no evidence of manufacturing or distribution problems, is considered a market withdrawal.
MDUFMA	Medical Device User Fee and Modernization Act.
MHLW	Ministry of Health, Labour and Welfare.
NCTR	National Center for Toxicological Research (United States)
Nonfiber releasing filter	Non-fiber releasing filter means any filter, which after appropriate pretreatment such as washing or flushing, will not release fibers into the component or drug product that is being filtered.
OTC	Over the counter.
Packaging	All operations, including filling and labeling, that a bulk product has to undergo to become a finished product. *Note*: Sterile filling would not normally be regarded as part of packaging.
Packaging material	Any material employed in the packaging of a medicinal product, excluding any outer packaging used for transportation or shipment. Packaging materials are referred to as primary or secondary according to whether or not they are intended to be in direct contact with a product.
PDCO	Pediatric Committee.
PDUFA	Prescription Drug User Fee Act.
Percentage of theoretical yield	The ratio of the actual yield (at any appropriate phase of manufacture, processing, or packing of a particular drug product) to the theoretical yield (at the same phase), stated as a percentage.
Procedures	Description of the operations to be carried out, the precautions to be taken and measures to be applied directly or indirectly related to the manufacture of a medicinal product.

Production	All operations involved in the preparation of a medicinal product, from receipt of materials, through processing and packaging, to its completion as a finished product.
QCU	Quality control unit.
QA	Quality assurance.
Qualification	Action of proving that any equipment works correctly and actually leads to the results expected. The word *validation* is sometimes widened to incorporate the concept of qualification.
Quality control unit	Any person or organizational element designated by the firm to be responsible for the duties relating to quality control.
Quarantine	The status of starting or packaging materials, intermediate, bulk, or finished products isolated physically or by other effective means while awaiting a decision on their release or refusal.
Recall	*FDA-ordered recall*: a recall initiated by a firm in response to an order for such action. Examples: device recalls ordered under Sec. 518(e), infant formula recalls ordered under Sec. 412(e)(1), and human tissue for transplantation ordered under 21 CFR Part 1270.

FDA-requested recall: a recall initiated by a firm in response to a formal request for such action by the associate commissioner for regulatory affairs, or the appropriate center director when the authority has been delegated.

Firm-initiated recall: a recall that is initiated by a firm on its own volition without a formal request from FDA.

Recall: a firm's removal or correction of marketed products, including its labeling and/or promotional materials, that FDA considers to be in violation of the laws it administers. The agency would initiate legal action (e.g., seizure or other administrative or civil actions available to the agency) if the product was not recalled. Recall does not include market withdrawal or a stock recovery.

Recall classification: a numerical designation assigned by the FDA to a particular product recall that indicates the relative degree of health hazard. Class I is a situation in which there is a reasonable probability (strong likelihood) that the use of, or exposure to, a violative product will cause serious adverse health consequences or death. Class II is a situation in which the use of, or exposure to, a violative product may cause temporary or medically reversible adverse health consequences or where the probability of serious adverse health consequences is remote. Class III is a situation in which the use of, or exposure to, a violative product is not likely to cause adverse health consequences.

Recall completed/recall terminated: Recall completed is the classification status used for monitoring purposes when the recall action reaches the point at which the firm has actually retrieved and impounded all outstanding product that could reasonably be expected to be recovered, or has completed all product corrections. Recall terminated is the monitoring classification used to indicate that FDA has determined that all reasonable efforts have been made to remove or correct the violative product in accordance with the recall strategy, and proper disposition has been made according to the degree of hazard.

Recalling firm: the firm that initiates a recall or, in the case of an FDA-requested recall, the firm that has primary responsibility for the manufacture and/or marketing of the product to be recalled.

Recall number: number assigned to the recall of one product regardless of package size, lot numbers, or private buyers' labels, provided that the labels are otherwise identical. If a manufacturer requests a wholesaler, distributor, or relabeler to extend the recall to a lower level (assuming that no change of the product has occurred), the same recall number assigned to the manufacturer will be used. If the recalled product has undergone a change due to further processing or use as an ingredient or component in a new product, or has had the directions or indications for use changed, it will be considered a different product and its recall will be the responsibility of the firm responsible for the change. FDA will then assign a separate recall number.

Recall strategy: the planned course of action to be carried out by a firm in the achievement of its recall goals. The strategy will normally be developed by the recalling firm following 21 CFR Part 7, Subpart C, Recalls. For FDA-requested recalls, the agency will recommend a strategy to the recalling firm.

Reconciliation
A comparison, making due allowance for normal variation, between the theoretical and actual amount of product or materials produced or used.

Recovery
The introduction of all or part of previous batches of the required quality into another batch at a defined stage of manufacture.

Representative sample
A sample that consists of a number of units that are drawn based on rational criteria such as random sampling and intended to assure that the sample accurately portrays the material being sampled.

Reprocessing
The reworking of all or part of a batch of product of an unacceptable quality from a defined stage of production so that its quality may be rendered acceptable by one or more additional operations.

Return
Sending back to the manufacturer or distributor of a medicinal product that may or may not present a quality defect.

Seizure
Removal of goods through court order by a U.S. marshal pursuant to Section 304 of the FD&C Act.

Lot-specific seizure: seizure of all units in a specific lot or batch of a product.

Open-ended seizure: seizure of all units of a specific product, regardless of lot or batch.

Mass seizure: seizure of all products and equipment at an establishment/facility.

Multiple seizures: seizure of the same product in more than one district court.

SISQP
Safety, identity, strength, quality, and purity.

Starting material
Any substance used in the production of a medicinal product but excluding packaging materials.

Sterility
The absence of living organisms.

Strength
The concentration of a drug substance (e.g., weight/weight, weight/volume, or unit dose/volume basis), and/or the potency, that is, the therapeutic activity of the drug product as indicated by appropriate laboratory tests or by adequately developed and controlled clinical data (expressed, for example, in terms of units by reference to a standard).

System
Used in the sense of a regulated pattern of interacting activities and techniques that are united to form an organized whole.

Theoretical yield	The quantity that would be produced at any appropriate phase of manufacture, processing, or packing of a particular drug product, based on the quantity of components to be used, in the absence of any loss or error in actual production.
U.S.C.	United States Code.
Validation	Action of proving, in accordance with the principles of good manufacturing practice, that any procedure, process, equipment, material, activity, or system actually leads to the results expected (*see also* Qualification).
WHO	World Health Organization.

CHAPTER 22

REGULATIONS FOR COMPOUNDING PHARMACIES

22.1 INTRODUCTION

Pharmacy *compounding* is an age-old practice. "The Food and Drug Administration (FDA) views traditional pharmacy compounding as combining, mixing, or altering of ingredients to create a customized medication for an individual patient in response to a licensed practitioner's prescription. In its simplest form, it may involve taking an approved drug substance and making a new formulation to meet the medical needs of a specific patient" (Galston, 2003). The safety and efficacy of compounded drugs are not verified by the FDA and, as such, they are not approved by the FDA. In its traditional form, pharmacy compounding is a vital service that helps many people, including (Galston, 2003; FDA, 2007):

- Those who are allergic to inactive ingredients (e.g., a dye or a preservative) in the FDA-approved medicines and formulating a product without those allergens
- Patients needing medications marketed previously now not available commercially (i.e., the prescriber believes that the discontinued drug is still the best solution for the patient)
- Children or elderly patients who may have difficulty swallowing a tablet or a capsule who need a suspension or suppository dosage form
- Children needing diluted dosage of a drug for adults
- Those children who are simply unwilling to take bad-tasting medicines
- Traditional roles of compounding also include (NABP, 2011):
  - Preparation of drugs or devices in anticipation of prescription drug or device orders based on routine, regularly observed prescribing patterns
  - Preparation of drugs or devices for the purpose of, or as an incident to, research, teaching, or chemical analysis

Pharmacists are legally allowed to compound a wide variety of dosage forms, including sterile, nonsterile, intravenous admixtures, parenteral nutrition solutions,

Integrated Pharmaceutics: Applied Preformulation, Product Design, and Regulatory Science,
First Edition. Antoine Al-Achi, Mali Ram Gupta, William Craig Stagner.
© 2013 John Wiley & Sons, Inc. Published 2013 by John Wiley & Sons, Inc.

radiopharmaceuticals, and veterinary pharmaceuticals. The scope of compounding is limited to processing only those prescription orders, for individual patient use, that cannot be met by manufactured products and would not infringe the manufacturing process. A difference in dosage size is not a meaningful difference between a compounded and an FDA-approved product if they contain the same amount of active drug substance. (The FDA took an enforcement action in 2006 by issuing a warning letter (WL: CIN-07-28792-06) to a Kentucky compounding pharmacy for compounding such a preparation.)

Section 510 (g) of the Federal Food, Drug, and Cosmetic (FD&C) Act provides, in part, that pharmacies are exempt from registering as manufacturers and therefore do not have to comply to current good manufacturing practice [21 CFR Parts 210 (FDA, 2011a) and 211 (FDA, 2011b)] provided that they do not manufacture, prepare, propagate, compound, or process drugs or devices for sale other than the regular course of their business of dispensing or selling drugs or devices at retail (FDA, 2011a). Pharmacies are required to register as a manufacturer if they:

- Repackage over-the-counter products or in any way change the container, wrapper, or labeling of these products for resale [FD&C Act 510 (a)(1)]

- Repackage prescription drug products for sale to other health care providers.

Pharmacists or compounders engaged in the preparation of compounded formulations for dispensing and/or administration to humans or animals are expected to conform to applicable state and federal compounding laws, regulations, or guidelines. Differentiating between state and federal laws governing the practice of pharmaceutical compounding may be difficult at times, as some of the areas are reserved exclusively for state governments, whereas others are governed exclusively by the federal government. If there is a conflict between state and federal laws, it is usually safer to follow the stricter law. As a general rule, state laws regulate practice of pharmacy, and the federal government regulates pharmaceuticals, including their marketing, production, and distribution.

22.2 COMPOUNDING GUIDELINES

Governmental agencies and professional organizations have published several "global" guidelines covering all aspects of compounding, ranging from the definition of compounding to the specifications for specialized compounding equipment for various dosage forms for human and animal use. The following is a partial list of those guidelines:

- The Model State Pharmacy Act and Model Rules of the National Association of Boards of Pharmacy, Appendix B: Good Compounding Practices Applicable to State Licensed Pharmacies (NABP, 2011)

- The Food and Drug Administration Compliance Policy Guide: Sec. 460.200, Pharmacy Compounding of Human Drugs (FDA, 2002)

- The FDA Compliance Policy Guide: Sec. 608.400, Compounding of Drugs for Use in Animals (FDA, 2003)
- The *United States Pharmacopeia* (U.S.P)–*National Formulary* (N.F.): compliance with the information published in the U.S.P.–N.F. is enforced by the FDA and the state boards of pharmacy. The four main U.S.P. chapters that pertain to compounding are:
  - <795>: Pharmaceutical Compounding—Nonsterile Preparations
  - <797>: Pharmaceutical Compounding—Sterile Preparations
  - <1075>: Good Compounding Practices
  - <1191>: Stability Considerations in Dispensing Practice

The U.S.P.–N.F. contains many other chapters (a partial list is provided in Section 22.4.1.2 of this chapter) that influence the practice of compounding and testing of pharmaceutical products either directly or indirectly. Compounding pharmacies are routinely audited against these guidelines and standards by the state boards of pharmacy, the FDA, and accreditation organizations such as the Joint Commission on Accreditation of Healthcare Organizations, the Accreditation Commission for Health Care, and the Community Health Accreditation Program.

22.3   FDA COMPLIANCE POLICY GUIDES

The following two guides represent the FDA's current thinking on this topic and describe the types of compounding activities that might be subject to enforcement action under the law. An alternative approach may be used, provided that it is in compliance with the requirements of the applicable statues and regulations. These and other guidelines may be obtained from the FDA's website, http://www.fda.gov/drugs/guidancecomplianceregulatoryinformation/guidances/default.htm.

22.3.1   Pharmacy Compounding of Human Drugs
[Sec. 460.200] (FDA, 2002)

Compounding and manipulating reasonable quantities of human drugs by pharmacists, upon receipt of a valid prescription for an individually identified patient from a licensed practitioner, is well recognized and accepted by the FDA. This traditional activity is not the subject of this guidance, which was issued on May 29, 2002. The FDA will take serious enforcement actions, however, if pharmacy activities are similar to those of a drug manufacturer and result in significant violations of the new drug, or adulteration or misbranding provisions of the FD&C Act (Galston, 2003). The agency may take necessary enforcement actions if a pharmacy is engaged in any of the following activities:

- Compounding of drugs in anticipation of receiving prescriptions, except in very limited quantities in relation to the amounts of drugs compounded after receiving valid prescriptions. [Preparation of drugs or devices in anticipation of a prescription drug or device order based on routine, regularly observed prescribing pattern is permitted by the National Association of Boards of Pharmacy (NABP).]

- Compounding drugs that were withdrawn or removed from the market for safety reasons (Appendix 22.1) [or on the FDA list of drugs products that present demonstrable difficulties in compounding unless approved by the state board of pharmacy. In extreme situations, the state board of pharmacy may permit compounding of those products that present demonstrable difficulties in compounding provided that (NABP, 2011):

  ○ A documented assessment has been made that other FDA-approved drugs would not treat the patient successfully.

  ○ The clinical assessments, benefits of risk analysis, etc., is justified.

  ○ The patient and practitioner are informed and aware of the benefits or risks].

- Compounding of finished drugs from bulk active ingredients that are not present in the current FDA-approved drug list (this information may be obtained from the FDA's Orange Book, http://www.accessdata.fda.gov/scripts/cder/ob/default.cfm) without an FDA-sanctioned investigational new drug application in accordance with 21 *United States Code* (U.S.C.) 355(i) and 21 CFR 312.

- Receiving, storing, or using a drug substance without first obtaining written assurance from the supplier that each lot of the drug substance has been made in an FDA-registered facility.

- Receiving, storing, or using drug components neither guaranteed nor determined to meet official compendia requirements.

- Using commercial-scale manufacturing or testing equipment for compounding drug products. [In September 2002, the FDA issued a warning letter to a California pharmacy for using commercial-scale manufacturing equipment and making large quantities of drugs for shipment across California and to patients in other states (Galston, 2003).]

- Compounding drugs for third parties who resell to individual patients.

- As a wholesaler, offering compounded drug products to other state-licensed persons or commercial entities for resale.

- Compounding drug products that are identical or essentially copies of commercially available FDA-approved drug products. In certain circumstances a small quantity that is only slightly different from a commercially available FDA-approved drug may be compounded provided that documented evidence exists that shows the medical need for the particular variation of the compound for the particular patient. [Following are examples of such exceptions:

  ○ The FDA issued a guidance in December 2009 (FDA, 2009) to pharmacies for compounding Tamiflu oral suspension from Tamiflu capsules in advance of receiving prescriptions (anticipated to be received within the next 24 hours) for multiple preparations provided that the Tamiflu for oral suspension is not readily available in a timely manner because of an actual shortage of the product.

- ○ On March 30, 2011, the FDA indicated that it would not prevent (for the time being) pharmacies from producing "compounded" versions of the FDA-approved drug Makena (active ingredient, hydroxyprogesterone capo-rate) based on a valid prescription for an individually identified patient unless the compounded products are unsafe, of substandard quality, or not being compounded in accordance with appropriate standards for compound-ing sterile products (FDA, 2011c).]

- Failing to operate in conformance with applicable state laws regulating the practice of pharmacies.

In cooperation with the state authorities, the FDA encourages its district offices to identify compounding establishments that are operating outside the tradi-tional practice of pharmacy. A noncompliant establishment may receive an FDA-initiated regulatory action such as a warning letter, seizure, injunction, and/or prosecution. Charges may include, but not be limited to, violations of the 21 U.S.C. Secs. 351(a) (2) (B), 352(a), 352(f) (1), 352(o), and 355(a) of the act.

The following are a few of examples of warning letters that the FDA has recently issued to those pharmacies that have extended their operations beyond their traditionally accepted compounding activities:

- The FDA issued warning letters to seven pharmacies for compounding hor-mone therapy drugs that contained estriol as well as progesterone and estro-gen. No drug containing estriol has been approved by the FDA, and the safety and effectiveness of the estriol is unknown. Additionally, these phar-macies improperly claim that their drugs are superior to the FDA-approved menopausal hormone therapy, and they prevent or treat serious diseases, including Alzheimer's disease, stroke, and various forms of other actions (FDA, 2008).

- The FDA warned five pharmacies to stop compounding and distributing stan-dardized versions of topical anesthetic creams that were marketed for general distribution rather than responding to the unique medical needs of individ-ual patients. According to an FDA news release (FDA, 2006), two deaths have been connected to compounded topical anesthetic creams made by two of these five pharmacies. The FDA is very concerned that exposure to high concentrations of local anesthetics, like those in compounded topical anes-thetic creams, can cause grave reactions, including seizures and an irregular heartbeat. Hence, the FDA stated further in the news release that this action "serves as a general warning to firms that produce standardized versions of these drugs."

- In 2006, the FDA warned three firms to stop manufacturing and distributing thousands of doses of compounded unapproved inhalation drugs nationwide (FDA, 2007).

- In 2006, the FDA issued a warning letter to a compounding firm in Kentucky for compounding several prescription drugs which were considered as "new drugs," and one drug was essentially a copy of an FDA-approved capsule (WL: CIN-07-28792-06).

- In 2006, the FDA issued a warning letter to a pharmacy in Utah for compounding a "new drug" which was found to be associated with the death of a 25-year-old female (WL: DEN-07-02).

22.3.2 Compounding of Drugs for Use in Animals
[Sec. 608.400] (FDA, 2003)

Section 608.400 provides guidance to drug compounders, veterinarians, and the FDA staff on how the agency intends to address compounding of drugs intended for use in animals. This guidance describes the FDA's current thinking on the types of compounding that might be subject to enforcement action. An alternative approach may be used that meets the requirements of the applicable statute and regulations.

The FD&C Act does not distinguish compounding from manufacturing or other processing of drugs for use in animals. According to the FDA, compounding of veterinary medicines to treat many conditions in a number of different species, some of which are known to have unique physiological characteristics, is acceptable. Additionally, the FDA permits the compounding of products from approved animal or human drugs under the conditions set forth in 21 CFR 530.13. This activity is not the subject of this guidance. However, the FDA is greatly concerned about veterinarians and pharmacies that are engaged in manufacturing and distributing unapproved new animal drugs in a manner that is clearly outside the bounds of traditional pharmacy practice, and that violates the FD&C Act (e.g., compounding that is intended to circumvent the drug approval process and provide for the mass marketing of products that have been produced with little or no quality control or manufacturing standards to ensure the purity, potency, and stability of the product). These activities are the focus of this guidance. Pharmacies and veterinarians who engage in activities analogous to manufacturing and distributing drugs for use in animals may be held to the same provisions of the act as manufacturers.

The act does not exempt veterinarians or pharmacists from the approval requirements in the new animal drug provisions of the act, 21 U.S.C. 360b. In the absence of an approved new animal drug application, the compounding of a new animal drug from any unapproved drug or from bulk drug substances results in an adulterated new animal drug in violation of 21 U.S.C. 351(a)(5). The compounding of a new animal drug from an approved human or animal drug also results in an adulterated new animal drug in violation of 21 U.S.C. 351(a)(5) unless the conditions set forth in 21 CFR 530.13(b) are met.

The FDA will take serious enforcement action if the scope and nature of activities of veterinarians and pharmacists raise the types of concerns normally associated with a drug manufacturer and result in significant violations of the new animal drug, adulteration, or misbranding provisions of the act. In determining whether to initiate such an action, the agency will consider whether the veterinarian or pharmacist engages in any of the following activities:

- Compounding of drugs for use in situations (a) where the health of the animal is not threatened; and (b) where suffering or death of the animal is not likely to result from failure to treat.
- Compounding of drugs in anticipation of receiving prescriptions except in very limited quantities in relation to the amounts of drugs compounded after receiving prescriptions issued within the confines of a valid veterinarian–client–patient relationship.
- Compounding of drugs that are prohibited for extra label use in food-producing or non-food-producing animals [under 21 CFR 530.41(b)], as these drugs present a risk to the public health.
- Compounding finished drugs from unapproved human or animal drugs or drug substances other than those specifically addressed for regulatory discretion (e.g., antidotes listed in Appendix 22.2) by FDA's Center for Veterinary Medicine. Contact the Division of Compliance for information about compounding from unapproved drugs or bulk drug substances.
- Compounding from approved human drugs for which the FDA has implemented a restricted distribution system.
- Using commercial-scale manufacturing equipment for compounding drug products.
- Compounding drugs for third parties who resell to individual patients.
- Offer compounded drug products at wholesale level to other state-licensed persons or commercial entities for resale.
- Failing to operate in conformance with applicable state law regulating the practice of pharmacies.
- Compounding of drugs for use in animals where an identical approved new animal drug or approved new human drug is available and if used as labeled (or in conformity with 21 CFR 530) to treat the diagnosed condition.
- Compounding from a human drug for use in food-producing animals if an approved animal drug can be used for the compounding.
- Instances where illegal residues occur in meat, milk, eggs, honey, aquaculture, or other food-producing animal products, and such residues were caused by the use of a compounded drug.
- Labeling a compounded drug with a withdrawal time established by the pharmacist instead of the prescribing veterinarian.

- Labeling of compounded drugs without sufficient information, such as withdrawal times for drugs for food-producing animals or other categories of information that are described in 21 CFR 530.12.

The FDA encourages its district offices to consult with state regulatory authorities when taking a regulatory action against an establishment that is operating outside the traditional practice of pharmacy. An FDA-initiated regulatory action may include issuing a warning letter, seizure, injunction, and/or prosecution.

Examples of warning letters issued by the FDA:

- On July 6, 2010, the FDA issued a warning letter (WL: CIN-10-107809-16) to a veterinarian who was using a drug that did not conform with its approved uses or with the regulations for Extralabel Drug Use in Animals; the FDA determined this new drug to be unsafe and adulterated.

- On January 7, 2005, the FDA issued a warning letter (WL: KAN-2005-04) to a veterinary compounding pharmacy in Nebraska for:

  ○ Compounding (manufacturing) and distributing several drugs for use in animals in a manner that was clearly outside the bounds of traditional pharmacy practice and that violated the FD&C Act. The drugs were not compounded for individual patients in the context of a valid veterinarian–client–patient relationship. Instead, they appeared to be sales to veterinarian for use as office stock and/or for general distribution.

  ○ The pharmacy was compounding drugs for use when an approved drug was available in the dosage form and concentration that would treat the animals.

- On December 8, 2004, the FDA issued a warning letter (WL: 2005-DAL-WL-06) to a veterinary compounding pharmacy in Oklahoma for:

  ○ Compounding (manufacturing) and distributing several drugs for use in animals in a manner that was clearly outside the bounds of traditional pharmacy practice and that violated the FD&C Act. The firm was not compounding for individual patients (animals) in context of a valid veterinarian–client–patient relationship for administration by an end user, but were compounding for third party for resell to individual patients.

  ○ The pharmacy prescription drug labeling for clinical use did not identify the animals to receive treatment, provide the dosage frequency, or provide the duration of treatment.

  ○ The pharmacy was compounding drugs for use when an approved drug was available in the dosage form and concentration that would treat the animal.

  ○ The drug compounded could be used in food-producing animals as the product labeling did not specify the target animal species and therefore could result in unsafe drug residues in edible tissue. For example, at least three compounded drugs (nitrofurarzone, chloramphenicol, and diethylstilbestrol) are not permitted for extralabel use in food-producing animals because they present a risk to public health.

22.4   GOOD COMPOUNDING PRACTICES
(FDA, 2007; NABP, 2011)

In this section we cover the minimum current good compounding practices (GCPs) for the preparation of drug products by state-licensed pharmacies for dispensing and/or administration to humans and animals. Compounding pharmacists and pharmacies are expected to practice in accordance with these good compounding practices and with applicable state and federal compounding laws, regulations, or guidelines. Poor practices on the part of drug compounders can result in contamination or in products that would not possess the required strength, quality, and purity. Several tragedies have been reported where supervising compounding pharmacists may have been lax in following these guidelines. Through voluntary reporting, the media, and other sources, the FDA knows of more than 200 adverse events involving 71 compounded products since 1990 (through 2007); some of these instances were devastating. A few examples of the tragedies reported include:

- Twenty-one polo horses died in April 2009 within 3 hours of receiving a vitamin and mineral supplement compounded by a Florida pharmacy. Based on reported findings of the investigation, this incident was a result of the incorrect strength of an ingredient in the supplement. This supplement (similar to a French-made supplement, Biodyl) was not approved for use in the United States (Skoloff, 2009).

- In 2009, the FDA issued a warning letter to a compounding facility in New Jersey for producing and distributing various strengths of an injection solution which were found to be contaminated with diethylene glycol monoethyl ether, which has not been studied for use in injectable drugs, so there are no approved drugs for injection that contain this material (WL: 2009).

- In July 2008, two infants died, possibly after receiving a dose of heparin 100 times as strong as recommended, due to mixing errors (www.medpagetoday. com/publichealthpolicy/publichealth/1008/).

- In March 2006, a child died after a hospital technician compounded a chemotherapy drug with a saline solution that had up to 26 times more salt than that needed for the treatment (http:/www.ivteam.com/saline-error-child-dies/).

- In March 2007, two persons died after receiving IV colchicine which contained eight times greater than the standard dose recognized (JAMA, 2007;298(20):2364–2366; jama.ama-assn.org/cgi/content/full/298/20/2364).

- Two patients were blinded and several others had their eyesight damaged by a compounded ophthalmic solution used in cataract surgery. The product was contaminated with bacteria. In August 2007 the FDA announced a nationwide recall of this contaminated product, which had been distributed to hospitals and clinics in eight states (FDA, 2007).

- The FDA issued a nationwide alert concerning a contaminated compounded solution that caused five cases of bacterial infections, and a patient treated with the product developed sepsis and died (FDA, 2007).

22.4.1 Organization and Personnel

22.4.1.1 Responsibilities of a Compounder (OSBP, 2009; NABP, 2011; U.S.P. <795>; U.S.P. <1075>)   A compounder engaged in drug compounding has the responsibility and authority to:

- Ensure the validity of all prescriptions.
- Certify all compounding orders.
- Ensure that all preparations are compounded in accordance with good compounding practices, official standards, and relevant scientific data and information, and that the quality is built into the compounding preparations.
- Ensure that compounding preparations are packaged and labeled appropriately and have acceptable strength, quality, and purity.
- Purchase components following the quality criteria listed in Section 22.4.4.1.
- Ensure that the drug product and components of drug products have not been withdrawn or removed from the market for public health reasons (see Appendix 22.1 for this list) or are difficult to compound (Davidson, 2011).
- Inspect and approve (or reject) all components, drug product containers, closures, in-process materials, and labeling to ensure that all materials used in compounding have expected identity, quality, and purity.
- Prepare all compounding records accurately and timely.
- Maintain adequate procedures and records for investigating and correcting failures or problems in compounding, testing, or in the preparation of the compounded orders.
- Review all compounding records to ensure that no errors have occurred in the compounding process and, if errors have occurred, conduct a full investigation. Prepare a written record of the investigation, including its conclusions and follow-up. Maintain the investigation reports for a required period, which may vary from state to state.
- Assure proper maintenance and cleanliness of all facilities and equipment used in compounding.
- Assure that the compounding equipment, facilities, and environment are suitable for their intended purposes.
- Implement procedures to prevent cross-contamination when compounding with drugs (e.g., penicillin, and other allergens) that require special precaution to prevent cross-contamination.
- Validate critical processes to ensure that procedures, when used, will result consistently in the expected qualities in the finished preparation.

- Ensure that all personnel (including compounders themselves) engaged in compounding are:
  - Proficient in compounding and maintain that proficiency with ongoing education/training at sufficient frequency; or
  - Nonpharmacists become certified by a compounding certification program approved by the state boards of pharmacy and are familiar with all details of the GCPs and the U.S.P.–N.F. compounding standards.
- Ensure that all personnel engaged in the compounding don clean clothing appropriate to the operation being performed; where necessary, employees must wear protective apparel, such as coat/jacket, apron, hand, arm, beard, or head coverings to protect drug products from contamination
- For sterile compounding operations involving one or more aseptic manipulations, refer to U.S.P. <797> and the NABP Model Rules for Sterile Pharmaceuticals.
- Limit access to the immediate vicinity of the drug-compounding operation to authorized personnel only.
- Instruct all personnel engaged in compounding to:
  - Report if they are sick or have open lesion(s) that may adversely affect the safety or quality of a drug product being compounded.
  - Refrain from direct contact with components, drug product containers, closures, in-process materials, and drug products until the condition is corrected or determined by a competent medical personnel not to jeopardize the safety or quality of the products(s) being compounded.
- Establish reliable beyond-use dating to ensure that the finished preparations have their expected potency, purity, quality, and characteristics, at least until the labeled beyond-use date. The beyond-use date may be established from literature or from appropriate stability evaluations or as described in Section 22.5.2.
- Assure that each compounding process is always carried out as intended or specified and is under control.
- Perform the final check of preparations prior to their release from the pharmacy.

22.4.1.2 *Personnel Training* (U.S.P. <795>; U.S.P. <1075>)

- All personnel involved in the compounding, evaluation, packaging, and dispensing of compounded preparations are well trained for the type of compounding (nonsterile, sterile compounding with varying degrees of risks, veterinary, or radiopharmaceuticals) conducted and their job functions.
- All training activities are covered by appropriate standard operating procedures and are properly documented (NDBOPH, 2011).
- The compounder must ensure that a training program has been implemented and that it is ongoing.

- The following steps should be included in a training procedure:
  - All employees involved in compounding are familiar with the applicable U.S.P.–N.F. chapters:

    <795>: Pharmaceutical Compounding—Nonsterile Preparations

    <797>: Pharmaceutical Compounding—Sterile Preparations

    <1075>: Good Compounding Practices

    <1151>: Pharmaceutical Dosage Forms

    <1160>: Pharmaceutical Calculations in Prescription Compounding

    <1163>: Quality Assurance in Pharmaceutical Compounding

    <1176>: Prescription Balances and Volumetric Apparatus

    <1191>: Stability Considerations in Dispensing Practice

    <1265>: Written Prescription Drug Interaction Guidelines
  - All employees are familiar with each of the compounding procedures, including those involving the facility, equipment, actual compounding, evaluation, packaging, storage, and dispensing.
  - The compounder meets with employees regularly to review their work and answer any questions that employees may have concerning compounding related procedures.
  - The compounder should demonstrate the procedures, observe, and guide the employee throughout the training process. The employee should be able to repeat the procedures without any assistance from, but under the direct supervision of, the compounder.
  - An employee is not permitted to perform a procedure without direct supervision provided that the employee has demonstrated both verbal and functional knowledge of the compounding procedure. Upon satisfactory completion of the training, the supervisor will sign-off on formal records to show that the employee is trained properly.
  - The compounder must continually monitor to assure that the employee's calculations and work are accurate and performed adequately.
  - The compounder is ultimately responsible for the finished preparation and must approve all final preparations.

22.4.2 Drug Compounding Facilities (OSBP, 2009; NABP, 2011; U.S.P. <795>; U.S.P. <1075>)

The requirements of a drug compounding facility are:

- It should be Suitably sized for orderly placement and flow of equipment and materials to prevent mix-ups or contamination between components, containers, labels, in-process materials, and finished products.
- Aseptic processes used in compounding sterile preparation must be conducted in an area separate and distinct from areas that are used for the compounding of nonsterile products.

- The facility should be constructed appropriately and be situated so as to facilitate its cleaning, maintenance, and proper operation.
- The compounding work surface must be smooth, impervious, free of cracks and crevices (preferably seamless), and nonshedding.
- The facility must be maintained in a good state of repair.
- The plumbing system must be defect-free.
- Adequate washing facilities should include, but not be limited to, hot and cold water, soap or detergent, and an air drier or single-service towels.
- Potable water for hand and equipment washing is to be supplied under continuous positive pressure and must meet the standards prescribed in the Environmental Protection agency's National Drinking Water Regulations (40 CFR 141).
  - Use purified water, if required, for compounding a nonsterile drug preparation.
  - Use purified water for the final rinse of equipment and utensils.
  - For sterile preparations, use water for injection, sterile water for injection, or bacteriostatic water for injection.
- The facility must have adequate lighting, heating, ventilation, and air conditioning to prevent contamination or decomposition of components.
- The compounding area must be free of infestation by insects, rodents, and other vermin.
- Sewage, trash, and other refuse in and from pharmacy and immediate drug compounding area(s) are to be held and disposed of in a safe, sanitary, and timely manner.
- The components used in the compounding of drugs must be stored in a clean, dry area and under appropriate temperature conditions (controlled room temperature, refrigerator, or freezer) as directed by the manufacturer, or according to applicable U.S.P. monograph requirements.

22.4.3 Drug Compounding Equipment (NABP, 2011; U.S.P. <795>; U.S.P. <1075>)

- Equipment is to be designed appropriately, sized adequately, and located suitably to facilitate operations for its intended use and for its cleaning and maintenance.
- Equipment is to be made of suitable material so that the surfaces that come in contact with components, in-process materials, or drug products are not reactive, additive, or absorptive, so as to alter the safety, identity, strength, quality, or purity of the drug products.
- The equipment should be thoroughly cleaned promptly after use to avoid cross-contamination of ingredients and preparations. Take extra care when cleaning equipment used in compounding preparations requiring special

precaution, such as antibiotics, cytotoxins, cancer drugs, preparations containing allergenic ingredients (e.g., penicillins or sulfonamides), and other hazardous materials. Where possible, equipment should be dedicated for such use. If the same equipment is used for all drug products, necessary cleaning procedures must be written and followed to ensure that the equipment is meticulously cleaned and would not carry contamination to the next drug.

- Store cleaned equipment and utensils properly to protect them from contamination prior to use.

- Immediately prior to the initiation of compounding operations, the compounder must inspect and determine if the equipment is suitable for use. If necessary, clean and sanitize equipment immediately prior to use to prevent contamination of the drug product.

- The equipment, utensils, and containers/closures that are used in the compounding of sterile drug products must be cleaned, sterilized, and maintained according to the procedures set forth in the NABP Model Rules for Sterile Pharmaceuticals.

- Maintain equipment properly to prevent malfunctions that could alter a drug product's safety, identity, strength, quality, or purity.

- Develop cleaning schedules and cleaning methods used in the cleaning and maintenance operations.

- Automatic, mechanical, electronic, or other types of equipment used in compounding or testing of compounded preparations are routinely inspected, calibrated, and checked to ensure proper performance according to the manufacturer's instructions.

- To ensure accuracy and performance reliability, equipment and accessories used in compounding must be inspected, maintained, cleaned, and validated at appropriate intervals.

- Defective equipment must be labeled clearly and, preferably, stored separately from functional equipment.

22.4.4 Control of Components and Drug Product Containers and Closures

22.4.4.1 *Component Selection and Storage* (OSBP, 2009; NABP, 2011; U.S.P. <795>; U.S.P. <1075>)   Criteria for selecting and storing components:

- The order of selecting a component is as follows:
  - U.S.P.–N.F. drug substances (active ingredients) manufactured in an FDA-registered facility.
  - Inactive components are manufactured in an FDA-registered facility.
  - Components must be of high quality and chemically pure, analytical reagent grade, American Chemical Society–certified, or Food Chemicals Codex grade.

- ○ Obtain components from sources deemed acceptable and reliable in the professional judgment of the compounder.
- When components are derived from ruminant animals (e.g., bovine, caprine), the supplier must provide written assurance that these animals were born, raised, and slaughtered in countries where bovine spongiform encephalopathy (BSE) and scrapie does not exist.
- Do not use components if withdrawn or removed from the market by the FDA for public health reasons. Use only those inactive ingredients that are either generally recognized as safe (GRAS) by the FDA (http://www.fda.gov/ Food/FoodIngredientsPackaging/GenerallyRecognizedasSafeGRAS/ GRASSubstancesSCOGSDatabase/ucm084104.htm) or approved by the FDA (www.accessdata.fda.gov/scripts/cder/iig/index.cfm).
- All components are to be labeled properly and stored off the floor in a clean dry area under appropriate temperature conditions (controlled room temperature, refrigerator, or freezer) as directed by the manufacturer, or according to applicable U.S.P. monograph requirements.
- Components are rotated so that the oldest acceptable stock is used first.

22.4.4.2 *Packaging Containers and Closures* (OSBP, 2009; NABP, 2011; U.S.P. <795>; U.S.P. <1075>)

- Containers and closures used in packaging compounded preparations must meet the U.S.P.–N.F. requirements described in several general chapters, such as ⟨660⟩, Containers—Glass; ⟨661⟩, Containers—Plastics; and ⟨671⟩, Containers—Performance Testing. A written record must be obtained from the supplier to show that the containers meet appropriate U.S.P. requirements.
- Containers and closures must be handled, sterilized (where appropriate), and stored as described the U.S.P.–N.F. general chapters: ⟨797⟩, Pharmaceutical Compounding—Sterile Preparations; and ⟨795⟩, Pharmaceutical Compounding—Nonsterile Preparations. The use of commercially available presterilized containers is encouraged for sterile preparations.
- The containers and closures must be made of materials that are not reactive, additive, and absorptive so as to alter the quality, strength, or purity of the compounded drug.
- The containers and closures must be handled properly and stored off the floor to prevent contamination and to permit inspection and cleaning of the work area.
- Components are rotated so that the oldest acceptable stock is used first.

22.4.5 **Drug Compounding Controls** (OSBP, 2009; NABP, 2011; U.S.P. <795>; U.S.P. <1075>)

- All significant procedures performed in the compounding area (inclusive of the facility, equipment, personnel, preparation, packaging, and storage of compounded preparations) must be written as standard operating procedures and approved by a responsible compounder.

- These procedures should:
  - Ensure accountability, accuracy, quality, safety, and uniformity in compounding
  - Provide consistency in the orientation and training of personnel
- Establish appropriate control procedures to monitor the output and to validate the performance of those compounding processes that may be responsible for causing variability in the final compounded preparations. Factors that may cause variability include:
  - Particle size distribution of ingredients
  - Weight variation
  - Adequacy of mixing to assure uniformity and homogeneity
  - Clarity, completeness, or pH of solutions
- The compounding procedures should include a description of:
  - The components, their amounts, the order of component addition, and the compounding process
  - The equipment and utensils required
  - The drug product container and closure system
  - The product must be formulated with the intent to provide 100% of the labeled or established amount of the active ingredient.
- The written procedures described above are followed in the execution of the compounding process.
- The compounder must:
  - Check and recheck each procedure at each stage of the process to ensure that each weight or measure is correct as stated in the written compounding procedures.
  - Describe the tests or examinations of the compounded preparation (e.g., the degree of weight variation among capsules) to assure uniformity and integrity of drug preparations.
  - Observe the finished drug product to ensure that it has the appearance expected.
  - Record the various compounding steps completed at the time of performance.
  - Investigate any discrepancies and take appropriate corrective actions before the drug product is dispensed to the patient.
- If a component is transferred from the original container to another container (e.g., a powder is taken from the original container, weighed, placed in a container, and stored in that other container), identify the new container with:
  - The component name (and grade, if available)

- ○ The weight or measure
- ○ The lot or control number
- ○ The expiration or beyond-use date
- ○ The transfer date
- ○ The manufacturer's name, where available
- To assure the sterility of sterile compounded drug products, appropriate written procedures, including validation of the sterilization processes, should be designed and followed to prevent microbiological contamination of these preparations. Refer to the U.S.P. <797> for further details.
- The compounder should establish appropriate beyond-use dates based on the information available from the U.S.P.–N.F. monographs, appropriate testing, or from peer-reviewed literature or as described in Section 22.5.2.
- The compounder should establish appropriate storage requirements according to the Preservation, Packaging, Storage, and Labeling chapter under the U.S.P.–N.F. General Notices.
- Rejected in-process and finished materials must not be used for compounding operations for which they are unsuitable, and should be stored separately from the acceptable materials to prevent their inadvertent use in compounding.
- Documentation of all significant steps should enable a compounder to trace, evaluate, and replicate systematically the steps included throughout the preparation process of a compounded product.

22.4.6 Labeling Controls (NABP, 2011; U.S.P. <795>; U.S.P. <1075>)

- The label for the compounded preparation should contain information required by state and federal laws:
  - ○ Use of established name or distinct common name (do not use the trade name of a manufactured product).
  - ○ Do not indicate that the compounded product is therapeutically equivalent to a manufactured product.
  - ○ The label should state that this is a compounded preparation.
  - ○ Do not use a National Drug Code (NDC) number assigned to another product; an NDC number is not required for a compounded preparation.
  - ○ A complete list of ingredients or preparation name and reference or established name or distinct common name.
  - ○ Dosage form.
  - ○ Strength and/or concentration.
  - ○ Weight or measure.
  - ○ Preparation date.
  - ○ Name and address of compounder.

- ○ Inactive ingredients.
- ○ Batch or lot number.
- ○ Assigned beyond-use date, based on published data, appropriate testing, or U.S.P.–N.F. standards.
- ○ Affix an auxiliary label if special storage requirement is required to ensure their strength, quality, and purity.

- Preparations compounded in anticipation of a prescription prior to receiving a valid prescription should be prepared in amounts that can be justified by the history of prescriptions filled by the pharmacy and should be labeled with the information listed above.

- The compounder must examine the preparation for correct labeling after completion of the compounding process before dispensing.

22.4.7 Records and Reports (NABP, 2011; U.S.P. <795>; U.S.P. <1075>)

- Comply with the record-keeping requirements of state and federal laws governing compounding. If a drug is compounded in accordance with the manufacturer's labeling instructions, further documentation is not required.

- The objective of the documentation is to allow another compounder to reproduce the identical prescription at a future date.

- Keep adequate records of controlled drug substances (scheduled drugs) used in compounding.

- A formulation record (Appendix 22.3) and a compounding record (Appendix 22.4) are the most important documents and must be maintained as original records or true copies by all compounding pharmacies for a duration required by the state and/or federal laws, whichever is longer.

- Make sure that these documents are readily available for authorized inspection during the retention period at the establishment where the activities described in such records occurred.

- These records are readily accessible to all employees working with drug substances or bulk chemicals located on the compounding facility premises.

- Train employees on how to retrieve and interpret needed information.

22.4.8 Quality Control and Continuous Quality Improvement Program (FDA, 2007)

As reported by the FDA, the following three examples of contaminated products causing devastating repercussions would have been avoided if the responsible compounding pharmacies had employed effective quality control systems:

- Three patients died of infection stemming from contaminated compounded solutions (which were used to paralyze the heart during open-heart surgery).

- Two patients at a Washington, DC Veterans' Affairs hospital were blinded, and several others had their eyesight damaged, by a bacterially contaminated compounded ophthalmic solution used in cataract surgery.

- A patient treated with a contaminated compounded magnesium sulfate solution developed sepsis and died; an additional five cases of bacterial infection were reported in a New Jersey hospital.

22.4.8.1 *Quality Control* (U.S.P. <795>; U.S.P. <1075>)

The safety, quality, purity, and performance of compounded preparations depend on correct ingredients and calculations, accurate and precise measurements, appropriate formulation conditions and procedures, and prudent pharmaceutical judgment. In analyzing 29 compounded products from 12 compounding pharmacies, the FDA found that 34% of the products failed one or more standard quality tests. Nine of the 10 failed samples had low potency (59% to 89% of the values expected) (Galston, 2003). To achieve the desired goals of quality, safety, and performance, the compounder must:

- Verify the accuracy and completeness of each step performed in compounding.

- Visually inspect the finished preparation to ensure that the final preparation appears as expected.

- Investigate any discrepancies and take appropriate corrective action before the prescription is dispensed to the patient (refer to the Checklist for Acceptable Strength, Quality, and Purity for various pharmaceutical dosage forms under the compounded preparations in U.S.P. <795>).

22.4.8.2 *Continuous Quality Improvement Program* (NABP, 2011)

- Implement a continuious quality improvement (CQI) program for:
  - Monitoring, evaluating, correcting, and improving the activities and compounding processes
  - Placing emphasis on maintaining and improving the quality of systems and the provision of patient care
  - Ensuring that the plans aimed at correcting identified problems includes appropriate follow-up to make certain that effective corrective actions are performed
  - Adhering to the provisions set out in the NABP Model Rules for the Practice of Pharmacy

- A CQI program is documented through written policies and procedures; it should include the following:
  - All steps involved in the preparation and dispensing of compounded products
  - A description of specific monitoring and evaluation activities and how the results are reported and evaluated

    ○ A collection of complaints, returns, or recalls related to the identity, strength, quality, and/or purity of the compounded drug products

    ○ Identification of appropriate follow-up when action levels or thresholds are exceeded

    ○ The people responsible for each aspect of the CQI program

- In developing a specific plan, establish objectives and measurable indicators for monitoring activities and processes that are deemed high risk, high volume, or problem prone.

- Assure proper evaluation of environmental monitoring (e.g., trending of an indicator, such as settling plate counts for sterile products).

- Reassess the selection of measurable indicators and the effectiveness of the overall CQI program as needed or on an annual basis.

22.4.9 Compounding for a Prescriber's Office Use (NABP, 2011; U.S.P. <1075>)

- Compounders may compound drug preparations for a prescriber's office use if permitted by federal and state requirements. The office use does not include the resale of the compounded medications by the prescriber.

- Where compounding for office use is permitted, the preparation must be compounded for the sole purpose of administration by or for the prescriber. The preparation label should state "For Office Use Only—Not for Resale."

- Maintain a record of the compounding process.

22.4.10 Compounding Veterinary Products (OSBP, 2009; U.S.P. <795>)

- Compound prescriptions for animals on the basis of the prescription orders from a licensed prescriber. Compounding for office use for administration (but not to be dispensed to the patient) by veterinarians is allowed; the label of such a preparation should state "For Office Use Only—Not For Resale."

- Compounders must assess the intended use of the animal patient (e.g., companion, performance, food) before compounding for that patient.

- Handle and fill these prescriptions according to the guidelines available for compounding of veterinarian products.

- Compounding with bulk substances for food-producing animals is not permitted. The veterinarian must provide an accurate length of time that treated animal tissues (e.g., milk, eggs, meat) are to be withheld from the human food supply, and this withdrawal time must appear on the label of every compounded preparation for food-producing animals (U.S.P. <795>).

- Do not violate federal patent laws by duplicating a commercially available drug in inordinate quantities.

22.4.11 Compounding Herbal Products

If a compounded (pursuant to prescription) herbal product is intended to treat a disease, it could be considered a drug. In such cases, the bulk ingredients must conform to U.S.P.–N.F. standards; otherwise, the compounded drug could be considered a new drug.

22.4.12 Compounding Nonprescription Drugs

Technically, a nonprescription compounded drug product may be considered a new product and must be labeled according to the act.

22.5 STABILITY CRITERIA AND BEYOND-USE DATING OF COMPOUNDED PREPARATIONS

22.5.1 Stability Criteria Considerations (Allen, 2008; U.S.P. <795>; U.S.P. <1191>)

U.S.P. <1191> defines *stability* "as the extent to which a product retains, within the specified limits, and throughout its period of storage and use (i.e., its shelf life), the same properties and characteristics [chemical, physical, microbiological, therapeutic, and toxicological] that it possessed at the time of its preparation."

22.5.1.1 Factors Affecting Stability   One or more of the following sources could affect the stability (physical, chemical, microbiological, therapeutic, and/or toxicological characteristics) of a compounded preparation and makes it unsuitable for its intended use:

- Ingredients (active/excipient)
- Environmental factors (temperature, humidity, light, oxygen, etc.)
- Particle size (in emulsions and suspensions)
- pH
- Solvent-system composition: aqueous/nonaqueous (% of free water and polarity)
- Compatibility of anions and cations
- Solution ionic strength
- Primary container
- Specific chemical additives
- Molecular binding and diffusion of drugs and excipients

22.5.1.2 Signs of Instability   Patients should be advised to stop using a preparation if the following signs of instability are observed:

- Solutions, elixirs, syrups: precipitation, discoloration, haziness, gas formation due to microbial growth

- Suspensions: caking, difficulty in resuspending, crystal growth
- Gels: shrinkage, separation, discoloration, microbial growth
- Emulsions: breaking, creaming, separation of oil phase
- Creams: emulsion break, crystal growth, shrinkage, gross microbial contamination
- Ointments: consistency, separation, grainy/gritty, drying
- Capsules: physical appearance, softer/harder shells, discoloration, expansion/distortion of capsules
- Powders: caking into hard masses, discoloration, presence of objectionable odor
- Suppositories: excessive softening/hardening/drying/shriveling
- Troches: softening/hardening, crystallization, microbial growth, discoloration

22.5.2 Guidelines to Assign BUD (Allen, 2008; OSBP, 2009; NABP, 2011; U.S.P. <795>)

The *beyond-use date* (BUD; also referred as the *discard-by date*) is defined as "that date and time, as appropriate, after which a compounded preparation is not to be used and the date is determined from the date the prescription was compounded." Beyond-use dates are different from the expiration dates displayed on commercially available manufactured drug products. The expiration dates on commercial products are determined based on extensive controlled stability studies and are typically in months or years. Beyond-use dates are assigned to compounded preparations based on the criteria listed in this section and are generally in days or months. The main difficulty in determining the BUD is the extemporaneous nature of the compounded preparations, where the manufacturer's expirations dates cannot be used to estimate and/or extrapolate the BUD, due to their differences in drug concentrations, diluents, fill volumes, and/or packaged in a different container type.

When assigning a beyond-use date:

- Compounders should consult and apply drug-specific and general stability documentation and literature when available.
- Compounders should consider the nature of the drug and its degradation mechanism.
- Compounders should consider the container in which it is packaged, the expected storage conditions, and the intended duration of therapy (see the U.S.P., General Notices: Preservation, Packaging, Storage Labeling for Expiration Date and Beyond-Use Date section, for further details).
- If the compounded formulation has an official U.S.P.–N.F. monograph, the BUD given in the monograph can be used provided that the procedure listed in the monograph is used.
- In the absence of a U.S.P.–N.F. monograph, extrapolation of a manufacturer's expiration date or the availability of stability information from other

sources (reference books or primary sources) that is applicable to the specific preparation being compounded, the following maximum (but be conservative) beyond-use dates are recommended for nonsterile compounded drug preparations that are packaged in tight, light-resistant containers and stored at controlled room temperature, unless otherwise indicated (see Preservation, Packaging, Storage, and Labeling in the General Notices and Requirements in the U.S.P.–N.F.)

○ Solids and nonaqueous liquids prepared from commercially available dosage forms:

■ Twenty-five percent of the remaining expiration date of the commercial product, or

■ six months, whichever is earlier

○ Solids and nonaqueous liquids from U.S.P.–N.F. grade active ingredients:

■ Up to six months provided that all the active pharmaceutical ingredients present in the preparation have expiration periods beyond this period

○ Water containing liquid formulations (prepared from ingredients in solid form):

■ Up to 14 days (when refrigerated)

○ All other formulations:

■ Up to 30 days or the intended duration of therapy, whichever is earlier

These beyond-use date limits may be extended if supported by valid scientific stability information applicable directly to the specific preparation (i.e., the same drug concentration range, pH, excipients, vehicle, water content, etc.). See also the beyond-use dating information in the labeling section in U.S.P. <681>, Repackaging Into Single-Unit Containers and Unit-Dose Containers for Nonsterile and Liquid Dosage Forms. The labels on the containers or packages of all preparations, including official compounded preparations, must bear beyond-use dates.

22.6   VERIFICATION (U.S.P. <795>)

Perform routine verification of compounding procedures, including batch compounding documentation, according to written procedures. Verification ensures that calculations, weighing and measuring, order of mixing, and compounding techniques were appropriate and performed accurately.

22.7   PATIENT COUNSELING (U.S.P. <795>)

Counsel patients, at the time of dispensing, on the proper use, storage, and evidence of instability in compounded preparations.

22.8 PHARMACY COMPOUNDING ACCREDITATION

A compounding pharmacy may apply for accreditation by the Pharmacy Compounding Accreditation Board (PCAB). The accreditation is voluntary and is not required by state boards of pharmacy as a prerequisite to provide compounding services. The PCAB is a nonprofit organization formed by eight national pharmacy organizations for the development of policies and comprehensive standards for the practice of compounding pharmacy as well as to establish PCAB accreditation criteria and processes. The member organizations are:

> American Colleges of Apothecaries
>
> American Pharmacists Association
>
> International Academy of Compounding Pharmacists
>
> National Association of Boards of Pharmacy
>
> National Community Pharmacists Association
>
> National Council of State Pharmacy Association Executives
>
> National Home Infusion Association
>
> United States Pharmacopeia Convention

To earn the PCAB accreditation, a pharmacy must meet strict criteria, including completion of an extensive application, documentation for written policies, and on-site inspection of its quality procedures. The PCAB performs a "review and survey" every three years after the initial accreditation. For further details, refer to the PCAB website.

REFERENCES

Allen LV Jr. *The Art, Science, and Technology of Pharmaceutical Compounding* 3rd ed.. Washington, DC: American Pharmacists Association. 2008. pp. 43–54.

Davidson G. Should I compound itraconazol? North Carolina Board of Pharmacy. http://www.ncbop.org/faqs/Pharmacist/faq_CompoundedProducts.htm. Accessed Aug. 2011.

FDA. 2002. CPG Sec. 460.200: Pharmacy compounding (reissued May 29, 2002). Updated Jan. 20, 2010. http://www.fda.gov/ICECI/ComplianceManuals/CompliancePolicyGuidanceManual/ucm074398.htm. Accessed Dec. 2011.

———. 2003. CPG Sec. 608.400: Compounding of drugs for use in animals. Updated Feb. 23, 2010. http://www.fda.gov/ICECI/ComplianceManuals/CompliancePolicyGuidanceManual/ucm074656.htm. Accessed Dec. 2011.

———. 2006. FDA news release: FDA warns five firms to stop compounding topical anesthetic creams. Dec. 5, 2006. Updated Mar. 2, 2010. http://www.fda.gov/NewsEvents/Newsroom/PressAnnouncements/2006/ucm108793.htm. Accessed Dec. 2011.

———. 2007. FDA consumer health information: The special risk of pharmacy compounding. May 31, 2007. http://www.fda.gov/ForConsumers/ConsumerUpdates/ucm107836.htm. Accessed Dec. 2011.

———. 2008. FDA news release: FDA takes action against compounded menopause hormone therapy drugs. News & Events. Jan. 9, 2008. Updated June 18, 2009. http://www.fda.gov/NewsEvents/Newsroom/PressAnnouncements/2008/ucm116832.htm. Accessed Dec. 2011.

———. 2009. Guidance to Pharmacies on Advance Compounding of Tamiflu Oral Suspension to Provide for Multiple Prescriptions. Dec. 23, 2009. http://www.fda.gov/Drugs/DrugSafety/InformationbyDrugClass/ucm188629.htm. Accessed Dec. 2011

———. 2011a. *Code of Federal Regulations.* 21 CFR 210. Current good manufacturing practice in manufacturing, processing, packing, or holding of drugs; general. Updated Apr. 1, 2011. http://www.accessdata.fda.gov/scripts/cdrh/cfdocs/cfcfr/CFRSearch.cfm?CFRPart = 210&showFR = 1. Accessed Dec. 2011.

———. 2011b. *Code of Federal Regulations.* 21 CFR 211. Current good manufacturing practice for finished pharmaceuticals. Updated Apr. 1, 2011. http://www.accessdata.fda.gov/scripts/cdrh/cfdocs/cfcfr/CFRSearch.cfm?CFRPart = 211&showFR = 1. Accessed Dec. 2011.

———. 2011c. FDA statement on Makena. Mar. 30, 2011. Updated Mar. 30, 2011. http://www.fda.gov/NewsEvents/Newsroom/PressAnnouncements/ucm249025.htm. Accessed Dec. 2011.

Galston SK. Federal and state role in pharmacy compounding and reconstitution: exploring the right mix to protect patients. Testimony before the Senate Committee on Health, Education, Labor, and Pensions, Oct. 23, 2003. Updated July 24, 2009. http://www.fda.gov/NewsEvents/Testimony/ucm115010.htm. Accessed Dec. 2011

NABP. 2011. Appendix B: Good compounding practices applicable to state licensed pharmacies. Model State Pharmacy Act and Model Rules of the National Association of Boards of Pharmacy. Aug. 2011. http://www.nabp.net/publications/model-act/. Accessed Dec. 2011.

NDBOPH. 2011. Pharmaceutical compounding standards. 61-02-01-03. North Dakota Board Of Pharmacy. http://www.nodakpharmacy.com/pdfs/compoundingStandards.pdf. Accessed Dec. 2011.

OSBP. 2009. Title 535, Oklahoma State Board of Pharmacy; Chapter 15, Pharmacies; Subchapter 10, Good compounding practices; Part 1, Good compounding practices for non-sterile products. May 6, 2009. http://www.ok.gov/OSBP/documents/2009_Compounding_Draft_Rules.pdf. Accessed Dec. 2011.

Skoloff B. AP newsbreak: Pharmacy made mistake in horse drug. Yahoo! News. Apr. 23, 2009. http://news.yahoo.com/s/ap/2009423/ap_on_re_us/us_dead_polo_horses/print. Accessed Apr. 2009.

U.S.P. <795>. Pharmaceutical compounding: nonsterile preparations. USPC Official General Chapters online. Accessed Dec. 2011. http://www.pharmacopeia.cn/v29240/usp29nf24s0_c795.html

U.S.P. <1075>. Good compounding practices. USPC Official General Chapters. http://www.pharmacopeia.cn/v29240/usp29nf24s0_c1075.html#usp29nf24s0_c1075. Accessed Jan. 2012.

U.S.P. <1191>. Stability considerations in dispensing practice. USPC Official General Chapters. online. Accessed Dec. 2011.

WL: 2005-DAL-WL-06. Veterinary Enterprises of Tomorrow, Inc. Dec. 8, 2004. Updated July 8, 2009. http://www.fda.gov/ICECI/EnforcementActions/WarningLetters/2004/ucm146698.htm. Accessed Dec. 2011.

WL: 2009. Drugs Are Us, Inc. Sep. 28, 2009. Updated Nov. 3, 2009. http://www.fda.gov/ICECI/EnforcementActions/WarningLetters/2009/ucm188449.htm. Accessed Dec. 2011.

WL: CIN-07-28792-06. Customs Crips Pharmacy. Dec. 1, 2006. Updated July 8, 2009. http://www.fda.gov/ICECI/EnforcementActions/WarningLetters/2006/ucm076191.htm. Accessed Dec. 2011.

WL: CIN-10-107809-16. Country Road Veterinary Services LLC. July 6, 2010. Updated Sep. 1, 2010. http://www.fda.gov/ICECI/EnforcementActions/WarningLetters/2010/ucm218111.htm. Accessed Dec. 2011.

WL: DEN-07-02. University Pharmacy. Dec. 4, 2006. Updated Dec. 8, 2009. http://www.fda.gov/ICECI/EnforcementActions/WarningLetters/2006/ucm076192.htm. Accessed Dec. 2011.

WL: KAN-2005-04. Essential Pharmacy Compounding. Jan. 7, 2005. Updated July 8, 2009. http://www.fda.gov/ICECI/EnforcementActions/WarningLetters/2005/ucm075242.htm. Accessed Dec. 2011.

GLOSSARY

Beyond-use date	Date and time, as appropriate and imprinted on the medication label, after which a compounded preparation is not to be used. This date is determined from the date the preparation is compounded.
BUD	Beyond-use date/Discard Date.
Bulk drug substance	A substance that is used in a drug and that when used in the manufacturing, processing, or packaging of a drug becomes an active ingredient or a finished dosage form for the drug; this term does not include intermediates used in the synthesis of such substances.

CFR	*Code of Federal Regulations.*
Component	Any ingredient intended for use in compounding a drug product, including those that may not appear in such a product.
Compounder	A pharmacist or other licensed professional authorized by the appropriate jurisdiction to perform compounding pursuant to a prescription order by a licensed prescriber.
CQI	Continuous quality improvement.
Extra-label use	Actual use or intended use of a drug in an animal in a manner that is not in accordance with the approved labeling.
FDA	U.S. Food and Drug Administration.
FD&C	Food, Drug, and Cosmetic Act.
GCPs	Good compounding practices.
GRAS	Generally recognized as safe.
NABP	National Association of Boards of Pharmacy.
NDC	National Drug Code.
N.F.	*National Formulary.*
PCAB	Pharmacy Compounding Accreditation Board.
U.S.C	*United States Code.*
U.S.P.	*United States Pharmacopeia.*
Veterinarian–client–patient relationship	A valid relationship is one in which (1) a veterinarian has assumed the responsibility for making medical judgments regarding the health of the animal patient and the need for medical treatment, and the owner of the animal or caretaker has agreed to follow the veterinarian's instructions; (2) the veterinarian has sufficient knowledge of the animal to initiate at least a general or preliminary diagnosis of the medical condition of the animal; and (3) the veterinarian is readily available for follow-up in case of adverse reactions or failure of the regimen of the therapy. For this, the veterinarian should have recently examined and be personally acquainted with the keeping and care of the animal and/or by medically appropriate and timely visits to the premises where the animal is kept.

APPENDIXES

APPENDIX 22.1   List of Compounding Drugs Withdrawn or Removed from the Market for Safety Reasons

The following drug products may not be compounded under the exemptions provided by Sec. 503A(a) of the FD&C Act.

Adenosine phosphate: All drug products containing adenosine phosphate.

Adrenal cortex: All drug products containing adrenal cortex.

Aminopyrine: All drug products containing aminopyrine.

Astemizole: All drug products containing astemizole.

Azaribine: All drug products containing azaribine.

Benoxaprofen: All drug products containing benoxaprofen.

Bithionol: All drug products containing bithionol.

Bromfenac sodium: All drug products containing bromfenac sodium.

Butamben: All parenteral drug products containing butamben.

Camphorated oil: All drug products containing camphorated oil.

Carbetapentane citrate: All oral gel drug products containing carbetapentane citrate.

APPENDIX 22.1 (*Continued*)

Casein, iodinated: All drug products containing iodinated casein.

Chlorhexidine gluconate: All tinctures of chlorhexidine gluconate formulated for use as a patient preoperative skin preparation.

Chlormadinone acetate: All drug products containing chlormadinone acetate.

Chloroform: All drug products containing chloroform.

Cisapride: All drug products containing cisapride.

Cobalt: All drug products containing cobalt salts (except radioactive forms of cobalt and its salts and cobalamin and its derivatives).

Dexfenfluramine hydrochloride: All drug products containing dexfenfluramine hydrochloride.

Diamthazole dihydrochloride: All drug products containing diamthazole dihydrochloride.

Dibromsalan: All drug products containing dibromsalan.

Diethylstilbestrol: All oral and parenteral drug products containing 25 mg or more of diethylstilbestrol per unit dose.

Dihydrostreptomycin sulfate: All drug products containing dihydrostreptomycin sulfate.

Dipyrone: All drug products containing dipyrone.

Encainide hydrochloride: All drug products containing encainide hydrochloride.

Fenfluramine hydrochloride: All drug products containing fenfluramine hydrochloride.

Flosequinan: All drug products containing flosequinan.

Gelatin: All intravenous drug products containing gelatin.

Glycerol, iodinated: All drug products containing iodinated glycerol.

Gonadotropin, chorionic: All drug products containing chorionic gonadotropins of animal origin.

Grepafloxacin: All drug products containing grepafloxacin.

Mepazine: All drug products containing mepazine hydrochloride or mepazine acetate.

Metabromsalan: All drug products containing metabromsalan.

Methamphetamine hydrochloride: All parenteral drug products containing methamphetamine hydrochloride.

Methapyrilene: All drug products containing methapyrilene.

Methopholine: All drug products containing methopholine.

Mibefradil dihydrochloride: All drug products containing mibefradil dihydrochloride.

Nitrofurazone: All drug products containing nitrofurazone (except topical drug products formulated for dermatologic application).

Nomifensine maleate: All drug products containing nomifensine maleate.

Oxyphenisatin: All drug products containing oxyphenisatin.

Oxyphenisatin acetate: All drug products containing oxyphenisatin acetate.

Phenacetin: All drug products containing phenacetin.

Phenformin hydrochloride: All drug products containing phenformin hydrochloride.

Pipamazine: All drug products containing pipamazine.

Potassium arsenite: All drug products containing potassium arsenite.

Potassium chloride: All solid oral dosage form drug products containing potassium chloride that supply 100 mg or more of potassium per dosage unit (except for controlled-release dosage forms and those products formulated for preparation of solution prior to ingestion).

Povidone: All intravenous drug products containing povidone.

Reserpine: All oral dosage-form drug products containing more than 1 mg of reserpine.

Sparteine sulfate: All drug products containing sparteine sulfate.

Sulfadimethoxine: All drug products containing sulfadimethoxine.

Sulfathiazole: All drug products containing sulfathiazole (except those formulated for vaginal use).

Suprofen: All drug products containing suprofen (except ophthalmic solutions).

(*continued*)

APPENDIX 22.1 *(Continued)*

Sweet spirits of nitre: All drug products containing sweet spirits of nitre.

Temafloxacin hydrochloride: All drug products containing temafloxacin.

Terfenadine: All drug products containing terfenadine.

3,3′,4′,5-Tetrachlorosalicylanilide: All drug products containing 3,3′,4′,5-tetrachlorosalicylanilide.

Tetracycline: All liquid oral drug products formulated for pediatric use containing tetracycline in a concentration greater than 25 mg/mL.

Ticrynafen: All drug products containing ticrynafen.

Tribromsalan: All drug products containing tribromsalan.

Trichloroethane: All aerosol drug products intended for inhalation containing trichloroethane.

Troglitazone: All drug products containing troglitazone.

Urethane: All drug products containing urethane.

Vinyl chloride: All aerosol drug products containing vinyl chloride.

Zirconium: All aerosol drug products containing zirconium.

Zomepirac sodium: All drug products containing zomepirac sodium.

Source: FDA (2002).

APPENDIX 22.2 List of Bulk Drug Substances for Compounding and Subsequent Use in Animals to Which CVM Would Not Ordinarily Object

Ammonium molybdate

Ammonium tetrathiomolybdate

Ferric ferrocyanide

Methylene blue

Picrotoxin

Pilocarpine

Sodium nitrite

Sodium thiosulfate

Tannic acid

Source: FDA (2003).

APPENDIX 22.3 Formulation Record

Name: _____

Strength: _____

Dosage form: _____

Route of administration: _____

Date of last review or revision _____

Person completing last review or revision _____

This review approved by: _____ Date: _____

APPENDIX 22.3   (*Continued*)

Formula:

Serial Number	Ingredient	Quantity Required	Manufacturer	Quality Grade

Examples of calculations that must be made each time the formula is compounded:

Equipment required:

Method of preparation:

The method is a step-by-step sequence in the correct order of mixing, including mixing temperature and other environmental controls; also duration of mixing and list of in-process checks (during compounding), where appropriate.

Description of finished product:

Quality ontrol rocedures (QA tests/evaluation to be performed on the finished product):

Packaging ontainer:

Storage requirements:

Beyond-use date assignment (criteria for assigning BUD):

Label information (include auxiliary labels where appropriate):

References (source of the formula):

APPENDIX 22.4 Compounding Record

Name: _____

Strength: _____

Dosage form: _____

Route of administration: _____

Formulation record reference: _____

Quantity prepared: _____

Date of preparation: _____

Internal identification number and/or prescription number: _____

Person preparing formulation: _____

Person checking formulation: _____

Formula:

Serial Number	Ingredient	Manufacturer/ Lot No.	Quantity Required	Actual Quantity Used

Examples of calculations that must be made each time the formula is compounded:

Record all deviations from the reference formulation record and state reasons for each deviation:

Calculation: (to be performed at the time of compounding):

Equipment used:

Method of preparation:

Description of finished products:

Quality control test results and data:

Beyond-use date assignment:

CHAPTER 23

IND AND NDA PHASE-APPROPRIATE NEW DRUG DEVELOPMENT PROCESS

23.1 INTRODUCTION

The process of developing a new drug and ensuring its safety and efficacy is complicated, time consuming, costly, and the end result is never guaranteed. For every 5000 to 10,000 compounds that enter the research and development pipeline, ultimately only one receives approval. It is estimated that it takes an average of 10 to 15 years to develop one new medicine, from discovery to when it is available for treating patients, at a cost in excess of $800 million (this could be as high as $1.7 billion). It is further estimated that only about one-third of the drugs that are approved and marketed generate sufficient revenue to recover development cost (Grabowski et al., 2002; DiMasi et al., 2003; Pharma, 2007).

The Food, Drug, and Cosmetic (FD&C) Act does not allow the FDA to develop new drugs. The agency's role is small during the early stages of drug research but becomes of major importance when a new drug sponsor moves beyond these early stages and into animal and human testing, with the ultimate goal of marketing the new drug. During the later stages, the U.S. Food and Drug Administration (FDA) reviews the test result data submitted by the sponsor and determines whether the drug is safe to test in humans. If so, after human testing is complete, the FDA decides whether the drug can be sold to the public and what its label should say. The following are the major steps involved in a new drug review process (FDA Centennial, 1906–2006):

- Preclinical (animal) testing
- An investigational new drug (IND) application outlining the new drug sponsor's proposal for human testing in clinical trials
- Phase 1 studies (typically involve 20 to 80 people)
- Phase 2 studies (typically involve a few dozens to about 300 people)
- Phase 3 studies (typically involve several hundred to about 3000 people)

Integrated Pharmaceutics: Applied Preformulation, Product Design, and Regulatory Science,
First Edition. Antoine Al-Achi, Mali Ram Gupta, William Craig Stagner.
© 2013 John Wiley & Sons, Inc. Published 2013 by John Wiley & Sons, Inc.

- The pre–new drug application (NDA) period, just before the NDA is submitted, is commonly viewed as a time for the FDA and the drug sponsors to meet.
- Submission of an NDA is the formal step asking the FDA to consider a drug for marketing approval.
- After the NDA is received, the FDA has 60 days to decide whether to file it so that it can be reviewed.
- If the FDA files the NDA, an FDA review team is assigned to evaluate the sponsor's research on the drug's safety and effectiveness.
- The FDA reviews information that goes on a drug's professional labeling (information on how to use the drug).
- The FDA inspects the facilities where the drug will be manufactured as part of the approval process.
- The FDA reviews find the application either approvable or not approvable.

23.2 PRECLINICAL DEVELOPMENT OVERVIEW
(FDA, 1998a)

To satisfy FDA's minimum requirements, the sponsors must:

- Develop a pharmacological profile of the drug
- Determine the acute toxicity of the drug in at least two species of animals [usually in a rodent (e.g., rat), and a nonrodent (e.g., dog)]
- Conduct short-term (from 2 weeks to three months) animal toxicity studies, depending on the proposed duration of use of the substance in the clinical studies proposed

Before initiating clinical studies in humans, a sponsor must first submit data to the FDA demonstrating that the drug is reasonably safe for initial small-scale studies in humans. The sponsor has several ways to fulfill this key requirement:

- Compiling existing nonclinical data from past in vitro laboratory or animal studies on the compound
- Compiling data from previous clinical testing or marketing of the drug in the United States or another country with a population comparable to that of the United States
- Undertaking new preclinical studies designed to provide the evidence necessary to support the safety of administering the compound to humans

During this phase, the sponsor's goals are to develop a stable bioavailable active ingredient and dosage form, and to evaluate the drug (based on in vitro and in vivo animal pharmacology/toxicology studies) for genotoxicity screening, pharmacological effects, absorption, metabolism, rate of excretion, toxicity of metabolites, and rate of metabolite excretion.

23.2.1   Synthesis and Purification

There are no set rules through which drugs are developed. A new drug may be developed by targeting a broad disease category (e.g., cancer or cardiovascular diseases), a specific disease state (e.g., breast cancer or hypertension), or a specific disease site at the gene or cellular level. New drug research starts with an in-depth understanding of both normal and abnormal body functions and how a drug might be used to prevent, cure, or treat a disease or medical condition. This provides the researcher with a target. On rare occasions, scientists find the right compound quickly. Usually, however, discovery of a biologically beneficial lead compound requires hundreds or thousands of compounds (including their structural analogs) to be synthesized, purified, screened, assayed, and evaluated for their biological performance. Ongoing advances in computer modeling and other technologies provide clues to select a compound in the shortest possible time, but they all require testing in living beings to prove their effectiveness. Another approach, used primarily in the biotechnology industry, involves testing compounds made naturally by microscopic organisms (i.e., fungi, viruses, and molds) which are grown in what is known as a "fermentation broth," with one type of organism per broth. Sometimes, 100,000 or more broth samples are tested to determine whether any microorganism-derived compound has a desirable effect.

23.2.2   Animal Testing

Drug companies make every effort to use as few animals as possible and to ensure their humane and proper care. Generally, two or more species (one rodent, one nonrodent) are tested because one animal species may be affected by a drug differently than another animal species, due to physiological or metabolic differences (e.g., acetaminophen is toxic to dogs but not to rodents or humans). Animal testing is conducted using the same administrative route as that intended for human use. Chemical assays are used to measure:

- How much of a drug is absorbed into the blood
- How the drug is broken down chemically in the body
- The toxicity of the drug and its breakdown products (metabolites)
- How quickly the drug and its metabolites are excreted from the body

Prior to initiating animal studies, the sponsor should schedule a meeting (a pre-IND meeting) with the FDA to discuss and agree upon the design of the animal studies needed to provide evidence that the compound is biologically active and is reasonably safe for initial human testing. Data acquired during animal testing are used to determine the appropriate magnitude (e.g., $\mu g/kg$) for a human dose for the first-time-in-human testing.

23.2.2.1 *Short-Term Testing*   Short-term testing in animals ranges in duration from 2 weeks to three months, depending on the proposed use of the substance.

23.2.2.2 Long-Term Testing Long-term testing in animals ranges in duration from a few weeks to several years. Some animal testing continues after human tests have begun to learn whether or not long-term use of a drug causes cancer or birth defects. Much of this preliminary information is submitted to the FDA when a sponsor requests proceeding with human clinical trials. The FDA reviews the preclinical research data and makes a decision whether or not to allow clinical trials in humans to proceed (see Section 23.3).

23.2.3 Institutional Review Boards

An institutional review board (IRB) ensures the rights and welfare of people participating in clinical trials both before and during their trial participation. An IRB must be composed of at least five members, including scientists, doctors, and laypeople with varying backgrounds. The purpose of an IRB is to ensure a complete and adequate review of activities commonly conducted by research institutions. It includes a review of compliance to applicable laws and professional competence of the investigators, including professional and ethical conduct. Ethical and community attitudes are also evaluated by the IRB. IRBs verify that study participants are fully informed and have given their written consent before studies begin. IRBs are monitored by the FDA to protect and ensure the safety of study participants in medical research. For more information, refer to the IRB Operations and Clinical Requirements list provided by the FDA's Office of Health Affairs and the Technical Amendments concerning protection of human subjects (45 CFR 46).

23.3 PHASE-APPROPRIATE CLINICAL TRIALS OVERVIEW (FDA, 1998a)

While the purpose of preclinical work (animal pharmacology and toxicology testing) is to develop adequate data to undergird a decision that it is reasonably safe to proceed with human trials of the drug, the clinical trials process represents the ultimate premarket testing ground for unapproved drugs. During these trials, an investigational compound is administered to humans and is evaluated for its safety and effectiveness in treating, preventing, or diagnosing a specific disease or condition. The results of this testing comprise the single most important factor in the approval or disapproval of a new drug. (FDA, 1998a)

Although the goal of clinical trials is to obtain safety and effectiveness data, the overriding consideration in these studies is the safety of those in the trials. The Center for Drug Evaluation and Research (CDER) monitors the study design and conduct of clinical trials to ensure that people in the trials are not exposed to unnecessary risks. To conduct these clinical trials, the sponsor will probably ship the drug to clinical investigators in many states. The sponsor must seek exemption from the federal statue which prohibits an unapproved drug from being shipped in interstate commerce. An IND application is the means through which the sponsor technically obtains this exemption from the FDA, whereas an NDA is the vehicle through which drug sponsors formally propose that the FDA approve a new pharmaceutical for sale in the United States. To obtain this authorization, a drug manufacturer submits:

1. Nonclinical (animal) and clinical (human) test data and analyses
2. Drug information
3. A description of manufacturing processes and controls
4. Information as to labeling proposed

An NDA must provide sufficient information, data, and analyses to allow the FDA reviewers to reach several key decisions, including:

- Whether the drug is safe and effective for its proposed use(s) and whether the benefits of the drug outweigh its risks
- Whether the drug's proposed labeling is appropriate, and if not, what the drug's labeling should contain
- Whether the methods used in manufacturing the drug and the controls used to maintain the drug's quality are adequate to preserve the drug's identity, strength, quality, and purity

Clinical trials are sponsored or funded by a variety of organizations or individuals, such as physicians, medical institutions, foundations, voluntary groups, and pharmaceutical companies in addition to federal agencies such as the National Institutes of Health, the Department of Defense, and the Department of Veterans' Affairs. Trials can take place at a variety of international locations, such as hospitals, universities, doctors' offices, and community clinics. There are various types of clinical trials, and most of them are conducted in four phases.

23.3.1 Types of Clinical Trials (NIH, 2007)

The clinical trials may be classified into the following five types:

1. *Treatment trials:* involve experimental treatments, new combinations of drugs, or new approaches to surgery or radiation therapy.
2. *Prevention trials:* look for better ways to prevent disease in people who have never had the disease or to prevent a disease from returning. These approaches may include medicines, vaccines, vitamins, minerals, or lifestyle changes.
3. *Diagnostic trials:* are conducted to find better tests or procedures for diagnosing a particular disease or condition.
4. *Screening trials:* find the best way to detect certain diseases or health conditions.
5. *Quality-of-life trials* (or *supportive care trials*): explore ways to improve comfort and quality of life for persons with a chronic illness.

23.3.2 Phases of Clinical Trials (NIH, 2007)

Clinical trials are conducted (typically as single- or double-blinded randomized trials) in phases. The trials at each phase have a different purpose and help answer different questions.

- In *phase 1 trials*, researchers test an experimental drug or treatment in a small group of people (20 to 80) for the first time to evaluate its safety, determine a safe dosage range, and identify side effects.
- In *phase 2 trials*, the experimental study drug or treatment is given to a larger group of people (100 to 300) to see if it is effective and to evaluate its safety further.
- In *phase 3 trials*, the experimental study drug or treatment is given to large groups of people (1000 to 3000) to confirm its effectiveness, monitor side effects, compare it to commonly used treatments, and collect information that will allow the experimental drug or treatment to be used safely.
- In *phase 4 trials*, postmarketing studies delineate additional information, including the drug's risks, benefits, and optimal use.

23.3.2.1 *Phase 1 Clinical Studies* (FDA, 1998a)
Phase 1 includes the introduction (if an exploratory IND, described in Section 23.4.2, was not conducted) of an investigational new drug into humans. These studies are closely monitored and may be conducted in patients but are usually conducted in healthy volunteer subjects. The total number of subjects included in phase 1 studies varies with the drug but is generally in the range of 20 to 80. During phase 1, sufficient information about the drug's pharmacokinetics and pharmacological effects should be obtained to permit the design of well-controlled scientifically valid phase 2 studies.

Phase 1 studies are, therefore, designed to determine the following in humans:

- Drug metabolism
- Pharmacological actions
- Mechanism of action
- Chemical structure–activity relationships
- Side effects associated with increasing doses
- Early evidence on effectiveness, where possible
- If an investigational drug can be used as a research tool to explore biological phenomena or a disease process

In phase 1 studies, the Center for Disease Evaluation and Research (CDER) can impose a clinical hold (i.e., prohibit the study from proceeding or stop a trial that has started) for reasons of safety or because the sponsor failed to accurately disclose the risk of study to investigators. Although CDER routinely provides advice in such cases, investigators may choose to ignore any advice regarding the design of phase 1 studies in areas other than patient safety.

23.3.2.2   Phase 2 clinical Studies (FDA, 1998a)   Phase 2 includes the early controlled clinical studies conducted to obtain some preliminary data on the effectiveness of a drug for a particular indication or indications in patients with a disease or condition. This phase of testing also helps determine the common short-term side effects and risks associated with the drug. Phase 2 studies are typically well controlled, closely monitored, and conducted in a relatively small number of patients, usually involving several hundred people. This phase may be subdivided into phase 2a (pilot clinical trials) and phase 2b (most rigorous pivotal trials to demonstrate a drug's efficacy).

23.3.2.3   Phase 3 Clinical Studies (FDA, 1998a)   Phase 3 studies are expanded controlled and uncontrolled trials. These studies can be large and costly. They are performed after preliminary evidence suggesting that the effectiveness of the drug has been obtained in phase 2, and are intended to gather the additional information about effectiveness and safety that is needed to evaluate the overall benefit–risk relationship of the drug. Phase 3 studies also provide an adequate basis for extrapolating the results to the general population and transmitting that information in the physician labeling. Phase 3 studies are very costly and involve several hundred to several thousand people. The phase 3 clinical trials are comprised of phases 3a and 3b clinical trials. In phase 3a, trials are conducted after the efficacy of the drug has been demonstrated (after phase 2b), but prior to regulatory submission of a new drug application. Phase 3b clinical trials are conducted after NDA submission but prior to drug approval and launch. In both, phases 2 and 3, the CDER can impose a clinical hold if a study is unsafe (as in phase 1) or if the protocol is clearly deficient in design in meeting its stated objectives. The CDER takes great care to ensure that the determination is not made in isolation, but reflects current scientific knowledge, agency experience with the design of clinical trials, and experience with the class of drugs under investigation.

23.3.2.4   Phase 4 Studies   Phase 4 trials, often called *postmarketing studies*, are conducted after a drug has been approved for sale. These long-term studies may be required by the FDA as a condition of approving the drug, or the sponsor company may undertake these studies to assess the drug's risks, benefits, optimal use, and for other competitive reasons. Depending on the findings of the study, a drug may be withdrawn from the market or restricted to a certain use. For further details, refer to Chapter 26. If a marketed drug is evaluated for a new indication, those clinical trials are considered phase 2 clinical trials.

23.3.3   Sponsor–FDA milestone meetings (FDA, 1998a)

23.3.3.1   Pre-IND Meeting   Prior to clinical studies, the sponsor needs evidence that the compound is biologically active and that the drug is reasonably safe for initial administration to humans. Preclinical meetings, typically requested by sponsors, are conducted with the appropriate division that would review the drug marketing application. Meetings at this early stage in the process are useful opportunities for open discussion about testing phases, data requirements, and any scientific issues that may need to be resolved prior to an IND submission. At these meetings, the sponsor and the FDA discuss and agree upon the design of the animal studies needed to provide evidence that the compound is biologically active and the drug is reasonably safe to initiate human testing (refer to 21 CFR 312.47 and 312.82 for more details).

23.3.3.2 End of Phase 2 meeting The primary focus of "end of phase 2" meetings is to determine whether it is safe to begin phase 3 testing. It is also intended to establish an agreement between the agency and the sponsor of the overall plan (objectives and design of studies) for phase 3 and any additional information that may be required to support the submission of a new drug application. By clarifying data requirements, these meetings avoid unnecessary expenditures of time and money.

One month prior to the meeting, the sponsor should submit the background information and protocols for phase 3 studies. This information should include data supporting the claim of the new drug product, chemistry data, animal data and proposed additional animal data, results of phases 1 and 2 studies, statistical methods being used, specific protocols for phase 3 studies, as well as a copy of the proposed labeling for a drug, if available. This summary provides the review team with information needed to prepare for a productive meeting.

23.3.3.3 Pre-NDA Meeting The purpose of a pre-NDA meeting is to discuss the presentation of data (both paper and electronic) in support of the application. The information provided at the meeting by the sponsor includes:

- A summary of clinical studies to be submitted in the NDA
- The format proposed for organizing the submission, including methods for presenting the data
- Other information that needs to be discussed

The meeting is conducted to uncover any major unresolved problems or issues, to identify studies the sponsor is relying on as adequate and well controlled in establishing the effectiveness of the drug, to help the reviewers to become acquainted with the general information to be submitted, and to discuss the presentation of the data in the NDA to facilitate its review. Once the NDA is filed, a meeting may also occur 90 days after the initial submission of the application to discuss issues that are uncovered in the initial review.

23.4 INVESTIGATIONAL NEW DRUGS

The FDA's role in the development of a new drug begins when the drug's sponsor (usually, the manufacturer or potential marketer) has screened the new molecule for pharmacological activity and acute toxicity potential in animals, and wants to test its diagnostic or therapeutic potential in humans. At that point, the molecule changes in legal status (under the FD&C Act) and becomes a new drug subject to specific requirements of the drug regulatory system. The IND is not an application for marketing approval; rather, it is a request for an exemption from the federal statute that prohibits an unapproved drug from being shipped in interstate commerce. A sponsor will probably ship the investigational drug to clinical investigators in many states; hence, it must seek an exemption from that legal requirement to ship the drug across state lines. The IND is the means through which the sponsor technically obtains this exemption from the FDA to ship in interstate commerce.

23.4.1   Types of INDs (FDA, 1998a, 2011a)

INDs may be classified in two broad categories: commercial or noncommercial. Commercial INDs (e.g., exploratory INDs) are submitted primarily by companies whose ultimate goal is to obtain marketing approval for a new product. The vast majority of INDs, including Investigator, emergency use, and treatment INDs, are non commercial INDs filed for noncommercial research.

An *investigator IND* is submitted by a physician who both initiates and conducts an investigation; under whose immediate direction the investigational drug is administered or dispensed. A physician might submit a research IND to propose studying an unapproved drug, or an approved product for a new indication or in a new patient population.

An *emergency use IND* allows the FDA to authorize use of an experimental drug in an emergency situation that does not allow time for submission of an IND in accordance with 21 CFR 312.23 or 312.34. It is also used for patients who do not meet the criteria of an existing study protocol, or if an approved study protocol does not exist.

A *treatment IND* is submitted for experimental drugs showing promise in clinical testing for serious or immediately life-threatening conditions while the final clinical work is conducted and the FDA review is taking place.

23.4.2   Exploratory (or Phase 0) IND Studies (FDA, 2006)

"A new medical compound entering phase 1 testing, often represents the culmination of upwards of a decade of preclinical screening and evaluation and has been estimated to have only an 8 percent chance of reaching the market," according to the critical path report in March 2004 (FDA, 2004). In this report the FDA explains that new tools need to be employed earlier in the process to distinguish those drug candidates that hold promise from those that do not. These new tools include exploratory IND (early phase 1 exploratory or phase 0) approaches that are consistent with regulatory requirements while maintaining needed human subject protection. Such studies are conducted prior to traditional dose escalation, safety, and tolerance studies that ordinarily initiate a full-blown clinical drug development program. The duration of dosing in an exploratory IND study is expected to be limited in time and scope (e.g., 7 days). Existing regulations provide more flexibility in regard to the preclinical testing requirements for an exploratory IND compared with traditional phase 1 studies. Exploratory studies usually involve fewer resources and limited human exposure than are involved in customary phase 1 studies.

There is only an evaluation of safety assessment and no therapeutic or diagnostic intent. An early phase 1 approach should help sponsors to move ahead efficiently with the development of promising candidates and to determine the following:

- Determine if the proposed mechanism of action would be observed in humans (e.g., a binding property or inhibition of an enzyme).
- Provide information on the pharmacokinetics (PK) and pharmacodynamics (PD).
- Select the most promising lead product (drug product and/or drug substance) from the candidates based on the pharmacokinetic and pharmacodynamic properties.
- Find the products' biodistribution using imaging techniques.
- Exploratory IND studies can help reduce the number of human subjects early in the study by identifying the promising candidates early in the study.
- Potential risk of toxicity is less in exploratory IND studies than in a traditional phase 1 study, as only perceived subpharmacological doses or doses expected to produce pharmacologic effect are administered in these studies.
- Exploratory IND studies present fewer potential risks because they are designed differently. Exploratory IND investigations in humans can be initiated with less, or different, preclinical support than is required for traditional IND studies. Traditional phase 1 studies look for dose-limiting toxicities. Generally, exploratory studies would not be carried out in pediatric patients or in pregnant or lactating women.
- It is important for sponsors to consider this exploratory IND approach to develop drugs that treat serious or life-threatening illnesses (refer to Chapter 25 for details on the accelerated new drug approval process).

23.4.3 Applicant (Drug Sponsor) (FDA, 1998a)

An applicant, or drug sponsor, is a person or entity who assumes responsibility for the investigation of a new drug, including responsibility for compliance with applicable provisions of the Federal Food, Drug, and Cosmetic Act and related regulations. The sponsor is usually an individual, partnership, corporation, government agency, manufacturer, or scientific institution.

23.4.4 Investigational New Drug Application
(FDA, 1998a, 2011a)

The IND application is the result of a successful preclinical development program and the vehicle through which a sponsor advances to the next stage of drug development, known as clinical trials (human trials). During the preclinical phase, if a product is identified as a feasible candidate for further development, the sponsor then focuses on collecting the data and information necessary to establish that the product will not expose humans to unreasonable risks when used in limited, early-stage clinical studies. The sponsor must wait 30 days from the IND submission date before initiating any clinical trials; the FDA needs this period to review the application for any safety concerns to make sure that the research subjects are not subjected to unreasonable risks. As detailed in the 21 CFR 312.23, IND Content and Format, an IND includes the data and information in three broad areas, including previous experience on drug use in humans (possibly available from foreign countries):

- Clinical protocols and investigator information: detailed protocols for proposed clinical studies to assess whether the initial-phase trials will expose subjects to unnecessary risks. Also, information on the qualifications of the clinical investigators or professionals (generally, physicians) who oversee the administration of the experimental compound and to assess whether they are qualified to fulfill their clinical trial duties. Finally, the sponsor must commit to obtaining an IRB review of the study protocol, informed consent from the research subjects, and to adhere to the investigational new drug regulations.

- Chemistry, manufacturing, and controls (CMC) information: information pertaining to the composition, manufacture, stability, and controls used for manufacturing the drug substance and the drug product. This information is assessed to ensure that the company can produce and supply consistent batches of the drug.

- Animal pharmacology and toxicology studies: preclinical data to permit an assessment as to whether the product is reasonably safe for initial testing in humans.

23.4.4.1 *Clinical Protocol and Investigator Information* (FDA, 2006, 2011b)
Information on clinical development plan including rational and types of clinical studies proposed; the types of studies could be (with limited duration of dosing):

- Single- and multiple-dose studies
  - In single-dose studies, a subpharmacologic (of radiolabeled candidate) or pharmacologic dose is administered to a limited number of subjects (healthy volunteers or patients) to collect PK information and/or performing imaging studies, or both.
  - Repeated-dose studies can be designed with the pharmacologic or pharmacodynamic endpoints.
- In these studies the duration of dosing is limited (7 days).
- For escalating studies, doses should be designed to investigate a PD endpoint but not to determine the limits of tolerability.
- Investigator information: If an investigator's brochure is required under 21 CRF 312.55, it should contain summaries of the following information [312.23(a)(5)]:
  - Description of the drug substance and the formulation, including the structural formula
  - Pharmacological and toxicological effects of the drugs in animals

- ○ Pharmacokinetics and biological disposition of the drug in animals
- ○ Safety and effectiveness in humans obtained from prior clinical studies
- ○ Anticipated risks and side effects based on prior experience with the drug under investigation or with related drugs
- ○ Precautions and special monitoring needed as part of the investigational use of the drug

23.4.4.2 *Chemistry, Manufacturing, and Controls Information* (FDA, 2006)

The regulations, 21 CFR 312.23(a)(7)(i), emphasize that the CMC information is updated as the IND application progresses. At each phase of clinical investigational program, sufficient information should be submitted to ensure the proper identification, strength, quality, purity, and potency of the investigational compound. For the purpose of an exploratory IND application, the CMC information indicated below can be provided in a summary report to enable the FDA to make necessary safety assessments.

23.4.4.2.1 General Information for the Candidate Product The extent and type of CMC information submitted in an exploratory IND application is similar to that described in the current guidance for use of investigational products [Content and Format of Investigational New Drug Applications Phase 1 Studies of Drugs, Including Well-Characterized, Therapeutic, Biotechnology-Derived Products; http://www.fda.gov/downloads/Drugs/GuidanceComplianceRegulatoryInformation/Guidances/UCM071597.pdf]. Information on each candidate product [e.g., the active pharmaceutical ingredient (API)] can be submitted in a summary report containing the following items:

- A discussion when a sponsor believes that the chemistry or manufacturing of the candidate product may present any potential human risk (21 CFR 312.23), along with steps proposed to monitor for such risk
- Information on the candidate product (on the API), including physical, chemical, and/or biological characteristics as well as its source or synthetic route and therapeutic class
- Description of dosage form
- Description of formulation and route of administration
- Grade and quality of excipients used in the product; all excipients should be one of the following:
  - ○ Generally recognized as safe (GRAS) [the GRAS list may be accessed: http://www.accessdata.fda.gov/scripts/fcn/fcnNavigation.cfm?rpt=gras Listing]
  - ○ Part of a formulation that is approved or licensed in the United States for the same route of administration and amount
  - ○ Adequately qualified through appropriate animal studies [refer to the FDA guidance document Nonclinical Studies for the Safety Evaluation

of   Pharmaceutical   Excipients;   http://www.fda.gov/downloads/Drugs/
GuidanceComplianceRegulatoryInformation/Guidances/ucm079250.pdf]

- Name and address of manufacturers
- Method of preparation of candidate product
- Quantitative composition of product
- Information demonstrating product stability and methods for analyzing stability during clinical studies
- For ophthalmic and parenteral preparations, results from sterility and pyrogenicity studies

23.4.4.2.2 Analytical Characterization of Candidate Product

- There are two scenarios for this:
  - The first scenario involves using the same batch of the candidate product in both the toxicology studies and clinical trials.
  - The second scenario involves where the batch used in toxicological studies is different from that used in clinical trial studies; in such cases the sponsor should demonstrate by analytical testing that the batch to be used is representative of batches used in the nonclinical toxicology studies
    - Tests to accomplish the second scenario include identity, structure, assay for purity, impurity profile, potency assay, physical characteristics, and microbiological characteristics.
- Impurities (e.g., chemical and microbiological) should be characterized in accordance with recommendations in the agency guidance if, and when, the sponsor files a traditional IND for further clinical investigation [FDA guidance document INDs for Phase 2 and Phase 3 Studies: Chemistry, Manufacturing, and Controls Information—http://www.fda.gov/downloads/Drugs/GuidanceComplianceRegulatoryInformation/Guidances/UCM070567.pdf].

23.4.4.3 Animal Pharmacology and Toxicology Studies (FDA, 2006)

The toxicology evaluation recommended for an exploratory IND application is more limited than a traditional IND application [Internationals Conference on Harmonization (ICH) guidance for industry: M3 Nonclinical Safety Studies for the Conduct of Human Clinical Trials and Marketing Authorization for Pharmaceuticals describes what is expected for a traditional IND http://www.ich.org/fileadmin/Public_Web_Site/ICH_Products/Guidelines/Multidisciplinary/M3_R2/Step4/M3_R2__Guideline.pdf]. The level of preclinical testing performed to ensure safety will depend on the scope and intended goal of the clinical trials. Examples of such tailored preclinical safety study designs include (1) confirming that an expected mechanism of action (MOA) can be observed in humans; (2) measuring binding affinity/or of localization of drug; (3) assessing PK and metabolism; and (4) comparing the effect on a potential therapeutic target with other therapies. Three of these examples are described below.

23.4.4.3.1   Clinical Studies of Pharmacokinetics or Imaging with Radio-labeled Drugs

- Microdose studies (less than 1/1000 of the dose that yields a pharmacological effect) are carried out to see the pharmacokinetic properties but not to include the pharmacological effects.

- Extended single-dose toxicity studies in animals that support single-dose studies in humans are accepted by the FDA.

- For microdose studies, a single mammalian species (both genders) can be used if justified by in vitro metabolism data and by comparative data on in vitro pharmacodynamic effects.

- The route of exposure in animals should be the same as that intended for the clinical route; animals should be observed for 14 days postdosing with an interim necropsy, and endpoints evaluated should include body weights, clinical signs, clinical chemistries, hematology, and histopathology.

- To establish a margin of safety, the sponsor should demonstrate that a large multiple (e.g., 100-fold) of the proposed human dose would not induce adverse effects in experimental animals.

- Scaling from animals to humans based on body surface area (or body weight) can be used to select the dose for use in the clinical trial.

- Scaling based on pharmacokinetic and pharmacodynamic modeling would also be appropriate if such data are available.

23.4.4.3.2   Clinical Trials to Study Pharmacologically Relevant Doses

- Clinical trials are designed to study the pharmacological effect of candidate products; more extensive safety data, however, would be needed to support the safety of such studies.

- Rats are a usual species chosen for this comparative confirmatory purpose; additional studies in nonrodents, most often dogs, can be used to confirm that the rodent is an appropriately sensitive species.

- Repeat dose clinical trials lasting up to 7 days can be supported by 2-week repeat dose toxicology study in a sensitive species; the goal of such a study is to select the safe and maximum doses for clinical trials.

- The number of animals used in the confirmatory study could be fewer than normally used to attain statistically significant comparisons but of sufficient number to rule out any toxicologically significant difference in sensitivity compared with the rodent. The confirmatory study could be a dedicated study involving repeat administration of a single-dose level approximating the rat NOAEL (non-observed-adverse-effect level) calculated on the basis of body surface area.

- The route of administration should be same as that selected for clinical trial, and toxicokinetic measurements should be used to assess exposure.

- If confirmatory tests suggest that the rodent is not the more sensitive species, a 2-week repeated-dose toxicity study must be performed in a second species to select the doses for human trials.

- This study should include measurements of body weight, clinical signs, clinical chemistries, hematology, and histopathology.

- Evaluation of the central nervous system and respiratory systems can be performed by rodent studies, and evaluation of cardiovascular system can be performed by nonrodent studies.

- The results from preclinical trials can be used to select the starting and maximum doses for clinical trials.

- The starting dose is anticipated to be no greater than 1/50 of the NOAEL from the 2-week toxicology study in the sensitive species on an mg/m^2 basis. The maximum clinical dose would be the lowest of the following :

  ○ One-fourth of the 2-week rodent NOAEL on a mg/m^2 basis

  ○ Up to one-half of the area under the curve at the NOAEL in the 2-week rodent study, or the area under the curve in the dog at the rat NOAEL, whichever is lower

  ○ The dose that produces a pharmacological and/or pharmacodynamic response or at which target modulation is observed in the clinical trial

  ○ Observation of an adverse clinical response

23.4.4.3.3 Clinical Studies of MOAs Related to Efficacy (Phase 2a, Also Called Proof of Concept)   The clinical trials are conducted to evaluate MOAs. To support this approach, the FDA will accept alternative, or modified, pharmacological and toxicological studies to select clinical starting dose and dose escalation schemes. The animal studies would incorporate endpoints that are based mechanically on the pharmacology of the new chemical entity and thought to be important to clinical effectiveness. The degree of saturation of a receptor or the inhibition of an enzyme would be characterized and determined in the animal study and used as endpoint in a subsequent clinical investigation. The clinical investigation would be designed to assess this binding mechanism as it relates to effectiveness.

23.4.4.4   *Good Laboratory Practice Compliance*   All preclinical safety studies supporting the safety of an exploratory IND application are expected to comply with good laboratory practices (21 CFR 58). The GLP provisions apply to a broad variety of studies, test articles, and test systems. Sponsors should discuss with the FDA (preferably in a pre-IND meeting) any need or exemption from these provisions prior to conducting safety-related studies. Sponsors must justify any nonconformance with GLP provisions [21 CFR 312.23(a)(8)(iii)].

The full application (refer to the 21 CFR 314.50 and 314.54) should be filed with:

Center for Drug Evaluation and Research

Food and Drug Administration

Document and Records Section

5901-B Ammendale Road

Beltsville, MD 20705-1266

Correspondence not associated with a particular application should be addressed specifically to the intended office or division and to the person as follows:

Center for Drug Evaluation and Research

Food and Drug Administration

Attn: [insert name of person]

HFD-[insert mail code of office or division]

5600 Fishers Lane

Rockville, MD 20857

23.4.5 IND Application Review Process (FDA, 1998a)

23.4.5.1 Medical Review A medical reviewer evaluates the clinical trial protocol to determine (1) if the participants will be protected from unnecessary risks; and (2) if the study design will provide data relevant to the safety and effectiveness of the drug. Under federal regulations, proposed phase 1 studies are evaluated almost exclusively for safety reasons. In evaluating the safety and efficacy of phase 2 and 3 investigations, FDA reviewers must also ensure that these studies are of sufficient scientific quality to yield data that can support marketing approval.

23.4.5.2 Chemistry Review The chemistry reviewers are responsible for reviewing the chemistry and manufacturing control sections of drug applications. The reviewing chemist evaluates the manufacturing and processing procedures for a drug to ensure that the compound is adequately reproducible and stable. If the drug is unstable and/or not reproducible, it could undermine the validity of any clinical testing, and more important, the studies may pose significant risks to participants. The sponsor should state these human risks along with the steps to monitor such risk at the beginning of the "chemistry and manufacturing" section of the submission. Additionally, sponsors should describe any chemistry and manufacturing differences between the drug product proposed for clinical use and the drug product used in the animal toxicology trials, and how these differences might affect the safety profile of the drug product. If there are no differences in the products, that should be stated.

23.4.5.3   *Pharmacology/Toxicology Review* (FDA, 2011b)   The pharmacology/toxicology review team evaluates the results of animal testing and attempts to relate animal drug effects to potential effects in humans.

23.4.5.3.1   Pharmacology and Drug Distribution   This section of the application should contain, if known, (1) a description of the pharmacological effects and mechanism(s) of action of the drug in animals, and (2) information on the absorption, distribution, metabolism, and excretion of the drug. A summary report without individual animal records or individual study results would suffice for submission. To the extent that such studies may be important to address safety issues, or to assist in the evaluation of toxicology data, their inclusion may be necessary; however, lack of this potential effectiveness should generally not be a reason for a phase 1 IND to be placed on clinical hold.

23.4.5.3.2   Toxicology Data   Present regulations [21 CFR 312.23(a)(8) (ii)(a)] require an integrated summary of the toxicological effects of the drug in animals and in vitro. The particular studies need to depend on the nature of the drug and the phase of human investigation. When species specificity, immunogenicity, or other considerations appear to make many or all toxicological models irrelevant, sponsors are encouraged to contact the agency to discuss toxicological testing.

23.4.5.4   *Safety Reviews*   Generally, drug review divisions do not contact the sponsor if there are no concerns with drug safety and the clinical trials proposed. If the sponsor hears nothing from the CDER after 30 days, the study may proceed as submitted on day 31 after submission of the IND.

23.4.6   Clinical Hold Decision (FDA, 1998b)

The CDER will contact the sponsor within the 30-day initial review period to stop the clinical trial if it does not believe, or cannot confirm, that the study can be conducted without unreasonable risk to the subjects or patients. Prior to their issuance, all clinical holds are reviewed by upper management of the CDER to assure consistency and scientific quality in the center's clinical hold decisions. The CDER may either delay the start of an early-phase trial on the basis of information submitted in the IND, or stop an ongoing study based on a review of newly submitted clinical protocols, safety reports, protocol amendments, or other information. When a clinical hold is issued, a sponsor must address the issue that is the basis of the hold before the order is removed.

23.4.7   Notification of Sponsor (FDA, 1998a)

Once a clinical hold is placed by the CDER, the sponsor is notified immediately via telephone by the division director, and a letter is sent within 5 working days following the telephone call describing the reasons for the clinical hold. The letter must bear the signature of the division director (or an acting division director).

If a response to the clinical hold is received from the sponsor, the division reviews the sponsor's response and decides within 30 days whether the hold should be lifted. Otherwise, the division director will telephone the sponsor and discuss what is being done to facilitate completion of the review.

If the reviewer decision is not to lift the clinical hold, the office director must decide within 14 calendar days whether or not to sustain the division's decision to maintain the clinical hold. If the decision is made to lift the hold, the division telephones the sponsor and informs them of the decision, and sends a letter confirming that the hold has been lifted. The letter will be sent within 5 working days of the telephone call. However, the trial may begin once the decision has been relayed to the sponsor by telephone.

23.4.8 Sponsor Notified of Deficiencies (FDA, 1998a)

If other deficiencies (but not serious enough to justify delaying clinical studies) are found in an IND, the division may either telephone or forward a deficiency letter to the sponsor. In either case, the sponsor may proceed with the planned clinical trials, but that additional information is necessary to complete or correct the IND file, or the deficiencies need to be addressed prior to a marketing application (NDA) submission.

23.4.9 Study Ongoing (FDA, 1998a)

Once CDER's 30-day initial review period expires and no clinical hold is issued, clinical studies can begin immediately upon submission of the clinical protocol to the IND. If the sponsor was notified of deficiencies that were not serious enough to warrant a clinical hold, the sponsor addresses these deficiencies while the study proceeds.

23.5 NDA REVIEW PROCESS (FDA, 1998a)

Since 1938, the regulations and controls to commercialize new drugs in the United States have been based on the NDA. The NDA process has evolved considerably during its history from requiring information pertaining to investigational drug's safety (the FD&C Act was passed in 1938) to contain evidence that a new drug is effective for its intended use and that the benefits of the new drug outweighed its known risks (as required by the Kefauver–Harris Amendment to the FD&C Act, passed in 1962). The last revision in 1985 (the NDA rewrite) restructured the ways in which information and data are organized and presented in the NDA to expedite (FDA, 2010) reviewers to reach the following decisions:

- Whether the drug is safe and effective in its proposed use(s), and whether the benefits of the drug outweigh the risks
- Whether the drugs proposed labeling (package insert) is appropriate, and what it should contain

- Whether the methods used in manufacturing the drug and the controls used to maintain the drug's quality are adequate to preserve the drug's identity, strength, quality, and purity

23.5.1 NDA Classifications (FDA, 1998a)

The CDER classifies new drug applications with a code that reflects both the type of drug being submitted and its intended uses. The numbers 1 through 7 are used to describe the type of drug:

1. New Molecular Entity
2. New Salt of Previously Approved Drug (not a new molecular entity)
3. New Formulation of Previously Approved Drug (not a new salt *or* a new molecular entity)
4. New Combination of Two or More Drugs
5. Already Marketed Drug Product—Duplication (i.e., new manufacturer)
6. New Indication (claim) for Already Marketed Drug (includes a switch in marketing status from prescription to over the counter)
7. Already Marketed Drug Product—No Previously Approved NDA

23.5.2 Documentation Requirements (FDA, 1998a)

The documentation required in an NDA is supposed to tell the drug's entire story, including what happened during the clinical tests, the ingredients of the drug, the results of the animal studies, how the drug behaves in the body, and how it is manufactured, processed, and packaged. The components of any NDA are, in part, a function of the nature of the subject drug and the information available to the applicant at the time of submission. Hence, the quantity of information and data submitted in NDAs can vary significantly, as outlined in Form FDA-356h, Application to Market a New Drug for Human Use or as an Antibiotic Drug for Human Use NDAs can consist of as many as 15 different sections and should be submitted at the address listed under the IND section:

- Index
- Summary
- Chemistry, Manufacturing, and Control
- Samples, Methods Validation Package, and Labeling
- Nonclinical Pharmacology and Toxicology
- Human Pharmacokinetics and Bioavailability
- Microbiology (for antimicrobial drugs only)
- Clinical Data
- Safety Update Report (typically submitted 120 days after the NDA's submission)

- Statistical Results
- Case Report Tabulations
- Case Report Forms
- Patent Information
- Patent Certification

23.5.3 NDA Review Process (FDA, 1998a)

The following letter codes describe the review priority of the drug:

S: Standard review for drugs similar to currently available drugs

P: Priority review for drugs that represent significant advances over existing treatments

23.5.3.1 Is the Application Filable? After an NDA is received by the agency, it is screened to ensure that sufficient data and information have been submitted in each area to justify filing the application for initiating CDER's formal review of the NDA.

23.5.3.2 Refuse-to-File Letter Issued If an NDA is incomplete, the applicant receives a letter (refuse-to-file) detailing the decision and the deficiencies that form its basis. This decision must be forwarded within 60 calendar days after the NDA is received by the CDER.

23.5.3.3 Medical Review Medical reviewers are responsible for evaluating the clinical sections of submissions and to formulate the overall basis for a recommended agency action on the application. These sections should provide information on the safety of the clinical protocols in an IND or the results of this testing as submitted in the NDA and for synthesizing the results of the animal toxicology, human pharmacology, and clinical reviews.

23.5.3.4 Biopharmaceutical Review Pharmacokineticists evaluate the rate and extent to which a drug's active ingredient is made available to the body and the way it is distributed in, metabolized by, and eliminated from the human body.

23.5.3.5 Statistical Review Statisticians evaluate the statistical relevance of the data and various methods used to analyze the data in the NDA. Through this review, the medical officers should get a better idea of the statistical significance and clinical significance of the findings to be extrapolated to the larger patient population in the country.

23.5.3.6 Microbiology Review Clinical microbiology information is required only in NDAs for anti-infective drugs. Since these drugs affect microbial rather than human physiology, reports on the drug's in vivo and in vitro effects on the target microorganisms are critical for establishing product effectiveness. An NDA's microbiology section usually includes data describing:

- The biochemical basis of the drug's action on microbial physiology
- The drug's antimicrobial spectra, including results of in vitro preclinical studies demonstrating concentrations of the drug required for effective use
- Any known mechanisms of resistance to the drug, including results of any known epidemiologic studies demonstrating prevalence of resistance factors
- Clinical microbiology laboratory methods needed to evaluate the effective use of the drug

More specific guidance on developing the microbiology component of the NDA is available from the FDA's Guideline for the Format and Content of the Microbiology Section of an Application (FDA, 1987) (http://www. fda.gov/downloads/Drugs/GuidanceComplianceRegulatoryInformation/Guidances/ UCM075101.pdf).

23.5.4   Field Office Preapproval Inspection Review

An FDA field office conducts, at its discretion, preapproval inspections of the sponsor's facilities in which the product will be manufactured to assure that the site is capable of producing a product meeting the current good manufacturing practice (CGMP) requirements and the commitments made in the sponsor's documentation filed with the FDA.

23.5.5   Is Preapproval Inspection Acceptable? (FDA, 1998a)

The CDER review division initiates a request for a preapproval inspection of the sponsor's manufacturing facilities and clinical trial sites. During such inspections, the agency:

- Verifies the accuracy and completeness of the manufacturing-related information submitted in the NDA
- Evaluates the manufacturing controls for the preapproval batches upon which information provided in the NDA is based
- Evaluates the manufacturer's compliance with the CGMP requirements and manufacturing-related commitments made in the NDA
- Collects a variety of drug samples for analysis by the FDA field and CDER laboratories

According to the CDER's policy, product-specific preapproval inspections generally are conducted for products (1) that are new chemical or molecular entities, (2) that have narrow therapeutic ranges, (3) that represent the first approval for the applicant, or (4) that are sponsored by a company with a history of CGMP problems or that has not been the subject of a CGMP inspection over a considerable period. More specific guidance on CDER's preapproval inspection program is available from CDER's Compliance Program Guide 7346.832 (http://www.fda. gov/downloads/Drugs/DevelopmentApprovalProcess/Manufacturing/Questionsand AnswersonCurrentGoodManufacturingPracticesCGMPforDrugs/UCM071871.pdf).

The results of the preapproval inspection may also affect the final approval decision. When such inspections discover significant CGMP problems or other issues, the reviewing division may withhold approval until these issues are addressed and corrected.

23.5.6   Advisory Committees (FDA, 1998a)

As is the case with INDs, the CDER may seek advice and opinions from outside national experts (advisory committee) on various issues described in the NDA; their recommendations are not binding, but the agency considers them carefully when deciding drug issues.

The CDER may request a committee to provide opinions on:

- A new drug
  - Whether to conduct proposed studies for an experimental drug
  - Whether the safety and effectiveness information submitted for a new drug is adequate for marketing approval
- A major indication for an already approved drug
- A special regulatory requirement being considered, such as a boxed warning in a drug's labeling
- Information
  - Labeling desired
  - Help with guidelines for developing particular types of drugs

23.5.7   Are Reviews Complete and Acceptable? (FDA, 1998a)

Upon completion of technical reviews, each reviewer develops a written evaluation of the NDA that presents their conclusions and their recommendations on the application. The division director or office director then evaluates the reviews and recommendations and decides the action that the division will take on the application. The result is an action letter that provides an approval, approvable or nonapprovable decision, and a justification for that recommendation.

23.5.8   Meetings with Sponsor (FDA, 1998a)

The CDER usually communicates often with sponsors about scientific, medical, and procedural issues (e.g., easily correctable deficiencies, need more data/information, and/or need for technical changes) that arise during the review process. Major scientific issues, however, are usually addressed in an action letter at the end of the initial review process.

23.5.8.1   End-of-Review Conference   At the conclusion of the CDER's review of an application, one of three possible action letters can be sent to the sponsor:

1. Not approvable letter: lists the deficiencies in the application and explains why the application cannot be approved.

2. Approvable letter: signals that, ultimately, the drug can be approved; lists minor deficiencies that can be corrected, often involves labeling changes, and possibly requests commitment to do postapproval studies.

3. Approval letter: states that the drug is approved; this may follow an approvable letter but can also be issued directly.

If the action taken is either an approvable or a not approvable (as opposed to an approval action), the CDER provides applicants with an opportunity to meet with agency officials and discuss the deficiencies and further steps necessary before the application can be approved. This meeting is available on all applications; however, the priority is given to applications for "priority review" drugs and major new indications for marketed drugs. Requests for such meetings are directed to the director of the division responsible for reviewing the application.

23.5.8.2   Other Meetings   Other meetings between the CDER and applicants may be held to discuss scientific, medical, and other important issues that arise during the review process. Refer to the FDA guidance for more information on meetings between the CDER and applicants: Guidance for Industry—Formal Meetings with Sponsors and Applicants for PDUFA (Prescription Drug User Fee Act) Products (http://www.fda.gov/downloads/Drugs/GuidanceCompliance RegulatoryInformation/Guidances/UCM079744.pdf).

23.5.9   Additional Information (Amendment) (FDA, 1998a)

In some cases an applicant may seek to augment the information provided in the original NDA during the review process; such information provided for an unapproved application is considered an NDA amendment and may result in an extension of the FDA's time line for application review.

23.5.10   Is Labeling Review Acceptable? (FDA, 1998a)

Each statement proposed for drug labeling must be justified by data and results submitted in the NDA. The *Code of Federal Regulations* (CFR) describes labeling requirements in 21 CFR 201, Labeling. The labeling is organized in the following sections:

- Description: proprietary and established name of drug; dosage form; ingredients; chemical name; and structural formula
- Clinical pharmacology: summary of the actions of the drug in humans; in vitro and in vivo actions in animals if pertinent to human therapeutics; pharmacokinetics
- Indications and usage: description of use of drug in the treatment, prevention, or diagnosis of recognized disease or condition

- Contraindications: description of situations in which the drug should not be used because the risk of use clearly outweighs any possible benefit
- Warnings: description of serious adverse reactions and potential safety hazards, subsequent limitations in use, and steps that should be taken if they occur
- Precautions: information regarding any special care to be exercised for safe and effective use of the drug. This section includes general precautions and information for patients on drug interactions, carcinogenesis/mutagenesis, pregnancy rating, labor and delivery, nursing mothers, and pediatric use
- Adverse reactions: description of undesirable effect(s) reasonably associated with proper use of the drug
- Drug abuse/dependence: description of the types of abuse that can occur with the drug and the adverse reactions pertinent to them
- Overdosage: description of the signs, symptoms, and laboratory findings of acute overdosage and the general principles of treatment
- Dosage/administration: recommendation for usage dose, usual dosage range, and, if appropriate, upper limit beyond which safety and effectiveness have not been established
- How supplied: information on the available dosage forms to which the labeling applies

23.5.11 Sponsor Revisions (FDA, 1998a)

When an NDA nears approval, agency reviewers evaluate each element of the draft package labeling for accuracy and consistency with the regulatory requirements for the applicable prescription or over-the-counter drugs. If the CDER has concerns about the draft labeling, the center will contact the sponsor, detailing suggested revisions. The sponsor may submit several revisions until an agreement is reached with the FDA.

23.5.12 NDA Actions (FDA, 1998a)

Once an approval, approvable, or nonapprovable recommendation is reached by the reviewers and their supervisors, the decision is evaluated and agreed to by the director of the applicable drug review division or office who has sign-off authority for such a drug. Once the division director (or office director, as appropriate) signs an approval action letter, the product can be legally marketed in the United States as of that date.

REFERENCES

DiMasi JA, Hansen RW, Grabowski HG. The price of innovation: new estimates of drug development costs. J. Health Econ. 2003;22:151–185.

FDA. 1987. Guideline for the format and content of the microbiology section of an application. Feb. 1987. http://www.bcg-usa.com/regulatory/docs/1987/FDA198702D.pdf. Accessed Dec. 2011.

———. 1998a. New drug development and review process. The CDER Handbook. Revised Mar. 16, 1998. http://www.fda.gov/downloads/AboutFDA/CentersOffices/CDER/UCM198415.pdf. Accessed Dec. 2011.

———. 1998b. CDER MAPP 6030.1: IND process and review procedures(including clinical holds). May 1, 1998. http://www.fda.gov/downloads/AboutFDA/CentersOffices/CDER/ManualofPolicies Procedures/UCM082022.pdf. Accessed Dec. 2011.

———. 2004. Challenges and opportunities report. Mar. 2004. http://www.fda.gov/ScienceResearch/SpecialTopics/CriticalPathInitiative/CriticalPathOpportunitiesReports/ucm077262.htm#execsummary. Accessed Jan. 2012.

———. 2006. FDA/CDER guidance for industry, investigators, and reviewers:—Exploratory IND studies. Jan. 2006. Pharmacology/Toxicology. http://www.fda.gov/downloads/Drugs/Guidance ComplianceRegulatoryInformation/Guidances/ucm078933.pdf. Accessed Dec. 2011.

———. 2010. New drug application (NDA). Updated Aug. 20, 2010. http://www.fda.gov/Drugs/DevelopmentApprovalProcess/HowDrugsareDevelopedandApproved/ApprovalApplications/NewDrug ApplicationNDA/default.htm. Accessed Dec. 2011.

———. 2011a. Investigational new drug (IND) Application. Updated June 6, 2011. http://www.fda.gov/drugs/developmentapprovalprocess/howdrugsaredevelopedandapproved/approvalapplications/investigationalnewdrugindapplication/default.htm. Accessed Dec. 2011.

———. 2011b. *Code of Federal Regulations*. 21 CFR 312.23. IND content and format: investigational new drug application. Revised Apr. 1, 2011. http://www.accessdata.fda.gov/scripts/cdrh/cfdocs/cfcfr/CFRSearch.cfm?fr=312.23. Accessed Jan. 2012.

FDA Centennial, 1906–2006. From test tube to patient:—protecting America's health through human drugs. A special report from FDA Consumer Magazine, pp. 12–21.

Grabowski HG, Vernon J, DiMasi JA. Returns on research and development for 1990s new drug introductions. J. Pharmacoecon. 2002;20(Suppl.3): 11–29.

NCI. 2007. New approaches to cancer drug development and clinical trials: questions and answers. National Cancer Institute/National Institute of Health. June 4, 2007. http://www.cancer.gov/newscenter/qa/2007/phasezeronextqa. Accessed Dec. 2011.

NIH. 2007. Understanding clinical trials. Updated Sep. 20, 2007. http://clinicaltrials.gov/ct2/info/understand. Accessed Dec. 2011.

Pharma. 2007. Drug discovery and development: understanding the R&D process. Feb. 2007. http://www.pharma.org/sites/default/files/159/rd_brochure_022307.pdf. Accessed January 2012.

GLOSSARY

API	Active pharmaceutical ingredient.
CDER	Center for Drug Evaluation and Research.
CFR	*Code of Federal Regulations*.
CGMP	Current good manufacturing practice.
CMC	Chemistry, manufacturing, and controls.
FD&C Act	Food, Drug, and Cosmetic Act.
FDA	U.S. Food and Drug Administration.
Genotoxicity tests	In vitro and in vivo tests designed to detect compounds that induce genetic damage directly or indirectly by various mechanisms. These tests should enable hazard identification with respect to damage to DNA and its fixation.
GLP	Good laboratory practice.
GRAS	Generally recognized as safe.
ICH	International Conference on Harmonization of Technical Requirements for Registration of Pharmaceuticals for Human Use.
IND	Investigational new drug.
IRB	Institutional review boards.

MOA	Mechanism of action.
NDA	New drug application.
No observed adverse effect level	The highest dose tested in an animal species that does not produce a significant increase in adverse effects compared to the control group. Adverse effects that are biologically significant, even if not statistically significant, should be considered in determining a NOAEL.
NOAEL	No observed adverse effect level.
PD	Pharmacodynamics.
Pharmacodynamics	Describes the biochemical and physiological effects of a drug on the body, including how long a drug is absorbed, moves throughout the body, and interacts with certain molecules within target tissues.
Pharmacokinetics	Describes the effects of the body on a drug that includes the process by which the drug is absorbed, distributed in the body, localized in the tissues, metabolized, and excreted.
PK	Pharmacokinetics.

CHAPTER 24

GENERICS, BIOSIMILARS, AND OTCS

24.1 GENERIC DRUGS (FDA, 1998a, 2005a, 2009a, 2011a–c)

A generic drug is chemically identical and bioequivalent to a brand name (innovator) drug in dosage form, safety, strength, route of administration, quality, performance characteristics, and intended use. A brand name or innovator drug is also known as the *reference listed drug* as identified in the U.S. Food and Drug Administration (FDA) list of Approved Drug Products with Therapeutic Equivalence Evaluations; this list is commonly referred to as the *Orange Book* (FDA, 2010a). Generic drugs are available in both over-the-counter (OTC) and prescription forms. For example, ibuprofen is the generic version of the OTC pain medicine Advil, and gabapentin is the generic form of the prescription antiseizure drug Neurontin. According to an IMS Health press release (Gatyas, 2011), in 2010 generics accounted for 78% of total retail prescriptions dispensed in the United States as a result of brand product patents expiring and patients choosing lower-cost generic options. Adoption of generic equivalents is occurring at a much faster rate on average; more than 80% of a brand's prescription volume is replaced by generics within six months of patent loss.

Drug companies must submit an *abbreviated new drug application* (ANDA) to the Center for Drug Evaluation and Research (CDER)/Office of Generic Drugs (OGD) for approval to market a generic product. Generic drug applications are termed "abbreviated" in that there are no requirements to provide clinical data to establish safety and efficacy, since these parameters have already been established by the approval of the innovator drug product. Instead, a generic applicant must demonstrate scientifically that a product is bioequivalent (i.e., performs in the same manner as the innovator drug) by delivering the same amount of active ingredients into a patient's bloodstream in the same amount of time as the innovator drug from a statistically valid number of subjects: typically, 24 to 36 healthy volunteers.

According to Gary Buehler, director of the Food and Drug Administration's OGD: "The FDA ensures a rigorous review of all drugs, and consumers can be

Integrated Pharmaceutics: Applied Preformulation, Product Design, and Regulatory Science, First Edition. Antoine Al-Achi, Mali Ram Gupta, William Craig Stagner.

assured that generic drugs are as safe and effective as brand-name drug products" and meet the same rigid standards as the innovator drug (FDA, 2005a).

24.1.1 FDA Requirements for Approval of Generic Drugs
(FDA, 2003, 2005a, 2011a)

The Drug Price Competition and Patent Term Restoration Act of 1984, more commonly known as the Hatch–Waxman Act, made ANDAs possible by creating a compromise in the drug industry where by generic drug companies gained greater access to the market for prescription drugs, and innovator companies gained restoration of patent life (up to an additional five years) of their products which was lost during the FDA's approval process. A generic drug can enter the market only after the brand-name patent or other marketing exclusivities have expired and only after FDA approval has been granted. The FDA requirements for approving a generic drug are as follows:

- Generic drugs must have the same active ingredients (inactive ingredients may vary) and the same labeled strength of the active moiety and salt as the brand-name drug product.

- Generic drugs must have the same dosage form (e.g., tablets, liquids) and must be administered in the same way.

- Generic drug manufacturers must show that a generic drug is bioequivalent to the brand-name drug, which means that the generic version delivers the same amount of active ingredients into a patient's bloodstream in the same amount of time as the brand-name drug. As noted in 21 CFR 320.24 (FDA, 2011d), several in vivo and in vitro methods can be used to establish bioequivalence (BE). These include, in descending order of preference, pharmacokinetics, pharmacodynamics, clinical, and in vitro studies. A pilot study in a small number of subjects (e.g., 12) may be carried out to validate analytical methodology, assess variability, optimize sample collections time intervals, and provide other relevant information before proceeding with a full BE study.

  ○ In situations where a pharmacokinetic approach is not feasible, suitable pharmacodynamic methods may be used to demonstrate BE.

  ○ Where there are no other means, well-controlled clinical trials in humans may be useful to provide supportive evidence of BE.

  ○ Under certain circumstances, BE can be documented using in vitro approaches [(21 CFR 320.24(b)(5) and 21 CFR 320.22(d)(3)]; for example, for a highly soluble, highly permeable, rapidly dissolving, orally administered drug product, documentation of BE using an in vitro approach (dissolution studies) is appropriate for class 1 drug compounds based on the Biopharmaceutical classification system (FDA, 2009b).

- Generic drug labeling must be essentially the same as the labeling of the brand-name drug.

- Generic drug manufacturers must fully document the generic drug's chemistry, manufacturing steps, and quality control measures.

- Firms must assure the FDA that the raw materials and finished product meet specifications of the *United States Pharmacopeia* (U.S.P.), which sets standards for drug purity in the United States.

- Generic drugs meet the same batch requirements for identity, strength, purity, and quality.

- Firms must show that a generic drug will remain potent and unchanged until the expiration date on the label.

- Firms must comply with federal regulations for good manufacturing practices and provide the FDA with a full description of the facilities they use to manufacture, process, test, package, and label the drug; these facilities are inspected by the FDA to ensure compliance.

  - As part of an ANDA application, the applicant must include a patent certification as described in section 505(j)(2)(A)(vii) of the Food, Drug, and Cosmetic (FD&C) Act; the certification must make one of the following statements:

    - Such patent information has not been filed.

    - Such patent has expired.

    - The date on which such patent expires.

    - Such patent is invalid or will not be infringed through manufacture, use, or sale of the drug product for which the ANDA is submitted. This fourth certification is known as a paragraph IV certification. Section 505(j)(5)(B)(iv) of the FD&C Act established an incentive for generic manufacturers to file paragraph IV certifications and to challenge listed patents as invalid, or not infringed, by providing for a 180-day period of marketing exclusivity (commonly known as *180-day exclusivity*); this applies to drugs first marketed after 1962.

24.1.2 Generic Drug Review Process
(FDA, 1998a, 2003, 2011a)

The CDER, through its office of Generic Drugs (OGD), assures that safe and effective generic drugs are available to the American public. The Generic Drug Review Process provides an overview of CDER's abbreviated new drug application (ANDA) and abbreviated antibiotic drug application (AADA) review process, and how CDER determines the safety and bioequivalence of generic drug products prior to approval for marketing.

24.1.2.1 Applicant An applicant submits an ANDA or AADA (original applications as well as all amendments, supplements, and resubmissions) to the FDA's Office of Generic Drugs at the following address for review and approval to market a generic drug product.

Office of Generic Drugs (HFD-600)

Center for Drug Evaluation and Research

Food and Drug Administration

Metro Park North II, Room 150

7500 Standish Place

Rockville, MD 20855

24.1.2.2 Accept/Refuse to File Letter Issued The OGD performs an initial review and documents if the application contains all the necessary components and is therefore acceptable for filing and review. If the application is missing one or more essential components, a "refuse to file" letter is issued to the applicant. This letter identifies the missing component(s) and informs the applicant that the application will not be filed until it is complete.

24.1.2.3 Bioequivalence Review The FDA requires an applicant to provide the following information to establish bioequivalency:

- A formulation comparison for products whose bioavailability is self-evident: for example, oral solutions, injectables, or ophthalmic solutions where the formulations are identical:
- Comparative dissolution testing where there is a known correlation between in vitro and in vivo effects (e.g., the BCS class I drug compounds, which have high permeability and high solubility).
- In Vivo bioequivalence testing comparing the rate and extent of absorption of the generic to the reference product.
- For nonclassically absorbed products, a head-to-head evaluation of comparative effectiveness based on clinical endpoints. The use of comparative clinical trials is generally considered insensitive and should be avoided (21 CFR 320.24: FDA, 2011d) except when both pharmacokinetics and pharmacodynamics approaches are infeasible.

The Division of Bioequivalence has developed new data summary tables so that data can be submitted to the Office of Generic Drugs in a concise format consistent with the Common Technical Document (CTD) (CTD, 2009; OGD, 2011). ANDA applicants should complete these tables and send the completed tables along with the rest of the bioequivalence submission of their ANDA. If the review determines that the bioequivalence portion of the application is acceptable, a letter indicating that there are no further questions at that time will be issued. Otherwise, a bioequivalence deficiency letter is issued to the applicant detailing the deficiencies found and requesting the information and data needed to resolve the deficiencies.

24.1.2.4 CMC/Microbiology Review The chemistry, manufacturing, and controls (CMC)/microbiology review provides assurance that the generic drug will be manufactured in a controlled, consistent manner meeting all applicable CGMP requirements. If deficiencies are found in one or more portions, these deficiencies, along with the regulatory direction on how to amend the application, are communicated to the applicant in a not approvable letter.

24.1.2.4.1 Question-Based Review for CMC Evaluations of ANDAs

(FDA, 2010b) In early 2005, the OGD began developing a question-based review (QbR) for evaluation of an ANDA. The QbR assessment process was designed with the expectation that the ANDA applications would be organized according to the Common Technical Document (CTD), a submission format adopted by multiple international regulatory bodies, including the FDA. Additionally, this process would transform the CMC review into a modern, scientific, and risk-based pharmaceutical quality assessment based on the FDA's Pharmaceutical CGMPs for the 21st Century: A Risk-Based Approach and Process Analytical Technology (FDA, 2009c). The main objectives of this enhanced review are to:

- Assure product quality through design and performance-based specifications
- Facilitate continuous improvement and reduce CMC supplements through risk assessment
- Enhance the quality of reviews through standardized review questions
- Reduce CMC review time when sponsors submit a quality overall summary that addresses the QbR

Appendix 24.1 provides a list of questions to be completed by the ANDA sponsors for the preparation of a QbR quality overall summary (FDA, 2009d). An ANDA checklist for CTD or eCTD format for completeness and acceptability of an application for filing may be obtained from the FDA website, http://www.fda.gov/downloads/Drugs/DevelopmentApprovalProcess/HowDrugsareDeveloped andApproved/ApprovalApplications/AbbreviatedNewDrugApplicationANDA Generics/UCM151259.pdf (FDA, 2011e). An example of a completed application, prepared using this new format, on ersatzine tablets, U.S.P. may be found at the FDA website, http://www.fda.gov/downloads/Drugs/DevelopmentApprovalProcess/HowDrugsareDevelopedandApproved/ApprovalApplications/AbbreviatedNewDrugApplicationANDAGenerics/ucm120979.pdf.

24.1.2.5 Labeling Review
The labeling review ensures that the proposed generic drug labeling is identical to that of the reference listed drug except for differences due to a change in manufacturer, patent or exclusivity issues, and to avoid drug mix-ups and prevent medication errors.

24.1.2.6 Preapproval/Request for Plant Inspection
Upon filing an ANDA or AADA, the OGD requests the CDER's Office of Compliance (OC) to determine whether the product manufacturer, the bulk drug substance manufacturer, and any outside testing or packaging facilities are operating in compliance with the current GMP regulations as outlined in 21 CFR 211 (FDA, 2011f). A preapproval, product-specific inspection may be performed by the OC on certain applications to assure data integrity. If the preapproval inspection is found unsatisfactory, a not approvable letter may be issued, and the approval of the generic drug product is deferred pending a satisfactory reinspection and recommendation by the OC.

24.1.2.7 ANDA/AADA Approval If all components of the application are found acceptable, an approval or tentative approval letter is issued to the applicant to market the generic drug product. If the approval occurs prior to the expiration of any patents or exclusivities accorded to the reference listed drug product, a tentative approval letter is issued to the applicant. A tentative approval does not allow the applicant to market the generic drug product until the patent/exclusivity condition has expired.

24.1.2.8 Changes to Approved ANDA (FDA, 2004) According to the FDA guidance document Changes to an Approved NDA or ANDA, and in accordance with Sec. 506A of the FD&C Act and 21 CFR 314.70, postapproval changes in components and composition, manufacturing sites, manufacturing process, specifications, container closure system, labeling, miscellaneous changes, and multiple related changes are classified and processed based on the four categories described below.

24.1.2.8.1 Major Change Requiring a Prior Approval Supplement A major change that has a substantial adverse effect on the identity, strength, quality, purity, or potency of a drug product, potentially affecting its safety or effectiveness, requires submission of a supplement and approval by the FDA prior to distribution of the drug product made using the change. A major change may include:

- A move to a different manufacturing site that has never been inspected by the FDA for this type of operation
- Addition or deletion of major processes or equipment
- Relaxing or deleting specifications or analytical methodologies
- A change in primary packaging component materials
- Labeling changes associated with new indications or usage, claims, etc.

24.1.2.8.2 Moderate Change

24.1.2.8.2.1 Changes Being Effective in 30 Days This is a change that has a moderate potential to affect the drug product characteristics listed above and on the safety or effectiveness of the product. For this type of moderate change, the supplement is submitted to the FDA at least 30 days before the distribution of the drug product affected. The supplement should be clearly labeled "Supplement—Changes Effective in 30 Days". This type of change may include:

- A move to a different manufacturing site
- Some selective changes in process, process parameters and/or equipment, analytical procedures, and container closure system
- A labeling change that adds or strengthens a contraindication, warning, precaution, or adverse reaction, etc.

24.1.2.8.2.2 Changes Being Affected The FDA may identify certain moderate changes for which distribution can occur when the FDA receives the supplement.

For either type of moderate change, the FDA may disapprove requested changes and may order the manufacturer to cease distribution of the drug products made using the disapproved changes [21 CFR 314.70(c) (7)].

24.1.2.8.3 Minor Change: Annual Report A change that has minimal potential to adversely affect the identity, strength, purity, potency, safety, and effectiveness of the drug product is considered a minor change. Minor changes must be described in the next annual report [21 CFR 314.70(d)]. A minor change may include:

- Moving to a different manufacturing site for secondary packaging or labeling
- Changes to equipment of the same design or operating principle
- A change in specification to comply to an official compendium
- A change in the size or shape of a container for a nonsterile solid dosage form, etc.

24.2 BIOSIMILAR DRUGS (AAPS, 2011; FDA, 2011g,h; Ledford, 2010)

Biosimilars differ from generic drugs because their active ingredients are huge molecules with intricate structures (a collection of large protein isoforms; not a single entity). Unlike small-molecule active pharmaceutical ingredients (APIs) in generic drugs, these huge molecules are impossible to replicate in every detail, and there are currently no analytical techniques to establish biopharmaceutical equivalence. Definitions, based on the terminology used by the European Medicines Agency (EMA, 2010), for generic and biosimilar medicinal products are provided in the Glossary.

The Biologics Price Competition and Innovation Act (BPCI) of 2009 (a "biosimilar statue") established an abbreviated approval pathway [under Sec. 351(k)] for biologic drugs that have been demonstrated to be highly similar (biosimilar) to, or interchangeable with, FDA-licensed biological products. This new legislation provides 12 years (vs. 10 years in Europe) of data exclusivity and prohibits the FDA from allowing another manufacturer of a highly similar biologic to rely on the agency's prior findings of safety, purity, and potency from the innovator product. This statue, however, does not prohibit another manufacturer from developing its own data to justify the FDA approval of a full biologics license application rather than an abbreviated application that relies on the prior approval of a reference product. This legislation further prohibits any product from being granted more than one period of data exclusivity unless a manufacturer modifies an approved product to produce a change in safety, purity, or potency. Under such conditions, the modified product would be considered a new product and would be entitled to new data exclusivity of 12 years. The BPCI became part of the Patient Protection and Affordable Care Act, which was signed into law in March 2010.

24.3 OVER-THE-COUNTER DRUGS (FDA, 1998b, 2005b, 2010c–e, 2011a,i)

Over-the-counter (OTC) drug products are those drugs that are available to consumers without a prescription. There are more than 300,000 marketed OTC drug products from 80 different therapeutic categories of OTC drugs, ranging from drug products for acne control to drug products for weight control. As with prescription drugs, the CDER (through its Office of Drug Evaluation IV) oversees OTC drugs to ensure that they are safe, effective, properly labeled, and that their benefits outweigh their risks. It is estimated that six of every 10 medications bought by American consumers are OTC drugs. A Nonprescription Advisory Committee, which meets regularly, assists the agency in evaluating issues surrounding these products. Most OTC drug products have been marketed for many years prior to the laws that required proof of safety and effectiveness before marketing. For this reason, the FDA has been evaluating the ingredients and labeling of these products as part of its OTC Drug Review Program. The goal of this program is to establish OTC drug monographs providing information on acceptable ingredients, doses, formulations, and labeling for each class of products. Monographs are updated continually, adding additional ingredients and labeling as needed. Products conforming to a monograph may be marketed without further FDA approval. Many of these monographs are found in Sec. 300 of the *Code of Federal Regulations*. Those drugs that do not conform to the OTC monographs must undergo separate review and approval through the new drug approval system. Under the new application process, some drugs may be approved initially as OTC drugs, but most are approved for prescription use and later switched to OTC. Following the establishment of a final monograph, any related OTC drug that fails to meet the requirements of the monograph and 21 CFR 330.1 will be recognized as being misbranded (Sec. 502 of the FD&C Act) or as a new drug requiring an approved NDA before it can be marketed (Sec. 505 of the FD&C Act).

Section 501(a)(2)(B) of the FD&C Act requires drugs to be manufactured in conformance with current good manufacturing practice. This section does not differentiate between OTC and prescription products; hence, the CGMP regulations apply to all drug products, whether OTC or prescriptions. An OTC drug listed in 21 CFR 330 is considered or being recognized as safe and effective and is not misbranded if it meets each of the conditions contained in specific final monographs and each of the conditions of 21 CFR 330.1, which are:

- The product is manufactured in compliance with the CGMP (21 CFR 210 and 211) requirements.
- The product manufacturer is registered with the FDA and the product is listed in compliance with 21 CFR 207.
- The product is labeled according to Chapter V of the FD&C Act and the format and contents of the label meet 21 CFR 201.66 requirements.
- Advertisement for use of the product is limited to the conditions stated in the labeling.

- The product contains only those inactive ingredients which:
  - Are safe in the amounts administered
  - Do not interfere with the product effectiveness or its suitable tests or assays to determine if the product meets its professed standards of identity, strength, quality, and purity
  - Incorporate color additives used in accordance with Sec. 721 [21 *United States Code* (U.S.C.) 379e] of the FD&C Act and Subchapter A of 21 CFR 70, Color Additives.
- The container and closure components of the product meet 21 CFR 211.94 requirements.
- The labeling of all drugs contains the general warning: "Keep out of reach of children" (highlighted in bold type).
- Based on the route of administration, the labeling of drugs should also include the following statements:
  - For oral administration: "In case of overdose, get medical help or contact a Poison Control Center right away."
  - For topical, rectal, or vaginal administration: "If swallowed, get medical help or contact a Poison Control Center right away."
- If the maximum daily dose limit of the active ingredient is not established, it should be used in a product at a level that does not exceed the amount reasonably required to achieve its intended effect.
- The 21 CFR 330.1 provides a list of terms that may be used interchangeably (e.g., "administer" or "give") or connecting terms that may be deleted (e.g., "discontinue use") from the labeling of the OTC products provided that such use does not alter the meaning of the labeling that has been established and identified in an appropriate monograph or regulation.

24.3.1   OTC Drug Monograph Review Process
(FDA, 1998b, 2011a)

24.3.1.1   Data Submitted by a Drug Sponsor   Data regarding OTC monographs can be submitted by anyone as a request to amend an existing drug monograph or is an opinion regarding a drug monograph. The request is submitted in the form of a citizen petition or as correspondence to an established monograph docket. If a monograph does not exist, data must be submitted in the format as outlined in CFR 330.1. Data are submitted to the Dockets Management Branch, where they are logged in and a copy is made for the public files. The data are then forwarded to the Division of Over-the-Counter Drug Products for review and action.

24.3.1.2   Review by the CDER   A project manager of the division conducts an initial review to determine the type of drug being referenced and then forwards the package to the appropriate team (Advisory Review Panel/Committee, appointed by the FDA commissioner) for a more detailed review. The team leader determines if the package will need to be reviewed by other discipline areas in the review divisions, such as chemists or statisticians, or by other consultants, such as those from other centers or agency offices. For all petitions and requests to amend a monograph, the OTC division has 180 days to review the data and respond to the sponsor. The drug is then categorized through the monograph rule-making process as follows:

- Category I: generally recognized as safe and effective and not misbranded.
- Category II: not generally recognized as safe and effective or is misbranded.
- Category III: insufficient data available to permit classification. This category allows a manufacturer an opportunity to show that the ingredients in a product are effective, and, if they are not, to reformulate or relabel the product appropriately.

The OTC division also oversees OTC drug labeling, as the safety and effectiveness of OTC drug products depend not only on the ingredients but also on clear and truthful labeling that can be understood by consumers. Once the initial review is complete, a feedback letter outlining the CDER's recommendations is prepared for the sponsor. If the sponsor is not satisfied with the recommendations made by the division, the applicant may request a meeting to discuss any concerns.

24.3.1.3 OTC Advisory Committee Meeting (FDA, 1998b, 2011j) Advisory committee meetings are usually held to discuss specific safety or efficacy concerns, or the appropriateness of a switch from prescription to OTC marketing status for a product. Usually, the OTC advisory committee meets jointly with the advisory committee that has specific expertise in the use of the product. The OTC Advisory Committee consists of a core of 14 voting members, including the chair, selected by the FDA commissioner or designee from authorities knowledgeable in the fields of internal medicine, family practice, clinical toxicology, clinical pharmacology, pharmacy, dentistry, and related specialties. The core of voting members may include one technically qualified member who is identified with consumer interests and is recommended by either a consortium of consumer-oriented organizations or other interested persons. In addition to the voting members, the committee may include one nonvoting member who is identified with industry interests.

24.3.1.4 Consultants Review Depending on the type of data submitted, the OTC team may request that the information be reviewed by consultants from other review divisions. These consultants include chemists, statisticians, experts in other FDA centers or agency offices, or by advisory committee members selected for their specific scientific expertise on a critical issue. The comments from these experts and consultant are returned to the OTC review team when their review is complete.

24.3.1.5 Preparation of the Feedback Letter A feedback letter explaining the CDER's actions or recommendations is prepared and forwarded to the sponsor after the OTC reviewers complete their review. A copy is filed at the FDA's Dockets Management Branch.

***24.3.1.6 Proposed Monograph/Amendment Published in the* Federal Register**   If the CDER supports the recommendation of the sponsor to either amend an existing monograph or to create a new monograph, a notice is published in the *Federal Register*. If the sponsor's recommendation is not accepted, a letter is sent to the sponsor explaining the decision for not accepting the petition.

24.3.1.7 Meeting Between the OTC Division and the Sponsor   If the sponsor is not satisfied with the CDER's feedback letter, the sponsor can request a meeting with the division to provide more information and/or respond to the center's concerns.

24.3.1.8 Public Comment   The public usually has 30 to 90 days to respond to the proposal after it is published in the *Federal Register*. This deadline depends on the controversial nature of the notice and can be extended if a request to do so is made. Anyone can request an extension to the deadline. All comments are sent to the Dockets Management Branch and then are forwarded to the Division of Over-the-Counter Drug Products. The comments are reviewed and evaluated by the appropriate team and sent to other discipline areas for further review, if needed.

24.3.1.9 Final Monograph or Final Amendment Prepared   After the public comments have been reviewed, the final monograph is prepared. The final monograph sets final standards that specify ingredients, dosage, indications for use, and certain labeling. The final monograph is sent out for approval through the appropriate review channels, including Division, Office, Center, Office of General Counsel, Deputy Commissioner for Policy, and Regulations Editorial Staff.

24.3.1.10 Return to OTC Division for Revisions/Further Discussion   The package is returned to the Division of Over-the-Counter Drug Products for revision if any office does not concur and is then rerouted to the appropriate sources.

***24.3.1.11 Final Monograph/Amendment Published in the* Federal Register *and* Code of Federal Regulations**   The final monograph is published in the *Federal Register* once all revisions are made and the package receives all the appropriate final approval and concurrence form the review offices. All final monographs and amendments that have been published in the FR are forwarded to the Regulations Editorial Staff for publication in the *Code of Federal Regulations*.

24.3.2   OTC Combination Drug Products (FDA, 2010c)

The OTC combination drug products (where two or more APIs are included together into one dosage form; for example, acetaminophen and caffeine, U.S.P. tablets) not marketed on or before May 11, 1972, are subject to the policy described below.

24.3.2.1 Interim Marketing Permitted (At Risk)   Marketing of a combination product not marketed on or before May 11, 1972, may begin (but taking a risk in case if the agency subsequently adopts a different position that may require relabeling, recall, or other regulatory action) if all of the following conditions are met:

- Each of the active ingredients in the combination has been marketed commercially in the United States on or before May 11, 1972 and is subject to the OTC Drug Review.

- Each of the active ingredients in the combination product has been classified by an OTC advisory review panel in category I (generally recognized as safe and effective and not misbranded) in a published advance notice of proposed rule-making.

- The combination of ingredients has been classified in category I by an OTC advisory review panel in a published advance notice of proposed rule-making.

- The agency has not dissented from these panel recommendations.

24.3.2.2 *Interim Marketing Not Permitted (Pending Subsequent Notice)*

A combination product, not marketed on or before May 11, 1972, is considered a new drug and/or misbranded and is subject to regulatory action if any one of the following applies:

- A panel has recommended in a published advance notice of proposed rule-making that the combination of ingredients or categories of ingredients be classified in category I, and the agency has dissented from this recommendation.

- A panel has recommended that the combination be classified as category II (not generally recognized as safe and effective or misbranded) or category III (available data insufficient to classify as either category I or II).

- A panel has recommended that one of the active ingredients in the combination be classified as category II or III).

- No OTC advisory review panel has considered the combination.

- It's possible to market such a combination product but only after:

  ○ The FDA commissioner tentatively determines that the ingredients and the combination are generally recognized as safe and effective and states by notice in the *Federal Register* (separately or as part of another document) that marketing of the combination under specified conditions will be permitted.

  ○ The commissioner determines that the combination is generally recognized as safe and effective and the combination is included in a published OTC drug final monograph.

  ○ A new drug application for the product has been approved.

24.3.3 OTC Drug Facts Label (FDA, 2005c, 2009e)

The FDA published the OTC Drug Facts Label regulation in the *Federal Register* in March 1999. This regulation required most OTC drug products to comply with the new format and content requirements by May 2002. The OTC labeling rule applies to more than 100,000 OTC drug products. Patterned after the nutrition facts food label, the drug facts label (Figure 24.1) uses simple language and an

DRUG FACTS

Active ingredients (in each tablet)   **Purpose**

Chlorpheniramine maleate 2 mg.......................Antihistamine

Uses: Temporarily relieves these symptoms due to hay fever or other upper respiratory allergies: Sneezing, Runny nose, Itchy and watery eyes, Itchy throat.

Warnings: Ask a doctor before use if you have

- Glaucoma
- A breathing problem such as emphysema or chronic bronchitis
- Trouble urinating due to an enlarged prosta te gland

Ask a doctor or pharmacist before use if you are taking tranquilizers or sedatives.

When using this product:

- You may get drowsy
- Alcohol, sedatives, and tranquilizers may increase drowsiness
- Avoid alcoholic drinks

- Be careful when driving a motor vehicle or operating machinery
- Excitability may occur, especially in children

- **If pregnant or breast-feeding,** ask a health professional before use.
- **Keep out of reach of children**. In case of overdose, get medical help or contact a Poison Control Center right away.

Directions:

Adults and children 12 years and over	Take 2 tablets every 4–6 hours Not more than 12 tablets in 24 hours
Children 6 years to under 12 years	Take 1 tablet every 4–6 hours Not more than 6 tablets in 24 hours
Children under 6 years	Ask a doctor

Other information :

- Store at 20–25 °C (68–77 °F)
- Protect from excessive moisture

Inactive ingredients : D & C yellow no. 10, lactose, magnesium stearate, microcrystalline cellulose, pregeletinized starch.

FIGURE 24.1   OTC drug facts label.

easy-to-read format to help people compare and select OTC medicines and follow dosage instructions. The following information must appear in this order:

- (Title) "Drug Facts"
- The product's active ingredients, including the amount in each dosage unit.
- The purpose of the product.
- The uses (indications) for the product.
- Specific warnings, including when the product should not be used under any circumstances and when it is appropriate to consult with a doctor or pharmacist; this section also describes side effects that could occur and other drugs, substances, or activities to avoid.
- Dosage instructions: when, how, and how often to take the product.
- The product's inactive ingredients: important information to help consumers avoid ingredients that may cause an allergic reaction.

Along with the standardized format, the label uses plain-speaking terms to describe the facts about each OTC drug. For example, *uses* replaces *indications*, while other technical words, such as *precautions* and *contraindications*, have been replaced by more easily understood words and phrases. The label also requires a type size large enough to be read easily and specific layout details. Additionally, bullets, spacing between lines, and clearly marked sections are added to improve readability.

24.3.4 Tamper-Resistant Packaging Requirements for Certain Over-the-Counter Human Drug Products (CPG 7132a.17/CPG Sec. 450.500) (FDA, 2009f)

The regulations, published in the *Federal Register* on February 2, 1989, require that all OTC human drug products (except dermatologic products, dentifrices, insulin, and throat lozenges: 21 CFR 211.132), cosmetic liquid oral hygiene products and vaginal products (21 CFR 700.25), and contact lens solutions and tablets used to make these solutions (21 CFR 800.12) must be packaged in tamper-resistant packaging. Manufacturers and packagers are free to use any packaging system as long as the tamper-resistant standards listed in the regulations are met. The TRP requirements are intended to assure that the product's packaging "can reasonably be expected to provide visible evidence to consumers that tampering has occurred." For additional information, refer to the section "Tamper-Evident Packaging Requirements for Over-the-Counter Human Drug Products" in Section 21.2.1.

The TRP requirements are part of the current good manufacturing practice regulations. Regulatory actions for deviations from these requirements should be handled in the same manner as any other deviation from CGMP regulations.

REFERENCES

AAPS. 2011. AAPS News. Senators seek clarity from FDA on biosimilar exclusivity provisions. AAPS 2011;3:6–7.

CTD. 2009. Common technical document modules/sections corresponding to summary data tables in bioequivalence submissions to ANDAs. Updated Apr. 30, 2009. http://www.fda.gov/Drugs/ DevelopmentApprovalProcess/HowDrugsareDevelopedandApproved/ApprovalApplications/AbbreviatedNewDrugApplicationANDAGenerics/ucm120962.htm. Accessed Dec. 2011.

EMA. 2010. Glossary (terms and abbreviations). Dec. 9, 2010. http://www.ema.europa.eu/docs/en _GB/document_library/Other/2010/12/WC500099907.pdf. Accessed Dec. 2011.

FD&C Act, Sec. 505. 21 USC 355. New drugs. Revisions posted Feb. 2008. http://www.fda. gov/RegulatoryInformation/Legislation/FederalFoodDrugandCosmeticActFDCAct/FDCActChapterV DrugsandDevices/ucm108125.htm. Accessed Dec. 2011.

FDA. 1998a. Generic drug review process. The CDER Handbook. Revised Mar. 16, 1998, pp. 29–34. http://druganddevicelaw.net/CDER. Accessed Dec. 2011.

FDA. 1998b. Over-the-counter drug review process. The CDER Handbook. Revised Mar. 16, 1998; pp. 35–40. http://druganddevicelaw.net/CDER. Accessed Dec. 2011.

———. 2003. FDA/CDER guidance for industry: bioavailability and bioequivalence studies for orally administered drug products—general considerations. Mar. 2003. BP, Rev. 1. http://www.fda.gov/ downloads/Drugs/GuidanceComplianceRegulatoryInformation/Guidances/ucm070124.pdf. Accessed Dec. 2011.

———. 2004. FDA/CDER guidance for industry: Changes to an approved NDA or ANDA. Apr. 2004. CMC, Rev. 1. http://www.fda.gov/OHRMS/DOCKETS/98fr/1999d-0529-gdl0003.pdf. Accessed Dec. 2011.

———. 2005a. Greater access to generic drugs. From Test Tube to Patient—FDA Consumer Magazine 2005;4(Dec.):25–27.

———. 2005b. Now available without a prescription. From Test Tube to Patient—FDA Consumer Magazine 2005;4(Dec.):28–30.

———. 2005c. OTC drug facts label. From Test Tube to Patient—FDA Consumer Magazine 2005;4(Dec.):31.

———. 2009a. The therapeutic equivalence of generic drugs. Updated May 4, 2009. http://www. fda.gov/Drugs/DevelopmentApprovalProcess/HowDrugsareDevelopedandApproved/ApprovalApplications/AbbreviatedNewDrugApplicationANDAGenerics/ucm147161.htm. Accessed Dec. 2011.

———. 2009b. The biopharmaceutics classification system guidance. Updated Apr. 30, 2009. http://www.fda.gov/AboutFDA/CentersOffices/CDER_ucm128219.htm. Accessed Dec. 2011.

———. 2009c. Pharmaceutical CGMPs for the 21st century—a risk-based approach: second progress report and implementation plan. Updated Aug. 17, 2009. http://www.fda.gov/Drugs/Development ApprovalProcess/Manufacturing/QuestionsandAnswersonCurrentGoodManufacturingPracticesCGMP forDrugs/UCM071836. Accessed Dec. 2011.

———. 2009d. Office of Generic Drugs QbR quality overall summary outline. Updated Apr. 30, 2009. http://www.fda.gov/Drugs/DevelopmentApprovalProcess/HowDrugsareDevelopedand Approved/ApprovalApplications/AbbreviatedNewDrugApplicationANDAGenerics/ucm120974.htm. Accessed Dec. 2011.

———. 2009e. OTC drug facts label. Updated May 28, 2009. http://www.fda.gov/Drugs/ ResourcesForYou/Consumers/ucm143551.htm. Accessed Dec. 2011.

———. 2009f. CPG Sec. 450.500: Tamper-resistant packaging requirements for certain over-the-counter human drug products. Updated Nov. 05, 2009. http://www.fda.gov/ICECI/Compliance Manuals/CompliancePolicyGuidanceManual/ucm074391.htm. Accessed Dec. 2011.

———. 2010a. Orange book: Approved drug products with therapeutic equivalence evaluations. Updated Dec. 21, 2010. http://www.accessdata.fda.gov/scripts/cder/ob/default.cfm. Accessed Dec. 2011.

——. 2010b. Question-based review for CMC evaluations of ANDAs. Updated Aug. 27, 2010. http://www.fda.gov/Drugs/DevelopmentApprovalProcess/HowDrugsareDevelopedandApproved/ApprovalApplications/AbbreviatedNewDrugApplicationANDAGenerics/ucm120971.htm. Accessed Dec. 2011.

——. 2010c. CPG Sec. 450.300: OTC drugs—general provisions and administrative procedures for marketing combination products. Revised Mar. 1995. Updated Jan. 20, 2010. http://www.fda.gov/ICECI/ComplianceManuals/CompliancePolicyGuidanceManual/ucm074389.htm. Accessed Dec. 2011

——. 2010d. *Code of Federal Regulations*. 21 CFR 330. Over-the-counter (OTC) human drugs which are generally recognized as safe and effective and not misbranded. Revised Apr. 1, 2010. http://www.accessdata.fda.gov/scripts/cdrh/cfdocs/cfcfr/CFRSearch.cfm?fr=330.1. Accessed Dec. 2011.

——. 2010e. Drug applications for over-the-counter drugs. Updated Apr. 26, 2010. http://www.fda.gov/Drugs/DevelopmentApprovalProcess/HowDrugsareDevelopedandApproved/ApprovalApplications/Over-the-CounterDrugs/default.htm. Accessed Dec. 2011.

——. 2011a. Generic drugs: questions and answers. Updated Aug. 24, 2011. http://www.fda.gov/Drugs/ResourcesForYou/Consumers/QuestionsAnswers/ucm100100.htm. Accessed Dec. 2011.

——. 2011b. Generic drugs: information for industry—news and announcements. Updated May 26, 2011. http://www.fda.gov/Drugs/DevelopmentApprovalProcess/HowDrugsareDevelopedandApproved/ApprovalApplications/AbbreviatedNewDrugApplicationANDAGenerics/ucm142112.htm. Accessed Dec. 2011.

——. 2011c. Abbreviated new drug application (ANDA): generics. Updated Mar. 21, 2011. http://www.fda.gov/Drugs/DevelopmentApprovalProcess/HowDrugsareDevelopedandApproved/ApprovalApplications/AbbreviatedNewDrugApplicationANDAGenerics/default.htm. Accessed Dec. 2011.

——. 2011d. *Code of Federal Regulations*. 21 CFR 320.24. Types of evidence to measure bioavailability or establish bioequivalence. Revised Apr. 1, 2011. http://www.accessdata.fda.gov/scripts/cdrh/cfdocs/cfcfr/CFRSearch.cfm?fr=320.24. Accessed Dec. 2011.

——. 2011e. ANDA filing checklist (CTD or eCTD format) for completeness and accessibility of an application. Updated Aug. 30, 2011. http://www.fda.gov/downloads/Drugs/DevelopmentApprovalProcess/HowDrugsareDevelopedandApproved/ApprovalApplications/AbbreviatedNewDrugApplicationANDAGenerics/UCM151259.pdf. Accessed Dec. 2011.

——. 2011f. *Code of Federal Regulations*. 21 CFR 211. Current good manufacturing practice for finished pharmaceuticals. Apr. 1, 2011. http://www.accessdata.fda.gov/scripts/cdrh/cfdocs/cfcfr/CFRSearch.cfm?CFRPart=211&showFR=1. Accessed Dec. 2011.

——. 2011g. Biologics Price Competition and Innovation Act of 2009. Updated May 10, 2011. http://www.fda.gov/Drugs/GuidanceComplianceRegulatoryInformation/ucm215031.htm. Accessed Dec. 2011.

——. 2011h. Implementation of the Biologics Price Competition and Innovation Act of 2009. Updated May 10, 2011. http://www.fda.gov/Drugs/GuidanceComplianceRegulatory Information/ucm215089.htm. Accessed Dec. 2011.

——. 2011i. Manual of Compliance Policy Guides. Revision/Update History. Updated May 25, 2011. http://www.fda.gov/ICECI/ComplianceManuals/CompliancePolicyGuidanceManual/default.htm. Accessed Dec. 2011.

——. 2011j. Nonprescription Drugs Advisory Committee. Updated May 27, 2011. http://www.fda.gov/AdvisoryCommittees/CommitteesMeetingMaterials/Drugs/NonprescriptionDrugsAdvisory Committee/default.htm. Accessed Dec. 2011.

Gatyas G. IMS Institute reports U.S. spending on medicines grew 2.3 percent in 2010, to $307.4 billion. http://www.imshealth.com/portal/site/ims/menuitem.d248e29c86589c9c30e81c033208c22a/?vgnextoid=24b2f14cddc40310VgnVCM10000071812ca2RCRD&vgnextfmt=default. Accessed Apr. 2011.

Ledford H. Biosimilar drugs poised to penetrate market. Nature 2010(Nov. 4);468:18.

OGD. 2011. Office of Generic Drugs model bioequivalence data summary tables: formatting of bioequivalence summary tables. http://www.fda.gov/downloads/Drugs/DevelopmentApprovalProcess/HowDrugsareDevelopedandApproved/ApprovalApplications/AbbreviatedNewDrugApplicationANDAGenerics/UCM120957.pdf. Accessed Dec. 2011.

GLOSSARY

AADA	Abbreviated antibiotic drug application
ANDA	Abbreviated new drug application.
API	Active pharmacentical ingredient.
BCS	Biopharmaceutical classification system.
BE	Bioequivalence.
Bioequivalence [European Medicines Agency (EMA)]	The absence of a significant difference in the rate and extent to which the active ingredient or active moiety in pharmaceutical equivalents or pharmaceutical alternatives becomes available at the site of action when administered at the same molar dose under similar conditions in an appropriately designed study.
Biosimilar medicinal product (EMA)	A medicinal product similar to a biological medicinal product that has already been authorized (i.e., the biological reference medicinal product). The active substance of a biosimilar medicinal product is similar to that of the biological reference medicinal product. The name, appearance, and packaging of a biosimilar medicinal product may differ from those of the biological reference medicinal product.
BPCI	Biologics Price Competition and Innovation Act of 2009.
CDER	Center for Drug Evaluation and Research.
CMC	Chemistry, manufacturing, and controls.
CTD	Common technical document.
FDA	U.S. Food and Drug Administration.
FD&C Act	Food, Drug, and Cosmetics Act.
Generic medicinal product (EMA)	A medicinal product that has the same qualitative and quantitative composition in active substances and the same pharmaceutical form as the reference medicinal product, and whose bioequivalence with the reference medicinal product has been demonstrated by appropriate bioavailability studies. (EMA Reg. 726/2004, Art. 10, 2b) Generic "copies" can be marketed only after the originator's patent protection and/or marketing exclusivity has expired.
OC	Office of Compliance, FDA.
OGD	Office of Generic Drugs, CDER.
OTC	Over the counter.
QbR	Question-based review.
U.S.P.	*United States Pharmacopeia*.

APPENDIX

APPENDIX 24.1   Office of Generic Drugs QbR Quality Overall Summary Outline

Questions to be completed by ANDA Sponsors for the preparation of a QbR Quality Overall Summary Pharmaceutical product quality: QbR for ANDAs

Definition: simple dosage form—either a solution or an immediate release (IR) solid oral dosage form

[The following numbering system is taken directly from the reference cited above]

2.3 Introduction to the Quality Overall Summary

Proprietary name of drug product:_____

Nonproprietary name of drug product:_____

Nonproprietary name of drug substance:_____

(*continued*)

APPENDIX 24.1 (*Continued*)

Company name:_____

Dosage form:_____

Strength(s):_____

Route of administration_____

Proposed indication(s):_____

2.3.S DRUG SUBSTANCE

2.3.S.1 General Information

What are the nomenclature, molecular structure, molecular formula, and molecular mass of the API?
What are the physicochemical properties including physical description, pK_a, polymorphism, aqueous solubility (as function of pH), hygroscopicity, melting points, and partition coefficient?

2.3.S.2 Manufacture

Who manufactures the drug substance?
How do the manufacturing processes and controls ensure consistent production of drug substance?

2.3.S.3 Characterization

How was the drug substance structure elucidated and characterized? How were potential impurities identified and characterized?

2.3.S.4 Control of Drug Substance

What are the drug substance(s) specification? Does it include all the critical drug substance attributes that affect the manufacturing and quality of the drug product?
For each test in the specification, is the analytical method(s) suitable for its intended use and, if necessary, validated? What is the justification for the acceptance criterion?

2.3.S.5 Reference Standards

How were the primary reference standards certified?

2.3.S6 Container Closure System

What container closure system is used for packaging and storage of the drug substance?

2.3.S.7 Stability

What drug substance stability studies support the retest or expiration date and storage conditions for the drug substance?

2.3.P DRUG PRODUCT

2.3.P.1 Description and Composition

What are the components and composition of the final product? What is the function(s) of each excipient?
Does any excipient exceed the IIG limit for this route of administration?
Do the differences between this formulation and the reference listed drug present potential concerns with respect to therapeutic equivalence?

2.3.P.2 Pharmaceutical Development

2.3.P.2.1 Components of the Product

2.3.P.2.1.1 Drug Substance

Which properties or physical chemical characteristics of the drug substance affect drug product development, manufacture, or performance?

2.3.P.2.1.2 Excipients

What evidence supports compatibility between the excipients and the drug substance?

APPENDIX 24.1 *(Continued)*

2.3.P2.2 Drug Product

What attributes should the drug product possess?

How was the drug product designed to have these attributes?

Were alternative formulations or mechanisms investigated?

How were the excipients and their grades selected?

How was the final formulation optimized?

2.3.P.2.3 Manufacturing Process Development

(If the Product Is an NTI Drug or a Nonsimple Dosage Form)

Why was the manufacturing process described in 2.3.P.3 selected for this drug product?

How are the manufacturing steps (unit operations) related to the drug product quality?

How were the critical processes parameters identified, monitored, and/or controlled?

What is the scale-up experience with the unit operations in this process?

2.3.P.2.4 Container Closure System

What specific container closure attributes are necessary to ensure product performance?

2.3.P.3 Manufacture

(For All Products)

Who manufactures the drug product?

What are the unit operations in the drug product manufacturing process?

What is the reconciliation (reconciliation limits) of the exhibit batch?

Does the batch formula accurately reflect the drug product composition? If not, what are the differences and the justifications?

What are the in-process tests and controls which ensure that each step is successful?

(If Product Is Not a Solution)

What is the difference or comparison in size between commercial scale and the exhibit batch? Does the equipment use the same design and operating principles?

(If the Product Is a NTI Drug or a Nonsimple Dosage Form)

In the proposed scale-up plan, what operating parameters will be adjusted to ensure that the product meets all in-process and final product specifications?

What evidence supports the plan to scale-up the process to commercial scale?

2.3.P.4 Control of Excipients

What are the specifications for the inactive ingredients, and are they suitable for their intended function?

2.3.P.5 Control of Drug Product

What is the drug product specification? Does it include all the critical drug product attributes?

For each test in the specification, is the analytical method(s) suitable for its intended use and, if necessary, validated? What is the justification for the acceptance criterion?

2.3.P.6 Reference Standards and Materials

How were the primary reference standards certified?

2.3.P.7 Container Closure System

What container closure system(s) is proposed for packaging and storage of the drug product? Has the container closure system been qualified as safe for use with this dosage form?

APPENDIX 24.1 *(Continued)*

2.3.P.8 Stability

What are the specifications for stability studies, including justification of acceptance criteria that differ from the drug product release specification?

What drug product stability studies support the proposed shelf life and storage conditions? What are the proposed shelf life and storage conditions?

What is the postapproval stability protocol?

Source: FDA (2009c).

CHAPTER 25

ACCELERATED NEW DRUG APPROVAL AND EXPEDITED ACCESS OF NEW THERAPIES

25.1 INTRODUCTION

Prior to the 1980s, the U.S. Food and Drug Administration's (FDA's) new drug approval process [from investigational new drug (IND) to several phases of clinical studies as described in Chapter 23] was extremely laborious, costly, and lengthy for all drugs, including the drugs for life-threating diseases such as AIDS (Naeger et al., 2010). However, the AIDS crisis in the late 1980s brought an awareness of the new disease, which according to Naeger et al., "resulted in an almost certain death sentence," as there was very little known about this disease. In 1987, the FDA approved Retrovir (zidovudine, azidothymidine), which soon became ineffective as AIDS patients developed resistance to the drug. In response to an increasing demand for effective AIDS drugs and to the pressure of AIDS activists, the FDA started a policy to expedite the process so that the public could access drugs still in clinical trials. During the years from 1993 to 2003, the median time to review a priority review drug was reduced from 13.9 months to 6.7 months. According to a recent FDA list (FDA, 2011a), the time to approve antiretroviral drugs on a priority review process ranged from 10.9 months to as low as 1.4 months. This also affected other disease states. Since 1996, sixty-eight drugs for cancer therapies also received priority review and approval (FDA, 2010a). The several topics relevant to the accelerated new drug approval process and expanded access to new therapies included in this chapter are expedited review and approval of new therapies, expanded access to new therapies, orphan drugs, pediatric drugs, pediatric drug development, and the orphan drug act incentives, including the common European Medicines Agency and FDA applications for orphan medicinal product designation.

Integrated Pharmaceutics: Applied Preformulation, Product Design, and Regulatory Science,
First Edition. Antoine Al-Achi, Mali Ram Gupta, William Craig Stagner.
© 2013 John Wiley & Sons, Inc. Published 2013 by John Wiley & Sons, Inc.

25.2 EXPEDITED REVIEW AND APPROVAL OF NEW THERAPIES (FDA, 2009a, 2010a, 2011b)

After the HIV epidemic in late 1980s, the FDA has taken significant steps to make experimental drugs, intended to treat life-threatening diseases (e.g., HIV/AIDS), widely available to severely ill patients, as well as toward speeding the review and approval of the applications for these products. The Center for Drug Evaluation and Research (CDER) and Center for Biologics Evaluation and Research (CBER) have established classifications, policies, and procedures for the development and expedited review of new drug applications (NDAs) and biologics license applications for innovative drug products. Expedited approaches (i.e., AA priority, Subpart E regulations, accelerated approval regulations, fast track drug development programs, and priority review policies) described in this section are intended to make therapeutically important drugs available at an earlier time without compromising the standards for safety and effectiveness.

25.2.1 AA Priority (1987)

The FDA created an AA priority category to classify all applications for potential AIDS therapies to ensure that these products receive the highest priority in the review process.

25.2.2 Subpart E Regulations (1988)

Subpart E of the *Code of Federal Regulations*, 21 CFR Part 312 (the investigational new drug application regulations), allows FDA to make a medical risk–benefit judgment in deciding whether to approve a drug or biological product. The "Subpart E regulations asserted the importance of flexibility in the FDA's application of the statutory (FD&C Act—federal Food, Drug, and Cosmetic Act) standards of safety and effectiveness, particularly in recognition of the increased risk-tolerance of seriously ill patients for whom no satisfactory treatment alternative existed" (Greenberg, 2000). The regulations emphasize the importance of early communication between the agency and a sponsor for improving the efficiency of preclinical and clinical development, and reaching early agreement on the design of the major clinical efficacy studies that will be needed to support approval. "The regulations allow early access to medications and biologics for such life-threatening diseases by permitting a 'treatment protocol' before getting marketing approvals if the preliminary analysis of phase 2 trials is promising" (Naeger et al., 2010). In addition, the early consultation between the FDA and a drug company serves as a predicate to eliminate phase 3 studies if enough safety and efficacy data have been established with phase 2 study. Moreover, so-called "phase 4" postmarketing trials have also been established to postpone some of the research burden until after marketing of the drug. This, in turn, expedites the access where the initial clinical trials provided positive results. According to Greenberg, "Taken together, early consultation, consolidation of phase 2 and phase 3 clinical testing, and the possibility of post-approval testing, served to radically alter the new drug approval process with regard to life-threatening illnesses, particularly for AIDS" (Greenberg, 2000).

25.2.3 Accelerated Approval Regulations (1992)

The FDA's accelerated approval procedures and restricted distribution provisions in 21 CFR 314 Subpart H are available for new drug and biological products (1) that have been studied to treat serious or life-threatening illnesses and (2) that provide meaningful therapeutic benefit to patients over existing treatments (e.g., ability to treat patients unresponsive to, or intolerant of, available therapy, or improved patient response over available therapy). The FDA may approve drugs based on surrogate endpoints (21 CFR 314.510 and 601.41) that reasonably predict that a drug provides clinical benefit, thus considerably shortening the FDA approval time. This clinical benefit is then confirmed through additional human studies that will be completed after marketing approval (during postmarketing clinical trials, also known as phase 4 confirmatory trials). The accelerated approval approach provides for removal of the drug from the market if further studies do not confirm the clinical benefit of the therapy.

25.2.4 Fast Track Drug Development Programs (1997)

The FDA's fast track drug development programs are designed to facilitate the development and expedite the review of drug and biological products that are intended to treat serious or life-threatening conditions and that demonstrate the potential to address unmet medical needs. A drug that receives fast track designation is eligible for some or all of the following benefits:

- More frequent meetings with the FDA to discuss the drug's development plan and ensure collection of appropriate data needed to support drug approval
- More frequent written correspondence from the FDA about such things as the design of the proposed clinical trials
- Eligibility for *priority review*
- Eligibility for *accelerated approval* (i.e., approval on an effect on a surrogate, or substitute endpoint reasonably likely to predict clinical benefit)
- *Rolling review*, which means that a drug company can submit completed sections of its NDA for review by the FDA rather than waiting until every section of the application is completed before the entire application can be reviewed
- Dispute resolution if the drug company is not satisfied with an FDA decision not to grant fast track status

25.2.5 Priority Review Policies (1997)

Section 112 of the Food and Drug Administration (P.L. 105-115) amended the FD&C Act by adding a new section 506(21 U.S.C. 356) authorizing the FDA to facilitate the development and expedite the review of new drugs that are intended to treat serious or life-threatening conditions and periodically to address unmet medical needs (fast track products). The goal for completing a priority review is six months. The agency has made significant strides, in that the median time for FDA approval period has been reduced from 13.9 months to 6.7 months in the 10-year span from 1993 to 2003. During fiscal year 2011, the FDA approved 16 of the 35 new drugs within the priority review period of six months (FDA, 2011d). The FDA is devoting additional resources to drugs that have the potential to provide significant advances in treatments, such as:

- Evidence of increased effectiveness in treatment, prevention, or diagnosis of disease
- Elimination or substantial reduction of a treatment-limiting drug reaction
- Documented enhancement of patient willingness or ability to take the drug according to the required schedule and dose
- Evidence of safety and effectiveness in a new subpopulation, such as children

A request for a priority review must be made by the drug company. The FDA determines within 45 days of the drug company's request whether a priority designation will be assigned. Products regulated by the CDER are eligible for priority review if they provide a significant improvement compared to marketed products in the treatment, diagnosis, or prevention of a disease. Products regulated by the CBER are eligible for priority review if they provide a significant improvement in the safety or effectiveness of the treatment, diagnosis, or prevention of a serious or life-threatening disease. Designation of a drug as priority does not alter the scientific or medical standard for approval or the quality of evidence necessary.

25.3 EXPANDED ACCESS TO NEW THERAPIES
(FDA, 2009a)

Expanded access (sometimes referred to as *compassionate use*) mechanisms are designed to make promising products available as early in the drug evaluation process as possible to patients without therapeutic options, either because they have exhausted or are intolerant of approved therapies.

25.3.1 Treatment IND (1987/revised 2009)

Treatment IND regulations allow the use of an investigational drug for treatment under a treatment protocol, or treatment IND application, outside the clinical trial. Treatment INDs have generally been granted when the product is well into clinical trials, or when clinical trials have been completed, after the accumulation of adequate safety data, and there is evidence of probable effectiveness.

According to regulations [21 CFR 312.305(a)], the investigational drug can be used for this purpose only if all of the following criteria are met:

- The patient or patients to be treated have a serious or immediately life-threatening disease or condition and there is no comparable or satisfactory alternative therapy available to diagnose, monitor, or treat the disease or condition.

- The potential patient benefit justifies the potential risks of the treatment use, and those potential risks are not unreasonable in the context of the disease or condition to be treated.

- Providing the investigational drug for the use requested will not interfere with the initiation, conduct, or completion of clinical investigations that could support marketing approval of the expanded access use or otherwise compromise the potential development of the expanded access use.

In accordance with 21 CFR 312.320, the FDA must further determine that:

- The drug is being investigated in a controlled clinical trial under an IND designed to support a marketing application for expanded access use, or all clinical trials of the drug have been completed.

- The sponsor is actively pursuing marketing approval of the drug for expanded access use with due diligence.

- When the expanded access use is for a serious disease or condition, there is sufficient clinical evidence of safety and effectiveness to support the expanded use. Such evidence would ordinarily consist of data from phase 3 trials, but could consist of compelling data from completed phase 2 trials.

- When the expanded access use is for an immediately life-threatening disease or condition, the available scientific evidence, taken as a whole, provides a reasonable basis to conclude that the investigational drug may be effective for expanded access use and would not expose patients to an unreasonable and significant risk of illness or injury. This evidence would ordinarily consist of clinical data from phase 3 or phase 2 trials, but could be based on more preliminary clinical evidence.

25.3.2   Parallel Track (1992)

The purpose of the parallel track policy (published in the *Federal Register* on April 15, 1992) was to expand the availability of promising investigational drugs to those persons with AIDS and HIV-related diseases (and not for all diseases as in a treatment IND) who are without satisfactory alternative therapy and who cannot participate in controlled clinical trials. Although it is still technically available, it has not been used since stavudine (D4T) was approved in 1994. Treatment INDs have proven to be a more practical mechanism to provide treatment access.

25.4 ORPHAN DRUGS (EMEA, 2007, FDA, 1998a, 2005a, 2011c)

The Orphan Drug Act (ODA) was signed into law on January 4, 1983. The intent of the ODA is to stimulate the research, development, and approval of products that treat rare diseases. According to Coté (FDA, 2010b), there are over 7000 rare diseases; so far, 357 new therapies have received full marketing approval and another 2100 promising compounds have been labeled as orphan drugs as of June 2010. An *orphan disease* is defined as a condition that affects fewer than 200,000 people in the United States or fewer than 1 in 20,000 in the European Union; some diseases have patient populations of fewer than 100. It is estimated that about 50 to 60 million people in these two regions are affected by these rare diseases. A few examples of these rare diseases are cystic fibrosis, ALS (Lou Gehrig's disease), Huntington's disease, Tourette syndrome, Hamburger disease, Job syndrome, acromegaly (i.e., gigantism), and Jumping Frenchmen of Maine disorder. Symptoms of rare diseases may be apparent at birth or in childhood. These symptoms include infantile spinal muscular atrophy, lysosomal storage disorders, patent ductus arteriosus, familial adenomatous polyposis, and cystic fibrosis. But according to Sharma et al. (2010) more than 50% of rare diseases (such as renal cell carcinoma, glioma, and acute myeloid leukemia) are present during adulthood. Eighty percent of the rare diseases have been identified as of genetic origin, whereas others are the result of bacterial or viral infections, allergies, or are due to degenerative and proliferative causes. These diseases collectively affect as many as 25 million Americans (and about 25 to 36 million in the European Union), according to the National Institutes of Health (NIH), which makes finding treatments for them a serious public health concern. With the aim of encouraging research into rare diseases, the United States, the European Union, and other regional governments have begun offering incentives to sponsors to develop and market orphan drugs.

In the United States, the ODA created financial incentives for drug and biologics manufacturers, including tax credits for costs of clinical research, government grant funding, assistance for clinical research, and a seven-year period of exclusive marketing given to the first sponsor of an orphan-designated product who obtains FDA market approval for the same indication. At the same time, federal programs at the FDA and the NIH began encouraging product development as well as clinical research for products targeting rare diseases. Since 1983, the ODA has resulted in the development of more than 250 orphan drugs, which now are available to treat a potential patient population of more than 13 million Americans. In contrast, the decade before 1983 saw fewer than 10 such products developed without government assistance. As a result of the ODA, treatments are available to people with rare diseases who once had no hope for survival.

Within the European Union, the sponsor of a designated orphan drug benefits from a number of incentives. Financial incentives include 100% fee reductions for protocol assistance and follow-up, for preauthorized inspections and 50% fee reductions for application fee, and for postauthorization activities and 10-year market exclusivity. Similar incentives are offered by other regional governments. The requirements for establishing the quality, safety, and efficacy are applied equally to designated orphan drugs as for drugs not designated as such.

According to Wellman-Labadie and Zhou (2010), the ODA has gained a lot of popularity over the years, as is evident by designating over 2000 as orphan drugs and approval of more than 300 as orphan drugs. Manufacturers have gained interest in orphan drugs, as they are less likely to face any competition. In addition, companies' ethical profiles are also strengthened with the production of an orphan drug for a rare disease. Thus, orphan drugs are likely to occupy greater proportions of annually approved pharmaceutical drugs and health care budgets. According to Wellman-Labadie and Zhou, "The sales of biopharmaceutical drugs including many orphan drugs, have increased by over 100% in U.S. and over 200% in most European nations during 2001–2005."

25.4.1   Orphan Drug Act (FDA, 2011c)

25.4.1.1   Recommendations for Investigations of Drugs for Rare Disease or Conditions [Sec. 525 (21 United States Code (U.S.C.) 360aa)] (FDAa)
The sponsor of a drug for a disease or condition that is rare in the United States may request the Secretary of Health and Human Services (HHS) to provide written recommendations for the nonclinical and clinical investigations that must be conducted with the drug before:

1. It may be approved for such disease or condition under Sec. 505
2. It may be certified for such disease or condition under Sec. 507 if the drug is an antibiotic
3. It may be licensed for such disease or condition under Sec. 351 of the Public Health Service Act if the drug is a biological product

25.4.1.2   Designation of Drugs for Rare Diseases or Conditions [Sec. 526 (21 U.S.C. 360bb)(a)(1), (b), (c), (d)] (FDAb)
The manufacturer or the sponsor of a drug may request that the HHS secretary designate the drug as a drug for a rare disease or condition. If the secretary finds that a drug for which a request is submitted under this subsection is being or will be investigated for a rare disease or condition and (a) if an application for such a drug is approved under Sec. 505, (b) if a certification for such a drug is issued under Sec. 507, or (c) if a license for such a drug is issued under Sec. 351 of the Public Health Service Act and the approval, certification, or license would be for use for such a disease or condition, the secretary will designate the drug as a drug for such a disease or condition.

25.4.1.3   Grants and Contracts for Development of Drugs for Rare Diseases and Conditions [Sec. 527 (21 U.S.C. 36022)(a)] (FDAc)
The secretary may make grants to and enter into contracts with public and private entities and individuals to assist in (1) defraying the costs of qualified clinical testing expenses incurred in connection with the development of drugs for rare diseases and conditions, (2) defraying the costs of developing medical devices for rare diseases or conditions, and (3) defraying the costs of developing medical foods for rare diseases or conditions.

25.5 PEDIATRIC DRUGS (FDA, 1998b, 2005b)

25.5.1 Drug Research and Children

According to Bowen et al. (2008), "The first clinical trials of antiretroviral (ARV) use in HIV-infected children were published in 1988." Since then, many studies have been addressed; however, the questions associated with pediatric antiretroviral drug development, such as dosing and child-friendly formulations, have not been answered thoroughly. "Compared with 28% coverage for adults meeting international criteria for ARV treatment, only 15% of children who needed ARVs were receiving it by the end of 2006." In addition, statistics from the Joint United Nations Programme on HIV/AIDS (http://www.unaids.org) show that there were approximately 2.1 million children under the age of 15 who were living with HIV/AIDS in 2007. Moreover, approximately 40% of all children with the disease are under the age of 18 months, with a mean age of 2.3 years. These statistics show urgency in developing safe and effective HIV/AIDS treatments for children.

A common approach has been to use data from adults and to adjust the dose according to a child's weight. Experimenting over the years has taught doctors to use many drugs in children safely and effectively. But this trial-and-error approach has also resulted in tragedy. Adult experiences with a drug are not always a reliable predictor of how children will react. The antibiotic chloramphenicol was widely used in the 1950s to treat infections in adults resistant to penicillin. But many newborn babies died after receiving the drug because their immature livers could not break down the antibiotic effectively, so the drug became toxic in children. "Experience has shown us that we need to study drugs in children because they aren't small adults," says Ralph Kauffman, director of medical research at Children's Mercy Hospital in Kansas City, Missouri. "It's not just about smaller weight," he says. "There are dynamics of growth and maturation of organs, changes in metabolism throughout infancy and childhood, changes in body proportion, and other developmental changes that affect how drugs are metabolized." Drug testing for pediatrics has been sparse historically for a combination of several reasons, including:

- The pharmaceutical companies generally viewed drugs for children as bringing only limited financial benefits.
- "It's harder to carry out studies in children," says Dianne Murphy, director of the FDA's Office of Pediatric Therapeutics. "You need child-friendly environments in every sense, from age-appropriate equipment and medical techniques to pediatric specialists who are sensitive to a child's fear." "[D]rawing blood or getting a urine sample can be difficult with children" (Jeffrey Blumer, chief of pediatric pharmacology at Case Western Reserve University in Cleveland).

The ethical issues are also stickier. For example, whereas adults can give informed consent to participate in a clinical trial, children cannot, as the "consent" implies full understanding of potential risks and other considerations. Parents are involved in the decision to enroll children in a study, and children ages 7 or older can "assent" or "dissent," meaning that they can agree or disagree to participate in a study. Because of these challenges, the experts say that it is more important to build the foundation and resources needed to conduct the studies rather than avoiding pediatric research; without these studies, children face significant risks.

Fortunately, "there have been more studies conducted in children in the last five years than in the previous 30 years combined," Kauffman says. New discoveries have revealed underdosing, overdosing, ineffectiveness, and safety problems. The information coming out of these studies has added pediatric information to the drug labeling for more than 80 drugs, and more changes are coming. For example, ibuprofen, one of the most common over-the-counter drugs on which parents rely to reduce children's fever, carried no dosing information for children younger than 2 years old until recently. Now, because of studies in thousands of young infants, the safe and effective dosing for over-the-counter use has been established for children ages 6 months to 2 years.

25.5.2   Pediatric Research Initiatives (FDA, 2005b)

As a result of the two initiatives discussed here, more than 100 drugs have been granted exclusivity so far and, as of December 31, 2004, 691 studies had been requested and 298 written requests issued. The FDA estimates that about 80% of the studies outlined in written requests will be conducted.

- The voluntary pediatric exclusivity provision of the Food and Drug Administration Modernization Act of 1997 (FDAMA), which was reauthorized in January 2002 and extended through 2007 as the Best Pharmaceuticals for Children Act (BPCA)
- The Pediatric Research Equity Act (PREA), made into law on in 2003, which allows the FDA to require pediatric studies
- The Food and Drug Administration Amendment Act of 2007, which reauthorized both the BPCA and the PREA

25.5.2.1   *The Pediatric Exclusivity Provision of the BPCA*   The pediatric exclusivity provision has done more to spur pediatric studies than any other regulatory or legislative initiative so far. The provision allows companies to qualify for an additional six months of marketing exclusivity if they do the studies in children as requested by the FDA, thus delaying the availability of low-cost generic drugs. The FDA is authorizing six months of exclusivity to all dosage forms marketed by the firm provided that they all contain the same active ingredient(s) and that at least one of the dosage forms is studied in the pediatric population.

The process can be initiated either by a drug company or by the FDA. A drug company may submit a proposal to the FDA to conduct pediatric studies. If the FDA agrees that studying a drug may produce health benefits for children, the agency will issue a written request addressing the type of studies to be conducted, study design and goals, and the age groups to be studied. The agency may issue a written request on its own initiative when it identifies a need for pediatric data. No matter how the studies are initiated, if the FDA determines that the data submitted fairly respond to the written request, the company will be granted six months of pediatric exclusivity.

Since the incentive under the FDAMA did not apply to older antibiotics and other drugs that lack marketing exclusivity or patent protection, some categories of drugs have remained inadequately studied. For these products, the BPCA provides a contract mechanism through the NIH to fund pediatric studies. In addition, if a company that has a drug with existing exclusivity or patent protection chooses not to conduct the pediatric studies requested, the FDA can refer the written request to the Foundation for the National Institutes of Health to award grants so that third parties can conduct the needed studies. The NIH, in consultation with the FDA and other pediatric experts, publishes an annual List of Drugs for Which Additional Pediatric Studies Are Needed in the *Federal Register*. Since the BPCA went into effect, the FDA has issued 11 written requests for off-patent drugs and the NIH has published four requests for contracts (FDA, 2005b).

25.5.2.2 The Pediatric Research Equity Act Under the PREA, the FDA can require pediatric studies of a drug submitted in a new drug application if the FDA determines that the product is likely to be used in a substantial number of pediatric patients, or if the product would provide a meaningful benefit in the pediatric population over existing treatments. At the same time, the PREA does not delay the availability of drugs for adults. "The BPCA and PREA have worked in tandem," says Murphy. "We have told sponsors who submit a new drug application and who are required under PREA to conduct pediatric studies that they also may qualify for pediatric exclusivity."

25.5.2.3 Changing Drug Labels (FDA, 2005b) The FDA works with the American Academy of Pediatrics to educate pediatricians about new physician labeling changes through an online continuing medical education program called PediaLink. Recent pediatric drug studies have resulted in the addition of pediatric information to the labeling for more than 80 drugs. Here are a few examples of several changes that are considered significant for dosing and risk:

- Claritin (loratadine) syrup: used to treat allergy and hives. Patients' aged 2 to 5 years require a lower dose (5 mg) than the 10-mg dose recommended for older children and adolescents.
- Duragesic (fentanyl) transdermal patch: used to manage chronic pain. It is now only to be used in patients older than 2 years who have been on opioids and have tolerated them well. This medication is administered through a patch placed on the skin.
- Luvox (fluvoxamine maleate) tablets: treats obsessive-compulsive disorder. The dose of the drug may need to be increased to the recommended adult dose in adolescents, but girls ages 8 to 11 years may need a lower does than that recommended.

- Midazolam hydrochloride syrup and injection: used as a sedative. The drug was shown to have a higher risk of serious and life-threatening adverse events for children with congenital heart disease and pulmonary hypertension. Research identified the need to begin therapy with doses at the lower end of the dosing range to prevent respiratory problems in this special pediatric population.

- Neurontin (gabapentin) capsules, tablets, and oral solution: used as adjunctive therapy in the treatment of partial seizures in pediatric patients ages 3 to 12 years. Neuropsychiatric adverse events were identified in 3- to 12-year-olds.

- Pepcid (famotidine) tablets, injection, and oral suspension: used to treat gastroesophageal reflux disease. Patients up to 3 months of age require a lower dose because their ability to get rid of the drug is less than that of older children and adults.

- Ultane (sevoflurane) volatile liquid for inhalation: used in general anesthesia. Pediatric studies revealed rare reports of seizures in pediatric patients given this drug.

25.6 PEDIATRIC DRUG DEVELOPMENT AND THE ORPHAN DRUG ACT INCENTIVES (FDA, 2010c)

The Orphan Drug Regulations [Final Rule, 57 F.R. 62076 (1992)] stipulate that a drug sponsor may request orphan drug designation for a drug or a biological product under development for only a subset of persons with a particular disease or condition if the subset is "medically plausible." In the pediatric population, growth and developmental changes can influence the way drugs are absorbed, distributed, metabolized, and excreted, which are vastly different from these actions in an adult. Based on these unique pharmacokinetic properties, the Office of Orphan Products Development has determined that pediatric patients are "therapeutic orphans" or constitute a medically plausible subset of a patient population. Therefore, a sponsor of a new drug or biological product (including currently marketed drugs with no pediatric indication) may seek orphan drug designation for treatment of a disease or condition in the relevant pediatric subset of a patient population. The sponsors are awarded multiple economic incentives for their efforts in this area:

- Tax credits for clinical research
- Waiver of Prescription Drug User Fee Act application fee
- Seven years of orphan drug marketing exclusivity
- Research grants to defray clinical development cost (authorized under the Orphan Products Grant Program)

25.7 COMMON EMEA/FDA APPLICATION FOR ORPHAN MEDICINAL PRODUCT DESIGNATION
(FDA, 2009b)

The sponsor of a medicinal product for human use may desire to seek orphan designation of its medicinal product for use to diagnose, treat, or prevent a rare disease or condition from the European Commission (EC) in accordance with EC Regulation No. 141/2000 of December 16, 1999 and EC Commission Regulation No. 847/2000, and from the U.S. Food and Drug Administration in accordance with Sec. 526 of the FD&C Act (21 U.S.C. 360bb). In such a case, the sponsor may apply for orphan designation of the same medicinal product for the same use in both jurisdictions by using this common application for its submissions to the European Medicines Agency and the FDA.

REFERENCES

Bowen A, Palasanthiran P, Sohn AH. Global challenges in the development and delivery of paediatric antiretroviral. Drug Discov. Today. 2008;13(11–12):530–535.

EMEA. 2007. Orphan drugs and rare disease at a glance. Doc. Ref. EMEA/290072/2007. July 3, 2007. http://www.ema.europa.eu/pdfs/human/comp/29007207en.pdf. Accessed Dec. 2011

FDAa. *U.S. Code*, Sec. 525 [21 USC 360aa]. Recommendations for investigations of drugs for rare diseases or conditions. http://www.fda.gov/RegulatoryInformation/Legislation/FederalFood DrugandCosmeticActFDCAct/FDCActChapterVDrugsandDevices/ucm110318.htm. Accessed Dec. 2011.

FDAb. *U.S. Code*, Sec. 526 [21 USC 360bb]. Designation of drugs for rare diseases or conditions. http://www.fda.gov/RegulatoryInformation/Legislation/FederalFoodDrugandCosmeticActFDCAct/FD CActChapterVDrugsandDevices/ucm110321.htm. Accessed Dec. 2011.

FDAc. *U.S. Code*, Sec. 527 [21 USC 360cc]. Protection for drugs for rare diseases or conditions. http://www.fda.gov/RegulatoryInformation/Legislation/FederalFoodDrugandCosmeticActFDCAct/FD CActChapterVDrugsandDevices/ucm110326.htm. Accessed Dec. 2011

FDA. 1998a. Orphan drugs. The CDER Handbook. Revised Mar. 16, 1998; p. 75. http://drugand devicelaw.net/CDER_handbook.pdf. Accessed Dec. 2011.

————. 1998b. CDER pediatric initiatives. The CDER Handbook. Revised Mar. 16, 1998; pp. 81–82. http://druganddevicelaw.net/CDER_handbook.pdf. Accessed Dec. 2011.

————. 2005a. Orphan products: hope for people with rare diseases. From Test Tube to Patient—FDA Consumer Magazine 2005; 4 (Dec.): 60–62.

————. 2005b. Drug research and children. From Test Tube to Patient—FDA Consumer Magazine 2005; 4(Dec.): 63–68.

————. 2009a. Expanded access and expedited approval of new therapies related to HIV/AIDS. Updated Sept. 11, 2009. http://www.fda.gov/ForConsumers/ByAudience/ForPatient Advocates/HIVandAIDSActivities/ucm134331.htm. Accessed Dec. 2011.

————. 2009b. Common EMEA/FDA application for orphan medicinal product designation. Updated June 18, 2009. http://www.fda.gov/ForIndustry/DevelopingProductsforRareDiseasesConditions/How toapplyforOrphanProductDesignation/ucm124795.htm. Accessed Dec. 2011.

————. 2010a. Fast track accelerated approval and priority review: accelerating availability of new drugs for patients with serious diseases. Updated May 28, 2010. http://www.fda.gov/ForConsumers /ByAudience/ForPatientAdvocates/SpeedingAccesstoImportantNewTherapies/ucm128291.htm. Accessed Dec. 2011.

_____. 2010b. FDA basics videos: Tim Coté on orphan drugs—How does FDA's orphan drugs program help in the fight against rare diseases? Updated June 16, 2010. http://www.fda.gov/AboutFDA/Transparency/Basics/ucm213762.htm. Accessed Dec. 2011.

_____. 2010c. Pediatric drug development and the Orphan Drug Act incentives. Updated Dec. 16, 2010. http://www.fda.gov/ForIndustry/DevelopingProductsforRareDiseasesConditions/HowtoapplyforOrphanProductDesignation/ucm135125.htm. Accessed Dec. 2011.

_____. 2011a. Antiretroviral drugs used in the treatment of HIV infection. Updated Aug. 15, 2011. http://www.fda.gov/ForConsumers/ByAudience/ForPatientAdvocates/HIVandAIDSActivities/ucm118915.htm. Accessed Dec. 2011.

_____. 2011b. Guidance for industry: Available therapy July 2004—clinical medical. Updated Jan. 6, 2011. http://www.fda.gov/RegulatoryInformation/Guidances/ucm126586.htm. Accessed Dec. 2011.

_____. 2011c. 21 CFR 316. Orphan Drug Regulation. Subpart C, Designation of an Orphan Drug. Updated Feb. 22, 2011. http://www.fda.gov/ForIndustry/DevelopingProductsforRareDiseasesConditions/HowtoapplyforOrphanProductDesignation/ucm124562.htm. Accessed Dec. 2011.

_____. 2011d. FDA news release. Nov. 3, 2011. http://www.fda.gov/NewsEvents/Newsroom/PressAnnouncements/ucm278383.htm. Accessed Mar. 2012.

Greenberg MD. AIDs, experimental drug approval, and the FDA new drug screening process. Legislation Public Policy 2000;3(295):295–350.

Naeger LK, Struble KA, Murray JS, Birnkrant DB. Running a tightrope: regulatory challenges in the development of antiretrovirals. Sci. Direct Antiviral Res. 2010;85:232–240.

Sharma A, Jacob A, Tandon M, Kumar D. Orphan drug: development trends and strategies. J. Pharm. Bioallied Sci. 2010;2(4):290–299.

Wellman-Labadie O, Zhou Y. The US orphan drug act: rare disease research stimulator or commercial opportunity? Sci. Direct Health Polity 2010;95:216–228.

GLOSSARY

BPCA	Best Pharmaceuticals for Children Act.
CBER	Center for Biologics Evaluation and Research.
CDER	Center for Drug Evaluation and Research.
CFR	*Code of Federal Regulations*.
EC	European Commission.
EMA	European Medicines Agency.
Fast track	A process designed to facilitate the development and expedite the review of drugs to treat serious diseases and fill an unmet medical need. The purpose is to get important new drugs to the patient earlier. Fast track addresses a broad range of serious diseases.
FDA	U.S. Food and Drug Administration.
FDAMA	Food and Drug Administration Modernization Act of 2007.
FD&C Act	Food, Drug, and Cosmetic Act
HHS	U.S. Department of Health and Human Services.
IND	Investigational new drug.
NDA	New drug application.
NIH	National Institutes of Health.
ODA	Orphan Drug Act.
Orphan drug	United States: Any drug developed under the Orphan Drug Act of January 1983; a disease that affects fewer than 200,000 people is defined as an orphan disease.

Europe: A drug for a disease or disorder that affects fewer than 5 in 10,000 citizens (regulation 141/2000).

Japan: A drug that meets the following three conditions: (1) used to treat a rare disease, which is a disease with fewer than 50,000 prevalent cases (0.4%); (2) treats a disease or condition for which there are no other drugs already available in Japan, or if the drug's superiority is proven to the drugs that is already available in the Japanese market; and (3) the applicants have a clear product development plan and scientific rationale to support the necessity of the drug in Japan. Japan also has orphan drug legislation, which has room for interpretation.

Australia: A drug not intended for use in more than 2000 patients per year if it is a vaccine or in vivo diagnostic.

Canada: Although the country has no official "orphan disease" status, based on international standards it could be defined a drug that treats a disease with a potential patient population numbering between 3300 (Australian standard) and 22,500 (U.S. standard).

PREA Pediatric Research Equity Act.

Priority review A designation given to drugs that offer major advances in treatment or provide treatment where no adequate therapy exists. A priority review results in less time for an FDA review of a new drug application. The goal for completing a priority review is six months. Priority review status can apply both to drugs that are used to treat serious diseases and to drugs for less serious illnesses. A request for a priority review must be made by a drug company, but the length of the clinical trial period is not affected. The FDA determines within 45 days of the drug company's request whether a priority or a standard review designation will be assigned. Designation of a drug as "priority" does not alter the scientific and medical standards for approval or the quality of evidence necessary.

Serious disease A disease may be classified as serious if patient survival or day-to-day functioning are threatened, or if the likelihood that if left untreated, the disease will progress from a less severe condition to a more serious one. AIDS, Alzheimer's, heart failure, and cancer are obvious examples of serious diseases; however, diseases such as epilepsy, depression, and diabetes are also considered to be serious diseases.

Surrogate endpoint A marker (e.g., a laboratory measurement or physical sign) used in clinical trials as an indirect or substitute measurement that represents a clinically meaningful outcome, such as survival or symptom improvement, and the endpoint studies are adequate and well controlled. The use of a surrogate endpoint can shorten considerably the time required prior to receiving FDA approval, but approval is conditional based on verifying the anticipated clinical benefits during postmarketing clinical trials (also known as phase 4 confirmatory trials). For example, the FDA could accelerate approval of a new drug based on evidence that the drug shrinks tumors, rather than waiting to learn whether a cancer patient actually survived longer, because tumor shrinkage is considered reasonably likely to predict a real clinical benefit.

Unmet medical needs   Filling an unmet medical need is defined as providing a therapy where none exists or providing a therapy that may be potentially superior to existing therapy: for example, showing superior effectiveness, avoiding serious side effects of an available treatment, improving the diagnosis of a serious disease where early diagnosis results in an improved outcome, or decreasing the clinically significant toxicity of an accepted treatment.

U.S.C.   *United States Code.*

CHAPTER 26

POST–DRUG APPROVAL ACTIVITIES

26.1 POSTMARKET REQUIREMENTS AND COMMITMENTS (FDAa,b; FDA, 2008, 2011e)

The phrase *postmarketing requirements and commitments* refers to studies and clinical trials that sponsors conduct after approval to gather additional information about a product's safety, efficacy, or optimal use. Some of the studies and clinical trials may be required [referred to as *postmarket requirements* (PMRs)] or committed to be conducted by a sponsor [referred to as *postmarket commitments* (PMCs)]. The U.S. Food and Drug Administration (FDA) could require the following studies or clinical trials:

- Demonstration of benefits for drugs under the accelerated approval requirements in the *Code of Federal Regulations*, 21 CFR Subpart H (21 CFR 314.510 and 601.41)
- Deferred pediatric studies [21 CFR 314.55(b) and 601.27(b)], where studies are required under the Pediatric Research Equity Act (PREA)
- Demonstration of safety and efficacy in humans to be conducted at the time of use of products approved under the Animal Efficacy Rule [21 CFR 314.610(b)(1) and 601.91(b)(1)]

Section 130(a) under Title I of the FDA Modernization Act of 1977 became law on November 21, 1997, and added Sec. 506B (Reports of Postmarketing Studies) to the act [21 *United States Code* (U.S.C.) 356b]. This provision requires sponsors of new drug applications (NDA) and abbreviated NDAs (ANDAs) to report to the FDA annually on the progress of PMCs and requires the FDA to make certain information available to the public.

Section 901 in Title IX of the FDAAA (Food and Drug Administration Amendments Act of 2007) created Sec. 505(o) of the Food, Drug, and Cosmetic (FD&C) Act, which authorizes (effective as of March 25, 2008) the FDA to require certain studies and clinical trials for prescription drugs and biological products approved under Sec. 505 of the FD&C Act or Sec. 351 of the Public Health Services Act. Under Sec. 505(o)(3)(D)(i), the FDA must find that adverse event

Integrated Pharmaceutics: Applied Preformulation, Product Design, and Regulatory Science,
First Edition. Antoine Al-Achi, Mali Ram Gupta, William Craig Stagner.
© 2013 John Wiley & Sons, Inc. Published 2013 by John Wiley & Sons, Inc.

reporting under Sec. 505(k)(1) of the FD&C Act and the new pharmacovigilance system that will be established under Sec. 505(k)(3) of the FD&C Act will not be sufficient to meet the purposes described in Sec. 505(o)(3)(B) before requiring a postmarketing study. Similarly, under Sec. 505(o)(3)(D)(ii), the FDA must find that a postmarketing study will not be sufficient to meet the purposes described in Sec. 505(o)(3)(B) before requiring a postmarketing clinical trial. Thus, the postmarket studies or clinical trials may be required to:

- Assess a known risk related to the use of a drug
- Assess signals of serious risk related to the use of the drug
- Identify an unexpected serious risk when available data indicate the potential for a serious risk
- Demonstrate clinical benefits for drugs under the accelerated approval requirements in 21 CFR Subpart H (21 CFR 314.510 and 601.41)
- Complete deferred pediatric studies [21 CFR 314.55(b) and 601.27(b)] where studies are required under the Pediatric Research Equity Act
- Demonstrate safety and efficacy in humans, to be conducted at the time of use of products approved under the Animal Efficacy Rule [21 CFR 314.610(b)(1) and 601.91(b)(1)]

Postmarketing study commitments or the final report may be requested either directly from the applicant or from FDA under the Freedom of Information Act (5 U.S.C. 552). 21 CFR 314.81 (FDA, 2011a) requires an applicant to submit an annual report, within 60 days after the anniversary date of U.S. approval, for every approved NDA and ANDA; the status of all open commitments, including postmarket study commitments, must be included in this annual report. Section 505(o)(3)(E)(ii) of the FD&C Act requires an applicant to report periodically on the status of any study or clinical trial required under this section. This section also requires an applicant to report periodically to the FDA on the status of any study of clinical trial otherwise undertaken to investigate a safety issue. The FDA will consider the submission of an annual report as specified by Sec. 506B and 21 CFR 314.81(b)(2)(vii) to satisfy the periodic reporting requirement under Sec.505(o)(3)(E)(ii), provided that the report includes the elements listed in Sec. 505(o) and 21 CFR 314.81(b)(2)(vii). To comply with Sec. 505(o), an annual report must include a report on the status of any study or clinical trial otherwise under-taken to investigate a safety issue. Failure to submit an annual report for studies or clinical trials required under Sec. 505(o) by the date required will be considered a violation of Sec. 505(o)(3)(E)(ii) and could result in enforcement action.

26.2 POSTAPPROVAL MANUFACTURING CHANGES
(FDA, 2009a, 2010)

After a drug is approved and marketed, the FDA uses different mechanisms to assure that manufacturers (1) adhere to the terms and conditions of approval described in the application, and (2) produce the drug in a consistent and controlled manner. This is done by periodic unannounced inspections of drug production and control facilities by the FDA's field investigators and analysts. An applicant must notify the FDA of a change to an approved NDA or ANDA application in accordance with the statutory and regulatory requirements, including Sec. 506A of the FD&C Act (21 U.S.C. 356a), which was added to by Sec. 116 of the Food and Drug Modernization Act (Public Law 105–115) and 21 CFR 314.70. These manufacturing changes may be characterized as major, moderate, or minor, based on the potential effect that a given change may have on product quality (product identity, strength, quality, purity, and potency) as it relates to the safety or effectiveness of a product.

- If a change is considered to be major, an applicant must submit and receive FDA approval of a supplement before the product made with the manufacturing change is distributed.
- If a change is considered to be moderate, an applicant must submit a supplement at least 30 days before the product is distributed or, in some cases, submit a supplement at the time of distribution.
- If a change is considered to be minor, an applicant may proceed with the change but must notify the FDA of the change in an annual report.

For any change an applicant must assess how the change affects the product's safety and effectiveness through appropriate studies to determine the change's category and the agency reporting requirements (as a supplement or an annual report). Regardless of the category reported, applicants continue to be required to comply with the current good manufacturing practice for finished pharmaceutical regulations (21 CFR 210 and 211) and, if applicable, to Q7A, Good Manufacturing Practice Guidance for Active Pharmaceutical Ingredients (FDA, 2001). For further details on the characterization of manufacturing changes, refer to the following documents:

- FDA Guidance for Industry: Changes to an Approved NDA or ANDA, April 2004 (http://www.fda.gov/downloads/Drugs/GuidanceComplianceRegulatory Information/Guidances/ucm077097.pdf)
- FDA Guidance for Industry: Chemistry, Manufacturing, and Controls—Changes to an Approved NDA or ANDA, May 30, 2007 (http://www. fda.gov/downloads/AnimalVeterinary/GuidanceComplianceEnforcement/ GuidanceforIndustry/UCM052415.pdf)

26.2.1   Contents of Annual Report Notification (FDA, 2010)

In accordance with 21 CFR 314.81(b)(2)(iv)(b) and 314.70(d)(3), the applicant must include the following in a notification of a change in an annual report:

- A full description of the Chemistry, manufacturing, and controls) (CMC) changes that were made, where the applicant believes the changes to be minor in nature and not requiring a supplement application under Secs. 314.70(b) and (c), including:

- ○ A list of each change by the date and the change that was made
- ○ The relevant summary of data from studies and tests performed to evaluate the effects of the change, including cross-references to validate protocols and standard operating practices and policies
- ○ A list of all drug products involved

- • A description of each change in enough detail to allow the agency to determine quickly whether the appropriate category was assigned
- • A list of all changes in the summary section of the report [see 21 CFR 314.81(b)(2)(i)]

For further details, refer to the primary reference used in preparing this section, FDA Guidance for Industry: CMC Post Approval Manufacturing Changes Reportable in Annual Reports, June 2010.

26.3   CLINICAL PHASE 4 STUDIES: POSTMARKETING SURVEILLANCE AND RISK ASSESSMENT
(FDA, 2009a, 2011b)

The goal of the Center for Drug Evaluation and Research's (CDER's) postmarketing surveillance system is to monitor the ongoing safety and effectiveness of marketed drugs for unexpected adverse events and to recommend ways to manage potential risks to public health. In accordance with Title IX, Sec. 915 of the Food and Drug Administration Amendments Act of 2007, which created a new section, 505(r), of the FD&C Act [21 U.S.C. 355(r)], these postmarketing evaluations are performed 18 months after approval of the drug or after its use by 10,000 individuals, whichever is later. The FDA posts this information to provide patients and health care providers with better access to information about drug safety. Despite CDER's vigilant premarket review, all possible side effects of a drug cannot be anticipated based on preapproval studies involving only several hundred to several thousand patients. Hence, the FDA maintains a system of postmarketing surveillance and risk assessment programs. This is done through a variety of activities, including identification of changes in product labeling, reevaluation of approval decision, and the use of other tools that are outlined below. This work is accomplished primarily through the CDER's Division of Pharmacovigilance and Epidemiology.

26.3.1   Adverse Event Experience Reporting
(FDA, 1999, 2009a, 2011b)

Manufacturers of prescription medical products, both applicant and nonapplicant holders, prescription drug packers, and self-distributors of prescription drugs are required by regulations (21 CFR 310.305, 314.80, and 314.98) to submit postmarketing adverse event experience (ADE) reports to the FDA. These regulations also apply to over-the-counter (OTC) drugs that have approved applications, including those initially marketed as prescription drugs under approved applications (i.e., prescription-to-OTC switched drugs). In addition, drug manufacturers must submit either error and accident reports or drug quality reports when deviations from current good manufacturing practice regulations occur. All serious, unexpected (not listed in the drug product's current labeling) ADEs must be reported (MedWatch FDA Form 3500A is used for this mandatory reporting of ADEs) to the FDA within 15 calendar days. All others nonserious ADEs are reported to the FDA periodically (quarterly during the first three years following approval of the drug and yearly thereafter). A serious ADE is one that is fatal, life-threatening, persistent, disabling or incapacitating, requires inpatient hospitalization, prolongs existing hospitalization, or involves a patient experiencing cancer (refer to the Glossary for conditions referred to as serious ADEs). The 15-day time frame begins on the day the manufacturer or any of its affiliates received the ADE information or the date when the report is reclassified as "serious and unexpected." During site inspection, an FDA investigator will check if the firm has written procedures describing how ADEs are investigated, evaluated, and submitted to the agency and whether the procedures are followed. Clear deviations, such as failure to submit ADE reports, failure to investigate an ADE event promptly, inaccurate information, incomplete disclosure of available information, lack of written procedures, or failing to adhere to reporting requirements, are cited on FDA Form 483. This form is issued to the firm management at the conclusion of an inspection and describes conditions found to be in violation of the FD&C and related acts. Up to 25 serious unexpected ADE reports are submitted with the establishment inspection report to the HFD-332 (the Surveillance and Data Analysis Branch, CDER). FDA considers the following violations significant enough to warrant issuance of a warning letter:

- Failure to submit ADE reports for serious and unexpected adverse drug experience events [21 CFR 314.80(c)(1) and 310.305(c)]
- Fifteen-day alert reports, submitted as part of a periodic report and not otherwise submitted under separate cover as 15-day alert reports (this applies to foreign and domestic ADE information from scientific literature and postmarketing studies as well as spontaneous reports) [21 CFR 314.80(c)(1) and 310.305(c)]
- Fifteen-day alert reports that are inaccurate and/or not complete
- Fifteen-day alert reports that are not submitted on time
- Repeated or deliberate failure to maintain or submit periodic reports in accordance with the reporting requirements [21 CFR 314.80(c)(2)]
- Failure to conduct a prompt and adequate follow-up investigation of the outcome of ADEs that are serious and unexpected [21 CFR 314.80(c)(1) and 310.305(c)(3)]
- Failure to maintain ADE records for marketed prescription drugs

- Failure to have written procedures for investigating ADEs for marketed prescription drugs without approved applications [21 CFR 314.80(i) and 211.198]
- Failure to submit 15-day reports derived from a postmarketing study where there is a reasonable possibility that the drug caused the adverse drug experience

The FDA may take the following enforcement actions if the firm fails, in a proper and timely manner, to address all the existing violations and violations reported in the warning letter:

- Withdrawal of NDA or ANDA approval: warranted only when agency evaluation of ADEs shows that a drug is no longer safe and should be removed from the market, or the drug product requires labeling changes that the applicant is unwilling to make [FD&C Act, Sec. 505(e)].

- Injunction: considered appropriate when follow-up inspection or investigation shows that the firm has a continued pattern of significant and substantial deviations despite previous attempts by the agency to obtain compliance. Repeated failures on the part of the firm to submit required serious ADEs or failure to take steps to ensure that required serious ADE reports are complete and accurate may warrant injunctive action.

- Citation or prosecution: may be considered (in consultation with the HFD-322) when it is evident that a firm is not submitting reports for required serious ADEs, or is withholding important information, or submitting false information (the submission of which might have resulted in the agency requiring labeling changes or withdrawal of an application).

26.3.2   Adverse Event Reporting System (FDA, 2009a,b)

The adverse event reporting system (AERS) is a computerized information database designed to support the FDA's postmarketing safety surveillance program for all approved drug and therapeutic biologic products. The safety reports in AERS are evaluated by multidisciplinary staff safety personnel, epidemiologists, and other scientists in the CDER's Office of Surveillance and Epidemiology to detect safety signals, especially unlabeled serious reactions, and to monitor drug safety. As a result of AERS safety reports, the FDA may take regulatory actions to improve product safety and protect the public health by updating a product's labeling information, sending out a "Dear Health Care Professional" letter, or reevaluating an approval decision.

The AERS contains over 4 million reports of adverse events and reflects data from 1969 to the present and summary statistics of data since 2000. The AERS does have limitations; for example, there is no certainty that the event reported was actually due to the product. The FDA does not require that a causal relationship between a product and event be proven, and reports do not always contain enough detail to evaluate an event properly. Further, the FDA does not receive all adverse event reports that occur with a product. Many factors can influence whether or not an event will be reported, such as the time a product has been marketed and the publicity that an event receives. Therefore, the AERS cannot be used to accurately calculate the incidence of an adverse event in the U.S. population.

26.3.3   MedWatch Program (FDAc; FDA, 1998a; 2009c)

MedWatch, initiated in June 1993, is the FDA's safety information and adverse event reporting program and plays a critical role in the agency's postmarketing surveillance. Although this is a voluntary program; health professionals report adverse reactions, product problems, and use errors related to drugs, biologics, medical devices, dietary supplements, cosmetics, and infant formula. MedWatch has several goals:

- To make it easier for health care providers to report serious events and to show health care providers that the FDA is interested in learning about serious events involving deaths, life threatening conditions, hospitalization, disability, a congenital anomaly, or intervention in one of these serious conditions
- To accept reports of product quality problems with drugs or devices, suspected product tampering, or counterfeit drugs, medical errors, and any other unexpected problems with the product that could pose a safety risk

Appendix 26.1 provides examples of the four types of problems with human health care products that are reported to MedWatch (FDA, 2009c). MedWatch receives about 25,000 voluntary reports, each year. These reports, along with the mandatory reporting of the ADEs, are entered into the AERS database and evaluated. Once the FDA receives early signals of possible safety issues, the agency can rapidly identify problems and take appropriate actions. These actions include broadcasting new safety information, such as MedWatch Alerts, letters to health care professionals, labeling changes, product withdrawals, or further postmarketing research. Over the past few years, MedWatch postmarketing drug safety alerts have increased. In 2007, 161 safety warnings about drugs, medical biological, medical devices, and nutritional products were issued. One example of a widely publicized alert occurred on May 21, 2007; this alert reported that people receiving the antibiotic drug rosiglitazone showed increased risk of ischemic myocardial infraction and cardiac death (FDA, 2007).

26.3.4   The Office of Prescription Drug Promotion
Surveillance Program (FDA, 2011c)

The Office of Prescription Drug Promotion (OPDP; formerly, the Division of Drug Marketing Advertising and Communications its stated mission "to protect the public health by assuring prescription drug information is truthful, balanced, and accurately communicated through a comprehensive surveillance, enforcement and education program, and by fostering better communication of labeling and promotional information to both healthcare professionals and consumers." Refer to Section 26.4 for more details on the prescription drug advertising and promotional labeling direct-to-consumers program.

26.3.5 Medication Errors (FDAd)

The FDA receives medication error reports on marketed human drugs, including prescription, generic and over-the-counter drugs, nonvaccine biological products, and devices. The National Coordinating Council for Medication Error Reporting and Prevention defines a *medication error* as "any preventable event that may cause or lead to inappropriate medication use or patient harm while the medication is in the control of the health care professional, patient, or consumer." Such events may be related to:

- Professional practice, health care products, procedures, and systems, including prescribing
- Order communication
- Product labeling, packaging, and nomenclature
- Compounding
- Dispensing
- Distribution
- Administration
- Education
- Monitoring
- Use

The CDER medication errors program staff reviews medication error reports from several sources, including the *United States Pharmacopeia*, the Institute for Safe Medication Practices, the Medication Errors Reporting Program, and Med-Watch. The CDER staff classifies them to determine the cause and type of error. The FDA can change the way it labels, names, or package a drug product, depending on the findings. The FDA educates the public on an ongoing basis to prevent repeat errors. In subsequent sections we describe areas the FDA is working on to reduce medication errors.

26.3.5.1 Bar Code Label Rule (FDA, 2009d)

This rule became effective on April 26, 2004. The bar code rule applies to prescription drugs, biologics (other than blood, blood components, and devices regulated by the CBER), and OTC drugs that are commonly used in hospitals. Manufacturers, repackers, relabelers, and private label distributors of prescription and OTC drugs are subject to the bar code requirements. The bar codes provide unique identification information about drugs given at a patient's bedside. According to Lottie Lockert, a nursing administrator at the Houston Veterans' Affairs Medical Center, before giving medications, nurses use a scanner to pull up a patient's full name and Social Security number (or some other unique identification number assigned to the patient) by scanning the patient's wristband along with the medication, label(s); if there is not a match between the patient and the medication, or if there is another problem, a warning box appears on the nurse's computer screen.

26.3.5.2   Drug Name Confusion (FDA, 2009d)   The FDA reviews and rejects names if they look or sound alike, to minimize confusion between drug names. The FDA tracks reports of errors due to drug name confusion and spreads the word to health professionals along with recommendations for avoiding future problems. For example, the FDA has reported errors involving the inadvertent administration of methadone, a drug used to treat opiate dependence, rather than the intended Metadate ER (methylphenidate) for the treatment of attention-deficit/hyperactivity disorder.

26.3.6   Product Problems (FDAe)

Product problems should be reported to the FDA when there is a concern about the quality, authenticity, performance, or safety of any medication or device. Problems with product quality may occur during manufacturing, shipping, or storage. They include:

- Suspect counterfeit product
- Product contamination
- Defective components
- Poor packaging or product mix-up
- Questionable stability
- Device malfunctions
- Labeling concerns

A pharmacist is often the first to recognize a product quality problem with drugs, and nurses are often the first to recognize a problem with a medical device. Suspected product or device quality problems can be reported to the FDA through MedWatch.

26.3.7   Drug Shortages (FDA, 2009a, 2011d)

It is the FDA's policy to attempt to prevent or alleviate shortages of medically necessary products. Drug shortages may arise from various causes, such as the unavailability of raw materials or packaging components, marketing decisions, and enforcement issues. Refer to the FDA website Drug Shortages (FDA, 2011d) and the Manual of Policies and Procedures for an overview of CDER's drug shortage management responsibilities, and how drug shortage reports are processed.

26.3.8   Therapeutic Inequivalence Reporting
(FDA, 1998b, 2009a)

In recent years, the CDER has been receiving an increased number of reports of drug products that fail to work for patients because the product simply has no effect or is toxic. These problems could be attributed to switching brands of drugs. The FDA created the Therapeutic Inequivalence Action Coordinating Committee as part of the CDER on September 14, 1988 to identify and evaluate reports of therapeutic failures and toxicity that could indicate that one product is not equivalent to a similar product.

26.4 PRESCRIPTION DRUG ADVERTISING AND PROMOTIONAL LABELING DIRECT TO CONSUMERS (FDA, 1998c)

Part of CDER's mission is to assure that prescription drug information provided by drug firms is truthful, balanced, and communicated accurately. This mission extends to direct-to-consumer advertising. The mission is accomplished through a comprehensive surveillance, enforcement, and education program, and by fostering better communications of labeling and promotional information (refer to the Glossary for information on the prescription drug advertising and labeling requirements) to both health professionals and consumers. This work is accomplished primarily through CDER's OPDP. This office administers an FDA education outreach program (the Truthful Prescription Drug Advertising and Promotion, sometimes called the "bad ad program") to help health care providers to recognize misleading prescription drug promotions and to provide them an easy way to report misleading material to the agency.

Manufacturers voluntarily submit promotional pieces, including introductory (launch) promotional campaign material, to OPDP for comments [as specified in 21 CFR 202.1(g)(4)] before dissemination. Under 21 CFR 10.85, companies may request an advisory opinion on promotional pieces before those pieces are used by the company. Review of launch campaigns is the highest priority of the OPDP because these campaigns create the initial and often lasting impression to prescribers regarding the product's safety and efficacy. Effective March 25, 2008, Section 906 of the FDAAA made it mandatory that the published direct-to-consumer advertisements for prescription drugs include the following text: "You are encouraged to report negative side effects of prescription drugs to the FDA. Visit www.fda.gov/medwatch, or call 1-800-FDA-1088." The prescription drug advertising must be accurate, balanced in terms of the risk and benefit information, consistent with the prescribing information approved by the FDA, and include information only if supported by strong evidence from clinical studies.

The OPDP monitors all types of prescription drug promotions (e.g., sales representative presentations, speaker program presentations, TV and radio advertisements, and all other written or printed prescription drug promotional materials) for compliance with the law. A promotion cannot be false or misleading and must be presented with fair balance. The OPDP conducts surveillance of submissions from drug applicants. FDA's regulations, 21 CFR 314.81(b)(3)(I), require drug applicants to:

- Submit specimens of mailing pieces and any other labeling devised for promotion of the drug product at the time of initial dissemination of the labeling and at the time of initial publication of the advertisement for a prescription drug product

A completed transmittal Form FDA-2253 (Transmittal of Advertisements and Promotional Labeling for Drugs for Human Use) must be submitted and a copy of the product's current professional labeling must be included. Form FDA 2253 may be obtained from the following address:

PHS Forms and Publications Distribution Center

12100 Parklawn Drive

Rockville, MD 20857

The OPDP staff also attends medical professional conferences, where they observe company exhibition booths and collect promotional materials to review.

The OPDP reviewers evaluate whether the materials meet the requirements for advertising or promotional labeling and may recommend one of the following enforcement actions:

- Untitled letter: address promotion violations that are less serious than those addressed in warning letters. A reviewer's untitled letter is peer-reviewed and has the concurrence of the branch chief. In such letters, the OPDP usually requests that a company take specific action to bring a company into compliance within a certain amount of time, usually 10 working days. There is no requirement that the agency take enforcement action. Letters may serve as a basis for additional regulatory action.

- Warning letter: written communications from the OPDP notifying a company that one or more promotional pieces or practices are in violation of the law. If the company does not take appropriate and prompt action to correct the violations, as requested in the warning letter, there may be further enforcement action without further notice. Warning letters are issued by the OPDP division director and receive concurrence from appropriate officials in the Center for Drug Evaluation and Research. Companies have 15 working days to respond to a warning letter. Warning letters are put on display at the time of issuance in the FDA's Freedom of Information Office.

- Other enforcement actions: recalls, seizures, injunctions, administrative detention, and criminal prosecution are also possible.

REFERENCES

FDAa. Postmarket requirements and commitments: legislative background. http://www.fda.gov/Drugs/ GuidanceComplianceRegulatoryInformation/Post-marketingPhaseIVCommitments/ucm064633.htm. Accessed Dec. 2011.

FDAb. Postmarket requirements and commitments: introduction. http://www.fda.gov/Drugs/ GuidanceComplianceRegulatoryInformation/Post-marketingPhaseIVCommitments/default.htm. Accessed Dec. 2011.

FDAc. MedWatch: the FDA safety information and adverse event reporting program. http://www.fda. gov/Safety/MedWatch/default.htm. Accessed Jan. 2012.

FDAd. Medication errors. http://www.fda.gov/Drugs/DrugSafety/MedicationErrors/default.htm. Accessed Jan. 2012.

FDAe. Product problems. http://www.fda.gov/Safety Accessed Jan. 2012.

FDA. 1998a. MedWatch program. CDER Handbook. Revised Mar. 16, 1998, p. 44. http://www.fda.gov/downloads/AboutFDA/CentersOffices/CDER/UCM198415.pdf. Accessed Dec. 2011.

———. 1998b. Therapeutic inequivalence reporting. CDER Handbook. Revised Mar. 16, 1998, p. 67. http://www.fda.gov/downloads/AboutFDA/CentersOffices/CDER/UCM198415.pdf. Accessed Dec. 2011.

———. 1998c. Prescription drug advertising and promotional labeling. CDER Handbook. Revised Mar. 16, 1998, pp. 49–61. http://www.fda.gov/downloads/AboutFDA/CentersOffices/CDER/UCM198415.pdf. Accessed Dec. 2011.

———. 1999. Staff Manual Guide; Chapter 53: Postmarketing surveillance and epidemiology—human drugs adverse drug effects. 7353.001. Updated May 12, 2010. http://www.fda.gov/Drugs/GuidanceComplianceRegulatoryInformation/Surveillance/ucm129115.htm. Accessed Dec. 2011.

———. 2001. Guidance for industry: Q7A—Good manufacturing practice guidance for active pharmaceutical ingredients. Aug. 25, 2001. http://www.fda.gov/downloads/RegulatoryInformation/Guidances/UCM129098.pdf. Accessed Dec. 2011.

———. 2006. Drug shortage management. Manual of Policies and Procedures. MAPP 6003.1. Office of New Drugs, CDER. Effective date Sept. 26, 2006. http://www.fda.gov/downloads/AboutFDA/CentersOffices/CDER/ManualofPoliciesProcedures/UCM079936.pdf. Accessed Dec. 2011.

———. 2007. Avandia (rosiglitazone). Posted May 21, 2007. http://www.fda.gov/Safety Accessed Jan. 2012.

———. 2008. Postmarketing surveillance programs. Updated Aug. 19, 2008. http://www.fda.gov/Drugs/GuidanceComplianceRegulatoryInformation/Surveillance/ucm090385.htm. Accessed Dec. 2011.

———. 2009a. Surveillance: post drug-approval activities. Updated Nov. 10, 2009. http://www.fda.gov/Drugs/GuidanceComplianceRegulatoryInformation/Surveillance/default.htm. Accessed Dec. 2011.

———. 2009b. Protecting America's health through human drugs. FDA Consumer Magazine and FDA CDER, 4th ed. Jan. 2006. Updated May 22, 2009. http://www.fda.gov/Drugs/ResourcesForYou/Consumers/ucm143455.htm. Accessed Dec. 2011.

———. 2009c. FDA 101: How to use the consumer complaint system and MedWatch. Feb. 27, 2009. http://www.fda.gov/ForConsumers/ConsumerUpdates/ucm049087.htm. Accessed Jan. 2012.

———. 2009d. Strategies to reduce medication errors: working to improve medication safety. Updated Aug. 12, 2011. http://www.fda.gov/Drugs/ResourcesForYou/Consumers/ucm143553.htm. Accessed Dec. 2011.

———. 2010. Guidance for industry: CMC post approval manufacturing changes reportable in annual reports. June 2010. http://www.fda.gov/downloads/Drugs/GuidanceCompliance RegulatoryInformation/Guidances/UCM217043.pdf. Accessed Dec. 2011.

———. 2011a. *Code of Federal Regulations*. 21 CFR 314. Applications for FDA approval to market a new drug. Title 21, Food and drugs; Chapter I, Food and Drug Administration/Department of Health and Human Services; Subchapter D, Drugs for human use. Updated Apr. 1, 2011. http://www.accessdata.fda.gov/scripts/cdrh/cfdocs/cfcfr/CFRsearch.cfm?CFRPart=314. Accessed Dec. 2011.

———. 2011b. Postmarketing drug and biologic safety evaluations. Updated Nov. 10, 2011. http://www.fda.gov/Drugs/GuidanceComplianceRegulatoryInformation/Surveillance/ucm204091.htm. Accessed Dec. 2011.

———. 2011c. The Office of Prescription Drug Promotion (OPDP) [formerly, Division of Drug Marketing, Advertising and Communications (DDMAC)]. http://www.fda.gov/AboutFDA/CentersOffices/OfficeofMedicalProductsandTobacco/CDER/ucm090142.htm. Accessed Dec. 2011.

———. 2011d. Drug shortages. Updated Dec. 2, 2011. http://www.fda.gov/drugs/drugsafety/drugshortages/default.htm. Accessed Dec. 2011.

_____. 2011e. Guidance for industry: Postmarketing studies and clinical trials—Implementation of Section 505(o)(3) of the Federal Food, Drug, and Cosmetic Act. Apr. 2011. http://www.fda.gov/downloads/drugs/guidancecomplianceregulatoryinformation/guidances/ucm172001.pdf. Accessed Dec. 2011.

GLOSSARY

ADE	Adverse event experience.
Adverse drug experience	Adverse drug experience includes an adverse event occurring in the course of the use of a drug in professional practice.
AERS	Adverse event reporting system.
ANDA	Abbreviated new drug application.
Animal efficacy rule requirement	On May 31, 2002, the FDA published the final rule to allow use of animal data for evidence of a drug's effectiveness for certain conditions when the drug cannot, ethically or feasibly, be tested in humans. In these situations, certain new drug and biological products used to reduce or prevent the toxicity of chemical, biological, radiological, or nuclear substances may be approved for marketing based on evidence of effectiveness derived from appropriate studies in animals and any additional supporting data. Under this rule, the applicant must conduct postmarket studies or clinical trials to verify and describe the drug's clinical benefit when such studies or clinical trials are feasible and ethical. Such postmarket studies or clinical trials may not be feasible until an emergency arises that necessitates use of the product.
CDER	Center for Drug Evaluation and Research.
CFR	*Code of Federal Regulations*.
Clinical trials	Clinical trials are any prospective investigations in which the applicant or investigator determines the method of assigning a drug product(s) or other interventions to one or more human subjects.
CMC	Chemistry, manufacturing, and controls.
Epidemiology	Study of the distribution and determinants of health-related states and events in specified populations, and the application of this study to the control of health problems.
FD&C Act	Food, Drug, and Cosmetic Act.
FDA	U.S. Food and Drug Administration.
FDA Form 483	An FDA Form 483 is issued to firm management at the conclusion of an inspection when an investigator(s) has observed any conditions that in their judgment may constitute violations of the FD&C and related acts. Companies are responsible to take corrective action to address objectionable conditions cited and any related noncited objectionable conditions that might exist and are encouraged to respond to the FDA Form 483 in writing with their corrective action plan and then implement that corrective action plan expeditiously. FDA Form 483 does not constitute a final agency determination of whether any condition is in violation of the FD&C Act or any of its relevant regulations. FDA Form 483 is considered, together with a written report called an establishment inspection report, all evidence or documentation collected on-site, and any responses made by the company. The agency considers all of this information and then determines what further action, if any, is appropriate to protect the public health.

FDAAA

Food and Drug Administration Amendments Act.

Labeling (product/ promotional)

Section 201(m) of the FD&C Act defines labeling as all labels and other written, printed, or graphic matters (1) upon any article or any of its containers or wrappers, or (2) accompanying such an article. The regulations provide examples of labeling under 21 CFR 202.1(l)(2): "Brochures, booklets, mailing pieces, detailing pieces, file cards, bulletins, calendars, price lists, catalogs, house organs, letters, motion picture films, film strips, lantern slides, sound recordings, exhibits, literature, and reprints and similar pieces of printed, audio or visual matter descriptive of a drug and references published (for example, the *Physician's Desk Reference*) for use by medical practitioners, pharmacists, or nurses, containing drug information supplied by the manufacturer, packer, or distributor of the drug and which are disseminated by or on behalf of its manufacturer, packer, or distributor are hereby determined to be labeling as defined in section 201(m) of the FD&C Act."

Labeling contents

Labeling must include the established name, proprietary name (if any), adequate directions for use, and adequate warnings. The agency considers the approved product labeling, sometimes called the *full prescribing information*, to be adequate directions for use and adequate warning.

Labeling requirements: Exceptions

Reminder labeling is exempt from the requirements for adequate directions for use and adequate warnings. Reminder labeling, as defined in 21 CFR 201.100(f), is exempted. Reminder labeling calls attention to the name of the drug product but does not include indications or dosage recommendations for use. Reminder labeling may contain only the proprietary name of the drug, the established name of each active ingredient, and optionally, information relating to quantitative ingredient statements, dosage form, and quantity of package contents, price, and other limited information. The exemption does not apply to products with "black box" warnings in their approved product labeling.

NDA

New drug application.

OPDP

Office of Prescription Drug Promotion.

OTC

Over the counter.

Prescription drug advertising (PDA)

21 CFR 202.1(l)1 states that advertisements subject to Sec. 502(n) of the FD&C Act include advertisements published in journals, magazines, other periodicals, and newspapers; and broadcast through media such as radio, television, and telephone communications systems. This is not a comprehensive list of advertising media subject to regulation. For example, FDA also regulates advertising conducted by sales representatives, on computer programs, through fax machines, or on electronic bulletin boards.

PDA contents

Under Sec. 502(n) of the FD&C Act, advertisements must include the established name, the brand name (if any), the formula showing quantitatively each ingredient, and information in brief summary that discusses side effects, contraindications, and effectiveness. The brief summary is discussed further in 21 CFR 202.1(e)(1).

PDA regulation exceptions	There are a few exceptions but only to the requirement to provide a true statement of information in brief summary as required under 21 CFR 202.1(e)(1). 21 CFR 202.1(e)(2) describes which ads are exempt:

1. *Reminder advertisements*: advertisements that call attention to the name of a drug product but do not include indications or dosage recommendations for use of the product, or any other representation. Reminder ads contain the proprietary name of the drug and the established name of each active ingredient. They may also contain additional limited information, such as the name of the company, price, or dosage form. The exception does not apply to products with "black box" warnings in their approved product labeling.
2. *Advertisements of bulk-sale drugs*: promote sale of a drug in bulk packages to be processed, manufactured, labeled, or repackaged and contain no claims for the therapeutic safety or effectiveness of the drug.
3. *Advertisements of prescription-compounding drugs*: promote sale of a drug for use as a prescription chemical or other compound for use by registered pharmacists.

PMC	Postmarket commitment
Postmarket commitment	Studies or clinical trials to which sponsors commit but that they are not required to conduct.
PMR	Postmarket requirement.
Postmarket requirements and commitments	Studies and clinical trials that sponsors conduct after approval to gather additional information about a product's safety, efficacy, or optimal use. Some may be required; others may be studies or clinical trials that a sponsor has committed to conduct.
Serious adverse drug experience	An adverse drug experience that:

(A) Results in (1) death; (2) an adverse drug experience that places the patient at immediate risk of death from the adverse drug experience as it occurred (not including an adverse drug experience that might have caused death had it occurred in a more severe form); (3) inpatient hospitalization or prolongation of existing hospitalization; (4) a persistent or significant incapacity or substantial disruption of the ability to conduct normal life functions; or (5) a congenital anomaly or birth defect; or

(B) Based on appropriate medical judgment, may jeopardize the patient and may require a medical or surgical intervention to prevent an outcome described under subparagraph (A).

Serious risk	Risk of a serious adverse drug experience.
Serious risk—unexpected	A serious adverse drug experience that is not listed in the labeling of a drug or that may be symptomatically or pathophysiologically related to an adverse drug experience identified in the labeling, but differs because of greater severity, specificity, or prevalence.
Studies	All other investigations, such as investigations with humans that are not clinical trials as defined above (e.g., observational epidemiologic studies), animal studies, and laboratory experiments.
U.S.C.	*United States Code*.

APPENDIX

APPENDIX 26.1 Problems to Report to MedWatch

1. Serious Adverse Event	2. Product Quality Problem	3. Product Use Error	4. Problem with Different Manufacturer of the Same Medicine
Death	Suspected counterfeit product	Mixing up products with similar drug names or packaging	Not getting the same results from a generic drug as from a brand zname drug or from another generic
Life-threatening situation	Potentially contaminated product indicated by suspicious odor or unusual color	Taking the wrong dose of a drug because of confusing dosing instructions on the label	
Requires admission to hospital or longer-than-expected hospital stay			
Permanent disability	Inaccurate or unreadable product labeling		
Birth defect, miscarriage, stillbirth, or birth with serious disease			
Requires medical care to prevent permanent damage			

CHAPTER 27

DRUG MASTER FILES AND EU DOSSIERS

27.1 DRUG MASTER FILES (FDA, 2001, 2011a–e)

A drug master file (DMF) is a submission of confidential information to the U.S. Food and Drug Administration (FDA) that may be used to support an investigational new drug (IND) application, a new drug application (NDA), an abbreviated new drug application (ANDA), another DMF, an export application (EA), or amendments and supplements to any of these applications. Confidentiality of information in a DMF is covered by the *Code of Federal Regulations*, 21 CFR 314.430(g) (FDA, 2011a), and is the same as for other types of submissions. Usually, a DMF contains information related to the chemistry, manufacturing, and controls (CMC) of a component of a drug product. Drug product information or other non-CMC information may also be filed in a DMF but is preferred to be filed as part of an IND, NDA, or ANDA. Here are some key facts about DMFs:

- There is no legal or regulatory requirement to file a DMF.
- A DMF is submitted solely at the discretion of the holder.
- Separate DMFs are submitted if multiple sites are involved in performing different processes.
- There are no fees for submitting DMFs.
- A DMF is not a substitute for an IND, NDA, ANDA, or EA.
- A DMF is neither approved nor disapproved.
- The technical contents of a DMF are reviewed only in connection with the review of an IND, NDA, ANDA, or EA. The CMC for a compendial [*United States Pharmacopeia–National Formulary* (U.S.P.–N.F.)] excipient or a drug substance used in some over-the-Counter (OTC) drug products may not be reviewed.
- A DMF can be referenced in several applications (of IND, NDA, ANDA, and/or EA). To review DMF information in support of an application, the FDA must receive a letter of authorization from the DMF holder in advance of such a review.

Integrated Pharmaceutics: Applied Preformulation, Product Design, and Regulatory Science,
First Edition. Antoine Al-Achi, Mali Ram Gupta, William Craig Stagner.

- A DMF is generally created to allow an applicant (who files any one of the applications listed above), other than the DMF holder, to reference the material without disclosing the contents of the DMF file to that applicant.

- If an applicant references its own material, the applicant should reference the information contained in its own IND, NDA, or ANDA directly rather than establishing a new DMF.

- There is no requirement to file DMFs in an electronic format, but once filed as such, all future submissions related to those DMFs must be in an electronic format.

The key differences between applications and DMFs are listed in Table 27.1.

27.1.1 Types of Drug Master Files (FDA, 2011b)

DMF types II through V are current at present; type I is no longer accepted but the numbering system remains the same.

- Type I: Manufacturing site, facilities, operating procedures, and personnel (no longer applicable; to avoid confusion the numbering remains the same as the original five types). Holders of DMF types II, III, and IV should not include information regarding facilities, personnel, or general operating procedures in these DMFs. Only the addresses of the DMF holder and manufacturing site and contact personnel should be submitted.

- Type II: Drug substance, drug substance intermediate, and material used in their preparation, or drug product

- Type III: Packaging material

- Type IV: Excipient, colorant, flavor, essence, or material used in their preparation

- Type V: Other: sterile manufacturing plants, biotech contract facilities, clinical, toxicology, etc.

TABLE 27.1 Key Differences Between Applications and DMFs

Applications	DMFs
Submitted to a particular review division [e.g., the Center for Drug Evaluation and Research (CDER)]	Submitted to the Office of Business Process Support.
Each submission (including supplement) is entered in the application database.	Each submission is entered into a database (different from the application database).
Each submission is assigned to a reviewer.	Reviewed only when referenced in an Application; not assigned to a reviewer.
Each submission has a due date based on review clocks set by CDER, the Prescription Drug Fee User Act, etc.	No due date is assigned.
An acknowledgment letter is sent.	No acknowledgment letter is sent.

- ○ Submission of clinical and toxicology DMFs require preclearance from the FDA.
- ○ DMFs for sterile processing and biotech manufacturing facilities may be filed without preclearance.

Each DMF should contain information and all supporting data on only one type of DMF; the DMF content can be cross-referenced to any other DMF.

27.1.2   Submissions to Drug Master Files (FDA, 2011e)

The following information should be supplied, as applicable, to a DMF:

- Transmittal (cover) letter
- Administrative and technical information
- Letter of authorization
- General information

The DMF must be in the English language. Whenever a submission contains information in another language, an accurate and complete English translation must be included. Each page of each copy of the DMF should be dated and numbered consecutively. An updated table of contents should be included with each submission. Two copies (original and duplicate copy) of all DMF submissions (original, amendments, annual reports, letter of authorization, and all other subsequent submissions) are submitted to the following address:

Central Document Room

Center for Drug Evaluation and Research

Food and Drug Administration

5901-B Ammendale Road

Beltsville, MD 20705-1266

DMF holders and their agents or representatives should retain a complete reference copy that is identical to, and maintained in the same chronological order as, their submissions to the FDA. DMFs are submitted in the FDA-specified binders (two-piece prong fasteners, 81/2 in. center to center, 31/2 in. capacity). Refer to the FDA guidelines (FDA, 2010, 2011b) for further details on submissions and binder recommendations.

27.1.2.1   Transmittal (Cover) Letter
27.1.2.1.1   Original Submissions
- Identification of submission: original, type of DMF (listed previously), and its subject
- Identification of the applications, if known, that the DMF is intended to support, including the name and address of each sponsor, applicant, or holder, and all relevant document numbers

- Signature of the holder or the authorized representative
- Typewritten name and title of the signer

27.1.2.1.2 Amendments

- Identification of submission: amendment, the DMF number, type of DMF, and the subject of the amendment
- Descriptions of the purpose of submission (e.g., update, revised formula, or revised process)
- Signature of the holder or the authorized representative
- Typewritten name and title of the signer

27.1.2.2 Administrative Information Administrative information should include the following:

27.1.2.2.1 Original Submissions

- DMF holder name and addresses of the corporate headquarter and the manufacturing or processing facility
- Name, telephone and fax numbers, and e-mail address of the DMF holder's contact person for FDA correspondence
- Agent(s), if any
- Statement of commitment
- Signed statement by the holder certifying that the DMF is current and that the DMF holder will comply with the statements made in it

27.1.2.2.2 Amendments   Submit each change (identify the change) as a separate amendment to include the following:

- Name of DMF holder; change in holder name and/or address
- DMF number
- Name and address for correspondence; change of contact person or responsible official
- Agent appointment/termination
- Request for closure
- Affected section and/or page numbers of the DMF
- Name and address of each person whose IND, NDA, ANDA, DMF, or EA relies on the subject of the amendment for support
- Number of each IND, NDA, ANDA, DMF, and EA that relies on the subject of the amendment for support, if known
- Particular items within the IND, NDA, ANDA, DMF, and EA that are affected, if known

27.1.2.3 Technical Information and Content by DMF Type   The public availability of data and information in a DMF, including the availability of data and information in the file to a person authorized to reference the file, is determined under 21 CFR 20, 21 CFR 314.420(e) (FDA, 2011c), and 21 CFR 314.430. Since DMFs cover manufacturing information, they are not usually considered releasable via a Freedom of Information Act request.

27.1.2.3.1 Type I DMF Type I DMF is no longer applicable.

27.1.2.3.2 Type II DMF: Drug Substance, Drug Substance Interme-diate, and Material Used in Their Preparation, or Drug Product A type II DMF should, in general, be limited to a single drug intermediate, drug substance, drug product, or type of material used in their preparation. All items covered by a type II DMF are expected to be manufactured under current good manufacturing practice (CGMP), and a statement of such compliance is required. Completed batch records are expected to be filed for all type II DMFs. The CMC information for some drug substances used in some OTC drug products is not reviewed by the FDA.

27.1.2.3.2.1 Drug Intermediates, Substances, and Material Used in Their Preparation Summarize all significant steps in the manufacturing and controls of the drug intermediate or substance. The FDA guidance (FDA, 1987a) provides detailed information on what should be included in a type II DMF for drug substances and intermediates. Starting material is not expected to be manufactured under CGMP. All changes, including a change from intermediate to starting material, are reported to the DMF and to all applications supported by the DMF.

27.1.2.3.2.2 Drug Product Manufacturing procedures and controls for finished dosage forms should ordinarily be submitted in an IND, NDA, ANDA, or EA. If this information cannot be submitted in these applications, it should be submitted in a DMF and the applicant should follow the guideline for submitting documentation for the manufacture of and controls for drug products (FDA, 1987b) and the guideline for submitting samples and analytical data for method validation (FDA, 1987c).

27.1.2.3.3 Type III: Packaging Material A type III DMF should provide sufficient CMC information for the FDA reviewers to assess whether the packaging components intended for use with drug substances and/or drug products are safe. Assessment of the safety of a packaging component or material is generally based on the potential or observed interaction between the material of construction (MOC) and the drug substance or drug product during storage and usage. Typically, the supplier of the packaging component or material provides the composition information. The applicant provides the remaining safety information (including the list of leachable materials). Refer to relevant U.S.P. monographs as well as U.S.P. extraction tests on plastics (U.S.P. <661>, Containers—Plastics) and the U.S.P. biological reactivity tests (U.S.P. <87>, in vitro Tests, and <88>, in vivo Tests).

The following information is typically included in a type III DMF:

- Description of the packaging component or material of construction [e.g., high-density polyethylene (HDPE) bottles, polypropylene (PP) caps, or HDPE or PP resin]
- Intended use of the component
- Names of the suppliers or fabricators of the components used in preparing the packaging material
- Quantitative or qualitative statement of the composition [chemical composition is identified corresponding to the appropriate *Code of Federal Regulations* (CFR) citation, and not simply the trade name]
- Release specification or quality attributes for the packaging component or material
- Results of certain types of qualification testing (e.g., some of the tests are described in U.S.P. <661>)
- Data supporting the acceptability of the packaging material for its intended use should also be submitted as outlined in the guidance document Container Closure Systems for Packaging Human Drugs and Biologics: Chemistry, Manufacturing, and Controls (FDA, 1999)
- Toxicological data on these materials is either included under this type of DMF or cross-referenced to another document

The following information is usually provided in the application:

- Information that addresses a container closure system's protection, compatibility, and performance attributes
- Fundamental or basic physical attributes (e.g., component dimensions); these attributes are usually addressed in the applicant's acceptance specifications
- Child resistance attributes

Information regarding packaging material can be submitted directly to the applicant for inclusion in an NDA and other applications; a DMF is not required. Refer to the table provided in MAPP 5015.5 (FDA, 2011b) to determine if the information provided in the application is adequate. This table lists the dosage forms, routes of administration, MOCs, and summarizes the information typically accepted to support the safety assessment of the MOCs used in the packaging component.

27.1.2.3.4 Type IV Excipient, Colorant, Flavor, Essence, or Material Used in Their Preparation Each additive should be identified and characterized by its method of manufacture, release specifications, testing methods, and safety. The following sources may be used for this information:

- Official compendia (e.g., U.S.P.–N.F.)
- FDA regulations for color additives (21 CFR 70 to 82)
- Direct food additives (21 CFR 170 to 173)

- Indirect food additives (21 CFR 174 to 178)
- Food substances (21 CFR 181 to 186)
- List of acceptable or generally recognized as safe excipients (FDA, 2011f)

Toxicological and other supporting information on these materials should be included under this type of DMF or cross-referenced to another document. Guidelines for a type II DMF may be helpful for preparing a type IV DMF. Except for new and novel excipients, a new route of administration or total dosing that may affect safety and efficacy, the CMC information for a compendial excipient is usually not reviewed, and therefore a DMF is not necessary. The CMC requirements for a novel excipient (one not used in an approved drug product) are the same as those for a new drug substance.

27.1.2.3.5 Type V: Other   The FDA discourages the use of type V DMFs for miscellaneous information, duplicate information, or information that should be included in one of the other types of DMFs. As required by the 21 CFR 314.410(a)(5), if a holder wishes to submit information and supporting data in a DMF that are not covered by types II through IV, it must first submit a letter of intent to the DMF staff or send e-mail to dmfquestion@cder.fda.gov to obtain clearance prior to submitting this information except for the following DMFs, which may be submitted without preclearance:

- Manufacturing site, facilities, operating procedures, and personnel for sterile manufacturing plants; see the guidance document Submission Documentation for Sterilization Process Validation in Application for Human and Veterinary Drug Products (FDA, 1994).
- Facilities for the manufacture of biotech products; see the Draft guidance document Submitting Type V Drug Master Files to the Center for Biologies Evaluation and Research (FDA, 2009).

27.1.2.4 *Letter of Authorization to FDA*   The FDA will review a DMF only when it is referenced in an application (IND, NDA, ANDA, or EA) or another DMF. A letter of authorization (LOA) does not permit anyone except the FDA to access and review the DMF. The DMF holder must submit two copies of a LOA to the DMF to permit the FDA to reference the DMF. The LOA should include the following:

- Date
- Name of DMF holder
- DMF number
- Name of person(s) authorized to incorporate information in the DMF by reference
- Specific product(s) covered by the DMF
- Submission date(s) of the product(s) covered by the DMF
- Section numbers and/or page numbers to be referenced

- Statement of commitment that the DMF is current and that the DMF holder will comply with the statements made in it
- Signature of authorizing official
- Typed name and title of official authorizing reference to the DMF

The holder should also send a copy of the LOA to the affected applicant, sponsor, or other holder who is authorized to incorporate by reference the specific information contained in the DMF. The applicant, sponsor, or other holder referencing the DMF is required to include a copy of the DMF holder's LOA in the application.

27.1.2.5 General Information and Suggestions

27.1.2.5.1 Environmental Assessment An environmental assessment usually applies to the impact of the drug's use on the environment, not on production. The National Environmental Policy Act requires that all government agencies prepare an Environmental Impact Statement (EIS) or a Finding of No Significant Impact (FONSI) when they take an action (e.g., approving a drug application). Companies submitting an application are required to submit an environmental assessment (or a waiver request) to permit the FDA to determine whether an EIS or a FONSI is needed. The FDA guidance (FDA, 1998) provides further information on the environmental assessment of human drug and biologics applications.

Since the FDA does not approve a DMF, its holder does not have to file an environmental assessment but provides sufficient information to the customer (applicant) to permit the customer to file an environmental assessment. It is expected, however, that the type II, III, and IV DMFs contain a commitment by the firm that its facilities will be operated in compliance with applicable (including local) environmental laws. Refer to 21 CFR 25 for details on environmental assessment.

27.1.2.5.2 Stability Stability study design, data, interpretation, and other information should be submitted, when applicable, as outlined in the Guideline for Submitting Documentation for the Stability of Human Drugs and Biologics (FDA, 1987d).

27.1.2.5.3 Quality by Design The principles of quality by design (QbD) can be applied to drug substance manufacture. Implementation of QbD, including establishment of design space and control strategy, in a DMF could reduce frequency for reporting changes to the DMF.

27.1.2.5.4 Inspections Inspections of drug substance manufacturers are usually triggered when there is an application under review that references a DMF for the manufacture of that drug substance. When inspecting an active pharmaceutical ingredient (API) manufacturing facility, the FDA follows compliance requirements similar to those used for the inspection of a drug product manufacturing facility (21 CFR 210 and 211). The agency will issue an observation FDA form

(FDA-483) if deficiencies are found in the inspection. Along with the typical deficiencies listed in a warning letter but without citing violation to a specific CGMP regulations, the agency makes standard opening and closing statements in a warning letter (WL: 320-09-09).

A typical opening statement: "These CGMP deviations cause your APIs to be adulterated within the meaning of Section 501(a)(2)(B) [21 *United States Code* (U.S.C.) 351(a)(2)(B)] of the FD&C Act. Section 501(a)(2)(B) states that drugs are adulterated when they are not manufactured, processed, packed, and held according to CGMP. Failure to comply with CGMP constitutes a failure to comply with the requirements of the Act."

A typical closing statement: "Until all corrections have been completed and the FDA has confirmed corrections of the violations and your firm's compliance with CGMP, this office may recommend withholding approval of any new applications or supplements listing your firm as an API manufacturer. In addition, failure to correct these deficiencies may result in FDA denying entry of articles manufactured by your firm into the United States. The articles could be subject to refusal of admission pursuant to the Section 801(a)(3) of the Act [21 U.S.C. 381(a)(3)], in that the methods and controls used in their manufacture do not appear to conform to Current Good Manufacturing Practice within the meaning of the Section 501(a)(2)(B) of the Act [21 U.S.C. 351(a)(2)(B)]."

27.1.3   DMF Formats and Review Process

27.1.3.1   DMF (CTD/eCTD) Formats   An original and duplicate copies (properly collated, fully assembled, and individually jacketed) are submitted for all DMF submissions. DMF holders and their agents or representatives should retain a complete reference copy that is identical to, and maintained in the same chronological order as, their submissions to the FDA. DMF submissions and correspondence should be sent to the address provided in Section 27.1.2. Refer to the FDA guidelines (FDA, 2010) for details on the specifics for the paper size, margins, jacketed types and colors, volume identification, pagination, and so on.

27.1.3.1.1   Common Technical Document for DMFs   The quality section of a common technical document (CTD) is acceptable to regulatory agencies of all three regions (Europe, Japan, and United States). CTDs are not intended to indicate what studies are required. The guidance (FDA, 2001) indicates an appropriate format for the data that have been acquired. Applicants should not modify the overall organizational format of a CDT except for the nonclinical and clinical summaries sections, in order to provide the best possible presentation of the technical information to facilitate the understanding and evaluation of the results.

The key CTD sections applicable to DMFs are (FDA, 2001, 2011b,e; Shaw, 2011):

- Module 1: Administrative information, to include (there are no forms for DMFs):

- ○ Sec. 1.2: Cover letter and statement of commitment
- ○ Sec. 1.3: Administrative information
  - ■ 1.3.1 Contact/sponsor/applicant information
    - ▫ 1.3.1.1 Change of address or corporate name
      - ✦ Can be used to supply addresses of DMF holder and manufacturing and testing facilities
    - ▫ 1.3.1.2 Change in contact or agent
      - ✦ Can be used to supply the name and address of contact persons and/or agents, including the agent appointment letter
  - ■ 1.4.1 Letter of authorization
    - ▫ Submission by the owner of information, giving authorization for the information to be used by another person or agency
  - ■ 1.4.2 Statement of right of reference
    - ▫ Submission by the recipient of a letter of authorization with a copy of the LOA and statement of the right of reference
  - ■ 1.4.3 List of authorized persons to incorporate by reference
    - ▫ Generally submitted in DMF annual reports
- ○ Sec. 1.12.14: Environmental Analysis
- • Module 2: A Quality overall summary is expected to be submitted.
- • Module 3: Format of the quality section
  - ○ Sec. 3.2.S: Body of data for drug substance
  - ○ Sec. 3.2.R: Regional information
    - ■ Executed batch records (United States only)
    - ■ Method (analytical) validation package: not usually submitted for DMFs but can be submitted as complete methods validation information in Sec. 3.2.S.4.3
    - ■ Comparability protocols: not usually submitted for DMFs

27.1.3.1.2 Electronic Filing of DMFs and CTD   There is no legal requirement to submit any type of application (including a DMF) in electronic format. However, a DMF originally submitted in paper may be resubmitted as an electronic DMF, but it would require that the entire DMF be resubmitted in electronic format (using the eCTD structure format). Once a DMF has been submitted in electronic form, paper documents (including LOAs) are no longer accepted.

27.1.3.2 DMF Review Process   An original DMF submission is examined upon receipt by the Office of Business Information (OBI) staff to determine whether it meets minimum requirements for format and content. If the submission is administratively acceptable, the FDA will acknowledge its receipt, assign it a DMF number, and enter it into the DMF database. This process takes about 2 to 3 weeks. This database is updated quarterly and can be accessed via the FDA website

(FDA, 2011b). The DMF database includes the DMF assigned number and type, title (subject), and the holder's name.

If the submission is administratively incomplete or inadequate, it will be returned to the submitter with a letter of explanation from the OBI staff, and it will not be assigned a DMF number. (Note that a DMF is neither approved nor disapproved.) When an FDA reviewer receives an application that refers to a DMF, the reviewer requests the DMF and reviews it following the same regulatory and scientific criteria as those applied to the application.

- If deficiencies are not found:
  - No letter is sent to the holder.
  - The applicant is not notified.
- If deficiencies are found:
  - The detailed deficiencies are communicated to the DMF holder.
  - The applicant is notified that deficiencies exist in either an information request (IR) or a complete response (CR) letter, but the nature of the deficiencies is not communicated to the applicant.

If the applicant was sent an IR letter, the review clock for the application is generally not affected and the response to an IR letter may be reviewed at the reviewer's discretion, depending on timing relative to the application review due date. If the applicant was sent a CR letter, the review clock is stopped. The application (and supporting DMFs) will be reviewed only when all issues in the CR letter (including DMF deficiencies) have been addressed. Therefore, it is important that the DMF holder addresses all deficiencies and submits an amendment to the FDA review division that identified the deficiencies as well as sending copies to all applicants who are relying on that DMF. The CR letter may have a significant impact on the sponsor's business.

27.1.4 Holder Obligations

27.1.4.1 *Changes to a DMF* For any change, addition, or deletions of information in the file, the DMF holder must notify, in writing, each applicant authorized to reference that information. The notification should be provided well in advance of making the change in order for the applicant(s) to supplement or amend any affected application(s) as needed. The DMF holder must submit two copies and describe by name, reference number, volume, and page number the information affected in the DMF (FDA, 2011c).

27.1.4.2 *Listing of Persons Authorized to Refer to a Drug Master File* A DMF is required to contain a complete list of each person currently authorized to incorporate by reference (identifying name, reference number, volume, and page number) any information in the file (FDA, 2011c). The holder should update the list in the annual update. The updated list should contain:

- Holder's name
- DMF number

- Date of the update
- Identification by name (or code) of the information that each person is authorized to incorporate
- Location of that information by date, volume, and page number
- Personal identification if an authorization was withdrawn during the preceding year

27.1.4.3 Annual Update/Annual Reports The holder should provide an annual report on the anniversary date of the original submission. This report should contain:

- Holder's name
- DMF number
- Date of the update
- Identification of all changes and additional information incorporated into the DMF since the preceding annual report on the subject matter of the DMF
- Statement that the subject matter of the DMF is current if the subject matter of the DMF is unchanged
- List of parties authorized to reference information
- Date of the LOA
- List of changes reported during the past year

The FDA will not send a reminder if the anniversary date is missed. If the list is unchanged on the anniversary date, the DMF holder should still submit a statement that the list is current. Failure to update or to keep DMF content current may cause delays in an FDA review of a pending IND, NDA, ANDA, EA, or any amendment or supplement to such an application.

27.1.4.4 Appointment of an Agent When an agent is appointed, the holder should submit a signed letter of appointment (not to be confused with the letter of authorization) to the DMF giving the agent's name, address, and scope of responsibility (administrative and/or scientific). Domestic DMF holders do not need to appoint an agent or representative, although foreign DMF holders are encouraged to engage a U.S. agent.

27.1.4.5 Transfer of Ownership/Agents as Holders To transfer ownership of a DMF to another party, the holder should so notify the FDA and authorized persons in writing. The letter should include the following:

- Name of transferee
- Address of transferee
- Name of responsible official of transferee
- Effective date of transfer

- Signature of the transferring official
- Typewritten name and title of the transferring official

The new holder must submit a letter of acceptance of the transfer and an update of the information contained in the DMF, where appropriate. Any change relating to the new ownership (e.g., plant location and methods) should be included. The new owner must take full responsibility for the accuracy of all information in the DMF and for all processes and testing performed by the manufacturer. The new owner must submit all changes as required under 21 CFR 314.420(c) (FDA, 2011c).

27.1.4.6 Major Reorganization of a DMF A holder who plans a major reorganization of a DMF is encouraged to submit a detailed plan of the proposed changes and request its review by the OBI staff. The staff should be given sufficient time to comment and provide suggestions before a major reorganization is undertaken.

27.1.4.7 DMF Retirement and Closure A holder who wishes to close a DMF should submit a request to the OBI staff stating the reasons for the closure. If there has been no activity (amendment and/or annual reports to a DMF) in three years, the FDA will initiate a retirement procedure. (Note: An LOA does not count as activity.) The agency has started issuing an overdue notice letter (ONL) to the holder and/or agent using the most recent address; this highlights the importance of keeping the holder or agent name and address up to date. The holder could respond to close the file or keep it open by submitting an annual report. If a response from the holder/agent is not received in 90 days, the DMF is designated as unavailable for review; the FDA will shred the circulatory copy and send the archival copy to the Federal Records Center.

27.2 EUROPEAN MARKETING AUTHORIZATION DOSSIERS

An API quality documentation for a European marketing authorization dossier may be submitted using any one of several suitability procedures. Regardless of the option chosen, for a pharmaceutical API, its manufacturer has legal obligations to comply with the *European Pharmacopoeia* (E.P.) monograph. The following procedures may be used:

- Certificate of suitability (CEP)
- European drug master file (EDMF)/active substance master file (ASMF)
- Full details of manufacture submitted by a marketing authorization applicant
- Other evidence of suitability of the pharmacopeial monograph

27.2.1 Certificate of Suitability (EDQMa,b; EMA, 1998; EDQM, 2010a)

In accordance with Resolution AP-CSP(07)1 and Directives 2001/83/EC and 2001/82/EC as amended by the European Council and the Parliament, "manufacturers or suppliers [regardless of their geographical locations] of active substances or excipients ..., products with transmissible spongiform encephalopathy (TSE) risk, ... used in ... the preparation of pharmaceutical products can apply for a certificate concerning (EDQMa):

- the evaluation of the suitability of the [E.P.] monograph for the control of the chemical purity and microbiological quality of their substance; or,

- the evaluation of the reduction of TSE risk according to the general monograph; or,

- both of the above; ...."

This certificate of suitability procedure ensures the quality of substances and compliance to the E.P. monographs and the requirements of the European Union directives for medicines. A CEP can therefore be used in a marketing authorization application (MAA) to demonstrate that the active substance (listed in the MAA) complies with the E.P. monograph and with Directives 2001/83/EC and 2001/82/EC. The MA applicant should include a copy of the CEP (which was obtained from the CEP holder) in the MAA dossier, together with a written assurance that no significant changes have been made to the active substance manufacturing process since the date of certification (EMA, 1998). A paper CEP can be converted into electronic format at any time during the life cycle of a CEP application or a CEP that has been granted provided that all future information is sent electronically; a mixture of paper and electronic formats is not allowed (EDQM, 2010a). Submissions in electronic format must be in accordance with the ICH M2 EWG-eCTD and EMEA requirements. The eCTD CEP dossier should be a stand-alone document and is distinct from any marketing authorization eCTD dossier and life cycle (EDQM, 2010a). The following documents should assist in deciding how to comply with the EDQM procedures related to paper and electronic submissions for CEP applications (EDQMb; EDQM, 2010a):

- ICH eCTD specification (ICH, 2011)
- EU M1 eCTD specification (EMA, 2011)
- Guidance for submission of electronic applications for CEP: revised procedures (http://www.edqm.eu/medias/fichiers/cep_guidance_for_electronic_applications.pdf)
- Explanatory note: updated EDQM procedures related to paper and electronic submission for CEP applications (http://www.edqm.eu/en/News-and-General-Information-164.html)

The list of CEPs granted by the European Directorate for the Quality of Medicines and Healthcare (EDQM) is updated daily in the certification database (EDQMe).

The CEPs are recognized by all signatory states of the European Pharmacopoeia Convention and by the European Union, including the health authorities of Canada, Australia, New Zealand, Tunisia, and Morocco (EDQMa).

27.2.1.1 Revisions and Renewals

27.2.1.1.1 Revisions The EDQM does not accept changes or additions to documents submitted under assessment unless requested. Exceptions to this policy include administrative updates such as change in the company name and/or the details of the contact person for the dossier (EDQMb). Each time a revised or renewed CEP is granted, the CEP holder must provide a copy of the CEP to all customers for them to update their MAAs. The CEP holders are obliged to inform their customers in case of CEP revision, CEP suspension, withdrawal, or negative outcome of an EDQM inspection. To enforce this obligation, the EDQM requires the CEP holders to include a commitment document signed by the applicant in the CEP application forms for all new applications and revisions/renewals. For further details, refer to the Procedures for Management of Revisions/Renewals of Certificates: PA/PH/Exp CEP/T (04) 18 2R (EDQMd).

The revision application is submitted [electronic format is preferred] to the certification division of the EDQM. It is validated or rejected and then listed for assessment. After it is assessed, the EDQM may send queries to the applicant. When the queries have been resolved, the EDQM sends the applicant a CEP (EDQMb).

27.2.1.1.2 Renewals (EDQMd) A CEP is valid for five years from the date of first issuing and is renewed only once. Once renewed, the CEP is valid for an unlimited period unless the EDQM decides to request one additional renewal (by Directives 2004/27/EC and 2004/28/EC). The request for renewal (electronic format preferred) should be submitted six months before the expiry date. Any requests sent later may lead to a gap between the expiry date of the certificate and the approval of the request for renewal, during which no valid certificate would be available.

27.2.2   The EDQM Inspection Program (EDQMc)

The inspections are normally carried out as team inspections by official inspectors from the EDQM and other agencies; no appointed auditors are used for these inspections. The aim of the inspection program is to check compliance with both the GMP (Vol. 4 of the Rules Governing Medicinal Products in EU: EC, 2011) and the CEP application dossier (and updates) at the manufacturing/distribution sites covered by the CEPs.

If inspection findings are favorable, an attestation of compliance with good manufacturing practice and a CEP dossier are issued by the EDQM. If major/critical deficiencies are found, the corresponding CEPs may be suspended. In such cases the suspension is reported on the EDQM's website and to all other concerned authorities in order for them to take necessary actions related to the affected marketing authorization(s); the holder must inform their customer(s) of this decision. The suspension will end only when the company takes satisfactory corrective actions and the implementation of the corrective actions is confirmed by reinspection.

27.2.3 European Drug Master File/Active Substance Master File (EMA, 1998, 2005)

The main objective of the ASMF procedure, commonly known as the EDMF procedure, is to keep AS manufacturing know-how confidential while allowing the MA applicant to take full responsibility for the medicinal product, including the quality of the active substance. The European Medicines Agency (EMA) and other component authorities will access the EDMF information in order to evaluate the suitability of the use of the active substance in the medical product. For this review, the EDMF must contain detailed scientific information and be submitted in the CTD format. The EDMF information is typically submitted in two parts: the open/applicant part (AP) and the restricted part (RP). The AP contains the information considered nonconfidential by the applicant/MA holder, whereas the RP is confidential. Exception to the RP may be to include sterilization process validation data in the applicant's part when the final product does not require terminal sterilization. In the dossier and/or in the expert report, a thorough discussion should be included to demonstrate whether the pharmacopeial monograph tests are in fact able to control all recurring impurities at or above 0.1%, and resulting toxicological implications are addressed (EMA, 1998).

Full details of manufacture (or minimally, the restricted part) are submitted as a DMF by the AS manufacturer as outlined in the Committee for Proprietary Medicinal Products (CPMP) guideline: Europe Drug Master File Procedure for Active Substances (EMA, 2005). In such a case, the MA applicant's part (which includes the Open part) should be included in the MA applicant's dossier as well as being submitted by the AS manufacturer. The ASMF procedure is limited to APIs (and is not for excipients, starting materials, intermediates, or packaging materials) and is accepted only in context with a marketing authorization application (MAA).

An EDMF holder may procure both the EDMF and CEP for a single active substance. Both documents are not necessary and generally are not accepted, except in cases where the component authorities/EMA may decide that additional information (not included in the CEP: e.g., stability data) should be provided in the dossier. In such cases it may be acceptable to refer to both EDMF and CEP.

The EDMF holder should give permission to the authorities to access the data in the EDMF in relation to a specific MAA, in the form of a letter of access. The EDMF holder should submit to the applicant/MA holder the following:

- A copy of the latest version of the AP
- A copy of the quality overall summary (QOS)/ER on the latest version of the AP
- The letter of access when this letter has not been submitted earlier for the product concerned; it is the MAA/applicant's responsibility to submit this to the authorities

In addition, the EDMF holder should submit the following to the authorities:

- The EDMF accompanied by a cover letter
- The letter of access if not submitted previously for the product concerned

The submission of the relevant documentation to the authorities should arrive at approximately the same time as the MAA. The EDMF holder must notify MA holders and the authorities of all changes to AP and/or RP so that the MA holder can update all affected MAs accordingly. The applicant/MA holder is responsible for ensuring access to all relevant information concerning:

- Current manufacturing of the AS
- Specifications to control the quality of the AS
- A copy of the most recent version of the AP in the MA dossier and is identical to the AP as supplied by the EDMF holder to the authorities as part of EDMF
- A single compiled specification that is identical for each supplier in cases where there is more than one supplier

27.2.3.1 *Changes and Updates to the ASMF* The EDMF holder should:

- Keep the content of the EDMFs updated with respect to the actual synthesis or manufacturing process
- Ensure that the quality control methods are kept in line with the current regulatory and scientific requirements
- Verify that modifications of the EDMF content (e.g., manufacturing process and/or specifications) are not made without informing the applicant/MA holder and the authorities
- Provide a cover letter to the authorities containing the following information:
  - A tabular list summarizing the changes carried out since the first compilation of the EDMF
  - An overview comparing the old and new content of the EDMF
  - Information as to whether the change has already been accepted, rejected, or withdrawn by another member state
  - The names of the relevant applicants, MA holders, and MAs
  - An updated QOS/ER if relevant

At the five-year renewal anniversary of a medical product, the MA holders (in consultations with the EDMF holders, where applicable) are required to declare that:

- The quality of the product, with respect to the methods of preparation and control, has been updated regularly
- The product conforms with current CPMP quality guidelines
- No changes have been made to the product particulars other than those approved by the authorities

27.2.4 Full Details of Manufacture (EMA, 1998)

The MA applicant may submit full details on the active substance, its manufacture and control, and to demonstrate that potential impurities (at or above 0.1%

level) can be controlled by the pharmacopeial monograph. For further details, refer to the guideline Chemistry of the Active Substance (EMA, 1987, 2003). This procedure is open for chemicals or herbal API, new or compendial, but is not open to biological. For biological API, the manufacturing process is usually confidential, and its manufacturer would prefer not to share it with the MA applicant.

27.2.5 Other Evidence of Suitability of the Pharmacopoeial Monograph (EMA, 1998)

The MA applicant may provide other evidence obtained from the active ingredient manufacturer. This may include the following evidence:

- Information as to the length of time that the particular named source has been on sale in the European Union and elsewhere; and

- A statement that during the sale period, there had been no significant change in the method of manufacture, leading to a change in the impurity profile of the active substance; and

- Evidence that samples from the source named had been supplied to the European Pharmacopoeia Commission or National Pharmacopoeia Commission and have been taken into account in the development of their monograph; and

- A statement that no additional tests arising from the use of the manufacturing route were necessary in order to identify and limit additional impurities at or above 0.1% (for toxic impurities at even lower levels, if appropriate) not specifically controlled by the pharmacopoeial monograph

The approach described above is one possible way to provide reassurance to the authorities of the suitability of the pharmacopeial monograph to control a well-defined active substance with long and safe patient exposure from the named source. It is noted, however, that even when a monograph has been in force for many years, it will not necessarily be sufficient in relation to a new route of synthesis.

REFERENCES

EC. 2011. EudraLex, Vol. 4, Good Manufacturing Practice (GMP) Guidelines. European Commission. Updated July 16, 2011. http://ec.europa.eu/health/documents/eudralex/vol-4/index_en.htm. Accessed Dec. 2011.

EDQMa. Background and legal framework. Certification of Suitability. http://www.edqm.eu/en/Background-Legal-Framework-77.html. Accessed Dec. 2011.

EDQMb. Certification of Suitability to the Monographs of the European Pharmacopoeia: News & General Information. http://www.edqm.eu/en/News-and-General-Information-164.html. Accessed Dec. 2011.

EDQMc. The Inspection Programme. http://www.edqm.eu/en/The-Inspection-Programme-159.html. Accessed Dec. 2011.

EDQMd. Revisions and Renewals. http://www.edqm.eu/en/Revisions-Renewals-663.html. Accessed Dec. 2011.

EDQMe. Search database online: certification. https://extranet.edqm.eu/publications/recherches_CEP.shtml. Accessed Dec. 2011.

EDQM 2010a. Guidance for submission of electronic applications for Certificates of Suitability (CEPs), revised procedures. Certification of Suitability to Monographs of the European Pharmacopoeia. PA/PH/CEP(09)108, 1R. Mar. 2010. http://www.edqm.eu/medias/fichiers/cep_guidance_for_electronic_applications.pdf. Accessed Dec. 2011.

———. 2010b Explanatory note: Updated EDQM procedures related to paper and electronic submissions for CEP applications. Certification of Suitability to Monographs of the European Pharmacopoeia. PA/PH/CEP(09)109, 1R. Mar. 2010. http://www.edqm.eu/medias/fichiers/cep_procedures_for_cep_applications.pdf. Accessed Dec. 2011.

EMA. 1998. Note for guidance on summary of requirements for active substances in Part II of the dossier. European Agency for the Evaluation of Medicinal Products, Human Medicines Evaluation Unit. CMPM/QWP/297/97. Jan. 28, 1998. http://www.ema.europa.eu/docs/en_GB/document_library/Scientific_guideline/2009/09/WC500002812.pdf. Accessed Dec. 2011.

———. 2002. Guideline on summary of requirements for active substances in the quality part of the dossier. CHMP/CVMP. CHMP/QWP/297/97 Rev. 1 Corr. EMEA/CVMP/1069/02. http://www.ema.europa.eu/docs/en_GB/document_library/Scientific_guideline/2009/09/WC500002813.pdf. Accessed Dec. 2011.

———. 2003. Guideline on the chemistry of new active substances. European Agency for the Evaluation of Medicinal Products. Evaluation of Medicines for Human Use. CPMP/QWP/130/96, Rev. 1. Dec. 17, 2003. http://www.ema.europa.eu/docs/en_GB/document_library/Scientific_guideline/2009/09/WC500002815.pdf. Accessed Dec. 2011.

———. 2005. Guideline on active substance master file procedure. European Medicines Agency Inspections. EMEA/CVMP/134/02 Rev. 2 Consultation. CPMP/QWP/227/02 Rev. 2 Consultation. Apr. 27, 2005. http://www.ema.europa.eu/docs/en_GB/document_library/Scientific_guideline/2009/09/WC500002811.pdf. Accessed Dec. 2011.

———. 2011. eSubmission. Updated Apr. 14, 2011. http://esubmission.ema.europa.eu/eumodule1/index.htm. Accessed Dec. 2011

EME. 1987. Chemistry of active substances. Directive 75/318/EEC as amended. Revised 1987. http://www.ema.europa.eu/docs/en_GB/document_library/Scientific_guideline/2009/09/WC500002817.pdf. Accessed Dec. 2011.

FDA. 1987a. Guideline for submitting supporting documentation in drug applications for the manufacture of drug substances. CDER/FDA. Feb. 1987. http://www.fda.gov/Drugs/GuidanceComplianceRegulatoryInformation/Guidances/ucm149499.htm. Accessed Dec. 2011.

———. 1987b. Guideline for submitting documentation for the manufacture of and controls for drug products. CDER/FDA. Feb. 1987. http://www.fda.gov/Drugs/GuidanceComplianceRegulatoryInformation/Guidances/ucm149499.htm). Accessed Dec. 2011.

———. 1987c. Guidelines for submitting samples and analytical data for methods validation. Feb. 1987. http://www.fda.gov/Drugs/GuidanceComplianceRegulatoryInformation/Guidances/ucm123124.htm. Accessed Dec. 2011.

———. 1987d. Guideline for submitting documentation for the stability of human drugs and biologics. CDER/FDA. Feb. 1987. http://www.seoho.biz/GMP_Quick_Search/Data/1.%20FDA%20Documents/1.6.31.pdf. Accessed Dec. 2011.

———. 1994. Guidance for industry for the submission documentation for sterilization process validation in applications for human and veterinary drug products. CDER, CVM. Nov. 1994. http://www.fda.gov/downloads/Drugs/GuidanceComplianceRegulatoryInformation/Guidances/ucm072171.pdf. Accessed Dec. 2011.

———. 1998. Guidance for industry: environmental assessment of human drug and biologics applications. CDER/CBER. July 1998. CMC 6, Rev. 1. http://www.fda.gov./downloads/Drugs/GuidanceComplianceRegulatoryInformation/Guidances/ucm070561.pdf. Accessed Dec. 2011.

———. 1999. Guidance for industry: Container closure systems for packaging human drugs and biologics—chemistry, manufacturing, and controls documentation. FDA/CDER/CBER. May

1999. http://www.fda.gov/downloads/Drugs/GuidanceComplianceRegulatoryInformation/Guidances/UCM070551.pdf. Accessed Dec. 2011.

———. 2001. M4Q: Guidance for industry: The CTD—quality. CDER/FDA. Aug. 2001. ICH. http://www.fda.gov/downloads/Drugs/GuidanceComplianceRegulatoryInformation/Guidances/UCM073280.pdf. Accessed Dec. 2011.

———. 2002. Guidance for industry: Container closure systems for packaging human drugs and biologics—questions and answers. May 2002. CMC. http://www.fda.gov/downloads/Drugs/GuidanceComplianceRegulatoryInformation/Guidances/ucm070553.pdf. Accessed Dec. 2011.

———. 2009. Draft guidance for industry: Submitting type V drug master files to the Center for Biologics Evaluation and Research. Updated July 14, 2009. http://www.fda.gov/Biologics BloodVaccines/GuidanceComplianceRegulatoryInformation/Guidances/General/ucm171812.htm. Accessed Dec. 2011.

———. 2010. FDA IND, NDA, ANDA, or drug master file binders. Effective Apr. 1, 1998. Updated Feb. 17, 2010. http://www.fda.gov/Drugs/DevelopmentApprovalProcess/FormsSubmission Requirements/DrugMasterFilesDMFs/ucm073080.htm. Accessed Dec. 2011.

———. 2011a. *Code of Federal Regulations*. 21 CFR 314.430. Availability for public disclosure of data and information in an application or abbreviated application. Updated Apr. 1, 2011. http://www.accessdata.fda.gov/scripts/cdrh/cfdocs/cfCFR/CFRSearch.cfm?fr=314.430. Accessed Dec. 2011.

———. 2011b. Drug master files (DMFs). Updated Dec. 12, 2011. http://www.fda.gov/Drugs/DevelopmentApprovalProcess/FormsSubmissionRequirements/DrugMasterFilesDMFs/default.htm. Accessed Dec. 2011.

———. 2011c. *Code of Federal Regulations*. 21 CFR 314.420 Drug master files. FDA. Updated Apr. 1, 2011. http://www.accessdata.fda.gov/scripts/cdrh/cfdocs/cfCFR/CFRSearch.cfm?fr=314.420. Accessed Dec. 2011.

———. 2011d. CMC reviews of type III DMFs for packaging materials. Manual of Policies and Procedures. MAPP 5015.5, Rev. 1. Effective date Mar. 22, 2010, Aug. 12, 2011. http://www.fda.gov/downloads/AboutFDA/CentersOffices/CDER/ManualofPoliciesProcedures/UCM205259.pdf. Accessed Dec. 2011.

———. 2011e. Drug master files: Guidelines. FDA Center for Drug Evaluation and Research. Sep. 1989; updated Mar. 11, 2005, Oct. 11, 2011. http://www.fda.gov/Drugs/GuidanceCompliance RegulatoryInformation/Guidances/ucm122886.htm. Accessed Dec. 2011.

———. 2011f. Generally recognized as safe (GRAS). Updated June 10, 2011. http://www.fda.gov/Food/FoodIngredientsPackaging/GenerallyRecognizedasSafeGRAS/default.htm. Accessed Dec. 2011.

ICH. 2011. eCTD specification and related files. Updated Nov. 29, 2011. http://estri.ich.org/eCTD. Accessed Dec. 2011.

Shaw AB. Drug master files. FDA Small Business Office Webinar, Nov. 14, 2011. http://www.fda.gov/downloads/Drugs/DevelopmentApprovalProcess/SmallBusinessAssistance/UCM279666.pdf?utm_source=fdaSearch&utm_medium=website&utm_term=CTD%20for%20DMFs&utm_content=6. Accessed Apr. 2012

GLOSSARY

Agent or representative	Any person who is appointed by a DMF holder to serve as the contact for the holder.
Amendment to an application	Additional information to an existing IND, a pending ANDA, or a pending ANDA supplement.
ANDA	Abbreviated new drug application.
AP	Applicant part of an EDMF.
API	Active pharmaceutical ingredient.

Applicant (or Customer)	Any person who submits an application or abbreviated application or an amendment or supplement to them to obtain FDA approval of a new drug or an antibiotic drug and any other person who owns an approved application.
Applicant's part of a DMF	Section of the European DMF given to the applicant to include in the application for a product marketing authorization.
Application	An IND, NDA, ANDA, another DMF, or export application that references the DMF.
ASM restricted part of DMF	Section of the European DMF given by the ASM only to the authorities.
ASMF	Active substance master file.
CDER	Center for Drug Evaluation and Research.
CEP	European procedure for a certificate of suitability of monographs of the *European Pharmacopoeia*.
Certificate of analysis	A document listing the test methods, specifications, and results of testing a representative sample from the batch to be delivered.
Certificate of suitability to *European Pharmacopoeia*	Certification granted to individual manufacturers by the EDQM when a specific excipient or API is judged to be in conformity with a Ph. Eur. monograph.
CFR	*Code of Federal Regulations*.
CGMP	Current good manufacturing practice.
CMC	Chemistry, manufacturing, and controls.
Contact person	A person to whom correspondence to the DMF holder should be addressed.
CPMP	Committee for proprietary medicinal products.
CR	Complete response.
CTD	Common technical document.
Drug master file	A submission of information to the FDA intended to provide confidential information on the CMC of an API or excipient in support of an application, amendment, or supplement; a DMF is also submitted to the Japanese Pharmaceutical and Medical Device Agency and to Health Canada.
DMF	Drug master file.
Drug product	A finished dosage form (e.g., tablet, capsule, or solution) that contains a drug substance, generally, but not necessarily, in association with one or more other ingredients.
Drug substance, active substance or active pharmaceutical ingredient	An active ingredient that is intended to furnish pharmacological activity or other direct effect in the diagnosis, cure, mitigation, treatment, or prevention of disease or to affect the structure or any function of the human body; does not include intermediates used in the synthesis of such an ingredient.
EA	Export application.
EC	Council of Europe.
eCTD	Electronic common technical document.
EDMF	European drug master file.
EDQM	European Directorate for the Quality of Medicines.
EIS	Environmental impact statement.
EMA	European Medicines Agency [formerly, the European Medicines Evaluation Agency (EMEA)].
E.P.	*European Pharmacopoeia*.

European drug master file	EC procedure where information can be provided to the authorities and the applicant, where the active substance manufacturer is not the applicant for a product marketing authorization, with a view to protecting valuable manufacturing know-how.
Export application	An application submitted under Sec. 802 of the Food, Drug, and Cosmetic Act to export a drug that is not approved for marketing in the United States.
FDA	U.S. Food and Drug Administration.
FONSI	Finding of no significant impact.
Holder	A person who owns a DMF.
IND	Investigational new drug.
IR	Information request.
Letter of access	Letter of authority from the ASM to allow the competent authorities to assess a EDMF on behalf of a specified applicant.
Letter of authorization	A written statement by the holder or designated agent or representative permitting FDA to refer to information in a DMF in support of another person's submission.
LOA	Letter of authorization.
MAA	Marketing authorization application.
MOC	Material of construction.
NDA	New drug application.
NEPA	National Environmental Policy Act.
New or novel excipients	A new chemical entity, physically or chemically modified existing excipient, coprocessed mixtures of existing excipients, or existing food additive or GRAS substance used in the United States for the first time in a human drug product or by a new route of administration. The CMC information for these excipients should be provided in the same level of detail and in the same format as the information provided for a new drug.
N.F.	*National Formulary*.
OBI	Office of Business Information.
Person	Includes a person, partnership, corporation, or an association.
QbD	Quality by design.
QOS	Quality overall summary.
Quality agreement	A formal agreement between API or excipient manufacturers and their customers that stipulates the responsibilities of each party in meeting regulatory requirements for sale and use of the substance (API/excipient) in a dosage form.
Representative	A third party (usually, a company) that acts on behalf of the holder in its interactions with the agency. For a foreign firm this party is often referred to as the U.S. agent.
RP	Restricted part of an EDMF.
Secondary DMF	A DMF that is referenced by another DMF.
Significant change	Any change that alters an API or excipient physical or chemical property from the norm or that is likely to alter the API/excipient performance in the dosage form.
Sponsor	A person who takes responsibility for and initiates a clinical investigation. The sponsor may be an individual or a pharmaceutical company, governmental agency, academic institution, private organization, or other organization.
Supplement	A report of a change in an approved ANDA. Since a DMF is not approved, there can be no supplement to a DMF, only amendments.

Supplement to an ANDA	A report of a change in an approved ANDA. Since a DMF is not approved, there can be no supplements to a DMF, only amendments.
Supplier	Used here to donate the company providing the excipient ingredient to a pharmaceutical customer; may be either the excipient manufacturer or the distributor.
U.S. agent	A company or agent resident in the United States appointed by a foreign firm to act as its representative.
U.S.C.	*United States Code*.
U.S.P.	*United States Pharmacopeia*.

CHAPTER *28*

COMMISSIONING AND QUALIFICATION

28.1 REGULATORY REQUIREMENTS

To quote the U.S. Food and Drug Administration (FDA): "Validation of manu-facturing processes is a requirement of the Current Good Manufacturing Practice (CGMP) regulations for finished pharmaceuticals (21 CFR 211.100 and 211.110), and is considered an enforceable element of Current Good Manufacturing Practice for active pharmaceutical ingredients (APIs) under the broader statutory CGMP provision of section 501(a)(2)(B) of the Federal Food, Drug, and Cosmetic Act. A validated manufacturing process has a high level of scientific assurance that it will reliably produce acceptable product. The proof of validation is obtained through rational experimental design and the evaluation of data, preferably beginning from the process development phase and continuing through the commercial production phase" (FDA CPG 490.100: FDA, 2010a).

The term *commissioning*, *qualification*, or *validation* is not defined in the Food, Drug, and Cosmetic Act or in the FDA's CGMP regulations, but the CGMP expectations and requirements (and accordig to FDA Compliance Policy Guide 490.100, stated above) are that all processes (including the building, utilities, equip-ment) will perform as expected and the proof documented. Similar requirements are found in good manufacturing practice guidelines and regulations of most other international health agencies; for example, "... manufacturers [are required to] identify what validation work is needed to prove control of the critical aspects of their particular operations ... a risk assessment approach should be used to deter-mine the scope and extent of validation" (EC, 2001); and "all critical production processes be validated" (Health Canada, 2009).

Following is a partial list of the CGMP (FDA, 2011a) requirements, examples of selected warning letters (paraphrased) for noncompliance, and the commissioning and qualification (C&Q) elements, if executed properly, that would address these requirements.

- Buildings and facilities used in the manufacture, processing, packing, or holding of a drug product are of suitable size, construction, and location to

Integrated Pharmaceutics: Applied Preformulation, Product Design, and Regulatory Science,
First Edition. Antoine Al-Achi, Mali Ram Gupta, William Craig Stagner.

facilitate cleaning, maintenance, and proper operations [21 *Code of Federal Regulations* (CFR) 211C].

- ○ FDA warning letters : (1) failure to maintain necessary separation, defined areas, or such control systems to prevent contamination during aseptic processing, cleaning, and for disinfecting rooms and equipment to produce aseptic conditions (WL: OEWL-08-01); (2) operations related to the manufacturing, processing, or packaging of penicillin are not adequately separated from nonpenicillin products (WL: 320-08-02).

- ○ C&Q elements: basis of design, commissioning, facility controls, utilities qualification, standard operating procedures (SOPs, ongoing training, and other appropriate qualification elements.

- Equipment used in drug processing is of appropriate design, adequate size, and suitably located to facilitate operations for its intended use and for its cleaning and maintenance (21 CFR 211D).

  - ○ FDA warning letters: (1) equipment not cleaned and maintained at appropriate intervals to prevent contamination that would alter the safety, identity, or quality of the drug product; the firm did not conduct adequate cleaning validation or provide scientific justification for several tablet and capsule drug products (WL: 45-10); (2) failure to follow written procedures for cleaning and maintenance of equipment, including utensils, used in the manufacture, processing, packing, or holding of a drug product (WL: SJN-2009-07).

  - ○ C&Q elements: equipment qualifications (design, installation, operational, qnd performance qualifications, and process validation), calibration, preventive maintenance, cleaning validation/verification, validation of sterilization and depyrogenation process, SOPs, ongoing training.

- Automatic, mechanical, or electronic equipment or related systems must be calibrated, inspected, or checked routinely according to a written program designed to assure proper performance. Inputs to and outputs from the computer are checked for accuracy; the degree and frequency of verification is based on the complexity and reliability of the computer or related system (Sec. 211.68).

  - ○ FDA warning letter: failure to routinely calibrate, inspect, or check according to a written program designed to assure proper performance of automatic, mechanical, or electrical equipment, including computers, used in the manufacture, processing and holding of a drug (WL: 320-09-05).

  - ○ C&Q elements: validation/qualification of these systems, SOPs, ongoing training.

- Assure batch uniformity and integrity of drug products, establish and follow written procedures to monitor the output and to validate the performance of manufacturing processes responsible to cause variability in the characteristics of in-process material and the drug product [Sec. 211.110(a)]. Validate in-process specifications as being consistent with drug product final specifications [Sec. 211.110(b)]. Section 211.180(e) requires that the information and data about the product quality and manufacturing experience be reviewed periodically to determine if any changes to the established process are warranted.

○ FDA warning letters: (i) failure to establish valid in-process specifications derived from previous acceptable process average and process variability estimates where possible (WL: 2008-DT-05); (ii) failure to establish control procedures that monitor the output and validate performance of those manufacturing processes that may be responsible for causing variability in characteristics of in-process materials and the drug product (WL: SJN-2009-07); (iii) failure to conduct adequate annual review as required (WL: NYK 2009-03); (iv) a high number (554 batches) of rejected batches demonstrates a lack of adequate process controls and raises significant concerns regarding capability and reliability of processes (WL: 320-09-06); (v) process validation batches were not prepared using the routine process parameters and production batch size (WL: 320-10-07).

○ C&Q elements: validate/revalidate, ongoing verification, change controls.

• Samples for validation must represent the batch being analyzed [Sec. 211.160(b)(3)]; the sampling plan must result in statistical confidence [Sec. 211.165(c) and (d)]; and the batch must meet its predetermined specification [Sec. 211.165(a)]. The accuracy, sensitivity, specificity, and reproducibility of test methods employed by the firm must be established and documented in accordance with Sec. 211.194(a)(2), [Sec. 211.165(e)].

○ FDA warning letters: (i) the firm has not established scientifically sound and appropriate sampling plans (e.g., sampling size used for the 5-day retain sample inspection is not scientifically sound) designed to assure that the drug products conform to appropriate standards of identity, strength, quality, and purity (WL: 10-ATL-12);(ii) laboratory records did not include a record of all calculations performed in connection with the required laboratory tests (WL: FLA 09-13); (iii) laboratory records did not include a complete record of all data secured in the course of each test, including graphs, charts, and spectra from laboratory instrumentation, properly identified to show the specific drug product and lots tested (WL: 320-06-03); (iv) the specificity test, included in the method validation reports, is inadequate; it has not shown to be capable of detecting potential impurities (WL: 320-10-07).

○ C&Q elements: analytical instrument qualification, test method validation, calibration, requalification, SOPs, ongoing training.

In summary, the CGMP regulations require that the manufacturing process (including analytical instruments, test methods, and documentation) be designed and controlled to assure that in-process materials and the finished product meet predetermined quality requirements consistently and reliably.

28.2 PRELIMINARY C&Q ACTIVITIES

The following two primary documents are prepared to guide the entire process of commissioning and qualification of facilities, utilities, equipment, and/or systems.

28.2.1 Validation Master Plan (EC, 2001; PIC/S, 2007; Health Canada, 2009)

Based on the complexity of operations or projects requiring qualification, several different plans (e.g., commissioning plan, validation master plan, instrument qualification plan) should be prepared to serve intended purposes. These high-level but living documents provide a life-cycle approach starting from the basis of design (or user requirements) through construction, implementation, commissioning, qualification, to the endpoint process of decommissioning of all systems. This should include all major elements that are considered when commissioning and qualifying a facility, system, equipment, or a utility.

The purpose of the validation master plan (VMP) is to define the base requirements and general guidelines to assure that the facilities, systems, equipment, and processes are capable of operating as specified and in a repeated and reliable fashion. The VMP should include an overview of the entire validation process, focusing on its organizational structure, content, and planning.

- All validation activities should be planned, clearly defined, and documented in a validation master plan or equivalent document. For large projects it may be necessary to create separate validation master plans.
- The VMP should be a brief, concise, and clearly stated summary document; do not repeat information such as that in policy documents, SOPs, validation protocols, and reports that has already been documented.
- The VMP should contain the following:
  ○ Validation policy and scope
  ○ Organizational structure of validation activities and personnel responsibilities
  ○ Rationale for inclusion or exclusion of the plant/process/product description and the extent of validation
  ○ Critical process considerations requiring extra attention
  ○ List of facilities, systems, equipment, and processes to be validated and validation approaches
  ○ Revalidation activities, actual status, and future planning
  ○ Key acceptance criteria
  ○ Documentation format for protocols and reports
  ○ Planning and scheduling of each validation project and subproject
  ○ Change control
  ○ Reference to existing documents

28.2.2   Basis of Design

Basis of design is the primary design document that establishes what is to be designed to meet the firm's requirements (user requirements) for the manufacture of its products.

28.2.2.1   Design Considerations

- Product type
  - Finished dosage form (nonsterile/sterile)
  - Biopharmaceutical product
  - Active pharmaceutical ingredient
- Single product or multiple products
  - Dedicated (separate facility by product: e.g., penicillin, viral vaccines)
  - Multiuse (multiple products but dedicated facility and equipment for each product)
  - Multipurpose (multipurpose facility and equipment for multiple products)
- Exposure risk and protection levels
  - Isolated rooms
  - Inside vs. outside
  - Contamination and cross-contamination
  - Controlled vs. critical environment
  - One vs. multiple batches at a time
  - Open vs. closed system
- Production volume
  - Batch size and number of batches per year
- Utilities
  - Direct impact [*United States Pharmacopeia* (U.S.P.) purified water, water for injection, clean steam, process air/gases, process vacuum, etc.]: these systems are both commissioned and qualified as they come in contact or potential contact with product or processing equipment
  - Indirect/no impact (potable water, instrument air, plant steam, waste treatment, etc.): these nonproduct or process equipment–contacting utilities are commissioned only
- Other design considerations:
  - Architectural finishes (based on usage requirements or room classification: floor—seamless or not, drain covers or no drains, sinks or no sinks; walls—porous or nonporous, rounded or curved corners; ceiling—sealed or not, no or suspended ceiling, curved or rounded corners, flush-mounted fixtures, fire detectors or not)
  - Pressure differentials

- ○ Air classification/zones (unclassified to class 100 or comparable International Standards Organization classification)
- ○ Temperature and humidity
- ○ Personnel flow
- ○ Material/product/waste flow
- ○ Equipment flow
- • Space (footprints and number of stories)
  - ○ Manufacturing or bioprocessing
  - ○ Aseptic filling
  - ○ Labeling and packaging
  - ○ Utilities and mechanical space
  - ○ Quality control labs
  - ○ Warehouse
  - ○ Offices
  - ○ Employee lockers, cafeteria, restrooms, parking, etc.
  - ○ Future expansion
- • Cost (this is a business, not a regulatory consideration)

28.2.2.2 *Impact Assessment* (ISPE, 2001)   The level of commissioning and qualification needed is determined based on the function and impact of the facility, equipment, or utility on product quality. This impact should be assessed as early as possible in the C&Q phase (preferably during the basis of design stage) and employed during the design phase to gain full cost benefits. If a system is expected to have an impact on product quality (or patient safety), it is categorized as a *direct impact system*. Conversely, *indirect impact systems* are not expected to affect product quality (or patient safety). Direct impact systems are both commissioned and qualified, whereas indirect impact systems would be commissioned only to meet regulatory expectations of the FDA and other regulatory authorities. Some possible examples of direct and indirect systems:

- • Direct systems (U.S.P. purified water, water for injection, autoclave, etc.)
- • Indirect systems (chilled water, office air conditioner, etc.)

28.3   COMMISSIONING

Commissioning is defined as "a well planned, documented, and managed engineering approach to the start-up and turnover of facilities, systems, and equipment to the End-User that results in a safe and functional environment that meets established design requirements and stakeholder expectations" (ISPE, 2001). Commissioning supports validation and streamlines qualification. Commissioning ensures that all facilities, equipment, utilities, and systems are designed, installed, tested, and fully operational in accordance with the owner's design intent. Starting at the design phase, commissioning can serve as a precursor to process validation for direct impact systems. For indirect impact or nonimpact systems, commissioning may serve as the final activity prior to routine operation. The direct impact systems are the only ones that require qualification (in addition to commissioning) to be in full compliance with pharmaceutical regulatory requirements. To streamline and reduce qualification cost, it is advised that all systems (and their components, where necessary) be categorized at the earliest possible stages of qualification as having direct, indirect, or no impact on product quality. Overall qualification cost will be reduced by proper execution of the elements in this phase:

- Equipment or system is verified to be as specified and is working as it should be.
- Commissioning is integrated with qualification to ensure that the first attempt to qualify will be successful (hopefully, no failures or deviations are discovered during qualification protocol execution).
- To streamline and reduce qualification cost, the commissioning protocol for a direct impact system should be preapproved by the quality control group before execution.
- Commissioning is to be documented properly for traceability.

28.3.1 Principles and Elements of Commissioning

A well-planned commissioning program will:

- Accelerate project startup
- Improve project completion time and on-time startup
- Reduce the validation effort by incorporating commission information into the qualification
- Verify that the equipment or system meets design specifications and is working as intended
- Result in more streamlined and thorough documentation
- Ensure a GMP-compliant facility

    Commissioning elements include (Tyree, 2005):

- All planning and documentation are verified.
- Equipment is prepared for startup and is started to confirm power and motor rotation.
- Piping and equipment installed are compared to the piping and Instrumentation diagram (P&ID).
- Utilities connections are confirmed.
- As-built drawings are prepared.
- Instruments are calibrated.
- Control loops are tuned.

- Operators are trained.
- Spare parts list is prepared.

28.3.2 Documentation Requirements

Depending on the project complexity, commissioning documents may include:

- A commissioning plan: a separate master commissioning plan for a large and complex project should be prepared, but for a smaller project or single equipment, requirements may be included in the SOPs.
- Precommissioning: includes predelivery inspections and test activities [factory acceptance test (FAT) and site acceptance test (SAT)]
- Commissioning test and inspection plans: these could be stand-alone plans for individual equipment and systems and should supplement areas not covered by FATs or SATs.

28.3.3 Commissioning Plan

The commissioning plan should define boundaries of a system (facility, utility, equipment, or process) to be commissioned. This plan should coordinate all life-cycle activities starting from the construction to process validation or through operation and maintenance. Changes to the plan should be reviewed and approved by the original approvers, and a revision history of significant changes to the plan should be maintained.

28.3.4 Facility Commissioning

The scope of facility commissioning is to identify critical processes (e.g., temperature control, humidity) and verify the following:

- Floor plan
  - The floor plan comprises documentation of walls, ceilings, floor finishes, room dimensions, and lighting.
  - Verification is based on comparing the drawings and design specifications.
- Calibration program
  - The calibration program is documented.
- Maintenance program
  - Maintenance and routine testing are documented.
- Security and alarm systems
  - Alarm systems are verified for entry as well as out-of-specified-range environmental conditions; the alarm and response systems must be verified and documented on a scheduled basis.
- Backup systems
  - The backup systems must be documented and tested for response time.

28.3.5   Utilities Commissioning

- All utility systems are commissioned.
- Use points for water systems, compressed air systems, and heating/ventilation/air conditioning systems are verified.
- "As-found" information is compared against design specifications.
- All system parameters are tested.
- All system alarms are verified.
- Safety alarms and interlocks are verified.
- Commissioning will ensure that qualification of direct impact systems would be successful.

28.3.6   Equipment Commissioning

28.3.6.1   *Predelivery Inspection and Factory Acceptance Test*   When possible, these tests should be performed on systems (and including their major components) at the supplier's (manufacturer's or fabricator's) site prior to delivery to the end-user site. This effort will identify and remedy possible problems and will avoid delays in the startup schedules. For systems that have a direct impact on product quality, the test protocols should be preapproved by quality assurance personnel so that the test results will be incorporated in the qualification package.

28.3.6.2   *Postdelivery/On-Site Commissioning Activities*
- Piping and instrumentation diagram
  - Verify all piping, construction materials, slopes, and so on, by comparing them to the P&ID and design specifications.
- Verify that the electrical supply and other utilities are correct and connected.
- Verify motor rotation (if applicable).
- Test and verify all system instrumentation parameters.

28.3.7   Inspection (Site Inspection/Site Acceptance Test)

Inspection of a facility, utility, and/or equipment is verified against detailed design, specified construction standards and material, and any other relevant legal or regulatory requirements. Inspection is primarily a visual comparison activity, but tests may be performed to verify that the construction material or installation is correct. For example, it would be very difficult to prove that ductwork had been constructed to a low-leakage specification through visual inspection.

28.3.8   Commissioning Plan Closeout

At closing, a formal turnover report, with proper approvals, should be prepared summarizing all aspects of the commissioning plan, from approval through execution and documentation, including any areas of concerns, deviation, or changes from the original plan.

28.4 QUALIFICATION AND VALIDATION

Although the terms *qualification* and *validation* are used interchangeably, it would be more appropriate to use *qualification* for physical items (such as equipment qualification, analytical instrument qualification) and *validation* for the systems (e.g., process validation, test method validation).

28.4.1 Principle of Qualification (What/Why/How?)
(EC, 2001; PIC/S, 2007; FDA, 2011a,b)

The principle of qualification:

- Consists of a series of tests that verify or qualify that equipment or a system is installed and operates properly and is fit for use
- Establishes key parameters and operating limits
- Ensures that procedures, training, and maintenance program are in place
- Ensures product quality and process reproducibility
- Develops documentation for life-cycle support
- Is a regulatory expectation to validate processes, sterilization, computers, and thorough testing of manufacturing equipment and controls as stated earlier in the chapter

The CGMP regulations for validating pharmaceutical (drug) manufacturing require that drug products be produced with a high degree of assurance of meeting all the attributes they are intended to possess [21 CFR 211.100(a) and 211.110(a)]. Effective process validation contributes significantly to assuring drug quality. The basic principle of quality assurance is that a drug should be produced that is fit for its intended use; this principle incorporates the following:

- Quality, safety, and efficacy are designed or built into the product. This requires careful attention in the selection of quality materials and components, product and process design, control of processes, in-process controls, and end-product testing.
- Quality cannot be adequately assured merely by in-process and finished-product inspection or end-product testing.
- Each step of a manufacturing process is controlled to assure that the finished product meets all design characteristics and quality attributes, including specifications.
- Significant changes to the facilities, equipment, and processes that may affect product quality should be validated.
- The manufacturers should make a risk assessment to determine the scope and extent of validation.

A successful validation program depends on the degree of information and knowledge gained from the product and process development. This knowledge and understanding helps in establishing controls appropriate for the manufacturing process. Manufacturers should:

- Understand the source of variation
- Detect the presence and degree of variation
- Understand the impact of variation on the process and ultimately on product attributes
- Control the variation in a manner commensurate with the risk that it represents to the process and product.
- After establishing and confirming a process, maintain it in a state of control over the life of the process, even as the materials, equipment, production environment, personnel, and manufacturing procedures change.

28.4.2 Commissioning vs. Qualification

- Direct impact systems are both commissioned and qualified, whereas indirect impact systems are commissioned only.
- Both types of systems require a demonstration of system functionality:
  - ○ Commissioning requires a demonstration of full system functionally.
  - ○ Qualification requires a demonstration of functionality determined to affect patient safety and/or product quality.
- Test both systems against a design specification:
  - ○ Commissioning is based on user requirements.
  - ○ Qualification is based on both user and regulatory requirements.
- Different approvals are needed:
  - ○ Commissioning documents are reviewed and approved by all stakeholders (engineering, users, purchasing, and vendors), but approval is not needed from the quality control group except for those commissioning documents or protocols that are planned to be used to support qualification.
  - ○ In addition to the stakeholders listed above, the qualification protocols must be reviewed and approved by the quality control group.

28.4.3 Current Trends in Qualification
(ASTM, 2007; FDA, 2010a)

- Critical process parameters: focus on critical areas (e.g., critical process parameters, critical functions, and critical design parameters) that affect product quality and ensure patient safety.
- Quality by design (QbD): move from a compliance mind-set to QbD; do not rely solely on the verification that the manufacturing systems are fit for

use after installation but apply planned and structured verification approach throughout the system life cycle.

- Design space: fitness for use and understanding of the design space must be key drivers.

- Value-added approaches: focus on value-added approaches to the C&Q process, eliminate unnecessary or duplicate efforts and costs (e.g., do not repeat qualification steps during process validation, and by not qualifying indirect impact systems that only require commissioning).

- Risk- and science-based approaches: are employed for those specifications, design, and verification of manufacturing systems and equipment that potentially affect product quality and patient safety (the ISPE guide describes these systems as direct impact systems).

- Process analytical technology (PAT): a methodology that strives for continuous process capability improvements through the implementation of PAT.

- Good engineering practice: engineering methods and standards should be applied throughout the life cycle to deliver appropriate and cost-effective solutions.

- Subject-matter experts (people with specific expertise and responsibility in a particular field): should take the lead role in the verification of manufacturing systems, including planning and defining verification strategies, defining acceptance criteria, selecting of appropriate test methods, executing of verification tests, and reviewing results.

- Use of vendor documentation: vendor documentation, including test documents, should be used as part of verification documentation provided that the vendor has an acceptable quality system, technical capability, and employs good engineering practices. The decision and justification to use vendor documentation should be documented and approved by both subject-matter experts and the quality control unit.

- Implementation of continuous process improvement: experience gained in commercial production and effective implementation of change management should provide technically sound continuous improvement based on periodic review and evaluation, operational and performance data, and root-cause failure analysis.

28.4.4 Documentation

- Prepare a written protocol specifying critical steps and acceptance criteria on planned qualification and validation projects.

- This protocol is reviewed and approved by appropriate organizational units prior to its implementation.

- Execute the approved protocol and document execution:
  - Prepare a summary report of the results obtained and conclusions drawn.

- ○ Comment on any deviations observed, including recommended changes necessary to correct deficiencies.
- ○ Document any and all changes to the approved protocol with appropriate justification.
- After satisfactory completion of a qualification step, provide a written authorization to proceed with the next step.
- Thus, the sequence of steps followed for all qualification and validation is:
  - ○ Protocol writing
  - ○ Protocol approval
  - ○ Protocol execution
  - ○ Data review
  - ○ Deviations resolved, if any
  - ○ Summary reports written
  - ○ Summary reports approved

28.4.5 Qualification Terminology

- The qualification and validation process should establish and provide documented evidence that a premise, supporting utility, equipment, or process:
  - ○ Has been designed in accordance with the user and functional specifications and CGMP requirements; this normally constitutes a design qualification/detailed design review.
  - ○ Has been built and installed in compliance with design specifications and the activities performed prior to powering the equipment or system and constitutes installation qualification.
  - ○ Operates in accordance with their design specifications; activities done after the equipment or system is powered to determine the limits of operating parameters and to demonstrate that the equipment or systems work as designed; confirmation of these activities constitutes Operational Qualification.
  - ○ Performs within lower and upper operating limits when subjected to the stresses of everyday use; this constitutes performance qualification.
  - ○ The process will consistently produce a product meeting its predetermined specifications and quality attributes; this constitutes process validation PIC/S, 2007).
- Carefully designed and successfully validated system and process control should provide a high degree of confidence that all lots or batches produced will meet their intended specifications. Elements of a successful validation plan are as follows (Health Canada, 2009):
- Validate all critical production processes.
- Conduct validation studies in accordance with predefined protocols.

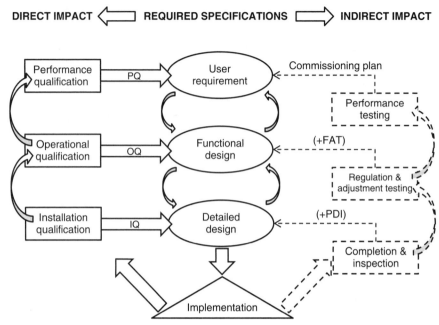

DIRECT IMPACT ⟨▭ REQUIRED SPECIFICATIONS ▭⟩ INDIRECT IMPACT

FIGURE 28.1 Qualification model: qualification-required specifications along with the direct and indirect impacting qualification protocols.

- Write summary reports on the results and conclusions recorded; review, approve, and archive these summary reports.
- Validate changes to production processes, operating parameters, equipment, or materials that may affect product quality and/or the reproducibility of the process prior to implementation.

28.4.6 Qualification Model (ISPE, 2001)

The relationship between the required specifications (in the center of Figure 28.1) and their verification of meeting those requirements by executing various direct and/or indirect impact qualification protocols are depicted on the right and left of the figure, respectively.

28.5 QUALIFICATION PROTOCOLS
(ISPE, 2001; PIC/S, 2007; Health Canada, 2009)

The scope and level of detail required for a qualification depends on the complexity of the system involved and the critical nature of that system with respect to the quality of the final product. To streamline qualification, properly executed and documented predelivery inspection and factory acceptance test protocols could be used to supplement installation qualification, operational qualification, and/or performance qualification of direct impact systems.

Qualification protocols or documents must be preapproved prior to execution. If test findings or results are different from the values expected or predicted, they should be noted as deviations. If a deviation is of critical nature (affects the operational status), it must be addressed before proceeding to the next qualification protocol. All deviations must be reviewed and approved by appropriate personnel. The results of a fully executed protocol (with all deviations addressed) are summarized by comparing the protocol goals against the results, listing all deviations encountered and how these were addressed. A formal (written authorization) release is required in order to proceed with the next step in qualification or validation.

28.5.1 User Requirement Specifications

A successful project depends on clear definition, communication, and understanding of the project scope and objectives, stakeholder's intent (both business and compliance), and flexibility and/or constraints of the project. The user requirement specifications (URSs) provide the objectives for the design and describe the requirements in terms of products to be manufactured, required throughput, and conditions in which products should be made. The URSs should focus on what is needed without being overly prescriptive as to how the requirements should be met, but should provide sufficient information to the engineering group to develop and assess specific options. Some of the information that should be considered includes:

- Product types
- Product forecast and ranges
- Batch sizes (and ranges) and potential restrictions
- Process description and estimated processing times
- Operating ranges and process parameters and their tolerances
- Safety information related to materials and processes
- Known compatibility or incompatibility for materials of construction (including seals)
- Cleaning methods, agents, and limits
- Past experiences with proposed products and processes
- Preferred vendors

URSs may be referenced in the qualification documentations and thus should be subject to a formal quality assurance change control process.

28.5.2 Functional Specification/Functional Design Specification

The functional specification (FS) and functional design specification (FDS) should state what functions a system will perform under both planned and emergency situations. The FS outlines, in detail, how the equipment or system is to accomplish the requirements outlined in the URS. The FS should be signed by the owner and by quality assurance personnel, including those key end users, developers, or engineers who were involved in developing functional requirements. Depending on the size and complexity of a given project, the functional requirements document may be combined with either the URS or the DS. Requirements outlined in the FS are usually tested in the operational qualification.

28.5.3 Design Qualification/Detailed Design Review

- The first element of the validation of new facilities, systems, or equipment could be a design qualification incorporating a system's intended use, specifications for all critical design parameters, and design parameters for future process flexibility.

- The compliance of the design to the CGMP requirements should be demonstrated and documented.

- During the design stage, an effective change management procedure should be in place, ensuring that all changes to the original criteria have been documented and required changes to equipment specifications, plant functional specifications, and piping and instrumentation diagrams have been made.

- During the final phase of the design stage, calibration requirements for facilities and equipment are identified where applicable.

- Review of design at appropriate stages for conformance to operational and regulatory expectations must be documented; in the ISPE Guide (ISPE, 2001) this review is called an enhanced design review.

28.5.4 Installation Qualification

Typically, an installation qualification (IQ) verifies and documents that the equipment, utilities, and systems have been properly built, supplied, and installed in accordance with the detailed design specifications, manufacturer's criteria, and user requirement specifications. IQs are implemented concurrently with the construction of each system and may integrate commissioning information into qualification process, where appropriate.

- Perform the installation qualification process on new or modified facilities, systems, and equipment.

- The IQ should include, but not be limited to, the following:
  - Document and verify correct installation of equipment, piping, services, and instrumentation.
  - Check against approved engineering drawings and plant functional specifications developed during the project planning stage.
  - Collect and collate supplier operating and working instructions and maintenance requirements.

○ Decide and document the calibration requirements.

○ Verify the materials of construction.

For existing equipment and systems that are not modified, a legacy approach may be followed. In such cases, the IQ will verify the documentation required for future setup, operation, cleaning, maintenance, and change control. For redesigned and modified equipment and systems, the IQ will verify and document that the equipment has been supplied and installed properly based on the available design specifications, manufacturer's criteria, and user requirement specifications.

Documents required for writing and executing an IQ protocol:

- Validation master plan
- User requirements and specifications
- Supplier drawings and specifications
- Purchase orders and contracts
- Equipment and instrument lists
- P&ID and drawings
- System hardware and software specifications
- Process description and manuals
- Material certifications
- Spare parts list
- Installation check sheets
- Lubrication schedule
- Calibration check (optional during the IQ phase, but must be done prior to operational testing)

28.5.4.1 IQ Protocol   A typical IQ protocol may include the following verifications and testing items:

- Protocol preexecution approval signatures
- Objective
- Scope
- Introduction
- System description
- Responsibilities for the protocol writing, review, approval, and execution
- References
- Installation qualification checklist
- Equipment and component specification listing and specifications
- A list of equipment that, when operating, has the potential to affect product quality or process performance (critical equipment and components)
- Auxiliary equipment listing and specifications

- Instrument and controls listing and specifications
- Instrument loop diagram verification
- Instrument wiring verification
- Test instrumentation listing and calibration verification
- List of utilities required, specifications, and verification
- System and utility points of use and labeling
- Materials of Construction of product contact materials and indirect product contact materials
- Lubricant list for lubricants having product contact and indirect product contact
- Drawings list (design, construction, red-lined and as-built) and verification
- As-built diagrams and specifications for all purchased equipment, new or used
- Verification that all equipment purchased and its installation meet the original intent (functional specifications and design parameters), including applicable building, electrical, plumbing, and other such codes
- Miscellaneous documentation (vendor correspondence, submittals, internal correspondence, etc.)
- Spare parts and change parts lists and verification
- Preventive maintenance and schedule
- Filter list and verification
- Testing, adjusting, and balancing review
- Reference document list
- Validation test instrument list
- System software verification
- Documentation of set points and configurable parameters
- Acceptance criteria for all IQ protocol parameters listed above
- Signature/initials log
- Deviation handling, deviation summary page, and deviation form
- Protocol postexecution approval signatures

Note: The IQ for any equipment- system-related programmable logic controllers (PLCs) are best addressed in separate validation protocols. This will be critical in administrating and managing project costs and in being able to review contractor quotes in an equitable manner. The validation of PLCs is complex and best treated under separate protocols and quotes.

28.5.4.2 *Change Control During IQ*   Changes should be assessed for potential impact and approved prior to implementation of the change. Following are a few examples of such changes:

- A fundamental change in a previously approved design
- A change in the user requirements or specifications
- When an indirect impact system is reclassified as a direct impact system

28.5.4.3 IQ Summary Report Upon completion, a report summarizing the results and data analysis should be prepared, reviewed, and approved by all stakeholders, including the quality assurance unit.

28.5.5 Operational Qualification

- Operational qualification (OQ) should follow IQ to verify and provide documented evidence that the equipment, systems, and supporting utilities operate in accordance with their design specifications. Sometimes IQ and OQ protocols are combined in a single document.
- During OQ, individual functional aspects of the equipment or system is tested against user requirements and/or sequence of operation.
- These tests should include a condition or set of conditions encompassing upper and lower processing or operating limits and circumstances commonly referred to as worst-case conditions.
- Where applicable, a simulated product may be used to conduct the OQ.
- Finally, issue a report summarizing the results and data analysis. Approval of this report signifies a formal release of the facilities, systems, and equipment to the next stage in validation.
- The completion of a successful OQ should allow the finalization and approval of standard operating procedures (SOPs), which include calibration, operating and cleaning procedures, operator training, and preventive maintenance requirements.

28.5.5.1 OQ Protocol A typical OQ protocol may include the following:

- Protocol preexecution approval signatures
- Objective
- Scope
- Introduction
- System description
- Responsibilities
- References
- Equipment and utility instrumentation listing and calibration verification
- Test instrumentation listing and calibration verification
- SOP listing and verification
- Data collection and documentation procedures
- Drawings (as applicable to protocol execution)

- Test functions (specific and individualized for each piece of equipment and utility)
- Alarm functionality and verification
- Control system functionality and verification
- Required utility system functionality and verification
- Sequence of operations verification
- Smoke profiling around doors to show proper directional airflow (if relevant)
- Three days of static testing of temperature, relative humidity, and differential pressure
- Process operating parameters for each module, including those designated as critical, to demonstrate that each module operates as intended throughout each process operating parameter range
- OQ protocol acceptance criteria for each parameter listed above
- Signature/initials log
- Deviation handling, deviation summary page, and deviation form
- Protocol postexecution approval signatures
- Task reports describing the successful execution of each OQ protocol
- A list identifying each module (step, unit of operation, or stages) of the process

28.5.5.2 *Change Control* Changes to direct impact systems should be assessed and approved.

28.5.5.3 *OQ Summary Report* At the conclusion of the qualification exercise, a report summarizing the OQ-related results and data analysis should be issued.

28.5.6 Performance Qualification

- Performance qualification (PQ) should follow successful completion of installation qualification and operational qualification. For utilities this could be the final qualification step, but for other systems and processes, PQ could be a precursor to process validation.
- PQ verifies that a process will consistently produce a product meeting its predetermined specifications and quality attributes as the user requirements.
- All direct impact systems are subjected to PQ and PQ protocols (original and all subsequent changes) for these systems require quality assurance (QA) review and approval.

28.5.6.1 *PQ Protocol*

- PQ should include, but not be limited to, the following tests (EC, 2001):
  - Employ actual production materials where feasible.

○ Qualified substitutes or simulated products may be used, if justified, based on the knowledge gained during development of the process and the facilities, systems, or equipment.

○ Include a condition or set of conditions encompassing upper and lower operating limits.

28.5.6.2 Change Control Changes must be implemented with quality assurance (QA) change control SOPs, and all changes must be reviewed and approved by QA.

28.5.6.3 PQ Summary Report Similar to the summary reports issued for IQ and OQ, this report includes the PQ-related results and data analysis outcomes.

28.5.7 Qualification of Established (In-Use) Facilities, Systems, and Equipment

To perform detailed installation and operational qualification of established facilities, systems, or equipment is usually not possible. Nonetheless, it is required that documented evidence be available to support and verify the operating parameters and limits for the critical variables of the operating equipment. Additionally, the calibration, cleaning, preventive maintenance, operating procedures, and operator training procedures and records should also be documented. In summary, the outcomes of the qualification are:

- Completed product specifications
- Scientific rationale or basis for criteria—thoroughly documented
- IQ and OQ steps completed and reports written, reviewed, and approved
- Operating personnel trained and qualified
- Change control procedures in place

28.6   PROCESS VALIDATION (PIC/S, 2007; FDA, 2011b)

The process validation, as defined by the FDA guidance document (FDA, 2011b), is the collection and evaluation of data from the process design stage through commercial production, which establishes scientific evidence that a process is capable of consistently delivering high-quality product that is fit for its intended use; it provides a structured way to assess factors that affect the final product.

- Commissioning and qualification (IQ/OQ/PQ) serve as the foundation for process validation.
- The facilities, systems, and equipment that are planned to be used for a given process should have already been qualified, analytical testing methods validated, and all personnel appropriately trained for the tasks they are to perform.

- Facilities, systems, equipment, and processes should be evaluated periodically to verify that they are still operating in a valid manner and are reliably producing finished products of required quality.
- The company's policy and procedures for process validation should be clearly defined.

An effective process validation incorporates understanding that the following conditions exist:

- Quality, safety, and efficacy are designed or built into the product.
- Quality cannot be adequately assured merely by in-process and finished inspection or testing.
- Each manufacturing process step is controlled to assure that the finished product meets all quality attributes, including specifications.

Success of a validation program depends on applying the knowledge and understanding gained from the product and process development in controlling a manufacturing process that results in products of desired quality. Manufacturers should:

- Understand sources of variation
- Detect the presence and degree of variation
- Understand the impact of variation on the process and ultimately on product attributes
- Control the variation in a manner commensurate with the risk it represents to the process and product

The FDA guidance describes process validation activities over the life cycle of the process in three stages:

- Stage 1: Process design. The commercial manufacturing process that can consistently deliver a quality product is defined in this stage based on knowledge gained through development and scale-up activities. The functionality and limitations of commercial manufacturing equipment, including the contributions of variability predicted for different component lots, production operators, environmental conditions, and measurement systems in the production setting, should be considered in the process design. Generally, early process design experiments (for noncommercially distributed batches) should be conducted in accordance with sound scientific methods and principles, including good documentation practices, but do not need to be performed under the CGMP conditions. Decisions and justification of the process controls should be documented sufficiently and reviewed internally to verify and preserve their values to be useful later in the life cycle of the process and product. The FDA expects controls to include both examination of material quality and equipment monitoring and encourages use of more advanced strategies such as process analytical technology (PAT) for timely analysis and control loops to adjust the processing conditions so that the output remains constant. These controls are incorporated in the master production and control records, and should be carried forward to the next stage.

- Stage 2: Process qualification. During this stage, the process design is evaluated to determine if it is capable of reproducible commercial manufacture. This stage has two elements: (1) facility design and qualification of the equipment and utilities to properly design and commission facility and qualify equipment and utilities to demonstrate that they are suitable for their intended use, and (2) process performance qualification (PPQ), which combines the qualified facility, utilities, equipment, and trained personnel with the commercial manufacturing process, control procedures, and components to produce commercial batches. A successful PPQ will confirm the process design and demonstrate that the commercial manufacturing process performs as expected. Completion of a successful PPQ is a prerequisite to beginning commercial distribution of a drug product. The PPQ lots should be manufactured under normal conditions (of utilities, environment, equipment, and manufacturing procedures) by the personnel routinely expected to perform each unit operation of the process. A report documenting and assessing adherence to the written PPQ protocol should be prepared, reviewed, and approved by all appropriate departments and the QA unit in a timely manner after completion of the protocol.

- Stage 3: Continued process verification. The goal of this stage is continual assurance that the process remains in a state of control (the validated state) during commercial manufacture. An ongoing program to collect and analyze product and process data related to product quality and to detect unplanned departures from the desired process is essential to accomplish this goal. Variation can also be detected by timely assessment of defect complaints, out-of-specification findings, process deviation reports, process yield variations, batch records, incoming raw material records, and adverse event reports. Based on the data from the variation reports and depending on how the proposed changes might affect product quality, additional process design and process qualification could be warranted. All changes proposed must have a documented well-justified rationale, implementation plan, and approval from the QA unit before implementation. The facility, utilities, and equipment must be properly maintained, calibrated, and requalified to ensure that the process remains in control. Maintenance and calibration frequency should be adjusted based on feedback from these activities.

In all stages of the product life cycle, the following practices should ensure uniform collection and assessment of process information and enhance accessibility of such information later in the product life cycle:

- An integrated team approach, consisting of experts from various disciplines, should be implemented.
- All studies are planned, documented, approved, and conducted according to sound scientific principles appropriate for the stage of the life cycle.

- All attributes and parameters are evaluated in terms of their risk to the process and process output; a higher degree of control is appropriate for attributes and parameters that pose a higher risk.

- Process validation should normally be completed prior to the distribution and sale of the medicinal product (prospective validation). In exceptional circumstances, where this is not possible, it may be necessary to validate processes during routine production (concurrent validation); processes in use for some time should also be validated (retrospective validation).

28.6.1 Prospective Validation (Health Canada, 2009)

Prospective validation is conducted before a new product is released for distribution or before a product made under a revised manufacturing process (which may affect the product's characteristics) is released for distribution (FDA, 2009).

- Prospective validation protocol should minimally include the following (EU, 2001):
    - A description of the process
    - A description of the experiment, including the investigation of critical processing steps
    - A list of the equipment/facilities to be used
    - A record of the calibration status of the measuring/monitoring/recording equipment
    - Finished product specifications for release
    - A list of analytical methods
    - Proposed in-process controls with acceptance criteria
    - Additional testing that must be carried out, with appropriate acceptance criteria and analytical validation samples to be taken—where, when, how, and how many
    - Methods for recording and evaluating results including statistical analysis,
    - Personnel responsibilities
    - Time table proposed
- All equipment, the production environment, and analytical testing methods planned for use should have been fully validated and validation staff appropriately trained.
- Prepare the master batch documentation after the critical process parameters have been identified and the machine settings, component specifications, and environmental conditions have been determined.
- Using the final approved parameters (including specified components), a series of batches (typically, three consecutive batches or runs) of the final product should be produced to observe the normal extent of variation to establish trends and to provide sufficient data for evaluation. This would constitute validation of the process.

- Batches made for process validation should be the same size as that of the intended full production scale. When full production starts, the validity of any assumptions made should be demonstrated. During the processing of batches, extensive testing is performed on the product at various process stages and on the final product and its package.

- Upon completion of the review of full-scale validation batches, recommendations should be made as to the extent of monitoring and the in-process controls necessary for routine production. Incorporate these recommendations into a batch manufacturing or packaging record or into appropriate SOPs.

- If planning to sell the validation batches, the entire processing conditions and validation outcomes must comply fully with the requirements of good manufacturing practice and with the marketing authorization (where applicable).

28.6.2   Concurrent Validation

According to FDA guidance on process validation (FDA, 2011b), a process performance qualification (PPQ) study must be completed successfully and a high degree of assurance achieved in the process before the PPQ batch is distributed commercially. In special but rare situations, the PPQ protocol can be designed to release the resulting batch for distribution before fully executing the PPQ protocol (i.e., releasing the batch concurrent to PPQ protocol execution):

- Manufacture of drugs with limited distribution (such as orphan drugs)
- Drugs that have a short lifetime (e.g., radiopharmaceuticals)
- Drugs that are necessary medically but are in short supply

Circumstances and rational for concurrent release should be fully described in the PPQ protocol; the concurrently released batch must comply with all CGMPs, regulatory approval requirements, PPQ protocol release criteria, and appropriate quality attributes. Lots released Concurrently must be included in the formal stability program and a rapid detection program established to correct any problems arising from stability failures, customer feedback and complaints, or from any negative PPQ study findings.

28.6.3   Retrospective Validation (FDA, 2009)

Retrospective validation is the validation of a process based on accumulated historical production, testing, control, and other information for a product already in production and distribution. Historical data must contain enough information (e.g., process capability studies results, trend charts) to provide an in-depth picture of how the process operated and whether the product consistently met its specifications; this analysis must be documented. Incomplete historical information mitigates against conducting a successful retrospective validation. Some examples of incomplete information are:

- Customer complaints that have not been investigated fully to determine the cause of the problem, including the identification of complaints that are due to process failures
- Complaints investigated but for which corrective action was not taken
- Scrap and rework decisions that are not recorded, investigated, and/or explained
- Excessive rework
- Records that do not show the degree of process variability and/or whether process variability is within the range of variation that is normal for that process
- Gaps in batch records for which there are no explanations (retrospective validation cannot be initiated until the gaps in records can be filled or explained)

Retrospective validation is not appropriate where changes have been made recently in the composition of the product, operating procedures, or equipment.

28.7 CLEANING VALIDATION (FDA, 1998, 2010b; PIC/S, 2007; Health Canada, 2008)

- Perform and document cleaning validation to confirm the effectiveness of a cleaning procedure.
- Employ a logical rationale for selecting limits of carryover of product residues, cleaning agents, and microbial contamination (and contamination from other processing materials, such as airborne particles, dust, lubricants, raw materials, and intermediates) based on the materials involved.
- Ensure that these limits are achievable and verifiable and pose no associated risk of cross-contamination of active and/or nonactive ingredients. The approach to setting limits can be:
  - Product specific
  - Grouped into product families and a worst-case product chosen
  - Grouped by properties (e.g., solubility, potency, toxicity, or formulation ingredients known to be difficult to clean)
  - Based on maximum carryover primarily of the active pharmaceutical ingredient [e.g., most stringent of no more than (NTM) 0.1% of the normal therapeutic dose of any product to appear in the maximum daily dose of the following product or NMT 10 ppm of any product to appear in another product]
- Analytical methods with necessary specificity and sensitivity to detect residues or contaminants at the established acceptable levels should be validated prior to use for verifying the effectiveness of a cleaning process.
- Develop cleaning procedures for all product-contact equipment surfaces as well as those noncontact parts into which a product may migrate (e.g., seals, flanges, mixing shafts, fans of ovens, heating elements). Product-contact

equipment should be dedicated until validation of the cleaning procedure has been completed.

- Cleaning procedures for products and processes that are very similar do not need to be validated individually. A single validation study using the worst-case scenario may be carried out which takes all relevant criteria into account.

- If automated cleaning procedures (e.g., clean-in-place) are used, monitor the critical control points and the parameters with appropriate sensors and alarm points to ensure that the process is highly controlled.

- During a campaign (production of several batches of the same product), cleaning between batches may be reduced, if justified.

- Validation of cleaning processes should be based on a worst-case scenario, including:

  - Challenge of the cleaning process to ensure that the process is indeed removing "challenge" soil to required levels (demonstrates log removal or recovers in sufficient quantity)

  - The use of reduced cleaning parameters (e.g., lowered or minimal concentration and/or contact time of cleaning agents and detergents)

- To prove that the cleaning method is valid, prove its effectiveness in a minimum of three consecutive applications.

- The intervals between use and cleaning as well as cleaning and reuse should be validated. Cleaning intervals and methods should be determined.

- "Test until clean" is not considered an appropriate alternative to cleaning validation.

- Document detailed cleaning procedures and SOPs.

- A cleaning validation protocol should include the following:

  - The objective of the cleaning validation process

  - Responsibilities for performing and approving the validation studies

  - A description of the equipment to be used

  - The interval between the end of production and the beginning of the cleaning cycle

  - A number of consecutive lots of the same product manufactured before a full cleaning is done

  - The detailed cleaning procedures to be used for each product, each manufacturing system, or each piece of equipment

  - Any routine monitoring equipment

  - Sampling procedures, including sampling locations and the rationale for selecting a certain sampling methodology

  - Data on recovery studies, where appropriate

  - A validated analytical method, including the limits of detection and the limit of quantitation of those methods

- ○ The acceptance criteria, including the rationale for setting the specific limits
- ○ Other products, processes, and equipment for which the planned validation is valid according to a "bracketing" concept
- ○ Change control
- ○ Revalidation
- Prepare a final validation report to conclude whether the cleaning process has been validated successfully and the limitations applied to the use of this method, including the analytical limits at which cleanliness can be determined. The report should be approved by management.

28.8 COMPUTER SYSTEMS VALIDATION (ISPE, 2001)

Commissioning and qualification of most systems would be incomplete without considering the level of computerization of the systems that they contain.

There are two basis approaches:

1. Stand-alone systems: these systems (e.g., distributed control system, building management system systems control and data acquisition) link or control multiple systems and should be qualified as independent systems.

2. Integrated control systems: these systems (e.g., computers, programmable logic controllers) control a single system and are typically qualified within the qualification protocols of the system.

28.9 CHANGE CONTROL (EC, 2001; PIC/S, 2007)

- A change control system is an important element in any quality assurance system and should ensure that all changes reported or requested are investigated, documented, and authorized satisfactorily.
- Written procedures should be in place to describe the actions to be taken if a change is proposed to any of the following:
  - ○ Changes in raw materials (including physical properties) and in packaging components
  - ○ Changes in the equipment and/or processes used for manufacture of raw ingredients, finished products, or both
  - ○ Changes in equipment (e.g., addition of automatic detection systems)
  - ○ Production area and support system changes (e.g., rearrangement of areas, new water treatment method)
  - ○ Change in the method of production or testing
  - ○ Transfer of processes to another site

  ○ Unexpected changes observed during self-inspection or during routine analysis of process trend data that may affect product quality or reproducibility of the process

- All changes that may affect product quality or reproducibility of the process should be formally requested, documented, and accepted. The probable impact of a change of facilities, systems, or equipment on a product should be evaluated, including the risk assessment and the need for, and the extent of, requalification and revalidation.

- Products made by changed processes should not be released for sale without properly authorized approvals.

28.10   REVALIDATION (EC, 2001; PIC/S, 2007; FDA, 2009)

- Revalidation provides documented evidence that changes to facilities, systems, equipment, and processes, including cleaning, introduced either intentionally or unintentionally, do not adversely affect process characteristics and product quality. The changes listed in Section 28.9 and the following planned or unplanned changes may require revalidation:

  ○ Changes in the source of active raw material manufacturer

  ○ Changes in equipment unless the replacement is based on a "like for like" basis

  ○ Changes in the plant or facility

  ○ Variations revealed by trend analysis (e.g., process drift)

- Revalidation review should be carried out periodically at scheduled intervals. A decision not to perform revalidation must be fully justified and documented.

- Documentation requirements are the same as for the initial validation, and in many cases similar protocols can be employed.

REFERENCES

ASTM. 2007. Standard Guide for Specification, Design, and Verification of Pharmaceutical and Biopharmaceutical Manufacturing Systems and Equipment. ASTM 2005-07.

EC. 2001. Qualification and Validation: Final Version of Annex 15 to the EU Guide to Good Manufacturing Practice. July 2001. http://ec.europa.eu/health/files/eudralex/vol-4/pdfs-en/v4an15_en.pdf. Accessed Dec. 2011.

FDA. 1998. Guidance for industry: Manufacturing, processing, or holding active pharmaceutical ingredients. Mar. 17, 1998. http://www.fda.gov/downloads/Drugs/GuidanceComplianceRegulatory Information/Guidances/UCM070289.pdf. Accessed Dec. 2011.

FDA. 2009. Process validation. Medical Device Quality Systems Manual, A Small Entity Compliance Guide (supersedes the Medical Device Good Manufacturing Practices Manual). Updated Apr. 28,

2009. http://www.fda.gov/MedicalDevices/DeviceRegulationandGuidance/PostmarketRequirements/QualitySystemsRegulations/MedicalDeviceQualitySystemsManual/default.htm. Accessed Dec. 2011.

FDA. 2010a. CPG Sec. 490.100. Process validation requirements for drug products and active pharmaceutical ingredients subject to premarket approval. Feb. 24, 2010. http://www.fda.gov/ICECI/ComplianceManuals/CompliancePolicyGuidanceManual/ucm074411.htm. Accessed Dec. 2011.

FDA. 2010b. Validation of cleaning process (7/93). Updated June 14, 2010. http://www.fda.gov/ICECI/Inspections/InspectionGuides/ucm074922.htm. Accessed Dec. 2011.

FDA. 2011a. *Code of Federal Regulations*. 21 CFR 211. Current good manufacturing practice for finished pharmaceuticals. Apr. 1, 2011. http://www.accessdata.fda.gov/scripts/cdrh/cfdocs/cfcfr/CFRSearch.cfm?CFRPart = 211&showFR = 1. Accessed Dec. 2011.

FDA. 2011b. Guidance for industry: Process validation—general principles and practices. Jan. 2011. http://www.fda.gov/downloads/Drugs/GuidanceComplianceRegulatoryInformation/Guidances/UCM070336.pdf. Accessed Jan. 2012.

Health Canada. 2008. Cleaning validation guidance. GUIDE-0028. Modified Jan. 1, 2008. http://www.hc-sc.gc.ca/dhp-mps/compli-conform/gmp-bpf/validation/gui-0028_cleaning-nettoyage_ltr-doc-eng.php. Accessed Dec. 2011

Health Canada. 2009. Validation guidelines for pharmaceutical dosage forms. GUI-0029. Aug. 7, 2009. http://www.hc-sc.gc.ca/dhp-mps/alt_formats/pdf/compli-conform/gmp-bpf/validation/GUI-0029-eng.pdf. Accessed Dec. 2011.

ISPE. 2001. Commissioning and qualification. Pharmaceutical Engineering Guide for New and Renovated Facilities, Vol. 5. Mar. 2001.

PIC/S. 2007. PI006-3: Recommendations on validation master plan, installation and operational qualification, non-sterile process validation, cleaning validation. Pharmaceutical Inspection Convention/Pharmaceutical Inspection Co-operation Scheme. Sept. 25, 2007. http://www.picscheme.org/publication.php?download&file = cGktMDA2LTMtcmVjb21tZW5kYXRpb24tddmFsaWRhdGlvbi1tYXN0ZXItcGxhbi5wZGY_. Accessed Dec. 2011.

Tyree R. Design and commissioning of a contract API facility: a contract API manufacturer explains how to build with CGMP in mind. Contract Pharma 2005 (Aug. 22). http://www.contractpharma.com/issues/2000-06/view_features/design-and-commissioning-of-a-contract-api-facilit/. Accessed Dec. 2011.

WL: 10-ATL-12. Hospira, Inc. Apr. 12, 2010. Updated Apr. 15, 2010. http://www.fda.gov/ICECI/EnforcementActions/WarningLetters/ucm208691.htm. Accessed Dec. 2011.

WL: 2008-DT-05. Caraco Pharmaceutical Laboratories, Ltd. Oct. 31, 2008. Updated July 8, 2009. http://www.fda.gov/ICECI/EnforcementActions/WarningLetters/2008/ucm1048080.htm. Accessed Dec. 2011.

WL: 320-06-03. Ranbaxy Laboratories Limited. June 15, 2006. Updated July 8, 2009. http://www.fda.gov/ICECI/EnforcementActions/WarningLetters/2006/ucm075947.htm. Accessed Dec. 2011.

WL: 320-08-02. Ranbaxy Laboratories, Ltd., Paonta Sahib, India. Sep. 16, 2008. Updated July 8, 2009. http://www.fda.gov/ICECI/EnforcementActions/WarningLetters/2008/ucm1048133.htm. Accessed Dec. 2011.

WL: 320-09-05. Lupin Limited. May 7, 2009. Updated June 18, 2009. http://www.fda.gov/ICECI/EnforcementActions/WarningLetters/2009/ucm162745.htm. Accessed Dec. 2011.

WL: 320-09-06. Apotex Inc. Updated June 23, 2011. http://www.fda.gov/ICECI/EnforcementActions/WarningLetters/2009/ucm170912.htm. Accessed Dec. 2011.

WL: 320-10-07. Haw Par Healthcare Limited. July 20, 2010. Updated Aug. 9, 2010. http://www.fda.gov/ICECI/EnforcementActions/WarningLetters/ucm221006.htm. Accessed Dec. 2011.

WL: 45-10. Nexgen Pharma, Inc. Sept. 27, 2010. Updated Nov. 10, 2010. http://www.fda.gov/ICECI/EnforcementActions/WarningLetters/ucm233276.htm. Accessed Dec. 2011

WL: FLA 09–13. Hill Dermaceuticals, Inc. Apr. 27, 2009. Updated Aug. 4, 2009. http://www.fda.gov/ICECI/EnforcementActions/WarningLetters/2009/ucm174100.htm. Accessed Dec. 2011.

WL: NYK 2009-03. Jerome Stevens Pharmaceuticals, Inc. Oct. 23, 2008. Updated July 8, 2009. http://www.fda.gov/ICECI/EnforcementActions/WarningLetters/2008/ucm162951.htm. Accessed Dec. 2011.

WL: OEWL-08-01. Catalent Pharma Solutions. Mar. 28, 2008. Updated July 8, 2009. http://www.fda.gov/ICECI/EnforcementActions/WarningLetters/2008/ucm1048356.htm. Accessed Dec. 2011.
WL: SJN-2009-07. Olay LLC. Apr. 24, 2009. Updated June 18, 2009. http://www.fda.gov/ICECI/EnforcementActions/WarningLetters/2009/ucm161902.htm. Accessed Dec. 2011.

GLOSSARY

C&Q	Commissioning and Qualification.
CFR	*Code of Federal Regulation.*
CGMP	Current good manufacturing practice.
Change control	A formal system by which qualified representatives of appropriate disciplines review proposed or actual changes that might affect the validated status of facilities, systems, equipment, or processes. The intent is to determine the need for action that would ensure and document that the system is maintained in a validated state.
Cleaning validation	Documented evidence that an approved cleaning procedure will provide equipment suitable for processing medicinal products.
Commercial manufacturing process	Manufacturing process resulting in commercial product (i.e., drug that is marketed, distributed, and sold or intended to be sold). For the purposes of this guidance, the term does not include clinical trial or treatment IND material.
Concurrent release	Releasing for distribution a lot of finished product manufactured following a qualification protocol, which meets the lot release criteria established in the protocol, but before the entire study protocol has been executed.
Concurrent validation	Validation carried out during routine production of products intended for sale.
Continued process verification	Assuring that during routine production a process remains in a state of control.
Critical process parameter	A parameter which if not controlled will contribute to the variability of the end product.
Design qualification (DQ)	The documented verification that the proposed design of the facilities, systems, and equipment is suitable for the purpose intended.
Equipment qualification	Studies to establish with confidence that the process equipment and ancillary systems used for the manufacture of commercial batches are capable of operating consistently within established limits and tolerances. These studies must include equipment specifications, installation qualification, and operational qualification of all major equipment and should simulate actual production conditions, including "worst case" or stressed conditions.
FAT	Factory acceptance test.
FDA	U.S. Food and Drug Administration.
FDS	Functional design specification.
FS	Functional specification.
Installation qualification	The documented verification that facilities, systems, and equipment, as installed or modified, comply with the approved design and the manufacturer's recommendations.
Major equipment	A piece of equipment that performs significant processing steps in the sequence of operations required for the manufacture or packaging of drug products. Some examples of major equipment include tablet compression machines, mills, blenders, fluid-bed dryers, heaters, drying ovens, tablet coaters, and encapsulators.

Master production document	A document that includes specifications for raw material, packaging material, and packaged dosage form, master formula, sampling procedures, and critical processing related standard operating procedures, whether or not these are specifically referenced in the master formula.
MOC	Material of construction.
NMT	No more than.
Operational qualification	The documented verification that the facilities, systems, and equipment, as installed or modified, perform as intended throughout the operating range santicipated.
OQ	Operational qualification.
P&ID	Pipeline and instrumentation diagram.
PAT	Process analytical technology.
Performance indicators	Measurable values used to quantify quality objectives to reflect the performance of an organization, process or system, also known as performance metrics in some regions. (ICH Q10).
Performance qualification	The documented verification that the facilities, systems and equipment, as connected together, can perform effectively and reproducibly, based on the approved process method and product specification.
PLC	Programmable Logic Controller.
PPQ	Product Performance Qualification.
PQ	Performance Qualification.
Process capability	Studies conducted to identify the critical process parameters that yield a resultant quality, and their acceptable specification ranges, based on the established $+/-3$ sigma deviations of the process, under stressed conditions but when free of any assignable causes.
Process design	Defining the commercial manufacturing process based on knowledge gained through development and scale-up activities.
Process qualification	Confirming that the manufacturing process as designed is capable of reproducible commercial manufacturing.
Process validation	The documented evidence that a process, operated within established parameters, can perform effectively and reproducibly to produce a medicinal product meeting its predetermined specifications and quality attributes. The collection and evaluation of data, from the process design stage through commercial production, which establishes scientific evidence that a process is capable of delivering consistently high-quality products.
Prospective validation	Validation carried out before routine production of products intended for sale.
QA	Quality assurance.
QbD	Quality by design.
Quality	The degree to which a set of inherent properties of a product, system, or process fulfills requirements.
Retrospective validation	Validation of a process for a product that has been marketed based on accumulated manufacturing, testing, and control batch data.
Revalidation	A repeat of the process validation to provide an assurance that changes in the process or equipment introduced in accordance with change control procedures do not adversely affect process characteristics and product quality.
Risk analysis	Method to assess and characterize the critical parameters in the functionality of equipment or process.
SAT	Site acceptance test.

Simulated product	A material that closely approximates the physical and, where practical, the chemical characteristics (e.g., viscosity, particle size, pH) of a product under validation. In many cases, these characteristics may be satisfied by a placebo product batch.
SOP	Standard operating procedure.
State of control	Condition in which the set of controls consistently provides assurance of continued process performance and product quality.
System	A group of pieces of equipment with a common purpose.
URS	User required specification.
U.S.P.	*United States Pharmacopoeia*.
Validation	The documented evidence that any procedure, process, or activity will lead consistently to the results expected.
Validation master plan	An approved written plan of objectives and actions stating how and when a company will achieve compliance with the GMP requirements regarding validation.
Validation protocol	A written plan stating how validation will be conducted, including a minimum number of batches needed for validation, test parameters, product characteristics, production equipment, decision points on what constitute acceptable test results, and who will sign, approve, or disapprove of the conclusions derived from such a scientific study.
Validation team	A multidisciplinary team of personnel primarily responsible for conducting and/or supervising validation studies.
VMP	Validation Master Plan.
Worst case	A condition or set of conditions encompassing upper and lower processing limits and circumstances, within standard operating procedures, which pose the greatest chance of product or process failure compared to ideal conditions. Such conditions do not necessarily induce product or process failure.

CHAPTER 29

QUALITY SYSTEMS AND CONTROLS

29.1 PHARMACEUTICAL QUALITY SYSTEM
(ICH, 2008; FDA, 2009a)

An effective pharmaceutical quality system (PQS), as described in the U.S. Food and Drug Administration (FDA) Guidance for Industry Q10 (FDA, 2009a)/International Conference on Harmonization (ICH) Pharmaceutical Quality Systems Q10 (ICH, 2008), based on International Organization for Standardization quality concepts, should include applicable good manufacturing practice (GMP) regulations and should complement ICH Q8 Pharmaceutical Development (ICH, 2009) and ICH Q9 Quality Risk Management (ICH, 2005a). This system should be implemented throughout the various stages of a pharmaceutical product life cycle to enhance the quality and availability of medicines around the world in the interest of public health. Additionally, implementation throughout the product life cycle should facilitate innovation and continual improvement and strengthen the link between pharmaceutical development and manufacturing activities. The product life cycle includes the following technical activities for new and existing products:

- Pharmaceutical development
  - Drug substance development
  - Formulation development (including the container closure system)
  - Manufacture of investigational products
  - Delivery system development (where relevant)
  - Manufacturing process development and scale-up
  - Analytical method development
- Technology transfer
  - New product transfers during development through manufacturing
  - Transfers within or between manufacturing and testing sites for marketed products

Integrated Pharmaceutics: Applied Preformulation, Product Design, and Regulatory Science,
First Edition. Antoine Al-Achi, Mali Ram Gupta, William Craig Stagner.

- Commercial manufacturing
  - Acquisition and control of materials
  - Provision of facilities, utilities, and equipment
  - Production (including packaging and labeling)
  - Quality control and assurance
  - Release
  - Storage
  - Distribution (excluding wholesaler activities)
- Product discontinuation
  - Retention of documentation
  - Sample retention
  - Continued product assessment and reporting

29.1.1 Objectives and Enablers for Implementing a PQS

Implementation of the PQS, based on the Q10 model, should result in achievement of three main objectives that complement or enhance GMP requirements.

- Achieve product realization: to deliver products of appropriate quality attributes that would meet the needs of all stakeholders, including patients, health care professionals, and regulatory authorities
- Establish and maintain a state of control: to develop and use effective monitoring and control for process performance and product quality through employment of quality risk management concepts
- Facility continual improvements: to identify and implement product quality and process improvements, variability reduction, innovations, and pharmaceutical quality system enhancements; these improvements should allow manufacturers to meet their own quality needs consistently

Use of knowledge management and quality risk management will enable a company to implement the PQS effectively and successfully by providing the means for science- and risk-based decisions related to product quality. Enhanced science- and risk-based regulatory approaches could potentially be beneficial:

- To comply with the requirements of regional GMP regulations
- To ensure regulatory inspections
- To facilitate pharmaceutical quality assessment
- To optimize postapproval change processes
- To maximize benefits from innovation and continual improvement
- To enable innovative approaches to process validation
- To establish real-time release mechanism

29.1.2   Design and Content Considerations of the PQS

The design, organization, and documentation of a PQS should be:

- Well structured and clear, to facilitate common understanding and consistent application
- Appropriate and proportionate to each of the product life-cycle stages
- Appropriate to incorporate risk management principles
- Responsive to provide assurance of the quality of outsourced activities and purchased materials
- Aimed at identifying management responsibilities
- Inclusive of elements such as process performance and product quality monitoring, corrective and preventive action, change management, and management review
- Focused on identifying performance indicators to monitor the effectiveness of processes

29.1.3   Quality Manual

The PQS should be described in a quality manual or equivalent documentation. This document should include:

- The quality policy
- The scope of the pharmaceutical quality system
- Identification of the pharmaceutical quality system processes, as well as their sequences, linkages, and interdependencies, including process maps and flowcharts
- Management responsibilities

29.1.4   Pharmaceutical Quality System Elements

The following four elements may be required in part under regional GMP regulations; however, the intent of the Q10 model is to enhance these elements to promote the life-cycle approach to product quality. Applications of each of these elements are provided in Appendixes 29.1 to 29.4.

- Process performance and product quality monitoring system (Appendix 29.1)
- Corrective action and preventive action (CAPA) system (Appendix 29.2)
- Change management system (Appendix 29.3)
- Management responsibilities (Appendix 29.4)

29.1.5   Continual Improvement of the PQS

The following activities should be conducted to manage and continually improve the PQS:

- Management should formally review the PQS on a periodic basis for:
  - Measurement of achievement of pharmaceutical quality system objectives
  - Assessment of performance indicators used to monitor the effectiveness of processes within the pharmaceutical quality system, such as:
    - Complaint, Deviation, CAPA, and change management processes
    - Feedback on outsourced activities
    - Self-assessment processes, including risk assessments, trending, and audits
    - External assessments such as regulatory inspections and findings and customer audits
- Monitoring of internal and external factors that can have an impact on the PQS: factors monitored by management may include:
  - Impact of emerging regulations, guidance, and quality issues on the PQS
  - Innovations to enhance the PQS
  - Changes in business environment and objectives
  - Changes in product ownership
- The outcome of management review of the PQS and monitoring of internal and external factors should include:
  - Improvements to the PQS and related processes
  - Allocation or reallocation of resources and/or personnel training
  - Revisions to the quality policy and quality objectives
  - Documentation and timely and effective communication of the results of the management review and actions, including escalation of appropriate issues to senior management

The major features of this PQS model are illustrated in Figure 29.1, and discussed below.

- The PQS covers the entire life cycle of a product, including pharmaceutical development, technology transfer, commercial manufacturing, and product discontinuation (as illustrated in the upper portion of the diagram). The PQS augments regional GMPs, including as they apply to the manufacture of investigational drugs.
- The next bar illustrates the importance of management responsibilities during all stages of a product life cycle.
- The next bar lists the PQS elements which serve as pillars for the model and should be applied to each life-cycle stage to identify areas for continual improvement.
- The bottom set of horizontal lines illustrates the two enablers: knowledge management and quality risk management; as illustrated, they are applicable throughout life-cycle stages and support the PQS goals of achieving product

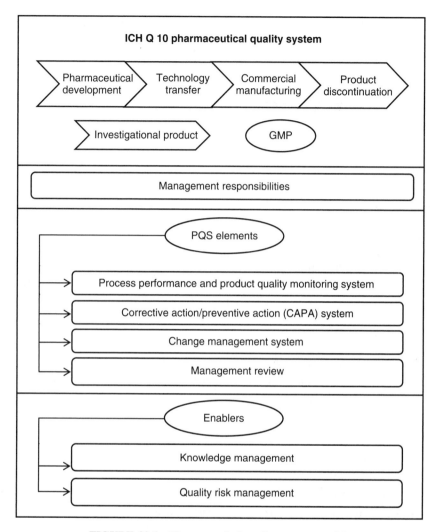

FIGURE 29.1   Pharmaceutical quality system model.

realization, establishing and maintaining a state of control, and facilitating continual improvement.

○ Knowledge management: management of product and process knowledge from development through the commercial life of the product up to and including product discontinuation. Knowledge management is a systematic approach to acquiring, analyzing, storing, and disseminating information related to products, manufacturing processes, and components.

○ Quality risk management: a proactive approach to identifying, scientifically evaluating, and controlling potential risks to quality. It facilitates continual improvement of process performance and product quality throughout the product life cycle. The ICH Q9 (ICH, 2005a) provides principles and examples of tools for quality risk management that can be applied to different aspects of pharmaceutical quality.

29.2 QUALITY SYSTEMS APPROACH TO CGMP REGULATIONS

This section is based on the FDA guidance document Quality Systems Approach to Pharmaceutical CGMP Regulations, September 2006 Pharmaceutical CGMPs (FDA, 2006). This guidance is intended to help manufacturers implement modern comprehensive quality systems and risk management approaches to meet Current good manufacturing practice (CGMP) regulations [21 *Code of Federal Regulations* (CFR) Parts 210 and 211].

Good manufacturing practice and good business practice require a robust quality system. When fully developed and effectively managed, a quality system will lead to consistent, predictable processes which ensure that pharmaceuticals are safe, effective, and available for consumers. Successful quality systems share the following characteristics:

- Science-based approaches
- Decisions based on an understanding of the intended use of a product
- Proper identification and control of areas of potential process weakness
- Responsive deviation and investigation systems that lead to timely remediation
- Scientifically sound methods for assessing and reducing risk
- Well-defined processes and products, starting from development and extending throughout the product life cycle
- Systems for careful analysis of product quality
- Supportive management (philosophically and financially)

The overarching philosophy articulated in both the CGMP regulations and a robust modern quality system is: "Quality should be built into the product, and testing alone cannot be relied on to ensure product quality."

29.2.1 CGMP and the Concepts of Modern Quality Systems

The following key concepts are critical for modern quality systems as they relate to the manufacture of pharmaceutical products.

- Quality. Preestablish identity, strength, purity, and other quality characteristics to ensure that each product meets the required levels of safety and effectiveness.
- Quality by design and product development. Design and develop a product and associated manufacturing processes that will be used during product

development to ensure that the product consistently attains a predefined quality at the end of the manufacturing process. This should provide a sound framework to transfer product knowledge and process understanding from drug development to the commercial manufacturing processes and for postdevelopment changes and optimization.

- Quality risk management. This process should guide:
  - In setting specifications and process parameters for drug manufacturing
  - In assessing and mitigating the risk of changing a process or specification
  - In determining the extent of discrepancy investigations and corrective actions

- Corrective and preventive action (CAPA). CAPA focuses on investigating, understanding, and correcting discrepancies with the goal of preventing their recurrence. Quality system models discuss CAPA as three separate concepts:
  - Remedial correction of a problem that has been identified
  - Root-cause analysis with corrective action to understand the cause of the deviation and potentially, to prevent recurrence of a similar problem
  - Preventive action to avert recurrence of a similar potential problem

- Change control. Change control focuses on managing changes to prevent unintended consequences. By utilizing the knowledge gained during a product's life cycle, a manufacturer is empowered to make certain changes (e.g., variability of materials used in manufacturing). Some other major manufacturing changes (e.g., changes that alter specifications, a critical product attribute, or bioavailability), however, would continue to require regulatory filings and prior regulatory approval (21 CFR 314.70, 514.8, and 601.12).

- Quality unit. The concept of a quality unit is also consistent with modern quality systems in ensuring that the various operations associated with all systems are planned, approved, conducted, and monitored appropriately. Current industry practice generally divides the responsibilities of the quality control unit (QCU), as defined in the CGMP regulations, between quality control (QC) and quality assurance (QA) functions.
  - QC usually involves (1) assessing the suitability of incoming components, containers, closures, labeling, in-process materials, and finished products; (2) evaluating the performance of manufacturing processes to ensure adherence to proper specifications and limits; and (3) determining the acceptability of each batch for release.
  - QA involves primarily (1) review and approval of all procedures related to production and maintenance, (2) review of associated records, and (3) auditing and performing/evaluating trend analyses.

- Six-system inspection model. The FDA's Drug Manufacturing Inspection Compliance Program is a systems-based approach to inspection and is consistent with the robust quality system model. Figure 29.2 shows the relationship among the six systems: the quality system and the five manufacturing systems.

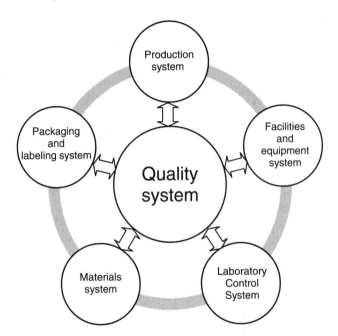

FIGURE 29.2 Six-system inspection model (From FDA, 2006.)

The quality system provides the foundation for the manufacturing systems that are linked and function within it. One of the important themes of the systems-based inspection compliance program is its ability to assess whether each of the systems is in a state of control.

29.2.2 Quality Systems Model

A properly implemented robust quality systems model can provide the controls to consistently produce a product of acceptable quality. Conversely, the lack of this consistent implementation of the robust quality system (QS) is the root cause of all compliance-related issues (FDA 483 observations, warning letters, drug product recalls, product seizures, etc.). The CGMP regulations related to each of the four major factors in implementing a robust QS model are listed in Appendixes 29.5 to 29.8.

- Management responsibilities (Appendix 29.5)
- Resources (Appendix 29.6)
- Manufacturing operations (Appendix 29.7)
- Evaluation activities (Appendix 29.8)

29.2.2.1 Management Responsibilities Management has ultimate responsibility to provide the leadership needed for the design, implementation, and successful functioning of a quality system.

- Provide leadership and commit necessary resource at all levels of management to support the quality system by:
  - Actively participating in system design, implementation, and monitoring, including system review
  - Advocating continual improvement of operations of the quality system
  - Committing necessary resources
- Structure the organization with assigned authorities and responsibilities that will support the production, quality, and management activities needed to produce quality products.
- Implement robust QS to meet CGMP regulations to ensure drug safety, identity, strength, quality, and purity.
- Establish policies, objectives, and plans to provide the means by which senior managers articulate their vision of, and commitment to, quality to all levels of the organization.
- Review the system to ensure its continuing suitability, adequacy, and effectiveness under a quality systems approach. A review should consider at least the following:
  - The appropriateness of the quality policy and objectives
  - The results of audits and other assessments
  - Customer feedback, including complaints
  - The analysis of data-trending results
  - The status of actions to prevent a potential problem or a recurrence
  - Any follow-up actions from previous management reviews
  - Any changes in business practices or environment that may affect the quality system (such as the volume or type of operations)
  - Product characteristics that meet customer needs
    - The review outcomes typically include:
    - Improvements to the quality system and related quality processes
    - Improvements to manufacturing processes and products
    - Realignment of resources

29.2.2.2 Resources
Appropriate allocation of resources is the key to creating a robust quality system and compliance to CGMP regulations.

- Senior management is responsible for providing adequate resources for the following:
  - To supply and maintain the appropriate facilities and equipment to manufacture a quality product consistently
  - To acquire and receive materials that are suitable for their intended purpose
  - For processing the materials to produce the finished drug product

- For laboratory analysis of the finished drug product, including collection, storage, and examination of in-process, stability, and reserve samples
- Personnel development. Senior management should support a problem-solving and communicative organizational culture and develop cross-functional groups to share ideas to improve procedures and processes. When operating in a robust quality system environment, it is important that managers verify that skills gained from training are implemented in day-to-day performance. To establish an effective training program, managers should include the following:
  - Evaluation of training needs: define appropriate qualifications (education, training, and experience of any combination thereof) for each position, including understanding of CGMP regulations
  - Provision of training to satisfy these needs
  - Evaluation of effectiveness of training
  - Documentation of training and/or retraining
- Facilities and equipment under a quality system and the CGMP regulations:
  - The technical experts (e.g., engineers, development scientists) are responsible for defining specific facility and equipment requirements
  - The QCU has the responsibility of reviewing and approving all initial design criteria and procedures pertaining to facilities and equipment and any subsequent changes [Sec. 211.22(c)]
  - Equipment must be qualified, calibrated, cleaned, and maintained to prevent contamination and mix-ups (Secs. 211.63, 211.67 and 211.68)
- Control outsourced operations to ensure that a contract firm is qualified before signing a contract with that firm.

29.2.2.3 Manufacturing Operations

- Design, develop, and document product and processes: quality and manufacturing processes and procedures and changes to them must be defined, approved, and controlled (Sec. 211.100); document processes, associated controls, and changes to these processes to ensure that sources of variability are identified. Documentation includes:
  - Resources and facilities used
  - Procedures to carry out the process
  - Identification of the process owner who will maintain and update the process as needed
  - Identification and control of important variables
  - Quality control measures, necessary data collection, monitoring, and appropriate controls for the product and process

- ○ Any validation activities, including operating ranges and acceptance criteria
- ○ Effects on related process, functions, or personnel

- Examine inputs. A robust quality system will ensure that all inputs (materials) to the manufacturing process are reliable because the quality controls have been established for the receipt, production, storage, and use of all inputs. The quality systems approach calls for periodic auditing of suppliers based on risk assessment; if full analytical testing is not done, the audit should cover the supplier's analysis. The CGMP regulations require either testing or use of a certificate of analysis (COA) plus an identity analysis (Sec. 211.84) for the release of materials for manufacturing.

- Perform and monitor operations. Implementing a robust QS approach is to enable a manufacturer to perform and monitor operations more efficiently and effectively, and to ensure that controls are scientifically sound and appropriate to manufacture finished products that have the required identity, strength, quality, and purity. In accordance with both the CGMP regulations (Sec. 211.110) and the quality systems approach:
  - ○ Process validation activity should be continued throughout a product's lifetime with a goal of making continuous improvements.
  - ○ Critical processes that may be responsible for causing variability during production are monitored.
  - ○ Written procedures are followed and deviations from them are justified and documented [Sec. 211.100(b)] to trace the history of the product, concerning personnel, materials, equipment, and chronology.
  - ○ Processes for product release are completed and recorded.
  - ○ Changes to an established process and/or data analysis reveal an area for improvements; they must be controlled and documented to ensure that the attributes desired in the finished products would be met [Sec. 211.100(a)].
  - ○ When implementing data collection procedures, consider the following:
    - Are data collection methods documented?
    - When in the product life cycle will the data be collected?
    - How and to whom will measurement and monitoring activities be assigned?
    - When should the analysis and evaluation (e.g., trending) of laboratory data be performed?
    - What records should be collected?

- Address nonconformities. A key component in any quality system is the handling of nonconformities and/or deviations. The investigation, conclusion, and follow-up must be documented (Sec. 211.192). Remedial action can include any of the following:
  - ○ Correct the nonconformity.

- With proper authorization, allow the product to proceed with justification of the conclusions regarding the problem's impact.
- Use the product for another application where the deficiency does not affect the product's quality.
- Reject the product.

29.2.2.4 Evaluation Activities

- Analyze data for trends. Quality systems call for continually monitoring trends and improving systems. This can be achieved by monitoring data and information, identifying and resolving problems, and anticipating and preventing problems. Trend analyses can help focus internal audits.
- Conduct internal audits at planned intervals to evaluate effective implementation and maintenance of the quality system to determine if processes and products meet established parameters and specifications. Managers responsible for the areas audited should take timely action to resolve audit findings and ensure that follow-up actions are completed, verified, and recorded. [Following Compliance Policy Guide 130.300 (FDA, 2010a), internal audit reports or records are not inspected by the FDA].
- Quality risk management. Elements of risk should be considered when developing product specifications, critical manufacturing process parameters as relative to patient safety, and ensuring the availability of medically necessary drug products. Since risk management is an iterative process, it should be repeated if new information is developed that changes the need for, or nature of, risk management.
- Corrective action. This is a reactive tool for system improvement to ensure that significant problems do not recur; determine what actions will reduce the likelihood of a problem recurring and document the entire CAPA process as required by Sec. 211.192. Examples of sources that can be used to gather such information include the following:
  - Nonconformance reports and rejections
  - Returns
  - Complaints
  - Internal and external audits
  - Data and risk assessment related to operations and quality system processes
  - Management review decisions
- Preventive actions. This proactive tool will help ensure that potential problems and root causes are identified, possible consequences assessed, and appropriate actions considered.
- Promote improvement. The effectiveness and efficiency of a quality system can be improved through the quality activities described in this guidance provided that senior management is involved in evaluation of the improvement process.

29.3 INSPECTION OF PHARMACEUTICAL QUALITY CONTROL LABORATORIES (FDA, 2009b)

The pharmaceutical quality control laboratory serves an important function in pharmaceutical production and control, as a significant portion of CGMP regulations (21 CFR 211) deal with the quality control laboratory and product testing. A laboratory inspection may include:

- Specific methodology that will be used to test a new product
- A complete assessment of laboratory's conformance with all applicable GMP requirements
- A specific aspect of laboratory operations

29.3.1 FDA Approach to Inspecting QC Laboratories

Following FDA guidance (FDA, 2009), comprehensive inspections of pharmaceutical quality control laboratories cover all aspects of the laboratory operations, including compliance with CGMP regulations and to the commitments made in a new drug application or abbreviated new drug application or drug master file. Following is a list of activities that may be audited during a comprehensive inspection.

- Deficiency letters issued by the Center for Drug Evaluation and Research (CDER). Evaluate responses to assure that the data are accurate and authentic.
- Laboratory records and logs. Inspect to determine the overall quality of the lab operation and the ability to comply with all applicable CGMP regulations:
  - Standard operating procedures are complete and adequate; laboratory operation conforms to the written procedures.
  - Specifications and analytical procedures are suitable and conform with applicable commitments and compendium requirements.
  - Raw laboratory data (preferably recorded directly in bound notebooks), laboratory procedures and methods, laboratory equipment (including maintenance and calibration), and methods validation data are accurate and authentic.
  - Examine chromatograms and spectra for evidence of impurities, poor technique, or lack of instrument calibration.
  - A system or standard operating procedure (SOP) that provides for the investigation of laboratory test failures to identify the cause of out-of-specification (OOS):
    - Analyses results of product lots that have failed to meet specification and decisions to release and who released them
    - Analyses of product lots that have been retested, rejected, or reworked
    - OOS investigation process and decision to conduct retests; an OOS result may fall into one of three categories:

□ Laboratory error

□ Non-process-related or operator error

□ Process-related or manufacturing process error

- If laboratory investigation is inconclusive (the reason for the error is not identified), the company:

  □ Cannot conduct two retests and base release on the average of three tests; the number of retests performed, where appropriate, is a matter of scientific judgment, but retesting cannot continue ad infinitum. If retesting is done:

    ◆ It must be done on the same sample, not on a different sample.

    ◆ It may be done on a second aliquot from the same portion of the sample that was the source of the first aliquot.

    ◆ It may be done on a portion of the same larger sample collected previously for laboratory purposes.

  □ Cannot rely on resampling; a successful resample result alone cannot invalidate an initial OOS result unless the failure investigation provides evidence that the original sample is not representative or prepared improperly

  □ Cannot use an outlier test in chemical tests

  □ Cannot use a resample to assume a sampling or preparation error

  □ Can conduct a retest of different tablets from the same sample when a retest is considered appropriate

- All failure investigations are conducted formally, including writing of an investigation report within 20 business days of a problem's occurrence

- An OOS is invalidated only if a laboratory error responsible for the OOS condition is identified; non-process- and process-related errors would represent product failure.

- The practice of averaging analytical results should be avoided, as it hides the variability among individual results, particularly if testing generates both OOS and passing individual results which when averaged are within specifications.

- Investigate to assure that suitable standards are used which are current, stored properly, and identified appropriately.

○ Review method validation records for completeness, accuracy, and reliability.

- If a firm decides to use an alternative method instead of the available compendium method, the firm must compare the two methods and demonstrate that the in-house method is equivalent or superior to the official procedure.

- If a firm is using a compendium method, the firm must demonstrate that the method works under the actual conditions of use.

○ Examine laboratory equipment use, maintenance, calibration logs, repair records, and maintenance SOPs and determine whether equipment is being used properly.

○ Determine if the company has a program to audit the certificate of analyses of the bulk product component, and if so, check the results of these tests.

○ Review results from in-process testing for conformance with established sampling and testing protocols, analytical methods, and specifications.

○ Evaluate the manufacturer's validation report for stability testing and verify that stability-indicating methods were employed; evaluate the raw data that was used in establishing the stability-indicating method and level of impurities.

○ The use of computerized laboratory data acquisition systems must comply to 21 CFR 11 for security and authenticity where

■ Only authorized persons can enter data.

■ Data entries cannot be deleted.

■ Changes are made as amendments.

■ The database is made tamperproof; authority to delete files and override computer systems is established

■ Data acquisition and system controls are validated initially and revalidated periodically.

■ The validity of the data and other security controls must be described in an SOP.

○ Evaluate the span of laboratory supervisory control, personnel qualification, the active training program, and documented evaluation of the training of analysts, analyst turnover, and the scope of the laboratory responsibilities.

○ Review any FDA citations (Form 483s and warning letters).

29.4 PHARMACOPEIAS (U.S.P., 2008)

In addition to the *United States Pharmacopeia–National Formulary* (U.S.P.–N.F.) (http://www.uspnf.com), there are three other major pharmacopeias published in the world: the *European Pharmacopoeia* (E.P.) (www.pheur.org), the *British Pharmacopoeia* (www.pharmacopeia.org), and the *Japanese Pharmacopoeia* (J.P.) (http://jpdb.nihs.go.jp/jp14e/index.html); they all share the goal of publishing and producing standards for pharmaceuticals. Except for the U.S.P., which remaines a practitioner-based non-governmental-standard-setting organization, all of its global counterparts are part of the ministries of health in their countries. All pharmacopeias have the same goal, which is to seek advanced public health by ensuring the quality and consistency of medicines and promoting the safe and proper use of medications. Brief overviews of the U.S.P.–N.F., E.P. and J.P. are presented in this section along with a comparison of specification parameters listed in the aspirin monographs from these three pharmacopeias.

29.4.1 United States Pharmacopeia
(U.S.P., 2008, 2011a,b, 2012)

The *United States Pharmacopeia* was first published in 1820 and contained formulas for the preparation of 217 drugs considered to be the "most fully established and best understood." The *National Formulary*, which included formulations and unofficial preparations for widely sold products, was created in 1888 by the American Pharmaceutical Association. The two publications were combined when the U.S.P. purchased the N.F. in 1975 and created the *United States Pharmacopeia–National Formulary* under one cover. It was not until 1906, when the federal Food and Drugs Act was passed, that the standards in the U.S.P. were recognized as official standards for the United States. The federal Food, Drug, and Cosmetic (FD&C) Act of 1938 further solidified the U.S.P.'s role by designating the U.S.P. as well as the N.F. and the *Homeopathic Pharmacopeia* standards, as enforceable by the U.S. Food and Drug Administration (FDA). If a drug's name is recognized in the U.S.P.–N.F., the drug must comply with the compendial standards for identity, strength, quality, and purity, and for the packaging and labeling requirements or be deemed adulterated and/or misbranded [FD&C Act Secs. 501(b), 502(e)(3)(b), and 502(g); 21 CFR 299.5(a,b,c)]. The FDA has issued citations when nonconformance to U.S.P. requirements is found during inspections. For example, the FDA issued a warning letter (WL: 320-11-01) to an active pharmaceutical ingredient (API) manufacturer: "Failure to thoroughly investigate complaints for API batches do not meet the *United States Pharmacopeia* (U.S.P.) compendia requirements that may have associated with the specific failure or discrepancy. . . . " The FDA states further in the warning letter that the manufacturer certified (through COAs) batches to meet the U.S.P. quality specification when one of the tests was performed without meeting U.S.P. environmental conditions.

The U.S.P.–N.F. contains more than 4000 monographs for prescription and over-the-counter products, dosage forms, drug substance excipients, dietary supplements, medical devices, and other health care products. The monographs for drug substances and preparations are featured in the U.S.P. and those for dietary supplements and ingredients appear in a separate section of the U.S.P.; excipient monographs are in the N.F. section. A typical monograph includes the following sections:

- Name of the ingredient or preparation
- Its definition
- Packaging, storage, and labeling requirements
- Specification composed of a series of tests, procedures for the tests, and acceptance criteria
- Requirements for the use of official U.S.P. reference standards during testing

The U.S.P. invites qualified candidates (pharmaceutical scientists, academicians, regulatory professionals, health care professionals, and others) to serve as scientific decision makers in the Council of Experts and on its expert committees. To seek expert candidates, the U.S.P. publishes formalized information detailing the focus areas of each committee, roles and responsibilities, and expertise required of potential candidates to serve in these committees. The 2010–2015 Council of Experts includes 20 expert committees focusing on 20 different topics, ranging from general chapters to monographs, nomenclature, reference standards, statistics, and toxicology.

The U.S.P.–N.F. is revised annually as standard revisions, which are supplemented by twice-yearly supplements. Revisions may also occur more frequently through accelerated revision processes such as errata, interim revision announcements, and revision bulletins and posted on the U.S.P. website in the new official text section; these bulletins supersede U.S.P.–N.F. or supplements. The content of U.S.P. 34-NF29 through the second supplement (effective until April 30, 2012) and the U.S.P. 35-NF30 and its supplements (official from May 1, 2012) is as follows:

- My U.S.P.–N.F., Bookmarks, Searches
- Front Matter, General Notices, Revision Bulletins
- General Chapters, Dietary Supplement Chapters, Reagents, References, Tables, Dietary Supplements
- N.F. Monographs, U.S.P. Monographs
- Chromatographic Columns, Glossary, Contact U.S.P.

29.4.2 European Pharmacopoeia (EP, 2008, 2011)

The *European Pharmacopoeia* (E.P.) was inaugurated in 1964 through the Convention on the Elaboration of a European Pharmacopoeia under the auspices of the Council of Europe. The E.P. Commission is responsible for its preparation. The functions of the commission are:

- To determine the general principles applicable to the elaboration of the E.P.
- To decide on the methods of analysis
- To prepare and adopt monographs for the E.P.
- To recommend time limits for implementation of the E.P. within the territories of the contracting parties

The purpose of E.P. is to promote public health by recognizing common standards for use by health professionals and others as a basis for the safe use of medicines by patients and consumers. The existence of these common standards:

- Facilitates the free movement of medicinal products in Europe
- Ensures the quality of medicinal products and their components imported into or exported from Europe

The E.P. monographs and other texts are mandatory throughout the 36 member states and the European Union who are signatory to the European Pharmacopoeia Convention, and are designed for the needs of:

- Regulatory authorities

- Controlling the quality of medicinal products and their constituents
- Manufacturers of starting materials and medicinal products

The E.P. is published by the Directorate for the Quality of Medicines and Healthcare of the Council of Europe in accordance with the convention on the elaboration of a European Pharmacopoeia (European Treaty Series No. 50). The 7th edition (in two volumes) was published on July 15, 2010 and became effective as of January 1, 2011. This edition will be complemented by several (a total of eight—two in 2010, three in 2011, and three in 2012) noncumulative supplements. Users are required to comply with the latest versions of the monographs and general chapters published in this edition as well as in the supplements. The key contents of the edition are:

- Volume 1 (General Notices, Methods of Analysis, Materials for Containers and Closures, Reagents, General Texts, Monographs on various dosage forms, including vaccines, immunosera, radiopharmaceuticals, herbal drugs, and homoeopathic preparations)
- Volume 2 (Monographs—individual, Index)

29.4.3 Japanese Pharmacopoeia (JP, 2011)

The 16th edition of the *Japanese Pharmacopoeia* (J.P.) which became official on April 1, 2011, has 1764 articles comprised of the following:

- Notification of the Ministry of Health, Labour and Welfare, Contents, Preface, General Notices
- General Rules for Crude Drugs, General Rules for Preparations
- General Tests, Processes, and Apparatus
- Official Monographs
- Infrared Reference Spectra, Ultraviolet–Visual Reference Spectra
- General Information, Table of Atomic Mass as an Appendix, and Cumulative Index

  The *Japanese Pharmacopoeia*:

- Provides an official standard to assure that the quality of medicines in Japan responds to the progress of science and technology and medical demands at the time
- Defines the standards for specifications and methods for testing
- Clarifies the criteria for quality assurance of drugs that are essential for public health and medical treatments

29.4.4 Pharmacopeia Comparison

A comparison (Appendix 29.9) of the contents of aspirin monographs from these three pharmacopeias (U.S.P., E.P., and J.P.) shows a significant disparity. This one example shows significant variations (conflicting, in some cases) in official requirements from these pharmacopeias, thus creating undue burden on pharmaceutical manufacturers. To minimize (or possibly eliminate) such disparity, efforts have been ongoing for the last couple of decades to develop a single set of specifications. In November 2003, the ICH established the Q4B (ICH, 2007) EWG (Expert Working Group) to evaluate and recommend pharmacopeial texts and to facilitate their recognition by regulatory authorities for use interchangeably in the three ICH regions (United States, European Union, and Japan). The Q4B EWG anticipates that the Pharmacopoeial Discussion Group will be the principal source of pharmacopeial test proposals. The PDG was founded in 1990 and consists of representatives from the European Directorate for the Quality of Medicines, the Council of Europe, the Ministry of Health, Labour and Welfare of Japan, and the United States Pharmacopoeial Convention, Inc. Following favorable evaluations, the ICH will issue topic specific annexes for their implementation to avoid redundant testing by industry. The ICH has issued two comprehensive guidelines on specifications [Q6A (ICH, 1999a) and Q6B (ICH, 1999b)] to assist in the establishment of a single set of global specifications for new drug substances and new products.

29.5 ANALYTICAL INSTRUMENT QUALIFICATION
(FDA, 2010b; U.S.P. <1058>)

The pharmaceutical industry relies on the precision and accuracy of analytical instruments to obtain reliable and valid data for research, development, manufacturing, and quality control. There are a large number of laboratory instruments and computerized analytical systems used by the industry to analyze samples and acquire data to help ensure that products meet required quality attributes and that the products are suitable for their intended use. Depending on the applications, users validate test procedures, qualify and calibrate instruments, and perform system suitability tests and analyses of in-process quality control check samples to ensure that the data acquired are reliable. The FDA routinely inspects firms and cites if nonconformance to instrument qualification is found. One such example is when FDA issued a warning letter (WL: 320-11-04) to an API manufacturer citing "failure of your quality control unit/laboratory to ensure that analytical instrumentation and test equipment used to assure the quality of your APIs has been appropriately qualified and calibrated for their intended use."

Qualification of analytical instruments is one of four components of data quality. These components may be depicted as layered activities within a quality triangle where each layer adds to the overall product quality. The analytical instrument qualification (AIQ) forms the base of this triangle followed by the analytical methods validation, system suitability, and quality control checks as we move up the ladder to generate quality data. The AIQ and analytical method validation assure the quality of analysis before the samples are tested, whereas the system suitability tests and quality control checks assure quality of analytical results immediately before or during sample testing. These quality components are described briefly below.

- Analytical instrument qualification: provides documented evidence that an instrument performs its intended purpose and that it is properly calibrated and maintained to generate reliable data.

- Analytical methods validation: documented evidence that an analytical method does what it purports to do and instills confidence that the method will generate data of acceptable quality.

- System suitability tests: tests verifying that a system works according to the performance expectations and criteria set forth in a test method and assuring that the system met an acceptable performance standard at the time these tests were conducted; typically, these tests are conducted during sample analysis.

- Quality control checks: most analyses are performed using reference or calibration standards; some analyses require inclusion of quality control check samples to provide inprocess assurance of a test's performance suitability.

29.5.1 Instrument Categories

It would be scientifically inappropriate to apply similar criteria when qualifying various types (from simple to complex) of instruments. On the basis of complexity, instruments may be categorized into three groups (U.S.P. <1058>).

- *Group A*. These are standard instruments with no measurement capabilities or required calibration. Examples of the equipment in this group are magnetic stirrers, vortex mixers, centrifuges, and other instruments where user requirements may be verified and documented through visual observation of their operation. Manufacturers' specifications of basic functionality are accepted as users' requirements.

- *Group B*. These instruments provide measured values as well as equipment controlling parameters (e.g., temperature, pressure, flow). Example of the instrument and equipment included in this group are balances, pH meters, thermometers, melting-point apparatus, variable pipettes, refractometers, titrators, viscometers, muffle furnaces, ovens, refrigerators, freezers, water baths, pumps, and dilutors. The instruments in this group are calibrated, and conformance to the user's requirements is determined according to the instrument SOP (or to the manufacturer's specification of functionality and operational limits) and is documented during installation and operational qualification.

- *Group C*. Installation of these more complex instruments can be difficult at times and may require assistance from specialists. These instruments should be qualified using the process outlined in Section 29.5.2. Examples of the instruments in this group include spectrophotometers and spectrometers (e.g., infrared, Raman, near infrared, atomic absorption, mass, x-ray fluorescence, flame absorption, ultraviolet–visible, etc.), chromatographs (high-pressure liquid, gas, electron microscopes, dissolution apparatus, differential scanning calorimeters, etc.)

Laboratory equipment may also be categorized based on the FDA Office of Regulatory Affairs (ORA) Lab Procedure 5.5 (FDA, 2010b):

- General service equipment (e.g., blenders, ovens, hotplates, furnaces, stirrers)
- Volumetric equipment (e.g., class A glassware, mechanical and automatic pipettes, burets).
- Measuring instruments (e.g., balances, chromatographs, spectrometers, thermometers)
- Physical standards (e.g., reference weights, reference standards)

After an initial required qualification, all critical equipment is calibrated and performance verified based on a preestablished documented schedule; one such schedule for the minimum calibration or verification of the most common laboratory equipment is provided in the ORA laboratory procedure.

29.5.2 Analytical Instrument Qualification Process

29.5.2.1 Roles and Responsibilities

- Users responsibility
  - Ultimately responsible for instrument operations and data quality
  - To be trained adequately in the instrument's use and maintain training records
  - Responsible for instrument qualification
  - Maintaining the instrument in a qualified state by performing preventive maintenance and recalibration/recertification routinely

- Quality unit responsibility—to assure that:
  - The AIQ process meets compliance requirements
  - The process is being followed
  - The intended use of equipment is supported by valid and documented data

- Manufacturers and developers are responsible for:
  - Design qualification, where applicable
  - Validation of relevant processes used in manufacturing and assembly of the instrument
  - Testing the assembled instruments before shipping them to users
  - Notifying users about hardware defects discovered after a product's release
  - Offering user training, service, repair, and installation support
  - Inviting user audits as necessary

29.5.2.2 Qualification Phases   AIQ activities may be grouped into one of four phases: design qualification, installation qualification, operational qualification, and performance qualification. It's more important that a given activity is

performed where it is logically suited rather than forcing the activity to be performed in a specific phase.

29.5.2.2.1 Design Qualification During the design qualification (DQ) phase, purchasing design criteria are defined based on the desired functional and operational specifications of the instrument, and vendors are selected. For most instruments and for commercial-off-the-shelf (COTS) instruments in particular, the vendor (developer/manufacturer) would probably be performing the DQ and the user does not need to repeat all aspects of DQ. The users should assure that:

- The COTS instruments are suitable for their intended use.

- The vendor performs DQ on instrument controls, data acquisition, and processing software, validates this software, and provides a validation summary to the user.

- The vendor has acceptable quality control systems for developing, manufacturing, and testing.

- The vendor provides necessary support, service, and training.

- The vendor allows site audits.

Vendor audits and vendor-supplied documentation may satisfy DQ requirements.

29.5.2.2.2 Installation Qualification For an installation qualification (IQ), a user establishes and documents that an instrument is delivered as designed and specified and is installed properly in the environment selected. An IQ applies to a new, preowned, or existing on-site instrument that has not been qualified previously. The activities and documentation typically associated with an IQ are:

- Description. Provide a description of the instrument or the collection of instrument components, including its manufacturer, model, serial number, software version, and location; use drawings and flowcharts where appropriate.

- Instrument delivery. Verify that the instrument, software, manuals, supplies, and all other instrument accessories, as specified in the purchase order, arrived undamaged.

- Fixed parameters. These nonchanging parameters (such as length, height, weight, voltage inputs, acceptable pressures, and loads) may be accepted from specifications supplied by the manufacturer; the user may confirm these parameters by testing at the user's site.

- Utilities/facility/environment. Verify that the installation site meets manufacturer-specified environmental requirements satisfactorily; it is not necessary to measure voltage for a standard voltage instrument and/or humidity measurement for an instrument that would operate at ambient conditions.

- Assemble and installation verification.: Assemble and install the instrument; perform all initial diagnostics and testing of the instrument according to the manufacturer-established installation tests and guides to assess instrument acceptability; firmware (integrated chips with low-level software) is considered a component of the instrument itself and is treated as such during validation; no separate qualification is needed for firmware; document if there is any abnormal event observed during assembly and installation.

- Network and data storage. If the network connections and data storage capabilities are required at the installation site, connect the instrument to the network, and check its functionality.

After a successful IQ, the instrument is ready for OQ testing.

29.5.2.2.3   Operational Qualification   During operational qualification (OQ), a user demonstrates and documents that the instrument functions according to its operational specification in the environment selected. The OQ tests can be modular (testing of individual components) or holistic (entire system). Testing of individual components may facilitate interchanging of these components without requalification. Testing activities in the OQ phase may consist of the following parameters:

- Secure data storage, backup, and archiving. When applicable, test secure data handling, such as storage, backup, audit trails, and archiving at the user's site according to written procedures.

- Test instrument functions in the user's environment to verify that the instrument operates as intended by the manufacturer and required by the user; refer to the manufacturer-supplied information for identifying specifications for these parameters and in designing tests to evaluate the parameters identified.

- Repeating OQ functional testing may not be required unless the instrument is moved to a new location.

29.5.2.2.4   Performance Qualification   Performance qualification (PQ) begins after successful completion of IQ and OQ phases. During PQ, the user demonstrates and documents that an instrument performs according to the specifications defined by the user and is appropriate for its intended use. As is the case with OQ testing, PQ tests may be modular or holistic. The PQ phase may include the following parameters:

- Performance checks. Set up a test or series of tests by analyzing known components and standards to verify the performance of the instrument for its intended use at the user's site; some system suitability tests or quality control checks that are performed concurrently with the test samples can be used to demonstrate that the instrument is performing suitably; the PQ tests may resemble those performed during the OQ phase, but the specifications for PQ tests should demonstrate that the instrument is operating perfectly for the applications intended.

- Retesting frequency depends on the ruggedness of the instrument environment and the criticality of the tests performed; generally, the same PQ tests are repeated each time so that a history of the instrument's performance can be complied; some system suitability tests or quality control checks that are performed concurrently with the test samples also imply that the instrument is performing suitably.

- Standard operating procedures for operation, calibration, maintenance, and change control. Establish procedures for all these activities, and document each calibration and maintenance activity.

- Preventive maintenance and repairs. After needed maintenance or repair, the relevant PQ test(s) should be repeated after the needed maintenance or repair to ensure that the instrument remains qualified; refer to the FDA's/ORA Laboratory Procedure ORA-LAB 5.5 (FDA, 2010b) for equipment maintenance and performance requirements.

29.5.2.3 Change Control The impact of changes to instruments, including software, should be assessed and a determination made if requalification is required. This documented process requires proper approval prior to implementing changes.

29.5.2.4 Major Repairs or Modifications Relevant OQ and/or PQ tests should be repeated to verify whether the instrument continues to operate satisfactorily after major repairs or modifications.

29.5.3 Equipment Maintenance and Performance Checks

A schedule for the laboratory equipment and performance checks should be established and executed routinely to identify and eliminate potential sources of problems. One such schedule is provided in the ORA procedure (FDA, 2010b).

29.5.4 Out-of-Service Equipment

- Out-of-service equipment must be calibrated or verified prior to use.
- Equipment must be clearly tagged "out of service" if it is not operating properly.
- Equipment that has not been calibrated or verified and not in use should be clearly tagged "Out of service."
- Equipment is not returned to service until performance checks and verification have been performed and documented.

29.6 VALIDATION OF ANALYTICAL PROCEDURES
(FDA, 2000; ICH, 2005b; U.S.P.<1225>)

The FDA, International Organization for Standardization, European, and other health regulators have requirements and expectations regarding documentation of an analytical method validation. Method validation is the process of demonstrating that analytical procedures are suitable for their intended use. The FDA investigators would review the analytical procedures used for product release and stability testing to ensure compliance with current good manufacturing practice (21 CFR 211) or good laboratory practices (21 CFR 58), as appropriate. Method validation is required when a new method is developed, an existing method is modified significantly, or an existing validated method is applied to a different sample matrix.

Each new drug application and abbreviated drug application must include the analytical procedures necessary to ensure the identity, strength, quality, purity, and potency of the drug substance and drug product, including the bioavailability of the drug product [21 CFR 314.50(d)(1) and 314.94(a)(9)(i)]. Data must be available to establish that the analytical procedures used in testing meet proper standards of accuracy and reliability [21 CFR 211.165(e) and 211.194(a)(2)].

29.6.1 Types of Analytical Procedures

29.6.1.1 *Regulatory Analytical Procedures*   A regulatory analytical procedure is used to evaluate a defined characteristic of the drug substance or drug product. The analytical procedures from the U.S.P.–N.F. are legally recognized as the regulatory procedures for the compendia items under Sec. 501(b) of the Food, Drug, and Cosmetic Act. Users of U.S.P.–N.F. methods are not required to validate the accuracy and reliability of test methods, but only to verify their suitability under actual conditions of use.

29.6.1.2 *Alternative Analytical Procedure*   An alternative validated analytical procedure may be used provided that its performance is equal or better than the regulatory analytical procedure. If an alternative analytical procedure is submitted, the applicant should provide a rationale for its inclusion and identify its use (e.g., release, stability testing), validation data, and comparative data to the regulatory analytical procedure.

29.6.1.3 *Stability-Indicating Assay*   A stability-indicating assay is a validated quantitative analytical procedure that can detect changes with time in the pertinent properties of the drug substance and drug product. A stability-indicating assay accurately measures the active ingredients without interference from degradation products, process impurities, excipients, or other potential impurities. Analytical assay procedures for stability studies should be stability-indicating unless scientifically justified.

A stability-indicating assay method is typically validated by forcibly degrading the API or the drug product and using the degraded samples to verify the separation of the impurities from the API or the drug product. Forced degrading involves exposing representative samples of drug substance or drug product to relevant stress conditions of acid–base hydrolysis, oxidation, light, heat, and humidity. Forced degradation plays an important role in understanding the chemistry of both the drug substance and the drug product and facilitates the development of stability-indicating analytical methodology. The desired target for the degradation observed is approximately 5 to 20%; beyond this, overstressing may lead to degradation of primary degradants and may generate irrelevant degradants that may not be generated in formal stability studies.

A stability-indicating method should separate process-related impurities and degradants generated in the stressed samples. A stress study protocol suggested in Appendix 29.10 may be used, as there are no specific guidelines with respect to the scope and timings provided by the FDA, ICH, or U.S.P..

29.6.2 Types of Analytical Procedures To Be Validated

The following four most commonly employed analytical procedures (and others e.g., dissolution testing) used in assessing safety, purity, identity, strength, and quality of a drug product must be validated:

- Identification tests
- Quantitative tests of the active moiety in samples of drug substance or drug product or other component(s) in the drug product
- Quantitative tests for impurities' content
- Limit tests for the control of impurities

29.6.3 Validation Protocols

29.6.3.1 *Method Validation* Method validation is the process of demonstrating that analytical procedures are suitable for their intended use. All relevant data collected during validation and formulas used for calculating validation characteristics should be submitted in regulatory filings and discussed as appropriate. Well-characterized reference materials, with documented purity, should be used throughout the validation study; the degree of purity necessary depends on the use intended. The following criteria should be considered when validating a test method:

- Accuracy
- Precision
  - Repeatability
  - Intermediate precision
- Specificity and selectivity
- Detection limit
- Quantitation limit
- Linearity
- Range
- Robustness and ruggedness

Each of these validation characteristics is defined in the Glossary. Appendix 29.10 lists those validation characteristics regarded as most important for the validation of different types of analytical procedures.

29.6.3.2 Revalidation Revalidation may be necessary in the following circumstances:

- Changes in the synthesis of the drug substance
- Changes in the composition of the finished product
- Changes in the analytical procedure
- Other changes and method used for validation
- Other changes or modifications that may require revalidation

The degree of revalidation required depends on the nature of the changes.

29.6.3.3 Other Validation Considerations

29.6.3.3.1 System Suitability Testing Each analytical procedure should include the appropriate number of system suitability tests defining critical characteristics of that system. As defined in the CDER's Guidance on Validation of Chromatographic Methods (FDA, 1994), the system suitability tests include the following parameters to verify resolution and reproducibility of a chromatographic system:

- Tailing factor
- Relative retention
- Resolution
- Relative standard deviation of retention time and area under the curve of the analyte
- Capacity factor
- Number of theoretical plates

29.6.3.3.2 Stress Testing As indicated in the next section, the stress testing demonstrates the specificity of the analytical procedures to ensure that the quantitation of the active ingredient and/or excipients is not affected by the impurities and degradants.

29.7 STABILITY TESTING OF NEW DRUG SUBSTANCES AND PRODUCTS (ICH, 1996a,b, 2003a,b; U.S.P. <1150>)

A systematic approach should be adopted in the presentation and evaluation of the stability information. This approach should include results from the physical, chemical, biological, and microbiological tests, including attributes of the dosage form (e.g., dissolution rate for solid oral dosage forms). The purpose of stability testing is to provide evidence:

- On how the quality of a drug substance or drug product varies with time under the influence of a variety of environmental factors, such as temperature, humidity, and light

- To establish a retest period for the drug substance or a shelf life for the drug product to recommend storage conditions

A minimum of three primary batches of a drug substance or a drug product should be used for stability studies; the degree of variability among these three batches should provide enough confidence that a future production batch will remain within acceptance criteria throughout its retest period or shelf life. For convenience in planning for packaging and storage, and for stability testing, the world may be classified into four zones based on observed temperatures and average relative humidity.

- Climatic zone I: Japan, United Kingdom, Northern Europe, Canada, Russia, United States. 21°C/45% relative humidity (RH).
- Climatic zone II. Mediterranean, subtropical (United States, Japan, Southern Europe, Portugal, Greece. 25°C/60% RH.
- Climatic zone III. hot, dry (Iran, Iraq, Sudan). 30°C/35% RH.
- Climatic zone IV. hot, humid (Brazil, Ghana, Indonesia, Nicaragua, Philippines). 30°C/70% RH.

The United States, Europe, and Japan are characterized by zones I and II. The stability information generated in any one of the three regions of the EU, Japan, and the United States would be mutually acceptable to the other two regions, provided that the information is consistent with the ICH Q1A (ICH, 2003a) guidance, and the labeling is in accord with national and regional requirements.

29.7.1   Stress Testing of a Drug Substance

Stress testing of the drug substance can help:

- To identify the likely degradation products.
- To establish degradation pathways.
- To establish intrinsic stability of the molecule.
- To validate the stability-indicating power of the analytical procedures

Stress testing is likely to be carried out on a single batch of the drug substance. The testing should include the effect of temperatures [in 10°C increments (e.g., 50°C, 60°C) above that for accelerated testing], humidity (e.g., 75 % RH or greater) where appropriate, oxidation, and photolysis on the drug substance. The testing should also evaluate the susceptibility of the drug substance to hydrolysis across a wide range of pH values when in solution or suspension. Photostability testing should be an integral part of stress testing. The standard conditions for photostability testing are described in the ICH Q1B (ICH, 1996a): photostability testing (1 million lux hours of visible light and 200 watthours of near ultraviolet) of new drug substances and products. Refer to the Appendix 29.11 for a suggested stress stability protocol.

Examining degradation products under stress conditions is useful in establishing degradation pathways and developing and validating suitable analytical procedures. However, such examination may not be necessary for certain degradation products if it has been demonstrated that they are not formed under accelerated or long-term storage conditions. According to the FDA guidance on investigation new drugs for phase 2 and phase 3 studies (FDA, 2003), if a stress study is not performed earlier, it should be conducted during phase 3 to demonstrate the inherent stability of the drug substance, potential degradation pathways, and the capability and suitability of the analytical procedures proposed. Results from these studies will form an integral part of the information provided to regulatory authorities in the chemistry, manufacturing, and controls section. Refer to the ICH Q1A(R2) (ICH, 2003a) for detailed information on the stability protocols for a drug substance.

29.7.2 Stability Testing of a Drug Product

Design of the formal stability studies for the drug product should be based on knowledge of the behavior and properties of the drug substance, results from stability studies on the drug substance, and experience gained from clinical formulation studies. The likely changes on storage and the rationale for the selection of attributes to be tested in the formal stability studies should be stated. Stability protocols for new dosage forms should follow the parent (API) stability guideline; however, a reduced stability database at submission time (i.e., six months of accelerated and six months of long-term data from ongoing studies) may be acceptable in certain justified cases. Refer to the ICH Q1C (ICH, 1996b) guidelines for further details.

29.7.2.1 Photostability Testing (ICH, 1996a) Photostability testing should be conducted on at least one primary batch of the drug product if appropriate. According to the decision flow chart (Appendix 29.12), the stability should be carried out in the following sequential stages:

1. Tests on the exposed drug product outside the immediate pack, and if necessary
2. Tests on the drug product in the immediate pack, and if necessary
3. Tests on the drug product in the marketed pack

29.7.2.2 Selection of Batches Data from stability studies should be provided on at least three primary batches of the drug product. Two of the three batches should be at least pilot-scale batches, and the third one can be smaller if justified. Where possible, batches of the drug product should be manufactured by using different batches of the drug substance. Stability studies should be performed on each individual strength and container size of the drug product unless bracketing or matrixing is applied. Refer to ICH Q1A (ICH, 2003a) for details on the bracketing or matrixing approaches.

29.7.2.3 Specification The testing should cover, as appropriate, the physical, chemical, biological, and microbiological attributes, preservative content (e.g., antioxidant, antimicrobial preservative), and functionality tests (e.g., for a dose delivery system). Analytical procedures should be fully validated and stability indicating. One primary stability batch of the drug product should be tested for antimicrobial preservative effectiveness (in addition to preservative content) at the proposed shelf life for verification purposes if there is a preservative. Refer to the ICH Q6A (ICH, 1999a) and Q6B (ICH, 1999b) guidelines for further details.

29.7.2.4 Stability Storage Conditions

The storage conditions and lengths of studies chosen should be sufficient to cover storage, shipment, and subsequent use. The long-term testing should cover a minimum of 12 months' duration on at least three primary batches at the time of submission and should be continued for a period of time sufficient to cover the proposed shelf life. Long-term, accelerated, and, where appropriate, intermediate-term storage conditions for drug products are listed in Appendix 29.13. Alternative storage conditions may be used if justified.

29.7.2.5 Data Evaluation and Extrapolation

Data from formal stability studies and, as appropriate, supporting data should be evaluated to determine the critical quality attributes likely to influence the quality and performance of the drug substance or product. In general, certain quantitative chemical attributes (e.g., assay, degradation products, and preservative content) of a drug substance or drug product can be assumed to follow zero-order kinetics (refer to the Chapter 8 for further details) during long-term storage. Data for these attributes are therefore amenable to various types of statistical analyses, including linear regression and poolability testing (refer to the Chapter 10 for further details). Although the kinetics of other quantitative attributes (e.g., pH, dissolution) is generally not known, the same statistical analysis can be applied, if appropriate. Qualitative attributes and microbiological attributes are not amenable to this type of statistical analysis. Appendix 29.14 provides a decision tree for the data evaluation for the retest period or shelf-life estimation for drug substances or products. Refer to ICH Q1A (ICH, 2003a) and Q1E (ICH, 2003b) for further details on data evaluation and extrapolation considerations. If the data (would be used to grant shelf life) show little degradation and little variability that is apparent from looking at the data, it is normally unnecessary to go through the formal statistical analysis; provide a justification for such omission.

29.7.2.6 Stability Commitments

- If the available long-term stability data on three primary batches (preferably, the first three production batches) do not cover the proposed shelf life granted at the time of approval, commitments should be made to continue the long-term studies (and six months of accelerated studies where necessary) postapproval to firmly establish the shelf life.

- If the submission includes data from stability studies on fewer than three production batches: a commitment should be made to continue the long-term studies through the proposed shelf life and the accelerated studies for 6 months and, to place additional production batches, to a total of at least three, on long-term stability studies through the proposed shelf life and on accelerated studies for 6 months.

- If the submission does not include stability data on production batches: a commitment should be made to place the first three production batches on long-term stability studies through the proposed shelf life and on accelerated studies for 6 months.

29.7.2.7 Storage Statements/Labeling
A storage statement should be established for the labeling in accordance with relevant national/regional requirements. The statement should be based on the stability evaluation of the drug product. Specific instructions should be provided in particular for drug products that cannot tolerate freezing. Terms such as ambient conditions or room temperature should be avoided. There should be a direct link between the label storage statement and the demonstrated stability of the drug product. An expiration date should be displayed on the container label.

REFERENCES

EP. 2008. European Pharmacopoeia . 6th ed., Vols. 1 (pp. i–v) and 2. Jan. 2008. European Directorate for the Quality of Medicines and Health Care.

——. 2011. Contents of the 7th edition, European Pharmacopoeia, pp. xvii–xxii. https://www.edqm.eu/store/images/majbdd/201006160933570.Contents%207-0_E.PDF. Accessed Dec. 2011.

FDA. 1994. Reviewer guidance: Validation of chromatographic methods. Center for Drug Evaluation and Research. Nov. 1994. http://www.fda.gov/downloads/Drugs/GuidanceCompliance RegulatoryInformation/Guidances/UCM134409.pdf. Accessed Dec. 2011.

——. 2000. Guidance: Analytical procedures and methods validation, chemistry, manufacturing, and controls documentation. Aug. 8, 2000. CMC. http://www.fda.gov/downloads/Drugs/ GuidanceComplianceRegulatoryInformation/Guidances/UCM122858.pdf. Accessed Dec. 2011.

——. 2003. Guidance for industry: INDs for Phase 2 and Phase 3 studies—chemistry, manufacturing, and controls information. May 2003. CMC. http://www.fda.gov/downloads/ Drugs/GuidanceComplianceRegulatoryInformation/Guidances/ucm070567.pdf. Accessed Dec. 2011

——. 2006. Guidance: Quality systems approach to pharmaceutical CGMP regulations. Sept. 2006. http://www.fda.gov/downloads/Drugs/GuidanceComplianceRegulatoryInformation/Guidances/ucm 070337.pdf. Accessed Dec. 2011

——. 2009a. Guidance: Q10—Pharmaceutical quality system. Apr. 2009. ICH. http://www.fda.gov/ downloads/Drugs/GuidanceComplianceRegulatoryInformation/Guidances/UCM073517.pdf. Accessed Dec. 2011.

——. 2009b. Guide to inspections of pharmaceutical quality Control laboratories. Updated Apr. 30, 2009. http://www.fda.gov/ICECI/Inspections/InspectionGuides/ucm074918.htm. Accessed Dec. 2011.

——. 2010a. CPG Sec. 130.300: FDA Access to Results of Quality Assurance Program Audits and Inspections. Document release date June 2, 2007. Updated Feb. 26, 2010. http://www.fda.gov/ICECI/ComplianceManuals/CompliancePolicyGuidanceManual/ucm073841.htm. Accessed Dec. 2011.

——. FDA 2010b. FDA/ORA laboratory procedure: Food and Drug Administration—equipment. ORA-LAB.5.5, Version 1:6. Revised May 25, 2010. http://www.fda.gov/ downloads/ScienceResearch/FieldScience/UCM092154.pdf. Accessed Dec. 2011.

ICH. 1996a. Guideline, Q1B: Photostability testing of new drug substances and products. Current Step 4 version. Nov. 6, 1996. http://www.ich.org/fileadmin/Public_Web_Site/ ICH_Products/Guidelines/Quality/Q1B/Step4/Q1B_Guideline.pdf. Accessed Dec. 2011.

—— 1996b. Guideline, Q1C: Stability testing for new dosage forms. Current Step 4 version. Nov. 6, 1996. http://www.ich.org/fileadmin/Public_Web_Site/ICH_Products/Guidelines/Quality /Q1C/Step4/Q1C_Guideline.pdf. Accessed Dec. 2011.

_____. 1999a. Guideline, Q6A: Specifications—Test procedures and acceptance criteria for new drug substances and new drug products; chemical substances. Current Step 4 version, Oct. 6, 1999. http://www.ich.org/fileadmin/Public_Web_Site/ICH_Products/Guidelines/Quality/Q6A/Step4/Q6A step4.pdf. Accessed Dec. 2011.

_____. 1999b. Guideline, Q6B: Specifications—Test procedures and acceptance criteria for biotechnology/biological products. Current Step 4 version. March 10, 1999. http://www.ich.org/fileadmin/Public_Web_Site/ICH_Products/Guidelines/Quality/Q6B/Step4/Q6B_Guideline.pdf. Accessed Dec. 2011.

_____ICH 2003a. Guideline, Q1A(R2): Stability testing of new drug substances and products. Current Step 4 version. Feb. 6, 2003. http://www.ich.org/fileadmin/Public_Web_Site/ICH_Products/Guidelines/Quality/Q1A_R2/Step4/Q1A_R2__Guideline.pdf. Accessed Dec. 2011.

_____. 2003b. Guideline, Q1E: Evaluation for stability data. Current Step 4 version. Feb. 6, 2003. http://www.ich.org/fileadmin/Public_Web_Site/ICH_Products/Guidelines/Quality/Q1E/Step4/Q1E _Guideline.pdf. Accessed Dec. 2011.

_____. 2005a. Guideline, Q9: Quality risk management. Current Step 4 version. Nov. 9, 2005. http://www.ich.org/fileadmin/Public_Web_Site/ICH_Products/Guidelines/Quality/Q9/Step4/Q9_ Guideline.pdf. Accessed Dec. 2011.

_____. 2005b. Guideline, Q2(R1): Validation of analytical procedures: text and methodology. Current Step 4 version. Parent guideline, Oct. 27, 1994 (complementary guideline on methodology, Nov. 6, 1996, incorporated in Nov. 2005). http://www.ich.org/fileadmin/Public_Web_Site/ICH_Products/Guidelines/Quality/Q2_R1/Step4/Q2_R1__Guideline.pdf. Accessed Dec. 2011.

_____. 2007. Guideline, Q4B: Evaluation and recommendation of pharmacopoeial texts for use in the ICH regions. Current Step 4 version. Nov. 1, 2007. http://www.ich.org/fileadmin/Public_Web_Site/ICH_Products/Guidelines/Quality/Q4B/Step4/Q4B_Guideline.pdf. Accessed Dec. 2011.

_____. 2008. Guideline, Q10: Pharmaceutical quality system. Current Step 4 version. June 4, 2008. http://www.ich.org/fileadmin/Public_Web_Site/ICH_Products/Guidelines/Quality/Q10/Step4/Q10 _Guideline.pdf. Accessed Dec. 2011.

_____. 2009. Guideline, Q8(R2): Pharmaceutical development. Current Step 4 version. Aug. 2009. http://www.ich.org/fileadmin/Public_Web_Site/ICH_Products/Guidelines/Quality/Q8_R1/Step4/Q8_ R2_Guideline.pdf. Accessed Dec. 2011.

JP. 2011. Overview _JP16th, Introduction. http://www.pmda.go.jp/english/pharmacopoeia/pdf/jpdata/ Overview_20111101.pdf. Accessed Dec. 2011.

Ngwa G. Forced degradation studies. Forced degradation as an integral part of HPLC stability-indicating method development. June 2010. http://www.particlesciences.com/docs/Forced_Degradation_Studies-DDT_June2010-rd3.pdf. Accessed Dec. 2010.

Reynolds D, Facchine K, Mullaney J, Alsante K, Hatajik T, Motto M. Available guidance and best practices for conducting forced degradation studies. Feb. 2002. http://www.pharmalytik.com/images/stories/PDF/forced%20degradation%20feb02.pdf. Accessed Dec 2010.

U.S.P. <1058>. Analytical Instrument Qualification. U.S.P. 34–N.F. 29 online. Accessed Dec. 2011

U.S.P. <1150>. Pharmaceutical Stability. U.S.P. 34–N.F. 29 online. Accessed Dec. 2011.

U.S.P. <1225>. Validation of Compendial Procedures. U.S.P. 34–N.F. 29 online. Accessed Dec. 2011.

U.S.P. 2008. What is a Phar•ma•co•pe•ia? /??/ *U.S. Pharmacopeia*. Sept. 2008. http://www.usp .org/pdf/EN/aboutUSP/pressRoom/pharmacopeiaBackgrounder.pdf?&⟨=en_;us&output=json. Accessed Dec. 2011.

_____. 2011a. Understanding USP–NF. http://www.usp.org/USPNF/understandingUSPNF.html. Accessed Dec. 2011.

_____ 2011b. U.S.P. 34–N.F. 29 online, through Second Supplement. Official Dec. 1, 2011 to Apr. 30, 2012. Accessed Dec. 2011.http://www.uspnf.com/uspnf/pub/index? usp=34&nf=29&s=1&officialOn=August%201,%202011

U.S.P. 2012. U.S.P. 35–N.F. 30. Official May 1, 2012 to July 31, 2012. http://www.uspnf.com/ uspnf/pub/index?usp=35&nf=30&s=0&officialOn=May%201,%202012. Accessed Dec. 2011.

WL: 320-11-01. Yunnan Hande Bio-Tech. Co. Ltd. Oct. 15, 2010. Updated Nov. 2011. http:// www. fda.gov/ICECI/EnforcementActions/WarningLetters/2010/ucm230829.htm. Accessed Dec. 2011.
WL: 320-11-04. Yuki Gosei Kogyo Co., Ltd. Dec. 9, 2010. Updated Dec. 23, 2010. http://www. fda.gov/ICECI/EnforcementActions/WarningLetters/2010/ucm236841.htm. Accessed Dec. 2011.

GLOSSARY

Accelerated testing	Studies designed to increase the rate of chemical degradation or physical change of an active drug product using exaggerated storage conditions as part of a formal, definitive storage program.
Acceptance criteria	Numerical limits, ranges, or other suitable measures for acceptance of the results of analytical procedures.
Accuracy	The accuracy of an analytical procedure expresses the closeness of agreement between the value that is accepted either as a conventional true value or an accepted reference value and the value found; sometimes termed *trueness*.
AIQ	Analytical instrument qualification.
Analytical procedure	The way of performing an analysis. It should describe in detail the steps necessary to perform each analytical test. This may include, but is not limited to, the sample, the reference standard and the reagents preparations, use of the apparatus, generation of the calibration curve, and the use of formulas for the calculations.
Annual review	An evaluation, conducted at least annually, that assesses the quality standards of each drug product to determine the need for changes in drug product specifications or manufacturing or control procedures.
API	Active pharmaceutical ingredient.
BP	*British Pharmacopeia*.
Bracketing	The design of a stability schedule so that at any time point only the sample on the extremes, for example, of container size and/or dosage strength, is tested. The design assumes that the stability of the intermediate condition sample is represented by those at the extremes.
CAPA	Corrective action and preventive action.
Corrective action and preventive action	*Corrective action*: action taken to eliminate the causes of an existing discrepancy or other undesirable situation to prevent recurrence. *Preventive action*: action taken to eliminate the causes of an existing discrepancy or other undesirable situation to prevent such an occurrence (a proactive approach).
CDER	Center for Drug Evaluation and Research.
CGMP	Current good manufacturing practice.
Change management	A systematic approach to proposing, evaluating, approving, implementing, and reviewing changes.
Climatic zone	The concept of dividing the world into four zones based on defining the prevalent annual climatic condition.
COA	certificate of analysis.
COTS	Commercial-off-the-shelf.
Critical process parameter	A process parameter whose variability has an impact on a critical quality attribute and therefore should be monitored or controlled to ensure that the process produces the desired quality.
Critical quality attribute	A physical, chemical, biological, or microbiological property or characteristic that should be within an appropriate limit, range, or distribution to ensure the desired product quality.

Design space	The multidimensional combination and interaction of input variables (e.g., material attributes) and process parameters that have been demonstrated to provide assurance of quality. Working within the design space is not considered a change. Movement out of the design space is considered to be a change and would normally initiate a regulatory postapproval change process. Design space is proposed by the applicant and is subject to regulatory assessment and approval.
Detection limit	The detection limit of an individual analytical procedure is the lowest amount of analyte in a sample that can be detected but not necessarily quantitated as an exact value.
	Drug master file.
DQ	Design qualification.
Drug product	Finished product.
E.P.	*European Pharmacopoeia*.
Excipient	Anything other than a drug substance in dosage form.
Expiry/Expiration date	The date placed on the labels and/or container of a drug product designating the time during which the batch of the product is expected to remain within the approved shelf-life specification if stored under defined conditions and after which it must not be used.
FDA	U.S. Food and Drug Administration.
FD&C Act	Food, Drug, and Cosmetic Act.
Forced degradation	Testing studies are those undertaken to degrade a sample deliberately. These studies, which may be undertaken in the development phase normally on the drug substances, are used to evaluate the overall photosensitivity of the material for method development purposes and/or degradation pathway elucidation.
GMP	Good manufacturing practice.
ICH	International Conference on Harmonization.
Immediate (primary) pack	That constituent of the packaging that is in direct contact with a drug substance or drug product, and includes any appropriate label.
Intermediate precision	Expresses within-laboratory variations: different days, different analysts, different equipment, etc.
IQ	Installation qualification.
J.P.	*Japanese Pharmacopoeia*.
Knowledge management	Systematic approach to acquiring, analyzing, storing, and disseminating information related to products, manufacturing processes, and components.
Lifecycle	All phases in the life of a product from the initial development through marketing until the product's discontinuation.
Linearity	The linearity of an analytical procedure is its ability (within a given range) to obtain test results that are directly proportional to the concentration (amount) of analyte in the sample.
Long-term (real-time) testing	Stability evaluation of the physical, chemical, biological, and microbiological characteristics of a drug product and a drug substance, covering the expected duration of the shelf life and retest period, which are claimed in the submission and will appear on the labeling.
Marketing pack	A combination of immediate pack and other secondary packaging, such as a carton.
Matrixing	The statistical design of a stability schedule so that only a fraction of the total number of samples is tested at any specified sampling point. At a subsequent sampling point, different sets of samples of the total number would be tested. The design assumes that the stability of the samples tested represents the stability of all samples.

	Ministry of Health, Labour and Welfare of Japan.
N.F.	*National Formulary*.
OOS	Out of specification.
OQ	Operational qualification. Office of Regulatory Affairs of the FDA.
Outsourced activities	Activities conducted by a contract acceptor under a written agreement with a contract giver.
Performance indicators	Measurable values used to quantify quality objectives to reflect the performance of an organization, process, or system, also known as *performance metrics*.
Pharmaceutical quality system	Management system to direct and control pharmaceutical company quality.
Pilot-plant scale	The manufacture of either a drug substance or drug product by a procedure fully representative of and simulating that to be applied on a full manufacturing scale.
Placebo (or blank)	A dosage form that is identical to the drug product except that the drug substance is absent or replaced by an inert ingredient or a mixture of the drug product excipients quantitatively equivalent to those found in the drug product dosage form.
PQ	Performance qualification.
PQS	Pharmaceutical quality system.
Precision	The precision of an analytical procedure expresses the closeness of agreement (degree of scatter) between a series of measurements obtained from multiple sampling of the same homogeneous sample under the conditions prescribed. Precision may be considered at three levels: repeatability, intermediate precision, and reproducibility. Precision should be investigated using authentic, homogeneous samples. However, if it is not possible to obtain a homogeneous sample, it may be investigated using artificially prepared samples or a sample solution. The precision of an analytical procedure is usually expressed as the variance, standard deviation, or coefficient of variation of a series of measurements.
Preventive action	Action to eliminate the cause of a potential nonconformity or other undesirable potential situation. *Note:* Preventive action is taken to prevent occurrence, whereas corrective action is taken to prevent recurrence.
Product realization	Achievement of a product with the quality attributes appropriate to meet the needs of patients, health care professionals, and regulatory authorities (including compliance with marketing authorization) and internal customers' requirements.
Production batch	A batch of a drug substance or drug product manufactured at production scale by using production equipment in a production facility as specified in the application.
QA	Quality assurance.
QC	Quality control.
QCU	Quality control unit.
QS	Quality system.
Quality	A measure of a product's or service's ability to satisfy a customer's stated or implied needs. The degree to which a set of inherent properties of a product, system, or process fulfills requirements.
Quality assurance	Proactive and retrospective activities that provide confidence that requirements are fulfilled.

Quality control	The steps taken during the generation of a product or service to ensure that it meets requirements and that the product or service is reproducible.
Quality management	Accountability for the successful implementation of the quality system.
Quality manual	Document specifying the quality management system of an organization.
Quality objectives	Specific measurable activities or processes to meet the intentions and directions as defined in the quality policy. A means to translate the quality policy and strategies into measurable activities.
Quality plan	The documented result of quality planning that is disseminated to all relevant levels of an organization.
Quality planning	A management activity that sets quality objectives and defines the operational and/or quality system processes and the resources needed to fulfill the objectives.
Quality policy	A statement of intentions and direction issued by the highest level of the organization related to satisfying customer needs. It is similar to a strategic direction that communicates quality expectations that the organization is striving to achieve.
Quality system	Formalized business practices that define management responsibilities for organizational structure, processes, procedures, and resources needed to fulfill product and service requirements, customer satisfaction, and continual improvement. In the CGMP regulatory context, the quality system establishes the foundation to promote the effective functioning of the five other major systems.
Quality unit	A group organized within an organization to promote quality in general practice.
Quantitation limit	The quantitation limit of an individual analytical procedure is the lowest amount of analyte in a sample that can be determined quantitatively with suitable precision and accuracy. The quantitation limit is a parameter of quantitative assays for low levels of compounds in sample matrices and is used particularly for the determination of impurities and/or degradation products.
Range	The range of an analytical procedure is the interval between the upper and lower concentration (amounts) of analyte in the sample (including these concentrations) for which it has been demonstrated that the analytical procedure has a suitable level of precision, accuracy, and linearity.
Repeatability	Expresses the precision under the same operating conditions over a short interval of time. Repeatability is also termed *intra-assay precision*.
Reproducibility	Expresses the precision between laboratories (collaborative studies, usually applied to standardization of methodology).
RH	Relative humidity.
Risk	The combination of the probability of occurrence of harm and the severity of that harm.
Risk assessment	A systematic evaluation of the risk of a process by determining what can go wrong (risk identification), how likely it is to occur (risk estimation), and the consequences. A systematic process for organizing information to support a risk decision that is made within a risk management process. The process consists of the identification of hazards and the analysis and evaluation of risks associated with exposure to those hazards.

Risk management	The systematic application of quality management policies, procedures, and practices to the tasks of assessing, controlling, communicating, and reviewing risk.
Robustness	The robustness of an analytical procedure is a measure of its capacity to remain unaffected by small but deliberate variations in method parameters and provides an indication of its reliability during normal use.
SOP	Standard operating procedure.
Specification	The quality standards (i.e., tests, analytical procedures, and acceptance criteria) provided in an approved application to confirm the quality of drug substances, drug products, intermediates, raw materials, reagents, and other components, including container closure systems, and in-process materials.
Specification—release	The combination of physical, chemical, biological, and microbiological test requirements that determine that a drug product is suitable for release at the time of its manufacture.
Specificity	The ability to assess unequivocally an analyte in the presence of components that may be expected to be present. Typically, these might include impurities, degradants, matrix, etc. Lack of specificity of an individual analytical procedure may be compensated by other supporting analytical procedure(s). This definition has the following implications: *Identification:* to ensure the identity of an analyte. *Purity Tests:* to ensure that all the analytical procedures performed allow an accurate statement of the content of impurities of an analyte (i.e. related substances test, heavy metals, residual solvents content, etc.). *Assay* (content or potency): to provide an exact result that allows an accurate statement on the content or potency of the analyte in a sample.
Stability-indicating assay	A validated quantitative analytical procedure that can detect the changes with time in the pertinent properties (e.g., active ingredient, preservative level) of a drug substance and drug product. A stability-indicating assay accurately measures the active ingredients without interference from degradation products, process impurities, excipients, or other potential impurities.
Storage conditions tolerances	The acceptable variation in temperature and relative humidity of storage facilities. The equipment should be capable of controlling temperature to a range of $\pm 2°C$ and relative humidity to $\pm 5\%$. The actual temperatures and humidities should be monitored during stability storage. Short-term spikes due to opening of doors of the storage facility are accepted as unavoidable. The effect of excursions due to equipment failure should be addressed by the applicant and reported if judged to affect stability results. Excursions that exceed these ranges for more than 24 hours should be described in the study report and their impact assessed.
Stress testing (drug substance)	Studies undertaken to elucidate intrinsic stability characteristics. Such testing is part of the development strategy and is normally carried out under more severe conditions than those used for accelerated tests. Stress testing is conducted to provide data on forced decomposition products and decomposition mechanisms for the drug substance. The severe conditions that may be encountered during distribution can be covered by stress testing of definitive batches of drug substance.

U.S.P.	*United States Pharmacopeia.*
Validation	Confirmation, through the provision of objective evidence, that the requirements for a specific intended use or application have been fulfilled.
Verification	Confirmation, through the provision of objective evidence, that specified requirements have been fulfilled.

APPENDIXES

APPENDIX 29.1 Application of Process Performance and Product Quality Monitoring System Throughout a Product Life Cycle

Pharmaceutical Development	Technology Transfer	Commercial Manufacturing	Product Discontinuation
Gain process and product knowledge Monitor development process Control strategy for manufacturing	Monitor during scale-up activities Knowledge gained during transfer and scale-up activities can be useful in developing the control strategy further	A well-defined system of quality monitoring should assure performance within a state of control and identify improvements	Once manufacturing ceases, monitoring such as stability testing should continue Appropriate action on marketed product should continue according to regional regulations

Source: FDA (2009a).

APPENDIX 29.2 Application of Corrective Action and Preventive Action System Throughout a Product Life Cycle

Pharmaceutical Development	Technology Transfer	Commercial Manufacturing	Product Discontinuation
Explore product for process vulnerability Incorporate CAPA methodology into the interactive design and development process	CAPA is an effective methodology for continual improvement	CAPA should be used and the effectiveness of the actions should be evaluated	CAPA should be continued after the product is discontinued to assess the impact on product remaining on the market as well as other products that might be affected

Source: FDA (2009a).

APPENDIX 29.3   Application of Change Management Systems Throughout a Product Life Cycle

Pharmaceutical Development	Technology Transfer	Commercial Manufacturing	Product Discontinuation
Formally document all changes Change management process should be consistent with the stage of pharmaceutical development	Manage and document adjustments made to all process activities during technology transfer	Employ a formal change management system with an oversight by the QU to provide assurance of appropriate science- and risk-based assessments	Any changes after product discontinuation should go through an appropriate change management system

Source: FDA (2009a).

APPENDIX 29.4   Applications of Management Review of Process Performance and Product Quality Throughout a Product Life Cycle

Pharmaceutical Development	Technology Transfer	Commercial Manufacturing	Product Discontinuation
Ensure adequacy of product and process design	Ensure that the the developed product and process can be manufactured at a commercial scale	Support continual improvement	The review should include items such as product stability and product quality complaints

Source: FDA (2009a).

APPENDIX 29.5   21 CFR CGMP Regulations Related to Management Responsibilities

Quality System Element	Regulatory Citations
Leadership Structure	Establish quality function: Sec. 211.22(a) [see definition Sec. 210.3(b)(15)] Notification: Sec. 211.180(f)
Build QS	QU procedures: Sec. 211.22(d) QU procedures, specifications: Sec. 211.22(c), with reinforcement in Secs. 211.100(a) and 211.160(a) QU control steps: Sec. 211.22(a), with reinforcement in Secs. 211.42(c), 211.84(a), 211.87, 211.101(c)(1), 211.110(c), 211.115(b), 211.142, 211.165(d), and 211.192

(continued)

APPENDIX 29.5 *(Continued)*

Quality System Element	Regulatory Citations
	QU quality assurance: review/investigations: Secs. 211.22(a), 211.100(a,b), 211.180(f),
	211.192, and 211.198(a)
	Record control: Secs. 211.180(a–e), 211.186,
	211.192, 211.194, and 211.198(b)
Establish policies, objectives, and plans	Procedures: Secs. 211.22 (c,d), 211.100(a)
System review	Record Review: Secs. 211.100, 211.180(e),
	211.192 and 211.198(b)(2)
Establish policies, objectives, and plans	Procedures: Secs. 211.22 (c,d), 211.100(a)
System review	Record Review: Secs. 211.100, 211.180(e),
	211.192 and 211.198(b)(2)

Source: FDA (2006).

APPENDIX 29.6   21 CFR CGMP Regulations Related to Resources

Quality System Element	Regulatory Citations
General arrangements	
Develop personnel	Qualification: Sec. 211.25(a)
	Staff number: Sec. 211.25(c)
	Staff training: Sec. 211.25(a,b)
Facilities and equipment	Buildings and facilities: Secs. 211.22(b), 211.28(c), 211.42–211.58, and 211.173
	Equipment: Secs. 211.63–211.72, 211.105, 211.160(b)(4), and 211.182
	Lab facilities: Secs. 211.22(b)
Control outsourced operations	Consultants: Sec. 211.34
	Outsourcing: Sec. 211.22(a)

Source: FDA (2006).

APPENDIX 29.7   21 CFR CGMP Regulations Related to Manufacturing Operations

Quality System Element	Regulatory Citations
Design and Develop product and processes	Production: Sec. 211.100(a)
Examine inputs	Materials: Secs. 210.3(b), 211.80–211.94, 211.101, 211.122, and 211.125

APPENDIX 29.7 (*Continued*)

Quality System Element	Regulatory Citations
Perform and monitor operations	Production: Secs. 211.100, 211.103, 211.111, and 211.113
	QC criteria: Secs. 211.22(a–c), 211.115(b), 211.160(a), 211.165(d), and 211.188
	QC checkpoints: Secs. 211.22(a), 211.84(a), 211.87, and 211.110(c)
Address nonconformities	Discrepancy investigation: Secs. 211.22(a), 211.100, 211.115, 211.192, and 122.198
	Recalls: 21 CFR 7

Source: FDA (2006).

APPENDIX 29.8   21 CFR CGMP Regulations Related to Evaluation Activities

Quality System Element	Regulatory Citations
Analyze data for trends	Annual review: Sec. 211.180(e)
Conduct internal audits	
Risk Assessment	
Corrective action	Discrepancy investigation: Secs. 211.22(a) and 211.192
Preventive Action	
Promote Improvement	Sec. 211.110

Source: FDA (2006).

APPENDIX 29.9   Pharmacopeial Comparison of Aspirin

Monograph Contents	U.S.P. 34–N.F. 29 Second Supplement	E.P. 7.0	J.P.
Title	Aspirin	Acetyl salicylic acid	Aspirin
Content on dried basis	= 99.5 to <100.5%	99.5% to 101.0%	<99.5%
Appearance/description	NA[a]	White, or almost white, crystalline powder or colorless crystals	White crystals, granules, or powder;
			odorless and has slight acidic taste
Solubility	NA	Slightly soluble in water; Freely soluble in ethanol (96%)	Slightly soluble in water; freely soluble in ethanol (95%); soluble in diethyl ether; dissolves in sodium hydroxide and in sodium carbonate

(*continued*)

APPENDIX 29.9 (*Continued*)

Monograph Contents	U.S.P. 34–N.F. 29 Second Supplement	E.P. 7.0	J.P.
Melting point	NA	About 143°C (instantaneous method)	About 136°C (bath fluid is heated at 130°C previously)
Identification Ferric chloride	Yes	NA	Yes
Infrared absorption	Yes	Yes	Yes
Sodium hydroxide test	NA	Yes	NA
Calcium hydroxide test	NA	Yes	NA
Heating of the sodium hydroxide test solution	No	Yes	N/A
Sodium carbonate test	NA	NA	Yes
Loss on drying	Yes (0.5 % max, dry on silica gel for 5 hrs.)	Yes (Max 0.5% determined on 1.000 g sample by drying in vacuum oven)	Yes
Readily carbonizable substances	Yes	NA	Yes
Residue on ignition	= 0.05%	NA	=0.1%
Substances soluble in sodium carbonate/clearing of solution	Yes	NA	Yes
Chloride content	Yes	NA	Yes
Heavy metals	= 10 ppm	= 20 ppm	< 10 ppm
Sulfate content (methodologies differ)	Yes	Sulfate ash max. 0.1%	Yes
Limit of free salicylic acid	Yes	NA	Yes
Assay	Yes	Yes	Yes
Storage–airtight container	Yes	Yes	Yes
Related substances	NA	Yes	NA
Impurities	NA	Yes (5 specified)	NA

Source: E.P. (2011), J.P., (2011), U.S.P. (2011b).

[a] NA, not available

APPENDIX 29.10   Recommended Validation Characteristics of the Various Types of Tests (FDA, 2000)

Types of Tests/ Characteristics	Identification	Testing for Impurities Quantitative	Limit	Assay Dissolution (Measurement Only), Content/Potency	Specific Test
Accuracy	$-^a$	$+^b$	$-$	$+$	$+^c$
Precision–repeatability	$-$	$+$	$-$	$+$	$+^c$
Precision–intermediate precision	$-$	$+^a$	$-$	$+^a$	$+^c$
Specificity	$+^d$	$+$	$+$	$+^f$	$+^c$
Detection limit	$-$	$-^e$	$+$	$-$	$-$
Quantitation limit	$-$	$+$	$-$	$-$	$-$
Linearity	$-$	$+$	$-$	$+$	$-$
Range	$-$	$+$	$-$	$+$	$-$
Robustness	$-$	$+$	$-^e$	$+$	$+^c$

Source: FDA (2000).

^{a} A minus sign signifies that this characteristic is not normally evaluated.

^{b} A plus sign signifies that this characteristic is normally evaluated.

^{c} May not be needed in some cases.

^{a} Where reproducibility has been performed, intermediate precision is not needed.

^{d} Lack of specificity for an analytical procedure may be compensated for by the addition of a second analytical procedure.

^{e} May be needed in some cases.

^{f} Lack of specificity for an assay for release may be compensated for by impurities testing.

APPENDIX 29.11   Forced Degradation Parameters for Drug Substances

Condition	Description	Range of Testing	Time Period(s)
Solid API–thermal	Employing a calibrated oven	60° C, 80° C	$T_0, T_1, T_2, \ldots$
Photolysis in solution and solid	UV light—365 nm, fluorescent light	1.5 × ICH Guidelines; typically 4 days in UV and 14 days fluorescent	Same as above
pH (hydrolysis) at 5°C, 20°C, 60°C, 80°C	Employing 0.1 N HCl, PBS, 0.01 N NaOH, and water; Use of co-solvents as needed for solubility	pH 1.0, pH 7.4, pH 12.0, water	Same as above

(continued)

APPENDIX 29.11 (*Continued*)

Condition	Description	Range of Testing	Time Period(s)
Oxidation	30% v/v H_2O_2	10 Meq at room temperature	T_1

Source: FDA (2003), ICH (2003a), ICH (1996a), Ngwa (2010), Reynolds (2002), T_0, initial time period or day 0; T_1, day 1; T_2, ..., day 2 up day to 14.

a) Prepare 30 to 50 mL for each of the degradation conditions at 1 mg/mL; pH condition are acid (0.1 HCl), base (0.01 N NaOH), neutral (PBS), and natural (water). Measure the pH of the bulk solution before aliquoting out into vials.

b) Where there is limited solubility of the compound in aqueous conditions, cosolvents can be used. Attempt to use the minimum amount of cosolvent that allows for a solution of the compound.

c) Once the solution is prepared, aliquot 5 mL into six clear vials, sealed with septa and crimped, for each condition. Place one set in the refrigerator, one at room temperature, one at $60°C$, and one at $80°C$. The Time required at the elevated temperature depends on the amount of degradation seen. The goal is to have the sample with 10 to 20% degradation. Samples can be pulled on a daily basis and analyzed to determine the extent of degradation and returned to the ovens.

d) Two samples from each of the two conditions are used for photostability. The visible light exposure should be no less than the ICH photostability guideline of $1,200,000$ lux $\cdot$ h/m^2 and, preferably, 1.5 to 2 times this guideline.

e) The UV light exposure is for the long wavelength UV of 364 nm and should be no less than the ICH photostability guideline of 200 Wh/m^2 and, preferably, 1.5 to 2 times this guideline.

f) With one of the natural samples, add 10 Meq of 30% H_2O_2 and store at room temperature for 24 hrs.

Forced degradation conditions guidelines for solids

a) Place powdered drug substances or drug product in a petri dish and place in a $60°C$ oven for 14 days, UV light chamber, and fluorescent light chamber for 1.5 to 2 times the *ICH guideline*, which is: Samples should be exposed to fluorescent light, providing an overall illumination of not less than 1.2 million lux $\cdot$ h and an integrated near-UV energy of not less than 200 Wh/m^2.

APPENDIX 29.12   Decision Flowchart for Photostability Testing of Drug Products

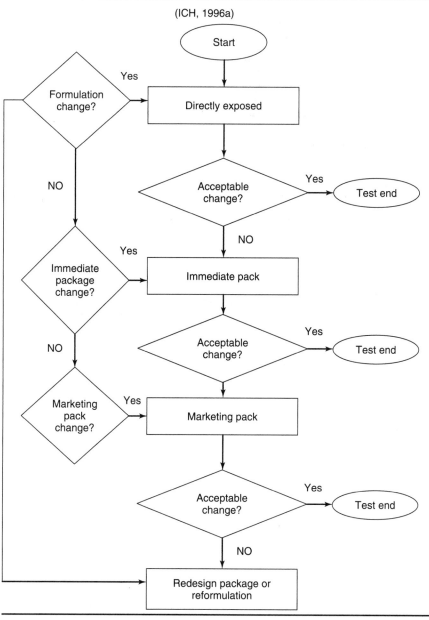

(ICH, 1996a)

Source: ICH (1996a).

APPENDIX 29.13   Stability Studies Test

Study	Storage Conditions	Time Period (months)	Suggested Test Time Period (months)
Long term[a]	25 ± 2°C/60 ± 5% RH or 30 ± 2°C/65 ± 5% RH	12	0, 6, 9, 12, 18, 24, 36
Intermediate[b]	30 ± 2°C/65 ± 5% RH	6	0, 3, 6
Accelerated	40 ± 2°C/75 ± 5% RH	6	0, 3, 6

Source: ICH (2003a).

[a] It is up to the applicant to decide whether the long-term stabilities studies are performed at 25 ± 2°C/60 ± 5% RH or 30°C ± 2°C/65 ± 5% RH.

[b] If 30° C ± 2°C/65 ± 5% RH is the long-term condition, there is no intermediate condition. If long-term studies are conducted at 25 ± 2°C/60 ± 5% RH and "significant change" occurs at any time during six months of testing at the accelerated storage condition, additional testing at the intermediate storage condition should be conducted and evaluated against significant change criteria. The initial application should include a minimum of six months' data from a 12-month study at the intermediate storage condition. In general, *significant change* for a drug product is defined as:

1. A 5% change in assay from its initial value.

2. Does not meet the acceptance criteria for potency when using biological or immunological procedures.

3. Any degradation product's exceeding its acceptance criterion.

4. Failure to meet the unexpected acceptance criteria for appearance, physical attributes, and functionality test (e.g. color, phase separation, resuspendability, caking, hardness, dose delivery per actuation). And, as appropriate for a dosage form:

5. Failure to meet the acceptance criteria for pH; or

6. Failure to meet the acceptance criteria for dissolution for 12 dosage units.

APPENDIX 29.14   Decision Tree for Data Evaluation for Retest Period or Shelf Life Estimation for Drug Substances or Product (Excluding Frozen Products)

Estimation for Drug Substances or Product (Excluding Frozen Products) (ICH 2003b)

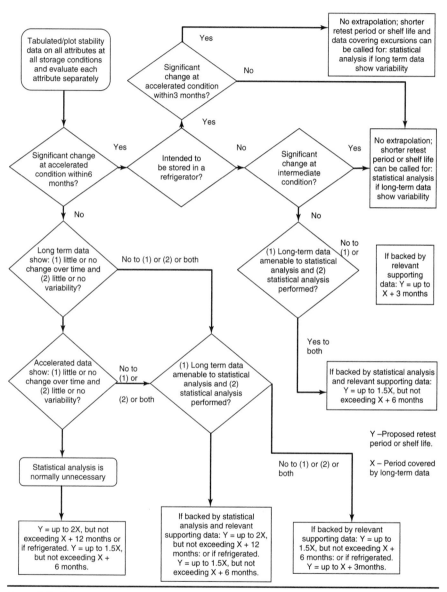

Source: ICH (2003b).

CHAPTER 30

SAFETY, TOXICOLOGY, AND PHARMACOGENOMICS

30.1 NONCLINICAL SAFETY STUDIES (ICH, 2009)

30.1.1 Planning and Design

Nonclinical safety studies and human clinical trials should be planned and designed to be ethical and scientific. The International Conference on Harmonization (ICH) guidance Nonclinical Safety Studies M3 (R2) (ICH, 2009) promotes the safety, ethical development, and availability of new pharmaceuticals and harmonizes nonclinical safety studies among the three regions (the United States, the European Union, and Japan). This harmonized guidance should facilitate timely conduct of clinical trials, reduce the use of animals in accordance with the 3R (reduce/refine/replace) principles, and the use of other drug development resources; where possible, consideration should be given to use of new in vitro alternative methods for safety evaluation.

- The nonclinical safety assessment for marketing approval of a pharmaceutical usually includes:
  - Pharmacology studies, general toxicity studies, toxicokinetic and nonclinical pharmacokinetic studies, reproduction toxicity studies, genotoxicity studies, and assessment of carcinogenic potential for drugs that have special cause for concern or are intended for a long duration of use
  - Case-by-case-based studies for phototoxicity, immunotoxicity, juvenile animal toxicity and abuse liability, and for
  - Pharmaceuticals under development for indications in life-threatening or serious diseases (e.g., advanced cancer, resistant HIV infection, congenital enzyme deficiency diseases) without current effective therapy also warrant a case-by-case approach for both toxicological evaluation and clinical development
  - For biotechnology-derived products, appropriate safety studies should be determined in accordance with the ICH S6 guidance, as these studies differ from those used for small molecules

Integrated Pharmaceutics: Applied Preformulation, Product Design, and Regulatory Science,
First Edition. Antoine Al-Achi, Mali Ram Gupta, William Craig Stagner.
© 2013 John Wiley & Sons, Inc. Published 2013 by John Wiley & Sons, Inc.

- The development of a pharmaceutical involves:
  - Stepwise evaluation of animal and human efficacy with safety information
  - Characterization of toxic effects with respect to target organs, dose dependence, relationship to exposure, and potential reversibility
  - Use this information:
    - To estimate a safe initial starting dose and dose range for the human trials
    - To identify parameters for clinical monitoring for potential adverse effects
  - The decision to expose for an increasing duration and/or the number of patients should be based on the demonstration of adequate safety in previous clinical trial(s), and on nonclinical safety information that becomes available as clinical development proceeds
- High-dose selection for general toxicity studies may include one of the following criteria:
  - In toxicity studies, clinically relevant effects can be adequately characterized as using doses up to the maximum tolerated dose (MTD); however, it is not necessary to demonstrate MTD in every study.
  - Other equally appropriate limiting doses that achieve large exposure multiples or saturation of exposure, or use the maximum feasible dose (MFD).
  - Limit doses for acute, subchronic, and chronic toxicity studies to 1000 mg/kg per day for rodents and nonrodents are appropriate provided that this dose provides a 10-fold exposure margin, or a dose of 2000 mg/kg per day, or the MFD, whichever is lower.
  - To support phase 3 clinical trials in the United States, identify dose-limiting toxicity in at least one species when using a 50-fold margin of exposure as the limit dose, or based on the study for one month or longer duration in one species conducted at a 1000-mg/kg limit dose, MFD, or MTD, whichever is the lowest.
  - The appropriate maximum dose should be selected based on MFD, MTD, or a limit dose of 1000 mg/kg per day if genotoxicity endpoints are to be incorporated in general toxicity studies.
- Estimating the maximum recommended starting dose (MRSD) for first-time use in clinical trials. For such estimation, consider all relevant nonclinical data, including the pharmacological dose response, the pharmacological/toxicological profile, pharmacokinetics, and the "no observed adverse effect level" (NOAEL) determined in nonclinical safety studies. The process recommended for selecting the MRSD is presented in Appendix 30.1 (FDA, 2005a). The NOAEL for each animal species tested should be identified and then converted to the human equivalent dose (HED) using appropriate factors.

One such factor is based on body surface area normalization and extrapolation of the animal dose to the human dose (Appendix 30.2). The species that generated the lowest HED is called the *most sensitive species* (FDA, 2005a).

30.1.2 Toxicokinetic and Pharmacokinetic Studies

In vitro metabolic and plasma protein binding data and systemic exposure data (ICH, 1994) for repeated-dose toxicity studies generally should be evaluated before initiating human clinical trials. Further information on pharmacokinetics (PK) and in vitro biochemical information relevant to potential drug interactions should be available before exposing large numbers of human subjects or treatment for a long duration (generally before phase 3). The nonclinical characterization of human metabolite(s) is required when exposure from that metabolite(s) exceeds 10% of total drug-related exposure and is at significantly greater levels in humans than the maximum exposure seen in toxicity studies.

30.1.3 Acute Toxicity Studies

Acute toxicity information may be obtained from single-dose toxicity studies in two mammalian species using both the clinical and a parenteral route of administration. This information may be obtained from appropriately conducted dose-escalation studies or from short-duration dose-ranging studies that define an MTD in the general toxicity test species (NC3Rs, 2007; Robinson et al., 2008). An earlier assessment of acute toxicity could be important for therapeutic indications for which patient populations are at higher risk for overdosing (e.g., depression, pain, dementia) in outpatient clinical trials.

30.1.4 Repeated-Dose Toxicity Studies

30.1.4.1 Clinical Development Trials The duration of the animal toxicity studies conducted in two mammalian species (one nonrodent) should be equal to or exceed the duration of the human clinical trials up to the maximum duration recommended for repeated-dose toxicity studies (Appendix 30.3). For example, repeated-dose toxicity studies in two species (one nonrodent) for a minimum duration of 2 weeks would generally support any clinical development trial up to 2 weeks in duration. Six-month rodent and nine-month nonrodent studies generally support dosing for longer than six months in clinical trials (see Appendix 30.3 footnotes for exceptions).

30.1.4.2 Marketing Authorization The durations of repeated-dose toxicity studies to support marketing are based on the duration of indicated treatments presented in Appendix 30.4; for example:

- One month repeated-dose studies in both rodents and nonrodents is recommended to support the treatment duration for less than 2 weeks.

- Longer durations of nonclinical testing, beyond three months, may be valuable for a small number of conditions (e.g., anxiety, sensual allergic rhinitis, pain) for which there is extensive clinical experience, suggesting widespread and long-term use beyond recommended periods between 2 weeks and three months.

30.1.5 Estimation of the First-Dose-in-Humans

The NOAEL determination from nonclinical safety studies performed in most appropriate species gives the most important information. The clinical starting dose proposed will also depend on various factors, including pharmacodynamic (PD) properties, drug molecules, and the design of the clinical trials. The regional guidelines should be used for specific approaches in estimating the first-dose-in-humans.

30.2 SAFETY PHARMACOLOGY STUDIES (ICH, 2000)

The purpose of safety pharmacology studies (SPS) is to investigate potentially undesirable pharmacodynamic effects of a substance on physiological functions in relation to exposure in the therapeutic range and above. Studies on the mode of action and/or the effects of a substance in relation to its desired therapeutic target are considered as the primary pharmacodynamic studies. Studies on the mode of action and/or effects of a substance not related to its desired therapeutic target are secondary pharmacodynamic studies (these have sometimes been referred to as part of general pharmacology studies).

30.2.1 Objectives and Design of SP Studies

- Objectives. The following three objectives of safety pharmacology studies should be clearly identified and delineated:
  - To identify undesirable PD properties of a substance that may have relevance to its human safety
  - To evaluate adverse pharmacodynamic and/or pathophysiological effects of a substance observed in toxicology and/or clinical studies
  - To investigate the mechanism of the adverse pharmacodynamic effects observed and/or suspected
- Design. These studies should be designed to address varying pharmacological effects, depending on specific properties of each test substance. The following factors (a partial list) should be considered when selecting and designing a SPS:
  - Effects related to the therapeutic class of the test substance for similar mechanism of action, which may suggest specific adverse effects

- Adverse effects associated with members of the chemical or therapeutic class but independent of the primary PD effects (e.g., antipsychotics and QT wave prolongation)
- Ligand-binding or enzyme assay data, suggesting a potential for adverse effects
- Findings from all previous safety pharmacology-related studies relevant to potential adverse effects in humans

30.2.2   Animal Models and Other Test Systems

- General considerations on selecting test systems:
  - Adopt a rational approach when selecting animal models or other test systems.
  - Use scientific methods to derive valid information.
  - Use new technologies and methodologies.
  - Incorporate SPs endpoints during the design of toxicology, kinetic, and clinical studies, or evaluate endpoints in other specific studies.
- Use of in vivo and in vitro studies as test systems
  - Animal (preferably, unanesthetized) models as well as ex vivo and in vitro preparations can be used as test systems; during conducting these experiments, consideration should be given to avoid discomfort or pain in unanesthetized animals; data from unrestrained or laboratory-conditioned animals are preferred to those obtained from restrained or unconditioned animals.
    - These test systems may include (but are not limited to) isolated organs and tissues, cell cultures, cellular fragments subcellular organelles, receptors, ion channels, transporters, and enzymes.
  - In vitro systems may be used to conduct supportive studies (e.g., to obtain a drug substance activity profile or to investigate the mechanism of effects observed in vivo).
- Experimental design
  - Sample size and use of controls. The size of the groups should be sufficient to allow meaningful scientific interpretation of the data generated; appropriate negative and positive control groups should be included in the experimental design.
  - Route of administration. In general:
    - The expected clinical route of administration is preferred.
    - Exposure to the parent substance and its major metabolites should be similar to or greater than that achieved in humans.
    - If the test substance is intended for clinical use by multiple routes of administration, consider more than one route.

30.2.3 Dose Levels or Concentrations of the Test Substance

- In vivo studies. If feasible, the highest dose tested should produce moderate adverse effects in this or in other studies of similar route and duration; testing of a single group at the limiting dose may be sufficient in the absence of an adverse effect on safety pharmacology endpoints in the test species.
- In vitro studies. These studies should be designed to establish a concentration–effect relationship; in the absence of such an effect, the range of concentrations selected should be justified.

30.2.4 Duration of Studies

Safety pharmacology studies (SPSs) are generally performed by administering a single dose. The duration of an SPS developed to address the following effects should be rationally based:

- When pharmacodynamic effects occur only after a certain duration of treatment, or
- When concerns about safety pharmacological effects are raised from repeat-dose nonclinical studies or results from use in humans

30.2.5 Studies on Metabolites, Isomers, and Finished Products

Evaluate SPSs when:

- Any parent component or its major metabolite(s) achieve, or are expected to achieve, systemic exposure in humans
- The major human metabolites are found to be absent or present only at relatively low concentrations in animals
- Metabolites contribute to the pharmacological actions of the therapeutic agent
- The product contains an isomeric matrix; then consider in vivo and in vitro testing of an individual isomer
- Finished product formulations substantially alter the PK and/or PD of the active substance in comparison to the formulation tested previously

30.2.6 Safety Pharmacology Core Battery

The purpose of the safety pharmacology core battery is to investigate the effects of the test substance on such vital functions or systems as:

- Cardiovascular (evaluate: blood pressure, heart rate, and electrocardiogram)
- Respiratory (measure: respiratory rate and other respiratory functions, such as tidal volume or hemoglobin oxygen saturation)

- Central nervous system (evaluate: effects on motor activity, behavioral changes, coordination, sensory/motor reflex responses, and body temperature)

30.2.7 Follow-up and Supplemental Safety Pharmacology Studies

- Consider these studies when:
  - Adverse effects are suspected based on the pharmacological properties or chemical class of the test substance
  - Additional safety concerns arise from the safety pharmacology core battery, clinical trials, pharmacovigilance, experimental in vitro or in vivo studies, or from literature reports
- Follow-up studies for safety pharmacology core battery are meant:
  - To provide a greater depth of understanding on a case-by-case basis
  - In some cases it may be appropriate to address these adverse effects while conducting other nonclinical and/or clinical studies
- Supplemental studies are meant to:
  - Evaluate potential adverse PD effects on the functions of various organ systems (renal/urinary, autonomic nervous, and/or gastrointestinal) if not addressed previously by the core battery or repeated dose-toxicity studies

30.2.8 Conditions Under Which Studies Are Not Necessary

- For locally applied agents (e.g., dermal or ocular), where the pharmacology of the test substance is well characterized, and where systemic exposure or distribution to other organs or tissues is demonstrated to be low
- Prior to the first administration in humans of cytotoxic agents for the treatment of end-stage cancer patients
- For biotechnology derived products that achieve highly specific receptor targeting
- A new salt with similar PK and PD

30.2.9 Timing of Safety Pharmacology Studies in Relation to Clinical Development

Assess if a given study is recommended or not when planning a safety pharmacology program:

- Studies conducted prior to first administration in humans
  - To investigate the effects of a test substance on the functions listed in the safety pharmacology core battery
  - Any follow-up or supplemental studies identified as appropriate, based on a cause for concern

- Studies during clinical development may be warranted to clarify adverse effects observed or those suspected
- Studies before approval: should be assessed; justify if not warranted

30.2.10 Application of Good Laboratory Practice

It is important to ensure the quality and reliability of nonclinical safety studies. This is normally accomplished, where required, through conducting these studies in compliance with good laboratory practice (GLP):

- Studies generally conducted in compliance with GLP
  - Safety pharmacology core battery studies
  - Follow-up and supplemental studies
  - Safety pharmacology investigation studies
  - Primary pharmacodynamic studies
- Studies generally not conducted in compliance with GLP
  - Secondary pharmacodynamic studies; in certain circumstances, when the results of secondary pharmacodynamic studies could make a pivotal contribution to the safety evaluation for potential adverse effects in humans, conduct these studies in compliance with GLP

For detailed information on good laboratory practice for nonclinical laboratory studies, refer to 21 CFR (*Code of Federal Regulations*) Part 58 at the U.S. Food and Drug Administration (FDA) website, http://www.accessdata. fda.gov/scripts/cdrh/cfdocs/cfcfr/CFRSearch.cfm?CFRPart=58&showFR=1&utm _campaign=Google2&utm_source=fdaSearch&utm_medium=website&utm_term =21%20cfr%2058&utm_content=1.

30.3 CARCINOGENICITY STUDIES OF PHARMACEUTICALS (ICH, 1995)

The objectives of carcinogenicity studies are to identify a tumorigenic potential in animals and to assess the relevant risk in humans. The fundamental considerations in assessing the need for carcinogenicity studies are:

- Maximization of the duration of patient treatment
- The intended patient population
- Prior assessment of carcinogenic potential
- The extent of systemic exposure
- The (dis)similarity to endogenous substances
- The appropriate study design
- The timing of study performance relative to clinical development

30.3.1   Factors to Consider for Carcinogenicity Testing

- Duration and exposure. Carcinogenicity studies should be performed for any pharmaceutical whose expected clinical use is continuous for at least six months or if used frequently in an intermittent manner in the treatment of chronic or recurrent conditions (e.g., allergic rhinitis, depression, anxiety).

- As a cause for concern, the following factors should be considered when recommending carcinogenicity studies:
  - Previous demonstration of carcinogenic potential in the product class that is considered relevant to humans
  - Structure–activity relationship suggesting carcinogenic risk
  - Evidence of preneoplastic lesions in repeated dose-toxicity studies
  - Long-term tissue retention of parent compound or metabolite(s), resulting in local tissue reactions or other pathophysiological responses

- Genotoxicity. Unequivocally genotoxic compounds presumed to be trans-species carcinogens should not be subjected to long-term carcinogenicity studies unless such compounds are intended to be administered chemically to humans. A single positive result in any assay for genotoxicity does not necessarily mean that the test compound poses a genotoxic hazard to humans (ICH, 2008).

- Indication and patient population. When carcinogenicity studies are required they need to be completed before applying for marketing approval. If pharmaceuticals are developed to treat certain serious diseases (where no satisfactory alternative therapy exists), carcinogenicity testing may not be conducted before market approval but needs to be conducted postapproval. In instances where the life expectancy in the indicated population is short (i.e., less than two to three years), long-term carcinogenicity studies may not be required.

- Route of exposure. The route of exposure in animals should be the same as the intended clinical route when feasible. For further details, refer to·the ICH guideline Dose Selection for Carcinogenicity Studies of Pharmaceuticals (ICH, 2005a).

- Extent of systemic exposure. Pharmaceuticals applied topically (e.g., dermal and ocular routes of administration) may need carcinogenicity studies; however, pharmaceuticals showing poor systemic exposure from topical routes may not need carcinogenicity studies by the oral route to assess the carcinogenic potential to internal organs.

- Endogenous peptides and protein substances or their analogs may require special considerations. Carcinogenicity studies are not generally required for endogenous substances given essentially as replacement therapy (i.e., physiological levels), in particular where previous clinical experience exists with similar products (e.g., animal insulin, pituitary-derived growth hormone, calcitonin). Carcinogenicity studies may, however, be important in the following circumstances:

○ For products where there are significant differences in biological effects to the natural counterpart(s)

○ For products where modification leads to significant changes in structure compared to the natural counterpart

○ For products resulting in humans showing a significant increase over the existing local or systemic concentration (i.e., pharmacological levels)

30.3.2 Need for Additional Testing

In cases where the relevance of the results obtained from animal carcinogenicity studies to assess human safety are in doubt, further research may be necessary to confirm the presence or lack of carcinogenic potential for humans.

30.4 GENOTOXICITY TESTING (ICH, 2008)

Genotoxicity tests can be defined as in vitro and in vivo tests designed to detect compounds that induce genetic damage by various mechanisms. These tests enable hazard identification with respect to damage to DNA and its fixation. Compounds that show positive test results in inducing genetic damage have the potential to be human carcinogens and/or mutagens.

30.4.1 Standard Test Battery for Genotoxicity

A battery approach is reasonable because no single test is capable of detecting all genotoxic mechanisms relevant in tumorigenesis. Optimize the standard genetic toxicology battery for the prediction of potential human risks. The general features of a standard test battery are:

- Assessment of mutagenicity in a bacterial reverse mutation test; this test has been shown to detect relevant genetic changes and the majority of genotoxic rodent and human carcinogens.

- Genotoxicity should also be evaluated in mammalian cells in vitro and/or in vivo.

The following three assays are currently considered equally appropriate, sufficiently validated, and therefore are interchangeable when used together with other genotoxicity tests in a standard battery for testing of pharmaceuticals: (1) the in vitro metaphase chromosome aberration assay, (2) the in vitro micronucleus assay, and (3) the mouse lymphoma L5178Y cell mutation assay. In vivo test(s) for genetic damage should provide additional relevant factors (absorption, distribution metabolism, excretion) that can influence the genotoxic activity of a compound and permit the detection of some additional genotoxic agents. An in vivo test for chromosomal damage in rodent cells largely fulfills this need.

There are several additional in vivo assays that can be used in the battery or as follow-up tests to develop the weight of evidence in assessing results of in vitro or in vivo assays. Negative results in appropriate in vivo assays (usually two), with adequate justification for the endpoints measured, and demonstration of exposure are sufficient to demonstrate the absence of significant genotoxic risk. The following two options for the standard battery are considered equally suitable:

- Option 1:
  - A test for gene mutation in bacteria
  - A cytogenetic test for chromosomal damage (the in vitro metaphase chromosome aberration test or in vitro micronucleus test), or an in vitro mouse lymphoma *tk* gene mutation assay
  - An in vivo test for genotoxicity, generally a test for chromosomal damage using rodent hematopoietic cells, either for micronuclei or for chromosomal aberrations in metaphase cells
- Option 2:
  - A test for gene mutation in bacteria
  - An in vivo assessment of genotoxicity with two tissues, usually an assay for micronuclei using rodent hematopoietic cells and a second in vivo assay

Under both standard battery options, the in vivo genotoxicity assays can often be integrated into repeat-dose toxicity studies when the doses are sufficient. For compounds that give negative results, the completion of either test battery will usually provide sufficient assurance of the absence of genotoxic activity, and no additional tests will be needed. Compounds that give positive results in the standard test battery may need to be tested more extensively.

30.4.2 Recommendations for In Vitro Tests

- Test repetition and interpretation. Routine testing of drugs with standards, which are scientifically well characterized and have sufficient controls, often does not need replication. Test results should clearly be declared as positive, negative, or equivocal. Statistical application may aid in data interpretation, but the interpretation of biological data is more important. An equivocal test that is repeated may result in:
  - A clearly positive outcome, and thus an overall positive result
  - A negative outcome, so that the result is not reproducible and overall negative, or
  - Another equivocal result, with a final conclusion that remains equivocal
- Recommended protocol for the bacterial mutation assays:
  - Selection of maximum dose level. The maximum dose level recommended is 5000μg/plate when not limited by solubility or cytotoxicity.
  - Limit of solubility. For a bacterial culture, precipitating doses are scored provided that the precipitate does not interfere with scoring, toxicity is not limiting, and the top concentration does not exceed 5000μg/plate.

○ Limit of cytotoxicity. In the bacterial reverse mutation test, the doses scored should show evidence of significant toxicity, but without exceeding a top dose of 5000μg/plate.

• Recommended protocols for the mammalian cell assays:

○ Maximum concentration. The maximum top concentration recommended is 1 mM or 0.5 mg/mL, whichever is lower, when not limited by solubility or culture medium or by cytotoxicity.

○ Cytotoxicity. It is not necessary to exceed a reduction of about 50% in cell growth for in vitro cytogenetic assays for metaphase chromosome aberrations or for micronuclei, or a reduction of about 80% in relative total growth for the mouse lymphoma *tk* mutation assay.

○ When an in vitro genotoxicity test is positive (or not done). Assessments of in vivo exposure should be made at the top dose or other relevant doses using the same species, strain, and the dosing route used in the genotoxicity assay. Demonstration of in vivo exposure should be made by any of the following measurements:

▪ Cytotoxicity (cytogenetic or other in vivo genotoxicity assays)

▪ Bioavailability (measurements in blood/plasma and/or direct measurement of drug-related materials in target tissues)

○ When in vitro genotoxicity tests are negative. If in vitro tests do not show genotoxic potential, in vivo (systemic) exposure can be assessed by any of the methods above, or can be assumed from the results of standard absorption, distribution, metabolism, and excretion studies in rodents done for other purposes.

30.4.3 Recommendations for In Vivo Tests

• Tests for the detection of chromosome damage in vivo. Either the analysis of chromosomal aberrations or the measurement of micronucleated polychromatic erythrocytes in bone marrow cells in vivo is appropriate for the detection of clastogens. Both rats and mice are appropriate for use in the bone marrow micronucleus test.

• Automated analysis of micronuclei. Systems for automated analysis (image analysis and flow cytometry) can be used if validated appropriately (Hayashi et al., 2007)

• Other in vivo genotoxicity tests. The same in vivo tests described as the second test in the standard battery (option 2) can be used as follow-up tests to develop the weight of evidence in assessing results of in vitro or in vivo assays.

• Use of male/female rodents in in vivo genotoxicity tests. If gender-specific drugs are to be tested, the assay can be done in the appropriate gender.

- Dose selection for short-term in vivo studies. For short term (usually 1 to 2 administrations) protocols, the top dose recommended for genotoxicity assays is a limit dose of 2000 mg/kg, if this is tolerated, or maximum tolerated dose
- Recommendations for determining whether the top dose in a toxicology study (typically, in rats) is appropriate for micronucleus analysis and for other genotoxicity evaluations (use any one of the following):
  - Maximum feasible dose (MFD) based on physicochemical properties of the drug in the vehicle (provided that the MFD in that vehicle is similar to that achievable with acute administration);
    - For common vehicles such as aqueous methylcellulose, this dose would usually be appropriate, but
    - For vehicles such as Tween 80, the volume that can be administered could be as much as 30-fold lower than that given acutely
  - A limit dose of 1000 mg/kg for studies of 14 days or longer, if this is tolerated
  - Exposure. Maximum possible exposure is demonstrated by:
    - Reaching a plateau/saturation in exposure, or
    - By compound accumulation
- A substantial reduction in exposure to the parent drug with time (e.g., a 50% reduction from the initial exposure) would usually disqualify the study.
- Use of positive controls for in vivo studies. For in vivo studies, periodically (and not concurrently with every assay) treat animals with a positive control after a laboratory has established competence in use of the assay.

30.5 IMMUNOTOXICITY STUDIES (ICH, 2005b)

Immunotoxicity is defined as unintended immunosuppression or enhancement, excluding drug-induced hypersensitivity and autoimmunity. Evaluation of potential adverse effects of human pharmaceuticals on the immune system (suppression or enhancements) should be incorporated into standard drug development. Suppression of the immune response can lead to decreased host resistance to infectious agents or tumor cells. Enhancing the immune response can exaggerate autoimmune diseases or hypersensitivity. Subsequent exposures to the drug can lead to hypersensitivity (allergic) reactions. Immunosuppression or enhancement can be associated with two distinct groups:

- Drugs intended to modulate immune function for therapeutic purposes (e.g., to prevent organ transplant rejection) where adverse immunosuppression can be considered exaggerated pharmacodynamics;
- Drugs not intended to affect immune function but cause immunotoxicity due, for example, to necrosis or apoptosis of immune cells or interaction with cellular receptors shared by both target tissues and nontarget immune system cells

The general principles applicable to immunotoxicity studies are:

- All new human pharmaceuticals should be evaluated for the potential to produce immunotoxicity;
- Include standard toxicity studies (STS) methods, and where appropriate, additional immunotoxicity studies may be conducted if necessary based on the weight of evidence reviewed from the following factors:
  - Findings from STS
  - Pharmacological properties of the drug
  - Intended patient population
  - Structural similarities of known immunomodulators
  - Disposition of the drug
  - Clinical information

The objectives of the immunotoxicity studies are to provide:

- Recommendations on nonclinical testing approaches to identify compounds that have the potential to be immunotoxic
- Guidance on a weight-of-evidence decision-making approach for immunotoxicity testing

The following inconsistencies that exist between the guidelines on immunotoxicity studies for human pharmaceuticals from the three health agencies [The Center for Drug Evaluation and Research (CDER)/FDA, European Committee for Proprietary Medicinal Products (CPMP), and Japanese Ministry of Health, Labor and Welfare (MHLW)] complicate mutual acceptance of immunological data among the regions and countries (ICH, 2003):

- According to CDER/FDA guidance, additional studies to determine potential drug effects on immune function are warranted based on the observations from nonclinical general toxicology studies
- Following CPMP guidance, all new chemical entities applied for marketing authorization should be screened for immunotoxic potential by distribution of lymphocyte subsets and NK-cell activity or the primary antibody response to a T-cell-dependent antigen in addition to the standard toxicological parameters, even when no toxicological findings suggestive of immunotoxicity are observed in the repeated dose-toxicity study
- The MHLW draft guidance indicates that when an abnormal finding is observed in a repeated dose-toxicity study, the antibody response should be examined before phase 1.

30.5.1 Factors to Consider in the Evaluation of Potential Immunotoxicity

Factors that might prompt additional immunotoxicity studies can be identified in the following areas:

- Signs of immunotoxicity potential from the STS data:
  - Hematological changes
  - Alterations in immune system organ weights and/or histology
  - Changes in serum immunoglobulin
  - Increased incidence of infections
  - Increased occurrence of tumors as a sign of immunosuppression
- The pharmacological properties of a test compound: signal that additional immunotoxicity testing should be considered when indicating a potential to affect immune function (e.g., anti-inflammatory drugs)
- The intended patient population: when the majority of the patient population is immune-compromised by a disease state or concurrent therapy
- Structural similarities to known immune modulators
- The disposition of drugs: when the compound or its metabolites are retained at high concentrations in cells of immune systems
- Clinical information: when clinical findings suggestive immunotoxicity in patients exposed to the drug

A weight-of-evidence review should be performed to determine the existence of a cause for concern. A finding of sufficient magnitude in a single area should trigger additional immunotoxicity studies. Findings from multiple factors, each of which would not be sufficient on its own, could trigger additional studies. If additional immunotoxicity studies are not performed, the sponsor should provide justification.

30.5.2 Selection and Design of Additional Immunotoxicity Studies

Additional immunotoxicity studies should also help determine the cell type affected reversibility, the mechanism of action, and possibly lead to biomarker selection for clinical studies.

- Selection of assays. An immune function study, such as a T-cell-dependent antibody response is recommended.
- Study design. To assess drug-induced immunotoxicity, conduct a 28-day study with consecutive daily dosing in rodents.
  - The high dose should be above the NOAEL but below a level inducing changes secondary to stress; multiple dose levels are recommended to determine dose–response relationships and the dose at which no immunotoxicity is observed.
  - The species, strain, dose, duration, and route of administration used in additional immunotoxicity studies should be consistent, where possible, with the standard toxicity study in which an adverse immune effect was observed.

○ Usually, both genders should be used in these studies,

- Evaluation of additional immunotoxicity studies and need for further studies. Results from additional immunotoxicity studies should be evaluated as to whether sufficient data are available to reasonably determine the risk of immunotoxicity. If the risk of immunotoxicity is acceptable and/or can be addressed in a risk management plan (ICH, 2004), no further animal testing is necessary.

30.5.3 Timing of Immunotoxicity Testing in Relation to Clinical Studies

- If the weight-of-evidence review indicates that additional immunotoxicity studies are appropriate, these studies should be completed before exposing a large population of patients (or before beginning phase 3 clinical studies).
- The timing of the additional immunotoxicity testing might be determined by:
  ○ The nature of the effect by the test compound
  ○ The type of the clinical testing that would be called for if a positive finding is observed with an additional immunotoxicity testing
- If the target patient population is immune-compromised, immunotoxicity testing can be initiated at an earlier time point in the development of the drug.

30.6   SAFETY REPORTING REQUIREMENTS

The sponsor investigating a drug under an investigational new drug (IND) application is required to notify the FDA and all participating investigators of potentially serious risks from clinical trials or from any other sources, including the findings from tests in animals. This IND safety report must be submitted as soon as possible but no later than 15 calendar days after the sponsor receives the safety information and determines that the information qualifies for reporting as required under 21 CFR 312.32(c) (FDA, 2011a). The sponsor must also submit information from the clinical study as prescribed by the postmarketing safety reporting requirements under Sec. 310.305 (FDA, 2011b) and Sec. 314.80 (FDA, 2011c). Appendix 30.5 (FDA, 2010) summarizes reporting requirements for submitting safety report from clinical studies. The sponsor should submit a development safety update report annually to the FDA and other interested parties (e.g., ethical committees). This report should provide a comprehensive but concise thoughtful annual review and evaluation of pertinent safety information collected from all clinical trials and other studies, ongoing or completed, during the reporting period related to a drug under investigation, whether or not the drug is marketed. This overall safety assessment should be integrated evaluation of all new relevant clinical, nonclinical, and epidemiologic information relative to previous knowledge of the investigational drug. When evaluating risks, particular emphasis is placed on newly identified safety concerns or new significant information relative to safety concerns identified previously (FDA, 2011d).

30.7   PHARMACOGENOMICS (FDA, 2005b; NIGMS, 2005; ORLN, 2010)

Each person is genetically unique; the genes, contained in a person's DNA, determine the makeup of a body's proteins. Medicines travel through the body, interacting with many of these proteins. Small, but normal variations (in makeup and amount) in a person's genes can produce proteins that would interact with a medicine differently than they would in other people, including close relatives. Variations in genes could be responsible for the following:

- Some people get no pain relief from certain prescription painkillers: in particular, those painkillers that work only when they are converted from an inactive to an active form by body protein

- Certain allergy and asthma medicines work well for some people and not at all for others.

- Nearly 3 million people in the Unites States are at risk for overdose (causing annoying, sometimes life-threatening side effects) when given the standard amount of a medicine commonly used to prevent blood clot.

- A normally safe dose of leukemia treatment can, in rare cases, lead to death in a child with an unusual change in just one gene.

Pharmacogenomics is a science that examines how an individual's genetic inheritance (or inherited variations in genes) affects the body's response to drugs and explores ways in which these variations can be used to predict whether a patient will have a good response to a drug, a bad response to a drug, or no response at all. In recent years, scientists have found genetic variations affecting patient responses to cholesterol-lowering medicines, cancer treatments, AIDS medicines, and many other commonly used medicines. Pharmacogenomics holds the promise that drugs might one day be tailormade for individuals and adapted to each patient's genetic makeup and the key to creating personalized drugs with greater efficacy and safety. *Pharmacogenomics* comes from two words, *pharmacology* and *genomics*, and combines traditional pharmaceutical sciences with annotated knowledge of genes, proteins, and single-nucleotide polymorphism.

30.7.1   Anticipated Benefits of Pharmacogenomics

- More powerful medicines. Pharmaceutical companies will be able to create drugs based on the molecules associated with genes and diseases. This will facilitate drug discovery targeted to specific diseases, thus maximizing therapeutic effects, but would also decrease damage to nearby healthy cells.

- Better, safer drugs the first time. Instead of the standard trial-and-error method of matching patients with the right drugs, doctors will be able to analyze a patient's genetic profile and prescribe the best available drug therapy from the beginning. This will take the guesswork out of finding the right drug, speed up recovery time, increase safety, and possibly eliminate adverse reactions to drugs. Pharmacogenomics has the potential to dramatically reduce the estimated 100,000 deaths and 2 million hospitalizations that occur each year in the United States as a result of adverse drug response (Lazarou et al., 1998)

- More accurate methods of determining appropriate drug dosages. Current methods of basing dosages on weight and age or "one-size-fits-all" medicine dosing will be replaced with dosages based on a person's genetics: how well the body processes the medicine and the time it takes to metabolize it. This will maximize the therapy's value and decrease the likelihood of overdose.

- Advanced screening for disease. Knowing one's genetic code will allow a person to make adequate lifestyle and environmental changes at an early age so as to avoid or lessen the severity of a genetic disease. Similarly, advance knowledge of particular disease susceptibility will allow careful monitoring, and treatments can be introduced at the most appropriate stage to maximize their therapy.

- Better vaccines. Vaccines made of genetic material, either DNA or RNA, promise all the benefits of existing vaccines without all the risks. They will activate the immune system but will be unable to cause infections. They will be inexpensive, stable, easy to store, and capable of being engineered to carry several strains of a pathogen at once.

- Improvements in the drug discovery and approval process. Pharmaceutical companies will be able to discover potential therapies more easily using genome targets. Previously failed drug candidates may be revived as they are matched with the niche population they serve. The drug approval process should be facilitated as trials targeted for specific genetic population groups providing greater degrees of success. The cost and risk of clinical trials will be reduced by targeting only those persons capable of responding to a drug.

- Decrease in the overall cost of health care. Decreases in the number of adverse drug reactions, the number of failed drug trials, the time it takes to get a drug approved, the length of time that patients are on medication, the number of medications that patients must take to find an effective therapy, the effects of a disease on the body (through early detection), and an increase in the range of possible drug targets will promote a net decrease in the cost of health care.

30.7.2 Barriers to Pharmacogenomics Progress

Pharmacogenomics is a developing research field that is still in its infancy. Several of the following barriers will have to be overcome before many pharmacogenomics benefits can be realized.

- Complexity of finding gene variations that affect drug response. Single-nucleotide polymorphisms (SNPs) are DNA-sequence variations that occur when a single nucleotide (A, T, C, or G) in the genome sequence is altered. SNPs occur every 100 to 300 bases along the 3-billion-base human genome; therefore, millions of SNPs must be identified and analyzed to determine their involvement (if any) in drug response. Further complicating the process is our limited knowledge of which genes are involved with each drug response. Since many genes are likely to influence responses, obtaining the big picture on the impact of gene variations is highly time consuming and complicated.

- Limited drug alternatives. Only one or two approved drugs may be available for treatment of a particular condition. If patients have gene variations that prevent them from using these drugs, they may be left without alternatives for treatment.

- Disincentives for drug companies to make multiple pharmacogenomic products. Most pharmaceutical companies have been successful with their one-size-fits-all approach to drug development. Since it costs hundreds of millions of dollars to bring a drug to market, will these companies be willing to develop alternative drugs that serve only a small portion of the population?

- Educating health care providers. Introducing multiple pharmacogenomics products to treat the same condition for different population subsets undoubtedly will complicate the process of prescribing and dispensing drugs. Physicians must execute an extra diagnostic step to determine which drug is best suited to each patient. To interpret the diagnostic accurately and recommend the best course of treatment for each patient, all prescribing physicians, regardless of specialty, will need a better understanding of genetics.

30.7.3   Submission of Pharmacogenomics Data

- Pharmacogenomics data submission for an IND application is required under Sec. 312.23 (FDA, 2011a) if any of the following apply:
  - The test results affect dose selection, entry criteria into a clinical trial safety monitoring, or subject stratification.
  - A sponsor is using the test results to support scientific justification with respect to pharmacological mechanism of action, the selection of drug dosing, or the safety and effectiveness of a drug.
  - The test results constitute a known, valid biomarker for physiologic, pharmacologic, pathophysiologic, toxicologic, and clinical states, or outcomes in humans.
  - The test results constitute a known, valid biomarker for a safety outcome in animal studies.
  - The test results can serve as a probable valid biomarker in human safety studies.

- Pharmacogenomics data submission for an IND application is not required, but voluntary submission is encouraged (i.e., information does not meet the criteria of Sec. 312.23), if:
  - The information is from exploratory studies or is research data, and/or
  - The results are from test systems where the validity of the marker has not been established.

- Steps in PG data submission for a new drug application (NDA), biologic licence application (BLA), or supplement:
  - Submit a full report (including the information about test procedures and data) in the relevant sections of the NDA or BLA if the sponsor plans to use the test results on the drug label or as part of the scientific database being used to support approval. The following examples would fit this category:
    - Pharmacogenomic test results that are being used to support scientific arguments made by the sponsor about drug dosing, safety, patient selection, or effectiveness
    - Pharmacogenomic test results that the sponsor proposes to describe on the drug label
    - Pharmacogenomic tests that are essential to achieving the dosing, safety, or effectiveness described on the drug label
  - Submit as an abbreviated report (not as a synopsis or voluntary genomic data submission) if the test results are known valid biomarkers (or represent probable valid biomarkers) for physiologic, pathophysiologic, pharmacologic, toxicologic, or clinical states or outcomes in the relevant species but the sponsor is not relying on or mentioning this on the label.
  - Voluntary submission of the data from general exploratory or research studies is encouraged; the agency does not view these studies as germane in determining the safety or effectiveness of a drug.
- Steps in PG data submission to an approved NDA, BLA, or supplement:
  - Submit as an abbreviated report or a synopsis if the test results are known valid biomarkers (or represent probable valid biomarkers) for physiologic, pathophysiologic, pharmacologic, toxicologic, or clinical states or outcomes in the relevant species
  - Voluntary submission of the results from test systems where the validity of the marker has not been established is encouraged but not required

REFERENCES

FDA. 2005a. Guidance for industry: Estimating the maximum safe starting dose in initial clinical trials for therapeutics in adult healthy volunteers. June 2005. http://www.fda.gov/downloads/Drugs/GuidanceComplianceRegulatoryInformation/Guidances/ucm078932.pdf. Accessed Dec. 2011.

———. 2005b. Guidance for industry: pharmacogenomics data submissions. Mar. 2005. http://www.fda.gov/downloads/Drugs/GuidanceComplianceRegulatoryInformation/Guidances/ucm079849.pdf. Accessed Dec. 2011.

———. 2010. Guidance for industry and investigators: Safety reporting requirements for INDs and BA/BE studies. Sept. 2010. http://www.fda.gov/downloads/Drugs/GuidanceCompliance RegulatoryInformation/Guidances/UCM227351.pdf. Accessed Dec. 2011.

———. Code of Federal Regulations. 2011a. 21 CFR 312. Investigational new drug application. Apr. 1, 2011. http://www.accessdata.fda.gov/scripts/cdrh/cfdocs/cfcfr/CFRSearch.cfm?CFRPart=312&show FR=1. Accessed Dec. 2011.

———. Code of Federal Regulations. 2011b 21 CFR 310.305. Subchapter D, Drug for human use. Apr. 2011. http://www.accessdata.fda.gov/scripts/cdrh/cfdocs/cfcfr/CFRSearch.cfm?fr=310.305. Accessed Dec. 2011.

_____. *Code of Federal Regulations*. 2011c. 21 CFR 314. Applications for FDA approval to market a new drug. Apr. 1, 2011. http://www.accessdata.fda.gov/scripts/cdrh/cfdocs/cfcfr/CFR Search.cfm?CFRPart=314&showFR=1. Accessed Dec. 2011.

_____. 2011d. Guidance for industry: E2F development safety update report. Aug. 2011. http://www.fda.gov/downloads/Drugs/GuidanceComplianceRegulatoryInformation/Guidances/UCM073109.pdf. Accessed Dec. 2011.

Hayashi M, MacGregor JT, Gatehouse DG, Blakey DH, Dertinger SD, Abramsson-Zetterberg L, et al. In vivo erythrocyte micronucleus assay: III. Validation and regulatory acceptance of automated scoring and the use of rat peripheral blood reticulocytes, with discussion of non-hematopoietic target cells and a single dose-level limit test. Mutat. Res. 2007; 627: 10–30.

ICH. 1994. Guideline; S3A: Note for guidance on toxicokinetics—the assessment of systemic exposure in toxicity studies. Oct. 1994. http://www.ich.org/fileadmin/Public_Web_Site/ICH_Products/Guidelines/Safety/S3A/Step4/S3A_Guideline.pdf. Accessed Dec. 2011.

_____. 1995. S1A: Guideline on the need for carcinogenicity studies of pharmaceuticals. Nov. 29, 1995. http://www.ich.org/fileadmin/Public_Web_Site/ICH_Products/Guidelines/Safety/S1A/Step4/S1A_Guideline.pdf. Accessed Dec. 2011.

_____. 1997. Guideline, S2B: Genotoxicity—: a standard battery for genotoxicity testing for pharmaceuticals. July 1997. http://www.ich.org/fileadmin/Public_Web_Site/ICH_Products/Guidelines/Safety/S2_R1/Concept_papers/S2_R1__Concept_Paper.pdf. Accessed Dec. 2011.

_____. 2000. S7A: Safety pharmacology studies for human pharmaceuticals. Nov. 8, 2000. http://www.ich.org/fileadmin/Public_Web_Site/ICH_Products/Guidelines/Safety/S7A/Step4/S7A_Guideline.pdf. Accessed Dec. 2011.

_____. 2003. S8: Immunotoxicology studies for human pharmaceuticals—final concept paper. Nov. 11, 2003. http://www.ich.org/fileadmin/Public_Web_Site/ICH_Products/Guidelines/Safety/S8/Concept_papers/S8_Concept_Paper.pdf. Accessed Feb. 2012.

_____. 2004. E2E: Pharmacovigilance planning. Nov. 14, 2004. http://www.ich.org/fileadmin/Public_Web_Site/ICH_Products/Guidelines/Efficacy/E2E/Step4/E2E_Guideline.pdf. Accessed Dec. 2011.

_____. 2005a. Guideline; S7B: The nonclinical evaluation of the potential for delayed ventricular repolarization (QT interval prolongation) by human pharmaceuticals. May 2005. http://www.ich.org/fileadmin/Public_Web_Site/ICH_Products/Guidelines/Safety/S7B/Step4/S7B_Guideline.pdf. Accessed Dec. 2011.

_____. 2005b. S8: Immunotoxicity studies for human pharmaceuticals. Sept. 15, 2005. http://www.ich.org/fileadmin/Public_Web_Site/ICH_Products/Guidelines/Safety/S8/Step4/S8_Guideline.pdf. Accessed Dec. 29, 2010.

_____. 2008. S2(R1): Guidance on genotoxicity testing and data interpretation for pharmaceuticals intended for human use. Mar. 6, 2008. http://www.ich.org/fileadmin/Public_Web_Site/ICH_Products/Guidelines/Safety/S2_R1/Step2/S2_R1__Guideline.pdf. Accessed Dec. 2011.

_____. 2009. M3(R2): Guidance on nonclinical safety studies for the conduct of human clinical trials and marketing authorization for pharmaceuticals. June 11, 2009. http://www.ich.org/fileadmin/Public_Web_Site/ICH_Products/Guidelines/Multidisciplinary/M3_R2/Step4/M3_R2__Guideline.pdf. Accessed Jan. 2010.

Lazarou J, Pomerouz BH, Corey PN. Incidence of adverse drug reactions in hospitalized patients: a meta-analysis of prospective studies. JAMA 1998; 279(15): 1200–1205.

NC3Rs. 2007. National Centre for the Replacements, Refinement and Reduction of Animals in Research. Challenging requirements for acute toxicity studies: workshop report. May 2007.

NIGMS. 2005. Medicines for you: Studying how your genes can make a difference. HHS/NIH/NIGMS. NIH Publication 05–4657. Feb. 2005. http://publications.nigms.nih.gov/medsforyou/MedsForYou.pdf. Accessed Dec. 27, 2010.

ORNL 2010 Human Genome Project Information — Pharmacogenomics. US Department of Energy Office of Science, Office of Biological and Environmental Research, Human Genome Program. Sep. 2010. Accessed Dec. 2010. http://www.ornl.gov/sci/techresources/Human_Genome/medicine/pharma.shtml

Robinson S, Delongeas J, Donald E, Dreher D, Festag M, Kervyn S, Lampo A, Nahas K, Nogues V, Ockert D, Quinn K, Old S, Pickersgill N, Somers K, Stark C, Stei P, Waterson L, Chapman K. A European pharmaceutical company initiative challenging the regulatory requirement for acute toxicity studies in pharmaceutical drug development. Regul. Toxicol. Pharmacol. 2008; 50(3): 345–352.

GLOSSARY

BLA	Biologic license application.
Body surface area conversion factor	A factor that converts a dose (mg/kg) in an animal species to the equivalent dose in humans (also known as the *human equivalent dose*), based on differences in body surface area. The ratio of the body surface areas in the species tested to that of an average human.
CDER	Center for Drug Evaluation and Research.
CFR	*Code of Federal Regulation.*
Gene endpoint	The precise type or class of genetic change investigated (e.g., gene mutations, chromosomal aberrations, DNA strand breaks, DNA repair, DNA adduct formation)
Gene mutation	A detectable permanent change within a single gene or its regulating sequences. The changes may be point mutations, insertions, or deletions.
Genotoxicity	A broad term that refers to any deleterious change in genetic material, regardless of the mechanism by which the change is induced.
GLP	Good laboratory practice.
HED	Human equivalent dose.
Human equivalent dose	A dose in humans anticipated to provide the same degree of effect as that observed in animals at a given dose. In this guidance, as in many communications from sponsors, the term is generally used to refer to the human equivalent dose of the no observed adverse effects level (NOAEL). When reference is made to the human equivalent of a dose other than the NOAEL (e.g., the PAD), sponsors should note this usage explicitly and prominently.
ICH	International Conference on Harmonization of Technical Requirements for Registration of Pharmaceuticals for Human Use.
IND	Investigational new drug application.
Known valid biomarker	A biomarker that is measured in an analytical test system with well-established performance characteristics and for which there is widespread agreement in the medical or scientific community about the physiologic, toxicologic, pharmacologic, or clinical significance of the results (FDA Guidance on Pharmacogenomics, 2005)
Lowest observed adverse effect level	The lowest dose tested in an animal species with adverse effects.
Maximum recommended starting dose	The highest dose recommended as the initial dose in a clinical trial. In clinical trials of adult healthy volunteers, the MRSD is predicted to cause no adverse reactions. The units of the dose (e.g., mg/kg or mg/m^2) may vary, depending on practices employed in the area being investigated.
Maximum tolerance dose	In a toxicity study, the highest dose that does not produce unacceptable toxicity.
MFD	Maximal feasible dose.
MHLW	Ministry of Health, Labour and Welfare in Japan.
MRSD	Maximum recommended starting dose.

MTD	Maximum Tolerance Dose
NDA	New drug application.
No observed adverse effects level	The highest dose tested in an animal species that does not produce a significant increase in adverse effects compared to the control group. Adverse effects that are biologically significant, even if not statistically significant, should be considered in determining this level.
No observed effect's level	The highest dose tested in an animal species with no detected effects.
NOAEL	No observed adverse effects level.
PD	Pharmacodynamics.
PG	Pharmacogenomics.
Pharmacodynamics	Describes the biochemical and physiological effects of a drug on the body, including how long a drug is absorbed, moves throughout the body, and interacts with certain molecules within target tissues.
Pharmacogenetic test	An assay intended to study interindividual variations in DNA sequence related to drug absorption and disposition (pharmacokinetics) or drug action (pharmacodynamics), including polymorphic variation in the genes that encode the functions of transporters, metabolizing enzymes, receptors, and other proteins.
Pharmacokinetics	Describes the effects of the body on a drug that includes the process by which the drug is absorbed, distributed in the body, localized in the tissues, metabolized, and excreted.
Pharmacologically active dose	The lowest dose tested in an animal species with the intended pharmacologic activity.
PK	Pharmacokinetics.
Probable valid biomarker	A biomarker measured in an analytical test system with well-established performance characteristics and for which there is a scientific framework or body of evidence that appears to elucidate the physiologic, toxicologic, pharmacologic, or clinical significance of the test results. A probable valid biomarker may not have reached the status of a known valid marker because, for example, of any one of the following reasons: (1) the data elucidating its significance may have been generated within a single company and may not be available for public scientific scrutiny; (2) the data elucidating its significance, although highly suggestive, may not be conclusive; or (3) independent verification of the results may not have occurred.
Relative growth treatment	A measure of cytotoxicity that takes the relative suspension growth (based on cell loss and cell growth from the beginning of treatment to the second day posttreatment) and multiplies it by the relative plating efficiency at the time of cloning for mutant quantization.
Safety factor	A number by which the human equivalent dose (HED) is divided to introduce a margin of safety between the HED and the maximum recommended starting dose.
SNP	Single-nucleotide polymorphism.
SPS	Safety pharmacology studies.
STS	Standard toxicity studies.
Valid biomarker	A biomarker that is measured in an analytical test system with well-established performance characteristics and for which there is an established scientific framework or body of evidence that elucidates the physiologic, toxicologic, pharmacologic, or clinical significance of the test results. The classification of biomarkers is context-specific.

Validation of a biomarker

Validation of a biomarker is context-specific, and the criteria for validation will vary with the intended use of the biomarker. The clinical utility (e.g., predict toxicity, effectiveness, or dosing) and use of epidemiology or population data (e.g., strength of genotype–phenotype associations) are examples of approaches that can be used to determine the specific context and the criteria necessary for validation.

Voluntary genomic data submission

The designation for pharmacogenomic data submitted voluntarily to the FDA.

APPENDIXES

APPENDIX 30.1   Selection of Maximum Recommended Starting Dose for Drugs Administered Systematically to Normal Volunteers

Administered Systematically to Normal Volunteers (FDA, 2005a)

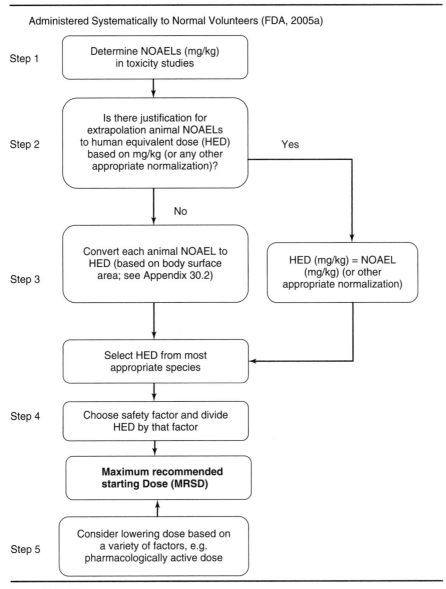

Source: FDA (2005a).

APPENDIX 30.2   Conversion of Animal Doses to Human Equivalent Doses Based on Body Surface Area

Species	To Convert Animal Dose in mg/kg to Dose in mg/m², Multiply by k_m	To Convert Animal Dose in mg/kg to HED[a] in mg/kg, Either:	
		Divide Animal Dose by	Multiply Animal Dose by
Human	37		
Child (20 kg)[b]	25		
Mouse	3	12.3	0.08
Hamster	5	7.4	0.13
Rat	6	6.2	0.16
Ferret	7	5.3	0.19
Guinea pig	8	4.6	0.22
Rabbit	12	3.1	0.32
Dog	20	1.8	0.54
Primates			
Monkeys[c]	12	3.1	0.32
Marmoset	6	6.2	0.16
Squirrel monkey	7	5.3	0.19
Baboon	20	1.8	0.54
Micro-pig	27	1.4	0.73
Mini-pig	35	1.1	0.95

Source: FDA (2005a)

[a] Assumes 60 kg human. For species not listed or for weights outside the standard ranges, HED can be calculated from the following formula: HED = animal dose in mg/kg x (animal weight in kg/human weight in kg)$^{0.33}$

[b] This k_m value is provided for reference only since healthy children will rarely be volunteers for phase 1 trials

[c] For example, cynomolgus, rhesus, and stumptail.

APPENDIX 30.3 Recommended Duration of Repeated-Dose Toxicity Studies to Support the Conduct of Clinical Trials

Maximum Duration of Clinical Trial	Recommended Minimum Duration of Repeated-Dose Toxicity Studies to Support Clinical Trials	
	Rodents	Non-rodents
Up to 2 weeks	2 weeks[a]	2 weeks[a]
Between 2 weeks and 6 months	Same as clinical trial[b]	Same as clinical trial[b]
>6 months	6 months[b,c]	9 months[b,c,d]

Source: ICH (2009).

[a] In the United States, as an alternative to 2 week studies, extended single-dose toxicity studies can support single-dose human trials. Clinical studies of less than 14 days can be supported with toxicity studies of the same duration as the proposed clinical study.

[b] Clinical trials for duration longer than 3-months can be initiated, provided that

- the data are available from a 3-month rodent and a 3-month non-rodent study, and
- the complete data from the chronic rodent and non-rodent study are made available, and
- it is consistent with local clinical trial regulatory procedures

For serious or life-threatening indications or on a case-by-case basis, this extension can be supported by complete chronic rodent data and in-life and necropsy data for the non-rodent study. Complete histopathology data from the non-rodent should be available within an additional 3 months.

[c] There can be cases where a pediatric population is the primary population, and existing animal studies (toxicology or pharmacology) have identified potential developmental concerns for target organs. In these cases, long-term toxicity testing starting in juvenile animals can be appropriate in some circumstances.

[d] In the EU, studies of 6 months duration in non-rodents are considered acceptable. However, where studies with a longer duration have been conducted, it is not appropriate to conduct an additional study of 6 months.

APPENDIX 30.4 Recommended Duration of Repeated-Dose Toxicity Studies to Support Marketing

Duration of Indicated Treatment	Rodent	Non-rodent
Up to 2 weeks	1 month	1 month
>2 weeks to 1 month	3 months	3 months
>1 month to 3 months	6 months	6 months
>3 months	6 months[c]	9 months[c,d]

Source: ICH (2009).

N.B., See footnotes c and d in the Appendix 30.3

CHAPTER 31

REGULATORY SCIENCE INITIATIVES FOR ADVANCING PUBLIC HEALTH

31.1 INTRODUCTION

On February 24, 2010, the U.S. Food and Drug Administration (FDA) and the National Institutes of Health (NIH) unveiled an initiative designed to accelerate the process from scientific breakthrough to availability of new and innovative medical therapies for patients. This initiative involves two interrelated scientific disciplines that which are needed to turn biomedical discoveries into products that benefit people (FDA–NIH, 2010):

- *Translational science*: the shaping of basic scientific discoveries into treatments

- *Regulatory science*: the science of developing and using new tools, standards, and approaches to develop products and assess the safety, efficacy, quality, and performance of FDA-regulated products more efficiently

The FDA launched in October 2010 its Advancing Regulatory Science Initiative (ARS) to build upon the achievements of existing agency programs, such as the critical path initiative (FDA, 2011a), and to expand its scope to encompass every dimension of regulatory science. The goals of the ARS span the breadth of FDA's activities as highlighted in the following two documents (FDA, 2012):

- Advancing Regulatory Science for Public Health: The Promise of Regulatory Science (October 2010)

- Advancing Regulatory Science at FDA: Strategic Plan for Regulatory Science (August 2011)

Integrated Pharmaceutics: Applied Preformulation, Product Design, and Regulatory Science,
First Edition. Antoine Al-Achi, Mali Ram Gupta, William Craig Stagner.
© 2013 John Wiley & Sons, Inc. Published 2013 by John Wiley & Sons, Inc.

31.2 ADVANCING REGULATORY SCIENCE FOR PUBLIC HEALTH: THE PROMISE OF REGULATORY SCIENCE (FDA, 2010)

Recent advances in science and technology, ranging from sequencing of the human genome to the application of nanotechnology to new medical products, have the potential to transform our abilities to prevent, diagnose, and treat diseases, and to form the bridge to critical twenty-first-century advances in public health. With a focused agenda and targeted investment, the FDA will continue to work with its partners on current activities and future opportunities for regulatory science to tackle some of the most important and pressing health challenges facing Americans. These challenges include:

1. Accelerating the delivery of new medical treatments to patients
2. Improving pediatric and child health
3. Protecting against emerging infectious diseases and terrorism
4. Enhancing safety and health through informatics
5. Protecting the food supply
6. Modernize safety testing
7. Meeting the challenges for regulating tobacco

For these advances to reach their full potential, the FDA is playing an increasing integral role as an agency, beyond its traditional role of ensuring safe and effective products, to promote public health and participate more actively in the scientific research directed toward new treatments and interventions.

31.3 ADVANCING REGULATORY SCIENCE AT FDA: STRATEGIC PLAN FOR REGULATORY SCIENCE (FDA, 2011b)

The FDA, a science-based regulatory agency, protects and promotes the health and safety:

- Of all Americans through enhancing the availability of safe medical products and foods and promoting innovations to address unmet medical and public health needs
- Of animals through assuring the availability of safe animal drug products and foods
- By helping consumers and health care providers get the accurate and science-based information needed to make the best possible decisions about the use of products for human and nonhuman animal use

To meet these challenges during this century, the FDA has developed a strategic plan for regulatory science to speed innovation, improve regulatory decision-making processes, and get safe and effective products. To implement this strategic plan, the FDA is working with diverse partners to protect and promote the health of all Americans and the global community. This strategy plan includes the following eight priority areas:

1. *Modernize toxicology to enhance product safety*. The goal here is to address critical gaps that exist between the patient response and preclinical toxicology findings due to (a) uncertainty in predicting the accuracy of many toxicology safety assays, and (b) the need for the most rigorous validation against actual human and animal adverse event data to define the reliability and possible limitations. Modernizing toxicology and continually improving the ability of nonclinical tests, models, and measurements to predict product safety issues will increase the likelihood of identifying toxicity risks earlier in product development, assuring patient safety, and mitigating the need to withdraw products approved previously. To implement this strategy, the FDA is addressing the following needs:

   a. Developing better models of adverse human response

   b. Identifying and evaluating biomarkers and endpoints that can be used in nonclinical and clinical evaluations

   c. Using and developing computational methods and computer modeling

2. *Stimulate innovation in clinical evaluations and personalized medicine to improve product development and patient outcomes*. Despite the progress made in understanding how genomic variations alter a person's response and the tests that can be used to tailor treatment to individual patients (personalized medicine), the process of translating new scientific findings into safe and effective use of medical products and optimizing the use of existing products for all populations remain a major challenge. To address these challenges, the FDA is focusing on the following new tools and approaches:

   a. Developing and refining clinical trial designs, endpoints, and analysis methods

   b. Developing quantitative models and leveraging existing and future clinical trial data

   c. Identifying and qualifying biomarkers and study endpoints

   d. Increasing accuracy and consistency, and reducing interplatform variability of analytical methods to measure biomarkers

   e. Developing a virtual physiologic patients

3. *Support new approaches to improve product manufacturing and quality*. The goal here is to assess how new and novel science and technologies (manufacturing innovations and advances in analytical technologies) affect product safety, efficacy, and quality, and to use this information to enhance regulatory policy relevant to these innovations. The FDA will support the application of these innovations by addressing each of the following needs:

   a. Enabling development and evaluation of novel and improved manufacturing methods

b. Developing new analytical methods

c. Reducing risk of microbial contamination of products

4. *Ensure FDA's readiness to evaluate innovative emerging technologies.* Groundbreaking discoveries in complex chemistry and biosynthesis, and emerging fields such as gene therapy, and tissue engineering, are yielding innovative approaches to improving our health. To bring the rewards of these discoveries safely forward to benefit patients, the FDA must be a step ahead of these new scientific developments by addressing each of the following needs:

   a. Stimulating the development of innovative medical products while concurrently developing novel assessment tools and methodologies

   b. Developing assessment tools for novel therapies

   c. Assuring safe and effective medical innovation

   d. Enhancing readiness for new applications of information technology

5. *Harness diverse data through information sciences to improve health outcomes.* The FDA receives a vast amount of information from a variety of sources. Successful integration of these data would provide knowledge and insight not possible from any single source alone. The FDA is in early stages of constructing the information technology to address the following needs:

   a. Enhancing information technology infrastructure development and data mining

   b. Developing and applying simulation models for product life cycles, risk assessment, and other regulatory science uses

   c. Analyzing large-scale clinical and preclinical data sets

   d. Incorporating knowledge from the FDA regulatory files into a database integrating a broad array of data types to facilitate the development of predictive toxicology models and model validation

   e. Developing new data sources and innovative analytical methods and approaches to advance the regulatory science and surveillance of medical products and devices throughout their life cycle

6. *Implement a new prevention-focused food safety system to protect public health.* The strategy here is to emphasize prevention and risk-based priority setting and resource allocation to address the challenges of the food safety environment in the twenty-first century. Specifically, the FDA will focus on the following:

   a. Establishing and implementing centralized planning and performance measurement processes

   b. Improving information sharing internally and externally

   c. Maintaining mission-critical science capabilities

   d. Cultivating expert institutional knowledge

7. *Facilitate development of medical countermeasures* (MCMs) *to protect against threats to U.S. and global health and security*. MCMs are drugs, biologics, devices, and other equipment and supplies for response to public health emergencies involving chemical, biological, radiological, or nuclear threat agents or naturally occurring infectious disease outbreaks. In close alignment with priorities identified by the Public Health Emergency Medical Countermeasures Enterprise, the FDA will facilitate development of safe and effective MCMs and will address the following needs:

   a. Developing, characterizing, and qualifying animal models for MCM development

   b. Modernizing tools to evaluate MCM product safety, efficacy, and quality

   c. Developing and qualifying biomarkers of diseases or conditions

   d. Enhancing emergency communication

8. *Strengthen social and behavioral science to help consumers and professionals make informed decisions about regulated products*. The focus here is on setting and enforcing high standards for product information and quality to ensure that labels are accurate, that products are truly what their labels claim, and that advertising about these products is clear, truthful, and in no way misleading. To facilitate the translation of science-based regulatory decisions and to provide the public with easy access to sound information, the FDA is addressing the following needs:

   a. Knowing the audience

   b. Reaching the audience

   c. Ensuring audience understanding

   d. Evaluating the effectiveness of communication about regulated products

The two documents described above provide examples of numerous innovations and resultant positive benefits to public health (FDA, 2012). Each of the scientific plans presented in these documents represents opportunities to further drive new and important research agenda using a variety of means, from FDA's internal research program to participation in consortium models.

31.4 COLLABORATIVE IMPLEMENTATION FRAMEWORK (FDA, 2010, 2011b)

The FDA has outlined a four-part strategic framework to advance regulatory science within the agency and around the nation:

- Leadership, coordination, strategic planning, and transparency to support science and innovation. The key inputs would be provided by
  - the Science and Innovation Strategic Advisory Council, and
  - the FDA Science Board
- Support for mission-critical applied research, both at the FDA and collaboratively. Support within the FDA is critical in expanding the field of regulatory

science. Some of the following projects are already under way:

- ○ A joint leadership council (FDA and NIH)
- ○ Centers of excellence in regulatory science (FDA and academic scientists)
- ○ Enhanced strategic collaboration and coordination with other governmental agencies
- ○ Enhanced support and focus for the critical path initiative
- ○ Partnership with the Reagan–Udall Foundation

- • Support for scientific excellence, professional development, and a learning organization. In support of this goal, the FDA will explore several approaches:

- ○ Scientific exchange programs through active participation in scientific and professional meetings and conferences
- ○ Access to cutting-edge experiences presented by talented scientists through its Commissioner's Fellowship Program, continuing education and professional development for the FDA staff

- • Recruitment and retention of outstanding scientists. The FDA will seek to recruit and retain an outstanding scientific workforce trained in new and emerging technologies. The FDA will consider other approaches:

- ○ Proposed merit awards to the most accomplished and productive FDA scientists
- ○ Proposed FDA expert physician program to support joint part-time academic faculty/FDA positions
- ○ Appointment of scientists expert in emerging technologies to work as researcher-reviewers throughout the FDA

Successful implementation of the Regulatory science program, including engagement with diverse stakeholders, will allow the FDA to bridge the gap from basic science discoveries to safe, effective, and high-quality products that help patients and protect and promote public health and security.

REFERENCES

FDA. 2010. Advancing regulatory science for public health: a framework for FDA's Regulatory Science Initiative. Office of the Chief Scientist/U.S. Food and Drug Administration. Oct. 2010. http://www.fda.gov/ScienceResearch/SpecialTopics/RegulatoryScience/ucm228131.htm. Accessed Mar. 2012.

———. 2011a. FDA's Critical Path Initiative. Updated Jan. 20, 2011. http://www.fda.gov/ScienceResearch/SpecialTopics/CriticalPathInitiative/ucm076689.htm. Accessed Mar. 2012.

———. 2011b. Strategic plan for regulatory science: Advancing regulatory science at FDA—a strategic plan. Aug. 2011. http://www.fda.gov/ScienceResearch/SpecialTopics/RegulatoryScience/ucm267719.htm. Accessed Mar. 2012.

———. 2012. Advancing Regulatory Science: Moving regulatory science into the 21st century. Updated Feb. 12, 2012. http://www.fda.gov/ScienceResearch/SpecialTopics/RegulatoryScience/default.htm. Accessed Mar. 2012.

FDA–NIH. 2010. News release: NIH and FDA announce collaborative initiative to fast-track innovations to the public. Feb. 24, 2010. http://www.fda.gov/NewsEvents/Newsroom/PressAnnouncements/2010/ucm201706.htm. Accessed Mar. 2012.

GLOSSARY

ARS	Advancing regulatory science initiative.
FDA	U.S. Food and Drug Administration.
MCM	Medical countermeasure.
NIH	National Institutes of Health.

INDEX

Note: Page numbers in *italics* refer to Figures; those in **bold** to Tables; those with (a) refer to appendices

Integrated Pharmaceutics: Applied Preformulation, Product Design, and Regulatory Science, First Edition. Antoine Al-Achi, Mali Ram Gupta, William Craig Stagner.
© 2013 John Wiley & Sons, Inc. Published 2013 by John Wiley & Sons, Inc.